Y0-BCD-786

THE GREAT
CONTEMPORARY
ISSUES

ETHNIC GROUPS IN AMERICAN LIFE

OTHER BOOKS IN THE SERIES

DRUGS
Introduction by J. Anthony Lukas
THE MASS MEDIA AND POLITICS
Introduction by Walter Cronkite
CHINA
O. Edmund Clubb, *Advisory Editor*
LABOR AND MANAGEMENT
Richard B. Morris, *Advisory Editor*
WOMEN: THEIR CHANGING ROLES
Elizabeth Janeway, *Advisory Editor*
BLACK AFRICA
Hollis Lynch, *Advisory Editor*
EDUCATION, U.S.A.
James Cass, *Advisory Editor*
VALUES AMERICANS LIVE BY
Garry Wills, *Advisory Editor*
CRIME AND JUSTICE
Ramsey Clark, *Advisory Editor*
JAPAN
Edwin O. Reischauer, *Advisory Editor*
THE PRESIDENCY
George E. Reedy, *Advisory Editor*
POPULAR CULTURE
David Manning White, *Advisory Editor*
FOOD AND POPULATION: THE WORLD IN CRISIS
George S. McGovern, *Advisory Editor*
THE U.S. AND THE WORLD ECONOMY
Leonard Silk, *Advisory Editor*
SCIENCE IN THE TWENTIETH CENTURY
Walter Sullivan, *Advisory Editor*
THE MIDDLE EAST
John C. Campbell, *Advisory Editor*
THE CITIES
Richard C. Wade, *Advisory Editor*
MEDICINE AND HEALTH CARE
Saul Jarcho, *Advisory Editor*
LOYALTY AND SECURITY IN A DEMOCRATIC STATE
Richard H. Rovere, *Advisory Editor*
RELIGION IN AMERICA
Gillian Lindt, *Advisory Editor*
POLITICAL PARTIES
William E. Leuchtenburg, *Advisory Editor*

THE GREAT
CONTEMPORARY
ISSUES

ETHNIC GROUPS IN AMERICAN LIFE

The New York Times

ARNO PRESS
NEW YORK/1978

JAMES P. SHENTON
Advisory Editor

GENE BROWN
Editor

Library of Congress Cataloging in Publication Data

Main entry under title:

Ethnic groups in American life.

(The Great contemporary issues)
Composed of articles taken from The New York times.
Bibliography: p.
Includes index.
SUMMARY: A collection of articles illustrating the diversity of origins of the American people.

1. Minorities—United States—History—Addresses, essays, lectures. 2. United States—Emigration and immigration—History—Addresses, essays, lectures. 3. United States—Foreign population—Addresses, essays, lectures. [1. Minorities—Addresses, essays, lectures. 2. United States—Emigration and immigration—History—Addresses, essays, lectures. 3. United States—Foreign population—Addresses, essays, lectures.
I. Shenton, James Patrick, 1925- II. Brown, Gene. III. New York times. IV. Series.
E184.A1E83 301.45'0973 77-11053
ISBN 0-405-10747-1

Manufactured in the United States of America

The editors express special thanks to The Associated Press, United Press International, and Reuters for permission to include in this series of books a number of dispatches originally distributed by those news services.

Book design by Stuart David

Contents

Publisher's Note About the Series

It would take even an accomplished speed-reader, moving at full throttle, some three and a half solid hours a day to work his way through all the news The New York Times prints. The sad irony, of course, is that even such indefatigable devotion to life's carnival would scarcely assure a decent understanding of what it was really all about. For even the most dutiful reader might easily overlook an occasional long-range trend of importance, or perhaps some of the fragile, elusive relationships between events that sometimes turn out to be more significant than the events themselves.

This is why "The Great Contemporary Issues" was created—to help make sense out of some of the major forces and counterforces at large in today's world. The philosophical conviction behind the series is a simple one: that the past not only can illuminate the present but must. ("Continuity with the past," declared Oliver Wendell Holmes, "is a necessity, not a duty.") Each book in the series, therefore has as its subject some central issue of our time that needs to be viewed in the context of its antecedents if it is to be fully understood. By showing, through a substantial selection of contemporary accounts from The New York Times, the evolution of a subject and its significance, each book in the series offers a perspective that is available in no other way. For while most books on contemporary affairs specialize, for excellent reasons, in predigested facts and neatly drawn conclusions, the books in this series allow the reader to draw his own conclusions on the basis of the facts as they appeared at virtually the moment of their occurrence. This is not to argue that there is no place for events recollected in tranquility; it is simply to say that when fresh, raw truths are allowed to speak for themselves, some quite distinct values often emerge.

For this reason, most of the articles in "The Great Contemporary Issues" are reprinted in their entirety, even in those cases where portions are not central to a given book's theme. Editing has been done only rarely, and in all such cases it is clearly indicated. (Such an excision occasionally occurs, for example, in the case of a Presidential State of the Union Message, where only brief portions are germane to a particular volume, and in the case of some names, where for legal reasons or reasons of taste it is preferable not to republish specific identifications.) Similarly, typographical errors, where they occur, have been allowed to stand as originally printed.

"The Great Contemporary Issues" inevitably encompasses a substantial amount of history. In order to explore their subjects fully, some of the books go back a century or more. Yet their fundamental theme is not the past but the present. In this series the past is of significance insofar as it suggests how we got where we are today. These books, therefore, do not always treat a subject in a purely chronological way. Rather, their material is arranged to point up trends and interrelationships that the editors believe are more illuminating than a chronological listing would be.

"The Great Contemporary Issues" series will ultimately constitute an encyclopedic library of today's major issues. Long before editorial work on the first volume had even begun, some fifty specific titles had already been either scheduled for definite publication or listed as candidates. Since then, events have prompted the inclusion of a number of additional titles, and the editors are, moreover, alert not only for new issues as they emerge but also for issues whose development may call for the publication of sequel volumes. We will, of course, also welcome readers' suggestions for future topics.

Introduction

Ethnicity and race are central to an understanding of the American people. Drawn from all corners of the world and bringing diverse cultures, religious beliefs, politics, and social mores, Americans are preeminently a heterogeneous people. Their diversity of origins has accentuated the importance of assimilation. A unity to bridge this range of differences has been a continuous concern of Americans and has frequently resulted in deep tensions that have disrupted social peace and domestic tranquillity. A particularly sharp edge has been the coexistence both of ethnic and racial diversity. The fact, as John F. Kennedy noted, that the United States was "a nation of immigrants" was the circumstance that set the American experience apart from that of other people.

The sheer size of immigration to the United States is difficult to imagine. All but a handful of Americans are descendants of immigrants. Federal immigration statistics record that between 1820 and 1924 thirty-six million made the journey to the New World. Most certainly this figure is an understatement since much immigration went unrecorded. For instance, although the entry of slaves was to have terminated in 1808, their importation continued illegally until the outbreak of the Civil War. Furthermore there are no reliable figures for Mexican immigration. At the very least, the official figures leave out ten million immigrants. Still this official count documents a demographic impact of staggering proportions. Between 1850 and 1920 the foreign born population averaged about fourteen per cent of the total population. The tendency of immigrants to concentrate in the northeastern and middle western industrial-urban areas of the country resulted in an uneven impact. For example, in 1860 the populations of New York, Chicago, Cincinnati, Milwaukee, Detroit and San Francisco were about one half foreign born. In New Orleans, Baltimore and Boston, the foreign born made up a third of the population, while St. Louis counted an astonishing sixty per cent of foreigners. The proportions in the future declined but still remained substantially higher than the national average.

The impact of immigration was further complicated by the tendency of certain groups, most particularly the Italians and Chinese, to emigrate only temporarily and then to return to their homelands. These "birds of passage" came to accumulate a nest egg with which to improve their lot when they returned to the old country. The transitory nature of their interest gave them little incentive to assimilate.

The conditions of assimilation, however, were defined by the "native" Americans, who, themselves, were the descendants of earlier immigrants. They differentiated between the "Old" and "New" immigration. By implication the first wave of immigrants drawn from Ireland, Germany, Scandinavia and Great Britain were more readily assimilable than the second wave whose roots were in eastern, central and southern Europe. But all immigrants with the exception of the British were viewed with misgivings when they first began to arrive in significant numbers. The grim comment of one nativist that "we allow every nation to pour its pestilential sewage into our reservoir" was typical of thousands of similar observations. White ethnics soon learned that anglicized whites were more equal than other whites. And all ethnic and racial minorities were subjected to a period in which their differences were labeled as evidence of deviant or criminal behavior.

The "native" Americans were generally of British origin or those who had assimilated to the Anglo-American system of values. Primarily Protestant, holding to the seventeenth-century political values of the Puritan and Glorious Revolutions, and the eighteenth-century economic teachings of Adam Smith, they spoke English and claimed the British intellectual and cultural milieu as their own. Their rebellion against the British monarch had originally been legitimated by an appeal to the rights of Englishmen which had then been extended to the rights of all humanity. An essentially conservative people, they had, in the words of Thomas Jefferson, justified their revolution as an expression of the commonly accepted values inherited from the British political philosophers, Locke, Harrington and Sidney. But the natives had split in the first half of the nineteenth-century along sectional lines on the fundamental question of whether humans could be held in slavery. Northerners rejected the institution as antithetical to a free labor society while Southerners made it the cornerstone of their society. But there was an irony in the division. Neither North nor South was prepared to accept non-whites as equals; where the South enslaved, the North segregated. Racism was the bond that united North and South. The Civil War excluded slavery as a solution to the question of racial inferiority, but it did not end racism.

In the aftermath of the Civil War, the triumphant North shaped a racial policy which consigned large numbers of Americans to a caste relationship with the dominant Anglo-Americans. Red, brown, black and yellow were set apart from the main stream. The black and brown were burdened with segregation; the Asiatics were either by law or "Gentlemen's Agreement" denied permanent entry; and Indians were confined to reservations. Between 1871 when the Indians were denied further recognition as a sovereign people and 1914 when Woodrow Wilson segregated federal employment and Washington, D.C., an American "apartheid" was instituted.

No one felt the implications of this new policy more fully than the American Indians. From the moment that the white man first set foot on the North American continent, the native population had been driven remorselessly westward. As one congressman noted, "...the Indian receded as the white man advanced." The eastern Indian became an immigrant to the west, slowly retreating, caught between the upper and nether millstones of a relentless white advance. Time and again, they adjusted to a new environment, accepting modifications of their original culture, only to discover in the decades of the Civil War that they faced extermination as an autonomous, distinct culture. Their resistance flared in a series of wars that gave them an occasional fleeting victory which was followed by a crushing defeat. By 1876, in the aftermath of Custer's massacre at Little Big Horn, the white American agreed that the Indian "needed... a stable rule of action to restrain him from committing injuries on the rights and property of others," which was to be accomplished by punishing and coercing him "until he submits to be guided by that rule." Submission did not bring citizenship to the Indian, a right denied by the Supreme Court. The general view was summed up by an 1883 Conference which proposed that "the Indians be admitted to United States citizenship so soon, and only so soon, as they are fitted for its responsibilities." The Indians were left in an anomalous position; although native born, unlike aliens, they were denied access to citizenship. Not until the passage of the Snyder Act of 1924 were they finally granted citizenship, even as the National Origins Act of 1924 established quotas that effectively terminated unlimited immigration into the United States.

Restrictions on immigration had first been instituted with the Chinese Exclusion Act of 1882. The measure reflected in part the growing fears of California white workers, largely of Irish extraction, that Chinese laborers would drive them from competition. Underscoring these fears was a pervasive racism. As one magazine noted, "We are accustomed to think of the Chinese as belonging to a degraded race, ignorant of civilized life..." The same restrictive principle was extended to the Japanese in 1907-08 by "the Gentlemen's Agreement." Tokyo assuaged its bruised pride by agreeing to halt the emigration of further laborers to the United States, while leaving the right of emigration of Japanese in the abstract still in existence. The 1924 National Origins Act specifically excluded Japanese immigrants as "aliens ineligible to citizenship." The Supreme Court in 1923 upheld the right of states, particularly California, to deny Japanese residents the ownership or leasing of farmlands. At the same time, violence was directed against both groups. Chinese in mining camps and railroad work gangs were frequently the target for lynchings. The Japanese suffered the climactic humiliation of being confined to relocation centers during World War II. On the floor of the House of Representatives, John Rankin of Mississippi, barely a week after Pearl Harbor, justified the impending confinement when he declared, "I'm for catching every Japanese in America, Alaska, and Hawaii now and putting them in concentration camps..."

Less overt, at least in law, was the treatment of white ethnics. Hostility directed toward them was often in the organization of nativist movements. In the 1850's, the "Know Nothings" rose up to "check foreign influence...and to purify the ballot box." More exactly, they were reacting to the massive influx of Irish immigrants who fled the catastrophe of the Potato Famine. Between 1846 and 1855, about 1,300,000 Irish arrived, and in approximately the same period, more than 1,100,000 Germans also came. Within a decade, the United States was called upon to absorb newcomers who totalled more than ten per cent of the resident population. "To the Americans who... looked upon the immigrants as they landed, or saw them in the city streets,'' as one perceptive historian has noted, ''the Famine emigration looked like a plague let loose in a fair land—bundles of disease, destitution, squalor, and hopelessness.''

Since almost sixty per cent of all immigrants were male, most of whom were young adults, their political impact was quickly felt. The ''Know Nothing'' politicians described the new voter with scurrilous precision as ''the refuse and scum of other nations, the thief and cut-purse, the incendiary, the cast-off pauper...(who) make our ballot-box a saturnalia of vulgar villainy.'' Their distaste was reinforced by their Protestant fears of the Catholic Church to which many of the immigrants owed at least nominal allegiance. The Roman Church was seen as a subversive institution bent upon ''usurping our government, destroying our liberties and shaping everything to the standard of priestly ambition.'' These messages of the ''Know Nothings'' were repeated in the 1880's by the American Protective Association and in the 1920's by the revived Ku Klux Klan.

Although in theory the ''old Immigration'' of Irish and Germans was more readily absorbed than the ''new Immigration'' of Jews, Slavs, Hungarians, Italians, and Greeks, the Irish and Germans remained targets of discrimination well into the twentieth-Century. In part, this arose out of the speed with which Irish and German Catholics had organized a parochial school system in reaction to the Protestantism of the public schools. Bluntly, as one Catholic churchman asked, ''How can we think of sending our children to those schools in which every artifice is resorted to in order to reduce them from their religion?'' The ''old Immigrants'' had consciously set about separating themselves from the dominant natives. The natives reacted by obliging German-Americans to make public displays of loyalty during World War I. The nomination of Al Smith by the Democrats in 1928 provoked slanderous attacks on Catholicism and the Irish. The fact that Smith would have lost under any circumstance was overshadowed by the Irish and Catholic realization that they were perceived as second-class citizens. The impact of 1928 was so demoralizing that Chicago's Cardinal Mundelein pleaded with Jim Farley to abandon his presidential aspirations in 1940 lest Catholics be faced with a renewed onslaught of bigotry. Even in 1960 when John F. Kennedy eked out a razor-thin victory, his Catholicism was an issue he had to face.

Despite the open prejudice they so frequently faced, the immigrants continued to come. The economic opportunities within the country were an irresistable lure. Without the constant influx of new labor, American industrial development would have been significantly retarded. The implication was not lost on native workers who realized that their wage scales were being undercut by cheaper immigrant labor. Understandably, organized labor agitated for restrictions on immigration. In contrast, American industrialists generally opposed restrictions, fearing they would undermine industrial expansion.

Generally, emigration reflected economic and social conditions both in Europe and the United States. When a recession or depression disrupted the European economy, heavy emigration abroad was the rule, and when the reverse was true in the United States, emigration from Europe declined. Disruptive events such as war also affected emigration. During the American Civil War, the flow of immigrants declined significantly, while the wars of German unification triggered an upsurge of emigration, not only from Germany but from a defeated Denmark. The push and pull factor was also reflected in the emigration, particularly from Britain, of highly skilled workers. The inability of the mid-nineteenth-century American educational system to provide the required industrial and engineering skills encouraged American employers to recruit workers from Britain's steel and iron manufacturers, engineering enterprises, tool and cutlery factories, as well as textile and mining operations.

Despite the recruitment of skilled workers, the mass of European and Asiatic immigration, as is presently true of Latin-American and Caribbean emigration, was made up of unskilled or semi-skilled workers. Their influx often had the effect of pushing native workers into positions of greater skill, and consequent security. But frequently the newcomers adversely affected the economic position of ethnic workers who had arrived at an earlier date. They and especially their children perceived of themselves as natives. Occupying the next to the bottom rung of the social and economic ladder, they not only felt economically threatened, but as often they were faced with competition for a place to live. Typically the Irish masses of the Lower East Side of New York were forced at the turn of the century to give way to incoming Jews and Italians, as these groups subsequently gave way to the Chinese, Blacks and Hispanics. The result was acute inter-ethnic tensions.

Ethnic divisions fragmented class consciousness, although within specific ethnic groups, it created an ethnic solidarity that was used effectively to advance the interest of these groups. Ironically, many of the ethnic elements arrived, not with a national identity, but with a regional identity. Newly arrived immigrants from Italy, for example, thought of themselves as Piedmontese, Lombards, Sicilians, Calabrese, Apulians, Neapolitans; but native Americans, including other ethnics, lumped them together as Italians, creating the hyphenated phenomenon of Italian-Americans. A peculiar result was that the Mafia—indigenous to Sicily—became the purported instrument of a pervasive Italian-American criminality. It hardly mattered that the evidence in no way supported such a conclusion for Sicilian emigrants, let alone Italians from other regions; the stereotype once set became as fixed in the American mind-frame as the cowboy and Indian version of the West.

In fact, criminal behavior knew neither ethnicity nor race. One need only recollect the thoroughly native antecedents of "Bonnie and Clyde" and John Dillinger to make the point. It is useful, however, to note that so-called criminal behavior is often a reflection of poverty and deprivation. Unsurprisingly, in the mid-nineteenth-century, when the Five Point's slum of New York was an Irish ghetto, it was widely perceived by outsiders as swarming with Irish criminals. Later, in 1906, the director of the New York YMHA reported that "between 28 and 30 per cent of all children brought to the children's court in New York are Jewish." The disruptive factor of criminality was but one reflection of the profound difficulties that surrounded life in every ethnic or racial ghetto.

Pressured by the dominant culture to anglicize in language and culture, the newly arrived ethnic was plunged into a crisis that affected their perceptions of family, law, government, authority, religion, to be exact, the whole sub-structure on which their previous life styles had been based. In addition, they were frequently trapped in an economic relationship that made them vulnerable to the grossest of exploitation, such as the sweat shop, the speed-up, or, as in the steel industry, the eighty-four-hour week. The hyphenated American straddled two worlds, one that they had abandoned, and a new one which only grudgingly made a place for them.

The ghetto, whatever its flaws, was the halfway house in which the immigrant worked out resolutions of the conflict between the old and the new value systems. Within the ghetto, the old language and the old customs and mores still served a purpose. The Roman Catholic parish organized on the basis of ethnicity and the Jewish synagogue around the shtetl. The result was a familiar God in familiar surroundings. No matter how dismaying the confrontation with the new and alien world, the ghetto was a secure base in which familiar values persisted. But to the native American, it reinforced a perception of ethnic separateness as an insidious threat to American unity. Nonetheless, the ghetto contained the seeds of its own destruction. Within it, the second, or American generation completed the journey of their parents. To the American born, the old country was a homeland never seen, an abandoned land recalled in second hand memories. To the second generation, America was an all embracing homeland which would tolerate nothing less than a 100 per cent Americanism. Those who had known the harshness of nativist rejection embraced the very values that had tormented them. Often the new Americans were in the forefront of those who would direct similar prejudices against the latest wave of newcomers. The patterns of the past persisted into the present and beyond into the future.

The end of unrestricted immigration in 1924 did not halt immigration. Since then, at least twelve million immigrants have legally entered the country. No one seems to have an accurate count of the illegal entries, but they too number in the millions. Simultaneously, a massive influx of Puerto Ricans and migration of Blacks from the South northward has repeated many of the patterns that have characterized previous immigrant groups. They have experienced the same concentrations in ghettos, the same exploitative economic patterns, the same social dislocations, but there has been a significant difference. Both groups have had to deal not only with traditional prejudices against ethnic groups, but, particularly the Blacks, with the added ingredient of racism.

The Black American has been the alien in our midst from the dimmest beginnings of the American experience. Except for the recent past, Blacks have endured either slavery or segregation. For them American history has been a crushing burden. Their flight northward brought not emancipation but the poverty and dislocation of the ghetto. Nevertheless, within the black ghetto, a process of self-discovery has taken place. Blacks have become hyphenated into Afro-Americans, and they have begun to rediscover their lost roots.

Simultaneously, the Spanish speaking immigrants from Latin-America and the Caribbean are being redefined as Hispanics as the native Americans perceive their collective rather than parochial identities. The processes of past assimilation are reasserting themselves.

The Afro-American's assertion of their identity has stimulated among white ethnics a renewed awareness of the abandoned past. The implications of this renewed interest are difficult to ascertain. At its worse, it could fragment American society into a balkanized pattern of petty, self-preoccupied groups devoid of a unifying commitment to a larger America. At its best, it may be an infusion into the American identity of the rich diversity that is the heritage of a people drawn from many ethnicities and races. It may mark the beginning of a process that makes assimilation a two-way street in which the best of the past becomes the heartbeat of the future American.

James P. Shenton

CHAPTER 1

The Native Americans

Painter Frederic Remington's version of "Custer's Last Stand."

Missionary Intelligence.

Some interesting items of Missionary intelligence communicated at a public meeting in Boston, on Sunday last, are reported in the *Traveller:* Rev. Mr. Treat, one of the Secretaries of the A. B. C. F. M. was present, and introduced the exercises by a very valuable statement respecting the present number, condition and prospects of the North American Indians; in part the result of his own observations during a recent extended tour among the different tribes. It appears that the whole number of Indians within the bounds of the United States is supposed to be 400,000. More than one-half of these, from their wandering habits and from other causes, are in circumstances which, for the present, discourage and repel the efforts of missionaries in their behalf. Our different Societies, however, have their eye upon them constantly: and as soon as a door is fairly opened, they are ready to enter.

Among the Indians who occupy the Indian Territory, or who reside within our States and Territories, missions are prosecuted with encouragement and success. If we commence with the Choctaws on Red River, and travel north through the Creeks, Cherokees, &c., &c., till we come to the Sioux and Ojibwas of Minnesota, and then survey the various tribes in Wisconsin, Michigan, New-York, &c., we shall find that there are about one hundred ordained preachers of the Gospel, who devote their time and strength to our red brethren. A few of these preachers are themselves Indians; and Mr. Treat bore his testimony, from personal observation, to the earnestness and propriety with which some of them deliver their message. A distinct and emphatic reference was made to certain heresies which have prevailed in this country in regard to the Indians. It was formerly said, that but little fruit could be expected from any attempt to give them the Gospel. But now we have the gratifying fact, ascertained after much careful inquiry, that the mission churches among the Indians have (including a few whites and negros,) about 10,000 communicants! And though it may be true that some of these have been received to Christian fellowship on lower evidence than we should approve, it is also true that all have been admitted in accordance with the principles and practice of the denominations to which they belong. And Mr. Treat mentioned one fact of a very significant character. To the church among the Tuscaroras forty-five persons have been admitted since Jan. 1, 1852. And this in a population of three hundred souls!

But it has been said that the Indians, even if they become converted men and women, cannot be *civilized.* Facts were stated to show the groundlessness of this allegation. Mr. Treat has visited several tribes of Indians, and he declared from his personal knowledge that the red man could be civilized. They are giving up the hunter's life, and becoming agriculturists. Many of them have comfortable houses, good farms, strong fences, horses and cattle, gardens, &c., &c. One of the Tuscaroras raises about 2,000 bushels of grain annually. One of the Cherokees is said to own 1,500 head of cattle. In the Choctaw country, farms may be seen that would greatly surprise any New Englander. The government of the Cherokees, modeled after our own, and successfully administered, was adduced as additional evidence that the Indians can be civilized.

And it has been also said that, do what we may for the Indian race, *it must waste away;* it is "doomed," and nothing can save it. Mr. Treat remarked that if the people in this country settle down in this belief, our forebodings will doubtless be verified; for in this way we insure the catastrophe. Still the history of the Indians shows that there is no necessity for such a result. Whenever and wherever they are in favorable circumstances they *increase.* At this very moment the Choctaws, Creeks and Cherokees, three large and interesting tribes, are advancing in population; and this, notwithstanding all which, they have suffered from the white man within a few years. The Senecas, also, though surrounded by a white population, are manifestly increasing.

On the whole, we are much gratified with the picture given us of Indian missions. There is hope for the red man. Let us not forget his claims upon us. Our Missionary Societies should continue their efforts, with the cordial support of the churches, that so a remnant may be saved.

April 8, 1852

The Red Man—What Shall be Done With Him?

Correspondence of the New-York Times.

GREAT SALT LAKE CITY, Friday, June 2, 1865.

What to do with the red men is still a problem which, it appears, cannot be satisfactorily solved. For this Spring there seems to be as much chance of difficulties with them, all around, as ever. We hear of Indian troubles from every quarter nearly. They are rampageous, in a limited way, and sometimes not so limited in California, Nevada, Arizona, New-Mexico, Colorado, Nebraska, Minnesota, Montana, Oregon and Utah, and probably will soon be, if they are not, in Dakotah.

In all the States and Territories afflicted, except this, the sentiment of extermination grows stronger and stronger, and it is generally conceded, whateve may be the ground of quarrel, that there will be no "good peace," sure and certain, while there is a red skin above ground. How far this may be true I shall not determine, but certainly experience goes in favor of that idea largely.

The original cause of the multitudinous quarrels and difficulties it is not easy to learn, but nobody is, by any means, anxious to bear the blame. There are a few notions and habits connected with Mr. Indian which are not exactly favorable to good neighborship with the whites. For instance, with the Indians, stealing is an accomplishment, and they will do it from friend or foe when they get a little "mad," particularly when their larders runs low.

If an Indian gets seriously "huffed" with his "father," the agent, which is very apt to be the case, and not always the redskin's fault, he must have his revenge, not particularly upon the agent always, as his native sagacity teaches him of a sort of sacredness of official character, or at least of greater danger in attacking the representatives of the government than private individuals, consequently innocent persons have to suffer frequently from the imprudence of officials. The Indian's wrath is poured out, with indiscriminate discrimination, upon the passing emigrant, or the industrious settler, and thus a general character is given to a murderous struggle which commenced with a few. Further, as in white society, so among Indians, there are a number of reckless fellows who occasionally will not be controlled by the more prudent, but will have their own wild way, and go on a "bust" sometimes. They will do a little stealing, get saucy, impudent, presuming, and when very "mad" will be cruel and kill. The whites, irritated and provoked, even when the Indians do not murder, but steal only, shoot at the marauders, if a sight can be obtained of them. Then the matter becomes serious, and the band or tribe take it up, for the Indians believe in the doctrine of Moses—blood for blood.

A few weeks ago there were some difficulties with the Aborigines in San Pete Valley, a hundred miles south, and two or three whites were killed or seriously hurt. The Indians in that case are reported to have come up from New-Mexico, pressed out of that Territory by the troops.

Recently a still more serious affair occurred in that county. The perpetrators are said to be a roving band from Grand River, Colorado, with three or four Utah Indians mixed up with them. About a dozen persons have been killed in that county this Spring, besides helping themselves liberally to stock.

The most recent difficulty is one of the most serious that has occurred in the Territory, at least very lately. The beginning was on the 25th ult., in the murder of JEUS LARSEN, who was tending a large flock of sheep, four miles north of Fairview. He was shot through the neck and right side.

Next morning, twelve miles north of Fairview, a whole family, occupying a temporary shanty, were swept away. They are as follows: JOHN GIVEN, Sen., aged 45; ELIZA GIVEN, 40; JOHN GIVEN, Jr., 19; MARY, 9; ANNA, 5, and MARTHA, 3.

The father, mother and son were first dispatched. The father was shot through the chest and tomahawked in the forehead; the mother was shot through the head, under the ear, and tomahawked in the mouth; the son was shot through the chest, left arm and left leg, and a finger of the left hand was almost cut off.

Two men—CHARLES BROWN and CHARLES W. LEAH—were sleeping in a wagon-box in the shanty. They say that four Indians did the first shooting. After the three elder GIVENS were shot, BROWN and LEAH leaped out of the wagon-box, when the Indians ran away. The two men then ran to the willows, and were shot at nearly a score times by the Indians. Being so hard pressed, BROWN and LEAH could not return to the shanty. The Indians entered it again, and then it was that Mrs. GIVEN was finished with the tomahawk. The little girls, crying over their parents, were next dispatched—one being shot in the face, another in the lower part of the body, and the third tomahawked in the face.

The Indians took the flour, axes, guns, &c., from the shanty, and drove off 100 to 200 head of stock, horses and the best of the cattle.

After the Indians had withdrawn, one of the men in the willows ran to the nearest settlement and gave the alarm. BLACK HAWK, it was reported, had been in the neighborhood of the catastrophe the night before.

On the 25th DAVID H. JONES, while hunting horses three miles northwest of Fairview, was killed by the savages.

Last Monday night ninety of the settlers started after the Indians.

Operations in connection with the Pacific Railroad are being commenced. Mr. S. B. REED, Division Superintendent of this portion of the line, has arrived from the east, with assistants SCHIMONOSKY and BISSELL. On Monday they start northward to make a more thorough examination of the Weber Valley, the head of Echo Cañon, and on to the South Pass. Afterward they go southward for Spanish Fork Cañon and Uintah Valley to survey that route.

A party are also going westward by Black Rock, and through Tooele and Rush Valleys, to examine the country in that direction.

Gen. B. M. HUGHES and party left the city, several days since, for Hobble Creek or Spanish Fork Cañon, Uintah Valley, and on to Denver, exploring for a wagon-road. Lieut.-Col. JOHN has also started on the same route, with 180 of the Third Infantry C. V., for escort and assistance to Mr. HUGHES and party. If the party are successful, the overland mail will soon pass over that route in preference to the Bridger route.

June 30, 1865

THE INDIANS.—There is one war in this country to which time brings no surcease—that of the white man against the red man. It has been waged for over two centuries—the Indian always defeated, yet never defeated. From all their vast hunting-grounds on the Atlantic seaboard, in the Valley of the Ohio, and in the Valley of the Mississippi, they have been driven, till now, in their last refuge on the great plains and under the shadow of the Rocky Mountains, they find themselves confronted and surrounded by the old white enemy, and no possibility of further retreat. The Indians are at bay. They are bewildered; they are helpless. It is not in their nature to adopt the habits and follow the pursuits of the white man; and it is not in the nature or destiny of the white man to permit the Indian to follow his hereditary habitudes. Feuds and fights, cruelties and hatreds, wretchedness and despair, exile and extermination, constitute the present, as the past, history of the poor Indian, in presence of the white settler.

It is very hard upon the Indian. And the worst of the matter is that we can see no good way in which his career is likely to close. There is to be a great convention of all the tribes, wild and civilized, held next month at Fort Smith; and our government is to be represented by an excellent delegation, among whom will be the distinguished descendant of RED JACKET, Col. PARKER, of Gen. GRANT'S staff. But what is it possible for the convention to do? It can neither guarantee the Indians their lands, their rights, or their lives. They are scattered over an immense surface, and will everywhere be in enmity with the contiguous settlers, and we fear it is impossible to get them all to live together. If they would take it into their heads to move southward to *Mexico*, where the bulk of their race is now concentrated, it would be a happy thing for them and us.

August 12, 1865

The Policy of Extermination as Applied to the Indians.

A gentleman somewhat known in connection with adventure and speculation in our Southwestern Territories, has written a letter to Gen. GRANT, proposing to take command of a regiment of cavalry, for the purpose of punishing the Apaches of Arizona. He says:

"I desire neither rank nor pay, only the absolute handling of this force without restrictions. I will then undertake to clear Arizona of Apaches in twelve months."

We do not suppose that Gen. GRANT would ever entertain the least thought of acceding to any such proposition, any more than we doubt that, if he did, it would lead to the extermination of this wildest of all the wild tribes of the Western plains.

It is now unquestionably in the power of our Government to exterminate not only one obnoxious tribe of Indians, but the whole Indian race, from the savage Sioux of the far North to the untamable Apaches of the far South, from the wild Arapajos and Cheyennes of the central regions to the wretched tribes on the Pacific coast. And such a policy of extermination would receive almost universal support from the white settlers on the Western plains. At all events, those living within range of the Sioux would like *that* tribe extirpated, those living within range of the Apaches would like *that* tribe extirpated, those living within range of the Cheyennes would like *their* extermination, and so on. The white settlements that are now planted here and there all over the Plains, and on both sides of the Rocky Mountains, would eagerly cooperate with the military forces in the work of extermination; and these settlements are so numerous, so widely scattered, and so advantageously located, that they could operate to most sanguinary purpose. They would not be rendered the more averse to do so, in that they would acquire the coveted lands and the "plunder" of the Indians, and would secure the expenditure among themselves of vast sums of Government money. Under this policy, the red man, once so powerful and populous on this continent, now so lean and circumscribed in numbers and domain, would quickly be relegated forever to those shadowy "hunting grounds" where the "pale face" will never dispute his possession.

We think, however, that even those most covetous of the Indian's heritage might be satisfied at the rate with which he is exchanging the soil for the sod. Looking at the restricted and disconnected territorial allotments now left him, at the tenure of his possession and the conditions of his toleration—looking at the rapidity and the steadiness with which his range is ever being limited, and the swiftness with which he himself is fading away, one might think that such furious desires of extermination as are expressed in the aforementioned letter to Gen. GRANT, and which, as we have said, are so widely entertained in the far West, might well be somewhat modified.

As for the Indian "outrages" of which we are continually hearing and of which the Apaches in Arizona furnish the latest illustration, it must be said that we only hear one side. The Apaches print no newspapers, and have no means of communicating to the world a knowledge of the outrages committed upon *them.* But it is on record, as the experience of all those most familiar with Indian affairs, that the aggressors in nearly all quarrels, and the beginners of nearly all outrages, are the whites—rarely indeed the Indians.

Our permanent policy toward the Indians should be, in brief and comprehensive terms, to exercise a steady pressure and influence toward securing the *aggregation* of the entire race and all the scattered tribes as far as possible; and secondly, to appoint military officers to the various Western posts who have a deep appreciation of justice, and who will be more anxious to avoid a quarrel than secure a fight; in the meantime spurning with proper contempt, as inhuman and unchristian, any and all propositions for exterminating any tribe in twelve months "with a regiment of cavalry."

May 16, 1866

The Civilization of the Indians.

In affirming that the civilization of the Indian is the easiest and cheapest, as well as the only honorable way of securing peace, the *Nation* brings forward a fact that is too generally ignored. It shows that whenever practical, civilized settlement has been made possible, the Indians have been ready for it, while the whites have repressed and excluded them—as in denying them citizenship in Minnesota, in 1857 8. That is the key to this difficulty, as to all others where dealing with human nature is concerned. Like begets like. Treat the Indians as irresponsible savages and they will realize the character. Treat them like men, subject to the laws, the responsibilities and the privileges of that condition and they will soon learn to accept it. The most hostile and lawless of the tribes are now under treaty again. When they settle on their reservations, as they will, to get the subsidies, let them be brought under the State laws if they are in Kansas, or the Federal Government if in a Territory. And when the individuals are held responsible, their tribal organizations will soon disintegrate, personal tastes and associations will take hold upon them, their fixed habitation will, as heretofore, stimulate them to agriculture; education can be sent among them, and the Indian may be made a citizen.

November 3, 1867

The Duty of the Churches to the Indians.

The letter addressed by the Secretary of the Interior to the Indian Convention which recently adjourned in this City, contained some very frank statements which our religious bodies would do well to consider. Mr. COX urged very justly that President GRANT and the public expected, on the appointment of a commission of Friends and well-known philanthropic citizens upon Indian affairs, that the powerful religious bodies of the nation would co-operate with these Commissioners and the Government to educate and Christianize the Indians; but that thus far the churches had done almost nothing, being so much occupied with the foreign heathen that they have almost utterly neglected these wards of the nation. The Secretary and the President esteemed these private philanthropic and religious efforts so highly, that the former asserted the alternative before the public to be, a vigorous attempt by the Christianity of the country to educate the Indian, or to witness the consummation of extermination in a general plundering and massacre of the wild tribes. Those officials both attach the highest value to the combined efforts of the churches and humane bodies of the nation in rendering justice to our own barbarians.

We may have overlooked the report, but we cannot recall in any missionary convention or church synod for the year, any important action originating new missionary and Christianizing efforts in harmony with the new Commission for our heathen at home. Action enough there has been about the Zulus, the Sandwich Islanders, and the Hindoos: but the American Indians, for whose miseries and crimes we are so largely responsible, have been mainly forgotten.

And yet any ordinary mission operation, such as might be started in South Africa or India, is not what is wanted. A solitary missionary, with Bible and prayer-book, going among the Comanches or the Sioux, can accomplish little in solving this great problem. Sufficient means should be raised to enable a missionary or band of missionaries to purchase and convey to some reservation all the material necessary for simple agriculture. There the mission should commence—as the old Spanish Catholic missions did, which were so successful with the Indians of the Pacific Coast—with teaching the rudiments of civilization, without which even the truths of Christianity will be of little avail to save them in this world. Beginning thus with instructions in planting crops, using farming tools, and tilling the ground, accompanying those lessons with the teachings of morality and the Divine truths of the Christian faith, material advancement will go hand in hand with moral and spiritual conversion. The wild nomadic Indian will get the rudiments of the great lesson of civilization; he will learn to labor for a result beyond the immediate moment; he will acquire the art of providing for his wants when game has disappeared or is scarce; he will possess a settled home and become accustomed to a steady occupation. This wise plan has already been tried with eminent success by an experienced Indian superintendent, Mr. DENT, Gen. GRANT'S brother-in-law, among the wild tribes near the Lower Colorado. He has proved what an advance in civilization can be made among the most nomadic tribes by a system of reservations, and the teaching of agriculture.

But the Church missionary would have an advantage in these labors possessed by no agent of the Government. The heart of every wild people is especially open to the sentiment of religion. A missionary comes as if from above. He has no selfish aims. He teaches the truth for all men and all times. No doubt agriculture will be easier learnt from a religious teacher, and religion from one who has the physical means to raise the neophyte above starvation and barbarism. No other system will at all succeed with aborigines, except this combination. And our Indian Commission, with its members, so influential among the churches, is especially the one to recommend and carry out such a suggestion. As fast as the wandering tribes are placed on reservations, the religious bodies should establish agricultural missions, with chapels, schools, plows and seed, with farmer missionaries, and teachers of the Bible who know how to plant corn. We fear, however, that the routine of the churches is too much fixed in the old method to allow of this innovation. And we dread that the fatal apathy of our religious bodies, wherever the heathen at home are concerned, may disappoint all hopes of any widespread missionary work for the Indians.

May 30, 1870

THE GREAT CHIEF.

Red Cloud Meets His White Brethren at Cooper Institute.

A "Big Talk" There in the Interests of Peace.

VAST GATHERING OF PALE FACES.

Speeches by the Indian Chieftains and Their White Advocates.

The Wrongs of the Past and Redress for the Future.

Notwithstanding the short notice which was given of the public reception of RED CLOUD and his companions by the Indian Commission, at the Cooper Institute, yesterday, and the unseasonable hour at which it occurred, such was the curiosity of the people to see the most distinguished living representatives of the race who originally possessed the American soil, and whose history contains so much romantic interest, albeit little to the credit of the white race, that long before the hour of 12 an eager throng besieged the entrance to the great hall and waited impatiently for the opening of the doors. At first only those who had taken the trouble to secure tickets, which were distributed gratis at the reading-room of the Institute and at the St. Nicholas Hotel, were admitted, but as soon as those persons had been accommodated with seats the crowd was let in and rapidly filled up every available foot of space. The platform had been considerably enlarged for the occasion, and many invited persons took seats thereon. Promptly at 12 several of the large carriages of the St. Nicholas drove up, and the twenty chiefs and braves, with Gen. SMITH, Mr. BEAUVAIS, and the interpreters, alighted and made their way through the throng in the passage-way to the committee-room. Their appearance soon after on the stand was the signal for tumultuous applause, and some little confusion occurred as to the assignment of the seat of honor to RED CLOUD, that dignified personage apparently refusing to accept the position proffered in the middle of the stage, with the other braves on either side of him. He speedily carried his point with the indulgent Commissioner, and took a seat near the edge of the platform, at one end of the semicircle. On the stand about him were Mr. PETER COOPER and other members of the Indian Commission, Judge DALY, Hon. WILSON G. HUNT, Rev. Drs. BELLOWS and FROTHINGHAM, and many other prominent citizens.

The meeting was opened with prayer by Rev. Dr. HOWARD CROSBY, at the close of which Mr. COOPER, the Chairman, spoke as follows:

PETER COOPER'S ADDRESS.

It is, my friends, but a few weeks since the country was filled with reports of an inevitable Indian war. All expectation of peace was abolished by the authorities in Washington; troops were hurried forward to the frontier, and the minds of tax-payers already familiar with the odious income tax were prepared to expect fresh burdens. During those dark hours a suggestion was made to the Secretary of the Interior and the Commissioner of Indian Affairs that, if they would invite the hostile Indians to come to Washington, and discuss their grievances, they might all be redressed on the principles of justice, and peace might thereby be preserved. In reply, it was said that those Indians would not trust themselves in our power. Within twenty-four hours from this conversation, official assurances were received that RED CLOUD and his principal Chiefs would come on this errand. Today we have before us the very men of whom but yesterday we were assured that nothing could be expected but merciless war! In the interviews between RED CLOUD and the Secretary of the Interior, the Indian has shown himself equal to the occasion, and his speeches must have given our honored Secretary, in common with every honest man in the country, a painful illustration that "thrice is he armed who "Thrice is he armed who hath his quarrel just." We have recognized in solemn treaties the Indians' claims to the hunting grounds upon which they have from time immemorial enjoyed the rights of "life, liberty and the pursuit of happiness;" and it is too late to deny his title now, while we profess to be a Christian nation. As banditti or free-booters we could claim the right of might, but on no other ground. The Indians' land yields them a plentiful supply for all their necessities. If we want their land, let us, as honest men, give to them a fair consideration for it—one that will yield them a full supply for all their wants. They do not question the right of eminent domain vested in our Government; but where is the white man on earth who can substantiate a just legal claim to one foot of the Indians' land except by voluntary cession from them? If you refuse to pay this equivalent amicably, and seek by force to wrest it from them, you will inaugurate a war that will cost so many millions that the interest alone will more than equal the price at which you can now purchase it, and find too late that honesty would have been the best policy. If members of Congress cannot understand this; if, while making to their constituents professions of economy, they take a course which can only end in war, and subject the industry of the country to millions of unnecessary taxation, the people will see the necessity of sending men to Congress who will in good faith carry out the policy of Gen. GRANT, the announcement of which was indorsed by his election to the Presidential chair. We, the people, want peace—not only with those who were lately in rebellion, but with all mankind.

REV. DR. CROSBY'S ADDRESS.

Rev. Dr. CROSBY then, as the mouthpiece of the Indian Commission and of the assembled multitude, addressed RED CLOUD, the great Chief of the Sioux nation, as follows:

"There are good white men and bad white men. There are also good Indians and bad Indians. The good white men want justice to be done to the Indians. The Great Father thinks with the good men, and we hold up the hands of the Great Father. The Indian has no newspaper; we want to be his newspaper and tell the Indian's story. If the Indian fight, then our power to help him is gone. Nobody will listen to us. We are the tree to shield the Indian. Do not cut the tree down. If there are troubles, settle them by talk, and not with guns. War means hatred, we want to make love. Bad men say the Indian is cruel and won't be peaceable. Show the bad men to be liars and so strengthen the hands of the Indians' friends. White men are happy and rich because they work. If the Indian work he will be rich and happy too. If he fight he will be poor and wretched. As you, RED CLOUD, said, the Great Spirit made us all and put us both on this land, therefore let us be brothers. And now our people want to hear RED CLOUD and RED DOG and others speak for the Indians."

The foregoing speech was delivered, or rather read in detached sentences and translated to the Chief by the interpreter. At its conclusion RED CLOUD arose and faced the audience, drawing his blanket around him majestically. He was greeted with an outburst of applause and waving of handkerchiefs. As soon as the tumult had subsided, he began on a somewhat high key and with a rapid utterance, his "talk." At the end of each sentence he paused, and stood calmly surveying the audience, while the interpreter explained his words to Dr. CROSBY, and he again in stentorian tones, gave them to the immense audience. Almost every sentence was received with loud applause by the assembly, while the other chieftains and warriors signified their assent by a guttural "Ugh."

RED CLOUD'S SPEECH.

My Brothers and my Friends who are before me today: God Almighty has made us all, and He is here to hear what I have to say to you today. The Great Spirit made us both. He gave us lands and He gave you lands. You came here and we received you as brothers. When the Almighty made you, He made you all white and clothed you. When He made us He made us with red skins and poor. When you first came we were very many and you were few. Now you are many and we are few. You do not know who appears before you to speak. He is a representative of the original American race, the first people of this continent. We are good, and not bad. The reports which you get about us are all on one side. You hear of us only as murderers and thieves. We are not so. If we had more lands to give to you we would give them, but we have no more. We are driven into a very little island, and we want you, our dear friends, to help us with the Government of the United States. The Great Spirit made us poor and ignorant. He made you rich and wise and skillful in things which we know nothing about. The good Father made you to eat tame game and us to eat wild game. Ask any one who has gone through to California. They will tell you we have treated them well. You have children. We, too, have children, and we wish to bring them up well. We ask you to help us do it. At the mouth of Horse Creek, in 1852, the Great Father made a treaty with us. We agreed to let him pass through our territory unharmed for fifty-five years. We kept our word. We committed no murders, no depredations, until the troops came there. When the troops were sent there trouble and disturbance arose. Since that time there have been various goods sent from time to time to us, but only once did they reach us, and soon the Great Father took away the only good man he had sent us, Col. FITZPATRICK. The Great Father said we must go to farming, and some of our men went to farming near Fort Laramie, and were treated very badly indeed. We came to Washington to see our Great Father that peace might be continued. The Great Father that made us both wishes peace to be kept; we want to keep peace. Will you help us? In 1868 men came out and brought papers. We could not read them, and they did not tell us truly what was in them. We thought the treaty was to remove the forts, and that we should then cease from fighting. But they wanted to send us traders on the Missouri. We did not want to go on the Missouri, but wanted traders where we were. When I reached Washington the Great Father explained to me what the treaty was, and showed me that the interpreters had deceived me. All I want is right and justice. I have tried to get from the Great Father what is right and just. I have not altogether succeeded. I want you to help me to get what is right and just. I represent the whole Sioux nation, and they will be bound by what I say. I am no SPOTTED TAIL, to say one thing one day and be bought for a pin the next. Look at me. I am poor and naked, but I am the Chief of the nation. We do not want riches, but we want to train our children right. Riches would do us no good. We could not take them with us to the other world. We do not want riches, we want peace and love.

The riches that we have in this world, Secretary COX said truly, we cannot take with us to the next world. Then I wish to know why Commissioners are sent out to us who do nothing but rob us and get the riches of this world away from us! I was brought up among the traders, and those who came out there in the early times treated me well and I had a good time with them. They taught us to wear clothes and to use tobacco and ammunition. But, by and by, the Great Father sent out a different kind of men; men who cheated and drank whisky; men who were so bad that the Great Father could not keep them at home and so sent them out there. I have sent a great many words to the Great Father but they never reached him. They were drowned on the way, and I was afraid the words I spoke lately to the Great Father would not reach you, so I came to speak to you myself; and now I am going away to my home. I want to have men sent out to my people whom we know and can trust. I am glad I have come here. You belong in the East and I belong in the West, and I am glad I have come here and that we could understand one another. I am very much obliged to you for listening to me. I go home this afternoon. I hope you will think of what I have said to you. I bid you all an affectionate farewell.

When the prolonged applause with which this speech was greeted had subsided, Dr. CROSBY said: "You have heard from the great warrior Chief RED CLOUD, I now introduce to you the orator Chief RED DOG."

RED DOG'S SPEECH.

RED DOG, who is a short, thick-set Indian, with a broad, open face and a pair of cogged wheels in his swarthy ears, spoke briefly as follows:

I have but a few words to say to you, my friends. When the good Great Spirit raised us, he raised us with good men for counsels, and he raised you with good men for counsels. But yours are all the time getting bad, while ours remain good. These are my young men. I am their Chief. Look among them and see if you can find any among them who are rich. They are all poor because they are all honest. Whenever I call my young men together in counsel, they all listen to what I say. Now you have come together in counsel, I want you and your children to listen to what I say. When the Great Father first sent out men to our people, I was poor and thin; now I am large and stout and fat. It is because so many liars have been sent out there, and I have been stuffed full with their lies. I know all of you to be men of sense and men of respect, and I therefore ask you confidently to see that when men are sent out to our country, they shall be right men and just men, and will not do us harm. I don't want any more men sent out there who are so poor that they think only of filling their pockets. We want those who will help to protect us on our reservations, and save us from those who are viciously disposed toward us.

REV. DR. WASHBURNE'S REMARKS.

After the speech of RED DOG was finished, Rev. Dr. WASHBURNE, of the Indian Commission, was introduced. He asked what response this vast assembly of the people of New-York would make to the appeal of these Indian Chiefs. They had not been invited there to gratify an idle curiosity by allowing an opportunity of gazing upon these dusky children of the forest in their blankets and their moccasins. They were a savage race, unlike us in their modes of life, but who that had heard their words could doubt that they were men endowed with all the noblest feelings of manhood? They had a strong love for their families and for their people. The question of the treatment of these people was now pressing upon our attention. Hitherto it had seemed somewhat remote, but now it is the great question before the country. Soon all the lands of the country will be occupied to a greater or less extent by white men. How, then, shall the original possessors of the soil be treated? Shall we wage against them a war of extermination? [Loud cries of "No, no."] Let the interpreters tell these Chiefs, continued the speaker, that this great audience, representing the people of New-York and the people of the country, say that there shall be no war of extermination, but justice and peace. [This sentiment was interpreted to the Chiefs, and received with expressions of satisfaction.]

The speaker continued by saying that the Indian question was a difficult one. He did not understand all its diplomatic complications, but one policy there was which overrides all others, and which is simple and easy in its application. That policy must be applied to this question. It was the policy of simple Christian justice. Treat these men as men. Give those who can be induced to accept the customs and habits of civilized life, reservations of land, and afford them the means of cultivating the industrial arts and of acquiring education, and on those reservations and in the enjoyment of the institutions thus afforded them, give them the protection of the Government. If there are any who have not the capacity for civilization, still give them justice and fair treatment. The policy of the Government heretofore had been the policy of neglect, followed by the policy of violence. The Indians have been left in the hands of land sharks to be deceived, cheated and abused, and when they, in their rude conceptions of justice, have resorted to warfare as their only means of redress, they have been treated with severity. Under the present Administration there had been hope of better days for the Indians. President GRANT had foreshadowed a policy of peace and justice, and in his efforts and wishes he had been seconded by his large-hearted Secretary of the Interior; but the hopes which had been raised in the friends of the red man had been of late somewhat overclouded. This meeting was called to demand the inauguration of a new policy, and the Administration should be made to feel that the people of this country were beginning to take an interest in this subject, and that they were determined that there should be a new condition of things.

REV. DR. BELLOWS' REMARKS.

Rev. Dr. BELLOWS was next introduced, and said that the red men showed that they had the same sensibilities and the same feelings which white men have. They know when they are wronged, and demand redress. If, when the scales were held between the justice of the Indian and the white man, it preponderated in favor of the former, where was the value of our Christianity and our civilization? The Indian should be made to understand that the Pacific Railroad, which he regards as his enemy, is indeed his greatest friend, since it brings his treatment nearer to the knowledge of the people, so that they will know what is done among them and require the Government to correct abuses. Tell them we cannot take the ruffians who cheat them by the throat, but we can take the Government by the throat and force it to do right, and not allow the Indian to be trodden upon any longer. The great people of New-York should say to the Government that there must be no more of this cheating and lying. Let us have no more Indian orators stuffed with our lies. Secretary COX meant well by the Indians. He told them just what he could do, and what he could not do, and that was far better than making promises which he could not fulfill. The Indians were a doomed race. Before our children's children can look upon their faces, in all probability they will have faded from the earth. Their history has afforded the material for the only national novels which we have, and for some of our best poetry. While this remnant of the race remains they should surely receive fair treatment at our hands. Let their affairs be no longer hid away in a secret bureau, but let us know everything that is done, and require that nothing shall be done but justice and right.

At this point Mr. COOPER announced that before the letter which he had written to Washington, offering to furnish the seventeen horses asked for by RED CLOUD, had reached its destination, the authorities there had already given orders that the horses be supplied at the expense of the Government.

JUDGE DALY'S REMARKS.

Judge DALY was then introduced, and remarked that this was the first time there had been so great a council of white men and Indians, and it was well that it should take place at the great Metropolis of the country. The only thing he had to say now was to call to mind a remark which he heard made by one who had no superior in his knowledge of the Indian character and of our relations with that race—old Gov. HOUSTON, of Texas. He said that in every instance of a contest with the Indians which he had ever known the white men had been the aggressor. Judge DALY closed his remarks with an appeal that the experience of the past fifty years be no longer continued in the treatment of these men, and that if ever again their chiefs should meet the people of New-York in a great council, it should be to express their gratitude, and not to set forth their grievances.

At the close of Judge DALY's remarks, Dr. GRENVILLE M. WEEKS explained the leading points of the plan for the settlement of the Indian question which has been prepared on behalf of the Indian Commission, and his remarks were interpreted to the Chiefs. The meeting then adjourned, and there was a general rush for the stage, the people being eager to get a nearer view of the distinguished aborigines. With considerable difficulty the crowd was induced to leave the hall in the usual way, and, in the meantime, the Indians were quickly and rapidly conducted to their carriages and driven away, amid the shouts and hurrahs of the populace.

Late in the afternoon they took their departure from the City, and will make no further stop until they reach Omaha, where the seventeen horses will await them to take them to their homes on the distant plains.

THE PLAN OF THE INDIAN COMMISSION.

The Committee of the Indian Commission, appointed at the recent Convention in this City to prepare a plan for the settlement of all our Indian difficulties, consisting of Peter Cooper, Grenville M. Weeks, Cornelius Du Bois, Lewis Masquerier, Rev. E. L. Janes, Col. C. N. Vann and S. W. Perryman, the two latter belonging to the Cherokee delegation, have prepared a small pamphlet, addressed to Hon. J. D. Cox, Secretary of the Interior, in which, after reciting the facts with regard to the past treatment of the Indians and their character and capabilities, they submit the outline of their plan in twenty-six articles, "drawn up for presentation to Congress." The material features of the scheme are: The concentration of all the tribes west and north of the Mississippi on not less than four nor more than seven reservations, allowing about eighty acres of good land to each individual, due regard being had to the relations of the different tribes in locating them; no white men to be allowed to locate on the reservation except those who are sent to benefit the Indians, and those to be married and of "well attested philanthropic proclivities" and "Christian character," and when railroads are built through these lands, only such stations to be established as are necessary, and no white settlements to be allowed near them; men to be sent and money employed to civilize the Indians and built up industries and educational institutions among them; and all necessary means to be used to protect the reservations and their occupants in all the rights, titles and interest guaranteed to them under this arrangment.

June 17, 1870

The Indian Chiefs at Washington.

To the Editor of the New-York Times:

Spotted Tail, the famous Ogalalla Chief, and other Sioux Chiefs and warriors, are expected in this City Monday night from Philadelphia and Washington. These frequent visits of the red man to the capitals of the East are probably not understood in their true significance by the majority of our people. It will, perhaps, enlighten the minds of many to state that the untutored children of the Western prairies have no means of forming an adequate conception of the numbers and strength of the white nation, and it is found to be the surest way of disposing the tribes to abandon the nomadic life and become "like the whites" (in all but their vices, let us hope!) to allow them to see with their own eyes the power and resources of the nation. The Western Indian hears of the East as the home of the white man, but imagines that it is as thinly settled as the Western States and Territories, and the few chiefs who have visited our Eastern cities, and told him of the density of their population, are distrusted, and accused of having accepted bribes. So large delegations of principal chiefs are now moved East, from time to time, and their return is found to have a very quieting effect upon their followers, especially where there is discontent and opposition to the course of the Government in regard to locating them on reservations.

Spotted Tail was formerly a refractory chief, but is now a good friend to the whites, and very anxious that his fellow-chiefs should see with their own eyes what he has before beheld in the East, and so confirm his words to his people.

The delegation will remain in town only two or three days, and it would be well if they could be shown the resources of the Empire City for peace and war, but especially for the cultivation of peace and the protection of the people. Their agent is Dr. R. Risley. About twenty persons compose the party. The Ogalialas and Brules represented by them are Sioux tribes living in Dakota. The Episcopal Church is establishing missions among them. K.

NEW-YORK, Saturday, Aug. 3, 1872.

August 4, 1872

REFORMING THE INDIANS.

INDIAN AGENCY, CHIPPEWA, WHITE EARTH, Wednesday, July 16, 1873.

From a Special Correspondent.

On my way back from the steam saw-mill worked by the Indians, spoken of in my last, I was struck with the practical way the Indian agents have here of giving lessons to their wild scholars. In one place we passed an Indian garden where the beets had been sown too thickly. The interpreter jumped out, and with his own hands weeded out the bed, telling the Indians that they would get nothing if they left them so thick. They immediately fell to work, and I suppose will understand, after this, how a beet-bed should be kept.

Mrs. Smith, the wife of the Indian Commissioner, was in one house, to show them how to make soap; in another, to teach how the rushes must be cut for the new trade of mat-weaving they are starting. We examined, also, the willow-patches, and my companions were already instructing the savages as to the kind of baskets which must be made. In one place one of the teachers had the day before gone into an Indian's inclosure and milked his cow for him, finding he knew nothing about it. These are "object lessons" not easily forgotten.

We met an Indian driving a team of oxen. "There," said my friend, "you see that fellow. Six months ago he was only a blanket Indian. He came to me and said he wanted to be something more. I set him at digging post-holes. That's pretty stiff work, you know. He kept at it though and saved enough to buy those good clothes you see, and now he has a team and is making money."

"Ah!" said Mrs. M., but look at poor Katy (she was crossing the field). She will wear that blanket over her head. We cannot get her to the bonnet; but she's a good girl." I fear I did not show enthusiasm enough on the subject of the bonnet and blanket question. It is a sad pity that the first steps to civilization should be the exchange of the graceful costume of the barbarians for our stiff modes. But the missionaries are right. Costume is a symbol of a whole history and condition, whether of barbarism or civilization.

"How do you prevent the use of the fire-water?" I asked of one of the agents.

"We follow up the sellers," he answered. It is a $500 fine you know. We have sent up fourteen or fifteen of the rascals, and one man in Oak Lake was fairly ruined and driven out of the country by the fines. They never get drunk now."

On our return to the village we went over the "Industrial House," where the various trades are to be taught, and where Indian sewing circles will be held, and evening meetings come together. I also visited the school, where sixty or seventy Indian children are taught, and a night school is held for those at work during the day.

These teachers are, I believe, employed by the American Missionary Association. The school was not thoroughly satisfactory. The teachers seemed faithful and conscientious; but there was too much routine. The children understood but little English, and what they need is thorough and active "object teaching."

A well-trained object teacher, here, at a good salary, would do an invaluable work. The only white child in the school did not equal the Indian children in their recitation. The great difficulty with them, the teacher says, is their shyness. Out of doors they are boiling over with fun and frolic. The experiment of the co-education of the sexes—at least, when the pupils are over fourteen years—does not work well here, and will probably be abandoned when more ample buildings are furnished. On one evening of my stay I took tea with the scholars in a large fine dining-room. The room, however, was not at all clean, and did not smell clean. We all had the same fare, which was very good and healthy: a brown-colored wheat bread, molasses, no butter, some preserved fruit and tea without milk. There seems to be an unaccountable want of milk and cream about an establishment which could feed a hundred cows without expense. Indeed, the Indian children ought to be instructed in the care of cows.

The manners of the Indian boys and girls at table were quite as good as those of the same number of white children would be.

In the front of the school is a farm which the lads cultivate, and the potatoes and vegetables looked beautifully in it.

THE TALK OF A CHIEF.

On the Fourth of July a grand festival was held here, and the ladies feasted 800 Indians in a grove. At the close a chief—the White Fisher—made a characteristic speech, which I only heard in fragments, but which would have been well worth preserving. In substance, as reported, or rather interpreted, it was:

"Friends and brothers, this medicine day among the people of our Great Father is the greatest day in the year, except Sunday, which is the greatest of all: the day of rest of the Master of Life. We are asked here to meet Oyemah ('Father,' or Indian Agent,) who comes from the great great Father (Gen. Grant). We are wards of the Great Father, and we would keep the day as our white friends keep it. I have often warned you to follow close in the trail of our Christian friends, and what a day this is for those to begin who still hold all the foolish ways of their fathers. I believe in the coming time. In a few years that great paper, (Declaration of Independence) which is read to-day at every council-fire of all the great white people, will be read by our own children. They will hold it as sacred as the whites. Oh, what a change! Wards of our Great Father, from one day old to ninety years; wards, as old as the Republic, which was born ninety-seven years ago, let us try, my friends, while a little of our fathers' land is left, to show to the world that we are men. Let us try to be rid of the bonds which bind the ward to his guardian, and no longer live, as we do now, without law, but let us soon be able to hold to the Declaration of Independence, and live under it as the whites do, and then we will defend it with our lives, for we know that law belongs to order, and that there is no order without law."

To the Indian Agent Wau-bon-o-quod (White Fisher) turned and said, after some introductory remarks: "Before this country was discovered by the whites we were free and independent. This great land, its lakes and rivers, its game and fish, its corn-fields and woods, were all ours. The red man held it all. Now, like a lake on a hill drained by a sudden outlet, so have our possessions passed away from us, leaving nothing but the dry roots of a marsh—the mark of the lost fields of our forefathers. Be gentle to us then, ye American people, ye have kept us as wards ever since the beginning of your Government, and you have kept us in ignorance till this late day. And now, when you come into the possession of all our fields, you ask for the first time the great question, 'What is to become of the Indians?' Then remember, oh Oyemah, while you respect the great great father and his flag, and report to him what happens every day among his red children, that there is a still greater father, even the good Spirit of all to whom you must give an account of your work, and how you helped the poor placed under your care. To-day we have here a company of Indian soldiers who fought side by side with their white brethren during the rebellion for more than three years. Are these red men not worthy of being American citizens, and of having the right to vote like many others who are more ignorant than they?"

When Major Smith first came here the chiefs demanded the whole of the appropriation which had been made to the Indians of this reserve. Mr. Smith replied that it would come to them in "improvements" and similar ways. They demanded money again, when he replied again decisively that they must either take the money in the mode he selected or not at all. They at once submitted.

It was proposed one day, during my visit, that we should call upon an Indian named Agoss, who had rendered great service to the whites during the massacres of 1862. We passed among the little farms and met the women with their "papooses" in the little wooden cases upon their backs, just as they have always carried them.

By some cottages were curious little wooden structures in the shape of dog-kennels. They were tombs. "Those women where you see at those three graves kept their faces black for weeks after the death." "This fine-looking fellow, who is coming across the field, was one of the worst characters in the whole reservation, a drunken blanket Indian. We consider him now a good man; he comes to church faithfully. He can speak English."

I was introduced to him—and after some talk he said, "I been twenty years in this country and never see such changes as in two year last. The Ogemah, (father) Major Smith, he done everything. We used to be so poor and always cheated. Now see them houses and the cattle, and those fields. Everything be different. Major Smith good man!" Taking him with us we passed through a very rich potato patch and garden of the Indian we would visit, the wild strawberries growing luxuriantly on the edges and under the trees, and entered the cabin.

It was neat and comfortable, with two bedsteads or raised stands for beds, all in one room. Clouds of mosquitoes filled it, and seemed to trouble our host as much as ourselves. We were welcomed cheerily, and we told our host that we had come to hear the story of his saving of the prisoners. He offered us chairs, and an admiring circle sat about on the beds and stools to hear the narrative. I had not fully reckoned on the Indian's love of long stories when I made the request, and I soon found that I was in for a protracted sitting. Agoss began at the beginning, and set forth each detail with dramatic power, and this, with the necessary translation by the interpreter, kept us an unconscionable time.

His service, however, was a most useful one. He was at the north, I think, near Leech Lake, when command came from the great chief, Hole-in-the-Day, to plunder all the whites and murder the heads of the families, and then to go south and join in the universal massacre and plunder which was about to take place through the State. They accordingly stripped the houses in that neighborhood, and killed the cattle, and gathered together a band of white men, unarmed fathers of families, to consult what should be done with them. The majority were for braining them at once, and the Indians were about carrying out this purpose, when this man arose, and, at the risk of his life, made a very skillful and persuasive speech. He told them that they had no certainty that all the whites in the south had been murdered, and that Hole-in-the-Day might not have wished these men to be killed at once, and that the best policy would be to take them as prisoners to the south, and let the chief decide. They had their goods, why should they take their lives? Perhaps the whites would recover and kill them. It was safer to send them off as prisoners. He said, very graphically, that as he made this speech he expected to feel a tomahawk crushing in his skull behind; but he would not look round. This passage appeared especially to touch the impassive squaws who were listening, and two obstreperous papooses were summarily spanked by one of the fathers for interrupting the speaker.

Agoss' appeal, however, succeeded, and the white men were carried captive toward the south, and subsequently released, as it was discovered that the white population was by no means "wiped out."

Agoss himself was never rewarded in any shape or form for his loyalty, which was a great mistake on the part of the authorities.

In listening to this animated speaker, and to the others, and looking at the cabin, these Indians seemed to me quite equal in mental capacities and mode of life to many of the Transylvanian peasantry whose cabins I visited last Summer. Who shall say that Indian reform is a delusion? C. L. B.

August 3, 1873

THE INDIANS.

Seminole Interview with the President.

From the National Intelligencer.

According to an arrangement made by the Commissioner of Indian Affairs, the Florida Indians were yesterday admitted to an interview with the President of the United States. They were accompanied by the Commissioner and Gen. LUTHER BLAKE. The Secretaries of War and the Navy were also present.

Gen BLAKE commenced the business of the interview by remarking that the Seminole Indians had encountered many troubles and difficulties in various ways, and that their object now was to call upon their Great Father, the President, himself, and learn from him what they might expect.

To this the President replied, that he should be happy to hear whatever they might wish to say, and would in return give them any information they might desire.

Col. LEA, Commissioner of Indian Affairs, explained the particulars of what had occurred at the meeting on Thursday, in the course of which he said he had assured the Seminoles that their Great Father, the President, would in like manner with himself listen patiently to all they might desire to communicate. This was the first time BILLY BOWLEGS had been to Washington, or indeed out of Florida.

To the President's inquiry if BILLY BOWLEGS had anything to say, BILLY replied that he had come here to learn from his Great Father the whole truth respecting their affairs. His father was a warrior; so was he; and he came here not to ask for favors, but for justice. He came not to pay a mere visit of compliment, but to seek for justice; and whatever his Great Father decided to give, he would be satisfied with. The white people are his friends and brothers, and it was with these feelings that he asked for justice. He said he had no ill-feeling against the whites; none whatever.

Finding BILLY BOWLEGS somewhat slow in coming to the real subject for consideration, the Commissioner of Indian Affairs suggested that the motive which brought the Indians to see the President was to hear from him his sentiments respecting the removal of Seminoles remaining in Florida to the west of the Mississippi.

Gen. BLAKE remarked that he had often advised the Seminoles that the arrangement with Gen. WORTH was merely temporary, and did not amount to the force of a treaty.

BILLY BOWLEGS said he had not yet mentioned anything respecting the treaty with Gen. WORTH. General WORTH told him he had authority from the President to make a treaty. The General said he had orders to make a friendly treaty; that he had come among the Seminoles to put a stop to bloodshed; and that there should be no more fighting. All the Seminoles (continued Gen. WORTH) that were left in Florida, must gather together, draw a line, and live within it. When the line was run, the Seminoles might live south of it, and could remain in the country. "This," repeated Gen. WORTH, "I came to tell you on the authority of the President. I can do nothing without his authority, and I am telling you the truth." After this, Gen. WORTH said: "We have made a treaty; there is to be no more fighting between us; war is all over; you have now nothing to do but to go and raise your children." Gen. WORTH again stated that he said all this by authority of the President. "He had," he said, "made a treaty of peace with the Seminoles, and if it were ever denied, the Seminoles might call him (Gen. WORTH) a liar forever."

To a question here put by the PRESIDENT asking if this was done in writing or in mere verbal conversation, BILLY replied that Gen. WORTH had a paper before him.

To another question from the PRESIDENT inquiring if BILLY himself had ever signed any paper to this effect, BILLY replied that he had not; he was so glad at the peace that he did not think of such a thing. Gen. WORTH advised him how he and his people should conduct themselves: he told them to go and raise their children, and keep hold of the country; and if they saved the lives of any of the whites who might be shipwrecked on the coast, or should stray into their country, they should be paid for them. In consequence of this, BILLY said that he and his people had conducted themselves accordingly, and, by way of instance, gave four cases in which white persons had been rescued from impending death by starvation and exposure.

One of these was the case of a young man who had drifted to a rock on a piece of wreck, and had been seven days without food or fresh water; another was that of a man who had strayed four days' travel into the Indian country; a third was an insane person whose track betrayed him; and a fourth was a drummer-boy, who having lost himself whilst gunning, and becoming so reduced as to be compelled to feed on cypress leaves, had laid him down to die, but was discovered by BILLY's brother SIMON, who, by judicious management in the very gradual administration of nutriment, restored the boy to life and health. It was such things as this that Gen. WORTH told him to do, and said that he would mention such acts to you, (meaning the President for the time being;) and he obeyed Gen. WORTH in generally righting everything that went wrong. Such was his (BILLY BOWLEGS's) anxiety to remain in the country that he would deliver up for punishment any of his own people who should commit a wrong. He would always obey the whites' orders, and give up malefactors. He loved his home very much; yes, if it were only a little place with a pine stump upon it, he should wish to stay there. He would do anything at all so as to stay. In fine, he would willingly listen to his Great Father, for he had heard of him for a long time, and had wished to see him.

The President answered, that he was happy to see BILLY BOWLEGS and the rest of his red children from Florida. I have, said the President, heard of him long before I saw him here. I know he is a great man among his people, and I am glad to hear that he has done so many good things to the whites. I fell a great regard for all my red children wherever they may be, but for none more than those living in Florida. I have lived many years of my life close by the Seneca Nation, and I am therefore acquainted with Indian habits. I am happy to hear that you come here simply to ask for justice. I am anxious to do everything I can to make your people happy, and to do justice to them as far as the laws and treaties of the country will permit me. Treaties made between the whites and Indians, or between different nations of whites, are *laws*, and all must obey them. Twenty years ago a treaty was made between the whites and the Seminole Indians in Florida. By that treaty the Seminoles granted all the land in Florida to the whites, and agreed to remove west of the Mississippi, and settle by the side of the Creeks. That treaty has never been abrogated or set aside, and no new treaty has ever been made. This treaty, made twenty years ago, is made binding on me and the people of the United States, and on all the Seminole Indians, whether in Florida or west of the Mississippi. Such treaties are always made in writing, and are signed by Indians and the Commissioner, who make them. They are then printed in our books, so that we may look at them and see what we have agreed to do, and also see what the Indians have agreed to do. Such treaties as these have no end. Sometimes our Commissioner or General carrying on a war, makes a truce, or agrees to stop fighting, but that is not a treaty. Such an arrangement or truce as that by which the parties stop fighting and keep apart, is not always made in writing, but by word of mouth, as BILLY says he made the arrangement with Gen. WORTH. We understand that Gen. WORTH made a truce or agreement by which fighting was to be stopped, and the Indians were to go south of a certain line. But this arrangement was not permanent; it was understood as temporary, and that the Indians were to go south of a certain line. But this arrangement was not permanent; it was understood as temporary, and that the Indians there were to go west of the Mississippi.

I am, continued the President, anxious only to do what is for the Indian's good, and for the good of the people of Florida. I never was in Florida; I do not know the people of Florida more than the red men now present, and am as anxious to do justice to one as to the other. But the inhabitants of Florida are increasing and will crowd on the settlements where the Indians live. The people of Florida have a right to ask that I see this treaty performed. What I fear is, that as the whites get on, the Indians and they will get into a fight. Now, if a war should break out again, I fear the Indians would be destroyed. We have prepared a place for them west of the Mississippi. That is where the rest of their brethren have gone, and if these should remain in Florida and die there, their children would want to remain their also; but it is impossible. I know it is a painful thing to remove from the place where one was born and brought up, but we all do it in case of necessity. I have removed from the place where I was born, and never expect to go there again.

If the Indians go West, they will go under the protection of this Government, and I shall be happy to make their journey as comfortable to themselves, their wives and children, as possible. When they get there they will soon find a good country and their old friends, and they will soon feel as if they were at home. There will be no danger of their leaving their children where there will be war with the whites, and I shall do everything in my power to render their new home happy and comfortable.

But I *must* say to them, and they must understand that this treaty which they made in 1832 to go west of the Mississippi *must be performed*; there is no way I can avoid it, and the only way for them to do it for their own benefit is to do it peaceably, quietly; and in that they will have the love and affection of this Government instead of having us for their enemies, and compelling us to go to war with them

We have had a great deal of trouble already to prevent the people of Florida from injuring them. We cannot do this much longer; we cannot keep those cowboys from getting up a fight much longer. The only way to do this is for the Seminoles to go quietly west of the Mississippi. Gen. BLAKE is authorized by me to make arrangements to take them there, and I have no doubt that all he tells them in reference to it is true. If he should tell them anything not true, and it comes to my knowledge, I would let them know it, and would immediately remove him from office.

This is all I have to say, except to return you my very sincere thanks for your kindness to the whites who have strayed in among you, and to express to you the hope that you will feel the importance of what I have said to you. I have told you the truth; and for everything else you wish to know, I refer you to the Commissioner of Indian Affairs, who will provide for all your wants while you remain with us. You are our friends and will be treated as such.

The President here announced that he had concluded, and wished to hear BILLY, or any other Indian who might wish to say anything. The Indians having expressed their desire to say no more at the time, after shaking hands, withdrew, two or three of them appearing by their countenances, to take the remarks of the President rather hard.

September 21, 1852

Indian Hostilities on the Plains—An Emigrant Train Attacked and Destroyed—Four Men Killed, two Men and one Woman Wounded—The Indians in rear of Col. Sumner's Command.

From the St. Louis Republican, 16th inst.

We have just received a letter from Fort Riley, which we publish below, informing us that an emigrant train has been attacked and destroyed by the Cheyenne Indians, about 80 miles west of that post.

It may be recollected that Col. SUMNER left Fort Leavenworth some three or four weeks ago, with six companies of cavalry and two companies of infantry, for the purpose of making war upon the Cheyennes. He divided his force, sending one portion up the Arkansas, under command of Major SEDGWICK, and leading the other himself up the Platte. It would appear that the Indians have slipped down between these two columns, and commenced murdering and robbing in their rear.

The Cheyennes are amongst the boldest and most warlike Indians on our Western prairies, and if once fairly aroused, will be difficult to subdue. They number about 1,000 warriors, and will, doubtlessly, be joined by many young Sioux braves, with whom they are closely connected.

As there are a great many emigrants crossing the plains this season, it is very much to be feared that we shall soon hear of other catastrophes similar to that described below:

NEAR FORT RILEY, Tuesday, June 9, 1857.

MR. EDITOR: Our quiet community has just been thrown into considerable excitement by the news that the Indians, supposed to be the Cheyennes, have attacked a small party of emigrants about 80 miles West of Fort Riley, and killed four men and wounded two men and one woman. One of the survivors, Mr. A. P. WEAVER, has reached this place, and makes the following statement:

"About eighty miles from the Post on the Republican fork of Kansas River, my party had just left camp on the morning of Saturday, the 6th of June, 1857, about 9 o'clock, A. M. About 150 Indians, mounted, charged on our train and surrounded it; they commenced firing on our men; they killed four men of our party. After their guns were discharged, the Indians retired to a creek close by, and continued their fire until we had left the wagons. Before we had got out of sight they had emptied the wagons; a part of them pursued us. Our party consisted of ten men, eight women, and ten children. I left the party coming down in this direction, with two men and one woman wounded, all on foot, and out of provisions. One of the four men killed was endeavoring to escape, but was overtaken, and the last that was seen of him the Indians were dragging him by a lariat.

The names of three of the men killed are S. D. WEAVER, M. LEWIS and SAM SMITH. The wounded are J. HOUSTON, J. SMITH, and a woman, name unknown. Capt. HENDRICKSON, with two companies of the 6th Infantry, who had just arrived here from Leavenworth, has gone out to bring in the survivors. As his command is on foot it will be impossible for him to pursue the Indians, who are all well mounted.

This may be looked upon as the commencement of the Cheyenne war. Col. SUMNER has gone out after this tribe; but one portion of his command is on the Arkansas and the other on the Platte, two hundred miles apart, so that the Indians have a fine chance of slipping in between and getting in his rear, which, it appears, they have done. As the emigration crossing the plains this year is very large, there will be a great loss of life and property unless the Government promptly sends an additional mounted force in that direction. Instead of sending such an unnecessarily large number of troops to Utah, a portion should be sent to chastise the Indians who are murdering and robbing out citizens at our very doors.

Yours, &c., A. B."

June 22, 1857

Our Indian Relations.—The Indian has made himself more conspicuous during the last year than he had been for a long time previous. The rebel Indians in the territory west of Arkansas, the wild Indians who again rule in Arizona and who have figured extensively in New-Mexico, and the murderous Indians who have carried desolation and slaughter over a large section of Minnesota, have successively attracted attention. There has been an idea widely prevailing among those best informed on the subject, that attempts have been made by white and red rebels to form a general league of all the tribes this side of the Rocky Mountains against the settlers; and of this there is a great deal of confirmatory proof. *That* has failed, but the horrid deeds of the red men have induced a stronger feeling of animosity than ever against them in the West. The people of Minnesota threaten to exterminate all those in that State, and along the whole line, from Minnesota to Texas, there is a strong desire to have the Indian question quickly brought to a definite solution. Most of the settlers are in favor of at least "driving them out"—no matter where they may be driven to. This, however, like "putting down the rebellion," is a thing easier spoken of than accomplished.

Commissioner Dole, in his able report on Indian affairs, gives the country a large amount of very valuable information concerning the "noble savage," and offers some valuable, though not new suggestions, concerning him. He states that hostilities may be apprehended in the Spring with the Sioux of Dacotah, the most powerful tribe on the continent; and, if these hostilities break out, there is danger that they may extend over a very wide surface. This must be avoided, if at all possible, as the Indians are yet powerful enough to inflict vast damage on our frontier settlements. We earnestly hope that the proper authorities will take hold of the matter in time, as these difficulties can be as easily prevented as they would be difficult to quell.

Mr. Dole is in favor of aggregating the tribes as much as possible in one region. He would mass the middle tribes in what we call the "Indian Territory"—the region west of Arkansas; and the northern tribes he would mass in the far Northwest. He recommends that a line of forts be erected along the Red River of the North, and that they be sent to that region. This plan we approve of. It seems the best solution of a difficult question.

Mr. Dole is opposed to the hanging of the three hundred warriors in Minnesota who surrendered themselves and have been condemned to death. He favors only the execution of the chiefs who instigated the slaughter. These chiefs and head-men, he says, wield an influence over their subordinate savages which it is difficult for us to understand or appreciate; and we shall never be justified in judging the latter by our Christian standard of morals. For the ordinary mutineers, he urges a milder punishment than death—which punishment, we think, should take the shape of banishing the entire Sioux tribe from Minnesota. Mr. Dole discusses these and cognate questions with a lucidity of statement and accuracy of style that is comforting to the reader who has been rendered uneasy by the prolix paragraphs and shocking syntax of some other of our official documents.

December 3, 1862

Execution of Indians in Minnesota.

St. Paul, Minn., Satutday, Dec. 27.

Thirty-eight condemned Indians were hung at Mankato, at 10 o'clock A. M. yesterday. The gallows was so constructed that all fell at once. Several thousand spectators were in attendance. All passed off quietly.

December 29, 1862

The American Aborigines—The Indian Territory.

We spoke yesterday of the necessity of some arrangement being made advantageous to the Indians, as well as to the prospective white settlers, in the splendid region designated the Indian Territory. The necessities of our Western growth demand the useful occupation of this region, and we believe it is possible to secure it. We propose to state the present position of the Indian nations who occupy the important territory lying between Kansas and Texas, Arkansas and New-Mexico.

The region near Chattanooga, over which Sherman's victorious legions were lately operating, was formerly the home of the nations now living in the Indian Territory. These consist of the Cherokees, Creeks, Choctaws, Chickasaws, Seminoles, a small body of Delawares, Kickapoos, Shawnees, Senecas, Quapaws, and a number of nomadic tribes, known as the Ionies, Kechees, Kiowas, Washitas, etc. According to an estimate made in 1860, the census not having been regularly taken owing to the disturbed condition of the frontier, the population of the territory was as follows: Indians, 65,680; colored—slaves, 7,869; free, 404; whites, 1,988; a total of 75,431. Of this population, the Cherokees had about of Indians, 26,000; Creeks, 15,000; Choctaws, 10,000; Chickasaws, 3,000; other civilized tribes, 1,000; and the others about 10,000. The Cherokees had 2,504 slaves among 384 owners. The Choctaws, 2,297, among 384 owners. The Creeks, 1,651, among 267 owners. The Chickasaws, 917, among 118 owners. The average of slave population was about 12½ per cent., or one to eight Indians. The Choctaws had 802 resident whites, and 67 free colored; the Cherokees 713 whites, and 17 free colored; the Creeks, 819 whites, and 227 free colored; the Chickasaws, 146 whites and 13 free colored; the Seminoles owned no slaves, and intermarry with the negroes. They were 8 whites and 80 free colored. It is a noteworthy fact that this tribe were entirely loyal, with the exception of about forty warriors. It is also noteworthy that the Choctaw Nation, which had the largest slave and white population, was the bitterest supporter of the rebellion. From it came the principal portion of the Indian regiments in the rebel Gen. Cooper's command, who for more than a score of years was the agent of this tribe. We believe there is not a full-blooded Indian left among the Choctaws. Almost exclusively the full-blood members of the different nations remained loyal, while the half and quarter bloods are in sympathy with the Confederacy. There are noteable exceptions to this, as in the case of the Cherokee Chief, John Ross, and his influential family, who have but little Indian blood in them; and of Stand Waitie, the rebel Cherokee leader, who is a full blood.

The territory occupied by these nations is one of the most desirable regions on the continent, situated between latitude 33° 30′ and 37° north, and longitude 94° 20′ and 100° west; with a length, east and west, of 320 miles; a breadth, north and south, of 220 miles; and a total area of 74,127 square miles, or 47,441,480 acres; its climate is salûbrious and inviting, and the soil is fertile in the extreme, producing all the cereals of the temperate zone in abundance. The vallies of the Spring, Grand, Neosho, Verdigris and Arkansas Rivers, are inexhaustible in fertility. Under the impetus of emigration their prairies would become one of the great corn granaries of the world. The vallies of the Canadian and Red Rivers are among the most productive cotton growing sections in the South. In every respect it is a delightful and inviting region. Well watered and timbered, with magnificent prairies, it is scarcely possible that the intrusive and transforming feet of the white men can be kept from its occupation. What is to become of the Indian nations who now occupy it? and what are to be the future relations the territory will hold to the movements of white emigration and labor, are questions which may properly be mooted?

When secession culminated in open treason, the necessities of the hour temporarily severed the control of the National Government over these nations. The Federal posts in their midst were surrendered to the rebels. Capt. (now General) STURGIS retreated from Fort Smith in the night, and escaped with his command to Kansas. The South had been preparing for years to carry the Indians with them. A large portion of the Cherokees and Creeks, and all of the Choctaws and Chickasaws sided with the rebels.' All of the nations were compelled or induced to form treaties with the Richmond Government. ALBERT PIKE negotiated them. In the original draft of these treaties, the annuities, &c., given by the National Government, were reaffirmed by the rebels, some additions made, and an agreement entered into to admit a delegate to the Confederate Congress who should be jointly elected by the various nations. These people were then very prosperous. The Cherokees were probably the richest community in the world. The different nations had reached quite a high state of civilization. The land was held in common, each cultivating what amount they wished. Their school funds were very large. They have written constitutions, and limited representatives and elective governments. The female seminaries at Tahlahquah, the chief Cherokee town, were splendid institutions, occupying handsome brick buildings.

All of this is changed. The slaves have in great part become free. Both armies, by turns, have fed from the vast herds of cattle belonging to the Indians, their towns and dwellings were destroyed or sacked; while the mortality among them has, indeed, been fearful. In the Winter of '61'–2, a large body of Creeks and Cherokees declared for the Union, and after fighting three battles, were defeated in the third and compelled to flee precipitously to Kansas. Their number was swelled by various accessions to at least eight thousand, most of whom remained for a considerable time in that State. At least one-third died from exposure and famine. Two regiments were raised among the warriors, and placed under command of white officers. A third was afterward raised from a Cherokee regiment in the rebel service, which joined us in a body, during the first campaign for the reoccupation of the territory in the Summer of '62. This Indian brigade has served with gallantry and efficiency in all the campaigns of the army under Gen. BLUNT. It was at Newtonia, Cane Hill, Prairie Grove, Honey Springs, Cabin Creek, the defeat of COBELL near Fort Smith, and in a large number of engagements of lesser importance. For four months during last Summer, the brigade under Col. WM. A. PHILLIPS, held the line of the Arkansas, keeping the Cherokee country clear, and a line of communication and supply of nearly two hundred miles open. He never had more than three thousand men, twenty-five hundred of whom were Indians, while the rebels had from five to seven thousand, about one-third of whom were Indians, the remainder being Texas and Arkansas troops. The rebel Indian regiments consisted of one Cherokee, two Creek and three Choctaw, none of which were more than half full.

Since our occupation, important changes have been recognized by the Cherokees. Their Legislature has abolished Slavery; declared *all*, without distinction of color, born in the nation, to be citizens; disfranchised the rebels; and otherwise legislated against them. Treaties have been made with the loyal Creeks to the same purpose. Both these nations agree to give lands to their freed people. A large number of their able bodied slaves are serving in the Kansas colored regiments.

Predicating their claims upon this state of affairs, the Cherokee Chief, JOHN ROSS, visited Washington and asked, for his own and the other nations, that the original treaty stipulations be fulfilled, and the Indians maintained inviolate in occupation of the territory they now hold. The question is one fraught with difficulty, and involves the whole subject of the future disposition of the Indian tribes. The present abnormal relation of these nations to the General Government demands a settlement. The question, whether, if advisable, it be possible to keep their fertile territory uncoveted by the white man, and free from the incursions of the emigrant army finding its way westward from the seaboard, deserve careful consideration? They can be more readily met and solved at this than at any future time.

August 14, 1864

THE RED MEN.

The Indian Population.

STATEMENT SHOWING THE TRIBES OF INDIANS WITHIN THE LIMITS OF THE UNITED STATES TERRITORY, NUMBER OF SOULS, AND PLACE OF RESIDENCE OF EACH TRIBE, MADE UP FROM THE BEST DATA IN THE POSSESSION OF THE INDIAN OFFICE.

Name of Tribe.	No. of Souls.	Place of Residence.
Apaches	7,000	New-Mexico Ter'y.
Apaches	—	Texas.
Apaches	320	Arkansas River.
Assinaboines	3,360	Upper Mo. River.
Arickarees	800	Upper Mo. River.
Arrapahoes	3,000	Ark. & Platte River.
Anadahkoes, Caddoes, and Ionies	600	Texas.
Blackfeet	7,500	Upper Mo. River.
Cherokees	17,530	West of Arkansas.
Cherokees	2,200	Nor. Carolina, Tennessee, Georgia and Alabama.
Choctows	16,000	West of Arkansas.
Choctows	1,000	Mississippi.
Chickasaws	4,787	West of Arkansas.
Creeks	25,000	West of Arkansas.
Creeks	100	Alabama.
Chippewas of Lake Sup	4,940	Michigan.
Chippewas of Lake Sup	4,940	Wisconsin.
Chippewas of Lake Sup	4,940	Minnesota Ter'y.
Chippewas of the Miss	2,206	Minnesota Ter'y.
Chippewas and Ottawas	5,152	Michigan.
Chippewas of Saginaw	1,[illegible]	Michigan.
Chippewas of Swan Creek	138	Michigan.
Chippewas of Swan Creek	33	Kansas Territory.
Cayugas	143	New-York.
Catawbas	200	N. and S. Carolina.
Christians, or Munsees	44	Kansas Territory.
Crows	3,360	Up. Missouri River.
Crees	800	Up. Missouri River.
Caddoes	—	Texas.
Comanches and Kioways	20,000	Texas.
Comanches	—	New-Mexico T.
Comanches	3,600	Arkansas River.
Cheyennes	2,800	Ark. & Platte Riv.
California Tribes	*33,539	California.
Delawares	902	Kansas Territory.
Gros Ventres	750	Up. Missouri River.
Ionies	—	Texas.
Iowas	433	Kansas Territory.
Kickapoos	344	Kansas Territory.
Kickapoos	—	Texas Border.
Kioways	—	Texas.
Kioways	2,800	Arkansas River.
Kansas	1,375	Kansas Territory.
Keechies, Wacoes and Towacarros	300	Texas.
Kaskaskias	—	Kansas Territory.
Lipans	560	Texas.
Miamies	207	Kansas Territory.
Miamies	353	Indiana.
Manoans	259	U. Missouri River.
Minatare	2,500	U. Missouri River.
Menomonees	1,930	Wisconsin.
Missourians	—	Nebraska Territory.
Nunsees	—	Kansas Territory.
Muscaleros, or Apaches	400	Texas.
Navajoes	7,500	New-Mexico Ter'ty.
Oneidas	249	New-York.
Oneidas	978	Wisconsin.
Onondagas	470	New-York.
Ottawas	—	Michigan.
Ottawas	249	Kansas Territory.
Omahas	800	Nebraska Territory.
Ottoes and Missourians	600	Nebraska Territory.
Osages	4,098	West of Arkansas.
Oregon Territory tribes	13,000	Oregon Territory.
Poncas	700	Nebraska Territory
Pottawatomies	236	Michigan.
Pottawatomies of Huron	45	Michigan.
Pottawatomies	3,440	Kansas Territory.
Pawnees	4,000	Nebraska Territory.
Piankeshaws, Weas, Peorias, and Kaskaskias	220	Kansas Territory.
Puebla Indians	10,000	N. Mexico Territory.
Quapaws	314	West of Arkansas.
Stockbridges	13	Kansas Territory.
Stockbridges	240	Wisconsin.
Sioux of the Mississippi	6,383	Minnesota Territ'y.
Sioux of the Missouri	15,440	Upper Mo. River.
Sioux of the Plains	5,600	Platte and Ark. Riv.
St. Regis Indians	450	New-York.
Senecas	2,557	New-York.
Senecas (Sandusky)	180	West of Arkansas.
Senecas and Shawnees (Lewistown)	271	West of Arkansas.
Shawnees	851	Kansas Territory.
Sacs and Foxes of Mississippi	1,626	Kansas Territory.
Sacs and Foxes of Missouri	180	Kansas Territory.
Seminoles	2,500	West of Arkansas.
Seminoles	500	Florida.
Tuscaroras	280	New-York.
Towaccaros	—	Texas.
Tonkawas	400	Texas.
Utah Territory Tribes	12,000	Utah Territory.
Utahs	2,500	N. Mexico Territory.
Wacoes	—	Texas.
Wichitas	950	Texas.
Weas	—	Kansas Territory.
Winnebagoes	2,546	Minnesota Territ'y,
Wyandots	208	Kansas Territory.
Washington Territ'y Tribes	544	Kansas Territory.
Wandering Indians of Comanches, Cheyenne and other tribes	14,000	Washington Ter'y.
	17,000	N. Mexico Territory.
†Total number	314,622	

* Obtained from a report of the Secretary of State of California, on the census of 1852, in which they are designated as "domesticated Indians." Superintendent BEALE, in November, 1852, estimated the Indian population of California at from 75,000 to 100,000. Commissioners BARBOUR and WOZENCRAFT, in March, 1851, 200,000 to 300,000, though their colleague, REDICK MCKEE, Esq., at the same time stated that he had information which would greatly reduce that number. And the Spanish missionary authorities reported it to be, in 1802, 32,231. The census of the State of California is believed to be the most reliable.

† Possibly some of the tribes embraced in this statement, especially those inhabiting the mountainous regions and the plains, are not correctly reported; their number may exceed or fall short of the estimate here made of them. The Indian population within the limits of the United States territory, exclusive of a few in several of the States, who have lost their tribal character or amalgamated with whites or blacks, may be estimated at from 320,000 to 350,000.

August 3, 1865

THE INDIAN WAR.

The following order, purporting to have been issued by the Superintendent of the United States Express Company at St. Louis to its employes and agents, reaches us through a St. Louis paper:

UNITED STATES EXPRESS COMPANY, (DENVER DIVISION,) ST. LOUIS, May 18, 1867.

To Agents and Employes:

Gen. HANCOCK, commanding Department of the Missouri, will at once place a sufficient number of soldiers at each station, from Lookout to Lake Station inclusive, for your and our mutual protection. You are instructed to carry their mail matter and communications from the officers and men to the several posts and commanders, to deliver them promptly and with certainty. You will also carry to the amount of two hundred pounds of rations for soldiers between stations, when required; furnish transportation for all soldiers on regular coaches, on proper authority in writing from the commanding officers of the several posts, and will not furnish without, under any circumstances; neither encourage nor countenance desertions in any form.

You will hold no communication with Indians whatsoever; if they wish "talk," they must go to the regular posts. If Indians come within shooting distance, shoot them; show them no mercy, for they will show you none.

You must be watchful; when you have so many at each station, one must constantly be on guard, day and night, and at all hours. In this watchfulness you will save your stations from being fired, and can prepare yourselves for any danger.

You will also report to either Fort Hayes, Fort Wallace, or posts that will be established, whichever is nearest to you, all depredations as they may occur, and to your division agent without exaggeration.

Gen. HANCOCK will protect you and our property, but requires, as we do, your vigilance and hearty co-operation. W. H. COTTRILL, Superintendent.

June 1, 1867

THE INDIAN WAR.

The expedition of Gen. HANCOCK against the Indians of Kansas and Colorado, and the movements of the Indian Commission under Gen. SULLY along the Platte River route to California, have served to draw general attention to the war which has been going on for the past two or three years with the Cheyennes, Arrapahoes, Kiowas, Sioux and Comanches, and indeed with all the Indians of the Plains. The announcement that this war has been progressing for three years, will doubtless surprise many readers who have supposed that hostilities were precipitated by the horrible massacre in December last of a portion of the Fort Philip Kearny garrison. The first hostile demonstration was really made on April 11, 1864. It will, doubtless, also surprise many to learn that the first breach of the peace was committed by United States volunteer troops, and that from the first the whites have been the belligerents, and that the Indians have repeatedly appealed and negotiated for peace on any terms that would enable them to continue their former mode of living in idleness and dissipation on the bounty of the Government. We do not mean to say the United States Government has been the belligerent party, but certainly the war has been precipitated by the demands of the white settlers of the disputed territories for "more room"—"more land" probably expresses it more clearly and forcibly. It is the old fight of industry against idleness, civilization against barbarism, and a thrifty race against a non-producing one. The morality of the invasion is not taken into consideration in this article; it is a very nice question to decide upon the moral right of an idle and debauched race, with every vice and with no virtues, to have and to hold a rich country like that of our Western plains. Similarly debased white people are daily arrested as "disorderly persons" in New-York City, and sent to the reservations for such on Blackwell's and Ward's Islands. The whites of Colorado and Nebraska have demanded more room, and are determined to have it. The Indians must give way before the demands of commerce, and the white settlers have decided—cruelly it may be, but inexorably decided—to take Gen. SHERMAN'S advice and "exterminate the Indians of the Plains." The Indians of the Mountains are now our allies, but this same question will eventually present itself with regard to them, and be answered in the same words. The affiliation of the two races appears to be impossible.

THE INDIAN RESERVATIONS.

Until 1857 the Indians of the Plains were a very belligerent race, and had made constant warfare on the white settlers of Kansas, Nebraska and Arkansas. This somewhat retarded, but did not stop, emigration; white settlers filled up these Territories, pushing the Indians back toward the mountains, and building their forts on the frontier. Forts Kearny in Nebraska, Riley in Kansas, and Gibson in Arkansas, represented our frontier in 1857. At this period, badly beaten, greatly reduced in numbers, and forced so far westward as to come into conflict with their old time enemies, the Indians of the Mountains, the Indians of the Plains were glad to make peace, and a peace was concluded in that year by which reservations and certain annuities were granted them. The Cheyennes and Arrapahoes were given a large tract in the south-eastern part of what is now known as Colorado, (then called Kansas) but their domain really extended from the Arkansas to the Platte Rivers, and embraced all the splendid country about the headwaters of Smoky Hill fork of the Kansas River. The Sioux or Dacotahs were located north of the Platte in Western Nebraska; the Pawnees in Dacotah, and the Kiowas and Comanches south of the Arkansas River in Southwestern Arkansas and the eastern part of New-Mexico. These tribes thus located, are the principal tribes of the Indians of the Plains living east of the Rocky Mountains. There any number of other tribes with unpronounceable names and of insignificant numbers, but as they live on the reservations of the tribes named, and depend on them for protection and guidance, they are considered herein as practically a part of the same.

THEIR NUMBERS AND CHARACTER.

The various tribes on the plains occupy, or rather roam over, a vast territory, but their numbers are rather insignificant. The following estimates are made from reliable data.

SAMUEL G. COLLEY, agent of the Cheyennes and Arrapahoes, in 1865 estimated the number living between the Platte and Arkansas Rivers as follows:

	Warriors	Total pop
Arrapahoes	1,000	5,000
Cheyennes	1,500	7,500

Col. JESSE H. LEAVENWORTH, agent of the Kiowas, Comanches and Apaches, is intimately acquainted with the tribes named above, and estimated their forces in 1865 as follows:

	Warriors.	Total pop.
Arapahoes	1,700	8,500
Cheyennes (including Dog Cheyennes)	2,100	10,500

The same officer thus estimated the tribes of which he had charge, *i. e.*, those living south of the Arkansas:

	Warriors.	Total pop.
Kiowas (including the Apaches)	2,000	10,000
Comanches	3,500	17,500

The estimates of the forces of the Sioux Indians are made up from incomplete returns of several small bands, including the Crows, (estimated by MALHON WILKINSON, Indian agent, at 1,000 warrriors,) the Assineboines, (1,000 warriors,) the Arickarees, Gros Ventrees and Mandans, (400 warriors.) The Sioux and Pawnees, and their adherents, can muster, it is believed, about 7,000 warriors, out of a total population of 40,000. This would give, as the total of the several tribes involved in the present war, the number of 70,000, men, women and children, and a fighting force of 15,000 warriors, all mounted.

Col. FORD, Second Colorado Cavalry, who commanded at Fort Larned, writing May, 31, 1865, of a portion of these warlike tribes, says: "From the best of my information all the tribes of Indians are hostile. The Kioways, Cheyennes, Comanches, Arrapahoes and parts of other tribes, with their families, are now south of the Arkansas, on the Red River, which is one of its tributaries. In February last a large number of them were about 150 miles west of south of this point. From the best information I can get, there are about 7,000 warriors, well mounted, some on fleet Texan horses. On horseback they are the finest skirmishers I ever saw." Col. LEAVENWORTH says of the Comanches: "The Comanches have a great many horses. A Comanche never moves except on a horse. There are no better horsemen in the world. They ride from the moment they can sit up straight. They are tied on the horse by the mother and the mother leads the horse, and that is the way they move from place to place. The Comanches are the most warlike Indians we have on the Continent, I think."

All the evidence which can be collected shows that these small forces are rapidly growing weaker. Gen. JOHN POPE says: "They are rapidly decreasing in numbers from various causes—by disease; by wars; by cruel treatment on the part of the whites, and by steady and resistless encroachments of the white emigration toward the West, which is every day confining the Indians to narrower limits, and driving off or killing the game, their only means of subsistence." The same authority states that the most fatal diseases among them are those resulting from the "universal prostitution prevailing," and adds that not one out of every five, men, women and children, are free from the effects of loathsome venereal diseases. Col. JOHN T. SPRAGUE, Seventh United States Infantry, attributes the rapid decrease of the Indians in numbers to "their proximity to the white man, whose vices they adopt and exclude all his virtues," Col. SPRAGUE also alludes to the prevalence of venereal diseases among them, and forcibly adds: "In this, striking at the very basis of procreation, is to be found the active cause of the destruction of the Indian race."

REVIVAL OF THE EMIGRATION MANIA.

Soon after the peace of 1857 was declared, and the frontier became safe, white emigration developed rapidly; so that while the Indians were growing weaker the whites were growing stronger. The frontier line of 1857 was pushed still further West, and further encroachments made on the Indian territories. The country lying between the Platte and Arkansas rivers was the richest, and this fact, and the other advantages of salubrious climate, fine streams, &c., &c., led to its being more rapidly occupied than other districts by the emigrants. Kansas was rapidly filled up and became a State; then Colorado was settled, and in a few years its capital, Denver City, became a prosperous place. Through this district the overland routes to California were located, surveys for two railroads were also made through it, and these served to incite further emigration. Three well-beaten routes of travel soon became marked out. One of these was along the Platte River. The army followed emigration, and for its protection built along this route Forts Kearny, McPherson, Gratton, and Laramie, and finally the Union Pacific Railroad line, of which we gave a full account in the TIMES of May 28, was located along the same route. Another route, chiefly used by emigrants to New-Mexico and traders to Santa Fé, was opened along the Arkansas River and Forts Larned, Dodge, Aubrey and Lyon were built. A third, more direct and favorite route was opened from Kansas City along the Smoky Hill River to Denver City. It has become known as the "Smoky Hill Route," has been long used as the route of Butterfield's Overland Express, and has lately been adopted as the line of the Kansas Pacific Railroad. Forts Harker, Hays, Ellsworth and Wallace have been built for its protection, and the important towns of Topeka, Manhattan, Saline City and Downer Station have sprung up as if by magic along it. The Smoky Hill route runs through the heart of the seat of war, and it is about the head waters of the river of that name that Gen. HANCOCK is now operating, and where the conflict will doubtless end in the removal of the Indians either to the north of the Platte or south of the Arkansas Rivers. It is important that they should leave the immediate vicinity of the two lines of railroad now rapidly building along the Platte and Smoky Hill Rivers. Suggestions to this effect were made by Senator J. R. DOOLITTLE and others in 1865, and the fixed purpose now seems to be to drive them out of this part of the Plains. The people of Colorado want the land and are aggressive. Commerce demands it also, and will have it. With this disposition prevailing, and this motive as an incentive, it did not take the people of Colorado long to find a cause for a quarrel.

OPENING OF HOSTILITIES.

The opportunity was found on April 11, 1864. On that day a white man reported to Capt. SANBORN, of the First Colorado Cavalry, then commanding at Fort Lyons, that three of his cattle had been stolen by the Cheyennes. Capt. SANBORN ordered Lieut. CLARK DUNN to pursue with a detail of twenty men, rescue the cattle and disarm the Indians. A band of Indians were overtaken, but no cattle found. The Indians were friendly, awaiting for their pursuers to come up, and shook hands with them when they did so. Although satisfied of the innocence of the Indians, Lieut. DUNN attempted to carry out his other orders and disarm the Indians. This attempt was resisted; a fight followed, DUNN losing one man killed and three wounded; the loss of the Indians was not ascertained.

This affair created great excitement among both the whites and Indians. Col. J. M. CHIVINGTON was at that time commanding, as an officer of United States volunteers, the District of Colorado. He at once issued orders for a series of regular operations against the Indians. The Cheyennes and Arrapahoes were much frightened at this demonstration, and at once made overtures for peace.

MAJOR DOWNING'S DESTRUCTION OF A CHEYENNE VILLAGE.

Soon after the Dunn affray, an unprovoked attack was made by the forces at Fort Lyon, under Major DOWNING, upon an Indian village at Cedar canyon, in the Cheyenne Reservation. This resulted in a fight, in which forty warriors, squaws and children were killed, and the village burned.

RETALIATORY MEASURES OF THE INDIANS.

Up to the time of this occurrence no retaliatory acts had been committed by the Indians. Shortly after the Downing massacre a band of Arrapahoes, under a chief called LITTLE RAVEN, made their appearance at Fort Lyon and proposed a peace to Major E. W. WYNKOOP, who had relieved Capt. SANBORN in the command of that fort. Major WYNKOOP was desirous of peace, and when LITTLE RAVEN had explained that a band of Cheyennes, under their chief, LEFT HAND, had gone to Fort Larned to propose the same terms there, Major WYNKOOP advised LITTLE RAVEN to remain in camp near Fort Lyon until he could hear from Fort Larned, telling him that Col. CHIVINGTON was there, that he would see LEFT HAND and conclude a peace with him. Under assurances of protection LITTLE RAVEN went into camp near Fort Lyon. In the meantime LEFT HAND made his appearance under a flag of truce near Fort Larned and explained his mission, at the same time showing his good faith by warning the commandant of a design on the part of a band of Kiowas to run off some animals grazing near the fort. No notice was taken of the warning or peace propositions by Col. CHIVINGTON. The robbery was committed and LEFT HAND made his appearance and offered to aid in the pursuit of the thieves. Col. CHIVINGTON ordered the artillery of the fort to be opened upon LEFT HAND's band, and at the first fire the Indians left the vicinity of the fort and retired to their reservation. As soon as the news of this reached LITTLE RAVEN he broke up camp and started north. Satisfied that no peace was to be obtained, he at once began to depredate on the Smoky Hill route, stealing stock and killing a teamster and carrying off his wife. The majority of the bands of Arrapahoes, and all of those of the Cheyennes refused to engage in these depredations, and continued to make overtures for peace.

LIEUT. AYRES' FIGHT.

The Fort Larned affair was followed up by the destruction of another village by Lieut. AYRES, of the First Colorado Battery. This was done under orders from Col. CHIVINGTON, "to kill all the Indians he came across." He marched from Fort Larned about forty miles westward to Beaver Creek, and encountered a band under a chief named LEAN BEAR, engaged in hunting buffalo. LEAN BEAR was invited by Lieut. AYRES, through a Sergt. FRIBLEY, to enter his camp. He did so, unaccompanied, and was soon after shot dead, and his band fired upon. A running fight ensued for two hours, AYRES being driven from the field, with a loss of several killed. Subsequently the Indians abandoned their village, which was burned by AYRES.

MORE PEACE OVERTURES.

Subsequent to this affair, the Indians appear to have again sued for peace. In August they wrote to their agent, Major S. G. COLLEY, the following proposition in reply to a letter from Col. BENT, urging on them to continue their efforts to conclude a peace:

CHEYENNE VILLAGE, Aug. 29, 1864.

MAJOR COLLEY: We have received a letter from BENT, wishing us to make peace. We held a council in regard to it. All come to the conclusion to make peace with you, providing you make peace with the Kiowas, Comanches, Arrapahoes, Apaches and Sioux. We are going to send a messenger to the Kiowas and to the other nations about our going to make peace with you. We heard that you have some (prisoners) in Denver. We have seven prisoners of yours which we are willing to give up, providing you give up yours. There are three war parties out yet, and two of Arrapahoes. They have been out some time and expected in soon. When we held this council there were few Arrapahoes and Sioux present.

We want true news from you in return—that is, a letter. BLACK KETTLE, and other Chiefs.

Subsequently, on Sept. 4, Major WYNKOOP held an interview with three Cheyenne Indians near his post, Fort Lyon, in which he became convinced of their desire for peace. These Indians stated that the Arrapahoes and Cheyennes had congregated to the number of two thousand, and were anxious to conclude a peace. A force was ordered under arms, Major WYNKOOP marched to the Indian rendezvous, and held a consultation with them. He told them he had no authority to conclude a peace with them, but invited them to remain in camp, under his protection, while a deputation of their principal chiefs went with him to Denver City, and made a peace with Gov. EVANS. Several prisoners were delivered up to Major WYNKOOP, and BLACK KETTLE and seven other Cheyenne and Arrapahoe chiefs went under his escort to Denver City, the rest accepting WYNKOOP's promise of protection. At Denver Gov. EVANS referred the deputation to Col. CHIVINGTON, who was the military commander. Col. CHIVINGTON informed them that he would have to consult his superiors, but told them that they would remain at the disposal of Major WYNKOOP until higher authorities had acted in their case. The Indians appeared to be perfectly satisfied, presuming that they would eventually be all right, as soon as those authorities could be heard from, and expressed themselves so. BLACK KETTLE embraced the Governor and Major WYNKOOP, and shook hands with all the officers present, perfectly contented, deeming the matter was settled. On returning to Fort Lyon Major WYNKOOP requested the chiefs to remove their camp to the immediate vicinity of Fort Lyon, and they unhesitatingly obeyed, satisfied that they were in safety. Shortly after this removal of the camp Major WYNKOOP was, by orders of Col. J. M. CHIVINGTON, suddenly relieved from the command by Major SCOTT J. ANTHONY. The Indians were somewhat dissatisfied with this change of commanders, but Major ANTHONY repeated the assurance of protection which WYNKOOP had given, and they remained quietly in camp awaiting Col. CHIVINGTON's consultation with his superior officers. Here they remained for nearly two months under ANTHONY's protection.

THE CHIVINGTON MASSACRE.

In the meantime Col. CHIVINGTON had left Denver, repaired to Fort Larned and begun the concentration of all the troops he could collect.

On Nov. 19 he left Fort Larned with the Third Colorado Cavalry and a battalion of the First Colorado Cavalry, and on Nov. 28 arrived at Fort Lyon. On the next morning this command, under Col. CHIVINGTON in person, and against the earnest protests of Major ANTHONY and his subordinates, marched out of the fort, and about sunrise surrounded the Indian camp. One battalion was interposed between the camp and the ponies of the Indians, while the rest were ordered to charge upon the village. The Indians were taken by surprise, and made little or no resistance. BLACK KETTLE, the chief, hoisted the American flag and a white one over his lodge, but no regard was paid to either. A wholesale slaughter ensued, and men, women and children were killed, and the camp burned. Most horrible outrages were committed by the troops. First Lieut. JAMES D. CONNER, of the First New-Mexico Infantry, reports that "in going over the battle-ground next day he did not see the body of man, woman or child but had been scalped, and in many instances their bodies mutilated in the most horrible manner." "I heard a man say that he had cut off the fingers of an Indian to get the rings on his hand." "I heard of one instance of a child a few months old being thrown into a feed-box of a wagon, and after being carried some distance left on the ground to perish." JOHN T. SMITH and his son, interpreters at Fort Lyon, had been given permission on the day previous to the massacre to go to the Indian camp; they were there when the attack was made. Both were taken prisoners and held as such by CHIVINGTON'S troops; during the night the son was shot and killed by Lieut. DUNN. The elder SMITH in his testimony corroborated the statements made as to the atrocities committed. The same is affirmed by other testimony before the Congressional Committee, showing without doubt that the massacre was one of the most horrible ever committed by white men or savages. Half the horrors cannot be given—they were of such a character as renders a recital of them unfit for publication here.

THE ALLIANCE OF THE INDIANS.

All hopes of peace were now abandoned, and depredations were begun by the Indians with savage earnestness. From the day of this massacre until January, 1865, they had possession of the chief routes of travel, the country was laid desolate, communication between the east and Denver City, Fort Lyon and other points was cut off, and, according to the estimates of Major WYNKOOP, "over one hundred whites fell victims to the fearful vengeance of these betrayed Indians." Col. LEAVENWORTH and Major WYNKOOP made some ineffectual efforts to bring about a peace, and at last they were discouraged from any further efforts by an order from Gen. G. M. DODGE, to the effect that the duty of the military was not to treat with Indians, but make them keep the peace by punishing them for their hostility. Satisfied of the impossibility of providing for their safety by peaceful measures, the Cheyennes and Arrapahoes made the now existing alliance with the Kiowas, Comanches and Sioux, and their adherents, by which they hoped to defend themselves against the settlers. This powerful alliance placed the settlers on the defensive; the overland routes were interrupted, daily depredations were made on the various stations, and at last the Chivington massacre was avenged by that of Fort Philip Kearny, committed by the Sioux on Dec. 21, 1866. This condition of affairs has shown the Government that more troops are necessary to the maintenance of the communication with the West, and hence the expedition of Gen. HANCOCK, whose operations in opening the "Smoky Hill Route," are now going on. This statement of the origin of the war will, perhaps, give the reader a clear understanding of the issues involved and operations as they are reported from day to day.

June 2, 1867

Our New Indian Policy.

The vigorous measures of Gen. SHERMAN, and the prompt action of Gen. SHERIDAN in executing them, appear to have put a stop for the present to the outrages of the Indian tribes in Kansas and Colorado. At any rate, we have had no reports of additional massacres since that prevented by the stubborn resistance of Col. FORSYTHE and party. But there is little reason to hope that this cessation of cruelties on the part of the Indians will be more than temporary if these measures are not maintained. This, it is gratifying to learn, is to be done; the policy of Gen. SHERMAN and the vigilance of Gen. SHERIDAN promise an early and final settlement of the Indian question, as far as the important vicinity of the two Pacific Railroads are concerned.

The last Indian war had our unqualified condemnation. It was begun by a horrible massacre by the white settlers organized in a military body, and was prosecuted by them for nearly two years with more than Indian cruelty and ingenuity, and hardly more than Indian discretion. It was avowedly a war of extermination; two races of diametrically opposite natures, habits, and education fighting for the same land. The present hostilities were begun by the Indians under circumstances that lead us reluctantly to the belief that it must be prosecuted and concluded by the adoption of the most stringent measures.

The last Congress passed appropriations of over $7,500,000 to purchase presents and pay annuities and interest to the Indians, and the Peace Commission of last Spring agreed in most instances to repeat these presents and continue the annuities. The actual close of hostilities last Winter was followed by a distribution of clothing, trinkets, guns, &c., which only served to increase the Indian appetite for more; and when the circumlocution of Congress and the Indian Bureau delayed indefinitely the issue of the last promised presents, the Indians began their depredations, as Gen. SHERIDAN says, in the hope of forcing a new distribution of gifts. Their new apparent bad faith and atrocity must disarm the most humane apologists among the whites, utterly change our Indian policy, and will authorize and justify the most stringent measures which our military commanders in the disturbed district may inaugurate and execute.

Our information is to the effect that the military authorities intend to effect the removal of the Indians of Colorado and Kansas to reservations beyond the Arkansas and Platte Rivers, and remote from that favored district through which the Pacific Railroads and the Santa Fé trains are now running, and which promises in the next decade to become the most thickly settled of all our Trans-Mississippi Territories. The operations, whose good effects are already visible in the sudden stop which has been put to Indian depredations, are to be continued with all the humanity consistent with energy, until the Indians are again collected on their reservations in the southeast corner of Kansas. A little energy on Gen. SHERIDAN'S part, and an early touch of the coming Winter, both of which may be looked for with equal confidence, will soon drive the Indians home. In the Winter season the Indian is a very tractable and sluggish animal, and the whole of the eight or ten thousand Cheyennes and Arrapahoes located in Colorado and Kansas can be removed by one-fourth as many soldiers to the remote western part of the Indian Territory, or even to Northern Texas, without calling on any of the forces so judiciously distributed throughout the South. There is no reason why, thus removed from the temptations for robbery and spoliation to which they are now subjected, the Cheyennes and Arrapahoes may not immediately become as docile as the Cherokees of the South or the Assiniboines of the North, and eventually die out as gracefully and peacefully as those tribes are slowly but inevitably dying.

If he possesses the power and feels authorized, we can trust Gen. SHERMAN to carry out this policy. The arming of the settlers, under a late order issued by Gen. SHERMAN, did not at the time seem to many as either advisable or necessary, but it should be remembered that this action was taken under very different circumstances from those of 1864. Then the armed settlers had no controlling or directing power other than their own passions and their hatred of the Indian race; now that they are under the direction of a wise and experienced commander, they will be guilty of few excesses like that of the "Chivington massacre." The arming of the settlers at this time and under these circumstances is an effective without being necessarily a cruel measure. Gen. SHERMAN has the confidence and will have the support of the public in carrying out any measures which he may conceive for the removal of the Indians.

Undoubtedly Congress, at its next session, will be urged to adopt this policy. It is the only alternative which remains for us to do. Our commerce as well as our emigration demands the removal of all hostile Indians from their dangerous vicinity to the Platte River, Smoky Hill and Santa Fé lines of traffic and travel. The Indians must give up their reservations in that region for habitations and hunting-grounds more remote from the great highway of the West and nearer the buffalo range. If the consent of Congress is necessary to the accomplishment of this policy, we trust that no dilatory action on its part will let loose these predatory tribes of Indians again next Spring; if its authority is not requisite, Gen. SHERMAN will add to the great debt which the country already owes him if he will promptly use his power to effect this now absolutely necessary disposition of the Indians.

October 4, 1868

Mismanagement of Indian Affairs—Necessity for a Reform.

To the Editor of the New-York Times:

I have noticed with a good degree of satisfaction the reasoning advanced in the TIMES, to show that the management of Indian affairs should be transferred to the War Department. I cannot advance new arguments but I have a few hard facts which may serve to illustrate those already put forth touching the worthlessness and dishonesty of those in whose hands the management is at present. These facts have been acquired by several years' experience in the Indian country, an experience which has rendered me capable of appreciating this whole subject, and made me familiar particularly with the question of these Indian agencies. I am sure it is capable of being shown that the Indians would be vastly better off, if the agents were wholly dispensed with, even though there should be no transfer of their management. This is so from the fact that nine-tenths of these agents are either practiced scoundrels when appointed, or readily become such through the allurements of money-getting, which beset them in most instances after they have been appointed. Among the score or more that I have known, I cannot recall but a solitary exception. The ostensible duties of an agent are to protect the rights of the tribe to which he may be accredited, as well, also, as the rights of the white people who may be residing in or sojourning in the territory over which his jurisdiction extends. These duties they never perform. In the eyes of the "noble red man" the agent is a vulture; in the estimation of the whites he is the very scavenger of the "flesh pots," and both classes dread about equally coming in contact with him. The Indian, in all cases which have come under my observation, has a universal distrust of him, and when aggrieved will apply for redress to the commanding officer of the nearest military post rather than to him.

Senator HENDERSON, of the Committee on Indian Affairs, is reported to be in favor of a transfer to the War Department of the management of the wild tribes, but opposed to a transfer of the half-dozen half-civilized tribes occupying the Indian Territory proper. If any doubt can exist at all with reference to the transfer, it can only exist in the case of the wild tribes. Those civilized nations in the Indian Territory, among which may be mentioned the Cherokees, Choctaws, Chickasaws, Creeks and Seminoles, no more require the services of a Government agent than do the people of the contiguous States. They have long since ceased to be the especial wards of the Government in the sense that the wild bands are, the Government annuities are no longer kept up, and each clan has organized a government of its own. They treat through committees and delegations direct with the authorities at Washington. The agents of these tribes are as useless as a cat in a cupboard. They have absolutely nothing to do but to stir up mischief and draw their pay monthly. These half-dozen agencies, with a single exception, are filled at the present time with as many contemptibly dishonest men. There are a thousand Indians in the Territory, who are as much their superiors morally and intellectually as one can imagine. They do not in any sense share the respect of the Indians or the whites living among them, but are held in utter contempt by both classes. The salary they draw would be better applied if thrust in the fire. They each have a hoard of special agents and hangers-on, and all are notoriously the tools of the Indian jobbers and contractors.

Most of these agents are indirectly engaged in trade with the tribe to which they are accredited, and so use their official position to promote their private interests. I have known one agent to draw pay for his brother as a Choctaw interpreter, at a round rate of compensation, when the fact was patent that he knew less of the Choctaw language than the "learned pig." To a large extent these agents control the question of trade with their tribes. When they have the power of granting licenses, they generally farm them out to the highest bidder as an auctioneer would his wares, except a little more slyly, and in cases where this privilege has been taken from them, they still claim that they are a power in the premises, and any one seeking to do business with the tribe for which they are agent, must first pay their price or be content to have every imaginable obstacle cast in his way. One instance of the kind, which came under my personal observation, may be taken as a fair sample of the whole.

Some parties were buying cattle and carrying on trade in the Chickasaw country, near Fort Arbuckle, under a permission received from the Governor of the Chickasaws, the authority having been delegated to him by the treaty with that tribe of 1866. The agent allowed these parties to proceed in business for the period of six months without intimating that he had any option in the premises; but when they had secured a large stock of goods he sent his "man Friday" (they all have one) to Fort Arbuckle, seized the goods and closed up the store, at the same time declaring it to be his purpose to have the goods shipped to Van Buren, Ark., two hundred and fifty miles distant, where they would be libeled in the United States District Court, and the parties prosecuted as violators of the Intercourse laws. Even though the agent should be beaten in Court, the transfer of the goods and the injury to business would involve an expense of many thousands of dollars. Being somewhat familiar with the conduct of this agent in similar cases, I assured the parties directly affected that he only wanted to extort money. I was delegated, therefore, "to take soundings." I started in quest of the agent, who was supposed to be at Fort Smith, some two hundred miles from his proper station. I had the fortune to meet him and his lackey at a wayside ranche, when I had only accomplished one-half of the journey. We immediately assumed the *diplomatic*. The agent pronounced many maledictions upon all violators of the Intercourse laws, and spoke in most feeling terms of the manner in which his authority had been trampled in the dust. But as I spoke of the known liberality of my friends and their readiness to appreciate a service, he assumed a milder and more congenial tone, and at the same time begged to refer me to his bottle-holder, assuring me he would sanction any arrangement we might consummate. I commenced to talk business with this fellow at once. He said to me in so many words, "You know we fellows are on the make." I replied that I faintly expected as much, but that I hoped in this instance he would not be too exacting. Nothing was now left to be done but to arrange the slate, and we both commenced to make figures. He assured me that in case the goods were confiscated, his own and the agent's share would be one-half the proceeds, which, in this case, would amount to at least $9,000; but that if they could have the money without further risk or trouble, they would sign a release for $2,000. I succeeded in cutting these figures down considerably by hard effort, and when the price had been fixed upon, I paid him the money, and received in lieu thereof the release of the goods and the written permission of the agent to trade in the Chickasaw country.

To add to the perfidy of this man's conduct, I would state that several months after this transaction occurred, he again seized what was left of this stock of goods, therefore violating his purchased pledge and promise in the first instance, and this must be my excuse for entering upon a theme that would seem to involve confidential features under ordinary circumstances.

I left the place of our interview with my confidence in human kind slightly impaired, and feeling that the half has never been told with reference to the dishonor and peculation of Indian agents.

I have reported these facts to the Commissioner of Indian Affairs in the form of a sworn statement, yet I believe this same agent is still guarding the interests of the Chickasaws.

NEW-YORK, Dec. 17, 1868. E. H. H.

December 22, 1868

The President's Policy Toward the Indians.

Probably most intelligent readers in different countries, who look over our annual public documents, turn with a peculiar and sad interest to what is said of the only native American race—the Indians. The public has long since ceased to have any sentiment about "the noble savage;" it knows him to be a wild, half-brutalized creature, given to many vices, and hating industry. The frontiersman looks upon him as he does on the wolf and the bear, and is quite as ready to exterminate him. And many of our States, bordering on his territories, share in these barbarous feelings. But the nation, though it has no love for the Indian and the sentiment about him, has at length become aroused to its duty toward this race, whose main crime has been that it stood in the pathway of civilization on this continent. The people feel now that in the past the Indians have been the victims of incessant fraud and oppression. The sense that our conduct toward these weak barbarians has been a disgrace to our civilization and Christianity, has become burned into the national conscience. There is at length a widespread determination to do justice to these tribes, to put a stop to the extortions of agents, and the wrongs done by treaties, and to really attempt to settle the "Indian problem."

One of the fortunate occurrences in the treatment of this question is that we have a soldier at the head of affairs. The army officers who have been most familiar with the Indians, it is notorious, have always taken an entirely different view of the mode of permanently settling the question from the civilians or frontiersmen who had transactions with these tribes. They have always held that honesty and good faith in public dealings, and an application of some of the instrumentalities of civilization, could make of these barbarians a docile, peaceful and even useful people. The frontiersmen, on the other hand, hold that all that can be done with an Indian is to shoot him. General GRANT has been long personally familiar with the frauds and cheating perpetrated by the Indian Agents; he has known enough of the bad faith practiced toward the savages, under the guise of treaties, and how steadily white progress first demoralizes and then crushes them. He understands how the anger and desire of revenge of one wronged Indian will light a fierce war on the whole border. General GRANT has had personal cognizance, too, of what can be done for the elevation and civilizing of the

Indians. His own brother-in-law, Mr. DENT, has long had charge of the wild tribes of Arizona and Southern California. He has put them on reservations and has shown what remarkable progress in agriculture and settled habits these wild nomads may make under proper direction. The President, too, has lived among the Pah Utahs—formerly the most savage of the tribes bepond the Rocky Mountains—and has beheld what a quiet, peaceable and industrious folk they may become under the system of "reservations" and suitable instruction. This experience, and the characteristically humane views of an army officer, have led the President to one of the most original and enlightened of his measures—the remitting of the selection of the Indian Agents to the Society of Quakers. Congress, too, has appointed a special Indian Commission, composed of private citizens of the highest character and well-known benevolence, to inspect the action of these agents.

The policy which both the President and the Secretary of the Interior recommend, to govern the action of these Agents and of Congress toward the Indians, is of settlement on large reservations, and instruction in agriculture, and thus of gradual preparation for self-government. By such a policy, the pioneers will be freed from the terrors of wandering hostile tribes, and these "wards of the nation," as the President justly terms them, may be elevated in the scale of humanity.

Congress has already passed an act to enble the civilized Indians of the Indian Territory to form an organization and prepare for a Territorial Government; but no money has been appropriated to carry out this legislation. The Secretary of the Interior believes that the Cherokees and their associate tribes are already sufficiently advanced to dispense with the tutelage of Agents and Superintendents, and to be represented on the floor of Congress by a Delegate able to speak for them. He says very justly that the white constituencies nearest to the Indians are precisely those who are least qualified to represent the tribes in Congress, as they are uniformly hostile to them, and their interests are opposite.

The Secretary thinks that the tribal organization could easily merge into the county, and that their present "grand councils" could be readily changed into territorial legislatures. The tribes north of the Platte River he does not, however, believe to be sufficiently prepared for this development and concentration, though in time two or three Indian territories may embrace all the tribes east of the Rocky Mountains. The same policy is recommended for the tribes west of the Rocky Mountains, though many of these are probably much more difficult to manage, and will need a longer time to civilize.

We confess that from considerable study of the subject, we see no other solution of the Indian problem. As the President remarks, it would be a disgrace to our civilization to adopt any system which simply looked to an extinction of this unfortunate race. We are bound by all considerations to give them a chance for improvement and elevation. But we cannot permit them to rove at will over the vast territories which are now opened by the Pacific Railroad, and which will soon be dotted by thriving towns. They must be confined to large districts, where they can still hunt if they choose, but where the conveniences for agriculture are supplied them, and where trustworthy agents can instruct them in the arts of civilization. If, under the stern law which condemns so many savage races, they perish from contact with civilization, we shall have at least done our duty. We shall have held out to them an opportunity for improvement, and have made some slight atonement for the unnumbered wrongs inflicted on their race in the past.

December 10, 1869

THE INDIANS.

Action of the Government in Relation to the Apaches—White Encroachments on the Reservations.

Special Dispatch to the New-York Times.

WASHINGTON, Nov. 9.—The formal orders necessary to carry out the policy agreed upon at the conference between the President, the Secretary of the Interior and the Secretary of War in regard to the treatment of the Arizona Indians have been prepared. The Secretary of the Interior has written a letter to the President which recites the details of the policy to be pursued, and asks to have orders sent to the military commanders instructing them as to the duties which will fall to them to perform. These instructions go from the President through the War Department, and were prepared by Gen. SHERMAN. Today the outline of the plan arranged for the suppression of Indian disorders in Arizona, may be briefly drawn. It is believed that putting a stop to outrages by the Apaches is a matter of good administration alone. The orders given look to harmonious and united action by the military forces and the agents of the Indian Office. Gen. CROOK is, in the first place, to circulate among both whites and Indians, a notice that the latter must go on their reservation; that the Government cannot permit them longer to send their old men and women and children on to the reservations, to be fed and taken care of, while their young men go on the war path; that all who go upon reservations shall be protected and fed by the Government to the full extent of its power, but that those who stay away from the reservations, and commit murders and depredations, shall be punished by the soldiers. The soldiers and agents will disseminate this notice as widely as possible, and give the Indians an opportunity to comply with its requirements. It is sent among the whites also, that they may understand what the Government is doing, and be deterred from attempting to imitate the Camp Grant massacre. The operations of all the Government officers will be directed to enforcing on both parties compliance with the terms of the notice. An army officer will be appointed Superintendent at each reservation until the Interior Department is ready to take charge of the Indians. Superintendent BENDELL is directed by the Secretary of the Interior to have his head-quarters with Gen. CROOK, and the two are expected to confer and work together. If both persuasion and threats fail to induce the warriors to accept the terms of the Government, then Gen. CROOK will have an opportunity to engage in a vigorous campaign, in which he will not be hampered and restrained and his efforts thwarted by the agents of the Peace Commission, because they have consented in advance to the policy. It is hoped that, with the exercise of good judgment and a wise discretion by the officers of the Government in the execution of their orders, Indian outrages in Arizona, at least, may be hereafter suspended.

The white settlers near the Indian territory seem determined to steal the Indian lands. They seem possessed with the idea that speedy extermination surely awaits the red men, and that those who have secured lodgment on their territory will do so to their profit. The various warnings given them by the Secretary of the Interior have not had the desired effect—to cause them to leave. It has come to this, that the preservation of the faith of the Government with the Indians in the interests of peace requires that those who settle upon Indian lands contrary to treaty stipulations shall be removed, and it is probable that a call will soon be made upon the War Department to see that their removal is made.

November 10, 1871

MINOR TOPICS.

The land speculators appear to be already casting covetous glances at the Indian Territory. One gentleman, just returned from a tour of observation, reports that among the tribes the question of selling off all the surplus land in the territory, and making the fund a general one, is fast gaining ground. This brilliant idea proceeds upon the assumption that the 44,000,000 of acres, comprising what is known as the Indian Territory, are held in fee-simple by the sixty thousand Cherokees, Creeks and others, who are at present its sole occupants. As pointed out in the report of the Secretary of the Interior, the present population of the Territory is but one to every 630 acres. The purpose of the Government in making this huge allotment for the benefit of the Indians, was to gather within it all the tribes who at present either roam on the plains, or hold large reservations elsewhere. Were the entire Indian population of the country gathered within the territory, there would still be 180 acres for every man, woman, and child they could muster. This is doubtless more than sufficient, but it will be time enough to talk of admitting white settlers within the Indian territory when the country has induced the 172,000 Indians, who occupy some 96,000,000 of acres in reservations outside of the territory, to migrate within its borders.

December 21, 1871

THE OSAGE INDIANS.

The Indians' Side of the Story—Agricultural Filibusters—A Magnificent Reserve Squatted Upon—The Troubles of Poor Lo—Justice Vindicated at Last.

From Our Own Correspondent.

PORT GIBSON, I. T., Thursday, June 6, 1862.

An interesting letter remains to be written about the Osage tribe of Indians, and if I fail to engage the attention of the reader in the performance of this task, it will be through no lack of suitable materials, but from a want of narrative power in the writer.

A RETROSPECT.

In my report of the last General Council held in Okmulgee, published in the TIMES last June, I made special mention of the Osage delegation which took part in the proceedings. It consisted of PAH-E-NO-PO-SHA, Governor of the Great and Little Osages; CHITOPA, HARD-ROPE, SAL-BA-TEA, delegates; a Chief or head man, who attended as invited guest, (whose name I have forgotten,) and a half-breed interpreter, named JAMES BIGHEART. With the exception of the interpreter, who was a tall, gaunt man, with a beaming black eye, a weasel face and a contracted lower jaw, this delegation was composed of noble-looking chieftains, of splendid physical development. The tribe has a princely fund in the hands of the Government, derived from the sale of their lands; and these chiefs presented themselves in all the bravery of war paint, with massive nose and ear jewelry, immense necklaces of bear's claws, (a highly valued ornament,) elaborate embroidery, and scarlet blankets rivaling the Syrian dye. Their entrance into the Council Chamber always produced a profound impression; and the beholder in running his eye over the motley array of one hundred Indians, costumed in every variety of style, from the ignorant Keechie, with his faded breech cloth and tomahawk, to the educated Cherokee, having the dress and manners of a gentleman, would find his gaze most attracted by this magnificent group of Osage warriors. And when his mind had taken in the whole scene, these painted and bespangled braves would arrest his final glance, and the eye would dwell upon them as objects of commanding artistic interest. Their former homes were in Southern Kansas, and comprised an area of seven million acres. This reservation had been assigned them by Government, when population began to flock into Kansas, and the stereotyped promise was made them that they should be protected from intrusion by the whites. But what border settler was ever known to pay regard to Indian treaties? Millions of acres of fine land was open to the homesteader in other portions of the State, and land offices were accessible at every requisite point, where he could file his claim and acquire right to a [illegible]m under the sanction of laws made and provided. But stolen waters are sweet, and the rude frontiersman takes especial delight in staking out his quarter-section when he knows it has been stolen from an Indian.

AGRICULTURAL FILIBUSTERING.

This agricultural filibustering was allowed to go on until the Osage reservation swarmed with thousands of settlers. The Indians complained to the Superintendent that they were being stolen poor by their white neighbors. When the Osages went west on their annual buffalo hunt, the squatters captured their ponies, raided on their stores of provisions, cleaned out their lodges and destroyed what they they could not carry away. The matter was maturely considered by the Indian Department, and it was found that the evil had grown to such extensive proportions that it would now be difficult to deal with. To attempt to eject the settlers would require a small army, and would raise such a howl through the whole western country that the eventual success of the "Quaker policy" would be seriously jeopardized. The old story was repeated. The weaker had to give way to the stronger. Superintendent HOAG called a council of the chief men of the tribe, and by skillful diplomacy, backed by his unbounded influence over the red race, induced them to part with their lands in trust to the Government, and take up a more confined home within the Territory, upon a tract of land ceded them by the Cherokee nation.

A CHIEF'S STORY.

When I made acquaintance with the delegation last June they had been living in the Territory some eight months. JOSEPH PAH-E-NO-PA-SHA, the Governor, I found to be an educated man, having a fair knowledge of the English tongue, and possessing a good share of natural astuteness. He retained the blanket, and spoke through an interpreter, as he was unwilling to lose influence with his tribe through adopting alien habits. I met him in Col. W. P. Ross' room (a Cherokee delegate) one day, and he departed from his unapproachable stolidity by extending his hand to me. I received the civility with intense gratification, and further placated the haughty chieftain by tendering him a cigar. An interesting conversation ensued. In answer to my inquiries, he said his people had not planted any corn that Summer, nor settled down to a permanent life. The western limit of their reservation was in dispute by the Cherokee nation. The preliminary survey of the ninety-sixth principal meridian, which formed the western boundary of their tract, gave them the Caney River, with a strip of five or six miles of rich bottom land. This the Cherokees claimed as belonging to them, and to settle the dispute an official survey had been ordered. If the resurvey should deprive the Osages of this strip, the reservation was not worth retaining, as it was rough, broken and uncultivable. They were again harassed with white intruders, he complained. Hundreds of settlers were swarming upon their land, and while Government showed such bad faith with its wards, he should make no effort to reclaim his people.

JUSTICE FOR THE RED MAN.

Every case has two sides. It is popular to paint the Indian race as thieving, treacherous and bloodthirsty, irreclaimable in their habits and an impediment to the advance of civilization. I have no desire to create any sympathy in their behalf, and in choosing a hero I should never imitate FENNIMORE COOPER. But I believe in simple justice. I approve the humane and wise policy of President GRANT in seeking to solve the Indian problem by just dealings, and an effort at civilising the aborigines instead of abandoning them to the tender mercy of the unscrupulous frontiersman, and then harrying them with fire and sword in their retributive attempts to gain wild justice.

THE BOUNDARY SETTLED.

The official survey has since been made, and the Caney River, with its strip of rich bottom land, is found to belong to the Cherokees. From Mr. HOAG I learn that the causes of dissatisfaction which rankled in the Osage Governor's mind at the time of his conversation with me, have been removed, and a more contented feeling is now manifested by the whole tribe. Last February a Commission, consisting of Messrs. THOMAS WISTAR, of Philadelphia, JOHN B. GARRETT, Philadelphia, and GEO. HOWLAND, New-Bedford, Mass., all members of the Society of Friends, visited the Territory, under the appointment of Secretary DELANO, to hold a council with the Osages and effect some equitable adjustment of their differences. In this they have happily succeeded. Their reservation has been extended from the ninety-sixth meridian to the Arkansas River east and west, and from the Kansas border to the Creek nation, north and south. Within this tract the Kaw Indians, a small tribe of about 700 persons, lately removed from Kansas, have been affiliated with the Osages.

THE SQUATTERS REMOVED.

The settlers upon the Osage lands have also been removed. A notice or manifesto was issued by the Superintendent after the convening of this council, in which he warned all persons illegally trespassing upon Indian lands to pull up stakes and depart. But hard words break no bones, and the settlers paid no attention to the summons. They are mainly ignorant and unscrupulous, and they listened to the counsels of persons (having a land speculation in view) who assured them that Government would not persist; that when the direct issue was between the white and the red races, the powers in Washington would never proceed against a community of hardy pioneers in the interest of a few thousand lazy, thieving Indian devils. But their predictions were falsified this time. An order issued from the War Department to Gen. SHERIDAN and through that officer to Gen. POPE, commanding the Department of the Missouri, at Fort Leavenworth, brought Major UPHAM upon the ground with a detachment of United States cavalry, who quietly notified the recalcitrant settlers that they must stand not upon the order of their going, but go at once. No violence was used in the ejection. The misguided squatters, seeing their case was a hopeless one, sullenly folded up their tents, loaded their wagons with their few calamities and took their solitary way to new homes.

KANSAS.

June 15, 1872

THE INDIANS.

The Sioux Reservation Demanded by Civilization.

WASHINGTON, May 15.—It has heretofore been stated that a commission has been appointed to the Sioux. The objects are stated in the letter of the Secretary of the Interior to the Commissioner of Indian Affairs, which has just been written. Referring to the alleged violations of the sixteenth article of the treaty concluded with the various bands of Sioux Indians April 29, 1868, he says the progress of population westward has already rendered it desirable that the territory embraced in this article be no longer considered to be unceded Indian territory, and that it be surveyed according to our system of Government surveys, and made accessible to homestead and pre-emption settlement, as well as to sale to persons desiring to settle upon it. To continue much longer the exclusion of white people from the settlement of this country will necessarily occasion great complaint and provoke much feeling among the inhabitants of the State of Nebraska. To prevent the difficulties here referred to, and to remove complaints already made upon this subject, I have concluded to appoint commissioners charged with the duties of treating with the various bands of Sioux Indians, parties to the treaty of 1868, for a relinquishment of all the privileges reserved to said Indians by the sixteenth article of the treaty aforesaid, and for the restoration of the territory therein embraced, without incumbrance to the public domain.

The Secretary says that, in consideration of the eleventh article, I have to remark that white settlements near to if not within this territory, and upon which there was reserved to the Indians the right to hunt so long as the buffalo may range thereon in such numbers as to justify the chase, in my opinion render it exceedingly hazardous to permit this privilege to be longer exercised. Large herds of cattle owned by settlers upon our frontier are now fed, if not upon, yet very near to the territory assigned to the Indians as hunting-grounds. To permit this privilege of hunting on the part of the Indians is likely to cause difficulties and outrage between white settlers and those engaged in the hunting expeditions.

I deem it, therefore, of very great importance to the peace of the country, to the security of settlements on the frontier, and to the welfare of the Indians, that the latter should be induced to relinquish these hunting privileges. The commission will therefore be charged with the duty of endeavoring to procure the agreement of the Indians to relinquish the privilege here referred to. It is probably important that the present agency selected for Red Cloud and his band should be located elsewhere. The commission will, therefore, be charged with the duty of inquiring whether such changes should be made, and if so, where the new location should be made. Specific instructions have been given to carry the above programme into effect, and also with a view to acquire part of the Crow lands.

May 16, 1873

WASHINGTON.

Special Dispatch to the New-York Times.

WASHINGTON, May 18.—The Interior Department has on its hands at the present time the most difficult and embarrassing question of governmental affairs—that of dealing with, controlling, and pacifying the Indian tribes, in the face of the ever-encroaching demands of railroad construction, and the never-ceasing pressure of the settlers. The field which now really presents the most difficult position of Indian affairs is that of Montana and Dakota. It is now less than five years since a solemn treaty was made with the Sioux nation, particularly the tribes represented by Red Cloud and Spotted Tail, guaranteeing them settled reservations covering portions of both Nebraska and Dakota, and the right to hunt on the Republican River in Kansas, so long as the buffalo shall range there in sufficient numbers to justify the chase. In the meantime the settlers of Nebraska are pushing up into this reservation, and meet all expostulation with the retort that the Government had no right to locate a reservation within the State limits without the consent of the State. The settlers in Kansas object to the ranging of the Indians on the Republican because they fear outrage and loss of stock. On the other hand, the Northern Pacific Railroad is crossing the Missouri into Northern Dakota, and with its progress the Sioux see that their hunting grounds disappear forever. If, therefore, the Summer passes without trouble, it will be because of the exercise of great patience and discretion on the part of the Secretary of the Interior and his Commissioner of Indian Affairs.

Pertinent to this point is the result of a council recently held at Red Cloud's agency. Late in March a commission was appointed, consisting of Jno. P. Williamson, Rev. S. D. Hinman, and Agent J. W. Daniels, to sound the various bands of Teton Sioux as to their feeling in regard to the progress of the railroad. The council was held April 9, with representatives of bands of Teton Sioux numbering 400 to 500 lodges. The Indians said they were glad to hear from their Great Father, but did not want the Northern Pacific Railroad built nor white men in their country. All the white man they wanted to see was the trader, and wanted him to sell guns and ammunition. They promised, however, to take the words of the Great Father home with them, and talk to their people, and when they agreed to let the commissioners know. The commissioners say in conclusion:

Your commission does not see the way open to prosecute the work further at present. They feel no hesitancy in assuring the department that there will be no combined resistance to the construction of the Northern Pacific Railroad. The Indians have neither ammunition nor subsistence to undertake any general war; neither do they manifest any such spirit in council. Small raiding parties will doubtless visit the Northern Pacific Road, and perhaps the border settlements. It is probable that a majority of the Indians with whom we consulted will remain in the vicinity of this agency, and in view of the scarcity of buffalo in the Sioux country it is believed that all Northern hostile Sioux will ultimately be compelled to come to the different agencies for subsistence.

May 19, 1873

THE INDIANS.

A BRUSH WITH A RAIDING PARTY—RECAPTURE OF STOLEN STOCK—ONE OF THE INDIANS KILLED.

WASHINGTON, May 4.—The following telegram has been forwarded here:

FORT ABRAHAM LINCOLN, via BISMARCK, D. T., April 23.

To the Assistant Adjutant General of the Department of Dakota, at St. Paul:

About 11 o'clock to-day a band of Indians attacked a party of citizen herders a short distance below this post, and drove off about eighty head of mules belonging to the citizens. Within ten minutes after receiving the notice of this, I had six companies of cavalry in the saddle, and began a vigorous pursuit, the Indians having several miles the start. For twenty miles we kept up the pursuit, almost at a continuous gallop. As the result, I report the recapture of every animal taken by the Indians, and the capture of a portion of the stock belonging to the Indians. I suffered no loss in men, while the Indians had at least one of their number wounded, besides being compelled to abandon saddles and other property in their flight. My command behaved handsomely, and reached Fort Lincoln on its return at 11 o'clock the same night. No portion of the stock captured by the Indians and recaptured by my command belonged to the Government.

(Signed,) G. A. CUSTER, Brevet Major General.

The commanding officer at Fort Fetterman, Wyoming Territory, reports that two Sioux fired into the camp of his scouts on the night of April 26.

Capt. Carleton, of the Tenth Cavalry, commanding Camp Augur, Texas, shows that Indians from the Reservation fired into his camp on the night of April 9, at a distance of about eighty-five paces. This is the second time they have done so within three months. Camp Augur is situated just outside the Reservation, across the Red River.

May 5, 1874

THE INDIANS.

THE DEATH OF COCHISE—HIS ADVICE TO HIS PEOPLE.

WASHINGTON, June 26.—A letter from Indian Agent Saffold to the Commissioner of Indian Affairs, reporting the death of Cochise, stated that there was no danger of trouble resulting therefrom. His last words to his people, as reported by Agent Saffold, were to come to the agencies, men, women, and children, and forever live at peace with our people. Cochise's eldest son Taza had been proclaimed as his successor, and acknowledged as such by all his tribe.

THE SIOUX IN DAKOTA AND CUSTER'S CAVALRY.

ST. PAUL, Minn., June 26.—From all information thus far obtained, it does not look as if the Sioux in Dakota are particularly anxious for a fight with the Seventh Cavalry, which they can have any day. Gen. Custer's expedition to the Black Hills would have started last Sunday if there had been any military necessity therefor. Its departure is now postponed until the 30th inst, or 1st prox.

June 27, 1874

WASHINGTON.

OWNERSHIP OF INDIAN RESERVATIONS.

A decision has recently been rendered by the Supreme Court which has an important effect upon the tribal relations of Indians. Chief Justice Waite read the opinion, which decides that the Indian tribes do not own the fee to the lands within their respective reservations, and are no more than tenants of the United States, with whom alone the fee is vested. The decision also holds that timber and minerals are a portion of the realty, and that the sale of timber, or a lease of the mineral lands, cannot be made by the Indian tribes. The case which evoked this decision is one growing out of the sale by two Indians on a reservation near Green Bay, in Wisconsin, of a quantity of pine logs. The Indian Agent brought a suit in behalf of the tribe to recover the logs, on the ground that the lands and the timber belonged to the Indians in their tribal relations. The decision is as noted above, and it makes the United States the only owner of the Indian lands. One of the effects of this decision is to invalidate the contract for timber made by E. P. Smith, the present Commissioner of Indian Affairs, when he was an Indian Agent in Wisconsin, which is known as the "Wilder contract." It will also deprive the Indians on several important reservations of what has been supposed to be their tribal possessions, the valuable pine lands. In all the cases where reservations have been ceded to tribes by treaty, this decision makes the Government the Trustee of the Indians, and the money for the sales of lands belongs to the United States, though under existing laws in most cases it must be disbursed for the benefit of the Indians.

November 28, 1874

Dispatch to the Associated Press

NEW FEATURE IN THE INDIAN POLICY.

A new feature is added to the Indian policy, namely: In distributing the supplies and annuities the agent is directed to require all able-bodied male Indians between the ages of eighteen and forty-five to perform service upon the reservation for the benefit of themselves or of the tribe at a reasonable rate to be fixed by the agent in charge, and to an amount equal in value to the supplies to be delivered. The allowances to be provided for such Indians is to be distributed to them only upon condition of the performance of labor. The object is to make the Indians self-supporting. Congress incorporated in the Indian Appropriation bill a section providing that hereafter no purchase of goods, supplies, or farming implements, or any other article whatsoever, the cost of which shall exceed $1,000, shall be paid for from the money appropriated unless the same shall have been previously advertised and contracted for.

March 15, 1875

INDIANS' RIGHTS.

A LETTER FROM BISHOP HARE.

To the Editor of the New-York Times:

The recent visit of a large delegation of Sioux chiefs and head men to New-York will, perhaps, gain for a few words of appeal a hearing which, under other circumstances, I might not hope to get. I should not presume to seek it but that residence among the Sioux, and frequent trips through the Indian country as a Missionary Bishop of the Episcopal Church, and two journeys into the vicinity of that part of the country known as the Black Hills, have made me familiar with some facts that may not be generally known.

These Sioux chiefs represent a people numbering about 35,000, who in the past have been among the fiercest of the meat-eating tribes of the Northwest. The chiefs are all of them famous themselves for prowess in past years upon the war-path. They accepted some years ago, however, treaty relations with our Government, and may be said, all things considered, to have observed in a commendable degree the obligations then assumed. They stand, therefore, midway between the Northern Sioux, some 10,000 in number, who yet maintain an attitude of utter independence, and those Sioux on the Missouri River who have begun to erect houses, till the ground, and wear the white man's dress, and who have been gathered into schools and churches, and have learned to read and write.

No one who has mingled among the Sioux Indians can doubt that, however far short of his wishes their present condition may be, it is strikingly advanced to what it was a few years ago. Civilization has been effecting slow but real victories; missions have been advancing; children are being gathered into schools; and Christian women engaged in the mission work are to-day living undisturbed in districts where, but five years ago, few white men, except squaw men, would have dared to show their faces.

But, unfortunately for the trust and quiet which such efforts engender, and in which they flourish, the Sioux, like Naboth in sacred history, and the poor man in Nathan's parable, own something very dear to them which a more powerful neighbor covets. It is the country known as the Black Hills of Dakota. The Indians' attachment to it is a passion. And well it may be, for this district is the kernel of their nut, the yelk of their egg. While the rest of their reservation is "a dry and thirsty land where no water is," this hill country is reported as abounding in "fountains and wells that spring out of valleys and hills." While many of the streams outside of these hills I know to be in Summer nauseously tepid, turbid, and alkaline, the streams in the Hills are said to be even in the hottest weather deliciously cool and always sweet and crystal clear. While most of the rest of their land is sun-baked and blasted by scorching siroccos, these hills are reported as attracting frequent showers. While much of the rest of their land is utterly denuded of all soil and the famous "*mauvaises terres*" or bad lands of Dakota occupy large stretches of it, the soil of the valleys in these hills is reported to be rich and deep and carpeted with grass and flowers; and while much of the rest of their reserve is utterly treeless, and the traveler seeks in vain, as I know by experience, for wood enough to heat water to make a cup of coffee with, these hills are well covered with elm and oak and pine. In the opinion of four gentlemen, all of them familiar with the country, with whom I conversed a few days ago, the timber (the only fuel for the Indian in a climate where the Winters are long and the mercury ranges from ten degrees to forty degrees below zero for weeks together) will, at the present rate of consumption, have all disappeared in less than ten or fifteen years. Manifestly, no one needs this tract of land so much as its present possessor, its rightful owner.

But what rights has an Indian? Three years ago an expedition was projected and partially organized in Dakota for the purpose of seizing upon this territory. Happily the Executive, in this instance, acted with great decision. A proclamation was issued warning evil-disposed persons of the determination of the Government to prevent the outrage, and troops were set in motion to deal effectively with the marauders. Thus checked, rapacity slumbered until a year ago, when a military expedition having penetrated the Black Hills and inflamed our cupidity with stories of its wealth, bands of reckless adventurers began to invade the Sioux country—thus far, thank God, only to be captured by the military. Many others hover on its borders; our cities, from the Missouri to the Atlantic, are placarded with "Gold! Gold! Gold! Ho! for the Black Hills." And the excitement and the pressure are so great that even the President of the United States, to whom I would pay a grateful tribute as the tried friend of the Indian, thinks it wise to succumb and to try and direct the storm which he cannot resist, and is obliged, with other friends of the Indians, in the effort to see that the evil is done as gently as possible, to play what may seem to some the rôle of *particeps criminis.*

The only plea which the proposed effort to obtain this country from the unwilling Indians can make with any force is this, that civilization having driven the game from the plains, the Indians have become dependent for their food on the bounty of the white man's Government, and that being "beggars" they must not be "choosers." Whether this apology will avail in view of all that the white man has already taken away from the Indian, let the kind and just determine.

The other plea under which the proposed effort seeks cover, viz, that barbarism has no right to hold back vast areas of land from the tillage of the needy settler, is, in this case, entirely without point. The chief sinners in this line are not Indians, but white speculators, who have bought up land and hold it by the ten thousand acres, to the exclusion of the needy. And, in the next place, it is an entire mistake to suppose that the area occupied by these Sioux Indians is vast. Their reservation proper, adding in the neutral lands which bound it on the south and west, is only about four hundred miles square. Their reservation proper is not two hundred and fifty miles square. Of this, as I have shown, a large portion is an utterly inhospitable waste. In God's name, I ask, may not the Sioux, who number some 35,000 souls, enjoy the occupation of the pitiful remainder? That they will willingly surrender it, driven to the wall as they already are, is, I fear, a fond hope. But if they will, manifestly we should be prepared to pay them so liberally and judiciously that the loss of this part of their land shall redound to their good. As, however, these Indians are a brave and warlike people, as they love their homes passionately, and as all the past has revealed to them that the white man has no pity, we should not be surprised if, insisting now on buying with money what the Indian does not wish to sell, we drive him to frenzy, our covetous enterprise end in massacre, and we pay for the Indian's land less in money than in blood.

WILLIAM H. HARE,
Missionary Bishop of Niobrara.

PHILADELPHIA, Friday, June 18, 1875.

June 21, 1875

MASSACRE OF OUR TROOPS.

FIVE COMPANIES KILLED BY INDIANS.

GEN. CUSTER AND SEVENTEEN COMMISSIONED OFFICERS BUTCHERED IN A BATTLE ON THE LITTLE HORN-ATTACK ON AN OVERWHELMINGLY LARGE CAMP OF SAVAGES-THREE HUNDRED AND FIFTEEN MEN KILLED AND THIRTY-ONE WOUNDED—TWO BROTHERS, TWO NEPHEWS, AND A BROTHER-IN-LAW OF CUSTER AMONG THE KILLED—THE BATTLE-FIELD LIKE A SLAUGHTER-PEN.

SALT LAKE, July 5.—The special correspondent of the Helena (Montana) *Herald* writes from Stillwater, Montana, under date of july 2, as follows:

Muggins Taylor, a scout for Gen. Gibbon, arrived here last night direct from Little Horn River, and reports that Gen. Custer found the Indian camp of 2,000 lodges on the Little Horn, and immediately attacked it. He charged the thickest portion of the camp with five companies. Nothing is known of the operations of this detachment, except their course as traced by the dead. Major Reno commanded the other seven companies, and attacked the lower portion of the camp. The Indians poured a murderous fire from all directions. Gen. Custer his two brothers, his nephew, and brother-in-law were all killed, and not one of his detachment escaped. Two hundred and seven men were buried in one place. The number of killed is estimated at 300, and the wounded at thirty-one.

The Indians surrounded Major Reno's command and held them one day in the hills cut off from water, untill Gibbon's command came in sight, when they broke camp in the night and left. The seventh fought like tigers, and were overcome by mere brute force.

The Indian loss cannot be estimated as they bore off and *cached* most of their killed. The remnant of the Seventh Cavalry and Gibbon's command are returning to the mouth of the Little Horn, where a steam-boat lies. The Indians got all the arms of the killed soldiers. There were seventeen commissioned officers killed. The whole Custer family died at the head of their column.

The exact loss is not known as both Adjutants and the Sergeant-major were killed. The Indian camp was from three to four miles long, and was twenty miles up the Little Horn from its mouth.

The Indians actually pulled men off their horses, in some instances.

This report is given as Taylor told it, as he was over the field after the battle. The above is confirmed by other letters, which say Custer has met with a fearful disaster.

ANOTHER ACCOUNT.

SALT LAKE CITY, July 5.—The *Times* publishes a dispatch from Boseman, Montana Territory, dated July 3, 7 P. M.

Mr. Taylor, bearer of dispatches from Little Horn to Fort Ellis, arrived this evening, and reports the following:

The battle was fought on the 25th of June, thirty or forty miles below the Little Horn. Gen. Custer attacked an Indian village of from 2,500 to 4,000 warriors on one side, and Col. Reno was to attack it on the other side. Three companies were placed on a hill as a reserve.

Gen. Custer and fifteen officers and every man belonging to the five companies were killed. Reno retreated under the protection of the reserve. The whole number killed was 315. Gen. Gibbon joined Reno.

When the Indians left, the battle-field looked like a slaughter-pen, as it really was, being in a narrow ravine. The dead were much mutilated.

The situation now looks serious. Gen. Terry arrived at Gibbon's Camp on a steamboat, and crossed the command over and accompanied it to join Custer, who knew it was coming before the fight occurred. Lieut. Crittenden, son of Gen. Crittenden, was among the killed.

July 6, 1876

THE INDIAN SLAUGHTER.

THE BATTLE AND THE BATTLE-FIELD

NARROW ESCAPE OF RENO'S COMMAND FROM ANNIHILATION—FORTY-EIGHT HOURS' FIGHTING WITH INDIANS—FRIGHTFUL SUFFERING FROM THIRST—A BOLD AND SUCCESSFUL DASH FOR WATER—THE RELIEF BY GEN. TERRY—MEN WEEPING FOR JOY—THE FRIGHTFUL FATE OF CUSTER'S FORCE GRADUALLY DISCOVERED—NO LIVING MAN LEFT—THE FIVE COMPANIES DESTROYED SUCCESSIVELY—TOTAL LOSS 261 KILLED IN BOTH COMMANDS—FUTURE OPERATIONS.

Special Dispatch to the New-York Times.

CHICAGO, July 7.—Further particulars of the desperate encounter which Col. Reno had with the Sioux Indians on the 25th of June teil of the sufferings which his command experienced while it was so completely hemmed in by the red devils. I have already forwarded you a hasty sketch of the encounter. For thirty-six hours the troops were without a drop of water. The appeals of the wounded for drink were heartrending, while the others were almost exhausted, in many instances their tongues protruding from their mouths, and few of them able to speak aloud. They tried to eat crackers, but could not moisten them. Others attempted to chew and swallow blades of grass to secure relief, but these clung to their parched lips and intensified their agony. It was while thus suffering that they determined at all hazards to gain the water from which they were cut off, and made the desperate dash which, while it cost them a number of lives and many wounded, secured that which they so much needed. It was then early night, and when firing ceased Col. Reno at once took steps to relieve his animals, which, like the men, were completely exhausted. He knew full well that the Indians would resume the attack in the morning.

It was in this position that Gen. Terry, with Gibbon's command, consisting of five companies of infantry, four of cavalry, and the Gatlin battery, found Reno. Terry had started to ascend the Big Horn to attack the Indians in rear, while Custer attacked them from his point of contact. The march of the two columns was so planned as to bring Gibbons' forces within co-operating distance of the anticipated scene of action by the evening of the 26th. In this way only could the infantry be made available, as it would not do to encumber Custer's march with foot soldiers. On the evening of the 24th Gibbon's command was landed on the south bank of the Yellowstone, near the mouth of the Big Horn, and on the 25th was pushed twenty-three miles over a country so rugged that the endurance of the men was taxed to the uttermost. The infantry then halted for the night, but the department commander with the cavalry advanced twelve miles further to the mouth of the Little Big Horn, marching until midnight in the hope of opening communication with Custer. The morning of the 26th brought the intelligence, communicated by three badly frightened Crow scouts, of the battle of the previous day and its results. The story was not credited, because it was not expected that an attack would be made earlier than the 27th, and chiefly because no one could believe that a force such as Custer commanded could have met with disaster. Still the report was in no way disregarded. All day long the toilsome march was kept up, and every eye bent upon a cloud of smoke resting over the southern horizon, which was hailed as a sign that Custer was successful, and had fired the village. It was only when night was falling that the weary troops lay down upon their arms. The infantry had marched twenty-nine miles. The march of the next morning revealed at every step some evidence of the conflict which had taken place two days before. At an early hour the head of the column entered a plain half a mile wide, bordering the left bank of the Little Big Horn, where had recently been an immense Indian village, extending three miles along the stream, and where were still standing funeral lodges with horses slaughtered around them, and containing the bodies of nine chiefs. The ground was strewn everywhere with carcasses of horses. Their camp was strewn with robes, gaudily painted, with finely-dressed hides and interesting and valuable trinkets. The ground was covered everywhere with carcasses of horses, besides buffalo robes, packages of dried meat, and weapons and utensils belonging to the Indians. On this part of the field was found the clothing of Lieuts. Sturgis and Porter, pierced with bullets, and a blood-stained gauntlet belonging to Capt. Yates. Further on were found the bodies, among whom were recognized Lieut. McIntosh, the interpreter from Fort Rice, and Reynolds, the guide. It was evident also that the Indians had shown little solicitude for their wounded, as the ravines were covered with dead.

While making these gloomy discoveries a scout came up in breathless haste with the announcement that Col. Reno, with a remnant of the Seventh Cavalry, was intrenched on a bluff near by waiting for relief. The command pushed rapidly on and soon came in sight of a group surrounding a cavalry guard upon a lofty eminence on the right bank of the river. Gen. Terry forded the stream, accompanied by a small party, and rode to the spot. All the way the slopes were dotted with the bodies of men and horses. The General approached, and the men swarmed out of the works and greeted him with hearty and repeated cheers. Within was found Reno, with the remains of seven companies of the regiment, with the following named officers, all of whom are unhurt: Capts. Benteen, Weir, and Moylan, and Lieuts. McDougal, Godfrey, Mathey, Gibson, De Rudio, Edgerly, Wallace, Varnum, and Hare. In the centre of the inclosure was a depression in the surface, in which the wounded were sheltered, covered with canvas. Reno's command had been fighting from Sunday noon, the 25th, until the night of the 26th, when Terry's arrival caused the Indians to retire.

Up to this time Reno and those with him were in complete ignorance of the fate of the other five companies which had been separated from them on the 25th to make an attack under Custer on the village at another point. While preparations were being made for the removal of the wounded, a party was sent on Custer's trail to look for traces of his command. They met a sight to appall they stoutest heart. At a point about three miles down the right bank of the stream, Gen. Custer had evidently attempted to ford and attack the village. From the ford the trail was found to lead back up to the Bluffs and to the northward, as if he had been repulsed and compelled to retreat, and at the same time had been cut off from rejoining the forces under Reno. The bluffs are cut into by numerous ravines, and all along these slopes and ridges, and in the ravines lay the dead, lying in the order of battle as they had fought. Line behind line showed where defensive positions had been successively taken up and held, till at last few were left to fight, and then, huddled in a narrow compass, horses and men were piled promiscuously. At the highest point of the ridge lay Custer, surrounded by his chosen band. Here were his two brothers and his nephew, Mr. Reed, Capt. Yates, Lieuts. Cooke and Smith, all lying within a circle of a few yards, their horses beside them. Here behind Yates' company the last stand had been made, and here one after another of these last survivors of Custer's five companies had met their death. The companies had successively thrown themselves across the path of the advancing enemy, and had been annihilated. Not a man had escaped to tell the tale, but it was inscribed on the surface of these barren hills in a language more eloquent than words.

Two hundred and sixty-one bodies have been buried from Custer's and Reno's commands. The last one found was that of Mr. Kellogg, correspondent of the Bismark *Tribune*.

The following are the names of the officers whose remains are recognized: Gen. Custer, Capts. Keogh, Yates, and Custer, Lieuts. Cooke, Smith, McIntosh, Calhoun, Hodgeson, and Reilly. All of these belonged to the Seventh Cavalry. Lieut. Crittenden, of the Twentieth Infantry, was serving temporarily with the regiment. Lieuts. Porter, Sturges, and Harrington and Assistant Surgeon Lord are reported missing, as their remains were not recognized; but there is small ground to hope that any of them survived, as it is obvious that the troops were completely surrounded by a force of ten times their number. Charley Reynolds, a scout, was also killed. Reynolds, with Kellogg, Dewolf, Reed, and Boston Custer were the only citizens killed.

The history of Reno's operations comprises all that is now known of this sanguinary affair. It seems that Custer, with eight companies, reached the river in the forenoon of the 25th, having marched continuously all the previous day and night. Seeing the upper or southern extremity of the village, and probably underestimating its extent, he ordered Reno to ford the river and charge the village with three companies, while he, with five companies, moved down the right bank and behind the bluff, to make a similar attack at the other end. Reno made his charge, but finding that he was dealing with a force many times his own numbers, dismounted his men, and sought shelter in the timber which fringed the river bank. The position appearing to him untenable, he remounted and cut his way to the river, forded under a murderous fire, and gained the bluff where he was subsequently found. Here he was afterward joined by Capt. Benteen with three companies which had just reached the field, and by Capt. McDougall with his company and the pack mules. The position was immediately after completely invested by the Indians, who for more than twenty-four hours allowed the garrison no rest and inflicted severe loss. But for the timely arrival of relief the command would have been cut off to a man. The number saved with Reno was 329, including fifty-one wounded. The loss among the Indians was probably considerable, as bodies have been found in every direction, and they left behind only a small portion of their dead.

In closing this narrative of this affair, in certain respects the most remarkable in modern history, I purposely refrain from comment. The naked facts, so far as they are known, must guide your readers to a conclusion as to the cause of the calamity. Information derived from many sources, including, of course, the observations of officers engaged in the battle, lead to the conclusion that 2,500 or 3,000 Indians composed the fighting force arrayed against Custer and his 600. Still these were odds which any officer of the Seventh Cavalry would have unhesitatingly accepted for his regiment under any ordinary circumstances of Indian warfare. The force under Gen. Terry's immediate command was designed not only to cut off the retreat of the Indians, but to afford support to Custer if needed. Its march was made in accurate accordance with the plan communicated to each of the subordinate leaders before the movement commenced. It reached the point where the battle was expected at the time proposed, and had not the action been precipitated for reasons which are as yet unknown, a force would have been present on the field sufficient to retrieve any repulse of the attacking column.

Col. Gibbon's cavalry followed the Indians for about ten miles, and ascertained that they had moved to the south and west by several trails. A good deal of property had been thrown away by them to lighten their march, and was found scattered for many miles over the prairie. Many of their dead were also discovered secreted in ravines a great distance from the battlefield. Among them were Arapahoes and Cheyennes as well as Sioux.

A correspondent who was on the scene with Terry's column writes that they remained there nearly two days, burying the dead and preparing to transport the wounded to a place of safety. The neighboring country was still full of scattered bands of Indians, watching the movements of the troops, and doubtless prepared to take advantage of any want of vigilance to add to the number of their victims. A species of rude horse-litter was constructed of poles and strips of hide, and on these the disabled were carried twenty miles, to the forks of the Big Horn, where they were placed on board the steamer.

Thirty-five wounded men were sent forward to Fort Lincoln, where Mrs. Custer was at the time the battle was fought. The wounded were in charge of Capt. Smith, who also bears dispatches to Gen. Sheridan. Three of the wounded died on the way. The remainder of the wounded were cared for in the field. Capt. Smith proceeded by the steamer Far West, which made 900 miles in less than thirty-six hours, the current of the Yellowstone and the Missouri Rivers aiding her in this quick transit.

Gen. Terry has, through Capt. Smith, who was also to proceed with dispatches, submitted his plan of campaign to Gen. Sheridan, and the action to be taken will depend upon the answer given. Meanwhile Terry will await supplies of provisions and clothing, of which his command is nearly destitute.

The latest intelligence at head-quarters here to-day is that the official report of Custer's encounter has been sent via Fort Ellis, communication with which point is now closed owing to the fact that the wires are down. A detachment of cavalry was sent out a week ago to repair the lines, and it is expected that they will be working again within a few days. It will, therefore, be several days probably before Terry's report is received giving the full details. This report, however, can contain but little that has not been published beyond the names of the soldiers killed. The terrible nature of the action is already known beyond doubt.

There is general interest in knowing what action will be taken by the Government and in the matter and the plan of operations which will be adopted by Gen. Sheridan in any further operations against the savages. Yesterday, in obedience to orders from the Lieutenant General, Col. Miles, with six companies of the Fifth Infantry, was ordered to the relief of Gen. Terry's command, and will leave Fort Leavenworth immediately via Yankton, Dakota Territory. The Fifth Regiment of Cavalry has been withdrawn from its advanced position in front of the Red Cloud and Spotted Tail agencies, and ordered to concentrate at those agencies. No further intelligence has been received at head-quarters this forenoon concerning any movement of troops of Crook's or Terry's commands.

A special dispatch to the Chicago *Evening Telegraph* received to-day from St. Louis says: "Lieut. James Garland Sturgis, killed with Custer, was out on his first campaign. He was the son of the Colonel of the Seventh Cavalry. He was assigned to the Seventh, his father's regiment, and joined the command last September at Fort Rice. The feeling against Custer is very strong among army men here. Nearly all the officers killed have spent more or less time on duty here. The greater number af the privates came from about St. Louis, Desmoines, Iowa, and from Central Illinois. "I think," says Gen. Sturgis, "that Custer pursued much the same tactics as he did during the war when fighting white men, and was heedlessly drawn into an ambush resulting in the loss of his own life and the destruction of his commrnd. We who have lost sons by the rash act feel deeply and can scarcely refrain from speaking out. Somebody is very much to blame, and it is the duty of the War Department to ascertain who is responsible for this loss of life." Col. Sturgis said that the very flower of his regiment had been cut down, and that more gallant and chivalrous men never went into battle; and while he grieved for the loss of his own son, he also felt deeply for the widows and orphans of the brave officers and soldiers who had so often followed him into battle.

July 8, 1876

SITTING BULL AND THE SIOUX.

PERSONAL SKETCH OF THE SAVAGE CHIEF —PECULIARITIES OF TRIBE WHEN ON THE WAR PATH—CONVERSATION WITH A FORMER AGENT'S CLERK.

The St. Louis *Globe-Democrat* of Saturday, gives the following sketch of the Sioux Chief in a conversation with Mr. J. D. Keller, of that city:

Mr. Keller was from 1868 to 1873 clerk of the agent at Standing Rock, and had ample opportunities to get acquainted with this tribe of bloodthirsty savages. In fact he lived among them so long that he learned to speak their language "like a native," and was a great favorite of the big chiefs who came to the agency. They called him "Minnehua Ochila" [the Writing Boy.] The word Sioux means "cut-throat." According to Mr. Keller's statement, the various bands of Sioux number from 35,000 to 45,000, and are divided into the following different tribes: Unkapapas, Black Feet, Sans Arcs, Two Kettles, Upper Yanktonais, Lower Yanktonais, Santee Sioux, Burgklys, Minneconjous, and Guikas.' Part of these live east and part west of the Missouri River. Tatonka Otanka (Sitting Bull.) who led the savages in the fight against Custer, belongs to the Unkapapas (dried beef eaters.) Mr. Keller knows him well, and describes him to be about five feet in height. He has a large head, eyes, and nose, high cheek bones; one of his legs is shorter than the other, from a gun-shot wound in the left knee. His countenance is of an extremely savage type, betraying that bloodthirstiness and brutality for which he has been so long notorious. He has the name of being one of the most successful scalpers in the Indian country. There has been a standing reward of $1,000 offered for his head for the last eight years, by the Montana people, who have special cause to know his ferocious nature, some of his worst deeds having been perpetrated in that Territory. The Sioux, when on the war path, black their faces from the eyes down, the forehead being colored a bright red. When in mourning, and very eager to revenge the death of friends or relations, they cut their hair short and daub their faces with white earth. Their feats of horsemanship are wonderful. They consider the greatest act of valor to be the striking of their enemy with some hand instrument while alive, and, whether alive or dead, it is the first one that strikes the fallen foe that "counts the *coup*," and not the one that shoots him. They do not always scalp. Their object in scalping is to furnish a proof of their deed, and give them to their women to dance over. They always attack in a sweeping, circling line, eagle-like, give a volley, pass on, circle, and return on a different angle. When they kill one of the enemy there is always a rush to get the first crack at him, so as to "count the *coup*," and then some Indian who was disappointed in getting a cut at the victim while alive scalps him. The Sioux always camp with tepes (lodges) in a circle, making, as it were, a stockade, and when on dangerous ground they picket their ponies in the centre. Mr. Keller is familiar with the ground where the disastrous engagement of Custer occurred. Concerning this he said: "My idea of the Custer slaughter is that the Indians had no women and children in their lodges, and had parapets dug under the lodges out of sight. Custer, thinking it was a family camp, rushed in the centre of their fort, where resistance would necessarily prove fatal. His only means of escape was, after finding himself in this fix, to run right through and out, and not stop to fight, but join Reno's command and retreat."

July 10, 1876

EXTERMINATION.

The repulse of CROOK, the defeat of RENO, and the slaughter of CUSTER and his men have profoundly stirred the nation. It is natural that we should smart under the victories of a foe whom we had despised. It is right that we should mourn with the sincerest sorrow the gallant men who died with the dauntless CUSTER. It is even desirable that our defeats should impel us to wage war in the sharp, vigorous manner which is the truest mercy to friend and foe. But it is neither just nor decent that a Christian nation should yield itself to homicidal frenzy, and clamor for the instant extermination of the savages by whose unexpected bravery we have been so sadly baffled.

All through the West there is manifested a wild desire for vengeance against the so-called murderers of our soldiers. The press echoes with more or less shamelessness the frontier theory that the only use to which an Indian can be put is to kill him. From all sides come denunciations of what is called in terms of ascending sarcasm, "the peace policy," "the Quaker policy," and "the Sunday-school policy." Volunteers are eagerly offering their services "to avenge CUSTER and exterminate the Sioux," and public opinion, not only in the West, but to some extent in the East, has apparently decided that the Indians have exhausted the forbearance of heaven and earth, and must now be exterminated as though they were so many mad dogs.

What is meant by "the Quaker policy" which is thus bitterly assailed? If it means anything, it means the policy of justice and humanity. Whatever may have been the faults of the present Administration, history will credit it with having at least made the attempt to treat the Indians fairly. Where we have become involved in war with the Modocs or the Sioux, the cause is to be found not in the maligned "Quaker policy," but in occasional acts of willful or ignorant injustice, which were so many deviations from the very policy to which the present Indian war is falsely imputed. We ordered the Modocs to remove to a barren reservation in Oregon, where they were in absolute danger of starvation. They left the reservation, preferring, as they expressed it, to die in battle in their own country rather than to starve in a strange and sterile land. The facts in the case have been fully set forth by Mr. MEACHAM, a man who bears the scars of Modoc rifle-bullets, but who still advocates the "Quaker policy," and who maintains that the Modoc war was in no possible sense the result of that policy. Neither was the Sioux war brought about by peace men, or Quakers, or Sunday-school sentimentalists. We bound ourselves by treaty with the Sioux to prevent white men from entering the Black Hills country, which we had ceded to them forever. We then sent CUSTER to explore the country at the head of a column of troops, and his report of the discovery of gold mines was followed by a rush of reckless gold-hunters. These acts, which were in direct violation of our solemn treaty obligations, were surely not the outgrowth of a peace policy. CUSTER'S troopers were not Quakers, nor were the Black Hills miners Sunday-school superintendents. The one bright feature in this miserable business was the long forbearance of the savages to attack either the exploring expedition or the miners, and the loyal bearing of YOUNG MAN AFRAID OF HIS HORSES, who, at the council where the Sioux declined to sell a territory as large as the State of Michigan for fifty thousand dollars, saved the Commissioners from the massacre meditated by the wilder tribes. It was not until after we had failed to cajole the Sioux into a sale, and had openly abandoned all pretense of observing our treaty obligations, that the Indians attacked the miners, and with the aid of outlying clans like the band of SITTING BULL, renewed the fight of centuries against white aggression. This is the true and shameful origin of the Sioux war; and had the Quaker policy of justice been faithfully and intelligently carried out, neither the Modocs nor the Sioux would have been provoked into hostility.

If it is unreasonable to lay at the door of the peace policy results due strictly to deviations from it, there is a like lack of reason in the anger which styles SITTING BULL'S recent victory a "fiendish massacre." CUSTER went out to beat the Sioux. Had he succeeded, would he have been guilty of a "fiendish massacre"? The soldier has blows to take as well as to give, and there is no justice in styling the defeat of an attacking force "a fiendish massacre," when its success would have been called a glorious victory. We did not fancy that the Southern people deserved extermination because we were beaten at Bull Run, nor did the rebels call the defeat at Gettysburg a "fiendish massacre."

Over the border the Indians and the colonists live in peace. The peace policy which we have tried as a new thing—dropping it now and then through weariness or inadvertence—has there proved so complete a success that its wisdom is conclusively demonstrated. Is there a strange and baleful magic in the invisible boundary line, whereby the Indians who, on the other side of it are peaceable and trustworthy, become on this side utterly treacherous and bloodthirsty? If not, there must be some mistake in the theory that extermination is the only policy which should be pursued toward the Indians of the United States.

We are now at war, and there is nothing left but to prosecute the struggle with the utmost rigor. Peace can be obtained only by the thorough defeat of the enemy, and for the moment any peace policy must necessarily be suspended. But we can wage war without displaying an unnecessary ferocity, more worthy of savages than of civilized men. We must beat the Sioux, but we need not exterminate them. And when the war is over it will be the duty of every humane man to demand that the peace policy which has made border warfare unknown in the vast Hudson Bay territory shall be made successful on our side of the line, in spite of the men who believe that the laws of GOD should be suspended in our relations with the Indians, and the dictates of humanity trampled under foot.

July 12, 1876

A TALK WITH SITTING BULL

HE RELUCTANTLY GIVES SOME ACCOUNT OF HIS LIFE.

HOW HE WON HIS CHIEFTAINSHIP—HIS FIGHT WITH CUSTER AND EXCUSE FOR IT—HIS FAMILY.

Fort Yates (Dakota) Correspondence of the St. Paul (Minn.) Pioneer-Press.

The conversation, omitting the delays, side remarks, and questions to which no response would be given, run as follows:

Interpreter—Where were you born, and when?

Sitting Bull—I don't know where I was born and cannot remember. I know that I was born, though, or would not be here. I was born of a woman; I know this is a fact, because I exist.

Sitting Bull here held a long conversation with his uncle, Chief Four Horns, and after pointing at different fingers for some time, said:

"I was born near old Fort George, on Willow Creek, below the mouth of the Cheyenne River. I am 44 years old, as near as I can tell; we count our years from the moons between great events. The event from which I date my birth is the year in which Thunder Hawks was born. I am as old as he. I have always been running around. Indians that remain on the same hunting-grounds all the time can remember years better."

Reporter—How many wives and children have you?

Sitting Bull (running over his fingers and then with thumb and forefinger of one hand pinching and holding together two fingers of his other hand)—I have nine children and two living wives, and one wife that has gone to the great spirit. I have two pairs of twins.

Lieut. Dowdy—Tell Sitting Bull he is more fortunate than I am. I can't get one wife.

At this interruption Sitting Bull laughed.

Reporter—Which is your favorite wife?

Sitting Bull—I think as much of one as the other. If I did not I would not keep them. I think if I had a white wife I would think more of her than the other two.

Reporter—What are the names of your wives?

Sitting Bull (raising the side of the tent and calling a squaw to him; evidently he asked her)—Was-Seen-By-The-Nation is the name of the old one. The One-That-Had-Four-Robes is the name of the other.

Reporter—Are you a chief by inheritance, and if not, what deeds of bravery gave you the title?

Sitting Bull—My father and two uncles were chiefs. My father's name was The Jumping Bull. My uncle that is in the teepee is called Four-Horns and my other uncle was called Hunting-His-Lodge. My father was a very rich man, and owned a great many good ponies in four colors. In ponies he took much pride. They were roan, white, and gray. He had great numbers, and I never wanted for a horse to ride. When I was 10 years old I was famous as a hunter. My specialty was buffalo calves. [Here Bull indicated with his arms how he killed the buffalo.] I gave the calves I killed to the poor that had no horses. I was considered a good man. [Here Bull again counted on his fingers and joints.] My father died 21 years ago. For four years after I was 10 years old I killed buffalo and fed his people, and thus became one of the fathers of the tribe. At the age of 14 I killed an enemy, and began to make myself great in battle, and became a chief. Before this, from 10 to 14, my people had named me The Sacred Standshoty. After killing an enemy they called me Ta-Tan-Ka I-You-Tan-Ka, or Sitting Bull. An Indian may be an inherited chief, but he has to make himself a chief by his bravery. [Although several efforts were made and much tact used, Sitting Bull would not speak of his life beyond the age of 14.]

Reporter—Besides yourself, whom do you think the greatest and bravest chief of the Sioux nation?

Before answering this question Sitting Bull took a long smoke, then handed his pipe around, and played with a knife in his sheath. Withdrawing it, he said: "When I came in Buford I gave up everything. I even gave up all my knives but this. This is the only weapon I have. It is not sharp. I keep it to fix pipes. [Meditatively again he recalled the last question and said:] There are five great chiefs of the Sioux nation before me. They were: He-to, (meaning Four Horns,) Ce-su-ho-tan-ka, (meaning Loud Voiced Hawk,) Helo-ta, (meaning Scarlet Horn,) Can-te-tanka, (meaning Big Heart,) and Ta-to-ka-en-yan-ka, (meaning Running Antelope.) All are dead but Running Antelope and Four Horns. He is the bravest chief besides myself. Antelope is witko, (meaning a Fool.) He has been among the whites, and asked all of us to surrender.

At the conclusion of this sentence the interpreters turned to the reporter and told him that this statement of Sitting Bull was made from a spirit of hatred and jealousy, and that Running Antelope was a great chief, and had done more than any other to get the Indians to surrender. Sitting Bull, it was ascertained, had in some manner interpreted or understood this conversation, for he shortly interfered, and said he had no hatred or jealousy in his heart when speaking of other chiefs.

"What induced you to surrender and what wrongs have you suffered at the hands of the Government?" These and other questions were put in all sorts of ways, and at first the chief refused to answer them, saying this was an ordinary talk, and these were questions of great importance to him. After hesitating a long time, and being assured by the interpreter that it was best to speak, he finally spoke in an excited and rambling way as follows: "Already have I told my reasons. I was not raised to be an enemy of the white. These five chiefs that I have named were not enemies of the white man. The pale faces had things that we needed in order to hunt. We needed ammunition. Our interests were in peace. I never sold that much land. [Here Sitting Bull picked up with his thumb and forefinger a little of the pulverized dirt in the tent, and holding it up let it fall and blow away.] I never made or sold a treaty with the United States. I came in to claim my rights and the rights of my people. I was driven in force from my land and I now come back to claim it for my people. I never made war on the United States Government. I never stood in the white man's country. I never committed any depredations in the white man's country. I never made the white man's heart bleed. The white man came on to my land and followed me. The white man made me fight for my hunting grounds. The white man made me kill him or he would kill my friends, my women, and my children."

Reporter—The white man admires your conduct in a battle. You showed yourself to be a great chief in the Custer fight.

Sitting Bull—There was a Great Spirit who guided and controlled that battle. I could do nothing. I was sustained by the great mysterious One. [Here Sitting Bull pointed upward with his forefinger.]

Reporter—You conducted the battle well; so well that many thought that you were not an Indian, but that you were a white man and knew the white man's ways.

Sitting Bull, (pointing to his wrist)—I was not a white man, for the Great Spirit did not make me a white skin. I did not fight the white man's back. I came out and met him on the grass. When I say Running Antelope is a fool I mean he made treaties and allowed the white man to come in and occupy our land. Ever since that time there has been trouble. I do not want aid or assistance from the whites or any one else. I want them to stay from my country and allow me to hunt on my own land. I want no blood spilled in my land except the blood of the buffalo. I want to hunt and trade for many moons. You have asked me to come in. I wanted the white man to provide for me for several years if I came in. You have never offered me any inducements to come in. I did not want to come. My friends that come got soap and axe-handles, but not enough to eat. I have come in, and want the white man to allow me to hunt in my own country. That is the way I live. I want to keep my ponies. I can't hunt without ponies. The buffalo runs fast. The white man wanted me to give up everything.

Reporter—What treatment do you expect from the Government? If not satisfied, what shall you do?

Sitting Bull—I expected to stay but a few days at Buford. When I came in I did not surrender. I want the Government to let me occupy the Little Missouri country. There is plenty of game there. I have damages against the Government for holding my land and game. I want the Great Father to pay me for it.

The reporter here asked the interpreter to get an idea of what Sitting Bull meant by the Little Missouri country?

Sitting Bull—My hunting ground is from the bad lands to the end of the Little Missouri, and I want it extended down here where some of my people are, so that I can trade.

Reporter—What do Chiefs Gall and Running Antelope say about their treatment here?

Sitting Bull—Antelope is a fool. I have seen Gall. He can't tell me anything. He is not a chief of my people.

Reporter—Don't you think the Indians here are treated well?

Sitting Bull—I have not had a chance to talk with them. They are waiting for me to speak. They want to give me a feast and hold a council. I am not jealous of them. I don't know whether we will hold a council or not.

Reporter—Tell us all about the Custer battle. How did it happen? Did you direct the main forces?

Sitting Bull (after a long silence)—I am not afraid to talk about that. It all happened; it is passed and gone. I do not lie, but do not want to talk about it. Low Dog says I can't fight until some one lends me a heart. Gall says my heart is no bigger than that, [placing one forefinger at the base of the nail of another finger.] We have all fought hard. We did not know Custer. When we saw him we threw up our hands and I cried: "Follow me and do as I do." We whipped each other's horses, and it was all over.

Reporter—Custer's men were all killed. There is no one to tell us about the battle but you. We keep a record of our battles and study them. We write histories of brave men. We will never fight the Sioux again. Tell us more about it.

Sitting Bull—There was not as many Indians as the white man says. They are all warriors. There was not more than 2,000.

Reporter—Crow King says, on the second day, when you were fighting Reno's men, that you asked your warriors not to kill any more, that you had already killed enough.

Sitting Bull—Crow king speaks the truth; I did not want to kill any more men. I did not like that kind of work. I only defended my camp. When we had killed enough, that was all that was necessary.

All further efforts by a series of a dozen questions failed to induce Sitting Bull to say anything further about this matter.

Reporter—Do you understand what this is all for? Do you know how a great newspaper like the *Pioneer-Press* goes out to thousands of people every morning?

Sitting Bull (making a mark on the ground and placing his finger on one side of it)—Yes, I have seen the great newspaper over there. [Meaning across the British lines.]

Reporter—Have you ever been interviewed before by a newspaper man upon these subjects?

Sitting Bull—I have never talked about these things to a reporter before. None of them ever before paid me money. My words are worth dollars. If the Great Father gives me a reservation I do not want to be confined to any part of it. I want no restraint. I will keep on the reservation, but want to go where I please. I don't want a white man over me. I don't want an Agent. I want to have the white man with me, but not to be my chief. I ask this because I want to do right by my people and can't trust any one else to trade with them or talk to them. I want interpreters to talk to the white man for me and transact my business, but I want it to be seen and known that I have my rights. I want my people to have light wagons to work with. They do not know how to handle heavy wagons with cattle. We want light wagons and ponies. I don't want to give up game as long as there is any game. I will be half civilized till the game is gone. Then I will be all civilized. I want peace and no trouble. I want to raise my children that they may have peace and prosperity. I like the way the white brother keeps his children. Miss Fanny Culbertson, of Poplar River, was the first person I shook hands with when I came over the line. My daughter came to see me last night. We both cried. I was happy to see her. The soldiers would not let her come into my camp at first. She came here before I did, and I listened a long time to hear word from her, and for the winds to tell me how she was treated. I did not hear. I came down to see her. She seems to be doing well, but I saw she had no respect from the whites. The soldiers would not spread down a blanket for her to walk into my camp. She is well dressed, but she says her relatives at the Agency gave her her clothes.

Reporter (to the interpreter)—Explain to Sitting Bull how President Garfield has been shot by a coward, and ask him what he thinks about the act; also, what the Indians would do if a coward would shoot their chief.

Sitting Bull—It was a cowardly act. If the warrior had been there he would have gone to the Great Father's face and looked him in the eyes, and then shot him. I heard when way up there about the Great Father being shot, but had no one to tell me all about it. I don't know whether the warrior was wise in doing it or not. He might have shot the Great Father because he was not treating the Indians right. If that was so it was not a bad thing to do. If a coward should shoot one of our chiefs or warriors without looking him in the eyes our friends would go and kill him. If he was a very rich coward he could pay the damages in many ponies and we would let him leave.

Reporter—How many scalps have you taken, not counting those taken by your people, which are always credited to the chiefs?

Sitting Bull (spreading out both hands and putting his two thumbs together, and pointing to his joints, and thinking for fully five minutes)—I have killed 16 enemies. I never killed a white man. I have made raids upon the Crees, Gros Ventres, and Northern Blackfeet, and stolen horses 22 times. I never stole horses from the whites.

Reporter—The President takes the *Pioneer-Press*. He will read your words. What message do you wish to send to him?

Sitting Bull—I have told you all I want. I would like to have the Great Father listen to what I have said and help me accomplish what I ask.

August 7, 1881

THE DEATH OF CRAZY HORSE.

HOW THE ONCE DREADED CHIEFTAIN RECEIVED HIS FATAL WOUND—SOME FACTS CONCERNING HIS LIFE.

The Cheyenne *Sun* says: "The notorious Sioux Chieftain, Crazy Horse, died at Camp Robinson at midnight of the 5th inst. It is a coincidence worthy of note that the unconquered and atrocious chief received his death wound—a terrible bowie-knife cut in the side—at the hands of one of his former followers, a soldier Indian, who had stood side by side with the chief in the battle of the Rosebud, in the Custer massacre, and in other engagements of the Sioux war. When Crazy Horse essayed his escape, a few days before his death, the Indian soldiers who had been foremost in acts of brutality one year ago hunted him down like blood-hounds. Relationship, tribal ties, and savage pride have all lost their hold before the keen manipulation of Gen. Crook and his junior officers, and the words of the General a year ago regarding these hostiles, that he could 'make brother hunt brother,' are strikingly verified in this instance. The scene around and inside the Adjutant's office as the chief died, was in perfect keeping with his long years of thrilling adventure and unstinted revenge. His father, with one or two faithful followers, watched by his side, while crowds of savages hung around the building, giving vent to their feelings in the curdling whoop and lending utter drear to the murky darkness with the ominous death chant. Indian soldiers and disarmed hostiles were there alike, the first glorying in their delegated peace-keeping power, and the others only harmless because helpless. When the chief was dead he was quickly borne down by the side of White Earth River, where four months ago he first knelt to the sway of the military, and there cared for with the customary savage ceremony by his few remaining friends. Crazy Horse was about 35 years old, a relative of Red Cloud, and an aborigine from head to foot. He was of medium height, slender and wiry build, and as mean a looking Indian as ever knocked for admission to the happy hunting ground. However, a bullet wound through his face, received in battle, added much to his brutal and treacherous look. He has known nothing but fight and marauding since boyhood, and was last seen at the agencies about 12 years ago. Among the hostiles he has been considered the peer of Sitting Bull in influence, and vastly that old chieftain's superior in out and out bravery, devilishness, and unrest. He has done more to retard the settlement of this frontier, by his deeds of blood and pillage, than all other war chiefs combined. The taking off of such a savage will result in more real good than a decisive battle, for it is he who has sown continued dissension and inspired atrocity in others, even on the heels of defeat."

September 14, 1877

A LESSON FROM THE NEZ PERCES.

Now that the Nez Percé war has ended in victory, thanks to the energy and courage of our much-enduring Army, it is worth while, before it passes out of mind, to ask why it was fought. We freely express the opinion that the Nez Percé war was, on the part of our Government, an unpardonable and frightful blunder—a crime, whose victims are alike the hundreds of our gallant officers and men who fell a prey to Nez Percé bullets and the peaceful bands who were goaded by injustice and wrong to the war-path. It is greatly to be regretted that the immediate responsibility for its occurrence is so obscurely distributed that it is difficult to bring anybody to account for it at the bar of public opinion.

The Nez Percé comes into history as the white man's friend. The famous exploring party sent out by President JEFFERSON said of this tribe: "The Pierced-Nose nation are among the most amiable men we have seen—stout, well-formed, well-looking, active, their character placid and gentle, rarely moved into passion, yet not often enlivened by gayety." From their first warm welcome of our explorers in 1805, up to the present war, no full-blooded Nez Percé is known to have murdered a white man—an extraordinary fact which is on record in the Government archives at Washington. With the Nez Percés we have always been at peace; and when we have had wars with other neighboring tribes, the Nez Percés have invariably been the allies of our Army. While other tribes have been roving and hard to control, a large part of the Nez Percés have taken to grazing and farming. Most of them live in houses like white men, and build fences around their lands. The brother of Chief LAWYER is a Presbyterian minister in Oregon. These harmless and peaceful neighbors, these faithful allies in every war, were the nation that we drove to desperation and deeds of blood.

When white men first found them, the Nez Percés had bands, but no general chief—a system often, if not always, found among Indians west of the Rocky Mountains. Thirty-five years ago the United States Indian Agent for that region undertook to remedy this, which in his wisdom he conceived to be a defect, by giving them a grand chief named ELLIS, whose main recommendation was that he had learned English at a mission-school and so could talk to the agent. Less peaceful Indians than the Nes Percés might have gone to war rather than agree to the agent's labor-saving contrivance. The Nez Percés protested, but waited patiently for ELLIS' death, when, however, they were pressed to choose a new general chief. The rivalry lay between JOSEPH, a scion of the most illustrious Nez Percés, the father of the present JOSEPH, and an Indian named LAWYER. As LAWYER, like ELLIS, knew English, he received the powerful support of the Government agents, who gave him an enormous advantage by conducting all their business with the Nez Percés through him. JOSEPH's father at length withdrew in disgust from the councils of his tribe, still claiming the headship, if there were to be any.

The dwelling-place of the Nez Percés, as far back as their tradition goes, has been the Wallowa Valley, prized by them for its roots and its fishing, and now for its grazing. The whites at length began to increase in numbers, and, of course, took measures to dispossess the Indians of the valley. For this purpose they framed the successive treaties of 1855, 1863, and 1868, providing the Nez Percé Reservation and annuities instead of their lands. Old Chief JOSEPH, however, refused to go upon the reservation, and remained, with his band and the other non-treaty Nez Percés, so called, in the Wallowa Valley, rightly claiming that the rest of the tribe had no right to give it up to the United States and white settlers as against the non-treaty Indians, because it was held in common. It should be said here that the present Chief JOSEPH insists that his father never signed any of these treaties; that this was his father's instruction to him, and that the fact that their band remained away from the reservation shows it. But the Government is said to be able to show that JOSEPH reluctantly signed the treaty of 1855. Admitting this latter statement to be correct, as we have no doubt that it is, yet the treaty which definitely undertook to give up the Wallowa Valley was that of 1863, and this most unquestionably old JOSEPH did not sign, as he also did not sign the treaty of 1868. He died in 1871.

The claim of JOSEPH to his ancestral homestead was, therefore, good, *primâ facie*—so good, at any rate, that there was no case for driving him out with the bayonet. Even the treaty Indians could almost have claimed to have the Wallowa Valley restored to them, for it is proved beyond any doubt that the Government never carried out its stipulations for land partition, which formed an essential part of the treaties. The Government, however, holds with reason that the continued acceptance of the benefits of the treaties partly meets that objection as regards them; and for the non-treaty bands, off the reservation, another means has been found—coercion, violence, and a bloody war.

Now, these are not fancy sketches or rumors; they are officially-ascertained facts, well known to the Government, and they are to be found in two admirably clear reports, drawn up and presented to the Government more than a year ago, by no less an authority than Col. H. CLAY WOOD, a staff officer of Gen. HOWARD himself. They are the result of most careful investigations, and one of Col. WOOD's conclusions in a later report was that the present JOSEPH and his band have in law an undivided and individual interest in *all* lands ceded to the United States by the treaty of 1863, though he only claimed the Wallowa Valley, or rather the tract of land set apart by President GRANT's order of 1873.

A brief reference to this last order must close the story. Settlers having begun to encroach on the non-treaty Indian lands, in June, 1873, President GRANT ordered that these possessions should be "withheld from entry and settlement as public lands, and that the same be set apart as a reservation for the roaming Nez Percé Indians, as recommended by the Secretary of the Interior and the Commissioner of Indian Affairs." The decision took the form of an order, because no treaties are allowed since the act of 1871. JOSEPH continued, therefore, with his band to peacefully occupy his ancestral home in the Wallowa Valley under an order which warned off others from interfering with his rights. Two years ago President GRANT issued another order summarily revoking the former, and saying that "the said described tract of country is hereby restored to the public domain." The settlers at once encroached; the Government ordered the non-treaty Nez Percés to go upon the reservation; to their pleadings against the injustice, the menacing reply was the gathering of troops which the Indian Department called for, in order, if necessary, to put them on by force. Just before the time set for executing this scheme, JOSEPH, who had held back for months, and to the last moment, from a resort to which the peaceful Nez Percés were repugnant, at length, counseling with his brother non-treaty chiefs, and seeing soldiers assembled to drive him from his home, desperately plunged into war—a war which, on our part, was in its origin and motive nothing short of a gigantic blunder and a crime.

October 15, 1877

GEN. HOWARD AND CHIEF JOSEPH.

SPEECH OF THE INDIAN CHIEF IN SURRENDERING TO GEN. MILES.

WASHINGTON, Nov. 15.—Lieut. Wood, Assistant Adjutant-General to Gen. O. O. Howard, is in this city. At the time of the surrender of Chief Joseph to Gen. Miles, Lieut. Wood made a verbatim report of the speech delivered by the chief, which was as follows: "Tell Gen. Howard I know his heart. What he told me before I have it in my heart. I am tired of fighting. Our chiefs are killed. Looking Glass is dead. The old men are all dead. It is the young men who say yes or no. He who led on the young men is dead. It is cold and we have no blankets. The little children are freezing to death. My people, some of them, have run away to the hills and have no blankets, no food. No one knows where they are—perhaps freezing to death. I want to have time to look for my children and see how many of them I can find. Maybe I shall find them among the dead. Hear me, my chiefs, I am tired. My heart is sick and sad. From where the sun now stands I will fight no more forever."

November 16, 1877

INDIANS GOING TO WORK.

WASHINGTON, June 16.—A letter received at the Bureau of Indian Affairs, mentions that Chief "Howling Wolf," who was one of the prisoners at St. Augustine, has induced 70 Cheyenne Indians to cut off their scalp locks and assume the garb of civilized Americans. They have also, under the direction of the school contractor of their reservation, gone actively at work cutting wood. The Arrapahoes, watching the Cheyennes at work, have also decided to become industrious, and want to go to work in the same manner.

June 17, 1878

THE NEZ PERCES AND THE MODOCS.

The Leavenworth (Kan.) *Times* prints the following private letter from Interpreter Chapman, of the Nez Percés:

NEZ PERCES CAMP, QUAWPAW AGENCY,
Indian Territory, Aug. 29, 1878.

DEAR SIR: Chief Joseph is very much dissatisfied, and, I think, with very good reasons, having lost 39 of his people since our arrival here. We have been very poorly provided with medicines and other necessaries for a sick camp, especially when we have had as many as 265 on the sick list at one time. I have never heard of so much suffering among the same number of people in all my life as we have here, and nothing for them to eat but beef and bread. Mothers dying, leaving children 6 and 8 months old to be taken care of, and no milk or anything else to feed them on except what I send off and buy and pay for out of my own pocket. I am living in my tent near the Indian camp, which is situated about two miles from the agency. My duties are acting agent, hospital steward, Commissary Sergeant, interpreter, and Superintendent of Farming, and I am looking for an appointment as agent of a life insurance company. The Modoc Indians, who are our next-door neighbors, numbered 155 when they landed here, and out of that number 87 have died. So I am told by their head chief, Steam-boat Charley. A pretty good country to get rid of Indians in. Yours truly, A. I. CHAPMAN.

Interpreter and Superintendent of Farming.

September 8, 1878

PROTECTING INDIAN RIGHTS.

A PROCLAMATION BY THE PRESIDENT WARNING UNAUTHORIZED PERSONS TO KEEP AWAY FROM INDIAN TERRITORY.

WASHINGTON, April 26.—The following proclamation was issued this afternoon by the President:

Whereas, It has become known to me that certain evil-disposed persons have, within the territory and jurisdiction of the United States, begun, and set on foot, preparations for an organized and forcible possession of and settlement upon the lands of what is known as the Indian Territory, west of the State of Arkansas, which Territory is designated, recognized, and described by the treaties and laws of the United States and by the executive authorities as Indian country, and as such is only subject to occupation by Indian tribes, officers of the Indian Department, military posts, and such persons as may be privileged to reside and trade therein under the intercourse laws of the United States; and

Whereas, Those laws provide for the removal of all persons residing and trading therein without express permission of the Indian Department and Agents, and also of all persons whom such Agents may deem to be improper persons to reside in the Indian country;

Now, therefore, for the purpose of properly protecting the interests of the Indian nations and tribes, as well as of the United States, in said Indian Territory, and of duly enforcing the laws governing the same, I, Rutherford B. Hayes, President of the United States, do admonish and warn all such persons so intending, or preparing to remove upon said lands, or into said Territory, without permission of the proper Agent of the Indian Department, against any attempt to so remove or settle upon any of the lands of said Territory, and I do further warn and notify any and all such persons who may so offend that they will be speedily and immediately removed therefrom by the Agent, according to the laws made and provided, and if necessary the aid and assistance of the military forces of the United States will be invoked to carry into proper execution the laws of the United States herein referred to.

April 27, 1879

WHITE MEN ROBBING INDIANS.

OUTRAGES NEAR THE RED CLOUD AGENCY—THE ARMY POWERLESS TO INTERFERE.

WASHINGTON, July 3.—The Interior Department is to-day in receipt of official information that white men, during the past few weeks, have stolen about 700 horses from the Indians at the Red Cloud Agency, and run them across the Nebraska line. The State authorities are doing nothing to prevent similar raids upon the Indians' property, and the military authorities, on account of the Posse Comitatus law of last year, stand by without intercepting or pursuing the marauders, although the stolen horses are driven right past Camp Sheridan and Camp Robinson, on the way to market or to the horse-thieves' corrals. The Indian Agent, having no armed force at his command, is powerless to stop the depredations, and for the present they seem likely to continue. The Indians, notwithstanding their keen sense of injury, arising from this want of adequate protection, manifest no symptoms of insubordination, but remain entirely peaceable and are beginning to devote themselves to farming. The Spotted Tail Indians, within the past two years, have lost several thousand horses in the same way.

July 4, 1879

IN THE WAY.

The oratory that followed the luxurious indulgence in creature comforts, wherewith the New-England Society celebrated the stern virtues and simple tastes of the Puritans, was somewhat varied from the rhetorical monotony of congratulation and self-admiration customary on such occasions, by the diversion of Gen. SHERMAN against the Indians. After reminding the company that we had a vast, unpeopled domain that would have to wait many years yet for the plowshare, he declared that the little surviving remnant of the original possessors of the continent would have to get out of the way for advancing civilization. We have millions of acres of fertile soil yet untilled, and vast wildernesses whose primal solitude is still undisturbed, save by the screeching wild-fowl, the howling beasts, and the roaring cataracts. We are urging the surplus population of other countries to come and take a share in this vast domain. We tell them there is room for them for generations to come, and yet we have no place to be allotted in peace and security to the few thousand aborigines who still linger about our borders.

They are "entitled to fair consideration," says Gen. SHERMAN, but they are continually getting in the way, and must move on whenever the superior white man takes a fancy to the particular spot they chance to occupy. We have always treated them with "fair consideration." We have a peculiar theory, which they obstinately refuse to understand, but which we find wonderfully convenient for making them get out of the way. All the territory within the boundaries of the United States is subject to the jurisdiction of its Government, which may be exerted whenever there is any object to be promoted by it. Every human creature within those limits must yield obedience to the behests of the Government, whenever it sees fit to put its authority in exercise over them. And yet we have always dealt with the Indians as if they were an independent people, and fostered the delusion that within such domain as they were permitted to occupy they were subject to no Government but their own. We have made treaties and agreements with them, such as no national authority ever before made with its own subjects. They have been led to believe that they were not subjects, and so they have entered into bargains with a perverse expectation that these would be respected. Having thus flattered them with the notion that they were independent within certain territorial limits, we have proceeded to establish agents, under the pretense of carrying out our agreements with them, and have insisted that they should be submissive to these representatives of national authority. This they cannot be made to understand; but they are perverse barbarians. We permit them to be robbed and exercise no jurisdiction for their protection; but when white men suffer from collision with them, we remember that these are subjects of the Government and must be protected and even revenged. Our theory has two entirely different faces, like the shield in the fable; one side is presented to the Indians, the other to the whites, and it is no wonder that they are continually quarreling as to what it is made of.

When we negotiate and make treaties with Indians, and induce them to accept certain reservations of land, we deal with them as an independent people, and they so understand it. The land is theirs, and under the agreement the authority of the United States is withdrawn from a certain circumscription. That is the side of the theory presented to them, and yet the other side is kept in view, turned toward the advancing white man. The authority of the United States may still be exerted over the Indians at the will of the Government and against the will of the other party to the agreement. Settlers and miners may still go on their way into the plains and forests in any direction, for these are under the jurisdiction of the Government, according to this side of the theory, and the pioneers are its subjects and entitled to protection. When the Indian possessions are reached, the Indians are in the way. They must move on. Civilization and progress are coming, and cannot be impeded in their course. Indians are not independent peoples, and their lands are not their own. The jurisdiction of the Government is over them and their territory, in spite of treaties and agreements. They must move on. This is Gen. SHERMAN'S view. It is the view on which our Indian policy has all along been based, and it has led to continued misunderstanding and a settled sense of wrong and injustice on the part of the Indians. It can have no consummation but the final extermination of all the aboriginal tribes. Civilization must ultimately overrun and surround them, and gradually extinguish their existence as completely as it has that of the Mohawks and Senecas. Can there not be enough humanity injected into our statesmanship to give us an Indian policy that shall look to the preservation and protection of this race, instead of its destruction? It can only be by adopting a theory consistent with itself and understood on both sides, and acting upon it.

December 24, 1879

THE WARDS OF THE NATION.

The annual report of Secretary SCHURZ is naturally occupied more largely by a discussion of the Indian policy of the present Administration than with any other topic which falls within the purview of the Department of the Interior. During the past year there have occurred several events which attract attention to what is known as the Indian policy of the Government. In his exceedingly able report, Secretary SCHURZ summarizes that policy in the following language:

"To respect such rights as the Indians have in the land they occupy; to make changes only when such lands were found to be unsuitable for agriculture and herding; to acquaint the Indians with the requirements of civilized life by education; to introduce among them various kinds of work by practical impulse and instruction; gradually to inspire them with a sense of responsibility through the ownership of private property and a growing dependence for their support upon their own efforts; to afford to them all facilities for trade consistent with their safety as to the disposition of the product of their labor and industry for their own advantage; to allot to them lands in severalty, with individual ownership, and a fee simple title inalienable for a certain period; then with their consent and for their benefit to dispose of such lands as they cannot cultivate and use themselves to the white settlers; to dissolve, by gradual steps, their tribal cohesion, and merge them in the body politic as independent and self-relying men, invested with all the rights which other inhabitants of the country possess."

No humanitarian, no philanthropist, could ask anything more fair than this. Unhappily, however, the Secretary has been hampered by what may be termed bequests from a previous Administration, and he has not been able, in all cases, to carry out his wise and generous designs. Several of these complications, derived by inheritance, are referred to in the report of the Secretary. Take, for example, the case of the Ponca Indians, which Mr. SCHURZ doubtless had in his mind when he referred to the difficulties which he encountered at the very outset of his administration. When he entered upon the duties of his office he found that it had been ordered that this small tribe, domesticated on their reservation in the North, and living in that condition of happiness which the Secretary fondly pictures as the ideal toward which all good Indians must struggle, had been ordered to give up their farms and homesteads and move southward into the Indian Territory. At first, we assume, the Secretary did not realize how deeply the Indians had been wronged by this order. And when, after he had become more familiar with the business of the office, he learned the truth, he felt that things had gone too far for him to reverse the official drift of things. The Ponca Indians had been bereft of homes in which they were working out the problem of self-support, so ably sketched in the policy of the Secretary of the Department of the Interior, and just now quoted. While the Secretary, and many others conversant with the matter, deeply deplored the injustice and wastefulness of the removal of the Poncas from their reservation, it seems to have been considered that no mandate could be recalled, and no backward steps taken, for fear that the Indians would consider this as a concession to their demands and would conclude that the Government of the Republic of the United States is a weak concern that could be bullied into making any sacrifice for the sake of peace. In the face of official and popular condemnation of the act, the Poncas were driven from their homes in order that the majesty of the law might be vindicated.

In fact, the dealings of the Government with almost every tribe of American Indians are irreconcilably in contest with the admirable but apparently impracticable theory propounded by the Secretary. **It is proposed that the right of the individual Indian to a certain tract of land, to be held in perpetuity, shall be held and recognized. But this right, as enjoyed by the tribe or by the individual, is not now held to exist. For example, it is just now reported that the Indian Agent at the Crow Creek Agency is on his way to Washington with a delegation of Brulé Sioux chiefs, their tribe having conceded to the Chicago, Milwaukee and** St. Paul Railroad Company the right of way through their reservation. The chiefs visit Washington to settle the details of this concession. It appears, then, that the Sioux have conceded a right of way through their reservation to a railroad corporation. The railroad corporation has thought it worth while to ask this concession at the hands of the Indians, and not of the United States Government. Will the Government allow this bargain to be made between the Brulé Sioux and the railroad company as between owners and buyers of lands? Do Indian tribes, then, own the right of way through their reservations? If this is so, how did it happen that the Indian Bureau, last Summer, issued orders to the agent in charge of the Ponca reservation in the Indian Territory to arrest any member of the Omaha Ponca Relief Committee who should set foot on that reservation? If Indians may concede a "right of way" through their reservation to a railroad corporation, how does it happen that the agent in charge of such a reservation may, without consultation with said Indians, and presumably against their wish, arrest and deport an agent of a relief committee, as was done in the case of Mr. TIBBALS, who was forbidden to return to the Ponca reservation at the peril of his life?

These are some of the inconsistencies which beset our exceedingly variegated and ramshackle Indian policy at the present time. Secretary SCHURZ, who propounds his admirable theories in a truly admirable manner, does not attempt to reduce this confusion to any semblance of order. It is a gratifying fact, as shown in the report of the Secretary, that the number of Indians who are able to sustain themselves, wholly or in part, is slowly and surely increasing. Notwithstanding the fact that the policy of the Government has been fickle and capricious, and notwithstanding the discouragements which those who wish well for the Indian encounter from mere theorists in power, it may be said that the so-called Indian problem is steadily solving itself. If it were possible for the Government to pursue a uniform policy toward the Indians, so far as fundamental rights are concerned, without being afraid to right a wrong whose existence is once admitted, the philosopher's dream and the humanitarian's hope of seeing the red men merged in the body politic will not long remain unfulfilled.

December 2, 1880

NEW PENAL CODE FOR THE INDIANS.

Indian Commissioner Price has sent a circular letter to all the Indian Agents containing new rules for their guidance. The rules establish a tribunal to be called "the Court of Indian Offenses," and to consist of three persons, not one to be a polygamist, at each agency, to be nominated by the agent after consultation with the Indians. The "Judges of Indian offenses" are to be paid $20 per month; their term is a year, and they are removable at any time by the Commissioner. The court is to hold two sessions a month, and special sessions. The agent is empowered to enforce attendance of witnesses and the orders of the court. The "Indian offenses" to be tried by this court are the sun-dance, scalp-dance, war-dance, and "sports assimilating thereto." Unlucky Indians found guilty of such festivities are to lose their rations for 15 days. Any plural marriage is to be punished by a fine of $20 or hard work for 20 days or both. "Medicine men," for using their influence against civilization, are to be shut up in prison for from 10 to 30 days and to be fed on bread and water. Stolen property must be returned and the thief shut up for 30 days. When an able-bodied Indian is convicted of failing to support his wife and children no rations are to be issued to him. An appeal lies from the "Court of Indian Offenses" to the Commissioner of Indian Affairs.

December 28, 1882

DEPREDATIONS BY INDIANS

RUMOR OF AN ENCOUNTER BETWEEN APACHES AND TROOPS.

GERONIMO SAID TO HAVE BEEN DEFEATED BY CAVALRY—MURDERS OF DAILY OCCURRENCE—TROOPS IN PURSUIT.

SILVER CITY, New-Mexico, May 28.—A report has reached here that a small detachment of the Tenth Cavalry encountered Chief Geronimo and his band of Apaches in Cook's Cañon, through which the Indians were trying to escape to Mexico. The hostiles were driven back with a loss of four killed and eight wounded. Two soldiers were killed and eight wounded. Owing to the small number of troops the Indians were not pursued. This detachment is trying to effect a junction with two companies of the Fourth Cavalry, when an active pursuit will be made. The Apaches have been joined by a number of Utes and Navajos, and the band is composed of nearly 200 warriors. The Indians retreated toward Diamond Creek, where their women, numbering nearly 100, preceded them.

A dispatch from Fort Bayard reports that Indians are leaving the reservation daily. The number of Indians who have been killing the whites during the last 10 days is said by the military authorities to be only 134, 34 bucks, 8 half-grown boys, and 92 squaws. Outside reports indicate that there are many more. News from the north says that several men have been killed in the Black Mountain country.

Capt. Smith, of the Fourth Cavalry, who followed the Indians from the reservation, passed through Silver City yesterday on his way to Fort Bayard. In the fight at Devil's Park one Indian was killed. One soldier and one Indian scout were wounded. Capt. Smith routed the Indians and captured 200 rounds of Government ammunition and nine ponies. Gen. Bradley, who is now at Fort Bayard, has ordered two troops of the Tenth Cavalry after the Indians who are reported to be on the Upper Gila River. Later advices are to the effect that the Indians are breaking up into small parties and scattering through the country in this direction from the Gila River. Forty-five armed men left here this evening to protect the families now surrounded on Bear Creek.

A courier from Juniper Spring, 10 miles from here, states that 30 Indians, including squaws and children, are camped there. One ranch has been captured. A man from a ranch near Negro Creek, four miles from here, reports fighting there. One man and a child were killed and one man was wounded. Parties are organizing to go out. Arms are scarce. A courier from Capt. Madden's command has arrived here with a request for supplies. He reports finding the bodies of two more murdered prospectors and a hot trail.

May 29, 1885

MONEY FOR INDIAN SCALPS.

ARIZONA AND NEW-MEXICO SETTLERS PROPOSE TO DESTROY THE SAVAGES.

DEMING, New-Mexico, Oct. 11.—It has been recently telegraphed that the pioneer settlers in the border counties of Arizona have brought to light an old law in several counties offering a reward of $250 each for Indian scalps. Under this law, which is nothing more than an order made by the County Commissioners, the ranchmen and cowboys in Cochise, Pima, and Yavapai Counties are organizing in armed bodies for the purpose of going on a real old-fashioned Indian hunt, and they propose to bring back the scalps and obtain the reward. Word now comes from Tombstone, the county seat of Cochise County, that the reward in that county has been increased to $500 for a buck Indian's scalp. The authorities of Pima and Yavapai Counties have taken steps to increase the reward to $500, and it is said Yuma, Apache, and Maricopa Counties will follow suit.

This reward system, while it may seem savage and brutal to the Northern and Eastern sentimentalist, is looked upon in this section as the only means possible of ridding Arizona of the murderous Apaches. The settlers of New-Mexico and Arizona are aroused on this question, and propose to act henceforth independent of the military authorities. From time immemorial all border countries have offered rewards for bear and wolf scalps and other animals that destroyed the pioneer's stock or molested his family. Why, therefore, asks the Arizona settler, should not the authorities place a reward upon the head of the terrible Apache, who murders the white man's family and steals his stock like the wolves? "Extermination" is the battle cry now, and the coming Winter will witness bloody work in this section.

Public sentiment in this part of New-Mexico and Arizona is strongly in favor of the immediate removal of Gen. Crook, who, it is declared, has always been overrated as an Indian fighter. It is the general belief here among old white scouts that Gen. Crook has been duped by his Indian scouts, and that the latter have always had secret understandings with Geronimo, and have repeatedly sent him word of movements of troops.

October 12, 1885

PLUNDERING THE INDIANS

AND ALSO SWINDLING THE GOVERNMENT.

HOW RINGS GO ABOUT THE WORK OF SECURING POSSESSION OF IMPROVED RESERVATIONS.

SANTA FÉ, New-Mexico, Oct. 12th.—The removal of the Chiricahua Apaches to Florida opens up another eligible opportunity for reservation speculation, and the set of demagogues ruling this Territory will not be slothful in hiving the spoil. For 30 years the Government has been neatly practiced upon by a method unfamiliar to any one outside of this Territory and as simple as it is ingenious. The first step in the scheme is to demand a removal of a certain tribe from a certain locality for reasons of "public" advantage. This is done by pressure brought to bear by the Representative, voluntarily or under compulsion. Simultaneously reports come up from the Indian Agent that the present place is unsuitable; that the tribe is restless, and would be benefited by a change. Then a discovery is made that a certain remote and fertile spot is just the panacea for all these ills together. The order for removal comes, and willing or unwilling the tribe leaves its reservation and goes to pastures new. Now, of course the abandoned reservation is useless to Government, and is offered at public auction, the ring buys it in, and thus the improved property, with roads, streams, dams, cultivated fields, fences, storehouses, and all that is needed for a great ranch, upon which the Government spent millions and the Indians lavished years of toil is acquired for a song.

If in the investigation of the Maxwell grant frauds the Government will rigidly inquire into the means by which the old Navajo Reservation, at Bosque Redondo, after having been made one of the most fertile and desirable tracts in the Territory, was abandoned and sold for a pittance considerable light will be thrown not only upon Maxwell's career, but perhaps upon that of officials then high in the Indian Department and of Representatives affecting to serve the best interests of New-Mexico and really serving their bank account at Government expense. The facts in the case are these: At the close of the Navajo war, in 1863, Gen. James H. Carleton, then commanding the Department of New-Mexico, ordered the conquered tribe from their almost inaccessible stronghold in the Cañon de Chelle to the Bosque Redondo, where a large reservation was set apart for their use. The Navajoes are a diligent and pastoral people, and under careful supervision set about improving the property that was to be their future home; dug long *acequias* to convey water for irrigating purposes, built roads, cleared and fenced fields in the fertile bottom, built substantial storehouses, planted orchards, and in fine made what was a wilderness a vast and prosperous farm. The Navajoes were contented and fast regaining prosperity when the fine Italian hand of the ring began to assert itself. One day an emissary from the Indian Department arrived from Washington to make a "report." Gen. Carleton was shortly afterward informed that this gentleman, instead of examining into the condition of affairs at the Bosque, was holding pow-wows with the chiefs, urging to infractions of the rules prescribed by the military authorities, inflaming their temper by bugaboo threats to cut off their supplies, and making promises of what the Great Father would do in their favor if they would make a unanimous demand for a new reservation. So flagrant indeed was his conduct and language that Gen. Carleton felt compelled to issue a peremptory order to the effect that if the ambassador did not attend strictly to his business and cease to excite mutiny he was to be summarily ejected from the reservation. A council was then held with the chiefs, and they unanimously expressed their satisfaction and desire to remain at the Bosque. A few months later, however, agitation of the subject began in Washington, and the Indians were ordered to be sent to their old stamping ground, where they now are. It was in vain that the military authorities opposed it or that the Indians themselves clamored to remain. The fiat was final, the Indians were removed, the Bosque came under the hammer and passed to Maxwell's ring for $200,000.

The case of the Chiricahua Apaches was parallel. In 1870 they were placed on a reservation on the Tulerosa River, in a fertile and well-timbered tract 20 miles north of the San Augustine plain. There are very few streams of sweet water in New-Mexico, and this was an especially desirable possession for many reasons, but not until roads had been built, timber cleared, fields plowed, dams and mills built, &c. Government of course wrought all these improvements, and just as soon as the reservation was valuable and in full swing the Apaches were removed to Ojos Calientes, near the San Mateo Mountain—one of the worst places which could be selected, because near the settlements in the heart of the Miembres Mountains, where renegades could not be followed, and within easy reach of the rendezvous at Cañon del Alamo, used by Victorio, Juh, Loco, and Natchez as a resting place during their subsequent raids. It is a noteworthy fact that while these Indians were at Tulerosa they gave little or no trouble. Victorio made an occasional raid into Sonora, Mexico, but no domestic troubles of any consequence occurred. Capt. George Chilson, of the Eighth Cavalry, had a little fight with Haralche, Juan, and a few braves at the foot of the San Mateo, and there was one petty case of cattle stealing on the Cuchillo Negro River, afterward traced to some renegade Navajoes and Mescaleros, who had settled in the Magdalena Mountains; but the Territory enjoyed several years of absolute immunity from Apache troubles, and the famous Cook's Cañon and the Jornada del Muerte, or "Journey of Death," were as safe thoroughfares as Broadway.

When I visited the Tulerosa Reservation in 1874 Loco, Natchez, Haralche, Juh, Chosito, and Juan were living peaceably and contentedly, playing monte and the Apache lance game, with no thought of murder or rapine. On the contrary, Loco, who was head chief, a fat and good-natured polygamist, was daily lecturing his young men on the moral impropriety of going off the reservation for either hair or cattle. Immediately on his removal to Ojos Calientes, however, Loco changed his tune, and then began the series of raids just happily ended. It was openly whispered at the time that the removal was not for military reasons or to advance the prosperity of the Chiricahuas, but to the end that the Tulerosa property, as improved by Indian labor and Government money, might come cheaply into the hands of the ring.

Now is the turn of the Jicarilla and Mascalero Apaches to be moved from their property near Fort Stanton. It has been converted from a desert into quite a valuable property, and it will be amazing if a strong effort is not soon made to get possession of it.

Another system, which is less skillful and more knavish, is that of practicing on the Indian's ignorance of relative value to "buy" lands from the Pueblos, Zunis, and other Indians holding them. The white buys an acre of arable land and goes subsequently into court with a signed and witnessed deed conveying to him one, two, or six square miles. Nearly all the arable land of the Zuni's at Nutria Springs has gone in this fashion, and unless Government interferes the tribe will be homeless, for they have no patent to their lands.

There is a fertile field here for investigation in this line, and the industrious pathfinder will find rich treasures of surprise awaiting his shrewd diligence.

October 18, 1886

THE DEFENSELESS INDIAN

CONSTANT PREY OF AGENTS AND CONTRACTORS.

HOW THE GOVERNMENT IS SWINDLED AND THE RED MEN ARE GOADED ON TO REVENGE ON THE WHITES.

SANTA FÉ, New-Mexico, Oct. 19.—The statement made by Geronimo to Gen. Miles that the causes of his leaving the reservation were the tyranny and abuse of the officials in charge is generally believed by those who are acquainted with the system usually employed on these fiefs of the United States. The Indian agent, especially in districts far removed from civilization, is Czar-like in the exercise of his power, and can exact slavish submission to his whims and caprices or make the Indian's life a burden to him. Once on the reservation the Indian is a dependent of the agent, through whom the supplies of Government, which are absolutely necessary to his life, are doled. If he will the agent can withhold the rations. If the Indian grows restive under the deprivation the troops are appealed to, and the "insurrection" is quieted by a few well administered doses of lead.

It is a recognized fact in human nature that the consciousness of autocratic power leads almost inevitably to tyranny, and I know no autocrat possessed of a keener appreciation of his powers or more cruelly prompt in their exercise than the average agent of the Interior Department on the reservation. He is the Government's representative; his "reports" are accepted at Washington as Gospel truth; it is always taken for granted that he is a large-souled man, with an enormous philanthropic energy, entirely devoted to lifting the veil of ignorance from the untutored savages, teaching them the ways of civilization, and wooing them gently and with oily gammon from the red paint and feathers of their wayward nature, and only upon his committing some crime that smells to heaven is he removed or investigated. It is generally supposed that the "agent" is a meek and lowly minion who merely and perfunctorily carries out the instructions he receives from his chief; but in fact the agent, once firmly established in the reservation, does exactly what he pleases. His report is usually a pleasant piece of rhetoric, detailing his vast labors and their beautiful results, and this one-sided statement is all the public receives; for the Indian is not called upon for a report, and the only knowledge of his discontent comes when he breaks away and becomes an outlaw. Sometimes, at long intervals, an inspector comes around. The time of his coming is known in advance; an extra ration or two and a few smooth promises restore the simple wards of Government to good-humor, the agency is put in apple-pie order, a score of boys and girls are borrowed to organize a "school," and the inspector finds everything lovely and in exact accord with the agent's report. This is not a fancy sketch, but drawn from personal knowledge and observation.

In the issue of rations the agent is irresponsible to any one. If he happens to be honest the Indians will get what is due them. If he chooses to be dishonest there is no hindrance in his way. See how easy and delightfully simple the peculation. Say there are 761 Indians on the reservation—at least the agent says there are 761, though there may be only 500, and Government is never the wiser. Upon the agent's requisition for supplies rations for 761 Indians are sent him—flour, beef on the hoof, blankets, coffee, &c. On Monday of each week, say, he "issues." There may be only 300 Indians present—200 being off on a hunt, gathering mescal or piñons or engaged in other ways. To the 300 present he issues what really would be full allowance for only 250, and, satisfied or not, the Indians must take quietly what is given or be fined their rations next time. In his report the agent claims to have issued to the full number, or to, say, 680, and by a frugal arrangement between himself and the contractor gets the benefit of the extensive saving. An agent some years ago in this Territory—a Presbyterian parson at that—came down to serve the Indians at a nominal salary of $1,200 a year and retired after two years worth nearly $150,000, but the reservation meanwhile got to look like a home for indigent consumptives.

No man of business would place a twelve-hundred-dollar-a-year man in a position where he could if he wished embezzle $50,000 a year without detection, yet this is practically what the Government is doing under the so-called peace policy. The agent has no resident supervisor, and his accounts must be accepted as rendered. The Indian does not know when he is being cheated. The agent receives so many tons of supplies, and if he claims to have distributed an equal amount to that borne on his receipt, the balance shows up as correct on the annual report of the Interior Department.

Another point is conspicuous. The agent employs a certain number of Indians to cultivate arable lands, ostensibly for their own advantage. Say 12,000 bushels of grain result. There is nothing to prevent the agent's reporting only 6,000 and selling the difference. It is well known in the Territories that the contractors carry from the reservations in many cases quite as large loads as they bring, to whose advantage no one is allowed to know.

A record of our Indian troubles for 20 years will show that nine-tenths of them began on the reservation. The problem often presented to the Indian is whether it is better to remain on the reservation and be slowly starved to death to glut the agent's rapacity or make raids upon neighboring settlements for stock and, perhaps, incidentally, for fun and hair. The Nez Percé war was directly traceable to this cause and no other, while the flagrant robbery of the Yankton Sioux, which was carried on for years before the Government discovered any fraud, is too well known to need more than reference. The United States agreed through its Commissioner, in consideration of the Indians relinquishing certain valuable territory, to expend for their benefit $65,000 annually for 10 years, $40,000 annually for the succeeding decade, and $15,000 each year for 20 years thereafter. This treaty was made in 1859, and an investigation made in 1884 showed that of the $1,135,000 which the Yanktons should have received probably not $90,000 had really been expended for their benefit. A "school" was built and much boasted of by the agent, which cost the Government $10,000. Subsequent investigation showed the "school" to be a log house 39 by 19 by 8, which could not have cost more than $200 and was used most of the time as a stable—the "school" having been on permanent vacation. A second "school" was built by contract in 1881 at a cost of $8,000. At the end of two years the foundations crumbled, and the special agent sent out to investigate found it to have been constructed as a mere shell, and on plans which made it a fire trap.

These swindles are well known to the Indians, but are viewed by them as dishonesty on the part of the Government, which obtained their lands under false pretenses, and it is small wonder if at last they take to the warpath again for revenge.

The wresting of lands from the Indians by squatters or fraudulent purchasers is in many cases done with the full knowledge of the agent and by his influence. The number of these squatters on all Indian reservations is now 2,401, and the number of acres unlawfully occupied by them is 117,000. The land thus taken is invariably the choicest, and under the present system the Indian has no recourse. If the agent chose he could easily prevent this robbery, and why he does not choose is among the many financial mysteries which hedge his career.

There are some curious features in the "school" system as at present conducted. The job of teaching the young Indian idea how to shoot is for the most part farmed out at figures which must make it very lucrative. Of the 23 contract schools not connected with agencies the Rev. Joseph A. Stephan is contractor for 7, scattered through Dakota, Minnesota, and Montana. There were 263 pupils at these schools last year, and Mr. Stephan received $21,083 93 as his share. Of the 35 other contract schools the Rev. Joseph A. Stephan has 21, claiming a total of about 1,500 pupils, for the education of which he receives $157,800, making his educational farm embrace 28 schools, whose annual income is, in round numbers, $178,884 for the education of 1,700 little Indians. Next to Mr. Stephan the Rev. Henry Kendal shows up prominently, and other gentlemen are rapidly acquiring the knack of running these aboriginal baby farms, which are well paid for by the Government. A glance at the report of the Indian Commissioner will show a singular mathematical feature in the attendance at these schools. Bear in mind that the contractor is paid $25 a quarter, or $10 a month in some cases, for each little Indian. One would naturally suppose that sometimes an odd number of little Indians would be at a school. Not so. The table of attendance runs thus: 125—50—50—40—50—60—70—40—30—showing that never less than 10 little Indians go together—probably being actuated by a desire to make the contractor's accounts come out in nice round numbers.

The report sets forth solemnly that at these schools "the following industries are taught: Carpentry, shoemaking, farming, sewing, and housework." Do you remember the class in natural history at Dotheboys Hall? W-i-n-d-e-r spelled window and c-l-e-a-n spelled clean, and when a boy knew this he got soap and water and acquired a practical knowledge of the two under Mrs. Squeers's direction. B-o-t-t-i-n-e-y spelled botany, and the pupil learning this went out to hoe potatoes for Mr. Squeers.

There are too many peculiarities in the Indian service to be analyzed in a short space, but the present Administration will find it a fruitful field. Nowhere in the whole range of public affairs is wholesale peculation or mismanagement as easy. It is significant that all observant persons living near reservations are of one opinion—that the care and protection of the Indians should rest with the army and not with civilians. The system of isolating the Indians under charge of a small-salaried man, who is a disburser of large sums and is supervised by no one, is that which has led to nearly every trouble the Government has had with its red children, and in many cases the pillage has amounted to nearly all the money and stores provided for the Indians' relief.

October 25, 1886

THE INDIAN SEVERALTY LAW.

A few weeks hence the Interior Department will begin to enforce a law whose enactment marked the adoption of a new policy concerning the Indians and the reservations which they occupy. Probably the importance of the Severalty act is not fully comprehended by a majority of the people. The principle which it embodies had been set forth with more or less clearness in bills which two or three Congresses failed to pass, and when at last the Forty-ninth Congress accepted it the measure received less attention from the public than it deserved. There are now upon the reservations about 260,000 Indians, and they occupy 135,000,000 acres of land, a very small part of which they use. The purpose of the law is to place these Indians (the members of the five so-called civilized tribes in the Indian Territory excepted) upon farms of reasonable size; to secure these farms to them in fee simple in such a way that they shall be unable to sell or give away the land until the expiration of a period of 25 years; to open the surplus lands to white settlers under the homestead laws and for the pecuniary benefit of the Indians, and to make every Indian who takes a farm so allotted in severalty a citizen of the United States.

The reservations in which the department will begin its work are small ones—the Devil's Lake reservation in Northern Dakota, the Lake Traverse in Eastern Dakota, and the Siletz, on the Pacific coast in Northern Oregon. It is reported that in these, as well as in thirteen other reservations, a majority of the occupants not only approve the allotment plan but are also anxious that the allotments shall be made without delay.

The execution of the law in certain small reservations where surveys have already been made will be followed by its enforcement in the great reservations, where millions of acres are now of no value to those who hold them, except so far as the leases procured by cattlemen yield small sums to the tribes. It is of great importance that at the beginning, as well as throughout the entire work of making allotments, the law shall be enforced with a scrupulous regard for the interests of the Indians, and that the Government's agents shall be honest and capable men. It is fortunate, therefore, that these agents are to be appointed by the President, who will doubtless pay special attention to their qualifications. The Government will permit representatives of the Indian Rights Association to be present when the allotments are made. This association warmly supports the law.

The department will be opposed either openly or secretly by the corporations which hold very profitable leases of the Indians' surplus lands and by an organization called the National Indian Defense Association of Washington. The leaseholders naturally are unwilling to be deprived of the use of millions of acres of grazing land for which they pay an annual rent of two or three cents per acre. They exert considerable influence among the Indians who receive the money. But these leases were made without authority of law, and they will all be swept away. The loss of the annual rent for their surplus lands will tend to convert the tribes to the support of a policy that will make these lands again a source of income. The National Indian Defense Association holds that the tribal organizations should not be broken up, but that the Indians should continue to hold their vast estates in their present condition. Its attitude is fairly shown by the expressed wish of its Vice-President, the Rev. Dr. BYRON SUNDERLAND, of Washington, "that a wall of adamant high as the stars and permanent as heaven might be erected around the Sioux Reservation," thus making a quiet and secluded wilderness out of a tract of 32,000,000 acres in Dakota. These opposing forces will not prevail.

Fortunately, the surplus land to be released can be taken by settlers only under the homestead laws. If the same restriction could be enforced with reference to the lands released to settlement by the opening of the railroad indemnity belts there would be insured a more equitable distribution of those lands than can be made under the other laws which have been so extensively used by land grabbers.

May 27, 1887

THE EDUCATION OF INDIANS.

PRESIDENT CLEVELAND EXPRESSES HIS VIEWS.

REPLY TO RESOLUTIONS ADOPTED BY THE PHILADELPHIA METHODIST EPISCOPAL CONFERENCE.

WASHINGTON, April 3.—The following letter was written by the President in response to a resolution adopted at a session of the Philadelphia annual Conference of the Methodist Episcopal Church, held at Philadelphia on March 20:

WASHINGTON, March 29, 1888.

To the Rev. James Morrow, D. D., 71 Walnut-street, Philadelphia, Penn.:

MY DEAR SIR: I have received from you certain resolutions passed at the annual Conference of the Methodist Episcopal Church, held at Philadelphia on the 20th inst. I am not informed how to address a response to the officers of the Conference who have signed these resolutions, and for that reason I transmit my reply to you.

The action taken by this assemblage of Christian men has greatly surprised and disappointed me. They declare:

"That this Conference earnestly protests against the recent action of the Government in excluding the use of native languages in the education of the Indians, and especially the exclusion of the Dakota Bible among those tribes where it was formerly used. That while admitting that there are advantages in teaching English to the Indians, to compel them to receive all religious instruction in that language would practically hinder their receiving it in the most effective way. The line of power travels with the human heart, and the heart of the Indian is in his language. That it is in harmony with the genius of our country, a free church in a free State. That the operations of all missionary societies should be untrammeled by State interferences."

The rules of the Indian Bureau upon the subject referred to are as follows:

I. No text books in the vernacular will be allowed in any school where children are placed under contract, or where the Government contributes in any manner whatever to the support of the school. No oral instruction in the vernacular will be allowed in such schools. The entire curriculum must be in the English language.

II. The vernacular may be used in missionary schools only for oral instruction in morals and religion where it is deemed to be an auxiliary to the English language in conveying such instruction, and only native Indian teachers will be permitted to otherwise teach in any Indian vernacular, and these native teachers will only be allowed so to teach in schools not supported in whole or in part by the Government and at remote points, where there are no Government or contract schools where the English language is taught. These native teachers are only allowed to teach in the vernacular with a view of reaching those Indians who cannot have the advantages of instruction in English, and such instruction must give way to the English-teaching schools as soon as they are established where the Indians can have access to them.

III. A limited theological class of Indian young men man be trained in the vernacular at any purely missionary school, supported exclusively by missionary societies, the object being to prepare them for the ministry, whose subsequent work shall be confined to preaching, unless they are employed as teachers in remote settlements where English schools are inaccessible.

IV. These rules are not intended to prevent the possession or use by any Indian of the Bible published in the vernacular, but such possession or use shall not interfere with the teaching of the English language to the extent and the manner hereinbefore directed.

The Government seeks in its management of the Indians to civilize them and to prepare them for that contact with the world which necessarily accompanies civilization. Manifestly nothing is more important to the Indian from this point of view than a knowledge of the English language. All the efforts of those having the matter in charge tend to the ultimate mixture of the Indians with our other people, thus making one community, equal in all those things which pertain to American citizenship. But this ought not to be done while the Indians are entirely ignorant of the English language. It seems to me it would be a cruel mockery to send them out into the world without this shield from imposition, and without this weapon to force their way to self-support and independence.

Nothing can be more consistent, then, than to insist upon the teaching of English in our Indian schools. It will not do to permit these wards of the Nation, in their preparation to become their own masters, to indulge in their barbarous language because it is easier for them or because it pleases them. The action of the Conference, therefore, surprises me if by it they mean to protest against such exclusion as is prescribed in the order. It will be observed that "text books in the vernacular" are what are prohibited and "oral instruction;" the "entire curriculum" must be in English. These are the terms used to define the elements of an ordinary secular education, and do not refer to religious or moral teaching. Secular teaching is the object of the ordinary Government schools, but surely there can be no objection to reading a chapter in the Bible in English, or in Dakota if English could not be understood, at the daily opening of those schools, as is done in very many other well-regulated secular schools.

It may be, too, that the use of words in the vernacular may be sometimes necessary to aid in communicating a knowledge of the English language, but the use of the vernacular should not be encouraged or continued beyond the limit of such necessity, and the "text books," the "oral instruction" in a general sense, and the curriculum certainly should be in English. In missionary schools moral and religious instruction may be given in the vernacular as an auxiliary to English in conveying such instruction. Here, while the desirability of some instruction in morals and religion is recognized, the extreme value of learning the English language is not lost sight of. And the provision which follows, that only native teachers shall "otherwise" (that is, except for moral or religious instruction) teach the vernacular and only in remote places and until Government or contract schools are established, is in exact keeping with the purpose of the Government to exclude the Indian languages from the schools so far as is consistent with a due regard for the continuance of moral and religious teaching in the missionary schools, and except in such cases as the exclusion would result in the entire neglect of secular or other instruction. Provision is made in the rules for the theological training of young men in missionary schools to fit them as Indian preachers, and the possession and use of the Bible so far as it does not interfere with the secular English teaching insisted upon is especially secured.

I cannot believe that these rules of the Indian Bureau were at hand when the resolutions before me were adopted. If they were, I think they were strangely misunderstood, though the mild admission that "there are advantages in teaching English to the Indians" indicates that there is a wide difference between those who appear cautiously to make such an admission and the many others interested in Indian improvement who deem such teaching the paramount object of immediate effort. The rules referred to have been modified and changed in their phraseology to meet the views of good men who seek to aid the Government in its benevolent intention, until it was supposed their meaning was quite plain and their purpose satisfactory. There need be no fear that in their execution they will at all interfere with the plans of those who sensibly desire the improvement and welfare of the Indians. At any rate, until it is demonstrated that these rules operate as impediments to Indian advancement they will be adhered to, while the Government will continue to invoke assistance of all Christian people and organizations in this very important and interesting part of the labor intrusted to it. Yours very truly, GROVER CLEVELAND.

April 4, 1888

GALL SIGNS THE TREATY

SO DO THE BLACKFEET AND THE YANKTONIANS.

SITTING BULL'S BAND GRUNT DISAPPROVAL—A STAMPEDE WORSE THAN THAT TO OKLAHOMA PREDICTED.

STANDING ROCK AGENCY, Dakota, Aug. 6.—The requisite number of signatures for the opening of the great Sioux Reservation has finally been secured. The sensation of the day was the signing by Chief Gall. Gall made no speech, as was expected, but with his faithful followers around him and in the presence of those whom for years he had influenced against giving up the lands, he marched silently to the roll, touched the pen, and amid the applause of the friendly Indians and the disapproving grunts of the Sitting Bull band it was announced that Gall had signed.

This settled it. The Blackfeet and Upper and Lower Yanktonians followed Gall and signed with a rapidity and eagerness that proved the wonderful influence of this powerful chief. Now 11,000,000 acres of land to which the whites have been looking longingly for years are theirs. The Commissioners are rejoiced over their success.

Hundreds of settlers have been camped on the eastern bank of the Missouri during the last two months, awaiting the success of the commission, and although for several weeks they were despondent, they are now jubilant and are receiving telegrams from friends all over the country giving notice of probable reinforcements. It is predicted that the rush to the reservation will be greater than the Oklahoma stampede, as the land is of much better quality and the prospects for prosperity better.

Gen. Crook said the Commissioners hope to close their reports within a few weeks and place everything in readiness for formal opening of the reservation. There is some fear that a premature rush of whites to the lands will cause much trouble between the Indians and the settlers, as there will be for some time many points of dispute as to boundary lines and survey. Sitting Bull, although in the minority, has a sufficient number of followers to make a vast amount of trouble, and will require close watching until they resign themselves to the order of things.

Gall was sought out by a correspondent. "I have given my consent," said he; "my Indians have signed because I told them to after learning the Government could take our lands for nothing if it wanted to. The whites have now got our lands, and I hope they will be satisfied and let us live in peace in the future."

John Grass said he had been holding out for better terms, but when he found the Indians at the lower agency were signing he thought it best to do so. "We hope," he added, "to receive such help from the Government and the white people as to help us to become like them, to become civilized. There is one big log in our camp, though, and that is Sitting Bull. He is utterly worthless to us and keeps us back more than he helps us. He is of no consequence to us, and if the whites think so much of him they had better come and get him. He never was a chief and is always noisy and making trouble. He has gathered around him a band of bad men having no idea of civilization, and all he seems to want is notoriety. He has not only opposed the present treaty, but always has opposed civilization and always will be a nasty man to get along with."

"What is the general idea of the Indians regarding civilization—do they want it or not?"

"Yes; we want to be civilized and live like the white people. We want to earn our living and be as the whites; but so long as we have the disturber and his bad influence to contend with we cannot accomplish much."

"What do you think of Gov. Foster's idea of having industrial schools at the agency?"

"It is a good one. If the Government would build big houses where they could teach our young men to make wagons, harness, shoes, and clothes for ourselves, and teach them the white people's languages besides, it would be good, and we would like it. I hope they will do this. It would help us along and give our young men employment, instead of leaving them to loaf around their camps making mischief."

Sitting Bull, when asked what he believed the effect of opening the reservation on the Indians would be, exclaimed: "Don't talk to me about Indians; there are no Indians left except those in my band. They are all dead, and those still wearing the clothes of warriors are only squaws. I am sorry for my followers, who have been defeated and their land taken from them."

It is learned that the Indians have all along doubted whether the Government had the power to take their lands without paying them what they asked, and not until John Grass became satisfied of this point by consultation with lawyers in Bismarck did he consent to accept the treaty.

August 7, 1889

THE LAST OF SITTING BULL

THE OLD CHIEF KILLED WHILE RESISTING ARREST.

A DESPERATE FIGHT BETWEEN HIS FOLLOWERS AND THE INDIAN POLICE—THIRTEEN SAID TO HAVE BEEN KILLED ON BOTH SIDES.

ST. PAUL, Minn., Dec. 15.—A report has been received here to the effect that Sitting Bull, the Sioux chief, has been killed. It is stated that the Indian police started out this morning to arrest him, and meeting him three miles from camp tried to effect his capture. A fight ensued in which Sitting Bull was killed.

The news of the killing has been confirmed by advices received by Gen. Miles at the military headquarters in this city. He received two dispatches this evening, the first from Pierre, S. D., stating that Sitting Bull and his son had been killed, but giving no further particulars. The other dispatch was from Standing Rock Agency, and stated that the Indian police started out this morning to arrest Sitting Bull, having understood that he proposed starting for the Bad Lands at once. The police were followed by a troop of cavalry under Capt. Souchet and infantry under Col. Drum.

When the police reached Sitting Bull's camp on the Grand River, about forty miles from Standing Rock, they found arrangements being made for departure. The cavalry had not yet reached the camp, when the police arrested Bull and started back with him. His followers quickly rallied to his rescue and tried to retake him. In the mêlée that ensued, the wily old chief is said to have been killed, and five of the best of the Indian police were also killed.

A *Pioneer Press* (Dickinson, S. D.) specia says reliable dispatches received are to the effect that Sitting Bull's camp was attacked by troops, and himself and seven warriors were killed. The remainder of the band are now in retreat up the Grand River, but it is not yet known definitely along which fork their trail will lie. Information of the most reliable nature was received to-day that a band of eight wagons were encamped on the Little Missouri opposite Pretty Buttes. It is therefore probable that the fugitives will make this camp their objective point. They will not be able without great exertion to reach the forks of Grand River to-day. It is estimated that 150 warriors are in the band, and this number is likely to be increased by other bands.

Lieut. Casey, with a troop of Cheyenne scouts and Capt. Adams's troop of the First Cavalry, are headed for the north end of the Powder River range, opposite the mouth of the Box Alder Creek. Capt. Fountain's troop of the Eighth cavalry, with pack transportation, which will leave here in the morning for White Buttes, will probably intercept the band before it reaches the Little Missouri. If not, Lieut. Casey and Capt. Adams will do so. Settlers who are aware of the movements of the troops are little alarmed, as the weather is such that intelligence of disturbances and of movements travels rapidly, and it is well known that the troops are so distributed as to have the situation in hand. A general outbreak on the Sioux reservation is not feared, and those disaffected bands which are now giving trouble will soon be placed where they will cease to be a cause of alarm for the settlers.

The Sioux reservation is surrounded by troops, thoroughly equipped for a Winter campaign in the most difficult country. All are in communication with each other and department headquarters. No outbreak can become general in the face of the precaution already taken, and the wild rumors which have caused the population of entire valleys to fly for their lives are malicious and groundless. The arch villain is dead, and his followers will soon lose the enthusiasm necessary to follow his teachings. Troops are now hot on their trail and before another sun has set Sitting Bull's celebrated chorus of dancers will be good Indians or prisoners.

CHICAGO, Dec. 15.—At 9 o'clock to-night Assistant Adjt. Gen. Corbin of Gen. Miles's staff received an official dispatch from St. Paul saying that Sitting Bull, five of Sitting Bull's men, and seven of the Indian police have been killed. The thirteen casualties were the result of an attempt by the police to arrest Sitting Bull.

Gen. Brooke, in command of the troops at Pine Ridge, telegraphed the situation to Assistant Adjt. Gen. Corbin at army headquarters to-night as follows: "All the Indians who can be brought in are now here, leaving about 200 bucks in the Bad Lands who refuse to listen to any one or anything. Against these I will send a sufficient force to capture or fight them. All has been done that can be done. The Indians now out have a great many stolen horses and cattle with them. I hope to be able to end this matter now."

The following official telegram was also received:

ST. PAUL, Minn.

To Col. Corbin, Assistant Adjutant General, Chicago:

Sitting Bull was arrested this morning at daylight by the Indian police. Friends attempted his rescue and a fight ensued. Sitting Bull, his son Black Bird, Catch Bear, and four others were killed; also seven Indian police. Capt. Fechet arrived just in time with his two troops, Hotchkiss and Gatling guns, and secured the body of Sitting Bull.

By command of Gen. Miles. MAUS, A. D. C.

The story of the last visit paid by a white man to Sitting Bull's camp prior to the tragic events of to-day is told in a report received this afternoon by Assistant Adjt. Gen. Corbin. The narrative throws a flood of light on the old chief's wily character, and strongly depicts the circumstances existing in the isolated camp. The document is addressed to Commissioner of Indian Affairs Morgan by United States Indian Agent James McLaughlin of Standing Rock, and reads in full as follows:

"Having just returned from Grand River District, and referring to my former communication regarding to the ghost-dance craze among the Indians, I have the honor to report that on Saturday evening last I learned that such a dance was in progress in Sitting Bull's camp, and that a large number of Indians of the Grand River settlements were participators. Sitting Bull's camp is on the Grand River, forty miles southwest from the agency, in a section of country outside of the line of travel, only visited by those connected with the Indian service, and is therefore a secluded place for these scenes. I concluded to take them by surprise, and on Sunday morning left for that settlement, accompanied by Louis Primeau, arriving there about 3 P. M. Having left the road usually traveled by me in visiting the settlement, I got upon them unexpectedly and found a 'ghost' dance in its height. There were about 45 men, 25 women, 25 boys, and 10 girls participating. A majority of the latter (boys and girls) were until a few weeks ago pupils of the day schools of the Grand River settlements. Approximately 200 persons were lookers-on, who had come to witness the ceremony either from curiosity or sympathy, most of whom had their families with them and encamped in the neighborhood.

"I did not attempt to stop the dance then going on, as, in their crazed condition under the excitement, it would have been useless to attempt it, but, after remaining some time talking with a number of the spectators, I went on to the house of Henry Bull Head, three miles distant, where I remained over night and returned to Sitting Bull's house next morning, where I had a long talk with Sitting Bull and a number of his followers. I spoke very plainly to them, pointing out what had been done by the Government for the Sioux people, and how this faction by their present conduct were abusing the confidence that had been reposed in them by the Government in its magnanimity in granting them full amnesty for all past offenses, when from destitution and imminent starvation they were compelled to surrender as prisoners of war in 1880 and 1881; and I dwelt at length upon what was being done in the way of education of their children and for their own industrial advancement, and assured them of what this absurd craze would lead to, and the chastisement that would certainly follow if these demoralizing dances and disregard of department orders were not soon discontinued. I spoke with feeling and earnestness, and my talk was well received, and I am convinced that it had a good effect.

Sitting Bull, while being very obstinate, and at first inclined to assume the rôle of "Big Chief" before his followers, finally admitted the truth of my reasoning and said that he believed me to be a friend to the Indians as a people, but that I did not like him personally, but that when in doubt in any matter in following my advice he had always found it well, and that now he had a proposition to make to me which if I agreed to and would carry out would allay all further excitement among the Sioux over this ghost-dance, or else convince me of the truth of the belief of the Indians in this new doctrine. He then stated his proposition, which was that I should accompany him on a journey from this agency to each of the other tribes of Indians through which the story of the Indian Messiah had been brought, and when we reached the last tribe, or where it originated, if they could not produce the man who started the story, and we did not find the new Messiah, as described, upon the earth, together with the dead Indians returning to reinhabit this country, he would return convinced that they (the Indians) had been too credulous and imposed upon, which report from him would satisfy the Sioux, and all practices of the ghost societies would cease; but that, if found to be as professed by the Indians, they be permitted to continue their medicine practices, and organize as they are now endeavoring to do.

"I told him that this proposition was a novel one, but that the attempt to carry it out would be similar to the attempt to catch up the wind that blew last year; that I wished him to come to my house, where I would give him a whole night or day and night, in which time I thought I would convince him of the absurdity of this foolish craze, and the fact of his making me the proposition that he did was a convincing proof that he did not fully believe in what he was professing and endeavoring so hard to make others believe. He did not, however, promise fully to come into the agency to discuss the matter, but said he would consider my talk and decide after deliberation.

"I consumed three days in making this trip, and feel well repaid by what I accomplished, as my presence in their midst encouraged the weaker and doubting, and set those who are believers to thinking of the advisability of continuing the nonsensical practices they are now engaged in. I also found that the active members in the dance were not more than half the number of the earlier dancers, and believe that it is losing ground among the Indians, and while there are many who are half-believers I am fully satisfied that I can keep the dance confined to the Grand River district.

"Desiring to use every reasonable means to bring Sitting Bull and his followers to abandon this dance and to look upon its practice as detrimental to their individual interests and the welfare of their children, I made the trip herein reported to ascertain the extent of the disaffection and the best means of securing its discontinuance. From close observation I am convinced that the dance can be broken up, and after due reflection would respectfully suggest that in case my visit to Sitting Bull fails to bring him in to see me in regard to the matter, as invited to do, all Indians living on Grand River be notified that those wishing to be known as opposed to the ghost doctrine, friendly to the Government, and desiring the support provided in the treaty must report to the agency for such enrollment and be required to camp near the agency for a few weeks, and those selecting their medecine practices, in violation of department orders, to remain on Grand River, from whom subsistence will be withheld.

"Something looking toward breaking up this craze must be done, and now, that cold weather is approaching, is the proper time. Such a step as here suggested would leave Sitting Bull with but few followers, as all, or nearly all, would report for enrollment, and thus he would be forced in himself.

"There are not many firearms among these Indians, still there are a few, and as a pledge of good faith on their part they should be required to turn in all their arms to the agent and get a memorandum receipt for the same. Knowing the Indians as I do, I am confident that I can, by such a course, settle the Messiah craze at this agency, and also thus break up the power of Sitting Bull, without trouble and with but little excitement. This will be sustained by public sentiment and conform to the discipline approved by the better-disposed Indians. It is true that it would unsettle the Indians of that district in their home life for a few weeks, but after that all worry and uneasiness would cease, while with the ghost practices continued, all the participants being Indians regularly rationed by the Government, without any appearance of withdrawal of this support, anxiety among well disposed and the greater temptation for others to join is increased.

WASHINGTON, Dec. 15.—Indian Commissioner Morgan this evening received from Indian Agent McLaughlin the following dispatch, dated Fort Yates, North Dakota, Dec. 15:

"Indian police arrested Sitting Bull at his camp, forty miles northwest of the agency, this morning at daylight. His followers attempted his rescue and fighting commenced. Four policemen were killed and three wounded. Eight Indians were killed, including Sitting Bull and his son, Crowfoot, and several others were wounded. The police were surrounded for some time, but maintained their gr und until relieved by United States troops, who now have possession of Sitting Bull's camp, with all the women, children, and property. Sitting Bull's followers, probably 100 men, deserted their families and fled West up the Grand River. The police behaved nobly, and great credit is due them. Particulars by mail."

Commissioner Morgan showed this telegram to the President late this evening. The President said that he had regarded Sitting Bull as the great disturbing element in his tribe, and now that he was out of the way he hoped that a settlement of the difficulties could be reached without further bloodshed.

Gen. Schofield was asked for his opinion of the effect on the other Indians of the killing of Sitting Bull, but he was disinclined to discuss the matter, saying that it was not possible to predict the result. He indulged the hope expressed by others that this would hasten the settlement of the Indian trouble. He thought it would make more definite the line of division between the friendly Indians and those determined to be hostile, but just how numerous the latter might be could not be told at this time. He had from the start of the troubles in the Northwest hoped the matter would be settled without conflict and regretted that blood had been shed, but he hoped for favorable results. Further than this, Gen. Schofield declined to be interviewed.

When Secretary Proctor was asked concerning the effect of the killing he said he did not think it would have any bad effect on friendly Indians. They had not been kindly disposed toward Sitting Bull and had no love for him. It was only with the disaffected Indians that he held any influence.

THE DESPERATE CHIEF'S CAREER.

A BITTER FOE OF THE WHITES, SAGACIOUS, CRUEL, AND BLOODTHIRSTY.

Sitting Bull, of all the Indians, was the most unrelenting, the most hostile, the most sagacious, the most cruel, and the most desperate foe of the whites of any chief of modern times. He never assented to the control of the United States Government over his people, but persistently fought the troops whenever they came in his way.

He claimed that the country belonged to the Indians; that they had a right to hunt or fish wherever they pleased; that the white man had wronged them, and thus appealing to the feelings of the younger portion of his race, he induced a large number to follow him, and for twenty years carried on his war of murder and rapine, until he finally surrendered to the United States forces on the 19th of July, 1881.

Sitting Bull began his career as a "medicine man." He was the son of Four Horns, one of the four supreme chiefs elected by the Sioux Nation more than a century ago. The Custer episode marked the zenith of Sitting Bull's fame and the beginning of his downfall. He was undoubtedly the most wily and astute Indian in the Sioux Nation. With no reputation as a warrior to aid him, he gained ascendancy as a "medicine man," a prophet, a preacher, a teacher, a politician.

He divided the chieftainship with Gall. Ever since the Little Big Horn fight there has been a bitter rivalry between the men. He commanded the Indians on that occasion, Sitting Bull remaining in his tent performing incantations, distilling "medicines," and indulging in the foolery that was supposed to exercise great influence over the fortunes of the battle. His influence had become so great and was working such deleterious effects among the agency Indians that finally the department at Washington was obliged to take prompt measures to prevent a general outbreak among the friendly Indian tribes, and to this end Sitting Bull and his followers were ordered to come into the reservation, or they would fall under the control of the military power. Sitting Bull laughed at these commands of the Indian Department and still continued his raids, and then followed the campaign of 1876, inaugurated by Gen. Sheridan, wherein three powerful columns of troops were to move simultaneously upon the enemy and force them either into civilization or extermination. Gen. Crook was repulsed, and Gen. Custer, in his eagerness to make an attack, without proper support, which could have been obtained, brought on a fight in which he and his men were all killed, leaving Sitting Bull and his warriors complete masters of the field.

After the battle the savages crossed into Canada. Then the old Chieftan began again to commit depredations among the Americans, and finally a commission was appointed by the United States Government to cross the line and try to effect by diplomacy what had failed by force of arms. Sitting Bull sneeringly rejected every overature for peace.

Matters now continued quiet with Sitting Bull for nearly one year and a half, when, in 1879, he broke out again, and commenced depredations upon the settlers. He was met by Gen. Miles, and a battle ensued, Bull being in command in person. Then followed the surrender of Rain in the Face, Crow Wing, Chief Gall, and many thousand Indians. For endeavoring to stir up the Indians at Standing Rock against his rival, Gall, Sitting Bull with his relatives were made prisoners and confined at Fort Randall. He was afterward released, and has since made his home in Grand River Valley.

December 16, 1890

A FIGHT WITH THE HOSTILES

BIG FOOT'S TREACHERY PRECIPITATES A BATTLE.

CAPT. WALLACE OF THE SEVENTH CAVALRY KILLED, WITH MANY INDIANS AND SOLDIERS—LIEUT. GARLINGTON SERIOUSLY WOUNDED.

PINE RIDGE AGENCY, S. D., Dec. 29.—Big Foot's braves turned upon their captors this morning and a bloody fight ensued. The trouble came when the soldiers attempted to disarm the Indians, who had surrendered to Major Whiteside. This move on the part of the troops was resisted, and a bloody and desperate battle at close quarters followed, in which the Indians were shot down ruthlessly and in which the lives of several soldiers were sacrificed.

Capt. George W. Wallace was killed, and Lieut. Garlington and fifty troopers were wounded.

Big Foot's band, numbering 150 warriors, surrendered yesterday to Major Whiteside, who, with the Seventh Cavalry, had been in readiness to intercept the hostiles who were making for the Bad Lands. This party of warriors had previously escaped while being taken under escort by Col. Sumner to Fort Bennett. They made no show of resistance when offered by Major Whiteside the alternative to surrender or engage in battle, and were marched to the old camp on Wounded Knee Creek, where they were surrounded by the troops, while couriers were sent to Gen. Brooke for reinforcements.

This camp is about twenty miles from Pine Ridge Agency. With the Indians thus captured were the immediate followers of Sitting Bull, and all were in a sullen and ugly mood. They had been closely pressed by Col. Sumner's troopers, and were harassed on every side by the cavalry. They had made a forced march for the Bad Lands, accompanied by their squaws and children, who were suffering for food.

Col. Forsythe arrived at the camp early this morning, bearing orders from Gen. Brooke to disarm Big Foot's band. Col. Forsythe assumed command of the regulars, which comprised two battalions of 500 men, with Hotchkiss guns. It was feared that the Indians would offer resistance, and every precaution was taken to prevent an escape and to render the movement successful. Col. Forsythe threw his force around the Indian camp and mounted the Hotchkiss guns so as to command the camp, and at 8 o'clock issued the order to disarm the redskins.

The preparations were quickly made. The command was given to the Indians to come forward from the tents. This was done, the squaws and children remaining behind the tepees. The braves advanced a short distance from the camp to the place designated and were placed in a half circle, the warriors squatting on the ground. A body of troops were then dismounted and thrown around the Indians, this force, comprising Company K, Capt. Wallace, and Company B, Capt. Varnum. The order was then given to twenty Indians to go to the tents and get their guns. Upon returning it was seen that only two guns were brought. A detachment at once began to search the village, resulting in thirty-eight guns being found.

As this task was about completed, the Indians, surrounded by Companies K and B, began to move. All of a sudden they threw their hands to the ground and began firing rapidly at the troops not twenty feet away. The troops were at a great disadvantage, fearing to shoot their own comrades. The Indians, women, and children then ran to the south, the battery firing rapidly at them as they ran. Soon the mounted troops were after them, shooting them down on the wing on every side.

The engagement lasted fully an hour and a half. To the south many took refuge in a ravine, from which it was difficult to dislodge them. The Indians from cover kept up a constant fire on the soldiers, who replied, picking off the redskins at every opportunity. The Hotchkiss gun was also run up so as to command the ravine, and a withering fire was poured upon the reds. It is estimated that the soldiers killed and wounded number about fifty. Just now it is impossible to state the exact number of dead Indians. There are more than fifty, however, killed outright.

The Indians were shot down wherever found, no quarter being given by any one. Capt. Wallace was killed by a blow of a club on the head, and Lieut. Garlington of arctic fame was shot through the arm at the elbow.

The soldiers pursued the red skins who attempted to escape from the r vine, and few of the band of 150 who surrendered yesterday escaped. To say that it was a most daring feat, 120 Indians attacking 500 cavalry, expresses the situation but faintly.

It could onl have been insanity which prompted such a deed. It is doubted if by night either a buck or a squaw out of all Big Foot's band is left to tell the tale of this day's treachery.

The members of the Seventh Cavalry have once more shown themselves to be heroes in deeds of daring. Single-handed conflicts were seen all over the field.

The Indians were not all armed with guns, many of them having only pistols or knives and clubs. They fought with desperation, and after the first surprise were greatly at a disadvantage. After breaking through the line that surrounded them they were at the mercy of the mounted troopers, the ground for some distance being unbroken.

After the first volley the Indians threw themselves upon the troopers who surrounded them, and who were so completely taken by surprise that they were unable to return the first fire, and could only fight with their clubbed guns or small arms.

Before the battle was over another skirmish occurred near the agency this afternoon. One of Col. Forsythe's troopers of the Seventh Cavalry was fired on by some Indians who went out from the Rosebud Camp, near Pine Ridge, and on their return they fired into the agency. This caused a skirmish, in which two soldiers were wounded. Owing to the absence of the cavalry there is great trepidation here. Indian scouts who have just come in say that but few of Big Foot's men are left alive.

WASHINGTON, Dec. 29.—Official dispatches from Gen. Miles dated Rapid City, S. D., were received to-night by Gen. Schofield telling of the fight in the Bad Lands to-day between the Indian hostiles and the white troops. The dispatches were first sent by Gen. Brooke to Gen. Miles. The first was as follows:

"Whiteside had four troops of cavalry and held the Indians till Forsythe reached him with four more troops last night. At 8:30 o'clock this morning, while disarming the Indians, a fight commenced. I think very few Indians have escaped. I think we will have this matter in hand as soon as all are in position. There was no precaution omitted. The fight occurred near the head of Wounded Knee Creek. I have just seen many of the Indians who went out toward Forsythe this morning come back."

The next dispatch was: "Gen. Brooke telegraphs Forsythe reports that while disarming Big Foot's band this morning a fight occurred. Capt. Wallace and five soldiers were killed. Lieut. Garlington and fifteen men were wounded. The Indians are being hunted up in all directions. None are known to have gotten their ponies. General Brooke also reports that many of the young warriors that were going out from the camp in the Bad Lands to the agency have gone toward Forsythe. All troops have been notified. Col. Forsythe had two battalions of the Seventh Cavalry and Hotchkiss guns. Other troops are in close proximity."

A later dispatch says: "Gen. Brooke reports that two shots were fired near the agency (Pine Ridge) by some one and several were fired in return. Quite a large number of Two Strikes's band ran away and all at the agency are greatly excited. All this makes matters look more serious."

Gen. Schofield, though deeply regretting the occurrence, was not greatly surprised when he learned of the treachery displayed by the Indians in the fight referred to above. He had been on the look-out for treachery all the time; it was almost inevitable. That the trouble would end without a conflict of this kind was almost too much to hope for. So far as he could see just now there appeared to be no further danger at hand, except that to be feared from the disarmament of the band of Indians that is still out, though the excitement following the fight of to-day might be the means of leading to further trouble.

Secretary Proctor also expressed regret at the occurrence, as he had hoped for the settlement of the trouble without further bloodshed. He supposed that inasmuch as Big Foot was connected with Sitting Bull's band it was a case where the Indians wanted revenge for the killing of their friend. Both Secretary Proctor and Gen. Schofield felt disinclined to talk at length in the absence of detailed iformation.

CHICAGO, Dec. 29.—The following was received at Army Headquarters to-night at a late hour:

RAPID CITY, S. D., Dec. 29.

To Col. H. C. Corbin, Army Headquarters, Chicago, Ill.:

Col. Forsythe reports that while disarming Big Foot's band a fight occurred. Capt. Wallace and a few soldiers were killed. Lieut. Garlington and fifteen men were wounded. This again complicates the surrender of all the Indians which would have taken place in short time had this not occurred. Forsythe had two battalions and Hotchkiss guns. Quite a large number of young warriors have been away from the camp, that were going from the Bad Lands; also quite a number of Two Strikes' band going towards Forsythe. The troops are in close proximity. MILES, Commanding."

December 30, 1890

INDIANS TELL THEIR STORY.

A PATHETIC RECITAL OF THE KILLING OF WOMEN AND CHILDREN.

WASHINGTON, Feb. 11.—The Sioux Indian conference was concluded to-day and the Indians will to-morrow or Friday start for home, going by way of Philadelphia and Carlisle. The feature of to-day's meeting was the story of the fight at Wounded Knee, which was told by Turning Hawk and American Horse.

Turning Hawk said that a certain falsehood came to his agency from the West, which had the effect of fire upon the Indians. "When the fire came upon our people," he said, "those who had a certain farsightedness and could see into the matter made up their minds to stand up against it and fight it. The reason we took this hostile attitude to this fire was because we believed that you yourself would not be in favor of this particular mischief-making thing; but, just as we expected, the people in authority did not like this thing, and we were quietly told that we must give up or have nothing to do with this certain movement. Though this is the advice from our good friends of the East, there were, of course, many silly young men who were longing to become identified with the movement, although they knew that there was absolutely nothing bad nor did they know there was anything absolutely good in connection with the movement, and in the course of time we heard that the soldiers were moving toward the scene of the trouble."

Turning Hawk then told how, frightened at the approach of the soldiers and hearing all manner of rumors as to what the soldiers were going to do with them, they fled into the Bad Lands. Their friends and relatives left behind at the agency became very anxious about them and sent parties to them to try and induce them to return. Finally they succeeded. Turning Hawk then continued:

"When our people, who had been frightened away, were returning to Pine Ridge, and when they had almost reached the agency, they were met by the soldiers and surrounded, and finally taken to the Wounded Knee Creek, and there, at a given time, their guns were demanded, and when they had delivered them up the men were separated from their families, from their tepees, and taken to a certain spot, their guns having been given up. When the guns were thus taken and the men thus separated, there was a crazy man, a young man of very bad influence, and in fact a nobody, among that bunch of Indians who fired his gun, and, of course, the firing of a gun must have been the breaking of a military rule of some sort, for immediately the soldiers returned the fire and the indiscriminate killing followed."

Spotted Horse—As soon as the first shot was fired, the Indians immediately began drawing their knives, and they were exhorted from all sides to desist, but this was not obeyed; consequently the firing began immediately on the part of the soldiers.

Turning Hawk—All the men who were in the bunch were killed right there, and those who escaped that first fire got into the ravine, and as they went along up the ravine for a long distance they were pursued on both sides by the soldiers and shot down, as the dead bodies showed afterward.

Turning Hawk said the women had no firearms to fight with, and thought it very improbable that the women were shot because of a similarity in dress to the men. Then American Horse took up the story.

"The men were separated," he said, "from the women, and they were surrounded by the soldiers, and then came next the village of the Indians, and that was entirely surrounded by the soldiers also. When the firing began, of course the people who were standing immediately around the young man who fired the first shot were killed right together, and then they turned their guns, Hotchkiss guns, &c., upon the women who were in the lodges, standing there under a flag of truce, and, of course, as soon as they were fired upon they fled, the men fleeing in one direction and the women running in two different directions. So that there were three general directions in which they took flight."

Commissioner—Do you mean to say that there was a white flag in sight over the women when they were fired upon?

American Horse—Yes, Sir; they were fired upon, and there was a woman with her infant in her arms who was killed as she almost touched the flag of truce; and the women and children of course were strewn all along the circular village until they were dispatched. Right near the flag of truce another was shot down with her infant. The child, not knowing that its mother was dead, was still nursing, and that was especially a very sad sight. The women as they were fleeing with their babes on their backs were killed together, shot right through, and the women who were very heavy with child were also killed. All the Indians fled in these three directions. After most of them had all been killed, a cry was made that all those who were not killed or wounded should come forth and they would be safe, and little boys who were not wounded came out of their places of refuge, and as soon as they came in sight a number of soldiers surrounded them and butchered them there.

"Of course we all feel very sad about this affair. I stood very loyal to the Government all through those troublesome days, and believing so much in the Government and being so loyal to it my disappointment was very strong, and I have come to Washington with a very great blame against the Government on my heart. Of course it would have been all right if only the men were killed; we should feel almost grateful for it. But the fact of the killing of the women, and more especially the killing of the young boys and girls who are to go to make up the future strength of the Indian people—those being killed is the saddest part of the whole affair, and we feel it very sorely."

The Rev. Mr. Cook, a Sioux half-breed, pastor of an Episcopal church at Pine Ridge, said, in reference to the statements as to the good spirit with which the members of the Seventh Cavalry went to that scene of action, that one of Gen. Miles's scouts, who was among the soldiers after the fight, told him that "an officer of high rank, he did not know whom, came to him and said with much gluttonous thought in his voice, 'Now we have avenged Custer's death.'"

February 12, 1891

COL. CODY GETS HIS INDIANS.

MR. MORGAN FINDS HE IS NOT A BIGGER MAN THAN MR. NOBLE.

WASHINGTON, March 6.—Indian Commissioner Morgan has discovered once more that he is not a bigger man than the Secretary of the Interior. Somebody started the story last Fall that the Indians who had been taken abroad by Col. William F. Cody to take part in the "Wild West" had been badly treated and were surrounded by degrading influences.

As soon as Col. Cody, better known as "Buffalo Bill," heard of this talk he rounded up his Indians, who were in Winter quarters at Strasburg, and brought them to Washington to give Commissioner Morgan an ocular demonstration of the falsity of the report. It was an expensive undertaking. When the Indians reached Washington, Col. Cody sought an interview with Morgan. To his surprise, the Commissioner kept him waiting from day to day, and it was not until Buffalo Bill had secured the active help of a prominent Cabinet officer that he was able to bring his troop before the Commissioner. Then the Indian troubles broke out in the West, and, at the solicitation of the authorities, Cody and his friendly Indians went to Pine Ridge and did good service for the Government.

But when the Colonel got ready to take his Indians back to Europe, he found a large-sized snag in his way. Commissioner Morgan flatly refused to permit a single red man to leave the reservation. He had been told that the Indians did not have proper surroundings abroad, and he considered it much better that they should remain on the reservation than be demoralized by foreign travel.

No amount of evidence that Col. Cody could produce, including that of many leading Americans who had seen the show in Europe, could affect Commissioner Morgan, nor would he listen to the recommendations of Indian agents and Gen. Miles and Col. Forsythe that it would be the best way to prevent a renewal of troubles in the Spring to let Cody take a hundred of the Sioux out of the country. The Nebraska Senators and Representatives joined in trying to induce Morgan to issue the permits, but the only result was to call out an order from Morgan for the arrest of any agent of Cody who tried to take an Indian away from the reservation.

It had cost Col. Cody a good many thousand dollars to bring his hundred Indians to the United States and take care of them here, and then to have them suddenly corralled by the Indian Commissioner after their return passage had been paid for was a little disheartening. Col. Cody has been here a week working hard to prevent the destruction of his show by Morgan's arbitrary act. The matter was finally laid before Secretary Noble, and to-day the Secretary overruled the Commissioner, and issued an order directing that Col. Cody be given liberty to take to Europe as many of the Sioux Indians as he wished. A fortnight hence 100 of the redskins will sail with Buffalo Bill, and Commissioner Morgan, who boasted that he never attended a theatre or a circus in his life, will have to give them up to the demoralizing and degrading influences of foreign travel and contact with the civilization of the white man.

March 7, 1891

AN IMPORTANT DECISION.

DEPARTMENT OFFICIALS AT WASHINGTON ACTING CONTRARY TO LAW.

ST. PAUL, July 30.—A Pierre, (S. D.) special to the *Pioneer Press* says: "Considerable comment has been aroused by the decision of the United States Court in regard to the status of children born of Indian women and white or citizen husbands. The case was that of the United States against Ward, on the charge of selling liquor to a half-breed. The evidence was that the half-breed in question had a negro father, who was a citizen, and an Indian mother. The decision of the court is that the children follow the status of their father and hence are citizens of the United States and amenable only to its laws.

"If the decision holds good, it will affect the ownership of the greatest part of the lands taken up in the vicinity of Fort Pierre and Stanley, across the river from Pierre, as it is nearly all held by squaw men's children or their wives. The department officials at Washington have always held the contrary, and the alloting agents who have been and are now at work are performing their duties under instructions to give all people of Indian blood a preference for lands under the allotment law."

July 31, 1891

THE FIVE CIVILIZED TRIBES

Not Fit to Govern the Territory Ceded to Them.

UNBEARABLE LAWLESSNESS THERE

The Commission Charged with Treating with the Indians Says the Government Must Take Control.

WASHINGTON, Nov. 21.—The anomalous condition of affairs now existing between the United States and the perturbed Indian Territory makes intensely interesting the report submitted to-day to the Secretary of the Interior by the commission sent to investigate matters concerning the five civilized tribes of Indians.

If the recommendations of the commission shall be adopted, the Federal Government will recover possession of the great domain owned by these peoples and revoke the right given them to govern themselves.

Charges that the tribal Governments have perverted the trust conferred by the United States, and also shown their inability to care for their own interests, are preferred by the commission, which concludes its report with these straightforward sentences:

The United States put the title to a domain of countless wealth and unmeasured resources in these several tribes, or nationalities, but it was a conveyance in trust for specific uses, clearly indicated in the treaties themselves, and for no other purpose. It was for the use and enjoyment in common of each and every citizen of his tribe of each and every part of the Territory, thus tersely expressed in one of the treaties, "to be held in common, so that each and every member of either tribe shall have an equal undivided interest in the whole."

The tribes can make no other use of it. They have no power to grant it to any one, or to grant to any one an exclusive use of any portion of it. These tribal Governments have wholly perverted their high trust, and it is the plain duty of the United States to enforce the trust it has so created, and recover, for its original uses, the domain, and all the gains derived from the perversion of the trust, or discharge the trustees.

The United States also granted to these tribes the power of self-government, not to conflict with the Constitution. They have demonstrated their incapacity to govern themselves, and no higher duty can rest upon the Government that granted this authority than to revoke it when it has so lamentably failed.

The commission consists of ex-Senator Henry L. Dawes of Massachusetts, Meredith H. Kidd of Indiana, and Archibald S. McKennon of Arkansas. They went to the Indian Territory early in the present year, and in February addressed a convention of all the civilized tribes except the Seminoles, explaining fully the policy of the Government and the reasons for desiring a change.

A strong inclination was manifested at first toward taking steps looking to negotiations, but dispatches from Washington, representing that the Government would hold to the treaty provisions and make no change unless the Indians desired it, resulted in the adoption of resolutions to resist any change and to decline to negotiate.

At the invitation of the various tribes, the members of the commission went among the people and made addresses on the objects of their mission, but the councils of all the tribes except the Cherokees passed resolutions refusing all negotiations.

Propositions were made during the Summer by the commission to divide all lands among the Indians, except town sites and coal and mineral deposits, which were to be sold, and the proceeds divided among the people. Each citizen was to receive sufficient land for a good home, and all intruders were to be removed.

A final adjustment of all claims against the United States was also promised, and it was proposed that after these and other propositions should have been carried into effect Congress should form a Territorial Government. An answer to these propositions was requested by the 1st of October, but no answers were received then, and none has been received since then. The Cherokee Council alone asked further time.

The commission says the Indians refuse to sell any portion of their land. It also states that the full bloods are less fit for citizenship than they were twenty years ago. All progress with them has been arrested. The commission thinks the Indians deserve little consideration in their demands that the National Government remove white people from their territory, as the whites were induced to settle by the Indians, and have lived up to their agreement.

Shrewd whites, through intermarriage with the Indians, have obtained valuable lands for pasturage and cultivation in violation of the agreements with the United States, and some of these have secured from 30,000 to 60,000 acres. This has resulted in preventing the real Indian from obtaining possession of any part of his common property.

In one tribe, with a total territory of 3,040,000 acres, 61 citizens have enclosed and hold 1,237,000 acres, more than a third of the property belonging to 14,632 citizens. This is a violation of the plain terms of the treaty and a perversion of the uses and purposes for which the territory was conveyed to the Indians.

The influx of white citizens and the failure of the tribal governments to observe and enforce the treaty stipulations for the protection of citizens, and the lamentable corruption of these governments in all their branches, have brought the commission to the conclusion that is is impossible to enforce the executory provisions of the treaties.

All the functions of the tribal governments, the members of the commission say, have become powerless to protect the life and property of the citizen. The courts of justice have become helpless and paralyzed. Violence, robbery and murder are almost daily occurrence, and no effective measures of restraint or punishment are put forth to suppress crime. Railroad trains are stopped, and their passengers robbed within a few miles of populous towns.

"A reign of terror exists," the commission adds, "and barbarous outrages, almost impossible of belief, are enacted, and their perpetrators hardly find it necessary to shun daily intercourse with their victims."

The commission heard that fifty-three murders occurred in one of the tribes in September and October, and no one was brought to justice. The tribal governments, they contend, have fallen into the hands of cunning politicians, while the real Indians have little to do in their management.

THE SENATE COMMITTEE'S VIEWS.

Its Report of Last May in Line with That of the Commission.

WASHINGTON, Nov. 21.—The Senate last March, instructed its Committee on the Five Civilized Tribes of Indians to inquire into the present condition of the Indians and of the white people dwelling among them, and to make recommendations as to the legislation required to put an end to existing difficulties. Mr. Teller is the Chairman of this committee, and he went to the Territory, with Senators Platt and Roach, spending nine days there, and getting a good idea of the situation. The report which he submitted last May was thoroughly in line with the prevailing belief that no time should be lost in changing the conditions which exist in the Territory. The committee ascertained that, according to the census of 1890, the Indian population, including the colored claimants to Indian citizenship, was only about [illegible], and the white population over 109,000. It is well known that since the census of 1890 was taken thou-sands of whites have gone into the Territory, and it is now estimated that there are at least 300,000 whites there, making the proportion about 5 to 1 in favor of the whites.

In some agricultural sections the whites outnumber the Indians 10 to 1. This is especially true of the Chickasaw country, where the Indians number only about 3,500, while the white population is variously estimated at from 50,000 to 70,000. Flourishing towns have grown up along the lines of the railroads composed wholly of white people. The town of Ardmore, in the Chickasaw country, is said to contain 5,000 white people, and not to exceed 25 Indians. Duncan and Purcell contain populations of from 1,000 to 1,500, composed of white people. The town of Muskogee, in the Creek country, contains a population of from 1,200 to 1,500 white people, and many other towns of from 500 to 1,500 people are known as "white towns." It is rare to see an Indian in any of these towns, except those that come in from their farms to dispose of their produce or purchase goods of the white trader.

Mr. Teller's committee uncovered the root of the existing troubles in the following paragraph, which appears in its report:

This section of country was set apart to the Indian with the avowed purpose of maintaining an Indian community beyond and away from the influence of white people. We stipulated that they should have unrestricted self-government and full jurisdiction over persons and property within their respected limits, and that we would protect them against intrusion of white people, and that we would not incorporate them in political organizations without their consent. Every treaty from 1828 to and including the treaty of 1866 was based on this idea of exclusion of the Indians from the whites and non-participation by the whites in their political and industrial affairs.

We made it possible for the Indians of that section of the country to maintain their tribal relations and their Indian polity, laws, and civilization if they wished so to do. And if now the isolation and exclusiveness sought to be given to them by our solemn treaties is destroyed and they are overrun by a population of strangers five times in number to their own it is not the fault of the Government of the United States, but comes from their own acts in admitting whites to citizenship under their laws and by inviting white people to come within their jurisdiction, to become traders, farmers, and to follow professional pursuits.

The committee assumed by reason of its discoveries regarding the population of the Territory that the Indians had determined to abandon the policy of exclusiveness and freely to admit white people to the region. While it was there the committee did not receive from any Indian the suggestion that the white people in the Territory with the consent of the Indians should be removed. Considerable space is devoted in the report to a description of the difficulties surrounding the detection and punishment of persons who commit offenses against any of the laws of the United States. Prior to the redistricting of the Territory with reference to the United States courts it was necessary sometimes to take prisoners 600 miles or more to reach a United States court. Even now it is necessary to convey prisoners 200 or 300 miles. The courts are so burdened with business that prompt disposition of the cases brought before them is impossible. The court established for the Indian Territory, having cognizance of all minor offenses and of the smallest civil controversies, is naturally the only court having police powers within the Territory, and is absolutely the only court of final jurisdiction administering justice, in matters large of small, in a territory as large as the State of Indiana, for a people numbering now at least 300,000 and rapidly increasing.

The committee reported further that the criminal business of the Territory was transacted at enormous expense. Cases of the smallest importance, like ordinary assaults, often cost the Government from $200 to $300 each, by reason of the distance traveled by the Deputy Marshals and the fees of witnesses for travel and attendance. The committee made a number of suggestions regarding the disposition of court business, which it is not necessary to explain here. It called attention to the startling fact that, although the Indians practically had invited the whites to enter their Territory, the large white population was without the means of maintaining schools. No public schools were possible for this class, it was argued, while the present condition of affairs continues in the Indian Territory. Upon this point the committee reported:

It may be said that these people went to the Indian Territory with the knowledge that the education of their children would be left to their individual effort, and, therefore, they ought not to complain. We do not stop to inquire whether the parents of these children complain or not—the Nation at large has the right to protest against a condition that deprives the children of 200,000 or 300,000 white and several thousand colored people of the opportunity to acquire an education that will fit them for the discharge of the duties of citizenship, which they have the right to exercise in other parts of the country, if not in the Indian Territory. It is not the concern of the parents alone, nor of the children alone, but of all the people of the United States, and it is a matter of concern to the citizens of those States contiguous to the Indian Territory. Common humanity demands that we take steps to secure to the people the advantages of education, even if they do not appreciate such advantages.

The committee found further that while the theory of the Government in giving to the Indian tribes as bodies politic title to the lands in the Territory was that the title was held for all of the Indians, and all were to participate in the benefits to be derived from such holding, a few enterprising citizens of the tribe, frequently not Indians by blood, but by intermarriage, had in fact become the practical owners of the best of these lands, while the title still remained in the tribe, so that the great body of the Indians derived no benefit from the title. There are many shrewd Indians who under existing Indian law appropriate large bodies of the public domain and lease them to white men, who pay rental either in crops or in cash, as they may agree with their landlords. Instances came to the committee's notice of Indians who had as high as 100 tenants, and one case was found where a single Indian had 400 holdings, amounting to 20,000 acres. The monopoly is so great that in the most wealthy and progressive tribes 100 persons are said to have appropriated fully one-half of the best lands. The evidence is only too strong that while the titles to these lands are held by the tribe in trust for the people, this trust is not being properly executed, nor will it be if left to the Indians, and the question arises, what is the duty of the Government with reference to this trust?

Mr. Teller and his associates were convinced that the system of government which prevailed in the Indian Territory could not continue long. They put the case in this way:

It is not only non-American, but it is radically wrong, and a change is imperatively demanded in the interest of the Indians and whites alike, and such change cannot be much longer delayed. The situation grows worse, and will continue to grow worse. There can be no modification of the system. It must be abandoned, and a better one substituted.

Mr. Teller's committee has not completed its work, and promises to make further recommendations regarding necessary legislation. By the time Congress convenes there will be sufficient information at its disposal to enable it to take up the question intelligently and endeavor to find a solution. Although the outlawry which exists in the Indian Territory is to be deplored, it probably will result in a more speedy settlement of the questions growing the occupation of the Territory by the five civilized tribes than would be possible if thieves and cutthroats were not so plentiful there. This is the opinion of persons who have watched the growth of lawlessness in the Territory, and chafed under the restrictions which long have operated to prevent the United States Government from taking a hand in the internal affairs of the tribes. The depredations of the Cook gang of outlaws and the general tendency to disorder in the Territory are due entirely to the form of government which enables the Indians to maintain an empire in the very heart of the United States, in which, according to Attorney General Olney, United States troops may not enter except in case of trouble between the tribes. It is plain that the Congress which authorized the first treaty with these Indians in 1828 had no conception of the future growth of the United States, and Congresses which accepted later treaties were equally lacking in perception.

The Cherokees secured the first foothold in the Territory, and later the Choctaws, Chickasaws, Creeks, and Seminoles were settled there under treaties which secured to them the unrestricted right of self-government and full jurisdiction over persons and property within certain limits, excepting all white persons, with their property, who are not by adoption or otherwise members of the tribes.

Under these treaties, according to Attorney General Olney's construction, the United States Government cannot send troops into the Territory unless the tribes engage in war among themselves. The Indians have the power, under the treaties, to regulate their internal affairs, but the events of the last two weeks have shown that they are unable to do this, and that the Territoy to-day more than at any time in its history is overrun by desperadoes. It is, in fact, the only region in which lawless characters like the Cooks and the Daltons can prosecute their murderous calling with comparative immunity. The country abounds in hiding places, and large areas are as wild as they were when the Territory was first given over to the Indians.

Nobody with knowledge of the facts doubts now that a great mistake was made in permitting the five tribes to have a government of their own, and the question now to be settled is, How shall existing conditions be changed? The opinion is freely expressed by prominent men that the interests of civilization demand that the treaties shall be held to have been disregarded by the Indians, and that the tribes shall be forced to agree to terms which would extinguish the tribal form of government and throw the Territory open to settlement, giving the Indians lands in severalty and the whites opportunity to acquire lands and hold them.

Mr. Jones of Arkansas, who is Chairman of the Senate Committee on Indian Affairs, has just returned from the Indian Territory, where he remained about a fortnight. He said to-day to a correspondent of The New-York Times that the reports of outrages in that region failed to give an accurate idea of the lawlessness which prevailed there. In all parts of the Territory, he said, there were many desperate characters who had been driven from the States where law and order prevailed, and these were continually engaged in criminal pursuits. The Senator is convinced that the United States Government will be obliged before long to enter the Territory and take charge of its affairs. In his opinion, the leaders of the five civilized tribes soon will see the necessity of effecting a change, and as soon as they come to an understanding with the United States the better it will be for all concerned.

November 22, 1894

THE PRESIDENT TO THE INDIANS

He Suggests that Full Citizenship, Reached by Easy Stages, Is the Ideal Future for the Redskin.

MUSKOGEE, I. T., May 14.—A uniform letter has been sent by Chairman Dawes of the commission to the Chief of each of the five civilized tribes. In substance it states that the commission has been directed to present to the several nations, for their consideration, a letter from the Secretary of the Interior, in which he incloses one from the President, disclosing his interest in the success of the commission in coming to some agreement which will sanction all their just rights and promote their highest welfare. He asks the Chiefs to lay the matter before their people for favorable consideration. The letter from President Cleveland is as follows:

Executive Mansion,
Washington, D. C., May 4, 1895.
To Hoke Smith, Secretary of the Interior.

My Dear Sir: As the commission to negotiate and treat with the five civilized tribes of Indians are about to resume their labors, my interest in the subject they have in charge induces me to write you a few words concerning their work. As I said to the Commissioners when they were first appointed, I am especially desirous that there shall be no reason, in all time to come, to charge the commission with any unfair dealing with the Indians, and that whatever the results of their efforts may be the Indians will not be led into any action which they do not thoroughly understand or which is not clearly for their benefit.

At the same time, I still believe, as I have always believed, that the best interests of the Indians will be found in American citizenship, with all the rights and privileges which belong to that condition. The approach to this relation should be carefully made, and at every step the good and welfare of the Indian should constantly be kept in view, so that when the end is reached, citizenship may be to them a real advantage, instead of an empty name. I hope the commission will inspire such confidence in these Indians with whom they have to deal that they will be listened to, and that the Indians will see the wisdom and advantage of moving in the direction I have indicated. If they are seen willing to go immediately, so far as we may think desirable, whatever steps are taken should be such as to point out the way and the results of which will encourage these people in future progress. A slow movement of that kind, fully understood and approved by the Indians, is infinitely better than swifter results gained by broken pledges and false promises. Yours very truly,
GROVER CLEVELAND.

Secretary Smith says in his letter to Mr. Dawes: "The impossibility of permanently continuing their present form of government must be apparent to those who consider the great difficulty already experienced, even by an Administration favorable to the enforcement of treaties in preserving for them the rights guaranteed by the Government."

May 15, 1895

TREATY WITH THE INDIANS

Dawes Commission Reaches an Agreement with the Choctaws and Chickasaws at Anoka, I. T.

TRIBAL GOVERNMENTS TO END.

Citizenship to be Assumed upon Admission to Statehood—The Senate Needs to Ratify the Terms to Make Them Effective.

WASHINGTON, April 28.—The representatives of both the Dawes Indian Commission and the Choctaw and Chickasaw tribes, who five days ago at Anoka, Indian Territory, jointly executed an agreement, or treaty, for abolishing tribal organization and allotting lands in severalty, have reached here, and have announced formally the results of their conferences.

The agreement now has to be ratified by the Senate, and there is little likelihood of material delay in securing that approval. The substance of the agreement, brief announcement of which was wired to the Interior Department last week follows, a large part of the body of the text being similar to that executed with the Choctaws last year, but to which the Chickasaws then refused to agree:

The tribal governments are to continue for eight years from March 4, 1898, on the ground that no further change will be needed till the lands shall, in the opinion of Congress, be prepared for admission to Statehood. Provisions practically identical with those in the former unsanctioned agreements are made as to direct payment of per-capita funds; as to the Choctaw and and Chickasaw trust funds and their payment to the Indians; as to Choctaw orphan lands in Mississippi; the assuming of citizenship on the expiration of tribal existence, and fixing the Senate as the arbitration tribunal for claims between the United States and the two tribes; forty-acre shares are to be allotted Choctaw freedmen; 640 acres instead of ten each are given certain eleemosynary institutions designated; the Federal Government agrees to maintain strict laws in the territory of the two nations against the liquor traffic in any form; lots not exceeding 50 feet front and 100 feet deep for churches and parsonages in the towns are set apart and exempted from sale, with reversion to the tribes; all coal and asphalt mines are to be controlled by two Trustees appointed by the President and recommended by the heads of the tribes; past agreements for operating coal or asphalt mines are declared void, but all contracts hereafter made by the National agents thus authorized are ratified by the agreement. Coal and asphalt leases are to include 960 acres and to run 30 years. A royalty of 15 cents per ton on all coal mined and 60 cents on asphalt is provided for, subject to changes by the Indian Legislatures.

April 29, 1897

TRYING TO RECONCILE INDIANS TO HARD WORK

And the Government Agents Have a Serious Time of It.

TIRE AFTER TEN DAYS' LABOR

Will Wait for Months for $4 Interest—Squaws Take Prizes in an Annual Fair.

Special to The New York Times.

WASHINGTON, Nov. 18.—The Commissioner of Indian Affairs is engaged in serious efforts to persuade the Indian that there is dignity in labor. The annual report of the department, just submitted to the Secretary of the Interior, summarizes the results hopefully, though the officials engaged in the task admit its gigantic proportions.

An employment bureau has been in operation throughout the year. As a result some 600 Indian men and boys have developed into wage earners in the Southwest, and some other hundreds in other sections. They have labored away from reservation influences on railroad construction, irrigation ditches, and as beet farmers. No persuasion has induced most of them to keep at it for more than ten days at a time, and then they have gone back to the old business of loafing around reservations, doing only what they pleased, and that only when they pleased to do it.

"It must be said for them, however," says the report, "that for such times as they do work under contract without cessation they are the steadiest and most conscientious workers known in their part of the country.

The experiment, it is believed, would have proved more successful but for eighteen months of prosperity on farms and cattle ranches which promised a living on terms more agreeable to the labor-scorning red man.

In the North the fact that in the Spring, when labor is in greatest demand, the Government pays to each Indian his interest money has also interfered. The sum averages $4 to the individual, but, says the report, "the average Indian will hang around for months, doing nothing, waiting for his $4." On some reservations land allotments are being made. This also kept the Indians close to home, despite the allurements of wage earning at a distance, as "they are afraid the land they have selected will not be allotted to them or that some one will jump their claims."

Despite constitutional opposition on the part of the Indians as a class and these other drawbacks, the experiment has proved successful enough to prompt further effort along the same line, and in teaching him powers of sustained effort the sponsors for "Poor Lo" hope to get his face turned permanently toward civilization.

The beet-sugar field and sugar factory are regarded as particularly effective civilizers. Further effort will be made to get Congress to lengthen the present five-year limit on leases of Indian lands, with the hope of encouraging this industry. The Indian himself is given up as hopeless in the matter of farming.

"An Indian owning eighty acres of land," says the report, "has at least sixty acres more than he knows what to do with, and in saying this I am giving the Indian the benefit of a very liberal estimate of his competency."

The department, therefore, wants the Indian owner put in position to lease tribal lands in bulk for long periods to capitalists. It is argued that he would profit by the leases and could be improved by the opportunity to learn beet growing and sugarmaking, and thus become a valuable member of society.

"The Indian," declares the report, "takes to beet farming as naturally as the Italian takes to art or the German to science. Even the little papoose can be taught to weed the rows, just as the pickaninny in the South can be used as a cotton picker." A strong argument is made for the industry as a developer of the West, "and," says the report, "in view of these facts, I cannot think that the campaign for sound economics in the training of the Indian has been doomed a failure by one session's repulse."

The report places blame for much of the thriftlessness of the Indians on the ration system. "Although," it says, "as a broad principle, the ration system has already ceased to exist, it is still in limited practice here and there, owing to the unfortunate language of the treaty pledge that the Government will aid the Indians until they are self-sustaining. The Sioux cling to this pledge. To push even a part of these Indians into actually earning money in a big, competitive world, where there are no rations, will do more than any other one thing could do to awaken the spirit of self-respect in which alone lies the doom of the ration system. About 2,000 Indians are now self-supporting six months of the year. It is to go on from this point, and also invade the ranks of the 4,100 who still draw rations all the year round that an employment bureau is needed."

The report tells at length of the annual fair of the Crow Indians in Montana, which promises to be a big success. It is built up on the tendency of the Indians to gather for feasts, horse races, and dances. The first fair was held two years ago. It was planned to be an exhibition of farm products. The Indians arrived in droves, ready for horse racing, gambling, or carousing, but not a vegetable. They had not taken the project seriously. But the agents went to work again, stirred up a spirit of competition, got a part of the nation interested, and last Fall held a genuine Indian exposition akin to the county fair of the East. There were cash prizes for horses, cattle, pigs, and garden and field products to interest the braves, and prizes for jellies, sewing, and the like to capture the attention of the squaws. There was horse racing on the side, but no gambling or drinking was allowed.

After keen competition in most of the classes nearly $700 was awarded as prizes, a feature being the victories of uneducated women over Indian girls and women who had been educated in the ways of civilization. The squaws generally distinguished themselves, demonstrating their right to a place in the ranks of "new women."

"Chief Plenty-Coos, the most eminent Indian on the reservation," says the report, "competed for the prize for the best driving team, but was beaten by a team owned and driven by a squaw, and the other Indians had a great deal of amusement at his expense, which he took without offense, as became a father of his people, saying that the award was good."

The fair is to be an annual event, and, says the report, "proved that many of these Crows are anxious by their own example to show their neighbors that it is not impossible for an Indian to make a living for himself and family from the farm the Government has provided for him."

November 19, 1906

JOCKEY TO U. S. SENATOR AN INDIAN'S ACHIEVEMENT

Charles Curtis of Kansas Is a Member of the Kaw Tribe.

STUDIED LAW AS CAB DRIVER

Goes to the Upper House After Eight Elections as a Representative—First Indian in the Senate.

Special to The New York Times.

WASHINGTON, Jan. 12.—Congressman Charles Curtis of Kansas, named last night for Senator by the Republican caucus, is a sure-enough native son. His mother was a full-blooded Kaw Indian, and he will be the first of his race to sit in the United States Senate. He is strictly a self-made man. In two weeks he will be 47 years old.

Curtis began life in Shawnee County, Kan., where North Topeka now stands, and earned his first money in the days of the old Kansas City Inter-State Fair Association, when spider-web tidies, embroidered table covers, and pumpkins divided interest with the $10,000 in cash prizes in the speed ring.

Floral Hall, Machinery Hall, and the Art Gallery were deserted one afternoon for the white rail that marked the race course. One of the entries was known to most of the crowd as a crazy horse, who bolted at a certain spot on the track. A new rider was handling the bolter that afternoon—a little fellow with coal-black, straight hair, flashing eyes, and the high cheekbones of an Indian. The rider had been borrowed from another stable.

Off in a bunch they went at the crack of the pistol. When the dark-skinned boy and his crazy mount reached the dangerous point on the course, the bolting place, there was a short, sharp struggle. The horse was conquered, and sped on with the others, under the lash. Half around and then there was another struggle and a spill. Boy and horse landed in a thundering heap against the high board fence. When the dust cleared away the little fellow was picked up unconscious, covered with dust and blood. A long gash lay across his head. That boy was Charley Curtis. To-day he bears the scars.

As he grew up he ran a peanut stand for a time, and the ntook to hack driving. Most of the schooling he got he gave himself, studying at home at such odd moments as he could find. It was while working as a cab driver that he began the study of law. He read in the office of a Topeka lawyer, and was admitted to the bar when he was only 21. Immediately he was taken into partnership with Mr. Case, the man in whose office he had studied, and continued the partnership for three years, until he was elected County Attorney.

After two terms in that office Mr. Curtis formed the Congressional habit and has kept it up ever since. He was elected to the Sixtieth House last Fall, that being his eighth straight election. In the House his most conspicuous service has been on the Committee on Indian Affairs. He was recognized as the House authority on Indian matters, although not the Chairman of the committee. His bill, known as the Curtis act, for the allotment in severalty of the lands and moneys of the five civilized tribes wound up the communal affairs of 97,000 Indians. As a member of the Kaw tribe he obtained allotments for himself and children aggregating more than 3,000 acres in Oklahoma. He is a sturdy, well-built man, his Indian blood showing in his straightness of figure as well as in his black eyes and swarthy complexion. He has a fine voice and is a ready speaker.

To-day 117 of his tribe live in Indian Territory, a few miles below Arkansas City. Every September, during the season of their tribal festivities, Curtis goes to visit them. Much ceremony, much rejoicing, greets him. Feasting and dancing are indulged in. He is always admitted to the council chamber, and his voice is listened to with great respect by the older members of the tribe. They are proud of him. To-day their affection has advanced many marks.

January 13, 1907

STANDS UP FOR REDSKIN.

Real Mohawk Tells Germans Dime Novels Malign American Indian.

Special Cable to THE NEW YORK TIMES.

BERLIN, June 30.—Brant Sero, who calls himself Ojijatheka and is a full-blood Mohawk, has declared war on the publishers of the penny dreadful literature in Germany, which depicts the American Indians exclusively as a race of blood-thirsty scalpers and horse thieves.

He is furnishing the Berlin newspapers with vivacious interviews describing his fellow-redskins as a maligned, misunderstood people. At their forthcoming congress at Muskogee, Brant Sero says, the modern generation of American Indians intends to take vigorous action in the direction of clearing up the world's dime novel conception of the noble red man.

July 1, 1910

SEEK RICHES OF INDIANS.

Oil Promoters in Oklahoma Resorting to Many Schemes to Get Lands.

Special to The New York Times.

MUSCOGEE, Okla., April 19.—Grabbing of Indian lands has led to many scandals in what was formerly Indian Territory, but never were the white promoters more active than now, when the tribesmen have come into possession of fortunes in money and land rich in minerals and lumber. Persons who wish to exploit these lands are scattered among the Indians in profusion, and in the scramble for wealth desperate measures have been taken. Tales of bribery, of fraud in securing deeds from ignorant Indians, and violence toward those who will not accept proffers are heard on all sides.

So bad are the conditions that Federal agents in many cases have stepped in to protect heirs of Indians. Secret Service men are in the field and are watching developments, but they cannot see everything that is going on. They have to deal also in many cases with persons who have strong financial and political backing and who cannot be balked unless most conclusive evidence is obtained.

Carrie Cochran, a Cherokee girl, who will come into possession of oil lands worth about $500,000 when she reaches her majority, is only three-eighths Indian, and so not under the protection of the tribes. Her father is a hotelkeeper in this city. According to her own story she came of age last month, although the census rolls show that she will not until next September. The courts have ruled that the census roll must be taken as conclusive as to the age of Indians, but oil exploiters took no heed of this, and set out to get her signature to papers giving them control of her property.

Indians related to her learned of the pressure being brought to bear, and in court charged that her father was in league with the exploiters. In consequence, Houston Tahee, an Indian of Tahlequah, was appointed as her guardian, and the girl went to his home. Later she was kidnapped and taken to Hot Springs, where, under the authority of her father, she was kept under restraint. An injunction had to be sued out to prevent her father obtaining a lease from her, and the guardian had to go to Hot Springs to rescue her. The matter is in the courts, and probably will drag along until after the girl has become independent.

Marcus Covey, a Cherokee boy, who owns oil land now valued at $75,000, was induced to make a lease to one promoter, but immediately a rival firm entered the field and declared that Covey was not of age. The matter was taken to court, and it was decided that the boy would not reach his majority until next July. His father, who had agreed to use his influence to get a lease for the second firm, was appointed his guardian.

Covey disappeared last Thanksgiving Day. It was learned that persons interested in oil lands had plied him with liquor and had invited him to see the world. He was traced to New York, and it was found that he had broken a leg there and was treated for a time. As soon as possible he was put on a steamer and taken to London, where he continued to have a good time. His companions intended to keep him there until next July, when he could sign the lease they wanted. Government agents brought such pressure to bear that he was returned to his father.

April 20, 1912

INDIANS ON THE WARPATH

Redskins Are Taking Part in Struggle, and One Nation, Under Its Treaty Rights, Has Prepared a Formal Declaration of Hostilities.

When dispatches from Syracuse, N. Y., told last week that the Onondaga Indians were drawing up a declaration of war against Germany, under their treaty with George Washington, which made them a separate nation, it was a reminder of the part the redskins are taking in this struggle. The Onondagas took action because of indignities visited by the Germans upon some of their number taken prisoner while traveling with a circus abroad. The following account of the activities of Indians since the United States was drawn into the war was prepared by the Committee on Public Information:

IN South Dakota, nestled in a picturesque valley of the Grand River, there is a little settlement called Bullhead. It is not a great way from the scene of Custer's last fight against the circling Sioux, and only a few miles from the spot where Marcellus Redtomahawk of the Indian police slew Sitting Bull in single combat. This grassy valley was once the very hotbed of hostile Indian plots against the United States Government.

Last December there took place at Bullhead a ceremony full of significance for the whole race of red men—full of meaning, indeed, for all Americans. Seven young full-blooded braves were about to volunteer for the military service of the United States in our war against the Kaiser. Even their names were redolent of the wild—Eugene Younghawk, James Weaselbear, Samuel Bravecrow, James Villagecenter, John Ironthunder, Joseph Leaf, and Thomas Pheasant.

Over in France not long ago John Peters, a Menominee Indian serving with Company A, First Engineers, died of wounds received in a fight with the Germans. Back home in Wisconsin, at the Keshena Indian School, the American flag flew at half-mast. Through Shawano County antique women of the victim's tribe revived the age-old custom of wailing for the dead, which lasts for days. In the case of John Peters it is worth remembering, for he was probably the first Indian to enlist in the army after war came and was undoubtedly among the first Americans to cross the ocean in transports.

But there are plenty of Indians waiting to avenge him. Down at Camp Bowie, near Fort Worth, Texas, Captain Walter Veach commands Company E, 142d Infantry. It is composed wholly of Choctaws, all volunteers. Through this camp alone there are scattered more than a thousand Indians. In Nebraska the Winnebagoes formed a company early in the war. For the most part separate Indian units are frowned upon, as it is the wish of the Government to merge the aborigines upon an equal footing with our white soldiers. But wherever Indian soldiers are found they are reported as earnest, efficient, silently observant, and equal to the best. Above all, they are anxious to fight.

The total Indian population of the United States is only 335,998. Of these just about half are citizens, 50,000 still wear skins and blankets, while only 30 per cent. read and write English. There are less than 33,000 male Indians of military age. Yet there are over 6,000 Indians in the United States Army, 85 per cent. of them volunteers, and several hundred more in the navy, every one a full citizen.

Fourteen tribes are represented in the service, and when young enough the chiefs themselves have enlisted. In rank our Indian soldiers scale down from Major to private, and almost every branch has lured some Indians. One Indian helps run a flock of balloons, and there are many in the Aviation Corps. Some have become proficient in wireless telegraphy, and there are others scattered through various technical divisions of the army. A large percentage of the civilized Indians have received military training at the Government schools and so enter the army with a certain advantage over raw recruits.

There are about twenty-five big Indian schools. Every one of them is an automatic recruiting station.

None of the many races which mingle in the American melting pot have a better Liberty bond record than the Indians. They are not the wealthiest people of the world, but on the three loans they have managed to subscribe more than $13,000,000—between $30 and $40 per capita. Jackson Barnett, a member of the Five Civilized Tribes of Oklahoma, individually took $660,000 of the second issue and $157,500 of the third issue. The Osages, with a population of only 2,180, are the richest Indian tribe in the country. To the last loan they subscribed $226,000. At Camp Travis in the 358th Infantry Regiment it is said that every company has its Indian noncommissioned officers. One of these, Otis Russell, owns some valuable oil lands yielding him an income of anywhere from $500 to $1,000 a month. He puts it regularly into Liberty bonds.

Here and there the Government had trouble with a few of the more remote tribes over the draft. In every case it was due to lack of knowledge and misinformation among the Indians. In but one instance was a show of force necessary. That was in Utah last Spring, when a troop of cavalry was thrown into the mountains. In Washington and northern Oregon the Nisqualli, Puyallups, Kayapullas, and several other tribes appealed to a clause in the treaties of 1854-55, by which they agreed to lay down their arms and never "make war against any other tribe except in self-defense." But when the Government explained that this really was a war of self-defense they decided readily enough that those savage tribes over in Germany needed the Indian sign more than the Iron Cross.

The war attitude of the Indians at large has been a revelation of patriotism. The Indian insists that he is merely following his traditions—that from the old days back in 1609, when kindly Indians relieved the starving settlers at Jamestown, he has always been a friend of the white man whenever reasonably encouraged. Indians have helped the United States in every war it has fought. Since 1831 they have been accepted as volunteers in the army and have written a soldierly, honorable record.

But in this war the spirit and blood of the race seem to have risen in one hot compound of militant Americanism.

August 4, 1918

MAKING INDIANS CITIZENS.

No little sentimental and historical interest attaches to the fact that the President has signed the bill conferring American citizenship upon all Indians born within the territory of the United States. Heretofore only certain classes and individuals have had this standing. Most of the Indians living on reservations have been considered simply as wards of the nation, without the legal status of citizenship. Thus in the one hundred and forty-eighth year of the independence of these United States it has pleased Congress to admit the descendants of the original American people to the same legal status as aliens who have gone through the necessary procedure after five years of continuous residence here. In granting citizenship to the Indians it is expressly provided that they shall not thereby be deprived of any right in tribal or other property. Nor is their new status considered inconsistent with wardship, so that the Federal Government is not relieved of its obligations to those residing under its care on reservations.

In the past the Indian tribes have held a special relationship to the Government, which JOHN MARSHALL described as that of "domestic dependent nations." In the early history of the Republic, relations with them were largely conducted through treaties, which our State and Federal Governments shamelessly disregarded whenever it was convenient. Not even the system of reservations with resident agents of the Indian Bureau has insured them fair treatment. In the Western States the white settlers are so near the frontier days in spirit that they have not shaken off the attitude toward the Indian that all frontiersmen have had since the westward march began. To them he is a creature to be ignored, cheated or dispossessed. He is, in fact, little more than an interloper who occupies good lands (in some few instances) which should be turned over to the whites.

What material advantages the Indians will obtain from their new status as citizens it is hard to [illegible] Presumably they will become subject to local laws, at least when off the reservations, although there is no reason to suppose that their tribal form of government will be interfered with. In due time we may expect to hear that the "Indian-American" vote has been mobilized behind a candidate pledged to this or that issue. This might strengthen their position technically, and make it easier for them to maintain their rights. But not even the new legal status affords them practical protection from the jealousy and ignorance of white men determined to "civilize" them by gradual extermination.

If there are cynics among the Indians, they may receive the news of their new citizenship with wry smiles. The white race, having robbed them of a continent, and having sought to deprive them of freedom of action, freedom of social custom and freedom of worship, now at last gives them the same legal status as their conquerors. If the Indians are thankful for this new privilege, it is to be hoped that their gratitude is what TALLEYRAND described as "a keen sense of favors to come." Certainly the Indians are entitled to much if they are ever to be compensated for the evil which has been done them.

June 7, 1924

FORWARD · BVILDING
175
紐約
靈糧堂
NEW YORK
LING LIANG CHURCH
靈糧堂
主恩
托兒所

CHAPTER **2**

Immigrants From Europe and Asia

For over half a century this building housed *The Jewish Daily Forward. The Forward,* edited by Abraham Cahan, reflected the problems, aspirations and culture of the Jewish immigrants on the Lower East Side. The present status of the building is evident from signs at either side of the entrance.

Courtesy Gene Brown

'A Nation of Immigrants'

Five years ago the Junior Senator from Massachusetts wrote an eloquent plea for a change in the restrictive immigration policy of the United States. In a pamphlet called "A Nation of Immigrants," he sharply criticized the quota system—the system of admitting immigrants on the basis of fixed annual quotas for different races and nationalities. The system is heavily weighted in favor of Northern Europeans.*

Now, as President, John F. Kennedy has acted to carry out the ideas of "A Nation of Immigrants." He has called on Congress to abolish the quota system. Under his proposed law, 165,000 immigrants (the present ceiling is 156,700) would be allowed in each year without regard to racial or national origins.

Here are excerpts from Mr. Kennedy's story of immigration to America.

By JOHN F. KENNEDY

SINCE the first settlers reached the New World, some 40 million people have migrated to America. This is the greatest migration of people in all recorded history. It is hard to imagine how many people 40 million is. It is all of the people in Arizona, Arkansas, Colorado, Delaware, Idaho, Kansas, Maine, Montana, Nevada, New Hampshire, New Mexico, North Dakota, Oregon, Rhode Island, South Dakota, Utah, Vermont, Wyoming—two and one-half times over!

Another way of measuring the importance of immigration to America is to say that every American who ever lived, with the exception of one group, was either an immigrant himself, or a descendant of immigrants.

The exception? Will Rogers, part Cherokee Indian, said that his ancestors were at the dock to meet the Mayflower.

This means that in just over 300 years, a nation of 175 million people has grown up, populated by persons who came from other lands and their descendants. It was the literal truth when President Franklin Delano Roosevelt greeted a convention of the Daughters of the American Revolution with the words, "Fellow immigrants."

Any great social movement must leave its mark. The great migration of peoples to the New World did just that. It made America a nation different from all others. The effects of immigration—to put it another way—the contributions of immigrants—can be seen in every aspect of our national life. We see it in religion, in politics, in business, in the arts, in education, in athletics and in entertainment. There is no part of America that has not been touched by our immigrant background.

Immigration to America can be pictured as ocean waves breaking on the shoreline. We can clearly see the crest of each successive wave, but if we look closer we can see that the waves are not really separate but continuous. Even at the moment that one wave is reaching its crest, the next is gathering force and building momentum.

THE first wave began with the [Virginia] settlers in 1607 and the Puritans and Pilgrims in 1620. The first wave was predominantly, but not solely, English in origin. The urge for greater economic opportunity mixed with the desire for religious freedom impelled these people to leave their homes. Of all of the groups that have come to America, these English settlers had perhaps the most difficult physical environment to master, but the easiest social adjustment to make. They mastered a rugged land and that was hard, but they built a society in their own image and never knew the hostility of old toward new that succeeding groups would meet.

By 1820, Ireland began to replace England as the chief source of new settlers. Indeed, in the century between 1820 and 1920, some four and a quarter million Irish came to America—most of these between 1820 and 1860. Most of the Irish settled in the great cities of the North and provided

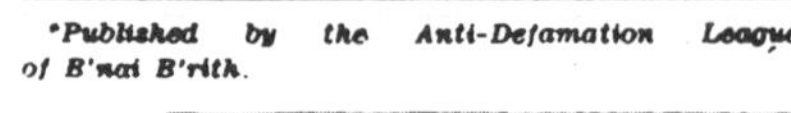
*Published by the Anti-Defamation League of B'nai B'rith.

THREE WAVES OF IMMIGRATION—They brought 40,000,000 to America.
The Irish wave (emigrants await departure) reached its peak with the famine of the eighteen-forties.

a laboring force to meet the needs of a rapidly expanding industrial economy. Many worked in factories and many others left Ireland when offered jobs by the builders of the new American railroads and canals.

The Irish were among the first to meet the hostility of an already established group of "Americans." It was not long before employment circulars included the phrase "No Irish need apply."

It is not unusual for people to fear and distrust that which they are not familiar with. Every new group coming to America found this fear and suspicion facing them. And, in their turn, members of these groups met their successors with more of the same. The Irish are perhaps the only people in our history with the distinction of having a political party, the Know-Nothings, formed against them. No party based on bigotry and hatred could be successful in America and the Know-Nothings, after a vigorous start, died an ignominious death in the mid-1800's.

The wave of German immigration began to rise in the middle of the 19th century as Irish immigration began slowly to recede. For the rest of the century the two overlapped. By 1910, there were eight million Americans either themselves born in Germany or whose parents were born there.

FOR the most part, these new Americans were farmers and artisans. They were instrumental in opening the West to settlement. Lured by the promise of free land, these people were among the first to cultivate the fertile soil of the Mississippi Valley. Their skill as craftsmen, too, found ready acceptance in the developing American industry.

Toward the end of the 19th century, immigration to America underwent a significant change. For the first time the major sources of settlers became Southern and Eastern Europe rather than Northern Europe and the British Isles. Large numbers of Italians, Russians and Poles came to this country and their coming created new problems and gave rise to new tensions.

In the 1930 census, New York City had more people of Italian birth or parentage than did Rome, Italy. Most large cities had well defined "Little Italys," or "Little Polands" by 1910, and walking through them one might well imagine himself in Italy or Poland.

The history of cities shows that when conditions become overcrowded, when people are poor, and when living conditions are bad, tensions run high and crime flourishes. This is a situation that feeds on itself—poverty and crime in one group breed fear and hostility in others and this, in turn, impedes the acceptance and progress of the first group, thus perpetuating its depressed condition. This was the dismal situation that faced many of the Southern and Eastern European immigrants just as it had faced some of the earlier waves of immigrants. Indeed, one New York newspaper had these intemperate words for the newly arrived Italians: "The floodgates are open. The bars are down. The sallyports are unguarded. The dam is washed away. The sewer is choked . . . the scum of immigration is viscerating upon our shores. The horde of $9.60 steerage slime is being siphoned upon us from Continental mud tanks."

AS it had been with their predecessors, the struggle to establish themselves in the New World was a hard one for the newcomers from Southern and Eastern Europe. Indeed, for many, the struggle continues to this day. Fear, bigotry, hatred—these do not die easily and since they are not based on fact and logic, they do not yield to the evidence of fact and logic. The history of new peoples in America shows clearly, however, that given time and opportunity, virtually every group has found its way up the economic and social ladder—if not the original settlers then their children or grandchildren. There is no reason to believe that this process has ended now.

Each new group was met by the groups already in America and adjustment was often difficult and painful. The early English settlers had to find ways to get along with the Indians; the Irish who followed were met by these "Yankees"; German immigrants faced both Yankees and Irish; and so it has gone down to the latest group of Hungarian refugees. Somehow, the diffi-

Then came Germans (shown embarking at Hamburg) and Scandinavians.
The third wave, late in the 19th century, flowed from Eastern and Southern Europe.

Above, newly arrived Italians.

cult adjustments are made and people get down to the tasks of earning a living, raising a family, living with their new neighbors, and in the process, building a nation.

THERE has always been public sentiment against immigration, or, more accurately, against immigrants. At times this sentiment was only latent, at times, it has been manifest, indeed, crudely so. Most often it has been unorganized, but in some periods it has been most effectively organized. The usual term for this sentiment is "nativism," which has been defined as "the fear of and hostility toward new immigrant groups."

Yet it is a remarkable fact that in spite of this agitation there was at first no official governmental response. The forces favoring free and open immigration were clearly dominant. The sense of America as a refuge for oppressed and down-trodden people was never far from the consciousness of Americans. Thus, for almost 100 years of the Republic's history, even through the period of Know-Nothingism, there were no Federal laws of any consequence dealing with immigration. Not only were new settlers allowed to enter freely, but they were positively sought after in some periods.

Inevitably, though, this mass movement of people presented problems which the Federal Government was forced to recognize. In 1882, recognizing the need for a national immigration policy, Congress enacted the first general legislation on the subject. The most important aspect of this law was that, for the first time, the Government undertook to exclude certain classes of undesirables, such as lunatics, idiots, convicts and people likely to become public charges. In 1891, certain health standards were added as well as a provision excluding polygamists.

By the turn of the 20th century the opinion was becoming widespread that the amount of new immigration should be limited. Those who believed sincerely, and with some basis in fact, that America's capacity to absorb immigration was limited were joined by those who were opposed to all immigration and to all "foreigners." Anti-immigration sentiment was heightened by World War I and the aftermath of disillusion with the way peace was settled, which brought on a strong wave of isolationism. In 1921, Congress passed and the President signed the first major law in our country's history severely limiting new immigration. An era in American history had ended and we were committed to a radically new policy toward the peopling of the nation.

The [restriction] was based on . . . the so-called "national-origin" system [which] limited numbers of each nationality to a certain percentage of the number of foreign born individuals of such nationality residing in the United States. . . . The effect was to cut drastically the amount of immigration from Eastern and Southern Europe and from Asia.

The famous words of Emma Lazarus on the pedestal of the Statue of Liberty read: "Give me your tired, your poor, your huddled masses yearning to breathe free." Under present law it is suggested that there should be added: "as long as they come from Northern Europe, are not too tired or too poor or slightly ill, never stole a loaf of bread, never joined any questionable organization, and can document their activities for the past two years."

A new, enlightened policy of immigration need not provide for unlimited immigration but simply for so much immigration as our country could absorb and which would be in the national interest—the most serious defect in the present law is not that it is restrictive but that many of the restrictions are based on false or unjust premises. We must avoid what the Massachusetts poet John Boyle O'Reilly once called:

Organized charity, scrimped and iced,
In the name of a cautious, statistical Christ.

Such a policy should be generous; it should be fair; it should be flexible. With such a policy we could turn to the world with clean hands and a clear conscience.

August 4, 1963

THE EXPERIENCE OF EMIGRATION

Condition and Care of Emigrants on board Ship

To the Editors of The New-York Daily Times.

I think it would be rendering a service to humanity, if those who ship so many poor Irish emigrants from Liverpool, &c., to this country, could be induced to take more pains, and examine properly the conditions of the passengers, instead of driving them on board ship like so many cattle at the last day of embarkation!

I have crossed the ocean many a time and oft, and have since late years, acted as Doctor on board of several large American ships bound from Liverpool for New-York, and find it absolutely necessary that some sort of reform must be made to better the condition of the poorer classes of emigrants.

In the health office at Liverpool where sometimes 1000 emigrants are examined per day from 10 in the morning till 3 o'clock in the afternoon, they are merely required to show their tongue; but no other notice is taken of their filthy condition nor is the body which may be full of ulcers, itch, small pox, or other disagreeable diseases properly examined, and I have sent back many a one, when the ship lay at anchor in the river Mersey, who had an infectious disease or who was maimed, lame or blind in one eye; but every captain or surgeon does not take that trouble especially when the ship is going to sea in a hurry, hence so many deaths occur at sea and infectious diseases are spread throughout the vessel, which might be avoided by proper care being taken to examine each passenger thoroughly, before the ship leaves the docks or the river at Liverpool. And then more attention should be paid to the baggage of the passengers, for it is well known that many a one goes on board without a shirt and a bed, clothed in rags and full of vermin!

It has been, and is now, the practice in Ireland to send away a pauper (nay, hundreds of them,) in the fall of the year to better his condition in the United States of America: for it costs the almshouse in Ireland £5 to feed a pauper through the winter, and a passage to New-York can be obtained for from £2 to £2 10s. in the steerage, hence the almshouse is the gainer by this operation. I have asked many a one at sea what, in their advanced age, they intended doing in the States, and they answered: "Why, we came out of a poorhouse and must go into it again at New-York."

Now if the present commutation money of $1 50 for each passenger was raised to $5 or $10 per head, I think the American vessels would obtain a better set of passengers at Liverpool, and the poorer classes and the paupers would be shipped to Quebec or St. Johns; and after all it is but fair that her Majesty's poor subjects, who are starving in Great Britain, and particularly in Ireland, should first be sent to her Majesty's own colonies, than to fill the poorhouses of New-York and other American Atlantic cities.

Another important point is, that each passenger should at least have three shirts, an extra jacket and trowsers, and a bed; for cleanliness is undoubtedly the best prevention against sickness. And too much attention cannot be paid to cleaning the berths and the between-decks, for it is inconceivable how much filth and dirt will accumulate between the boxes, and in and under the berths of three, four, nay, sometimes nine hundred passengers, crowded together in the steerage. It is really sickening and revolting to humanity to see the misery existing on board of some vessels, where the captain makes his own tyrannical laws, and the mates and crew treat the poor passengers more like common brutes than human beings.

From Liverpool each passenger receives weekly 5 lbs. of oatmeal, 2½ lbs. biscuit, 1 lb. flour, 2 lbs. rice, ½ lb. sugar, ½ lb. molasses, and 2 ounces of tea. He is obliged to cook it the best way he can in a cook-shop 12 feet by 6! This is the cause of so many quarrels and hard fightings for might makes right there. Of course many a poor woman with her children can get but one meal done, and sometimes they get nothing warm for days and nights when a gale of wind is blowing and the sea is mountains high and breaking over the ship in all directions. On such days we have been obliged to go with buckets of water and a bag of biscuits in the steerage and feed the passengers to save them from starvation.

On board of the German vessels from Bremen and Hamburg, the ship's cook gives at breakfast, dinner, and supper time, to the head of a mess of 8 passengers their share, and there is no trouble, confusion, nor quarrelling and fighting; but where passengers are obliged to cook for themselves there can be no order nor regulations enforced, except by having the rod of iron constantly above their heads, and this of course creates ill-feeling, ill-will, and bad behavior.

There is many a worthy man and woman amongst the poor Irish—though on the other hand there is many a one who leaves his country for his country's good!—and a strict regulation is necessary to keep them in order; but it is revolting to humanity to see how shamefully they are treated by some captains, mates and crew, without the slightest provocation, merely because they are poor steerage passengers!

I leave to wiser heads than mine to devise the laws that should be established, and have made this statement merely to show the necessity that some sort of Reform must take place to avoid the paupers of Great Britain filling the Alms-houses of this country, and to insure more comfort on board ships to passengers in general.

If you find the foregoing interesting enough, be pleased to have it inserted in your valuable Journal, and in case it should find support I might suggest some remedies which would, I think, do a vast deal of good to emigrants' health and comfort whilst crossing the Ocean.

I have the honor to be, sir, your obedient servant,

DOCTOR.

NEW YORK, Sept. 25, 1851.

October 15, 1851

NEW-YORK CITY.

Fixed.

No more Operas, or Kossuth Receptions, or Promenade Concerts in Castle Garden—that is settled! Yesterday the Commissioners of Emigration took possession of these beautiful and very historical quarters, on a lease for four years; and hereafter the Garden will be used as a sort of a dépôt, *vulgarite*, pen, into which the paupers, convicts, honest men and reputable of Europe shall be tunneled from shipboard through the city to their places of destination.

The Commissioners intend to let an office on their new premises to each Railroad office that engages to sell tickets at no other place in the City to emigrants, requiring each Railroad Company, so provided with an office, to send a steamer to the Garden, and thence directly to their depots, with passengers and baggage—thus relieving the emigrants who prefer it from the necessity of even going into the City for any purpose, if they have any further point of destination in view. The economy of the thing to the Board, on the terms stated, cannot be doubted. The humanity of it is unquestionable, for the thousand runners that make their living out of emigrants would thus doubtless be effectually headed off from much of their plunder. There can be no objection to the measure on the score of public health, for the Health Officer has not any new reason for relaxing the vigilance that prevents sick emigrants from landing; and if they are allowed to do so by inadvertence, it certainly is no worse to confine them together in an airy place like Castle Garden than to let them be distributed over the City as they now are.

And yet it comes hard to surrender this favorite old resort of citizens to such purposes, fitting as it is. The City Hall would be a capital hospital ground, and the Fifth-avenue, from its width and freedom from travel, a delightful thoroughfare to drive pigs through to market. Still some people would object to such innovations.

May 8, 1855

HOW EMIGRANTS ARE FLEECED.

Overcharges by the Railroad Agents-Emigrants Charged More than First-Class Passengers-Estimate of the Agent's Profits-Proposed Remedy-An Interesting Affidavit.

In February last the Castle Garden Committee of the Commissioners of Emigration appointed Mr. JOHN P. CUMMING, Superintendent of Emigration, a Committee of One to communicate with the ticket agents and others in relation to the irregularities of charges for the transportation of emigrant passengers West and South, and to report a plan of arrangements.

In March last Mr. CUMMING presented his report, stating that he had made as thorough an investigation of the same as the time would permit, and finds:

That, on the 13th of April, 1853, the Legislature of this State passed an act for the better protection of emigrant passengers arriving at the port of New-York. Section seven of this act imposes a penalty for any person to charge an emigrant more than one and one quarter cent per mile for passage tickets to the interior.

At the time this act was passed all the railroads and transportation companies were in open competition for this business, and of course each advertised their lowest rates. In 1853 advertised prices were, from New-York to

Buffalo, by N.Y.C.R.R.	$4 00	Erie	$4 00
Cleveland, O., by N.Y.C.R.R.	4 85	Erie	5 00
Cincinnati, O., by N.Y.C.R.R.	7 25	Erie	7 50
Chicago, Ill., by N.Y.C.R.R.	6 50	Erie	7 00
Dunkirk, by N.Y.C.R.R.	4 00	Erie	4 00
Toledo, O., by N.Y.C.R.R.	4 84	Erie	5 00
Detroit, Mich., by N.Y.C.R.R.	4 84	Erie	5 00
Milwaukee, Wis., by N.Y.C.R.R.	6 50	Erie	7 00

And to all other points accessible to railroad communication in proportion to the foregoing.

Having in view the better protection of emigrants, the Legislature, in 1855, gave power to the Commissioners of Emigration to establish a general landing depot for all emigrants arriving to the Port of New-York, and making the landing of them at such designated depot obligatory upon the owner, master or consignee of the vessel bringing them to this port. Castle Garden was selected as such depot, and the New-York and Erie and New-York Central (Hudson River) Railroad were admitted thereto, thus securing all western bound emigrants to the exclusion of all other railroads and transportation companies, and at that time a tariff was settled upon, making the fare to all points much less than rates allowed by act of 1853.

At a subsequent period, the Pennsylvania Railroad (Camden and Amboy) made application to enter Castle Garden and solicit business over their line.

For a long time the roads already in possession retained the sole control of this business, forcing the Pennsylvania Railroad in order to secure a portion of the same, to establish agencies in the principal ports of Great Britain and the Continent of Europe, where, by means of low published rates and orders on their home office in this City, a portion of this traffic was diverted from the routes whose headquarters were held at the Garden, and ultimately resulted in the Pennsylvania Railroad being admitted as one of the parties in interest.

The before-named three roads still continue therein, and divide the business by giving to the Erie and New-York Central thirty-five per cent. each, and the Pennsylvania Railroad thirty per cent. of all the passengers to western competing points; but, on account of the Camden and Amboy line being the feeder from the City of New-York for the Pennsylvania Railroad, this latter road demanded, and was allowed, forty miles more than the actual length of their road, thus increasing the price of emigrants fifty cents each for the benefit of the Camden and Amboy line.

In this connection, I compare the prices as now charged emigrants with those then established, when the requirements of law regulating prices were in no instance exceeded:

N. Y. to Buffalo	423 miles,	now $6 50,	then $5 50
N. Y. to Dunkirk	460 miles,	now 6 60,	then 5 00
N. Y. to Cleveland	596 miles,	now 8 50,	then 6 50
N. Y. to Cincinnati	770 miles,	now 11 50,	then 9 00
N. Y. to Detroit	678 miles,	now 10 50,	then 7 50
N. Y. to Chicago	899 miles,	now 13 00,	then 10 00
N. Y. to Milwaukee	952 miles,	now 15 50,	then 10 00

And to all other points reached by, or quite accessible to, railroad communication, in proportion to above rates.

After stating that the accomodations provided for emigrants at the depots, with one exception, are of the most shabby and insufficient character. Mr. CUMMING gives a comparison of the rates charged between Chicago and Janesville, Wis., showing that between these two places emigrants are actually charged $6 50—being $3 more than a first-class, and $3 50 more than a second-class passenger pays over the same road, yet one of the roads admitted to the privileges of the Garden, advertises the second-class time to be less than half that of the emigrant train.

Mr. CUMMING says the above abuse, after having existed the past year, has only been corrected the first of the present month, and then only after the parties practicing the same had knowledge that the Commissioners of Emigration were investigating this abuse.

There are so many instances of the emigrant being overcharged at Castle Garden for all or a portion of his ride that to note them all would be superfluous; one more only will be given:

Milwaukee, Wis., first-class fare is	$27 85
Milwaukee, Wis., second-class fare is	21 85
Milwaukee, Wis., emigrant fare is	15 50

The New-York Central advertises to deduct from the first class $3, and from the second class $2 50, of the above, whenever the passenger buys tickets via Detroit and Milwaukee road. There is, however, no deduction advertised or allowed the emigrant when he buys tickets via Detroit and Milwaukee Railroad; and yet much the largest proportion, in fact, almost all of them, are sent by the Detroit and Milwaukee Railroad through all the summer months. Distance from New-York to Milwaukee via Chicago is 984 miles; distance to same place via Detroit and Milwaukee Railroad is 952 miles; fare to Chicago, at one and a quarter cents per mile, is $11 24, to which add $2 50, arbitrary rate from Chicago to Milwaukee, would give $13 74 as the fare by this route, while the fare by Detroit and Milwaukee Railroad, at one and a quarter cents per mile, would be only $11 90 to Milwaukee, yet the Castle Garden agency, for the past year, has charged the emigrant $15 50 by either route, and still continue the same.

The representatives of the roads now admitted to the Garden have adopted a new tarrif of prices, which took effect on the 1st of March. This late tariff differs but little from the previous one, except in those cases where the emigrant was charged more than first or second-class rate for a portion of his ride. In other respects the new one is fully as objectionable as the one of last year. In this connection, I will state that I have had interviews with JAMES H. BANKER, Esq., one of the Directors of the New-York Central, and with ROBERT H. BERDELL, Esq., President of the Erie Railway, each of whom have assured me that they will heartily cooperate with the Commissioners of Emigration in reforming any and all abuses practiced on emigrants. I have also had an interview with Messrs. JOHNSTON and BALDWIN, President and General Ticket Agent of the Central Railroad of New-Jersey. These gentlemen inform me that they will convey emigrant passengers over their line to all points West, on the basis as agreed upon and recommended by the railroad ticket agents, that is, one cent per mile not from New-York to destination, except to those points where arbitrary rates are exacted, such rates to be added from the nearest point where the cent per mile rate ceases, to destination.

In regard to baggage the report says:

Emigrants are charged for their extra baggage still higher in proportion than for their passage, in many instances exceeding $1 per 100 pounds more than the price as agreed upon and recommended by the Committee heretofore mentioned. In fact, there are numerous instances where the excess added to the regular price will swell the profits of the Castle Garden agency to a larger sum on each hundred pounds than the railroads receive for the transportation of such baggage from New-York to destination. To illustrate this take a single instance—Cleveland, Ohio. The price reckoned on the basis as agreed upon by said Committee, would be $1 49 for each 100 pounds, out of which the roads retain $1 19 for transportation, and allow the agency 30 cents for weighing and handling the same; yet they charge the emigrant at the Garden $2 55 per 100 pounds, thus swelling their profits to $1 30 on each 100 pounds, or 17 cents more than the roads receive for transportation the whole distance.

Mr. CUMMING says he has been unable to obtain from the railroad agents at Castle Garden an account of their receipts and expenditures for the past year; but from the report of the Commissioners of Emigration he has made an estimate of the profits accruing to the agents during 1865, which he believes to amount to $273, 708 81.

In concluding his report, Mr. CUMMING says that it is apparent that when the various lines of railroad were in open competition the fare was a great deal less than at present.

That the accomodations furnished emigrants at the various depots are insufficient.

That there have existed for the past year gross abuses in charging emigrants more than first or second class passengers' pay.

That the emigrant is much longer delayed at certain points of his transit than is necessary.

That the Castle Garden agency demand and receive a greater amount from the emigrant than is allowed by the report of the Committee having this business in charge.

That the booking of emigrant passengers in Europe is pregnant of great and continued abuses.

That in the present tariff emigrants are still charged more than allowed by report of Railroad Committee.

That emigrants are excessively charged on all their extra baggage.

That the profits of Castle Garden agency are more than double what they should be for 1865.

That the ten cents collected from emigrants on packages by City delivery go only to swell the profits above noticed.

That the commissions paid to outside parties are in the end just so much money taken from the emigrant.

Mr. CUMMING further says:

To show, once for all, the protection that the emigrant has received from the Castle Garden agency, I will compare the prices to a few places charged emigrants at Castle Garden, New-York, with the prices charged them in Philadelphia, where they have no Castle Garden and no Commissioners of Emigration:

New-York		Philadelphia	
To Alliance, Ohio	$9 00	To Alliance	$6 15
To Adrian, M.	12 00	To Adrian	8 80
To Alton, Ill	17 00	To Alton	14 40
To Chicago, Ill.	13 00	To Chicago	10 20
To Mattoon, Ill	16 00	To Mattoon	11 65

Now, if the fare of emigrants was computed pro rata per mile on the actual distance from New-York by the short line, as charged by the Pennsylvania Central from Philadelphia to the following poinnts, viz.:

New-York to Alliance, Ohio, $7.88, in place of $9, as charged in Castle Garden.

New-York to Adrian, Mich., $9 31, in place of $12, as charged in Castle Garden.

New-York to Alton, Ill., $15 43, in place of $17, as charged in Castle Garden.

New-York to Chicago, Ill., $11 09, in place of $13, as charged in Castle Garden.

New-York to Mattoon, Ill., $13 04, in place of $16, as charged in Castle Garden.

In view of the foregoing statement, I respectfully recommend that the Presidents of all the trunk lines be invited to meet with the Castle Garden Committee, for the purpose of exchanging views, and, if possible, to agree upon some plan by which the abuse before mentioned may be corrected, either by the officers of the railroads appointing one person as receiver of all moneys from emigrants paying to the roads the net price now paid to them, and divide the business equally between them, or on any other basis they may agree upon, he paying to the Commissioners of Emigration the twenty per cent, now allowed the Castle Garden Agency, this money to be expended first in the payment of the salaries and contingent expenses necessary to transact this business, and also rent of Castle Garden, balance to be expended in the forwarding of indigent emigrants at once to their destination. Any surplus yet remaining to be expended in employing suitable agents at the different railroad centers, whose duty it will be to look after the interests of the emigrants while in transit.

Should the foregoing not be satisfactory, that then the Commissioners of Emigration respectfully request the United States Commissioners of Immigration to ask for proposals, under section 4 of Act of Congress approved July 4, 1864, from all the trunk lines running from the City of New-York, for the lowest price for emigrants and their luggage to all points West and South, and be requested to make a contract with the party or parties offering the best terms. Should these proposals be uniform from all the trunk lines, then the business to be equally divided between them, or on any basis they shall agree upon.

The statements contained in Mr. CUMMING'S report in regard to the over-charges of emigrants are supported by the following affidavits:

State, City and County of New-York, as.

GEORGE W. DALEY, being duly sworn, doth depose and say that he is a resident of Edgewater, Richmond County, State of New-York; that he has for the past year been employed as an officer under the United States Superintendent of Emigration; that he knows, of his own knowledge, that overcharges have been made on emigrants in the Castle Garden Emigrant Depot. From the first day of July, 1865, to the last of February, 1866, emigrants were charged from Chicago to Beloit, one dollar each; from Chicago to Davenport, two dollars and five cents each; from Chicago to Janesville, three dollars and fifty cents each; from Chicago to Rock Island, one dollar and fiftyfive cents each more than second-class passengers. He further swears that there were numerous other points, as many as thirty-five, where the same overcharges were made on the emigrants, varying in amounts from ten cents to five dollars and fifty cents on each ticket; that in numerous instances emigrants were ticketed to Davenport, Iowa, and then only furnished tickets to Rock Island, they charging the emigrant fifty cents more than the price to Rock Island on each ticket; that from the first of July aforesaid until the closing of Lake Michigan, emigrants were charged in Castle Garden fifteen dollars and fifty cents each from New-York to Milwaukee, when the tickets could be purchased for twelve dollars each outside the Garden. He further swears that on the fifteenth day of the present month, he purchased a ticket from New-York to Milwaukee direct, from the offices of the railroads in this City forming that line for which he paid the sum of twelve dollars, and that this was all that he was asked or charged for such ticket, and that there was not one cent of commissions or drawback asked for or received by this deponent; that on the morning of the sixteenth of the present month he applied for and purchased a ticket to Milwaukee at the emigrant agency in Castle Garden, for which he paid the sum of fifteen dollars and fifty cents; that he neither asked for nor received any commissions or drawback on said ticket; that upon comparing said two tickets, they were each by the same route, as follows: New-York to Albany by Hudson River Railroad, from Albany to Suspension Bridge by N.Y.C.R.R., from Suspension Bridge to Detroit by Great Western Railroad, from Detroit to Grand Haven by D & M.R.R., and from Grand Haven to Milwaukee by steamers. He further swears that emigrants were charged $5 each in Castle Garden during all the year 1865, for tickets from New-York to Baltimore, and that the emigrant fare from New-York to Philadelphia was but $1 75, and that the second-class fare from Philadelphia to Baltimore was but $2 15 during all that time; or a total of $1 10 less than the price charged emigrants in Castle Garden. And further this deponent saith not. G. W. DALEY.

U.S. Int. Rev. stamp canceled.

Sworn to and subscribed before me at the City of New-York this 19th day of May, 1863.

BERNARD CASSERLY, Notary Public.

To the report is appended a letter from Messrs. ROBERT CHRISTIE and JOHN McDONALD, agents, denying that the profits amount to Mr. CUMMING's estimate. They admit that more than the legal rates are charged to emigrants, because they say, the legal rates are not remunerative. In regard to the charges between Chicago and Janesville, they say that this was caused by an error in the tariff, which has been corrected. As to detentions of trains, with this they disclaim having anything to do.

May 29, 1866

LOCAL INTELLIGENCE.

EMIGRANT LIFE IN NEW-YORK.

How Thousands Sleep and Eat-Boarding-Houses and Runners-London Lodging-Houses.

Every great and overcrowded city has its unsupplied wants, and the City of New-York affords no exception to the rule. The trite axiom, so often quoted that one-half the world does not know how the other half lives, is peculiarly verified in crowded localities, and it is in just such places that we find the least concern upon the subject. Were the underground population of New-York ferreted out of their pest holes, and the swarming dens of reeking filth and vice in the Five Points turned inside out for the view of the aristocratic residents of Fifth-avenue and similar localities, it would afford a spectacle at which the dainty sensibilities of these favored ones would be grossly shocked. The subject has oftentimes been taken up by the public Press, and gigantic, sweeping reforms of the most prodigious character have been proposed more than once, and have as often failed to be put in practice.

One of the most serious wants existing in this City is cheap and clean lodging-houses for men of limited means. As matters stand at present, a clean bed cannot be secured for less than fifty cents, and a great many dirty ones are let out for the price. New-York has necessarily a surplus population of emigrants seeking employment, of persons out of employment temporarily, and of others working for almost nominal pay. All these require clean lodgings just as much as their more aristocratic fellow creatures residing in the regions of wealth and luxury, and there can be no reason why they should not have them, and that, too for ten cents per night. To illustrate the present shifts and inconveniences with which an emigrant has to put up, we will imagine the case of one who has just arrived from the old sod, with only a few dollars in his pocket. First of all he lands at Castle Garden, and unless he is extremely lucky, he leaves it quickly to wander around a great and strange City, without a friendly hand to guide him, a prey for all the rascally land sharks hanging around various corners, seeking whom they may devour. In all probability, bewildered by the strangeness of all about him and the persistent importunities of the "runners," he allows himself to be conducted to a strange boardinghouse, where he is unmercifully fleeced, and out of which, when all his money has gone he is turned unceremoniously.

BOARDING-HOUSES AND RUNNERS.

These houses as a rule, though happily there are some one or two exceptions, are little better, if any, than houses of prostitution, and are the resort of thieves, loafers and swindlers. The first floor is set apart as a sitting and bar room, and the upper floors for sleeping apartments. Three or four beds are placed in a single room, and in each bed two, and sometimes three, persons sleep together, while the rooms themselves are almost choked up with trunks, valises, clothing, & c., presenting anything but an inviting or tidy appearance. Most of these houses employ "runners," who hang around Castle Garden and waylay emigrants as they leave the building. Inside the Emigration Depot a few of these "runners" are allowed, but these for the most part represent decent houses, and emigrants would do well, if they intend to accept the services of "runners" at all, to employ those found within the building and to disregard the importunities of those outside, however emphatic their assurances. Owing to the exertions of the Commissioners of Emigration great improvements have been made recently in respect to runners. Only a few years ago, hardly a week passed without furnishing instances of young girls being entrapped by these fellows into houses of ill-fame, where their ruin was effected. In a similar manner, young men have been induced to live a dishonest life by the artful incentives put forth to induce them to follow a career of vice and crime. It is in this respect that these boarding houses are a curse. The inevitable bar-room, which is ever to be found in these places, is to the emigrant what the candle-flame is to the moth. He hangs round it until his money is all gone, drinking, drinking, and what with the incessant drink, the many "friends with whom he is sure to meet, who are always "dry" and often "dead broke," and the thieves who will rifle him of what he does not spend, he soon finds himself a penniless stranger, and wakes up to the bitter truth that all this time he has been losing opportunities of employment. He has spent each day in eating, drinking, swearing, and probably gambling. Thus it is with hundreds and thousands of emigrants who land at Castle Garden, and thus it is the record of the Criminal Court of this City shows so large a percentage of criminals of foreign birth. Having first been robbed themselves they turn to robbing others, driven to it by the presence of starvation, and no prospect of immediate relief. The bar, the dice-box, the cards, and the insinuating voice of the tempter, all combine to demoralize the mind of a young emigrant, away from home and friends. If he does not turn to stealing, then is he left to seek employment under the most straightened circumstances, often having to make the City Hall Park his nightly resting-place, and a crust of dry bread with a drink of cold water his daily food. Perhaps, after enduring hardships that none can imagine but those who have drank of the bitter cup of its very dregs, he finds employment on a paltry salary, with his clothes all worn and seedy, and carrying his wardrobe on his back. Situated as he is, a respectable boarding-house is out of the question, and if he goes to anything like a decent lodging-house, he has to pay each week the better part of his insignificant stipend. He has either to do this or sleep in some miserably dirty hole, with disgusting vermin for bedfellows, and drunken, brutal, debased and criminal men for companions. An emigrant who has been forced to the later says that, even on this plan, living had cost him *per diem* as follows:

Breakfast, 15 cents; dinner, 15 cents; supper, 10 cents; bed, 15 cents, thus costing $3 85 per week on the very lowest fare that it is possible for a working man to live. His description of the beds in these houses is truly revolting, and his remarks about his room-mates equally so. In a letter to a friend he says: "The bed was a tumble-down ricketty old thing with an old straw mattrass swarming with lice, fleas and bugs, and covered with an old army blanket, evidently innocent for several months of any acquaintance with soap and water. My companions comprised some of the viles. They swore and drank, while many were common thieves and vagrants, entertaining as supreme a contempt for honest labor as any pampered scion of European aristocracy. "There was no accommodation for self abrution; in fact such a thing among a crowd like that would have bordered on the absurd. They would have scorned it."

He says, however, that there were decent men among them and as he writes, "like myself forced to this sort of thing."

Thus it is that thousands live in the City of New-York. Many of them who have simply been driven to their condition by a stroke of illfortune, soon rise out of the mire into which they have fallen, and in a short time they leave for more congenial, and cleanlier resting places. Others there are who are constant lodgers from year's end to year's end at these miserable dens, until they begin to regard them as a home, and become confirmed, irredeemable loafers from whom no better things can be expected. There is little, of anything, to be hoped of men when they become safeguided as to be satished and happy under such circumstances.

LONDON LODGING—HOUSES.

In this matter of cheap lodging-houses London is ahead of New-York, though the system there is by no means as perfect as it might be. Still the plan there adopted is a good one, and it can be easily improved upon. The lodgings are provided at four-pence (8 cents) per night, or two shillings (50 cents) per week. The building to which we refer is a large six-story one, with six rooms on each floor, and a window in each room. It is all well ventilated, and each man has a separate bed. The large rooms contain five iron bedsteads, and the small rooms three. The rooms and beds are kept scrupulously clean, no dirty person being admitted. The whole is under the supervision of the Inspector of Lodging-houses, and a code of rules, prohibiting drinking, smoking, gambling, &c., in the building hangs on the walls. The place is closed at midnight, so that all persons must be in by that hour. Should any lodger return in an intoxicated condition, he is refused admittance. In the basement is a large kitchen, supplied with ranges, and all appuntenances for cooking. Several tables and benches are in various parts of the room, besides several closets for dishes, plates, &c. Here those of the weekly lodgers who desire to do so can cook and provide for themselves. A strange and not altogether unpleasing sight is to see them, strong, stalwart men, as they cook their beefsteaks, or any other provision they may have chosen. Not the least pleasant spectacle is the evident good-feeling that exists among the inmates. Notwithstanding their comparative poverty they seem happy, and hold up their heads with a free and independent air, not seen in men who content themselves with the filth and misery of the unlicensed cheap lodging-house.

The necessity for cheap lodging-houses cannot be more forcibly exemplified than by the statistics of the number of station-house lodgers. There are, in all, in this City 33 station-houses, including sub-stations, and each of these affords accommodation for homeless wanderers. The Sixth Precinct records the following as the number of lodgers who sought shelter during the first week in August:

	Males.	Females.		Males.	Females.
Aug. 1	-	17	Aug. 5	—	10
Aug. 2	7	12	Aug. 6	—	10
Aug. 3	12	—	Aug. 7	4	10
Aug. 4	5	12		—	—
Total				28	71
Total males and females					99

And this in the Summer months. In the Winter time this number is at least quadrupled.

August 12, 1867

EMIGRATION.

Comparative Statements of the German and Irish Influx—The Irish in the Majority.

The erroneous impression has gone abroad that the number of immigrants arriving here from Ireland is far below that of the Germans. In order that the TIMES' readers may be properly informed as to the true state of immigration with regard to these two nationalities, we subjoin the following tables, made up from the books of the Commissioners of Emigration, beginning Jan. 1, 1847, and bringing the exact figures down to Saturday last. It will be noticed that in 1847 the numbers of immigrants arrived from Germany and Ireland nearly balanced each other, there being a difference of only 134 for that year in favor of the Germans, That was the terrible famine-year in Ireland, and we see that during the four years immediately following, 1848, '49, '50 and '51, the excess of immigration was largely in favor of the Irish. In 1852, '53, '54, '55, '56, '57 and '58, it will be noticed, the tide turned in favor of the German element, and from 1859 to 1864, excepting the year 1861, was again changed, bringing hither a greater proportion of Milesians than of Teutons. In 1865, '66, '67, '68 and '69, however, the Germans regained the ascendency, and have retained it, the excess gradually diminishing down to the present time. In April, 1869, for instance, there were 1,194 more German than Irish immigrants; but during the month of April last past, the Irish outnumber the Germans by 1,523. Thus far during the present month the comparative excess of German arrivals is 470 only, while for the same period of time last year the excess of Germans was 4,993; so that it would appear that the Irish immigration is again "going to the 'fore." This appears still more palpable, too, when we compare the whole number of arrivals for the present year with the number for the same period of time last year. In 1869, the Germans, up to May 21, were in excess 12,406; this year, however, they lead the Irish thus far only 1,039.

From some cause which we will not attempt to divine, the German immigration during 1866, '67, '68, and '69, received an impetus which put it largely in excess of all others, and hence people argued that the aggregate for the past twelve years would show a much larger German element brought to these shores than has been brought from Ireland. The fact is just the reverse, however. If a difference be struck between the totals of the two classes from 1847 to 1869, (including both those years,) we find it to be 27,755 in favor of the Irish; and deducting the excess of 1,039 credited to the Germans up to Saturday last for the year 1870, we must still leave the Irish ahead by 26,716, since 1847. Doubtless, if the present ratio of Irish immigration be kept up, as the figures seem to imply it will be, the end of 1870 will find the Milesian element again in advance of the incomers from Faderland, and, possibly, they will hold that position for another five years; for by some unknown law which appears to govern this matter, the tide of immigration, as between these two classes of people, changes about every semi-decade. Those well informed on this subject say that the emigration from Ireland to the United States would be much greater this year than it is except for the fact that the efforts at Governmental reform in Ireland, which English statesmen are now making, keep many in the old country who otherwise would find a home in this. These are hoping (we trust not against hope) that a brighter future is opening up for Old Erin, and that under a revised and more just system of tenant-right and property-holding than has prevailed in the past, with the abolition of laws which now discriminate against those in Ireland who are "native and to the manner born," the time is fast approaching when Ireland will be ruled by Irishmen, and not by English stipendiaries. There may be force in this averment. It is a lamentable fact that of the Irish who come to this country, the far greater number prefer to risk the chance of getting a precarious subsistence in our large cities than to seek the almost certain competency which awaits them in the wide West, were they wise enough to settle there. It would be far better for them, as well as for the labor markets of our cities, already overstocked, were those who come in the future to imitate the foresight of the Germans who act upon a different principle, and who are, to their honor and credit, be it said, doing so much to build up the waste places of our extensive country, and are "making the wilderness to bloom as the rose." We subjoin the tables showing the comparative immigration of Irish and of Germans since the year 1847:

	From Ireland.	From Germany.		From Ireland.	From Germany.
1847....	52,946	53,180	1859....	32,652	28,270
1748....	98,061	51,973	1860....	47,330	37,899
1849....	112,591	55,705	1861....	25,784	27,139
1850....	117,038	45,535	1862....	32,217	27,740
1851....	163,306	69,919	1863....	92,157	35,002
1852....	118,131	118,611	1864....	89,399	57,446
1853....	113,164	119,644	1865....	70,462	83,451
1854....	82,302	176,986	1866....	68,017	106,716
1855....	43,043	52,892	1867....	65,134	117,591
1856....	44,276	56,113	1868....	47,571	101,989
1857....	57,119	80,974	1869....	66,204	99,605
1858....	25,075	31,874			
				1,611,009	1,633,254

Continuing the enumeration down to the present time we find that the relative proportion of emigrants, of the two classes (Irish and German) for 1870, up to May 21, Saturday last, is as follows, compared with the same period for 1869:

	1869.			1870.	
	Irel'd.	Ger.		Irel'd.	Ger.
January.......	745	3,233	January.......	1,012	2,140
February......	809	1,778	February......	1,403	1,634
March.........	2,969	5,727	March.........	3,109	4,142
April..........	8,258	9,456	April..........	9,799	8,276
May (to 21st)..	7,861	12,854	May (to 21st)..	8,838	9,308
Total......	20,642	33,048	Total......	24,461	25,500

The steam-ship *Smidt* arrived on Saturday from Bremen, with 633 emigrants. On the same day the steam-ship *Erin* arrived from Liverpool, with 1,290 emigrants, and the sailing ship *Cynosure*, from Liverpool, with 43 passengers, making the aggregate number of emigrants landed at Castle Garden in one day 1,966.

May 23, 1870

THE IMMIGRANT.

How He is Received, Protected and Sent on His Way Rejoicing—Who Welcome the Coming and Who Speed the Parting Guest.

Few, who have not made a personal examination, have any idea of the labor and attention which are bestowed upon the immigrant upon his arrival on these shores, and the care that is taken to see that he shall not be imposed upon during his brief sojourn in the City. When it is considered that almost the entire expense of the machinery necessary to accomplish these desirable ends is originally borne by the immigrant himself, and that in reality he is of comparatively little expense to the country which he intends to make his home, the public will admire the system by which these taxes—in the form of "head money"—are gathered and applied to the comfort and protection of the class who contribute them.

THE FIRST GREETING.

The first pressure of the hand of the Government whose protection the immigrant seeks, is extended to the wanderer as the vessel, which has been his home during the passage across the ocean, drops her anchor in front of the official residence of the Quarantine Officer. The wanderer has watched with eagerness every foot of the progress of the vessel from the time the pilot boarded her off Sandy Hook, and has continually looked forward to the time when he should be called upon to pass the ordeal of the final examination preparatory to his final discharge upon the shores of a free country. What his thoughts may have been during this interval those can best describe who have experienced the pleasure of first entering upon a future home. The ship is no sooner brought to an anchorage than the Health Officer's barge—known by carrying a yellow flag—is seen approaching, and presently that official boards the ship. The ceremony of visitation is a brief one, usually. The immigrants are all mustered upon the main deck, and as they emerge, pass in review before the Health Officer. He determines at a glance if any sick are being smuggled among the well passengers, and if any are there who are ill, they are examined with a view to learn the character of the disease. If the vessel has a "clean bill of health," she is suffered to pass up. If not, she is quarantined. If, at the general muster, there be some who are unable to respond, by reason of illness, the doctor examines the patients, and if the disease with which they are suffering is not contagious, the vessel is allowed to proceed to her destination.

QUARANTINED.

Should any of the passengers have a contagious disease, such as yellow fever or cholera or typhoid or typhus fever, the vessel must undergo fumigation, and for this purpose is sent to the Lower Quarantine. Meanwhile if no sickness exists among the cabin passengers they are transferred to a tug and sent up to the City. On arrival in the Lower Bay immigrant passengers are sent to the hospital-ship—to which duty the *Illinois* is now assigned—and there the sick are healed and the well are fumigated and prepared for association with their more fortunate fellow-beings in the City. The ship and baggage and cargo undergo thorough fumigation below, and after a time are discharged from Quarantine. Then the passengers come up and are treated at Castle Garden precisely as newly-arrived immigrants.

THE NEXT SALUTATION.

If the immigrant has passed the Health Officer immediately on his arrival, or if he has been forced to contemplate the promised land from the Lower Quarantine for a certain period, he is next greeted by the presence of the Custom-house Inspector, whose duty it is to examine his baggage, the process of which was graphically described in the TIMES a few weeks ago. The examination is necessarily thorough, for very often professional smugglers assume the garb of immigrants, or induce the ignorant steerage passengers to conceal dutiable goods in their baggage and in this way defraud the Government. The inspection is conducted with rapidity, and as fast as each piece of baggage is examined it is sent over the side into a barge, and is followed soon after by the owner. When the ship is discharged the barge proceeds to Castle Garden.

FIRST PLACING FOOT ON SHORE.

Here the immigrant's foot first touches the shore of a free country. He passes off the barge, and—presuming that he is not wanted by the officers of the customs for further examination—he is marshaled into the rotunda of Castle Garden, where once resounded the notes of the Swedish nightingale—herself an immigrant. This rotunda is divided into sections for each nationality, in case more than one ship-load should arrive at the same time. In the center of the room are desks, which are occupied by clerks under the charge of Col. COONAN, and here the immigrants have their names recorded, their destination, &c. If any have letters awaiting their arrival, they receive them. If there are friends expecting them they soon learn the fact by hearing their names called out from the desk. Those who have foreign coin which they wish to exchange can convert it into greenbacks at the desk and receive within half of one per cent. of the rate at which gold or silver is then selling in Wall-street. If any desire to communicate with friends at a distance, there is the telegraph office in the building; or, if, the slower process of the mail is desired, there are facilities at the desk, and clerks at hand whose business it is to assist the immigrant in this respect, even to the extent of writing the letter.

SPEEDING THE PARTING GUEST.

Leaving the sick and impecunious for awhile, we will follow the immigrant who is provided with means, and is anxious to reach his destination through the steps which he takes to accomplish his object. Having registered his name, and exchanged his money—if he desires to, though this is not compulsory—he looks after his baggage. He finds that it has been carefully landed, and is now in the weighing-room, where the weigh-master is ready, upon its being identified, to count the excess over eighty pounds allowed the emigrant over the railroads. The owner pays the freight charges, receives a check for his baggage, and departs in peace. If he is to travel on a steam-boat or railroad-car, his baggage is conveyed to the depot free of charge. If he is destined for a boarding-house, or other place in the City, Captain ALBERTSON, the baggage-master, will carry it for a reasonable tariff, which is regulated by the rules of the Garden. Should the owner desire to have his baggage retained for a short time, it is placed in the store-room, where it remains perfectly safe until called for, and no charge for storage is made. If the immigrant wishes to depart at once, and has an order for railroad tickets which he purchased on the other side of the ocean, he surrenders his order at the ticket office in the Garden and receives tickets in exchange, which carry him to his destination. If he has no order, and desires tickets, he purchases them at the office at rates which are established by the railroad companies and approved by the authorities at Castle Garden. When he is ready to go he is conveyed to the railroad or steam-boat depot on a barge—if there are a sufficient number of passengers to warrant the use of the steamer—or he climbs upon the express wagon with his baggage or goes up town on the cars—being cautioned meanwhile against the sharpers who congregate about the Garden on the watch for those whom they may take in and do for.

THE DESTITUTE IMMIGRANT.

If the stranger has no money with which to reach his destination, and has baggage, the latter is sometimes detained and the money advanced thereon; or the baggage is forwarded to the destination with the immigrant and the charges collected thereon when the stranger meets his friends. If he has neither money nor baggage, but has friends in the far West, or elsewhere, who might, if they knew of his circumstances, assist him, he is sent to Ward's Island for shelter, and letters are written for him, the response to which he awaits with anxiety. If he has neither money nor friends, and is entirely destitute, he is sent to the Labor Exchange, and what is done there will be shown presently. If he is sick, provision is made for him at the hospital on Ward's Island, and if he has met with an accident or is dangerously ill, he is immediately conveyed to the hospital at the Garden, where the best surgical attendance is given him. Though the masculine pronoun is used in this article, it applies equally to females, who are, if possible, treated with more consideration than males. Sometimes women are landed at the Garden who are about to become mothers, and in that case they are taken at once to the hospital in the building, and there cared for. One more noticeable feature of the Garden is one in which provision is made for the destitute, in the form of a lunch counter, where bread, fresh from the bakeries of the island, is set out, and those who are half famished can procure enough to sustain life until they can be conveyed to the island.

THE LABOR EXCHANGE.

This is an adjunct to the Garden, which has been in existence only a short time comparatively, but it operates beneficially both to the employer and the employed. Those immigrants who are destitute, and want to obtain work, are directed to this Exchange, where their names are recorded, together with the class of work they can perform, whether skilled or unskilled. The male department is under the control of a competent superintendent, and Mrs. MOODY presides over the female department. If there is a demand for labor—as there often is—the applicant has not long to wait, but is hurried off to his or her destination. Thus the Commissioners are relieved from the care of supporting the destitute immigrant in idleness, and the country is benefited by the labor of the workman just where it is most needed.

THE MACHINERY AND WHO OPERATE IT.

The vast machinery which moves so noiselessly yet so efficiently in the discharge of the multifarious duties above described, is under the direct control of Mr. BARNARD CASSERLY, who for years has been the general agent of the Commissioners of Emigration and charged with the management of every detail connected therewith. He has chosen able assistants who perform their duty well, and no complaint has ever been made that in the course of their administration they have been guilty of acts of injustice toward the immigrant. The constant aim is to make these strangers feel that they have protectors at a time when they most need them, and that none will be permitted to impose upon or defraud them. The cost of this machinery is, as was stated above, borne by the immigrant, who himself pays head money, or the consignees of the vessel pay commutation money, and thus a fund is created, and the supply is constant. The tens of thousands of immigrants who simply pass through the City en route for homes in the West or South, do not become chargeable to the Commissioners of Emigration, but they have to pay their head or commutation money as well as the hundreds of immigrants who arrive destitute, and are for a long time chargeable upon the authorities. Altogether, the institution at Castle Garden is one well worthy of the consideration of the public, and its benefits have only to be known to be thoroughly appreciated.

May 29, 1870

CASTLE GARDEN COMMISSION.

A Western View of Tammany Outrages on Immigrants and Extortions from Western Agents and Others.

From the Indianapolis (Ind.) Journal, July 1.

When Mr. WILLY WALLACH and his Castle Garden *confrères* came to Indianapolis last Fall and made the best showing they possibly could for the Immigration Board of New-York City, it was pretty generally conceded that that best was exceedingly poor. Before the caustic questions and under the tart replies of the Western delegates to that Convention, who had been there, and, to use a vulgarism, knew how it was themselves, the rhetoric and smooth sophisms of the New-Yorkers passed for naught. The Convention closed with a most decided expression of dissent from the conduct of the Castle Garden Commission, and an equally decided expression that the work of caring for the immigrant ought to be committed to the custody of the General Government, and taken from a single State or city, which might be, if it were not already, tempted to divert it to local and selfish ends. This had reference to the Commission as it had been constituted and operated—a Commission composed of men of both parties, some of them of high repute for honesty, interity, humanity and intelligence; a Commission which was confessed to be in many respects entitled to the confidence and gratitude of the people of this country and of the immigrants as well. The movement to make the management of immigration a national matter failed in the last Congress, as did many other measures of much greater importance than San Domingo and the Sumner decapitation, whereupon the Democratic chiefs of New-York took heart, and determined to so modify and make over the Immigration Commission that it would be entirely and servilely devoted to the interests of the Ring, which both rules and ruins that party-cursed City. The NEW-YORK TIMES of Thursday tells how it was done, and under what pretext. A little "riot" on Ward's Island, which in reality was nothing but a row between a few individuals, was made the pretense for the appointment of a Committee of Investigation from the State Legislature, who went down to the island, examined into the matter, and reported back to the Assembly that the food supplied to the emigrants by the existing Commission was not of that high and nutritive quality which should be spread out before the guests of a great nation. Not forgetting the prime object of their appointment, the committee coupled with their report a bill for the reorganization of the Commission. In this reorganization the Legislature did not trust Gov. HOFFMANN, Ex-Grand Sachem of Tammany, but reposed the appointment of all the Commissioners in the hands of "Boss" TWEED, who named his men, and with them inserted the measure passed, of course. The new Commission is to hold office for five years, fill all vacancies which may be created, and after the expiration of five years, the power of appointing the Board is to revert to the Governor, as under the old law. The five-year limitation is significant. In casting the probabilities of their political future, the managers of the Ring seem to have judged that within this time, at least, the conscience of the nation will have been so pricked that Congress will have relieved them of any further duty in this behalf; and also, that during the ensuing five years there will accrue the "richest pickings" off the immigrants from other countries. The five next immediate years are to be the golden period for Tammany, and Tammany was too smart and selfish not to avail itself of the opportunity. The TIMES gives the names of the Commission as follows: RICHARD O'GORMAN, Tammany Sachem; ISAAC BELL, Tammany Sachem; EMANUEL B. HART, Tammany Sachem; JAMES B. NICHOLSON, Tammany Sachem; ALEXANDER FREAR, Tammany legislative henchman; WM. B. BARR, General Ticket Agent of the Erie Railroad; JAMES W. HUSTED, ANDREAS WILLMAN and WILLY WALLACH.

We feel warranted in assuring the TIMES that its scathing exposures of the rascality of the Tammany Ring are warmly appreciated by the country, and its revelations of the abuses and excesses practised in and about Castle Garden will find a thousand tongues for repetition, until our Congressmen shall be forced into a remedy of the national scandal. The movement inaugurated at Indianapolis will not fail. It has obtained sufficient momentum to carry it to a successful consummation, over the combined self-interest of a few of the Eastern States, and the culpable negligence and apathy of others, who unwittingly contribute thereby to the continuance of this New-York corruption.

Pending the present Summer let the country be enlightened on this matter, for it is one of more interest—to the West and South particularly—than may at first be imagined; and then if President GRANT will call the attention of Congress to it at its next meeting, Tammany may not live to see the years the "Boss" has decreed for the plunder of the defenseless foreigners.

July 4, 1971

Statistics of Immigration at Boston.

From the Boston Traveller, July 24.

The number of immigrants that arrived in Boston by water during the quarter that ended June 30 was 11,442, 6,777 of whom were males and 4,665 females. Of this number 1,878 were males under twenty years of age; 1,899 females under twenty; between twenty and forty there were ,128 males, 2,284 females; over forty years, 710 males, 461 females; and 61 males and 21 females age not stated. Of the occupations of the males 3,559 were classed as laborers, 1,016 as mechanics, 347 mariners, 309 farmers, 230 fishermen, 100 merchants, 25 clerks, 15 miners, 15 shoemakers, 14 students, 11 engineers, 10 traders, and the balance divided among 14 occupations, with 1,074 not stated, mostly children. Of the females 76 were seamstresses, 7 dressmakers, 3 tailoresses, 1,343 spinsters, and 3,236 not stated. Of their nationalities Ireland takes the lead with 2,794 males, 2,343 females. England came next with 1,316 males, 767 females. Nova Scotia furnished 1,056 males, 531 females; Germany 433 males, 200 females; Scotland sent 150 males, 119 females; the Azores sent 155 males, 102 females; Italy 123 males, 44 females; Cape Breton 106 males, 38 females; Newfoundland 53 males, 81 females; Prince Edward Island 98 males, 100 females; Sweden 107 males, 87 females, and the United States 236 males, 133 females, while the balance was made up from twenty different places. Of the whole number all but 619 intend to reside under the broad flag of the United States, which alone should be allowed to wave over them. Only five of this large number died on the passage; 2,444 were cabin and 8,993 came in the steerage. The corresponding quarter last year gave 14,020, showing a decrease of 2,578. They came in 275 different vessels, that brought from one passenger to 891 each. In the corresponding quarter of 1869 12,747 arrived.

July 27, 1871

THE STEERAGE PASSAGE.

OFFICIAL REPORT ON THE TREATMENT OF STEERAGE PASSENGERS COMING TO THIS COUNTRY—RATES OF MORTALITY, &C.

From Our Own Correspondent.

WASHINGTON, Monday, Nov. 24, 1873.

Dr. John W. Woodworth, Supervising Surgeon of the Marine Hospital Service, has submitted to the Secretary of the Treasury his report of an investigation into the treatment and condition of steerage passengers, made during the months of July, August, and September, 1873, at the ports of New-York, Boston, Philadelphia, and Baltimore, with suggestions of needed legislation concerning the emigration service. The following is a synopsis of the more important portions of the report: Dr. Woodworth personally examined thirty passenger vessels, comprising twenty-one steam-ships and nine sailing vessels, and made an inspection of 8,488 steerage passengers carried on them. On account of the season of the year few of the compartments contained the maximum number of passengers allowed by law. The average space legally allowed on the vessels examined being 15 22-100 superficial feet, while the space actually occupied was 17 75-100 to each passenger. The official reports of the Collectors of Customs were found to be inaccurate. This inaccuracy consisted in incorrect reports as to the passenger capacity, and as to the number of decks, ventilation, and a variety of minor details. One result of these inaccuracies is that vessels frequently carry an excess of passengers. These errors, it is believed, originate in the loose and inoperative law under which examinations are made, and under which the duties of the examiners are merely nominal. The chief evil caused by these inaccuracies is in impairing the value of the statistics obtained from such sources, since no action has ever been maintained for a reported violation of any of the provisions of the act of March 3, 1855. The condition of the law in this respect is stated by Ex-Solicitor of the Treasury Jordan, who says that he is "of the opinion that there is too much doubt as to the application of the penalties to the cases of a vessel arriving from abroad to warrant any attempt to enforce them." It was found that the vessels took especial pains to have an appearance of cleanliness upon arriving in port, when they are to be examined by health officers and other officials of the public service. Consequently a chemical analysis of the atmosphere of the steerage compartments would have been entirely misleading for any purposes of sanitary knowledge or legislation. Instead of this, a careful study was made of the appliances for ventilation. There is no uniform system of ventilation, but the newer steam vessels are believed to be quite as well, if not better, ventilated than the average hospital, hotel, or other public buildings. It is believed that the combination of the present system of ventilation, "propulsion," with the "suction" system, would possess decided advantages over the use of either singly, and it is suggested that it is entirely feasible to utilize the escaping heat of the furnace-rooms and of the funnel, for creating the exhaust current from the steerage compartments through properly-located foul-air shafts, by which mode the "suction" of foul air from the steerage could be efficiently and economically combined with the "propulsion" of fresh air into the compartments by the usual ventilating-shafts and wind-sails.

The general treatment of emigrants on board ships was, from the manner in which the examinations were made, largely a matter of inference, rather than a matter of observation. The personal examination of a large number of steerage passengers, however, and a critical inspection of their quarters, together with a collection of the various reports and statistics for the last ten years, show that the ill-treatment of steerage passengers is, to a great extent, a matter of history. The substitution of steam vessels has greatly improved the condition of emigration. It has resulted in shorter voyages, better accommodations, and better food. The respective per millages of mortality being—

On sailing vessels for 1867, 11.67; 1872, 5.42. On steam-ships for 1867, 1.03; 1872, .45.

For these improvements the United States can take little credit, owing to the almost entire destruction of the American commercial marine. For, with the exception of collecting a penalty of $10 for every steerage passenger over eight years of age who dies on the voyage, we do not hold foreign vessels to any regulations whatever, and the examination of such vessels by our inspectors "to ascertain whether the requirements of the law have been complied with," (section 9, act of March 3, 1855,) are made as a matter of form only, so far as the exaction of any penalties for non-compliance with the requirements of the law is concerned. That the use of steam vessels instead of sailing vessels has much to do with the lessening of the death rate is shown by the fact that where the service is to any great extent performed by sailing vessels, the mortality rises in direct ratio to the proportion of passengers carried by them. Thus, during the five years above quoted, while the mortality on sailing vessels was reduced over fifty per cent., only about eight per cent. of the total number of emigrants was carried on them in 1872, while nearly twenty-five per cent. was so carried in 1867. A striking proof of the connection between these two facts, as cause and effect, was found in the statistical report of Baltimore during the Spring of 1873. At this port for the quarter ending March 31, 1871, there arrived 1,602 steerage emigrants, of whom 711 were brought by sail and 954 by steam. Of the former thirty-two died on the voyage, and only two of the latter, being in the ratio of one death to every forty-seven and seven-tenths steam passengers, and of one to every twenty-two and two-tenths sail passengers. The usual mode of stating the mortality on shipboard is misleading to those not accustomed to distinguish between per annum mortality and mortality per voyage. During the last six months of 1872 the per annum mortality of steam passengers was a little less than the per annum mortality of the United States, which is a most favorable showing for steam-ships compared with the mortality of sailing vessels for the same period—a mortality which equaled a per millage of forty-four and eighty-two hundredths. In connection with all ship mortality statistics, it should be remembered that emigrants are, to a certain extent, "selected lives," for they are all examined before leaving the port of embarkation, and are generally persons in the prime of life, and healthy.

With the exception of certain measures, looking to protecting passengers from contact with the crew, there would seem, from the general testimony, little to be desired in the treatment of emigrants on steam-ship lines. As to the sailing vessels, it is believed that competition is driving them out of the passenger-carrying trade so rapidly as to render it unnecessary to recommend any specific action with regard to them.

The report adds that as regards the subject of re-establishing the National Bureau of Emigration, concerning which inquiry was directed to be made by the Senate resolution, the conclusion is reached that such an agency is inexpedient and unnecessary, since it would be the occasion for renewed protests and hostilities on the part of State organizations, and would be contrary to the repeated decisions of the Supreme Court of the United States, to the effect that States have paramount authority in all matters for which such a bureau could be created. It is further urged that the bureau was formerly abolished on account of its actual failure.

November 27, 1873

LOCAL MISCELLANY.

ANXIETY AT CASTLE GARDEN.

EFFECT OF THE DECISION REGARDING EMIGRANT HEAD MONEY BY THE SUPREME COURT OF THE UNITED STATES.

The decision of the Supreme Court of the United States declaring the emigrant head money laws of this State unconstitutional has created consternation among the officials at Castle Garden. Commissioner Lynch, in conversation with a TIMES reporter yesterday, said that the decision would have the result of destroying the usefulness of Castle Garden, and that much suffering would accrue to the immigrants, who would in future, unless the State afforded some protection, be dumped on the wharves just like merchandise and left to manage for themselves. The protection afforded by Castle Garden was, he said, of great benefit to the immigrants, and it would also be advantageous to the State and City, as the department was frequently the means of preventing the landing of criminals from other countries, and had them sent back to their native land. The commission would still, he said, have the power to compel the steam-ship companies to give bonds for the support of such immigrants as may within a certain time become a burden on the State. The board, he said, did not know precisely what action they would take, as they had not yet seen the full text of Justice Miller's opinion. The steam-ship companies no doubt would decline to pay head money from this date, and under the circumstances he did not blame them. He said that the Supreme Court, on a former occasion, had given a decision adverse to that given on Monday, and that the effect of the recent decision would be to centralize power in the Federal Government and destroy State rights.

Commissioner Maujer said that the steam-ship companies paid the head money last year under protest, and it was questionable whether they would not now sue to recover the money so paid under protest. As soon as the subject shall have been thoroughly canvassed by the board they will bring the matter under the notice of the Governor and ask for instructions. Before the Emigration Commission was established in 1847 the owners of vessels gave bonds that the immigrants should not become a charge on the State. Under this system a good deal of trouble was created; hospitals and alms-houses had to be established on Long Island, and it was difficult to enforce the obligations of the bond. He thought that the steam-ship companies would regret the steps which they had taken to multiply the existing arrangements, as no persons were more concerned in the proper treatment of the immigrants than they were. If the Legislature declines to make a small appropriation for the maintenance of Castle Garden, the immigrants will undoubtedly be swindled and robbed of everything which they may possess. The Insane Asylum on Ward's Island, in which there are about two hundred inmates, should be transferred to the care of the Commissioners of Charities and Correction.

At the offices of the steam-ship companies the decision of the Supreme Court was received with satisfaction, and it was intimated that they would not in future pay the head money.

March 22, 1876

A PARTY OF COLONISTS FROM ICELAND.

Among the steerage passengers of the steam-ship Anchoria who landed at Castle Garden yesterday, were 76 natives of Iceland, who are on their way to form a colony in Minnesota. There are 14 families represented in the group, with a plentiful sprinkling of children and babies. The grown persons are mostly middle-aged, there being only three really old women in the party. The men are all vigorous and healthy-looking, and they appeared perfectly able to take care of themselves in any country. The most noticeable thing about the women was the national head-dress, which nearly every one wore. It is made of black cloth, and resembles nothing so much as the old-fashioned long net purses with a sliding ring in the centre. One end of the head-dress is pinned close to the head like a skull-cap. The rest hangs gracefully on one side, reaching the shoulder, and is ornamented with a shining metal ring from one to two inches wide. None of the party spoke English, and the Castle Garden officers talked a little with some of them in the language of Norway. This is the first large party of Icelanders that has come to America. It is said that this company of 76 has been preceded by leaders, who have selected a place for them in Minnesota. During the voyage one of the Icelandic women, named Kiersteum Ryansen, died. Her body was brought here to be buried on Ward's Island.

July 29, 1879

DESTITUTE IMMIGRANTS.

ARRIVAL OF NINETY HUNGARIANS WITHOUT A CENT OF MONEY.

The arrivals of immigrants at Castle Garden since Saturday, are as follows: Per City of Berlin, Liverpool, 354; Elysia, London, 53; Spain, Liverpool, 255; Maas, Rotterdam, 135; Anchoria, Glasgow, 177; total, 974. Among this number of immigrants were a large number of Hungarians, who arrived in a destitute condition. Thirty of these people came last week, and had immediatly to be taken charge of, by the Castle Garden officers. Superintendent Jackson finally found employment for them at Lenhardtsville, Bucks County, in the lumber region of Pennsylvania. They were all wood-choppers and hardy forest laborers, hence the reason for sending them there. Yesterday, the Spain discharged 90 more of these people, equally destitute, on the Emigration Department. The astonished officers became alarmed, and instituted inquiries. It was not easy to communicate with the men, as they could not speak English, French, or German; but an interpreter was found with some trouble. It was learned that they had come from the flooded districts of Hungary, where the crops have failed and long continued rains have caused the inundation of the country, sweeping away the subsistence of the people. The immigrants sold everything they had to get money enough to pay their passage to this country, which was necessarily expensive as they had to come from the interior of Hungary, through Germany, to Liverpool. The section they come from rejoices in the unpronounceable name of Sarosmegye. They are all lumbermen and farmers. It is exceedingly likely that the steerage fare on the Spain was the best feeding they ever had in their lives. They were landed looking strong and hearty, but without so much as a cent among the whole lot, and they had to get their breakfast from the department people in whose charge they for the present will remain. Superintendent Jackson says they are honest, hard-working fellows, and he thinks there will not be much difficulty to find work for them; but he says what he rather dreads is, that when they have got $15 or $20 together, each man will be sending for his family, and a long string of equally destitute Hungarian wives and children will trail through the department for a year or more.

December 16, 1879

EMIGRATION FROM EUROPE

ITS CHARACTER AND EXTENT.

A SETTLED PURPOSE ON THE PART OF THE GERMANS TO ESCAPE THE BURDEN OF IMPERIALISM—THE CURRENT FROM SWEDEN—A WIDE-SPREAD MOVEMENT IN ENGLAND.

LIVERPOOL, May 15.—The published statistics present the more obvious aspect of the emigration question. Everybody can see that the flow of emigration from the Old World is unusually great, and, in some instances, great beyond all precedent. But there are aspects of the question which mere figures cannot adequately exhibit. It is important to understand, for example, whether the outflow comes from sources hitherto closed, or whether it is merely an enlargement of streams which for an indefinite period have contributed to the industrial wealth of the United States. How far the acquisition of so many thousand souls per week is a proper ground of satisfaction, obviously depends on the general character of the emigrants, and the qualifications, habits, and tendencies which the classes they are divisible into respectively represent.

Of the German emigration a comparatively small proportion comes this way. The main body proceeds direct from German ports. It is not difficult, however, at this point to measure the forces that are at work and their probable duration. In some respects, German emigration is conducted under difficulties. Authority, local and imperial, is against it. You cannot actively promote it without subjecting yourself to an unpleasant surveillance and to contingencies still more distasteful. And yet nowhere, perhaps, is emigration pursued so methodically, or with so fixed a determination to seek new homes far away. The authorities would not tolerate the direct advocacy of emigration, but they cannot prevent the keen search for information which, outside of the chief cities, any one experiences who is in a position to talk practically of the different fields of emigration in the United States, and of their varying opportunities and attractions. Winter is the season during which this pursuit of knowledge under difficulties goes on most energetically. Clubs are formed. Reports received from friends on the other side of the Atlantic are handed round The cost of removal is carefully calculated and provision for meeting it is made gradually. So far as Germany is concerned, emigration signifies little that is spasmodic or evanescent. It represents a settled and a growing purpose to escape from conditions which the glories of German unification cannot conceal. The country people especially are wearied of burdens which the Bismarckian policy constantly increases, and of risks which that policy naturally entails. More than anywhere else in Europe, therefore, the volume of emigration indicates a steady yearning for deliverance from the yoke of imperialism, and though good times or bad times necessarily affect some of the points to be decided, they really do but affect temporarily the essential features of the emigration question in Germany. The drain which is gradually undermining the military strength of the Empire, may at periods be checked by the receipt of less favorable news from America, and, in not improbable contingencies, by sterner measures on the part of the Government; but flow it will, and, as heretofore, the United States will monoplize its benefits. We may yet discover that the disposition to extinguish the peculiar privileges of the Free Cities has a remote bearing upon the repression of German emigration, which this year proceeds mainly through these outlets.

Scandinavian emigration, again, is wholly free from political complications. The Swedish Government cannot look favorably upon the diminution of a population already scanty, but it resorts to none of the harsh measures for averting it which are among the features of Bismarck's rule. As regards the Swedes, poverty is the impelling cause of change. In Sweden, as in Norway and Denmark, labor has no hope. It sees no chance of improvement. Engaged in a hand-to-mouth struggle for tolerable existence, their eyes turn toward America with a somewhat extravagant conception of its benefits. To them, the North-west is the promised land. Letters and newspapers received thence do service throughout a neighborhood. One of themselves, returning on a visit from Wisconsin, Minnesota, or Dakota, becomes the centre of a local movement. People cluster around him, and he becomes their leader. This tendency to depend upon persons who profess to speak of their own knowledge is often turned to base account. Two or three months' observation at this point supplies illustrations of the manner in which popular credulity is sometimes abused. That there are truthful, upright men whom circumstances have converted into small leaders of emigrating groups, no one can doubt. They are made leaders in spite of themselves. On the other hand, it is equally certain that unworthy men are enabled to abuse an opportunity, and do abuse it most shamefully. Large clusters of emigrants follow individuals who, professedly disinterested, are intent upon making money at every step. They obtain free passage; they receive a commission on the passage-money, and if they deliver those who trust them into certain bonds at the end of the journey they receive another bonus. The ultimate fortunes of the emigrants may not be impaired by the sinister proceedings of their leaders, but the fact has something to do with the complaints and reproaches which come to the surface at the stopping place on the journey, which Liverpool really is. For Swedish emigrants are not more amiable than the emigrants sent forth by other nationalities. Their grumbling does not get into print, which, to certain parties, is a decided gain, but there is plenty of grumbling among them, nevertheless. Here and there, the trickery of self-constituted leaders is detected, and then the grumbling is legitimate. In other cases it is capricious and unreasonable. The steam-ship companies have more than their share of it. Their contract with the emigrant obliges them to provide for him during the period which elapses between arrival here and the sailing of the steam-ship; and, speaking generally, the obligation is discharged honorably. Each company has its own lodging-houses, where wholesome food is supplied in abundance. Even under these circumstances, an extra day's detention gives rise to vexatious complaints, and now and then to threats of litigation.

These are not peculiarities, however, and they do not detract from the remarkably high average of the Swedish emigration of the present year. All the information from the Swedish ports, and from Copenhagen, through whose agency so many of the Swedes come, is to the effect that the emigrants are nearly altogether of the peasant class. Of farmers, there are this season few. The great majority go out with the view of working for others. Germans have gone in large numbers to Michigan and Illinois. Swedes bend their footsteps in the direction of North-western Wisconsin, Minnesota, and Dakota. There are "returned emigrants" among them. One man I saw, who had already crossed the Atlantic four times, each time under the influence of discontent, and yielding at last to the feeling which so often proves irresistible in the minds of men who have grown familiar with American life. Nearly all tell of the hardships they have endured in Sweden, of ill-requited toil, and a lot that knows little sunshine. Keeping this fact in mind, the appearance of comfort possessed by the emigrants is remarkable. Men and women are well clad, clean, and bright. In physique they are superior to all other emigrants.

The contrast is most striking when the steerage passengers from the manufacturing districts of England are brought upon the scene. The English farm laborer is remarkable for stolidity, but at least he has bodily strength We see him rarely. His poverty makes emigration to the United States all but impossible When he does put in an appearance here it is in the capacity of an assisted emigrant, en route to Canada. Even in this character he is a rarity, as the Canadian authorities confess. The most familiar of the poorer classes of the English people are, then, from the towns, especially from manufacturing towns, and with regard to them the impressions formed are for the most part unfavorable. They have much of the ignorance and stupidity, with none of the physique, of the agricultural laborer. Many of them cannot read, and still more know nothing of their destination or of what they are going to do. They are pitiful specimens of manhood, and England suffers as little by losing them as America gains by the acquisition. Indeed, few things have struck me more painfully than the deterioration which has taken place, within my knowledge of England, in the character, stamina, and habits of the English working man. This is not the occasion for discussing the relative strength of the influences which have contributed to this result; but the result is undeniable, and its more palpable causes are working with unabated force. Hence it comes to pass that the great proportion of the English steerage passengers—the limp, listless, pale-faced young fellows, with pipe in mouth and hands in pockets, and the slatternly young women who accompany them—must be regarded as a curse rather than a blessing to the country which acquires them. Fortunately for the West, it will receive very few of them. And in judging of the emigration movement in this country, as it now appears, they must be left altogether out of the reckoning.

Perhaps there has been a certain degree of exaggeration in the statements put forward with reference to the movement among the tenant farmers, so far as immediate results are to be considered. There is not, and there is not likely to be, a heavy rush of farmers, or of other classes with corresponding means, to America or elsewhere. Various reasons account for this fact. The principal are (1) the habitual unfamiliarity with the associated effort which alone would challenge special attention, and (2) the difference in circumstances which compels farmers to proceed, each on his own account, both as to time and destination. Apart from these considerations, it may be safely said, I think, that the disposition to migrate has acquired unexampled strength among English and Scottish people to whom, until now, the idea of emigration has been repugnant. Instead of a sudden and violent movement, which in the nature of things would soon exhaust itself, there is a wide and settled purpose, not confined to any class, to transfer to other lands the knowledge, energy, and means which no longer command a fair field here. Among farmers particularly, the conviction prevails that the wiser course is to transfer effort to the country which will hereafter compete with nearly all the conditions of assured success on its side. A promise of a good crop does not materially affect the feeling. Intelligent farmers are persuaded that in this country they are overweighted, and though for the time unable to remove, they do not conceal their desire to go as soon as the opportunity presents itself. There may not be an enormous exodus of English or Scottish farmers this year, but enough will go to indicate the depth and extent to which the agriculturists are stirred. The movement may be slower than some have expected it to be, but it will be wider and more constant than others have anticipated. Moreover, it will embrace classes who have not the farmer's reasons for anxiety. From widely separated quarters—from Manchester and Glasgow, from Sheffield and Dundee, from Essex and Aberdeen, and from nearly all parts of Wales—evidences come of change on the part of men having more or less realized property, who recognize the coming supremacy of the United States and want to participate in their prosperity. Two instances are in my eye at this moment, one at Manchester, the other at Glasgow, in each of which the owner of £10,000 is arranging for a trip with the view of settling in the Western States. Other instances there are, where men high in official position—officers in the British Army, relatives of Cabinet Ministers, and so on—are determined to send out their sons to carve their fortunes in the trans-Mississippi country. All the efforts of the Canadian Government to turn the stream in the direction of its unpeopled territory do not perceptibly affect the current of thought in the circles contemplating emigration. Australia, too, is losing its power to charm. And New-Zealand suffers from the apprehension of financial collapse, occasioned by wild efforts of the kind to which Canada, with various modifications, is now committed.

G. S.

May 28, 1880

THE HEAD-MONEY TAX.

AGREEMENT BETWEEN THE EMIGRATION BOARD AND SECRETARY FOLGER.

The Commissioners of Emigration met at Castle Garden yesterday afternoon. The following agreement, signed by Secretary Folger, was submitted to the board, which gave its President authority to sign it:

"This agreement, made this 2d day of September, 1882, pursuant to an act of Congress entitled 'An act to regulate immigration,' approved Aug. 3, 1882, between the Secretary of the Treasury of the United States, of the first part, and the Commissioners of Emigration of the State of New-York, the party of the second part, witnesseth: That the party of the second part undertakes to examine into the condition of alien passengers arriving by vessel from a foreign port at the port of New-York; to ascertain who among them are convicts, lunatics, or unable to take care of himself or herself without becoming a public charge, and report the same in writing to the Collector of the Port of New-York. The party of the second part will also receive all alien immigrant passengers at Castle Garden or such other suitable place as may from time to time be secured and under their control for landing of immigrants, and there provide such means for their accommodation as are now provided, including necessary interpreters, and shall provide at the hospitals and other public buildings under the control of the party of the second part suitable accomodations for such alien immigrants as shall become sick, or in distress, or idiot, or lunatic, or a public charge, for a period not exceeding five years from the time such immigrant shall have arrived at the port of New-York. The party of the second part shall so far as possible keep a record of all alien immigrants arriving at the port of New-York by vessel from a foreign port and the place from whence they came. The party of the second part shall also carry out such regulations as the party of the first part shall from time to time prescribe, pursuant to law, so far as the same are applicable to the port of New-York. The party of the second part agrees to employ the necessary persons for carrying into effect this contract, and to render to the party of the first part on the 1st of each month a sworn statement, with vouchers for all items, of the necessary expenses of the preceding month incurred by the parties of the second part in executing this contract, which account, when audited, shall be paid on or before the 15th of the month. It is the intent and meaning of this contract that neither party shall be bound to execute its provisions or incur any liability beyond the amount of money properly applicable thereto under the act recited."

This will secure to the Commissioners the funds which they require to carry on the institutions under their charge. A resolution was offered to appoint a watchman to look after the baggage which is landed at the Castle Garden depot. Commissioner Ulrich proposed that each railway company doing business in the Garden should appoint a watchman for the baggage, who should be under the control of the Commissioners. The matter was referred to the Castle Garden Committee. Mayor Grace again brought up the subject of disposing of the Castle Garden franchises. Mr. Taintor, of the Castle Garden Committee, explained that the consideration of this subject had been postponed for the present. James W. Best was appointed apothecary of Ward's Island at a salary of $800.

September 6, 1882

FIRST STEPS IN AMERICA

CASTLE GARDEN A BUSY PLACE YESTERDAY.

THE ARRIVAL OF 1,637 IMMIGRANTS OF MANY DIFFERENT NATIONALITIES FILLING THE GREAT ROTUNDA.

Castle Garden on a Sunday is always an interesting spot, and yesterday was no exception to the rule. A busier place could not be discovered when the steerage passengers from three big ocean steamers pushed and crowded themselves out of the floating barges by which they are transferred from the steamers to the Garden. There were 1,637, nearly all of whom placed their feet for the first time on American soil. La Bretagne landed 965 men, women, and children of the Saxon and French type, and some from Southern Europe and the lower German provinces. The Etruria brought 639. They were mostly Scandinavians, with some stout boys and girls from old Ireland. The Alexandria furnished only a small quota of 33 from Leghorn and Mediterranean ports.

As they landed, chattering in their various tongues, they were huddled together like a great drove of sheep in a grove during a thunder shower. Then a great strong voice from within shouted, "Let 'em in." The big doors were thrown open and the procession moved into the amphitheatre. The procession divided on entering the Castle proper, one line swinging out to the right and one to the left. After a few feet the great river of humanity was again divided on each side, and four smaller streams coursed on into narrow passageways. The centre of the Garden is fenced in, and the labyrinth of passageways and wicket gates, clerks' desks, and railings is bewildering to the stranger, and the Irish lad, in trying to reach the quarters allotted to his countrymen, is as likely as not, by following the lanes and turnings, to bring up in a bevy of Germans squatted in a heap on the other side of the great room.

The system, however, is very simple. The procession is divided so that the immigrants may be registered quickly by the clerks. Each line passed before a desk where a few questions are put to each person. First the name is given, then the place from which he comes, his destination, and whether or not he has money. These clerks were facetious yesterday, and made parenthetical and supplementary remarks in connection with their questions.

"Your name?" was asked of a stout, wiry, red-bearded Irishman, whose lips made strenuous and ineffectual efforts to close over bulldog teeth.

"Michael Malone," responded the aspirant for American citizenship, with a grin.

"French or Italian?" nodded the clerk.

"Arrah, it's from the ould sod I hail," cried the Irishman, as the sides of his mouth reached away in an almost endless smile.

Some laughed and some did not know what to make of the sallies. A light-haired and florid-faced Swedish maiden, when asked whether she was Japanese or Chinese, looked up in silent wonder at the clerk and told her nationality without a smile or relaxation of feature. They all said they had money when the question was put to them.

"I have $15 in—" began a Donnybrook man.

"No, no, that won't do," cried the clerk; "you will have to go back."

But the poor man did not know he was doing any harm, and told a reporter that he was only going to say that he had $15 wrapped up in a stocking. Finally, he was allowed, with a warning never to associate $15 with any inside coat pockets, to pass happily on.

Right behind him happened to come a Finn, who was somewhat astonished to be asked "What was the matter with swimming over?" Of course, he did not understand, and said he didn't in his own language. These little jibes and remarks, however, were very pleasing to the clerks, and it didn't make much difference whether the poor immigrants understood them or not.

One of the questions asked was whether the steerage passenger had ever been in this country before. One Irishman said he had lived in Pittsburg.

"Alderman or policeman?" was the prompt question of the clerk.

"Neither," responded the Irishman seriously, "but oi 'ave a brother who is a night watchman."

So it went on until all the four streams had been filtered through the narrow passages and were turned loose in the rotunda. They were all treated fairly, and patiently were all questions answered. A big dark-haired interpreter, who would make even Dr. Dallon, of the Court of General Sessions, frown with envy, because of the dialects and languages he knows, was on hand ready to help out the poor men or women who had difficulty in making known their wants. After the registering the immigrants could do as they pleased. The most of them sat down on their baggage and camped out on the floor. The different nationalities, of course, formed little clans and groups by themselves, when they discussed who knows what.

The scene was a curious one. The Scandinavian men in dark colored caps, higher in front than in the back, sat on the queerest looking old boxes and trunks of every shape conceivable. Women in dark colored, substantial woolen dresses which were gathered in a bit at the waist and then allowed to take what course they pleased, heads tied up with scarfs or hats that had lost shape and color, sat like inhabitants of a silent city. Children looked from deep, big, solemn eyes at the lunch counter, but never a word did they peep or request did they hazard for some of the sandwiches and bolognas that were there set out. People of few words they were, but when voice did issue forth it was a big one and evidently to the point.

In the next corner a group of Italians nervously fidgeted about, keeping up an endless chatter. The women wore bright colored dresses of cheap material and with ugly rings dangling from their ears. Soon they were rushed out of the Garden and, piled up on rickety express wagons, were rattled off to the savory precincts of Crosby and Mulberry streets.

Interesting groups were those of the Irish people. Those who arrived yesterday were mostly stout, manly boys who were ready for hard work, and tall pleasant-eyed girls, whose soft rich brogue was charming as they greeted some friends. Sunday is a great day for meeting friends from over the water. A large crowd of waiting friends on the outside of the building was kept back by policemen. The Irish immigrants have the most girls and boys to meet them, who in turn carry them off, baggage and all, from the Garden.

There was a large number of Finns who came into port. Superintendent Jackson, as he saw them pass along, remarked that if many more came none would be left in their native land. "There are only 1,500,000 Finns," he observed, "and recently they have been coming over in large numbers. The reason is because they are wanted for the Russian Navy, as the Finns make the best sailors in that region."

"How do they get away?" was asked.

"Russia does not allow any agents to go among the people to make arrangements for coming over to this country, but the Finns get a leave of absence for six days. Then they hasten to a seaport town and embark for America."

The Finns who arrived yesterday were all good-looking fellows. They generally go West, and many drift into the lake region.

While the immigrants were encamped on the floor of the Garden the railroad agents were going among them and sorting them out for their respective roads, as most all of them had through tickets, especially the Scandinavians, who were bound for the Northwest. Others patronized the lunch counter, where substantial food, such as the foreigners liked, could be obtained. Toward night boats came up from the railroad companies and carried off loads to the stations, for the emigrant trains run Sunday nights as well as others, and leave New-York soon after sundown. Children, with placards attached and addresses pasted on their caps, were given rolls of bread and sausages and turned over to the care of railroad men. There were no old men without money to be sent back yesterday.

Some of the immigrants who were not ready to go away last night remained in the Garden. They had blankets with them and fixed up their beds in recesses and corners of the place. It was so bright a day that many of the new comers went out on the Battery Park and passed the afternoon in looking at the green and the elevated trains while they waited for the trains to take them West. Others left the Garden to wander through the city streets and hunt up boarding places in the city. Inside the Garden the usual other features of the week day were in full force. The rival telegraph companies shouted in their bidding for customers. The money exchangers plied their vocations, and the young man from a central desk shouted out the names of those immigrants for whom friends were waiting and those for whom telegrams had come. Twice during the afternoon the immigrants were changed to other sections in the building while the floor was carefully swept and made clean. The immigration has been very large this Spring and in the first half of the month more arrived than had been expected for all April.

April 25, 1887

WHY ITALIANS COME OVER

DEEP-SEATED CAUSES AND INEVITABLE EFFECTS.

OVERTAXATION AND THE CONSEQUENCES TO THE PEASANT — DISARMAMENT THE REMEDY AND WAR NEEDED.

NAPLES, Oct. 5.—Echoes of the great question of Italian emigration in North America, just now before the American public, have long since fallen upon the ears of Italians in their mother country in unmistakable accents of warning. During the discussion before the Congressional committee followed here with deepest interest, none save the most immediate causes and effects of Italian emigration have been considered. No speaker or writer among those self-exiled peoples, which, nevertheless, count among their number men of the analytical perspicacity of Vincenzo Botta, seems to have arisen to point out hidden first causes to the American legislator, unrolling before him the vivid panorama of Italian struggles, Italian heroism, and Italian misery. Yet herein lies the kernel of the whole question, the explanation, if not the solution, of the problem. To explain the reasons which induce the Italian laborer, and especially the Italian peasant, to abandon the soft air of his native land for the rigors of the North American climate and the fibrous tenacity of home and village ties for the charitable cold shoulder of the stranger one must remount the course of history to find in the tremendous drain of an emigration of over 150,000 souls yearly the logical sequence and completion of Italy's sacrifice to the great Italian idea. The humble exiles pushed away from the mother land by changed and inexorable conditions illustrate how the blood is still flowing from unhealed wounds, and are the extreme offering of the nation to its one persistent aim of liberty and unity.

It must be borne in mind that Italy, whose resources are mainly agricultural, with over 29,000,000 of hectares (nearly 72,000,000 English acres) of land, one-third of which is uncultivated, and the remainder lacking development, can only look upon emigration as an unmitigated evil. But, with the worm of Papal temporal power stirring uneasily at her heart, the Austrian double-headed eagle perching in the Trentino with one of his rapacious beaks constantly turned, despite of alliances, toward lost domains, and, above all, with the petulant French cock pecking at her fair borders, Italy must continue to spend annually 317,000,000f. on her army and 118,000,000f. on her navy. She must continue to purchase alliances with her standing army of 881,000 soldiers and her 150 superb ships of war. She must continue to oppress the groaning earth with taxes and to draft the reluctant peasant. She must continue to monopolize the sale of salt and the manufacture of tobacco. She must continue to run her demoralizing lotto, and put the screws on internal revenue. Nothing but disarmament can save her from the "body of this death." But, given the seething condition and the manifest hostility of France, war is considered inevitable. Therefore, the public yearnings of Kings and statesmen for the preservation of peace are to be read, as in a mirror, backward. Both Italy and Germany want war, the final settlement of vexed European questions, and subsequent disarmament to save them from financial ruin.

Let us now see how these questions, vital to the nation, become personal to the peasant. Contrary to the general belief, Italy, issuing from feudalism, was one of the European nations which entered most boldly on the way of progress, following the rules of fractional agriculture laid down by English economists. In fact, in few countries of Europe is landed property so largely subdivided. Even where large estates exist, save in some few known cases, they are mainly in the hands of beneficent institutions, and therefore, to a certain extent, held in trust for the poor. The overwhelming but unavoidable taxes of the State, the increasing importation at cheap rates from Australia, America, and the Indies, and the consequent depreciation in the value of home products involuntarily form the triple alliance that crushes the small holders. In numberless instances, these, unable to pay a tax of 24 per cent. on the gross income of their lands, abandon their property to the Government and themselves to the westward wave of emigration. Their small farms becoming ipse facto Crown lands, fall into immediate non-cultivation. I might describe in pages of eloquent narration the struggles of the Italian peasant under this subtraction of area and the grinding advance of living rates; the gradual and patient restriction of his domestic life to merest necessities; the elimination of meat from his daily fare, followed by the sterner suppression of condiments and of bread; the substitution therefor of *polenta* or cornmeal porridge, even, as in many parts of Calabria, of acorn and barley cakes, hardened by time, which constitute the sole food of the poorer classes. Add to this the interminable *giornata* or day's work of the Italian laborer, beginning at early daybreak and ending with the "twenty-fourth hour" night-fall, the poetic Angelus of mediæval times. Mark the miserable hovel where his nights are passed without ventilation and almost without repose; the dreadful *pellagra*, vindictive and incurable disease, whose first cause is insufficient and unvaried food and which reduces whom it attacks to madness or idiocy! Bound this leaden horizon with the impossibility of betterment, invert the order of things, and over the gates where his children enter life write the fateful sentence of Dante:

"Abandon all hope, oh ye, who enter here!"

Then, in this hotbed of ignorance, privation, and misery, sow the seeds of emigration; letters of the successful emigrant, (for here as elsewhere success is garrulous while failure is taciturn;) money sent to the home family; the improved condition of the returned wanderer; the exaggerated descriptions of the distant Eden whereby he enlarges and adorns his own importance. Let these fruitful germs quicken under the sun of the Southern imagination, and never did Persian fakir or Hindoo adept, bending over his mangoseed, evoke a speedier harvest—a harvest which the numerous societies of navigation press forward to reap. Let it, however, also be borne in mind that the Italian emigrant as we will consider him, one of the 13,000,000 of his compatriots who do not know the alphabet, is, in the words of the distinguished political economist, Alberto Errara, "the animal most refractory to persuasion on the face of the earth!" It would be easier to persuade a mule or a dog than an Italian peasant. The very attempt is fatal, arousing his natural or acquired distrust and closing the door to subsequent influence. The agents of navigation companies are careful not to fall into this error, reserving their efforts to directing the self-moving stream, each one seeking by concessions and promises to turn it to his own ends.

Here their responsibility, which is morally a grave one, begins. The majority of emigrants are bona fide paying passengers. It is inexact to say that they are in any large degree carried gratis on promises of payment on arrival. **The sale of their lowly cabin, of the domestic furniture, the gold ornaments of the wife and her meagre savings, all these go to make up the small sum of their passage money. In the recent collision and wreck of the Matteo Bruzzo the principal reason which deterred the terrified survivors from accepting the Government offer of free transportation to their homes was the fact that they no longer had any, having sold everything to embark. In former times, attracted by favorable conditions offered in South America (among which, I remember, a manifesto of the Emperor of Brazil offering free passage, remunerative employment and a bonus of public lands to the Italian emigrant)—in former times, I repeat, this emigration set mainly in the direction of Central and South America, and fortune nearly always favored the venture. The climate, propitious to the Italian; the language easily mastered, nay, both languages reciprocally intelligible from the very first; the religion and customs a connecting link, all served to foster and protect the growth of Italian prosperity, until now it may fairly be called the grain of mustardseed that 'grew and filled the land,' overtopping in wealth, position, and influence the native development. But a new element of emigration then entered. English and French companies, invading the Italian ports, began to rival each other in the transportation of emigrants to New-York and other ports of the United States. The great Italian company, (Navigazione Generale,) found itself obliged to enter the** lists and, at a heavy loss, run vessels regularly between New-York, Naples, and Genoa.

All the conditions which favored and favor emigration to South America diametrically oppose that to the north. The climate, with its wintery rigors and exhaustive Summer heats, is insupportable to the Italian used to the warm evenness of his own. Religion and customs of trade are in every sense diverse, and, most important of all, the language is a sealed book. In the way of private advice I cannot say how many capable and enterprising operatives I have dissuaded from going to New-York by simply giving them in English a series of orders such as they would be required to receive and understand in the exercise of their trade in America. The geographical ideas of the major part of Italian emigrants are but vague. Once in the seaport they become the defenseless prey of agents and manipulators, whose object is to fill the steerage quarters of the departing ship regardless of destination. As the peasant is apt to conclude that if you live in America you are bound to know whatever relative he may have there, so he is fain to believe that Buenos Ayres and New-York are sister cities to be indifferently chosen. The Italian Government, owing to the military surveillance it keeps, might add perhaps with benefit to the eminent, the moral influence of enlightened advice. But we all know the market value of that article to the receiver. The American cultured traveler who has suddenly had his linguistic ignorance assailed by the Babel of some French, German, or Italian Custom House, knows by experience the helpless dismay the bewildered exasperation of that moment. Let him make it permanent, dividing furthermore his own intelligence by that of the unlettered peasant; let him subtract the sustaining sense of money in one's pocket and add the misery of narrow or no means at all. He will then have some idea of the immense negative force contained in that one phrase—"I don't understand!" It is the snuffers clapped on the flame of intelligence; the straitjacket applied to human endeavor. I know of but one word —a French one—that expresses it, *ahurissement*, complete dumfoundedness, so to speak; and of but one figure in material things that illustrates it—the limp helplessness of a tree in process of transplantation. I once met in New-York an Italian street sweeper who had a schoolmaster's diploma and certificate in his pocket. Nor can I say how many poor organ grinders have poured with surprised delight into my ear their tale of insuperable displacement. Tillers of soil yearning with the peasant's homesickness for the wonted life of the fields and condemned to explore the filthy rag barrels of the city; able mechanics metamorphosed into petty peddlers; designers, workers in brass and iron, engravers, into common porters, keepers of fruit stalls, or hawkers of lemonade at 5 cents the glass.

Wherever the necessity of an intermediary arises, abuses follow in his train. Witness our own Indian Agency affairs. Thus we see explained and in a certain sense justified the existence of contractors or padroni. These, despite abuses, are a mouthpiece, the medium by which the dumb speak, the blind are led, and the naked clothed. They are a union of the rascally European guide whom Mark Twain flagellates, decoying his prey into unheard-of shops and impossible bargains, and the wily American Indian agent persuading the red man to sell his rifle for a drink of "firewater."

Still, "half a loaf is better than no bread," and even the bitter expedient of fattening one's natural enemy is preferable to being starved one's self. The Italian Government has announced its intention of proposing a new law on emigration at the opening of Parliament. An effort will be made to regulate the rivalry of agents, and even—so far as they come under Italian jurisdiction—of the padroni, by exacting solid guarantees in protection of the emigrant. But so long as the causes I have here set down exist, the flow of reaction, of adventure, and of reawakened hope will continue to sweep the Italian emigrant hence until such time as the ebb of disappointment and absolute failure may bring him back again and turn the tide.

There would seem to be a remedy, however, within the reach of American legislation, palliative if not radical, and worthy of America's great heart. The establishment of free evening schools in all Italian quarters or encampments for simple and exclusive instruction in the English language. Object and oral teaching—the application of the Froebel system to the adult, the ready Italian eye and ear aiding—would be of far more practical value than any attempt to teach the written and printed language. The spoken tongue and its use in the common ways of life are what the unlettered peasant needs and would readily acquire. A profounder remedy and one which time may eventually develop would lie in inducing Italian emigrants to renounce the beloved country that can no longer supply their needs and become citizens of the hospitable land that offers them not only subsistence, but also free manhood. The tenacity with which the Italian clings to his native land has no counterpart in other classes of emigrants. Every effort undertaken, every risk run, every success achieved, has a direct relation to his return. Still, the propaganda of liberal ideas, and above all the attainment of personal well-being, will in time obliterate even the glowing "picture on memory's walls," whose dark shades of suffering, privation, and woe only serve to enhance unforgotten beauties. The Italo-American may then become what the occult designs of fate may have perhaps already destined him to be, a factor in American progress, infusing into restless and exacting American labor something of his native spirit of large abnegation and accurate patience, as well as much of the splendid heritage of art which has come down to him in the uninterrupted traditions of his ancestors.

October 28, 1888

THE NEW IMMIGRATION LAW.

STEAMSHIP COMPANIES OBJECT TO RECEIVING PASSENGERS BACK.

Steamship lines affected by the new immigration law do not seem disposed to yield ready compliance with the provisions of that act. Col. Weber, the Superintendent of Immigration, refused admission to a number of passengers brought by the steamship Iniziativa. These were sent to that vessel to be returned to Italy, whence they came, and subsequently three of them made their escape. Yesterday Col. Weber received the following letter from Phelps Brothers & Co., who are the local agents of the Florio Line:

"We hereby protest against your having put on the steamship Iniziativa a number of persons that you have decided are to be returned to Italy on the ground that, under Section 10 of the last act, 'such persons are to be returned on the same ship, if practicable.' As this ship does not return to Italy, it is not practicable. We therefore decline any responsibility arising from their having been put on board, the same having been done, in the first instance, without notice to us. We request you to take and keep them in charge until such time as we can send them back, which we will do at the earliest moment."

Col. Weber stated in his reply that it might not be profitable to return the ship to Italy, but that in his opinion it was at least practicable. The Superintendent pointed out that if agents or owners of a vessel could evade the law by stating that it was inadvisable to return the ship to the port from which she sailed the law would nullify itself in every such case. Furthermore, the Colonel knew of no reason why the paupers of Europe should be foisted on America simply because the business interests of steamship companies made it necessary for the vessels to ply between other ports. Nor did he know of any regulation which required him to give notice of the return of persons who came in violation of the law.

The letter closed with an emphatic refusal to receive the debarred immigrants, as he did not consider it his duty as an official to take care of such persons merely to relieve the company of the inconvenience of caring for them.

A representative of the line subsequently called upon Col. Weber and made a more moderate presentation of the case, stating that they were willing to pay for the maintenance of the excluded passengers if the Barge Office authorities would take care of them. This Col. Weber declined to do.

The Iniziativa cleared yesterday for Lisbon and expects to sail to-day. The Superintendent telegraphed the facts of the case to the Secretary of the Treasury and asked instructions, but had received no reply at a late hour last night. It was expected that the Collector of the Port would withhold the papers of the vessel, but as no official notice was given him, the papers were issued.

A writ of habeas corpus issued by Judge Lacombe for Domico Marianno was served on the Iniziativa yesterday. Marianno is one of that vessel's passengers who was debarred by reason of the likelihood of his becoming a pauper. The writ was got out by the Italian Society, and the same organization sent a doctor to examine the passengers who were debarred on account of loathsome diseases.

Twenty-one of the twenty-four rejected Italian immigrants brought by the Burgundia, and who were returned to that vessel, have succeeded in escaping into the city. As soon as the report reached the Barge Office, Andre Palmeri, an interpreter connected with the Labor Bureau, was sent to inquire about it. The officers of the ship alleged that the men were on board, but the interpreter failed to find them. The Burgundia belongs to the Fabre Steamship Line, and was scheduled to sail yesterday. The vessel frequently visits this port, and Col. Weber said last night that proceedings would probably be instituted against her when she returns.

April 9, 1891

LANDED ON ELLIS ISLAND

NEW IMMIGRATION BUILDINGS OPENED YESTERDAY.

A ROSY-CHEEKED IRISH GIRL THE FIRST REGISTERED—ROOM ENOUGH FOR ALL ARRIVALS—ONLY RAILROAD PEOPLE FIND FAULT.

The new buildings on Ellis Island constructed for the use of the Immigration Bureau were yesterday formally occupied by the officials of that department. The employes reported at an early hour, and each was shown to his place by the Superintendent or his chief clerk. Col. Weber was on the island at 8 o'clock, and went on a tour of inspection to see that everything was in readiness for the reception of the first boatload of immigrants.

There were three big steamships in the harbor waiting to land their passengers, and there was much anxiety among the new-comers to be the first landed at the new station. The honor was reserved for a little rosy-cheeked Irish girl. She was Annie Moore, fifteen years of age, lately a resident of County Cork, and yesterday one of the 148 steerage passengers landed from the Guion steamship Nevada. Her name is now distinguished by being the first registered in the book of the new landing bureau.

The steamship that brought Annie Moore arrived late Thursday night. Early yesterday morning the passengers of that vessel were placed on board the immigrant transfer boat John E. Moore. The craft was gayly decorated with bunting and ranged alongside the wharf on Ellis Island amid a clang of bells and din of shrieking whistles.

As soon as the gangplank was run ashore, Annie tripped across it and was hurried into the big building that almost covers the entire island. By a prearranged plan she was escorted to a registry desk which was temporarily occupied by Mr. Charles M. Hendley, the former private secretary of Secretary Windom. He asked as a special favor the privilege of registering the first immigrant, and Col. Weber granted the request.

When the little voyager had been registered Col. Weber presented her with a ten-dollar gold piece and made a short address of congratulation and welcome. It was the first United States coin she had ever seen and the largest sum of money she had ever possessed. She says she will never part with it, but will always keep it as a pleasant memento of the occasion. She was accompanied by her two younger brothers. The trio came to join their parents, who live at 32 Monroe Street, this city.

Besides those of the Nevada, the passengers of the City of Paris and of the steamship Victoria were also landed at the new station. They numbered 700 in all, and the many conveniences of the mammoth structure for facilitating the work of landing were made manifest by the rapidity with which this number was registered and sent on to their various destinations. It was quite a populous little island about noon, when the steerage passengers from the three big steamships were being disembarked but within a very short time they had all been disposed of. Those destined for local points were placed on board the ferryboat Brinckerhoff and landed at the Barge Office. Those going to other places were taken to the various railroad stations by the immigrant transports.

The first ticket sold by the railroad agents in the new building was purchased by Ellen King, on her way from Waterford, Ireland, to a small town in Minnesota.

Col. John J. Toffey and Major Edward J. Andersen, who have succeeded to the contract for the supply of subsistence, signalized the day by entertaining Col. Weber, the Superintendent of Immigration; Major Hibbard, the Superintendent of Construction; Surgeon Toner and staff, and all the employes of the station at a New Year's Day spread. Capt. Charles W. Laws, their chief, had prepared the board for 300 guests, and the throng had a merry time at the tables.

Col. Toffey and Major Anderson had planned to have a pretentious opening and their friends were to have been invited, but the authorities at Washington directed that the opening be made without any ceremony.

All connected with the Immigration Bureau expressed themselves as exceedingly well pleased with the change from the cramped quarters at the Barge Office to the commodious building on its island site. The railroad people were the only ones who were heard to express any dissatisfaction. Their grievance is that the building is so large as to involve much running about on their part in getting their various passengers together. Others said that when the tremendous number of immigrants who had to be handled in this building was considered finding fault with its size was like complaining of a circle for being round.

"We can easily handle 7,000 immigrants in one day here," said Col. Weber. "We could not handle half that number at the Barge Office. At the old place the greatest delay was in the baggage department. All that is now done away with, as the baggage department has the entire first floor and the arrangement is perfect."

The building was erected by the Federal Government at a cost of $500,000. The wharves are so arranged that immigrants from two vessels can be landed at the same time. As soon as disembarked the passengers are shown up a broad stairway on the southern side of the building. Turning to the left they pass through ten aisles, where are stationed as many registry clerks. After being registered, those of the immigrants who have to be detained are placed in a wire-screened inclosure. The more fortunate ones pass on to a similar compartment, where those going to the West are separated from those bound for New-England or local points.

There is an information bureau in the building for the benefit of those seeking friends or relatives among the immigrants. There are also telegraph and railroad ticket offices and a money changer's office.

Except the surgeon, none of the officials will reside on the island. The surgeon occupies the quarters formerly used by the gunner when Ellis Island was a naval magazine.

January 2, 1892

NOTICE

The Russian Hebrews who were brought to this city by the *Massilia* came from Southern Russia. The reports of the Treasury Department already show a large increase in the number of immigrants from Russia. The figures for the calendar years 1890 and 1891 are as follows:

	1890.	1891.
Russia (except Poland).....	40,883	73,177
Poland........................	19,737	31,285

The report for the last six months of 1891 shows that the number coming from Russia (Poland excepted) in those months was 46,710, as against 20,934 in the corresponding months of 1890. The whole number of immigrants was greater in 1891 than in 1890 by about 100,000, and nearly half of this increase is ascribed in the reports to Russia and Poland. February 15, 1892

WORK FOR IMMIGRANTS

Interesting Stories of the Labor Bureau at the Battery.

REMINISCENCES BY MATRON BOYLE

Government's Care for Prospective Citizens and Citizenesses—The Average Wages $12 a Month.

"Immigrants' Free Employment Bureau" and "Freies Arbeits Bureau" are two signs, the first very large, and the second also conspicuous, at one side of Battery Park, but it is probable that very few of the many persons who wander through the lower part of the city ever give them a thought.

There are enough housewives in this city, however, who are so familiar with the place from frequent visits, that there are, on an average, 50 employers waiting for every woman who applies there for a position.

This is one of the ways in which the Government takes charge of immigrant girls. "Most of the girls who come to this country nowadays have friends," says Matron Boyle, who has been in the bureau for the past sixteen years, and it is very often through their influence that the newcomers find their way to positions.

The bureau is a dingy room filled with long settees, from which a glimpse can be had of the water and the boat from the island, bringing the latest immigrant arrivals.

Lillian Russell, wearing a coquettish little cap and smiling down from a gay-colored picture on the wall, is one of the most conspicuous objects of interest that the waiting applicants find to gaze upon.

Most of the immigrant seekers after work are young women. An occasional man breaks the monotony of the long lines of women, for the husbands are allowed to sit with their wives instead of joining the ranks of men in the outer room, and frequently a couple will obtain a position together.

The Irish predominate. A recent month's report shows 166 Irish, 34 German, 11 English, 2 Scandinavians, 1 Pole, and 2 Hungarians.

"That is not anything like what we used to have," Matron Boyle said to a reporter for The New-York Times. "I have known, say, five years ago, as many as that to come in in a day. The average wages the women receive are $12 a month. There are governesses and dressmakers, beside house servants. The greatest demand is for girls for general housework, and I have known dressmakers to take positions as domestic servants, these positions being so much easier to obtain.

"Good cooks are in demand, and there is no trouble in getting them into good positions when they don't want all this nonsense of a day off every week.

"Every employer who obtains a servant from us must have a letter of recommendation from some reliable business house in the city or we will not send a servant.

"No, we never have any trouble. We find out if the recommendations are genuine. If we think everything is not all right we investigate. Then if we find the family does not live in a locality which would be desirable for a girl ignorant of the ways of the world, we will not send one there.

"We register the name of each girl, her nationality, the place to which she is sent, the date of her arrival, and the ship upon which she came.

"We also take the name and address of the employer, and the rate of wages she is to pay, and make any marginal notes that may be necessary.

"The girls are engaged by the month, and, as a general thing, they turn out well in every way. The mistresses expect to take green servants. Some of the girls, however, have been carefully trained in the old country. A girl is an immigrant by law for a year, and during that time we assist her in finding a position. If she has no friends we are allowed to use a little discrimination and sometimes find positions for her after the first year.

"The wages range from $8 for a little girl up to $16 and $20. There we have a man and his wife registered for $20 a month together and full board. That is not bad.

"Some of the best people in the city come here to get their servants. Only, I am sorry to say, we never have enough for them.

"Yes, there is quite a demand for servants at this time in the year. Many people want them to go into the country, and are willing to pay a little more to have them go. The girls don't like it so well, and you can't blame them. They have just left friends in the old country, and it is hard to be obliged to leave the new ones, just made.

"I had one bright little Irish girl here once, waiting for a position, who said she had an aunt in this country, but her address had been lost and she did not know where she was.

"It happened that the girl's name was O'Conner. Well, there was a Mrs. O'Conner who had been getting servants of me for some years, who came back again just at that time.

"'I have a nice little girl who has the same name as your own,' I said, 'and I would like you to see her.'

"So I took the little girl to her.

"'Can you do so and so and so and so?' she asked. 'Well, I'll try,' said the little girl.

"Then she was asked about the county in Ireland from which she had come, and—well, the woman found she was talking with her own niece.

"'It will not be as a servant that I'll take you home,' she said.

"I saw the little girl afterward and she told me she had a lovely home.

"There was another girl I had here, and some way she never could find a place that was just right. She was always afraid of something. I had a woman at that time who was equally troubled to get a servant to suit her. I sent that girl to her one day. I knew she wouldn't stay, but I thought it might be better for her to try.

"Do you know it must be that those two were looking for one another, for neither of them came back.

"It was some time after when I next saw the girl, and then she drove up in a coach. She had married a rich man in Jersey, a widower, and she was going to France to be educated to take her position in his family and take care of his children.

"Yes, it is all true. We do a great deal of good here. We take care of the girls just as their mothers would, and sometimes better."

The German girls who patronize the bureau are in charge of Otillie Lenhard.

July 7, 1895

IN the good old days before New Amsterdam had become New York, and before New York had extended its northern boundary beyond what is now Chambers Street, indiscriminate immigration was favored by the authorities of the Colony of New York and of this city, and it mattered not whether the immigrants that reached this port came of their own volition or were kidnapped on the other side and brought here by men who made a business of supplying the colonists with laborers, white and black.

Those who purchased the laborers after they arrived here cared little or nothing about their pecuniary wealth, and paupers were then made welcome, whereas now efforts are made to keep persons of this class from coming to these shores.

If the immigrants of those days had not sufficient money with which to pay for a passage across the ocean, or had been kidnapped, an enterprising sea Captain was always to be found who was willing to risk the expense of bringing them here, he well knowing that if he once got them here he would receive money enough to compensate him for his trouble, and a handsome bonus besides, by selling their labor for a number of years to some colonist. And so the early pauper immigrants were usually sold and held in bondage until they had worked out the sum that had been paid for their passage. The working-out process was usually prolonged as much as possible, and until the period of bondage had expired the immigrants could be sold and resold any number of times. There was always a profit in the transaction for all concerned but the hapless immigrant—unless, soon after getting here, he was fortunate enough to run away and elude capture.

Since the early Colonial days immigration has undergone various phases, and, whereas, at times in the past any one was welcomed, great restrictions are now being placed on those who seek this country for a new home.

The Colonial period may be said to have ended and the period of immigration to have begun with the close of the Revolutionary War, in 1783. Decennial censuses were taken in 1790, 1800, and 1810, but no immigration statistics were gathered until 1820. The accepted estimate for the period is 250,000. From 1820 to 1856 no distinction was made in the returns between settlers and travelers.

From 1856 to 1868 settlers were distinguished from travelers, but the number of immigrants of each nationality was not displayed separately from the number of travelers of that nationality. Since 1885 immigrants from the British North American possessions and from Mexico have not been included, and in 1894 the Commissioner General of Immigration presented for the first time the number of European immigrants arriving in the Dominion of Canada destined for the United States.

The immigration records were never before kept as carefully and with as much detail as they are to-day, but, according to those which are in existence, it is shown that up to the end of the last Governmental fiscal year, June 30, 1896, the total immigration since 1783 was 17,813,750. By decades, from 1820 to 1890, and from 1890 to 1896, it was as follows: 1820-1830, 128,393; 1830-1840, 539,391; 1840-1850, 1,423,337; 1850-1860, 2,799,423; 1860-1870, 1,964,061; 1870-1880, 2,834,040; 1880-1890, 5,246,613; 1890-1896, 2,878,492. It will be noticed that from 1880 to 1890 it was nearly twice as much as it had been for any previous decade, and over two-thirds what it was for the entire period. In fact, nearly two-thirds of the entire immigration movement of the world in 1890 was directed to the United States.

The commercial depression of 1826-7 in England was accompanied by a decided increase in the number of immigrants to this country. In 1827 there were 18,875, almost twice as many as in any previous year, and in 1828, 27,382, a figure that was not reached again for some time. The American panic of 1837 registered itself in the immigration returns for 1838, when the number was less than it had been for any year since 1831, and was only about half of what it had been for the two years previous. The years 1848 to 1854 were marked by an enormous increase, culminating in the latter year, when the arrivals were 427,833, the largest number recorded until after the close of the civil war. Three principal things contributing to the change were the bad times in Germany, the famine in Ireland, and the discovery of gold in California. The financial depression of 1857 was followed by another decrease, the number for each of the years 1858 and 1859 being less than 120,000. The civil war naturally checked the incoming movement still further; then the immense business activity of the years following (when transportation facilities were being speedily improved and the West was being opened up) was coincident with another increase in immigration, as the hard times of the later seventies were with another falling away.

The return to prosperity after 1879 brought with it an enormous immigration, (788,992 in 1882 alone.) Finally, the recent depression (1893-5) was accompanied by a most significant decrease. For 1894 the immigrant arrivals were only 285,631, and for 1895 only 258,536, and for several months of this period the outflow of steerage passengers, almost exclusively foreign born, from American ports actually exceeded the inflow—an unprecedented condition. The gradual return to prosperity is already observable in the immigration figures. For the period from Jan. 1 to Sept. 1, 1895, for example, there was an increase of 36,270 at the Port of New York over the same period of 1894. Furthermore, returns show an increase of 51 per cent. for the month of July, 1895, over July, 1894; of a little over 60 per cent. for August, 1895, over August, 1894, and of more than 55 per cent. for September, 1895, over September, 1894.

From 1820 to 1840 the immigration was almost entirely from Great Britain and Ireland. During the decade ending with 1840 German immigration first became noticeable, and this steadily increased until, in the decade ending with 1890, it amounted to nearly 1,500,000, about the same as that from the whole United Kingdom. Norway and Sweden first made themselves felt in the returns for the decade ending 1870, with over 100,000 arrivals, a number which increased to over 500,000 during the decade ending with 1890. This last decade, however, is even more notable for an enormous increase in the immigration from Austria-Hungary, Italy, Russia, and Poland.

The first National immigration law was passed by Congress in July, 1864. It was for the purpose of encouraging immigration, all immigration theretofore having been cared for by the individual States. The passage of the act was due, in part, to the absence in the army of the large bodies of wage earners who had been employed in important indus-

IMMIGRANT INSPECTION ROOM, ELLIS ISLAND.

THE UNITED STATES IMMIGRATION BUREAU AT ELLIS ISLAND.
Commissioner J. H. Senner, M. D., Deputy Commissioner E. F. McSweeney, and Their Assistants.

INFORMATION BUREAU, ELLIS ISLAND.

tries, and it encouraged indiscriminate immigration. No safeguards were provided against the worst classes of foreign population, and idiots, criminals, paupers, &c., were admitted without inspection or examination. "Contract labor" was encouraged under this act.

The law of 1864 was repealed in 1868, and from that time until 1882 there was no United States statute bearing on immigration. The law of 1882, instead of encouraging immigration, was passed to restrict it. Since 1882 other laws and amendments to laws have been passed, and now the immigration officials exercise a power that permits them to stand between the objectionable persons in other countries and the people of their own. The laws declare who shall not be permitted to land, and their provisions are very broad.

A great percentage of the immigrants who enter this country enter by the way of New York, and consequently the history of immigration at this port is of much interest.

For many years the New York State officials had entire charge of immigration matters in this port. They used Castle Garden for their purposes, and there all immigrants were landed. For a time after the first Federal laws affecting immigration were passed, the Federal and the State officials worked together, the State officials doing the work under contract with the Nation. It was discovered, however, that the State officials were misusing their power, and were using the bureau for political purposes to the injury of the immigrants, and by the Federal act of 1891 the Immigration Bureau was placed exclusively in the charge of Federal officials. Thereafter Castle Garden was used for a time; then the Barge Office at the Battery was used, and then Ellis Island. Where the Commissioner of Immigration now has his office, and there the Federal Government has a valuable and almost perfect plant for the purpose of inspecting and caring for all immigrants who come to this port. Immigrants have been landed there since Jan. 1, 1892.

All passenger vessels are boarded at Quarantine by Inspectors from the Immigration Bureau. As the vessels proceed to their docks the passenger lists are examined by the Inspectors. Cabin passengers' tickets and declarations are scrutinized as well as steerage passengers', and if any cabin passenger is thought to be a person who comes within the restrictive clauses of the law he is compelled to go to Ellis Island and await investigation.

When the vessel has reached her dock the immigrants and their baggage are taken by barge to Ellis Island, and there they are all inspected and their baggage is examined.

The main building on the island has a great room on the ground floor into which the baggage is taken, and rooms on the upper floor into which the immigrants are sent. Every immigrant is numbered and tagged, and, 240 at a time, in groups of 30, are examined by the men and women Inspectors, before whom they are compelled to pass, and to whom they make their declarations. If any immigrant fails to pass an Inspector, he or she is at once sent before the Board of Special Inquiry for further examination, and if the board finds that the immigrant should not be allowed to land, he or she is put in the detention pen to await a re-examination, or his or her return to the place from which he or she came. Every immigrant who is found ineligible to land is detained on the island and returned to his or her home at the expense of the steamship company that brought him or her here.

The immigration officials are familiar with the ways of those who would prey upon the new-comers, and every effort is made to protect the men, women, and children. Women and children they are particularly solicitous about, and hedge them around with the safeguards of the law and of the Church to which they belong. If a woman or a child is not met by friends or relatives she is detained until some one whom she names can be communicated with, and until it is determined to be safe for her to land. In such instances the missionary societies aid the immigration officials.

As an instance of the care that is exercised to prevent improper persons from landing, statistics show that as many as 800 immigrants have been detained and returned to their homes in Italy in one month.

There is a well-conducted hospital on Ellis Island, and in it all ill immigrants are placed and kept until returned or allowed to land. The Immigration Bureau's care of an immigrant does not end with his landing, for every immigrant who becomes ill or unable to care for himself during the first year ashore must be cared for at Ellis Island. Those who are ill are cared for in the hospital and the others in the dormitories.

As it is possible to inspect and pass 5,000 immigrants in a day, it is seldom that there are many of them at the island over night. For those who are there, however, ample provision has been made, and wholesome food and clean beds are provided. The dormitories, which will accommodate 475 persons, are well lighted and ventilated, and are kept in an odorless condition. Separate rooms are provided for the men, and for the women and children. Every immigrant is permitted to land in this city as soon as possible after he has disembarked from his ship. All come ashore at the Barge Office, at the Battery.

Ellis Island, which was formerly a Government storage station for powder and shells, contains about fourteen acres and is being constantly added to by filling in the shallow waters that surround it. Upon it is the main building, which contains the Commissioner's offices, and in which the inspection, &c., is done; a hospital, a power house that supplies electricity for lighting, water for a fire and flushing system, heat for the buildings and for laundry and disinfecting purposes; a kitchen and storerooms in buildings in which powder was formerly stored, and great cisterns in which fresh water is gathered from the roofs of the buildings on the island. There are also other buildings, in which the physicians and minor officials live.

The bureau has its own steamboat for transportation between Ellis Island and this city, and on this boat all passengers, immigrants, and baggage are carried to and from the island. Dr. Joseph H. Senner is Immigration Commissioner, Edward F. McSweeney is his deputy, and upon them devolves the duty of properly conducting all that pertains to immigration in this port.

January 31, 1897

FOR MISTREATING IMMIGRANTS.

A Gateman and Two Messengers Dismissed from the Barge Office.

WASHIGTON, Oct. 4.—Assistant Secretary Taylor to-day took action in the case of four employes of the Immigration Office at New York who had been charged with various offenses of a more or less serious character. Four other cases now under consideration, it is expected, will be disposed of in a day or two. The action of the Assistant Secretary is the result of a recent investigation conducted at the Barge Office in New York.

Emil Auspitz, gateman, is dismissed for violent and profane language and generally rough treatment of immigrants. The Commissioner at New York recommended that he be reprimanded, but retained in the service. The Commissioner General recommended his dismissal. The appointment division referred the matter to Assistant Secretary Taylor for direction.

Karl E. Kumpf, messenger for the board of inquiry, is dismissed for accepting money from the friends of immigrants.

J. Ross Stewart, messenger, is dismissed for receiving money from the friends of immigrants.

The charges against John Lederhilger, alleging brutality toward female immigrants, were dismissed at the direction of Assistant Secretary Taylor.

October 5, 1900

THE RUSH TO AMERICA.

So far the year 1902 has broken the record of the past decade for immigrants landing at this port. January and February showed a large increase on the figures for the same months last year.

This March showed 23,000 more than March, 1901, and during the first two weeks of April there entered 40,000 souls, as against 28,000 in the same fortnight a year ago. For the rest of April the proportions are as large if not larger, and May bids fair to outdo April. None too soon have the new quarters of the Emigrant Palace on Ellis Island been made ready.

The flood of immigration, which subsided a little after the lean years of the nineties, is rising to unprecedented heights during the fat years which bear a rotund O auspiciously in the place of their penultimate numeral.

Rules that immigrants must have money in their pouch, that they must be healthy and free from suspicion of crime, have no deterrent effect. The Spanish war, the advance of American manufactures into Europe, the outcry of European papers against the American bugbear, and the evidence of their own senses, which show them how North America has become the land above all others which feeds Europe, have impressed the dullest and least imaginative.

No proclamations made by the press of lynching bees, and miners' strikes, and the bad treatment of immigrants in New York can offset the hope to better their condition. Would-be colonial kingdoms and empires spread their lures in vain. Russia is losing her most stalwart and trustworthy soldiers by the exodus from Finland. Hungary and Galicia pour their thousands through the German ports of Hamburg and Bremen. Germany herself, having for a term of years marked with complacency the great industrial advances and how the empire was absorbing her laboring men, is grown anxious again, now that things financial and commercial in the Fatherland are less propitious. Irish immigration is not so great, only because the population has already sunk so low by previous outgoings to America and the British colonies.

It is particularly Italy that sends a stream of immigrants, breaking for the peninsula all previous records. Next are the Poles and Slovaks of Austria, with Greeks and denizens of Turkey and the Balkans to swell the flood; South Russians, too, and a small but constant contingent of Irish, Scotch, English, and Germans.

The old bogy of illiteracy among the immigrants has lost much of its former force, since other nations are paying attention to education, although they have not attained that low percentage of illiterates we find in Scandinavia, Switzerland, and Northern Germany. Strange to say, it is this improvement in popular education which has done much to cause the rush to the land of dollars, since a reading people has the press and cheap mails to aid them in deciding where their chances of a livelihood are best. The situation is a serious one not only for the European nations who live in a constant state of menace of war, but for us, who may fairly feel anxiety lest the assimilative powers of the Great Republic shall not be equal to the task of weaving all these threads of diverse races into a homogeneous whole.

In the large cities the Germans are being driven out of the barbers' shops by Italians and the Irish out of rude laboring employments. Germans take to shopkeeping, beer selling, and other better-paying occupations. Irishmen turn to salaried places as janitors or through their singularly keen political sense obtain offices under municipal government in the great cities. The South Russian immigrants go largely to the West as farmers, and the Turkish subjects become peddlers of fruit and cheap Oriental goods, which are being made here in great quantities. Englishmen take to factory work and shopkeeping. The Italian immigrants are doing so well that they now import their families and settle down, generally in the large cities, instead of returning to Italy as they were formerly wont to do. The Slavs of Austria find employment in the mines.

So far it cannot be said that the immigration offers any very serious drawbacks, although many bad elements enter with the good. Our population is now so large that it may be depended upon to neutralize even so tremendous a foreign element, broken as it is into so many unconnected parts, separated yet further by diversity of language. Whether in the coming years we shall be forced to render immigration more difficult in order to winnow more thoroughly the desirable from the undesirable parts remains to be seen. A singular feature is the fact that, with the exception of certain lines of rough labor, this exodus of working people has not seriously affected the labor market. This perhaps is to be expected so long as prosperity reigns.

May 4, 1902

ELLIS ISLAND CHANGES HANDS.

To Be Controlled by Department of Commerce and Labor.

To-day Ellis Island passes from the control of the Treasury Department into that of the Department of Commerce and Labor.

Commissioner Williams does not expect there will be any formal procedings or that any ceremony will mark the change in control. The change has been so long expected that there will not be the slightest friction, and the work is expected to proceed to-day as in days past. Cases now pending, particularly those before boards of special inquiry, which arose during the Treasury Department's control, will necessitate two reports on their settlement to complete the report of the fiscal year of the older department.

July 1, 1903

NEW RULE FOR IMMIGRANTS.

Must Be Provided with More Funds Than Heretofore.

The unusual number of aliens who, as public charges, have been sent back to Ellis Island recently for deportation has led Commissioner of Immigration William Williams to prepare a set of rules governing the amount of money that immigrants arriving at New York must possess in order to be allowed to land. Under the law passed last year any immigrant who, within two years after his arrival, becomes a public charge may be sent back to the Immigration Bureau for deportation to his native country.

Mr. Williams has ordered that hereafter in reference to telegrams for funds, sent to friends by arriving immigrants, "each telegram shall request $10, and in proper cases $20 or $25. Only in exceptional cases shall the amount asked be as low as $5."

This order has created much discussion among the societies of the various nationalities who are represented at Ellis Island to look after the interests of their people. Many of the agents said that the reason the girls carried so little money was owing to a fear of losing it or being robbed.

By the new rules all immigrants must telegraph their friends or relatives on arrival and satisfy the authorities that they will be cared for when they reach their destination. It is expected that this rule also will furnish evidence of the good faith of those interested in the immigrant. The new rules will be sent to the various immigrant stations in Europe.

May 21, 1904

Refused an Entry to the Promised Land

One of the Tragic Phases of Ellis Island---Turned Back from Liberty's Gateway.

YOU can run the entire scale of human emotion in the deportation division in the big Federal building on Ellis Island. But for the asking you will hear stories of hardship and disappointment that have the power to seek out an unpetrified corner of a heart of flint and to cause that strange lump to rise in the throat.

The Government ferryboat Ellis Island leaves the Barge Office on the hour and soon glides into its slip in front of the big red building across from the Battery. A sea of eager, expectant faces peer down at you from the roof garden of the building. Friends and relatives are expected, and a great cry goes up from the roof as the passengers begin to walk ashore. Families are to be united and made happy after a separation of many years; sweethearts are to be joined who before had the great Atlantic between them.

Yet in this great sea of faces above are many who will never set foot on American soil, save to come and go from the little island controlled by Uncle Sam.

⁂ ⁂ ⁂

Those aliens who do not qualify physically and mentally—and often morally—are not permitted to remain in this country. They must go back to the land from which their long journey began. The Immigration Department presents a strict examination for the aliens, and they must pass letter-perfect in every subject if they would come under the protection of this Government. Those who do not satisfactorily pass the examinations are detained as "undesirables," and after a sojourn of a few days in the deportation division, are returned to Europe, at the expense of the steamship company responsible for their crossing the Atlantic.

A complete record is kept of this undesirable alien element, and it is stated that an average of 1,000 aliens are monthly refused admission to this country and are compelled to journey back to the land from which they came.

The study of immigration in all its varying departments is highly interesting; but to locate the human interest side of the subject you must climb the stairs in the big red building on Ellis Island and interview these poor unfortunates known as "undesirables."

Many of them had for years saved a few pennies each day, denying themselves even the bare necessities of life, in order to realize the price of a steerage ticket to this mystical land of wealth. The cost of the ticket meant many a supperless night. Then with light hearts and an abundance of ambition they traveled across the ocean, to find another cup of bitter disappointment awaiting them, filled to the brim by watchful immigration officials.

A little Russian boy, ill-shapen and pale from long suffering, was found packed in among a number of his fellow-countrymen in a large room on the upper floor of the immigration building. The bent piece of humanity could not have been more than 16 years old, and yet his pain-pinched features made him appear much older. He was eager to talk of his shattered hopes and to tell his story of bitter disappointment.

Less than two months ago this boy left the town of Kishinef in Russia. He could not make a living at home because he was a cripple and too weak to go into the fields. No one wanted to look on him, for his bent and twisted figure was unsightly to the superstitious peasants. Other children would cry after him and taunt him with his deformity. A deep sadness came into the life of the little fellow and he had given up all hope of ever wringing one moment of happiness out of life.

Then a letter written by an uncle arrived from America. This letter told of the great opportunities awaiting those who should hasten across the ocean. The hunchbacked boy read the letter over and over again, every word sinking deep into his memory. Through years of toil his parents had saved a sum sufficient to send the little fellow to America. They willingly gave up their all in the hope that their crippled son should find a place with his uncle. The uncle's address was written at the end of the letter, and with this the hunchbacked boy started on the long voyage.

When the big liner steamed slowly past the Statue of Liberty the boy stood on the forward deck, his eyes feasting on the beautiful scene. In a provincial dialect, hardly intelligible to one speaking Russian, the little fellow described his emotions on sighting the shores of America:

"I stood at the front of the ship with several of my fellow-countrymen, who, like me, were bound for America. As we came closer to the shore my joy knew no bounds. I was soon to be in a land where my race is not persecuted. I heard of this gold that could be had for the asking, and I longed to gather some of it and return to my old parents in Russia. Now they tell me I must return home, for they cannot find my uncle, and, furthermore, cripples like me are not wanted here."

Tears came into the cripple's eyes as he looked up appealingly to the knot of visitors who had gathered around him. An Inspector shambled up and after a casual glance at the little fellow turned to the crowd and vouchsafed the information that "Hunchy is clean loco."

But "Hunchy" isn't "loco," though the pain and sorrow that have been thrust upon this poor bent weakling have been enough to drive the strongest man along the straight road to the insane asylum.

⁂ ⁂ ⁂

Insanity among immigrants is daily increasing. The matron in charge of the women's quarters on Ellis Island explains that homesickness and lack of knowledge of the English language are the chief causes of mind derangement among female aliens. This is more so the case with servant girls who are permitted to land, but who later become insane and must be returned to their native lands. For three years after landing aliens are under the watchful eyes of the Government, and if during this period of probation they in any way step aside from the strict laws governing their conduct they are liable to be returned without notice. Matrons are kept busy traveling throughout the country and collecting female aliens who are found undesirable even after having been permitted to land.

A short time ago a mother and son arrived from Poland. Both expected to gain lucrative employment as factory hands. But during the voyage the son developed a severe type of tracoma—that dreaded disease of the eyes, which closes the gates on so many aliens. The mother gained the consent of the immigration authorities to land, but the son was ordered to immediately return to the port of departure. When the mother and son found they were to be separated a scene was enacted which brought tears to the eyes of every one standing about. The mother threw her arms about the lad, weeping all the while as if her heart would break. The son endeavored to hide his feelings of emotion and bitter disappointment, but like a mighty flood of water breaking through a dam his pent-up feelings passed beyond his strength and broke forth in a wail that sounded more like the cry of some hunted animal than of a man. He would have fallen to the floor had not his mother placed her loving arms around his neck. The boy continued to cry out at the top of his voice, while the mother's deep sobs sounded the accompaniment. It was a hard scene to look upon, and the visitors walked away.

When it is understood that an average of 5,000 aliens daily seek admission at the Port of New York, many incidents similar to the foregoing are bound to occur.

Up on the roof garden of the big building on Ellis Island the "undesirables" look out through a wire netting. Across the water they gaze longingly at the line of skyscrapers bordering the opposite shore. The strange sight of appalling greatness seems to taunt them and to take voice and whisper in their ears: "Look on me, so near and yet so far." Soon they shall take ship and travel back empty-handed to their homes. There they must again take up their old lives of wearisome toil, under slave-driving masters, in a land where corruption and oppression often run rampant. They return from the new world with only a view of that formidable wall of skyscrapers that seemed to bar their way.

Love plays strange pranks on Ellis Island and often causes the officials no end of worry and trouble. Young girls from Europe arrive on every ship to meet lovers who had gone before to prepare the nest. These girls from Europe are simple in their ways and modestly look forward to a simple life in a little home, and wish for nothing more than to be looked on as good housekeepers and mothers. The swain knows this and appreciates the girl's worth until he meets in this country girls of his own nationality who have been here for some time and have acquired somewhat the atmosphere of individuality that surrounds the real American girl. This is a novelty to the rustic fresh from the fields in Hungary or Russia, and with but little difficulty he soon forgets his promises made to the fair maid across the water. When the sweetheart arrives from Europe she often finds that her intended does not claim her at the island, so in tears and vowing vengeance on the fickle swain, she returns home at the expense of the steamship company by which she traveled to this country. Cases of this character are very common on the island.

⁂

A pen full of cattle crowding each other about before the entrance to a slaughter house is the comparison one makes while visiting the deportation division. And in truth Ellis Island is a slaughter house, for daily hundreds are sacrificed, their ambitions and hopes slain, to guard the interests of higher bred animals who fear contamination.

A feeling of depression comes over you while seeing these unfortunates and hearing their stories, and even after you have left the island, and Uncle Sam's ferryboat is gliding through the water, that depression is still with you. All the way over you say to yourself: "If I had my way I'd throw down the barriers to every one of these poor unfortunates with his hard-luck story." But this is your sentimental side asserting itself. Wait until you have taken the elevated; then look down through the crowded streets near the Battery; observe the congested state of the street traffic, and where thousands are battling for daily bread. This sight knocks out sentiment with one blow, and your better judgment says: "It is much better so."

August 12, 1906

TOPICS OF THE TIMES.

The steamship Wittekind arrived from Bremen on her historic voyage to Charleston, S. C., on Nov. 4, 1906, with 475 aliens aboard—the first European immigrants by direct route to the South. The State of South Carolina had induced them to migrate; their passage money had been paid by the State, at the discretion of State officials. The State distributed these aliens at various points within its borders, where they might freely accept or reject any offers of employment made thom. Tho quostion of the State's right so to encourage immigration was immediately raised. Secretary STRAUS of the Department of Commerce and Labor ruled on Tuesday that no violation of the Immigration law or of the law prohibiting the importation of alien contract laborers had been committed. Hereafter, if this ruling stands, a State may encourage immigration by paying the necessary expenses of its immigrants.

December 20, 1906

18,000 VOYAGERS ARRIVE.

Incoming Steamships Thronged—The Immigration Bureau Taxed.

Such a flood of arriving passengers has poured into the Port of New York in the last twenty-four hours that not only have the immigration officials been taxed to keep up with the work on hand, but the rather limited supply of Customs Inspectors has had all it could do to inspect the property brought in by passengers. The arrivals numbered 1,896 cabin and 16,053 steerage passengers.

The Kaiser Wilhelm der Grosse and the Grosser Kurfuerst o' the North German Lloyd Line; the Hamburg-American liner Hamburg and the Holland-American steamer Noodram arrived at almost the same time at their piers in Hoboken. The big Kaiser brought 427 cabin and 700 steerage; the Zeeland, from Antwerp, landed 229 cabin and 1,343 steerage; Roma, from Marseilles, 9 cabin and 1,406 steerage; Italia, Naples, 2 cabin and 1,689 steerage; Livonia, Russian ports, 4 cabin and 540 steerage; Indiana, Italian ports, 13 cabin, 1,501 steerage; Cretic, Naples, 260 cabin and 2,112 steerage; Mesaba, London, 34 cabin; Campania, Naples, 44 cabin, 2,032 steerage; Noordam, Rotterdam, 219 cabin, 1,810 steerage; Grosser Kurfuerst, Bremen, 285 cabin, 1,850 steerage; Hamburg, Naples, 130 cabin and 1,067 steerage.

It will be at least two days before the last of these arriving will have passed through the immigration station on Ellis Island, where they can handle 5,000 a day. In the meantime the steerage passengers must remain on the steamers and await their turn. Not only is there a shortage of customs Inspectors to do the work on the steamship piers, but Deputy Collector J. Castree Williams declared that instead of twelve Appraisers for the work there should be at least twenty.

Collector Williams has also increased the number of his deputies who go down the bay to take passenger declarations.

A record was made in the number of arrivals from Cuba in one day. The Havana of the Ward Line brought 200 cabin passengers, the largest number arriving on a single ship from that island.

March 28, 1907

GREEK CONSUL HELD.

Charged with Conspiracy in Importing Greek Boys.

Special to The New York Times.

BOSTON, Mass., April 8.—The Federal authorities were elated last night when Deputy Inspector Ruhl went to Lowell and returned with Michael Iathos, Greek Consul and United States interpreter, under arrest, charged with conspiracy against the United States Government. The charge is based on allegations that Iathos has for three years been a party to the importation of boys aged from 12 to 14 years, declaring that some Greek living in Lowell was their father.

The Government contends that the children's parents live in Greece, and that the children really weer imported for child labor purposes, many having been employed in the big cotton mills at Lowell and other cities.

Iathos was arraigned before the United States Commissioner and held in $2,000 for a hearing to-morrow morning. The Federal authorities profess to believe that they are on the verge of disclosing a gigantic plot to bring children here.

Iathos has been connected with various Greek publications, and is one of the leaders of his race in this country.

April 9, 1907

POWDERLY TELLS LABOR NEED

Immigration Bureau Can Provide Work for 256,400 Persons.

WASHINGTON, D. C., Sept. 18.—Terence D. Powderly, Chief of the Division of Information of the Bureau of Immigration, which was established two months ago, has made to Secretary Straus of the Department of Commerce and Labor an important report showing the nature and extent of the work of the division since its establishment. Efforts are being made by Mr. Powderly to obtain throughout the country exact information as to the labor needs of the various sections.

Precise and accurate information is desired in order that admitted aliens may be given practical information as to where and by whom labor is needed, the classes required, wages paid, house rents, transportation, and school facilities.

Already the division has information certifying that places can be provided for 256,400 men, women, and children at wages ranging from $3 per week to $3.50 per day. From individual employers of labor specific information has been received which will enable the division to place immediately 1,395 aliens at wages ranging from $1.25 to $3 per day. From Commissioners of Labor and State Boards of Agriculture have been received reports that 84,100 aliens can be employed at wages ranging from $18 a month to $3 per day. From the Commissioners of Agriculture of three States comes the information that an aggregate of 1,020,600 settlers on lands are needed in three States.

September 19, 1907

COUNTRY'S GATES SWING OUTWARD

And Out Go the Foreign Toilers in Thousands, Carrying Their Hoards with Them.

25,000 EVERY WEEK NOW

Special Trains Bring Them—Hundreds Left Behind by Every Ship—Fill Hoboken Park at Night.

The rush of foreign laborers back to their homes in Europe, which began in such noticeable volume about three weeks ago, has reached such proportions that the steamship companies at this and other ports are in serious trouble to provide accommodations for the homegoing. Hundreds who wanted to sail yesterday were unable to find berths, and many who wished to sail to-morrow will have to wait for next Wednesday's steamers.

President Harrison, in the course of an oratorical flight years ago, in describing the wonderful resources of this country and their capacity to accommodate all the toilers of Europe, said that the gates of Castle Garden always opened inward, never outward. Now, however, the gates are certainly swinging outward to their widest extent to accommodate the great stream of humanity returning eastward.

This week all the big transatlantic lines raised the steerage rate from $21 to $33, but this has not had a particle of effect on the demand for tickets. In fact, the demand has increased. A representative of the steerage department of the Hamburg-American Line told a TIMES reporter yesterday that 25,000 foreigners in this country were now leaving this port alone every week for Europe, and that there was not a sign anywhere of a let up in the tide of traffic. He estimated that $5,000,000 is being taken abroad weekly by these returning immigrants in cash or drafts on foreign countries.

Special Trains Bring Them In.

All the large railroad trunk lines in the East are now using special trains to bring the homegoers to port, some of the trains being ten and twelve cars long. Boston is sending here the overflow the steamers cannot accommodate. Outgoing ships, which sailed yesterday, took some of this overflow, although hundreds were left behind.

The Fabre Line found it necessary to provide accommodations in various parts of the city for 200 steerage passengers who wanted to sail on the Madonna, which left for Mediterranean ports on Wednesday with 1,480 in the steerage. Five hundred more stayed here, but 800 got aboard other ships. The 200 will have to remain over for the Venezia, which sails next Wednesday, with 1,890 steerage passengers already booked. The Arabic sailed yesterday with 1,100. The steamship Main of the North German Lloyd sailed yesterday, loaded to her full capacity. The steamer Batavia took 2,500, and the Pretoria, sailing for Hamburg to-morrow, will also carry 2,500.

The Red Star Line had an overflow of 400 steerage, and this overflow was taken by the steamship Main. The Oceanic went on Wednesday with 1,400 Italians, Servians, and Lithuanians, many of whom were left over by the Canopic, which sailed on Wednesday from Boston, the Philadelphia, and the North German Lloyd steamer Princess Irene.

3,200 on One Steamer.

The largest number to go on one steamer, so far, will be 3,200 on the President Grant, sailing on Wednesday next for Hamburg. Her steerage passengers will be composed mostly of Russians, Poles, and Hungarians. The St. Paul of the American Line will sail to-morrow with 900, her full capacity. Her steerage passengers will include 350 who have been waiting for a week to get accommodations. The Caronia and the Kaiser Wilhelm II. will take 700 and 1,000 steerage passengers respectively. On Wednesday next the Cambronian will carry out 1,000; 1,850 more are booked to sail on the Zeeland of the Red Star Line. The Republic will carry 2,300. She stopped booking three days ago.

Two trainloads of emigrants came in over the Erie yesterday from Western points. Eight hundred foreigners came from Sharon, Penn., where they were booked by one steamship agent. The Pennsylvania and Lehigh Valley also brought large trainloads. Three trainloads came in over the Baltimore & Ohio. One immigrant train over the New Jersey Central at midnight yesterday was made up of twelve cars. The passengers numbered 980. They were lined up ten deep for a block long in front of the Jersey Central Ferryhouse at the foot of Liberty Street. The crowd was so great that the ferry officials stretched a rope between the posts of the ferry entrance to keep a passage clear.

Most of these men, it was said, were either single or had left their wives at home in Europe. There were not half a dozen women in the whole multitude. Most of these people came from the districts in Pennsylvania where twenty large stone and cement works have shut down.

Where the Thousands Come From.

The representative of the Steerage Department of the Hamburg-American Line estimated yesterday that the steerage traffic abroad is fully 60 per cent. heavier this year than last. He attributed the increase wholly to the shutting down of industries caused by the financial disturbances. The traffic is noticeably heavy from the Pittsburg district and territory adjacent. The men come from the steel mills, coal and coke industries, railroads, mines, and large mills. The Hamburg-American Line's spokesman said it was estimated that each man had from $150 to $200 to take home. A great deal of money had, of course, been sent ahead already, he said, in the form of drafts to the men's families, so that the $150 or $200 by no means represented the total savings of each person. These men were Italians, Hungarians, Poles, Syrians, Russians, and even a few Turks.

The greatest congestion of returning immigrants is seen in Hoboken, where trains from the West land hundreds every few hours. Two train loads arrived in Hoboken yesterday afternoon and added to the hundreds who are awaiting passage to the other side on either the North German Lloyd or Hamburg-American liners. At least 300 men slept in Hudson Square Park, Hoboken, on Wednesday night until the cold drove many to exercising. Then they walked the streets until daybreak. Some of them got aboard outgoing ships yesterday, but most of them will not sail until to-morrow. They were a sorry-looking lot yesterday, having lost much sleep since they left the points at which they had been working.

Steamship agents united yesterday in expressing their belief that the outgo of foreigners was on the increase, and that it will continue until the end of the year.

November 22, 1907

WOMEN IN STEERAGE GROSSLY ILL USED

Government Inspectors Declare Men Passengers and Crew Invade Their Quarters.

NONE SAFE FROM INSULT

Report Which Has Reached Senate Asks Legislation to Improve Conditions Called Appalling.

WASHINGTON, Dec. 13.—A report on steerage conditions, based on information obtained by special agents of the Immigration Commission traveling as steerage passengers on different transatlantic steamers, was made public to-day through presentation to the Senate with recommendations for legislation to better conditions. Conditions found on many of these vessels are described as appalling, in spite of the fact that in some instances the letter of the law was obeyed implicitly.

The general report of the commission contains the reports of individual agents giving their experiences on board steamships where they posed as steerage passengers. Summing up one such trip, a woman agent of the Immigration Commission, who was herself miserably insulted, said:

"During these twelve days in the steerage I lived in a disorder and in surroundings that offended every sense. Only the fresh breeze from the sea overcame the sickening odors. The vile language of the men, the screams of the women defending themselves, the crying of children, wretched because of their surroundings, and practically every sound that reacheds the ears irritated beyond endurance. There was no sight before which the eye did not prefer to close.

"Everything was dirty, sticky, and disagreeable to the touch. Every impression was offensive. Worse than this was the general air of immorality. For fifteen hours each day I witnessed all around me this improper, indecent, and forced mingling of men and women who were total strangers, and often did not understand one word of the same language. People cannot live in such surroundings and not be influenced."

Agents of the Immigration Commission say that on many of the steamships men stewards and members of the crew as well as male steerage passengers crowd into the compartments set aside for women, and constantly pass through the passageways of such compartments, so that no woman in the steerage "has a moment's privacy."

The women agents of the commission say that the women's compartments in which they were quartered had but one entrance and exit, so that there could be no good excuse for the constant appearance of the men. It is stated that during the hour preceding the breakfast bell, while the women are rising and dressing, several men usually passed through and returned for no ostensible reason.

"If a woman were dressing," says one woman agent, "they always stopped to watch her, and frequently hit and handled her.

"One night, when I had retired very early with a severe cold, the chief steerage steward entered our compartment, but not noticing me, approached a Polish girl who was apparently the only occupant. She spoke in Polish, saying: 'My head aches; please go on and let me alone.' The girl, weakened by seasickness, defended herself as best she could, but soon was struggling to get out of the man's arms. Just then other passengers entered, and he released her. Such was the man who was our highest protector and court of appeal."

Describing further how women steerage passengers were compelled to submit to insults the report says of the passage just referred to that not one young woman in the steerage escaped such experiences. The writer herself was no exception, and tells of repelling advances on the part of the crew and stewards with a hard, unexpected blow in the offender's face.

Concerning other conditions in the old type steerage, which still exists on many of the steamships, the agents of the commission are just as severe. In the introduction to the report it is stated:

"The universal human needs of space, air, food, sleep, and privacy are recognized to the degree now made compulsory by law. Beyond that the persons carried are looked upon as so much freight, with mere transportation as their only due."

The sleeping quarters are described as being in many cases dirty, inadequate, and all that is bad. The average berth is six feet long and two feet wide, with only two and one-half feet of space above it, and that is all the space to which the passenger can assert a definite right. In that space he has to sleep and find room also for his baggage, all of his extra clothing, his eating utensils, his towels, and other toilet necessaries. The passageways between the berths are so narrow that none of the articles mentioned could be deposited there.

The floors, when iron, are continually damp, and when of wood they are not washed. The open deck available to the steerage is described as very limited. Much space is devoted to the lack of ventilation and to bad conditions in washrooms, which on some ships men and women steerage passengers are obliged to use in common.

Regular dining rooms are not a part of the old-type steerage. On such ships none of the men have places to sit when at their meals, and provision is not made for all of the women. The food served is described as fair in quality, but usually spoiled by being wretchedly prepared.

Good conditions are described in connection with investigations of some steamships, and it is declared that the competition was the most forceful influence that led to the development of an improved type of steerage. It is declared, however, that a division of the territory from which the several transportation lines draw their steerage passengers excludes the possibility of the use of this force for further extension of the new type of steerage to all immigrant-carrying lines. Legislation is advocated to complete what competition began.

The new statutes, which it was supposed would obviate dirt and provide ventilation for the steerage passengers, took effect Jan. 1 of the present year. The vessels fitted with the new steerage, in the opinion of agents of the Immigration Commission, fully comply with all that can be demanded under the law. On that account it is urged that a statute be immediately enacted providing for the placing of Government officials, both men and women, on vessels carrying third class and steerage passengers, the expense to be borne by the steamship companies.

Under existing conditions on some of the steamships where the old type steerage prevails, the report says, it is impossible for a woman to keep even reasonably clean. Of this condition one agent says:

"No woman with the smallest degree of modesty, and with no other conveniences than a washroom, used jointly with men, and faucet of cold salt water, can keep clean amid such surroundings for a period of twelve days and more. It was forbidden to bring water for washing purposes into the sleeping compartments, nor was there anything in which to bring it. On different occasions some of the women rose early, brought drinking water in their soup pails for washing, but were driven out when detected by a steward. No soap and no towels were supplied."

Senator Dillingham, Chairman of the Immigration Commission, introduced in the Senate to-day two bills intended to correct much of the evil of which complaint is made.

December 14, 1909

RUSSIAN JEWS BARRED OUT.

Contract Labor Law to Apply to Those Induced to Immigrate.

Special to The New York Times.

WASHINGTON, July 15.—The Department of Commerce and Labor announced to-day that Russian Jewish immigrants coming to this country in response to promises made by agents of American Jewish aid societies would be barred under the contract labor laws. It has been represented to the department that the aid organizations, which were originally designed to divert Jewish immigrants to the sparsely settled parts of the country and away from the Eastern cities, have developed into labor bureaus and conduct extensive advertising campaigns in Russia.

The decision was made in connection with the application of 280 Russian Jews to be admitted to the United States at Galveston. All but 34 were admitted, these being barred on the ground that they were destitute and would become public charges. It was announced, however, that the contract labor law would hereafter be applied to all such cases.

July 16, 1910

ALL ROB THE ALIEN, SAYS LABOR BUREAU

Private Bankers, Lawyers, and Real Estate Agents Accused in Legislative Report.

CONDEMNS PADRONE SYSTEM

Department Recommends a Camp School Bureau to Deal with Aliens' Educational Needs.

ALBANY, March 6.—The first annual report of the Bureau of Industries and Immigration of the State Department of Labor, which was presented to the Legislature this morning, outlines the State's policy with reference to admitted aliens, and deals with the transportation of aliens across the City of New York, labor and living conditions in labor camps and colonies throughout the State, industrial calamities and personal injuries borne by alien workmen, frauds and exploitation committed on aliens by private bankers, notaries public, lawyers, collection agents, insurance companies, real estate agents, employment agents, benevolent societies, and naturalization clubs.

The bureau finds that in the matter of transportation the combination of steamship agents, emigrant hotels, runners, porters, expressmen, and cabmen throughout the country, operating chiefly through New York City, forms one of the most stupendous systems for fleecing the alien, from the time he leaves his home country until he reaches his destination in America, and vice versa.

The report says:

"In labor camps aliens are discriminated against in regard to housing, sanitation, food supplies, and employment methods, being denied the ordinary decencies of life; aliens are checked and tagged, amounts ordered by the padroni are deducted from their wages without their knowledge or express sanction, and exploitations occur in hospital charges and the purchase of supplies.

"The private banking laws are affording only a small measure of protection owing to evasions of the law, and no protection whatever outside of cities of the first class; frauds in the sale of homes to aliens by means of the solving of puzzles or by means of excursions arranged to interest aliens in 'show' pieces of property, or by other means, are widespread, and the settlement of affairs in the old country, when an alien wishes to settle here, is in the hands of a most unscrupulous class of lawyers, notaries public, collection agents, information bureaus, and protective leagues."

Among other recommendations made in the report are the following relating to State laws:

That free employment bureaus be established in the various industrial centres in the State for the distribution of unskilled workers from congested centres.

That there be State supervision of both pay and free employment agencies.

That the one-and-a-quarter-cent immigrant fare law be repealed, as all such matters are now under the Public Service Commission and Inter-State Commerce Commission.

That legislation closing shoe-shining parlors one day or one-half day each week be enacted.

That the penal law be amended so as to prohibit use of puzzles and similar devices in advertising real estate, and prohibit the giving of clear titles when a purchase money mortgage exists.

That an investigation be made by the State of the insurance department of benevolent and fraternal societies, especially those doing an inter-State business.

That the lottery law be amended so as to cover raffles.

That there be created a Camp School Bureau or Commission to deal with the educational needs of immigrant colonies and camps.

That aliens marrying in this country be required to furnish satisfactory proof that there is no wife living in the home country.

Recommendations relating to the Federal laws and business are as follows:

That the Federal Government regulate and license labor bureaus which furnish labor to persons and corporations doing an Inter-State business.

That the Federal Steamboat Inspection Service be extended so as to include measures for the comfort and welfare and protection of passengers on board coastwise vessels.

That the discriminations in immigrant rates between steerage and second cabin west-bound passengers be removed.

That there be an investigation of the Naturalization Service, including a consideration of the restriction of naturalization to Federal courts and the appointment of Judges for purpose of naturalization who will serve throughout the State.

That the transit lines, rail and steamship, establish a transfer company for the safe transit of alien through passengers from rail to steamship lines, and vice versa, across the City of New York.

That the transportation lines and shippers abolish the present method of obtaining cattle attendants and coal passers and engage their employes directly through their own agencies.

It is urged that contractors and employers of unskilled labor, especially tranportation lines, abolish the padrone system, and that temporary quarters on all public contracts be erected and rented by the employer directly.

March 7, 1912

RECORD IMMIGRATION YEAR.

Balkan and Italo-Turkish Wars Help to Swell Total to 1,355,000.

Special to The New York Times.

WASHINGTON, July 16.—The highest record for immigration into the United States, which was made in 1907, has been broken by the fiscal year of 1914.

Up to June 1 last 1,254,548 foreigners had entered during the eleven months of the fiscal year to take up their residence here. This is close to the highest previous record, which was 1,285,349 in the year ended June 30, 1907. Reports received by Commissioner Caminetti since July 1 indicate that the total for the year will approximate at least 1,355,000.

Commissioner Caminetti finds that there has been a marked increase in the number of persons, principally men, coming in from every country taking part in the Balkan war— Turkey, Greece, Bulgaria, Servia, and Montenegro. However, there were a greater number of Italians than of any other nationality, many soldiers in the war against Turkey having returned to their homes at the close of the conflict to find they were without occupations and that conditions were generally impoverished. There were 312,818 Italians entered during the year. Admission was denied to a total of 30,000 immigrants. The percentage of persons with diseases from Italy and the Balkan States was found to be much higher than ever before. Of those denied admittance, 1,915 were Mexicans. A total of 8,886 Mexicans were admitted up to June 1.

July 17, 1914

TO AID WOMEN IMMIGRANTS.

New Rules for Better Care if Detained or Deported.

Special to The New York Times.

WASHINGTON, April 7.—Anthony Caminetti, the Commissioner General of Immigration, issued today new rules for the care of women and girl immigrants on their arrival at immigration stations. The purpose of the new regulations is to prevent the operation of white slave agents, who have been detected from time to time at New York and elsewhere. Hereafter immigration officials are to co-operate with charitable organizations here and in other countries in the effort to shield and assist female immigrants so that they may find good homes and opportunities for making an honest living.

No woman or girl is to be incarcerated in a jail or detention house if such a course can be avoided. Where it may be required to hold a woman or girl in custody of immigration officials she is to be placed in the care of some proper institution or organization which has facilities for detaining her under comfortable conditions. Philanthropic organizations of the same nationality as that of the immigrant are to be preferred. At all stations the custody of female immigrants is to be given to female immigrant officials.

Immigration officials have been instructed to co-operate with charitable societies and institutions abroad for the care of deported women, and to avoid giving such cases any undue publicity. Arrangements for this purpose have been made under the white slave traffic international agreement. As far as practicable, societies that undertake to help fallen women to reform will be called on in all such cases, every facility will be offered to shield the woman from publicity, get her employment, and look after her generally. Women who are deported will not be released to go their own way and possibly to sink deeper in immorality, but will be detained until philanthropic organizations can be communicated with and called in to undertake their work of helpfulness.

April 8, 1915

WAR CUTS IMMIGRATION.

Arrivals in Last Eight Months Fewer Than for Many Years.

Special to The New York Times.

WASHINGTON, April 8.—The effect of the war on immigration is indicated by the figures given out at the Bureau of Immigration for the fiscal year beginning July 1, 1914, and covering the eight months to Feb. 1. The total number of immigrants from all parts of the world, but chiefly from European countries since the war began, which is practically for the seven months beginning Aug. 1 last year and including February, is 173,861. For the period August, 1913, to February, 1914, the immigration was 688,206.

The total immigration this fiscal year is 234,238. Ths is the number classed as immigrant aliens. In addition there were admitted 81,044 non-immigrant aliens, making the total number of aliens admitted 315,282. Against this total must be placed the number of departing immigrant aliens, which, during the eight months, was 312,181, leaving as the net number of immigrants for the first eight months of the present fiscal year the insignificant grand total of 3,101, the smallest ever known in the history of the country.

An idea of the falling off of the volume of immigration as shown in the great number admitted, appears from a comparison of the returns for the ten years' period from 1905. The totals by years are as follows:

1905, 1,026,499; 1906, 1,100,735; 1907, 1,285,349; 1908, 782,870; 1909, 751,786; 1910, 1,041,570; 1911, 878,587; 1912, 838,172; 1913, 1,197,892; 1914, 1,218,480.

If the immigration for the four remaining months this year were to be the same as for the same four months last year, when the total was 392,030, the total for the complete fiscal year would be 626,268.

While immigration has been greatly reduced because of the war, the tide of non-immigrant aliens returning to Europe for military service has been very great, thus making the net addition to population from immigration very small. There is every indication that for the remaining four months of the fiscal year it will continue to be small, making the entire year phenomenally low in immigration increase of population.

The statistics by countries show some interesting facts. Just as many Englishmen are coming here now as before the war. The number for February was 2,137, and, for the eight months thus far this year 27,376. The annual number of English coming here runs from 40,000 to 50,000. In the eight months this year 8,674 French have come over, which is about the normal number. There have been 16,423 Germans coming into the country in the period named, which is much below the usual immigration from that country, which runs from 30,000 to 35,000.

The greatest number of immigrants coming from any country comes from Italy, which has sent over 38,087.

April 9, 1915

$7 A WEEK SERVANT COMING.

Influx of Foreign Girls Promises Relief to Housewife.

CHICAGO, Jan. 3.—The day of the $7 a week servant girl, who would cook, sweep, mind the baby, wash dishes, run the laundry, and do odd jobs of calcimining in her spare time is coming again, according to Miss Elizabeth Moynihan of the Travelers' Aid Society.

Every boat from Europe is bringing hundreds of Scandinavian, Irish, English, and Italian girls eager to do housework, Miss Moynihan says. The Travelers' Aid Society is assisting scores on their way to the West from New York.

"I expect that in three or four months," one employment agency head said, "we will have almost the old conditions back—girls willing to work for $7 or $8 a week instead of 'highty-tighty' dusters, willing to assist in housework for $15 a week."

January 4, 1920

NEW IMMIGRANT TYPE EXCELS PREDECESSOR

Ellis Island Finds Proof in Decreasing Numbers of Undesirable Arrivals.

Ellis Island admitted into this country 10,527 aliens during the week that ended with the close of the workday yesterday. Those who watch the flow of immigration into this port have agreed that a better class of immigrants are coming than in the days prior to the war. A good-looking colleen who landed yesterday from the White Star liner Baltic expressed the same thought when she said to an inspector: "It used to be true that most Irish girls who came to the United States came to be domestic servants. We are all stenographers now."

Another standard by which the quality of immigration can be judged is by the number of undesirable arrivals who are promptly deported. Deportations for a month used often to be more than 3 per cent. of the arrivals and were seldom below 2 per cent. Since January of this year deportations have not gone beyond two-thirds of 1 per cent. of the arrivals in any single month. Latest figures showed that during January, February, March, April and May the arrivals at Ellis Island numbered 152,987, while the total deportations were 885, or .58 of 1 per cent.

The right of many more arrivals to land was questioned by the local immigration authorities and they appealed to Washington, where a majority of such cases was admitted. This was strikingly illustrated in figures which show that out of about 500 detained as radicals only 23 have been ordered deported. The immigration figures do not show the numbers who were turned back at their home port.

The largest detachment of immigrants to land yesterday came on the White Star liner Baltic from Liverpool. There were 1,335 in this band, of whom a majority were young women. The over-topping number of women on the Baltic was counterbalanced by the arrival here during the week of nearly 2,000 reservists, of whom nearly 1,500 were Poles who left our mining districts two years ago and went to France to fight. They were inspected and sent to Camp Dix where they were discharged.

Forty-five undesirables recently brought here from the Far West for deportation were placed upon outgoing vessels yesterday and sent back to their native countries. About one-half of them were enemy aliens and radicals. Others were defectives or persons with criminal records. Forty more, the officials said, will be deported next week.

June 20, 1920

"PICTURE BRIDE" JILTS HIM.

Girl Chooses Deportation Rather Than Be Married to Miner.

Protesting that she wouldn't marry a coal miner and that she didn't like his looks anyway, Ruzica Romoivic, a "picture bride" who came to this country from Croatia to marry, refused to go through with the ceremony when she saw her prospective husband for the first time in the marriage license bureau in the Municipal Building yesterday. Rade Vuletich, a coal miner of Masontown, Pa., was the jilted bridegroom.

The young woman told the agent of the Travelers' Aid Society, in whose custody she was, that she was educated and was used to living in genteel and refined surroundings.

When told she would be deported unless she married Vuletich she consented to the wedding. Then she thought again and withdrew her consent.

The girl was taken back to Ellis Island. She will be deported unless she can get some friends or relatives to help her get in.

May 4, 1921

IMMIGRANT MOTHER LOST.

Leaves Ellis Island for Cincinnati and Disappears.

Special to The New York Times.

CINCINNATI, Oct. 18.—Aid of the Police Departments of Cincinnati and New York has been enlisted by Morris and Isidore Dunsky, merchants of Cincinnati, to find their mother, Mrs. Taube Dunsky, 62 years old, who left Ellis Island and New York last Friday, presumably on a Big Four train, for Cincinnati. She was expected to arrive here Saturday, but no trace of her has been found by her sons.

Morris Dunsky came to America from Lithuania and settled in Cincinnati seventeen years ago, and the following year brought his brother here. His father disappeared during the World War and has never since been heard from.

Several months ago Morris sent $200 to his mother to come to Cincinnati. He received a telegram from her Sunday before last, stating that she had arrived at Ellis Island, but was being detained until her sons should send affidavits that they were able and willing to care for her. The sons wired at once to the Hebrew Immigrant Aid Society in New York that they would send the affidavits and did so.

Hearing nothing definite by the following Friday morning, Morris called up Ellis Island and the Hebrew Immigrant Society, but could get no information, and his brother, Isidore, left Friday for New York. At 6:30 that evening Morris received a telegram ffrom the Immigrant society stating htta his mother had been released from Ellis Island and had taken a Big Four train for Cincinnati.

The woman cannot speak English, and it is feared that she may have taken a wrong train or got off the train at some place where she cannot be understood.

October 19, 1921

IRISH IMMIGRANT FAST DISAPPEARING

Effect of Establishing Irish Free State Already Noticeable, Says Congressman Siegel.

FEW EXPECTED TO COME

Picturesque and Important Factor in American Life Believed to Be a Thing of the Past.

The Irish immigrant, who has been a picturesque and important factor in American life since Colonial days has already become practically a thing of the past, although the Irish Free State is scarcely a month old, according to Congressman Isaac Siegel, a member of the House Committee on Immigration.

"The effect of the establishment of the Irish Free State is already noticeable," he said. "Irish immigration has already fallen off to practically nothing.

"There will probably be much going to and fro between this country and Ireland, but if the Irish Free State has the success that is expected, Irish immigration is probably at an end."

Irish immigration has been low at all times since 1914 and was practically cut off for considerable periods during the period of war when England permitted no able-bodied man to leave the island and after the war during the worst of the Irish troubles. One indication of the falling off of Irish immigration has been in the diminishing proportion of Irish names among appointees to the Police Department and the variegated cosmopolitanism of the lists of new candidates.

There have been a great number of inquiries for passage to Ireland during the last month, but the sailings have been comparatively few, according to an official of the White Star Line.

"The creation of the Irish Free State came so suddenly," he said, "that most of those who were anxious to return to Ireland have not had time to wind up their affairs and get started. Many do not care for transatlantic voyages in January and February. There are several big ships which now stop at Queenstown, but the travel has not been extremely heavy so far.

"There are reports that new lines are to be started for passenger ships to run directly between this country and Ireland without other stops, and there are preparations for a great movement to Ireland this Spring and Summer, but the matter is somewhat problematical. Many hope to obtain cheap passage to Ireland, believing that there will be low rates later, but this is doubtful, in view of the high cost of operating ships at present.

"There will probably be less and constantly diminishing Irish immigration, but for years many Irish will be coming over to join relatives in this country."

February 12, 1922

ADAPTING TO A NEW LIFE

Destitute Italians.

To the Editor of the New-York Times.

The members of the Italian School Young Men's Association, lately formed in connection with the Italian school, under charge of the Children's Aid Society, propose to procure employment for their poor countrymen, who, however deserving, willing and able to work, either as common laborers or mechanics, have no means, unaided, of obtaining such employment. Contractors, manufacturers, &c., needing a large number of men, can be supplied on short notice. Employers will please address Italian School Young Men's Association, No. 46 Franklin street, basement.

By kindly inserting this notice, you will greatly contribute to the success of the undertaking, and much oblige,

Respectfully yours,
A. E. CERQUA,
Superintendent Italian School.

NEW-YORK, Sunday, Dec. 8, 1872.

December 12, 1872

IRISH IMMIGRANTS IN CITIES.

If the Convention of Irishmen which has been in session at St. Louis for some days past had confined itself to a consideration of the best means for improving the condition of Irish immigrants, instead of rambling into the discussion of obnoxious subjects, the most feasible of the plans suggested would stand a very much better chance of being carried out. As it is, however, we fear that the main question is likely to be forgotten in the confusion of resolutions and fustian speeches. The overcrowding in the tenement-houses of the large cities has resulted, as is well known, from the inability of Irish immigrants, through poverty, to go westward immediately after landing. It would be perfectly safe to say that more than two-thirds of all those who have come here from Ireland during the past thirty years were born either on farms or in rural villages. Previous to emigration most of them had probably never been half a dozen times to a city. If, therefore, these people had had an opportunity to decide between city and country life, very few of them would have declared in favor of the former. In the absence of any benevolent scheme for furnishing them transportation, they were compelled to accept the only alternative. The process of settlement has always been rapid. A young Irishman no sooner touches the point of destination, and procures employment, than he begins to think of the folk at home. The first money earned by him is sent to pay the passage of a brother or sister. Next to his place of birth, he loves most the scene of his first struggles for existence. Thus it is that he is loth to leave it, no matter how fair the promises for pecuniary gain may be elsewhere. When his wife and children ultimately join him, and he is compelled to exchange the cheerful lodging for the tumble-down shanty or dingy tenement, he cannot go westward, no matter how eagerly he may desire to do so. There are hundreds of Irish laborers living in this City to-day who have had just such an experience, and who would, if they could see the way clearly, gladly agree to remove to farming districts. It is almost impossible to provide them all with work during the Summer months, much less to do it in Winter, and, as a consequence, the public charitable institutions are kept full, and the benevolent societies have more than they can do to provide for the necessities of those who do their utmost to exist from one year's end to another without breaking up their wretched homes and becoming paupers.

The Irish Convention at St. Louis began its labors with a most encouraging discussion of how best to put a stop to overcrowding and its attendant evils in large cities. It was proposed to establish an independent immigrant society, with headquarters in New-York, and branches judiciously distributed throughout the country, for the purpose of receiving the poorer immigrants, and sending them in charge of its officers to such States as shall offer the greatest inducements for settlers. It was also proposed to provide the means for building temporary houses for the protégés of the society, and, in short, put them in the way to become prosperous farmers. To accomplish a work of this magnitude would, of course, require a liberal expenditure. The society, according to the plan for its establishment, was to procure funds by popular subscription, the Irish element being expected to contribute generously. Now, we would have no doubt whatever as to the success of a scheme of this kind if it were started by any other nationality. The Germans would carry it to a success, for they would be sure to organize it upon a basis that would make it an object of interest to all classes. The Irish Convention would confine the benefits of the project to Roman Catholics. It has not said this in so many words, but it has led us to believe as much from the character of the resolutions and the debates which followed the introduction of the question. Under the circumstances, it could scarcely hope to gain the sympathy and contributions of Protestants, however much the latter should wish to see an end put to the overcrowding in large cities. It is to be regretted that the Convention could not have considered the scheme from a purely benevolent standpoint, without jumbling it up with the "Sovereign Pontiff," "Roman Catholic persecution in Germany," "The Bible in the public schools," and other topics with which it could have no possible connection. The idea in itself is an excellent one, and if carried out in a proper spirit—as we hope to see it some day—would be attended with benefit not only to the Irish immigrant, but to the people of the whole country.

October 20, 1873

THE COMING ITALIAN.

There was a time when people were inclined to believe that China would furnish us the coming man; when people felt convinced that the chivalrous Californians might so far forget their playful animosity as to at least neglect their favorite amusement of stoning innocent Chinamen to death, and allow the flood of tawny immigration to pour into the country through the Golden Gate. But we have latterly been convinced that the coming man will, perhaps, after all, not be a Celestial, but an Italian. There are many reasons for congratulating ourselves on a general emigration from Italy to this country; many reasons why we should do what we can to encourage the hardy peasants of the sunny historic land to find new homes upon our rich soil; and one of the most potent is their special adaptability to climates in the United States toward which the Germans, the Irish, the Swedes and Norwegians do not readily turn. We have for so many years been wont to associate an Italian in America with opera singers with a princely income, or with the political exile and organ-grinder with the incomes of mendicants, that we have not sufficiently considered the really admirable qualities of the race as workers, and the benefits which would accrue from their acquisition to our cosmopolitan roll of citizenship.

The United States census of 1870 informed us that there were but a few more than seventeen thousand native Italians in this country. But the number must have greatly increased since then, for the olive features under the black slouch hat of the man, and the low-browed, Madonna-eyed face of the peasant woman of Italy are fairly ubiquitous in our cities to-day. They flock into our Northern and Southern ports, and their neatly-arranged fruit-stalls make pretty spots of color against the dull brick or stone of our avenue corners. They cry our newspapers, they black our boots, they clean—or are supposed to clean—our streets; they sell us flowers, they cook us food, they carry burdens, they levy tribute on our coppers, their gracious gestures and musical language give animation and picturesqueness to our crowds, and their middle classes creep into the higher branches of trade when they are blest with education. One can hardly walk abroad in the New-York streets without recognizing that they are rapidly multiplying in our midst, and that we must speedily hail them as one of the important elements to be fused in our great crucible, out of which we hope to see arise the nation of the future.

The fact is that, although Columbus discovered America some centuries ago, the rest of the Italians have only just found it out. It is but natural to expect, as they have finally become convinced that it is a goodly land, and one eminently fit to emigrate to and prosper in, that great numbers of them will come hither, and gravitate to the climates most resembling those which they have left behind them. The cities are by no means the best places for them; they must be encouraged to find homes in the rich fields of the South and South-west, to plant vineyards and raise stock in the Alleghanies, to grow the orange and the cane in Florida and Louisiana, and to build up houses and fortunes for themselves in the fat lands which now lie waiting. They will be heartily welcome in the North and along the Pacific coast, but they will aid the country more and profit themselves better by turning to the southward. It is true that they have already been somewhat discouraged in their efforts to settle in the Southern States by the embarrassed condition of affairs there, and by the lack of knowledge of the proper status of the laboring man, so sadly manifest in some of the commonwealths south of MASON and DIXON's line. But there are immense inducements for them to settle in that section, and the Southern State which is enterprising enough to encourage their advent, and to receive them as they should be received, will find that there are plenty of them, and that they are industrious and well-disposed when placed where they have opportunities for enriching themselves by the labor of their hands. Let Italy send us a million men and women who can be nothing save beggars at home, and we will transform them into prosperous citizens. And let those States now so sadly in need of white immigration set to work courageously to study the coming Italian. He can be of infinite service to them, and they in turn can serve him by turning him from the overcrowded cities to the open country, where he can have room to grow rich as well as free and intelligent.

June 2, 1874

IMMIGRANTS IN THE SOUTH.

While there is by no means the great flood of immigration into the States of the South which the people of that section desire, there is yet a considerable movement which, in time, will result in modifying, and, in some respects, changing the character of the populations there. The presence of one hundred and fifty thousand Germans in Texas cannot fail, in due time, to work great influence in the South-west. This small army of hardy and frugal Teutons has entered the Lone Star State so silently, in such unceasing currents, and has scattered so widely up and down over the broad acres, that the Texans themselves were amazed, when, a few months since, the number of Germans in the State was announced. A dense stream of immigration has also been poured into Texas from the North-west, from the States of Iowa, Illinois, and Indiana. These various elements are, one and all, aiding in the upbuilding of a giant commonwealth. It is observed that the abominations of bad cooking and worse whisky fade away before the examples of the thrifty Germans, who show the native Texans what a variety of good things nature really intended them to have. Texas bids fair to overtake and, perhaps, distance Missouri in the number of immigrants received.

Virginia is constantly receiving immigrants from the Northern States. Many of them are prompted more by the cheapness of the lands in the "Old Dominion" than by any necessity of migrating. The Pennsylvanian finds that by moving into Maryland he can cheaply buy quantities of good land, slightly improved; and the Marylander is willing to sell, because, over the border, in Virginia, he finds land cheaper, as well as better, than that which he has just sold. So Maryland and Virginia are gradually changing the character of their populations. The Virginian is wont to say that the Northern immigrant would succeed better if he would make more allowances for the climate, and would not begin on a scale on which he can never finish. Virginia welcomes Northern immigration very heartily;

but is also glad to receive the hundreds of English families which have settled in some sections of the State. The arrivals from England are not so numerous as might be desired; the English generally bring goodly sums of money, and settle in fine farming or stock-raising counties, where they often accumulate large fortunes. As soon as Virginia has a sufficient tide of immigration to crowd the negro a little, and to force him to be more provident, the State's material prosperity will be fully assured, and all classes will work harmoniously together.

It is not a little singular that two of the remotest of the States south of MASON and DIXON'S line should be among the very first to secure large numbers of immigrants. Florida and Texas are each shining examples of what good advertising and proper representation of resources can do for a State. The character of Florida as a cane and orange and tropical fruit-bearing country having been carefully explained to the people of Italy, three hundred Italian families recently migrated to the Winter paradise, and will, doubtless, be followed by many more. A judicious crusade in Italy and Sicily in favor of the Floridian peninsula would probably result in the beginning of a flood of immigration which might attain the same proportions as that from Germany into Texas.

The South is beginning to give very earnest attention to this subject of an influx of population from the outer world. In some of the States the presence, or rather the omnipresence, of the negro is a bar to any very numerous entries of immigrants. But there is no Southern State where the native people feel anything save the most earnest desire to secure as many new citizens as possible by importation. Strong tides of immigration setting to South Carolina and Alabama would do much to solve the problem of the political situation there. The carpet-bag Governments never did much to encourage the coming of white Europeans into those States, because they felt that if they did so, they would be destroying the principal source of their own perverted and shameless power.

December 2, 1874

OUR SCANDINAVIAN ACCESSIONS.

THEIR NUMBER NEARLY ONE MILLION—NEARLY ALL HARD-MONEY REPUBLICANS.

From the St. Paul (Minn.) Folkebladet.

The Scandinavian population, or the population of Norwegian, Swedish, and Danish origin in the States is claimed now to amount to about 1,000,000. Unlike all other populations of foreign nationality, they nearly all live in the same part of the country, in the States and Territories north-west of the great lakes. There is a certain number living in Eastern ports, as New-York, Philadelphia, and Portland in Maine, also in some Eastern manufacturing towns, and there are a few scattered colonies in Southern States, Texas, Florida, Tennessee, Missouri. A greater number is to be found on the Pacific coast, in California, Oregon, and not least in Washington Territory—not to speak about Utah. But the greater majority lives in the North-west, including the lumbering towns and mines of Michigan near the lakes, in Wisconsin, Minnesota, Dakota, Northern Iowa, Nebraska, also in Illinois, and some in Kansas, until out in the Territories further west. In North-western Wisconsin, often one-half, in some counties the whole of the population, consists of Scandinavians. Through Southern Minnesota, most of the western part of the State, not least through the North-west, one can travel for hundreds of miles from one Scandinavian farm to another. It is said that not much less than one-third of the population in Minnesota is Scandinavian, and in Dakota even more. In several parts they own nearly all the land. Here they have their own clergy. There are more than 700 Scandinavian clergymen in the Union. They have their Representatives in the Legislatures—if we remember right, in the last session in Minnesota 19, and in Dakota 10, of 24 of the House; at present a Scandinavian is Secretary of State in Wisconsin and another State Treasurer in Minnesota; through the whole country Scandinavian county officers, usually honest and efficient men, are taking care of a large part of the public business.

The Scandinavians are the least clannish of all foreign nationalities. They learn English without difficulty, because it is so similar to their own tongue, and often not much more different from the local dialect of their home than the literary language of their Scandinavian capital. They know, which also is shown by experience, that they are most thriving in business where they are in the clearest connection with the American business life. The Scandinavian-American press has been unanimous for honest money, and the Greenbackers have never had much strength with our people. The Scandinavians want, as much as any class, a clean, honest administration. They don't appreciate the renowned civil service order of Hayes, understood as an interdict for officers to take any part in politics; but they are, as much as the Germans, in for a reform which could secure at the same time a greater capacity of the officers and fixity of tenure; they know from the old countries the importance of a good class of officers. We think, on the whole, our nationality as good reformers as any one. Except some few working men in the cities, nearly all the Scandinavians are Republicans; they have naturally been inclined to Republican politics as Protestants and as a northern race; they have early chosen the Republican Party, and they stick to their party. Two years ago, there was hardly any very strong Republican feeling among the Scandinavians; but the Democratic blunders of later years in the money question and in other matters have also here made the Republican Party stronger. Concerning the Presidential election, the majority of the Scandinavians do not have any marked personal preference.

February 23, 1880

MAINE'S SWEDISH COLONY

CELEBRATION OF ITS TENTH ANNIVERSARY.

HOW THE HARDY SETTLERS WERE BROUGHT INTO THE PRIMEVAL FORESTS OF AROOSTOOK.

The Hon. William W. Thomas, Jr., who was the agent and Commissioner of the State of Maine in bringing out the colony of Swedish immigrants for settlement in Aroostook County, 10 years ago, gave an interesting account of his labors at the recent decennial celebration of the event, and the dedication of a new Lutheran Church at New-Sweden. In it he said:

"On June 23 the colonists, who came from nearly every province of Sweden, were assembled at Gothenburg, where the Swedish festival of midsummer eve was observed, and in just 40 days after landing in Sweden, I sailed from Gothenburg with the first Swedish colonists of our State. The colony was composed of 22 men, 11 women, and 18 children—in all, 51 souls. All the men were farmers, but some were also skilled in trades and professions. There was a Pastor, a civil engineer, a blacksmith, two carpenters, a basket-maker, a wheelwright, a baker, a tailor, and a wooden-shoe maker. The women were neat and industrious, tidy housewives, and diligent workers at the spinning-wheel and loom. All were tall and stalwart, with blue eyes, light hair, and cheerful, honest countenances. There was not a physical defect or blemish among them.

"We arrived at Hull Monday evening, June 27, crossed England by rail to Liverpool, and on Saturday, July 2, sailed in the City of Antwerp, of the Inman Line, for America. On Wednesday, July 13, we landed at Halifax, where the agents of the Inman steam-ship lodged the colony in a vacant warehouse. The next day we proceeded to St. John, and on Friday, July 15, ascended the St. John River to Fredericton. Here steam navigation ceased. Two river boats were chartered for the colony, and on Saturday, July 16, we pushed forward. Each boat was towed by two horses. The progress was slow and toilsome, but the weather was fine, and the colonists caught fish from the river, and picked berries along the bank. Six days were spent in towing up from Fredericton to Tobique. The journey is now made by rail in as many hours. Near Florenceville, on Tuesday, July 19, the first misfortune befell the colony. Here died Hilma C. Clase, infant daughter of Capt. Nicholas P. Clase. Her little body was placed in a coffin, quickly constructed, and brought along with the colony. At Tobique Landing we debarked on the afternoon of July 21, and were met by the Hon. Parker P. Burleigh, Land Agent. The next morning, Friday, July 22, teams were provided by Mr. Joseph Fisher, of Fort Fairfield, and the Swedish immigrant train started for Maine and the United States. Mr. Burleigh and your historian drove ahead in a wagon. Then came a covered carriage drawn by four horses, with the women and children. Then two three-horse teams with the men, followed by two two-horse teams with the baggage.

"So we wound over the hills, and at 10 o'clock reached the iron post that marks the boundary between the dominions of the Queen of Great Britain and the United States. Beneath us lay the broad Valley of the Aroostook. The river glistened in the sun. The white houses of Fort Fairfield shone brightly among the green fields along the bank. As we crossed the line, the American flag was unfurled from the foremost carriage, and we were greeted by a salute of cannon from the village. Mr. Burleigh descended from the wagon and welcomed the colony to the State which was to be their new home. I translated the speech, and the train moved on. The people along the way greeted us with waving handkerchiefs, cheers, and every demonstration of enthusiasm. In ascending a hill the horses attached to one of the baggage-wagons became balky, backed into the ditch, and upset the wagon. The Swedes sprang lightly from their carriages, unhitched the horses, righted and reloaded the wagon, and ran it by hand to the top of the hill. This was their first act in Maine.

"At noon we reached the Town Hall at Fort Fairfield. A gun announced our arrival. Here we dismounted, and a multitude of people surrounded us. The Swedes clustered shyly apart. A public meeting was improvised in the American fashion. The Hon. Isaac Hacker was called to the chair, and introduced Judge William Small, who welcomed the strangers in a judicious and eloquent address. He was followed by the Rev. Daniel Stickney, of Presque Isle. The remarks of these gentlemen were repeated to the Swedes in their own tongue by your historian, who then returned thanks at their request. A bountiful collation was served in the Town Hall, and when, at 2 o'clock, the Swedes resumed their journey, it seemed as if half the population accompanied them in carriages and on foot. A Swedish youth of 20 and an American of about the same age ran at the head of the procession, with their arms about each other's waists, laughing and chattering, though neither understood a word of the other's language. Finally they crowned each other with garlands of green leaves.

"As we passed over a hill-top I pointed out the distant ridges of Township 15, rising against the sky, *Det lofvade Landet*, 'the promised land,' shout the Swedes, and a cheer goes along the line. Late in the afternoon we reached the bridge over the

Aroostook River. A salute of cannon announced our approach, and a concourse of 500 people turned out with a brass band to escort us to the picturesque village of Caribou. Here the Hon. John S. Arnold delivered an address of welcome, and supper was served in Arnold's Hall, where the settlers passed the night. At this supper one of the ladies of Caribou happened to wait upon our worthy Land Agent, and getting a reply from him in a language which she understood, exclaimed with delight and commendation, 'Why you speak very good English for a Swede!'

"Next morning the train was early in motion. A hundred and fifty Americans accompanied the march. We soon passed beyond the last clearings and entered the deep woods. Slowly the long line of wagons wound among the stumps of the newly-cut wood road and penetrated a forest now first opened for the abode of man. At noon, on Saturday, July 23, 1870, just four months after the passage of the act authorizing this undertaking, and four weeks after the departure of the immigrants from Sweden, the first Swedish colony of our State arrived at their new home. We called the spot New-Sweden, a name at once commemorative of the past and auspicious of the future. Here in behalf of the State of Maine I bade a welcome and God speed to these far travellers, our future citizens, and here under a camp of bark the colonists ate their first meal on this township.

"As soon as it appeared from my letters that a Swedish colony would surely come, the Board of Immigration had begun to make preparation for their reception. Under the direction of Mr. Burleigh, the township had been relotted, reducing the size of the lots from 160 to 100 acres. A road was cut into the township. Five acres were felled on each of the 25 lots, and 25 log houses would have been built, but the Swedes arrived much earlier than was anticipated, and only six houses were up and only two had glass in the windows. Necessary supplies and tools had also been provided.

"The next day was the Sabbath. The first religious service was the funeral of little Hilma Clase. The services were conducted up the Rev. James Withee, of Caribou.

"On Monday afternoon the farms were distributed. The only fair way of distribution appeared to be by lot; yet this method seemed likely to separate friends from the same province who preferred to be neighbors. This difficulty was finally overcome by dividing the settlers into groups of four friends each, and the farm into clusters of four, and letting each group draw a cluster, which was afterward distributed by lot among the members of the group. The division of the farms was thus left entirely to chance, and yet friends and neighbors were kept together. With two exceptions, every one was satisfied and these two were immediately made happy by exchanging with each other. It was determined to set the Swedes at work felling trees, cutting out roads, and building houses, allowing them $1 a day, payable in provisions and tools. Capt. N. P. Clase, a Swede, was placed in charge of the store-house. Every working party was under a foreman, who reported once a week to the storekeeper. The Swedes thus did the work which the State had intended to do for them, and were paid in the provisions which the State would have given them had they arrived later in the season. All through the Summer and Fall there was busy work in this wilderness. In nearly every instance the trees were felled on the contiguous corners of four lots, and a square chopping of 20 acres was thus made, letting the largest possible amount of air and light into each lot, and enabling the settlers to help one another in the clearing. The houses were placed in couples on opposite sides of the roads, so that every household had a near neighbor. It was too late for a crop, but it seemed best to give the Swedes ocular proof that something eatable would grow on this land. So on Tuesday, July 26, the prostrate trees on the public lot were piled; the next day they were burned, and two acres of land were sowed with English turnip-seed. The turnips were soon up, and grew luxuriantly, and in November we secured a large crop of fair size, some of them 15 inches in circumference. This was the first crop raised in New-Sweden. On July 28 we explored an old tote road running from one of the abandoned American farms in Woodland out toward Caribou. This road cut off three-quarters of a mile, saved a hard hill and a long pole-bridge, and was at once put in repair and used exclusively. The present turnpike follows this route substantially.

"On July 29 the first letters arrived from old Sweden, and the teamster brought word that a Swede was at Caribou on his way in. The next day Anders Westergren came in and joined the colony. He was a seaman, and had read an account of the colony in Bangor, and immediately decided to join us. On Sunday, July 31, Nils Olsson, the Swedish Pastor, held public religious services in the Swedish language. Tuesday, Aug. 2, the immigrants wrote a joint letter to Sweden declaring that the State of Maine had kept faith with them in every particular, that the land was fertile, the climate pleasant, the people friendly, and advising their countrymen emigrating to America to come to New-Sweden in Maine. The letter was published in all the leading journals in Sweden. On Friday, Aug. 12, the first child was born in New-Sweden to Korno, wife of Nils Persson. The youngster is alive to-day. He rejoices in the name of William Widgery Thomas Persson, and is happy in the contemplation of the constitutional fact that he is eligible to the office of President of the United States. On Friday, Aug. 19, Andrew Malmqvist arrived from Sweden via Quebec and Portland. He was a farmer and student, 22 years of age, and the first immigrant direct to us from the old country. Saturday afternoon, Aug. 21, Jœns Perssons was united in marriage to Hannah Persdotter by your historian, who luckily happened to be a Justice for the State. All the spoons at the wedding-dinner were of solid silver. Thus, within one month from the arrival of the colony, it experienced the three great events in the life of man—birth, marriage, death. The State extended a helping hand to the infant colony and guarded it with fostering care. But the State only helped those who helped themselves. The passage of the colony of 1870 cost over $4,000, every dollar of which was paid by the immigrants themselves. They also carried into New-Sweden over $3,000 in cash, and six tons of baggage. From 1870 until now, the State has never paid a dollar, directly or indirectly, for the passage of any Swede to Maine.

"In the matter of Government, New-Sweden presented an anomaly. It was an unorganized township, upon which there was not resident a single American citizen through whom the first step toward a legal organization could be taken. For two years the Commissioner found time to settle disputes between the colonists and arrange all matters of general concern. As the colony increased, the work became too great for one man, and a committee of ten was appointed to assist the Commissioner. Nine members of the committee were elected for terms of six months; the Pastor was the tenth ex officio. This decemvirate satisfactorily managed all municipal affairs, until the colonists had completed their term of residence, so as to become citizens and secure a legal organization. The Government of the United States promptly recognized the colony by establishing a Post Office at New-Sweden, and appointing N. P. Clase Postmaster. In the Fall of 1873, the settlement had outgrown the township of New-Sweden and spread over adjoining sections of Woodland, Caribou, and Perham. The little colony of 50 had increased to 600, and outside of the colony there were at least 600 more Swedes in Maine, drawn to us by our Swedish settlement. The colony was prosperous. The men had renounced their allegiance to the 'King of Sweden and Norway, the Goths and the Vandals.' Every child that talked at all could speak English. The experiment was an experiment no longer. I was able then to recommend that all special State aid should cease, and that the office which I had held should be discontinued. On the 19th of October, 1873, I laid down the work which for four years had occupied the better portion of my life and endeavor, and took leave of the people of New-Sweden. The colony continued to grow and thrive. Scandinavian immigrants were attracted to other portions of the State. The Swedish example gave a stimulus to the movement of our native population into this fertile region.

"And now, 10 years after the arrival of that little company of 50 in the heart of the forest, we meet in this Christian church. Around us lie pleasant fields, where the tall grain waves in the Summer wind. Sleek cattle and heavy-fleeced sheep graze in the pastures. Great clearings, dotted with cottages, open far into the woods on every hand. The Swedish colony to-day numbers 777 souls. These Swedes have cleared 4,406 acres of forest. They raised last year 971 tons of hay, 1,304 bushels of wheat, 5,287 bushels of rye, 1,605 bushels of buckwheat, 8,129 bushels of oats, 24,162 bushels of potatoes. They own 166 horses and 661 cattle, besides sheep and swine. In 1879 they made 2,000 pounds of cheese and 13,869 pounds of butter. The value of their farms, live stock, and farming implements and machinery is estimated at $120,000, and the value of their farm products last year at $24,000, where not a dollar was produced 10 years ago. The settlement numbers 163 dwellings and 151 barns. Besides the Capitol, there is a church and five school-houses. Eleven miles of road have been turnpiked, and 31½ miles have been grubbed, and are in a passable condition, with the swamps corduroyed. From the founding of the colony to January 1, 1880, there had been 65 deaths and 216 births in this community. The future of New-Sweden is assured. It will thrive and grow, and push out into the forest. It will continue to attract Scandinavian immigrants to Maine, and will supply a superior and needed class of labor to the older sections of the State. We have no better citizens than those countrymen of John Eriksson, the descendants of the Vikings, and the soldiers of Gustavus Adolphus."

July 27, 1880

AID FOR HEBREW IMMIGRANTS.

FORMING A SOCIETY TO ASSIST JEWISH REFUGEES FROM RUSSIA.

There was a large meeting of prominent Israelites yesterday in the Hebrew Orphan Asylum, at Seventy-seventh-street and Third-avenue. It was called in pursuance of a resolution passed on the 20th inst. at the same place. It was set forth in the call that the immigration of Russian Hebrew refugees to this country has assumed proportions which the means and appliances in the hands of the Russian Relief Fund Committee as a voluntary organization are inadequate to meet. It was resolved that at the meeting of yesterday those in attendance should take into consideration the exigencies of the case and the advisability of organizing a charitable corporation under the laws of the State, under the name of the Hebrew Emigrant Aid Society, for the purpose of aiding and advising Hebrew immigrants of every nationality in obtaining homes and employment, and otherwise providing means to prevent them from becoming burdens on the charity of the community. It was added in the call that it may safely be calculated that during the coming Winter and Spring several thousands of these refugees will arrive under the auspices of the Alliance Israélite Universelle of Paris. Among the gentlemen present yesterday were Charles L. Bernheim, ex-Judge Myer S. Isaacs, Isaac Isaacs, Jesse Seligman, Jacob Schiff, Julius Bien, ex-Judge Joachimsen, Commissioner Jacob Hess, Alderman-elect Ferdinand Levy, ex-Alderman William Bennett, Henry S. Allen, Lazarus Rosenfeldt, Henry Newman, Myer Stern, Simon A. Wolf, of the *Jewish Messenger;* B. L. Boas, Charles Minzesheimer, Fredrick Nathan, Julius Nathan, Nathan Menken, and the Rev. Dr. Mendes. Mr. Bien read an account of the sufferings of the Hebrews in Russia, and of the prospect of a large immigration to this country. He briefly mapped out a plan of the proposed society. It was announced by one of the gentlemen present that Baron Maurice de Hirsch had signified his intention to place 1,000,000f. at the disposal of the Hebrew Emigrant Aid Society of this country in order to assist in establishing Israelite colonies here, where the young men may engage in agricultural pursuits.

Addresses were delivered for and against the proposed incorporation of the society, it being urged by those who disapproved of the plan that the committee now having the matter of immigration under their care were fully able to cope with it, and that the Jewish people of this City and the United States would always be found ready to furnish the money required by the committee. It was urged by Mr. Schiff that the immigration was not such as was desired, and he asked whether any one had ever heard of any other race or nationality starting an emigration aid society. Many in the assemblage shouted, "I have," and a few freely expressed their disapproval of Mr. Schiff's remarks. The question was at length put to a vote, and it was almost unanimously agreed that a society should be incorporated. A special committee then prepared a long list of names of gentlemen to whom the matter of organization is to be intrusted. Among those selected were Charles L. Bernheim, who presided at the meeting; Commissioner Hess, ex-Judge Isaacs, Alderman Levy, ex-Alderman Bennett, Jacob Seligman, ex-Congressman Einstein, Fredrick Nathan, ex-Judge Joachimsen, Lazarus Rosenfeldt, and Moritz Ellinger. The meeting then adjourned.

November 28, 1881

THE POLISH JEWISH COLONY

SKETCH OF THE HEBREW RESIDENTS ON THE EAST SIDE.

THEIR CUSTOMS, HABITS, AND RELIGIOUS OBSERVANCES—STRICT FOLLOWERS OF THE MOSAIC LAW—OUTLINES OF A MARRIAGE CEREMONY.

In that section of the City lying between Division and Canal streets and in certain portions of Chatham, Centre, Pearl, Chrystie, Monroe, and Grand streets there lives and thrives in constantly increasing numbers a large Jewish colony. By far the larger number of the members of this colony came from Poland, which fact, however, they are somewhat loth to acknowledge. Ask any one of the brown-visaged, sharp-eyed men or of the women who are engaged in sorting rags, vending old clothes, or standing in or about the small, dirty-looking shops in the vicinity from what country they came, and the answer will be in broken German, "Von hinter Berlin," which, being interpreted, gives the somewhat vague information, "From behind Berlin." Even on further questioning the Polish Jew continues to give evasive answers and to display an evident reluctance to acknowledge his birthplace. For much thievishness and still more dirt have given the Jew from Poland an unenviable reputation, and he has grown to be looked down upon as a despised one among those of his race. However this may be, and in spite of the dirtiness of the neighborhood, the dinginess of the shops, and the miserable condition of the dwellings of the Polish Jewish colony, the fact still remains that they are distinguished as the most orthodox in religious matters of any of their race in this country. As far as outward form is concerned they adhere to the tenets of their religion with unswerving fidelity, and observe the customs of the Jewish faith with an exactness which is nowhere to be found surpassed. The women, when doing their marketing, it will be found, buy their meat only at the Jewish butchers' shops in order to offer compliance to the Mosaic law, which is very strict and explicit as to the manner in which animals shall be slaughtered, and exacts the complete removal of the blood. The law in this respect directs that the flesh of no animal shall be eaten that has not split hooffs or does not chew the cud. In slaughtering meat at Jewish slaughter-houses, in order to comply with this law, the throats of the animals are cut and the veins are removed from the carcase. The consumption of a dish of oysters or of eels is a forbidden enjoyment to the orthodox Jew, for the law regarding the eating of fish demands that all fish consumed shall have scales or fins. After meat has been purchased it is taken home and placed for at least half an hour in water, and then covered with salt for an hour, and washed again in order to insure the complete removal of the blood.

The strong religious fervor of the Polish Jew is still further indicated by the number of places of religious worship scattered through the section of the City chiefly inhabited by the colony. At repeated intervals, sometimes several times on the same block, the "Beth Hamedrasch," or "House of Interpretation," is to be discovered, the "Beth Hamedrasch" being a building or apartment where religious worship is held thrice daily. In passing through this quarter at any hour of the day members may also be seen reading the Talmud, for it is held that the reading of a passage from the Talmud each day for the benefit of the soul of a departed relative will assist the passage of that soul in the other world. It also frequently occurs that a dying person leaves behind sums of money to be expended in engaging people to read passages from the Talmud for the benefit of his soul. On the day when a parent dies the children fast and burn a light in the house. The former custom, which was not introduced until after the return from Babylon, drew its origin from the fact that David fasted on the day that Saul died. The burning of the light is symbolic of the brightness of the departed soul.

The leading place of religious worship of the Polish Jews in this City is the Church of the Beth Hamedrasch Hagodel, in Ludlow-street. The rabbi of this church acts in many instances as a species of magistrate among the Polish Jews of the City, for, although their religion exacts obedience to the laws of the nation in which they live, nevertheless many questions of dispute and matters of government are settled among themselves by appeal to the rabbi. The books of Jewish law are the Talmud, appertaining to religious matters, Cheschanmeschpot, relating to money matters: Aban Azar, relating to matrimonial matters; Joradiah, relating to the killing of animals and the kitchen, and the Megen Abram, relating to feasts and holidays. The Jewish courts are of four kinds. The Sanhedrim, or jury, of three in cases relating to money matters; the Sanhedrim of thirteen, in cases of divorce; the Sanhedrim of thirty-two in criminal cases, and the Sanhedrim of seventy-three in cases of murder. The last two courts, of course, never sit in this country. When a civil case is referred to the rabbi to be tried before a Sanhedrim of three, he selects one member of the court for the plaintiff, one for the defendant, and one to counsel and advise the rabbi himself, who acts as judge. In cases where the plaintiff is unable to bring sufficient proof to establish his claim, he can require the defendant to take an oath that he is not indebted as charged. This oath is administered in the synagogue on the tables of the law. It is of a most solemn character, and if the defendant takes it, the case is at once adjudged in his favor. In cases where application for divorce is made before the rabbi, a decree must first have been obtained from the courts of the State before the matter can be adjudicated from a religious stand-point. But even though the courts of the State had granted a decree, the divorce would not be considered valid according to the Jewish religion unless a decree had been obtained before the rabbi. If the necessary decree has, however, been obtained from the courts of the State the rabbi then summons a Sanhedrim of 13 and the matter comes to trial. In instances, as frequently occurs, when the wife of the applicant for divorce is in Europe, the rabbi here communicates with the rabbi of the district in which the wife is living. A Sanhedrim is then convoked by the rabbi there, and the two courts hold continual communication with each other as the case proceeds. In cases of this kind, especially as the grounds for the divorce are required to be very strong ones, the proceedings often drag on at tedious length, and the obtaining of a decree is a matter of much time and difficulty. Should, finally, a decree declaring the marriage ties dissolved be obtained, it is then necessary that the decree should be written on parchment, as are the books of the law, and that no blot or stain should appear upon the scroll. The utmost care must be exercised in this respect by members of the Sanhedrim and witnesses in signing their names, for should the slightest stain appear upon the parchment, the document would be held valueless and would have to be rewritten.

In matters relating to the marriage ceremony the Polish Jews are most strictly orthodox. The contracting parties are required to fast from the evening preceding the marriage day until after the completion of the ceremony. On the day of the marriage the bride and the bridegroom meet at the place where the ceremony is to be solemnized, and they are sprinkled by their friends with grain, as expressive of a hope that their marriage may be fruitful. In the larger number of cases, in Jewish communities, this wish is most amply fulfilled. After the sprinkling with grain the bridegroom is led before the rabbi by his father and mother, or in their absence by his nearest of kin. He is followed by the bride, who is also led forward by her nearest relatives or friends. The bride and bridegroom are then placed under what is known as the "chuppa," a canopy generally formed of scarlet and gold cloth, which is upheld by four of the near friends, usually young men or boys, of the bridal pair. The bride and the bridegroom are then enveloped together in the Talith, or prayer mantle, while the rabbi blesses the wine. He next holds up the wedding-ring and asks in a loud voice if it has been paid for, as, with commendable prudence, it is held that no man should enter the marriage state in debt. The rabbi, having handed the ring to the groom, the latter places it on the finger of the bride, saying: "Thou shalt be hallowed unto me according to the laws of Moses and Israel." The marriage contract by which the bridegroom agrees to take the bride as his wife and endow her with all his earthly possessions is read and the ceremony is completed. The dress which the bride wears on this occasion is kept by her to be used at her death as her shroud. The husband is also presented at a later period, generally one year after marriage, with a shroud made by his wife, which he wears in life once a year on Atonement Day.

When a man dies the Jewish law provides that his brother shall marry the widow and take charge of the children. In cases when the brother is already married or when neither parties desire to enter into such relationship, the brother whose duty it is to marry the widow appears before the rabbi in the synagogue, and she kneels and unlooses a light sandal from his foot, as a sign that she relieves him of his obligation to her, and grants him a release.

The Polish Jews of New-York have in recent years organized a mutual benefit society, known as the Kesher Shel Barzel, or Society of the Iron Ring, which has been of much benefit to its members and their families in cases of sickness and death. Chief among the other Jewish societies in this City are The Free Sons of Israel, which organization pays weekly installments in cases of sickness, and the B'nai B'rith, which furnishes a premium in cases of death.

June 19, 1882

THE HEBREWS FROM RUSSIA

A PROBLEM THAT PERPLEXES THE AMERICAN JEWS.

ENGLAND'S PROMPTNESS IN SENDING REFUGEES TO THIS COUNTRY — FIFTEEN THOUSAND HELPLESS PEOPLE ALREADY ARRIVED—BENEVOLENCE OF THE JEWS IN AMERICA.

The Jewish problem, which, in variously modified aspects—chiefly through brutal persecution—is seeking solution in different parts of Europe, in quite a unique form is now also added to our own stock of "burning" questions. The subject divides into two parts: One is philanthropic, and makes immediate claims upon the attention of the American people; the other is economic, and, while important in its bearing on the future of the country, does not now demand to be settled out of hand.

There are reasons for believing that the people of this country, except in special cases, are not fully alive to the importance of this Jewish influx. There is such a thing as getting too much even of the very best in this world. The American people are no doubt prepared to do their whole duty in this refugee matter, but circumstances — unless the case is kept well in hand—may compel them to do a great deal more than their legitimate share. John Bull, as represented by the Mansion House Committee, seems to have collected a large sum of money for the express purpose of increasing our population; however, what is to become of these people after they get here does not seem to concern him in the least.

The whole question should be discussed—and, if need be, settled—in the clearest lights, and without prejudice. Until now it has been left for solution with our Jewish population, as if the Nation at large had no concern in it. It is a mistake to suppose that the Jews are anxious to keep this matter in their own hands. It is rather true that they are quite willing to be relieved of all responsibilities in the case, and the most far-sighted among them view the future with alarm. They realize the peculiarity of their position most keenly. A leading Jewish merchant expressed himself to the reporter for THE TIMES to this effect: "We are a commercial people, and the force of circumstances, as well as our traditions, confine us to a limited range of occupations. Within this sphere we are liable to establish a controlling influence—to secure a monopoly, as it were—and thus incur the danger of fanning into a flame any smoldering fires of prejudice. We are deeply concerned in this Jewish immigration—our own future well-being seems at stake in it. A pressing demand is made upon us just now, and we are bound to do our duty in the emergency; but to throw the whole responsibility upon us, by treating it as a sect question, when it is a great internatianal problem, seems to me to be not only wrong and unjust but extremely short-sighted on the part of the American people."

The importance of this problem to the whole people is, perhaps, best brought into view by the aid of some statistics. The Jew, because of his special callings, as compared with average people, seems, at the least, to triplicate himself. That is to say, he appears to the casual observer at the lowest estimate three times more numerous than the census figures warrant. Any village containing only a dozen Jews to a new-comer will appear as almost an exclusively Jewish settlement. The reason is that all Jews are in business, their signs on the principal streets are ever the boldest and most conspicuous, and they will themselves appear where the stranger can not fail to take cognizance of them. In the larger cities they divide into two classes—on the one hand shop-keepers, on the other peddlers. In any case they are peculiarly conspicuous. Their shops are generally on leading thoroughfares, and not a small part of their business is "sidewalk trade." The peddler, of course, lives on the street, and a single individual in this line easily succeeds in making himself more "numerous" than 50 mortals having ordinary callings. A thousand street vendors readily lend themselves to give the appearance of 10,000.

Except he has given this illusive aspect of the subject his attention, it will surprise the reader to learn that the entire present Jewish population of the United States does not exceed 250,000 souls.

Since the 1st of January, 1882, Russian Jews have arrived at the rate of about 2,000 per month. There are said to be 3,000,000 Jews in Poland and Russia. Already 100,000 or more have fled the Czar's dominion in search of refuge. If America continues to remain an unrestricted asylum to receive whomsoever John Bull may see fit to send, the possibilities in the way of Semitic acquisitions seem almost unlimited.

Aside from the peculiar characteristics that distinguish Jews from the common run of people, and make them either desirable or undesirable acquisitions at given periods in a country, there is to be observed a wide distinction between refugees and ordinary immigrants. Thus a third and most important element is added to the problem. The average immigrant arrives on these shores with a fixed purpose, definite aims, and a disposition, as well as qualifications, that will enable him to assimilate with the body of our people. The refugee, on the other hand, is absolutely helpless when he arrives, and, what makes matters worse in this case, peculiar circumstances prevent these new arrivals from placing themselves in agreement with their new relations. In this we have still another amplification. It must be remembered that these Russian Jews are in the strictest sense orthodox. The story told by Col. Ingersoll, where a Jew imagined that the heavens thundered because he had eaten a small piece of pork, would have special application to these people, and that, too, without the addition in the case of Ingersoll's Jew, who really thought Jehovah had taken occasion to get angry over a very little matter. These people would consider the eating of pork as matter enough for a cosmic cataclysm. The average Jew in America is a Jew only from an ethnological point of view. His racial antecedents he holds an accident, and for the rest he does in New-York as New-Yorkers do. He believes no more than he is compelled to, and eats pork if he has a taste for it. From a religious point of view the Russian Jew is further from the American Jew than the American Jew is from a Christian or infidel. An orthodox Jew, who undertakes to earn his bread in a non-Jewish community, is literally confined to bread for his diet. Not only is he debarred by his creed from eating pork, but such other meats as he ordinarily permits himself to eat must be slaughtered according to certain forms, and by one specially consecrated to the office of shochet or butcher. Now it happens, unfortunately, that full-fledged shochets do not abound very numerously in American rural districts, and in consequence any isolated orthodox Jew finds it hard lines until he concludes to adapt himself to the customs of the country. "This is apparently a small, but a very serious, matter," remarked a prominent Hebrew to a reporter of THE TIMES. "One of the first batches to arrive we colonized in Colorado. They were without a shochet, and consequently without meat, until we sent them one. Few of us care for this form, but to these people it is sacred above almost anything else. Should we undertake to persuade them that this is an empty, or at any rate minor, form, they would be horror-stricken and lose all faith in our counsels, and besides it would be cruel to disabuse their minds. It is partly, at least, for conscience's sake that these Jews have been persecuted, and to tell them after losing their all that their sacrifices meet with no appreciation among their own race brethren would be the extreme of cruelty. It may be a superstition, but we cannot overcome it, though it adds immensely to our difficulties."

Up to the present time, since the beginning of the Russian persecutions, the arrivals of Russian Jews have aggregated about 15,000. At first all were embarked to arrive at New-York, and thus came under the care of the New-York Jewish Emigrant Society. This body had about $110,000 placed at its disposal—$35,000 of which was contributed by the Mansion House Committee and the rest raised among wealthy Jews in this City—and by judicious management hoped to carry forward a systematic work, estimating that they would be able to take care of about 300 arrivals per week. For a time all went fairly well, but all of a sudden, without previous warning, several thousand were thrown upon their hands, and then the whole machinery became clogged. Lately the English committee has made direct consignments to various inland localities—to many points previously supplied by New-York with the full number that can be maintained—and the result is that the whole work is thrown into confusion. For example, Milwaukee, it was thought, would be able to care for 300, and that number were shipped to that point from New-York. Within a fortnight a party of 350 have been consigned to the same point direct from England, via Montreal. The result is that Milwaukee suddenly finds herself paralyzed. It has more than it can assimilate in any practical way, and the consequences cannot fail to be disastrous. Other places are being served in the same way. It would appear that without the least regard to the practical expedients of such a course, the English committee have adopted a sort of cast-iron census rule, by which the number consigned to any place is entirely regulated by the general population as given in the last census report—and in pursuance of this method of distribution a party of nine have been sent direct from England to Newport. Whether Summer cottages go along with this assignment is left a matter of painful uncertainty.

Various colonization schemes have been tried with varying success. One in Louisiana is an acknowledged failure. Another in Colorado presents some promising features. However, whether inherently feasible or otherwise, a lack of funds for the carrying forward of schemes of this sort will prevent the colonization of these refugees from ever becoming anything beyond an interesting experiment. In the nature of the case these people must remain where they can be best provided for. This is obviously in the established Jewish centres, and this fact is being more and more recognized by those having this refugee problem to solve. Unless immense sums of money can be placed at the disposal of the local Jewish committee for transportation purposes they will be compelled to devote all their means to the procurement of the merest necessaries of life for the people already on their hands. Under such circumstances, but one result can be looked for: These people will remain at the places to which they are originally consigned, the larger cities, and New-York must necessarily absorb the bulk.

The number any city is able to assimilate being in a measure limited or regulated by the Jewish population within its borders, it becomes a question of importance how, under this law of aggregates, the new-comers will be apportioned, and in this connectin some statistics will be of interest.

Under the auspices of the United Jewish Associations, recent efforts have resulted in a very complete Jewish census. The figures obtained are, perhaps, a trifle under the mark, but in the main sufficiently correct for a basis of estimates and calculations. With regard to the leading cities of the Union the report of the association gives the following figures:

BY CITIES.

Cities.	Jewish Population.	Cities.	Jewish Population.
New-York	60,000	Cleveland	3,500
San Francisco	16,000	Newark	3,500
Brooklyn	14,000	Milwaukee	2,500
Philadelphia	13,000	Louisville	2,500
Chicago	12,000	Pittsburg	2,000
Baltimore	10,000	Detroit	2,000
Cincinnati	8,000	Washington	1,500
Boston	7,000	New-Haven	1,000
St. Louis	6,500	Rochester	1,000
New-Orleans	5,000		

BY STATES AND TERRITORIES.

The following are the returns by States and Territories, covering the congregations, their membership, and the population at large:

States and Territories.	Number of Congregations.	Number of Members.	To'l Jewish Population.
Alabama	8	254	2,045
Arizona	..	..	48
Arkansas	4	165	1,466
California	12	613	18,580
Colorado	1	31	422
Connecticut	3	169	1,492
Dakota	..	..	19
Delaware	..	..	585
District of Columbia	3	144	1,508
Florida	1	..	772
Georgia	7	313	2,704
Idaho	..	..	85
Illinois	10	567	12,625
Indiana	14	398	3,381
Iowa	3	91	1,245
Kansas	2	56	819
Kentucky	4	285	3,602
Louisiana	13	495	7,538
Maine	1	..	500
Maryland	14	600	10,337
Massachusetts	9	650	8,500
Michigan	6	263	3,233
Minnesota	1	28	414
Mississippi	8	239	2,262
Missouri	6	506	7,380
Montana	..	..	130
Nebraska	1	20	223
Nevada	1	29	780
New-Hampshire	..	..	150
New-Jersey	8	229	5,593
New-Mexico	..	..	108
New-York	52	3,371	80,565
North Carolina	3	65	820
Ohio	24	1,014	14,581
Oregon	2	60	868
Pennsylvania	26	1,969	20,000
Rhode Island	2	105	1,000
South Carolina	3	110	1,415
Tennessee	7	271	3,751
Texas	7	210	3,300
Utah	..	..	258
Vermont	1	19	120
Virginia	8	291	2,506
Washington Territory	..	..	145
West Virginia	2	58	511
Wisconsin	3	95	2,559
Wyoming	..	..	40
Totals	278	13,763	230,984

Taking 250,000 Jewish inhabitants as a basis of calculation, 15,000 refugees would allot 1 to every 17 settled Jews. According to an apportionment on this basis the quotas of the larger cities would be about as follows: New-York, 3,500; San Francisco, 950; Brooklyn, 800; Philadelphia, 750; Chicago, 700; Baltimore, 600; Cincinnati, 500; Boston, 450; St. Louis, 400; New-Orleans, 300; Cleveland, 200; Newark, 200; Milwaukee, 150, (it has already received 650;) Louisville, 150; Pittsburg, 125; Detroit, 125; Washington, 100; New-Haven, 60; Rochester, 60.

But any mere abstract arithmetical rule to cover a complex problem such as this of the Russian refugees, if too rigidly applied, cannot fail to work mischief. Besides, it will come to pass that the larger cities must do far more than their arithmetical proportion. In the first place, at least one-third of the Jewish population of the country is isolated in semi-rural communities in such a way that effective co-operation is out of the question. In the second place, these refugees are evidently best adapted for city life, and it is safe to assume that fully 75 per cent. of all now on these shores or to arrive will finally settle in the larger cities, and no doubt 50 per cent. of all new-comers will, in one way or another, make their home in or about this City. The whole tendency is now in this direction. Several large appropriations by wealthy Jews here have been made with special reference to the establishment of "shelters" in this vicinity, where the women and children may be taken care of, thus placing the men at liberty to seek and procure employment. Two thousand women and children are now on the hands of the committee, and most of these will eventually find their way to the "shelters" to be erected on Ward's Island. It has been proposed to the Mansion House Committee that they pay the local committee $10 per month for all refugees under their charge. Unless something definite in the way of assistance is soon devised, the local committee at an early day will be completely stranded. About $100,000 has already been expended. Now an additional sum of $75,000 has already been subscribed by the Hebrews of this City. It was expected that at least twice that sum would be contributed, but hopes in that direction have been disappointed. There is an evident unwillingness on the part of many Jews that this should be regarded as an exclusively Jewish matter, both because the burden is more than the Jewish population can bear and because of the consequences that such exclusive responsibility may entail. If too much is done by them they argue that the American people may eventually hold them accountable for any real or imaginary surplusage. The more is done for these refugees the more they will press for these shores, and sometimes sufficient for the day is the evil thereof.

It is for reasons detailed in the foregoing, and many others equally cogent that might be suggested, that the local Jewish committees deem it wise to relieve themselves as much as possible from at least a few of the responsibilities and duties hitherto devolved upon them. They contend that in a matter of such importance, involving, perchance, great national interests in the near future, the Nation should assume all responsibility. The State of New-York has appropriated $250,000 for immigration purposes. None of this sum will go toward caring for the refugees so long as the Jewish committee remains in control of this work, and the members of the committee are of opinion that at least a portion of this fund should be devoted to the needs of Jewish refugees. The Jews of this City have always taken care of their own immigrants. It was their intention to continue doing so, but now the work has grown beyond their means, and they feel compelled to abdicate in favor of the regularly constituted authorities.

The charge of clannishness has often been set up against the Jews. It is certainly their boast and pride that no Jew has ever been allowed to suffer want in the midst of their plenty. Their labors in behalf of the needy of their race, together with their contributions in the past, have been all that the most philanthropic could wish. It will be of interest to present some statistics bearing on this subject. Given a population of 250,000, there must in turn be subtracted from this an indefinite proportion, say 50,000, living too remotely from the centres of activity to participate in any concerted action. In order to bring the benevolent work of the Hebrews of this country clearly and concretely before the readers' mind, let him suppose a city of 200,000 people and then endow it with the institutions to be enumerated.

HOSPITALS.

There are six or more hospitals supported by the Jews of the United States. All of these are extensive institutions, and models in their arrangements and management. Although maintained by Jews they are by no means confined to Jewish needs. These institutions are situated in the following places: Mount Sinai Hospital, New-York; Jewish Hospital, Philadelphia; Hebrew Hospital, Baltimore; Jewish Hospital, Cincinnati; Touro Infirmary, New-Orleans; Michael Rees Hospital, Chicago.

ORPHAN ASYLUMS.

There are 11 orphan asylums and homes distributed among the following cities of the Union: Hebrew Benevolent and Orphan Asylum, Home for Aged and Infirm Hebrews, Deborah Nursery and Child's Protectory, Sheltering Guardian Society, New-York; Foster Home and Orphan Asylum, Philadelphia; B'nai B'rith Orphan Asylum, Cleveland; Jewish Orphan Asylum, Baltimore; Pacific Orphan Asylum, San Francisco; Home for Aged and Infirm, Family Orphan Society, Philadelphia; Home for Widows and Orphans, New-Orleans. Besides these there are other benevolent institutions in almost every city in the Union for dispensing charity, for free burial, &c.

BENEVOLENT ORDERS.

There are four Jewish orders or secret societies in the United States. Their object is the moral, social, and intellectual advancement of Israelites, the payment of pecuniary benefits to members in case of sickness, and, in case of death, an endowment of $1,000 to the family of the deceased member, as well as the promotion of all benevolent undertakings. These societies are as follows:

1. Independent Order of B'nai B'rith. This has 7 Grand Lodges, 302 subordinate lodges, and 22,814 members. For the five years ending December, 1878, there were paid for sick and endowment benefits $1,007,039. Funds on hand, $570,089.

2. Independent Order of Free Sons of Israel. This has 2 Grand Lodges, 86 subordinate lodges, and 8,604 members.

3. The Order of Kesher Shel Barzel. This has 5 Grand Lodges, 170 subordinate lodges, and 10,000 members. Paid in 1878 for endowment benefits, $129,808. Funds on hand, $112,698.

4. Improved Order Free Sons of Israel. This has 1 Grand Lodge, 44 subordinate lodges, and 2,849 members. Paid in 1879 for sick and endowment benefits, $39,038. Funds on hand, $21,964.

There are several female orders attached to the above, and others independent, which pay weekly benefits to their needy members, and specific amounts in case of death.

SCHOOLS AND COLLEGES.

There are 13 free schools maintained by the Hebrews of the United States devoted to Hebrew and religious instruction. They are distributed as follows: New-York, 5; Philadelphia, 4; Cincinnati, 1; St. Louis, 1; Chicago, 1; San Francisco, 1.

There has been established besides a Hebrew Union College. This is situated at Cincinnati. It has an efficient corps of Professors, and is under control of a Board of Managers of 24 members. This college affords gratuitous instruction in Hebrew, classical, and rabbinical departments, not only to Israelites, but to students of all denominations, and is authorized by law to confer degrees.

There are in all the larger cities Young Men's Hebrew Associations for mental, moral, and social culture by means of lectures on scientific and literary topics, Jewish history, &c. Reading-rooms and libraries are attached.

HEBREW CONGREGATIONS.

The Union of American Hebrew Congregations comprises 118 congregations, and has for its object the union of the Israelites of America in all that is beneficial to their interests, and especially to establish and maintain institutions for instruction in Hebrew literature and Jewish theology, and to establish relations with kindred organizations in other parts of the world for the relief of the Jews from political oppression, and rendering them aid in their efforts toward social, moral, and intellectual elevation.

The value of the property of these societies, chiefly represented by synagogues, by States is as follows:

States and Territories.	*Value of Property.*	*State and Territories.*	*Value of Property.*
Alabama	$50,000	Michigan	40,000
Arizona	25,000	Minnesota	35,000
California	450,000	Mississippi	30,000
Colorado	5,000	Missouri	200,000
Connecticut	30,000	New Jersey	75,000
Dist. of Columbia	35,000	New-York	2,750,000
Georgia	55,000	North Carolina	25,000
Illinois	400,000	Ohio	800,000
Indiana	85,000	Oregon	10,000
Iowa	20,000	Pennsylvania	825,000
Kansas	15,000	South Carolina	80,000
Kentucky	175,000	Tennessee	75,000
Louisiana	250,000	Texas	100,000
Maryland	75,000	Virginia	50,000
Massachusetts	50,000	Wisconsin	75,000
Total value			$6,900,000

RECAPITULATION.

Number of congregations	278
Number of members	13,763
Total Jewish population, (with recent accessions)	250,000
Hospitals	6
Orphan asylums and homes	11
Benevolent lodges	602
Funds held by above lodges	$704,646
Colleges and schools free	14
Value of synagogue, hospital, and church property	$6,900,000

"As rich as a Jew" has all the force of a proverb among the vulgar. The aggregate wealth of the Jews of New-York is supposed by many to be something quite fabulous, and the natural fondness on the part of Israelites for display goes far to give color to the supposition that every Jew has a gold mine in his back yard. Talking on this subject, a prominent Israelite remarked: "It is a mistake to credit us with extraordinary wealth as a class. A few among us possess considerable of this world's goods—are recognized millionaires—but in New-York there are not above a dozen of this class, and the rest are no richer than the average of the people among whom they reside."

Said another: "We are disappointed at the non-action, or what amounts to the same, of many of our money kings. I feel certain that the Rothschilds, with their influence at Courts, and especially at St. Petersburg, could have put a stop to these outrages. Perhaps we over-estimate their influence, but we are none the less disappointed that there has been no visible effort on their part in the direction indicated. Besides, they have not contributed from their abundance as they ought. According to the papers, several bankers at Paris and Frankfort have been large contributors—one is credited with having given 1,000,000f.—but we have reason to doubt these statements. However, sums as large as that would be none too large from people like the Rothschilds in an emergency so pressing as this."

The colonization scheme to any considerable extent being practically out of the question, the local committees are turning their attention to other means of supplying these refugees with employment. Unpleasant as the fact is to them, they are forced to admit that the bulk of these refugees must find employment in the larger cities. Many are intelligent, but few are adapted to manual labor. They lack the requisite physique. Under the inspiration of free institutions, a wholesome diet, and other improving forces, the native-born Jew of America fills out to excellent proportions; but the European Jew, especially of the lower order, is an animated deformity, and in Germany few are found qualified to perform military service. Taking all the circumstances detailed above into account, it is readily seen that the problem presented by the Russian Jew is by no means an easy one of solution, and at no distant day national action of some kind will no doubt be imperatively demanded by the situation and evoked.

July 16, 1882

IMMIGRANT COLONISTS

THE WORK OF ROMAN CATHOLIC BISHOPS IN THE WEST.

HOW TOWNS ARE CREATED BY THE IMMIGRANTS—THE COST OF PLACING A FAMILY IN THE WESTERN STATES.

Under the management of a few able men, among whom, perhaps, Bishop Ireland, of St. Paul, Minn., one of the most prominent Roman Catholic prelates of the North-west, may be reckoned as the most successful in the various enterprises he has conducted, an Irish colonization movement, intended to benefit the suffering peasant masses of the agitated island, has been quietly developed within the last few years, until its history has become one of considerable social and economical importance to the student of emigration statistics. Those who have carefully studied the history of the various communities that have been established in this country by avowed socialistic or religious enthusiasts—begun in fanaticism to end in desolation—and have investigated the causes of their failure, need not be informed that, however beautiful in theory, the communistic or co-operative plan of organizing colonies has invariably broken down under the strain of practical application, save when, as was the case with the Oneida Community, the pecuniary and industrial interests of the association have been guided and controlled by some able and sagacious leader, capable of subduing the restlessness of turbulent spirits and of stimulating the indolence habitual with all social dreamers. The special causes that have operated in special cases to produce disaster are fully and incisively discussed by John H. Noyes, the founder of the Oneida Community, in his work on communism in this country; by Mr. Charles Nordhoff, in his valuable book on the "Communistic Societies of the United States," and by the author of a concise manual on communism and communistic associations published by the Oneida Community several years ago and condensed from materials furnished by the preceding works, with some original observations on the part of the writer.

Dating from the vast, vague, and fantastic pagodas of socialist speculation, elaborated by the followers and immediate disciples of Charles Fourier—poetic dreams of human society—the shifting aspects of socialism have trodden so rapidly upon each other that permanent organization upon any settled basis has been impossible. Taking its hues and tints, and its special phraseologies, from the advanced free thought of the time, socialism has passed almost abruptly from the philosophical vagaries of Fourier to an attitude and a literature whose phrases are scientific, whatever may be said of their spirit, and under the manipulation of such leaders and organizing intellects as Karl Marx, Lassalle, Schultze Delitsch, and scores of others, has assumed the relation of a political party whose aims are declaratively subversive of the existing order of things, by peaceful or violent methods, depending upon the bias of the leader who happens to be foremost at the moment. More or less intimately interwoven everywhere with the struggles and aspirations of the working classes, it has thus become an element of terror and menace to the sovereignties of Europe, instead of busying itself as formerly with the foundation of colonies or co-operative societies—ideal republics on a small scale with a few hundred acres of domain. The attitude of the socialists of the present day being rather revolutionary than submissive, and their declared purpose the reorganization of the social system at home in preference to the realization of their doctrines on new territory under new conditions, the vast tide of followers which was so fondly expected by those who led small colonies originally into the unoccupied wilds of the West, turned back abruptly, in many cases, upon the eve of embarking for the new Eden. One of the largest and most unique of all the colonization schemes that have graced the annals of communism, having for its leader a gifted littérateur, whose social essays were eagerly sought by the *Revue des Deux Mondes*, failed miserably from the refusal of the larger proportion of the colonists to embark after a vast territory had been secured for the new republic in the most eligible and arable section of Texas. Its projecter died in New-Orleans after witnessing the wreck of his life-dream; and the few who came over with him as pioneers struggled back to France dispirited and disappointed.

The various societies, based upon communism or co-operation penetrated by some common idea of religious fanaticism, noticed in the works of Mr. Noyes and Mr. Nordhoff, have proved more permanent than their fellows founded in purely humanitarian theorizing. But the majority of the former have lost ground of late years, possibly through the attrition of the liberalism and tendency to free thought which cannot be altogether excluded even from an isolated community; and the date of their final extinction, according to their latest historian, is an event that cannot be long postponed; so that, whether regarded as an outcome of social speculation or of religious eccentricity such as gave rise to the "Berean" of the Rev. (or once Rev.) Mr. Noyes or the revelations of the first Mormon apostle, the communistic plan of founding and managing colonies may be said to have give way under the tension of practical experience and to have demonstrated its own impracticability by a series of most interesting experiments, some of them conceived and conducted by able and profound thinkers, who have thus left monumental evidence of the fatality of speculative social systems. Equally true, according to Bishop Ireland, is the principle that such enterprises cannot be successfully conducted as benevolent institutions, but in order to answer the evils for which they are intended—the placing of bodies of population composed of material such as are constantly arriving from Ireland or Germany—must be managed upon a strict business basis." "The utmost that can be done for the masses of immigrants who come to seek homes in the New World," says the Bishop, speaking from an experience of many years, "is to give the thrifty and industrious an opportunity; each must work out his own destiny in his own way, and pay back honestly, with business promptitude, every dollar advanced to secure his land and build his cabin." This is a rule whose expediency has been impressed upon him by long experience, and one from which he never departs.

His methods are very simple: and, according to Bishop Spalding, whose book on colonization is an authority on the subject, his mode of founding a colony may be described as follows:

He first selects an eligible tract of 50,000 or 100,000 acres, and takes from the railroad company the exclusive right to dispose of it for three years. The Bureau of Colonization is then set to work as a vehicle for bringing the enterprise before the immigrants he wishes to attract; circulars giving full details as to price, conditions of sale, &c., are issued and distributed among immigrant Catholics who wish to obtain homes for their families: and a priest is selected, with especial view to his previous training as a farmer and his acquaintance with farm life, to preside over the destinies of the future colony. The latter is on the ground at the advent of the first settler, receives him on his arrival, and renders any advice and assistance in his power. The first building erected is the little church about which the future settlement is to cluster, and the new-comers select their lands. Towns are laid out at convenient and eligible intervals along the railway route, and in a few weeks, ere the colony is opened for the reception of settlers, a little village containing a Post Office, a store, and a church, but no liquor saloon, has sprung up. Contracts are made with the railroad to deliver the lumber required to build houses for the settlers at reduced rates, farms are laid out in advance for those who declare a wish to become members of the new colony, and every precaution is taken to deal fairly and honestly with the colonists.

The tract is probably and preferably rolling prairie, so there are no trees to be felled, no roads to be made, and (owing to the herd law in force in the Western States) no fences to be built. But the beginning is rendered tedious and discouraging by the necessity of plowing the wild prairie a year in advance of sowing for the first crop—a necessity arising from the fact that the matted roots of the prairie-grass, which has been growing for ages, form a sod so compact that it requires to be frozen and thawed before it can be broken up. Vegetables and corn can be raised on this sod the first year; but, as a rule, settlers who have taken farms contract to have the soil turned for them the Summer before their arrival, so that they can begin to sow wheat, and in four months after they are actually settled reap an abundant harvest. The resident priest attends to all these details. The rule is to give the colonist six years in which to pay for his farm in annual or semi-annual installments at a low rate of interest; and it often happens, according to the testimony of one who has recently visited several of these colonies, that the first year's wheat sells for more than enough to meet the obligation and enable the settler to start on the second year free from debt. Bishop Ireland has been the instrument in founding a dozen such settlements, among which are Graceville, Avoca, and Minneola, and in placing hundreds of Irish immigrants in comfortable homes. The bureau is at St. Paul. The percentage of failures is stated to be about 20 per 100, mostly due to home-sickness and discouragement, which take effect during the first year of the colonist's residence in his new home. Of those who remain for a whole year, notwithstanding these drawbacks, and make a vigorous effort to succeed, the proportion of failures is scarcely 5 per cent. It is the rule of the bureau not to make any pecuniary advances to the settlers and not to assist them in any way beyond the aid and advice given by the resident priest, who concerns himself with their material as well as with their spiritual affairs.

Bishop Spalding is President of the Irish Catholic Colonization Association of the United States, a joint stock company which has been successful in placing a large number of immigrants. The company was established under the laws of the State of Illinois some years ago, and is operated upon methods differing in some degree from those of Bishop Ireland's celebrated Colonization Bureau in St. Paul. The corporate object of the association is the settlement of Irish Catholic immigrants on the unoccupied lands of the United States, and its capital is wholly employed in buying lands to be sold to colonists on the most advantageous terms to the latter, yet not with so little profit as to preclude paying fair dividends to the stockholders. Having settled upon the location of a colony, and obtained the refusal of a sufficient tract on favorable terms, the company builds a church, a residence for the priest, and a house for the reception of emigrants—a sort of general depot, where the settlers can find shelter until their own houses are finished. In cases where the security warrants the outlay money is also advanced to plow 30 or 40 acres before the arrival of the settler, and to put up a cottage for his reception at a cost of $150 to $200. With these improvements the farm is sold to the colonist on time, the association retaining as security the title to the property until the last payment is effected. The association operates exclusively in railroad lands, which, when taken in large tracts, may often be obtained at very low prices for cash. It is thus possible to sell to the colonist on time at exceptionally favorable rates. The only officer of the company who receives a salary is the Secretary, and the only outlay of money for which there is no return is the sum of $2,000 or $3,000 originally expended in building the church, the priest's residence, and the emigrant house. When warranted in doing so, the association advances the emigrant the amount required to build a house or to put in and harvest his first crop of wheat, but all the additional means needed for stock, implements, and maintenance during the first year must be supplied by the settler himself, and he must pay $50 down as an earnest of good faith.

Among the colonies established by the Irish Catholic Colonization Association, of which Bishop Spalding is President, are Adrien in Minnesota and O'Connor in Nebraska. The former was not, at last report, particularly flourishing, having encountered poor crops for several years; but the latter is among the most successful ventures in placing Irish immigrants which have been undertaken of late years. The $50 paid in earnest money holds the colonists to their work, and prevents the wholesale desertion that would else follow slight disasters.

There is a third colonization society engaged exclusively in placing Irish Catholics in the West whose operations are worthy of notice. It was originated by Mr. John Sweetman, of Dublin, Ireland, as a philanthropic project. Mr. Sweetman is a man of large means and enlightened philanthropy, but by no means a dreamer. Some years ago, in view of the suffering of Irish emigrants, he conceived the idea of founding a society in Ireland whose concern it should be to place Roman Catholics emigrating to the United States under more favorable conditions than could be done by the individual with the small means and limited facilities at his disposal. Embarking in the purpose with enthusiasm, but with due regard to business principles, he succeeded in enlisting other capitalists in his plan, and the result was the incorporation by special statute of the State of Minnesota of the Irish-American Colonization Company, with a capital of £30,000, since increased to nearly double the original amount. The Directors of the company are mostly resident in Great Britain, and Bishop Ireland is the American agent of the association—a position for which his experience and information peculiarly fit him.

The first attempt at colonizing undertaken by the company was in Minnesota, where a tract of 15,000 acres was secured in Murray County, on the Winona and St. Peter Railway. The land was bought for cash at a low price, and sold to the colonists at $6 per acre with a good profit. In addition to selling them their farms on time the association erects houses for the settlers, supplies them with the necessary stock and implements when necessary, and advances the money for their maintenance for a period of six months, or until the first year's crop of wheat is harvested. Personal property is sold to the colonists at the wholesale cash price, the obligation bearing interest at 8 per cent. per annum, while the rate of interest on the price of the land is 6 per cent. Experience has shown that it requires about $1,000 to settle a family of immigrants, and up to the present time about 50 destitute families have been settled, some of which had not money enough to pay their passage from Europe. A transcript of the distribution of the $1,000 thus advanced may be of interest to readers:

Furniture and housekeeping outfit	$38 45
Provisions	28 65
Farm implements	32 35
Fuel and seed	16 80
Live stock	132 50
Building house	110 00
Traveling expenses from Boston	68 25
Eighty acres of land at $6 per acre	480 00
Extra provisions and feed for cattle	50 00
Total required to place one family	$1,007 50

This will be found, according to Bishop Ireland, to represent about the average amount required to place a destitute Irish family; and a large capital or a rapid return of the money invested is naturally required to carry on the business of a company established upon the basis of such wholesale advances. About $60,000 has been expended in placing the 50 poor families under the care of the company, and latterly some colonists have been attracted, whose larger means did not necessitate such wholesale expenditure. The company holds the title to the land and personal property until the last payment is made, and the payments are thus arranged:

	Per Cent.
At the end of 2 years and 6 months	5
At the end of 3 years and 6 months	5
At the end of 4 years and 6 months	10
At the end of 5 years and 6 months	10
At the end of 6 years and 6 months	10
At the end of 7 years and 6 months	20
At the end of 8 years and 6 months	20
At the end of 9 years and 6 months	20
Total	100

The interest on each percentage of the principal is paid with it, and at the end of the first 18 months the colonist pays to date the interest on the whole indebtedness. The colonial interests are managed by a resident Superintendent, who is not, however, at liberty to interfere with the settlers in any way whatever, and Mr. Sweetman himself resides on the ground, where he has established a stock-raising farm and a creamery. The colonists have generally been able to meet accruing obligations without difficulty, but the system of payments seems needlessly, almost studiedly, complicated.

Other Roman Catholic prelates in the West have engaged extensively during the last 10 years in placing colonies of Irish immigrants with more or less success, and many thriving towns and villages have been established in this way, either as benevolent enterprises, or as business ventures; but by none have the economical and social aspects of colonization been so fully investigated and put in practice as by Bishop Ireland and by Bishop Spalding, who have been identified with the movement from its inception, and are fully informed as to its statistics.

Graceville was not founded until 1878, in May of which year the immigrant-house was finished. Some of the colonists had less than $300 to begin life with; some could not even afford to put up shanties to live in, and the first cottages of the best of them were cabins with two rooms, sodded without as a protection against the weather. A wagon cost $75, a yoke of oxen $65, a plow $25, a cow $25. Colonists have poured in, and Graceville is now an incorporated town, with three stores, a large flour mill, churches, schools, and a regular government. Exclusive of money for land, it has been the experience of Bishop Ireland that about $400 is required to begin life as a settler under fair conditions, but with industry, thrift, and a brave spirit many have done well with one-fourth of that sum—established permanent homes and became the owners of valuable farms within the short period of five years—a statement which should be carefully pondered by the tenement-house population of our overcrowded cities, as well as by the Irish peasant starving in his hut at home.

December 17, 1882

THE GERMANS IN AMERICA.

Among all the public anniversaries with which these latter years have been crowded for the people of this country none is more calculated to arrest attention than the second centenary of the German immigration to this continent which was celebrated yesterday. There were, indeed, many hundreds of persons of German descent or of German birth already living in this country, and especially in this State, before 1683. "New-York," says BANCROFT, "was always a city of the world." But all the immigrants by way of Holland were classed together by the English of that day as Dutch, including not only Germans but French and Piedmontese. It was reserved for PENN to found upon the Delaware what he declared to be "a free colony for all mankind," and to encourage immigration by means more resembling the modern system, which is so extensive and so quietly conducted that it seems to be absolutely automatic, than had ever before his time been employed. The first organization of an emigration specifically German was his organization, and it was the result of this organization that the German-Americans of New-Jersey and Pennsylvania so heartily celebrated yesterday.

The distinction of this first German immigration to America has continued to characterize German immigration ever since. A little more than two centuries ago the map of the Atlantic coast was divided into New-France, New-England, New-Netherland, New-Sweden, and New-Spain. The Dutch settlers conquered New-Sweden, and after being themselves conquered by England and reconquered by Holland, became part of the English power, and were secured for the English side in the greater conflict which was to form the history of America for the first half of the eighteenth century, the struggle between Great Britain and France, or, as we would now say, between the Teutonic and the Latin races for the control of the New World. In all this there is no trace of German influence. There was no New-Germany, in name or in fact, among the European "claims" that were "staked out" on this continent during the seventeenth century. Indeed, there was no old Germany then, and the great gamesters took no account of what has since become Germany, except as spoil, in the game that they were playing for the control of two worlds.

It followed from this helplessness of the German States and their isolation from what may be called the planetary politics of the seventeenth century that the Germans who came to this country left Europe more absolutely behind them than did the immigrants of any other race. And this has continued to be their distinction ever since. The first rill of German immigration was not destined steadily to broaden and deepen into a great river. On the contrary, throughout the eighteenth century it was of little account compared with that from the islands which determined the language, the laws, and the government of this country. It was not until after the revolts of 1848 had been put down, and the insurgents began to set their faces toward the Western Republic, that the immigration from Germany became in any way comparable to that from the British islands, swelled beyond all example as this latter was in the same years by the failure of the potato crop in Ireland. Not until 1854 did the German immigration take first place in our statistics, with 206,000 immigrants against 105,000 from Ireland and 160,000 from all the British islands, in a total immigration which was not exceeded until 1872. For these last thirty years it has pretty steadily kept this first place. And it has uniformly kept its characteristic of leaving Europe behind it. The German immigrant, indeed, cherishes as much as any other the social ties which bind him to the land of his birth. But his renunciation of a political allegiance not merely to any potentate but to any faction in European politics is absolute and unreserved. He never asks his fellow-citizens of American birth to take sides with him in German politics. His children are taught to look to the future of America and not to the past or even to the future of Europe. There are, to be sure, a certain number of German "Reds" who try to inoculate the working men of this country with notions which never could have originated here and which have no sort of applicability to the industrial or social circumstances of this country. But the rational German-American lets these foolish persons exhaust themselves without being at all affected by them. The good citizenship of the Germans comes from the readiness with which they leave European politics behind them, and their citizenship is so good that every rational American must give them welcome and allow that the German immigration has been an unmixed good to the United States. Whatever tends to make life better worth living they have not left behind them, and it would be difficult to compute the good that German immigration has done us in importing German music and German beer, and in the labors of the German immigrants as social missionaries, practically showing, what was practically unknown in this country before they came, that it is possible on occasion to be idle and innocent.

October 9, 1883

CHINESE TRADE UNIONS.

WHY IT IS THE POLICE NEVER FOUND MONEY IN MONGOLIAN GAMBLING HOUSES.

"Do the Chinese have trades unions?" said Qwong He Chong, the richest Chinese merchant in New-York. "Yes, and no. They have great organizations in all their working classes, and in other classes too, that correspond closely to the trades unions of the Americans. The gamblers in Mott-street have such organizations. The opium joint keepers, and even the proprietors of disorderly houses likewise. The Chinese perceive the advantages of co-operation and association, but they perceive it up to a certain point and then stop.

"Most of these societies, if not all, have a feature of practical benevolence about them that the Americans do not carry out. For example, a gambler who belongs to the Kai Von, or 'Street Opinion Society' of Mott-street, falls sick and is in want; he is alone in a strange land, and might die if left to himself. He sends word to the managing head or agent of his society. This man, instead of calling a meeting of the members of the organization to take action in the sick man's behalf, immediately makes out a subscription book in the name of the petitioner and dispatches agents to call upon the members in person. Thus the necessary funds are frequently acquired within a few hours. Then, in order that he may have trustworthy attendance, one of his fellow-members volunteers, or else is selected either by lot, by numerical order, or according to nearest kinship, to nurse him until he is recovered. Even in death he will not be forsaken by his fellow-members, who would embalm his body and send it back to his friends in China."

"How does a trades union benefit gamblers outside of sickness?"

"In more ways than one. A dishonest player robs the house or does not pay what he loses. After a reasonable time notice is sent out, and he can play nowhere. In fact he is almost sent to Coventry. For this reason most Chinese gambling is carried on without the money being seen. You may be at a fantan table (that's something like your faro) where hundreds and thousands of dollars change hands and never see a dollar displayed either to buy or redeem chips. If the player wins the bank will pay him in some anteroom or send it to his home; if he lose he will go away and bring or send the money in within 24 hours, but generally before sunset. That's the reason why when the New-York police have raided a Chinese gambling house they seldom if ever find any money. You needn't smile. I don't mean to insinuate they pocket it in the least.

"The laundrymen are generally well organized. This is especially so with the Chinamen of San Francisco, where now it is almost impossible for a Chinaman to open and run a laundry unless he is a member of the union; the fee to enter the organization is $25 per annum. Their head man is elected every three years. The proprietors of the ten hundred laundries of this city have an association, so have the ironers and washermen. The latter are very poorly organized, most of them being newcomers and serving in laundries as apprentices or contract labor men.

"By this means we have what you call strikes, lockouts, and boycottings. Last June the laundry business was very brisk and the ironers were nearly all overworked. Their leaders consulted, concluded that higher wages were in order, and notified the bosses and their constituents of the fact. In 48 hours daily wages were advanced from $3 to $4 50 and $5. In October trade fell off, the bosses consulted, and wages went back to about the former figures, or rather to $10 a week and expenses. A singular result of this endless organization is that the Chinese in New-York have a quasi-government of their own, similar somewhat to the town system of this country. Our Consul, Ah Ming Young, has a consular jurisdiction and considerable power, so that really there is no need of any further ruler. But the Chinese have a regular City Hall at No. 16 Mott-street, and every year elect a Chun Wah Whey Kwan Cean Sung, to govern them for 12 months. He has, of course, no legal status either in this country or at home, but through these organizations he exercises much authority."

December 8, 1884

A CHINESE CHARITABLE SOCIETY.

ALBANY, March 20.—The first Chinese society incorporated under the Societies act is the Chinese Charitable and Benevolent Association of the City of New-York, whose certificate has been filed with the Secretary of State. The objects set forth are to ameliorate the condition, care for the sick and distressed, give pecuniary assistance and advice as required by reputable and deserving Chinamen in New-York City, and to generally aid and succor all worthy Chinamen in need. The Trustees are: Leung Jum, 8 Mott-street; Mon Lee, 5½ Mott-street; Wong He Chong, 19 Bowery; Tom L. Lee, 4 Mott-street, and William A. Haug, 13 Pell-street. The signatures to the certificate are in English and are well executed.

March 21, 1890

WARRING ON THE PADRONES.

THE CARBONARI ASSOCIATION ENDEAVORING TO PROTECT ITALIANS.

A committee of the Italo-American Carbonari Association of Chicago has been holding a secret session in Eighth-avenue with the members in this city, and yesterday it completed its labors. The committee has been working in the principal cities and towns, and wherever there are colonies of Italian laborers in this country. Its object is to break up the padrone system and the Italian banking agencies which co-operate with the padrones. In spite of the laws against the importation of laborers on contract the Italian padrones have been able to bring over thousands of Italian laborers and keep them in a state of bondage here until they have made an immense profit out of them. A padrone goes to one of the banking agencies and asks for an advance of money to pay the passages of 200 or 300 Italian laborers for railroad or mining work. The bank furnishes the passage money—$17 to $21 a head—and then the padrone writes or telegraphs to his agent in Italy to send on the men. The agent goes among the laborers or small farmers, and offers to secure them work in America at 7f. a day and pay his passage. This amount to an Italian laborer, who is glad to get 2f. a day at home, although he can save money at that, is too tempting, and even if he has an acre or two of land he sells it and comes over. On his arrival here he is told that his passage cost $40 or $50, and besides that he is charged with a heavy commission for supplying him with work. He is a stranger and cannot help himself, and is compelled to go wherever the padrone sends him and work out the amount that the padrone says is due him.

Respectable Italians are indignant at these frauds practiced upon their ignorant countrymen, and are ashamed of the reproach cast upon them of bringing over pauper labor, and the carbonari branch in this country has already requested the parent society in Italy to bring the matter to the attention of the Italian Government.

The committee and New-York members of the Carbonari Society have decided to appoint three Italian private detectives in this city at $75 a month to ferret out the padrones and the Italian banks that are engaged in the business of trading in human flesh. The names of several padrones in this city were furnished to the Chicago committee, and they will be published in the *Mazzini Voice*, a monthly bulletin of the organization that has a secret circulation among the carbonari.

The meeting also condemned the action of District Assembly No. 49 of the Knights of Labor, after hearing the evidence given by Italian laborers in regard to their sufferings during the last longshoremen and coal handlers' strike. Fifty Italian families, it was said, are still in a destitute condition in consequence.

The committee will go to Lynn, Mass., and also to Boston to carry on its work in those places.

February 1, 1888

GENEROUS BARON HIRSCH

HE PROVIDES A CHARITY FUND OF OVER TWO MILLIONS.

IT IS TO BE USED IN IMPROVING THE CONDITION OF IMPECUNIOUS HEBREW IMMIGRANTS—HOW IT IS TO BE EXPENDED.

At the request of Baron Maurice de Hirsch of Paris several well-known Hebrew residents of this city have undertaken the management of a philanthropic scheme of great practical utility. They have consented to act under a deed of trust, by which Baron Hirsch places in their control the sum of $2,400,000 to be used in improving the condition of the poor Hebrew immigrants that are continually coming here from Russia and Roumania.

This scheme has been under way since last March, and the liberal Baron has been sending $10,000 a month to the New-York Trustees. All of the details have been arranged by correspondence, the trust deed was received here last week, and yesterday Mr. Jesse Seligman received from Baron Hirsch a cablegram stating that the full amount awaited his order.

The Trustees of this fund are Myer S. Isaacs, Jesse Seligman, Jacob H. Schiff, Oscar S. Straus, Henry Rice, James H. Hoffman, and Julius Goldman of New-York, and Mayer Sulsberger and William B. Hackenberg of Philadelphia. In a day or two the Trustees will take out articles of incorporation as the Baron de Hirsch fund Executive officers have already been chosen as follows: Myer S. Isaacs, President; Jacob H. Schiff, Vice President; Jesse Seligman, Treasurer; Julius Goldman, Secretary. The headquarters of the fund are at 45 Broadway, where Mr. Adolphus F. Solomons is stationed as the general agent.

It is not the design of the fund to dispense charity. Its purpose is mainly educational. Baron Hirsch's idea is to educate the poor Hebrew immigrants in the language and the ways of this country, teach them some trade, and put them in the way of taking care of themselves. It was the Baron's original intention to carry on this work in Russia. He proposed to colonize the unfortunate members of his race in Russia, give them land, build them schools and workshops, and help them until they became self-sustaining. The Russian Government, however, refused him permission to put his plan into operation, and he thereupon turned his attention to this country.

The trust deed provides that $240,000 shall be used in acquiring and improving land and in erecting and maintaining buildings and dwellings for the use and occupancy of such families of Hebrew immigrants as the Trustees may select. The remainder of the $2,400,000 is to be invested in such manner as the Trustees deem advisable, and the income is to be used in educating and improving the condition of ignorant and needy Hebrews. The objects of this trust fund are briefly stated in the articles of incorporation as follows:

The education and relief of Hebrew emigrants from Europe, chiefly from Russia and Roumania, and the education and relief of the children of such emigrants.

The training of emigrants and their children in a handicraft, contributing to their support while learning such handicraft and furnishing necessary tools, implements, and other assistance to enable them to earn a livelihood.

Instruction of emigrants and their children in the English language, in the duties and obligations of life and citizenship in the United States, in technical and trade education, and the establishment and subvention of schools, workshops, and other suitable agencies for such instruction.

The instruction of such emigrants and their children in agricultural work and in improved methods of farming and aid to agriculturists by way of loans on real or chattel security.

Co-operation with established agencies in the United States in furnishing aid, relief, and education to needy and deserving applicants embraced in the classes designated.

Sub-committees of the Trustees are preparing plans for training schools in this city, Philadelphia, and Baltimore, and for the establishment of rural settlements, where immigrants will be taught farming or other industry. It is designed to build factories as rapidly as possible on land to be purchased and held by the fund, and comfortable tenements will be erected for the homes of the operatives. It is probable that the first of these rural settlements will be situated in New-Jersey, not far from this city.

There are three or four schools now in operation under the direction of the Trustees. One is a technical training school in Ninth Street near Third Avenue, and there are two schools conducted for the purpose of preparing poor Hebrew children for the primary department of the public schools. There are about five hundred pupils in these two schools, one of which is in East Broadway near Jefferson Street, and the other is in Suffolk Street near Broome. Evening schools are being established for the benefit of girls who are employed in the daytime. Similar work will soon be started in Philadelphia and Baltimore.

President Myer S. Isaacs said last evening that since the Trustees had been in receipt of Baron Hirsch's generous contributions, they had found employment for 3,000 persons, mostly heads of families. Within the past few days thirty families of Hebrew immigrants were sent to Texas through this agency. In 1890 the average number of Hebrew immigrants that arrived in this port was 4,000 each month.

As soon as the necessary formalities have been observed the Baron de Hirsch fund will be transferred to this country and invested in first-class real estate mortgages.

February 8, 1891

THE ITALO-AMERICAN LEAGUE.

A MOVEMENT TO MAKE GOOD CITIZENS OF ITALIANS.

PHILADELPHIA, May 3.—That Italians shall not be a foreign horde, but become American citizens, is the purpose for which an important organization to be known as the Italo-American League is to be formed in this city.

It is estimated that there are upward of 20,000 Italians in Philadelphia. A score of secret societies exist, but there is more or less jealousy among them. The importance of the big league proposed is such that it is expected that an establishment of the organization here will be followed by a similar movement in New-York, Chicago, New-Orleans, and other large cities, the league having Philadelphia as its headquarters. Francis Tesoriere, who is connected with the law office of A. W. Horton and who is about to organize the league, gave the following outline of its purposes to-day:

"Our plan is to organize a society which shall be a purely philanthropic concern, calculated to advance the interests of Italian residents morally and materially. The league will be supported by monthly contributions of its members, and will admit any one, provided he will conform to the rules and spirit of the organization. The chief objects of the league will be to make American citizens of all Italians belonging to it; to secure through the medium of the Italo-American League the enactment of laws looking toward the suppression of low and degrading avocations carried on by Italians, such as organ grinding; to establish a bureau, and to encourage the full exercise of the rights of American citizenship, without in the least influencing the political opinions of the members who shall be deserving of kindly treatment at our hands."

May 4, 1891

OVER ON THE EAST SIDE

Customs and Habits of the Jewish Population of That Quarter of the City.

The man who was born in New-York and has not seen the city for fifteen or eighteen years would become bewildered if he were to take a walk through the great east side. The streets are much the same as they were in those days, except that several of them have since been paved with asphalt, which has vastly improved the sanitary condition of the district, and a number of high tenement houses have been built. But it is the population that has changed—a class entirely different in manners and customs and political ideas and beliefs from the old-timers, who have moved up town, east and west.

The district bounded by Catharine Street, the Bowery, Houston Street, and the East River, which was formerly divided up between the Irish in the lower and more eastern portion and the Germans in the upper part, and a fair sprinkling of Americans who were still able to support several churches there, has almost entirely changed hands. A certain proportion of the Irish element still holds the fort on the river front, where the rough work is done along shore, but the others have been steadily pushed upward and outward by the children of Israel in their new exodus, this time out of the Russian Egypt and house of bondage into the Canaan of the West.

A different language is now heard there. Neither German nor English nor yet Gaelic, but what is called the Yiddish, or Jewish, a jargon of old German, Hebrew, Polish, and Russian, with the addition of Hungarian, where the Jews come from Hungary. As an illustration of the mixture of this jargon may be taken the sentence, Goot shabes taty—good Sabbath to you, father. The first word is the German gut—good; the second is Hebrew, meaning Sabbath, and the third is Polish, or Slovak, tata, signifying father. These people have already adopted a number of English words.

In matters of dress, these new-comers have adapted themselves to the manners of the country as closely as the precepts of orthodox Judaism will permit. The long caftan, girdled or not, with the tsitses underneath, from which are suspended four cords made of twisted strings, signifying that Israel is to own the four ends of the earth, which is a distinguishing mark of the Jews in Russia and Poland, is hardly ever seen in the ghetto in this city, except, perhaps, on some newly arrived boys. Even the bushy, untrimmed beard which was characteristic of the Anarchists before they were tamed by Superintendent Byrnes and Inspector Williams, is generally disappearing. When the Jewish journeyman becomes a contractor he will trim and comb his beard, or he will shave it off and sport a mustache; and a good many of the young bloods shave off beard, mustache, and all. Of course, this is contrary to the ordinances, but they are minor ones in this age, and can stand suspension.

The dress of the women is much like that of the sex of other nationalities, except that the law requires married women to conceal their hair. According to strict orthodox custom, a girl must have her head sheared or shaved on her wedding day, and thereafter must keep it concealed from men's view by a turban, which in Poland is called a knoop, or by some other appropriate headgear. The more fashionable ones began wearing wigs over their growing hair, and this fashion is now prevalent here among the old-fashioned women of the orthodox faith. A very few women on the east side may still be seen wearing the hair-concealing headgear. The younger women, however, have rebelled against the laws that detract from their attractions and handicap them in their efforts to retain their husbands' affections or to capture new husbands when they become widows.

The separation of the sexes among the Jews on the east side in this city is now confined to the synagogue, where the women are relegated to the galleries, out of the sight of the men. But, as very many of these Jews are the descendants of Poles, Lithuanians, and Germans, who in times gone by were converted to Judaism, the feeling for the social as well as business intercourse of the sexes has perhaps proved too strong for the old Oriental prejudices. So they are seen together in the shops, stores, and at social gatherings. The Oriental seclusion of women, except in the synagogue, has been discontinued by even the most orthodox Polish and Russian Jews.

A custom which these people have preserved during their sojourn of centuries in Southwestern Russia, and which to a certain extent prevails in this country, is the divorce, which a man can give his wife, and marry another woman. Polygamy was prohibited among them many years ago, it is said, by an ancient rabbi in Poland, who pronounced a terrible ban upon all such as should marry more than one wife. The Russian Czars granted the Jews autonomy in their dealings among themselves, and the rabbis settled disputes and contracts among their coreligionists in their own courts. The Sultans of Turkey gave to the Jews, and also to the Christians, where they dwelt in small communities in Turkey, the same privileges.

The divorce of the woman was one of the ancient customs which has been maintained, and several orthodox rabbis in this city have believed that they were possessed with this authority in the United States. Appeals from these rabbinical divorces have on various occasions been made by wronged women, and the husbands, who were about to marry younger or richer wives, were made to understand that such practices were forbidden by law under severe penalties. The young women who have grown up in this city and who know their rights have generally set their faces against the custom, in spite of the influence of the orthodox rabbis from Europe, and there is no doubt that this custom will gradually die out.

August 27, 1895

LITTLE ITALY IN NEW-YORK

OVER ONE HUNDRED THOUSAND OF THE MODERN LATINS HERE.

Only About 25,000 are Voters—The Most Honored Citizen Who Might Be Boss—Growth of the Race and Spread of Its Institutions—Full List of Societies—Many Benefit Organizations—Many Chapels and Churches.

There are more Italians in New-York, Brooklyn, and Jersey City than there are in any city in Italy, if we except three or four of the largest. According to the very best authorities, there are in New-York City alone upward of 105,000 Italians and Sicilians, of whom at least 25,000 are voters, and all the adult males are willing to become such as soon as practicable. As explained by Commendatore Louis V. Fugazy, who enjoys the distinction of being President of at least fifty Italian societies in this city and vicinity, as well as being President of the body which has charge of the celebration of Italian unity next September, every male Italian immigrant declares his intention to become a citizen the moment he arrives, because he is informed by his compatriots here that only citizens can obtain employment in the street cleaning and other civic bureaus.

Commendatore Fugazy is the best authority on "Little Italy" in this vicinity, because he is the most honored and trusted among the masses who affectionately call him "Papa Fugazy." It must not be inferred from this that the Commendatore is an old man. He is an elderly man, acknowledging fifty-seven years, and really not looking any older. He is a banker and notary public, and, as he said to a reporter for THE NEW-YORK TIMES the other day, he had been a Captain with Garibaldi and had also served with Victor Emanuel both as a soldier and Chief of Secret Police during the troublous days of the struggle for Italian unity. That he acquitted himself in all his offices at home with credit and satisfaction to his King and country is attested by his many decorations and his great popularity with his fellow-countrymen here. He is without a doubt the one man to whom all Italians—that is to say, the majority of the simple, every-day Italians of New-York City—look for advice and aid in personal affairs. He is the leading spirit, if not actual President, of 132 mutual benefit associations, and in his homely, unostentatious way, sits in his office in Bleecker Street, in the very heart of "Little Italy," accessible to all and ready to aid by counsel and material those of his compatriots who seek him.

Some of the native-born citizens—that is, sons and daughters of Italian parents, but born in New-York—may not display that exalted reverence for the Commendatore visible at all times in the Simon-Pure Italian, and there may be among the Italo-Americans a spirit of independence, seeking a new régime and an up-to-date guidance in social and political affairs peculiar to the prevalent progressive impulse, which suggests a desire to break away from the Old World conservatism and patriarchal influence of the Commendatore; nevertheless, they yield a strict obedience to the vogue of the majority, which is in absolute accordance with the recognition of Fugazy and his claims upon the respect of "Little Italy." To his naturalized fellow-citizens he is a man of probity, patriotism, wisdom, and a representative of the best interests of Italians as expounded by Garibaldi and Victor Emanuel, and in America he has striven to serve his fellow-countrymen in their struggle for bread, education, maintenance of their families, and the establishment of new homes, while still keeping unbroken the chain of communication between the immigrants and those dear ones they have left behind in the vineyards, vales, and mountains of Italy and Sicily.

There have been local political padrones in New-York City, and names having been published as belonging to individuals who could deliver the Italian vote to this, that, or the other faction at election time, but the hearts and pockets of the Italians are closer to the Commendatore's hand than many people imagine, and if there were need of an Italian boss in "Little Italy," Papa Fugazy would make the running hot for any other candidate. But he disclaims any pretension to political power. He confines his efforts to benevolent and social purposes. He said on this point: "I do not bother my head about politics. My fellow-countrymen think for themselves and vote for their own best interests. Every political club has its own leader, who tells the members what, in his judgment, it is best to do, and if they agree with him and are convinced that he is right, they follow his advice. It is not easy, therefore, to tell in a word the trend of Italian politics in this city. My own observation would lead me to believe that the Italian vote, which is a little more than 25,000, would be largely Democratic this year. Still, I cannot say this positively. No one can tell at present. My own time is taken up with the work of the benevolent societies, a report of which I am now preparing for the Secretary of State."

There was a time when things were different in "Little Italy," as the Italian colony in New-York was and still is called by the people, although it is no longer little, but, on the contrary, large and important. In those days the Italian population was supposed to dwell, like the Chinese, in the Sixth and Fourteenth Wards, principally in Mott, Mulberry, and Crosby Streets and contiguous byways. It is true that then, say, twenty years ago, the bulk of the Italian colony found its habitations in that region. There the ragpickers and lazzaroni of Naples and the cutthroat refugees from Southern Italy found their lairs in common, it must be added, with the felons and outcasts of other races, who sought shelter and oblivion in the catacombs and rookeries of the Five Points and the streets before mentioned. Then the street musicians and little beggars earned money in this city for the rascally padrones, whose headquarters and barracks were situated as above described. Then were murders committed in secret, and societies of brigands from the Sicilian hills and the haunts of Calabrian bandits continued their organized warfare on law and order, even as they had conducted it in Europe. But this was not peculiar or singular to the Italians. Pirates from the Levant, freebooters from Central Europe, cutthroats from great cities, buccaneers from the Grecian Isles and the African coasts, footpads from the faubourgs of Paris, highbinders from China, and thugs from India lurked by day in these coverts like rats and reptiles, to wander forth at night to pilfer, rob, and slay, if need be.

That this condition then existed the writer can attest, for in company with Walls Mahay, the celebrated London artist; Stephen Adams, the composer of "Nancy Lee" and other famous songs; the eldest son of the Lord Mayor of London for that year, and a police Detective Sergeant who has since become great upon the force, he went forth one Winter's night to visit the Five Points and its haunts of vice and great criminals of vulgar and brutal grades. The party proceeded to White Street, between Centre and Baxter Streets, and descended into what appeared to be the basement of one of the little two-and-a-half-story brick houses that stood on the north side of the thoroughfare. Once below the sidewalk, the party found themselves at the opening to a chamber, which was only the anteroom to another cellar, through which descent was made to a still deeper subcellar, in what might have been the centre of the foundation. This place reached, opened the approaches to the labyrinth of catacombs which, like the cells of a honeycomb, were occupied for every purpose by every nationality under the sun. There were dance houses, in which Italian street musicians were playing, while low-browed, fierce-eyed men and women, with swarthy faces and jet-black, glossy hair, waltzed and pirouetted, regardless of the strangers; there were smaller dens, in which some men sat playing cards; there were opium joints of Chinese; there were little rooms, to which not a breath of air could penetrate save through the keyholes, yet in which were huddled together a score of nondescript men and women of the lowest and most depraved types of different nationalities, to whom were being served stale beer and a sort of liquor which the officer called "strike-me-blind," by a little, wizen-faced hag in a fragment of a calico gown, at 1 cent a drink.

Drinking, carousing, sleeping, gaming, eating, crouching and "hitting the pipe," singing, and enjoying themselves, the motley inhabitants of these catacombs in separate cells were seen by the writer and his friends, who finally emerged at a point in Mott Street through a cellar in a Chinese grocery store not far removed from Chatham Street. No one had spoken to the visitors, who were led by the Sergeant, to whom, however, certain persons nodded familiarly along the route, and whose approach it was evident had been announced and expected. There were more Italians and Chinese in this place than any other nationalities, and all the managers and controlling spirits were of the Latin race—either Italian, Spanish, or French.

But this was twenty years ago. Things have changed since then, in that region at any rate, for the better, it is presumed. Civic reforms and the will of the people have prevented the great cities of Europe from emptying their surplus desperadoes and criminals into this city. The padrones have been muzzled, if not entirely banished, and there are no Five Point cellars or catacombs to shelter the fugitive reptiles of the Old World.

The Italian brigands and Sicilian bandits who infested that region just described, giving it the name of "Little Italy," and covering the appellation with obloquy, can find no harbor here. Their own countrymen dwelling here detest and dread them. Peaceful peasants, artisans, and poor working men and women, for whom there is not bread enough at home, come here to get it by honest toil and honorable endeavor.

"Little Italy" of to-day is worthy the descendants of a noble people. It is not confined to one section of the city. The Sixth and Fourteenth Wards are still thickly peopled by the Italians; but they are not criminals any more than any of the other denizens of the vicinage. They are laborers; toilers in all grades of manual work; they are artisans; they are junkmen; and here, too, dwell the ragpickers, who practice an industry of which the Italians seem to have a monopoly. On the west side, from Grand Street up to Washington Square and west to Varick and Carmine Streets, there is a monster colony of Italians, who might be termed the commercial or shopkeeping community of the Latins. All sorts of stores, pensions, groceries, fruit emporiums, tailors, shoemakers, wine merchants, importers, musical instrument makers, toy and clay modelers, are found and abound here. There are notaries, lawyers, doctors, apothecaries, undertakers—followers of every profession and business, in fact found in a great city. There are more bankers among the Italians than among any other foreigners except the Germans in this city. In fact, some of the barbers are bankers. This is due to the illiteracy of the masses, who are not good at figures, and who also live in such frightfully crowded tenements that they fear robbery, and prefer to place the responsibility of their little savings in the hands of one man, who is of good repute in their little community, and can be found whenever or wherever sought by Italians on the habitable globe should he, in an evil hour, be tempted to decamp with his trusts, than to take the risk of losing them some night in their poorly guarded apartments.

Up town on the east side there is another large Italian setlement, from One Hundred and Third Street up to One Hundred and Fourteenth Street, from Second Avenue to the river. The people live in flats, and suffer like many others from overcrowding. They are for the most part a thrifty, industrious lot. Most of them are in fairly prosperous circumstances. They come from the southern part of Italy. In the Summer evenings it is not uncommon to see the young people strolling with the priests, who have striven to preserve their love for the old faith fresh and pure as it was in the parent country. Not long ago the writer saw a clergyman in the garb of his calling, shovel hat, soutane, cloak, bands, and girdle, strolling along the water front, accompanied by two striplings who had evidently returned from a hard day's work, and had sought the "padre" to learn orally from him, in the soft accents of their mother tongue, the stories of salvation and life as they are taught by the Church of Rome. The incident was absolutely Italian in semblance, and reminded one of those which can be witnessed daily at the same hour in any Italian or Sicilian village, in the Apennines, on the shores of the Adriatic, or outside of Palermo or Messina.

The "Little Italy" of to-day in New-York is full of the picturesque character and color so dear to tourists. There are many Italian chapels and churches in this city. The mission is full of priests, who can speak to the poor Calabrian and Neapolitan and Sicilian in their own patois. The organ grinder is no longer the sole type of the ancient Roman in America. The organ grinder is here, still very much in evidence, but his well-educated, cultured, the progressive countryman is likewise with us, and is doing his best to make New-Yorkers feel that the 105,000 Italians dwelling in the Empire City are striving to benefit and improve it, as well as themselves. They are proud of Columbus, to whom they have erected such grand testimonials everywhere. They are proud of the races which has nurtured and preserved the fine arts, and the learned men and priests among them, the bankers, merchants, and lawyers and doctors are bent upon proving the greatness of Italia proper by the virtues and progress of "Little Italy" in America. At least, so says Commendatore Fugazy, and he can speak for 145 Italian societies and clubs with a membership of over 30,000 people in this city.

These are the clubs:

Avetoma, Aurelio Saffi, Altavilla Cilentuna, Armentese, Avellinese, Abruzzo Citro G. Rossetti, Avigiimen Ailievi Balilla, Addolorata, Beneficenza Italiana, Bentivegno Corleonese, Barbieri Italiani, Biellese, Buontemporei, Camera di Commercio, Cilentuna, Corona di Italia Corleto Monforte, Cittadini Sassanesi, Cittadini Americani Second District, Catanzarese, Cittadini Balvanesi, Cittadini Teggianesi, Carraciolo, Citra Calabro-Americani, Cittadini Paduesi, Cristoforo Colombo di Colliano, Cittadini Calabro-Americano, Carlo Piscani, Castelmezzanesi, Duilia, Dante Alighieri, De Cristofaris, Esercito Italiano, Etna, Emiliana, Firenzi, Fraterna, Fruttivendoli Italiani, Fratellanza Bellesi, Fraterna Accetturese, Flavio Gioia, Fratelli Testa, Fratellanza Sanfelese, Fratellanza Calvellese, Fraterno Ainto, Guardia Savoia, G. P. Rives, Guardia Colombo, Giordano Bruno, Guardia G. Garibaldi, Giovanni Cristiani, Guadenti, Giuseppe Mazzini, Grande Unione Politica, Improved Order Red Men, Italo-Albanese, Italian Home, Indipendenti, Italian Rifle Guard, La Concordia, La Lega Ligure, Legione G. Garibaldi, La Lega Eolia, Lavoranti Calzolai, Liberty and Benefit Society, La Piemontese, La Lega Italiana, Liberi Pensatori, Lega Toscana F. D. Guerrazzi, Laurenzanese, Laurinese, Loggia Garibaldi, Loggia Italia, Mutuo Soccorso di Riciblìano, Musicale P. Mascagni, Militari Campagnesi Masaniello, Manuali Mosaicisti, Marineria Italiana F. Giova, Musicale Italiana di M. S., Montemurro, Marsico Nuovo, Massimo, d'Azeglio, Nizza Cavalleria, Nuova Italia, Novarese, Operaia Italiana di M. S. Operaia di M. S. di Montemurro, Ordine Colombiano Libera Italia, Politica Calveilese, Potenza Lucania, Principe Amedeo, Politica Italo-Americana, Petrucceli della Gattina, Progresso Olevitano, Patria e Lavoro, Primo Battaglie-Bersaglieri Africa, Politica Laurenzanese, Reduci Patrie Battaglie e Militari in Congedo, Reali Carabinieri, Roma Cavalleria, S. Antonio di Padova, Stella d'Italia, Stato Maggiore Aggregato, S. Argenio, Saati, S. Stefano d'Aveto, San Cono, San Rocco, Stato Magg. Bersaglieri Africa, Sarti Italiani, S. Antonio Martire, S. Giuseppe, Sciacca, S. Michele, Sarnese, Stato Magg. Duca d'Aosta, Stato Magg. Aggreg. Carabinieri, Tiratori Italiani, Torquato Tasso, Trinacria, Unione e Fratellanza, Unione Calzolai Italiani, Umberto I., Unione e Fratellanza Colliano, Vespro Sicilnano, Vittorio Emanuele II., Volontari Cientani, 1860; Venti Settembre, 1870; Velocipedisti Italiani.

These are benevolent societies. The following are simply clubs: Filodrammatico, Irredentista, Democratico, Eighth District; Italian-American Athletic, Politico S. Felese, Italia, Armonia, Democratico, Third District; G. Garibaldi, Young Italians' Indipendenti, Union, Republican Columbus, Roma.

The preceding list gives merely the names of the best known of the mutual benefit societies and great clubs in this city. No mention is made of secret societies, of which there are several. They are "secret societies" in every sense, and if there be one characteristic more prominent than another in the Italian nature it is secrecy. It is wonderful. One does hear of the Mafia and a dozen other organizations from time to time; but even the police know little or nothing concerning them. That such societies exist among the Sicilians, Calabrians, and others of the subdivisions of Italians in this city is indisputable, but they guard their secret—yes, even their existence—so well that none but the members knows aught about either, and they dare not tell.

Italy was the hotbed of secret societies—their nursery. It would be folly to suppose that there are none in "Little Italy." There are many; but there are also great societies, free and ably conducted Italian newspapers, and good men. These predominate, and are carving a glorious future for "Little Italy."

May 31, 1896

EQUAL RIGHTS FOR CHINESE.

Chicago Celestials Organize to Demand Citizenship.

CHICAGO, Nov. 23.—Americanized and native-born Chinese will appeal to Congress for the right of suffrage and ask that body to repeal the anti-Chinese law passed in 1882. A public mass meeting will be held at Central Music Hall next Saturday, at which prominent Chinese from all over the United States will speak.

This, it is said, is the first time that the Chinese have undertaken such a movement. They have organized the Chinese Equal Rights League of America, with offices in this city. Wong Chin Foo is President of the organization. In an interview, he said:

"We want Illinois, the place that Lincoln, Grant, and Logan called their home, to do for the Chinese what the North did for the negroes. Why should we not have a voice in the municipal and National affairs like other foreigners? There are 50,000 Chinese in this country who are desirous of becoming citizens."

Following the meeting at Central Music Hall, meetings will be held all over the United States. Among those who will address the meeting next Saturday are Wong Chin Foo and Wong Ock of Massachusetts, and Sam Ping Lee of New York.

November 24, 1897

DIED TOO FAR UP TOWN

SO A JEWISH BENEFIT SOCIETY WOULD NOT PAY HIS INSURANCE.

Chevre Chochmes Adam Merplinsk Will Pay No Death Benefit if a Person Dies Above Houston Street —When He Fails to Pay an Assessment, Because He is Dead Is No Excuse—Nathan Greenstein's Widow Sues the Chevre.

An interesting question in regard to downtown Hebrew benefit societies is to be decided by Justice Roesch in the Fourth Civil District Court, at Second Avenue and First Street. If a member of one of these societies should die above a certain street in this city, is he entitled to the benefits for which he has been paying dues or not? If that member failed to pay an assessment of 50 cents for a tombstone of a deceased member because he himself was also dead at the time, should he forfeit all the rights in the society? The President of such a society argues that he loses all rights.

Nathan Greenstein went into the clothing business in partnership with a certain Yellen, at 69 Hester Street, and contributed $800 to the capital of the firm, while Yellen put in $160. Greenstein lived with his wife and children at 26 Ludlow Street, and joined a society known as Chevre Chochmes Adam Merplinsk, which means Society of Human Wisdom of the City of Plinsk, and which has its headquarters at 61 Hester Street. He paid his dues, and in return the Chevre was to support him in case of sickness and give him a decent burial when he died, and his widow was to receive $1 from every member of the Chevre.

Greenstein was taken sick last month, and was removed to Mount Sinai Hospital, where he died on July 29. The Chevre refused to bury him or to aid his widow, and she did not have money enough to buy even a shroud. Another society to whose members he was known took charge of his body and buried it in its plot in the cemetery at Bayside, L. I.

Isaac Potashnik, the brother of Mrs. Greenstein, called on Harry Gutschick, President of the Chevre, and reminded him of the society's duties to the dead man. These were to pay the following expenses: For burial, $75; for "minyan," $9; for "shiva," $5; for "tachrichem," $4.40; for four weeks' sickness, $20, and $1 from each member for the widow.

Minyan is a prayer to be recited by ten men morning and evening for one week at the house of mourning. Shiva is the suspension of all work for one week by the family, who sit on the floor in stocking feet, and the Chevres allow them $5 for the enforced idleness. Tachrichem is a shroud, in which the corpse is enveloped.

President Gutschick looked at Potashnik with an expression of surprised horror.

"What!" he exclaimed. "We pay all these expenses? What for? Did I tell Greenstein to go and die above Houston Street, when our by-laws say he must die below that street? Never; we'll not pay one cent."

Supposing my brother-in-law had been run over by a cable car, and killed at Fourth Street?" asked Potashnik.

"Well, that would have been his own lookout," replied Gutschick. "He had no business to die above Houston Street."

"But that is not all," continued President Gutschick. "He was in arrears to the Chevre. Didn't we send him a ball ticket, and he never paid for it? That was $3.60. And when he was called upon to read the eliah in the synagogue, for which honor he was to pay 25 cents, he did not respond. And when he was fined $1 for not responding, did he pay up? Not at all."

"But you know very well," replied Potashnik, "that he was sick at the hospital when you sent him the ball ticket, and also when you called on him to read the eliah. I sent him the ticket, and the next day he died."

"Oh, there is still more!" cried Mr. Gutschick. "Wasn't he assessed 50 cents for another member's tombstone? Did he pay that? No! And when we fined him 20 cents for the neglect, did he pay up like a man? I leave it to yourself."

"He certainly couldn't pay up like a live man," retorted Potashnik, "because you well know that he was already dead at the time."

"Oh, well, now, what's the use of standing here arguing with both you!" exclaimed the President, gesticulating with both hands and shaking his head. "We didn't bury Greenstein, because he broke the by-laws of the Chevre, and went and died above Houston Street. And, as for the fines and assessments, the Chevre can't bother itself with such fine distinctions as to whether he died before or after they fell due."

Potashnik finding that he could do nothing with Gutschick, retained Lawyers Greenthal & Greenthal of 51 Chambers Street to sue the Chevre. They brought suit in the Fourth District Court, and obtained a summons for Gutschick. The latter expected a lawsuit and was on the lookout for subpoena servers. So it was a difficult matter to find him.

Abraham Jacobs, managing clerk of the firm, disguised himself as a countryman, and might be easily mistaken for an Italian farmer. He said yesterday: "I took a walk along Baxter Street, and the first barker or puller-in tackled me. This encouraged me, and I walked along toward Gutschick's store. Before I reached it half a dozen other barkers wanted to do business with me, and when I got there Gutschick's puller-in caught me by the arm, saying he would sell me the best outfit at the cheapest prices in the city.

"I allowed myself to be led in and Mr. Gutschick himself waited on me and showed me most of his stock.

"I pretended to take off my coat to try on one of his coats, and at the same time pulled out the summons and thrust it into his hand.

" 'What is this?' he asked, and I told him.

" 'May a devil take possession of your father and your father's father, you dog!' he shrieked, and seized a rule and wanted to brain me. I warned him against trying to assault me and then walked away."

One of the members of the firm said that it was very difficult to serve summonses on that class of people. In one case they could not possibly catch their man, who was as slippery as an eel. So at last they had to take his wife into their confidence. She was a big woman, while he was small, and she happened to have a grudge against him and actually kidnapped him, and brought him to the summons server.

Mr. Greenthal continued that the position of the Chevre and its President could not be sustained in court, as it was ridiculous. It was a fact that of all the members of the Chevre Greenstein was the only one who was in good standing, while the others were in arrears from $5 to $28, and even President Gutschick himself was in arrears.

Both Potashnik and a brother of his are also members of the Chevre, and Mr. Greenthal believed that they will both be expelled from that society for daring to sue it. In that case he will bring suit to compel the society to reinstate them.

August 6, 1896

MANY SOCIETIES OF ITALIAN COLONY

Their Uses and Their Abuses Discussed by an Italian Citizen.

No outsider can appreciate how very many mutual benefit societies exist in New York's Italian colony. The moral disunity of the old peninsula is transplanted here; each province, each town, each village has a society of its own, and although all on the same plan and serving essentially a like purpose, they never merge in action and seldom mix in their membership. They keep the colony divided, create petty jealousies and prevent beneficent activity on a large scale. Could they all be fused into one, they would probably have as large and effective a membership as that of the more prosperous German and Irish societies in this city. That such fusion is a present impossibility has been proved time and time again in various ways.

Comparatively large societies like the Fraterna have been split by internal dissensions, and even events of common interest which should have drawn all these bodies together only served to intensify the differences. The reason for this may, to a great extent, be explained in two ways, both of which would probably escape the foreign observer. The first is the real, though unexpressed, reason for the creation of many of such societies; the second is to be found in the quality of the men to whom is intrusted the leadership therein.

There are a number of societies which came into being at the earnest solicitation of those countless notai and avvocati (mostly Commissioners of Deeds who translate their title into that of the very important law functionary—the Notaio—of Italy) who are a terrible sore in the fair body of an honest and thrifty colony. In order to make money, they advise their countrymen to incorporate, undertaking at small cost to obtain a charter, "Incorporated!" Magic word to conjure with among these too trusting people. They have no clear idea what it means, and of course they do not know how simple a matter it is to incorporate such a society. But when a typewritten paper is shown them signed by a Judge of the Supreme Court, and countersigned by a "minister" in the Capital City of Albany—just think of it, from far away Albany—with a big red or gold seal—when they see such a wonderful document their pride at being "incorporated" is only surpassed by their admiration for the notaio whose skill and knowledge no less than his influence at the capital obtained for them such splendor of a charter.

What is the result? The notaio gains in importance and popularity, becomes a sort of counsel to the new society, is often elected an officer therein, or his "bank" may be made the deposito.y for the social funds. The papers praise "his noble initiative in founding a society to perpetuate the glorious name of Messina or Lipari or Avellino in this new land," and this increases the incorporating mania, and his orders grow and he becomes a power.

It is not only the notai that are responsible for many such societies. There are also the saloon keepers, whose "lodge-room" back of the bar would be useless without societies to hold meetings there. These, like those of the notai, are the product of what may be called the commercial spirit clothed under the mask of patriotic philanthropy. But there is another class, which gives the second reason for the existence of so many dissociated bodies. These are the product of the simon-pure, almost childish, vanity of certain self-deluded "leaders of men" in the colony, hiding it under such a very thin veil of beneficent intent that it is strange how their followers do not see the imposition.

THE DESIRE TO HOLD OFFICE.

They are men of little, if any, schooling, mad with the desire to be called "Presidents" and to wear an especially ornate badge at dances and funerals, and hoping for the joy of seeing their name in print. Perhaps they have laid aside a goodly sum, and feel that they can spend part of it for the pleasure of leadership. They may even dream of a future knightly cross from the King of Italy, or, in default thereof, a less honorific but more resplendent decoration from the Ruler of Venezuela or Colombia. These are the men who never miss an opportunity to cable felicitations to the King or the Queen or the Minister of Foreign Affairs for some happy national event, which naturally is acknowledged by the secretary of such dignitaries; or again they invite his Excellency the Ambassador of Italy at Washington to be honorary patron of the annual ball given by their society, and who, through his chancellor, is obliged perfunctorily to allow his name to be used. Those are t happiest days for the "President," for can show his friends such official an pressive correspondence.

If such men do little good they certainly do very little harm except, of course, as elements of disunion in movements for fusion and affiliation.

It is not to be implied that the countless societies in the Italian colony serve no good purpose. On the contrary, they are an aid to social regeneration. They are the beginnings, however frail, of a co-operative system and of organized labor; they are clubs with the advantages of social intercourse, and they are quite well developed in their mutual benefit features. Even if the critical will say that some of them exist only for the purpose of supplying a brass band at the funeral of its members, is not a society useful that makes the final struggle of these hardworking, big children easier by the assurance of an imposing procession to the grave?

Most, if not all, of these societies have approximately the same articles of incorporation, the same constitution and by-laws. Every detail is provided for, even to certain minor contingencies which would put Cushing's Manual to shame. A generous number of officers is the general rule, so as to satisfy many souls by the conferring of a title. The most striking official character is, as a rule, the Corresponding Secretary. He is the great factotum, being in many cases the one official who can write without laborious effort. You can pick him out at sight, for his ancient frock coat, the ever-present collar and cuffs, and the pale, thin face stamp him as the ex-village school teacher or quondam village journalist, who has sought a better fate in the land of gold.

In the colony he is always known and respected as the "professor," and is an invaluable adjunct as orator at all social functions. He is, generally, the only paid official (other than the doctor,) and works hard for his pay. The society physician is often chosen because of the geographical connection of his birth to the society's members, and is likewise a hard-worked individual.

There is nothing secretive about the deliberations of these societies, yet it is not easy for an outsider to be present at their sessions. They generally meet once a month in the back room of some saloon, giving the preference, whenever possible, to one run by a man from their own town or province. If a meeting is called for 8 P. M., it means that it will not begin until 9 o'clock. Promptness may be the politeness of Kings, but it certainly is not of Italians. And no matter how late the meeting begins, there are always laggards who accuse the Chairman of jamming measures through in their absence, even though there is a quorum and over. The by-laws provide disciplinary measures by fines and expulsion, but lack of discipline in such assemblies is so prevalent and general that they are not and cannot be enforced.

THE APPEAL TO HONOR.

These people are too imaginative to enjoy parliamentary routine or to subject themselves to parliamentary limitations. No matter what the constitution or the by-laws provide, the last resort is an appeal to "honor.'" If the majority can carry a motion, it cannot stop the minority from calling it names. What would be the fun of assembling if there were no opportunity for personalities? What if a man is out of order—isn't it always a matter of personal privilege to speak out one's rancor? As long as you are entertaining you are within the rules, for if the Chair ruled you out of order the assembly would overrule the Chair. And, by being entertaining is not meant that the orator must be funny, for whatever defects Italians may have they are not funny. No; it means to be dreadfully serious and talk grandly about one's honor, which the Chair or a fellow-member or an imaginary entity has outraged.

Members of the minority often feel that way and must speak out their mind. There is really something intrinsically pathetic in this idea of honor which comes to the surface so often. It sounds grandiose and comic, but it stands for something almost tragically strong underlying it. Sergeants-at-Arms are unnecessary, for an insulting epithet from the Chair will more effectively drive a member from the hall. But he will come back at the next meeting with a following to support him in his plan of action. He will arise and move that all action at the last meeting be declared null and void on some impossible ground. This is likely a ruse to try the Chairman's temper. The Chairman's temper is tried and he will likely call his oponent "pecoraro."

Now, although the literal translation of that epithet is the harmless title of shepherd, yet by proper inflection and a look of terrible scorn in its utterance, it can be made to convey a dreadful meaning. It doesn't mean merely ignorant and uncouth or dishonest and dishonorable, but all of these in a supreme degree—a social outcast and a dupe. It has a stigma all of its own which the imputation of no crime, however base, possesses. It is a social anathema under which one cowers and shrinks. It is on such an occasion that these assemblies become, as I said, entertaining. It creates divisions and ruptures, evokes oratory of the grandiose style followed by applause of impressive goings out. But it does not result in the drawing of knives, as the outsider might think; at most, a lawyer is consulted not on how to obtain one's rights but as to how to "break" the society.

THE CHARITABLE SIDE.

The existence of so many of these societies explains, in a measure, the reason why so few, if any, Italians apply to outside charitable organizations. There is probably no foreign colony in New York that depends less on outside help, for which the Bend should well be proud.

The dues are generally twenty-five cents a month in addition to a small initiation fee. The sick benefit includes doctor's care, the nursing by fellow-members and a weekly payment for a few months of from five to seven dollars. Chronic cases are sent to their native town, receiving, besides their fare, a lump sum, generally fifty dollars. The death benefit is never large, but always includes a "handsome funeral" and a wreath of flowers.

The annual balls given by these societies are family affairs. Between the dances, the little boys and girls have a good time making sliding ponds of the waxed floor. Even the babies are brought along, and Italian babies are so good that they submit quietly even to the music of "Professor Campobasso's Orchestra." And after the grand march everybody enjoys the buffet and the speeches, and there is a generally good time with perfect orderliness.

It will be seen, therefore, that these associations are a source of positive good, despite some decided drawbacks. But the colony is too large, its interests are now too important to confine itself to such divided and dividing work. It is capable of much better things and there is a large and growing element in it that needs only united action to make itself felt as a power for good within the colony and as a means of asserting its claims outside.

GINO CARLO SPERANZA.

March 8, 1903

CASH RETURNS TO EUROPE

Immigrants Send Millions Back to Homes.

ITALIANS IN THE LEAD

They Remit $62,000,000—Outflow Plays Important Part in Our Trade Balance.

How large a part the thrifty Italian plays in the outflow of money which turns from the credit to the debit side our annual balance of trade is indicated by the foreign money order statistics of the New York Post Office. All money sent beyond our borders by postal order is cleared through New York, and the figures subjoined represent the remittances from the entire United States. The purely commercial transfers of funds are usually accomplished by means of drafts or other forms of exchange; the money order recommends itself more especially to the saving immigrant with money to send home. The total, of course, is many times the sum sent through the Post Office, but, because of the many channels of shipment entering into a calculation of the total, the postal-order figures are the safest for the purpose of estimating the relative positions of the different nationalities in the army of remittance senders.

During last year the Post Office sent 295,221 money orders, representing $11,-092,446, to Italy, the amounts averaging $37 each, a fact which bears out the conclusion that these money orders represent almost entirely the homeward-bound savings of aliens. Next to Italy comes Austro-Hungary with $7,857,503, the United Kingdom with $7,506,932, Russia with $5,198,326, and Japan follows next in order with $2,144,175. Germany"s quota was little more than a million. In the same period the immigration figures were as follows: Italy, 193,296; Austro-Hungary, 177,156; Russia, 145,141, and the United Kingdom, 87,598.

This year everything points to vastly larger immigrant remittances to Italy than ever before. The Italian laborer ordinarily packs up his belongings early in the Fall, when railroad construction, building, and other forms of pick-and-shovel work shut down for the Winter, and goes back home for the season, taking with him his savings in the form of

Italian currency. The exodus this year has been away below the average because of the openness of the season which has permitted the continuance, almost without interruption, of the classes of work employing Italian labor. A natural consequence will be a corresponding increase in the sums sent home.

Lacking exact figures of the falling off in the Winter exodus, the experience of one Italian banking house in New York which does the greater part of the money changing business for returning laborers, furnishes an interesting clue. This firm imported early in the Fall nearly 5,000,000 lire to supply the first rush of homegoers with cash. Little more than half of this sum, which was not expected to last the season through, has so far been called for, and by now the Italian steerage traffic has turned from East to West, and is nearing its flood in our direction.

This firm's experience is a fair sample of the conditions which have prevailed at other ports.

Although New Orleans and Boston book a considerable part of the steerage traffic between the United States and Italy, New York gets the lion's share, and this city's participation in the business of money exchange is in proportion. The banking house referred to claims to issue about 7 per cent. of the money orders outside of those taken care of through the Post Office. This estimate, which seems a fair one in view of the house's steamship connections, would bring the total of cash carried to Italy, even in this past year of slack homegoing, to 12,500,000 lire, or $3,125,000. Almost all of the remittances sent to the other side by the laboring classes are in the form of money orders; drafts and cable orders are the forms of commercial transfers, and to arrive at the total of the laborers' annual drain on the Nation's capital it is necessary to add to the cash carried away the total of the money orders issued by banking houses and the Post Office. The latter sum has already been told; what the banking houses contribute to the total is problematical, and can only be guessed at. That it is no small factor in the final sum, and steadily growing, may be seen from the following table, which represents the money order business of one Italian banking firm in New York for four years:

Year.	Lire.	Dollars.
1905	15,155,611	3,788,902
1904	12,248,422	3,062,105
1903	11,924,847	2,981,212
1902	7,712,227	1,928,056

These figures, according to the bankers' own estimate, are about 7 per cent. of the total for the United States, exclusive of the Post Office figures. While this guess may be far from the exact facts, it is worth considering in the absence of definite data. It would indicate, for remittances of this class, a total of $51,044,042, to which should be added the postal orders, bringing the grand total up to $62,137,088. This sum is nearly twice the value of our exports to Italy, and seems at first glance to be highly extravagant, but it is borne out substantially by the figures available of the Italian banks themselves. The Bank of Naples alone, and it is credited with about 10 per cent. of Italy's foreign exchange, during the last fiscal year handled American exchange amounting to 22,022,394 lire, or $5,505,596. Then there is to be considered the not inconsiderable amount of Italian exchange transactions done through German and French banking houses and credited to those countries.

This immense movement of funds to Italy cannot but have a considerable effect on the nation's material prosperity.

In addition to her sons in America, Italy has another source of income in her emigrants to South America, who are more numerous there in proportion to population, and enjoy a more important position in the communities and hold higher place in the industrial scale.

Referring again to the Bank of Naples figures for the last fiscal year, Brazil sent 6,658,328 lire to Italy, and to the Argentine Republic 2,616,007. The Italian colonies in Africa contributed only 2,697 lire through the Bank of Naples.

An interesting attestation to the superior position of the Italian in South America is the higher average rate of the remittances from Brazil and Argentina. Our own transfers through the Bank of Naples averaged 176 lire, Argentine's 186 lire, and Brazil's 198. The average from the African colonies was 62.

February 18, 1906

Relieving Congestion in Russian Jewish Quarter

Immigrants Encouraged to Go West —Thousands Now Prospering in Interior Towns—Successful in Farming—Work of the Industrial Removal Office.

VERYBODY knows something about the overcrowded condition of the east side districts of New York City, especially in the quarters where the Russian Jews, fleeing from persecution in their own country, have been arriving of late in such large and ever-increasing numbers. But nothing is generally known about the efforts to relieve the congestion in these quarters and to open up to immigrants other sections of the country, which are not overcrowded, and which offer far better opportunities for earning a livelihood. Particularly interesting is the work of the Industrial Removal Office, created for the distribution over the country of Jewish immigrants.

The office was started in 1901. Since then it has sent away from this city 29,413 persons, almost all of them recent arrivals from Russia, who knew no English and were, in many cases, practically penniless. Of these about 16,600 were wage-earners; the remainder their wives and children.

When an immigrant, just arrived from Europe, is directed to the Removal Office by some acquaintance he gives a record of his antecedents to some employe there. This record is investigated by the office staff. If the applicant is adjudged to be a suitable person, he is sent away to some place in the United States where he will have an opportunity of earning his living. If he has friends or relatives in some particular town, he goes there; otherwise he goes to whatever place is determined upon by the Removal Office.

The success of this system has been very marked. To take the Russians sent to Toledo, Ohio, as an example, one Russian immigrant there is now earning $70 a month, another $60, another $16 a week. To show the diversity of occupations, one is employed in the Wheeling & Lake Erie Railroad shops, another in the Pope Motor Car Works, another in the American Can Company, another in the Gendron Wheel Company. These men had been in the United States only a few weeks at the time they were sent to Toledo by the Removal Office.

ꕥ ꕥ ꕥ

In Omaha, Neb., men sent away from New York City by the office are now working as upholsterers, plasterers, butchers, carpenters, tailors, and as mechanics in the Union Pacific Railroad shops, at wages ranging from $15 to $25 a week.

In short, estimating the average earnings of men sent away by the Removal Office at $500 per annum, which is far from a high figure, it may be said that they are earning $8,000,000 per annum, which is no mean addition to the industrial wealth of the country, when the fact is borne in mind that the great bulk of those earning this amount came to this country as helpless foreigners.

A recent investigation by agents of the office showed these facts: Six men sent to Columbus, Ohio,

Members of the Jewish Colony at South Bend, Ind., Working in the Field.

had savings in banks ranging from $75 to $300, and aggregating $960, notwithstanding the fact that three of them were sending money home to Russia and two of them had brought their families from there. Of nine men sent to Nashville, Tenn., three have their own stores already. Two carpenters sent to Minneapolis have savings of $800 and $700, respectively, and one shoemaker sent there owns his own house. A Pittsburg machinist is earning $20 a week; another $4 a day. Three of the four have bank accounts ranging from $300 to $500. Of those sent to Rochester, six have purchased houses and others have bank accounts as high as $500. Twenty-nine men sent to that city have aggregate bank savings and holdings in real estate amounting to $8,000.

It must be borne in mind that none of these men when sent away, had employment of any kind in New York. Every one was out a job when sent. It entails more responsibility than the Removal Office is willing to assume to take a man out of a paying job here and send him elsewhere, where he may or may not succeed. The bulk of the men whose cases are cited above came here as immigrants, many of them destitute.

In Milwaukee the local committee in charge of the work of removal has organized an agricultural society and bought 720 acres of land, dividing it into forty-acre tracts. They take some candidates, put them on these plots, advance the necessary supplies, permitting them to build their houses, and pay them wages during the first year of their work. At the close of the first year they give them a purchase contract on the property, and these men sent out originally as industrial workers are established as farmers.

Recently the Industrial Removal Office held an exhibition for the purpose of giving an idea to the Jews of New York of its work. The exhibition was held in conjunction with the Baron de Hirsch Farming Society and the Jewish Agricultural and Industrial Aid Society, both of which are engaged in encouraging Jews to leave the unhealthy, congested districts of the east side and become farmers.

The great majority of Jewish immigrants arriving in New York have a very vague idea of the extent and industries of this country and opportunities for earning a livelihood which it offers. It is this more than anything else that contributes to their remaining in New York. The exhibition was held in order to counteract this ignorance among immigrants concerning other parts of the United States. Large photographs and charts formed one of the main features of the exhibition. These were destined to bring home to the immigrants the real extent of the United States, and to show that there are hundreds of cities in the country in which there are Jewish communities, where the newcomers may find fellow-countrymen speaking their own tongue, and in which they may find schools for their children.

Photographs of streets and buildings in various cities were shown, to tempt Russian Jews away from New York. It is not uncommon that immigrants from Russia believe that, outside of three or four cities in the United States, there are no Jews in this country.

A special feature was made of the agricultural work being done by the Removal Office and the two societies mentioned above. Owing to the many photographs of farms and farming work exhibited, hundreds of inquiries have been made to the office regarding openings for farmers by recently arrived immigrants from Russia. One of the members of the Industrial Removal Office staff who visited South Bend, Ind., recently, to investigate the results obtained by Russian immigrants shipped to that place by the office, spoke enthusiastically of what he saw during his stay. He was in South Bend on New Year's Day, when the workmen were idle, so that his opportunities for seeing and observing them were particularly good. He saw them at home, in their clubs, and at their places of worship. He spoke with members of practically every one of the 200 families sent from New York to South Bend by the Removal Office. Not only were all earning a livelihood, but they were contented and full of gratitude to the Removal Office for sending them away from New York.

This is only one instance of the work done by the Removal Office providing openings for the thousands of destitute Russian Jews who flee from the persecutions in Russia.

January 20, 1907

Splendid Work Accomplished by the Educational Alliance

Its Aim Is to Overcome the Inherited Prejudices of Immigrants and Prepare the Newcomers for American Citizenship---How It Sets About to Achieve the Purpose.

LAST year 150,000 Jewish immigrants from Russia and other parts of Eastern Europe, fleeing from persecution in their own country, came to the United States. Of these, 90,000 stayed in New York City. When it is borne in mind that the average number of such immigrants for the last ten years has been one-half the total of last year, it will be seen what a problem confronts those who seek to meet the immediate needs of these immigrants and prepare them gradually for American citizenship.

Foremost among those engaged in this work is the Educational Alliance, whose large light-brick building is one of the landmarks of East Broadway. In view of the sudden influx of immigration from Eastern Europe, this institution has abandoned this year its classes in higher culture, such as music, painting, and singing, and turned its whole attention to the newly arrived immigrants. It is overcoming the prejudices brought by the immigrants from their native lands, and giving them that practical understanding of American government, institutions, and ideas which alone can make them real Americans.

The work of the Alliance is of four kinds—educational, religious, social, and moral. Each of these is applied to parents as well as children, to women as well as men.

The educational section of the work comprises evening schools for adults, open from April to October, when the evening schools of the Board of Education are closed, and courses of lectures, given in the language of the immigrants, on American history, industries, geography, and government. In this way the interest of the immigrant is aroused in his new home.

As for the children, a large majority of them arrive here totally unfit to enter the public schools. Some of them have never been to school in Russia; others know nothing but the Russian language and the methods of Russian schools. These children are at once taken in hand, taught English, and accustomed to American ideas of schooling. In six months, on an average, they are ready to leave the Alliance schoolroom and be entered in the public schools of their home district as regular scholars.

Other children, who have been in school in Europe and wish to go to work in their new country, but are not old enough or, according to our laws, sufficiently educated to give up school, are retained in the classes at the Alliance until they know English and are otherwise qualified to go to work.

For Jewish children to attend public schools in Russia is practically to estrange them from their families. Everything connected with the Russian Government is far away from the Russian Jew. Hence, while the children are at the Alliance schools, it is brought home to them that the fact of their attending an American public school will not in any way disrupt their family ties. Moreover, parents' meetings are held for the purpose of explaining this to the older immigrants.

Another important branch of the education work is the classes for women. In Russia women are almost nonentities. Education is practically unknown to them, especially to Jewish women. Hence, immigrants, at first, do not even think of sending their girls to school; to them it is a great discovery to find out that the latter are expected to attend the Alliance classes and the regular public schools. So true is it that women are educationally nonentities in their native lands in Eastern Europe that, whereas every male Russian Jew, no matter how ignorant he may be in general branches, has always a more or less complete religious education, their womenfolk have not even this, to the men, indispensable branch of knowledge.

But it is the tendency among female immigrants from Russia, as soon as they get their first taste of freedom—as soon as they find that they can go to restaurants when hungry and to department stores when they wish to buy anything—to neglect their home work. In Russia their activities were entirely domestic; in this country they suffer a reaction. They tend to unlearn their domestic virtues, owing to the many paths of new activity open to them.

To counteract this the Alliance has classes in domestic science, sewing, and cooking, where the pupils are encouraged in every possible way to continue their interest in their homes.

Another important branch is the encouragement of manual training among the immigrants. Jews have always tended to develop mentally, not physically. By its manual training classes the Alliance seeks to develop in the boys a taste and desire for manual labor. It does not teach them the technical side of any trade, for it is not a "bread-winning" industrial school. Its work does not encroach on that of schools which fit boys for trades. It confines itself to stimulating a desire for technical work among a people who have systematically neglected it.

The religious work is a particularly important part of the Alliance work. Underlying the system as pursued by the Alliance teachers is an intimate knowledge of the people to whom it is applied. The staff of Alliance teachers have borne in mind that of all the Russian Jews who come to this country nine out of ten come from small towns and villages. In them whatever talents a Jew may have are largely wasted. All the universities and many of the trades are closed to him. If he is artistic or musical, if he has dramatic ability, his progress is hampered at every step by prejudice and persecution. For him there is only one outlet for all such latent powers. It is the synagogue. There the musician becomes a reader; the man with a legal mind a rabbi. There the talents of the architect are utilized for the erection of the building, those of the painter for decorating it; there the man with social proclivities becomes active in forming societies of all kinds. The synagogue is not merely the Russian Jew's church; it is his arena, his club, his theatre, his opera house, everything.

Here he finds this no longer true. If he likes politics, he may enter them. If he has a legal mind, there are law schools open to him. If he is literary, he may write on an equality with other writers; if musical, he may compose to his heart's content. This sudden breaking down of the barriers frightens him. He fears there is no Jewish life in the new country.

"Here I shall disappear!" he cries. "In Russia I was persecuted, but I had my own life apart; here I am free, but my identity will soon be lost." His peace of mind is gone.

At this point the Educational Alliance steps in with its religious work. In its classes the mind of the pupil is gradually led away from the mere ceremonial of religion; he is taught to look upon it as a part of his existence, not, as he did in his native land, as the whole. Religion is used as a means for Americanization.

Special attention is also given to the stimulation

of interest in physical culture among the immigrants. For centuries Jews have systematically neglected this. They despised the Greeks, their rivals of ancient times, for their worship of external development; to the Jew internal development was the sole end to be pursued. Duty, not beauty, was their motto. "The Greeks worshipped the religion of beauty, the Jews the beauty of religion," said Ernest Renan.

Later Jews were massacred indiscriminately by Christian and Mohammedan armies. They came to look on physical strength with greater hatred than ever. Still later military service was required of them, but to enter the army was to ostracize themselves from their families. Hence to disqualify themselves for such survice they began to mortify their bodies. A philosophy of physical weakness grew up. And, as if to give it a greater hold on the Jewish mind, came the Russian massacres of the present time as another instance of what brute strength can do. To overcome gradually the prejudice of the immigrants against physical culture the Alliance has established a gymnasium, a Summer camp, a roof garden, and baths.

In the Alliance there are fifty-seven clubs of various kinds, in which the members play at democracy, elect officers, appoint committees, move, amend, and otherwise observe Parliamentary procedure. The clubs are thus practically classes in American government. In Russia the immigrant feared all connected with the Government; here he himself vests authority in a man. In Russia he was governed by the whims of others; here he learns that the will of the majority is what counts. He had an idea that the individual was not responsible, that everything Governmental was paternal. In developing his club spirit the Alliance develops communal spirit, the beginning of citizenship.

The newly arrived immigrant is easily the prey of the shyster lawyer and the quack physician; instead of freedom of the press, he too often sees only the yellow journal; instead of free government, only the ward boss; instead of free speech, only the ranting demagogue. He is cramped in unhealthily congested districts; he finds difficulty in getting employment; he misunderstands his rights and becomes quarrelsome and litigious. In many cases, even, his disappointment in the new land makes him abandon his family and turn to crime.

To counteract all this the Alliance brings the moral side of its work to bear. It includes courses of lectures, the employment of a Probation Officer at the Children's Court to take charge of delinquent immigrant children, and the maintenance of a Legal Aid Bureau, which endeavors to reconcile people and settle cases out of court. During the fifteen months that it has been at work this Legal Aid Bureau has handled 8,391 cases.

The large hall of the Alliance Building is the arena of the neighborhood. It is always open for meetings in the interest of clean streets, better tenement accommodations, eradication of the social evil, &c.

And last, but not least, the Alliance has a children's theatre, to counteract the tendency of the children to frequent the cheap shows of the neighborhood. Performances are given here once a week. All plays produced are carefully and adequately staged, and, serving as veritable object lessons for the children, these Alliance productions are regarded by educators as valued adjuncts in their daily work.

"In short, we understand the psychology of the immigrant," said the Superintendent of the Alliance, "and we meet his needs."

February 3, 1907

WANT THE SWEDES BACK.

King Oscar's Government to Try to Induce Immigrants to Return.

Acting on instructions from the Swedish Government, Alexander E. Johnson, who represents that country as acting Consul in this city, is making an effort to obtain information as to why Swedes emigrate to this country, with a view to inducing them to return home.

In this connection he has caused to be sent to the Swedish newspapers and clergy throughout the United States a circular requesting that they ascertain from as many as possible of their Swedish-American readers and parishioners their present condition, occupation, and earning power, whether they are engaged in agricultural, mechanical, or other pursuits, and whether they are unskilled laborers.

This information is sought for the benefit of the Royal Swedish Statistical Society, which, prompted by King Oscar, is anxious to induce many people of that country who have emigrated to the United States to return home.

It is hoped that through the information furnished in this country many of the Swedes may be influenced.

July 19, 1907

WHY SWEDES EMIGRATE.

Poor old King OSCAR is starting a statistical inquiry to ascertain why it is that so many of his lieges of the one kingdom that is left to him in his old age emigrate to America. The purpose of the investigation is to induce them to return to Sweden.

Some of the "by-products" of the investigation may be interesting and even useful. But the inquiry is evidently superfluous and the ultimate object unattainable. Swedes come here, as other immigrants come, in order to better themselves. They stay here because they have succeeded in bettering themselves. The attractions of the United States to industrious Swedes in the vigor of life overbalance any advantages they can enjoy at home. And we find them among the best and most useful of citizens. They are very welcome, and a short experience of American life disaffects them more and more from the social conditions of their native land, which is as essentially aristocratic as its neighbor Norway is essentially democratic.

A large proportion of the Swedes who come here find themselves, within a few years, able to return to Sweden for pleasure trips. Possibly many of them, when they have done their work and amassed a competency, may return to Sweden to end their days. It would be surprising if they did not. For, besides their sentimental attachment to their birthplace, a krona will go as far in Sweden as a dollar in this country. There they can be "prodigal within the circumference of a guinea." Only a krona is not so easily come by in Sweden as a dollar in this country. The hope of winning back to Sweden any considerable proportion of the Swedish emigrants to America while they are vigorous and active members of the community is illusory, and the attempt foredoomed to failure.

July 22, 1907

THE SERVANT QUESTION—IS IT TO BE ETERNAL?

IN A TYPICAL EMPLOYMENT BUREAU

"DESIRABLES" FROM ENGLAND.

THERE are problems a-plenty in this country, but there is one only that is spread all over the country, and can be guaranteed to start any gathering of men or women into enthusiastic discussion. It is the pet problem of the country—"Where are our servants coming from?"

Whenever two or three housekeepers are gathered together you can hear the watchword, "it gets worse every year," and even their harassed husbands remark to each other as they commute toward the city: "Our cook left yesterday; this business is certainly getting worse every minute."

Call—if you are so minded—on employment agencies and ask them what they think about it. They join in the chorus: "It is steadily getting worse." One of them declared there was a difference from week to week. When she was asked what the end would be, she looked like a sphinx and said:

"Don't ask me."

Go and see the women who deal with the newly-landed immigrant girls; they say that a girl, who by going into domestic service could make $16 or $18 a month the very moment she lands in this country with the prospect of rising quickly to $25 or $30, would rather go into a factory. The average wage of a factory hand is $5.32 a week; board costs her about $2.50, so she gets about $13 a month and has miserable quarters to live and work in. But she would rather do it than go into housework.

It is a very curious situation. The country is full of girls who might hold well-paying places as domestics, people are clamoring for them, and they go on their way rejoicing in their pittance gained in a shop. Social workers remark with that finality so admirable in the philosophic brain and so irritating to the average mind, "Girls will go into domestic service when it is made sufficiently attractive for them and not before." If a distracted housekeeper pleads that she does all she can to make the place agreeable for a girl, they only look askance and insinuate that they could undoubtedly find the fly in the ointment.

Well, perhaps they can. Anyway they try to. For the problem is still serious enough to attract more and more attention. The Woman's Municipal League has taken it up in this city and is sure that it will accomplish something by degrees. But it will take a long time to change the attitude toward domestic service. Then, too, further, fewer girls of domestic service traditions come over now than used to come. Probably not 10 per cent. of the old number of Irish girls now arrive annually. The Swedes, who form a large percentage of the servant class in this country, are coming less than formerly, and the Swedish Government is organizing a campaign to keep them at home. Plenty of Italians are coming, but these do not go in much for domestic service, nor do the Russian Jewish girls, who are arriving in great numbers. There is actually a deficiency in the supply of raw material.

Girls Want "Evenings Out."

It is, however, the general opinion that there are still servants to go around if the problem were properly met. Employment agencies, investigators, and social workers all agree in declaring that the housekeeper as she is to-day is largely responsible for the present condition of affairs. She doesn't understand the servants' point of view. The sphinxlike employment agent, referred to above, when asked to put her finger on the root of the matter, observed with Delphic succinctness: "It's the evenings." Most people who have studied the question will tell you that the reason the factory and shop girls go gayly on their way, never deviating to the more lucrative profession of housework, is that the average servant is required to give up practically all social life.

Girls in domestic service are remarkably like girls elsewhere, and they would rather have their freedom in the evenings than get higher wages and have no opportunity to "enjoy life." This is what all the immigrant girls say. Could they be guaranteed a certain amount of social freedom, they would consider the proposition of domestic service, for like most human beings they want to make all the money possible, but they think that you pay too dearly for your whistle if you give all your liberty in exchange.

Several employment agencies called attention to the fact that in a great many homes, even of well-to-do people, the servants are shoved into undesirable or even unhealthful rooms. "I don't wonder," said one employment agent, "that girls don't want to work in the country. I know lots of them who have no heat in their bedrooms. If anybody thinks that a girl is going to stand in the hot kitchen and then go to bed in a room with a temperature below freezing and keep either healthy or happy, they make a mistake. And you would be amazed at the number of houses where this is done."

Then the situation is complicated by certain popular prejudices. A good many employment agencies spoke up warmly in defense of the colored maid and the Italian. Many women positively refuse to take these as servants, but both, so the agencies agree, make capable and devoted domestics; better than many more popular races. The mistress has to get their point of view a little. That is all. Italians are good cooks by nature, so, for that matter, are negroes, and both have a willingness to stand by the mistress in an emergency which is often not found in the Northern races. But both negroes and Italians are unpopular as servants, and have to suffer from a prejudice which many people who have studied the question consider most unjust.

Jewish girls, who are coming over more and more, seldom go out to service, but a few of them do. They almost always give satisfaction, and bring great fidelity and intelligence to their work. One lady who found herself suddenly without a nursemaid took a Jewish girl, chiefly, it must be confessed, because she needed some one badly and the girl was equally anxious for a position. She had had no experience with a Jewish servant, was not a Jewess herself, and foresaw trouble. Asked a month or two after the girl was engaged how she liked her, she said: "Well, the first week she asked permission to reorganize the nursery entirely. When she had done this I found that everything ran twice as smoothly. She has hypnotized the baby, so that he cries to go from me to her, and she gazes at me like a jealous cat if I take my own child out with me. Then last week I went into the kitchen and found she had organized a night among the Irish girls. Of course, it's a little bewildering to know what she's going to do next, but I'm quite sure that whatever it is it would be to my advantage. I never knew a girl take such a pride in her work. My husband expects her to reorganize, his stock-broking business for him shortly." Something in this nature is the usual testimony from people who have tried Jewish servants.

When the Municipal League undertook an investigation into the employment agencies of this city it found a most disordered state of affairs. Laws have now been passed regulating the affairs of the agencies and protecting immigrant girls against "fake" places. Every immigrant girl who lands unaccompanied by her family is sought out at the home to which she has gone and help is offered her in the way of suggestions whenever she seems to need advice. Whenever possible the investigators try to place the girls in domestic service, firmly believing that it is the best place for them, and they are equally energetic in trying to persuade housekeepers that girls are social beings and must have some time to themselves.

The old idea of discipline is hard to eradicate. One mistress who lives in the country with six children, and has no trouble with her servants, has a standing order that one maid shall be in the house every evening, and she leaves them to settle among themselves the details of this arrangement. Most of her neighbors were shocked at the liberty thus given, and said that it was bad policy to give too much freedom, but she is the one woman about there who keeps her servants until they marry the ever-attractive coachman or grocery boy. And when they do leave they almost always send up a friend to take their place.

The general houseworker is disappearing; there is no doubt of that. Apartment hotels offer employment to numbers of servants who can specialize as chambermaids, waitresses, bathroom girls, &c., and have their evenings free. Always those evenings! A girl is hardly to be blamed for going into such work in preference to taking a place where she has to do all the work and the washing thrown in. The only girls who are likely to go into general housework are the "green" girls and the colored women The first needs training which the average housekeeper cannot give, and the second is unpopular. So there you are, with a fine economic deadlock!

The Woman's Municipal League thinks that the chief need for lightening the domestic problem is to furnish such elemental training both to negroes and immigrant girls as shall seem to be desirable to the woman herself; that is, there must be brief instruction in cooking, and in the case of the immigrant woman, some few lessons in English. The colored branch of the Young Women's Christian Association in Brooklyn is doing work of this kind already, and the Municipal League thinks that a school of training on the lines of the Educational Alliance is the great need for immigrant women. They must be in an elementary way Americanized before they can properly go into American homes.

Social Life and Model Agencies.

But the league states that adequate training is only a fraction of the problem; workers trained and untrained do not meet the demand. The supply will increase, says the league, only when employers have been educated to a thorough knowledge of domestic science and a fuller recognition of the social disadvantages of houseworkers. The young working woman must have a social life; if she cannot get it in domestic service she will get it in a factory.

In the meantime the league urges employers to consider the need of endowing model agencies. Such an agency should make a specialty of substitutes for the present system. For example, it should have a department to give trained service hourly or daily. Then it should make an effort to get into touch with gentlewomen who will help with housework for a small wage and the same social recognition that is accorded the trained nurse and the seamstress. It should try to get unemployed men to do cleaning work. In "effete Europe" house cleaning is considered too formidable a task for the average maid, but here in America the most delicate girl is supposed to be equal to any amount of sweeping. Such a model agency would co-operate with all organizations that come in touch with the newly landed immigrant and the Southern negro, and in a short time it should have appreciably lessened the difficulty of the servant problem.

Talking over this plan with several heads of employment agencies, there was universal agreement as to its common sense. "But," said one employment agent, "you have got to get out of people's minds the idea that a general houseworker must do the washing before you will get gentlewomen to help. They aren't strong enough. In fact, I say, and I don't care who hears me," she said, sitting up very straight, "that the main point in solving the servant problem in America is to train the mistresses."

In short, in this, as in most things else, it is largely a matter of the personal equation.

October 20, 1907

DINED AND DANCED IN A VAT.

Italian Visitor Tells of Unique Experience in California Wine Colony.

Prof. G. B. Penne, an Italian author and Secretary of the Theosophical Society of Rome, who has spent four months traveling through America as special correspondent of La Tribuna, the Government organ at Rome, returned to New York yesterday and will sail for home on the Liguria next Wednesday. In speaking of his experiences he said he had enjoyed his trip to the Pacific Coast and was much impressed by what he had seen, especially in California.

"The Italian colony in California," said Prof. Penne, "is the finest in America, and has the best class of immigrants who come from Lombardy, Piedmont, and Genoa. I was astounded at the enormous quantities of wine made in the colonies of Asti and Cucamonga. There they have the latest machinery, eclipsing anything we have in Italy. I saw one vat at Asti which contained 500,000 gallons of wine and was invited to a banquet served in an empty vat where 200 persons sat down and over 200 couples danced later in the evening.

"In Italy the wine is made in small quantities and not handled by electric pumps, which I find extracts its force and the distinctive taste that is the charm in the home product.

"In the colony at Cucamonga they have twenty-seven miles of vineyards surrounded by wire fences to keep out the jack rabbits. The Italian laborers there earn \$1.75 to \$2 per day, live in delightful houses under conditions that have never been realized in Italy. I thing that the prospects for the industrious Italian immigrant in America are fine."

May 31, 1908

FOREIGNERS EAGER TO LEARN.

Illiteracy Among Them Less Than Among Americans.

Dr. Leander Trowbridge Chamberlain, Vice Chairman of the Immigration Department of the National Civic Federation, has been studying lately the question of illiteracy among the children of foreign-born parents who have come to dwell in the United States. He says that the result of his investigations is rather astonishing. For instance, says Dr. Chamberlain, three-fifths of all the prizes given in public schools of this city are taken by children of the foreign born.

"The question of illiteracy is one of the distinct and decisive issues with regard to our immigration," said Dr. Chamberlain a few days ago in discussing the immigration problem. "Of the foreign born parents here, 28 per cent. of them are illiterate. But are they careless about their children? Do they let their children remain in such illiteracy as has been their misfortune in their own case? If so, then it is a peril which attacks the whole nation.

"But now listen to this: The census of 1900 showed us that of the children of white, native born parents the illiteracy was 5.6 per cent.; that is, there were nearly six children out of every 100 who at 12 years of age could not read and write. It also showed that the illiteracy of the children of the foreign born population of the United States was exactly 1.6 per cent.; one third of the illiteracy of our own children.

"Actually these children of the foreign born attend school more days than our own children do. They play truant less. This holds good not only in New York City, but in New York State, and in Massachusetts and in all the nearby and the frontier States, like Ohio, Illinois, Iowa, Wisconsin, and Minnesota; everywhere the foreign born are our superiors in the matter of the absence of illiteracy among their children."

August 23, 1908

JEWISH COMMUNITY OF NEW YORK FORMED

After a warm argument lasting more than four hours, in the course of which the Rev. Dr. J. L. Magnes of Temple Emanu-El, the presiding Chairman, broke his ivory gavel, the Jewish convention, in adjourned meeting at the Hebrew Charities Building yesterday afternoon, finally adopted a constitution.

The meeting had been adjourned from the night previous when at a late hour the first two articles of a constitution for the proposed Jewish Community of the City of New York were adopted. The first was passed on the motion of Lewis Marshall, as follows:

I. That it be the sense of this body that a Jewish community of the City of New York be formed.

Immediately upon its adoption Jacob H. Schiff moved the adoption of Article II., which was unanimously carried. This article read:

II. The purpose of the Jewish community of New York City shall be to further the cause of Judaism in New York City, and to represent the Jews of this city with respect to all local matters of Jewish interest. This organization shall not engage in any propaganda of a partisan political nature or interfere with the autonomy of a constituent organization.

Dr. Magnes called the adjourned meeting to order at 2.30 yesterday afternoon and after a resolution was passed on the recent death of Rabbi Adolph Radin, the Tombs Chaplain, the Credential Committee reported that 218 Jewish organizations were represented at the convention, among them being 74 synagogues.

The third article of the proposed constitution was then read in English and Yiddish. It dealt with the qualifications for membership in the Community and provoked a long discussion, in which Rabbi Joseph Silverman took an active part. The paragraphs that provoked the greatest argument were those stating that "No person shall be eligible as a delegate unless he be an American citizen." And that "No political organization shall be eligible for membership." Dr. Silverman had this to say on these points:

"I don't think any delegate should speak for uptown or downtown, because he really speaks for himself. We are neither up or down town Jews, but Jews of New York, so let us drop the other manner of speech. We all want to be Jews and Americans. Above all things, we want to be known not only as Jews, but as Americans, who are loyal to their country. We want those who are not citizens to take out papers at once, and until this is done they are not fit to be delegates or to represent us as a body. If this community is to send representatives to the Jewish National Committee they must be Americans to do American National Jewish business. As to political organizations not being eligible, we are professedly not a political body, and do not wish to recognize any Jewish political body. This is a Community devoted to Jewish interests alone."

No Aliens Wanted, Sulzberger Says.

Cyrus L. Sulzberger, President of the United Hebrew Charities, said: "We have no business to organize ourselves for political purposes under a religious banner. There are no Protestant or Catholic political bodies and there should be no Jewish political body. Everyman should be an American citizen. We want no aliens in our community. This body is not for men who care so little for their country that they are not citizens. It is their duty to become such, and let it not be said that we are a community of aliens but of Americans who know their needs and what they want."

Judge Leon Sanders moved the adoption of the article point by point, and this was done amid much discussion, in which Miss Sadie American, Dr. Silverman, H. P. Mendes, Louis Marshall, and Joseph Barondess joined. When the paragraph as to American citizenship being necessary for membership came up there was much excitement, and one man, who was not a delegate, was asked by the Chairman to leave the hall. Rabbi Rabinowitz shouted that the delegates did not know for what they were voting, but the article on membership was finally passed by a close vote.

Article IV., on meetings and officers, was then taken up and passed, and then there was a hot debate on Article V., which stated the relationship of the community to the National Jewish Committee.

Rabbi Rabinowitz declared that the community should be local and independent of the National Committee, and Albert Lucas denounced the National Committee for what he called "an illegal combination with Police Commissioner Bingham concerning the Jews' observance of the Sunday business laws." He said that the Jews could regulate their Sunday business affairs without the interference of the committee.

Judge Sanders protested that there had been no "illegal combination," and that the matter of the Jews who observed the Sabbath doing business on Sunday was in accord with the statutes of the State. Louis Marshall then said:

Disclaimer from Louis Marshall.

"As one of the criminals of the American Jewish Committee, some explanation is due you from me. This committee is not for trampling on the community of New York, and to show this the New York members have tendered their resignations to that body. When this community is organized, its quota of twenty-five members will practically control the American National Jewish Committee. Are you willing to take this power which the former New York members of the committee are offering you? New York Jews are suffering from too much individuality and independence, and it is high time that they surrendered some of this for the good of the whole.

"The Jewish National Committee had nothing to do with the Bingham arrangement. The Commissioner called certain Jews in conference and the law was legally interpreted that Jews should not have to work seven days in the week. Jacob Schiff called the committee into being because of a necessary head at the time of the Russion relief work to protect the interest of the Jews in any part of the world. Was it wrong that the committee helped Jews in distress in all parts of the world? Was it wrong to make a passport of worth and American citizenship respected in any country? Here is the organization if you please. We don't want to take you. You come and take us!"

After much applause and cheering, Rabbi Drachman said that it would be unwise to take this step at the present time. The committee was doing a different work than the community should do, at it merely belonged to the social, religious, and charitable work of the Jews. What was needed was a local city Jewish community to handle internal problems. Dr. Magnes here explained that all members of the National Jewish Committee would be elected by representative Jewish bodies in Chicago, Baltimore, and elsewhere if the New York community took the head. He declared it a matter of jurisdiction for the committee to decide what was local and what National.

Chairman Breaks His Gavel.

At this point many of the delegates began shouting at once, a dozen men rising to a point of order, and Dr. Magnes broke his gavel in an effort to restore order. Article V., in which a committee of Twenty-five of the community is provided for and made a part of the National Committee and which reserves the power over local Jewish matters to the Executive Committee of the community, was finally adopted. The other articles as to dues, quorum, and amendments were quickly passed, and then on the motion of Samuel Dorf the whole constitution was adopted as amended.

Edgar J. Nathan, Chairman of the Nominating Committee, then announced the list of officers and Executive Committee nominated. The reading aroused a storm of protest, as it was asserted that partisanship was shown. A motion to postpone the election of officers was carried.

The convention voted to meet again Saturday evening for the election of officers.

March 1, 1909

ALIENS FEAR CENSUS MAN.

Chicago Italians and Poles Think Their Object Is Deportation.

CHICAGO, April 21.—Fear of deportation among Italian and Polish residents constitutes the greatest obstacle met by the census enumerators here. In certain districts of Chicago neither cajolery nor threats have been of any avail, and in three or four districts the task was abandoned, and the bureau chiefs informed that no progress could be made.

An investigation of conditions in the outlying districts was made yesterday by Special Agent Edwin L. Thacker.

"Some one has circulated a report among many of the Italians, Poles, and Lithuanians that the enumerators are collecting information for the purpose of their deportation," said Mr. Thacker today. "This has terrified many of them and they refuse to talk at all. I found one section in Chicago Heights where an enumerator had collected only thirty-two names in three days."

April 22, 1910

ITALIAN SOCIETY UPLIFTS IMMIGRANTS

The tide of Italian immigration, temporarily reversed in the panic year, is now turned this way in still greater force, according to figures collected for the Society for the Protection of Italian Immigrants. In 1908 some 57,000 Italians entered at Ellis Island, while in 1909 there were more than 215,000. The most hopeful part of the increase lies in the fact that in 1908 the society had to send 89 persons to the Benevolent Institute, while last year, with four times the number of immigrants, none was a case for the institute. Furthermore, in 1908 more than twice the number of "charity tickets" were issued than in 1909.

These statistics are gathered for the annual report of the society, which looks after the new arrivals. The first protection afforded is in the Home "Casa per gli Italiani," at 129 Broad Street, a large five-story building, which was completed and opened a little more than a year ago. In 1909 it sheltered 12,909 travelers, housing them and giving them three meals a day for 50 cents each. When several of the foreign benevolent associations were investigated by the Ellis Island authorities it was one of those to receive a clean bill of health. Besides those staying overnight, there were also thousands who, wishing to go out of the city, would stay a few hours, until guided to the proper boat or railway for the further journey.

The "casa" is a picturesque sight on a steamer day. The immigrants come up from the Battery by hundreds, fathers, mothers, and families, or single straggling adventurers, with their bundles tied to their walking sticks. Most of them have been ticketed to the society, for the badges are distributed among 256 bankers and agents handling emigrants in Europe. Besides this, there are the society's agents, to keep the people from falling into the hands of "runners" from some of the disreputable rookeries near the water front. On the first floor all is commotion in the two large rooms, one for the incoming and the other for the outgoing parties. There is much shouting as the lists of names forwarded from Europe are read off or parties are made up to be sent with guides to their respective destinations. Doubtless, too, the room is clean as it can be under the circumstances, which include mud, orange peel, and refuse from luncheons.

When once the "sorting out" has been completed a sudden peace and quiet fall over the "casa," as the folk who are to stay all night go upstairs into the hotel quarters. There are nearly 200 beds in the various dormitories, which are so arranged that children are kept with their mothers, and, as often as possible, in a room by themselves. There are ample facilities for bathing, and, unlike the reception rooms, are scrupulously clean, quite up to the standard of a hospital ward.

The dining rooms are large enough to accommodate several hundred at once. The food is cooked in Italian style, and while the service is not elegant, the food is abundant. The earthenware is heavy enough to defy destruction, but the cook and his assistants are clad in neat white caps and aprons. All the cooking apparatus is of the best quality, and not a speck or a spot is to be seen anywhere. Not only is all this comfort provided for 50 cents a day, but the society has tended to improve the standard of the commercial lodging house.

When steerage passengers, through the fault of the steamship company, are detained in New York for one or more days, the companies are liable for their support at certain definite rates. Formerly the rate was as low as 30 cents a day, and the housing given was of the worst imaginable. Through the efforts of the society, supported by the Italian Consul here, the rate has been increased so that the unfortunate "leftovers" can obtain some decent degree of comfort.

Identifying Italians with European police records, when they first land, is another phase of the protection afforded. Last year some fifty undesirables were "spotted" and turned over to justice, entirely without the aid of the police, while the society is called upon by the city police to report on the character of applicants for "runners'" licenses. The society hopes to make a feature of its special detective work in the future. Serving as bankers for the foreigners, recovering anything from lost baggage to relatives, is also included in the programme.

For work outside the city the most important is the camp school movement, started and supported by the society under the guidance of Miss Sarah Wool Moore, the Field Secretary. Not only were the first schools for adult immigrants working throughout the country in construction gangs originated, but they have been so successful that there is now a bill before the New York State Legislature making such camp schools a part of the night school system. The very first of these were in Pennsylvania in the region around Latrobe, Aspinwall, and Ambridge. The Pennsylvania Legislature has just taken action in the matter. The New York State affair is still more striking. After two small schools had been started at Stoneco and Wappingers Falls a larger field was opened in the Ashokan Dam works, and a school was started in the camp of the MacArthur & Winston Contracting Company. Last Fall, when an investigating committee under Miss Lillian Wald made a tour of the construction camps in the State, only this one was excepted from the general condemnation for bad conditions.

Primarily the school is to teach English, but practically it has become the social centre for the 3,000 foreigners in this section of the work. The most convincing proof of the school influence is the decrease of police duty necessary. Compared with other camps, the policing there is very slight, and yet it is at the other places that the "terrible times" occur. There is the night school, the day school, the kindergarten, and the mothers' meetings, for the workmen, their children, and their wives.

Not more than fifty men generally attend the school at once, but within the year so many have acquired a reading knowledge of English that this Winter a library has been added, upon which there are so many demands that a librarian soon also had to be sent. Besides English books there are Italian, Polish, and Slavic for the different nationalities. A Greek of the classic name of Socrates has recently joined the school. His shift ends at 8 o'clock, after the session has already begun. Nevertheless Socrates comes straight from work and without any dinner at all goes to the school. He doesn't wait to clean up either, rather to the disapproval of the regulation students, who seem to regard the class as a gala occasion for collars and neckties.

Not only does the school decrease police duty, but it lessens the number of accident cases. Many of the blunders, which often end fatally, are made because some workman hasn't understood his orders. Accordingly, in the school the men are gathered and put through a pantomime drill. The teacher gives the orders in English, and the men are to show they have understood by following directions. "Start, men!" "Go faster!" "Stop where you are!" "Look behind you!" "Run for your life!" "Fire!" "Blast!" and at that the men are to run to cover behind or under desks, chairs, tables, or whatever is handy.

There is one very dangerous piece of work, building the concrete core wall of the dam, way below the river bed. This the men call "working in the hole," and Andy, a 14-year-old boy, is the signalman on whom the lives of the others depend. Using a boy instead of a man for this is not wanton carelessness, but a premeditated scheme. Experience has taught the contractors that the greatest danger is to have a signalman who drinks. Accordingly they prefer to use a young boy like Andy, who, by the way, is a diligent pupil at the school.

Down "in the hole," the concrete comes in huge steam shovels weighing a ton; fifteen men dump and spread the concrete, the emptied bucket goes overhead, and the call "heads up!" warns the men when the next load is coming. So, up in the schoolhouse, Miss Moore has rigged up a light box and derrick for the men who work "in the hole." It is hoisted to the ceiling, "gripped," "pushed," and dumped to the given orders, and if the box is loaded with bean bags at least the men have learned what may save their lives in some tight place.

February 20, 1910

AN ARMY OF A MILLION IMMIGRANTS CAME HERE LAST YEAR

Italians Outnumber Other Nationalities, Poles Next ---Where They Go, How Much They Bring with Them and Other Interesting Facts About the Great Population.

Two or three years ago it was prophesied that the total number of immigrants in the United States in one year would reach a round million. This statement was received with more or less incredulity, but this figure as given for the last official year has actually been passed, for during that period no fewer than 1,041,570 immigrants entered the United States. It is evident that the United States must be prepared to digest this enormous annual incursion of 1,000,000 human souls.

The following table gives some analysis of the various races which compose this huge invading horde:

Italians	223,453	Black Afric'ns	4,966
Poles	128,348	Japanese	2,798
Jews	84,260	East Indians	1,782
Germans	71,380	Chinese	1,770
English	58,498	Islanders	61
Irish	38,382	Koreans	19
Maygars	27,302		

The firm hand of the United States administration was shown in the 24,270 aliens debarred. These were 118 polygamists, 2 anarchists, 156 idiots, imbeciles, and feeble-minded, 160 insane, 9 professional beggars, 11 paupers, 2,471 persons with diseases, 12,632 persons likely to become public charges, and 1,365 contract laborers.

In the above view the artist of the London Sphere has endeavored to visualize a portion of this huge army. For the more effective arrangement of the whole the numerical order has not been followed. Italians are shown on the left, next come the Poles, a smaller group of Chinamen, Jews, and so on.

THE past ten years has marked an increase in immigration into the United States to an extent that not the wildest dreams of the optimist nor the shrewdest forecasts of the expert could have forethought at the beginning of the decade 1900-1910.

Saving a phenomenal flood of human imports that in 1882 raised the sum past the three-quarter-million mark, the half-million figure was not reached half a dozen times in any year from the beginning of the Republic to the year 1902, when the sum total of immigration for the twelvemonth shot up to 648,743. In 1905 it reached the million-a-year mark, hung there for two years more, dropped back under the influence of the panic of 1907, around the three-quarter-million mark in the two succeeding years, and rose again in the fiscal year ended June, 1910, above a million.

The following are the totals of immigration for the past decade:

NUMBER OF IMMIGRANTS COMING TO UNITED STATES.

1901	487,918
1902	648,743
1903	857,046
1904	812,870
1905	1,026,499
1906	1,100,735
1907	1,285,349
1908	782,870
1909	751,786
1910	1,041,570

This makes a total for the decade ending the fiscal year of 1910 about 8,705,000.

When the patriotic citizen recalls that in 1820, the first year any record of immigration was kept by the Government, the total immigration was only 8,385, and that five years thereafter it had only grown to 10,199, he will begin to appreciate by contrast what enormous strides we have been making since the beginning of this century.

The grand total of immigration from 1820 to the present time is nearly 28,000,000; this in the ninety years. Yet the past decade has furnished about 30 percent, of that grand total.

Should there be the same proportionate increase for another decade, it is highly probable that the total for the two decades of this century will equal that for the entire eighty years of the last.

It is to be noted in the above figures, taken from the official reports of the United States, that despite the ill-effects of the panic of 1907 the average annual immigration for the last six years has been within a few thousands of a million —figures that cause the old-line American to pause and wonder whether the fish will swallow the bait or the bait the fish; in other words, will America become Europeanized?

"The remarkable expansion in all branches of industry and the resulting need of additional labor forces wherewith to prosecute them, says the special Industrial Investigation Committee of Congress, "has resulted in a general, simultaneous, and constantly increasing demand. In the face of this extraordinary demand, the labor resources of the country, consisting of the native stock and the races of older immigration from Great Britain and Northern Europe, were found wholly inadequate and resource was had of necessity to the races of recent immigration from Southern and Eastern Europe."

As a consequence the racial composition of the industrial population of the country has quickly undergone a complete change, and the cities and industrial localities, as well as the farming communities, of the United States have received enormous additions to their population in the form of industrial workers far more alien in speech, manners, and customs than ever were the British and Northern Europe immigrant.

The extent of this tremendous influx of a new class of foreigner may be more quickly realized when it is stated that this commission recently attempted to study a purely American community for purposes of comparison with immigrant industrial communities.

Of course, the first thing necessary was to find the purely American community, but after an exhaustive search the Commission was unable to find an industrial community of any importance or activity north of the Potomac River and east of the Rocky Mountains that did not include within its limits such a considerable number of industrial workers of these races of recent importation as to make them an important factor in the community.

The foreign communities of recent immigration which have sprung into existence by reason of the industrial expansion are of two general types. The first is a community which by a gradual process of accretion, has affixed itself to the original population of a city or industrial center, already well established before the arrival of the races of recent immigration.

Foreign communities of this type are as numerous as the older cities and industrial centers of country, any one of which in New England, the Middle States, the Middle West or Southwest, will be found to have its section or colony of recent immigrants from the south or east of Europe.

The dividing line between these immigrant sections and the original community to which it is affixed, is much more sharply drawn than were the like sections of the older immigrants, who, being nearer to the American blood and more akin in manners and customs, readily mixed and blended with the original inhabitants. The immigrant of the south and east of Europe mixes not at all with the American, but seeks his kind as a sheep would leave a herd of cattle to flock with sheep.

The second type of the communities of the recent immigrants has sprung into existence of recent years by reason of the development of some natural resources such as coal, iron or copper, or by reason of some extension of great industrial enterprise, as the erection of mills or buildings of smelters and furnaces.

The communities are usually clustered around mines or industrial plants wherein their inhabitants are employed and their distinguishing feature is that the large majority, often practically all the population, is of foreign birth and racially composed of Slavs, Magyars (Hungarians,) Italians and other people of the recent immigration.

Illustrations of this type are common in the bituminous and anthracite regions of Pennsylvania and the coal-producing areas of Virginia, Ohio, Illinois, West Virginia and Oklahoma.

In the great Iron Ore Ranges in the Mesabi and Vermillion districts of Minnesota, many communities of this character are also found. They are frequently established in connection with mills in the manufacture of steel, glass, cotton and woolen goods, although the preference of this class of immigrant appears to be for the mine and the old mills of the East.

As representative types of this class in different sections of the country, there may be cited, Lackwanna City, near Buffalo, a steel town ten years old, with a total population of 20,000, more than eighty per cent. of which is of foreign birth. Hungary Hollow, Illinois, another steel producing locality, built up from the barren fields during the last seven years, is the centre of a Bulgarian colony of 15,000 persons.

It is these two types of settlements that have absorbed the majority of the new immigration. Of course, the large cities have also got their quota. New York State received as settlers about 2,750,000 immigrants during the last decade, and unquestionably the city got a large proportion, probably a large majority, of that number. In the present incomplete state of the census for 1910, the unfinished work of the Immigration Commission of Congress and the fact that the Bureau of Immigration has not yet fully compiled its figures for the last year, render exact figures as to the relative amount of immigrants going to various towns and cities impossible to obtain.

A study of the nationalities which make up the various classes of immigrants and the sudden rise in this respect of hitherto unheard-of countries reveals some startling results.

Ireland, which a generation ago was looked upon as the backbone of the immigration movement, sent to our shores in 1901 30,000 immigrants; in 1910 she sent 38,000, while little Greece, hardly heard of in America outside the grammar school geography class ten years ago in immigration or any other matters, and which sent in 1900 a paltry lot of 3,700, forges forward in 1910 with more than 39,000, outstripping Old Ireland and increasing her output—literally—more than tenfold in a little over ten years.

The following show the immigration from the leading nations for the years 1901 and 1910:

Race.	1901.	1910.
Armenian	982	5,508
Bulgarian, Montenegrin, and Servian	611	15,130
Chinese	2,462	1,770
Croatian and Slovenian	17,928	39,562
Dutch	3,299	13,120
English	13,488	53,498
Finnish	9,999	15,736
French	4,036	21,107
German	34,742	71,380
Greek	5,919	39,135
Hebrew	58,098	84,260
Irish	30,404	38,382
Italian, (North)	22,103	30,780
Italian, (South)	115,704	192,673
Japanese	5,249	2,798
Lithuanian	8,815	22,714
Magyar	13,311	27,302
Mexican	350	17,760
Polish	43,617	128,348
Portuguese	4,176	7,657
Roumanian	761	14,199
Russian	672	17,294
Ruthenian	5,288	27,907
Scandinavian	40,277	52,037
Scotch	2,004	24,612
Slovak	29,343	32,416
Spanish	1,202	5,837
Syrian	4,064	6,317
Turkish	136	1,283
Welsh	674	2,244
African, (Black)	594	4,966

Quite a topsy-turveydom in the relative importance of countries is an immigration viewpoint!

The figures for the intermediate years are too voluminous to be given here, but they disclose that these changes between the beginning and end of the decade are not gradual. On the contrary, the lines of change if laid down on a plat would rise and fall like the heights and hollows of a hilly road.

For example, Bulgaria and Servia in 1903 sent 6,479 immigrants; in 1904 this dropped to 4,577, shot up in 1907 to 27,174, fell back in 1909 to 6,214, and lifted again in 1910 to 15,130. Likewise Poland, starting in 1901 with 43,617, ran up in 1907 to 138,033, dropped like a sinker the following year to 68,105, or less than one-half that number, winding up in 1910 with 128,348.

Indeed, this vagary in the number of immigrants from year to year appears to be common to all countries, although the variations are greatest in number and degree of change in the southeastern countries of Europe.

The English and stolid Scotch, true to their natures, vary very little from a steady line of increase year by year.

Perhaps for extreme permutations in this regard the Pacific Islander, not mentioned above as being negligible on account of paucity of numbers, is first in the list. In 1901 there were 167 immigrants from these tropic isles; that fell away to 5 in 1907 and 2 in 1908, a falling off of more than one-half in a year. In 1910, however, the islands came forward with a total immigration of 61, a gain of numberless per cents. in two years!

The amount of money brought over by immigrants is a somewhat difficult matter to determine, since many times funds are brought over for the use of resident immigrants, the bearer acting merely as a friendly carrier.

The total amount brought over by the immigrants for the year ended June, 1909, was $17,331,828, or about $23 a head for the 751,786 immigrants arriving that year. This, as shown by the table quoted below, is only a negligible amount above the average brought by immigrants for the past ten years.

Of all the nations the English immigrant seems to provide himself most bountifully with funds wherewith to carve a future for himself in the New World. From certain representative statistics in the Immigration Bureau, it appears that out of about 25,000 Englishmen, 14,000 had more than $50 with them on landing, while only 11,000 had under that sum.

On the other hand, the Italian of the south verges on pauperism when he lands. Out of 133,000, only 6,000 had $50 or over; the remainder of 127,000 had less than that sum. The immigrant Pole, however is no millionaire, there being only 1,700 out of a lot of 60,000 that could boast as much wealth as $50.

In the statistics whence the above figures are quoted the Spanish-American excelled the Englishman in immigrant wealth, 564 out of a total immigration of 644 having $50 or more, but the smallness of the number detracts from its value as a comparison with the more numerous Englishmen.

Following is given a table, not yet published by the Government, showing the total immigration by countries for the last decade ended June 30, 1910, together with the average amount of money brought by each alien:

Race	Total Immigrants.	Average Amount Money in Dollars.
African	32,504	$22
Armenian	28,842	24
Bohemian and Moravian	94,603	27
Bulgarian, Servian, and Montenegrin	97,093	18
Chinese	19,702	29
Croatian and Slovenian	309,727	15
Cuban	40,115	32
Dalmatian, Bosnian, and Herzegovinian	30,654	20
Dutch and Flemish	83,096	42
East Indian	5,762	61
English	387,005	55
Finnish	133,065	20
French	111,410	54
German	698,061	39
Greek	210,794	22
Hebrew	976,263	13
Irish	371,772	26
Italian (North)	342,261	25
Italian (South)	1,761,948	14
Japanese	132,706	41
Korean	7,697	7
Lithuanian	158,089	11
Magyar	318,674	15
Mexican	41,490	11
Pacific Islander	661	53
Polish	873,660	12
Portuguese	66,560	15
Roumanian	82,210	16
Russian	80,602	19
Ruthenian	148,143	18
Scandinavian	580,105	24
Scotch	138,333	49
Slovak	332,446	14
Spanish	48,944	48
Spanish-American	10,572	104
Syrian	50,281	32
Turkish	12,742	31
Welsh	18,631	49
West Indian	11,347	55
Other peoples	10,772	32
Total for decade	8,795,386	Averge, $22

The above races or nationalities are divided scientifically into five grand divisions by the Immigration Bureau—the Slavic, contributing 3,178,490 immigrants to the total above; the Iberic, 2,188,527; the Teutonic, 1,831,332; the Celtic, 977,407, and the Mongolian, 166,528. All other peoples, 508,102.

Where has this vast horde of immigrants taken itself to abide? It has been shown above that they herd together, like cattle, having nothing to do with any other humans other than of their race. The following shows the States in which they have settled; the record, however, is not quite complete, since it lacks the last fiscal year. The table below, however, gives the ten years preceding that and is, for all practical purposes, correct.

Number of immigrants settling in the following States in the decade preceding 1909:

State	Immigrants
Alabama	9,155
Alaska	1,105
Arizona	8,648
Arkansas	3,122
California	197,439
Colorado	45,784
Connecticut	197,554
Delaware	10,522
District of Columbia	11,115
Florida	55,132
Georgia	4,290
Hawaii	91,984
Idaho	5,722
Illinois	565,340
Indiana	50,004
Iowa	48,105
Kansas	24,605
Kentucky	4,973
Louisiana	15,953
Maine	18,718
Maryland	61,831
Massachusetts	576,024
Michigan	175,017
Minnesota	145,823
Mississippi	4,893
Missouri	100,804
Montana	20,084
Nebraska	34,921
Nevada	6,352
New Hampshire	21,930
New Jersey	391,164
New Mexico	2,918
New York	2,492,613
North Carolina	1,624
North Dakota	47,904
Ohio	326,601
Oklahoma	6,728
Oregon	21,568
Pennsylvania	1,449,780
Philippine Islands	36
Porto Rico	9,751
Rhode Island	77,407
South Carolina	2,153
South Dakota	28,323
Tennessee	5,513
Texas	39,611
Utah	19,359
Vermont	15,468
Virginia	12,193
Washington	83,827
West Virginia	49,365
Wisconsin	119,240
Wyoming	8,275
Total	*7,762,817

*Being the total immigration for the decade.

It will be seen from these figures that the North, the Middle West, and the Pacific Coast gets the bulk of immigration from the last ten years. The proportion that drifted to the South is hardly worthy of mention, although the large majority of these immigrants came from warm climates.

It is clear that the new immigrant shuns the country, as contrasted with the life of the city, or rather his own little colony in the city or hard by the factory, where for the first time in his poverty-stricken life he gets living wages.

Unlike the German or the immigrant from the British Isles, he not only does not seek to learn, but positively abhors the customs of his adopted country. He has no confidence in any manner of business or enterprise conducted by Americans, or indeed any foreigner but his own race.

Their children are brought up in ignorance of the English language, and taught to look upon the country they never saw —and where their parents would have starved, had they remained—as the chosen of God. They inculcate in them dislike, even hatred for everything that is American.

That immigrants cherishing openly such sentiments, who will soon be by their numbers a political power in the land, themselves mere puppets in the hands of a few unscrupulous leaders, are a menace calls for no debate, and it is known that efforts, and strong ones, will be made at the forthcoming Congress to check the flood of undesirable aliens. The Commissioner General of Immigration in his last report urges most strenuously that steps be taken in this direction.

Authentic statistics from all the institutions of detention in the United States show that of the Kelts detained therein 15 per cent. are in penal institutions. Of the Teutonic 17 per cent. are thus held; of the Slavs detained 23 per cent. are held for crimes, and of the Iberic, (which embraces all Southern Europe,) 39 per cent. are detained for crime.

Thus it will be seen that the Iberic or South Europe division leads in crime, with the Slav of East Europe, second, the Teutonic, or German-Scandinavian, third, and the Keltic—embracing most of the English-speaking peoples of Europe—are the least criminal.

The same statistical reports show that there were at the time 809 aliens held for murder in penal institutions in this country, of whom 253 were Italians; that 373 were held for attempts to kill, of whom 139 were Italians. There was one Italian held for murderous crime for every two of all other races combined.

It is further shown that the States in which are located the large cities have the greatest proportion of aliens held for crime. Thus out of about 45,000 aliens held for crime in all the penal institutions in the country nearly 13,000 or 28 percent., are held in the State of New York; 5.601, or 12½ per cent., are held in Pennsylvania; 5.490, or 12 per cent., are held in Massachusetts, and 3.359, or 7½ per cent., held in Illinois, thus making nearly 27,000 in the four States mentioned, or 60 per cent. of all held in the entire country. The majority of these criminals belong to the South and East Europe immigrant classes.

In the schedule of occupations as classified by the Bureau of Immigration, there are fourteen professions, fifty occupations under the head of skilled labor, and a dozen under the head of miscellaneous, which include practically everything from a banker to an unskilled laborer.

The professional engineer is the most numerous class among the first. Among the second class of skilled labor it is surprising to find that the tailor is by far the most numerous among the immigrants. In the year ended June, 1910, there came into the country nearly 19,000 immigrant tailors, while the trade of carpenter and jointer only claimed about 13,000 and the year previous only 8,000. Bakers have been among the followers of some trade to make a notable increase during the last decade in joining the immigrant ranks, while barbers and blacksmiths both have more than doubled in immigration in the last few years.

Among the professions it should be mentioned that actors appear to have made a sudden inroad upon America, the number coming in last year being double of the year previous, while that, in turn, was twice in number of the year previous to it.

The greatest number in any single line was the common laborer, there being 174,000 immigrants of this line of business coming in during the last fiscal year; the farm laborer, however, was a close second with 171,000.

It is notable that while of the unskilled immigrant laborers 119,000 returned to Europe during the same period, less than 3,000 of the skilled laborers returned.

This vast difference in the number of returning immigrants in these two lines is said to be due to the fact that thousands of unskilled laborers in Europe are beguiled by the agents of steamship companies into coming to America by the rosy prospects set out before them. Arriving in one of the great cities the laborer finds that work is scarce and living high. Hence within a short time he returns, discouraged and perhaps not a little disgusted.

The farm laborer, on the other hand, not being dependent upon the city, finds congenial work and plenty of it, with ample food at the first farmhouse where he may inquire; hence he remains.

"That question has been the bane of the Immigration Bureau for years," said a prominent official, with a grimace between a smile and a frown, when asked what statistics he could furnish the inquirer as to the religious faith of the immigrants who came over during the last decade.

"We haven't got any statistics on the subject of the religion of one single immigrant. We carefully avoid asking them any question even remotely bearing on their beliefs. Many years ago we tried it, but they were of no particular value, such statistics, and it created a good deal of trouble in the inquiry. You see, many of these fellows have come from countries where they were persecuted on account of their religious beliefs, and it was quite a shock to a poor foreigner, on coming to the land of the free, to be confronted at the very door with official inquiry as to this very thing he fled to avoid. So we dropped the whole matter."

Some experiments and examinations recently made by Prof. Franz Boas of Columbia University under the invitation of the Immigration Commission have brought forth some startling theories, if indeed, they have not progressed sufficiently toward conclusive proof to be set down as scientific facts.

It has long been an admitted fact from general observation that the changed conditions, educational, political, and social, in whose atmosphere the immigrant lives, gradually changes his habits of life and modes of thinking, until the immigrant gradually becomes an American. But little or no thought has been given to the possible effect of the changed conditions on the physical type of the descendants of such immigrants.

It was thought by the commission that if accurate measurements were taken of the bodies of immigrants and their offspring at different ages valuable results might follow. Accordingly Prof. Boas was engaged to do the work and formulate his results. These results have so far been based entirely upon observations and measurements of Sicilians and East European Hebrews.

The results have been more far-reaching that was anticipated. From the report there is every reason to believe that the descendant "changes his type not gradually, but even in the first generation almost entirely. Children born a few years after the arrival of their immigrant parents in America develop in such a way that they differ in type essentially from their foreign-born parents. The differences seem to develop during the earliest childhood and persist through life."

Says Prof. Boas's report further:

"An attempt was made to solve the following questions: 1. Is there a change in the type of development of the immigrant and his descendants, due to his transfer from his home surroundings to the congested parts of New York? 2. Is there a change in the type of the adult descendant of the immigrant born in this country as compared to the adult immigrant arriving on the shores of our continent? The investigation has shown much more than anticipated, summarized as follows:

"1. The head form, which has always been considered one of the most stable and permanent characteristics of human races, undergoes far-reaching changes, due to the transfer of the races of Europe to American soil. The East European Hebrew, who has a very round head, becomes more long headed; the south Italian, who in Italy has an exceedingly long head, becomes more short headed; so that both approach a uniform type in this country so far as the roundness of the head is concerned. This fact is one of the most suggestive ones discovered in our investigation, because it shows that not even those characteristics of a race which have proved to be the most permanent in their old home remain the same under new surroundings; and we are compelled to believe that when these features of the body change the whole bodily and mental make-up of the immigrants may change.

"2. The influence of American environment upon the descendants of immigrants increases with the time that the immigrants have lived in this country before the birth of their children.

"3. The changes in the head form which the European races undergo here consist in the increase of some measurements and the decrease of others. The length of the Hebrew's head is increased. Among the Sicilians the changes are, on the whole, of an inverse nature. The length of the head is decreased, while the width of the head is increased.

"We are compelled to draw the conclusion," says Prof. Boas, after fully discussing the matter and the methods of his experiments in his report, "that if these forms and traits that are considered the most stable change under the influence of environment, presumably none of the characteristics of the human types that come to America remain stable. The adaptability of the immigrant seems, therefore, to be very much greater than we had a right to suppose before our investigations were instituted."

If all of which is true, then the fears of the native American that he may be swamped by this influx of Magyars, Slovaks, and other impossible peoples are utterly groundless. New York's climate will fix things all right.

October 16, 1910

TOPICS OF THE TIMES.

Bad Advice Given in Good French.

That New York, the gateway of a Nation, or, rather, of a continent, should have a newspaper for each group of strangers arriving, and only too often settling, here, is natural enough. Indeed, this has come to be, or at least to seem, rather a congeries of colonies, each with its own language, than a city in the older sense of a homogeneous community. Much more remarkable is it that Winnipeg, so new and so remote, should already have, according to a report of origin to make it credible, periodical publications in forty-two languages and dialects, including the Canadian variant of English.

To be really startled, however, one must have his attention called to the fact that even in the heart of New England foreigners not yet even lingually assimilated are come to be so numerous that they prefer, demand, and get the news printed for them in words that the waning representatives of the "old stock," so recently almost unalloyed by any alien admixture, cannot understand. Such a shock, with a lancinating little pain, is the result of receiving from Worcester, Mass., a chance copy of L'Opinion Publique, a French paper, well edited and better printed than most of those that come from Paris, and "carrying" enough advertising to indicate prosperity.

Of course we all know that Worcester is a big manufacturing town, containing many artisans of French Canadian origin, but why do they thus stand apart from their neighbors? And, as if to emphasize their separateness, L'Opinion Publique gives prominent editorial space to a vehement, almost passionate, defense of the language spoken by its readers as being, not as certain base detractors have said, a "français déformé," but the real thing, pure and uncorrupted. That contention is just a bit amusing, since the Canadian patois not only exists, with its innumerable and interesting reminders of the old French provinces whence came, to be thereafter isolated from later changes, the original settlers of New France, but that patois has in recent years won for itself a respectable, though small, place in dialectic literature, both prose and verse.

Of course L'Opinion Publique does not claim that all the Frenchmen in Canada and the United States speak "good" French, but it says that many of them do, and, under the impulse of a strange and lamentable misguiding, it urges its readers to see to it that their children devote more of their schooltime hours to the learning of correct speech—in French! A deadly enemy could not give worse advice.

October 22, 1910

MAKING AMERICAN FARMERS OF ITALIAN IMMIGRANTS

Successful Experiments in Building Up Colonies to Till the Soil, Though States Give Insufficient Encouragement.

Just Landed Immigrants Starting for a Farm

A Typical Italian Farm of Ten Acres

The Home of a Prosperous Italian Farmer.

A Land Clearing Force of Italians.

IN the recent discussion as to the manner of distributing the immigrant population that pours through Ellis Island every year and of prevailing on the new citizens not to herd together in large cities, but to go out and help solve the country's agricultural difficulties, the Labor Information Bureau for Italians took an interesting part.

The manager of the bureau, Mr. G. E. di Palma Castiglione, explained first that Italian immigrants were much more favorably disposed toward farm life than has been popularly supposed and, further, that they were met with little encouragement. The authorities in different States officially express a desire for immigrants who will settle on the soil, but practically they do little to attract this class, and do not even co-operate to any extent with the agencies that are trying to turn the incoming tide from the cities to the country.

The offices of the Labor Information Bureau at 59 Lafayette Street serve as a clearing house for Italians, whatever kind of work they are seeking. The bureau works in connection with the Bureau of the Department of Commerce and Labor, and coöperates as far as possible with the various State Departments of Agriculture, that is they co-operate as far as the States respond to their appeal.

Mr. di Palma Castiglione feels, however, that there is need of a more definite and far-reaching scheme of co-operation among agencies and of helping the immigrant to settle on farms than has yet been devised, and, while he tells of the success Italians have had, he points out some weak spots in the present system.

"There is a strong impression among Americans," said Mr. di Palma Castiglione, "that Italians do not settle on the land when they come over here. They see the floating crowds of laborers that go about working on railways and street improvements, and they come to the conclusion that the Italian farmer in this country is pretty nearly an unknown quantity.

"This is a mistaken impression. Italians always have taken up land when they could, and they are increasingly disposed to do so. It is not possible to give more than very roughly approximate figures as to the Italian agricultural population, and the list I have is, I know, incomplete. There are more colonies than I have noted.

"It is a matter of general knowledge among Italians, for instance, that nearly all the abandoned farms of Connecticut have been taken up by our immigrants in recent years. We Italians who travel a little about the country find more and more evidences of our fellow-countrymen wherever we go. We can speak with assurance of the increase in the Italian farming population, even though we can not give full details and figures.

"Roughly speaking, the Italian colonies run about as folows:

	Families.
Alabama	22
Arkansas	332
Louisiana	934
Mississippi	183
New York	500
New Jersey	1,093
North Carolina	50
Missouri	120
Tennessee	84
Texas	841
Wisconsin	50

"And let me assure you that the immigrant who takes up a farm in this country does so because he wants one very much. Few of the States give themselves any real trouble to attract immigrants to their vacant lands.

"When we opened this bureau we wrote to the various State Commissioners of Agriculture and Immigration Boards for co-operation, but we got very little. In Illinois, Virginia, and Missouri the authorities were kind enough to publish in the newspapers our requests for specific information as to farms that might be bought, and these newspaper notices brought us many letters from individuals who had land to sell. Only these three States did that for us.

"New York has prepared a book which gives descriptions of the farms to be had in this State. Maryland and Virginia have similar publications and Massachusetts is preparing one. These are very helpful to us in our work, but they would hardly be helpful if they came directly to immigrants, since they are written in English.

"From these books we take descriptions of farms that cost $1,000 or less and translate the items into Italian and publish them in little bulletins like this. You see on the one side is an account of conditions in cities where labor is wanted and on the other are the descriptions of the farms for sale, with careful mention of every detail connected with the place that we can get hold of. The prices run from about $600 to $1,000.

"It is difficult to reach the Italians, especially by literature, even if it is in their own tongue, for it must not be forgotten that 50 per cent. of the Italian immigrants are illiterate. Much information must be given by word of mouth. It is a tedious process, but an inevitable one. We send out bulletins to the parish houses of Italian priests and to Italian newspapers. We trust that those who can read will spread the news among those who cannot.

"Concerted effort on the part of the immigration and agricultural authorities would doubtless give excellent results if it was carried on with a full appreciation of the immigrant's mental processes, and if their bulletins were written in many languages and gave the most definite sort of information. Such efforts would be worth making, although in view of the fact that wages on farms are lower than wages on construction work, it is not likely that Italians will at first flock to the country.

"Little by little, however, as their families come over and join them, they will be inclined to buy farms with the money they have saved in construction work. Even working against all the difficulties that meet them now, many, you see, have turned to farming. Some of the colonies have been formed under the leadership of some man actuated by business or by philanthropic motives, and some have sprung up spontaneously, coming into being as Italians from the same part of the country drew together for companionship and mutual assistance.

"For instance, there is a very prosperous colony near Fredonia, in Chautauqua

County, N. Y., made up of Sicilians. A certain Constanzo Siracusa from the Valle Dolmo, in the Province of Palermo, went there about fifteen years ago and took a place in a jam factory in Fredonia. He had the spirit of the pioneer and wanted to own land. He saved from his wages at the factory and bought a little bit of land. Part of the year he worked as a day laborer and part he spent improving his farm—a practice that has been followed by many Italian farmers and that we recommend to immigrants.

"As far off as Arkansas there is an Italian colony, formed by a priest by the name of Tonti and called Tontitown. The first settlers were colonists who escaped from the fraud of Sunnyside—a colonizing scheme into which many Italians were led by misrepresentations and cheating. Tonti led 200 souls into the wilderness, and after a Winter or two of distressing hardships the colony began to prosper, until now it is one of the most prosperous we have in this country. These people are from the north of Italy.

"Vineland, N. J., is another most successful colony, of the kind that was deliberately planned. They have a beautiful stretch of country, and it is splendidly cultivated—a model of industry.

"There are many colonies of farmers on the outskirts of large cities. Any one who will take the trouble to go to the outlying districts of Brooklyn, or to Yonkers, or White Plains, or West Farms, will find colonies of Italians working as truck farmers. They raise their produce and bring it in on their own carts to their fellow-countrymen in the city, who sell it on their fruit and vegetable stalls.

"There is one point in which the Italian farmer is marked off from the American who works beside him. The Italian is, in this country as in Italy, primitive in his methods of farming. He does not use tools and machinery. He works with his hands. He is a long way from the point of mechanical efficiency that the up-to-date American farmer has reached.

"But in spite of this he manages to compete successfully with the American. The same primitiveness which makes him use his hands instead of machinery and which thus limits his output gives him fewer wants than the American knows. He is frugal in his habits and he works most industriously. His wife and children do not want luxuries, or even many of what the American would call necessities, and they all help the man in his work.

"In this way things are evened up. And no doubt, as the Italian acquires more knowledge of machinery and modern methods his wants, too, will increase, so that the difference will again be evened.

"There is no doubt that the Italians will turn more and more to the country, for various reasons. It must be remembered, to begin with, that the Italian immigration has been a matter of recent years; that is, the immigration in hordes that we see now. It dates back only to about 1900.

"Now, before a man can take up a farm, at least outside of the colonies, he must first know the language. The things that he grows he must sell, and he must deal with English-speaking people. In the towns he can get along perfectly well without speaking English, but in the country he is helpless without it.

"Then, secondly, before a man can wisely take up a farm he must understand climatic conditions. He must not only know about the seasons but he must know something of the nature of the soil and he must have observed. He cannot acquire this knowledge nor can he learn to speak the language properly without a fairly protracted stay. But as an increasingly large number of Italians learn to speak English and stay over here a year or two an increasingly large proportion of them will turn to agriculture.

"It would be an exceedingly useful thing if some association could be formed which would furnish the capital for buying abandoned farms and allow immigrants to pay for them on the installment plan. Many lots are now sold to immigrants on a basis of five dollars down and five dollars a month until paid for—lots that cost from $60 to $125. It would be perfectly possible to do the same thing with farms on a larger scale.

"Such an association would be national in character and would serve all immigrants alike. Some such work as the Jewish Agricultural Aid Association has carried on for Jews might easily be undertaken on a larger scale.

"Americans believe that it would be a good thing for industrious immigrants to settle in this country instead of merely working here to make money and then going back to Europe to spend it. They also know that agricultural conditions here are discouraging, that there is a constant tendency on the part of the young people of the farms to move to the cities.

"Would it not be possible to form some such society which would at one and the same time keep a worthy class of immigrants in this country and help to provide the agricultural population that is so much needed?" It seems to me that such a society could do a useful work at a very small cost."

December 4, 1910

PENNILESS IMMIGRANTS WHO HAVE MADE MILLIONS

HERMAN SIELCKEN got married the other day and took his bride to his country home, Mariahalden, at Baden Baden. Kaiser Wilhelm has thirty-three palaces and vast landed possessions, but no one of his estates approaches in beauty or extent that owned by Herman Sielcken. Kaiser Wilhelm is monarch of Germany, but his rule is limited to the German Empire. Herman Sielcken is a monarch of commerce and his rule extends the world over. Kaiser Wilhelm was born a Prince. Herman Sielcken was born poor. Perhaps 100,000,000 persons pay tribute to Emperor Wilhelm. Perhaps 500,000,000 persons pay tribute to Herman Sielcken.

Sielcken is but one of hundreds of immigrants who came to America with little more than energy and hope, and to-day are kings of industry, many of them with wealth greater than that of any hereditary monarch.

Of the tremendously rich and powerful men of the United States, Sielcken is one of the least known. He was born in Hamburg and before he was twenty-one went to Costa Rica to work for a German firm there. He didn't like the country and within a year left for California, where he got a job as shipping clerk. As soon as he learned to speak English with reasonable fluency he sought work that would give an opportunity to him to travel and get acquainted with people. A wool concern engaged him as buyer and for five or six years he traversed the territory between the Rockies and the Pacific buying wool. On one of these trips he was in a train wreck in Oregon and nearly lost his life. When he recovered from his injuries he came to New York seeking work. He got a clerical position in a concern that imported crockery and glassware.

It was in 1868 that Sielcken left Germany. It was in 1876 that he reached New York. In those eight years he made a fair living, nothing more. Then there came a remarkable change.

In Costa Rica he had learned to speak Spanish. Because of that fact he was able late in 1876 to obtain employment with the firm of W. H. Crossman & Son, which handled coffee on a commission basis. Sielcken went to South America to solicit consignments for the Crossmans. His success was surprising. For six or eight months every mail from the Southern continent brought business to the house. Then, as the story goes, his reports ceased suddenly. Weeks and months passed and the firm heard nothing from him.

Becomes a Partner.

What had become of him the Crossmans had no idea. They feared he had caught a fever and died. To trace him was difficult. He had no regular itinerary. It distressed them a good deal to lose so promising a representative. Giving up all hope of getting any information about him, they looked around for a man to take his place. Then one morning he walked into the office and said "How d'ye do," just as if he had departed only the evening before. The members of the firm questioned him eagerly. He answered some of the questions and some he didn't. Then he laid a package on the table.

"Gentlemen," he said, "I have given a large amount of business to you, far more than you expected, as the result of my trip. I have a lot more business which I can give to you. It's all in black and white in the papers in this package. I think any person who has worked as hard as I have and so well, deserves a partnership in this firm. If you want these orders you may have them. They represent a big profit to you. Good work deserves proper reward. Look these papers over and then tell me if you want me to continue with you as a member of the firm."

After the Crossmans looked those papers over they had no doubt of the advisability of taking Sielcken into partnership. He was only a junior for some years, but in 1894 the firm became Crossman & Sielcken. It prospered amazingly. For the last fifteen years it has been the leading coffee house of the world.

At various times Sielcken was credited with working corners in coffee. Because of this he got to be one of the most feared and hated men in the Coffee Exchange. After a while coffee didn't offer enough play for Sielcken's tremendous energy and ambition. He embarked in various enterprises, among them the steel industry and railroads. No one was too big for him to cross lances with. He and John W. Gates had a titanic fight in American Steel and Wire. Gates got the worst of it. Then Sielcken got in a row with E. H. Harriman and George J. Gould. This fight was for possession of the Kansas City, Pittsburgh & Gulf Railroad, now known as the Kansas City Southern. Harriman, Gould and Gates had taken it away from Arthur E. Stillwell. They had no particular regard for the Kansas City, Pittsburgh & Gulf. It was a north and south railroad and disturbed the east and west traffic on the trunk lines they controlled through the grain belt. The Kansas City, Pittsburgh & Gulf gave a short haul to the Gulf. The old established trunk lines gave long railroad haul from Kansas, Nebraska, and the Middle West generally to the Atlantic seaboard.

Harriman permitted the Kansas City, Pittsburgh & Gulf to droop. Very little money was spent on maintenance of road or equipment. Wrecks were frequent. Traffic fell off. So did receipts. Then Sielcken, representing a syndicate of Dutchmen who held a large block of bonds, got control. He had no practical railroad experience, yet what he did with that property within six months amazed transportation men. When he took charge there was an average of three wrecks a day. Within a few months the average was reduced to one wreck a day. Within a year he brought the property up to a fair state of efficiency. To-day the Kansas City Southern is getting back to what it was designed to be—a real railroad.

While busy with the Kansas City Southern, Sielcken found time to engineer one of the biggest deals in the world. Brazil produces 85 per cent. of the world's supply of coffee. Most of the Brazilian coffee is raised in the State of Sao Paulo and territory near by. Brazil made so much money out of coffee for some years that the planters thought there was no end to the world's demand for the bean. They increased their production so tremendously that they demoralized the market. The price of coffee declined to 6 cents a pound, but still they couldn't get rid of their stock. Each year the surplus was growing larger. The coffee trade was threatened with demoralization. Brazil, so far as its coffee industry was concerned, faced ruin.

About this time Sielcken conceived a scheme whereby the world would not get any more coffee than was necessary to maintain prices at what he thought was a proper level. His scheme is known the world over as the valorization plan. By it the Brazilian Government buys all the coffee that is produced in Brazil and regulates the production. The money for financing the Government in this operation was raised through the issue of $75,000,000 in bonds, which were taken by English, German, French, Dutch and American bankers. Sielcken has the marketing of all the coffee. He will not sell to a coffee broker to deliver on contract. All the coffee he sells goes to the jobber or is shipped abroad. He decides how much coffee the world is to have. He has saved Brazil, or rather the Brazilian coffee raisers, from ruin. But the coffee drinkers of the world pay the bill. As a result of his scheme coffee rose from 6 to 16 cents a pound, when the world had the largest amount of coffee in its history. The law of supply and demand cuts no figure with Herman Sielcken. He's above anything like that.

When his partner, Mr. Crossman, died, it was discovered that the two men had a remarkable contract. Each man had made a will giving one million dollars to the other. It was a sort of bet on which one would live the longer. Mr. Crossman died last January and Mr. Sielcken got one million dollars to add to his many other millions. How much money he has no one but Sielcken knows. In New York he lives at the Waldorf-Astoria. He bought the German estate known as Mariahalden some years ago and has made it a wonderland. He has one of the largest rose gardens there in the world and probably raises more orchids than any other one person on the globe.

Story of Cudahy.

The immigrants who have become kings in America came from all parts of Europe. Patrick Cudahy, who is the head of one of the greatest of pork packing firms, comes from that part of Ireland famous for its fighting cats. He was born in Callan, County Kilkenny, sixty-four years ago. They named him Patrick because he was born on St. Patrick's Day. He came across the ocean in a sailing vessel with the rest of the Cudahy family. His father had to work for a long time in New York as a common laborer. Then the elder Cudahy went to Wauwatosa, near Milwaukee, Wis., and started farming. One thing the father did was to raise pigs, and when little Patsy quit school at twelve years of age and took a job in a Milwaukee grocery store at a dollar a week the father told him he was a fool and that he wasn't beginning right. "Be a farmer, my boy," he said. "Raise pigs; stick to pigs."

Young Cudahy had no intention of being a farmer. He ran errands for the grocer for two years. The second year he got two dollars a week. Then pigs and three dollars a week proved too strong an attraction for him. He took a job with the Roddis Packing Company. The concern was not a big one, and the boy did a little of everything from slaughtering hogs to keeping books. He remained with the Roddis Company for six years and then went to a larger firm, with which he remained for four years. When he was twenty-four he became Superintendent of the slaughter house of Lyman & Wooley. He did so well for this concern that he attracted the attention of the Armours and was made Superintendent of the plant of Plankington & Armour at Milwaukee.

Up to the time he went with Armour he had not received more than $125 a month. Within a few years he was the highest salaried Superintendent in his line in America. He saved a fair share of his money and bought a small interest in the firm. In 1875 the main office was established in Chicago, but Cudahy was left in charge of the Milwaukee plant, and when Plankington died in 1888 Cudahy and his brother Michael bought the Plankington interest, took over the ownership of the Milwaukee business, and started on their own hook under the name of Cudahy Brothers. The growth of this establishment has been prodigious. It has spread until its product is sold all over the globe. The town which has grown up around the Cudahy plant near Milwaukee has the name of Cudahy. In that one plant a million hogs a year are slaughtered and the business amounts to nearly $15,000,000 a year.

Frederick Weyerhaeuser is the lumber king of America. He is past 70. He came from Neidersaulheim, Germany, when he was 18 years old, and went to Erie County, Penn. Four years later he moved to Rock Island, Ill., and went to work in a lumber yard. He rose to be foreman. He saved a little money and, with his brother-in-law, F. C. A. Denkman, bought a small mill. They did not have enough to pay for it, but had to give their notes. Weyerhaeuser did the buying for the mill. He was shrewd and prudent. The firm prospered. Its business broadened, and gradually the partners acquired pine land. Within fifteen years of the organization of the firm it was doing the largest lumber business in the Mississippi Valley. In 1896 it bought out the C. N. Nelson Lumber Company at Cloquet, Minn., and acquired not only a great lumber plant but 600,000,000 feet of standing timber.

To-day Weyerhaeuser controls not only a big share of the lumber business of Minnesota, Wisconsin, and Illinois, but through his purchases of timber land in the Appalachian country and various other parts of the United States, he owns more standing timber than any other man in the world. He makes his headquarters at St. Paul. He lives simply. No one would suppose from his quiet, modest manner that he is enormously rich. Ten years ago his wealth was estimated at $30,000,000. Since then the value of lumber has increased greatly. He may be worth forty million, fifty million, or sixty million dollars to-day. He has no fads. Work is his recreation. One of the queer things about Mr. Weyerhaeuser is that he never lost a dollar in a lumber deal and never made a dollar in any other business in which he invested money. One of the jokes he tells at his own expense has to do with his purchase of bank stock. Before he got to be very rich he was induced to take an interest in various small banking concerns. Not one of them succeeded. Since then he has stuck to lumber.

Comparatively few men know Jules Weber. He is the king of the kitchens. He came to America from France when he was a mere lad. He got work in the old Astor House. He was the egg boy. His duty was to keep track of the stock of eggs. After a while he became assistant cook. He had rare talent in culinary affairs and won a high reputation before he left the Astor. He saved his money and opened a restaurant in Thirty-fourth Street, between Broadway and Seventh Avenue. Incidentally, he imported French delicacies. The importing business grew so large that he gave up the restaurant and devoted all his attention to merchandising. To-day he is a great authority on cooking. Most of the French chefs who are installed in the big hotels of America were placed there through his influence.

Made Money in Real Estate.

He never has lost his French viewpoint. When he had a restaurant he lived over his establishment. When he built a great warehouse he followed the same rule and had his living quarters upstairs. He has made a lot of money in real estate. Years ago he made up his mind that there was one spot in New York where a man could not make a mistake in buying property. That was in the section between Thirty-fourth and Forty-second Streets and Fourth and Seventh Avenues. He put his surplus money into buildings in that part of the city, and to-day it is the richest district in Manhattan. Weber is the leading spirit of one of the quaintest organizations in America—the Thursday club. Its members are the leading French chefs of New York. Once a week the club has a luncheon. When the member in charge of the meal fails to win the approval of the critical persons who are at table he is disconsolate. When he gets the applause of the fellow-members and a vote of thanks he feels that he has achieved the highest honor attainable in his profession. Some of the chefs plan their luncheons a year ahead and give to them an amount of thought far greater than to the most important banquet they ever are called on to prepare.

Michael Idvorsky Pupin is the king of the telephone. He gets his middle name from the town where he was born. He is a Serb. His parents were peasants. He came to America as a stowaway, and did not have a dollar when he landed. He worked as a farmhand in Maryland. When he had learned a little of the English language he returned to New York and did many kinds of odd jobs. He worked in a factory, and as a rubber in a Turkish bath. While he worked in the factory he attended night school. He saved a little money and went to Columbia University. He not only worked his way through the university, but while a pupil there he earned $3,000 tutoring American youths. He had a great talent for electricity, and when he finished his course at Columbia he was made a teacher there. He has specialized on telephony and has invented some of the most important devices for the improvement of that great branch of communication. He devised the Pupin coil by which telephony over long distances has been made possible. For this invention he got $400,000 cash from the American Telephone and Telegraph Company.

He is a great business man and has made corporations pay handsomely for his inventions. He is the leading Serb in America and raised most of the money that was collected in America to finance the Serbs in their war with the Turks. He is at work now on ocean telephony and expects to make it possible for the voice to carry across the Atlantic. One of the big halls at Columbia is given over to him for his work. He is recognized to-day as one of the great scientists of the world. Whenever he feels fagged and in need of recreation he goes to his country estate in Connecticut and tills the soil as his father did before him.

United States Senator Knute Nelson is a prince of politics. He was only six years old when he came to America from Norway. He says humorously that he is descended from a long line of Norwegian pirates. He had a very hard time as a boy, his mother being widowed. He worked on a farm in Wisconsin and got his schooling as best he could. He went into the army, and at the close of the war studied law. He and John C. Spooner were partners. He was in the Wisconsin Legislature twice, in the Minnesota Legislature three times, a member of the House of Representatives three times, Governor of Minnesota twice, and has been a United States Senator from Minnesota eighteen years. He is one of the great Norsemen of America.

Oxnard a Sugar King.

Henry T. Oxnard is one of America's sugar kings. He is from Marseilles, France. He was only a youngster when he came to America. To him more than any other man is due the development of the beet sugar industry in America. To-day he is the President of the American Beet Sugar Company, and also President of the American Beet Sugar Association, which comprises all the beet sugar factories in the United States. He has cut quite a figure in American racing circles, having owned a large number of thoroughbreds which were contenders on the great tracks of the East in the days when racing was in better odor than it is now.

Andrew Carnegie, steel king, came here from Scotland as an immigrant. Most persons know his story. He worked as a messenger boy for three dollars a week in a telegraph office. He became a telegraph operator and drifted into the steel business. When he sold out the Carnegie Company to the United States Steel Corporation he got $250,000,000. There are half a dozen kings in Europe whose combined fortunes do not equal this amount.

The lemon king of America is Simone Saitta. He is from Palermo, Sicily. He had very little money when he landed in New York. He has built up a tremendous business and now handles nearly one-fourth of all the lemons imported into the United States, and one-fourth of all the grapes imported into the United States.

Joseph Di Gorgio is the banana king of America. He is only thirty-eight years old. He came here an immigrant lad and worked for $5 a week on Pier 20, North River. To-day he owns twenty-eight steamships engaged in bringing bananas from Central America and the West Indies to the United States. He gives 100,000 tons of freight a year to the American railroads.

Henry Siegel, dry goods prince, came to America from Germany when he was fifteen years old. The first work he got was as shop boy in a Washington store. It was four years before he rose to the dignity of earning $15 a week. He became a salesman, and after awhile went into business with two of his brothers in a little store in Parkersburg, Penn. In 1876 he moved to Chicago and became a manufacturer of cloaks. It was there he met Frank Cooper and went into partnership with him. Their start was very modest. What Henry Siegel has grown to be in the dry goods trade of Chicago, New York, and Boston is pretty well known.

Four men met at dinner in a New York hotel the other night. They were Carl Laemmle, one of the big men of the motion picture business; Julius Hilder, who has been a prominent figure as an importer of notions and fancy goods; Julius Klugman, a Fifth Avenue furrier; and Leo Hirschfield, Vice President of a great candy corporation. Twenty-nine years ago the four were companions in the steerage of the steamship Neckar, immigrants on their way to America. To-day they are millionaires. Each found fortune in the land of promise.

It looks from all this as if the immigrant boy, with the spur of poverty, does better in America than the native born with all his natural advantages.

November 2, 1913

NEW IMMIGRANTS STUDIOUS.

To demonstrate the serious character of the Jewish immigrants from Russia and Rumania and their eagerness for self-improvement and assimilation of American influences, the Hebrew Sheltering and Immigrant Aid Society made public last night figures compiled by Frank Goodell, librarian of the Seward Park branch of the New York Public Library. The society emphasized the importance of the figures, since they show the constant tendency of people to remove from the Seward Park district to the Bronx and to Brooklyn, leaving the newer immigrants and their families to fill their places.

During the year 1913 the branch circulated 395,231 volumes, of which about 40 per cent. were classed as juvenile. Less than 10 per cent of the library's circulation was of books in foreign languages. Only 41 per cent. of the books circulated were fiction, and this percentage included the juvenile fiction. Mr. Goodell said that the small percentage of fiction was the subject of incredulous wonder among librarians of other branches, since most branches circulate 65 per cent. fiction, and on the upper west side fiction represents 70 per cent. of the total circulation.

Of literature, poetry, drama, and essays, 51,120 volumes were circulated; of history, 31,000; of sociology, about 30,000; of science, 14,000, and of travel, 14,000. Of the literature circulated, 18,000 volumes were poetry. During the year 1913, 77,041 persons used the reference library. The anxiety of the new immigrants to keep abreast of the times led to the creation of a special file of pamphlets and clippings. Reporting on the keenness of the newer immigrants for up-to-the-minute information, Mr. Goodell said:

"When 'Rosenkeavalier' was first produced at the Metropolitan Opera House, the demand for the score and libretto could not be supplied, and when Rabindranath Tagore received the Noble Prize the first demand for his books came from this branch. The increased demand for opera scores and books bearing on the opera is another example of the varied interests of the neighborhood. Now that the Century Opera Company has placed grand opera within the reach of all, this branch circulates more than 200 librettoes and scores every month, as against 90 a month last year."

The English classes conducted by the library are crowded, Mr. Goodell says, and the demand for the use of the library is so great that it is kept open now until 10 P. M., whereas, formerly it was closed at 9 o'clock.

March 1, 1914

ENGLISH TONGUE PREVAILS.

The Language of the Greatest Number of Our Foreign Born.

WASHINGTON, May 16.—Of the 32,243,382 persons of foreign white stock in the United States in 1910, the English and Celtic, including Irish, Scotch, and Welsh, had the largest representation, according to the mother-tongue bulletin issued to-day by the Census Bureau. As reported to the Census Bureau, the total foreign white stock whose mother tongue was English and Celtic numbered 10,037,420. This number represented 12.3 per cent. of the total white population of the United States in 1910, which was 81,731,957.

The German group numbered 8,817,271, or 10.8 per cent.; Italian, 2,151,422, or 2.61 per cent.; Polish, 1,707,640, or 2.1 per cent.; Yiddish and Hebrew, 1,676,762, or 2.1 per cent.; Swedish, 1,445,869, or 1.8 per cent.; French, 1,357,169, or 1.7 per cent., and Norwegian, 1,009,854, or 1.2 per cent.

The number of persons in the United States of foreign white stock reporting other principal mother tongues were:

Bohemian and Moravian, 539,392; Spanish, 448,198; Danish, 446,473; Dutch and Frisian, 324,930; Magyar, 320,893; Slovak, 284,444; Lithuanian and Lettish, 211,235; Finnish, 200,688; Slovenian, 183,431; Portuguese, 141,268; Greek, 130,379; Serbo-Croatian, 129,254, (including Croatian, 93,036;) Servian, 26,752; Dalmatian, 5,505; Montenegrin, 3,961; Russian, 95,137; Rumanian, 51,124; Syrian and Arabic, 46,527; Flemish, 44,806; Ruthenian, 35,359; Slavic, (not specified,) 35,195; Armenian, 30,021; Bulgarian, 19,380; Turkish, 5,441; Albanian, 2,366; all other and those whose mother tongue was unknown, 313,834.

May 17, 1914

ASSERTS IMMIGRANT IS SCHOOL PROBLEM

Research Director Lays Stress on Education's Part in Work of Assimilation.

CITES OLD PREJUDICES

Albert Shiels Says Critics of the Foreign-Born Are Apt to Generalize Too Much.

The school and the immigrant are the subjects of a report made to the New York Board of Education by Albert Shiels, Director of Reference and Research. Within the last thirty years, he says, there has been considerable uneasiness over the continuous flow of immigration, and the question has been raised as to the power of this country to continue successfully to assimilate so many variant types that might imperil American institutions, customs, traditions, and standards of living.

Whatever justification there may be for this feeling, says Mr. Shiels, its existence is a fact. It has been formulated in magazine articles, books, public discussions, and Governmental reports. It is interesting to note, however, that the present protests affect to be less against immigration, as such, than against the present character of the immigrant laborer, usually referred to in terms of his geographic origin. In the period from 1810 to 1883, 95 per cent. of the total immigration was from countries west of the Russian boundary and north of the Mediterranean and the Balkan Peninsula. From 1883 to 1907 81 per cent. of the immigration was from Italy, Austria-Hungary, the countries of the Balkan Peninsula, Spain, and Asiatic Turkey. Out of the total immigration for the year 1913 the latter type of immigration represents 74 per cent. of the total.

In a discussion of present-day immigration, he declares, we should be mindful that our judgment is always likely to be affected by possible illusions, and that impressions and prejudices may be confounded with fact. One frequent illusion is due to a prejudice derived from familiarity with the newly arrived immigrant, as we may meet him in person—his physical condition and his habits. We are likely here to confound the accidental and temporary with the fundamental and permanent. An immigrant may be unclean in person, repellent in his habits and apparently content to live under offensive conditions. Yet in such matters he may, in many cases, be but a creature of circumstances, perhaps their victim. The real question is whether he is inherently vicious, criminal or indolent, and, on the other hand, whether he may not under favorable conditions develop into a desirable citizen.

Critics Prone to Generalize.

A second illusion, Mr. Shiels finds, is the acceptance of generalizations, which are themselves based on very hasty and superficial inferences. One axample is the tendency to repeat what appears on a printed page, without thoughtful reflection. Thus it is frequently stated that recent immigration implies a disproportionate increase in poverty, disease, and crime. Some of the opponents of the immigrant movement affirm this, others are very cautious. There do not seem to be many convincing statistics to prove the validity of all these assertions. Another charge is that immigration displaces native labor, reduces wages, and increases unemployment, or, at least, contributes to under-employment.

A third illusion, and a very real one, in Mr. Shiels's opinion, is a manifestation of what Lord Bacon called an idol of the tribe, meaning our own human tribe. It is the attitude which most persons have against any stranger. To the degree that he differs in nationality, or race, there is always apt to be a corresponding intensity of suspicion and dislike.

Our welcome to immigration, Mr. Shiels says, has been a splendid service to the world; it is necessary to remember, however, that our goodness has not been without reward. When we realize, for example, to what degree the development of mining industries depends upon foreign labor, or remember that three-quarters of all our laborers on railroad and construction work are foreign born, certainly we must hesitate to adopt policies of exclusion, even if we adopt methods so illogical as the imposition of a literary test.

The question, therefore, Mr. Shiels thinks, is not to be solved by personal prejudice, by unprofitable economic theories, nor altogether by the assumption of responsibility for the welfare of the unfortunate of other lands. What we need to know are the fundamental qualities of the immigrant as we are getting him, what promise he and his children hold for good citizenship, what influences, pro and con, he produces on wages, labor and opportunity, and—perhaps the most important question—what we are prepared to do to reap the richest advantage from his coming. What we are doing consciously and purposively is a question the answer to which would scarcely flatter our national pride. If we shall do no more than we have done, if we are content to depend upon the slow process of evolution that has worked fairly well under happier conditions, it is possible that, notwithstanding all the material contributions that the immigrant may make, immigration itself may be a bad thing.

Whatever may be done in the matter of future distribution, Mr. Shiels points out, the immediate problem for teachers is to consider the immigrant, not in the places where we might like to place him, but where he insists on going himself. It is true that many immigrants working in camps or as farmhands go to rural districts. Until State educational agencies evolve a more effective system, however, the main concern of public school teachers will be centred in the interests of immigrants in the cities.

Teachers Need Broad Culture.

It has been doubted, Mr. Shiels says, whether the regular public school teachers are best fitted for this special work. Whatever the merits of the contention for a special school system, the actual conditions are that for some time to come many day public school teachers will continue to do the work in evening classes. Many of them have already done it with conspicuous success. Whatever the system may be, every teacher should have an equipment greater than a mere command of special method, however necessary that is. He should have a real understanding of the questions other than curriculum and teaching technique. The broader understanding brings the deeper sympathy; the foreign pupil should not be a mere recipient for a certain mass of knowledge, or a lay subject for technical demonstration of teaching skill. The foreign pupil, appreciated in all his relations, becomes in the teacher's presence a man like himself, with hopes, aspirations, misadventures, misfortunes; a fellow-being to be aided and encouraged without contempt or patronage.

Our civilization, Mr. Shiels thinks, has indicated a somewhat naive confidence in the public school teachers' abilities, which, though flattering, is embarassing. The school is assumed to be the exclusive agency in education. Whatever faults the rising generation may have are calmly attributed to it. Such fundamental influences as the home, the street, the newspaper, the differences in individual instincts and experiences are all brushed aside, and the teacher is called upon to assume the rôle of mentor, parent, and inspired prophet. Omnipotence is the prerogative of divinity, and not even public sentiment can confer it on a corps of teachers. There is, however, one duty a teacher is required to perform and to perform well. That is the duty of effective instruction.

Now, in the case of the immigrant, Mr. Shiels says, the whole educational procedure is handicapped from the beginning. The evening school, in which the teaching of foreigners has so large a place, in the words of one teacher, is the stepdaughter of the educational organization. When school funds must be reduced, it is the evening schools that bear the reduction. When other branches of instruction come to require more and more rigid conditions for the training of teachers, it is only among evening classes for foreigners, that in so many communities a place remains for almost any kind of teacher with any kind of license. When elaborate courses in special methods have been evolved in training schools, extension courses, and teachers' colleges, consideration of the critical problem of instructing our immigrants has alone been omitted.

July 11, 1915

THE NEW IRISH ARRIVALS.

To the Editor of The New York Times:

In the interest of many of your readers who have written to me in the last few days, kindly mention in your columns that there are no girls at the Mission of Our Lady of the Rosary looking for employment. The girls lately arrived from Ireland were claimed by relatives who were more than willing to find them positions.

It would seem that the day of the Irish servant girl is waning, if not about to close, in the United States. The girls now landing are better educated, more intelligent and more awake to their opportunities than their sisters of past years. Many are determined to become saleswomen, stenographers, telephone and telegraph operators, nurses—positions for which some are already qualified.

I feel confident that they will make as good a record in these avocations as so many of their predecessors have made in the humble but honorable field of domestic service.

REV. M. J. HENRY.

New York, April 17, 1920.

April 18, 1920

More Old World in New York

By HELEN BULLITT LOWRY

NEW YORK, at our water's edge, has become a city of shadows—the shadows of vanished pomps of the Old World. The American tourist comes to the water's edge. He goes to the shows on Broadway, Charlie Chaplin and otherwise. He shops along Fifth Avenue, either by window or charge account. He "slums" in Greenwich Village. And he thinks that he knows New York, city of blatancy and gayety, obviousness and newness. New York is really old—as old as the racial history of the least of her adopted immigrant children. New York is really sad with the glory that was Poland and the grandeur that was China. Her immigrants have made her a city of memories and of experience of the centuries.

On the east side, in the foreign colonies, is where this glamour lingers. But do not think that all who come riding on a sightseeing bus may vulgarly peer into the traditions and memories of the centuries. The east side yields up her "local color" in exact proportion to the amount of understanding and knowledge of the old countries that you carry with you. The sage has said, "He who would bring back with him the gold of the Orient must carry with him the gold of the Orient. And he who would bring back knowledge from his travels must carry with him knowledge." What you would find of the vanished social history must be carried with you to the east side, since the memory of a vanished thing is as illusive as a butterfly.

One Summer, a Polish family lived near us on a Connecticut "abandoned farm." We literally did not exist for each other. They were nothing but a new brand of "Wops" to me. Only a little knowledge has since been added to my slender store—and yet—I met a Polish peasant, a peddler of brooms, just after talking with a musician about Polish folk songs. I had learned that the songs of the Polish mountaineers are so free that they sing of the Madonna herself as a peasant girl instead of a queen of heaven. I had learned that not one folk song had been added to the store of the nation since the partition of Poland a hundred years ago. And I knew that locked behind the lips of each native-born peasant on the east side were the songs of his native land.

"Won't you sing for me?" I asked him casually. He burst into song in a rollicking, care-free baritone, as naïvely as a canary when the cloth has been lifted from his cage. With his brooms swaying over his shoulders he walked swaggeringly away as he sang. I recognized a theme that Chopin had caught up into a polonnaise and made familiar to us mere dwellers of cities. On the east side you've got to know what you want before you get it.

Regular international jokes come sometimes as the reward of seeking. Now the Polish peasants had been thinking themselves the only Catholics in the world until they came to America. To the west of them the world was Lutheran. To the east of them was the Greek Orthodox Church, for miles and miles and miles. Indeed, it had just been an accident in 964 A. D. that it had been a German invader who "converted" the practical Polish Prince Mieszko to the Pope instead of a Slavic invader who converted him to a Patriach. By such a cast of the die of fate a millennium ago is it decided which church our immigrants shall attend in New York.

The Poles settle down in New York and they straightway seek out the Catholic Church—only to discover that there are Italian Catholics and Irish Catholics. They receive the same sort of shock that a New Yorker gets when he asks direction of a policeman in Minneapolis and is answered in a Swedish accent. I had thought that they all came from Ireland. Moreover, the Poles discovered that the deed for their church is lodged in a Bishop with an Irish name. And so there has been started a "National Polish Church" in America with a different administration from the American Catholic—simply to create Bishops with seemly names like Mieroslawski, for property to be vested in.

It is indeed the church life of the immigrant that yields up the richest memories of the centuries. There is the Ukrainian Catholic Church on Eighth Street east of Third Avenue, which has moved into an old disused Methodist Church. As the new tenant moves in a van load of furniture, they move in their Catholic dogmas and their ancient Greek forms. As I knelt in that church among those gentle, typical Ukrainians with their high cheek bones and slanting eyes—as the chanting rose and fell of unaccompanied human voices—as the deacon thundered out prayers a whole octave deeper than the Methodist brother before him—as the red lamps burned luridly before the icon of the Virgin, and as we lighted our tapers in the perpetual twilight—as we touched our heads to the floor and as we kissed the holy icon and breathed the medieval incense—well, I was one that hoped that the good, decent shades of those Methodist sisters hadn't returned to peep in at us.

The very walls of this Methodist Church (which had been calcimined pale green, as is the way with church calcimines) have been incrusted over with memories of mighty Byzantium. It is a poor little congregation that has moved in. They haven't the wealth that had St. Sophia for marble pillars and mosaics. So, very simply and naturally, they've painted over the calcimine a Byzantine stage setting of marble pillars and vistas and ninth century stern browed saints.

But if you would really see where the east meets the west and the fourth century mingles with the twentieth, go to the Russian Orthodox Church itself, with its ritual which received its last alteration in the fourth century. Since then not a gesture of the hand has been changed. Since then not a chanted word has been inserted. In the Cathedral, on East Ninety-first Street, just off Fifth Avenue, you breathe the very air of Russia before the revolution. Lighted tapers pierce the dim corridors of fifteen centuries ago. You stand without "pews" as when the cathedrals of Europe were new.

And before your very eyes the drama of the early Church is performed. It is the life and death and ascension of Christ performed in symbolism. The church is darkened, while a voice chants the prophesies from a balcony. The lights are lighted when the angel bears his tidings unto Mary. Round the altar pace the six priests, with their every gesture possessing an age-old meaning. When the bread is pierced by a sharp poinard we know that Christ's side is pierced by the spear. Those changes of sacerdotal dress, those crossings and prostrations, processions and genuflexions fairly absorb the senses.

I happen to be one of those who have scanned in great volumes the plans of the early Basilican churches and dreamed of the drama of High Mass when classic and Christian ritual met in the days before the Barbarians swept over Rome. I did not know that I could see such another drama right here in New York, preserved intact from the same centuries in Byzantium—the pages of history turned back. The Connecticut Yankee may have had the like sensation when he struck King Arthur's Court.

So, too, does Chinatown yield up her ancient religion to those who go truly seeking—not on a sightseeing bus. Much of Chinatown is stage setting—a nice business arrangement between the bus lines and the keepers of "opium dens" for the special benefit of tourists. But there is a perfectly real Chinese religious and domestic life which goes on upstairs, which the tourists do not see—where Chinese babies that look like slant-eyed yellow dolls jabber alternate Chinese and English words, and where Chinese women in their native costume keep the ancient home fires burning.

The restaurants are stage properties—but the grocery stores are not. They are filled with as great a variety of vegetables imported from China and California as has any American market—and not one of them is familiar. There are fringe-tailed green vegetables that must look down on our beans and spinach as the Cheops of Egypt would regard a tombstone in Woodlawn. They are vegetables with ancestors.

I went into a home of Chinatown, where I was received by a fascinating little Ming Toy sort of person in Chinese trousers, just able to lisp a few words of English.

Above me hung the tablets of her husband's ancestors. They were written on red silk, and before them burned the daily incense. The Chinese lady served me tea in a handleless cup. No Chinese hostess would allow you to leave her house otherwise. These Chinese women whom I met and conversed with in "pidjin English" live by conventions that belong to the outgrown Old World culture. The father-in-law still selects the bride—for the very good reason that in China for a few thousand years the patriarchal system prevailed, and the father-in-law supported his sons' wives under the ancestral roof. In just common justice he deserved a voice in the matter.

Yet these women of Chinatown, courted by a father and living a shut-in life, seem happy—happier than our standard-bearing suffragettes, happier than our upper west side New Yorkers, attending matinees and munching chocolates. Perhaps it is because the Chinese quarter is the one spot in America where the years hold no menace to woman—where her position grows in dignity with age and grandchildren. That at least is left her of the philosophy of Confucius, while all the new "liberties" that American women have wrested haven't been able to do away with woman's real problem, which is growing old and losing prestige by gray hairs.

Bohemia, too, has its fascinating New York quarter—in the Seventies stretching from Third Avenue to the River—more romantic than the would-be Greenwich Village Bohemia because these real Bohemians are incurably romantic. Serenely they refer to the glory of six hundred years ago—as if they were mentioning Bryan's defeat in 1896. They'll admit that it's ancient history—but they take for granted that ancient as it is, it is still fresh in everybody's memory.

They tell of a civilization founded on the culture of Byzantium, where baths really never quite went out. They tell of an aristocracy whose estates became Austrian crown lands before the dawn of the fifteenth century. A woman of this New York Bohemian colony goes home next month to claim such an estate, now that Czechoslovakia has again become a country and the crown lands are the property of the Czech Republic. Her work in New York has been to shampoo heads. Mine has been one of the proletariat heads that the daughter of an ancient nobility has laundered.

"The Bohemian Counts and Countesses that you've heard about belong to a new rich class," scornfully explains she who shampoos my head. "Why, they have been created by the Austrian Government in the last six hundred years! The old nobility, knighted for military service, ceased to use their titles when their estates were gone."

Yet each household taught the tradition of its ancient glory from father to son. That vanished nobility has largely become the professional class of Bohemia, since, after the aristocracy of landed estates was gone, they have consistently demanded the aristocracy of education as guarantee against annihilation. Each son was taught by his father that some day a son of that house would regain his rightful place in the world, and it behooved him to have table manners in readiness. When she who washed my head was a child, she was taken to an ancient castle just outside of Prague, a castle which was Government property.

"My daughter," had said the father in the beloved forbidden Bohemian tongue, "some day there will be a war in Europe. In that war Bohemia will regain her freedom. When that time comes, come claim the castle of our fathers. Here is proof of my right to the title. Put this record away in safety. Show it to no one till that day comes."

The record which he gave the child that day is at my elbow as I write. It is an official document issued by the Austrian Government, acknowledging his right to the crest and an ancient title conferred on the founder of his house in 1071 by "Kaiser Konrad" for the "invention of certain instruments of war." There was a Bohemian nationalistic revival in 1862. It was at that time the sons of the nobility "before last" went to the Austrian Government and demanded, in preparation for this very hour a record for their ancient heredity—and got it.

That father died leaving no moneys behind him, as is the way with national revivalists. And the little girl did not receive her rightful "aristocracy of a professional education." Life forced her to learn a trade, and by that trade she has lived in New York. If the Czech Government returns her her castle, the hair will not grow so well on my head.

New York's little Bohemia is vibrant with such strains of its ancient history. Because John Huss was martyred in 1415 and because Catholicism was enforced by the Austrian Government, New York's Bohemian colony is the only foreign group that has practically no religion. When they have to economize on a funeral, they choose the carriage and leave off the preacher. Yet it seems to make little difference ethically what we pay our devotion to, just so we pay it passionately to something. Instead of God or saints, the New York Bohemians have made a fetish of their national integrity. Because Austria forbade the use of the Bohemian language in schools, the second generation in America continues to study that once forbidden language as well as English.

Because of that national revival in 1862, a "Sokol Gymnasium" exists on New York's East Seventy-first Street. These sokols were secret patriotic societies that flourished for fifty-six years under Austria's very nose under the excuse of being just athletic gymnasia. It was a nation-wide scheme to prepare the Bohemians physically and emotionally for the hour of revolution. The very name sokol means falcon—reputed swiftest of the birds. Women were trained along with the men. There was no mid-Victorian type in Bohemia. Over on Second Avenue the butchers are burly women. She who washes my head tells of a mother who went out hunting in boots to her waist and was as cool a shot as a man. An epic grandeur lends itself to the great bare gymnasium in Seventy-first Street, for in the sokols history has been in the making.

Glamour of the Old World again lingers in a Polish dance hall in the Polish quarter—a glamour of Polish mazurkas mixed in with American toddles. These Polish peasants dance the American shimmy atrociously, but the minute a waltz is played a wild rhythm stirs the air—a rhythm that no American-built feet can follow. Middle-aged men seize middle-aged women about their thick waists and whirl them. They stamp their feet with the wild abandon of the Russian ballet—and it's a poor imagination that can't substitute shining black boots for tan oxfords. The Polish dance hall at 23 St. Mark's Place is very well worth seeing.

So is the needlework of Ukrainian women—if you happen to be looking for a real peasant art. The Ukrainians, more than any other of our foreign-born women, have been taught how to put their native needlework on the American market, through the help of the various organizations—among them the parish priest of the painted Byzantine church. There happens to be an exhibition this week of this Ukrainian peasant embroidery at the Arden Gallery, at 599 Fifth Avenue—peasant art tamed to run around a smock for Greenwich Village or an American sport hat—art that was the accumulation of the Slavic centuries.

Yet perhaps the best of all our foreigners, if the coin be vanished glory, are our Russian refugees. There hasn't been anything like them in New York since the ex-court of Marie Antoinette was taking in laundry in old New York and old Philadelphia by day and fiddling minuets by night—still trying to nibble at its poor crumbled bit of cake. They really are here, these princesses and Generals and Admirals, being errand boys and chauffeurs and milliners. Scratch the surface of any brownstone front on any east side street that runs under the elevated, and you may find one of them in the fourth floor back, washing out underwear and hanging it in front of the feeble gas burner. In the subcellar of your own apartment house there may be that Admiral who fled the Bolsheviki through Siberia, to work his way across the Pacific, by stoking the fires of a steamer. He was trained labor by the time he reached here, and quickly got himself a job.

Each of these refugees wears on his brow the memory of fearful happenings, before he made his escape. But you can't get them to talk of the black moment. Instead they will begin to tell you that in ten years from now they'll be back in Russia. Then a look of consecretion will come into—say the eyes of the long, lean Captain of the Royal Guard who is making lamp shades at Eightieth Street and Lexington Avenue, while he murmurs:

"Ours is a sacred trust. We must keep alive the brains and the education of Russia for the hour of rehabilitation." He pastes the gilt braid upon the parchment. "We have not the right to die. Therefore we must work." Put an ear to the east side of New York, and this is the refrain that you will hear instead of the minuet of a century ago.

Instinctively one ponders what an aristocracy this would be if they ever get back into power—an aristocracy that knows what work is. There is a Professor of Political Economy, for example, from a great Russian university, working in an Ontario quarry. He is learning a lot about political economy. He who was engineer of the Balkan Fleet is now "learning the business from the ground up" as a mechanic. A famous architect, who was wont to lay out garden cities, has joined our carpenters' union.

The lower walk of American life is the one that the refugees are fitting into. They do not seem interested in our American "social distinctions." All work being "declassing," they seem to conclude, why bother to choose your new friends, according to the brand of work they happen to be doing? If you come from a house of Princes what is the difference between a millionaire manufacturer of sparkling gelatine and his chauffeur? The Russians are good-natured people, and so take life as it comes.

Moreover, the Princes are usually not well dressed enough to be playing successfully the hungry boy game. Not one can be discovered who has managed to marry a living—much less an heiress. In the matrimonial market their titles seem about as negotiable as are notes of our late Southern Confederacy. So they are drifting into our chauffeur class. They associate with mechanics and go calling on one another.

There is the Princess of a famous Old World name in a lodging house of the O. Henry period, whose most intimate friend is the wife of a chauffeur. But stop—you don't know yet that she is a Princess any more than did the other lodgers before the limousine drove up to the door. A lady draped in sables stepped out and gingerly climbed the stairs to the fourth floor back, where the little old lady lived whose coat lining was so frayed and conspicuous that the landlady had thought of asking her to leave. She brought down the tone of the house.

The lodgers leaned their heads over the banisters, as the way with lodgers. Then, "Oh, my dear little Princess!" cried the lady in sables, seizing the lodger of the fourth floor back in her arms.

That is how they discovered that they had a princess among them, a princess whose most intimate friend is the wife of a chauffeur, who used to do modelling in the "suits and coats." When she and the princess go out to market on Third Avenue, the intimate friend still walks with the manequin gait She, too, has her pomps of yesteryear. Yes. Princes and Princesses have not proved snobbish. The real snobs will come next generation, when every Russian descendant of every immigrant will be announcing that his ancestors came over as refugees. So it used to be said by the malicious that every Southerner in New York had a family that "lost its money in the war." It is a familiar process known as making an American aristocrat in one generation.

But, for all that bleak prophecy, the east side of foreigners still has its echoes of vanished glories. If you don't believe me, stroll into any of the foreign clubs and watch how the red wine flows. It is an echo of the pre-1918 period. "We hope in several years from now to dispense with our bar," serenely explains the secretary of one such club. "That would make it pleasanter for the women and children. But we can't afford to for several years more."

April 17, 1921

Some Popular Misconceptions About Our Resident Chinese

BY Major S. P. R. de RODYENKO, Chinese Military Forces, Retired.

AN American friend of mine has a Chinese servant, honest, clean and laborious. Late at night this American called on me in great exitement. It appeared that the servant had asked for a day's leave, saying that he had to see some members of his tong down in Chinatown. My friend pictured his servant as implicated in some crime which was likely to get his and his master's name into the papers. Questioning the Chinese about his mission to his tong developed the fact that he wanted to see about an indorsement—a reference. He was buying a Victrola on the instalment plan and references were required. The various stories published about tongs, and tong wars, had never suggested that these institutions could be used in so pacific and practical a fashion.

Down in Chinatown, in Mott Street or in Doyers Street, are huge signboards indicating to the stranger where the headquarters of the various tongs are, and where he can meet people who come from his own province in China. There is no more secrecy about the tongs than there is—say—about the Kentucky Society in New York.

The tongs are organizations composed of natives of China who come from the same province or county, and in that way resemble the various State societies or clubs in this city. The difference is that the tongs have not been organized mainly for social purposes, but for mutual benefit and aid for new arrivals.

When toward the end of the last century a great number of Chinese laborers came to this country to work on railroads and other construction the Chinese, unable to speak English naturally formed little communities of their own in the various labor camps. The work completed, a great number of them preferred to remain here, but found the average American entirely ignorant of Oriental character and psychology, disposed to regard them with suspicion because of their different customs, their queues, their strange language and weird manner of writing. Drifting to the cities—to New York, to Baltimore, Chicago and San Francisco—the exiles began to congregate where rents were cheap, and thus the foundation of the various "Chinatowns" was laid.

Now, the Chinese, although generally regarded as one race, are by no means alike—either in looks or in traits of character. The northern Chinese has peculiarities which are strange to the Cantonese who live in the South of China, and vice versa. Their languages are so utterly different that they cannot understand each other at all without the help of an interpreter. The Cantonese refer to the northerners as "military barbarians," while the sturdy men from the north bestow upon the Cantonese epithets which I prefer not to translate.

Therefore each tribe, so to say, has its own tong, and when a stranger comes to this country he calls on the tong composed of people from his province. A helping hand is extended to him and if necessary he is assisted in finding work or a place to sleep. The Chinese have not failed to realize that a man in need is liable to become a criminal, and not entirely understanding or trusting the foreigners among whom they have come to live they have made successful efforts to forestall contact with the arm of the law. But a high official of the New York police says there are no "down and outers" to be found among the Chinese. That is at least partly because the tongs, which are now well known to the authorities, collect small fees from their members and use the fund to assist those in need.

The tong wars, which at an earlier period figured melodramatically in the public eye, were a by-product of transplanted local loyalties in a strange environment. In most cases business jealousy caused some friction between members of various tongs, till the précipitate act of some hot-headed partisan brought weapons into play. The Chinese love of quibbling about little things helped along the quarrels.

For instance, when a number of years ago many Chinese had quit their work on railways and were looking for some other means of making a living, a native of Sing Ying County, in the Province of Kuang-Tung, discovered that laundry work paid well and did not need much initial capital (laundry work, by the way, is by no means a man's job in China, where the women do the washing). Other natives of Sing Ying, members consequently of the same tong, followed his example and opened laundry shops. Presently the Chinese laundry became an American institution. Other Chinese, who did not belong to that particular tong, also opened laundries. And the men of Sing Ying, who had invented the idea, and regarded it as their rightful monopoly, did not like it.

When the interlopers refused to quit, it came to hard words and worse. For these Chinese had already become Americanized so far that they resorted to blows, when in their native country they might have been content with a contest of mutual abuse and expert vilification, including the ancestors of the adversary to the remotest generations.

First it came to fist-fights. Then any kind of weapon was brought into the fight by those who were not quite sure of their pugilistic ability—a brick, a hatchet or a knife. These were chosen because of their cheapness or immediate availability. It is a mistake to believe that there are professional hatchet-men in China.

The parties to these fights never took their troubles to the authorities of this country. They had discovered that the police can barely tell the difference between one Chinese and another. In most cases the officers of the law grabbed plaintiff as well as defendant and put them both behind the bars, hoping in vain to get at the bottom of all the trouble—hoping in vain because both parties out of excess of caution lied skillfully and fantastically whenever they were asked about anything that had happened. Police and the newspaper reporters, handicapped by ignorance of the language of the offenders, and unable to penetrate the thick web of lies, put the whole thing down as a mystery backed by black oaths of secrecy, and there was nobody to contradict the story. Thus material has been provided for many a fiction writer of blood-curdling yarns, based on a total misconception of what the tongs are and how the Chinese behave. Naturally, educated Chinese smile at this sort of literature.

Even so, stories of tong wars have become obsolete. The Chinese have ceased to wage private battle among themselves. Differences between the tongs are now mostly settled by arbitration. As an outlet for their feelings, they have gone back to the national sport of calling each other's ancestors bad names.

Americanization of the Chinese in New York's Chinatown has gone fairly far, in some cases not to their advantage. Many of them have forgot the high business ethics of the unspoiled Chinese. Some Chinese students have become psalm-singing hypocrites, while others have drunk at the fountain of folly in Greenwich Village and call themselves Socialists, even Bolsheviki. But the great majority of the Chinese in this country still stick to the sound principles taught to them by their forefathers. Their honesty is above reproach, they are courteous and loyal to their friends.

Many of them, of course, were born in this country, and quite a few of them cannot speak or write Chinese. They have formed organizations for Americanizing their co-nationals. One of the most interesting of these organizations is a society of young women in Chinatown who, co-operating with a sectarian organization of American origin, are doing work that is not negligible along educational lines. Incidentally, this organization, like the Chinese Boy Scouts, sold Liberty bonds during the war.

All this indicates that Chinatown never was as mysterious as many uninformed persons used to fancy. Of late the better sort of Chinese have discovered that they have been exploited and shown at their worst, and they are trying to overcome the prejudice created against them. It is partly because of energetic efforts this way that Chinatown is losing most of its "atmosphere."

The members of the Chinese colony have not forgot that the Government of this country was the first to recognize the young Republican Government which had emerged from the chaos following the overthrow of the Manchus. They deeply appreciate also the act of the United States in deciding that the Boxer indemnity should be used to send young Chinese to this country to be educated. They have the friendliest feelings for this country, and to promote friendship between the intellectuals of China and America is one of the great aims of the Chinese students here.

It still remains true, however, that many or most Chinese in America have scant knowledge of the English language and find it difficult to explain peculiar Chinese customs, all of which have a sound and logical reason behind them. For example, why do the Chinese in this country persist in using the "kwei-tsze," or "chop stick," instead of knives and forks? The fact is that the Chinese used metal forks many centuries ago till one of their physicians discovered that contact with metal mars the delicate flavor of many a dish. So as it is rather difficult to carve forks from the wood of the brittle bamboo, the Chinese adopted the "kwei-tsze." Besides, the use of forks hastens the process of taking food, while the kwei-tsze necessitates the gather ing of food in smaller quantities, thus enabling the eater to linger over the flavor of the dish to the benefit of his digestion.

The use of warm tea instead of cold water rests upon equally good reasons, according to their viewpoint. Ice water, they believe, along with many European nations, contributes much toward digestional troubles prevalent among Americans.

Again, the Chinese way of preparing food is equally well reasoned. Take one of the most popular Chinese dishes in this country, tsar-suey, the Cantonese pronunciation of which is "chop-suey," and the meaning in English "mixed fry."

It is simple enough to make. Peanut oil is poured into a deep frying pan, where it is heated till it smokes. Then chopped celery, onions, various kinds of meat, sprouted beans and other vegetables according to taste are added, as well as salt and seasoning. Cornstarch is used to add nutritive value, with a dish of Chinese cane syrup and of soy bean sauce for extra flavoring. Prolonged simmering over a slow fire makes the dish very easy to digest. The same process—chopping up of ingredients and cooking them over a slow fire-is used with practically every Chinese dish. It may be added that those who eat Chinese food rarely suffer from digestional or intestinal troubles.

Writers in America sometimes say that the Chinese never smile. It is true that most of the Chinese in this country have a rather sullen look. But it should not be forgotten that the Anglo-Saxons have more jovial faces than any other race and thus that cheery look is missed in others. Europeans generally look more stern, and the further one travels toward the East the more noticeable becomes the prevalence of the solemn countenance.

Chinese themselves say that the apparent surliness of the Chinese in this country can be accounted for by the ill-treatment many of them have received as newcomers. There is a disposition to regard every stranger as a prospective assailant or one from whom they may expect verbal insults or ridicule. This does not contribute toward making them look cheerful. In addition to this, many a student received his first "foreign" education from America or English missionaries, whose teachings tend not so much to cultivate the sense of humor and the jocund countenance as the serious and dignified demeanor. Western education has sometimes very peculiar results when applied to Orientals.

In fact, as those who really know them are well aware, the Chinese have a keen sense of humor, are fond of jokes and good stories and are excellent company and good sportsmen. They have a lot of personal dignity, they wait before they open up in conversation. But so does the Britisher, who is good company, too, once you know him.

The queue is no longer worn by the Chinese. In fact, either in this country or in China the queue never was a Chinese institution, but was ordered to be worn as a sign of submission when the Manchus had subjugated the "Sons of Han," as the Chinese like to call themselves. The queue—the sign of submission—was most thoroughly hated in China and one of the first things observed at the overthrow of the Manchus was how quickly the Chinese cut off their queues. I remember how, shortly after the republic had been proclaimed in China, scaffolds were erected in many cities upon which the people walked to have their queues shorn off by an individual wielding gigantic shears, while bands played and "a good time was had by all"—Including the commercial-minded chap who collected the cut-off queues and afterward sold them for good money to American and European hair exporters.

Today the Chinese are apt to resent being represented with queues, though artists find it hard to break themselves of the habit of drawing them that way. It was a picturesque and characteristic appendage. So, likewise, the citizens of the Republic of China like to be called "Chinese," rather than "Chinamen" or Chinks."

Perhaps the greatest fault of the race is their passion for quibbling about trifles. A Chinese is capable of wasting time, temper and money over the smallest things which give an opportunity for dispute. His provincial prejudices are not left behind him, and even here in New York the squabbles between Northern Chinese and the Cantonese, who come from the South, are carried on as vigorously as at home ten thousand miles away.

July 31, 1921

WANT ITALIAN TAUGHT.

New Rochelle School Board to Hear a Petition.

Special to The New York Times.

NEW ROCHELLE, N. Y., Dec. 29.—Representatives of Italian organizations will ask the New Rochelle Board of Education next week to establish Italian as a compulsory subject in the schools of this city.

Those behind the movement say the majority of New Rochelle public school pupils are of Italian extraction and that instruction of the language of their forefathers is imperative.

December 30, 1923

FOREIGN POPULATION CENTRE NOW MOVING FURTHER WEST

The centre of foreign-born population in America is moving westward for the first time in nearly half a century. It is generally supposed that a large proportion of our foreign population is to be found in New York and other Eastern cities. Most people would probably place the centre of this population not far west of the Atlantic coast. As a matter of fact, it has been in Michigan, Ohio and Indiana for some fifty years. Its exact position has been calculated by the Census Bureau.

While the centre of general population has been moving steadily westward, the centre of foreign population has zigzagged back and forth for more than a generation. In 1880 the centre of population was in Kentucky, some eight miles south of Cincinnati, Ohio. At the same time the centre of foreign population lay near the southeastern corner of Michigan. In forty years the centre of population has moved steadily westward until today it rests in Western Indiana.

The centre of foreign population, curiously enough, has not kept pace with this western movement. After 1890 it moved in a southwesterly direction, well into Indiana. It then sharply changed its course, moving east into Ohio. For another decade it roamed in a southeasterly direction. In the next ten years it moved steadily westward, once more entering Indiana, but still failing to keep pace with the movement of the general centre of population. In other words, the past ten years have witnessed the first westward movement of the centre of foreign population in forty years.

The present westward movement is due to an increase of the number of foreigners in the Far West, especially in California. In the past twenty years the native-born population of Los Angeles has increased 160 per cent., while the foreign population has grown 500 per cent. A large proportion of these are Greeks and Armenians. The Greeks are rapidly acquiring control of the fruit interests in general, while the Armenians are interested in orange growing.

Fewer Japs Coming.

Meanwhile the number of Japanese entering California is decreasing. It is generally supposed that the Japanese constitute the most serious immigration problem facing California, but if the present inflow of foreigners from the East continues the situation will soon be changed. Meanwhile the Japanese are flowing into Nebraska and other Western States which have no land laws against them.

The centre of foreign population has also been pulled westward by the situation in Chicago. Although it is common to speak of New York as a foreign city, Chicago rivals, if she does not surpass, it in this respect. There is not a ward in Chicago containing fewer than 30 per cent. foreign-born population. In some wards the foreign-born constitute 96 per cent. of the total population. Similar conditions are found in the great industrial centres throughout the Middle West, in Indiana and Michigan, which balance the heavy foreign element in the East. The foreign population of Wisconsin also tends to draw the centre westward.

The large proportion of foreign-born in the Far Western States is surprising. Although the total population of Arizona is comparatively small, the inflow of foreign-born has brought about a unique situation. Arizona is the only State without a single "white" county. A white county is one which contains not more than 5 per cent. foreign-born population. It is commonly supposed that the "great open spaces" of the Far West are populated largely by native-born whites. In some parts of Nebraska, again, the foreign communities are self-centered and little influenced by the native-born. There are instances of men living in Nebraska for thirty years without learning to speak a single word of English. Nebraska lies well west of the present centre of foreign population.

The influx of foreigners across the Mexican border is also a factor in determining the movement of the centre. In some communities in New Mexico English is not even spoken in the courts. An American facing trial would therefore be obliged to face a judge and jury speaking a foreign tongue. In North Dakota is found the largest percentage of naturalized foreigners of any State.

During the last thirty years the centre of foreign population has zigzagged to the southward. The movement is largely due to the influx of foreign population into the South. For years there has been a steady exodus of the negroes to the Northern States. This has in turn created, so to speak, a vacuum in many parts of the South which is being filled by foreign-born from many European countries.

Foreigners Replace Negroes.

The foreign-born are especially in evidence in great manufacturing centres such as Birmingham. If this double movement continues, as it promises to do, the next few years may witness the negro problem in part shifted to the North, while the South will face the necessity of assimilating a large proportion of foreign-born, hitherto a distinctly Northern and Western problem.

It is not anticipated that the centre of foreign population in its westward movement will pass the general population centre. It will be checked largely by the steady increase of the foreign-born throughout the North and East. Rhode Island has the largest native-born white population of any State, with 73 per cent. Throughout the New England States, however, the foreign and native born are about equally divided. In some of the mill towns of New England, such as Fall River, only some 15 per cent. are native-born.

The latest census returns show surprising conditions in many Eastern cities. Philadelphia has less than 50 per cent. of whites of native parentage. A number of cities have but 15 per cent. of native whites of native parentage. The highest percentage found in any of the larger cities is 70 per cent., which is reported in several parts of the South. The general average is about 50 per cent. The same conditions obtain in the Far West.

If the native and foreign born populations be compared in the States, the foreigner still remains in a minority. New York has 7,385,474 native whites to 2,786,613 foreign-born whites. In Massachusetts there are 2,725,990 native-born to 1,077,534 foreigners. Connecticut with 982,210 native-born whites has 376,513 foreign-born. The total for the entire country shows a population of 81,108,161 native whites to 13,712,754 foreign-born whites.

September 28, 1924

WHEN AL SMITH WAS AN EAST SIDE BOY

By R. L. DUFFUS

FOUR small boys are pelting along Water Street toward the shadowed angle where South Street swings in under the Brooklyn bridge. Legs, arms and heads are bare, and if their entire costume consists of more than pants and shirt, the onlooker misses his guess. They are on one of those mysterious and important errands of childhood—perhaps to elude a watchman and dive off a pier into the murky waters of the East River, perhaps merely to slip around a corner and flip pennies. A drayman trots his horses over the cobblestones. Two of the boys hop on behind for a free ride. He flicks his whip at them. "Get out of there!" They drop off unabashed, shouting after him some half understood remarks about his personal appearance and his ancestry. It is a warm afternoon. School is not keeping. Life for them has the same tingle of adventure that it has for country boys running barefoot across the stubble fields to the pool at the bend of the brook.

Look at these four boys carefully. They may be Irish, Italian, Jewish, Greek, even Spanish, for all these races are to be found in this side eddy of Manhattan. Forty-odd years from now one of them may be a famous scientist, an artist, a novelist, a millionaire business man, a governor, a senator, a candidate for the Presidency. If the possibility appears remote, try to imagine this same region, say in the year 1883, when the Brooklyn Bridge was opened.

At that time there was living on the third floor of a house at 174 South Street, beneath the thundering traffic of the newly dedicated structure which had supplanted the Brooklyn ferry of Walt Whitman's day, a family named Smith. Its circumstances were as humble as its name. The father, Alfred Emanuel Smith, was a truck driver who had been born near South and Oliver Streets, a few blocks from where he now lived. The mother, whose maiden name was Catherine Mulvihill, had been born at the corner of Dover and Water Streets, in a humble tenement over a store. Their home was, by every account, a happy one, but they had not risen much in the world.

On the first floor of the house at 174 South Street was a fruit and candy store. On the second floor was Herr Morgenweck's barber shop—a place of romance for small boys because here resorted the Sandy Hook pilots, a tough and wiry bearded set of men with stories of salt water. Sometimes Uncle Peter Mulvihill, the fireman, came to call, stopping at Morgenweck's on the way up to have his handsome mustache trimmed.

On their own third floor the Smiths had four rooms. In one of them, to the clatter of wagons rolling endlessly over the cobbles, on the last day but one of 1873 Catherine Mulvihill Smith gave birth to a son. Little Alfred Emanuel—for he was named after his father—had heard, behind the soothing murmur of his mother's voice, an accompaniment to the lullabies she crooned to him, the song of the growing city.

Above the Smith tenement was an attic, and here Alfred Emanuel, as he waxed in size and strength, played with the other children of his block on rainy days. The attic looked upon the river, and on the river schooners and fishing sloops and square-riggers from every port went up and down, making their last struggle against the steamboat. Even if he had had the wealth of the Comstocks under his hands, the elder Smith could not have purchased a view more stimulating to the imagination. And even if he had been able to send his son to Harvard or Yale he could not have bought him an education more fitting for the career he was to have than that which this noisy, dusty environment was to furnish.

The Fourth Ward, within whose narrow boundaries two generations of Smiths had been born, was mostly Irish in those days. But more important than the fact that a particular race inhabited it was the fact that this was an immigrant race. The Irish had poured into the country during the migrations of the hungry '40s. The "new immigration" had not yet made them seem like old settlers. Individuals among them had risen with startling rapidity and almost all of them had adapted themselves easily to American ways, but in the mass they had still to show what magnificent use they were to make of America's opportunities.

What this amounts to saying is that, unlikely as it may now seem that any one of four boys seen by chance skylarking on Water Street may become a Presidential possibility, it must have seemed far less likely in the early '80s that Alfred E. Smith Jr. should be picked out for the same prominent rôle. Presidents did not come from immigrant stock. A boy in the Fourth Ward might indeed become a politician. If he were an unusually bright boy and were not drawn into the service of the Church he could hardly help becoming a politician. But the gap between playing the political game in a Tammany ward on the east side and becoming a candidate for the Presidency of the United States was too appallingly wide for even the most active imagination to jump.

LITTLE Al Smith, galloping down to the water on a hot Summer day, selling papers with a starting capital of 20 cents, serving as altar boy at St. James's Parish Church, singing and dancing in the fire houses to which he had been introduced by his genial uncle, Peter Mulvihill, winning a medal for declamation in competition with the students of all the parochial schools of New York City—little Al Smith seemed much further from the Presidency than young Andrew Jackson or young Abraham Lincoln or young Herbert Hoover had seemed at a corresponding stage in their careers. The log cabin and the country farm had become standardized as cradles of greatness. The third-floor tenement house hadn't. It is only as we look back four decades and watch the growth of cities and the mingling of the new stocks with the old that we can see that there is something to be said for the tenement as a strategic point of departure for Al Smith's march toward eminence.

In fact, there is more than a little to be said. Governor Smith's biographers, Norman Hapgood and Henry Moskowitz, have pointed out that the activities of three generations of Smiths, up to the time that young Al went to Albany as a member of the Legislature, were mainly circumscribed within an area about a quarter of a mile square. This was Al Smith's university—and it covered a ground space only a fraction as large as the campus of Stanford University, where Herbert Hoover got his formal training. At an age when Mr. Hoover had become intimately acquainted with the Middle West, the Pacific Coast, the mining districts of Australia and a large part of Northern China, Al Smith's work, amusements and thoughts still centred about this village within a great city—this small though important corner of Manhattan Island. Al Smith was nearly 30 before he went to Albany for the first time. Since then he has grown and learned. But the essential Al had been hammered into shape by the Fourth Ward. A man doesn't forget his first thirty years.

Let us consider how the hammers of destiny in this clanging shop beside the East River beat upon the anvil of a young soul. They might, the casual observer is easily led to believe, have pounded out something ugly. It is but a short walk from South Street along Oliver Street to the Bowery, and the Bowery in 1873, 1883 and later had its venomously sinful aspects. On the other side, along the waterfront, where the tall ships came in, were other places where the children of darkness lay in wait for sailors. Prostitutes flaunted their pitiful profession, gangs fought each other to the death, thugs lay in wait in doorways to hit the passer-by over the head with a piece of lead pipe, men drank knockout drops with their beer and were dragged into back rooms and robbed, and all this was accepted as an unavoidable part of life on the east side of New York.

Whether city boys in the Fourth Ward learned more about human wickedness than country boys of the same generation were learning at the same time during their visits to general store and livery stable may be questioned. But there can be little doubt that sinfulness as a career was never more picturesquely presented than it was in Al Smith's juvenile days in the Fourth Ward. Nobody of normal intelligence—and Al's was a good deal more than normal—could grow up there and not know practically everything about what were then referred to in genteel circles as the Facts of Life.

But no large community can live by sin alone, any more than it can live by taking in its own washing. The sins of the east side received publicity because they were gaudy. The industry, the frugality, the piety and the good humor of most of the people living there were overlooked because there was, after all, nothing startling about them. Al Smith came, nevertheless, under the influence of these qualities, and they were the ones that rounded out his personality. He had to be hard-working and economical because there never was too much money in the house, even during his father's lifetime, and because after his father's death—which happened before Al was 13—he had to do his part in saving and earning.

The religious and moral teachings which he learned, literally, at his mother's knee and from a pastor whom he liked and respected were as strict as those handed down in any old-fashioned New England village. They differed in detail, of course. The east side could not look upon the saloon quite as Vermont did. The saloonkeeper at his best was what the east side considered a good neighbor. Anyhow, he was part of the social and political system.

THE east side's idea of goodness was of something practical and positive. A good friend was one who helped find one a job or helped get one out of jail. No doubt there are worse tests. Al Smith and every one else who ever got anywhere in the life of the east side, began by being friendly with his neighbors. That was his first lesson in politics, though as any one who studies his appointments knows he learned in time to be neighborly to a whole constituency rather than to a limited group of Tammany voters and heelers.

A boy could rise on the east side if he could make people both like him and trust him. And here the environment gave young Al exactly the conditions he needed. He was naturally gay—every one who knew him in childhood and early manhood testifies to that. He was also a natural leader. If four boys from South Street decided to go swimming, it may be taken for granted that Al Smith, if he were one of them, would be the first into the water. Perhaps the Smith attic helped to develop the quality of leadership. It was Al's attic, and when other boys came around it was naturally up to him to run things. He never had that shy and sensitive quality which was manifest in Herbert Hoover's Quakerish younger days. Shyness and sensitiveness didn't get far on South Street—and still do not. If a boy were to be listened to he had to make himself heard above the prevailing din.

Al always had confidence in him-

"The Bowery, the Bowery." Governor Smith Knew the Famous Street Well in His Early Days.

Photograph by Brown Bros.

self. There is no evidence that he indulged in those long, long thoughts of youth of which Longfellow wrote. Much in his surroundings must have appealed to the imagination—the ships, the sailors speaking foreign tongues, the salt water always lapping at South Street's front door, the bridge rising stone by stone and cable by cable, like a captured dream. In such a scene might have been produced a Masefield, a Melville, a Conrad. But Al did not run away to sea, nor is there record that he so much as wanted to. He turned away from that bright framework of Fourth Ward life to mingle exuberantly in the stir and tumult around him.

He left school at 15. A good school it must have been, so far as the groundwork of education was concerned. At least it gave him the mastery of the tool of language. It must have taught him logic, too, and how to use figures—as any one knows who has ever heard him make a speech about the State budget. But it was the dramatic society, the political club, and, above all, the market and the street that really gave him his education. This was life at first hand. It didn't give him a chance to build up those illusions which are cherished by youths with more carefully selected environments.

He was awkward and green enough when, at 30, he first went to Albany, but this wasn't because he didn't know human nature and politics like a book. It simply took him a little time to discover a fact of which many rural and small-town Americans are still unaware, that the east side, and specifically the Fourth Ward, was the world in miniature. He had been afraid that Albany was different. He needn't have been. Much later in his career he may have been afraid that Washington was different. He probably is so no longer.

This education of his was an unconscious process. He did not, like Herbert Hoover, deliberately seek education. He drifted, doing what came to hand, ambitious primarily to earn as good a living as possible. But one thing and another led him toward jobs in which he could make use of his genius for human contacts. Jobs and amusements, too, one might add. It was no accident that his favorite diversion was acting. He loved to express himself vocally, to feel the effect of his personality on other people. A secluded country youth might have put his adolescent emotions into verse. A smoother ability might have written popular songs, as Jimmy Walker did. But Al wrote his verse directly on the features of living auditors.

It is on record that he spent some months playing chaperon to the machinery of a steam pump plant in Brooklyn. But machinery did not please him. He wanted people. His hardest job in boyhood and young manhood was in the Fulton Fish Market, where he put in twelve lively hours a day, from 4 in the morning till 4 in the afternoon. Yet this is the job to which he seems to look back with the greatest satisfaction. To this day there are abundant traces of it in his anecdotes and similes. The fish market was a wonderful place for the study of humanity. Its personalities were as rich as its smells, which is saying a good deal.

The same frank interest in people and in his own effect upon them made young Al a success from the first in the political club over which Tom Foley presided. It is hard to believe that he went into this organization with a cold-blooded ambition to wrest a good job out of it. It is remembered that he was even more faithful in his attendance than his hard taskmaster demanded. Night after night he saw Tammany in operation in its simplest, most human form—doing things for people, holding its strength by craft, but also by genuine sympathy. Moralists could point out that it took in the long run far more than it gave. Young Al probably didn't concern himself about that. What the organization offered him was an opportunity to straighten out human problems.

HE liked that, as he has always liked it. He wasn't inclined to generalize about the effects of Tammany on city government. In fact, he has never been fond of generalizing about anything. Pioneers, whether on South Street or in Wyoming, can't generalize. A log cabin or a political organization either stands up or falls down, according to the pioneer way of thinking. What it ought to do or might have done is another matter, concerning which such temperaments as Al Smith's are rarely given to ponder.

So the east side, and especially Tammany, made young Al a practical-minded and conservative person. What is sometimes mistaken for liberalism or even radicalism in his attitudes does not come from a theory at all—it is just an effort to be as helpful to a large group of people in a large way as it was his practice to be, in the old days of the Seymour Club, to a small group of people in a small way. And possibly this liking for tangible and immediate things, as opposed to the remote and theoretical, is his only temperamental resemblance to Herbert Hoover.

The old house at 174 South Street is now unrecognizable. The home of Smith's early married life, in which his children were born and in which he lived until after he became Governor, was at 25 Oliver Street. He lives there no longer. The tides of change have swept all around the Fourth Ward, rearing new skylines, sending forth new and denser streams of traffic. Yet, though time has had its way, the visitor cannot help feeling that something of the old mood of the locality has been preserved.

It is a hodge-podge of buildings, of races, of traditions. Near the junction of Oliver Street and Chatham Square is the first cemetery of the Spanish and Portuguese Jews, opened in 1656, closed in 1833. St. James's Church still stands around the corner in the street to which it gave a name. Boys and girls come boiling out of the great public school almost across from 25 Oliver. On Cherry Street youngsters stand in line on hot Summer afternoons, waiting to be admitted to the public baths which have partially taken the place of the more adventurous East River. Men sit at a table on the sidewalk, playing cards. The influx of Greeks is shown in the signs and in the cafés, where patrons sit long hours over more or less innocuous drinks. The Spanish have come in, too. A drunken sailor argues with passers-by in broken Castilian. The Gospel Mission, with its motto, "Be sure your sin will find you out," announces its services in Spanish.

One wanders along under the rickety arcades of South Street, past a Fonda Española, past a warehouse smelling richly of hay, precisely as it might have done in the elder Smith's time. The Fulton Market is still there, although, with modern improvements in the technique and organization of the fishing business, some of its glamour has departed. A lunch car is stationed in front, horse-drawn vehicles as well as motor trucks still lumber up, and the shadowy interior is richly flavored. The long wharf runs out into the river. One can see three bridges, the heights of Brooklyn, Governors Island, the lower end of Manhattan, the band of haze above the open ocean. And the river itself, even in these degenerate days, is still a great parade ground.

Poverty in this environment might have been rigorous, it could never have been completely depressing. Always some lift of hope, some lusty determination to battle with the world on its own terms, would have come to those spirited enough to receive them. Standing on this pier and looking back toward South Street and Oliver Street, one can begin to understand Al Smith. Here was his campus.

And the Al Smith of 1968? Has the Fourth Ward still vitality enough to produce one? Is this child of the east side an accident, or are the crowded areas of our great cities at last beginning to justify themselves by producing leaders and thinkers? Let's wait forty years and see.

July 8, 1928

Pilgrims' progress

The Golden Door

Italian and Jewish Immigrant Mobility in New York City 1880-1915.
By Thomas Kessner.
256 pp. New York: Oxford University Press.
Cloth, $12.95. Paper, $3.50.

By DIANE RAVITCH

Legend has it that New York City was the golden door to the land of opportunity, that millions of poor European immigrants worked their way into the middle class thanks to the city's booming economy and its free schools. In recent years, the legend has come increasingly under attack, largely from Marxist-oriented historians who contend that oppression was more characteristic of the urban industrial economy than was opportunity.

In debates about whether America in general, and New York City in particular, was hospitable to the immigrant masses, each side has had its own stock of historical facts, memoirs and anecdotes. One necessary ingredient has heretofore been lacking: firm, broad-based evidence. Filling this lacuna is the purpose of Thomas Kessner's "The Golden Door," which is a fascinating, closely documented study of Italian and Jewish mobility in New York City between 1880 and 1915.

There have been many books about these groups before, but Kessner's is the first to employ quantitative techniques to measure their occupational progress over time. Kessner is one of a growing breed of urban historians who aspire to write history "from the bottom up," rather than from the perspective of great leaders and momentous events. The special concern of these historians is with the anonymous Americans, the poor, the workers, the immigrants and all those others who did not leave behind written accounts of their lives.

To his credit, Kessner goes beyond strictly quantitative sources to draw on literature, biography and contemporary reports for supplementary descriptive material. At one point, he demonstrates how straightforward census data from 1880 can be used to reconstruct the lives of working-class Italians and Jews. For a few brief paragraphs, these men and their families come to life as representatives of an experience that is as worthy of the historian's attention as the bustle of the nation's policymakers.

In his effort to identify what was typical and what was exceptional in the immigrant past, Kessner selected large samples from Federal and state censuses and from city directories. By comparing his samples over time and by tracing individual families, he is able to analyze the changing living conditions of these two large immigrant groups with greater precision than any previous study. Based on the vast evidence amassed by his research, Kessner concludes that the legend of New York City as "the golden door" is a fairly accurate portrait of what actually occurred.

Most significantly, Kessner finds that New York City offered extensive occupational mobility when judged by the standard of "rags to respectability." Thirty-seven percent of New York's new immigrants moved from blue-collar jobs to white-collar jobs in the decade from 1880 to 1890. Even at the floodtide of immigration, from 1905 to 1915, some 32 percent of Italians and Jews managed to rise out of the manual class. New immigrants in Brooklyn experienced even higher rates of occupational achievement. No other American city that has yet been studied can match this level of advancement.

The especial usefulness of "The Golden Door" is its detailed confirmation or refutation of specific elements of the conventional wisdom about this period in American history. It has long been said, for example, that the Italian immigrant was more transient than the Jew, that many single working men (called "birds of passage") returned to Italy after working here for a time. Kessner pinpoints the phenomenon, finding that in one four-year period, 73 Italians left for every 100 that arrived; by contrast, only 7 percent of Russian immigrants returned. On the other hand, Kessner rejects the common belief that Italian families were larger than Jewish families.

The data reveal how European background influenced career mobility in the new country. Jews, most of whom were literate and skilled craftsmen when they immigrated, moved up faster than Italians, most of whom arrived as illiterate and unskilled laborers. Unskilled construction work was "the great Italian beachhead" into the American economy, while many first-generation Jews rose from peddling and tailoring into business. Kessner points out that it was business, not medicine or law or education, that carried the Jewish immigrants forward occupationally.

One of Kessner's particularly provocative speculations is about the function of the "ghetto," the ethnic community. He sharply disagrees with those who criticize the ghetto and call it a hindrance to mobility, in a vein not unlike that of turn-of-the-century bigots who abhorred immigrant neighborhoods. Kessner argues that the ghetto was "a mobility launcher. It offered the immigrant hospitality of place . . . jobs, business, and political contacts as well as investment opportunities." Neighborhood institutions, like the schools and settlement houses, helped the immigrant adapt to the new culture and acquire essential skills. For the new arrival in New York City, the ethnic neighborhood was both "a buffer and a way-station."

Two points stand out as guideposts to the present situation of New York City. First, the successful absorption of the immigrant masses was more clearly related to the city's expanding economy than to any other single factor. Second, the remarkable progress of the city's poor Italians and Jews, so evident to the historian in 1977, was entirely obscure to those who lived in New York in the early twentieth century, and particularly to those who wailed, wrongly, that the new immigrants were dragging the city down.

Diane Ravitch is the author of "The Great School Wars: New York City, 1805-1973."

February 13, 1977

PRESIDENT SEES NEW PLAY.

Zangwill's "The Melting Pot" Produced in Washington.

WASHINGTON, Oct. 5.—Israel Zangwill's new play, "The Melting Pot," under the direction of Liebler & Co., with Walker Whiteside in the title part, had its first production here to-night in the Columbia Theatre before an audience, both from official and social life. The author was present and received a warm welcome. After playing here and in Baltimore the piece goes to New York.

"The Melting Pot" symbolizes America as a crucible wherein all nationalities are fused into an American type, a being with the dominant characteristics of all that is best the world over. Mr. Whiteside appeared as David Quixano, a young Russian Jew refugee, a musician, whose parents were murdered at Kishineff by Baron Ravendal, father of the girl, who herself fled as a Russian revolutionist, with whom David falls in love and eventually marries.

In the audience were President and Mrs. Roosevelt, Mr. and Mrs. Loeb, Gen. Crozier, Secretary Root, and Secretary and Mrs. Straus.

October 6, 1908

SAY ALIENS SOON GET AMERICAN PHYSIQUE

Immigration Commission Reports That Even Head Forms Assimilate in One Generation.

STATISTICS SECURED HERE

Show Children of Long-Headed Sicilians and Round-Headed Jews Approach a Type Between the Two.

WASHINGTON, Dec. 16.—That the physical form as well as the habits of living and ways of thinking of the descendants of foreigners who immigrate to America is different from that of their ancestors is the conclusion of the Immigration Commission as embodied in the preliminary report of that body which was presented to Congress to-day. The conclusion is regarded as of importance in anthropological science as indicating in persons of European descent a development of a distinct American physical type in a period which anthropologists have regarded as entirely too brief for such changes. The very head forms, it is said, are modified strikingly, and tend toward assimilation.

The investigation which has brought this fact to the surface was undertaken soon after the appointment of the commission, and it was conducted by the comparison of measurements of the bodies of such immigrants and their descendants at different ages and under differing circumstances. The investigation was placed in the hands of a committee composed of members of the commission, and Prof. Franz Boaz of Columbia University was engaged as expert.

The inquiry was confined to New York City, and so far as the present report shows was restricted to Sicilians and Jews from Eastern Europe. A later report will give the details of the investigation among Bohemians, Hungarians, and Scotch.

The present report indicates that the descendant of the European immigrant changes his type, even in the first generation, astonishingly. Children born even a few years after the arrival of the immigrant parents in America develop in such a way that they differ in type essentially from their foreign-born parents. These differences seem to develop during the earliest childhood and persist throughout life. Every part of the body is influenced in this way, and even the form of the head, which has always been considered as one of the most permanent hereditary features, undergoes considerable changes.

In an official synopsis of the commission's report attention is called to the fact that even the most stable racial characteristics seem to have changed under the new environment, and this fact is commented upon as follows:

"This would indicate the conclusion that racial, physical characteristics do not survive under the new social and climatic environment of America. The adaptability of the various races coming together on our shores seems, if these indications shall be fully borne out in later study, to be much greater than had been anticipated. If the American environment can bring about an assimilation of the head forms in the first generation may it not be that other characteristics may be as easily modified, and that there may be a rapid assimilation of widely varying nationalities and races to something that may well be called an American type?"

The commission feels that it is too early to pronounce absolutely upon this question. "The investigation is by no means complete, and moreover, considering the importance of the subject, it should clearly be conducted on a larger scale and in different surroundings in various parts of the country, and perhaps also be checked up by certain investigations made upon the same races elsewhere."

Among other results noted it is shown that the American-born children of the long-headed Sicilians and those of the round-headed Jews from the East of Europe have very nearly the same intermediate head form. The children of the long-headed Sicilians are more round-headed, the children of the round-headed Jews are more long-headed than their parents. Similar changes are traced in the development of the faces of these types.

The amalgamation is most rapid during the period immediately following the arrival of the immigrants. The difference in type between parents and children manifests itself almost immediately after their arrival here. Among individuals born a long time after the arrival of the parents in America the difference is increased, but only slightly as compared with the great difference that develops at once.

Up to this time the investigations have not been carried so far as to determine what happens in the second generation of immigrants; but it seems likely that the influences at work among the first generation born in America will be still further accentuated.

The commission also finds that as a rule there is a falling off in the size of families after their arrival in the United States, that as the number of children decrease the size of the individuals increase. This fact is discoverable among the children of the well-to-do as well as those of the poor.

Another result of the investigation is the development of the fact that while removal from Europe to New York has had a beneficial effect upon the physique of Jews from Eastern Europe the result has been just the opposite upon the Sicilians, the conclusion being that bad as they are the surroundings in New York are better for the Jews than their city homes in the Old World, while the cramped quarters which the Sicilians occupy in New York City are not so desirable as their rural surroundings in Southern Italy.

December 17, 1909

EMINENT CITIZENS JOIN PATRIOT BAND

Mrs. Vincent Astor's Guests at Dinner Launch "America First" Campaign.

SEEKING THE FOREIGN-BORN

Educational Work and Training Camps to Help Assimilate Land's Immigrants.

LABOR PEACE ONE OBJECT

Gen. Wood, Cardinal Gibbons, Edison, Schiff, and John Mitchell Among the Members.

A campaign with three slogans, "America First," "The English Language First," and "Efficiency," is beginning under the direction of the National Americanization Committee, a new oranization with offices at 20 West Thirty-fourth Street, and a membership including eminent men and women.

Among the workers of the committee are persons of such widely separated environments as Mary Antin, Thomas A. Edison, Mrs. Vincent Astor, Cardinal Gibbons, John Mitchell, Jacob H. Schiff, and General Leonard Wood; and its object has been summarized in the phrase, "internal preparedness." The preparedness aimed at is not only for war, but for peace, and the committee aims to promote the process of welding the immigrants who come in from all over Europe into a single nation.

The campaign was started formally Friday night at a dinner given by Mr. and Mrs. Vincent Astor, where the beginnings of the work were recounted by the leaders. The committee wants to get the immigrants assimilated, naturalized, and educated. One of its principal aims is to improve the labor situation—to promote the conservation of the labor supply and prevent labor wars, and to make the workmen who are doing the manual labor of the country realize that they are part of the country with duties, privileges, and responsibilities.

Training Camps for Citizens.

Citizens' training camps are one means by which the committee hopes to forward its work—training camps where native Americans and foreign-born Americans work together for a common ideal; and what is wanted is the effect of the camps on the spirit rather than on the actual military preparation of the men who take part in them.

Robert Bacon, one of the leaders in the Plattsburg camp and one of the speakers at the dinner, told the committee and a few guests that the flap tent was a great Americanizing institution. Some of the other weapons which are to be used in the campaign are night schools; posters urging the America First ideal, which are to be put up in post offices; committees to take hold of newly naturalized citizens and help them retain their patriotic interest in their new country, and slips promoting the ideals of the committee which are to be given to workmen in their pay envelopes.

The Americanization Committee announces its intention to co-operate with schools, colleges, Government agencies, and the foreign-language newspapers. Uniform laws are to be worked for which will reach a uniform interpretation of the meaning of American citizenship.

Frank Trumbull, Chairman of the committee, who is a Director of the Chesapeake & Ohio Railroad, presided at the meeting. Miss Frances A. Kellor, Vice Chairman, told how the work had begun with a study of conditions among immigrant workmen. The industries particularly studied were those relating to national defense—munitions plants, railroads, mines, quarries, and other lines of work where the presence of unassimilated foreigners in time of war might be a national danger.

Preparedness here is preparedness for peace also, for it is has been found that with increase in the proportion of citizens among the workingmen there has been an increase in patriotism, a decrease in sabotage and violence.

Eminent Men Describe Work.

Dr. P. P. Claxton, Federal Commissioner of Education, told of the spreading of the America-first posters, which indicate the night school as the first aid to the immigrant, and of the work which the Bureau of Education is doing to form civic classes. State Commissioner of Education John H. Finley told of the work done to reduce illiteracy among immigrants in New York; Commissioner of Immigration Frederic C. Howe pleaded for support of the present campaign to increase interest in the city's night schools and for the reception arranged by the Mayor's committee for new voters, which is to be held in the City College Stadium on Oct. 29. Presidents Arthur T. Hadley of Yale and Benjamin Ide Wheeler of the University of California spoke in support of the committee's work, and President Nicholas Murray Butler of Columbia sent a letter announcing that a course of training for adult immigrants in preparation for naturalization was being added to that institution's extension department.

John Fahey, President of the Chamber of Commerce of the United States, told of the realization among business men of the importance of Americanizing the workmen.

What one city had done was related by Charles B. Warren, President of the Detroit Board of Commerce, which this Fall in connection with the work of the National Americanization Committee has begun work to promote the "English language first" campaign in a city 75 per cent. of whose residents are foreign-born or of foreign parentage. One result of this campaign was an increase of 153 per cent. in the registration of night schools.

An art contest on the subject of "The Immigrant in America" is being conducted by the committee with the co-operation of Mrs. Harry Payne Whitney. Prizes totaling $1,100 have been offered and an exhibit of the works submitted is to be sent through the principal cities of the country.

Mrs. Stotesbury Aiding.

The dinner at the Astor home was the first of a series through which the movement is to gain a nation-wide character. The next will be held in Philadelphia, at the home of Mrs. Edward T. Stotesbury, on Jan. 7, and meantime Mrs. Stotesbury is working to get the Pennsylvania Department of Education and the University of Pennsylvania enlisted in the campaign and to extend the plan of the issuance of Americanization slips.

The members of the National Americanization Committee are Mary Antin, Robert Bacon, Nicholas Murray Butler, Richard Campbell, P. P. Claxton, R. T. Crane, 3d, Henry P. Davidson, Coleman du Pont, Thomas A. Edison, Howard Elliott, John H. Fahey, Maurice Fels, John H. Finley, David R. Francis, Elbert H. Gary, James Cardinal Gibbons, Myron T. Herrick, Clarence H. Goodwin, John Grier Hibben, Henry L. Higginson, Frederic C. Howe, Clarence H. Ingersoll, Dr. Abraham Jacobi, Judge Manuel Levine, Clarence H. Mackay, C. H. Markham, Alfred E. Marling, Wyndam Meredith, George von L. Meyer, John Mitchell, A. J. Montague, John H. Moore, U. S. N., retired; Joseph C. Pelletier, Samuel Rea, Julius Rosenwald, M. J. Sanders, Jacob H. Schiff, Bishop Thomas Shahan, Melville E. Stone, Mrs. William C. Story, William H. Truesdale, Rodman Wanamaker, S. Davies Warfield, Charles B. Warren, Benjamin Ide Wheeler, and General Leonard Wood.

The officers and Executive Committee are Frank Trumbull, Chairman; Percy R. Pyne, 2d, Mrs. Edward T. Stotesbury, William Sproule, Mrs. Vincent Astor, Miss A. Kellor, Peter Roberts, Mrs. Cornelius Vanderbilt, Felix M. Warburg, and William Fellowes Morgan.

October 17, 1915

Solving the Problem of Colonies.

One of our correspondents lamented, yesterday, the existence in so many of our cities of districts inhabited exclusively, or almost exclusively, by foreigners of a single race. These segregations are indeed highly undesirable, for they delay assimilation by making it unnecessary for those who thus live apart from their neighbors of other extraction. It is entirely natural, however, that immigrants of like language and habits should flock together in a strange land, and it is not surprising, either, that every such "colony" should have its own newspapers or that these journals, instead of trying to hasten the Americanization of their readers, should be led by self-interest to delay that process to the extent of their ability.

But the suggestion that legal measures should be taken to abolish both the colonies and their papers cannot be followed—is quite impracticable, and fortunately, as a remedy so drastic would be worse than the disease. And there are other ways of handling the question—ways not certainly successful, perhaps, but at least more promising because compatible with free institutions.

Nowhere, probably, has this problem of the herded foreigner made a more urgent demand for solution than in Detroit. At the opening of the century Detroit was a city with only 285,000 inhabitants, all except an almost negligible minority of them belonging to what is called the native stock. Now, chiefly because of the concentration there of no small part of the automobile industry, the population has been multiplied by three, only one resident in four is of American birth, and half the nations of Europe are represented, each in a district, a little city, of its own.

The Hungarians are the most numerous, the Italians hardly less, and there are considerable groups of Russians, Greeks, Rumanians, Serbians, Armenians, and others.

Detroit has met the situation—she hopes and believes—by establishing the best and most numerous night schools for adults to be found anywhere in the world. The advantages offered by these schools are appreciated, as is shown by the great attendance, and the students in them are taught to speak English and to be Americans, not to remain aliens and separate.

April 18, 1916

ROOSEVELT DEMANDS RACE FUSION HERE

With 38 Others He Issues Appeal to People of All Nationalities.

ONE FLAG, ONE LANGUAGE

Suggests That "Children of the Crucible" Support the Activities of the Vigilantes.

Colonel Roosevelt issued a signed statement yesterday under the title of "The Children of the Crucible," setting forth patriotic reasons for united action by all Americans of whatever national ancestry in active support of the Government and the prosecution of the war. This message, which was made public at his home in Oyster Bay, bears the names of thirty-eight other persons, representing ancestry of the allied nations, of the neutral countries, and of Germany.

The name of Oscar S. Straus follows the signature of Colonel Roosevelt. The other names include George von L. Meyer, who was Secretary of the Navy during the Administration of Presidents Roosevelt and Taft; Michael I. Pupin, a professor at Columbia University; William Loeb, Jr., former Collector of the Port; Anthony Fiala, and John D. Crimmins.

The appeal cites the example of Washington and his men, "who did not hesitate because of the ties of blood to resist and antagonize Great Britain when Great Britain wronged this nation," as one for Americans to follow today. The statement goes on with the suggestion that, "The best way to accomplish this fusing of Americanism is through the activities of the Vigilantes, a non-partisan, militant, anti-pacifist group of writers, artists, and other patriotic Children of the Crucible." The statement in part reads:

"We Americans are the children of the crucible. It has been our boast that out of the crucible, the melting pot of life in this free land, all the men and women of all the nations who come hither emerge as Americans and as nothing else; Americans who proudly challenge as a right, not as a favor, that they 'belong' just exactly as much as any other Americans and that they stand on a full and complete equality with them; Americans, therefore, who must, even more strongly, insist that they have renounced completely and without reserve, all allegiance to the lands from which they or their forefathers came, and that it is a binding duty on every citizen of this country in every important crisis to act solidly with all his fellow Americans, having regard only to the honor and interest of America and treating every other nation purely on its conduct in that crisis, without reference to his ancestral predilections or antipathies.

Pacifists Like Tories.

"All Americans of other race origin must act toward the countries from which their ancestors severally sprang as Washington and his associates in their day acted. Otherwise they are traitors to America. This applies especially today to all Americans of German blood who directly or indirectly in any manner support Germany as against the United States and the allies of the United States; it applies no less specifically to all American citizens of Irish blood who are led into following the same course not by their love of Germany but by their hatred of England. One motive is as inexcusable as the other, and in each case the action is treasonable to the United States.

"The professional pacifists have, during the last three years, proved themselves the evil enemies of their country. They now advocate an inconclusive peace. In so doing they have shown themselves to be the spiritual heirs of the Tories who in the name of peace opposed Washington, and of the Copperheads who in the name of peace opposed Lincoln. We regard these men and women as traitors to the Republic; we regard them as traitors to the great cause of justice and humanity.

"This war is a war for the vital interests of America. When we fight for America abroad we save our children from fighting for America at home beside their own ruined hearthstones. We believe that the large majority of Americans are proudly ready to fight to the last for the overthrow of the brutal German militarism which threatens America no less than every other civilized nation.

"We hold that the true test of loyal Americanism today is effective service against Germany. We should exert as speedily as possible every particle of our vast lazy strength to win the triumph over Germany. Therefore, we should demand that the Government act at once with unrelenting severity against the traitors here at home, whether their treasonable activity take the form of editing and publishing newspapers, of uttering speeches, or of intrigue and conspiracy.

One Flag, One Language.

"We must have but one flag. We must also have but one language. That must be the language of the Declaration of Independence, of Washington's Farewell address, of Lincoln's Gettysburg speech and second inaugural. We cannot tolerate any attempt to oppose or supplant the language and culture that has come down to us from the builders of this Republic with the language and culture of any European country. The greatness of this nation depends on the swift assimilation of the aliens she welcomes to her shores. Any force which attempts to retard that assimilative process is a force hostile to the highest interests of our country. We call upon all loyal and unadulterated Americans to man the trenches against this enemy within our gates.

"We believe that they can most effectively do this through some organization. The Vigilantes, a non-partisan, militant, anti-pacifist group of writers, artists, and other patriotic children of the crucible, who, from their headquarters in New York, have for months been conducting a vigorous pro-America campaign in the newspapers of the country, offer exactly such an organization.

"We ask that good Americans address themselves at once to The Vigilantes. We ask, moreover, that, whether through this organization, or independently, they uphold the hands of the Government at every point efficiently and resolutely against our foreign and domestic foes, and that they constantly spur the Government to speedier and more effective action.

"Of us who sign some are Protestants, some are Catholics, some are Jews. Most of us were born in this country of parents born in various countries of the old world in Germany, France, England, Ireland, Italy, the Slavonic and the Scandinavian lands; some of us were born abroad; some of us are of revolutionary stock. All of us are Americans, and nothing but Americans."

The following names are signed to the address: Theodore Roosevelt, Oscar S. Straus, John Quinn, Henry L. Slobodin, George von Lengerke Meyer, Michael I. Pupin, Owen Wister, William Loeb, Jr., Anthony Fiala, Henry Reuterdahl, Julius Kahn, Guy T. Visknisskl, Harvey J. O'Higgins, Nathaniel A. Elsberg, Gutzon Borglum, John D. Crimmins, Harry Olson, Isaac Adler, A. W. Erickson, Karl H. Behr, Gustavus Ohlinger, Jesse Isidor Straus, Einar Barfod, John J. Leary, Jr., Leo Wiener, James N. Beck, Charles J. Rosebault, Antonio Stella, F. Wellington Ruckstuhl, Roger M. Straus, Cornelius Rubner, Porter Emerson Browne, Philip Zoercher, Einar Hanson, Edwin Carty Banck, Lionel S. Marks, A. Toxen Worm, Frederick Hollman, and Hermann Hagedorn.

September 10, 1917

NEW CITIZENS, NEW NAMES.

Illinois Judge Americanizes Some Foreign Spellings.

CHICAGO, Nov. 9.—Americanism permeated the courtroom of Circuit Judge Oscar E. Heard, who came here yesterday from Freeport to aid the Cook County bench in clearing away a docket crowded with applications for citizenship. The Judge did not find the multiplicity of foreign names to his liking, so before admitting forty applicants to citizenship, he rechristened a number and curtailed the names of several.

"If you are going to be American citizens, you must have American names," Judge Heard said.

The following are some of the changes in names made by the court:

OLD NAME.	NEW NAME.
Vaclav Pisch.	James Pesc.
Gerліher Woralowski.	Harry Warsaw.
Leopold Szirtakiemcz.	Frank L. Franks.
Anastasius Latsis.	Harry H. Latsis.
Coceal Fidelman.	William Kaufman.
Jomini Belomsetz.	Joe White.
Sam Israel Wiloczszonski.	Sam Weil.
Aloyzy Walamender.	Louis Valender.

The changes were accepted with good grace in most instances, but at least two of the applicants protested and announced that they would lay the matter before the Federal authorities.

November 10, 1917

FOREIGN-LANGUAGE PAPERS IN AMERICA

FIGURES showing the circulation throughout the United States of foreign-language papers have just been compiled by the Bureau of Education in the Department of the Interior in its work in Americanization.

There were approximately 33,000,000 people in the country in 1910 who were either born abroad or under foreign home conditions and neighborhood environment. In all there are 38 different language groups in the United States, supporting publications which have a total circulation approximated at 10,982,000.

The foreign-language press consists of 1,575 publications, printed in 38 tongues. From this number, however, must be deducted many German-language papers which have been suspended or suppressed during the war, the total of which is doubtful. Prior to the suppression of German papers the number of publications in that language amounted to 483. The next in order was the Italian, with 190 publications. In the number of subscriptions the German papers headed the list with 3,000,000.

The Jewish newspapers stand third in number, but second in the aggregate number of subscribers. There are 156 Jewish publications, with a circulation of 1,500,000.

The Polish population numbers approximately 1,500,000, and the ninety-seven Polish papers have a circulation of 850,000.

The Scandinavian groups bulk large. There are approximately 600,000 persons in each group of Swedes, Norwegians, and Danes, the former being mostly in Minnesota, Wisconsin, Illinois, and New York, while the others are found mostly in Minnesota, Wisconsin, and Illinois. There are seventy-seven publications in the Swedish language, with a circulation of 700,000, and sixty in the Norwegian-Danish language, with a circulation of 440,000.

The circulation of papers in each group reaches, in most cases something over three-quarters of the population of the group. An anomaly appears in the case of the Spanish press, where a circulation of 250,000 is divided among as many as eighty-seven papers. This is explained by the fact that Mexicans and American business men are also among the readers of Spanish papers.

Other racial groups which have a press in their languages are the Albanians, Arabians, Armenians, Assyrians Bohemians, Bulgarians, Chinese, Croatians, Dutch, Finnish, Greeks, Japanese, Lithuanians, Magyars, Belgians, Portuguese, Rumanians, Russians, Serbians, Slovaks, Slovenians, Swiss, Turks, and Ukrainians.

February 2, 1919

AMERICANIZATION FALLACY

Danger of Upsetting Respect for All Order in Too Rapid Eliminations of Aliens' Native Customs

By L. P. EDWARDS.

THE United States is suffering from one of its periodic attacks of Know-Nothingism. It is seriously maintained in the public prints that our recent Eastern European, and particularly our Russian, immigration contains enormous numbers of murderers, thieves, counterfeiters, dynamiters, arsonists and other criminals of the most atrocious character. It is alleged that the lives and property of all of us are in imminent danger from these incredibly numerous blackguards, and that the only salvation lies in what is called the Americanization of the foreigner.

Now, it is known to every respectable sociologist in America that our recent Eastern European immigrants, including the Russians, are just as peaceable and law-abiding people as native Americans of native American ancestry. This is a fact about which there is not the slightest doubt in the mind of any competently informed person. It has been repeatedly established by careful studies made by the United States Bureau of the Census; by various State boards and by highly qualified private foundations.

Furthermore, the most honest, thrifty, industrious, upright, God fearing and conservative portion of our foreign population is precisely that portion which has clung most stubbornly to its native ways of life and has been least influenced by American customs. Our immigrants upon changing their foreign languages, customs, beliefs and ideals, upon becoming "Americanized," deteriorate profoundly in moral character; deteriorate to a degree that shows itself in the criminal statistics.

Futile Americanization Efforts.

It is very fortunate for the moral welfare of millions of our foreign population that the present furore for "Americanization" is destined to fail in its object. Its failure is in its own nature. The fundamental social virtues, honesty, industry, thrift, truthfulness and the rest, are the same for all societies on the same general level of development. They are not promoted by the custom of saluting any particular flag nor advanced by the ability to read any particular Constitution.

The very complete and profound change of character implied by the phrase "The Americanization of the Foreigner" can be wisely and safely accomplished only if spread out over at least three generations, while four or five would be better. Every year less than three generations that the progress is hastened means moral and spiritual breakdown for thousands—means domestic tragedy and congested criminal calendars. There is only one foreigner who is really a menace to American society. He is the foreigner who is in rapid process of "Americanization." The danger point is the foreign-born child and the American-born child of foreign parents.

The danger from these classes is real and serious, perhaps the most serious presented in the whole range of immigration questions. Here again we have very reliable statistics which leave no room for reasonable doubt. America needs protection, needs it urgently, against the foreigner of the second generation, particularly against the youthful foreigner who goes through our public school system. The father who stubbornly refuses to learn English or to adopt American ways is commonly a man of admirable moral character. The son, often quite as American as young men of our old stock, is equally commonly a youth of vicious and unprincipled character.

Fault of Public Opinion.

Public opinion in this matter is grievously at fault. There is danger to American institutions, and that danger is real, but it is just the opposite of what is popularly feared. The danger lies precisely in the process of Americanization itself, particularly in the endeavor to hasten that process. If, as is commonly maintained, the present need in America is peace and safety, security and conservatism, then the Americanization of the foreigner should be slowed down in every way possible. No encouragement should at this time be offered to the foreigner to abandon his native language or religion or to change his ethical or cultural standards.

On the other hand, every possible assistance should be given to Roman and Greek Catholic priests, Orthodox rabbis and other such leaders in maintaining and strengthening the traditional loyalties of their various groups. Our Mohammedans—no negligible element in recent immigration—should be encouraged to build mosques, to read the Koran and to obey the various other requirements of their faith. Our public libraries should provide themselves more liberally with books in foreign languages. Foreign language lectures and speakers of all sorts should be much encouraged. By such means, and only by such means can the spirit of unrest and disquiet be stilled, and the spirit of conservatism and contentment with the status quo be developed among our foreign population.

It is a most curious popular misconception that peace and quietness and respect for law and order can be developed in the foreigner by suddenly and violently disturbing his mental life. Changing a man's language, upsetting his moral and social conventions, altering his inherited traditions of conduct, unsettling his ancestral faith—these are the very best means possible for making him a disbeliever in all established institutions, including those of the United States. Yet this is precisely what "Americanization" aims to do with the best intentions.

Let us take a specific illustration. It may perhaps be theoretically desirable to bring our new immigrant to a realization of the crudity and superstition of his Eastern Orthodox faith, and to a lively recognition of the superiority of American Protestantism. Practically, it can be seldom done and the reason is simple. When a person has been brought to realize the faults, imperfections, and limitations of a traditional system of belief in religion, government or what not, he inevitably applies his new critical attitude toward whatever system of belief is offered to him as a substitute for the one he has been encouraged to cast aside.

Most commonly the alternative system, being human, has serious faults, imperfections and limitations of its own, which are easily enough discoverable. The net result of very much conscientious missionary work in America is that the foreigner ceases to believe his traditional faith, refuses allegiance to any American substitute and becomes an infidel, agnostic or atheist. The same thing is just as common in the realms of social, ethical and political faith as in that of religious belief.

Respect for government and law is not a natural instinct. It is an artificial attitude slowly built up in the individual by all sorts of direct and indirect social pressure. The breakdown of old habits of thought in any one of the great departments of social activity very rapidly affects the other phases of conduct. The whole moral life of the individual tends to become unsettled. Nothing is held firmly except the selfish determination to obtain material wealth. Ideas and ideals which stand in the way of this are cast aside. The Americanized foreigner possesses all the native American's ruthless greed without possessing his social, ethical, religious, or political idealism.

No man can learn a language perfectly who learns it deliberately, and social ideals are harder to learn than language. They can never be learned naturally and completely except when they are learned so gradually and imperceptibly that the process is unrecognized and largely unconscious. This can never be possible in the case of the foreign born and is only very partially attainable in the case of the children foreign born. Its complete realization is possible only in the case of children born and reared in an entirely American environment. That is to say it cannot be accomplished before the third generation at the earliest, and often not then.

May 16, 1920

The Melting Pot Problem

DEMOCRACY AND ASSIMILATION The Blending of Immigrant Heritages in America. By Julius Drachsler, Assistant Professor of Economics and Sociology in Smith College. The Macmillan Company.

THE fundamental mistake in much of the Americanization movement, so it would appear to a reader of Professor Drachsler's "Democracy and Assimilation," is that it expects the immigrant to shed his ancient culture as if it were a suit of clothes and accept in exchange a new suit of Anglo-Saxon pattern. The mistake is to the detriment of America even more than of the immigrant. The latter brings with him a culture different from that of the Thirteen Colonies and different from that of nineteenth century America. These various cultures have, however, various values which it is to the interest of America to seize upon and incorporate in her own system, to the end that a truly American race may be evolved containing the best that is to be found in all of them as well as in her own original culture. Such a race would be indeed great.

Against this belief stand two classes. One would stamp out all European cultures as soon as possible; they are not wanted, and the sooner the Slav is turned into a complete Yankee the better. The other would preserve the different race traditions and keep up the different foreign communities, to prevent any amalgamation or "mongrelization." Amalgamation, however, is taking place whether one likes it or not. Professor Drachsler has made a study of that subject in New York City, which he adopts as the best material for such an examination because it is the type of all cities having such a problem on their hands. His statistics, which there is not room to summarize in the limits of a review like this, show that the tendency to intermarry with other racial groups is notable even in the first generation of immigrants and becomes very great in the second.

His examination shows that intermarriage is more frequent among what he calls "the middle or mediocre culture group" than among the highest or lowest. The only way to get at this, the statistics on the subject being practically nonexistent, is to form one's conclusions by the occupations of the persons marrying. Professor Drachsler assumes that the highest "culture group" would be represented "by persons in professional service, the lowest by those in unskilled work." The middle or mediocre group he classes as including "persons in commerce and trade, manufacturing and mechanical pursuits and personal and domestic service." It is in this middle group that most of the intermarriages take place, 71.1 percent of the whole. Only 9.5 per cent were in professional service and 12 per cent, were unskilled workers.

The assumption he makes from this is that in this "middle group" there is not the high degree of cultural self-consciousness that there is among the highest group, nor are there the strong prejudices that mark the lowest, and that "constant contact in daily work levels differences." If this is the fact, it seems that the fusion of races is being largely carried on by those least likely to bring to it any of the peculiar culture of their respective tribes.

That Americanization by school education is a failure Professor Drachsler undertakes to prove by citations from the confessions of educational authorities. The rigidity of the American system of school education is apparently to blame for this, at least in the main. An adult Serbian workingman for instance, is taught in the same manner as a seven-year-old American child descended from a Pilgrim Father. As his whole mental makeup comes from a milieu worlds away, he does not understand it and, what is much more to the point, he is not interested.

A second cause of the failure is the school building itself, the arrangement of the classrooms, the stiff placing of grown men on children's chairs and before children's desks to study a lesson about "little drops of water, little grains of sand." If the missionary is disposed to give it up, Professor Drachsler points out the mighty response made by the foreign-born and their children to the call to arms. "The same fiery enthusiasm for freedom that animated the builders of this nation also stirred the hearts of the immigrant peoples." Freedom, however; not Americanism, which they did not understand. The trouble was that the right way to make them understand it had not been taken.

The communities in which the immigrant lives reproduce for him the old life at home. That, at any rate, he can understand and enjoy. Some enthusiasts, observing the intensity of this community life, draw conclusions from it concerning the future; they imagine it as persisting and setting up a lot of little nations in America. But they do not count on the powerful forces that work against the continuance of these communities, and Professor Drachsler declares that these centres would die out for lack of material if they were not constantly reinforced by immigration. Yet it is through them that the effort to Americanize the foreigner in a true sense should be made, since the intermarriages are chiefly among the mediocre" class, with no cultural contribution to offer to America. On this point he says:

> But what is far more to be regretted; the unique opportunity that America has of utilizing the rich cultural heritages of the immigrant groups and weaving them into the texture of its growing civilization—an opportunity such as no nation ever was offered under the same circumstances—will inevitably be lost. To be consoled by the thought that the new versatile nation resulting from the fusion of many peoples will soon replace, by the potency of its own genius, what may have been discarded or neglected or deliberately ignored in the culture of the immigrants groups, is very much like justifying the barbarities the invading Germanic tribes committed upon the civilization of ancient Rome, on the basis that they ruthlessly cleared the ground for the creation of a newer and more virile culture, irrespective of the high achievements already recorded in the Greco-Roman world. That a thousand years later the more civilized descendants of these empire-wreckers should rediscover the ruined remnants of a glorious past and cherish them as long-lost treasures is ample proof of the original sin and madness of the fathers.

Fusion comes forth from the racial community centre, and the race culture is lost. It will not become a part of American life unless we take the position that the immigrant has something to give us. Heretofore, perhaps, our attitude has been that it is we only who have something to give, and that the immigrant is no giver, only a taker. We cannot prevent the youngsters from marrying whom they please; but we can "safeguard and improve the social environment under which fusion is taking place." In this light the racial community centre becomes the ally of a true Americanization as contrasted with mere heterogeneity.

At present, in the absence of any recognition of this fact, the community centres seek to save the racial consciousness and transmit it to their children, instead of incorporating it in American life. Or, to put it in Professor Drachsler's own words. "Throughout all this work there is much that reminds one of frantic efforts to save the group from disintegration in a new environment, rather than a free blossoming of a healthy cultural life taking root in a new and rich soil."

Standing on the other side is the effort of the older American to educate the newer ones, and for that matter their own American children, without teaching them that this country owes anything in the formation of its character to the different and successive waves of immigration. It may be a fact that most children leave school with the idea that the American character was formed only by the ingredient put in it by the colonists, or even by those colonists who remained along the seaboard. That the subsequent immigrants have played any creative role whatever is not impressed on the child's mind in the public schools.

Nevertheless, Professor Drachsler holds that the public school must become the medium through which the coming American race of his vision shall be born. Only the school methods must be radically transformed and broadened. There must be a "conscious effort to marshal all the cultural contributions of the races and nations represented in the student body." The basis of study, English, civics, and history, must continue to be the basis for the instruction of the immigrant, but must be broadened and enriched in a manner to appeal to the active interests of the adult immigrant. "Instead of puerile language lessons having no intrinsic value for the adult mind, English must be taught to him as a living instrument for the expression of his daily needs." Or, to put it shortly, the coal miner who finds no interest or amusement in copying "Little drops of water" might be intensely interested in finding out how to understand the foreman's commands or the signs warning him of dangerous places.

Instead of expecting him to grasp what lies back of the American and English idea of freedom by merely teaching him our history (incomprehensible to him without a background), the immigrant should at every possible step be informed of the contrasts and similarities between American institutions and those of his home land. The school, too, should pursue the immigrant into his own community instead of asking him to come to it, when the demonstration is so ample that he will not come. As for the author's proposals for the alteration of the classrooms in which foreigners are to be taught, it is a subject beyond the space limits of this article. It is sufficient to say that he would have the classroom an immigrant's club instead of a children's task-room.

December 12, 1920

The Great Melting Pot and Its Problem

A Review by FRANZ BOAS

OLD WORLD TRAITS TRANSPLANTED. By Herbert A. Miller and Robert E. Park. Third Volume in a series of Americanization Studies. New York: Harper & Brothers. $2.50.

WE have become accustomed to ominous predictions of the future of the population of the United States. The great melting pot in which so many "inferior" European races contaminate the superior Northwest European elements must turn out at the end an alloy in which the sterling qualities of the great blond Europeans are only faintly discernible. Instead of steadiness, weakness; instead of individual initiative, collectivistic complacency; instead of rational restraint, emotional yielding; instead of superior bodily build, a gradual degeneracy—such are the dire predictions of our pessimistic race-conscious prophets.

It is interesting to follow the development of this idea. It was first developed on broad lines and brought to general attention by the Frenchman. Gobineau; it found fertile soil in the thought of ante-bellum days in America, when the institution of slavery required some rationalized justification that was most readily

based on the attempt to prove the innate inferiority of the negro; it was further developed in Europe by the Austrian German of English descent, Stewart Houston Chamberlain, and it has found its newest and most emotional representation in the United States.

It is curious to note that the reaction grew apace with the growth of the idea of race superiority. The German, Theodor Waitz, was the first to express clearly the opinion that race achievement is not due to racial superiority, but a result of cultural history; and his fundamental thesis became the basis of all modern anthropological research into the cultural history of mankind. The German, Bastian; the Englishman, Tylor; the Americans, Morgan, Brinton and Powell; the Italian, Montegazza; in fact, modern anthropologists almost without exception consider themselves justified in disregarding racial, innate inequalities as almost entirely irrelevant in the development of cultural history.

It is refreshing to find an author who looks at the problem of the future of the American people not from the emotional race-conscious point of view, but from the modern anthropological angle, who considers each immigrant element as impregnated with a certain traditional cultural heritage, and who investigates the processes by which a gradual amalgamation develops. This is the pre-eminent merit of the book, "Old World Traits Transplanted," by Herbert A. Miller and Robert E. Park, which has just been published. The authors frankly disregard innate differences that may perhaps exist, but press the purely cultural problem which is due to the transfer of the foreigner to American environment.

Before discussing the merits of their argument, it may not be amiss to make clear to ourselves the reason why the authors have the right to disclaim the importance of innate racial differences.

The opinion is quite untenable that racial mixture, such as occurs in America, is a unique historical phenomenon. As a matter of fact, the history of Europe, as well as of other continents, shows that race mixture is the rule, not the exception. First of all we see that there are no hard-and-fast lines between any two human types. When we migrate from Europe into Asia, we do not strike a line at which the people show all of a sudden the characteristic Mongol type, but the change is very gradual. When we migrate from Northern India to Southern India, there are no sudden breaks, but only gradual transitions. In North Africa changes from Mediterranean types to African types are bridged by many intermediate steps. Only in those places where, owing to recent migrations or to customary isolation, types come into contact that were in former times geographically far apart do we find sudden changes.

The history of various countries shows also clearly the progress of mixture. In Spain, for instance, which in modern times is one of the most secluded parts of Europe, we find in early times an aboriginal population called the Iberians. The Phoenicians, later on their North African congeners, settled on the coasts of Spain; Celtic tribes came into the peninsula from the north; the Romans from Italy colonized the whole country to such an extent that the people adopted Roman speech; the Goths and other Teutonic tribes whose home was in the region between the Black Sea and the Baltic swarmed southward and even crossed into North Africa; the Moors came from Africa—in short, the early history of the Spanish population has been such that the blood of practically all European types runs in the veins of the modern Spanish people. In the Balkan Peninsula we can follow waves of migration in even earlier times and continuing until the southward movement of the Slavs and the invasion of the Turks. Even England has not escaped this fate. In prehistoric times there was a sudden intrusion of a foreign type that came presumably from Central Europe; later we have Roman colonization, Teutonic migration from all around the North Sea, and finally the Norman invasion.

Setting aside the negro, Indian and Asiatic elements which present a separate problem, the elements which intermingle in the United States are in no way different from those that entered into the various European populations. The only difference that exists is that the rapidity of the process of mixture and the numbers involved are very great, while in early times the numbers were, comparatively speaking, small, and social barriers retarded the assimilation more than they do now.

Notwithstanding the large extent of ancient mixture, we think of European peoples as pure races. Evidently this thought is based primarily on the stability of population which developed in Europe with individual landholding and with the assignment of the serf to the soil. Since the Middle Ages, Northern Europe has had a stable population, and each nationality, particularly the landholding class in every village, is more or less an intensely inbred group. It is easy to substitute in thought the idea of purity for the idea of that kind of homogeneity which is due to long-continued inbreeding, and it is largely due to this condition that we consider European types as "pure" types. In reality, we may safely say that pure human types in the sense of types evolved by inner forces from single ancestral types are nonexistent.

From this standpoint the racial problem may well be put aside. It is, however, claimed that the racial types, no matter how they have originated, have not the same values. This question is also rightly disregarded by the authors. Our racial panegyrists take an attitude as though a racial type was a unit, as though every family conformed to what they please to describe as a racial type. As a matter of fact, every investigator knows that it is very difficult to find among a mass of individuals a single one who conforms to an ideal racial type, that the great majority rather deviate considerably in many directions—physically as well as mentally. What we call a racial type is merely an impression gathered from the traits of a great many individuals that cluster close together and which we call their type. The actual investigation shows always a wide range of variation in each group. This variation is not only individual. It may also be found in family lines. The members of a family, representing as they do the same kind of hereditary constitution, are alike among themselves, and the variety of forms represented by the family lines in each population is very great. When the populations of Europe are compared, it is found that both physically and mentally the differences between the "types" are very much less than the differences between individuals or family lines belonging to the same racial type, so that it always happens that the series representing district types overlap, and many individuals of the "superior" series are inferior to those of the "inferior" series. The same is true of hereditary lines. The consistent eugenist would, therefore, direct his attention not to the selection of the superior race, but to the selection of the superior strains of all races—provided he can attain the consensus of opinion as to which hereditary traits are the best. In regard to mental qualities we find the same kind of overlapping not only between European types, but between all races—whites, mongols and negroes; and for the whites particularly there is no evidence that racial types differ considerably in regard to their physical or mental makeup.

It is claimed, however, that of certain groups we obtain only those strains which are by heredity inferior. It may be agreed that in certain cases there has been on the part of European nations an unloading of undesirables which from the point of view of American society must be considered as objectionable. It is, however, quite a different question whether these undesirables are by heredity representatives of inferior strains. They may be inferior, due only to social or congenital causes, without being encumbered by hereditary defects. The history of the Australian convict colony is not in favor of the assumption that undesirables may not be the progenitors of a race physically and mentally sound. It is true that psychologists claim that the results of their test indicate mental inferiority — for instance, among Italians—but so far they have failed to prove to us that the difference in reaction is not largely culturally determined, and that the groups investigated may not be essentially the same, but that the tests to which they are subjected are of such a nature, and the attitudes of the people are so diverse, that they do not respond in the same manner. The differences found between Northern and Southern negroes point distinctly in this direction. It may be admitted that the question is an open one. From the standpoint of the anthropologist the more likely interpretation of the results obtained is that, to say the least, a large part of the differences that have been found are socially determined, not due to inferior heritage. Even should we admit that we find hereditary differences, we must not forget that there are excellent strains in the "inferior" groups which excel the poorer strains of the "superior" group.

If these facts were more widely recognized, books like those of Madison Grant and Lothrop Stoddard, written in praise of the Great Blond Race, would no longer succeed in stimulating an unsound race prejudice.

"Old World Traits Transplanted," which is a volume of the series of Americanization Studies, is as sound in its premises as the books just mentioned are unsound. Without committing themselves to any theory in regard to innate differences, the authors study impartially, and by means of excellently selected documents, the attitude of the immigrant as due to the cultural complex which he brings from his home to this country. It is immaterial whether we agree or disagree with the sociological theory underlying the summaries, or whether we accept the statement that the present hodgepodge of European States is an "attempt to make racial and political boundaries more nearly coincide" (whatever the term "racial" may mean), and not rather an attempt to tear away and subject to other nationalities as many Germans as possible—the facts are well marshaled and clearly put forward, setting forth the difficulties which the immigrant has to overcome in order to adjust himself to the new order in which he has come to live.

We might be inclined to point out that the difficulties which the authors discuss are not only one of the incidents of emigration, but that they occur in all periods in which there is a rapid change of cultural attitude. The experiences of the orthodox Jews who come to this country and who see their children break away from the faith of their fathers is merely a repetition of what happened in Western Europe in the period following the Napoleonic wars.

We should be inclined to emphasize more strongly than the authors do the effect of social antagonism which brings it about that groups inherently heterogeneous are welded into a social unit by outer pressure. Many members of the group may feel primarily not as such but as individuals. The American attitude toward the negro, the modern revival of anti-Semitism in Europe and partly in America, are cases in kind.

It seems somewhat curious that the Mexicans of New Mexico, Arizona and Colorado should be classed by the authors as immigrants, when in reality the Americans are the immigrants in their country. We have accustomed ourselves to rail against those European nations that try to assimilate the groups that live within their boundaries and which do not conform as much to the West European social, industrial and economic standards as do the rest of the people. Examples of this kind are the few Lithuanians in East Prussia; the Poles in the mixed lower valley of the Vistula. We forget entirely that we handle the same problem in our Spanish Southwest in exactly the same manner. There are schools in which ignorant teachers forbid children to speak Spanish, there is the successful political domination, and there is the strong resentment of the people themselves. For eighty years the Spaniards have preserved their cultural characteristics.

Notwithstanding minor differences of opinion we agree with all the main points of the book, which should be read by every one interested in the problems of immigration. and we may conclude with the authors:

Assimiliation is thus as inevitable as it is desirable; it is impossible for the immigrants we receive to remain permanently in separate groups. Through point after point of contact, as they find situations in America intelligible to them in the light of old knowledge and experience, they identify themselves with us. We can delay or hasten this development. We cannot stop it. If we give the immigrants a favorable milieu. If we tolerate their strangeness during their period of adjustment, if we give them freedom to make their own connections between old and new experiences, if we help them to find points of contact, then we hasten their assimilation.

February 6, 1921

WHERE MELTING POT MELTS

Remarkable Work Among Toilers of Mesaba Range —25,000 at Naturalization Night Schools

By R. K. DOE,
U. S. Naturalization Examiner, Department of Labor.

WHAT would you say to a country with a rural population of over 100,000, where more than 25 per cent. of the adult population goes to night school after a day of toil, not for a few days or weeks, but for months and months, and in a great many instances, for year after year; where the enrollment of the women outnumbers the men; where the parents of large families will leave their children at home and come to night school in order to learn the language of their adopted country; where the parents, living on farms, will walk miles after dark and after the day's toil in order to better themselves intellectually?

St. Louis County, Minn., is such a county. You may be interested in seeing what the melting pot is doing there.

This county is, roughly speaking, 100 miles square, or, in other words, as large as the States of Connecticut and Rhode Island combined. It has a population of 206,391, according to the census of 1920. Of these, 71,313 were born in foreign lands. Most of them come from Finland, Austria, Czechoslovakia, Jugoslavia, Poland, Russia and Italy—that is, they belong to that class of foreigners who are most difficult to reach, and are therefore often designated as unassimilable, if not undesirable.

This population is distributed as follows: 99,000 in the City of Duluth; 62,000 around the great iron mines on the Mesaba and Vermilion Iron Ranges, 46,000 on farms in the rural communities. Here, therefore, the alien is living under three different social conditions, and it may be interesting to see how the problem has been worked out in each instance.

Let us first visit the mining communities on the Mesaba Range.

The Federal Congress passed the present Naturalization law in 1906. Before that time no serious effort had been made to Americanize the alien. He was welcomed to the country because he furnished cheap labor. But very few saw anything else in him than a raw uncivilized foreigner who would stay here a few years until he had accumulated enough to return to his home and live there at leisure. As no one thought it worth while to speak to him, he was compelled to seek the companionship of his countrymen. He learned enough of our language to understand what he was told to do. That was all. When he came home to his wife and children they all spoke their own language. They remained 100 per cent. alien, no matter how many years they lived in this country.

Stringent Requirements.

It was the good fortune of St. Louis County to have Judges on the district bench who, from the very first, took a great and sympathetic interest in the new law and its proper enforcement. They insisted upon the highest possible qualifications before an alien could be granted citizenship. So stringent were their requirements that many an alien has been forced to come to court once or twice a year for seven long years before being found qualified. Every time he came he would hear the same: "Go home and go to night school and learn our language, and about our great men, and come back in " six months or a year, as the case might be. But the progress was very slow. It proved very hard to arouse the interest of the alien. He worked hard in the daytime. When he came home to a wife and a houseful of children there were no end of chores to do. There all spoke his native tongue. If he did gather up energy enough to go to evening class the work there seemed too impossible. Besides that, his wife resented his going.

The main reason for this slow progress lay in the fact that all Americanization work was directed toward the men from one point of the compass and toward the women from another point. The husband and father was urged to attend night school and learn English and become Americanized. The wife was either left entirely alone, or given lectures in the home on child hygiene. She was not asked to learn English. She was alone in a strange land. She naturally resented any move that would tend to separate her from her husband. Her only defense was to keep him as alien as herself.

Such was the situation up to 1917. If you could get the alien to attend night school for a few months in the Winter it was only for the purpose of getting past the Judge and getting his second papers. As soon as they were obtained he would quit. He wanted to be a citizen because in most communities there were laws that gave citizens preference in obtaining work during slack times in the Winter months. Whether he had been naturalized or not, he remained at heart the alien he was when he landed.

In 1917 the State Legislature of Minnesota passed an act giving the women the Presidential franchise. The District Court of St. Louis County at once took the position that, as the women had received the franchise, they must prove themselves in possession of the same qualifications as their husbands before he could become naturalized.

The court issued an order that a married man must bring his wife into court. She must be examined in the same way as he. If found qualified she must take the oath of allegiance with him. If not qualified, he cannot be granted citizenship papers before she is.

This changed the whole situation. The wife at once became vitally interested with her husband in learning the language and in becoming Americanized. Instead of being a brake upon the whole movement she became its very best support. She was generally the younger, and her mind more active and alert. For her the movement promised wider views, wider interests and a richer life. It was an escape from the narrow alien life she had been living since she came to this country. An escape from the contempt which she felt she was subject to from those who could speak the language.

Felt the Disgrace.

If her husband's papers had been held up because she was found wanting, she felt the disgrace very keenly. She would put forth all possible energy in order to remove the stain. Very often there arose a rivalry between husband and wife as to who would make the best progress.

The Honorable Martin Hughes is District Judge in charge of naturalization among the miners on the Mesaba Range. To him first of all should go the credit for the work accomplished. He is of Irish extraction. Both his parents came to this country after they had grown to manhood and womanhood. He takes the keenest possible interest in the work. While he is all the time raising the standard of admission, he shows such sympathy for the aliens that they do not become discouraged, no matter how often they are sent home for further preparation. That it is not easy to become naturalized on the Mesaba Range can be seen from the following figures:

During the year 1921 there were 1,082 petitions for citizenship filed before Judge Hughes. Of these 399 were granted, 95 dismissed and 588 were continued to make further preparations.

No one aside from Judge Hughes should be given more credit for the result obtained than Miss Martha O'Connor of Gilbert, Minnesota She has for years proven a veritable inspiration for the whole county. During the day she teaches a class in the normal department of the high school. That would prove a task big enough for most women; not so for Miss Martha.

The village of Gilbert is made up of five mining locations with a total population of 4,881, of which seventy-five per cent. must be foreign born. Miss O'Connor has thrown herself into this Americanization work with such ardor that she has acquired a working knowledge of the Finnish, Slovanian, Italian and Polish languages. There is not a birth, baptism, wedding or funeral where she is not asked to be present. She is consulted by all and she is generally the judge before whom most small scrapes and family difficulties are ironed out. She has a method of teaching all her own, the marvelous success of which is due to the fact that it consists of fifty per cent. real heartfelt sympathy with the whole human family, no matter where it comes from.

At the commencement exercises of her night school last year the Bureau of Naturalization's diplomas were given out to 106 graduates. Most of these had completed a course of two years' study. At these exercises the pupils staged an original pageant, "The Removal of the Hyphen." More than one hundred people took part in the performance. What an enormous amount of work such a performance involves. Those actors must have worked for months, and all after a hard day's work in the mines. There were more than one thousand people in the audience.

Two years ago she took a class of sixteen foreign-born women, all over thirty years of age. They had lived here for years, but had never spoken one word of English. In six weeks she had them conversing in English among themselves. "How many children have you?" "I have six children." "What do you raise in your garden?" "I raise carrots and peas." "What do you raise?" "I raise flowers," &c.

She has got the different nationalities to mingle in social intercourse. During the last presidential campaign she formed a Civic Woman's Club in every square of the city. The clubs meet twice a month for six months prior to the date of election. She further achieved the almost impossible of voting more women than men in a community in which 75 per cent. are foreigners.

At the present time she has 219 men and 204 women attending night school. Of these 323 have attended more than one year. Three men and one woman have attended for nine years.

While she led the remainder of the range were not slow in following. The Mesaba Range is made up of fifteen mining communities with an adult population of 40,000. At the present the enrollment in the evening schools is 5,000, which is 12½ per cent. of the whole adult population. As of necessity there are a great many native Americans who do not go to such classes, and also quite a few aliens above 60 years, the majority of whom do not attend, it is not an exaggeration to assume that 25 per cent. of the alien population between 21 and 50 years of age attend.

The City of Eveleth with 7,205 population has an enrollment of 753, or 22.3 per cent.

The City of Chisholm with 9,039 population has an enrollment of 899, or 21.2 per cent.

The City of Virginia with 14,022 population has an enrollment of 1,197, or 14.2 per cent.

The City of Buhl with 2,000 population has an enrollment of 210, or 26.2 per cent.

This attendance is not for a few weeks, but for a term of six months or more. It is a yearly occurrence to have the pupils petition the School Boards to have the classes continued longer than the term agreed upon. The average attendance runs from 70 to 95 per cent. of the total enrollment. Nurseries are kept during the evening to enable mothers with small children to attend.

Spirit of the Work.

To illustrate the spirit of the work the following examples may be given:

Here is a mother bringing five children under school age to the nursery. Consider the work that must involve to get five children under 7 years of age ready for school and put them to bed again after 9:30, when school is out, all after a day's work.

Here are two such cases in Virginia alone where the mothers have been doing this for three consecutive years.

Here are any number of cases where the parents will leave all the way from six to twelve children at home in charge of the oldest, still of school age, while the parents attend school.

Here is a case where three generations are attending. The children in day school, the parents and grandparents in night school.

Here is one woman who gets up at 4 A. M. to prepare the food for the family for the day. Then she goes to the mill to work. She does not return from work before 7 P. M. and only in time to dress for night school. She has no time to eat and gets her supper after 10 P. M. This is for a day, but three times a week for six months or more.

One woman had attended the night school for two years without missing a single day. The very last day of her second year her husband was killed in the mine and she was unable to come. She told her teacher, "Was it not too bad that if he had to be killed he could not have waited one day so I could have had a record of perfect attendance?"

Another student had to quit school to go away for his employer for two months. He returned to the city at 7 o'clock P. M. and was in night school in time for the opening. Supper could wait. It was not so important as the school.

Practically all of these farmers are foreign born as are their wives. They came to this country in an advanced age and found it difficult to earn a living in the mining centre, due to their age and inability to understand the language. Therefore they went out in the wilderness and took up homesteads. Now, homesteading is no easy matter in St. Louis County. The land is very heavily wooded with all kinds of evergreens. It takes years to clear even a few acres in such a way that the land can be seeded. Until you can raise grass or other feed for the stock no stock can be maintained.

It is therefore natural that these farmers are settled in very small communities far from each other and, as a rule, each farm quite a ways from its nearest neighbor.

Nearly all of these settlers come from Finland and Austria, and, due to the fact that each community consists of one nationality, there are many such where no English is spoken.

One would think that under such conditions no Americanization work would be attempted.

Over Forty Night Schools.

In spite of all these difficulties, be it stated that there are today over forty rural night schools in operation with a total enrollment of over five hundred. Thanks to the energetic work of County Superintendent of Schools L. A. Barnes, and his Supervisor of Night Schools, Miss Stella Smith.

The work among the foreigners in the city of Duluth may not depart very much from similar work in other large cities. Still there are some features of it that are not generally known and ought to be described.

The movement started in 1917 by the Commercial Club calling for a mass meeting to devise ways and means for beginning an energetic campaign of Americanization among the city's 30,000 foreign born. A committee of one hundred of the foremost men and women was formed. As the first result of their work the employers of foreign labor agreed to pay twenty cents to the committee for each alien employed. This furnished the capital necessary. The Hon. William A. Cant, Senior District Judge, was elected President for the first year. It is very largely due to the impetus given in this year that such results have been obtained. Judge Cant is of Scotch descent though he is fond of saying that he feels himself the embodiment of the melting pot. His stick-to-it-iveness is developed as it only can be in one emerging from the Highland race. As President he attended all meetings, supervised all publications and personally went into the large foreign boarding houses and formed night school classes. It is not too much to say that during that first year one-half of his time was given to this work.

The following year Judge Bert Fesler succeeded him as President and acts as such at this time. The committee is still the very active centre around which the work revolves. It was thought proper that such important work should not be dependent upon private subscriptions for support and the Board of Education of the city has assumed all obligations. In this it has the hearty support of the public. The first result of the work was an increase of enrollment from 250 to 1,800.

A Director of Night School was engaged who could devote his entire time to the work. H. J. Steel was appointed and his work cannot be praised too highly. The board has adopted the policy that any time ten people desire to be instructed in any conceivable subject the board is willing to furnish a classroom and a teacher. At the present time night school is held in seventeen places in the city, including four industrial plants, three afternoon schools and ten evening classes.

The following subjects are taught:

English	Machine Shop Work
Citizenship	Mechanical Drawing
Cooking	Manual Training
Sewing	Reed and Raffia Weaving
Millinery	
Domestic Science	Home Nursing
Woodwork	

Away From Evil Influences.

No matter how trivial some subjects may be, it is worth while keeping the people away from the evil influences of the street.

A department for foreign-born women was also organized, and Miss Margaret Quillard was engaged as Director.

With the exception of the Director, who is paid by the board, this department is entirely dependent upon volunteer service furnished by American women of insight, ability and a sense of the need for special and civic service. In 1921 Miss Quillard had 131 such volunteers who made over 1,500 calls and instructed between eighty and ninety women.

But the most important innovation, from the point of view of educating applicants for citizenship, is a course of eighteen lectures, which is given simultaneously the year round in four different places in the city—two under the direction of the public schools and two under the direction of John Samuel, Educational Director of the Y. M. C. A.

These eighteen lectures consist of two in Geography, four in History, and twelve in Government—Federal, State and local. The salient features of each lecture are mimeographed and one copy given to each pupil at the end of the lecture. The pupil is told to study the same at home. The first hour of the next lecture is then given over to a quiz upon the last lecture.

At the completion of such a course one of the District Judges and the Examiner from the Bureau of Naturalization examine the pupils. No one is allowed to pass who has not a working knowledge of these lectures. The requirements are so high that some of the less apt applicants may have to take the course twice and even three times before they can pass and receive their diplomas. This diploma is honored in court as evidence of satisfactory knowledge of the applicant.

These lectures have become so popular that out of the 447 petitions for citizenship admitted in the city in 1921, 384 had graduated from one of these courses.

It Surely Melts.

The melting pot surely melts in St. Louis County, Minnesota. Then why is it that so many people are discouraged about its efficiency in the rest of our country?

There are two main reasons:

First, the efforts in Americanization have mainly been directed towards the man, though he is the least malleable subject. He is older, more engrossed in manual work, less accustomed to mental effort. Even in his school years he is behind his sisters. How much harder does it come now after twenty or thirty years of manual labor.

Our success has been due to the fact that we teach him through his wife. If she is interested she will soon convince him. She has more at stake. Her life in this country is a great deal narrower. He has his work, his life among his comrades. She is practically a mental prisoner in her own home. Then she is also more sensitive to criticism. With her nimbler brain she grasps what Americanization means to herself and her children a great deal quicker.

Apply it to him alone and she will fight it for it tends to isolate her still more. Apply it to him through her and every ounce of her mental and physical power will be strained in order to accomplish the most.

The second reason is that not all of the Judges in the country take to the movement sympathetically enough. They are all for Americanization in the abstract. They are all willing once in a while to come down and make a speech about the flag and the immortal George and Abraham. But that is all.

March 12, 1922

PUBLIC SCHOOL BEST ROAD TO CITIZENSHIP

Best Point of Contact Between Native and Foreign-Born, Says Naturalization Bureau.

HOW WOMEN CAN HELP

By Co-operating With School Boards and Administrators Who Aid Americanization.

A number of suggestions have been issued by the Bureau of Naturalization of the United States Department of Labor to the women's organizations throughout the country, as to how they may assist the foreign-born to prepare for American citizenship. In its statement the bureau says that it has been its experience that the public schools offer the most universal point of contact between the native and foreign-born, and it considers, therefore, that the best effort that the women's organizations can make is along the line of broadening the work of the public schools and making possible results which might not otherwise be obtained.

Among the suggestions offered by the bureau are that public school classes for the foreign-born be made possible, this being brought about in a variety of ways; for instance, by influencing officials and official organizations, such as school boards and administrators, who may not yet have realized the need of this work; by organizing and conducting surveys to show the need; by financing the work, or by arousing public interest to help finance or support it; by furnishing volunteer teachers if paid ones cannot be secured; by organizing and financing classes to be turned over to public schools for teaching and supervision; by raising money to employ a home teacher to work under direct supervision of the public schools, and by arousing public opinion to see the need for this work. The last, it is suggested, may be accomplished in various ways—as the result of the action of individual club members; through going on record as an organization; through appeals made to various industries, and through mass meetings.

The bureau also suggests that the women's organizations assist in public school work directed to the foreign-born already started, by aiding in clerical work; by regularly visiting families and inviting the women to enroll; by following up absentees; by furnishing volunteer assistants for trained public school teachers to act also as substitutes when needed; by providing for the care of children while their mothers are attending classes; by visiting newcomers to America as soon after their arrival as possible, and inviting them to attend the classes, and by furnishing cloth and other material for home classes.

Women are also asked to direct their efforts to broadening and making more effective public school work already being conducted; and suggestions for means to accomplish this and include making possible the establishment of specially supervised home class work for foreign women; showing the existing need for classes throughout the year, and for day as well as evening classes; providing classes for special types of workers whose hours and duties make attendance at regular classes impossible, such as household workers, mothers, factory workers on night shifts, waiters and hotel employes; by making possible institutes for the training of teachers, and by encouraging adequate appropriations of public funds for the work.

Following these broad suggestions, the bureau points out many other detailed ways in which the work of helping the foreign-born to become loyal and intelligent American citizens can be furthered, such as, by appointing a city chairman of community work to cooperate with the local director of Americanization in seeking to unite racial groups; by arranging receptions for the newly naturalized; by spreading the idea of special ceremonies to welcome new citizens—American youths as well as naturalized foreigners; by arranging for special graduation ceremonies when certificates of graduation are presented, and furnishing appropriate souvenirs; by entertaining in American homes wives of foreign-born men who have been recently naturalized; by arranging for exhibits of handiwork of different nationalities, to show appreciation of the foreign-born and friendliness toward them.

In regard to such exhibits, it is pointed out that it is necessary to gain the confidence of the foreign-born and that this is actually the first step in Americanizing them, and it is recommended that their co-operation be sought in planning the details of the exhibitions. A suitable space in a suitable locality having been provided; and a neutral background and necessary fittings, such as tables and chairs put in, it is suggested that the space be divided into sections for the different nationalities represented, and that each day during the exhibition a typical entertainment be given by each nation; that the last day be an all-American day with an all-American program, every one pledging allegiance to the American flag; that helpful literature in different languages be provided for distribution together with invitations to join the public school classes.

Other suggestions cover the encouraging of art needlework, the providing of materials and arranging of a market for such handicraft as may be produced; providing free tickets to concerts and sending teachers into homes to give lessons to talented foreign-born children and young people whose parents cannot afford to pay for them; and the arranging for a course of lectures for the training of voluntter Americanization workers.

In carrying out these suggestions, says Community Service, which is furnishing this matter to its workers, it should be kept in mind that "appreciation begets confidence, confidence begets co-operation, co-operation means neighborliness and neighborliness means Americanization," as the Bureau of Naturalization itself points out.

July 9, 1922

FULL-BLOOD INDIANS GOING FAST, HE SAYS

Increasing Marriage With White Race Is Responsible, Declares Dr. Hrdlicka.

Increasing marriages with members of the white race is responsible for the rapid disappearance of the full-blooded Indian from North America, said Dr. Ales Hrdlicka, Curator in Anthropology of the Smithsonian Institution in Washington, and on the staff of the United States National Museum, speaking last week at a joint meeting of the New York Academy of Sciences and the American Ethnological Society, at Columbia University.

"There are about 335,000 Indians in the United States and its possessions," said Dr. Hrdlicka, "more than most people realize. As a body they have slowly but steadily increased during the last two or three generations, but it has been an increase in mixed-bloods. The full-blooded Indian is passing from existence. A recent investigation among the Chippewas and Ojibways of the Middle West showed that of 30,000 Indians only 800 claimed full Indian blood. Careful examination showed that of these 800 only seven had a valid claim, and they were all old people. Among the Shawnees in Oklahoma only three Indians were full-blooded.

"The pure Eskimo is also gradually passing. He intermarries with either whites or Indians, and as the invading white population pushes further and further northward the Eskimos slowly merge with it."

Passing to the race problems of Canada, Dr. Hrdlicka remarked that the French Canadian is deserving of more study than he has heretofore received. With a prolific birth rate, in marked contrast to their cousins across the sea in France, the French Canadians are exercising considerable "racial weight," he said, both in politics and other fields. The average family, he reported, has from seven to ten living children, and families of twelve are not in the least uncommon.

"In the United States," continued the speaker, "the white man's race problems have solved themselves in a very favorable way. The incoming Europeans have blended with the native stock in a way that many people call most successful, and there seems little doubt that the traditional 'melting pot' of this country is an established fact. The blending has been better on the whole in the cities than in the rural districts, where the same European strain persists even to the third and fourth generations."

Dr. Hrdlicka declared that there had been much unjustified outcry against the character of different elements in the European immigration to this country, which he said was not based upon real knowledge. "The time has come," he said "for us anthropologists to point out the real conditions in answer to the published stuff that has even influenced Congress in favor of legislation against certain groups of white people. There is little basis for excluding from the United States any part of the white race."

The strongest emphasis was laid by Dr. Hrdlicka upon the need of keeping from this country the physical and mental defectives, who, he remarked, were still being admitted in large numbers upon American soil.

April 1, 1923

Professor Kallen Proposes to Balkanize America

A Review by
NICHOLAS ROOSEVELT

CULTURE AND DEMOCRACY IN THE UNITED STATES. Studies in the Group Psychology of the American People. By Horace M. Kallen. 347 pp. New York: Boni & Liveright. $3.

THE thirty Jeremiahs who looked upon America two years ago and found it unlovely joyed in their intellectual muckraking. America had no past. It present was unspeakable. Its future was delightfully drab. They scratched and tore with almost sadistic glee, but they failed to penetrate very deeply.

Not so Professor Kallen. To be sure, he also finds the United States unlovely. But he has taken the plains to dissect this creature rather than simply to stick pins in it, and he believes it has a future. His dissection is in many instances penetrating. But any American-American (to distinguish him from the British-Americans, German-Americans, Jewish-Americans and other hyphenates with which alone, according to Mr. Kallen, the United States is populated) will be tempted to remark that the future which the professor foresees is, indeed, a thousand times more dismal than that foreseen by the thirty Jeremiahs.

Its form [says the professor] would be that of the Federal Republic; its substance a democracy of nationalities, cooperating voluntarily and autonomously through common institutions in the enterprise of self-realization through the perfection of men according to their kind. The common language of the Commonwealth, the language of its great tradition, would be English, but each nationality would have for its emotional and involuntary life its own peculiar dialect or speech, its own individual and inevitable esthetic and intellectual forms. The political and economic life of the Commonwealth is a single unit and serves as the foundation and background for the realization of the distinctive individuality of each nation that composes it and of the pooling of these in harmony above them all. Thus "American civilization" may come to mean the perfection of the cooperative harmonies of "European civilization" —the waste, the squalor and the distress of Europe being eliminated —a multiplicity in a unity, an orchestration of mankind.

In other words, Mr. Kallen hails the creation of an ethnic federation, in which the English language shall, reluctantly perforce, be the lingua franca, and in which the racial peculiarities and jealousies of Europe shall have unrestrained freedom of development, until the United States becomes a second Austria-Hungary. The hyphen is to be crowned king, and the native-American element (which Professor Kallen, like his friend Mr. Mencken, persists in referring to as "British-Americans") will gradually be relegated to that inferior place to which its lack of cultural distinction justly entitles it. The melting pot has failed, and there is no longer any "fear" (the word is Mr. Kallen's) of the formation of an "American race."

There is nothing new in these views. There is little that entitles them to serious attention except for the fact that Professor Kallen, unlike the thirty Jeremiahs, has obviously devoted serious thought to his studies and speaks with the authority of a philosopher who happens to be versed in the history of the American Continent. Furthermore, the conclusion which he seriously reaches is held by enough persons interested in perpetuating foreign cultures in the United States to make it the expression of an ever-growing challenge to those who are proud of America's past and who see America's future in the hands of an American race, different from, although a blend of, the races of Europe, and with a tradition that is distinctively American, having its roots not in the Old but in the New World. Professor Kallen's is the challenge of the dissatisfied newcomers—who came, as he points out, at the behest of our greed-obsessed capitalists of the last half century (though he fails to add that they came willingly enough and were proud of the opportunity). His is the challenge of those who prefer other cultures to that of America and who, rather than accept the American, seek to impose on us their own ways and traditions. It is a challenge which, as he says, the old stock must take up. America is at the parting of the ways. Is it to remain one in spirit, tradition and language, or is it to become a hodgepodge boarding house for alien groups?

Professor Kallen's book begins with a postscript—"to be read first"—in which he relieves his mind about the Ku Klux Klan, and advocates as an antidote to it what he calls "Cultural Pluralism," which, although he does not clearly define it, seems to mean the encouragement of all cultures other than the native American. The essay is penetrating and informing, and is a good introduction to the three main portions of his book—"A Meaning of Americanism," "Democracy Versus the Melting Pot" and " 'Americanization' and the Cultural Prospect."

"A Meaning of Americanism" elaborates the thesis contained in the title of the late Barrett Wendell's book "Liberty, Union and Democracy." "Liberty," as explained by Mr. Kallen, has become liberty of the individual, with the advantages which the term implies. "Union," however, is merely political, for "the American people are no longer one in the sense in which the people of Germany or the people of France are one, or in which the people of the American Revolution were one." "Democracy" implies freedom for ethnic groups as well as for other units. It involves "not the elimination of differences but the perfection and perpetuation of differences."

In "Democracy versus the Melting Pot" Professor Kallen exposes the failure of the melting pot. Incidentally he brings out clearly the blame which rests on the American people for the problem which they now have on their hands. At the same time, however, he expounds a number of fallacies which cannot go unchallenged. In the first place, he pretends that there is no American race and there are no American people. In their place he sees only hyphenates, of whom the largest group happens to be "British-Americans." Nothing more clearly shows that despite his penetration and his grasp of certain essentials, he has failed, in the final analysis, to understand the real America. There may be a small group of British-born subjects who can properly be called "British-Americans." But the old stock is neither in name nor in fact, nor, for that matter, even strictly ethnically speaking, British in origin. Rather is it a mixture of strains, English, Scotch, Welsh, Irish, French, Dutch, Swedish and Germans. Its members look upon England as a foreign country, just as they do upon Germany or France. In so far as they are conscious of race pride, it goes back not to ancestors in England but to ancestors who fought England. They are proud of their American rather than of their alien origin.

The next fallacy is that whatever else a man changes, "he cannot change his grandfather." This is used as an argument to prove that there can be no American race, inasmuch as ethnic unity only exists among Europeans and the children of these Europeans cannot escape their ethnic heritage. But what of the grandmothers and the other grandfather? How often do not all come from different stocks? To what "race" does the grandson of a French grandfather and New England grandmother on the one side and an Irish grandmother and Dutch grandfather on the other belong? To which of these precious alien ethnic groups in the new "Ethnic Federation of America" does this poor mongrel owe cultural allegiance?

But perhaps the greatest fallacy is the failure to see that by the perpetuation of alien ethnic groups and the encouragement of foreign traditions in these groups the inevitable outcome is a new Austria-Hungary on American soil. How this ideal can be desirable is incomprehensible to any one who realizes that in the old Austria-Hungary alien races lived side by side in groups for countless generations and still preserve against each other a hatred such as we poor native Americans cannot even conceive. Would he advocate the future of America as in the Banat of Batschka? If so, he would have a crazy-quilt of cultural groups such as even the Balkan States find intolerable.

And yet it is these very fallacies that make this book valuable to the Americans who (with due respect to Mr. Kallen's ethnic nomenclature) do not look upon themselves as members of some European national stock. It states forcibly, clearly and reasonably an interpretation of America's future which is, as he frankly points out, a challenge to the old American traditions. The melting pot has failed. This he repeats (and it cannot be repeated often enough). The result of this failure is that the country must choose between, on the one hand, the perpetuation of the old American ideals, developing and encouraging them, and using them as the basis of the America of tomorrow, and, on the other hand, Mr. Kallen's vision of "Cultural Pluralism," which, carried to its logical extremes, will result in the Balkanization of these United States.

April 20, 1924

"ABIE'S IRISH ROSE" FUNNY.

Anne Nichols's Little Human Comedy Heartily Received at Fulton.

ABIE'S IRISH ROSE. A Comedy in Three Acts. By Anne Nichols. At the Fulton Theatre.

Mrs. Isaac Cohen..........Mathilde Cottrelly
Isaac Cohen..................Bernard Gorcey
Dr. Jacob Samuels.............Howard Lang
Solomon Levy...............Alfred Weisman
Abraham Levy...........Robert B. Williams
Rosemary Murphy.............Marie Carroll
Patrick Murphy......................John Cope
Father Whalen................Harry Bradley
Flower Girl.....................Dorothy Grau

"Abie's Irish Rose" vindicated her middle name, at least, by coming to the Fulton last night after a merry war and temporary truce between the play's author, Anne Nichols, and its first producer elsewhere, Oliver Morosco. Judge Julian Mack in the Federal Court yesterday, at the request of Mr. Morosco as plaintiff, adjourned hearing on an injunction application until June 1.

Miss Nichols, whose play is still running with two companies in San Francisco and Los Angeles—ten weeks in the latter city alone—organized the New York production on her own account, contending that Mr. Morosco's option had been too long delayed. She feared, in effect, that "Abie's Irish Rose" might be mistaken for a mere California poppy if it were much longer withheld from Broadway.

An "all-star" cast was billed in the piece under Miss Nichols's own direction. The veteran Mathilde Cottrelly was among the players last night, as were also Marie Carroll, John Cope, Alfred Weisman, Bernard Gorcey, Howard Lang, Harry Bradley, and, as Abie himself, Robert Williams.

Why the play must sooner or later have been meant for New York was fairly easy to see when the curtain rose on old Solomon Levy's big apartment in the Bronx. It was far more clear when the last curtain fell on a Christmas Eve in the thrice-married Abie's and Rosemary's tiny flat "one year later." Fell, for instance, on Mme. Cottrelly as the kindly Jewish neighbor, cooking, against her will, a ham; on John Cope as the Irish grandfather and Alfred Weisman as the Jew who also "wanted grandchildren"—as Mr. Gorceley had said, he "always talked wholesale"—and finally on Harry Bradley and Howard Lang as the priest and the rabbi, and on "the family."

Abie was a Romeo, heir to riches in New York, but with the ghetto in his blood; Rose, a Juliet, with the blarney, an heiress in California. This Abie and Rose had met "over there" in France in the war; had been "merried good and tight by a nice little Methodist minister," and, when cast off as "unwelcome strangers," just lived and loved, "not wisely, but well." Perhaps this Irish Rose is a hybrid, but handsome Abie, too, was a bit of a Virginia rambler.

A highly sophisticated Summer audience took the little comedy very heartily, laughing uproariously at its juggling with some fundamental things in human life and at some others, not so fundamental, but deeply cherished, as lifelong feelings are wont to be. The New York scenes sagged on the lines in the play's first act as did the Brooklyn Bridge when the cables were being strung. The Irish, in the person of big John Cope, got the laughs going at the interrupted wedding at old Levy's.

Miss Marie Carroll's Rosemary Murphy-Levy, with a "Peg o' My Heart" brogue, was girlish and charming as she walked with her bridesmaids and flower girl, a picture from her forgotten days in "Oh Boy" or "Oh Lady, Lady" down at the Princess. Robert Williams, too, who told everybody "You'll like her; she's a great girl," was himself a fine, likable lad as Abie Levy, who "wasn't marrying a religion." It wasn't orange-blossom time, so the knot was tied under "real California navy oranges." Small wonder when Mr. Cope exclaimed, "Good God, is she marryin' an A. P. A.?"

The play has its little sermon that earned one of the heartiest bits of applause last night. Priest and rabbi, it appeared, also had met "over there." "I gave the last rites to many Jewish boys," said the fighting chaplain. "And I to many of your Catholic lads," the Jewish chaplain replied. "We're all on the same road, I guess, even though we do travel by different trains."

It takes two to make a quarrel, family or otherwise, fathers or sons. And to make that quarrel up, it takes two—but that would be telling. Rosebuds and ramblers never grow singly. And as the good priest said at last, "Sure, Abie's a great boy." Personally, we hope to be present at little Rebecca Rachel and Patrick Joseph Levy's second birthday, if not their Hudson-Fulton centennial.

May 24, 1922

DR. ELIOT STIRS IRE OF IRISH IN BOSTON

Special to The New York Times.

BOSTON, Dec. 14.—When Dr. Charles W. Eliot, President Emeritus of Harvard University, said the Irish and Jews have not been assimilated in America, at the same time indicating that for some reasons it was just as well that they were not, he stirred the ire of critics here.

Dr. Eliot's statements were made Friday before the Harvard Zionist Society, when he expressed his disapproval of Jews marrying Christians and of their discarding the traditions and customs of their parents.

Not all disapprove, however, for two prominent Jewish leaders, A. E. Pinanski, Secretary of the Federated Jewish Charities, and Alexander Brin, editor of The Jewish Advocate, find much that meets their approval in Dr. Eliot's sentiments.

Charles J. O'Malley, President of the Irish Charitable Society, says Dr. Eliot is wrong about marriages.

"There are at least fifty members of our society, all successful in business and happy at home, who have married outside the Irish race," he declares. "I fail to understand what Dr. Eliot means by saying that these men have not been assimilated into American life."

Michael J. Jordan, a leader in the Irish Historical Society and authority on Irish matters generally, says:

"If Dr. Eliot says the Irish have never been assimilated he is in error. If the second generation of Irish-born are not as American as George Washington I am mistaken in the race. The splendid Jews of New York are among our finest citizens."

December 15, 1924

THE JEWS IN AMERICA

Dr. Eliot's Opinion That They Should Retain Their Racial Individuality and Not Intermarry Is Attacked and Defended.

To the Editor of The New York Times:

I have read with great interest the report in THE NEW YORK TIMES of the remarks of Dr. Charles W. Eliot, President Emeritus of Harvard University, before the Harvard Zionist Society in Cambridge, Mass., a short time ago.

I have great regard for Dr. Eliot's opinion. He has on many occasions expressed himself very favorably toward the Jewish people, and only lately in a foreword that he wrote in a book entitled "Patriotism of the American Jew," written by the late Governor Samuel W. McCall of Massachusetts, his attitude has been most favorable toward the American Jews. But I think he is mistaken when he says that the Jews "should keep their race individuality in America just as the Irish have done," and that he has never known intermarriages between Jews and Christians to turn out well.

He goes on to say, "What we want in this country is a number of races with various gifts, each contributing its own peculiar qualities to the common welfare," and also that "the Irish have never been assimilated in America anywhere and it is not desirable that they should be." If it is not desirable that the Irish should be assimilated, the same is true of all the other nationalities. None of the others, then, should be assimilated—the English should always marry with the English, the Scotch with the Scotch, the French with the French, the Germans with the Germans, the Italians with the Italians, &c. Then we would have in America fifty or a hundred sets of nationalities, and no Americans. It is really an impossible idea and just the opposite has taken place and should have taken place.

Dr. Eliot is entirely mistaken when he complains about the tendency among young Jewish people to disregard the customs of their parents. He either forgets or does not know that the modern Jew does not believe in keeping the old dietary laws contained in the Old Testament any more than the Christians do. The dietary laws having been made more than 3,000 years ago were probably founded upon reasons of health, which was quite suitable at that time for the Eastern countries. Modern Judaism is a reform of the Old Testament, the same as is Christianity, leaving out such parts of the Old Testament as are not suitable for the present age. There are very few of the Jews whom we are apt to meet who know anything about these obsolete laws.

In my opinion there is no Jewish race. The Persians, Arabs and even the Armenian Christians are much more Semitic than the Jews whom we generally meet here. While their coming from Semitic ancestors and having generally intermarried among themselves has left its influence upon them, their racial condition is greatly influenced by the country and climate in which they have lived for many centuries. For instance, I have a family tree showing that my ancestors in 1609 came to Hamburg from Holland, where they no doubt lived for centuries before that time. The Disraeli family came from Portugal early in the sixteenth century. The Jews commenced to arrive in the United States about 300 years ago. I know of many happy marriages between Jews and Christians, and there is no good reason why that should not be so.

ADOLPH LEWISOHN.

New York, Dec. 30, 1924.

January 4, 1925

OUR LOST MINORITIES.

Despite the persistence of the "Pennsylvania Dutch" in a small section of Pennsylvania, and of a few French-speaking groups in Louisiana and in Northern New York and New England, America's "racial minorities" seem to be doomed. There are still colonies of aliens, many of which are in the cities. New England has a rural population of Polish, Italian and Portuguese origin. In portions of the Dakotas a few enclaves of Scandinavians still persist. Michigan has a colony of Hollanders. Elsewhere are other racial groups, cherishing the language of the Fatherland, or, as in the case of the Irish, still keenly interested in the current affairs of the "old country." New Mexico and Arizona have large Spanish-speaking populations, a portion of which trace their ancestry to families settled there 200 to 300 years ago.

Yet a glance at history shows how, with the exception of the Spanish, the French and the "Pennsylvania Dutch," the separate racial groups have gradually been amalgamated. New York City has long been a classic example. The descendants of the earlier settlers combine in their veins English, Dutch, German, French, Irish, Scotch, Scandinavian and Jewish blood. That this same process has occurred up-State has been less clearly recognized. A map published by Mr. W. PIERREPONT WHITE in the Rochester Historical Society publication fund series shows the racial groups in New York State in 1790. Although it omits modifying details, its main outlines indicate the then composition of this State's population. Long Island west of Brooklyn and Flushing had been peopled largely by the English, the Hudson Valley by the Dutch, the central Mohawk Valley by the German Palatinates, the region about Johnstown by the English. Already there were important settlements of New Englanders—along the New York side of Lake Champlain, at Troy, Whitestown, Binghamton, Geneva, Ithaca and elsewhere.

Most interesting today is not so much the distribution by racial origin as the amalgamation of all these people and their contribution to the American stock. There are traces—notably in family and place names—of the original nationality of the early settlers. But the traveler, and even the student, cannot but be impressed by the complete blending of Germans, Hollanders, English and New Englanders. In the process there has been a generous fusion of Irish and Scotch blood, and, in recent years, contributions from many other nations in Europe. In view of what has happened there and elsewhere, the surprising thing is not so much that racial ties with Europe have been lost as it is that the three language groups that are the exceptions—the Spanish New Mexicans, the "Pennsylvania Dutch" and the French of Canadian or Louisiana origin—have managed to maintain their separateness so long. The German settlers of Pennsylvania came over at the same time as those who took up lands in the Mohawk Valley. The original ancestors of the French- and Spanish-speaking stocks have been on this Continent fully as long. What would have happened if the other descendants of non-English speaking immigrants had succeeded in creating enduring colonies, perpetuating their own languages and customs, and reproducing on American soil the racial cleavages that are the bane of Europe?

March 17, 1929

RACIAL GESTURES DISAPPEAR

Effects of an American Environment on the Characteristic Habits of Certain Peoples

By VICTOR H. BERNSTEIN

AN anthropologist, an experimental psychologist and an artist, after an expedition in the jungles of human behavior, have just returned with a bagful of gestures as their sole prize. An odd capture, surely, but then their quest was a curious one: to determine whether immigrants, under the influence of an American environment, lose their gesture habits.

The quest involved the identification, classification and interpretation of commonplace gestures used by "race groups" in Little Italy, the lower East Side and on university campuses in New York City. Its conclusion reveals myriad hands pointing—for once—in the same direction: toward assimilability. Which means that gestures, like clothes, manners and language, are changed or discarded with change of environment.

The expedition was directed by Franz Boas, Professor of Anthropology at Columbia University, while the field work was done by Dr. David Efron, University of Buenos Aires psychologist, and Stuyvesant Van Veen, American artist. Their armaments were camera, sketch book and a deceiving nonchalance which allowed them to work, unsuspected, close to the subjects of their observation.

* * *

OF the groups studied, the "traditional" Italian of Little Italy was found to exceed in vigor and radius of gesture, while the Jew of the "ghetto" came first in frequency. Indeed a quantitative analysis of the "traditional" gesticulations of the Jew as he talks would defy the most rapid calculating machine—though it has not escaped the observant eye of the mimic. His forearm, his wrist, his fingers, his head, his shoulders are in almost constant motion, weaving into the air the subtle, rapid pattern characteristic of Jewish thought.

His gestures reflect the generations of "ghetto" environment that are behind him, and the prevailingly sedentary occupations that have been his for centuries. They are nervous, perky, of uneven tempo, lacking in balance and confined in radius. The axis of his arm motion has been transferred from the shoulder to the elbow, with the result that his upper arm seems tied to his side, while the area immediately in front of his chest is a battleground for his flying fingers.

But the Jew, despite the popular notion, does not and cannot "talk with his hands." His gestures are rarely pictorial, and—except in religious ritual—never symbolic. Dissociated from his speech they are practically meaningless. Dr. Etron puts it neatly: "Instead of using his hand and arm to describe, in space, the form of his ideas, the Jew uses them rather as a baton to relate one idea to another, or to trace the pattern of his thought."

The Italian—extrovert, emotionalist, realist, as contrasted with the introverted and rationalizing Jew—uses a freer, more vigorous and more balanced and rhythmic gesture. He motions with the freedom of one who is master of his environment; as, indeed, in comparison to the Jew, he has been.

Moreover, the Neapolitan (Southern Italian) gestures are emphatic and pictorial, and frequently symbolic. The Neapolitan can, literally, "talk with his hands"—like the deaf-mute or the North American Indian. He has a visual vocabulary of more than 100 gestures, understandable only by a fellow-Neapolitan, and sufficient to carry on a voiceless, pantomimic conversation at great length.

MANY of these symbolic gestures bear a discernible descriptive relationship to their meaning, but only after the meaning is known. The raising of the little finger, with the hand outstretched and the other fingers closed, means a thin man; the plucking of the collar away from the neck indicates gullibility (i. e., "that chap's esophagus is big enough to swallow anything"); a scissors-like movement of the index and middle fingers represents the interminable "jawing" of the gossip; open fingers held before the face, giving the effect of bars, means jail, or prisoner.

But in other symbolic gestures of the Neapolitan, the pictorialism, perhaps clear enough once, has been obscured by time and change of custom and thought. Thus the drawing of the thumb and forefinger down along the line of the jaw toward the chin means "beautiful," and may have been derived from the ancient Greek insistence on ovality as a principle of facial beauty. But explanation of the origins of such gestures as the rubbing of the nose between thumb and forefinger, indicating shrewdness, or the rapid outward brushing of the underside of the chin with the back of the fingertips, meaning "I don't know," or "I don't care," must now be guesswork at best.

So the streets of the "ghetto" and of Little Italy—the marketplaces, the shops, the restaurants, the homes—are streets of waving, swooping, stabbing hands. If the gesture patterns differ, it is because the streets lead back to different pasts; but ahead, they tend to converge into the broad highway of American life, and just so do the gesture patterns tend to converge upon, and eventually merge with, the pattern of the American gesture.

Like the Italian, the American gesture is broad, free, vigorous and explosive; but it is not nearly so frequent, nor often pictorial, and almost never symbolic. The American uses his hands principally for emphasis. If the gesture of the "traditional" Jew is primarily a baton connecting one thought with another, then the gesture of the Italian, like a musical overtone, adds richness to his language or serves as a phrase in itself, while the gesture of the American is primarily a pedal lending accent to the score which he is playing.

AS time passes "traditional" tendencies fall away. Governor Lehman, American for several generations, gestures even less than the average American of Protestant descent. Mayor La Guardia, tempestuous by nature, and subject to an American environment since early childhood, gestures frequently—but like an American. He has picked up all the old standbys of political rhetoric—the "physical" type of oratory used by Huey Long and former Governor Smith—and he uses it emotionally and forcefully. Only in private conversations do little Italianisms creep into the Mayor's gestures.

This assimilation does not take place, of course, in American-born children who grow up in a "traditional" environment. Americanization of gesture takes place only when the foreign-born, or those of foreign extraction, have been poured into the melting pot of the public schools, subjected to the influence of the American stage and screen, of American public speakers and newspapers.

Why does the Anglo-Saxon gesticulate less than the Latin or the Jew? It rather complicates the problem when we learn, through pictures and literature, that the Englishman of Elizabeth's time probably gestured as much as does the Latin today. The factors which have led him to increased restraint have not yet been clarified beyond dispute. But certainly today there is a kind of social stigma placed upon too profuse use of the hands.

Almost every conceivable antisocial phenomenon — criminality, insanity, mental and artistic inferiority—has been laid at the door of race. One by one, painstaking scientific investigations reveal them as more the result of environment than of hereditary strain. At a time when much is being made of these racial "barriers" in Germany and elsewhere it is doubly interesting to note how Professor Boas's study of gestures tends to show these "barriers" crumbling at a wave of the hand or a shrug of the shoulders.

October 6, 1935

THE CLOSING DOOR

Know-Nothing Convention in Philadelphia

"AMERICAN BANQUET."

SPEECH OF MAYOR CONRAD.

[FROM OUR SPECIAL REPORTER.]

The grand banquet got up by Mayor CONRAD and his political friends and fellow-citizens in the city of Philadelphia, in honor of the Convention which is now holding its sessions there, and to which its members were severally invited, took place on Thursday evening last, in the large hall in Sansom-street. The occasion excited an unusual interest, not only among those more intimately connected with the demonstration, either as hosts or guests, nor among that numerous and wide-spread class of inquirers vulgarly known as *quid nuncs*, but among all those who, having been so long shut out from a knowledge of the internal workings of the new party, looked anxiously forward to these proceedings with the hope that they might be furnished with a clue that should enable them to discover the aims, feelings, arrangements and principles by which that party seeks to characterize itself.

The card of invitation was as follows:

PHILADELPHIA, June 7, 1855.

DEAR SIR—You are cordially invited to attend a "BANQUET," to be given by the Citizens of Philadelphia, at Sansom-street Hall, this afternoon, at 4 o'clock. Please present this ticket at the door.

Respectfully.

J. M. CHURCH,	JOHN MECKE,
R. P. GILLINGHAM,	CHAS. D. FREEMAN,
J. L. GOSSLER,	E. WAMPOLE,
JOS. WOOD, Jr.,	GEO. P. HENSZEY,
C. S. PANCOAST,	JNO. FRY,
C. M. NEAL,	GEO. S. SHARP,
J. L. GIFFORD,	W. G. FLANIGAN,

Committee.

For gentlemen not specially invited, the price of admission was $10.

At the hour above-mentioned the ante-rooms and avenues leading to the dining apartment were filled with guests; and at half-past five they were conducted to the tables. The Hall was elegantly decorated. On the southern wall, at the further end of the room, the American flag was curtained over the window-tops, and looped up in the spaces between the window-frames with blue shields dotted with silver stars. At the lower end of the outer side-wall, between two of the windows, was hung a draped portrait purporting to be that of "GEORGE SCHIFFLER, Martyr; N. A. November, 1844," and containing in prominent letters the caution, "Beware of foreign influence." The "martyr" was represented with a book which, lest there should be a mistake as to its identity, was labelled "The Bible." The upper or northern end of the room was similarly decorated, save with the addition that in one corner of the raised platform, occupied by the Chairman and the more prominent guests, was piled a number of muskets and spears with flags, the whole surmounted by a stuffed eagle with outstretched wings, and bearing in front of him the American shield. At right angles to the platform set apart for the Chairman and the more noted of the gentlemen invited, four long tables were arranged parallel with each other, and placed longitudinally down the floor. About 450 gentlemen occupied seats. The tables were ornamented with patriotic designs in pastry, large candelabra, and immense vases filled with flowers. The viands were of the choicest description and the wines excellent. The whole affair was marked by a display so profuse, and a style of luxury so utterly regardless of cost, that few occasions of a similar public character in this city have exceeded it.

Behind the Chairman's seat in close, and as we should believe, unpleasant contiguity to him, the band was stationed, and performed a few airs during the evening. After the Mayor had taken his place he read the following list of Vice Presidents: Dillon Luther, Henry Simons, Wm. Whitney, Edward Gratz, R. R. Smith, John W. Ryan, A. Homerfelt, E. H. Butler, Joseph B. Myers, B. H. Brewster, W. S. Smith, John J. Tyson.

Messrs. C. GIBBONS, E. H. BUTLER, JOSEPH MYERS, and W. S. SMITH presided, one of them at each of the four tables appropriated by the mass of guests.

About 6½ o'clock the impatience of the assemblage to hear the speeches began to manifest itself by loud tramping of feet and jingling of glasses, which it was attempted to repress by calls of order from the Chairman's table. Eventually silence was restored, and the CHAIRMAN rose, amid loud applause, and said:

MAYOR CONRAD'S SPEECH.

There is a feast before you, to which I contribute no viand, richer than any that you have yet tasted, and in order to enjoy it is proper that you should receive it with decorous silence. Gentlemen—there is a parsimony peculiar to affluence. We have among us today such a wealth of eloquence should rob us of some portion of the rich feast that and patriotism that we are jealous of time, lest it is spread before us. As the banquet that we all have shared was such that an epicure would have lingered over a single delicacy, unwilling to part with it, yet still less willing to linger over it, lest he should lose that which was presented afterward, so in the rich succession of speakers that we have presented to you this evening, you must be content with a more brief entertainment than their merits and your own wishes would warrant. For my own part, my duty is not to speak but to forbear—[applause]—to strike the rock, and invoke the fountains of eloquence from others. Yet a higher law—would that all higher laws were as harmless—[applause]—requires that having been honored with the duties of guiding the festivities on this occasion. I should occupy your attention at least long enough to extend to the strangers—no, they are not strangers, they are the *brothers* of our country—[applause]—brothers "not in the fashion which the world puts on," but, in the communion of patriotism, "brothers of the heart,"—[renewed applause]—a welcome true as our cause and earnest as our devotion to it. Pennsylvania—and Philadelphia answers, throb for throb, to every pulsation of her loyal heart—Pennsylvania has never known, in relation to her kindred communities, even momentary sentiment false to the religious patriotism which links her to them. All she has, and all she is, whenever invoked, belongs to the Union. [Enthusiastic applause.] Her valleys are crowded with homes and hearts—[confusion]—ever open to the sons of her sisters and to their friends, and her hills will ever present their iron wealth, with sinewy arms to wield it, to their enemies. [Cheers.] I should be glad—it would be a luxury for me to address you, but I feel that I am encroaching on the duties of hospitality. [Cheers, shouts of "No," and "Go on,"] it is, therefore—[Pause, and renewed cries of "Go on"] I am a man of many duties. All day I have to sit in judgment upon my fellow citizens, (laughter) and in the night, it seems, my fellow citizens must sit in judgment upon me. [Applause and renewed laughter.] But to night, I am sorry to say, I find I have another engagement—a lecture at a church, which must be attended to; and it cuts short, necessarily, any remarks that I would desire to make to you. Hereafter, at a later period in the evening, perhaps, without entrenching upon our more eloquent friends from other sections, I would be happy to address you. [Applause.] But this much let me say, that now and ever, here and elsewhere, by word or by act, there is no crisis that can occur in this holy cause in which we are engaged, that I am not prepared to give every energy of my nature to, and to call the American who claims that title a traitor and no friend of mine, who hesitates or falters to come up to the full mark of duty. [Loud cheers.]

A VOICE—"Three cheers for Mayor CONRAD." [Enthusiastic cheering.]

We append the conclusion of the Chairman's address, as it was intended we presume to be spoken, and as we find it supplied to the Philadelphia papers. What follows, however, if uttered at all, must have been delivered between 12 and 1 o'clock on Friday (yesterday) morning.

"To meet our brothers of other States, upon any occasion, is esteemed a privilege, but to meet them gathered in such a duty—when, under the shadow of the Hall of Independence, they are engaged in the sacred labors that here employed their fathers, in that era hallowed to every American memory and every American hope—is something more,—it is an exulting joy that must seek expression, not in words, but in the alacrity of our coöperation, and in the fidelity that never has faltered, and never will.

Gentlemen, what a legacy of inestimable privileges, what a trust of noble duties, what a destiny of glorious hopes, did those fathers bequeath to us. For the protection, development and consummation of all this, nothing was needed but their virtues and the favor of that Providence which smiles upon virtue. It was not necessary, it was not well, that the thin and tainted blood of decayed communities should be poured into the pure and vigorous current of our youth and health. Enlarged population was and is here no necessity, and, if of a doubtful character, it is a doubtful good.

Even the tithe, nay, the tenth of a tithe of the years that ripened Rome, was certain to give us a sufficient population without external supplies—a population, too, unequalled in every moral trait—a population drawn from founts lofty and pure as the springs that burst from the the bosom of our Alleghenies—the undefiled children of the compatriots of Washington.

For protection, even against a world in arms, nature gave us defences sufficient; and virtue and valor made them impregnable. What more was wanted? Many a people have achieved liberty, but what people has retained it?

There was an exception here, under such influences, with a people self-protected and protected by a religion that acknowledges no Vicegerent of Him who reigns alone—a people incorrupt and inaccessible to foreign corruption—here liberty would have been immortal—a glory setting only with our last sun, not like that sun, to set forever, but to rise again in the "perfect freedom" which is eternal. When or where else had there, or has there been such a people, such a land, such a Government or such a hope? But even they, the sages and patriots of that illustrious era, as they gazed with pious gratitude upon the destiny which opened before their children, could not suppress the doubt—"can such blessings be upon earth, and be lasting?" How full of those mingled feelings of pride and apprehension are their injunctions to their posterity?

And can we compare the past with the present, and under that earth's greatest and best, the *Pater Patriæ*, uttered, in regard to foreign influence, his solemn BEWARE! or that JEFFERSON'S prophetic spirit prayed for a sea of fire between the old world and the new; or that, one and all, they, the lofty and devoted, the pure, the sage, and now the sainted, enjoined us—(do they not watch over and bless our efforts to obey that injunction?)—to guard, as the lioness guards her young, our inestimable heritage of virtue, liberty, and independence, from the foreign chains—a thousand fold heavier and more galling than the'rs—which the old world seeks to wrap around our hearths and hearts, our Senate-houses and our temples, making the land of the free the home of slaves—the basest because willing slaves, and therefore worthy only of fetters.

It is true that they, even they, offered a refuge to the virtuous martyrs of liberty—how could they otherwise? but solely as the home offered by freedom to virtue. They dreamed not of the commerce of emigration—the rival, with its annual half million, of the Slave trade at its worst—with all that is sordid in its motives and fatal in its consequences. But fearfully has that privilege been perverted! How has the portal, thus generously opened, been thronged? not by the excess of overcrowded Europe—for no land is overcrowded—but by acts and aids of hostile policy or unscrupulous commerce at home and abroad, which I need not describe.

But reverting to the times and men of our Revolution, how appalling is the contrast! They left us an united and homogeneous people; now mixed and mottled, our very original is doubted. I observe that the New-York press expresses bitter indignation because the late meeting of the Democracy was not addressed in high Dutch and low Irish; and you remember that KOSSUTH, in his speeches, appealed to his audiences as if they were foreigners—ignoring, upon American soil, the existence of an American people, and that thousands—thank heaven I was not one—waited upon and applauded him.

Nay, so confident was he of the foreign predominance, that he dared to claim the Government and the country for his foreign policy, nor was he rebuked until the spirit of WASHINGTON, entering the noble bosom of a dying American patriot, inspired his prophetic eloquence, and the sordid imposter received, and will, until his dying day, writhe under the fire (holy as that of the altar) which fell from the lips of the illustrious HENRY CLAY.

Our fathers left us religious liberty, unquestioned and unquestionable—a ballot-box venerated as the holy of holies of our freedom—an Union, that knew and feared no foe—and, above all, a public and private virtue that seemed about to realize the loftiest aspirations of man in his moral, religious, political and social condition. Where is that religious liberty? Under the feet of foreign archbishops, the mighty masters of many thousands of voting retainers, and of millions of property held with the dead hand—both ready to be used, at the bidding of a foreign potentate, whenever needed in the war against free education, an open Bible and Protestant liberty.

Where is that Union? Threatened by the violence of factions made violent by strangers who neither know nor love our Constitution nor our people. And, above all, where is that social virtue, which, in the childhood of the older among us, before prisons and almshouses numbered thousands on thousands of foreigners, shuddered at the darker crimes as a marvel and a horror? Where are the doors unlatched at night, and the streets where the shades of evening brought neither shame nor terror to unprotected innocence? The contrast is a fearful one, and extends to every political and every social interest; and is it strange that a moral degeneracy so ra[illegible]id—which cannot be native, for no people ever so [illegible]ted before they were ripe, no political body ever before fell into corpse-like putrefaction while it retained youth and vitality—should excite our wonder?

"Can such things be,
And overcome us like a Summer cloud,
Without our special wonder?"

Having wandered so far from our fathers, we would retrace our steps, and at least save what of hope is left to us. Their example—their creed—their virtues—these are our object. They were, in the revolution, forced by the necessities of self-protection, to have recourse to secret organization—so long and so far as the same cause requires it, the same policy is justifiable.

But the time will soon come when every change will be dispelled—when the world will know and admit that we are far from illiberal—far from proscriptive—far from disposed to regard the friends of Protestant liberty among us as other than brothers. Meanwhile our first duty is to our country and her children. When that country is true to herself, she cannot be unjust to any class of men.

To thine own self be true,
And it will follow as the night the day,
Thou canst not then be false to any man.

June 9, 1855

European Emigration to America Encouraged.

The Liverpool *Post* publishes the following circular of Secretary SEWARD, and accompanies it with approbatory comments:

TO THE DIPLOMATIC AND CONSULAR OFFICERS OF THE UNITED STATES IN FOREIGN COUNTRIES.

DEPARTMENT OF STATE, WASHINGTON, Aug. 3, 1862.

At no former period of our history have our agricultural, manufacturing, or mining interests been more prosperous than at this juncture. This fact may be deemed surprising in view of the enhanced price for labor, occasioned by the demand for the rank and file of the armies of the United States. It may, therefore, be confidently asserted that, even now, nowhere else can the industrious laboring man and artisan expect so liberal a recompense for his services as in the United States. You are authorized and directed to make these truths known in any quarter, which may lead to the migration of such persons to this country. It is believed that a knowledge of them will alone suffice to cause them to be acted upon. The Government has no legal authority to offer any pecuniary inducements to the advent of industrious foreigners.

WILLIAM H. SEWARD.

This is not, as may be supposed, an indirect means of getting recruits for the army. The object is a much more serious one—the providing of means to feed and pay the army. America, like all imperfectly peopled countries, has more land than hands. A large population is scant in reference to the territories needing inhabitants; and the United States, from the moment of its discovery to the breaking out of the unhappy civil war, had been hungry for emigrants. Every man choosing to work was certain of employment, and the result was a blessing to the old country and the new. Strong doubts go for nothing in presence of the fact that the Irish emigrants transmitted home to friends, annually, an average of £1,000,000. As there were, up to 1861, too few rather than too many mechanics and laborers, it necessarily followed that an abstraction of 600,000 men from the pursuits of industry left vacancies which it is the interest of the community to fill up as speedily as possible. Companies yield large dividends, and the manufacturers are every one realizing large fortunes. No matter what the cause, the facts are undoubted. The action is somewhat sudden, and may not be permanent, but that does not concern the emigrant. Peace may effect an alteration, but what then? He may stop in the country or return home. A country so full of progress he will hardly leave; but, even if he does, he will come to the old country a wealthier and a more experienced man.

How wealthier? In this way. In the manufactories he will get something like two dollars a day. For a mere laborer the wages are a dollar and a half a day or forty shillings a week. There is employment now in one direction alone for half a million of hands. The cost of living is comparatively small, as may be inferred from the prices of breadstuffs even in New-York. In the West ten pounds of corn meal costs little more than a penny, and pork can be bought for 2d. a pound. Any provision merchant will confirm this statement, and the rate of wages alone testifies to the want of hands. If other proof were wanted it can be had from the great employers of labor in America now in this country.

People in Liverpool, who twelve months ago professed to know a Yankee as well as if they had seen inside of him with a lighted taper, now confess that he is an anomaly hard to comprehend. The *Times* harks back, because compelled to acknowledge very serious mistakes. All, however, are not ignorant of the real circumstances which give force and vitality to America. Gentlemen on 'Change have smiled at the foolish inferences drawn, after every telegram, from the rates of the exchanges, because they knew full well that these rates were favorable to the Americans. An explanation will come by-and-by.

If the authorities in Lancashire were wise they would take instant action on the information contained in Mr. SEWARD'S circular. The able-bodied, at least, could be sent out, and from each a bond for repayment of the passage money could be taken. In nine cases in ten it would be honorably discharged. The Federals want laborers, not recruits. Recruits are to be had from other classes; and probably the opposition to the departure of people in the drafting panic originated in the desire to retain laborers, not militiamen. Emigrants would not be exposed even to the claims of a conscription: and although the war is an evil to the world, it would be a blessing to them. They will be in no fear of an invasion, and they will be far enough away from battle-fields.

September 11, 1862

MINOR TOPICS.

The Chinese labor question is still agitating the Californian mind. Meetings and organizations are continually under way to discourage and, if possible, choke off the employment of those cheap workers by the community; pledges are circulated binding the signers not to buy goods made by them, or to patronize their employers; and a hateful animosity is engendered toward them, which, in one case at least, culminated in actual violence. But there, as here, the lawless agitators are from the ignorant classes, who are moved by their leaders toward a supposed good, and attempt its accomplishment by instrumentalities foolish in themselves and ineffectual for their object. The last *Steamer Bulletin* publishes a very sensible note from "A Hibernian," in which the writer not only deprecates the style of turbulent agitation, but gives a point for consideration by declaring that it is the system of *Cooleeism*, and not voluntary Chinese immigration, that makes the evil. That Chinese labor has been of material benefit to California none can deny. Some complain that the fruit of their labor is entirely lost, because they hoard and send home their pay; but, in the case of the San Francisco Woolen Mills, instanced by the *Bulletin*, "while the Chinese receive $120,000 per annum their labor furnishes support for 1,000 whites, and puts into circulation for the benefit of the community at least $3,000,000 per annum." Still the suppression of the Coolie process will certainly bring a better and more independent class of men, who, in time, will be ambitious and capable of receiving the rights and duties of citizenship.

May 17, 1867

THE COMING COOLIE.

Opposition to His Advent on the Part of Certain Working Men.

Mass-Meeting in Tompkins-Square to Discuss the Chinese Labor Question—Speeches by Nelson W. Young, Mayor Hall, Alexander Troup and Others—Resolutions.

A mass-meeting of laboring men and trades' unions was held in Tompkins-square last evening, to oppose the introduction of Chinese labor into this country. Three stands were erected for speakers—two for English and one for German. The meeting was called for 8 o'clock, at which time but very few men had assembled around either of the stands, though there were plenty of women and children scattered over the grounds. NELSON W. YOUNG, President of the Labor Union, presided at the main stand, and at 8½ o'clock called the assembly to order, and made a brief address.

MR. YOUNG'S REMARKS.

He said the question which they had met to consider was one of great importance to the working men and working women of the country. He was glad that this work of introducing coolie labor had commenced in Massachusetts—the State of isms—a State that had always been ready to introduce anything of a disorganizing character. It was an outrage on the working men of the country, and was intended to degrade labor. The working men had made the country what it was, and had furnished it with some of the brightest intellects in the councils of the nation. The speaker then broke out in a tirade against Congress, and said it had spent its time in fruitless reconstruction instead of legislating for the benefit of working-men. Our duty now, he said, was to take political action and select Congressmen from the working classes, among whom was to be found ability amply equal to the highest office in the gift of the people. Mr. YOUNG closed by saying that the man who had introduced coolie labor into Massachusetts had never been a fair man, either to his employes or to his brother manufacturers. He was always undermining and underselling his neighbors, and never gave anybody a fair show. He did not believe he would get much of a show himself after death. [Laughter.]

RESOLUTIONS.

Mr. MASTERSON was then introduced and read the following resolutions, which were adopted without opposition:

Whereas, The agricultural and national prosperity of a country depends upon the following conditions: the freedom of its citizens, their intellectual condition and their productive labor, a proper appreciation of the comforts of life, and a disposition to secure a portion of the same; and

Whereas, Any attempt made by monopoly or capital to lower the condition of mankind in the social scale, by depriving him of a fair proportion and equitable share of the product of his labor, engenders in the mind of the workman first, a natural spirit of enmity to those who, by first diminishing and then destroying his limited means of support, compelling him to relinquish those simple but necessary enjoyments of comfort that go hand in hand with the intelligent citizen and well-paid workman, thereby destroying the power to purchase, through the want of means, those necessaries of life, the production of which renders the investment of capital remunerative, the workman happy, and the country prosperous; and

Whereas, Strenuous efforts are being put forth by a monopolizing class of manufacturers to demoralize and ultimately suppress all classes of working men by an arbitrary and forced importation of the lowest and most degraded of the Chinese barbaric race, whose labor, the value of which in this country they seem deprived of the powers of judging and appreciating, is to be put in competition with those whose more advanced intelligence and improved tastes have generated in them a proportionately greater number of wants and desires; and

Whereas, In opposition to the above facts, the opinion of all writers on political economy and national prosperity, the attempt is being made in the State of Massachusetts, and with the superabundant wealth wrung from the brain and body of the working man, to degrade still lower the toiling mass and by compelling them to either starve or associate with the degraded labor of Asia, and by such association and competition debase our manhood, destroy our morals, and prostitute our virtues by witnessing the horrors of servility, sodomy and social degradation, the natural result of labor without remuneration and licentiousness without limit; and

Whereas, Any arbitrary system of importation of mechanics, artisans and laborers from any country or clime, or under any pretence whatever, even though it be the flattering and plausible one of "freeing the hands of Americans from the meaner industries, to engage in the higher and better occupations," is reprehensible, and an unjust and an unholy attempt to covertly revive the slave trade and re-establish slavery in a country dedicated to liberty, human progress and civilization, and, as such, demands the condemnation of the American people and Government; therefore, be it

Resolved, That the working men of the City of New-York, in mass-meeting assembled, do hereby call on their representatives in Congress to put forth their utmost efforts for the immediate passage of the bill now pending in the Senate, or any other bill of like effect, that will prevent the importation of coolies, under such contract schemes as developed and practiced by such agencies as KOOPMANSCHAP & Co.; and further, that our State and Municipal authorities be and are hereby requested to use the moral influence at their hands to the same end.

Resolved, That in view of the fact that there is a superabundance of labor in the United States, and with the voluntary flow of civilized labor from Europe, and easily procured at a living price without proposing terms to employers (the assertions of Mr. SAMPSON to the contrary notwithstanding,) we take this public opportunity to denounce said assertions as unmitigated falsehoods, and proven to be such by the continual concessions made to capital through the necessities of labor, and the enormous wealth accumulated by the capitalist on the product of the same; the luxury and comfort of the one, the poverty and destitution of the other, stand forth as a sufficient refutation that labor is not exacting nor presuming in its demands; but that capital is avaricious, absorbing, ostentatious and aggressive, forcing us at all times, as in the present instance, to assert our rights as men. Therefore, we call upon all the working men in the United States of America, by public demonstration, by energetic and prompt measures, to make common cause against this matter, and to demand from their representatives in Congress and elsewhere the speedy passage of such laws as will prohibit this foreign contract system, and protect the American citizen and his family against servile labor from foreign shores.

Resolved, That a copy of these resolutions be sent to our Senators and Representatives in Congress, to the President of the United States, and to the Governors of the several States, and signed, on behalf of the working classes of this Metropolis and vicinity, by the Presidents and Secretaries of the Working Men's Unions and Arbeiter Union of the City of New-York.

SPEECH OF MAYOR HALL.

Mayor HALL was then introduced and said it afforded him pleasure to meet his hearers in the discussion of this coolie question. If it were a subject to joke about he might say that the coolie question was a good one to talk about on this hot June evening. But it was a subject of too serious a nature for joking. He had been introduced by the Chairman as the Chief Magistrate of this City. He felt that the Chief Magistrate ought to be here because the object of the meeting he was sure met the sympathies of at least four-fifths of his constituents. He was not here as a politician or a demagogue, but he believed every one holding public position should be willing to show his hand on this question and not hold his tongue. The State of Massachusetts was the first in the country to introduce the black slave trade two centuries ago, and it was now the first in the United States to introduce another kind of tawny slave labor. From the sudden introduction of this species of labor in one of the towns of Massachusetts, had sprung this movement. The American people, it had been said, would not and could not object to any kind of immigration into this country. The objection was to traders and man-speculators bringing men here to overrun labor. Men who were debased in morals were being brought here to compete with free white labor. If the white men of the North are to tolerate this new kind of slave labor, then it was in vain they had fought to break down black slavery in the South, for it was only to see a worse species of slavery introduced into the North. The trouble on this question commenced with the Burlingame treaty. It was the duty of the statesmen of the country to have seen to it at that time and prevented the difficulty. It was by virtue of that treaty that Mr. SAMPSON, of North Adams, claimed the right to introduce Chinese laborers to his work-shops. He held the statesmen of the country accountable for not providing in the treaty for this difficulty. It would be impossible for American free labor to compete with this new system of Chinese slave labor, and it was the duty of Congress to pass laws immediately regulating Chinese immigration to this country. It was a difficult question to deal with—and there was no use in objurgation or the elaboration of revolutionary ideas. He was not here to suggest a remedy; that lay with the men in power at Washington. It was not a question of cheap labor merely, but a question of moral and social influences in the United States. After some remarks about the vast reservoir of Chinese labor, and the necessity of regulating its introduction into this country, the Mayor closed amid the cheers of the audience.

LETTERS.

The Secretary then read letters from MANTON MARBLE and an agent of HORACE GREELEY, regretting their inability to attend.

SPEECH OF MR. TROUP.

Mr. ALEX. TROUP was then introduced, and told the audience how the laboring men of Massachusetts were protesting against the introduction of this new species of slave labor into that State. Much had been said about the tyranny of the Crispin organization. He was not there to defend them; they needed no defense. HORACE GREELEY tells the laboring men to go West; that there is too little labor here and too many laborers. If that was so, why introduce new laborers from China? It was the greed of men like SAMPSON that brought the Crispin organization into existence. SAMPSON has bragged in the streets of Boston that he would break up the Crispins. [A voice—"Shoot him."]

The speaker continued at length to defend trades' unions and labor combinations, and closed by exhorting his hearers to wake up to the dangers that threatened them; to refrain from violence and resort to the ballot for a redress of their grievances. Let them forget the Democratic Party and the Republican Party, and elect only workingmen to office. He also recommended co-operation among mechanics, and said that the thing was to be tried in Natick next week.

Mr. DRURY and Mr. PURDY were next introduced and addressed the audience briefly, after which the meeting adjourned. The audience around the main stand numbered at no time more than 300 persons, and was composed almost entirely of foreign-born citizens.

Stand No. 2.

The presiding officer at this stand was JAMES CONNOLLY, of the painters, and the Secretary was WM. FITZGERALD, of the coopers. The crowd was by no means numerous, and consisted mainly of small children who had assembled to witness the initiatory proceedings attendant upon the letting off of the fire-works.

The speakers who were announced for this stand were Mr. CASHMAN, of the tailors; Mr. ENNIS, of the plasterers; Mr. GRIFFIN, of the bricklayers; Mr. MASTERSON, of the Crispins; Mr. GRIFFIN, of the stair-builders; Mr. PURDY, of the molders, and Mr. TROUP and Mr. FAUCER.

Mr. TROUP gave some facts concerning the feeling at North Adams, Mass., regarding the coolie movement, stating that the working men of that section were fully aroused, and determined to vote for none but working men and the friends of that class for office in the future.

The proceedings at this stand were brief, and after the reading of the resolutions and a few remarks by the Chairman, the crowd, which was then gathered, went to the main stand.

Stand No. 3.

The third stand, from which addresses in German were delivered, was surrounded by about 2,000 persons. Mr. ALBERT KAMFRAD, the first speaker, stated as the cause of the meeting the circumstance that in Massachusetts seventy-five Chinese had been imported who were engaged to work for wages amounting to $23 a month. Other manufacturers, in order to compete with the boss of the coolies in Massachusetts, would be obliged to reduce the wages of their laborers, and thus by that coolie importation the laborers of this country, and especially of the East and of this City, were bitterly threatened in their own material welfare and that of their families. It was necessary to appeal to Congress for a law for the protection of the laboring classes in this country.

Dr. ADOLPHUS DONAI, editor of the German working men's paper of this City, was the next speaker. He said that this country had become what it was at present by the laborers that had come from Europe. When they first came there had been no capitalists in this country, which class of men was flourishing now. By their efforts these very Chinese were imported to this country. Formerly Chinese had been imported only to California and Oregon, but now they made their appearance also in the East, having been ordered by capitalists to this part of the country. China and East India were inhabited by about 600,000,000 people, who were accustomed for thousands of years to work for the lowest possible wages. By the importation of these coolies the civilization created by the white race in this country was endangered.

Mr. FREDERICK HOMRIGHAUSEN, who then spoke, said that of course no opposition could be made against the voluntary immigration of free Chinese to this country, but that the importation of coolies, which had been contracted for in China by capitalists in this country, ought to be forbidden. He dwelt at some length on the endeavors for cheap labor on the part of capital, while labor itself wanted good and fair wages. Mr. CONRAD KUHN, who followed him, spoke in a similar manner, as did also Mr. C. CARL. The above resolutions were then read in German, and adopted by the meeting. Mr. F. TOURELLE made the concluding speech, and the meeting then adjourned. The injustice of capital and its exactions from labor were bitterly opposed and attacked by the speakers from this stand.

July 1, 1870

THE CHINESE QUESTION.

A Calm Statement of the Case from the Chinese Stand-Point.

At a late meeting of the San Francisco Board of Supervisors, Rev. O. Gibson, Chinese Missionary, appeared with the following petition, signed by many leading Chinese merchants, which he had translated:

To the Honorable Board of Supervisors of San Francisco:

GENTLEMEN: Will you listen to a calm statement of the Chinese question from a Chinese stand-point? Public sentiment is strongly against us. Many rise up to curse us. Few there are who seem willing, or who dare to utter a word in our defense, or in defense of our treaty rights in this country. The daily papers teem with bitter invectives against us. All the evils and miseries of our people are constantly pictured in an exaggerated form to the public, and our presence in this country is held up as an evil, and only evil, and that continually. Laws, designed not to punish guilt and crime, nor yet to protect the lives and property of the innocent, have been enacted and executed, discriminating against the Chinese, and now the Honorable Board of Supervisors of this city proposes still further to afflict us by what seems to us most unjust, most oppressive, and more barbarous enactments. If these enactments are the legitimate offspring of the American civilization and of the Jesus religion, you can hardly wonder if the Chinese people are somewhat slow to embrace the one or adopt the other. Unfortunately for us, our civilization has not attained the use of the daily Press—that mighty engine for molding public sentiment in these lands—and we must even now appeal to the generosity of those who perhaps bear us no good will, to give us a place in their columns to present our cause.

THE POLICY OF CHINA.

1. We wish the American people to remember that the policy of the Chinese Government was strictly exclusive. She desired no treaty stipulations, no commercial relations, no interchange whatever with Europe or America. She was not willing that other people should come to reside in her limits, because she knew the antagonism of races. For the same reason she was unwilling that her subjects should go forth to other lands to reside. But the United States and other Christian nations held very different views, and advocated a very different policy. Treaty stipulations, commercial relations, friendly interchange of commodities and persons were demanded of the Chinese. To secure these with China, pretexts for war were sought and found, and as the result of defeat on the part of the Chinese, our Government was compelled to give up her traditional, time-honored policy, and to form treaties of friendship and interchange with our conquerors.

RESULT OF THIS POLICY.

2. Under these treaty stipulations dictated to China by Christian (?) governments, the people of Europe and America have freely entered China for the purposes of trade, travel, and Christian evangelization. Foreign residents in China are numerous, and many of them have amassed ample fortunes in that land. Their presence has ever been hateful to a large portion of the Chinese people. It is but fair to state this fact, that as much friction, if not more, is caused in China by the presence of foreigners, than the Chinese are creating in this land. The declaimers against us because we supplant white laborers in this country ought to know, what is well known to every intelligent Chinaman, that the introduction of American and English steamers upon the rivers and coast of China has thrown out of business a vast fleet of junks, and out of employment a whole army of men larger in number than all the Chinese now in America. And yet during these few years of commercial and friendly intercourse a large commerce has sprung up between China and America, creating a community of interests between the people of these two countries, and doing much to remove the strong prejudices of the Chinese against foreign intercourse, American merchants and American enterprise. American missionaries and Christian doctrine meet with far less opposition and much greater favor in China now than formerly. Great changes are taking place in the popular sentiments of the people, a striking feature of which change is a marked partiality for the American Government and American civilization.

The Chinese Government has already sent a score of youths to this country to learn your language, your customs and laws, and propose to send many more on the same errand. This fact of itself is significant.

DEMANDS OF AMERICA UPON CHINA.

3. We wish also to call the attention of the American public to the fatct that at the present time the American and European Governments are greatly embarrassing the Chinese Government by strenuously insisting upon these two points:

(a.) That Americans and other foreigners shall be permitted to travel, and trade, and preach in all parts of the Chinese Empire without being subject to Chinese law. The foreign Governments insist upon their right to take their code of laws with them into all parts of our country, thus humbling and disgracing our Government in the eyes of our people. How would that shoe fit the other foot? Or how can they claim to be reconciled to the "Gelden Rule," considering the present treatment of Chinese in America?

(b.) The audience question. Foreign Governments insist upon holding audience through their representatives with the Emperor of China without paying him the homage and respect which the throne of China has ever received from all who came before it.

4. We wish now also to ask the American people to remember that the Chinese in this country have been for the most part peaceable and industrious. They have kept no whisky saloons and have had no drunken brawls, resulting in manslaughter and murder. They have toiled patiently to build your railroads, to aid in harvesting your fruits and grain, and to reclaim your swamp lands. Our presence and labor on this coast, we believe, have made possible numerous manufacturing interests which, without us, could not exist on these shores. In the mining regions our people have been satisfied with claims deserted by the white miners. As a people we have the reputation, even here and now, of paying faithfully our rents, our taxes, and our debts.

In view of all these facts we are constrained to ask, why this bitter hostility against the few thousands of Chinese in America? Why these severe and barbarous enactments, discriminating against us in favor of other nationalities? From Europe you receive annually an immigration of four hundred thousand, among whom, judging from what we have observed, there are many—perhaps one-third—who are vagabonds and scoundrels, and plotters against your national and religious institutions. These, with all the evils they bring, you receive with open arms, and at once give them the right of suffrage, and not seldom elect them to office. Why, then, that fearful opposition to the immigration, of fifteen or twenty thousand Chinamen yearly. In the name of our country, in the name of justice and humanity, in the name of Christianity as we understand it, we protest against such severe and discriminating enactments against our people while living in this country under existing circumstances.

A PROPOSITION.

Finally, since our presence here is considered so detrimental to the country, and is so offensive to the American people, we propose and promise on our part to use all our influence to carry the proposition into effect. We propose a speedy and perfect abrogation and repeal of the present treaty relations between China and America, requiring the retirement of all Chinese people and trade from these United States, and the withdrawing of all American people and trade, and commercial intercourse whatever from China. This, perhaps, will give the American people the opportunity of preserving for a longer time their civil and religious institutions, which, it is said, the immigration of the Chinese is calculated to destroy. This arrangement will also, to some extent, relieve the Chinese people and Government from the serious embarrassments which now disturb them, and enable them, by so much, to return to the traditional policy of her sages and statesmen, *i. e.*, "Stay at home and mind our own business, and let all other people do the same." This is our proposition. Will the American people agree to it? Will the newspapers, which have lately said so many things against us and against our residence in this country, will they now aid us in bringing about this, to us, desirable state of affairs? In the meantime, as we are now here, under sacred treaty stipulation, we humbly pray that we may be treated according to these stipulations until such time as the treaty can be repealed, and commercial intercourse and friendly relations come to an end.

LAI YONG, portrait painter.
YANG KAY, 74 Commercial-street.
A YUP.
LAI FOON.
CHUNG LEONG.

Translated by Rev. O. Gibson.

June 16, 1873

IMMIGRANTS FROM CHINA.

ONE HUNDRED AND THIRTEEN THOUSAND RECEIVED IN THIRTEEN YEARS—LARGE INCREASE EXPECTED THIS YEAR.

From the San Francisco Bulletin, July 15.

The number of Chinese immigrants who have arrived during the past three years is largely in excess of the arrivals from the same source during any previous corresponding period, and, according to authentic advices, the number that will arrive within the ensuing year will far exceed that of any year yet past. Every steamer and sail vessel sailing for this port from the Orient is now taxed to its utmost capacity by this class of immigrants, and, although the accommodations are greater than at any time gone by, only an infinitesimal proportion of those desiring to come to these shores can be accommodated. The majority of these immigrants belong to the lowest classes, and their condition is little better than that of slaves. The steamer Great Republic arrived yesterday with 933 of these people. Two more steamers from China will arrive this month.

In addition to the arrivals by the Great Republic, the steamer China has arrived within a fortnight with 976 Chinese; the ship Avonmore, 438; ship Atlantic, 383; ship Her Royal Highness, 20, and the bark William H. Besse brought 417. The following tabular statement shows the number of Chinese immigrants who have arrived during the past thirteen years:

Year.	Males.	Females.	Totals.
1862-63	5,407	87	5,494
1863-64	5,182	...	5,182
1864-65	3,309	175	3,484
1865-66	1,495	1	1,496
1866-67	3,362	...	3,362
1867-68	6,607	43	6,650
1868-69	11,124	951	12,075
1869-70	13,023	1,085	14,108
1870-71	6,068	339	6,407
1871-72	6,422	146	6,568
1872-73	18,529	839	19,368
1873-74	12,941	132	13,073
1874-75	15,433	374	15,807
Totals	108,902	4,172	113,074

During the years 1863-64 and 1866-67 no Chinese females came to this coast, and only one arrived during 1865-66. The greatest number of Chinese arriving in any one year since 1862-63 was 19,368 in 1872-73. The number of Chinese who have departed during the same period is comparatively small, but cannot be definitely stated. The Chinese on their arrival here are carted to the Chinese quarters, where they are taken in charge by the six companies until they can be distributed in lots here and there, as the opportunity for contract labor may permit.

July 26, 1875

BLOODSHED IN SAN FRANCISCO.

SHARP ENCOUNTERS BETWEEN THE CITIZENS AND THE "HOODLUM" RIOTERS — A GREAT NUMBER WOUNDED, BUT ONLY ONE MAN KNOWN TO HAVE BEEN KILLED —DESTRUCTIVE INCENDIARY CONFLAGRATIONS—A STRONG BODY OF VIGILANTS GUARDING THE CITY.

SAN FRANCISCO, Cal., July 25—9:40 P. M. —The Citizens' Committee began to assemble at 7 P. M. at Horticultural Hall, and the building was soon filled. About 8 o'clock W. T. Coleman, President, called the meeting to order, and had scarcely done so when a message was received that a fire had broken out at the Pacific Mail dock. It was soon ascertained that the fire was in a large lumber-yard near the dock. One hundred of the committee, armed with clubs, were at once dispatched to the scene, followed soon after by 100 more. The remainder of the committee were then divided into companies by wards, and, with the exception of about 200, proceeded to the City Hall to await orders from the Chief of Police. Sixty were dispatched to Sixth and Howard streets to disperse the crowd collected there smashing Chinese houses. All the members of the committee were armed with clubs in addition to the pocket fire-arms carried by nearly all. Muskets will be issued if necessary. The fire at the lumber yard is now raging fiercely, and a crowd of several thousand people are collected there. A heavy force of Vigilantes and Police are standing guard. News has just been received of a man, detected in cutting the hose, being shot down. The Vigilantes have closed the streets leading to the scene of the fire.

A fire alarm has just been turned on from the corner of Stockton-street and Broadway, and a party of Vigilantes has been dispatched to keep order. All is quiet in the central portion of the city. No call has yet been made on the military.

A Vallejo dispatch says the corvette Pensacola grounded in leaving the harbor, but will probably be got off in an hour or so. President Coleman says he has 3,000 Vigilantes on hand, and just before reaching the City Hall he perfected certain important arrangements the nature of which has not yet been made public. The principal streets are full of people, but there are no signs of disturbance except as above mentioned.

July 27, 1877

NOTICE

The President has signed the new bill to suspend the immigration of Chinese laborers for a period of ten years. It is to be hoped that this will settle the much-vexed Chinese question for a time at least. The bill was drawn with special reference to the objections raised by the President in his Message disapproving the first bill passed by Congress. As it now stands, the law suspends the immigration of Chinese laborers, whether skilled or unskilled, or employed in mining. It provides for a system of certificates, to be issued on the identification of Chinese persons now living in this country or who may hereafter arrive here under provisions of the law authorizing them to come. The naturalization of all Chinese is expressly forbidden. Various fines and penalties are imposed upon the masters of vessels who shall bring unauthorized Chinese persons into this country, and upon any person who shall forge, alter, or make fraudulent use of the certificates to be issued to Chinese who are allowed residence in the United States. The bill as it has become a law does not infringe upon any of the rights of China as defined in existing treaties. The people of California will probably be satisfied with all its features, unless they may object to the shortness of the term during which immigration is to be suspended.

May 9, 1882

IMMIGRATION AND LABOR.

The recent arrival in American ports of considerable numbers of Irish, mostly from the west coast, has provoked much severe criticism on the course of the British Government, and this criticism is by no means without justification. It is difficult, from present information, to fix the proportion of these emigrants who remain in this country, or the proportion of those so remaining who are landed with no means of support. Many of them reach our shores *en route* for Canada, and of the rest some at least have enough means to keep them from absolute want until employment can be obtained. But there is reason to believe that the British Government has sent to this country a certain number of men and women who are practically paupers, and so far as it has done this, it has committed an offense for which weaker Governments—that of Switzerland, for instance—have been sharply called to account. The circumstances require careful investigation and firm treatment, and will, without doubt, receive it from our Government. The position of the President with reference to the demands made, or likely to be made, by the British Government is not so agreeable that he will fail to present any real grievances that we may have.

On the other hand, the outcry against the general immigration of laborers in this country is misdirected. The subject has two sides. Our Government is bound to discourage immigration of persons who cannot reasonably be expected to be self-supporting, and cannot tolerate any deportation by foreign Governments of such persons. But the voluntary immigration of able-bodied persons possessed of very little means or of no means at all is not necessarily undesirable. Our country enjoys very great absorbent and assimilative power. There is work here for a far greater population than we now have, and work which will add to rather than diminish the average comfort of the great body of workers. Taking one class with another, those who bring some means with them, or have a definite destination and opportunity for employment, and those who come at a venture, it is fair to say that in ten or fifteen years the net result of their stay here, and the labor they engage in, must be profitable to the community. With all the progress that we have made, there is still almost boundless room for the development of existing forms of production and for the introduction of new forms, and the greater the numbers of the population the more rapid and varied will be, on the whole, the advance of the country in these directions.

Special objection has been made by the various labor unions to the importation of laborers to compete with those who are struggling for higher wages or against reduction—such importation as was threatened in the Pittsburg railway mines, for example. Where this has taken place in industries protected by the tariff against foreign competition the complaint has a good deal of justice. It is obviously unfair for employers to ask that obstacles be placed in the way of imported goods made by cheap foreign labor on the plea that this will enable them to pay higher wages to American labor, and then to import the foreign labor itself in order to avoid paying higher wages. But there is no means of preventing or of punishing such unfairness. The employers can take advantage of the labor market abroad if they choose. Their course only serves to prove what every intelligent laborer ought to be able to see, that no matter what may be the profits of the employers or the special favors they obtain through the tariff, they will only pay the wages which the market for labor demands. But, apart from these particular instances, the general history of labor in the United States shows that the imported laborers are not long in learning what is a fair rate of wages, and obtain it. The true preventive of ill effects from importation or immigration of laborers lies not in trying to check it by law, which is impracticable, but in uniting to make living as cheap and employment as profitable as possible. For that purpose the working men should support for Congress those candidates, no matter what party they belong to, who are most to be relied on to secure the gradual abolition of duties on raw materials and the coarser forms of manufactures which are used in other industries, such as wool, metals, and lumber; who will confine taxation as far as possible to luxuries; who will avoid extravagant appropriations, and who will maintain a sound currency and a high standard of public credit. In State and municipal politics they should support men, without reference to party or to specious pledges, who are of established reputation for character and judgment and are not place-hunters or professional politicians, for such men only can keep down the burdens of taxation. It is in this direction and not in supporting the schemes of flattering demagogues that the laboring men can hope to use their political power for the improvement of their condition, and if they will turn their energies this way they need not fear the arrival on our shores of a few thousand, more or less, of British or European immigrants.

May 3, 1883

CHINESE RESTRICTION.

FEELING AND ACTION OF THE PEOPLE OF CALIFORNIA RESPECTING IT.

From the San Francisco Bulletin, Feb. 5.

The effort to solve partially the Chinese question by non-intercourse has assumed the proportions of a great State movement. It may be regarded as the first direct action on the part of the people at large following the vote of 1880 of 154,638 against Chinese immigration to 883 in favor of it. In the history of American politics no case of a like unanimity on any public question can be found. There was no such overwhelming majority in favor of independence in the war of the Revolution. This practically unanimous verdict having been recorded, our people looked to the National Government as the most effective agency to rid the State of the demoralizing servile element which had intruded upon them. As a partial response to their appeal the first Restriction act was passed. But it soon became apparent that that restriction was so inefficiently executed as to fail of accomplishing any relief. A second Restriction act was passed, but with scarcely better result.

From the beginning of restrictive legislation something very like a "Chinese sack" seems to have been in operation. Upon no other hypothesis can the extraordinary contradictions, usurpations, and subterfuges of many of those who had anything to do in the interpretation or execution of the law be explained. About this time it became apparent to the people of California that if they would preserve this, perhaps the fairest State in the confederation, to Christian civilization, they must take the matter in their own hands and do their share of the work. Legislation in several of the cities and towns was resorted to for the purpose of mitigating some of the grosser evils consequent upon the residence of a semi-barbarous race. Efforts of various kinds were made to protect the public health from the overcrowding of the Chinese, to save the cities themselves from conflagration by their careless use of fire, and to check petty crime among them by cutting off the cues of all convicted offenders against the law. But all these regulations were carried by the lawyers employed by the man dealers into the Federal courts, where they have been hung up for years at a time.

The series of mishaps and failures at length brought pretty nearly the whole State to its feet. Casting about for some mode of relief, the people of Eureka adopted the direct mode of action and sent the Chinese off bag and baggage. The citizens held the reins so firmly that while purging their town of its degrading and unassimilating element no rioting took place. Everything was done regularly and in order. But it was quite evident that that was not a mode of proceeding that could generally be adopted. Some other course which steered clear of the law had to be devised. The Chinese themselves pointed out the way by refusing to work in a cigar factory in this city unless the white operatives were discharged. Sacramento at once took the lead in the new movement. The mute appeal of 500 white workingmen, condemned to idleness by the Asiatic interlopers, marching through the streets, furnished a powerful stimulus to the movement.

The policy of non-intercourse so firmly put in operation in the capital of the State was soon followed in almost every city and town of consequence. Truckee had a hard fight to dislodge the intruders, but it won a glorious victory, though the odds appeared to be heavily against it at first. The movement, almost universal in its character, has disposed of many of the falsehoods which have clustered around the Chinese question. No efforts have been spared to impress the Eastern mind with the notion that it was only a certain class of our population that was opposed to the Chinese. In view of the facts which are now transpiring it will be useless for any one hereafter to make an assertion of that character. The index of the views of the people of California will always be found in the vote of 1880, to which reference is above made. There has been no change of opinion since then. On the contrary, the feeling of opposition to the Chinese has become intensified. It is not based on prejudice, but on moral, ethnological, economic, commercial, industrial, and sanitary grounds that cannot be successfully impeached. The movement is orderly and self-contained, and exhibits in a striking manner the California trait of firmness, the faculty of organization, and the adoption of the most practical methods to effect a given object.

Naturally a movement so general in its nature is in need of a State organization. The first gathering having that object in view is now in session at San José. Nine counties were represented on the first day—that is to say, Alameda, Nevada, San Francisco, Sacramento, Santa Clara, Santa Cruz, Sonoma, and Solano. The various sections of the State were fairly represented—that is to say, the mountains, the central valleys, and the commercial metropolis. But a larger and more complete assemblage is to be held at Sacramento at a later date. The San Jose convention wisely eschewed politics. It resolved that the "Chinese must go," but at the same time that all its methods must be within the law.

So far as the action of Congress is concerned, a disposition is manifesting itself not only to demand that restriction shall be made perfect, but that the Burlingame treaty shall be abrogated. When the argument comes up on the latter point it will not be difficult to demonstrate that in that compact we gave everything and got nothing whatever in return. It would be impossible to find an instance in which a nation was more grossly overreached than we were on that occasion. The instrument will be searched in vain for any right or privilege conceded to Americans which all other foreigners do not enjoy. The history of diplomacy does not reveal another instance of a bargain so entirely one sided.

But we want restriction made perfect before any advance is made upon the more important line of the abrogation of the Burlingame treaty. The popular movement in this State has so far been successful in detail. It will encounter its most serious perils in the efforts to join the various parts together. Much will depend on the final action of the convention at San José and the subsequent one at Sacramento. Dissension, hastily considered measures, or illegal methods may prove fatal. The opponents of free labor and a homogeneous population are counting on missteps of one kind or another. They are also calculating that some of our people will not endure the inconveniences which must necessarily result from so radical a reorganization of our industrial system as that which is now proposed. They are waiting for what they call the "reaction." But, if we are not wholly mistaken about the temper and disposition of the people, it is not likely to come.

February 15, 1886

CHINAMEN DRIVEN FROM THEIR HOMES.

PORTLAND, Oregon, March 1.—Between midnight and 2 o'clock this morning a mob of 80 men, divided into squads of 20 each, visited the Chinese working back of East Portland and Albina, an eastern suburb of this city, and drove them out. There were 180 Chinese in all, and all of them were engaged in wood chopping and grubbing on land lying from one to three miles back of the towns mentioned. Some of the men wore masks and some had their faces blackened, while others had sacks over their heads, with holes for their eyes. All were armed. They went to the camps where the Chinese were asleep, routed them out, and ordered them to pack up and leave at once. The Chinese offered no resistance, and allowed themselves to be driven to a ferryboat, which brought them to this city. The mob worked with great secrecy. The night was dark, and the officers of the law knew nothing of the raid until the Chinese arrived here and were marching up the street.

March 2, 1886

NOTICES.

The Commissioners of Emigration have ordered the return to Norway of a female immigrant who was about to augment the population of the United States with a fatherless child. There is no reason why this action should be denounced, or described as anything more than a proper exercise of the discretion of these officers. A few demonstrations that a real scrutiny is exercised into the character and circumstances of immigrants are urgently needed in order to dispel the belief prevalent in Europe that this country is a universal foundling asylum as well as a hospitable refuge for convicts and paupers.

August 4, 1887

NOTICES.

The people of Western Massachusetts ar beginning to inquire about the manner in which our Emigrant Commissioners do their work. Several parties of Arab beggars have recently appeared in the cities and country towns, to the great annoyance of residents. In Holyoke they built fires in the streets to cook food that had been given to them. In Springfield several of them were sent to jail. A party appeared in Belchertown, carrying a curly-headed boy of 2 years, who had probably been stolen. These wretched creatures came from Syria, and the Springfield authorities learned that the Emigration Commissioners had permitted them to enter the country at this port about three weeks ago. They had been told, possibly by the agents of steamship companies, that they could quickly make fortunes here. Do not the Commissioners think that the law provides for the exclusion of such immigrants?

August 9, 1887

DEGRADED LABOR FROM EUROPE.

It is well known that we are receiving from certain countries a very undesirable class of immigrants. The reports recently collected by the State Department and analyzed by Mr. WORTHINGTON C. FORD present some of the characteristics of these people as they are seen at their homes in Europe, and the thoughtful citizen who adds to the knowledge of their habits and character which he has obtained here these facts gathered by our Consuls abroad will be able to form sound opinions as to the merits or demerits of a national policy which allows them to come in.

Take the immigrants from Hungary, for example, whose brutish condition in the Pennsylvania mining regions has been the subject of so much writing. The number arriving in this country from Austria, Bohemia, and Hungary in the last fiscal year was 40,135, or 40 per cent. more than were admitted in the preceding year. The contract labor law has not checked immigration from that region. It is asserted that the contracts are now made secretly by agents stationed abroad or more openly after the arrival of the immigrants at this port. At all events, the law appears to be a weapon of no value for use with regard to the class of persons whom its authors had in mind. Of the immigrants who have come from England in the last thirteen years, from 13½ to 20 per cent. (the proportion varying from year to year) have been skilled workmen. The percentage of skilled labor in the cases of Scotland and France has been even greater. In the case of Austrian immigrants, in 1885 it was only 6⅔ per cent., and of the Hungarians less than 2½ per cent. were skilled. But the number of immigrants from Hungary has rapidly risen from only 373 in 1877 to 19,807 in 1887.

The report of Consul STERNE, of Buda-Pesth, shows that these immigrants are for the most part Slovacks. They do not come to America to remain and become citizens, but their aim is to accumulate here what is to them a fortune, and they usually obtain this fortune in about three years. And what is the sum which they strive to obtain? About $600; perhaps not more than $500. We recall a statement ascribed to one of them, who was about to return at the end of his term of service, that for four years he had lived here at a cost of only $5 a month. As his wages had been $18 a month he had saved $624, and was looking forward to a life of ease in a country where he could be comfortable, after his fashion, on $30 a year.

We are told by Consul STERNE that in Hungary their homes are huts of one room, wherein all the members of a family are huddled together by night as well as by day. "From all I can learn," he says, "their demand for water is very limited for the use of the outer body as well as the inner." Their diet is milk, potatoes, corn, and rye, with, occasionally, while they are laboring in cities, "the remnants or offal from the restaurant." Those who are familiar with scenes in the mining regions of Pennsylvania know how strong is their appetite for the worst liquor that is made. At home their favorite drink is potato brandy. "As many of them as can," says the Consul, "men and women alike, will pack themselves into a room or a cellar over night, without the least regard for cleanliness." In Buda-Pesth the health authorities are frequently compelled to dislodge them in the night from "these disease-breeding pest holes."

Concubinage is so general among them that in Hungary it is a publicly acknowledged evil. Consequently the number of illegitimate children is large. The mortality among children is appalling. It is due in large measure to the rude and barbarous treatment to which they are subjected by the mothers, whose ignorance of hygiene is dense, and who are prevented by superstition from using ordinary medical remedies. "I am of the opinion," says Consul STERNE, "that with the present condition of the labor market in the United States there is no room for this class of people." Nor does he think that under more favorable conditions for labor these Slovacks would be desirable acquisitions. "It will surely require generations to make them enlightened citizens." But it is unnecessary to regard them as citizens of any kind. They do not remain here. They do not become naturalized. And as fast as they save their money they send it back to Hungary. In many respects they closely resemble the Chinese immigrants, and the arguments that have been used so successfully against the admission of the Chinese can be used with almost equal force for the exclusion of these persons. Indeed, the court records in the mining districts indicate that they are less desirable than the Chinese because of their criminal tendencies.

The effect of their presence upon the labor market may be considered apart from their character and habits. Their way of living enables them to underbid the native laborer or the immigrant who becomes a citizen, uses proper food, and does not perpetually menace the health of civilized people who surround him. The workingmen of America who are not willing to live on garbage may with justice complain that, while the cost of the necessaries of life which they must have is raised by a protective tariff, their labor is not protected by any tax on imported labor of this kind. "The pauper labor of Europe" is at their very doors, and is gladly used by the very employers who are the foremost advocates of a protective policy for the benefit of the workingman. It is not surprising that HENRY GEORGE should favor the free admission of such immigrants, and should even oppose the enforcement of a law enacted for the exclusion of persons who are sure to become objects of charity. Growing taxes, the competition of a degraded and alien class, and the multiplication of incapables and shirks breed discontent and folly; and it is from the discontented and the foolish that he and his theories gain support.

September 12, 1887

"AMERICAN" ANARCHISTS.

Now that the trial of MOST has resulted in his conviction, some reflections suggested by it may be expressed without impropriety. The trial incidentally disclosed the nature and education of the people who go to hear him make speeches, whether he is in his usual ranting and roaring condition, or in the subdued and mournful state in which a number of them swear that he appeared when the Anarchists had been hanged. There is no reason to doubt, by the way, that he was quieter than usual, and one of the best results of the hanging was to induce quietude and depression among orators of his kind.

The striking thing about these witnesses is that residence in this country does not seem to have had the slightest effect upon them. Some of them have been in New-York for a few months, others for many years, but none of them have any more definite notions about the country than would be obtained by a sailor of a port of which he gathered his "impressions" in a week on shore, during which he was drunk all the time. They are opposed to the Government and propose to overthrow it by means of dynamite and massacre without having the least notion what it is. One of them, who had been in the country for four years and had declared his intention of becoming a citizen, was asked the question to which all applicants for naturalization are expected to make satisfactory answers. The question was whether he was attached to our form of Government, and the luminous answer was that "he had not been attached for anything yet." Of course the stupidity of this reply was due partly to his ignorance of the English language, but it is a question whether a man so ignorant of the language of the country he lives in can possibly discharge with intelligence the duties of a citizen. This particular witness declared that he was not an Anarchist, and did not seem to have any notion what Anarchy meant, though he was clear that the murderers in Chicago had been improperly executed. Another witness, who swore that he had been in the country seven years, but who had not become a citizen, knew so little English that an interpreter had to be summoned to translate his testimony. One of his answers illustrates the effect which his American residence has had upon him. He said, "I am no Anarchist, but I may be."

It is men of this kind who make up MOST's audiences. To all intents and purposes they are living in Europe yet, and are as far from being assimilated to the American people as Anglo-Indians are from having become Hindus. A curious illustration of their unconsciousness that they are living in the United States was furnished by one of the murderers in Chicago, who recited on the morning of the execution a poem of HEINE's about the sorrows of Germany which did not bear the least relation to the condition of this country. If he had been about to be hanged for taking part in the crime of the Niederwald, the poem expressed the sentiments he might reasonably be supposed to have entertained.

It is most unfortunate that the conditions of modern civilization should have done so much to nullify the expectations of the founders of the Government concerning the immigrants. Our statesmen supposed, when the naturalization laws were passed, and indeed for two generations afterward, that the new-comers would be really incorporated with the surrounding population. Their experience justified them in that expectation. Even to this day an immigrant who goes to the country soon rids himself of such of his traditions as are inapplicable to his new life, and comes to adopt the point of view of his neighbors. There are no Anarchists and no Socialists except in the cities. At the furthest the children of an immigrant farmer, brought up among Americans and taught in the public schools, become thoroughly Americanized. At the beginning of the century there were no cities of the first or even of the second class in the United States, and the Americanization of a townsman went on even more rapidly than that of a countryman. Now every great city has colonies of foreigners merely encamped within its borders. They may spend their lives in the United States and know as little about America when they die as they did when they arrived. It is even possible for their children to grow up as colonists among an alien race. The best preventive of this result is to be found in the public schools, and the Germans in some cities have unwisely resisted the application of this preventive by trying to secure the compulsory teaching of their own language, in order that their own ignorance of English might embarrass them less. The people of St. Louis, one of the most strongly German cities in the country, have just done an excellent work by defeating this attempt.

Of course we do not mean that Germans, or foreigners of any other nationality, are bad citizens by reason of their nationality. All Americans know and appreciate the enormous benefits that German immigration has bestowed upon the United States. It is nevertheless true that a German or any other foreigner becomes a valuable American citizen by becoming an American and not by remaining a foreigner. The political doctrines that have borne fruit in crime, as they could not have originated in this country, could neither be preached or put into practice by men who had not remained foreigners. It is a mere farce that such men as have appeared as witnesses in the trial of MOST should exercise American citizenship, and, in view of what has happened and what is threatened, it cannot be decently maintained that no restriction of American citizenship is needed.

November 30, 1887

THEY OPPOSE A CHANGE

IN THE LAWS GOVERNING IMMIGRATION AND NATURALIZATION.

WASHINGTON, March 20.—The committees of the House and Senate on immigration and naturalization, sitting jointly, gave a hearing to-day to persons opposed to changes in the laws on those subjects. Edward Rosewater, editor of the Omaha *Bee*, spoke for a number of German and other societies in the West. He said that the ratio of immigration to the population is decreasing and the time has not yet come to restrict immigration. He was born in Bohemia; his brothers in America. He was as good as they; the accident of birth in this country did not make them any better than he. This pride in birthplace is a survival of barbarous fanaticism. The pauper laborers of Europe transplanted to America are generally supposed to injure the interests of American laboring people. This was not the case. When the pauper laborers reach here their condition changes; their wants are larger, and in consequence they become greater consumers.

Native Americans have largely given over to foreigners the heavier kinds of labor, and these immigrants are needed to take their place. There had been a great deal of criticism against the coming in of Italians. For one he had a tender feeling for Italy and Italians. Christopher Columbus was an Italian; the people of that nationality were cultivated in the arts and sciences hundreds of years before America had been thought of. He believed in the application of the golden rule in our relations with peoples of other nations. People who believed there was something wrong with the machinery of Governments were classed as Socialists, and, therefore, under the terms of the bills before the committees denied admittance to the United States. Bellamy and Henry George are Socialists, holding opinions radically at variance with the principles of our form of government. The latter thinks and says that it is not wrong to hold these opinions. What would be the effect upon treaties we have in force now with other countries by a change in our naturalization laws? Would it not involve us unpleasantly with the Governments of those countries?

Richard Bartholdt, Chairman of the conference of delegates of German-American societies which met in Washington this week, read the protest adopted by that body against the passage of any and all measures now pending in Congress designed materially to change the present national laws on immigration and naturalization, as follows:

First—These proposed measures are fraught with the same mischief and breathe the same spirit which caused the founders of this Republic to rise in rebellion against a British tyrant and to hurl at him the following indictment: "He has endeavored to prevent the population of these States; for that purpose obstructing the laws for naturalization of foreigners; refusing to pass others to encourage their migration hither, and raising the conditions of new appropriations of lands."

Second—The industry, thrift, and honest intelligence of the immigrant have, for more than a hundred years past, been a chief factor in developing the mineral, agricultural, commercial, and industrial resources of this country, and in raising it to its present proud position among the nations of the world.

Third—The patriotic devotion of those who have in the last century emigrated to our land and acquired a right of citizenship among us has been such as to win a proud place in its history. What right have we to assume that the character of those who come under a continued liberal policy of immigration will be one particle lower than that of those who came before?

Fourth—The proposed change of the naturalization laws, as embodied in bills now pending in Congress, must be regarded as uncalled for and a mischievous deviation from the policy first established under Washington's Administration. Not merely should we encourage those who have cast their fortunes among us to soon become loyal members of our body politic, but the very existence of large bodies of unnaturalized residents would seem to constitute a menace to our institutions. Love of a free country can best be bred in men by securing to them the full and early enjoyment of its privileges and blessings.

Fifth—The scheme proposed of emigrant inquisition through our consular and Governmental representatives abroad is impracticable. It will aid the undesirable merely and deter the good. No European Government will assist in retaining its bad elements and forwarding the desirable. Besides, a system of espionage like this is odious and degrading, inasmuch as it promotes blackmailing and corruption.

Sixth—Our existing laws, if rigidly and honestly enforced, afford ample protection against all undesirable and criminal immigration; but no such system of laws as now proposed can be enacted without violating the fundamental principles of our national compact and darkening the brightest pages of our national history.

Mr. Simon Wolff of Washington, appointed by the German Delegates' Conference, declared that change in the present laws was neither necessary nor desirable.

The committees will leave Washington for New-York to-morrow morning. In the evening they will receive informally at the Fifth-Avenue Hotel those persons who desire to advise the committees respecting the location of the proposed immigrant landing station. On Saturday they will visit Governor's Island, Bedlow's Island, Ellis Island, and Castle Garden. On this trip the committees will be accompanied by Gen. Schofield, representing the War Department; Commander Folger, representing the Navy Department, and Solicitor Hepburn and Surgeon General Hamilton, representing the Treasury Department, all of them being interested in the settlement of the question of locating the landing station.

If the steamer Belgravia, now en route to this country with 1,100 Italians aboard, is likely to reach New-York Monday, the committees will remain in that city and observe the operation of debarkation at Castle Garden.

March 21, 1890

TOO MUCH IMMIGRATION.

RUSSIAN JEWS AND OTHERS REDUCING THE PRICES OF LABOR.

Many union workingmen have been complaining that the large immigration of Russian Jewish refugees into this country is interfering with their prosperity by stimulating keen competition in their trades. Among the trades affected, it is alleged, are certain ones connected with the clothing and the cotton weaving and spinning industries, out of which the Jewish operatives are said to be driving other operatives. It has been said that Congress would be asked to put a stop to the enormous tide of immigration.

Certain labor leaders said yesterday that protests of workingmen were directed, not against the Jews in particular, but against further immigration, as being hurtful to the welfare of the working classes. The great fight that the Knights of Labor were going to make would be against the Chinese; they would insist upon the passage of a law keeping Chinese out altogether.

William McNair, ex-Secretary of District Assembly No. 49, said that in this city the Jews had practically crowded the Germans out of the clothing industry by working for lower wages. Now, it was the Italians who were beginning to crowd the Jews out of that very same industry by working for still less than the Jews were willing to accept.

Mr. McNair said that in New-England the Canucks swooped down and crowded American and Irish operatives out of the mills by working cheaper and pinching themselves in their living. Now the Jewish refugees were making inroads into New-England towns and villages, and were offering their services for less than the Canucks get and were crowding them out. Among workingmen, Mr. McNair said, it was not a question of nationality. Immigration was too great and was hurting them. They want to have it stopped.

January 22, 1892

A CURIOUS ANALOGY.

We have no doubt that the statements of Mr. BOTKINE, Secretary of the Russian Legation at Washington, which we printed the other day, in regard to the real reason for the edicts driving the Jews from Russia, are substantially correct. "The Jews," he said, "are cleverer than the Russians. They are shrewder business men and keener in intellect, and if they were admitted to the schools freely they would outrank the Russians in all particulars. It is because of this need of protecting the Russian citizens from the Jews, who are really immigrants, that these edicts have been made."

This is a very frank statement from a representative Russian, and it affords a more reasonable explanation of the antipathy to Jews than mere race prejudice or difference of religion. It is a confession of superiority and a plea for protection against the competition of a people who are stronger industrially and commercially than the native population. Jews are hated because they push into occupations and trades and get the better of those with whom they compete. The distinction of race and religion serves to draw a line of segregation and enables the people and the Government to wage their warfare against a class that is feared and hated on account of its superior efficiency in the struggle for life. This same fear of competition and recognition of a certain kind of superiority is at the bottom of the anti-Semitic feeling in other countries than Russia.

There is a curious analogy between this feeling and the antipathy for Chinese laborers on our Pacific coast. It may not be admitted that the Chinese are "cleverer" than those who call themselves American workingmen, or "shrewder" or of "keener intellect," but they have certain characteristics which give them an advantage in competition. They are industrious, patient, and frugal, and they are not addicted to wasting their time and substance. They have what Poor Richard used to regard as industrial virtues of a high order. They do not demand shorter hours and higher pay or concern themselves about a "higher standard of living," but plod along and push their way, making and saving all they can. This is why they are hated by those who fear their competition, but care little about their race and less about their religion. The people wish to be protected against their competition and want them driven out or kept out for that reason. We do not say whether they are justified in this feeling or not, but they might apply Mr. BOTKINE'S excuse for the Russians to their own case, perhaps. He said: "From the broad standpoint of the world this is not the right spirit, but the edicts are made for Russians, to protect the real citizens of the empire."

June 18, 1893

NO MORE CHINESE LABORERS

IMMIGRANTS EXCLUDED FOR TEN YEARS BY NEW TREATY.

Chinese Teachers, Merchants, Students, and the Like May Enter the Country and Live Here—Under Certain Conditions Laborers May Go to China and Return to the United States—Protection to Life and Property Guaranteed.

WASHINGTON, Aug. 13.--In accordance with the understanding of Friday last, the Senate to-day immediately after some unimportant routine morning business went into executive session to take up the Chinese treaty. It had been agreed that the vote was to be had at once without further debate, and the roll call was taken shortly after the doors were closed.

A suggestion was made that the injunction of secrecy be removed from the vote, but no motion was made to that effect. The vote stood 47 to 20 in favor of the ratification of the convention, almost all of the Northwestern Senators opposing the treaty while the Eastern men, with the exception of Messrs. Lodge and Hoar of Massachusetts, voted for its ratification.

The treaty has been the recipient of much adverse criticism, and many petitions have been received both for and against it. The following is the full text of the treaty, it having been ratified without amendment, and as originally sent to the Senate by Secretary Gresham:

Whereas, On the 17th day of November, A. D. 1880, and of Kwanghsii the sixth year, tenth moon, fifteenth day, a treaty was concluded between the United States and China for the purpose of regulating, limiting, or suspending the coming of Chinese laborers to, and their residence in, the United States; and,

Whereas, The Government of China, in view of the antagonism and much deprecated and serious disorders to which the presence of Chinese laborers has given rise in certain parts of the United States, desires to prohibit the emigration of such laborers from China to the United States; and.

Whereas, The two Governments desire to co-operate in prohibiting such emigration and to strengthen in other ways the bonds of friendship between the two countries; and,

Whereas, The two Governments are desirous of adopting reciprocal measures for the better protection of the citizens or subjects of each within the jurisdiction of the other,

Now, therefore, the President of the United States has appointed Walter Q. Gresham, Secretary of State of the United States, as his plenipotentiary, and His Imperial Majesty, the Emperor of China, has appointed Yang Yu, officer of the second rank, sub-director of the Court of Sacrificial Worship, as his Envoy Extraordinary and Minister Plenipotentiary; and the said plenipotentiaries having exhibited their respective full powers, and having found them to be in due and good form, have agreed upon the following articles:

Article I. The high-contracting parties agree that for a period of ten years beginning with the date of the exchange of the ratifications of this convention, the coming, except under the conditions hereinafter specified, of Chinese laborers to the United States shall be absolutely prohibited.

Article II. The preceding article shall not apply to the return to the United States of any registered Chinese laborer who has a lawful wife, child, or parent in the United States, or property therein of the value of $1,000, or debts of like amount due him and pending settlement. Nevertheless, every such Chinese laborer shall, before leaving the United States, deposit, as a condition of his return, with the Collector of Customs of the district from which he departs, a full description in writing of his family, or property, or debts, as aforesaid, and shall be furnished by said Collector with such certificate of his right to return under this treaty as the laws of the United States may now or hereafter prescribe and not inconsistent with the provisions of this treaty, and should the written description aforesaid be proved to be false, the right of return thereunder, or of continued residence after return, shall in each case be forfeited. And such right of return to the United States shall be exercised within one year from the date of leaving the United States; but such right of return to the United States may be extended for an additional period, not to exceed one year in cases where, by reason of sickness or other cause of disability beyond his control, such Chinese laborer shall be rendered unable sooner to return, which facts shall be fully reported to the Chinese Consul at the port of departure, and by him certified, to the satisfaction of the Collector of the port at which such Chinese subjects shall land in the United States. And no such Chinese laborer shall be permitted to enter the United States by land or sea without producing to the proper officer of the customs the return certificate herein required.

Article III. The provisions of this convention shall not affect the right at present enjoyed of Chinese subjects, being officials, teachers, students, merchants, or travelers for curiosity or pleasure, but not laborers, of coming to the United States and residing therein. To entitle such Chinese subjects as are above described to admission into the United States, they may produce a certificate from their Government or the Government where they last resided, vizéd by the diplomatic or Consular representative of the United States in the country or port whence they depart. It is also agreed that Chinese laborers shall continue to enjoy the privilege of transit across the territory of the United States in the course of their journey to or from other countries, subject to such regulations by the Government of the United States as may be necessary to prevent said privilege of transit from being abused.

Article IV. In pursuance of Article III. of the immigration treaty between the United States and China, signed at Pekin on the 17th day of November, 1880, (the 15th day of the tenth moon of Kwanghsii, sixth year,) it is hereby understood and agreed that Chinese laborers or Chinese of any other class, either permanently or temporarily residing in the United States, shall have for the protection of their persons and property all rights that are given by the laws of the United States to citizens of the most favored nation, excepting the right to become naturalized citizens. And the Government of the United States reaffirms its obligations, as stated in said Article III, to exert all its power to secure protection to the person and property of all Chinese subjects in the United States.

Article V. The Government of the United States, having, by an act of Congress, approved May 5, 1892, as amended by an act approved Nov. 3, 1893, required all Chinese laborers lawfully within the limits of the United States before the passage of the first-named act to be registered as in said acts provided, with a view of affording them better protection, the Chinese Government will not object to the enforcement of such acts, and reciprocally the Government of the United States recognizes the right of the Government of China to enact and enforce similar laws or regulations for the registration, free of charge, of all laborers, skilled or unskilled, (not merchants as defined by said acts of Congress,) citizens of the United States in China, whether residing within or without the treaty ports. And the Government of the United States agrees that within twelve months from the date of the exchange of the ratifications of this convention, and anually thereafter, it will furnish to the Government of China registers or reports showing the full name, age, occupation, and number or place of residence of all other citizens of the United States, including missionaries, residing both within and without the treaty ports of China, not including, however, diplomatic and other officers of the United States residing or traveling in China upon official business, together with their body and household servants.

Article VI. This convention shall remain in force for a period of ten years, beginning with the date of the exchange of ratifications, and if six months before the expiration of the said period of ten years neither Government shall have formally given notice of its final termination to the other, it shall remain in full force for another like period of ten years.

"In faith whereof," the treaty concludes, "we, the respective plenipotentiaries, have signed this convention and have hereunto affixed our seals. Done, in duplicate, at Washington, the 17th day of March, A. D. 1894.

"WALTER Q. GRESHAM. [Seal.]
"YANG YU." [Seal.]

August 14, 1894

BOGUS AMERICANISM.

To the Editor of The New York Times:

A densely ignorant immigrant who reaches these shores learns in an incredibly short period that this is the freest of free countries. He scarce knows a dozen words of pidgin-English, but he has already absorbed enough of the spirit of so-called "Americanism" to impudently assert it to our oldest citizen. Fresh from the hard slavery of an obscure province, these folk arrive at a most hospitable shore inflamed with the exaggerated tales of the auriferous nature of our paving blocks, of a country whose only nobility is that of wealth, and where socialism and anarchy are punishable only in the deed.

In a generation, these crude opinions, tempered by a little education and some prosperity, are molded into what ultimately becomes the spirit of a very vicious kind of "Americanism," a spirit which, whether right or wrong, is clamorous for the one against the multitude, a spirit which rages against the uniform of the State Guard obeying its lawfully chosen magistrate, a spirit which howls for the freedom in a land for whose liberty neither they nor their ancestors fired a shot. It is oft a mob which is full of patron saints, "vaterlands," stilettos and unpronounceable names, headed by a mercenary with a small head and big voice whose waving arms is the only proceeding understood by the crowd. He talks "America, the land of the free." This is indeed a bogus "Americanism," which breeds cocoon-like in the seclusion of alien countries to finally emerge in full flight to these United States.

This blessed soil may be the promised abiding place, overflowing with milk and money to which the slaves of lands of oppression turn their steps, but with our unrestrained immigration, it will take many years to efface the flaming red paint with which the Goddess of Liberty is at present so plentifully bedaubed. IRVING E. DOOB.
New York, March 20, 1903.

March 22, 1903

IMMIGRATION.

Immigration at the rate of 100,000 per month, which it is expected will be the record for April, and which is likely to continue and possibly increase as long as present conditions last, is not at all an unmixed cause for congratulation. Not more so is the fact that the total immigration for the fiscal year ending with June will approximate a million. The largest immigration of any one year hitherto was in 1882, when a little less than 800,000 were received. The present inflow is chiefly from Italy, Austria-Hungary, Russia, Finland, and countries which lie on the borders of Asia. Very little of it is high-class immigration, and in spite of the laws for the exclusion of criminals, insane persons, incapables, and degenerates, it is probable that large numbers of such people manage to get in.

Public opinion respecting immigration is changing rapidly in this country, just as it is in England. The idea that it is our duty to open wide our doors to the oppressed of all the world, and that because a person is poor and unhappy in the country of his birth he may claim our National hospitality as a matter of right, irrespective of any advantage to ourselves from his coming, is no longer held by thoughtful persons. On the contrary, the conviction is gaining ground that our present immigration laws are altogether too lax, and that to make them much more stringent would be advantageous. The economic aspects of the question are somewhat obscure. If the labor of these hundreds of thousands of relatively unskilled immigrants is needed, and shall continue to be needed until the movement is checked and those already here are absorbed and assimilated, they will be of benefit to the country. If, on the contrary, it is giving us a supply of unskilled labor in excess of our needs, and to find work the late-comers must break down the established rate of wages, strikes and riots are to be expected, and Socialism in its more dangerous aspects will receive a great impulse. In a word, the only ground for an optimistic view of the present movement of the surplus population of Southern Europe to this country is in the hope that our National prosperity will continue for a long time, if not indefinitely. The next Congress might find it profitable to deal with this subject in a much broader way than has hitherto been attempted. The judicious restriction of immigration would be a popular measure, and it will be surprising if it is not seized as an issue in the next Presidential canvass.

April 28, 1903

WOULD USE NEGROES IN IMMIGRANTS' STEAD

Lawyer Young Wants Only Saxons to Come for Twenty Years.

New Rochelle People's Forum Hears ex-Senator Mills's Arguments Against Restriction of Immigrants.

Special to The New York Times.

NEW ROCHELLE, Feb. 7.—Lawyer Charles A. Young, during a debate on "Should Immigration Be Restricted?" at the People's Forum here to-day advanced the argument that in his opinion immigration should be shut off for twenty years for all but Saxons, and that the negro of the South should be encouraged to come here to do the work which it was being argued the immigrants did—dig the ditches, build the railroads, and do similar work. He saw in the carrying out of the latter plan also the solution of the servant girl problem, of which he spoke in the following way:

"I can see by the anxious faces of some of the women here that they have wrestled with that problem—they have learned that there is such a person as a dark-skinned Romeo. Now, if that Romeo happens to be from the South it is often difficult to keep the kitchen Juliet, for she invariably insists on following Romeo. All would be different if we could bring our Romeos up here from the South.

"During the past down to 1875," said Mr. Young, "there have been three large migrations to this country in addition to the always steady stream from England; one came from Ireland about the middle of the century, and somewhat later one from Germany, and Scandinavia, in which is included Sweden, Denmark, and Norway. During this century down to 1875, then, as in the two preceding it, there had been, save for the French Huguenots, scarcely any immigration except from kindred or allied races; and no other which was sufficiently numerous to have produced any effect on the National characteristics.

"Since 1875, however, there has been a great change. While the people who for 250 years have been migrating to this country have continued to furnish large numbers of immigrants to the United States, other races of totally different race origin, with whom the English-speaking people have never hitherto been assimilated or brought in contact, have suddenly begun to immigrate to the United States in large numbers. Russians, Hungarians, Poles, Armenians, Croatians, Bohemians, Italians, Slavs, and even other Asiatics, whose immigration to this country was almost unknown twenty years ago, have poured in in steadily increasing numbers, until now, for the past two years, they exceed nearly two to one the immigration of those races kindred in blood and speech, or both, by whom the United States has hitherto been built up and the American people formed.

"I ask for twenty years' restriction of all but Saxon immigration. It is born of expediency and founded upon common business sense; it needs no justification other than lowering of American wages, which follows hard upon unrestricted immigration."

Ex-Senator Isaac N. Mills took the ground that the immigration question should be looked at from a higher moral level than his adversary had placed it on. "In restricting immigration," he said, "it must be remembered that we would be violating the corner stone upon which this Republic rests — individual liberty. The Bible and our Constitution demand it, and all efforts to restrict immigration have met with disapproval by the masses whenever it has been brought up thus far. The question is, Is it right? We say in our Constitution that 'all men are born free and equal.' Can we now begin to discriminate as Mr. Young would have us do? This country cannot afford to be dishonest; it cannot afford to discriminate or classify. Thousands shed their blood for the negro, and we are not now, I think, going to discriminate against the poor immigrant who comes here to enjoy our liberty."

February 8, 1904

IMMIGRATION AND THE EAST SIDE.

Those engaged in the University Settlements have begun the publication of a little quarterly magazine under the title of University Settlement Studies, which is a valuable contribution to knowledge of conditions among a large population quite unfamiliar to the general public. The editors are JAMES H. HAMILTON and WALTER E. WEYL. The advantage enjoyed by the writers for this quarterly is that they at least write largely from personal observation, which with equal intelligence and good faith yields better results than book study. These writers seem sensible and very much in earnest.

The second number of the "Studies" is devoted to the question of immigration. The principal article is by Mr. WEYL on "Immigration and Industrial Saturation." There is an exceedingly informing article by Mr. DAVID BLAUSTEIN-WALLING, on "What the East Side Before Emigration and After Immigration," and another by Mr. WILLIAM E. WALLING, on "What the East Side People Do." Mr. WEYL argues very strongly against the present system of immigration, which he regards as simply one of police regulation which does not materially affect the inflow either as to character or numbers. He points out that only 1 out of 6,000 immigrants was excluded under the restrictions of the law, which is obviously an insignificant proportion. We think, however, he hardly gives the proper weight to the deterrent effect of our restrictions which may prevent a considerable number from trying to get in. Nor does he take into account sufficiently the definite tendency of European legislation of recent years to govern and restrain emigration, a tendency which is likely to increase, and to afford opportunity for co-operation between these Governments and our own. The fact remains that the influx of immigrants is greater than ever before, and that it must have definite and serious effect upon our industrial situation. Mr. WEYL thinks that that effect is injurious, cumulative, and ought to be prevented by stringent measures. He says:

Within the last quarter of a century there has occurred an event which is of as vital importance as was slavery in the Southern States, or the civil war which terminated it. We have struck our frontier. The western wave of migration has reached its limit, and the population has been obliged to recoil upon itself. From now on there will be no outlet for the unemployed and the discontented of our cities. The conditions of life will tend to become more and more similar to those in Western Europe. We shall be more and more an industrial Nation, and shall rapidly lose our preponderatingly agricultural character. The struggle for life in our large cities will become more intense, and will be carried out upon a more gigantic scale. The evils of unemployment, irregular employment, casual employment, the intensity of labor, the insecurity of labor, and the shortening of the trade life, will be more and more difficult to combat as the residuum of unemployed, but employable, men in our large cities increases.

This residuum must necessarily increase with increased immigration. The immigrant must enter American life through a job. It is not his ability to work, but his ability to secure work, that gives him a foothold. And to a large extent he must not only secure a job, but he must take it away from some one else by offering more work for the same sum of money or the same amount of work for a less sum of money.

Here, then, is the point of view of a man especially caring for the welfare of the American workingman. Of course Mr. WEYL disclaims the notion that the amount of employment is fixed and that what one gets another must lose or go without. But measuring the work to be had against the men to do it, and taking into account the organization and distribution of labor and of employment, he thinks that the admission of more labor from abroad will do more harm than good to the country. That is the trades union view, and it is probable that in the notion of faulty organization and distribution, Mr. WEYL has in the back of his head some scheme, or, at least some principle, of adjusting these things. We are not confident that that can be done, but it is interesting to have the subject of immigration studied from this general point of view.

August 21, 1905

GOMPERS CRIES FOR LESS IMMIGRATION

Stirs National Conference to a Hot Debate.

THEORISTS AGAINST LABOR

Immigrants Worth $1,000 a Head, Carnegie Says—Not the Two Beers and a Biscuit a Day Kind, Shouts Ohio.

Organized labor, speaking through the mouth of Samuel W. Gompers, President of the American Federation of Labor, had its say on the immigration question yesterday afternoon, taking issue squarely with those who would open wide the country's gates and precipitating a storm of debate that raged through the better part of the afternoon session of the National conference on immigration. The conference is in session in the Madison Square Garden Concert Hall.

"I take second place to no man," declared Mr. Gompers, "in recognizing the fundamental right of man to move where he pleases on the face of this earth of ours. The question which this conference has to solve, however, is not merely academic. It is concrete, practical.

"I would be the last man to deny in the abstract that a man has the right to go where he pleases. When by so doing, however, he jeopardizes the rights and well-being of other men, it is an entirely different proposition.

"I respectfully dissent from the assertion that the question of immigration is unimportant. I maintain with all the strength of my being that there is no more vital question to-day before the people of these United States. It may not be all absorbing to those whose positions are assured, and who can look forward in the knowledge that the future holds no menace for them, come what may. But to those who labor to live, if they do not live to labor, the immigrant presents a problem that concerns his very being.

"The Chinaman is a man. I have nothing against the Chinaman as a Chinaman, but there is a great deal of difference between a Chinaman and an American. I am not in favor of excluding the Chinaman from our shores because he is a Chinaman, but because his ideas and his civilization are absolutely opposed to the ideals and civilization of the American people. Never in the history of the world have the Chinese been admitted into a land save to dominate or to be driven out. The Chinaman is a cheap man—"

"I call the gentleman to order," interrupted a delegate. "To-morrow has been set aside for the consideration of immigration from an Asiatic standpoint."

"I admit the justice of the point raised," said Mr. Gompers, "and will leave the Chinaman out of the question. I merely brought him in to illustrate my attitude.

"I would like to ask this question, then: Is it not strange that there are a number of gentlemen in our country who favor a system of protection for American products, as against foreign-made products, and yet demand absolute free trade in the matter of American labor? Is it not inconsistent to impose a duty upon the product of the European workman when the European workman is free to come to this country and turn out that same product?"

Mr. Gompers was interrupted by an outburst of cheers which came from all parts of the hall. One delegate let out a whoop which would have done credit to a Wild West show. Mr. Gompers went on to charge that at certain stated periods workmen of a given calling were brought to this country and concentrated at a given point for the purpose of depressing labor of a certain kind. This brought out another demonstration.

"It is true now," he concluded, "as it has always been, that self-protection is the first law of nature, and it is of the utmost importance that the American workman should do all in his power to prohibit the importation of those who would still further press him down."

Eliot Attacks Gompers.

President Eliot of Harvard, introduced as the next speaker, proceeded without delay to attack the doctrines promulgated by Mr. Gompers. That sentiment was well divided was evidenced by the generous applause with which he was interrupted.

"I question the doctrine of self-protection," he said. "It is overworked. To a certain extent it is a good thing, but if it is to be followed at the expense of our fellow-men it is a dangerous thing. [Cheers.] It is not the nobler, more generous attitude to assume, and it will never commend itself to our people in regard to immigration. [Applause.]

"We need all the brain and sinew we can import to develop our resources. Every part of our country, North, East, West, and South, needs more labor this very minute. Keep out the criminals, the insane, and those afflicted with contagious diseases, but when it comes to excluding labor on the ground that it has been induced to come here we are treading on entirely different ground. There can be no reason for prohibiting sound, moral laborers from coming to our land, whether they are induced to come or not. The American people will welcome them, and will endure no legislation designed to keep them out.

"It is not a generous thought that American Labor wants to keep out all other labor because the wages of American labor might be lowered. The mobility of capital is becoming greater and greater. It is much more mobile than the working people. The question, as it presents itself to me, is whether we prefer to have foreign labor come here and help build up our country, or to have our capital go elsewhere to build up other lands. [Applause.]

"The industrial problem is transcending National bounds, and we would not do well to forget that capital can always take care of itself. National greatness after all depends on the brain and the brawn of the people."

"We don't want the protection of American labor at the expense of other people," said Joseph Lee of Boston. "All that we need to do is to keep out the less desirable immigrants that the more desirable may come more rapidly. Immigration is essential to the growth and prosperity of our country."

"It is not true," said Col. N. F. Thompson, editor of The Tradesman of Chattanooga, Tenn., "that immigration is absolutely essential to the development of our resources. In 1860 the property of the South had an assessed valuation of $5,0000,000,000. The South has had no immigration, yet the receipts of the South last year, for spending money, were $6,000,000,000. Good immigrants we will receive with open hands. We don't want the bad ones under any consideration."

The interest of the conference had been excited to concert pitch. When Col. Thompson took his seat a dozen men tried to catch the eye of Lieut. Gov. West of Georgia, who was presiding. Jesse Taylor, an employe of a bank at Jamestown, Ohio, was recognized, and with his very first words threw a bomb into the ranks of the delegates.

"I don't own any skyscrapers or big factories," he began, "and I'm not the President of a College, but, as one of the plain people of Ohio, I tell you that we don't want the worms and riff-raff from Southern Europe you people have been sending out to us."

A burst of hisses almost drowned the

speaker's voice, to be lost in turn in a storm of cheers. Mr. Taylor was all unmindful of the result of his words.

"If you want to protect the products of American manufacturers, then protect the American workman from alien competition. If you don't, I give warning to you people who have been kiting around the country on the tail end of parlor cars and talking about the full dinner pail, that the American workingman will flop over against you, dinner pail, shirt, breeches, and all." [Cheers.]

Even the women in the galleries were excited, and there was turmoil on the floor of the hall. Mr. Taylor waved his arms above his head and kept right on.

"I am in earnest. I tell you from my heart that we don't want people brought over here who can live on two beers and a hard biscuit a day. They are not our kind of people."

When Mr. Taylor took his seat a dozen men went over and shook hands with him and asked his name. He was followed by several others, some on the one side and some on the other. Half a dozen resolutions were introduced, all of which were referred to the Committee on Resolutions without action.

Plea for the Jewish Immigrant.

The afternoon session was devoted to the subject of the selection of immigrants. There was a full attendance. Congressman Gardiner of Massachusetts was the first speaker. Others heard during the afternoon included Judge Lynn Harrison of Connecticut and Herman Wolf. Mr. Wolf made an eloquent plea for the Russian Jew. A previous speaker had suggested that a good plan to keep out undesirable immigrants would be to increase the head tax on immigrants to $40. Mr. Wolf asked how such a payment could be expected from the oppressed people of the Jewish race, who were, he said, victims of the Greek Catholic Church. He dealt, too, with the matter of naturalization.

The features of the morning session were the addresses of Archbishop Ireland of St. Paul, Andrew Carnegie, and Robert Watchorn, Commissioner of Immigration at Ellis Island. August Belmont was in the chair when the meeting was called to order, but upon the adoption of the report of the Committee on Rules and Programme, which named the permanent officers of the conference, he retired.

Ex-Mayor Seth Low was named as President and John H. Holliday of Indiana, Gov. W. D. Jelks of Alabama, Lieut. Gov. West of Georgia, Anthony Higgins of Delaware, Warren S. Stone of Ohio, and Frederick F. Judson of Missouri as Vice Presidents. In the absence of ex-Mayor Low, Mr. Harrison presided.

Robert Watchorn, the first speaker, told at length of the operation of the immigration laws, and suggested such changes as in his judgment were advisable. He made a special plea for a law which would prohibit ineligibles from crossing the ocean only to be deported. Prescott F. Hall of the Immigration Restriction League followed. He declared that the immigration laws were too lax.

Mr. Hall was bitterly attacked by Prof. Morris Loeb of New York University, who said that the spirit which animated Mr. Hall's organization was the same that whipped the Quakers out of New England and drove Roger Williams and Ann Hutchinson into the wilderness. He charged the Restriction League with the misrepresentation of facts in attempting to bolster up its cause.

Joke on Mr. Carnegie.

C. P. Smith of North Carolina, mindful of the fact that a time limit had been placed on speakers, placed his watch on a table when he stepped upon the stage. Andrew Carnegie was sitting behind the table, and looked at the watch.

"Don't pay any attention to it, Mr. Carnegie," the Southerner warned. "It's a filled case."

The delegates fairly roared with laughter, in which Mr. Carnegie joined. Mr. Smith declared that the whole trouble with the immigration system was that it was worked from the wrong end.

"Ellis Island ought to be on the other side of the ocean," he declared; "and then there would be none of those deportations which are so pitiful."

The speaker then entered upon a dissertation upon the civil war, being interrupted by the time limit before he reached his conclusions, and giving Mr. Carnegie, who followed, a chance to get back at him.

"Our country has more than one serious problem," Mr. Carnegie began, "but immigration is not one of them. Our last speaker understood this fully and gave us a magnificent oration on the civil war. [Laughter.]

"We have solved the question already by our somewhat too drastic law. I would like to ask what this country would have been without immigration? And who is there among us who is not either an immigrant himself or has immigrant blood in his veins? The immigration problem is not for us, but for the poor, unfortunate countries from which we are draining the best blood. Prince Bismarck declared that he would pass laws to keep the United States from draining the most valuable blood of Germany, and he was a wise man. [Applause.]

"Consider what you are getting, and you will think with me that we are getting all the best of the bargain. Thousands of these people had to save the money to come to this country, as my father had to do, and the only tests we should employ is the ambition of the immigrant to enjoy the rights of an American citizen, and his possession of the sobriety and frugality necessary to save the sum for his passage.

"If I owned America and was running it as a business operation, I would not only look for such men, but I would give them premiums to come over, and consider it the best bargain I ever made. [Applause.]

"Taking the average value of a man, woman, or child in this Republic, and fixing it as low as $1,000, the value of a slave in ante-bellum days, you are getting $400,000,000 cash value in immigrants every year. Every immigrant is a consumer. The great mistake of labor, if labor ever makes mistakes, [laughter,] is that a man who comes to this country to work injures other working men by doing so. Labor is one undivided whole, and every laborer, being a consumer, employs other labor. Somebody said something about purity of blood. It is not purity of blood you want, for it is the mingling of the different bloods that makes the American.

"Why all this agitation? We have laws now. I approve of them. We had none before, and the American of to-day is the result of unrestricted immigration. Is he a failure? Answer that yourselves."

Archbishop Ireland said that immigrants had made this the greatest Nation on earth. Of course, it was right to exclude the idiot, the criminal, and the diseased, but to shut out the man who came, attracted by the insthitutions of the country, to put his hand to the plow, would be to turn back upon the best traditions of the past. It would be a great mistake.

He declared that there was much unfounded prejudice against certain races from Southern Carnegie. [Hear! hear! came loudly from Mr. Carnegie,] saying that they made good citizens and seldom became dependents or criminals.

"We need all elements," he concluded, "to make up the American. The question of when shall the immigrant vote is one to be considered after he has landed."

Congressman Adams of Pennsylvania, Chairman of the House Committee on Immigration, said that his bill restricting the immigration from one country would keep the overflow from Italy, Austro-Hungary, and Russia within proper limits and would not affect other and more desirable nationalities.

December 8, 1905

JAPAN GRIEVED BY US.

Surprise Over Expulsion of Children from San Francisco Schools.

LONDON TIMES—NEW YORK TIMES.
Special Cable. Copyright, 1906.

TOKIO, Oct. 21.—The Japanese people, having long endured patiently and silently the anti-Japanese attitude of the Americans on the Pacific Slope, are now beginning to express profound regret and surprise at recent instances of discrimination, especially the expulsion of Japanese children from the public schools of San Francisco.

Some journals note the strange contrast between America's attitude toward Japan in the days of Perry, who, with the cannon's voice, proclaimed the doctrines of universal brotherhood and the common right of all nations to nature's gifts, and the attitude of a section of the Americans to-day, who violently advocate the expulsion of all Orientals from the American continent.

The leading journals, however, decline to regard the action on the Pacific Slope as an index of the great heart of the American Nation and declare that such unworthy and unmanly incidents cannot shake Japan's steadfast faith in her proved and constant friend America.

Special to The New York Times.

SAN FRANCISCO, Oct. 21.—Last week the Board of Education passed an ordinance requiring that all children of Oriental parentage—Chinese, Japanese, and Koreans—attend one school. The ordinance was adopted because many white parents objected to Japanese children attending the regular schools, as many Orientals are afflicted with trachoma. Before the fire Chinese children only were segregated.

Local Japanese object strenuously to this segregation, declaring it is due to racial prejudice, and they refuse to send their children to the school designated.

October 22, 1906

The Little Man from Nippon

A Californian's View of the Issue Raised by the Presence and Practices of Japanese in This Country.

By L. O'Connell.

ON Oct. 23 a cable was received at the Department of State from the United States Ambassador to Japan, which was dated "Tokio," and called to the attention of our Government the anti-American popular agitation throughout Japan consequent upon alleged discrimination in San Francisco against attendance of Japanese pupils at public schools. To this dispatch the Department of State cabled in reply:

"Troubles mentioned in your cable of 21st inst. are so entirely local and confined to San Francisco that this Government was not aware of their existence until publication in our newspapers of what happened in Tokio. * * * Trouble about schools appears to have arisen from fact that schools which Japanese had attended were destroyed at time of earthquake and have not been replaced. * * * The purely local and occasional nature of San Francisco school question should be appreciated, when Japanese are welcomed at schools and colleges all over our country. ROOT."

For a long time admission of Japanese to the public schools of San Francisco has been so vexed a question that in 1904 mass meetings were held in the metropolis of the Pacific to concert some scheme to change existing conditions. But State and Municipal politics intervened, and action as to exclusion of Japanese from white schools was postponed. The acute sense of unsuitability of indiscriminate co-education of pupils, the majority of whom were of the lowest class of Asiatics, the coolies, with children of American citizens, continued up to the time of the catastrophe which desolated San Francisco, and which for a time paralyzed all public reforms.

The most responsible citizens of San Francisco hold that the public schools are supported by the taxpayers of the United States, on the principle that education is necessary to good citizenship. These schools are not intended to train citizens for other nations—aliens, who desire to learn English merely to use for the commercial advancement of their own country, and, incidentally, for their personal enrichment.

If Japanese immigrants were independent individuals striving for advancement and intending to remain members of American communities it would be one thing; but the vast majority of immigrants from Japan come into our Pacific ports ticketed in gangs as if they were bunches of cattle, and are completely subject to the native bosses to whom they have been consigned by enterprising Japanese capitalists.

Between the children of polyglot Europe who attend public schools on the Atlantic Coast and Japanese pupils who have in the past and who desire in the present to crowd children of American citizens from the schools in San Francisco there is enormous difference. Teachers who labor in the Eastern States to instruct in English the children of German, Italian, and Scandinavian families are manufacturing the good American citizens of the future. But to train Japanese pupils in our language, in our systems of bookkeeping, and in our customs is another matter. The national policy of the land of Nippon is one of assimilation of all that is most admirable and progressive in foreign nations, all that may promote the glory of the new empire, which, emerging from long isolation, now ranks with the great world powers. To the man of Nippon the industries, arts, and scientific results of the civilized world are fruit from which he sucks the juicy essence, then casts away.

There is reason to believe that the extent to which the Japanese have "garrisoned" the United States, especially the Pacific Coast, with an army of students, artisans, and laborers, all enthusiastically loyal to the Emperor of Japan, and all amenable to prompt mobilization, and the resultant gravity of any friction between the two nations is not known to nor realized by the American public.

Any visitor to Tokio can see Japanese artisans making "Remington" typewriters and "Domestic" and "Singer" sewing machines, and putting the name "Maxim" on field guns, and affixing to a dynamo of well known pattern the legend, "made in Schenectady," with list of dates of all patents under which the genuine machine has been produced.

On the Pacific Coast Oriental immigrant labor at wages of from 80 cents to $1 per day was at first a benefit to American land owners. But under conditions which have prevailed increasingly for the past five years the Japanese laborer who comes to this country is not for direct hire to white men; he is to be approached only through his Japanese boss; is, in truth, merely the instrument whereby a capitalist at home in Japan fills a contract to supply so many days' labor for a certain price, (reaching constantly to a higher figure.)

In the Oriental quarter of any city or town on the Pacific Coast, from the Canadian boundary to the Mexican line, one may go to a shop and buy Japanese labor as one would buy sausages. After the bargain is struck, the price paid, (in advance,) the boss delivers the goods, free on board the wagon bound for the place of industry, and expects his gang to do as little work as possible. These brown men from Nippon, who in crowds labor in fields, vineyards, shops, and factories throughout the Far Western States, are the shrewdest and cheapest of all Asiatics. Their patriotism is proved by the languor of their efforts in the interest of white employers, and their willingness to work eighteen hours out of the twenty-four under a Japanese master.

One serious result to the community at large when Japanese gangs of coolies are employed in rural districts in the "emergency" fruit and crop-gathering seasons is that only the lowest class of whites will be co-laborers with Asiatics. The appearance of coolie gangs in the fruit valleys of California is the signal for an invasion of such tramp families as are accustomed to wander through the country—men, women, babies, dogs, and ragged clothing loaded into old prairie schooners. Their work of a few days in fields or on ranches is characterized by disappearance of small farm stock. When the Japanese boss gives his coolie gang the word to move on, these vagrants also take the road and continue their predatory march through the farming districts.

It is easy to find basement rooms where forty or more Japanese are at work with sewing machines in the manufacture of silk and muslin and lace waists and undergarments for women and children. These garments are largely ordered by the high-class department stores in the East. The machines are set as close as possible together; the walls of the basement are curtained off; behind the curtains are tiers of bunks where the workers sleep. In one corner is a cooking stove and a table, from which meals of raw fish and pickled turnips are eaten. At least forty Japanese at a time work, eat, and sleep in many such dens.

❈ ❈ ❈

In the agricultural districts, if a white employer demands more faithful labor and longer hours, the boss proclaims a boycott against him. Not an open boycott; the Japanese insidiously cripples the small fruit rancher or farmer. When the white man with several hundred acres of ripened fruit staring him in the face or with a time order hot off the Eastern wires for several carloads of produce, approaches an Asiatic employment agency, the boss may announce with a cunning grin:

"Boys all busy to-day," or, "No time to talk to-day; better go on."

The upshot of this crippling system is that at this writing hundreds of fruit ranches in California are under lease, or are actually owned by Japanese capitalists who never have seen and never intend to see the shores of America. In the City of San Francisco at the time of its destruction small industries, such as laundries, curio shops, and shops for making and selling ladies' waists and light dresses, shoe repairing, and barbering were virtually monopolized by Japanese. This state of affairs had arisen in five years.

There is in Japan a certain capitalist named Nishomura, who, having a few years since made a fortune in the shoe trade, turned his attention to activities in the United States, and sent to this country a small army of cobblers, whose bosses promptly districted San Francisco and the outlying country, with the effect that any newly arrived Japanese artisan who desires to ply the trade of cobbler must apply to Nishomura's agents, (one of whom can be found in every town and village,) who will indicate to the stranger a location where his competition will injure only white men.

It is the Japanese money lender who is the factor most to be dreaded in American communities. Of late the Japanese who have been some years in this land are showing a desire to acquire farms, held, be it understood, for the ultimate benefit, not of the laborer, nor even of the boss, but for the enrichment of some absentee Japanese captain of industry. A Japanese can go to the Japanese Bank which exists in every Japanese settlement, get a note discounted, and be ready with cash rent for a ranch or a sum for cash purchase of any small business whose owner is in financial straits.

⁂

When American families are crowded off farms where shall we get the new blood that vivifies the population of the cities? When the time comes that the farms are worked by Japanese coolies, and white land owners are non-resident; when tracts of the Republic are owned in Japan; when these tracts have become Japanese settlements where the population bear enthusiastic allegiance to the Emperor of Japan, then real trouble will begin for the United States of America.

December 2, 1906

PRESIDENT BARS JAPANESE COOLIES

Uses the Powers Recently Conferred on Him by the Immigration Act.

CALLS OFF SCHOOL SUITS

WASHINGTON, March 14.—President Roosevelt to-day issued an executive order directing that Japanese or Korean laborers, skilled and unskilled, who have received passports to go to Mexico, Canada, or Hawaii, and who come therefrom, be not allowed to enter the continental territory of the United States. This is practically the final chapter, except so far as the question may be taken up in treaty negotiations with Japan, in the issue growing out of the differences with that country over the action of the San Francisco authorities in prohibiting Japanese school children from attending the schools provided for the whites.

Authority to refuse permission to the classes of persons cited by the President to enter the continental territory of the United States is contained in the immigration bill, approved Feb. 20. It was incorporated in that measure at the request of the President and in fulfillment of a promise he made to Mayor Schmitz and the School Board of San Francisco in the course of their negotiations at the White House. On their side the San Franciscans promised to rescind their action on the school question, which promise has now been fulfilled. The President's order is as follows:

Whereas, by the act entitled "An act to regulate the immigration of aliens into the United States," approved Feb. 20, 1907, whenever the President is satisfied that passports issued by any foreign Government to its citizens to go to any country other than the United States or to any insular possession of the United States or to the Canal Zone are being used for the purpose of enabling the holders to come to the continental territory of the United States to the detriment of labor conditions therein, it is made the duty of the President to refuse to permit such citizens of the country issuing such passports to enter the continental territory of the United States from such country or from such insular possession or from the Canal Zone;

And whereas, upon sufficient evidence produced before me by the Department of Commerce and Labor, I am satisfied that passports issued by the Government of Japan to citizens of that country or Korea, and who are laborers, skilled or unskilled, to go to Mexico, to Canada, and to Hawaii, are being used for the purpose of enabling the holders thereof to come to the continental territory of the United States to the detriment of labor conditions therein,

I hereby order that such citizens of Japan or Korea, to wit, Japanese or Korean laborers, skilled and unskilled, who have received passports to go to Mexico, Canada, or Hawaii, and come therefrom, be refused permission to enter the continental territory of the United States.

It is further ordered that the Secretary of Commerce and Labor be, and he hereby is, directed to take through the Bureau of Immigration and Naturalization, such measures and to make and enforce such rules and regulations as may be necessary to carry this order into effect. THEODORE ROOSEVELT.
The White House, Washington, March 14, 1907.

Coincident with this order the President has directed the dismissal of the two suits filed in San Francisco at the direction of the Department of Justice, which had in view the testing of the question of the treaty rights of Japanese children to enter the white schools. This step the President had promised to take when the School Board rescinded its original action barring Japanese children from the white schools.

It is understood that the President is now prepared to drop the negotiations for a treaty of exclusion, which have been pending for some time. He is satisfied to wait a while and see what develops in the working of the order of to-day. If there is to be exclusion, the Japanese would prefer to have it come through an arrangement, by treaty or otherwise, with their Government to having it result from act of Congress.

There is no question that such an act would arouse great resentment throughout Japan, although the Japanese Government recognize the fact that it would be within the terms of the present treaty. The Japanese people would regard it as distinct discrimination against them, and would be correspondingly angry. The President is aware of that fact, and for that reason, if it develops that there is need for further action than the order issued to-day, he will undoubtedly resume negotiations for a treaty.

The Japanese Embassy had no comment to make on the President's order this evening. Attention was called, however, to the pacific character of the response of Viscount Hayashi, Minister of Foreign Affairs, to an interpellation on the subject in the Diet recently.

SAN FRANCISCO, March 14.—Late today nine little Japanese girls, who had applied for admission to the Redding Primary School this morning, were admitted after an examination as to their knowledge of the English language.

In accordance with instructions from Washington, United States Attorney Devlin to-day had a formal order of dismissal of the Japanese cases entered in the Federal court. Similar action was taken in the State Supreme Court.

March 15, 1907

BLAMES IMMIGRATION FOR THE BLACK HAND

Lieut. Petrosino Says We Are the Dumping Ground for the Criminals of Italy.

NO CENTRAL ORGANIZATION

Lieut. Petrosino, the Italian specialist of the Detective Bureau, upon whose squad is placed the onus of ferreting out crimes among the Italian population, said yesterday that the so-called "Black Hand" outrages in the city would never be stamped out until some restrictions are made in the admission of Italian immigrants to this country.

"The United States has become the dumping ground for all the criminals and banditti of Italy, Sicily, Sardinia and Calabria," said Lieut. Petrosino. "A little over a year ago the Government officials of Tunis decided to clean out the Italian Quarter of that city on account of the great number of crimes that were being committed there. A rigid investigation was conducted by the French Government, and as a result, over 10,000 men were deported. Where did they go to? Uncle Sam received them with open arms. Nearly every one of them came to this country and are now thriving on the spoils of their blackmailing and other conspiracies.

"First, it must be understood," continued the Italian sleuth, "that there is no big central organization of criminals called the 'Black Hand.' What we call the 'Black Hand' is simply an organization of ignorant and unscrupulous immigrants who have put themselves under the leadership of a man who is a little more intelligent than they are and was probably a bandit or criminal in Italy or Sicily. There may be five and there may be a dozen in the band, and there may be a dozen different bands working in the city at the same time. They have no connection with each other, and are in all probability bitter enemies and warring against each other. The idea that there is a big criminal club in this city called the 'Black Hand' is all a myth. It has grown out of the custom of the newspapers of calling every crime committed by an Italian a 'Black Hand' outrage.

"The system under which these gangs work is peculiar. They select some Italian who has come to this country and become prosperous, but he is almost always some one against whom they have a grudge for something that happened in Italy. Then many of the crimes are committed against former members of the gang who have fallen out with their partners for some reason or other. Many of the crimes have been explained in this way.

"Take for instance the famous 'shotgun murders' of a year or so ago. One night Giuseppe Guiliano, a saloon keeper, was going to bed at his home at 16 First Street when he was killed by a heavy load of buckshot which was fired through the window of his bedroom, and had evidently been aimed at his shadow on the drawn curtain. His brother Pietro, who lived in the same house, while being questioned by my detectives, made the most solemn oaths to avenge his brother's death. A little later it was learned that he and his dead brother were at the time conducting a blackmailing scheme against an Italian merchant in Frankfort, N. Y., and Pietro was convicted and sent to State's Prison for seven years.

"A little later Salvatore Sferlozza was shot in the same way in his saloon at 204 Forsythe Street. The crime was, after several months, finally traced to Giovanni and Giuseppe Pellettieri, but it was absolutely impossible to fix it on them definitely enough for a conviction. Their records were sent for from the little town in Sicily from which they had come, and their deportation resulted.

"Vincenzao Cantone came to this city about a year ago and established himself as a butcher at 117 Elizabeth Street. My detectives learned that he was at the head of a gang of blackmailers, and just as we had the evidence warranting an arrest, he was shot and killed by one of the gang. Later we learned that he was one of the men deported from Tunis, and that he had been suspected there of nearly twenty murders.

"One of the plans which I advocate to abolish these crimes is the establishment of a special bureau of inspectors by the Government for the examination of all Italian immigrants. It would be an easy thing to obtain from the Italian Government a description and record of all criminals who are suspected of having left for America, and with the aid of this the inspectors could prevent a great many of these men from ever entering the country.

"Then that part of the Italian population that is law-abiding and honest could do a great deal of good if it wanted to. It is not that the law-abiding Italians are afraid, but there is no concerted action. They could give the police much information if they wished which would aid them greatly in running down the criminals. The trouble is that every one is waiting for every one else to act first. If they would only form some organization in the nature of a secret service they could do much good in stamping out Italian crime. One of the chief obstacles that I and my men run up against is the difficulty of getting one Italian to testify against another. In many instances where a man has been blackmailed and we have almost run down the criminal, we are held up at the last minute by his disinclination to assist us.

"Another thing that should be done is to create a law forbidding more than one family to live in the same apartment. This would split up these gangs in a way. Then the pushcarts should be done away with. They are conducted by the lowest class of Italians, who are for the most part only confederates and informers for the gangs. Also the sale of explosives by Italians should be restricted, and licenses issued only to persons who can establish their integrity and honesty.

"Our Penal Code should also be made stronger. The trouble with the immigrants that come over here from Italy, Sicily, Sardinia, and Calabria is that they don't know what to do with the freedom that is given them. In the country from which they came the Penal Code was designed particularly to deal with their ignorance and hotheadedness. There they continually felt the heavy hand of the law on their shoulder. When they reached America they found these restrictions gone and they let their freedom get away with them."

The record of Petrosino's department in the last year was 1,000 arrests and 550 convictions, including one execution, one life sentence, and many long terms of imprisonment. Petrosino has only twenty-three men, eight in Harlem and fifteen at Headquarters, to contend with an Italian population in the city of nearly 500,000.

Petrosino is warm in his praise of Commissioner Bingham. "Since he has been in office," he says, "I have had a much easier time of it than I had before. Now I do not have politicians bothering me. It would surprise the public if it knew the names of some of the men in both parties who have come to me to intercede for some Italian criminals."

January 6, 1908

FUTURE AMERICANS WILL BE SWARTHY

Prof. Ripley Thinks Race Intermixture May Reproduce Remote Ancestral Type.

TO INUNDATE ANGLO-SAXON

Special to The New York Times.

BOSTON, Nov. 28.—In an article on "Races in the United States," appearing in The Atlantic Monthly for December, Prof. William Z. Ripley of Harvard University declares that from the abnormal intermixture of alien peoples in this country may result a reversion to a swarthy and black-eyed primitive type "running back to a time before the separation of European varieties of men began." This may settle "the live issue as to whether the first Europeans were long-headed or broad-headed, that is to say, Negroid or Asiatic in derivation." Prof. Ripley speaks of the primary physical brotherhood, not only of all branches of the white race, but of all races of men, as a "scientific probability." That is, the white, yellow, black, and red races differ "only in their degree of physical and mental evolution."

Prof. Ripley remarks upon the tendency of intermixed stocks to increase the percentage of female births. In the blending which is "sure to assume tremendous proportions" is an opportunity to study men in relation to this "great natural law." Moreover, in the "powerful process of social selection," which is already apparent, the Anglo-Saxon or Teutonic element is threatened with complete submergence. But while he foresees that the Anglo-Saxon stock, with its attenuated birthrate, will be "physically inundated by the engulfing flood" of foreigners, he trusts that "the torch of its civilization and ideals may still continue to illuminate the way." To nourish, uplift, and inspire the immigrant peoples of Europe is henceforth the "Anglo-Saxon's burden." Prof. Ripley's article is in part as follows:

"Will the future American two hundred years hence be better or worse, as a physical being, because of his mongrel origin? What chance is there that, out of this forcible dislocation and abnormal intermixture of all the peoples of the civilized world, there may emerge a physical type tending to revert to an ancestral one, older than any of the present European varieties? The law seems to be well supported elsewhere, that crossing between highly evolved varieties or types tends to bring about reversion to the original stock. The greater the divergence between the crossed varieties, the more powerful does the reversionary tendency become.

"Many of us are familiar with the evidence: such as the reversion among sheep to the primary dark type; and the emergence of the old wild blue rock pigeon from blending of the fantail and pouter or other varieties. The same law is borne out in the vegetable world, the facts being well known to fruit growers and horticulturists. The more recently acquired characteristics, especially those which are less fundamentally useful, are sloughed off; and the ancestral features common to all varieties emerge from dormancy into prominence.

"We are familiar, in certain isolated spots in Europe, the Dordogne in France for example, with the persistence of certain physical types without change from prehistoric times. The modern peasant is the proved direct descendant of the man of the stone age. But here is another mode of access to that primitive type, or even an older one, running back to a time before the separation of European varieties of men began.

"Thus, to be more specific, there can be little doubt that the primitive type of European was brunette, probably with black eyes and hair and a swarthy skin. Teutonic blondness is certainly an acquired trait, not very recent, to be sure, judged by historic standards, but as certainly not old, measured by evolutionary time. What probability is there that in the unions of rufous Irish and dark Italian types a reversion in favor of brunetteness may result?

"Anthropologists have waged bitter warfare for years over the live issue as to whether the first Europeans were long-headed or broad-headed; that is to say, Negroid or Asiatic in derivation. May not an interesting and valuable bit of evidence be found in the results of racial intermixture, as it is bound to occur in the United States?

"* * * More females at birth is the response of nature to an increasingly favorable environment or condition. In-and-in breeding is undoubtedly injurious to the welfare of any species. As such, according to Westermarck, it is accompanied by a decline in the proportion of females born.

A powerful process of social selection is apparently at work among us. Racial heterogeneity, due to the direct influx of foreigners in large numbers, is aggravated by their relatively high rate of reproduction after arrival, and, in many instances, by their surprisingly sustained tenacity of life, greatly exceeding that of the native-born American. Relative submergence of the domestic Anglo-Saxon stock is strongly indicated for the future. Our great philosopher, Benjamin Franklin, estimated six children to a normal American family in his day. The average at the present time is slightly above two. Even among the Irish, who are characterized nowadays everywhere by a low birth rate, the fruitfulness of the women is 50 per cent. greater than for the Massachusetts native-born. Is it any wonder that serious students contemplate the racial future of Anglo-Saxon America with some concern. * * *

"On the other hand, evidence is not lacking to show that in the second generation of these immigrant peoples a sharp and considerable, nay, in some cases a truly alarming, decrease in fruitfulness occurs. The crucial time among all our newcomers from Europe has always been in this second generation. The old customary ties and usages have been abruptly sundered, and new associations, restraints, and responsibilities have not yet been formed. Particularly is this true of the forces of family discipline and religion, as has already been observed. Until the coming of the Hun, the Italian, and the Slav, at least, it has been among the second generation of foreigners in America, rather than among the raw immigrants, that criminality has been most prevalent, and it is now becoming evident that it is this second generation in which the influence of democracy and of novel opportunity makes itself apparent in the sharp decline of fecundity.

"In some communities the Irish-Americans have a lower birth rate even than the native born. Dr. Engelmann, on the basis of a large practice, has shown that among the St. Louis Germans, the proportion of barren marriages is almost unprecedentedly high. Corroborative, although technically inconclusive, evidence from the registration reports of the State of Michigan appears in the following suggestive table, showing the nativity of parents and the number of children per marriage annually in each class:

	Children.
German father; American-born mother	2.5
American-born father; German mother	2.3
German father; German mother	6.0
American-born father; American-born mother	1.8

"* * * For three of the most crowded wards in New York City the death rate of the Irish was 36 per thousand; for the Germans, 22; for natives of the United States, 45, while for the Jews it was only 17 per thousand. By actuarial computation at these relative rates, starting at birth with two groups of 1,000 Jews and Americans, respectively, the chances would be that the first half of the Americans would die within 47 years, while for the Jews this would not occur until after 71 years. Social selection at that rate would be bound to produce very positive results in a century or two. * * *

"Great Britain has its 'white man's burden' to bear in India and Africa; we have ours to bear with the American negro and the Filipino. But an even greater responsibility with us, and with the people of Canada, is that of the 'Anglo-Saxon's burden'—so to nourish, uplift, and inspire all these immigrant peoples of Europe that in due course of time, even if the Anglo-Saxon stock be physically inundated by the engulfing flood, the torch of its civilization and ideals may still continue to illuminate the way."

November 29, 1908

"FREE WHITE PERSONS."

A. RUSTEM Bey, Chargé d'Affaires of the Turkish Embassy at Washington, has appealed directly to the American people against the ruling of Chief CAMPBELL of the Division of Naturalization, that excludes Turks, Armenians, Syrians, and Arabs from citizenship in this country. They are not "free white persons," Mr. CAMPBELL says arbitrarily. It is as remarkable that the Turkish diplomatic representative should send his signed statement to the press, as it is remarkable that he should take up cudgels in behalf of subjects who would renounce allegiance to the Turkish rule. But the Campbell doctrine that provoked this protest is the strangest of all.

Ethnically these races are Caucasian. Only the Turks retain traces of their Tartar ancestry. It is certain that the fathers, in prescribing that "free white persons" be eligible for citizenship, did not intend to legislate against Asiatics. They knew little or nothing about them in 1790, when the original act was passed, therefore they could not have meant to exclude these future immigrants. The Department of Justice has instructed the Federal District Attorneys to take no further action in such cases until further advised. The Washington correspondent of The Boston Transcript reports that the State Department is not at all in sympathy with Mr. CAMPBELL'S ruling, which puts a stigma on the Ottoman peoples, and regards it as seriously interfering with our foreign trade relations. Doubtless the Turkish Chargé d'Affaires was acquainted with Secretary KNOX'S views before sending his statement to the press.

November 7, 1909

SAYS TIME HAS COME TO HALT IMMIGRATION

Commission, in Forty Volume Report, Declares It No Longer Is Necessary.

TENDS TO LOWER WAGES

Aliens Come to Better Their Condition, Not to Escape Tyranny, and Sentiment Should Be Waived, Is Asserted.

WASHINGTON, Dec. 5.—Sentimental considerations in restricting immigration should be waived in view of the economic problems arising from adverse effects on wages and living conditions which the large number of aliens have had in recent years by their entry into basic industries, according to the final report of the Immigration Commission transmitted to-day to Congress. The commission unanimously urges the restriction of unskilled labor immigration.

The commission, created under the Immigration act of 1907, expired to-day, and the results of its three-year investigation into practically all phases of the immigration question make about forty printed volumes. Says the report:

> The present immigration movement is in large measure due to economic causes, but emigration from Europe is not now an absolute economic necessity, and as a rule those who immigrate to the United States are impelled by a desire for better conditions rather than by the necessity of escaping from intolerable ones. This fact should largely modify the natural incentive to treat that immigration movement from the standpoint of sentiment and permit its consideration primarily as an economic problem.

The commission presents several proposals by which restriction of immigration might be affected, including a reading and writing test, the exclusion of unmarried unskilled laborers, limitations in the number arriving at any one port and from particular races, as well as in the amounts of money in their possession on arrival. All the members of the commission do not concur in the feasibility of the reading and writing test.

The immigrants now coming, it is agreed, do not furnish any more criminals or subects for charity than the native born, but the tendency toward industrial and city life remains unchanged, in spite of the fact that statistics show the immigrants to have been more successful in agricultural pursuits.

In connection with these findings, the report urges that the division of information in the Bureau of Immigration be so conducted as to co-operate with the States and various societies in a more beneficial distribution of immigrants among agricultural sections, where they especially are needed.

Efforts to exclude all British East Indians through an agreement with Great Britain, the continuance of the present Chinese exclusion laws as well as present regulations with regard to Japanese and Korean immigration, the passage of the House bill for the deportation of alien criminals, with changes in the immigration law to make it applicable to alien seamen, and the appointment of an additional Assistant Secretary of Commerce and Labor to determine immigration appeals are recommended.

The commission, which has been sitting continually since Nov. 15 last in reviewing the report, consisted of Senator Dillingham of Vermont, Chairman; Senators Henry Cabot Lodge of Massachusetts and Leroy Percy of Mississippi, Representatives Bennet of New York, Benjamin F. Howell of New Jersey, and John L. Burnett of Alabama; Prof. J. W. Jenks of Cornell University, and William R. Wheeler of California, formerly Assistant Secretary of Commerce and Labor.

December 6, 1910

IMMIGRATION BILL VETO AT LAST MINUTE

Taft Accepts Nagel's Objections to Reading Test—Unsuited to Our Institutions, He Says.

MAY OVERRIDE PRESIDENT

Lodge Proposes That the Senate Repass the Bill, Which Will Probably Be Done—Outlook in House.

Special to The New York Times.

WASHINGTON, Feb. 14.—In the last few hours allowed him by the Constitution, President Taft decided to-day against the Immigration bill and vetoed it in a brief message to the Senate. There, as soon as it was read, Senator Henry Cabot Lodge, Chairman of the Immigration Committee, after a conference with leaders of both parties in the House and Senate, announced that when the pending business was out of the way he would move to pass the bill, "the objections of the President of the United St..., to the contrary notwithstanding." It is expeced hat this effort will be made on Monday.

The reading test for would-be immigrants is the one reason mentioned by the President for his disapproval, and in the Senate the opinion was freely expressed this afternoon that the motion to override the veto would easily command the needed two-thirds vote. In the House there is greater doubt although Mr. Lodge's announcement was made with the approval of Congressman John L. Burnett of Alabama, Chairman of the House Committee on Immigration and Naturalization.

The measure had such strength in the Senate that it passed without roll calls. But in the House, at one stage of the proceedings, a motion to recommit the bill was defeated only by 149 to 70—or 9 more than a bare two-thirds, and a change of four votes on that motion would have shown the measure with less than two-thirds of the House behind it.

It has always been supposed that a motion to override the President's veto brings to the side of the Administration many members of the President's party, who otherwise would vote against him. But that is hardly true in this case. The President will retire into private life in eighteen days, and his influence is at low ebb. Besides, the circumstances of the present veto have greatly angered even the President's supporters in both Houses. Senator Lodge is a close friend of the President, as are many of the Senators who most cordially supported the Immigration bill and they will lead the fight against the veto.

Up to the very last they hoped that Mr. Taft would sign the measure. At the Cabinet meeting this morning the President was still undecided, and even when the Cabinet adjourned after discussing the bill for more than two hours, his mind was not made up.

Taft Swayed by Nagel.

The Cabinet itself appeared about equally divided. In the end the President accepted the reasoning presented to him in a letter from the Secretary of Commerce and Labor, Mr. Nagel, and that letter, appended to his brief message, constituted the real criticism of the measure.

The President's message contains little more than one hundred words, written on one sheet of paper in the President's own handwriting. This message the President wrote at his desk some time after the Cabinet meeting broke up. If the President had waited until after the Senate had adjourned this evening the bill would have become law without his signature.

His message, short as it is, indicates what a struggle there was in the President's mind as to the course he should follow. The large votes behind the measure in both houses, he said, and the able presentation of the bill's merits by delegations before him, had great weight, but he could not make up his mind to sign a bill out of harmony with American institutions. He explained that statement as having reference to the literacy test and referred to Mr. Nagel's letter in elaboration of his view.

The President's message reads:

I do this with great reluctance. The bill contains many valuable amendments to the present immigration law, which will insure greater certainty in excluding undesirable immigrants. The bill received strong support in both houses, and was recommended by an able commission after an extended investigation and carefully drawn conclusions.

But I cannot make up my mind to sign a bill which in its chief provision violates a principle that was, in my opinion, to be upheld in dealing with our immigration. I refer to the literacy test. For the reasons stated in Secretary Nagel's letter to me, I cannot approve that test. The Secretary's letter accompanies this.

Like the President, Mr. Nagel based his final objection to the bill on the sole ground of the literacy test. He mentions five other objections, of which a good deal has been made, but dismisses most of them with the remark that, while unsatisfactory, those clauses leave the law pretty much where it is. His minor objections have to do with the inclusion of Hawaii within the terms of the bill; the vague language of the provisions excluding aliens capable of assuming citizenship; the provision whereby the Secretary of Commerce and Labor's decision, admitting skilled labor under contract, may be submitted to a review by the courts in advance of affirmative action; the provision for detailing matrons and inspectors on immigrant vessels, and the drastic power in the hands of the Secretary for penalizing ship companies that solicit immigrants.

Objections to Literacy Test.

Mr. Nagel's objections to the literacy test are as follows:

In my opinion this is a provision of controlling importance, not only because of the immediate effect which it may have upon immigration, and the embarrassment and cost it may impose upon the service, but because it involves a principle of far-reaching consequences, with respect to which our attitude will be regarded with profound interest.

The provision as it now appears will require careful reading. In some measure the group system is adopted; that is, one qualified immigrant may bring in certain members of his family, but the effect seems to be that a qualified alien may bring in members of his family who may themselves be disqualified, whereas a disqualified member would exclude all dependent members of his family, no matter how well qualified they might otherwise be. In other words, a father who can read a dialect might bring in an entire family of absolutely illiterate people, barring his sons over 16 years of age, whereas, a father who can not read a dialect would bring about the exclusion of his entire family, although every one of them can read and write. Furthermore, the distinction in favor of the female members of the family as against the male members, does not seem to me to rest upon sound reasons. Sentimentally, of course, it appeals, but industrially considered, it does not appear to me that the distinction is sound. Furthermore, there is no provision for the admission of aliens who have been domiciled here and who have simply gone abroad for a visit. The test would absolutely exclude them upon return.

In the administration of this law very considerable embarrassment will be experienced. This at least is the judgment of members of the immigration force, upon whose recommendations I rely. Delay will necessarily ensue at all ports, but on the borders of Canada and Mexico that delay will almost necessarily result in great friction and constant complaint. Furthermore, the force will have to be very considerably increased, and the appropriation will probably be in excess of present sums expended by as much as a million dollars. The force of interpreters will have to be largely increased, and, practically speaking, the bureau will have to be in a position to have an interpreter for any kind of language or dialect of the world at any port at any time. Finally, the interpreters will necessarily be foreigners, and with respect to only a very few of the languages or dialects will it be possible for the officials in charge to exercise anything like supervision.

Provision Indefensive.

Apart from these considerations, I am of the opinion that this provision cannot be defended upon its merits. It was originally urged as a selective test. For some time recommendations in its support upon that ground have been brought to our attention. The matter has been considered from that point of view, and I became completely satisfied that upon that ground the test could not be sustained.

The older argument is now abandoned, and in the later conference, at least, the ground is taken that the provision is to be defended as a practical measure to exclude a large proportion of undesirable immigrants from certain countries. The measure proposes to reach this result by indirection, and is defended purely upon the ground of practical policy, the final purpose being to reduce the quantity of cheap labor in this country.

I cannot accept this argument. No doubt the law would exclude a considerable percentage of immigration from Southern Italy, among the Poles, the Mexicans, and the Greeks. This exclusion would embrace probably in large part undesirable but also a great many desirable people, and the embarrassment, expense, and distress to those who seek to enter would be out of all proportion to any good that can possibly be promised for this measure.

My observation leads me to the conclusion that, so far as the merits of the individual immigrant are concerned, the test is altogether overestimated. The people who come from the countries named are frequently illiterate because opportunities have been denied them. The oppression with which these people have to contend in modern times is not religious, but it consists of a denial of the opportunity to acquire reading and writing. Frequently the attempt to learn to read and write the language of the particular people is discouraged by the Government, and these immigrants in coming to our shores are really striving to free themselves from the conditions under which they have been compelled to live.

So far as the industrial conditions are concerned, I think the question has been superficially considered. We need labor in this country, and the natives are unwilling to do the work which the aliens come over to do. It is perfectly true that in a few cities and localities there are congested conditions. It is equally true that in very much larger areas we are practically without help.

Immigrant Distribution Bad.

In my judgment no sufficiently earnest and intelligent effort has been made to bring our wants and our supply together, and so far the same forces that give the chief support to this provision of the new bill have stubbornly resisted any effort looking to an intelligent distribution of new immigration to meet the needs of our vast country. In my judgment no such drastic measure, based upon a ground which is untrue and urged for a reason which we are unwilling to assert, should be adopted until we have at least exhausted the possibilities of a rational distribution of these new forces.

Furthermore, there is a misapprehension as to the character of the people who come over here to remain. It is true that in certain localities newly arrived aliens live under deplorable conditions. Just as much may be said of certain localities that have been inhabited for a hundred years by natives of this country. These are not the general conditions, but they are the exceptions. It is true that a very considerable portion of immigrants do not come to remain, but return after they have acquired some means, or because they find themselves unable to cope with the conditions of a new and aggressive country. Those who return for the latter reason relieve us, of their own volition, of a burden. Those who return after they have acquired some means certainly must be admitted to have left with us a consideration for the advantage which they have enjoyed. A careful examination of the character of the people who come to stay and of the employment in which a large part of the new immigration is engaged will, in my judgment, dispel the apprehension which many of our people entertain.

The census will disclose that with rapid strides the foreign-born citizen is acquiring the farm lands of this country. Even if the foreign-born alone is considered, the percentage of his ownership is assuming a proportion that ought to attract the attention of the native citizens. If the second generation is included it is safe to say that in the Middle West and West a majority of the farms are to-day owned by foreign-born people or they are descendants of the first generation. This does not embrace only the Germans and the Scandinavians, but is true in large measure, for illustration, of the Bohemians and the Poles. It is true in surprising measure of the Italians, not only of the northern Italians, but of the southern.

Aliens Who Return.

Again, an examination of the aliens who come to stay is of great significance. During the last fiscal year 838,172 aliens came to our shores, although the net immigration of the year was only a trifle above 400,000. But, while we received of skilled labor 127,016, and only 35,896 returned, we received servants, 116,529, and only 13,499 returned; we received farm laborers, 184,154, and only 3,978 returned, it appears that laborers came in the number, 135,726, while 209,279 returned. Those figures ought to demonstrate that we get substantially what we most need, and what we cannot ourselves supply, and that we get rid of what we least need and what seems to furnish, in the minds of many, the chief justification for the bill now under consideration.

The census returns show conclusively that the importance of illiteracy among aliens is overestimated, and that these people are prompt, after their arrival, to avail of the opportunities which this country affords. While, according to the reports of the Bureau of Immigration, about 25 per cent. of the incoming aliens are illiterate, the census shows that among the foreign-born people of such States as New York and Massachusetts, where most of the congestion complained of has taken place, the proportion of illiteracy represents only about 13 per cent.

Mr. Nagel was persuaded, he said in conclusion, that the provision was in principle of very great consequence, and that it was based upon a fallacy in undertaking to apply a test which was not calculated to reach the truth and to find relief from a danger which really did not exist. He did not think it admitted of compromise, and, much as he regretted to do so, because he deemed the other provisions of the measure as I most respects excellent and in no respect really objectionable, he was forced to advise the President not to approve the bill.

February 15, 1913

ARIZONA ALIEN BILL OUTDOES CALIFORNIA

Lower House Passes Act Directed at Foreigners Who Fail to Seek Citizenship.

Special to The New York Times.

PHOENIX, Ariz., May 5.—The lower house of the Arizona Legislature to-day passed an alien land ownership bill that is far more drastic than that just adopted by California. The bill prohibits any alien who has not declared his intention of becoming a citizen from acquiring title to land.

Opponents of the bill were able to muster only six votes. These members proposed to exempt aliens who, though eligible to citizenship, did not wish to forswear allegiance to their native land.

The bill was sent to the Senate and probably will be passed at once. It embodies in the code the law passed a year ago.

Japanese residents held a conference to-night and decided to file an unofficial protest with Secretary of State Osborn to-morrow against the adoption of the bill. At the same time a call was issued through the Acting President of the Japanese Association for a mass meeting of all the Japanese in Arizona, said to number upward of 1,400. It will be held next Sunday, at which time it is believed a formal protest will be drafted.

The Japanese truck farmers and small ranchers are indignant over the bill. N. Oyo, President of the Arizona Japanese Association, is in California on his way to Japan with the body of his wife, who died here last week, and Henry Ninifiki, one of the moving spirits in the organization, is in jail in Globe on a charge of killing an aged American woman there.

In the absence of these leaders the local Japanese were at sea in the matter until some of them got together to-night. They said that before Oyo started for Japan last Saturday night he was authorized to lay the sentiment of the Japanese in this State before the Government at Tokio. Whether this will take the form of a request for a revision of the existing law in Arizona or whether it will be made in behalf of their countrymen in California the members of the association here declined to say.

May 6, 1913

CENSUS FIGURES DISCLOSE GRAVE RACIAL PROBLEM

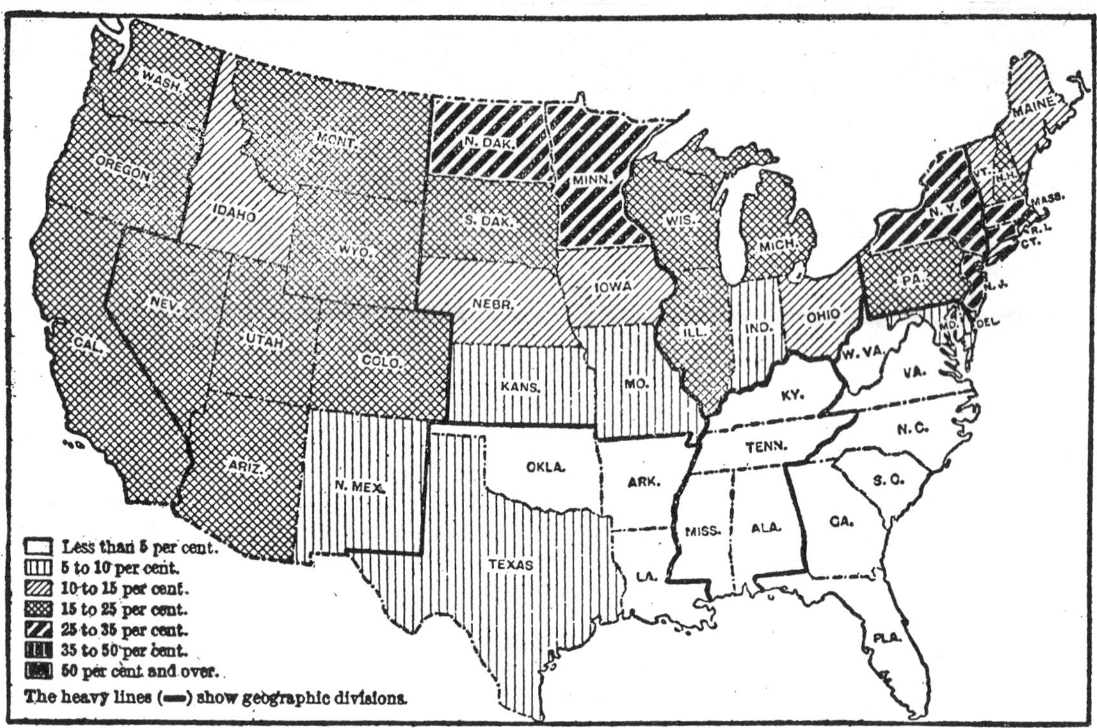

Percentage of Foreign-Born Whites in the Total Population, 1910.

By William Z. Ripley.

Professor of Economics, Harvard University.

THE racial composition of the population of the United States is a human phenomenon unique in history. Since 1820 about 30,000,000 people have come to America from all parts of Europe. This is a number about equal to the entire population of the British Isles at the time of our civil war. It is a number greater than the population of all Italy in the time of Garibaldi. It is a number, by itself alone, sufficient to populate, as it stood in 1910, all of New England and the Middle Atlantic States with Ohio thrown in to make good measure. The people of this great land, unlike those of the rest of the world, have thus been dropped upon this continent, if not from the skies, at least from every corner of Europe. In no sense are they indigenous to the soil. The population product is an artificial and exotic one. Never before has such an experiment in ethnic and social intermixture been attempted. Our future, not only as a race of men but as a nation, depends upon its outcome. Consequently no scrap of information throwing light upon the situation should escape the most careful scrutiny and the widest publicity. In this respect the mass of material concerning birthplace, nativity, and parentage in the recently published abstract of the Federal census of 1910 merits examination in some detail.

Some conception of the significance of the great influx of foreigners in recent years may be had by comparing it with the total immigration since 1819. According to the recent Federal Commission report on the subject 27,800,000 aliens were admitted to the United States during the ninety years to June 30, 1910. Our census report shows that 8,000,000 of these arrived during the last ten years of this period. Even allowing for the considerable number of transients, a phenomenon assuming such proportions, even had there been no change in the character of the immigrants, would occasion grave concern. Moreover, according to this census report, the second half of the last decade fully warranted the impression of the gravity of the situation. About one-fifth of our present total of foreign-born arrived in America between January, 1906, and April 15, 1910. The percentage of recent arrivals is much higher than this in all of the great cities and industrial centres. New York, for example, has derived almost one-fourth of its very high proportion of foreign-born from an equally recent date. Such facts as these are of the utmost importance to the political and social welfare of the communities concerned.

Menace to Our Civilization.

For the past decade all that we have known concerning changes in the racial composition of our population has been derived from the annual records of immigration. But this bulky Census volume, after an interval of ten years, affords an opportunity for once more taking account of our actual human stock in trade. The results should arouse Congress and the people at large to the menace to our Anglo-Saxon civilization presented by the horde of imperfectly Americanized human beings from whom our future citizenship in a large measure is to be derived. It is difficult to overestimate the significance for the future of democracy in America of the figures contained in this document.

Of about 92,000,000 inhabitants of the United States in 1910, 13,500,000, or 14.7 per cent., were born in some foreign country; that is to say, they are not native born American citizens. The corresponding proportion was 13.6 per cent. in 1900. There were then 10,300,000 foreign born in the United States. These figures, taken ten years apart, are not at first sight startling; nor do they seem to indicate any considerable change in conditions since our last official stock taking. We have so gradually become accustomed to them by a half century's experience with one-seventh of our people aliens by birth that we have lost sight of their significance. Mere percentages have, it is true, not greatly changed of late. But an important difference is to be noted, nevertheless, as the years go by. These latest immigrants, while relatively little more numerous, are now increasing absolutely at an alarming rate by comparison with our native population; and, of course, the children of these recent immigrants, themselves only half Americanized, are growing in numbers even more rapidly. Succinctly stated, the record of the last ten years is as follows: The total population of the country increased 21 per cent.; the native whites of native parentage by 20.8 per cent., while the number of foreign-born increased by 30.7 per cent. In other words, the rate of growth of our foreign-born has been about one and one-half times the rate of increase of the population claiming American citizenship by right of birth.

During the decade under consideration about 16,000,000 people in all were added to our numbers—this increment consisting of about 12,000,000 native whites, 1,000,000 negroes, and 3,175,000 foreigners.

This recent accession of aliens during the decade has brought the total foreign-born in 1910 to the formidable figure already stated. In other words, the foreign-born now number approximately double the population of all the New England States and exceed the entire population, white and negro, of that quarter of our territory east of the Mississippi and south of Mason and Dixon's line. Otherwise stated, there were more foreign-born in the United States in 1910 than the entire population of the country in 1830.

Had there been no change in the source and character of this immigration—a topic to be considered shortly —**the mere massiveness of the phenomenon, above described, would by itself constitute a grave source of danger. We were able in the past to assimilate the Irish immigrants of the potato-famine days and the Germans of a somewhat later date, not only because their antecedents were akin to our own, but also because their numbers were such that the influences of American life played directly upon each individual. In other words, they were not insulated by reason of their own great mass. But the present day condition is that of aliens congregated in such vast numbers that they promise, if not indefinitely at least for yet another generation, to preserve their former standards of living and of tradition in our midst, relatively untouched by immersion in the tide of American life.**

In other words, the facts of this Census make it clear that these alien colonies have become so large and, as we shall see, so compact that they threaten to remain quite undigested lumps of imperfect civilization in our midst.

Serious as the situation is, as revealed by the foregoing figures, further analysis is even more disquieting. According to the record of this census three-eighths of this army of 13,500,000 foreign-born came to us within the last ten-year period. The proportion of such newcomers in 1910, in other words, was half as large again as in 1900. Some 8,000,000 persons in all immigrated within the decade under review, but about three-eighths of that number either returned to Europe or died. The net gain from immigration for the ten years was thus 5,000,000. Yet, as we have already seen, the total addition to the number of foreign-born was only 3,175,000, according to the official count. How may we explain the apparent anomaly? Evidently we have to do with a case of substitution in a large measure of a new element of aliens for those who were already here in 1900. The ranks of the foreign-born ten years ago have, of course, been considerably reduced by death. But so large has been the immigration that the new aliens have not only filled up the ranks, but have, as we have seen, added some 3,000,000 more. We have to do, therefore, not with mere addition, but with a case of downright substitution. The significance of this substitution, when we consider the change in the sources of our immigration, will shortly appear.

Next to the size and recency of these foreign colonies their compactness is to be noted. The real menace to our civilization presented by the rude shock of this influx of aliens becomes more apparent when its concentration by States and cities is considered. The problem of assimilation would be difficult enough were they to be scattered broadcast throughout the land. But the census shows this to be far from the case. New England in this regard, with over one-fourth of its population foreign-born, is most seriously affected. In the States of Massachusetts and Rhode Island about one-third of the population is now foreign-born white. Bristol County, Massachusetts, stands at the head, with 37.6 per cent. of its population foreign-born. The accompanying map shows the distribution by States of the foreign-born whites. Two groups of concentration of aliens by birth exceeding one-quarter of the total population are discernible. These are, as we shall see, radically different in type. The eastern group, comprising the industrial centres between New York, Boston, and Buffalo, is recruited from an entirely different source from the western one centring in Minnesota and North Dakota. But it is apparent at a glance that a goodly proportion of the United States is characterized by an alien contingent which equals more than one-fifth of the total population.

Most imminent of all, of course, is the danger from concentration of these foreign-born in our urban centres and industrial districts. Almost three-fourths of the aliens in 1910 are found in cities, while less than half of the total population of the United States is urban in situs. Over five-sixths of our Russians, (Jewish,) Rumanians, Turks, and Irish; over three-fourths of our French Canadians, Italians, and Hungarians, and over seven-tenths of our English, Scotch, Austrians, and Greeks are located in cities. And of these urban centres the peculiarly industrial ones, large or small, are a focus of attraction.

The Situation in New York.

The City of New York (proper) naturally occupies first place as a foreign centre. Of its 4,770,000 inhabitants in 1910 1,930,000 were of foreign birth; 670,000 of these, or about one-third, having been added during the decade under review. Here is a colony comprising about one-seventh of the total foreign-born in the United States, resulting in a community of two-fifths foreign-born, about two-fifths native-born of foreign parents, and only one-fifth native whites of native parentage. Moreover, one-quarter of its approximately 2,000,000 aliens have immigrated since 1900, and about one-half of its total growth during the decade came from this source. Conditions in Manhattan Borough are even more extreme. Almost one-half (47.4 per cent.) of its inhabitants in 1910 were foreign-born whites, and less than 15 per cent. were native whites of native parentage. We shall soon present a table of cities with more than one-third of their inhabitants foreign-born. This is deeply significant. For it is evidently not a matter of size which determines the strength of this infusion of foreign blood. Amsterdam, N. Y.; Waterbury, Conn.; Lewiston, Me., or Woonsocket, R. I., present just as extreme instances of foreign birth as do the larger cities of the country. In fact, the great cities, with the exception of Manhattan, come along nearer the foot of the list, as will be seen.

The second significant feature of recent developments in respect of the racial composition of our popuplation is neither the absolute nor the relative number of foreigners, but the change in their character and quality. Even a big lump of aliens might be assimilated to American standards, although, perhaps, with difficulty, were it not that these later contingents are of an entirely different and inferior type.

This striking change may be expressed in another way by the statement that our foreign-born from Southern and Eastern Europe has more than doubled its proportion of the whole foreign-born during the decade. With upward of 3,000,000 more foreign-born in 1910, there was actually 250,000 fewer persons born in Northwestern Europe then resident within our borders. This decrease is wholly in the numbers from Germany, Ireland, and Wales. Fortunately, we still have a slightly larger number of English, Scotch, and Scandinavians—those sturdy and ever welcome additions to our host—than we had in 1900. **But whereas almost one-third of our** foreign-born in 1860 were Germans, they constituted only 18.5 per cent. of this class in 1910. In other words, the German contingent has been cut in halves. The Irish have shrunk in numbers proportionately within fifty years by about three-quarters. Fewer immigrants, in fact, were reported in 1910 as born in Ireland than at any census since 1860. Even Boston, close second to Dublin as a Celtic centre, had fewer Irish born than in 1900.

The disconcerting fact is the enormous increase in the foreign-born from the less favored parts of Europe: It is the shift of the centre of gravity of immigration from Northwestern to Southeastern Europe which is the striking feature. Down to 1883, 95 per cent. of our foreign-born came from north and west of the Alps; that is to say, even in 1900, about two-thirds (67.8 per cent.) of our foreign-born were from those parts of Europe closely allied to us by similarity of institutions and traditions. But now after ten years this last census shows as a result of the change in the sources of immigration less than one-half (49.9 per cent.) of this great horde of foreign-born originated in this eminently favored geographical and cultural region.

The phenomenon of change in racial composition may be best illustrated by the critical case of New York City proper. The following table brings out the sources of its recent growth. It is evident that almost one-half of its total increment of 670,000 aliens during the decade were of **Russian, that is to say, mainly Jewish,** origin. Add to this contingent 200,000 Italians, 100,000 Austrians, (mainly Slavs,) and about half as many Hungarians, then subtract a substantial number of Germans and Irish, and you have the net result:

	1900.	1910.	Inc.
Russian (Jewish)	180,000	484,000	304,000
Italian	145,000	340,000	195,000
German	324,000	278,000	—46,000
Irish	275,000	252,000	—23,000
Austrian	90,000	190,000	100,000
Hungarian	32,000	77,000	45,000

These substantial additions to the foreign colonies in New York serve to make it a larger Italian centre than even the Eternal City proper, and a bigger Jewish colony than any to be found elsewhere in the world. One has to turn to other great cities for similar instances of national prominence. Boston, of course, still remains pre-eminently Irish, although, measured by actual foreign birth, it has fewer Irish citizens than in 1900. Chicago still has a larger number of Germans than of any other nationality, although the Jewish and Slavic colonies are rapidly growing. During the decade it added about 200,000 foreign-born to produce an alien colony of 783,000 people. The Slavs, Jews, and Italians constituted practically all of this growth. The number of Germans, English, and Irish has substantially decreased.

The full effect of recent tendencies is revealed only when one considers not only the number of foreign-born, but of their immediate descendants as well. The census permits us to trace ancestry only through the first generation; that is to say, it distinguishes the native-born in two classes, one born of native parents and the other of foreign-born or mixed parentage. Aggregating the foreign-born and their children and setting them over against the native-born of native parentage, the results are striking. For the United States as a whole in 1910 these foreign-born with their children amounted to upward of one-third of our population (35 per cent.).

This proportion has not changed greatly for the United States as a whole since 1900. It is only when one considers the areas of concentration that the new situation develops. It then appears that Massachusetts and Rhode Island, for example, now have more than two-thirds of their population of this alien or recent American origin. And New York and Connecticut follow not far behind. But the most serious situation is presented, aside from the metropolis of New York, not in the great cities so much as in those of the second class. The facts in this regard are brought out in the following table showing the population allocated in three groups according to birth:

PER CENT. OF POPULATION BY BIRTH.

	Foreign Born White.	White. Foreign or Mixed Parent.	Native, Native Parent.
Passaic, N. J.	52.0	33.2	13.8
Lawrence, Mass.	48.1	37.9	13.6
Manhattan Borough	47.4	38.2	19.3
Perth Amboy, N. J.	44.5	39.1	15.9
New Bedford, Mass.	44.1	33.5	19.4
Woonsocket, R. I.	43.4	41.6	15.0
Fall River, Mass.	42.6	43.7	13.5
Manchester, N. H.	42.4	34.5	23.0
New Britain, Conn.	41.0	38.8	19.9
Lowell, Mass.	40.9	39.5	19.5
Shenandoah, Penn.	40.6	41.9	17.5
New York	40.4	38.2	19.3
Holyoke, Mass.	40.3	43.8	15.8
Hoboken, N. J.	39.3	41.3	19.1
Chicopee, Mass.	39.5	42.2	18.2
Duluth, Minn.	39.1	40.6	19.7
Chicago, Ill.	35.7	41.8	20.4
Boston, Mass.	35.9	38.3	23.5
San Francisco	31.4	36.9	27.7
Pittsburgh	26.3	35.9	33.0
Philadelphia	24.7	32.1	*37.7
Milwaukee	29.8	48.8	21.1

*5.5 negro.

The right-hand column in the foregoing table gives a list of cities in which, with only a few exceptions, *less than one-fifth of the population is native born of native parents,* and, of course, many of these would have foreign-born grandparents. In only two of these cities is the proportion of fully Americanized stock above one-quarter. And it is to this already heavily diluted American stock that we are adding raw immigrants from Southern and Eastern Europe at the rate already described. Is it any wonder that the Industrial Workers of the World prosper in Lawrence, Mass., and other industrial centres like Passaic and Paterson, N. J., or that politics seem almost incurably corrupt elsewhere? Or turn to the left-hand column of foreign-born whites and note in how many instances the proportion of pure aliens lies above 40 per cent. Indeed, New York's proportion is quite modest by comparison with those smaller places which precede it on the list. This table ought to be placarded on every bulletin board in the United States and shown continuously by stereopticon in the halls of Congress during the debates upon legislative measures for stemming the tide of immigration. Possibly it might arouse the nation to the real situation.

Are our different nationalities tending to coalesce in the production of what may turn out to be a hybrid physical type? The degree to which the sharp divisions of nationality are being broken down by intermarriage is extremely difficult to determine. The census throws but little light upon existing conditions. This arises primarily from the fact that after the first generation each foreign stock loses its identity in the great mass of native-born of foreign parentage. And of course it is through this class that amalgamation takes place. That is to say, the foreign-born male member of a colony predominately masculine most commonly finds a wife, if not among foreign-born women of his own type, then among the second generation of that same foreign colony born in the United States. And this choice at marriage is usually exerted upward in the social scale. That is to say, to be specific, the young Russian Jew, having made his way in the world, more commonly seeks a wife born in America of Jewish parentage than does the immigrant Jewish woman find a mate among the young men born in America of Russian parentage. In other words, the man being the active agent in matrimony, inclines to choose from a social station higher than his own; and it is a happy circumstance that Americanization and a rise in the social scale are synonymous terms.

Some Interesting Figures.

Seeking for light upon the degree of this intermixture, as indicated by statistics of mixed parentage, both the size of the colony and the length of time it has been established in America must be taken into account. Naturally, among the Germans, whose greatest immigration took place a number of years ago, the fringe of mixed parentage surrounding the central core of solid German origin is wider than in the case of Italians or Hungarians, who are both newcomers. And the relative proportion of the sexes must also be considered. A family immigration like the Jewish, as contrasted with a predominantly male influx like the Slavic, will materially change the current of marriages. Nevertheless, bearing these considerations in mind, it is of interest to compare the relative proportions of native whites born in America from the two classes of pure and mixed foreign parentage respectively. These proportions vary greatly as between different nationalities. Altogether there were almost 19,000,000 persons in 1910 born in the United States of foreign or mixed parentage. Of these approximately 13,000,000 had *both* parents from the same foreign country, while about 6,000,000 of this second generation born in America had one parent foreign born and the other born in the United States. In most cases, of course, this native-born parent—more often the wife than the husband, conformably to the social law above mentioned—was herself allied by birth to the same national stock. Real intermixture, that is to say, the breakdown of the sharp boundaries between different nationalities, is very much more rare.

NATIVE-BORN WHITES OF FOREIGN OR MIXED PARENTAGE IN 1910.

	No.	Per Cent.
Both parents from—		
Canada, (English)	307,000	39
England	592,000	49
Canada, (French)	331,000	69
Ireland	2,140,000	70
Germany	3,900,000	70
Sweden	546,000	78
Norway	410,000	79
Italy	695,000	80
Austria, (Bohemians, Poles)	709,000	90
Russia, (Jews)	949,000	93
Hungary	191,000	94
One parent born in U. S.—		
Canada, (English)	704,000	61
England	854,000	51
Canada, (French)	216,000	31
Ireland	1,010,000	30
Germany	1,870,000	30
Sweden	152,000	22
Norway	164,000	21
Italy	218,000	20
Austria, (Bohemians, Poles)	117,000	10
Russia, (Jews)	71,000	7
Hungary	13,500	6

Considering now the relative proportions of solid foreign-born parentage as distinct from a mixed type, one finds the sharpest differentiation naturally among the Russian Jews. The accompanying table sets forth the facts. There were 949,000 native-born whites with both parents Russian, while only 70,000 were born of the union of a Russian Jew and a native-born American, the latter, of course, being generally only one generation further removed from the pure Russian Jew. This solidarity of the Russian stock is undoubtedly due to a combination of recency of immigration and religious and social segregation. At the other extreme from the Russian Jew, at the top of the table, stand in order the English, either from Canada or the British Isles, and the Scotch. These people betray in the census a much larger number of persons of mixed parentage than of pure foreign origin. This is most marked among the English-speaking Canadians, among whom 704,000 persons were born in the United States with only one Canadian parent, while only 307,000 had both parents Canadian. The English from the British Isles betray the same tendency toward amalgamation, but to a somewhat less degree. There were 854,000 persons with one English and one American parent, while only 592,000 were born of two English parents. An odd circumstance to note is that France and Spain are the only other countries for which one finds mixed American marriages exceeding in number those between two parties of the same nationality.

Other Nationalities.

Between the two extremes of the isolated Russian Jew and the amalgamating English one finds the other nationalities ranging in series. First come the Irish, among whom about two-thirds of the second generation have both parents Irish born, while the remaining third is the offspring of combined Irish and American parentage. For the Germans the proportion of mixed parentage is somewhat less than the Irish. Then come the Norwegians and Swedes. The Swedes betray appreciably more clannishness, so to speak, than their Norwegian brothers. That is to say, the proportion of mixed parentage is distinctly lower. Otherwise stated, among the Swedes a larger proportion of Swedes marry Swedes than marry native-born Americans, even those of Swedish-American antecedents. And neither of them intermarry to the same degree as the Irish. The Germans are apparently similarly circumstanced, although, of course, they have been here longer than the Scandinavian people.

Recent immigrants of a decidedly lower social status than the Americans show relatively smaller proportions of intermarriage with native-born Americans even of their own type. Thus among the Italians 695,000 persons in 1910 were born in the United States of both Italian parents, while only 218,000 had one parent born in America. After the Italians come the Slavic peoples. Austrians, that is to say, Bohemians, Poles, &c., produced 709,000 native-born Americans of pure foreign stock, while only 117,000 had one American parent. The list may be closed by the Hungarians with 191,000 from pure Hungarian marriages, while only 13,000 represent the first stage of intermarriage with American-born. In these later cases, however, it is apparent that recency of immigration and a preponderance of male operate strongly to prevent intermarriage.

It would be of the utmost interest and scientific value to follow out a number of lines of inquiry suggested by the foregoing data. But, unfortunately, this is precluded by the fact that the statistical records close at this point. What a study it would be to go over complete genealogical records, not only of our chronologically "first families," but for all the subsequent immigrant and mixed ones as well! Years of time and mints of money would be needed to bring such an investigation to a head. But is it of any less importance than the calculation of the distance of the earth from the sun and allied inquiries in pure science? Some day, when it can be undertaken on a large scale, the general public and the generous patrons of sciences and the arts will be amazed at its economic and social significance.

June 22, 1913

IMMIGRATION VETO SENT TO CONGRESS

President Rejects Bill Because of Literacy Test and Denial of Asylum.

WANTS PEOPLE'S MANDATE

Won't Take Responsibility for Changing Traditional Policy—Veto Test Vote Next Thursday.

Special to The New York Times.

WASHINGTON, Jan. 28.—President Wilson today sent to the House his message vetoing the Immigration bill. His objections to the measure were that the literacy test provided for aliens desiring to enter the United States was not a fair test of prospective citizenship, and that the bill would precent this Government from granting asylum to political offenders, thus reversing a policy that had given the nation some of its most distinguished citizens. To the President's way of thinking the problem presented was too delicate to be adjusted without the sanction of the people, and he was unwilling to assume responsibility for changing the system.

Immediately after the veto message was read to the House Representative Burnett, Chairman of the Immigration Committee and the author of the bill, gave notice that he would make an effort to have it passed over the President's veto. His committee will report it to the House next Thursday and a vote will be taken on that day.

The bill was passed by the House originally—in February last year—by 252 to 126 and by the Senate on Jan. 2 last by 50 to 7, while the House vote on the conferees' report on Jan. 14 was 227 to 94. On the basis of the last vote only 214 members would constitute the necessary two-thirds required to overrule the President.

Following is the President's message in full:

Reasons for the Veto.

It is with unaffected regret that I find myself constrained by clear conviction to return this bill (H. R. 6,060, an Act to Regulate the Immigration of Aliens to and the Residence of Aliens in the United States) without my signature.

Not only do I feel it to be a serious matter to exercise the power of veto in any case, because it involves opposing the single judgment of the President to the judgment of a majority of both houses of the Congress, a step which no man who realizes his own liability to error can take without great hesitation, but also because this particular bill is in so many important respects admirable, well conceived, and desirable.

Its enactment into law would undoubtedly enhance the efficiency and improve the methods of handling the important branch of the public service to which it relates, but candor and a sense of duty with regard to the responsibility so clearly imposed upon me by the Constitution in matters of legislation leave me no choice but to dissent.

In two particulars of vital consequence this bill embodies a radical departure from the traditional and long-established policy of this country, a policy in which our people have conceived the very character of their Government to be expressed, the very mission and spirit of the nation in respect of its relations to the peoples of the world outside their borders. It seeks to all but close entirely the gates of asylum, which have always been open to those who could find nowhere else the right and opportunity of constitutional agitation for what they conceived to be the natural and inalienable rights of men, and it excludes those to whom the opportunities of elementary education have been denied without regard to their character, their purposes, or their natural capacity.

Restrictions like these adopted earlier in our history as a nation would very materially have altered the course and cooled the humane ardors of our politics. The right of political asylum has brought to this country many a man of noble character and elevated purpose who was marked as an outlaw in his own less fortunate land and who has yet become an onament to our citizenship and to our public councils.

The children and the compatriots of these illustrious Americans must stand amazed to see the representatives of their nation now resolved, in the fullness of our national strength and at the maturity of our great institutions, to risk turning such men back from our shores without test of quality or of purpose. It is difficult for me to believe that the full effect of this feature of the bill was realized when it was framed and adopted, and it is impossible for me to assent to it in the form in which it is here cast.

The literacy test and the tests and restrictions which accompany it constitute an even more radical change in the policy of the nation. Hitherto we have generously kept our doors open to all who were not unfitted by reason of disease or incapacity for self-support or such personal records and antecedents as were likely to make them a menace to our peace and order or to the wholesome and essential relationships of life. In this bill it is proposed to turn away from tests of character and of quality and to impose tests which exclude and restrict; for the new tests here embodied are not tests of quality or of character or of personal fitness, but tests of opportunity. Those who come seeking opportunity are not to be admitted unless they have already had one of the chief of the oportunities they seek—the opportunity of education. The object of such provision is restriction, not selection.

If the people of this country have made up their minds to limit the number of immigrants by arbitrary tests and so reverse the policy of all the generations of Americans that have gone before them, it is their right to do so. I am their servant, and have no license to stand in their way. But I do not believe that they have. I respectfully submit that on one can quote their mandate to that effect. Has any political party ever avowed a policy of restriction in this fundamental matter, gone to the country on it, and been commissioned to control its legislation? Does this bill rest upon the conscious and universal assent and desire of the American people? I doubt it. It is because I doubt it that I make bold to dissent from it. I am willing to abide by the verdict, but not until it has been rendered. Let the platforms of parties speak out upon this policy and the people pronounce their wish. The matter is too fundamental to be settled otherwise.

I have no pride of opinion on this question. I am not foolish enough to profess to know the wishes and ideals of America better than the body of her chosen representatives know them. I only want instruction direct from those whose fortunes with ours and all men's are involved.

A Week to Round p Absentees.

There was a vigorous round of applause at the close of the reading. The opponents of the literacy test manifested great satisfaction at the President's phrasing of his objections.

Representative Burnett at once asked that the message lie on the Speaker's table until Thursday or Friday of next week, when it could be taken up for consideration. There was prompt opposition to this request from several Republican members, and Representative Mann, the minority leader, demanded that the message be referred in regular order to the Committee on Immigration.

ne House voted to sustain this disposition of the document, and Mr. Burnett gave notice that his committee would ask the House to take up the veto next Thursday and vote to decide whether the bill should be passed over the President's objections.

"I do not desire," said Mr. Burnett, "to take snap judgment on anybody. I think no one who favors this bill had any tips as to when the veto would come in. I believe it should be fairly discussed and members should have an opportunity to arrange their affairs to be here. I think that the message itself, within its four corners, gives a good reason why this Immigration bill should become a law, regardless of the veto."

Representative Sabath of Illinois, who has fought the literacy test stubbornly, replied by saying that he resented the imputation that any of the enemies of the bill had received tips as to what the President intended to do.

"We received no tips," said Mr. Sabath. "We believed that the President would veto the bill because we thought the President would desire to be just and right."

The friends of the bill desired to have the interval of a week, so they could get back to Washington a large number of absentees. The opponents of the measure also intend to have their side present in full force next week, and "Hurry-home" telegrams were sent to them at once.

In viw of the large vote in the House and the Senate the advocates of the bill are very sanguine of ultimate success. Overwhelming votes, however, do not always display cohesion when subjected to the aoid test of party politics. With all the burdens that just now rest upon the shoulders of the majority party leaders in Congress, the proposition to reverse the President is a step that will necessitate thinking twice before acting.

The fact that the Imigration bill is not strictly a party measure affords the Democrats no advantage. If the veto is overridden they will have to assume responsibility for it before the country, and the foreign-born vote will be appealed to powerfully by the Republicans in the next general campaign.

Republican Senators today pointed out an apparent oversight in the President's message vetoing the bill. The President seemed to think the literacy test never had been advocated in a political platform, particularly one that had carried a party to victory. It was embodied in the Republican platform of 1896, when Mr. McKinley defeated Mr. Bryan for the Presidency.

APPREHENSION IN JAPAN.

Feared Immigration Bill Will Stimulate Anti-Asiatic Agitation.

TOKIO, Jan. 28, (Cable Dispatch to East and West News Bureau.)—The Jiji editorially says: "As one of the principles underlying the Immigration bill just passed by Congress is the granting of permission to enter America only to those aliens who are eligible to American citizenship, there is fear that the anti-Asiatic agitators in different States might construe it, if the bill is signed by the President, as the willingness of the Federal Government to recognize and even assist their propaganda. We cannot, therefore, but hope to see the bill vetoed by the President."

The Tokio Asahi, referring to the anti-Japanese bills likely to be introduced into the California Assembly, says: "We sincerely hope that the California Assembly will defeat those bills and thereby demonstrate in fact the sincerity of what the leaders of California have since last year said on the platform and in the press about the change of California's attitude toward Japan, which has been received with much satisfaction by the Japanese. We cannot view the recent utterances, favorable to Japan, of California's leaders as a temporary phenomenon brought about by their solicitude for the success of the Panama Exposition. We shall rely upon the good faith of the leaders and watch with interest their effort to do good for the benefit of both countries."

January 29, 1915

HOUSE UPHOLDS VETO OF THE ALIEN BILL

Friends of Measure Fail by Four Votes to Override President's Action.

ALL DAY GIVEN TO DEBATE

Changes of Front Shown by 13 Members in the Final Action on the Measure.

Special to The New York Times.

WASHINGTON, Feb. 4. — President Wilson's veto of the Burnett Immigration bill, sent to the House a week ago today, will stand. The House this afternoon, by a vetc of 261 to 136, refused to pass the bill over the veto by the constitutional two-thirds vote. It was, however, a close call for Mr. Wilson. A change of four votes would have given the friends of the bill a victory over the President.

It was the third time the House had undertaken to pass an immigration bill containing the literacy test over a President's veto. Two years ago today an attempt to get the necessary two-thirds to override the veto of President Taft failed. President Cleveland also vetoed a like measure. Today was notable as the anniversary of the passage of the Burnett bill by the House. It was a foregone conclusion that if the bill passed the House it would pass the Senate, the last vote there being 50 to 7.

Of the 425 members of the House, 399 were present. The entire day was given over to debate under the five-minute rule. At times the discussion arose to unusual heights of oratory. On one side the appeal was made for the oppressed and downtrodden of other lands, and on the other the argument for fair play to the American workman was urged.

Mr. Moore of Pennsylvania, who closed the debate for the Republicans in opposition to the measure, began his remarks by reading from an ancient document adopted once upon a time in his district in the City of Philadelphia, which declared that all men were born free and equal, and among other things complained that the King of England in that day was preventing immigrants from coming to this country.

Mr. Moore appealed for the principles of the Declaration of Independence. To the argument, that the Federation of Labor was asking for the passage of the Burnett bill, he said that there were only 2,700,000 members of the federation, while there were 30,000,000 laboring men outside that organization who made no complaint of competition from immigrant labor and were willing to keep the door open for those who wanted to come.

Mr. Burnett of Alabama, author of the bill and Chairman of the Immigration Committee, in closing the debate read from one of President Wilson's volumes of history an extract which set forth the argument against the admission of the southern European immigrant whose assimilation, it was asserted, was a task that was straining our institutions and putting undue burdens on our civilization. Mr. Burnett said that the pending bill with its literacy test had passed one house or the other seventeen times, and notwithstanding the President's demand in his veto message, that the question should be passed on by a party platform and a popular vote, it had been put in several platforms and public opinion was in favor of it.

As long ago as 1896, he said, the Democratic platform had asserted that there should be protection from excessive immigration, and the Republican platform had favored a literacy test.

It was several minutes before order could be restored when Mr. Burnett closed. The House broke out into cheers.

Mr. Sabath of Illinois led those Democrats who were opposed to the bill. Letters were read by him from the officers of the Garment Workers' Union, the Amalgamated Garment Makers, the Laundry Workers, the Bakers' Union, the United Neckwear Union, the United Trades of New York, the Cabinet Makers' Union, the Wood Turners' Union, the Shirtmakers' Union, and others in New York.

The whole number of members voting today was 397. When the bill passed a year ago the whole number voting was 378. The number voting on the conference report on Jan. 7 was 321. Of members who had gone on record heretofore on the measure, 13 changed their votes. Eight Democrats—Beakes of Michigan, Goeke of Ohio, Kindel of Colorado, Park of Georgia, Reed of New Hampshire, Taylor of Alabama, Whaley of South Carolina, and Williams of Illinois, who voted for the bill on its original passage and also for the conference report last month—voted today not to reverse the President.

Three Republicans, who previously voted against the bill, today voted to pass it over the veto. These were Cooper of Wisconsin and Drukker and Parker of New Jersey. Shreve, Republican, of Pennsylvania, changed from "Nay" to "Yea," and Steenerson of Minnesota, who had voted against the bill, answered "Present" when his name was called.

The 261 votes for passing the bill over the veto were cast as follows: Democrats, 166; Republicans, 78; Progressives and Progressive Republicans, 16; independent, 1.

Against the bill the vote was: Democrats, 101; "insurgent" Democrat, 1; Republicans, 32; Progressives and Progressive Republicans, 2.

February 5, 1915

PRESIDENT VETOES IMMIGRATION BILL

Asserts Literacy Test Constitutes a Change in Nation's Policy Which Is Not Justified.

Special to The New York Times.

WASHINGTON, Jan. 29. — President Wilson, as was expected, vetoed the Immigration bill today. In the statement of his reasons for refusing for the second time in his Administration to give his approval to the measure, the President stated that the provision incorporated in the bill to admit illiterates who may come to the United States seeking a refuge from religious persecution involved the responsibility on the part of United States immigration officials to pass on the laws and practices of foreign Governments, and this he regarded as an invidious function for any administrative officer of this Government to perform.

Singularly this feature of the bill was inserted to meet the objections raised by President Wilson in his first veto of the bill. The President's veto message follows:

"I very much regret to return this bill without my signature.

"In most of the provisions of the bill I should be very glad to concur, but I cannot rid myself of the conviction that the literacy test constitutes a radical change in the policy of the nation which is not justified in principle. It is not a test of character, of quality, or of personal fitness, but would operate in most cases merely as a penalty for lack of opportunity in the country from which the alien seeking admission came. The opportunity to gain an education is in many cases one of the chief opportunities sought by the immigrant in coming to the United States, and our experience in the past has not been that the illiterate immigrant is as such an undesirable immigrant. Tests of quality and of purpose cannot be objected to on principle, but tests of opportunity surely may be.

"Moreover, even if this test might be equitably insisted on, one of the exceptions proposed to its application involves a provision which might lead to very delicate and hazardous diplomatic situations.

"The bill exempts from the operation of the literacy test 'all aliens who shall prove to the satisfaction of the proper immigration officer or to the Secretary of Labor that they are seeking admission to the United States to avoid religious persecution in the country of their last permanent residence, whether such persecution be evidenced by overt acts or by laws or Governmental regulations that discriminate against the alien or the race to which he belongs because of his religious faith.'

"Such a provision, so applied and administered, would oblige the officer concerned in effect to pass judgment upon the laws and practices of a foreign Government, and declare that they did or did not constitute religious persecutions. This would, to say the least, be a most invidious function for any administrative officer of this Government to perform, and it is not only possible, but probable, that very serious questions of international justice and comity would arise between this Government and the Government or Governments thus officially condemned, should its exercise be adopted.

"I dare say that these consequences were not in the mind of the proponents of this provision, but the provision separately and in itself renders it unwise for me to give my assent to this legislation in its present form."

Representative Burnett, Chairman of the House Committee on Immigration, announced in the House late today that he would move to reconsider and pass the measure over the President's veto on Thursday. By unanimous consent it was agreed the veto message shall remain on the Speaker's table until that time.

The House vote on the motion to override the veto undoubtedly will be close. Mr. Burnett expressed confidence today that the required two-thirds majority will be obtained, but both sides are now engaged in telegraphing for absentees, and a half dozen votes may decide the issue. The House lacked four votes when it attempted to override a similar veto two years ago. When President Taft vetoed a similar measure the Senate succeeded in overriding him by more than a two-thirds majority, but the House fell short by about a dozen votes. The bill which the President vetoed today passed the House last March by a vote of 308 to 87, and the Senate in December by 64 to 7.

Representatives Burnett and Sabath, the latter representing the opponents of the Immigration bill, agreed today that the roll call on reconsideration shall be preceded by a debate of an hour and a half.

January 30, 1917

IMMIGRATION BILL ENACTED OVER VETO

Passed by Senate, 62 to 19, with Literacy Test, to Which President Objected.

Special to The New York Times.

WASHINGTON, Feb. 5.—By a vote of 62 to 19 the Senate today repassed the Immigration bill, containing the literacy test, thus enacting it into law despite the President's veto. The House repassed the bill over the veto last week by a vote of 287 to 106. By this action, the first time a veto by President Wilson has been overridden, Congress has at last ended a fight for the restriction of immigration by the literacy test which began in 1897, when President Cleveland vetoed the measure. President Taft also vetoed the provision, and President Wilson has done so twice. The first time Mr. Wilson refused his signature, the bill was repassed by the Senate with a sufficient majority to override the veto, but failed of the necessary two-thirds vote in the House. The law goes into effect May 1.

The literacy test, President Wilson declared in his recent veto message, is unjust in principle, constituting a test, not of character but of opportunity. He also objected that the provision of the bill which permits immigration officials to exempt from the operation of the test foreigners who, in their judgment, are fleeing from religious persecution would raise delicate questions which might involve this nation in international difficulties.

Vote on Repassage of the Bill.

The vote today was non-partisan, 34 Democrats and 28 Republicans voting for the bill and 11 Democrats and 8 Republicans voting to sustain the veto. The vote follows:

Ayes—Democrats: Ashurst, Bankhead, Beckham, Bryan, Chamberlain, Chilton, Culberson, Fletcher, Hardwick, Hughes, James, Johnson of Maine, Kern, Kirby, Lane, Lee of Maryland, Martin, Myers, Overman, Phelan, Pittman, Pomerene, Robinson, Shafroth, Sheppard, Shields, Simmons, Smith of Georgia, Smith of South Carolina, Thomas, Tillman, Underwood, Vardaman, and Williams. Republicans: Borah, Brady, Clapp, Cummins, Curtis, Dillingham, Fall, Gallinger, Gronna, Harding, Jones, Kenyon, La Follette, Lodge, McCumber, Nelson, Norris, Page, Penrose, Poindexter, Smoot, Sterling, Sutherland, Townsend, Wadsworth, Watson, Weeks, and Works.—62.

Nays—Democrats: Hollis, Husting, Johnson of South Dakota, Lewis, Martine, Ransdell, Reed, Saulsbury, Stone, Thompson, and Walsh. Republicans: Brandegee, Clark, Colt, du Pont, Lippitt, Sherman, Smith of Michigan, and Warren.—19.

The vote on the bill when it first passed the Senate at this session was 64 to 7. Senators Hollis, Johnson of South Dakota, Sherman, Smith of Michigan, and Thompson changed their votes, after the President's veto of the bill, from aye to nay. Senator Phelan of California changed his vote from nay to aye.

During the debate on the bill this afternoon Senator Reed, who has been one of its strongest opponents, objected not only to the literacy test, but also to the clause relating to the restriction of Asiatic immigration, asserting that he understood from an official of the State Department that the Japanese Government objected strongly to the present wording of this clause.

Says Japan Has Objected.

"I am authorized to say to the Senate," he went on, "that the Japanese Embassy has called attention to this language. The State Department feels that the clause may be the occasion of some misunderstanding, and is exceedingly desirous that nothing shall be done which will cause the Japanese Government to feel that we have in any way impinged upon the understanding that now exists."

Under the present wording, the bill excludes by geographical limitation Asiatics coming from certain countries. Japan was expressly omitted from this restriction at the request of the State Department because the immigration of Japanese labor is now forbidden by a "gentlemen's agreement" between the two nations, and the Japanese Embassy objected to having Japan discriminated against specifically while the Japanese Government was carrying out this agreement faithfully.

Senators from the Pacific Coast attempted to secure definite restriction of the Japanese in spite of this, but the Senate refused to permit it. The conference committee which framed the final draft of the bill finally agreed on a clause, to follow the geographical restriction paragraph, reading, "And no alien now in any way excluded from or prevented from entering the United States shall be admitted to the United States." This, Senator Reed asserted, was objectionable to the Japanese as referring to the existence of the agreement.

It was said by an official at the State Department today, however, that some misunderstanding must have occurred, as, so far as he knew, the Japanese Embassy had made no objection to the bill in its present form. The final draft of the bill was decided upon after Secretary Lansing had conferred with the committee on this particular point.

Senator Smith of South Carolina, Chairman of the Immigration Committee, answered Senator Reed with the declaration that the present state of international affairs emphasized the necessity for a pure, homogeneous American people, such as the bill was intended to protect.

What the Literacy Test Provides.

The literacy test provided for in the bill excludes from the United States all aliens over 16 years of age physically capable of reading, who cannot read the English language or some other language or dialect, including Hebrew or Yiddish. Any admissible alien, however, or any citizen of the United States, may bring in or send for his father or grandfather, over 55 years of age, his wife, mother, grandmother, or unmarried or widowed daughter, if otherwise admissible, regardless of whether such relatives can read.

Immediately after the Senate's action, Representative Gardner of Massachusetts introduced in the House a new immigration measure to limit the number of aliens coming into this country to a total of 200,000 a year, in excess of the outgoing aliens.

February 6, 1917

BIBLE FOR LITERACY TEST.

Government Will Employ It In Enforcing the New Immigration Law.

WASHINGTON, March 27.—Reading matter for a literacy test for aliens under the new immigration law will be taken from the Bible, the Department of Labor announced today. Passages will be selected in more than one hundred languages and dialects.

"This is not because the Bible is considered a sacred book by many people," said the department's announcement, "but because it is now the only book in virtually every tongue. Translations of the Bible were made by eminent scholars, and, what is more to the point, the translating was done by men whose purpose it was to put the Bible in such simple and idiomatic expressions in the various foreign languages as would make it possible for the common people of foreign countries to grasp the meaning readily and thoroughly."

March 28, 1917

AFTER-WAR IMMIGRATION CRISIS

Two Bills Before Congress—Need of Laborers vs. Desirability of Restrictions

By HENRY P. FAIRCHILD,
Professor of Social Economics at New York University.

THE war did to immigration what all the restrictionist agitation in the world could not have accomplished—it stopped it altogether. During the years of fighting the number of incoming foreigners dropped to negligible proportions, frequently exceeded by the number of those departing, so that many months showed a net loss in population through the movements of aliens.

Although hostilities ceased nearly a year ago, this condition has remained almost unchanged. The scarcity of shipping available for immigration purposes, added to other abnormal conditions, has prevented the immigration of such Europeans as would otherwise have come, while the refusal of the United States Government to grant permits to sail except for very urgent reasons has kept in this country most of those aliens who would be glad to return to the homeland.

There has resulted one of the most extraordinary immigration situations which the United States has ever experienced. The unique features of the situation are the following:

1. For nearly five years the industries of this country have been compelled to depend absolutely upon the labor forces already in the country. During that period we have weathered in a manner satisfactory and creditable on the whole the most severe strain ever put upon our productive capacity.

2. For the first time in half a century we are going through a period of intense business activity and vigorous industrial expansion without being able to recruit our labor supply to an unlimited extent from European sources. Under ordinary conditions the "boom" situation in this country would be felt in Europe and would be reflected in the arrival of foreign immigrants by the hundred every month.

3. In spite of the keen demand for labor and high wages in this country there are tens of thousands, perhaps hundreds of thousands, of aliens who would like to return to Europe, but who are prevented by the impossibility of securing passage. There are not wanting those who are ready to interpret this fact as evidence of the domination of our Government by capitalistic interests, who do not propose to suffer any diminution of their labor force if they can help it.

4. In the innumerable industrial disturbances which are taking place all over the land, notably in the steel strike, the foreign element is playing a leading part, with the result that many of the mills are "making Americanism the chief issue," and "have publicly announced that hereafter only American citizens or those with first papers will be employed."

There is a wonderful irony in this last feature of the situation. Here are the big producing interests of the country—again, notably the steel industry—who have been the chief influence in bringing foreign labor to the United States. For the last fifty years the large employers of this country have used every possible means to attract foreign labor to their plants, first by direct recruiting in European countries until this practice was forbidden by the contract labor law, and since then by less direct though hardly less effective methods of advertising their needs among the low paid laborers of Europe. Lower and lower economic strata in Europe have been tapped successively, and each new contingent of half-dazed, ox-like toilers has been welcomed as furnishing a cheaper and more tractable labor supply. This foreign labor has been used over and over again to turn the balance against dissatisfied and striking wage-earners in this country, sometimes consciously, sometimes unconsciously. The actual and potential effect upon the standard of living of the working people in the Unied States has been ignored. The importance of wealth production has been allowed completely to overshadow considerations of human happiness and welfare.

Thus the employers of labor have been sowing the wind, and now that they are reaping the whirlwind of discontent, resentment, and rebellion the only expedient that some of them, at least, seem to have is to turn about and bar from the mills the very workers whom half a decade ago they were so eager to receive. Is it to be wondered at that the seeds of cynicism, anarchy, and class hatred find fertile soil in the bosoms of some of these aliens?

These facts are not in any sense a justification of the position taken by the workers in many industrial disputes. They are simply an explanation. The great majority of immigrant laborers come to America wholly untrained in the ways of democracy, unfamiliar with our industrial methods and our social concepts, unused to the absence of authority and the postulated equality of this country. They cannot be expected to apply the same degree of reasoning to a complex social situation that the typical American workingman can command. They naturally respond to the emotional appeals of agitators and demagogues, and are easily moved to transform a discomfort into a grievance.

We, in this country, have all of us been too much inclined to treat the immigrant as if he were some unique and distinctive species of being. We do not apply to him the categories which we use for ourselves. In certain quarters he is exalted as the embodiment of idealism, loyalty, and love of liberty; in others he is treated like a clod, an animal, or at best an "economic" man. We forget that the immigrant, after all, is just a human being. The main thing that differentiates him from ourselves is that he happens to have been born in some other land, and to have grown up in a different social environment.

It is this difference in social heritage, coupled with whatever of differentiation there may be in racial inheritance, which makes the immigrant a "problem." And that is enough, forsooth, to constitute a very serious problem. But it must never be allowed to obscure the underlying similarities of desire, passion, hope, and ambition which are common to all humanity.

The lesson which the war seems to have impressed upon a larger part of our population than could probably have been made to see it in any other way is that while we can safely admit to this country aliens of such character and in such quantity as we can make real participators in our common life, we undermine our own national solidity if we throw open our doors to unlimited numbers of strangers who cannot possibly be treated as really members of our national body politic, hardly even as human beings.

This change in sentiment is reflected in the present attitude of Congress toward the control of immigration. If an industrial situation like the present had arisen before the war the remedy which would have seemed natural and almost a matter of course, to the employing class at least, would have been to facilitate in every way the migration of foreign labor to this country, even to the extent of easing up the various regulations intended to protect the interests of the workers in this country. In direct opposition to this, however, at the present time, we find Congress considering the expediency of imposing more stringent restrictions upon immigration than have ever been proposed before.

These restrictions take two forms and are embodied in two separate bills now before Congress. The most direct and drastic of these is the proposal to prohibit absolutely the immigration of all foreign labor for a period of five years. Of course exceptions are made in favor of highly skilled labor and such professional and special classes as are customarily exempted in our immigration laws. But in effect this measure, if it became a law, would shut off entirely the stream of foreign labor upon which we were accustomed to depend in busy seasons before the war. A bill very similar to this was introduced into the last session of Congress by the late Representative Burnett, then Chairman of the House Immigration Committee, who had great hopes of its passage. It was, however, killed in the filibuster which terminated that session so ingloriously.

The second method of restricting immigration is that proposed by the bill recently introduced by Senator Dillingham, Chairman of the Immigration Committee of the upper house. According to this plan the immigration of any foreign nationality in any given fiscal year shall be limited to 5 per cent. of the number of persons of such nationality resident in the United States at the time of the next preceding census. This type of restriction is not entirely a novelty, having been frequently considered by Congress ever since the report of the Immigration Commission. It is designed not only to fix a definite limit to the total immigration, but to differentiate in favor of immigrants from Northwestern Europe as against those from Southeastern Europe. The number of immigrants who could enter from the former section would be much larger than those customarily seeking admission, while the immigration from the latter region would be very materially reduced.

What gives special significance to Senator Dillingham's bill is that it also proposes to repeal the Chinese Exclusion acts and all other measures placing special restrictions upon Orientals.

The supporters of the measure claim for it the special advantage that, in addition to furnishing a good basis for general restriction, it also provides for an exclusion of Orientals practically as complete as the present, without wounding the pride of these ancient and dignified peoples by specifically designating them for debarment. The Chinese and Japanese have always regarded such discrimination as an imputation of inferiority. The proposed measure nominally treats all nations like, but in fact the number of Orientals resident in the United States are so small that it is contended that the permissible immigration would be almost negligible. This is evidently a matter of statistical computation, and convincing figures as to the probable future outcome have not yet been fully supplied. If it can be shown that the twofold aim of excluding Orientals without hurting their feelings can be secured in this way, a measure of this sort would be an almost ideal form of restriction.

As to the question of virtually suspending the immigration of common laborers during the acute period of reconstruction, there can hardly be two rational views. The injection of a new mass of undigested human material into the present turbulent and feverish situation would be immeasurably disastrous. As soon as the natural impediments to free movement begin to disappear artificial restrictions should certainly be imposed.

October 12, 1919

IMMIGRATION BILL STANDS.

Conferees Retain Essential Features of Measure Vetoed by Wilson.

WASHINGTON, May 11.—Conferees on the Immigration Restriction bill agreed today to retain the plan of limiting immigration inflow by nationalities next year to 3 per cent. of the number from each nation now in the United States, but struck out House proposals to exempt from these restrictions persons subject to religious persecution and those who had left this country temporarily. The House exemption in favor of children under 1 year of parents in this country was retained.

The conference report will be presented first to the House, with prospects of prompt ratification by both bodies. By conference agreement the restrictions would become effective fifteen days after signature by the President.

In essential features the conference measure is almost identical with that vetoed by former President Wilson.

May 12, 1921

IMMIGRATION ACT NOW LAW.

President Signs Measure Restricting Admission of Aliens.

WASHINGTON, May 19.—The immigration restriction bill was signed today by President Harding.

Under the terms of the measure as it was finally agreed to in conference and passed by both houses last Friday, its provisions will go into effect in fifteen days from today, the date of its approval by the President, and will remain effective until July 1, 1922.

The new law provides that the number of aliens admitted into the United States during that time shall not exceed 3 per cent. of the nationals of each country who were here in 1910

May 20, 1921

HARDING TAKES UP ALIENS' EXCLUSION

Consuls Ordered to Quit Issuing Passports When Quota Is Nearly Reached.

SIEGEL CHARGED CRUELTIES

Complained to President, Who, in Reply, Blames "Dishonest Steamship Agents."

Cruelty in the enforcement of the immigration laws has reached an extreme stage in the last two weeks, because of "the failure of the immigration officials to use common sense," according to Representative Isaac Siegel, who made public yesterday a correspondence between himself and President Harding on the subject. President Harding placed the blame largely on "dishonest steamship agents."

Representative Siegel said that because of the controversy over the interpretation of the immigration laws and the hardships worked in so many cases on individuals, the State Department sent out a general order last Friday to our Consuls in Europe, instructing them to suspend the issuance of passports to emigrants when the number of passports already issued approaches the quota of allowed immigrants. He said that in Poland and some other countries there was a conflict of opinion as to whether the quota had been reached or not, and that the Labor Department desired to stop emigration from such countries until the tangle is straightened out.

Mr. Siegel said that, according to the State Department figures, Poland had passed its quota and several other European countries had reached or approximated it. Mr. Siegel contended that the State Department figures regarding Poland were incorrect and that several thousand immigrants from that country might still be legally permitted to enter.

Citing an instance o what he regarus as cruelty inflicted on aliens through the new law's operation, Mr. Siegel said:

"Freda Berman is 10 years old. Her father, Jacob Berman, a tailor at 50 West Ninety-first Street, came to this country seven years ago, intending to bring his wife and child here as soon as he settled. The war broke out and most of the people of Poland became refugees. He was notified that his wife and child were dead. Nevertheless, he was successful a short time ago in finding them. He immediately sent the money to bring them to this country.

"His wife died on the eve of sailing and the ten-year-old daughter came over alone. She has not a relative in Poland. But she was ordered to be deported on the ground that the quota was filled.

"There have occurred sixty-seven similar cases of needless cruelty.

"The law allows the admission in any one month of as many as five-twelfths of the 3 per cent. permitted to enter in a year. But the immigration officials are permitting only one-twelfth of the 3 per cent. to enter each month, thus inflicting twelve times as much hardship and cruelty as is necessary."

Mr. Siegel's letter to President Harding was as follows:

Washington, Sept. 8.

Hon. Warren G. Harding,
President of the United States:

Dear Mr. President—It is exceedingly regrettable that the immigration laws, as now being enforced, are resulting in the grossest kind of injustice. There is no reason why the same attitude taken by the Labor Department heretofore should not be followed, in allowing members of families who arrive after the quota for each month has been filled to remain at Ellis Island or be allowed to land under bond and be counted in the next month's quota.

All I can say to you is that the separation of parents from children who arrive at Ellis Island — the parents being admitted and the children being sent back because they happen to arrive after the quota is filled—is something I don't believe you will approve.

The reason that I desire to bring this to your attention is because no one can picture to you in words the effect of the enforcement of this provision the way it is now being administered. Action on your part is absolutely necessary, if we are to uphold American traditions regarding humanity.

With kindest regard, yours very truly, ISAAC SIEGEL.

The reply of President Harding was as follows:

White House, Washington,
Sept. 9, 1921.

My Dear Mr. Siegel: I have your letter of Sept. 8. I haven't any doubt in the world but the enforcement of the immigration laws is working many a hardship. My own distress has been very great over some of the specific instances which have been reported to me.

If I have the situation correctly presented, the difficulty must be charged to dishonest steamship agents who have brought to this country innocent immigrants in spite of our continued warnings during a period of very great leniency. I know how very persistent have been the impositions which have been made on the Government agents who have been disposed to be sympathetic and more than generous in carrying out the law.

However, I am sending your letter to the Department of Labor for further information on the subject. I have great confidence in the Commissioner of Immigration, and I know the Secretary of Labor is one of the most humane and sympathetic men in the land. If there are conditions such as you suggest I feel pretty confident that they are unavoidable under the law. Very respectfully yours,

WARREN G. HARDING.

Plans have been made for a consultation of Secretary Davis of the Labor Department, Representative Siegel of the Immigration Committee and others to devise a more humane method of handling the immigration problem at the port of New York.

September 13, 1921

EUGENISTS DREAD TAINTED ALIENS

Believe Immigration Restriction Essential to Prevent Deterioration of Race Here.

MELTING POT FALSE THEORY

Racial Mixture Liable to Lower the Quality of the Stock—Prof. Osborn's Views.

THE LESSON OF EVOLUTION

Minute Scruting of Family History of Prospective Immigrants Is Advocated.

Severe restriction of immigration is essential to prevent the deterioration of American civilization, according to students of race and biology now taking part in the Second International Eugenics Congress at the American Museum of Natural History.

The "melting pot" theory is a complete fallacy, according to engenists, because it suggests that impurities and baser qualities are eliminated by the intermingling of races, whereas they are as likely to be increased, if not more likely to be increased. Speakers who touched on the subject were all on one side, holding that the mixture of poor stock with a good one does as much harm to the good stock as it does benefit to the poor.

The theory held by some eminent anthropologists that all races have an equal capacity for development and that all race questions, even the negro question, is to be solved in the long run by race mixture, was vigorously combatted. Denying that certain race stocks are poor entirely because of poor environment in the old world, eugenists asserted that education and better economic conditions in this country could only imperfetly overcome ingrained racial and family defects.

Stricter Immigration Guard.

Dr. Charles B. Davenport, Director of the Eugenics Record Office, urged stricter immigration laws than those now in force and suggested that amendments should eventually be made enabling researches to be made into the famly history of candidates for admission into the United States in order to bar tainted lines.'

One of the most outspoken addresses on the subject was by Professor Henry Fairfield Osborn, President of the congress, author of "Men of the Stone Age" and an authority on evolution.

"In the United States," he said, "we are slowly awaking to the consciousness that education and environment do not fundamentally alter racial values. We are engaged in a serious struggle to maintain our historic republican institutions through barring the entrance of those who are unfit to share the duties and responsibilities of our well-founded Government.

"The true spirit of American democracy, that all men are born with equal rights and duties, has been confused with the political sophistry that all men are born with equal character and ability to govern themselves and others, and with the educational sophistry that education and environment will offset the handicap of ancestry.

"South America is examining into the relative value of the pure Spanish and Portuguese and of various decrees of racial mixture of Indian and Negroid bloods in relation to the preservation of republican institutions."

The Lesson of Evolution.

Professor Osborn said that 500,000 years of evolution had impressed certain characteristics on the three great racial branches, the Caucasian, the Mongolian and the Negroid, and their variations. He said there was no for mof matter so stable as the germ plas mon which heredity depends, and that this accounted for the stubborn permanence to types and of the survival of their original qualities in admixtures.

"In the matter of racial virtues," he said, "my opinion is that from biological principles there is little promise in the melting-pot theory. Put three races together, and you are as likely to unite the vices of all three as the virtues."

He said that scientific proof or disproof of the "melting-pot" theory would be forthcoming as the result of researches such as those of Dr. Sullivan in the Hawaiian Islands.

"For the world's work, however," he said, "give me a pure-blooded negro, a pure-blooded Mongol, a pure-blooded Slav, a pure-blooded Nordic and ascertain through observation and experiment what each race is best fitted to accomplish in the world's economy."

Dr. Osborn argued that the State's right to prevent disease implied a similar right to prevent the multiplication of feeble-minded, idiocy and moral and intellectual diseases.

"The wisdom of British biologists," be continued, expressed by Tennyson in his memorable lines:

How careless of the individual,
How careful of the race?

has been translated into the fatal reverse,

How careful of the individual,
How careless of the race.

Sees Peril in Individualism.

"The closing decades of the nineteenth century and the opening decades of the twentieth have witnessed, what may be called a rampant individualism—not only in art and literature, but in all our social institutions—an individualism which threatens the very existence of the family; this is the motto of individualism, let each individual enjoy his own rights and privileges—for tomorrow, the race dies.

"The New England a century has witnessed the passage of a many-child family to a one-child family. The purest New England stock is not holding its own. The next stage is the no-child marriage and the extinction of the stock which laid the foundations of the Republican institutions of this country.

"It is questions of this kind which are being set forth before this Congress so that they may be disseminated among our people. Let us endeavor to discard all prejudices and to courageously face the facts. Recent works by Bury and Inge on human progress are regarded in some quarters as pessimistic. I do not regard them as pessimistic, because to my mind the pessimist is one who will not face the facts, and these writers, especially Inge, look at the worst as well as at the best. I regard an optimist as one who faces the facts but is never discouraged by them. The optimist in science is one who delves afresh into nature to restore disordered and shattered society. This was the constructive spirit of Francis Galton, founder of the science of eugenics. I trust it will be the keynote of this Congress. To know the worst as well as the best in heredity; to preserve and to select the best—these are the most essential forces in the future evolution of human society."

Professor Osborn, who was recently in Europe bringing together leaders in eugenics and biology from many European countries to attend the Congress, said that he had made a special study of parts of Belgium and France. Here he had been impressed, he said, with the manner in which the three main races of Frances, the Mediterranean, the Alpine and the Nordic, preserved their racial traits. He said that 12,000 years of similar environment and 1,000 years of similar education had caused only a slight divergence from the characteristics which were found in those races many thousands of years ago, as shown by evidences in the remains surviving from that period.

Effects of the War.

"To each of the countries of the world," he said, "racial betterment presents a different aspect. To the five countries most closely engaged in the recent fraticidal conflict, the financial and economic losses of which we hear so much are as nothing compared with the spiritual, intellectual and moral losses which each has sustained. In the Scandinavian countries, while kept out of th econflict, and to a large extent in the United States, the case is different. In Scandinavia, which I have recently visited, it is largely through the active efforts of leaders like Mjoen and Lundborg that there is a new appreciation of the spiritual, intellectual, moral and physical value of the Nordic race, and that a warning is being given that it must not be too severely depleted by emigration. Nearly half that race is now in the United States."

The difficulty in obtaining legislation to better the races, because of various prejudices and because of the fear on the part of politicians to give offense to any of their constituents, was emphasized by several speakers. Major Leonard Darwin said that it was very difficult to induce law-makers to pass laws for the benefit of the unborn who have no votes. Dr. Davenport said that the study of eugenics must progress until proofs of its contentions are piled high and have impressed the general community, before political action becomes a possibility.

September 25, 1921

To Restrict Immigration 2 Years More.

WASHINGTON, May 11.—President Harding today signed the bill extending for two years from June 30 the 3 per cent. immigration restriction act.

May 12, 1922

NEW IMMIGRATION LAW BARS FLOOD TO AMERICA

Southern Europe Uses Up Its Quotas, While Northern and Western Countries Fall Short.

WASHINGTON, Aug. 5.—The 3 per cent. limitation immigration law has proved an effectual bar against any overwhelming movement of immigrants from southern and eastern European countries to the United States, Secretary of Labor Davis announced today in making public a survey of immigration figures for the fiscal year just closed.

The figures made public show that northern and western European countries had fallen far short of filling their quotas, while the southern and eastern countries of Europe were sending just as many to America as would be accepted.

A summary of the immigrants admitted during the fiscal year which ended June 30 showed that the following countries sent to this country 100 per cent. of the quotas allowed them under the 3 per cent. law: Belgium, Greece, Hungary, Italy, Luxemburg, Poland, Rumania, Jugoslavia, Palestine, Turkey, Syria, the miscellaneous European and Asiatic countries, Africa, Australia and New Zealand.

In contrast with these nations the countries of northern and western Europe in some instances, it was explained, sent less than one-half of the number permitted under the quota law. Germany sent 28 per cent. of its allotment, Sweden 43 per cent., Norway 48 per cent., The Netherlands 66 per cent. and France 75 per cent.

"The figures clearly indicated," the Secretary said, "that the 3 per cent. limit has proved no bar to immigrants from the Nordic races, for it fixes the limitation well above the number of immigrants of this class normally coming to America, but it has operated effectually to check the stream from southern and eastern Europe, eliminating the widespread fear that the country would be flooded with immigrants from war-stricken Europe."

August 6, 1922

WON'T LIFT BARRIERS TO ADMIT REFUGEES

Administration Plans No Action on Appeals to Let In Armenians and Greeks.

Special to The New York Times.

WASHINGTON, Dec. 11.—The Administration, it was learned today, entertains no idea of raising the immigration restrictions now 'existing against the admission of Greek and Armenian refugees into the United States. It was pointed out that Congressional action would be necessary and it was stated that the Administration does not consider this a matter in which it should take the initiative.

There have been numerous appeals to the Administration to let down the bars to let in Greeks and Armenians now seeking admission but who cannot enter under the quotas fixed by law. The recent appeal of the Federal Council of Churches, backed up by other church organizations, was followed by a fresh appeal today from Rev. Samuel McCrea Cavert, one of the general secretaries of that body.

"At Ellis Island," said Mr. Cavert's appeal, "there are many Greeks and Armenians from Asia Minor and Thrace who are being denied admission because the annual quotas for these groups for the year 1922-23 are exhausted. Even in the case of those who have prosperous relatives in America ready to receive them no entrance is possible until some action is taken by Congress.

"Where are these refugees to go? Smyrna is only one incident, even if the most tragic of the Turkish policy. The roads from Thrace are reported as crowded with these people. Any day may record a similar exodus from Constantinople.

"Even with the help of Bulgaria and Serbia, Greece cannot possibly care for the boats pouring in upon her. More than half a million are reported as having arrived already, while the total number will probably be one million, which means an increase of 20 per cent. in the population of Greece. Obviously, Greece cannot bear this burden alone—America must help in some way."

December 12, 1922

75,000 KLANSMEN GATHER IN DALLAS TO IMPRESS NATION

But the Kloncilium Gets Only a Third of the Huge Number Summoned.

OBSERVERS SEE IT LOSING

Disintegration Has Already Set In, and Klan's Sway Is Lost, They Say.

EVANS READS ANATHEMAS

He Declares the Negro, Jew and Catholic Dangerous, and Would Bar Immigrants.

Special to The New York Times.

DALLAS, Texas, Oct. 24.—This, the second city of Texas, witnessed today the culmination of the long planned effort of the Ku Klux Klan to demonstrate that so far as the southwestern part of the United States is concerned the Klan is supreme. The hooded organization had published far and wide that between 200,000 and 250,000 masked men from Texas, Louisiana, Arkansas, Oklahoma and other States of the West and Southwest would invade Dallas, and that when the sun had set the absolute dominance of the Ku Klux Klan in this great section of the country could no longer be denied.

Hiram W. Evans, Imperial Wizard of all the Klans, accompanied by his staff of Grand Dragons, Klokards, Klabees, and others, arrived in state last night, and dawn found this Klan-dominated city ready to receive the masked thousands from near and far.

But a cog had slipped somewhere, and tonight it is admitted even by the Kloncilium that if the Klansmen now in Dallas were counted the total would not exceed 75,000. Leading citizens not members of the Klan say it is a sure sign that the crest of the Klan flood has passed and that Texas, where its control is undisputed, is at last ready to challenge the rule of the hooded knights.

"The spasm is passing and today proves it," said a Texan whose name is a household word, to THE TIMES correspondent tonight.

However, the day has been one Dallas will not soon forget. While the vast horde which the rulers of the Klan had boasted would concentrate in the city did not materialize, the fact remains that more Klansmen than ever before assembled at one time in any part of the South are tonight in Dallas. They are everywhere. There are thousands in hood and robe, and other thousands in ordinary civilian garb, their membership in the Klan disclosed by the little red "100 per cent. American" buttons and red ribbons with the number of the particular Klan to which they owe allegiance.

The Dragons, Klabees, Cyclopses and other high dignitaries of the organization are conspicuous in hotel lobbies, resplendent in robes of gold, royal purple or scarlet, the gorgeousness of the costume being in keeping with the rank of the wearer.

Blasts Catholics, Jews and Negroes.

The big feature of the day was the appearance at the State Fair Grounds this afternoon of the Imperial Wizard, who delivered a speech, the preparation of which is said to have taken five months. It was the official blast of the Klan, the first detailed statement by the Kloncilium of the things the masked horde stands for, a pronouncement in which Jews, Roman Catholics and negroes were grouped together as men and women incapable of attaining the "100 per cent. American standard"—people, the Wizard declared, who can never measure up to what he described as the Anglo-Saxon standard of patriotism.

Evans spoke to about 10,000 people, most of whom were members of the Klan. A loud-speaking apparatus had been installed to carry his words to all parts of the fair grounds.

Ku Klux brass bands were on hand to generate enthusiasm and "Onward, Christian Soldiers," the old hymn which the Klan has appropriated, was dinned into the ears of the thousands. The Wizard appeared on schedule time, which was 1 o'clock. He is a big man, with a rotund, clean shaven face. His mouth is large and he is always smiling. Incidentally he has a habit of waving his hand at the crowds in a kingly sort of fashion. On the stand with him were the chief officials of the Kloncilium. They were silent and undemonstrative, as befitted their positions as subordinates to the Imperial Wizard.

George K. Butcher, the Cyclops of Dallas introduced the Wizard. He lauded Evans as about the most 100 per cent. American of all the 100 per cent. Americans in the United States. Incidentally, Butcher referred to the impeachment of Governor Walton of Oklahoma, which he said was a great Klan victory. Then, with a deep bow, he presented the Imperial Wizard.

Very slowly, and with a broad smile, the Wizard advanced to the front of the stand. In his hand was a morocco bound case containing the manuscript of his speech. Before his elevation to the "throne" in Atlanta Evans was a dentist in Dallas. It was to be expected, therefore, that he would be accorded a great demonstration. But again something went wrong—a moment or two of cheering and handclapping. That was all. When he read that part of his speech in which he referred to the negroes there was a real outburst. But his references to the Jews and Catholics received only scattering applause.

The Wizard's Speech.

"Without being unkind or unjust," he said, "we can say that there are in New York City about two and one-tenth times as many unassimilable foreigners as native Americans, while the better inferior aliens, added together, reach a total almost three and one-half times that of the American stock.

"If this nation is to continue its charted course, and attain its destiny, there will have to be a harmonious assimilation of each and every alien element in complete accordance with these essential standards I have laid down. Such assimilation implies a common merging, which, to be successful, must bring about true social and political equality.

"It means intermarriage upon a basis of physical, mental and moral equality, with no menacing threat of uneugenic consequences to succeeding generations.

"It must mean, in a culminating sense, a spiritual wedding with the State, a kind and character of citizenship in which no higher temporal loyalty shall exist, or be permitted to exist.

"With respect to present and future immigration, this problem of the melting pot is made many times more difficult by reason of the fact that we already have at least three powerful and numerous elements that do now, and forever will, defy every fundamental requirement of assimilation. They cannot be merged because of insurmountable social, racial and religious barriers, they will always stand apart from our own people.

"First, there is the negro, ten and a half million in number, about a tenth of the whole population. They have not, they cannot, attain the Anglo-Saxon level. Both biology and anthropology prove it, and the experience of centuries confirms that conclusion. The low mentality of savage ancestors, of jungle environment, is inherent in the blood stream of the colored race in America. No new environment can more than superficially overcome this age-old hereditary handicap.

"In fact, with the present ever-increasing exodus from country to city, it is an undoubted fact that another generation will be marked by retrogression. The records, authoritative and unemotionally scientific show the negro to be specially susceptible to tuberculosis and alarmingly vitiated by venereal infections. Even though these eugenic considerations did not forbid such bodily merging, there could never be intermarriage between whites and blacks without God's curse upon our civilization.

Calls the Jew "Unblendable."

"Another absolutely unblendable element is the Jew. For ages, ever since his ejection from Judea, he has been a wanderer upon the face of the earth. In speaking of him there is never a reference to the country whence he came, because, throughout the centuries, there has been no country he would or could call his home. Into his life has come no national attachment. To him patriotism, as the Anglo-Saxon feels it, is impossible.

"Persecution has been his lot. With America, I am happy to say, the most tolerant of all the lands in which he has lived, an attitude reflected in the fact that already a fifth of the Jewish race is in this country. I would prefer to believe that all of this unceasing persecution has been unmerited. But, whether or not the Jew himself is largely to blame, its indelible impress is there, marked by generation after generation of unchanging and unchangeable racial characteristics.

"They are a people apart from all other peoples. They always will be. On the one side are their religious ceremonies, their social customs, their aversion to the Gentile, all as inflexible as fate; on the other, a racial and religious antipathy, unrelenting and unabating since the cross of Calvary. This should not be; it is.

"Would you have your daughter marry a Jew, were we willing? That could not solve the problem of assimilation, because the Jew would stand out against it. A merging through intermarriage can never be brought about. Were the melting pot to burn hundreds and hundreds of years, Jew and Gentile would each emerge as he is today. Only the Gentile would have been affected by many other elements and the Jew would not.

"As a race, the Jews are law abiding. They are of physically wholesome stock, for the most part untainted by immoralities among themselves. They are mentally alert. They are a family people, reverently and eugenically responsive to God's laws in the home. But their homes are not American, but Jewish homes, into which we cannot go and from which they will never emerge for a real intermingling with America.

"Not only because of the forbidden intermarriage, but also in an actual sense, is the Jew unmergable. By every patriotic test, he is an alien and unassimilable. Not in a thousand years of continuous residence would he form basic attachments comparable to those the older type of immigrant would form within a year. The evil influence of persecutions is upon him. It is as though he was here today and might be forced to flee tomorrow. He does not tie himself to the land. Jews owning farms are almost negligible; and it is largely only the Hebrew bankers and long established merchants that have their own homes.

"Rarely do you find him engaged even in the binding ties of constructive labor. Although charitable with each other, as a class the Jews are mercenary minded, money mad. 'To get, to have and to hold,' is their materialistic motto, and always, always do you find them seeking some tangible, quickly convertible, easily movable, kind of wealth which is in no vital way related to and dependent upon social and national values.

Declares "The Catholic a Danger."

"I must speak now of a third element among our people whose assimilation is impossible without the gravest danger to our institutions.

"No nation can long endure that permits a higher temporal allegiance than to its own Government. Throughout all history, medieval and modern, every churchly encroachment upon school and State, every attempt at temporal power by a religious organization, has been attended by sinister results both to Church and State. A divided allegiance is the deadliest, most menacing, of national dangers.

"The hierarchies of Roman and Greek Catholicism violate that principle. They demand, and increasingly seek to exercise, dominion outside the spiritual. To them the Presidency at Washington is subordinate to the priesthood in Rome. The parochial school alone is sufficient proof of a divided allegiance, a separatist instinct. They demand that our future citizens be trained not in public schools but under the control and influence of a priesthood that teaches supreme loyalty to a religious oligarchy that is not even of American domicile.

"I make no attempt to show cause and effect. In this connection you may make your own deductions. But do you realize, my friends, that the illiteracy of Europe is practically confined to Catholic countries?

"In Italy 37 per cent. of the population over ten years of age are without education and 38.7 per cent. of all marriages are by illiterates. Spain has 58.7 per cent. of illiterates; Portugal, 68.9 per cent.

"Consider the Greek Catholic countries and the comparison with our own standards is even more alarming. Greek illiteracy is 57.2 per cent.; that of Serbia, 78.9; while in Bulgaria and Rumania more than 68 in every hundred are illiterate.

"Is it unfair to suggest that Catholicism, if not actually desiring that condition, thrives upon ignorance?

"Since 1900 millions of Catholics from such sources have poured into America. They are coming now, under the quota law, in predominating numbers."

Tonight the Klan is holding a public initiation at the race track. It is announced that about 5,000 men will take the oath, while 2,000 women are to be initiated into the women's branch of the order.

At 11 P. M. the klansmen parade through the city. During the parade all traffic stops by order of the authorities.

October 25, 1923

1890 CENSUS URGED AS IMMIGRANT BASE

Eugenics Committee Report, Announced by Prof. Fisher, Would Limit South Europeans.

SELECTION ABROAD ASKED

Intelligence Test to Pick Aliens Superior to the American Average Is Suggested.

NEW HAVEN, Conn., Jan. 6.—An immigration law based on the census of 1890, a more effective system of inspection at American ports, and some sort of preliminary selection overseas are recommendations for a new immigration law made in the report of the Committee on Selective Immigration of the Eugenics Committee of the United States, given out today by Chairman Irving Fisher of Yale University, Chairman of the Eugenics Committee. Congressman Albert Johnson, Chairman of the House Committee on Immigration, is a member of the Eugenics Committee.

Numerical limitation and careful selection within the established limits are the two rules that should be followed in an immigration policy, the report states. American public opinion, it is asserted, is crystallizing around these conclusions:

"Never again is there to be an unlimited influx of cheap alien labor; a numerical limitation of immigration is here to stay, and there must be a careful selection of our immigrants within the fixed limits."

"The conviction that the census of 1890 should be used as the basis of any percentage law has been growing rapidly all over the country," the report continues.

"Since there were fewer Southeastern Europeans here in 1890 than in 1910, a percentage provision based on the former census would decidedly cut down the number of such immigrants. This provision would change the character of immigration, and hence of our future population, by bringing about a preponderance of immigration of the stock which originally settled this country.

Says Limitation Is Sound Policy.

"On the whole, immigrants from Northwestern Europe furnish us the best material for American citizenship and for the future upbuilding of the American race. They have higher living standards than the bulk of the immigrants from other lands; average higher in intelligence, are better educated, more skilled, and are, on the whole, better able to understand, appreciate and support our form of government.

"A percentage limitation based on the 1890 census is sound American policy, based on historical facts. It is not here a question of racial superiority of Northwestern Europeans or of racial inferiority of Southeastern Europeans. It is simply a question as to which of these two groups of aliens, as a whole, is best fitted by tradition, political background, customs, social organization, education and habits of thought to adjust itself to American institutions and to American economic and social conditions; to become, in short, an adaptable, homogeneous and helpful element in our American national life."

A percentage limitation based on the census of 1890, it was maintained, would not only reduce the inflow of "cheap" labor, but "would also greatly reduce the number of immigrants of the lower grades of intelligence, and of immigrants who are making excessive contribution to our feeble-minded, insane, criminal and other socially inadequate classes."

Would Examine Them Abroad.

"Consular certificates should be required of each intending immigrant before he starts on his voyage," the report continues. "This certificate should contain answers to questions essentially the same as are asked of the immigrant on his arrival at our ports, as well as full information about his health, civic record, political activities and character and the general standing and health of the immigrant's family. It should include a statement from the responsible police authorities of the immigrant's residential city or district that the applicant has not been convicted of crime (other than political), and should be verified by oath before a United States consular officer abroad.

"Certificates should be issued only up to the numbers allowed by the quotas, and should be good for six months, so that if the alien came at any time within that period he would not be denied admission as being in excess of the quota allowance.

"If our future population is to be prevented from deteriorating, physically and mentally, higher physical standards must be required of all immigrants. In addition, no alien should be admitted who has not an intellectual capacity superior to the American average. Aliens should be required to attain a passing score of, say, the median in the Alpha test, or the corresponding equivalent score in other approved tests, these tests to be given in the native tongue of the immigrant. Further, if possible, aliens whose family history indicates that they come of unsound stock should be debarred."

The members of the Committee on Selective Immigration include Madison Grant, Chairman; Robert De C. Ward, Vice Chairman; Charles W. Gould, Lucien Howe, Albert Johnson and Francis H. Kinnicutt.

January 7, 1924

NEW YORKERS FIGHT IMMIGRATION BILL AS RACIALLY UNFAIR

Special to The New York Times.

WASHINGTON, Feb. 24.—Organized opposition to the Johnson restrictive immigration bill developed today, when twenty of the twenty-two members of the New York State Democratic delegation in the House endorsed a declaration in which it was charged that the bill was deliberately framed to favor the Nordic races and discriminate against races from Southern and Eastern Europe.

The Johnson bill would base the quotas on the number of nationals of any country resident in the United States under the census of 1890, as opposed to the present basis of the 1910 census. The New York Congressmen hold that such racial and religious discrimination is a "new but perilous doctrine for democratic America," which was founded upon the principle that "all men are created equal."

The Administration leadership in the House has put immigration legislation next to tax legislation on the Congressional program, as the present law expires at the end of the fiscal year and a considerable majority of the members feel that some new restriction law is essential to the best interests of the country.

The New York Democrats do not openly take exception to this contention, but they hold that the basis for fixing the quotas should not be changed.

When the Johnson bill was framed it was forecast that strong opposition would appear against it on the part of delegations from large cities such as New York, Philadelphia, Boston and Chicago, which have big foreign communities. There is much speculation as to whether this fight will develop sufficient strength to force Congress to retain in the new law provisions making the 1910 census the basis of quota-fixing.

Text of the Protest.

The declaration by the New York Democrats follows:

"The undersigned, being Democratic members of the House of Representatives from the State of New York, are unalterably opposed to the rigidly restrictive immigration bill reported by the Committee on Immigration and Naturalization and known as 'the Johnson bill, H. R. 6540.'

"The foreign-born population of our country and those born here of a foreign parent comprise 33⅓ per cent. of the total population. Of these, at least 25 per cent. are recent immigrants and constitute the young men and women of today's laboring classes, so necessary to our industrial prosperity.

"We are underhoused, underconstructed and underdeveloped and are in sore need of those who are willing to do our work, both skilled and hard and laborious, and this bill would tend to keep out that class of immigrants best suited for such occupations.

"It would not, moreover, bring into this country a better class or a more assimilable body of immigrants.

"Our national policy, as expressed in the act of 1917, a distinctly selective measure, has been to welcome to our shores all immigrants who are desirable; that is, all who are mentally, morally and physically fit and friendly to our form of government.

"The proposed bill goes even further than the present law in fixing an arbitrary number of immigrants who can be admitted.

"It is the avowed purpose of the Immigration Committee to have this law embody our permanent policy of immigration, and bind us to a program which is inflexible, unscientific and unjust, and is, furthermore, an attempt to treat a human problem upon a cold, mathematical formula, since its basis is quantitative rather than qualitative.

Sees "So-Called Nordics" Favored.

"The 'Johnson bill' is particularly objectionable because it discriminates against certain nationalities already going to make up a great part of our population, and fans the flames of racial, religious and national hatreds and brands forever elements already here as of an inferior stock.

"It discriminates against Italy, who gave us the great Columbus; it discriminates against Poland, who gave us our revolutionary heroes, Kosciusko and Pulaski; it discriminates against Russia of the great Tolstoy; against Hungary, who gave us the great patriot Kossuth; against Greece, the land of Venizelos; against Czechoslovakia, from whence hails the distinguished Masaryk; against Yugoslavia, who sent us the great inventor Michael Pupin, and finally against France, from whence came the immortal Lafayette and Rochambeau.

"Have we so soon forgotten the World War, when the youth of those same nationalities, residents in the United States, joined hands with their relatives across the seas and brought victory to us and our allies in that great conflict? Shall we exclude those compatriots in arms by a mere mathematical formula? Is it fair? Is it American?

"This proposed law would adopt as a basis of entrance 2 per cent. of the foreign population of 1890. In its determined effort to be as unfair as possible, the committee, in addition to reducing the percentage from 3 per cent., adopts as a basis census figures thirty-four years old, instead of taking the census of 1920, now available, or even the census of 1910, the basis of the present law. This basis was deliberately selected to favor the so-called Nordic races and discriminate against races from Southern and Eastern Europe, which discrimination is, indeed, a new but perilous doctrine for democratic America, founded upon the declaration that 'all men are created equal.'

"Our great country is still big enough, geographically, politically and socially to receive those persons knocking at our doors, whether of brain or brawn, who answer our mental, moral and physical requirements and can contribute to our science, our art, our literature, our commerce or our industry."

The members of the New York Democratic delegation signing the statement are:

John F. Carew, John J. Kindred, Christopher D. Sullivan, Thomas H. Cullen, James M. Mead, Anthony J. Griffin, William E. Cleary, John F. Quayle, David J. O'Connell, Loring M. Black Jr., Sol Bloom, George W. Lindsay, Emanuel Celler, Parker Corning, Samuel Dickstein, John J. Boylan, John J. O'Connor, Frank Oliver, Anning S. Prall and Royal H. Weller.

President Coolidge and Secretary of Labor Davis both have endorsed immigration restriction legislation, but have not expressed an opinion concerning the basis on which quotas should be made.

February 25, 1924

COOLIDGE SIGNS THE IMMIGRATION BILL

EXCLUSION LAW IS SCORED

President Calls That Part of Act "Unnecessary and Deplorable."

WOULD VETO IT IF ALONE

Accepts Measure Because of Its Good Features and Because Comprehensive Law Is Needed.

SENATE LEADERS PLEASED

Alien Quotas of 2 Per Cent. Based on the 1890 Census Go Into Effect on July 1.

Special to The New York Times.

WASHINGTON, May 26.—The Immigration bill, with its provision for Japanese exclusion, was signed today by President Coolidge.

In a statement given out soon after he had signed the bill the President explained that he had approved the bill because it had many desirable features and a comprehensive act was necessary at this time, but that he would have vetoed it if Japanese exclusion had stood alone.

The President's Statement.

The President's statement follows:

In signing this bill, which in its main features I heartily approve, I regret the impossibility of severing from it the exclusion provision, which in the light of existing law affects especially the Japanese.

I gladly recognize that the enactment of this provision does not imply any change in our sentiment of admiration and cordial friendship for the Japanese people, a sentiment which has had and will continue to have abundant manifestation.

The bill rather expresses the determination of the Congress to exercise its prerogative in defining by legislation the control of immigration, instead of leaving it to international arrangements. It should be noted that the bill exempts from the exclusion provision Government officials, those coming to this country as tourists or temporarily for business or pleasure, those in transit, seamen, those already resident here and returning from temporary absences, professors, ministers of religion, students, and those who enter solely to carry on trade in pursuance of existing treaty provisions.

But we have had for many years an understanding with Japan by which the Japanese Government has voluntarily undertaken to prevent the emigration of laborers to the United States and in view of this historic relation and of the feeling which inspired it it would have been much better, in my judgment, and more effective in the actual control of immigration, if we had continued to invite that cooperation which Japan was ready to give and had thus avoided creating any ground for misapprehension by an unnecessary statutory enactment.

"Unnecessary and Deplorable."

That course would not have derogated from the authority of the Congress to deal with the question in any exigency requiring its action. There is scarcely any ground for disagreement as to the result we want, but this method of securing it is unnecessary and deplorable at this time.

If the exclusion provision stood alone, I should disapprove it without hesitation, if sought in this way at this time. But this bill is a comprehensive measure dealing with the whole subject of immigration and setting up the necessary administrative machinery. The present quota act of 1921 will terminate on June 30 next. It is of great importance that a comprehensive measure should take its place and that the arrangements for its administration should be provided at once in order to avoid hardship and confusion.

I must therefore consider the bill as a whole and the imperative need of the country for legislation of this general character. For this reason the bill is approved.

New Law in Effect July 1.

The new law will go into effect July 1, when the temporary immigration law which limited immigration to 3 per cent. of the national here under the 1910 census will expire. The new law is not only more rigid, in that it excludes Japanese laborers, but lowers the total immigration yearly to the United States from about 370,000 to approximately 160,000. The quota is based on 2 per cent. of the 1890 census.

Senators who led the fight for exclusion today commented favorably on the action of President Coolidge.

Senator Lodge, Chairman of the Foreign Relations Committee, said:

"I felt confident that the President would sign the Immigration bill, for I knew he sympathizes with the general purposes of the legislation and realizes fully its great importance. It is a very great measure, one of the most important if not the most important, that Congress has ever passed. It reaches far into the future. The President by his support of the bill will receive great and deserved credit from one end of the country to the other. It is a most important piece of constructive legislation."

Senator Hiram W. Johnson of California said:

"It is a matter of congratulation and rejoicing that California finally prevails in the long struggle for the protection and preservation of its own. With the last chapter closed now, I apprehend neither difficulties nor dangers.

"We have done what was clearly our right, without intention of offense, and, giving no cause for offense, offense cannot justly be taken. California's cherished policy is now the nation's maturely determined policy. The victory, after so many attempted checks, gives us great pleasure and increased rejoicing."

Senator Shortridge of California said:

"By approving the Immigration bill, the President has rendered a great service to our country and to civilization.

"We of California, who have urged the exclusion of aliens ineligible to citizenship are profoundly grateful to those from other sections of the country who have assisted us."

While the bill was pending in Congress efforts were made by President Coolidge to influence Congress to postpone the effective date of Japanese exclusion until March, 1925, so that the same object might be obtained by treaty. The conferees agreed to such postponement, but their report was rejected by the House, which fixed the date as July 1, 1924, and the Senate did the same thing over the protest of the President. In both houses the votes were far in excess of the two-thirds that would be necessary for passage over a veto.

After the original action by the House putting exclusion in the bill, Ambassador Hanihara, on April 11, sent a note to the State Department in which he spoke of "grave consequences" if the exclusion section became effective. This note greatly aroused the Senate, and some Senators say that the exclusion section might not have been accepted by the Senate if this note had not been written. Subsequent explanations by the Ambassador and Secretary Hughes failed to change the temper of the Legislature.

May 27, 1924

WARNS OF ALIEN INFLUENCE.

Dry League Links Foreign-Born to Fight on Prohibition.

"The growing social and political influence of the unnaturalized foreign-born in our population" harbors a menace to American institutions, according to a statement issued yesterday by the Anti-Saloon League of New Jersey.

Congressional representation and Presidential electors are apportioned on the basis of the total population, not of citizens, it is pointed out. The unnaturalized persons totaled 7,494,057, according to the census of 1920.

The system of Congressional apportionment allows thirty-one Congressmen to these aliens, according to the New Jersey organization, which argues that most of these Representatives come from large cities, where alien habits have great influence.

The organization says that the purpose of its discussion of these points is not to suggest a new Constitutional Amendment, but "to illuminate the wet and dry discussion, and other social and moral problems where Americanism meets alien habits, ignorance and prejudices."

"It may be taken as a truism," continues the statement, "that in Congress and the New Jersey House of Assembly, wherever there is ranting against the tyranny of the prohibitory law, and invasion of inalienable rights, it comes from the representatives of a region where there is great congestion of alien population who know little and care less for American ethical standards or the ideals that came over in the Mayflower.

"It will not do to say that, inasmuch as they have no votes, they have no influence. Then why should they have seven Congressmen from New York State, representing aliens chiefly from New York City, which has in its population 1,991,547 foreign born, including 203,450 Irish; 194,254 Germans; 145,679 Poles; 126,739 Austrians; 64,393 Hungarians; 479,797 Russians (mostly Jews) and 390,832 Italians, not to speak of Chinese, Japanese, Slavs, Greeks and other races and nationalities counted as inhabitants."

September 27, 1926

NOW "NATIONAL ORIGINS" FIX QUOTAS FOR ALIENS

By a Complicated Analysis of Census Figures, Experts Have Determined the Proportions of the Race Stocks Which Make Up the American Population

By S. A. MATHEWSON.

SURVIVING opposition from the White House and in Congress, the national origins immigration quotas, proclaimed by President Hoover only because the law is mandatory, go into effect tomorrow, displacing the 1890 census quotas used for the last five years.

The national origins provision seeks to preserve the existing racial proportions of the American people by distributing a total annual immigration of 150,000 among the quota countries in such proportions as to make this army of immigrants a reflection in miniature of the make-up of the white population derived from quota countries living in the United States in 1920—native and foreign-born alike.

First, of course, the make-up by national stocks here in 1920 had to be determined. This was found by estimating the number contributed by each country through descendants living in 1920, of early colonists and former immigrants, as well as immigrants themselves alive in 1920. The annual quota allotted to each country is then fixed by that percentage of the 150,000 which corresponds to its contribution to the population of 1920.

In contrast with national origins, the 1890 census basis is a foreign-born basis exclusively, since it fixed the quota for each nationality at 2 per cent of the foreign-born of that nationality resident in the country according to the census of 1890.

A comparison of the principal quotas under these two systems shows how widely they differ in results:

Country.	National Origins Quota.	Previous Quota.
Armenia	100	124
Australia (incl. islands)	100	121
Austria	1,413	785
Belgium	1,304	512
Czechoslovakia	2.874	3.073
Denmark	1,181	2,789
Estonia	116	124
Finland	569	471
France	3.086	3,954
Germany	25,957	51,227
Great Britain and Northern Ireland	65,721	34,007
Greece	307	100
Hungary	869	473
Irish Free State	17,853	28.567
Italy	5,802	3,842
Latvia	236	132
Lithuania	306	314
Netherlands	3.153	1,648
Norway	2,377	6.453
Poland	6,524	5.928
Portugal	440	503
Rumania	295	603
Russia (European and Asiatic)	2.784	2,248
Spain	252	131
Sweden	3,314	9.561
Switzerland	1,707	2,801
Syria and the Lebanon (French mandate)	123	100
Turkey	226	100
Yugoslavia	845	671

The new law allots a minimum of 100 immigrants each to countries (not included in the above list) which, upon the basis of national origins, would have quotas of less than 100.

The opponents of national origins maintained it was utterly impossible to determine the national origins of our vast white population; the supporters of that provision, on the other hand, insist that it has been done and with reasonable accuracy.

Though the immigration act of 1924 designated the Secretaries of State, Commerce and Labor to ascertain the national origins quotas and report them in 1927 to the President for proclamation, the actual work of computation was delegated to a subcommittee of experts, headed by Dr. Joseph A. Hill, Assistant Director of the Census.

The task confronting these experts was to ascertain the national origins of the 94,820,915 white population of the United States enumerated in the census of 1920.

After trying out various lines of approach they devised an ingenious method of estimating with substantial accuracy that part of the 94,-820,915 white population here in 1920 which was derived from the population here in 1790. This method was based upon the principle of taking the percentage of the white native-born population which in 1924 had native parents, then estimating what percentage of those native parents in turn had native parents, until the percentage of those having forebears living in the country in 1790 had been reached. Recent census classifications made this method possible.

Beginning with the census of 1890 up to that of 1920, inclusive, the population was classified into white and colored, these two divisions into native-born and foreign-born, and the native-born further classified into those who had native parents, those who had foreign-born parents and those who were of mixed parentage, native and foreign-born. What enabled the committee to follow out its computations was the fact that the entire population had been cross-sectioned by the census of 1890 and the following censuses into five-year age groups—0—5,—5—10, 10—15 and so on.

Considering only the white population, we know just how many in any age group were foreign-born and native-born and of the native-born how many in any age group had native parents, how many had foreign-born parents and how many were of mixed parentage.

To illustrate: Take the age group of the native-born 35-40 years of age, enumerated in the 1920 census. By comparing the number of native-born of native parents with the number of native-born of foreign-born parents we find that the native-born of native parents constitute 77 per cent of these native-born. Children of mixed parentage offer no difficulty, since one-half their children would be credited to the native-born of native parents.

But the persons 35-40 years of age in 1920 were born between 1880 and 1885, and they were first listed in the 1890 census in the 5-10 age group, where it appears that 76 per cent of the native-born children within that age group had native parents. These same children were similarly enumerated in the census of 1900, under the 15-20 age group, when they were 75 per cent of their age group; in 1910, when they were from 25-30, the percentage was 77; and again in 1920, when from 35-40 they were 77 per cent of their age group. The slight variations in percentage are due to difference in death rates between the children of native parents and the children of foreign-born parents. Now the average of these four percentages, or 76 per cent, is the average percentage of native white persons born in the United States between 1880 and 1885 who had native parents.

Determining the Activities.

The next step is to determine what proportion of the parents of the above 76 per cent were themselves

BASIS FOR NATIONAL ORIGINS QUOTAS

In its final report the Government's committee of experts made the following apportionment of the white population of the United States of 1920 by country of origin:

Country of Origin Quota Countries.	Total.	Colonial Stock.	Post-Colonial Stock.
Austria	843,051	14,100	828,951
Czechoslovakia	1,715,128	54,700	1,660,428
Denmark	704,783	93,200	611,583
France	1,841,689	767,100	1,074,589
Germany	15,488,615	3,036,800	12,451,815
Great Britain and North Ireland	39,216.333	31,803,900	7,412,433
Irish Free State	10,653,334	1,821,500	8,831,834
Italy	3,462,271	....	3,462,271
Netherlands	1,881,359	1,366,800	514,559
Norway	1,418,592	75,200	1,343,392
Poland	3,892,796	8,600	3,884,196
Russia	1,660,954	4,300	1,656,654
Sweden	1,977,234	217,100	1,760,134
All others	4,750,419	1.061.100	3,689,319
Total quota countries	89,506,558	40,324,400	49,182,158
Total non-quota countries	5.314,357	964,170	4,350,187
Total white population	94,820,915	41,288,570	53,532,345

the children of native parents. Since the parental age is usually between 20 to 35, it can be assumed that, with few exceptions, the parents of children born between 1880 and 1885 were themselves born between 1845 and 1860. Consulting the census tables, it apears that of the native population born between 1845 and 1850, 87 per cent had native parents; between 1850 and 1855, 82 per cent had native parents, and of those native-born between 1855 and 1860, 76 per cent had native-born parents. The average of these percentages, or 82 per cent, is therefore the average percentage of the parents born between 1845 and 1860 who were themselves the children of native parents.

Having started with the native-born population between 35 and 40 years of age in 1920, we found that 76 per cent of them had native parents, and we now find that 82 per cent of the parents of these children also had native parents. Now, 82 per cent of the 76 per cent gives 62 per cent, which is the percentage of native-born population between 35 and 40 years of age in 1920 who had native-born grandparents.

By continuing this process, the percentage of the population born between 1880 and 1885 having forebears living in 1790 can be found.

By applying this process to each age group, the percentage of the native-born white population living in 1920 who had forebears living in the United States in 1790 can be determined. Since the exact number of native-born in each age group living in 1920 is shown by the census of that year, the rest of the computation consists merely of applying percentages and adding the results.

The "Colonial Stock."

The 41,288,570 thus computed as that part of the population descended from the population of 1790 is called, for convenience, the "Colonial stock." Deducting this 41,-288,570 from the 94,820,915, we have 53,532,345, called the "post-Colonial stock," which includes the descendants living in 1920 of all immigrants who entered the country after 1790 as well as all foreign-born enumerated in the census of 1920.

It now remains to divide these two stocks according to their national origins and afterward combine the same national stocks to get the proportion of the 94,820,915 total.

Take first the 41,288,570 "Colonial stock" descended from the 3,172,444 white population in the United States in 1790, as enumerated by the existing schedules of the 1790 census. It is known that the 1790 population was mainly British, but to what extent has been a matter of dispute, since the census of 1790 did not classify the population by nativity or nationality. Using an official classification by nationality, based upon the apparent racial character of names in the census of 1790, whether Dutch, English, German, Irish, &c., the experts apportioned the "Colonial stock" accordingly. The quotas as reported in 1927 were based in part upon this apportionment which was admittedly provisional.

Immediately there was an outburst of protest. It was charged that many Irish names had been omitted from the 1790 census, while many persons of German and Scandinavian stock joined in condemning the national origins provision, and by vote of Congress its operation was postponed a year.

By 1928 the experts had revised their estimates, whereby the number attributed to the Irish of 1790 was substantially increased. But it was the eve of a Presidential election and national origins was again postponed. In 1929 the quotas reported were much the same as in 1928.

In consequence of the above revisions the Irish Free State quota jumped from 13,862 as reported in 1927 to 17,853 in this year's final report, while quotas for Gemany and certain other countries received slight increases, nearly all at the expense of the British quota, which was reduced from 73,039 as reported in 1927 to 65,721 in 1929.

As finally reported, the "Colonial stock" was apportioned as follows:

France	767,100
Germany	3,036,800
Great Britain and North Ireland	31,803,900
Irish Free State	1,821,500
Netherlands	1,366,800
All others	2,492,470
Total	41,288,570

Considering next the 53,532,345 "post-Colonial stock" living in 1920, we find more than three-fifths of this stock, comprising 13,712,754 foreign-born and 19,190,372 native-born children of foreign-born parents, already classified by national origins in the census of 1920. This is due to the fact that all censuses have classified the "foreign-born" by their countries of birth, including foreign-born children admitted as well as their foreign-born parents, and that, beginning with 1890, every census has classified also the "native-born children of foreign-born parents" according to the countries of birth of those parents.

Where father and mother were of different national stocks each stock was credited with half the number of the children, fractions being used where necessary, since "individuals are merely units of measure of blood stock."

The remaining 20,629,219 of the "post Colonial stock" necessarily consists of grandchildren and later generations of "post-Colonial" descent, who are, of course, children of native-born parents. Since even the recent censuses do not give the parent stock of children of native-born parents, the national origins of these 20,629,219 grandchildren and later generations can be determined only by finding in some other way the national origins of those who must have been their parents or earlier ancestors.

For statistical convenience the blood stock of these 20,629,219 will be considered as derived from two sources.

One source comprises the 5,752,578 grandchildren and later generations of "post-Colonial" descent here in 1890, their number having been computed with the aid of the 1890 census classifications in similar manner as were the 20,629,219 under consideration. The national origins of these 5,752,578 are determined from their grandparents, who must have been in the country a generation before 1890 to become grandparents by that date. Therefore, the composition of the "post-Colonial stock" of 1870 would determine the national origins of the grandparents and through them the national origins of the 5,752,578 and their contributions through descent to the 20,629,219 living in 1920. It would determine also the national origins of the survivors of the 5,752,578 still living in 1920, **who would likewise compose part of the 20,629,219, the number of such survivors being estimated from appropriate death-rate averages.**

The composition of the "post-Colonial stock" of 1870 was estimated from annual records of immigration from 1820 to 1870, showing the countries whence **immigrants** came; **decennial censuses from 1850 to 1870, which classified the foreign-born according to their countries of birth, and historical material covering 1790 to 1820, when immigration was very small.**

The rest of the blood stock of the 20,629,219 grandchildren and later generations living in 1920 would be derived through descent from the native-born children of parental ages between 1890 and 1920 whose parents were foreign-born and classified in the 1890 census and later censuses. This completes apportionment of the "post-Colonial" 53,532,345.

Combining now the same national stocks in the "Colonial" and "post-Colonial" parts we obtain the apportionment of the whole 94,820,915.

But we are concerned only with those countries which are subject to quota restrictions, whereas the 94,820,915 white population includes also the national stocks of certain countries not subject to quota restrictions, such as Canada, Mexico and all countries of Central and South America. Hence, in computing the national origins quotas the tabulators merely disregard the estimates of the blood stock of non-quota countries.

With respect to what degree any possible error in the estimate of a national stock would affect its quota, it should be noted that the total affected, 89,506,558, is to 150,000 practically as 600 to 1. Thus, an error of 600 in a national stock estimate would make a difference of only one immigrant in the quota, while an error of 60,000 would mean only 100 immigrants.

June 30, 1929

Alien Departures Exceeded Immigration In 1932 for First Time in Nation's History

Special to THE NEW YORK TIMES.

WASHINGTON, Aug. 17. — The number of aliens leaving the United States permanently in the last fiscal year was nearly three times that of the new immigrants admitted for permanent residence, according to a statement issued today by Secretary Doak.

"During the fiscal year ended June 30, 1932," the statement said, "only 35,576 permanent immigrants were admitted, and 103,295 alien residents left the United States with the expressed intention of making their homes in other countries. In other words, alien emigration exceeded immigration by 67,719.

"A comparison with corresponding figures for the preceding fiscal year is exceedingly interesting. In that year 97,139 immigrants were admitted, as against 61,882 who departed permanently from the United States. Thus for the fiscal year just ended we show an increase of departures over admissions of 67,719, as compared with an increase of 35,257 admissions over departures for the preceding year.

"This is the first year in the history of the country when the number of aliens permanently departing from the United States exceeded the arrivals. During this period 19,426 were formally deported, 2,637 aliens who had become destitute within three years after arrival were returned to their native countries by the department, and 10,750 aliens who were apprehended and found to be subject to deportation were permitted to depart at their own expense without formal deportation.

"Therefore, it will be seen that the department directly caused 32,813 aliens, who were here unlawfully or who had fallen into distress, to depart from the country. In addition, it has been variously estimated that thousands of other aliens left the country on their own initiative in preference to apprehension and subsequent deportation."

August 18, 1932

VEGA BAJA FOOD CENTER
COLD BEER • COLD CUTS • PRODUCTOS TROPICALES • VEGETABLES •
GOOD-O
KOLA
Vega Baja
Center
we accept food stamps
aceptamos...
CUPONES PARA COMIDA
EL DIARIO
LA PRENSA
SE VENDE AQUÍ
HOT SANDWICHES
White Rock
White Rock
Celebrate the Joys of Spring KODAK
INSIDE SPECIAL
CLEARANCE
SALE FINAL
Pampers
EL SOL
cheer

CHAPTER 3

Migration, Race and Poverty

The *bodega,* in addition to selling food, often serves as a social center for a Puerto Rican community.

Courtesy Gene Brown

RAW LABOR FROM THE SOUTH.

A telegraphic dispatch from Raleigh, N. C., a few days ago, stated that a railroad ticket agent at that place had sold 300 tickets to negroes leaving the State during the month of August, and the information was added that a large proportion of those leaving North Carolina were going to Massachusetts.

There was no suggestion that this movement of negroes was an indication of disapproval of the recent decision in the State to disfranchise the illiterate blacks. Indeed, such an explanation was not really necessary; for before the proposition to restrict the use of the ballot in North Carolina had taken definite shape a considerable movement of the black population toward the North had set in and had been kept up for several years.

Much of this North Carolina emigration has found its way to New York and New Jersey. In a sense it has been "promoted" immigration. The negro man or woman, particularly the woman, who had been able to reach a New Jersey city or town where domestic labor was in demand, was soon able, without other capital than a willingness to take service at $12 a month "and found," to send a message of good tidings to other negroes to "come North," where pay was good and where very few questions about qualifications were asked.

But in New Jersey and New York it is being discovered that while these negroes from the South soon learn to demand $12 a month to begin with, and speedily threaten to leave if their wages be not advanced to $15 or $20, they are expensive at any price. They are not domestic servants at all. They cannot build fires, but they understand the advantages and disregard the dangers of kerosene oil as a fuel. They permit kitchens to become as dirty as pig sties. Their unaccustomed hands cannot be trusted with crockery or glassware. They cannot cook. They are not ashamed to beg. They work when they choose; they quit without consulting the convenience of the employer.

If there was a great deal more of this sort of labor available in this section we should not be materially better off. An excess of supply might depress the rate of wages for this unskilled labor, and enable employers to hold on to servants a little longer. As it is, most of the housekeepers employing these colored immigrants from the South are engaged in an educational work, and are enduring all the pains of the teacher while paying the green pupil for unintelligent service, too often to find themselves abandoned upon the invitation of another housekeeper to assume the double relation of instructor and wage payer at a higher rate.

The Northern States that are receiving and assimilating this raw Southern negro labor ought to be ready to appreciate the efforts of some Booker Washingtons who will take upon themselves the task of preparing the negroes for domestic service. If such men will but teach the negro men to clean stables, care for horses, feed and harness and drive them, run lawn mowers, make and keep gardens, and also to keep engagements, they will deserve monuments and the gratitude of two races. If the colored women will but learn a few things before they come to us to demand salary, when they should pay our wives tuition, they will be more welcome, and our wives will be at greater liberty to devote their educational efforts to us and to our children.

September 9, 1900

NEGROES DESERTING SOUTH FOR NORTH

Considerable Discussion Caused by Migration of Laborers to Border Cities

INCREASE IS STRIKING

From 1900 to 1910 the Increase Was More Than Three Times That of the Total Negro Population.

New York, Oct. 28, 1916.

To the Editor of The New York Times:

The movement of negro laborers from the South in large numbers during the past few months has created considerable discussion in the public press, North and South, and not a little concern in parts of the South. A striking feature of most of this discussion is the absence of statements about the migration of negroes before the present movement. The migration of negroes northward in considerable numbers year by year for the last two or three decades has been quietly going on, although it may not have attracted much attention.

The indication of this movement since 1880 is shown by the percentage of increase of the negro population of the following nine Northern and border cities: Boston, Greater-New York, Philadelphia, Chicago, Cincinnati, Evansville, and Indianapolis, Ind.; Pittsburgh, and St. Louis. The census figures for these nine cities showed that between 1880 and 1890 the negro population increased about 36.2 per cent.; from 1890 to 1900 it increased about 74.4 per cent.; and from 1900 to 1910 about 37.4 per cent. In the first decade the increase was more than three times the increase of the total negro population; in the second period it was more than four times as large and shows the influence of the economic disturbances of the period. In the last period the increase was nearly three times larger than the increase of the total negro population.

The rate of increase in the Southern cities has been large, although less than that of the Northern cities during the same period, indication that similar causes were operating to draw negroes to Southern cities, although these causes were weaker than those operating in Northern cities. The percentage increase of negroes in fifteen Southern cities was, from 1880 to 1890, about 38.7; from 1890 to 1900, about 20.6; from 1900 to 1916, (sixteen cities with addition of Birmingham, Ala.,) 20.6 per cent. These percentages are based upon census figures for the following cities; Wilmington, Del.; Baltimore, Md.; Washington, D.C.; Norfolk and Richmond, Va.; Charleston, S. C.; Atlanta, Augusta, and Savannah, Ga.; Louisville, Ky.; Chattanooga, Memphis, and Nashville, Tenn.; Birmingham and Mobile, Ala., and New Orleans, La. It may be added in passing that from 1880 to 1910 the increase of White population in these Southern cities has been very similar to that of the negroes.

The causes of this movement during this longer period have been the same as those affecting the negro population in the last few months. The only difference has been the increase in the volume of the movement because of the increase in its influencing causes.

The newspaper discussion of the arrests, fines, and jail commitments, restlessness of the younger generation of negroes and political calculation may be given place as individual factors in the causes for such a movement. But a further shifting of the facts shows that at bottom the negro is reacting toward certain fundamental conditions in a similar manner to the response of other elements in our cosmopolitan population.

Economic, educational, and civic opportunities and conditions are having their influence. Among these forces economic conditions demand first consideration. The wages in all lines of occupations open to negroes in the North are considerably higher than those in the South. For instance, the writer found in New York City in 1910 that the majority of thirty-seven negro men who had recently come to the city had received weekly wages before they came of between $3 and $7, and after coming to the city the majority of the same group of men were receiving weekly wages of between $9 and $12 per week. Out of the twenty-six women who had recently come to the city the majority, before coming, had received weekly wages ranging from below $3 to $5, and after coming to the city the majority had advanced in weekly wages so as to range between $5 and $8 per week.

Even if these facts are not conclusive, they are corroborated by the replies to a further question which was asked of negro wage earners in 1909 and 1910. In those years 365 wage earners were asked their reason for coming to New York City. Nearly half of the answers showed a decided economic reason; some gave reasons that were probably economic. The answers showed that probably three-fourths of these wage earners had been drawn to the city through economic motives.

It has been frequently stated and generally believed that the negro has no economic opportunity in the North. And it cannot be denied that heretofore he has been practically shut out of most of the highly paid employments. He has also been crowded into very limited fields and mostly into occupations where white men in large numbers did not care to compete. This has been mainly in domestic and personal service. What has not often always been considered is that the Northern wage, in occupations where negroes might enter, coupled with board, lodging, and other perquisites, sometimes has been better than the wage in the more skilled occupations into which negroes could enter in the South, to say nothing of the unskilled.

The writer has found many cases in the North of semi-skilled or skilled negro workmen who were crowded out of occupations to which they aspired either because of color discrimination or because of an efficiency lower than that of competing white workmen. Instead of returning to the South with its open field in their line, these men have become janitors in office buildings, messengers, Pullman porters, porters in stores, &c. They remain partly because in such jobs they receive better wages than they were getting as skilled workmen in the South. It should be borne in mind that the average workman does not consider the higher cost of living and the greater congestion of residence in the northern centre. He makes a comparison of wages in terms of dollars and cents and does not think of the economists' distinction between money wages and real wages.

The present labor demand in the opening of wider fields of employment for negroes, therefore, has simply accelerated a movement which has been under way for a long time. What should be emphasized both North and South in considering the matter in connection with the welfare of the negro and the nation is the negroes response to a fundamental economic call in a way similar to that of other groups. The migration cannot be stopped by ordinances or statutes and will probably continue as long as the conditions that create it persist.

There is another phase of the matter which has not seemed to find much expression in the public press. The settlement of large numbers of negroes in Northern towns and cities has been discussed from the point of view of its effect upon opinion concerning the so called negro problem. Attentions should also be directed to the adjustment of these newcomers to the life of the Northern communities in such a way that the political and economic forces which have attempted to exploit other elements of the communities may not use the negro in the same way. For example, the Northern communities need to see that proper houses, sanitary, and sewerage regulations are carefully provided in negro neighborhoods. Otherwise, the whole community may suffer from official neglect and landlord exploitation. "Loan shark" agents, company stores, and sweat-shop conditions may suck new life from the blood of these labor recruits. Ward heelers and local political bosses may gain a new grip on the community by posing as friends and champions of the new citizens.

Naturally, the question of the best steps to take in these larger community matters arises. Two or three suggestions may not be out of place:

First-Among the negroes in each of these communities may be found individuals of character and intelligence who are eager to do everything they can for the advancement of their people. Public-spirited white citizens may get in touch with these persons through personal contact.

Second-These white and colored citizens can then best help the adjustment of the colored people by coming together in some form of joint organization with a definite purpose to benefit in the main the colored population.

Third-This organized effort may look over the field and agree upon a definite program of active work along a few lines such as the community most seems to need. This program may include a careful study of the living and working conditions of the colored people, a plan of publicity, and a plan to keep check on the conduct of public officials wherever their duties touch negro life. It takes public attention to see that the police department, the housing, the sanitary, the health, and other departments of the community government adequately serve the colored neighborhoods. So fundamental is the question of work and wages that a special part of any community program should provide some employment, finding agency or agencies, should seek methods of touch with employers, and should aim to set negro workers thinking and acting collectively with their employers for their own interest.

GEORGE E. HAYNES,
Professor of Social Science, Fisk University: Executive Secretary of the National League on Urban Conditions Among Negroes

November 12, 1916

MEXICAN ALIENS TO ENTER.

Special to The New York Times.

WASHINGTON, June 19.—To aid in meeting the present shortage in unskilled labor, Secretary of Labor Wilson has issued a departmental order temporarily removing restrictions on the importation of Mexican labor to be used in certain specified occupations.

The order goes into effect tomorrow, and provides that during the present emergency Mexicans entering the United States to engage in agricultural pursuits, in railroad section maintenance, and in lignite coal mining will be exempt from the head tax, literacy test, and contract labor provision imposed by previous rulings.

This step supplements the order by which the Department of Labor, through United States Employment Service and the Bureau of Immigration, has arranged to bring Porto Rican laborers to this country for work on Government contracts. It is estimated that 75,000 islanders can be brought in when transportation is available.

As additional insurance that an alien admitted under the new regulations will leave the country at the end of the war emergency, it is provided that all Mexican laborers at the time of their admission shall open a postal savings account at their port of entry. Employers shall withhold from the workman's wages 25 cents for each day's service, which will be deposited to his credit in the local postal savings bank, available to him, with interest, when he leaves the country. After the aggregate withheld for each workman totals $100, only $1 a month will be withheld and deposited.

June 20, 1918

PORTO RICO BILL SIGNED.

Gives Citizenship to Natives and Makes the Dependency "Dry."

Special to The New York Times.

WASHINGTON, March 2.—Porto Rico today obtained a new organic act when President Wilson signed the Jones-Shafroth bill to that end. The signing was marked by a small ceremony, at which were present Secretary of War Baker, Senator Shafroth, Chairman of the Committee on Pacific Islands and Porto Rico; Representative Jones, Chairman of the Committee on Insular Affairs; Brig. Gen. Frank McIntyre, Chief of the Bureau of Insular Affairs of the War Department; Gonzales Llamas, a leader of the Unionist Party of the dependency, and Samuel Gompers, President of the American Federation of Labor.

The bill provides that the residents of Porto Rico shall become American citizens in a body, with opportunity to renounce such citizenship at any time within one year; heretofore, they have not been citizens of the United States or any other country. While prohibition was imposed on the dependency in the bill, it was accompanied by a referendum provision, this, however, in a new form which, it is understood will be acceptable to the prohibitionists in Congress in most cases which may arise in the future. Instead of leaving the dependency "wet" and placing the burden of making it "dry" upon the prohibitionists, the bill makes the island "dry" and lays the burden of changing it upon the anti-prohibitionists.

The first election under the new law will take place in July, when a resident Commissioner will be chosen to take the place of the late Luis Munoz Rivera, years ago an untiring insurrectionist against Spanish rule, and the hero of Porto Rico on account of his frequent imprisonment and prosecution by Spain before 1898.

March 3, 1917

WELCOMED MEXICAN INVASION

Thousands of Families Crossing the Border to Till the Soil and Otherwise Build Up the Southwest

By GERALD B. BREITIGAM.

BROAD and brown the Rio Grande turned the corner of a high bluff on its northern bank and came ambling down between a sagebrush flat and a cactus desert, apparently the only thing alive and moving in the January morn.

But as the hot sun, leaping above the eastern horizon, began to dispel the early morning mists and the plain to the south of the river came into focus, a strange thing occurred. Here and there in the foreground little clumps that in the night had seemed but blotches of cactus or sage began to take form and shape and exhibit signs of animation.

Then these clumps separated into individual entities and women arising from them built little fires. Soon there was an appetizing smell of cookery. Other forms rolled and sat up, and were seen to be children and men. Only those in the foreground could have been noted by an observer on the deserted Texas bank, if such had been on the lookout. But beyond were other such groups and beyond still more, clear to the horizon, and then beyond again.

It was an invasion. They were all

headed for the Rio Grande with the one purpose of crossing into Texas.

Down at the river's edge a ferryman began getting his boat in shape to receive the first newcomers. He thought with pleasure of what the day's profits would be, for every Mexican whom he set on Texas soil would pay him a dollar for the trip. And hundreds would be transported during the day, while those who had to wait would cross on the morrow. There would be many tomorrows!

Mexicans by the thousands began coming into Texas as early as last Christmas. With their women and children, packing their lares and penates on their backs, they trudged hundreds of miles up from the interior of Old Mexico to the Rio Grande, and then were ferried over at desolate, unguarded spots. Not for them entrance through the few scattered immigration stations at Del Rio or Eagle Pass, Laredo or Brownsville. That would mean embarrassing questions as to literacy, worldly wealth and such like. Also, it would mean payment of a head tax of $8. Why bother, when one could be ferried over, at nominal cost, and free from the necessity of answering the strange questions of those brusque Gringoes?

Thus, beginning at Christmas and lasting until well along in March before the peak even was reached, began one of the strangest invasions of history. It was a peaceful invasion by a people sick of earthquake, sick of pestilence and famine, sick of the endless round of revolution. It was an invasion by an industrious, simple-minded agricultural people looking for work.

100,000 in the Northward Movement.

Since Christmas last, it has been estimated, more than 100,000 Mexicans with their families have thus entered Texas. In the Laredo district alone, a speaker at a business men's dinner recently estimated that between 10,000 and 15,000 of these "wetbacks," as they are called because of their method of entry, had crossed into Texas in that time. Perhaps an equally large number, moreover, have crossed the border since the new year into Southern California, Arizona and New Mexico.

No such tale is told, however, by the Immigration Department figures. Naturally, for these immigrants avoid official doorways into the Promised Land. And the immigration authorities are helpless to control the situation. When Congress cut down the departmental appropriations recently, the immigration men were forced to give up their border patrol and withdraw to the towns. They called on the army to patrol the border for them, but the army balked. Perhaps the fact that ranchers are eager to have these Mexican workers had something to do with that army obstinacy. Anyhow, there is no perceptible hindrance to Mexican entry by the unofficial route.

So these tens of thousands of Mexicans with their families, all looking for work, have streamed up into the Southwest in their strange sandals and serapes, some coming from far south of even Guadalajara, and they have been put to work.

Work? Why, they have put Texas on the map agriculturally. When practically every other part of the Union is shorthanded on the farms, Texas has an adequate labor supply. When virtually every other corner of the United States is cutting down on farm production because there aren't enough farm hands to go around, Texas is increasing her cultivated acreage. For these Mexicans, almost without exception, are farm hands, and the work they and their families find to do is not in Texas industries but on Texas farms.

In that connection, let me recall that this invasion, beginning last Christmas, while this year reaching unprecedented bounds, was not a new thing, but merely a sudden acceleration of the steady stream of Mexicans which has been flowing northward ever since the downfall of Madero. In that time it has been estimated that more than 500,000 Mexicans have entered the Southwest—not Texas alone, but Southern California, New Mexico and Arizona as well.

Thus, before the present invasion began, the Southwest for some time had been far better off in the matter of farm labor than other portions of the country. It was for that reason that Texas, once esteemed a joke agriculturally, was enabled last year to surpass every other State in the Union, even Iowa, in the value of farm products, being the only State whose crops totaled more than one billion dollars in value.

As for Southern California, Mexican labor has been a potent factor in putting her fruits and vegetables on the national market. Go up into Ventura and Santa Barbara counties, where most of our lima beans come from, and you will find Mexicanos picking them at harvest. Or take New Mexico and Arizona, and, both in their marvelously irrigated valleys and their great stock ranges, you will find the Mexican farm hand or vaquero at work.

Replacing the Drift to Cities.

Consider for the moment only what bearing this invasion of farm hands has upon the production of food. We are beginning to wake up to the perils of the situation caused by the shortage of farm labor. A steady drift of young men and women from the farms to the city which had been going on practically unnoticed for a decade was largely accelerated during the war years when the lure of high wages and shorter hours drew the young folks into industry. Today the rural population, it is estimated, is only 30 per cent. of the whole. While modern agricultural machinery has made it possible to farm more acres with a less number of hands than formerly, yet it comes a long way from replacing all the young folks who have left the farm. Decreasing acreage is the rule, and at the very time when there is a compelling necessity for increased food production to feed not only the hungry mouths of Europe, but, primarily, to take care of our own spawning cities.

At this moment, then, the Southwest comes to the fore—a territory so rich in soil fertility that one portion of it alone, namely, Texas, could feed the entire nation if it were all brought under cultivation.

I stood in front of the historic Alamo not long ago one bright hot day and noted the placard on a Chamber of Commerce auto cramped at the curb. Walking over I examined it. Here was a map of Texas—a State in which you can travel a thousand miles in a straight line, a State so large that when the Solons of the Panhandle set out for the legislative halls at Austin, in the middle of the State, they put on their ear muffs and overcoats, while the lawmakers from the South at the same time set forth in Panama and Palm Beach. The map was all broken out in feverishly colored areas like a smallpox patient, and these were marked: "Cotton," "Cattle," "Corn," "Sugar Cane," "Fruit." Above that map was tastefully lettered the modest assertion that "Texas can feed the nation."

When you recall that Texas already has leaped to premier position agriculturally, making such States as Iowa, Kansas, Indiana and many another usually associated with thought of farming take its dust, that assertion does not seem so boastful as it might. In the production of corn last year Texas ranked third among the States with 202,800,000 bushels to its credit. At the same time it raised 2,700,000 bales of cotton, more than one-fifth the total national crop.

When I asked John B. Carrington, the Secretary Director of the San Antonio Chamber of Commerce, how about it, he said, and said it emphatically:

"We couldn't do it if we didn't have the labor. Yes, Sir, we are dependent on the Mexican farm labor supply, and we know it. Mexican farm labor is rapidly proving the making of this State."

Nor does the value of these tens of thousands of Mexican laborers apply solely to the Southwest. Other employers besides the planters and cattle raisers have been quick to make bids for this labor supply, notably the railroads and the beet growers of Colorado and Michigan. Throughout the Southwest the railroads for years have been employing Mexicans as track repair workers, but today they are sending them on the same mission into other sections of the country as well. The box car home of the Mexican track worker is becoming a familiar sight in many a section of the Northern States which not so long ago did not know except by report that Mexicans existed. Take, for instance, the village of Burbank, Ohio, a typical rural American hamlet of small population. Dropping off there on business the other day I saw a Mexican colony beside the tracks. Its members had arrived so recently that they were still objects of curiosity to the villagers, and timid women would not go unescorted to the station near their camp. At Des Moines, Iowa, I found a growing colony on the outskirts. And in Kansas City, second railroad centre of the country and the "Gateway of the Southwest," I found a population of 10,000 Mexicans mainly down along the Southwest Boulevard near the great railroad yards.

Entering the Sugar Beet Field.

In the sugar beet fields, too, the Mexican is proving a veritable godsend. Carload after carload of Mexicans recruited by labor agents rolled out of San Antonio last Spring for Colorado and Michigan. Up in Michigan the beet growers are so eager to get them that they pay $35 an acre. Inasmuch as the Mexican takes his whole family with him when on the move and everybody works, he is able to cultivate as much as 25 acres a season. That means $800, no inconsiderable trifle.

Not all these peaceful invaders stay. They come across the Southern border in January, start working north when the harvests begin, turn back about the northern edge of the Southwestern States and then work back through the second harvest (Texas has two or three a year), and many of them cross into Mexico again to spend their wealth in easy living during the last quarter of the year. More and more do stay behind, however, so that the resident population is constantly increasing.

Naturally, such an invasion of aliens gives rise to many grave problems in the Southwest — political, social and economic.

There is, first of all, the need for understanding something of Mexican psychology and character, if American and Mexican are to live amiably together. The Mexican's racial pride is deep and wide, and it must be observed. The Mexican farm hands now invading the Southwest come almost entirely from that 60 per cent. of the population of Old Mexico which is unmixed Indian, descended from the Aztecs, Toltecs and subject races of the country who were conquered by the Spaniards but not wiped out. The term of Indian is a misnomer as applied to them, inasmuch as it creates the impression that they resemble the North American redskin. On the contrary, these people had a rather high type of civilization, and were village dwellers, not nomads. Their descendants—the people who go to make up the major part of the population of Old Mexico today—are industrious and make good farm workers.

They are sentimental and romantic. No Mexican peon trekking up into the Southwest to hunt work would dream of coming without his "old woman." The latter seldom has any claim to good looks, but, surrounded by a numerous brood of offspring, is either corpulent as a mediaeval friar or stringy and lean, yet her man would not dream of moving without her. Perhaps the fact that she can be counted on to find food for him when all his own efforts fail, has something to do with that. Nevertheless, the fact remains that family ties are strongly knit.

Yoked with this sentimental quality is the underlying trait of cruelty handed down from those ancestors who practiced human sacrifice to their gods. The same trait is answerable for much that goes on in the ubiquitous revolutions in Old Mexico. The sentimental and the cruel find their expression in the thrumming on mandolin or guitar of some hunting, plaintive air, while, as he sings with half-closed eyes, the Mexican dreams of just where he will plant his knife in his pet enemy.

Qualities to Be Reckoned With.

Because of this quality of irresponsibility, many employers have found it the policy of wisdom to give the Mexican a personal interest in his job in the hope of keeping him at it. Thus there are in the Southwest already thousands of acres of cotton plantations which are sub-rented in small lots to Mexican families, and the acreage is growing. The theory is that this will give the Mexican a feeling of independence and keep him on the job, and indications are that such reasoning is correct.

In getting at the facts in this matter, I found three men more conversant with the situation, more deeply versed in understanding of Mexican psychology than others, and each of them doing work in connection therewith. Of one I have already written, John B. Carrington. Mr. Carrington has occupied the position of Secretary-Director of the San Antonio Chamber of Commerce seventeen years, being longer on the same job than any man employed in similar capacity in the United States. His city is the Mecca of all those Mexican invaders who cross the Rio Grande, and it is from San Antonio that they are parceled out by labor agencies to the Texas farmers and the beet growers of the North.

The other two men are situated in the remaining "strategic centres" of this Mexican-held sector of the Southwest, namely Albuquerque, N. M., and Los Angeles. Both are Methodist ministers, the Rev. H. A. Bassett being President of Albuquerque College, an institution for Mexican youths, as well as Superintendent of all Methodist missions in New Mexico and Arizona, and the Rev. Vernon M. McCombs being Superintendent of the Methodist Latin-American Missions in Southern California. Mr. Bassett brings to his work an understanding of Mexicans built up during a stay in old Mexico lasting from 1897 to 1913, during part of which time he was Vice President of the Methodist College at Puebla. Mr. McCombs, who has been working among the Mexicans of Southern California ten years, was prior to that time in Peru.

It was from Mr. Carrington and other men similarly situated that I got the best leads toward obtaining data on the economic aspects of the peaceful invasion from old Mexico. But when I came to inquire what was being done to assimilate these aliens into American life, I was invariably referred to "church people." City, State and governmental agencies are doing very little to Americanize these Mexicans, to teach them our ways, customs and ideals, and the only agency at work along that line is the church organization.

Bilingual New Mexico.

In New Mexico, for instance, where more than half the population is of Mexican descent, having given the State its present Governor, the President of the lower house of the Legislature and other Government officials, many of the public schools are conducted in Spanish. And when the Legislature meets at Santa Fé one of the strangest sights to be seen in any legislative hall in America is presented. The Legislature enacts its laws bilingually, that is, so many of the legislators in the lower house speak only Spanish that discussions are carried on through the medium of interpreters. Thus, as soon as a legislator gains his feet, one of the watchful interpreters takes stand by his side. If the man speaks in English, the interpreter puts his words into Spanish for the benefit of those knowing only that tongue, and vice versa. All laws and the proclamations of the Governor are written, passed and put on the books in both languages.

"These Spanish-speaking citizens are loyal to the United States," Mr. Bassett told me. "Some time back, when there was all that talk of the possibility that Old Mexico might try to reclaim New Mexico and the Southwest, a magazine writer hinted strongly that the Spanish-speaking rural population of this State would favor such a move. I can deny that vigorously out of my own knowledge. These people are glad to be in the American fold."

On the site of the old Frémont 'dobe headquarters of late years, a notorious tenderloin place, is going up an institutional church which will provide ample space for the conduct of all sorts of classes and clubs in the Americanization program, as well as a gymnasium and swimming pool for the Mexicans. In all, I was given to understand the Methodists are putting in more than a quarter million dollars on their plaza projects.

Mr. McCombs, who supervises all these plaza projects as well as a Mexican boys' institute at Gardena and missions in various places throughout Southern California, believes the alien from the South can be made into a fine American citizen. He feels, however, that as a people we are not yet awake to the rapid increase in our Mexican population, and the necessity, both for their sake and our own, of helping them become Americanized.

"They hold fine possibilities of citizenship," he said, "being sturdy, independent and filled with racial pride. To the best of my belief, there isn't a Mexican tramp in the United States. But, despite their good qualities, they are, we must remember, illiterate and grossly misinformed about the United States. Accordingly, for instance, I. W. W. agitators find it comparatively easy to play upon their ignorance and to convince them the United States is a tyrannical country. We must remember that the only Government of which they have any knowledge is one of license and misrule. The church missions working among the Mexicans throughout the Southwest are striving valiantly to combat this sinister propaganda, but the Government should recognize the necessity for such efforts and tackle the problem on a big scale. The economic future of the entire Southwest depends more and more upon the Mexican and, if for no other reason than self-interest, we should not repeat past mistakes and let our incoming aliens go unassimilated."

June 20, 1920

THE NORTH AND THE NEGRO.

Remarks on the Hopeful Aspects of the Coming of Negroes to the Northern States.

By JULIUS ROSENWALD.

The so-called negro problem was confined largely to the South before the great war. Now it is a national problem. Prior to 1914 the negro usually was an agricultural worker. Today he is a big factor in America's industrial life.

The World War virtually put a stop to foreign immigration to the United States. Hundreds of thousands of Europeans living in this country were called to the colors. Under the stimulus of war conditions industries in the North expanded greatly.

These three factors caused a labor shortage that forced the Northern industrialists to scour the country for available workers. The greatest supply of such labor was found among the negroes of the South and this was drawn upon to a great extent.

The result was the beginning of the largest migration of negroes in the history of America—a migration that is still in progress.

The migration has been marked by two phases, that of 1916-1920 and that commencing anew in 1922. The first really began in 1915, reached its maximum in 1917 and continued at a slower pace up to 1920, when the economic depression brought it to a halt.

The revival of business prosperity in 1922 brought on a new exodus from the South. Thousands of negroes who had remained North during the period of depression wrote to their friends and relatives urging them to come North and in many cases sending them money for the journey.

The 1920 census reported the negro population of the nation to be about ten and one-half million, or 10 per cent. of the total population. The negro population in the North was shown to be more than one and one-half millions.

The essentially industrial implications of the northward movement in the last eight years is revealed by the fact that more than one million of the Northern negroes, or 73 per cent., live in ten industrial centres, as follows, using round figures:

Indianapolis District	47,500
Detroit-Toledo District	55,900
Cleveland-Youngstown District	58,800
Kansas City District	65,400
Pittsburgh District	88,300
Columbus-Cincinnati District	89,600
St. Louis District	102,600
Chicago District	131,600
Philadelphia District	248,300
New York District	251,300

The concentration in these ten Northern centres not only has projected the so-called negro problem into the North, but has presented it in new aspects. Eighty per cent. of the negroes in the South live in rural communities. The present status and future of the negro, therefore, are primarily linked with industry in the North and with agriculture in the South.

The present selective immigration law was passed by a Republican Administration and will probably remain in force for another four years. Even should the law be repealed, thousands of negroes have acquired skill in mechanical occupations that puts them beyond the likelihood of replacement by foreign labor. In short, Northern industrialists have come to look to the negro for the labor supply in their factories and workshops. For some of these tasks men and women who have had training at Hampton and Tuskegee are well fitted.

These industrialists are now carefully selecting their negro workers. Some have made special efforts to employ only married men, and then to provide such housing and working conditions as will keep them satisfied. One large iron foundry which pursued this policy reported that the turnover among its negro workers was only 10 per cent.

Apparently, the settling of the negro in the North is permanent. Many of the factors that brought him North operate to keep him there.

There is good ground for believing that the migration of the negro will have a beneficial effect on the nation. It will be a good thing for the South because the colored population will be more evenly distributed over the entire country and will lessen the Southern fear, real or alleged, of race domination, and will thus remove an outstanding factor that has hampered that section's development.

In this connection it might be well to quote the authority of a Southern white man, President Jacobs of Oglethorpe University, Atlanta, Ga., who has declared:

"The very finest effect of this exodus of negro laborers is its political effect. As long as there is a negro problem in America, the South is in political slavery, unable to vote her mind about matters of national and international importance. When the time comes that the negro problem is no longer a sectional problem, but in so far as it is a problem at all a national problem, then, indeed, will the Southern country be free.

"And it should be added that, from the political standpoint of the negro also, the change will be most highly advantageous. It is difficult for a white man to realize how it feels to be a 'problem,' and the negro will never be satisfied nor will the tension between the races be over until he ceases to be one."

The negro's rise in the scale of occupations has given him a greater purchasing power and a higher standard of living. To his credit it should be said that, for the most part, he tries sincerely to live up to his opportunities in the North. He is usually a law-abiding citizen, buys his own home when possible and gives his children the best schooling his income will permit.

One of the most hopeful signs for the future of the negro in the North is that the opportunities there are attracting young colored men and women trained in such schools as Hampton and Tuskegee. These two schools, and others of their type, not only give a thorough training in mechanical occupations, but their whole system of education tends to turn out young men and women who will be community teachers and leaders.

Those trained in what has become known as the "Hampton-Tuskegee" method strive to bring about cooperation between the white and colored races and to reduce interracial friction. Leaders of this type are bound to have a salutary influence on the negro communities everywhere.

March 9, 1925

NEGRO MIGRATION TO HARLEM HEAVY

Starvation Conditions in Some Sections of Deep South Cited as Reason

NATIONAL PROBLEM SEEN

Urban League Official Asserts Not Only City but U. S. Must Help Relieve Needy

During the hard times of the Thirties, in which relief and other social benefits here accentuated the higher living standards of Northern cities compared to the deep South, the non-white population of New York City increased by 135,130, according to census figures tabulated by the Department of Welfare. The non-white population increased from 343,216, or 4.95 per cent of the total, in 1930, to 478,346, or 6.42 per cent of the total, in 1940.

Not only did the non-white population of Manhattan, which includes Harlem, the city's largest Negro district, grow by more than 75,000, but an exodus from overcrowded Harlem helped increase that of Brooklyn's new Negro section by almost 40,000. There were smaller increases in other boroughs.

Welfare Department statistics show that the approximately 125,000 relief cases in New York City include about 32,000 Negro cases of 72,000 persons, or about 26 per cent of the total relief roll.

For a family of four the average home relief here amounts to $81.48 a month, including $59.60 cash and the rest consisting of food stamps, free milk, medical care and the like; for a family of five, $96.45 a month, including $69.10 cash; for a family of six, $110.62, including $76.80 cash. There are about 63,500 persons in WPA in New York City, of whom between 20 and 23 per cent are officially estimated as being Negroes.

In contrast to New York's comparatively high relief benefits, the latest figures available at the offices of the National Urban League here are contained in an FERA report showing that in certain Southern counties in 1936 the average monthly relief budget for Negro families was $8.31; for white families $12.65.

Comparison with Alabama

Whereas the Negro's wage list in the North is stated by the National Urban League to be $10 or $15 a week, a study of Negro tenant farmers in an Alabama rural county made by Charles S. Johnson and Will W. Alexander about five years ago showed an average cash income for Negro tenant families of $105 a year, including relief allowance where made.

Emphasizing that the "miserable existence" of Negroes in the North "can be riches compared with some of the wages paid in some sections of the South, not only to Negroes but to whites as well," Lester B. Granger, assistant secretary of the National Urban League, said yesterday that persons living in the conditions prevailing down South will manifestly "continue to escape where possible by moving to localities offering better opportunities."

He said that the problem of Harlem was clearly a national one, which should be solved not only by improving the condition of the Negro in New York, but also by improving the condition of the Negro in the South, so that he would not consider it necessary to migrate here.

The conditions that exist, he asserted, can be improved only by a national attack, including higher unemployment insurance and other social security standards, better schools, hospitals, housing and judicial processes.

Puts Question Up to City

"A conscious recognition by New York citizens of the interdependent nature of social problems in the North and in the South," he added, "will make them realize that it is imperative to relieve Harlem conditions, to be sure, but no local program can be completely effective in the long run unless it is backed up by national action."

Saying that the "much-advertised conditions in city Negro neighborhoods in New York are serious enough to engage the thoughtful attention of all New York residents," Mr. Granger added it should be pointed out that "the low economic status of Negroes and the unwholesome social conditions that continue to develop in Negro neighborhoods are an inevitable result of the attitude of New York City toward its Negro poplation."

"Much has been made," he continued, "of the migration of Negroes into the city from the South. Observers overlook entirely the fact that approximately as many white Southerners as Negro Southerners have left their homes and migrated to New York City and New York State in search of opportunity. No noticeable 'problem' has been created by the migration of whites because these have been accepted by their fellow New Yorkers as simply new residents, accorded all of the opportunity and privileges that older residents have enjoyed for themselves. Thus, white Southerners coming to New York have been speedily integrated into the population as a whole.

Different With the Negro

"It has been the contrary with Negroes. Instead of being accepted, they have been rejected by the older New York population, have been confined to certain residential areas, certain occupational callings and to a lower economic and social condition than that desired and enjoyed by the majority of old New Yorkers.

"Thus present conditions, which have been so much overadvertised and misadvertised in the daily press, have been inevitable. Some New Yorkers have felt that arbitrary steps might be taken to reduce migration, or at least that Negroes might be discouraged from moving to New York and other Northern cities if lack of opportunity were sufficiently advertised. This is a fallacious reasoning. As a matter of fact, New York City has actually had less migration of Southern Negroes than many other Northern cities, proportionately speaking.

"What has happened in New York City has happened in most important industrial and commercial centers where job opportunities are expected to be found. Even if New York were to reduce standard of living and opportunities for its Negro citizens to an even lower point than now exists, these standards could not possibly be reduced to the level of the isolated rural communities and city slums from which many of our Southern newcomers have escaped. As a matter of fact, it has been the finest kind of social instinct that has motivated these migrations of whites as well as Negroes; namely, the pioneer urge to escape from an unfavorable and unfruitful environment into a more fruitful one where opportunities are better.

Starvation Real Reason

"It is not that the 'riches' of the North attracts Negroes, but that the starvation of so many sections of the South ejects them. While we are taking steps to improve conditions in New York City, similar steps must be taken to improve conditions for Negroes in that part of the country where the majority are now located—the deep South. Such improvement would include larger employment opportunity, better housing conditions, better health and educational facilities, and more secure enjoyment of the civil liberties that are supposedly the possession of every American."

Welfare Commissioner William Hodson agreed with Mr. Granger that the Harlem problem was a national one, because of the migration from the South to New York and other northern cities.

"The problem can not be beaten," he added, "until there is a larger measure of economic justice for Negroes in all parts of the country."

He said the fact that it is basically a national problem, however, should not prevent New York from doing everything it can to improve conditions here.

November 19, 1941

WHY PUERTO RICANS FLOCK TO THE U. S.

Lack of Opportunities at Home and Easy Transportation Bring Many Islanders Here

By W. F. O'REILLY
Special to THE NEW YORK TIMES.

SAN JUAN, P. R., May 31—Puerto Ricans like other people yield to the lure of the United States as the Promised Land. But what distinguishes these islanders from other migrant groups is that the overwhelming majority — at least 75 per cent—over a period of years have been unable to adjust themselves to the new environment. Missing their families, finding the climate harsh, resenting American racial discrimination they return to Puerto Rico.

Emigration fluctuates with periods of prosperity in the United States and depressed economic conditions here.

This fact was pointed up in the report on New York City's relief problems, which was released a few days ago by Mayor O'Dwyer. In the report, New York City Welfare Commissioner Rhatigan stated that the 54 per cent rise in the non-resident case load in the last year reflects "the influx of Puerto Ricans into the city."

There has been an amazing increase in the number going out since 1941, when 1,000 left the island. In 1942 the number rose to 1,837, in 1943 to 2,600, in 1944 to 7,500 in 1945 to 14,800, in 1946 to 21,600, in 1947 to date 25,000 (tabulation is by the fiscal year ending June 30). Widespread unemployment—80,000 are jobless—stimulates the present migration.

University Study

The most comprehensive, accurate and revealing study of Puerto Rican emigration covering the years 1908-47, was recently completed by the University of Puerto Rico Social Science Research Center headed by Dr. Clarence Senior assisted by Dr. Harvey Perloff. The researchers find that from 1908 to March, 1947, 750,000 left the island, 95 per cent going to the United States. Only 127,000 remained in the United States as permanent residents.

During the war years, despite the island's lack of war industries, emigration was restricted. Transportation was monopolized by the war effort and only selected workers were sent north for the war industries. In the last two years expansion of plane travel and low passenger rates have permitted thousands of the island's poor to go to the United States. It is estimated that thirty companies are operating here with chartered planes. Some offer passage to the mainland at less than $50. An observer recently noted a chartered plane leaving Mayaguez for Miami with passengers in the low economic brackets packed like those on old-time immigrant trains.

The researchers submitted questionnaires to a sample 3,000 emigrants, 53 per cent of whom answered. Sixty-one per cent of the returned and unreturned found living more satisfactory in the United States than in Puerto Rico and the greatest sources of dissatisfaction were separation from family, climate and racial prejudice. Sixty-two per cent of those still in the United States gave better jobs as their reason for remaining on the mainland. The questionnaire showed that 64 per cent of those remaining were white.

The larger number of those permanently separating themselves from the island left good jobs here while 71 per cent of those unable to stick out the North were unemployed here when they left.

The Insular Legislature at the last session appropriated $25,000 for a Commissioner of Labor to inspect and approve contracts and survey conditions where Puerto Ricans are employed and another law requires agents to file fees and post bonds. A Puerto Rican official said this week that he had personnally inspected conditions of Puerto Rican workers in New Jersey and found that employers preferred Puerto Ricans to British West Indians.

June 1, 1947

Million a Year Flee Mexico Only to Find Peonage Here

By GLADWIN HILL
Special to THE NEW YORK TIMES.

SAN ANTONIO, Tex., March 24—A visitor to the Southwest today has to make a double mental adjustment to comprehend the conditions he finds.

First, he has to think back twenty years to prohibition days, when wholesale violation of the laws of the United States was being ignored, tacitly sanctioned or overtly encouraged by a large cross-section of the population.

Second, he has to project himself in imagination back a full century to the days of slavery, when the systematic exploitation of an underprivileged class of humanity as cheap labor was an accepted part of the American social and economic order.

Then the observer has to compound mentally these two diverse anomalies and conceive of their existence in 1951.

For that, in effect, is the situation today in the Southwest, and in parts of the deep South and the Far West as well, arising from the ceaseless and steadily increasing tide of illegal immigration from Mexico into the United States.

This illegal immigration now amounts to more than 1,000,000 individuals a year. They sneak across the thinly patrolled 1,600,-mile border between Brownsville, Tex., and San Diego, Calif., in an unending hegira from Mexican unemployment and wage levels as low as 40 cents a day, seeking farm work and any other labor available in this country.

Some 500,000 of these "wetbacks"—so called because of those who swim the Rio Grande, the Texas-Mexico boundary — were caught by the United States Immigration and Naturalization Services' Border Patrol last year and put back across the international line.

The estimate of a total traffic of 1,000,000 is conservative, based on the surmise that for every one caught there is one who isn't caught. Most immigration officers would concede a more likely average of five or ten to one, and for some areas responsible estimates go as high as 100 to one; in such a reckoning many of the total would be not different individuals, but repeaters who re-cross after having been deported.

There are fewer than 900 border patrol officers to guard the whole 1,600 miles of boundary, and sometimes the deportees get back into the United States before the deporting officers do. Some Mexicans have been deported as many as twenty times, only to return again and again.

Some Return, Some Stay

A large number of "wetbacks" regularly slip back into Mexico after seasonal farm work ends. But tens of thousands of others have remained in the United States, managing to settle down in semi-hiding in hundreds of communities along the border, or becoming transient laborers and filtering northward over the western half of the country all the way to the Canadian border.

Technically, every one of these "wetbacks" is an illegal alien, the same as a foreigner who slipped through the paradoxically tight immigration patrols at Ellis Island—a violator of the Federal law, subject to imprisonment.

But because the bulk of this im-

'WETBACKS' CROSSING THE RIO GRANDE

Mexican farm workers wading across the river near El Paso

Associated Press

migration is tidal, advancing and receding with the regularity of the seasons, it has come to be taken for granted, both by the United States border population and the nation at large.

Superficially, the "wetback" traffic is just a picturesque cat-and-mouse game on a grand scale between a mass of amiable Latins and an overwhelmed border patrol.

In actuality, however, a year's study of the situation by this correspondent, ending in a recent 5,000-mile tour of the border area, shows it to be a phenomenon whose real economic and social implications should shock the average American when fully known and comprehended.

The impression is widespread that these Mexicans simply come over as a regular essential and harmless supplement to the domestic harvest forces.

The fact is that they are attracted by the glittering prospect of high American wages, and in many cases actively recruited by large-scale ranchers or their representatives, only to work for wages that have been described by United States immigration officers as tantamount to peonage and under conditions compared on the same authority to ante-bellum slavery.

"This traffic in contraband labor actually is worse in its elements than either Prohibition or slavery," a nationally prominent figure in the Southwest commented a few days ago. "Prohibition involved only traffic in alcohol. This is traffic in human beings. And in slavery some responsibility, at least, was acknowledged for the release of the slaves. Here there is none."

The "wetback," a penniless fugitive, has to take whatever wages are offered, which, because of the multitude of "wetbacks" available, usually are at a bare existence level or below it. His traditional housing is the crudest form of shack or hovel, wherever he can find it, or just the open air. His status denies him access to regular community legal and welfare agencies. If he doesn't like his situation he can go back where he came from. There are thousands of other hopefuls south of the border ready to take his place.

Aside from the abuses to which these Mexicans themselves are subject, the traffic in "wetbacks" has manifestly baneful effects on the citizenry with which they mingle.

In many sections of Texas, Arizona, California and other areas they depress the general standards of wages and working conditions below the accepted American levels. Instead of supplementing the domestic labor force they undercut it, taking jobs from tens of thousands of native citizens, farm and urban workers alike.

In the agricultural field particularly the resulting native "displaced persons" have to join the great stream of migrant labor that fans out annually northward all the way from New York to Oregon, working a good deal of the time for substandard wages and working conditions and, in turn, taking jobs from resident local workers.

From a sociological standpoint the "wetback" traffic has obliterated the border completely, transforming the 3,000,000 Spanish-American citizens of the Southwest—many of whose families have been in the area for centuries—into a "cultural peninsula" of Mexico; retarding their assimilation, in the opinion of experts, by a generation or more, and perpetuating and aggravating some of the worst conditions of health, education and social, economic and political un-integration that can be found in the country.

Finally, the "wetback" traffic has engendered, in somewhat the way Prohibition did, an atmosphere of amorality and warped thinking which demonstrably extends from the farmer-exploiters of "wetback" labor, through their communities and local and state officials, to the highest levels of the Federal Government.

Although this may sound sensational, the remarkable thing is that there is no secrecy about the situation. Despite an elaborate legendry that has grown up to screen it, and occasional efforts to dissimulate the facts, most of the details are matters of formal public record or common knowledge in the region.

The main problem in inquiring into it is to preserve one's sense of proportion and avoid entanglement in the falsework of rationalization that has grown up over the years to justify the system and maintain it.

The "wetback" situation is of particular pertinence at this time. The defense emergency has evoked official proposals for the importation of hundreds of thousands of workers from outside the continental United States.

Before Congress is a draft extension of the 1949 agreement between the United States and Mexico, which expires in June, governing the so-called "importation" of Mexican farm workers under contracts, which in actuality have degenerated into the legalization of "wetbacks."

Before President Truman is the initial report of the special lay commission he appointed last summer to investigate the problems of migrant farm labor and the "wetback" evil.

Also, it has lately been realized that the "wetback" traffic offers a wide-open avenue for Communist spies to enter the country. There is no public record of any having been caught lately. But immigration officers acknowledge that every year at least a handful of Europeans, some of them with Communist backgrounds, are netted among the "wetbacks," and that in cold fact Joseph Stalin might adopt a perfunctory disguise and walk into the country this way.

Legal immigration into the United States from Mexico—on which there is no quota limit—involves such complications as fees, literacy standards, job assurance and consular approval. It is much easier to sail, wade or swim the Rio Grande, seldom more than 100 yards wide, or just to walk across the largely unfenced dry sections of the border.

Such an illegal crossing is a misdemeanor the first time and a felony after an initial deportation, subject to two years' imprisonment and $1,000 fine. But the "wetbacks'" numbers largely defy prosecution.

Detention facilities and court dockets are so crowded that anything beyond quick deportation of first offenders has had to be abandoned. The latest authoritative analysis indicates that only about twenty-five of every 10,000 caught are prosecuted.

So the influx has snowballed. In the San Antonio immigration district, the largest of three covering the border, the number of apprehensions—generally conceded to be in direct proportion to the total traffic—has jumped from 43,000 in 1945 to 215,000 last year.

In the Los Angeles district, covering California and western Arizona, they jumped from 4,000 in 1943 to 230,000 last year. And the traffic is mounting steadily. The Los Angeles district's apprehensions in January and February of last year totaled 19,500; for the same period this year they were 33,800.

The "wetback" traffic dwarfs the legal contracting of Mexican labor under the international agreement, and has become almost inextricably enmeshed with it. At the moment, according to the United States Employment Service, there are about 28,000 contract Mexicans in the country. Most of them were not "imported" at all, but are ex-"wetbacks" who were rounded up north of the border, processed at border immigration and consular stations, and given legal work permits under guarantees of somewhat better conditions than those accorded "wetbacks."

Once across the line, the "wetback" passes himself off as one of the 3,000,000 citizen Mexican-Americans. The main centers of "wetback" labor are the large-scale cotton, citrus and vegetable ranching areas of Arizona, and the Imperial and San Joaquin Valleys of California.

In lesser numbers "wetbacks" can be found seasonally on farms in Utah, Colorado, New Mexico and other parts of the West, and over in Arkansas, Mississippi and Missouri. Every week or two a planeload is shipped from Chicago back to the border.

Employment of "wetbacks," although from a realistic standpoint tantamount to harboring a fugitive, has been held by the Federal courts to be not a punishable offense because of the omission in the 1917 version of the immigration laws of any specific penalty.

March 25, 1951

NEGRO MOVEMENT FROM SOUTH SEEN

WASHINGTON, Oct. 30 (AP)—There was a definite trend of Negro population away from the South to industrial areas of the country during the 1940-1950 decade, Census Bureau figures showed today.

The agency has not issued its final figures on population by races or on movement of population from state to state, but a tabulation of its preliminary figures on white and non-white population for each state shows the trend.

The bureau classifies as non-white the country's Negro, Indian and Asiatic population, except in certain instances the non-white population is virtually all Negro. The picture presented by the figures was as follows:

From 1940 to 1950 the non-white population of the thirteen Southern states, commonly known as "the South," was virtually at a standstill, showing a net gain of only 55,637. During the same period, the white popuation in the thirteen states gained 4,453,354, or nearly 100 times that shown by the Negro population.

In the decade, the non-white population of eight major industrial states, California, Illinois, Michigan, Missouri, New Jersey, New York, Ohio and Pennsylvania, rose from 2,808,549 to 4,364,000, a gain of 1,555,451. In the same ten years the white population of those eight states rose 7,887,052.

In other words, in the South the white population gained about 16 per cent and the Negro population gained a ½ per cent. However, in the eight industrial states the white population gained about 14 per cent and the non-white population nearly 55 per cent.

Of the thirteen Southern states, seven showed declines in non-white population. They were Mississippi, with a drop of 87,000; Alabama, Arkansas, Georgia, Kentucky, Oklahoma and Texas. Southern states showing a gain in Negro population were Florida with 90,000 increase, Virginia with 75,000, North Carolina with 75,000, Louisiana, South Carolina and Tennessee.

The net gain in Negro population in the Southern states, experts say, dos not equal what the natural increase through births would be for the ten years. Thus obviously there was a migration of Negroes from the area sometime during the ten years.

Presumably this migration came during the war years when many Negroes left the South to take jobs elsewhere in the country. The big gains in Negro population of the industrial states support this theory.

The census figures show that in 1940 the Negro population was about 33 per cent of the white population in the thirteen Southern states. In 1950, it was down to about 30 per cent.

On the other hand in eight industrial states the Negro population was about 5 per cent of the white population in 1940. By 1950, it had increased to about 7 per cent.

October 31, 1951

BRACEROS SEEKING PERMANENT WORK

Special to The New York Times.

MEXICO CITY, Aug. 11—The influx of immigrant Mexican laborers into the United States this fiscal year probably will be the largest in many years if, in fact, it does not set a record.

Contrariwise, this probably will be a record low year for the entry of braceros, or temporary Mexican laborers into the United States. The two unusual situations are closely connected. Braceros who have spent repeated harvest seasons in the United States and who have been cut off this year are now trying to arrange for permanent residence there.

Braceros are Mexican laborers contracted for seasonally by United States planters. They are used primarily in border states for the harvesting of fruit and cotton crops.

The entry of braceros (from "brazo," or "arm") from 1956 through 1959 averaged about 400,000 a season. In 1960 the number dropped to 318,000. In 1961 it dropped to slightly less than 300,000. This year, according to the Mexican Government, less than 275,000 are expected to be taken across the line.

Immigrants Increasing

The immigrant, in contrast to the bracero, enters the United States on permanent status leading to naturalization and citizenship. In the calendar year of 1959 a total of 23,000 Mexicans were processed for immigration status. The following year 32,000 were processed. In the fiscal year 1961, when immigration figures were based on the fiscal rather than the calendar year, 41,632 immigrant laborers actually were accounted for in border crossings. The data has not been completed yet for the fiscal year of 1962 but the number of immigrants is well past 50,000.

A number of reasons have been advanced for the decline in bracero entries with the consequent increase in immigrant status labor. Probably the basic reason has been the establishment of a minimum wage for seasonally imported Mexican labor. In the principal states contracting for bracero labor this minimum wage is in: California, Michigan and Wisconsin $1.00 an hour, in Colorado, 90 cents an hour, in Texas, 70 cents an hour, and in Arkansas, the lowest hourly rate of all, 60 cents.

In some instances where the principal crops to be harvested are fruits and berries, there has been relatively little change in the entry of seasonal labor as a result of the minimum wage. In other instances, however—for example in certain areas where the seasonal crops are cotton or tomatoes—the establishment of a minimum hourly wage has stiffened an already latent opposition to the entry of bracero labor.

Large Farms Mechanized

The most significant result has been the mechanization of large farming operations. The data on this is necessarily arrived at haphazardly but it appears certain that last season in the cotton-growing areas of Texas, where bracero labor is imported for picking, more than 50 per cent of the crop was harvested by machines. Texas planters, here in Mexico for the purpose of arranging immigration papers for favored Mexican laborers, have maintained that this year upward of 80 per cent of the cotton in their areas will be machine-harvested.

According to news dispatches from these Texas cotton areas, there are instances in which the importation of braceros has dropped nearly 60 per cent this year under last.

The braceros suffering mostly from the turn of events are those who have attained status approaching that of overseer or of skilled laborer on the farms to which they have been going regularly. These, under the current United States-Mexican treaty governing braceros, can be ruled out by the United States Department of Labor as taking advantage of "wages, working conditions and employment opportunities" of United States workers.

Their employers to whom they have been going for years are now endeavoring to bring them in as immigrants. There are no quotas for Mexican immigrants to the United States, as there are none for any of the Western Hemisphere countries. The immigration is kept within certain limits by Immigration and Naturalization Service regulations and by facilities of the United States Consular Service.

Backlog of 85,000

The latter is working under severest strain in all the six processing offices in Mexico. Despite having processed nearly 175,000 immigration applications within less than five years, the consular service has a backlog of 85,000 to be processed. Some personnel have been added but the processing has been mainly speeded up by recent streamlining procedures in the service. The six processing offices in Mexico are in Mexico City, Monterrey, Tijuana, Ciudad Juarez, Guadalajara and Nogales.

The reaction here in Mexico to the decrease in bracero traffic is mixed. In some circles, particularly among nationalistic and Leftist groups, there has been constant criticism against the bracero system as undignified for Mexico. Among the braceros themselves, however, and generally in the Mexican Government the loss of revenue arising from the bracero decline is a matter of great preoccupation.

August 12, 1962

Negro Migration to North Found Steady Since '40's

By JACK ROSENTHAL
Special to The New York Times

WASHINGTON, March 3—The Census Bureau reported today that Southern blacks streamed north during the sixties at a rate nearly the same as the high level of the two previous decades.

The finding, drawn from newly computed 1970 census data, contradicts earlier, widely accepted reports that black migration north had tailed off sharply, falling to about half the earlier level.

Since large numbers of black migrants go from poor farms to urban slums and often require welfare assistance, the finding is likely to figure importantly in the present political debate about revenue sharing and Federal assumption of welfare costs.

More than three-fourths of the 1.4 million black migrants from the South went to five large states where the soaring cost of welfare is a heated public issue.

By far the largest gain was measured in New York, which gained 396,000 black migrants in the decade. California gained 272,000, and New Jersey, Illinois and Michigan each gained about 120,000.

Analysts said there were indications that such migration could well continue and even increase in the next decade.

The Census Bureau also released reports showing significant improvement in three indicators of the condition of housing units across the country.

"The quality of housing went up considerably," Maurice H. Stans, Secretary of Commerce, said at a news conference. He cited improvements in the number of units with basic plumbing facilities, in the age of the housing stock and in the degree of crowding.

The census report on migration showed that as Southern blacks moved north and west, Northern and Western whites moved south in historic numbers.

In total out-migration, New ork lost 638,000 white residents. Pennsylvania lost 423,-000. Illinois, Ohio, Iowa, Kansas, and Washington, D.C., each lost more than 100,000.

The result of this dual movement is a clear continuation of a long-term trend toward dispersion of the black population throughout the country.

The South still contains 53 per cent of the black population, but that compares with 77 per cent in 1940. Since that year, the population of the Northeast and North Central states has gone from about 11 to 20 per cent black.

The total black population increased from 18.9 to 22.7 million in the decade. Blacks now make up 11.2 per cent of the national population, up from 10.6 per cent in 1960.

Proportion to Be Steady

The black population growth rate continues to be higher than that of whites, but not enough to alter this proportion significantly in the future. Herman P. Miller, director of census population studies, said today he estimated that by the year 2000, blacks might account for no more than 13 per cent of the population.

More significant than the black population's size, analysts said today, is its distribution. Each of the 11 states of the Confederacy lost black population. Mississippi lost 279,000, and Alabama lost 231,000.

These losses continued strong trends evident since at least 1940.

In 1968, however, the Johnson Administration released a widely noted Census Bureau-Labor Department study reporting that these trends had at last begun to change dramatically.

"Average annual out-migration in recent years is about half of what it was in the forties," the report said.

Drop in Migration

This study, based on survey data, showed a drop in black migration from 159,000 a year in the forties to 80,300 in 1968.

The report released today, based on full census data, shows that the figure for the cities in fact was about 138,000 a year, close to the annual figure of 147,000 in the fifties.

George Hay Brown, director of the Census Bureau, took note of the 1968 findings at the news conference today.

Those findings and the figures released today were estimates, he said. The new data were computed by subtracting births and deaths from each state's population change. The result is defined as net migration.

He said this method was useful to provide interim data until precise migration studies were complete next year. But it also may encompass considerable error, he said.

'Migration Continuing'

In any event, Mr. Stans said, "migration of blacks to the North is continuing in the same degree as in the previous two decades."

The Commerce Secretary said, "I have no doubt that higher welfare benefits in the North are a factor." But, he added under questioning, "I would certainly assume that greater job opportunities [in the North] would be the chief motivating factor."

Other authorities have attributed continued migration as much to the push of poor rural conditions in the South as to the pull of opportunity in the big city.

Details of the 1970 census lent some support today both to these views and to the belief that heavy black migration north may continue.

These data show that the migration rates are highest among young adults in at least seven Southern states. In Mississippi, for example, there were 70,000 black men aged 15 to 24 in 1960. But now there are only 30,324 black men aged 15 to 24, a drop of 57 per cent.

The figures for young black women are somewhat lower.

Since there are few deaths in the 15-to-24 age group, analysts regard it as a significant measure. It suggests, they said today, that as farm mechanization continues to eliminate rural work, young blacks continue to look north.

Movement by Whites

Meanwhile, whites continued to migrate south and west. The West gained 2,855,000 migrants of all races. California accounted for 2,113,000 of these.

As Mr. Brown, the census director, reported last fall in a speech on the new South, that region as a whole gained about 1.8 million whites, offsetting the 1.4 million black outmigrants. The net gain is the first for the South since the eighteen-seventies. Florida accounted for 1,340,000 white migrants.

The North Central states lost 752,000 in migration while the Northeast gained 324,000, most of them black. Among the state figures are the following:

State	White	Black
New York	—638,000	+396,000
New Jersey	+336,000	+120,000
Conn.	+166,000	+38,000
Penn.	—423,000	+25,000
Mass.	+23,000	+33,000

Report on Housing

Today's Census Bureau housing report showed that the proportion of housing units without a toilet and tub or shower dropped in the sixties from 17 to 7 per cent. In 1950, the figure was 35 per cent. The total is now 4.7 million out of 68 million units.

The proportion of units that average more than one person per room dropped from 11.5 to 8.2 per cent of all units over the decade.

In both cases, the highest over-all proportions are in the South. But California, with 522,000, and New York, with 447,000, have by far the most housing units with more than one person per room.

Mr. Stans said that about half the present housing units in the country were built after World War II.

March 4, 1971

Distribution of U.S. Blacks

Special to The New York Times

WASHINGTON, March 3—The following tables, based on census data, show the increasing distribution of the black population throughout the major regions of the country.

The percentage of each region's population that is black:

	1970	1960	1950	1940
United States	11.2	10.6	10.0	9.8
Northeast	8.9	6.8	5.1	3.8
North Central	8.1	6.7	5.0	3.5
West	4.9	3.9	2.9	1.2
South	19.2	20.6	21.7	23.8

The amount and proportion of the black population living in each region:

	1970	Pct.	1960	Pct.
U. S.	22,672,570	(100)	18,871,831	(100)
Northeast	4,342,137	(19.2)	3,028,499	(16.0)
North Central	4,571,550	(20.2)	3,446,037	(18.3)
West	1,694,625	(7.5)	1,085,688	(5.8)
South	12,064,258	(53.2)	11,311,607	(59.9)

	1950	Pct.	1940	Pct.
U. S.	15,042,286	(100)	12,865,518	(100)
Northeast	2,018,182	(13.4)	1,369,875	(10.6)
North Central	2,227,876	(14.8)	1,420,318	(11.0)
West	570,821	(3.8)	170,706	(1.3)
South	10,225,407	(68.0)	9,904,619	(77.0)

Electronic Vigil Fails to Stem Mexican Alien Influx

By JAMES P. STERBA

Special to The New York Times

SAN YSIDRO, Calif., July 21 — Sensor No. 139 had been sending out a steady radio signal for two days—an obvious malfunction. But when No. 103 began beaming impulses to a magnetic tape at the Border Patrol sector headquarters here, the desk officer radioed the nearest patrol car.

"Got a reading on 103," he reported.

"Right, we'll check it out," came the response. Minutes later, three Mexicans, attempting to sneak into the United States, were tracked down and caught as a result of the electronic detection system operated by the United States along the Mexican border.

In the slower, morning hours, the sensors are quieter. But at night, especially on weekends, the electronic readout console becomes a Christmas tree, and stopping the swarm of illegal aliens crossing the border is an exercise in futility.

So heavy is the flood of aliens attempting to sneak across the border that the sensor program cannot keep up with it.

Federal immigration officials have told a special Federal grand jury in San Diego that they have "lost control of the situation" along the Mexican border, where the influx of illegal aliens has become a surging flood.

Criminal organizations formerly engaged in the narcotics traffic have turned to alien smuggling because the profits are large and involve far fewer risks, according to testimony heard by the grand jury, which is dealing with conditions in the Immigration and Naturalization Service.

The Federal authorities have reportedly gathered evidence establishing direct links between the smuggling ring and a number of Southern California business concerns, most of which depend largely upon unskilled labor.

"Our problem is that we get more signals than we have men to respond," said Richard Batchelor, deputy chief patrol agent at the Chula Vista sector headquarters of the immigration service's border patrol division.

"I could have 10 times as many patrol officers and we'd still be overworked," he said. "And even if we caught everyone, all we'd have is a huge people-shuffling operation that wouldn't solve the problem."

The Chula Vista sector is the busiest of nine Border Patrol regions along the Mexican border. It runs 70 miles from the coast, past Tijuana, into the interior. Last year, agents at the Chula Vista sector caught 94,220 people attempting to sneak into the country illegally. This year they expect to catch 150,000. But patrol officers here estimate that two or three times that many will sneak past them to begin lives and take jobs away from United States citizens or aliens who entered legally.

Statistics on seizures for this one section of the border illustrate the hugeness of the problem. In 1963, only 4,377 illegal aliens were caught here. In 1968, 26,206 were caught. By 1971, that figure jumped to 61,576.

Organized alien smuggling is increasing. Penalties are usually suspended jail sentences. The aliens themselves are simply trucked, bused or flown back across the border to try again.

Aliens sneaking across border paths represent only part of the problem. Thousands more simply drive through normal border highway checkpoints, flashing counterfeit border crossing cards or alien registration cards that have been stolen or sold.

A smuggler sometimes

In the field, an agent makes an arrest. Use of sensing devices to catch aliens entering U.S. illegally helps, but officials say volume of such aliens limits systems' effectiveness.

simply distributes the cards in Mexico to aliens who vaguely resemble the photograph of the original card holder. Then on this side of the border, the cards are collected for reuse.

Other aliens hide in car trunks of behind back seats.

Because of the volume of traffic, inspections are usually cursory. Nearly 24.5 million people, including 13.2 million aliens, crossed into and out of the United States at the San Ysidro border crossing point last year. At authorized entry points along the entire Mexican border, nearly 153 million people, including 91 million aliens, entered the country last year. They also come in and out by airplane and boat.

Last year the Border Patrol agents reported seizing 515,448 deportable aliens.

Without a Federal law making it illegal for Americans to hire aliens who enter illegally, the problem will grow and remain out of control, said Mr. Batchelor, who helped pioneer the use of electronic detectors along the Chula Vista sector starting in 1968 with a system of wired sensors some labeled "Batchelor's Folly."

Now the system is being extended along the 1,550-mile border except for certain areas regarded as inaccessible. The installation, for which $1.5-million has been budgeted this fiscal year, may be completed as early as mid-1974.

Some Congressmen have voiced criticism of the project and Mexico is asking the United States for a detailed explanation of what sort of devices are being installed.

According to Foreign Minister Emilio O. Rabassa, the Mexican Ambassador to the United States, Dr. José Juan de Olloqui, has been asked to make a formal request for the explanation from Secretary of State William P. Rogers.

The instruments are similar in nature to the electronic devices that were used in an effort to detect troops and supply movements along the Ho Chi Minh trail in Vietnam and Laos. Enemy troops defeated the system by sending through decoy detachments that drew off American troops and then dispatching larger forces at other points.

"Our stuff is basically military surplus but we have expanded it and constructed readout systems," Mr. Batchelor said.

The Pentagon, seeking appropriations for the use of the sensors along the Ho Chi Minh trail, was anxious to cooperate in return for data on sensor effectiveness.

In use are seismic sensors, buried with only a wire antenna above ground, which can pick up a footstep within a 38-pace radius. Infrared sensors can detect body heat, and metallic sensors can detect coins, guns or other metal equipment carried by border crossers.

But the sensors do not blanket the border. Rather, they are frequently moved and placed at likely entry points or along inland paths and trails frequently used by aliens. Sometimes simple traffic counters are used along dirt roads that are used at night to bring aliens in with cars and trucks.

Except near major towns and highway border check posts, there are no fences or walls along the border. And without an army of agents such as the 25,000-man military patrol that guarded the border before World War I, patrol agents here say fences or walls will not deter border crossers.

The Federal grand jury in San Diego has heard testimony that a dozen or more major smuggling rings are operating in the San Diego-Tijuana area alone, where the illicit traffic is heaviest.

These rings, according to William Duckhal, a border patrolman, are known to have connections with what he called "some big name businessmen" on the American side of the border.

There is evidence, officials said, that some employers, particularly in agriculture, food processing plants, hotels and a variety of small industrial concerns make advance contracts with a smuggling ring for delivery of a specific number of illegal aliens—men or women—on a certain date.

The employer pays a fee for delivery of this cheap labor, thereby enriching the smugglers beyond the $200 to $400 they collect from each Mexican for transporting him to a waiting job on this side.

Although reliable estimates are hard to come by, it appears that the illegal border traffic may be running at anywhere from 1.2 million to two million people a year since the emergence of the large-scale organized smuggling.

Once across the border and past the Immigration checkpoint along the main highways, the ones who are not hired—usually at substandard wages — as field hands in the Southwest and West often vanish into the protective barrios of Los Angeles and other cities.

Some, with the help of relatives or friends already in this country, move on to Denver, Kansas City, Mo., Chicago, Philadelphia and other locations more distant from the border where immigration surveillance is less rigid.

Once he gets a job in the United States, the illegal alien is still prey for the smugglers, who extort money from him—often nearly half his pay—under threat of exposing him to immigration authorities. If he refuses to pay, he may be killed.

Including the sale of entry and resident documents, the organized alien traffickers are said by Government investigators to be doing a billion-dollar-a-year business, besides taking away from American workers jobs with annual salaries totaling in the billions.

There is no effective way, and apparently no official or public desire, to prosecute illegal Mexican aliens. Much sympathy exists for their plight and for the manner in which they are underpaid, overworked and otherwise exploited by American employers.

Immigration officials are hoping for passage of a bill, already approved by the House and now before the Senate, which provides for the punishment of employers who knowingly hire illegal aliens.

July 22, 1973

Blacks Move South Again

The century-old migration of impoverished blacks from Southern farms to Northern cities is still continuing, but reverse migration has risen so significantly in recent years that there may now be as many blacks moving back South as moving North—and some demographers think there may now be slightly more.

Research into the subject is still meager and all that can be said with certainty is that the South is not losing its blacks the way it used to. From one admittedly limited Government sampling made in March, 1973, it is possible to conclude that some 166,000 blacks moved out of the South from 1970 to early 1973 while 247,000 moved in, for a net gain of 88,000. And one remark, by a 26-year-old black woman in Montgomery, Ala., summarized a great deal: "I grew up in Detroit. . . . Crime was everywhere. You couldn't save any money and people were rude. The South is better than that."

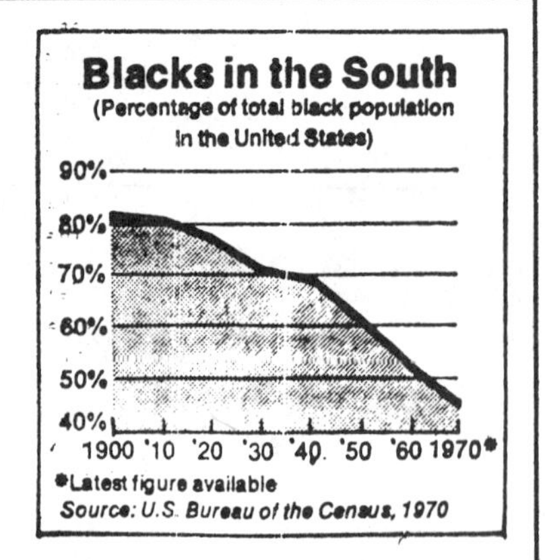

June 23, 1974

Official Urges National Assessment as Hispanic-American Population Rises Sharply

By ERNEST HOLSENDOLPH
Special to The New York Times

WASHINGTON, April 8—The nation's Hispanic-American population, increasing sharply in the last two decades, is approaching 20 million, raising broad and complex implications for the society that planners and policy makers have scarcely begun to assess, according to the Community Relations Service.

The service, a Justice Department agency, says that the ethnic group includes 11.2 million Hispanic-Americans accounted for by the Census Bureau last year, as well as almost all of more than eight million "illegal" aliens estimated by the Immigration and Naturalization Service—whose exact numbers may never be known.

Five years earlier, in 1970, the Census Bureau's estimate of the number of Hispanic-Americans was 10 million.

Benjamin F. Holman, director of the Community Relations Service, said that while there are no figures for the number of Hispanic-Americans two decades ago, reports from staff people in the agency's 10 regional offices have indicated a steady increase in communities across the country over the last 15 to 20 years.

The majority of the new residents are of Mexican origin, the service says; others are Puerto Ricans, Cubans and persons from Central and South America. The Census Bureau figures do not include residents of Puerto Rico itself.

The Hispanic-Americans are settling not only in such familiar areas as New York City, Miami and Los Angeles but also in a great tree-shaped sector with its thick trunk in Texas and its broad branches spreading up through the nation's center and outward along the Great Lakes.

In a speech prepared for delivery tomorrow in New York City, Mr. Holman describes the migration as a "blown tide" that resembles the movement of Southern blacks to Northern urban centers in the 1930's and 1940's, but with important differences.

Just as the huge migration of blacks was little-noticed by a preoccupied nation 40 years ago in the Depression, the Hispanic-American migration today constitutes "an evolving internal problem with both national and international consequences that gets little attention from the thoughtful concerned citizens and enlightens leadership sectors of our society," Mr. Holman says.

Mr. Holman's speech was prepared for a luncheon meeting of the American Immigration and Citizenship Conference at the Plaza Hotel.

The consequences of neglecting to monitor and sort out the new population phenomenon could have "unimagined implications for the maintenance of social order in our country," he states, "or [be] a golden opportunity for us to start planning, organizing and devising the future scheme of things."

Population Sampling

The Community Relations Service was created in 1964 to serve as the mediator to smooth over the process of the integration of public accommodations. Since then it has played a conciliation role in community conflicts arising out of school integration and easing local problems involving racial and ethnic groups.

The Census Bureau will conduct its first direct count of Hispanic-Americans in 1980. In 1970 an official estimate was made for the first time by taking a 5 percent population sampling. The estimate was updated last year.

The official Census Bureau estimates Mr. Holman says, indicate the following:

¶While 60 percent of the people of Spanish origin live in the five Southwestern states of Arizona, California, Colorado, New Mexico and Texas, New York City has the largest concentration of them, with estimates running from 1.5 million to 2.5 million.

¶In the Chicago-Great Lakes area alone, there are more Hispanic-Americans than in the states of Arizona, Colorado and New Mexico combined.

¶Among 80.8 percent of the native-born and foreign-born Hispanis-Americans combined, Spanish was the most common language spoken, indicating to the C.R.S. study that a growing number of communities are likely to become bilingual and bicultural. Underlining this development, Mr. Holman said are the increasing number of businesses in some cities, including New York, that display signs stating, "English spoken here."

¶Although some of the new migrants are professionals and businessmen, including many of the Cuban refugees in Miami and in Greater New York, most tend to be poorer and less educated than the national average.

¶Because of the youthfulness of the Hispanic-American population, its relatively high fertility rates and the prospect of continued population "pressure" from Mexico, the service concluded that this sector would outpace the rest of the nation.

Exception Is Cited

Generally, the nation, as in the case of the black migration, has done little to accommodate or adjust to the new migration, Mr. Holman says. One exception was legislation passed by Congress last year to protect the voting rights of citizens of Spanish origin.

Hispanic migrants, even more than the rest of the nation, tend to be concentrated in the cities. The 1970 Census estimates indicate that 83 percent live in urban areas.

With high unemployment nationally and a jobless rate among Hispanic - Americans that is double the national average, there have already been instances of friction between the newcomers and others, Mr. Holman indicates.

There have been a growing number of "confrontations" between the police and Hispanic-Americans, he says, and the police have been hampered by not having officers familiar with the new residents, their language and their cultures.

Mr. Holman emphasizes that the increase in Hispanic-Americans is a broad one.

"Cubans are now found in great numbers in Elizabeth, N.J., Puerto Ricans in Chicago are outnumbering white ethnics in what were once exclusive neighborhoods for Czechs, Poles and Lithuanians," he says. "Chicanos outnumber Indians in Utah. And our agency gets calls from Hispanics for assistance in such places as Burley, Idaho; Lancaster, Pa., and Omaha, Neb."

Mr. Holman believes that a number of changes are worth considering to avert problems.

He contends that school boards should insist that urban teachers be bilingual, that is "that they not only speak but also understand Spanish," and teachers colleges should require Spanish language training for all teaching personnel.

Urban planning should take into account the particular needs of Spanish-speaking citizens and should include Hispanic leaders among the planners where possible, he says.

And Mr. Holman, who is black, chided social planners by saying, "We should stop doing any more demographic studies of our cities that project the needs of blacks or poor whites only."

Aliens in New York Area

As much as 80 percent of the estimated 1.5 million illegal aliens in the New York metropolitan area are from South America, Central America and the Caribbean, according to officials of the Immigration and Naturalization Service here.

Other nationalities represented, they said, include Chinese, Greeks (mostly crew members who have jumped ship), Filipinos and Italians.

The considerable growth in illegal aliens here of Hispanic origin has occurred since 1965. Before that time, a citizen of a country in the Western Hemisphere could migrate to the United States by demonstrating to an American consul that he could support himself in this country.

Since 1965, however, there has been a ceiling of 120,000 immigrants each year from the Western Hemisphere, as well as other regulations that make emigration to the United States more difficult.

April 9, 1976

BLACKS STRUGGLE FOR SURVIVAL AND EQUALITY

THE TWO RACES.—We are informed by a gentleman from a region in the Southern States where society was greatly broken up, and the population, white and black, greatly scattered by the operations of the war, that, within the last year, the negroes who had fled or taken refuge in far away localities, have nearly all returned to the homesteads where, and to the employers by whom, they were brought up. This has been the case to a very great extent throughout the South. The poor darkeys find themselves completely lost in the wide world. Without the means of support, without habits of self-reliance and self-help, without knowledge of life and its ways, and with a sense only of its hard necessities, they have found that nothing but bitter experiences awaited them in their wanderings and struggles, and it has been a relief to them to return to the plantations and households where perhaps they had been only slaves, but where at least there was a state of things that approximately agreed with their characters and the measure of their development. We have often urged the idea that after all our legislation and laws, and regulations and orders, and bureaus and appropriations, it is to the hereditarily dominant white race in the South that the Southern blacks must mainly look for security and justice for a long time to come. Legislation can do something, and time and opportunity can do much more, but the present generation of negroes will never get over the influences of the terrible fact that they have been slaves. Hence, whatever legislation or policy tends to raise animosity or create dissensions and antagonism between the white and the black races of the South, must have a most baneful effect upon the Southern blacks.

June 28, 1866

EDUCATION FOR NEGROES.

Of all the experiments in education for the colored people, none come more prominently before us than the well-known institution at Hampton, Va. There are several somewhat similar establishments in the country, and from most of these we hear excellent reports. But the Hampton College is within easy reach from New-York, its patrons are so nearly allied to our own friends and neighbors, that we have a more active interest in its welfare; and we are possibly better informed of its progress. Last week the "Commencement" exercises of the Hampton College attracted many people from New-York and the Middle States, and our own reports show that the programme was not only admirably carried out, but the performances evinced a degree of careful culture for which the visitors were scarcely prepared. We observe three prominent features in the exercises, as reported by our correspondent The first of these was algebra, in which the students proved themselves wonderfully proficient. The second was a dissertation on the music and poetry of American slavery, of which many apt illustrations were given by a band of colored singers. The third was a scholarly disquisition on certain lectures on Man, prepared by that acute and subtle metaphysician, President HOPKINS. It should be remembered that these pupils have only been about three years in college, and yet they were required to give their own ideas on such topics as "The Difference between the Animal and the Vegetable," their objections to the Development Theory of DARWIN, the "Definition of the Higher and Lower Orders of Life," and an "Analysis of the Nature of Man."

It is very certain that the hope of the South must largely rest upon the education of the colored people. For better or for worse, four million blacks are now placed on equality with the white citizens of the Republic. Of these there are not a few whose blood is so far free from any African admixture that they have an exceptional interest in our eyes; they are inevitably bound to the so-called "inferior race," but their natural place would seem to be with the white people, from whose friendly and intimate society, however, they are debarred by a prejudice which is the growth of centuries. For the most part, however, the labors of educators and philanthropists, like those who have given so much success to the Hampton institution, are necessarily confined to the African, the ex-slave, or the direct descendant of those who have been born in slavery. Whatever may be said of the political vices of those colored people who have come to the front in public affairs, as in South Carolina, it cannot be denied that suffrage without education is a dangerous experiment. Our original idea of the common school in America was that the State, for self-protection, must give the voter an education sufficient to enable him to exercise the right of suffrage with intelligence. If we have gone beyond that—as we doubtless have in some instances—our duty to the unintelligent mass of voters is, nevertheless, plain. And to say that a few "smart" negroes in a Southern State have sharply competed with the sharpest of the whites in rascality, proves nothing against the general proposition that the freedmen must be educated before the experiment of universal manhood-suffrage can be fairly tried.

When we come to inquire into the details of the education proposed for the freedmen, there may be room for difference of opinion. There is a tendency toward over-education in these days. We have seen youths, whose time in training-schools was brief, put through a course of mental philosophy, rhetoric, and drawing, and then, with a taste of those studies, dismissed into the world to lay brick or hew timber—occupations for which their hurried education has given them small qualification. It is not fair to say that the colored youth now in course of training in the few collegiate institutions of the South are to be hewers of wood and drawers of water, and that their curriculum should be formed on such a basis. Nevertheless it is true that these forty graduates who go out from Hampton this year, and their comrades from similar institutions, are going out into the ranks of their former comrades in bondage as teachers. Temporarily they will take places as household servants; ultimately, they will be the leaders of those who have come up from slavery, and are to learn the way to take care of themselves. As our correspondent has already pointed out, some of the commoner duties of living are not so easily made plain in an institution where economy must be practiced by the use of labor-saving machinery. The scholar who has been educated in housekeeping in an establishment where steam, gas, and economical mechanism are employed, will be at a loss when he or she is set down in the primitive cabin or cottage of the South, where the housekeeper must know how to economize tallow candles, pine knots, fire-wood, and wash-tubs. We must confess that we should prefer that the black leader of his race should know how to put up a log house deftly and without loss of labor rather than be able to formulate the cosmic system of philosophy, or intelligently expound the theories of HERBERT SPENCER. Since we are obliged to take the colored people late in life, if we educate them at all, is it not better that they should be thoroughly grounded in the rudiments? We may trust the future to give them all the science which they may really require.

June 17, 1875

LESSONS OF THE EXODUS.

A LECTURE BY FREDERICK DOUGLASS—MIGRATION A MISTAKE.

BALTIMORE, May 4.—Marshal Frederick Douglass lectured before a large mixed audience to-night at the Centennial African M. E. Church, on the negro exodus. He said the condition of the race was not what it ought to be; that the colored race must rise to equality or fall lower than ever. He had great hope, but many in despair are leaving their old cabins and cotton patches in the far South, where they are doomed to oppression, hoping to improve their condition in the great West. He pitied these men. By violence, intimidation, and outrage they had been deprived of their rights as citizens. They had been maltreated, abused, and oppressed by the old slaveholders. Their lot had been a hard one, but it was a mistake for them to seek relief in migration. In the South they were acclimatized, they had a monopoly of labor, while in the West they must compete with the Irish and Germans. Their rights of franchise would be safer in Kansas, but they would have to pay too dearly for it. They would suffer more than ever, and only a few would succeed. They would be better off to stay where they are. The colored race must measure its condition not by what it hopes to attain, but what it has been. "Fifty years ago," the speaker said, "I landed here from my slave home on the Eastern Shore. The motto of the white men then was: "When you see a nigger's head hit it," and lucky was one that didn't get hit. If a "nigger" was killed nothing was thought of it. If a "nigger" struck a white man he was hanged for it. The church, the whipping-post, and the slave auction-block were side by side. Now we are citizens and better off, notwithstanding our wrongs, than the citizens of Europe. The earthquake of abolition sent some poor, uneducated negroes to higher positions than we could hold. We had more men in Congress once than now. We won't have any more for some time. Slavery was not a good school for statesmen. They will come later. We must not be discouraged by evils or sufferings. Each year we are better off than before." Mr. Douglass spoke nearly three hours, and was frequently applauded.

Dispatch from the Rebel Agent of the Associated Press.

VICKSBURG, Miss., May 4.—A telegram received last night from Leota Landing, 70 miles above Vicksburg, says that 150 negroes, bound for Kansas, have encamped on the bank of the river, and picketed the roads. No conflict between the whites and blacks has yet occurred.

The Mississippi Valley Labor Convention, which is to assemble here to-morrow, gives promise of being largely attended. Many delegates from the river counties of Mississippi have already arrived, and every train and steamer swells their numbers. Among the persons who have signified their purpose to be present are Gov. Stone and James Hill, of Mississippi, and T. Morris Chester and J. G. Lewis, two prominent colored men from Louisiana. There is diversity of opinion as to what the convention will recommend. Some delegates favor a better understanding between employer and employe, both as to wages and political rights, while others will introduce resolutions favoring a plan for replacing the lost labor by Chinese and white immgrants. Another conflict will arise if the resolution already prepared is introduced, which sets forth that the cause of the present exodus is not attributed to restriction of the political rights of colored men.

May 5, 1879

THE SOUTHERN SOCIAL QUESTION.

The evident candor and good faith of the communication that follows will commend it to those who are prepared to exercise these qualities in considering the social question at the South:

RICHMOND, Va., Dec. 4.

To the Editor of the New-York Times:

I inclose to you an advertisement which appeared in the Richmond *Daily Times* of yesterday:

WANTED—A washlady moving in the very best colored society desires a position in a fashionable family with full possession of the back building and privilege of daughter taking music and French. Address Mrs. L. C., P. O., city.

It gives you a glimpse of the state of affairs here since HARRISON'S election. The negroes have been completely demoralized by it, and are taking on most startling airs. It has become almost impossible to keep them at work, and their impudence and disobedience to employers are unbearable. I often see them riding in the fashionable streets with white coachmen and fine carriages. Ladies are constantly pushed off the sidewalks into the mud of the streets by negro women and men. They are often spoken to by negro men, forced to walk with them on the streets, and grossly insulted. Several of my lady friends have been so treated. I am a Northern man, but feel the strongest sympathy with the people of the South, especially the ladies, who have to endure all this. I wish you could do something to create a right state of feeling about it in the North. R. W.

We do not see how any fair-minded man or woman can read this without sympathy for the white people of the South and without seeing that the race question in that section is primarily a social and only secondarily a political question. That a laundress should describe herself as "a washlady moving in the very best colored society," and should insist upon "full possession of the back building," is on the face of it merely comic. Perhaps it is not more comic than that a white shopgirl should describe herself as a saleslady, though the white shopman has not yet insisted on being addressed as a salesgentleman. But the colored washlady's advertisement indicates a state of mind that would render her quite intolerable in any white household North or South. The necessity for earning a living forces people who take thought for the morrow to do their work, whatever it may be, in such a way as to establish a *modus vivendi* with their employers. But imprudence and improvidence are among the most characteristic traits of the negro. How far they are innate and how far they are due to the institution of slavery, under which the negro was neither forced nor permitted to provide for himself, does not affect the actual state of society in the South. It is eminently a condition and not a theory that confronts Southern employers, industrial and domestic. And the main fact of the condition is that the negro does not act with that view to his own interest which can be counted upon in dealing with white labor. The ambition of bettering their condition which is the incentive to industry and thrift in the progressive races affects only exceptional individuals among the African race. The testimony of housekeepers in Southern cities, including Washington, is uniformly to the effect that colored servants abandon their situations much more lightly than white servants, and what is true of domestic service is equally true of mechanical or agricultural labor. The standard of living is a mere subsistence, and that they can attain by begging or in other ways, with fitful intervals of work.

Evidently a region in which this is the temper of the laboring class, and which does not attract the immigration of steadier and more trustworthy laborers, cannot make rapid industrial progress. The wonder with regard to the South is not that there should be so few enterprising and advancing communities, but that there should be so many. Whoever puts himself in the place of the Southern whites will understand, if he does not wholly approve, the emphatic declaration made the other day by the Governor of South Carolina to the effect that the whites of that State meant to control its affairs without reference to the numerical superiority of the blacks.

But when all this is admitted, the admission does not do much toward settling the practical question raised by the letter of our Richmond correspondent—What is to be done about it? The experience of more than twenty years tends to show that nothing to the purpose can be done about it through the action of the General Government. The "reconstruction" was perhaps a necessary stage in the restoration of the Union, but nothing could have happened better adapted to strain the relations of the two races in the South than the appointment of political adventurers who used their places with the view of organizing and controlling the ignorant and newly-enfranchised negroes. So long as the blacks are politically "solid" the whites must be solid also. A Democratic term of the Presidency was especially to be desired in order to show the negroes that they had nothing to fear from Democratic success. The demonstration has been made. It is not pretended by anybody that their rights have not been as secure under CLEVELAND as they were under GARFIELD, HAYES, or GRANT. And yet it seems that they are encouraged into insolent behavior by the election of HARRISON. As they have not been molested or oppressed by a Democratic Administration, it is not the prospect of escape from tyranny that elates them now. It must be the hope that they are to be taken care of without any exertion of their own. There has apparently been a revival of the belief in "forty acres and a mule" that prevailed in the early days of the Freedmen's Bureau. Nothing could more effectually than such a belief check any tendency to help themselves they might otherwise show. The social question in the South must be left to be worked out by the people concerned. Congress cannot help them in this, but it may increase the difficulties of the question and postpone its settlement. That is the effect of every special committee of investigation that gives the negroes a hope that they are to be relieved from the necessity of supporting and protecting themselves.

December 9, 1888

THE NORTHERN COLOR LINE

STEADILY BECOMING MORE SHARPLY DRAWN.

ESPECIALLY IN NEW-YORK DO RACE PREJUDICES EXIST — EXPERIENCES OF TWO COLORED CLERGYMEN.

"Ever since the Civil Rights bill was declared unconstitutional by the United States Supreme Court, it appears to me that the spirit of intolerance of the colored people has been on the increase in the North," said the Rev. Hutchens C. Bishop, rector of St. Philip's Protestant Episcopal Church, (colored,) to a TIMES reporter. "Before that time, ten or a dozen years ago, the practice of the Northern white people bore some consistency with their professions and, for the most part, colored people were treated according to their personal merits as other human beings were. In the rural and suburban communities the old custom still continues, but in the larger cities the aspirations of the colored people for a better kind of existence, for a higher life, are hedged about by obstacles imposed by the white people and for which there is no excuse that I can find short of a spirit of prejudice that is altogether unworthy of them.

"This prejudice is most sharply defined and emphasized by the refusal of decent habitations to us, even though our people are willing to pay larger rentals than are required from white tenants. The agents and landlords are not altogether responsible for this condition of things. Several of the agents have said to me that they have advised the letting of houses to colored tenants, because they have uniformly found that colored people take better care of the premises than their social equals among the whites, and they are generally reported the best paying tenants a landlord can have. But the landlords are afraid to rent to colored people because of the objections of white people to living in the same building or in the same neighborhood with colored people. In the larger cities of the country, where this prejudice and spirit of intolerance is most pronounced, it springs apparently in greatest measure from the foreign and poorer elements of the population. They are the ones with whom, in the nature of things, the colored people are thrown most directly in contact, for the earnings of the colored people in general are among the lowest of the low, and not being able to buy their own homes they are compelled to accept whatever the white man will let them have in the shape of habitations."

Mr. Bishop and his wife are of fairer complexion than most white Americans, have light blue eyes and blonde hair. The rector told a story about his own experiences in renting a house.

"I called upon the agent," he said, and agreed with him quite readily for the letting of the house, for while I am in no way sensitive about my proper classification among the colored people, I do not care to hang a label to that effect around my neck. But when the agent came to look up my references he found that I was a colored man. He at once sent me word that, much as he regretted it, he could not rent me the house, the landlord had absolutely refused to rent his house to colored occupants. My wife called to see him about it, and was advised to see the landlord in person. We called upon him, and upon our assuring him that we wanted the house for our exclusive use and occupancy, and would not sublet it, or any part of it, we were allowed to have it.

"If I had not succeeded thus I would have been compelled to go into a noisy tenement house among a class of people that I do not care to be particularly identified with, be they white or black, or to take a house that had been previously occupied by disorderly white people who had been cleaned out by the police. Such environments as those are not suitable to the growth of a robust moral sentiment or the bringing up of children, but, good and bad alike, the colored people are closely restricted to them. And except by indirect means it would be very difficult for a colored man to even purchase a house in a respectable residence locality in New-York. So strong is this prejudice that many of our best people have left the city entirely and taken houses in the suburbs—in Brooklyn and New-Jersey.

"Instead of improving, the situation seems to be growing worse. Colored men would gladly learn mechanical trades, and wherever there have been openings for them the colored people have shown themselves the equals of their white competitors as mechanics. Only servile occupations are open to the colored men, and even in these the rates of pay are lower to colored people than to white men in the same occupation. Under such circumstances how can a rapid moral, mental, and material development be expected of the colored race? They put the little money they receive to better use than any other people I know of. But the margin above the actual necessaries of life is so very small that little show can be made of it in any direction.

"The forms this prejudice takes in the North are often more humiliating than the political antagonism the colored people encounter from their former masters in the South. There colored men work side by side with their white colleagues in all the mechanical trades, and no effort is made to prevent them from learning any branch of industrial employment they may choose. The material and moral and intellectual aspirations of the colored man are not interfered with in the South, but, on the contrary, are encouraged in many ways by the white people, who learned before the war to look to him for all the mechanical work that was wanted. For the colored man was the mechanic of the South before the war, as well as the field and plantation hand and day laborer. When, therefore, the colored artisan comes North and tries to find work at his trade and is met by a refusal of white men to work alongside of him, he encounters a kind of antipathy and prejudice that are unknown in the South. His development along the lines of morality and industry is interfered with and hampered, and there is small wonder if his progress is slow and precarious."

Another colored pastor who expressed similar views to the reporter was the Rev. H. A. Monroe of St. Mark's Methodist Episcopal Church. He maintains that the colored population of the city has been vastly increased of late by emigration from the South, and, in consequence, the difficulty that is encountered in the search for homes is greater than ever before. He declares there is a colored population in New-York City of over thirty-five thousand, and that it is rapidly on the increase.

"There are no more American Americans in New-York to-day," he said, "than the colored people. All their tastes and instincts, their ambitions and ideas, are purely American. They are not a dissipated race, nor given to brawls and disorder. For the most part they have acquired their ideas of life in its various phases by tuition of a superior class of white people. They have an inbred desire to appear as well as their white masters of slavery days, and to the extent their means will allow, they try to imitate them in the adornment of their homes and persons. But their means are limited, and they are forced to practice the most rigid economy. You will find, therefore, that a colored housekeeper is able to make a small money allowance go further in supplying a table than is general among the people of any other race or nation. And no matter how slender their incomes, they can always spare a little for house decoration. They make good housekeepers, pay their rents promptly, and keep their rooms and premises in good order. And yet they are a proscribed race.

"Before a neighborhood is opened up for colored settlement in New-York it must have been made untenantable for white people. Take this row that I live in—[one of five brick houses in West Forty-seventh-street]—every house had been one of the disorderly kind. This very house was one of the most notorious places of resort in the city, and at last the inmates became so disorderly that the police raided them several times in succession and drove them out of the street. The houses were advertised at $900 a year, and were vacant several months, because respectable white people would not live in them, and the landlord refused to rent them to colored occupants. At last, finding that he could not rent them to white people at all, the rent was raised to $1,000 and the houses were thrown open to colored people. We were compelled to sign leases containing waivers of any claims for damages we might have by reason of the uses to which the houses had been put by former occupants. The annoyances to which we were subjected for months after we moved in were something terrible, and yet it is to such houses and such neighborhoods that we are limited in our choice of homes. You can readily imagine what the moral effect of such an atmosphere upon the rising generation must be.

"I have contended against this proscription of my race upon every opportunity, but it seems as if the lines were being drawn closer all the while, and when such a representative Christian body as the Young Men's Christian Association closes its doors to our young men the prejudices of other people are scarcely to be wondered at."

"What excuse is given for refusing to admit colored youths to the Young Men's Christian Association?" he was asked.

"None, that I have ever heard. And we have college graduates, lawyers, physicians, and theological students among our applicants, men in every way worthy of equal social consideration with the better classes of white men. I have lived in the South, although, until I left school, I had never seen a colored person except the members of my father's family, for I was brought up in the East, went to the public school with white children, and had never been given to understand that my schoolmates or neighbors esteemed themselves superior to me because they were white and I was colored. But in the South there is none of this trade and neighborhood ostracism that we meet in the North. In Charleston or Wilmington—places that are considered the most hide-bound and storm-buttressed of Democratic strongholds—a colored man may have anything in the way of a residence that he is able to pay for. He may buy or he may rent in the most fashionable part of the city if he is willing and able to pay the price that would be asked of any white man.

"There, too, the trades are open to colored men, and they are not discriminated against in the payment of wages. I have seen white firemen working with colored engineers on Southern locomotives, and vice versa. And I have seen white bricklayers and laborers working under colored contractors on buildings in the South and nothing was thought of it. But after I came to New-York, when I went to attend an army reunion in Massachusetts—I was a member of the Fifty-fourth Massachusetts Volunteers—I heard a letter of Chief Arthur read in which he declared that no colored man would be received into the Brotherhood of Locomotive Engineers. The letter was written to Col. Hartwell of the Fifty-fifth Massachusetts Volunteers. This is only a fair example of the oppression under which the black man is expected to make progress in the North.

"Until the attitude of the white people shall change and a colored man shall be treated according to his just deserts, based upon his intelligence and moral conduct, it need not be surprising if he reflects more of the vices than of the virtues of his white brother. But right here I would like to say that among the bankers and brokers and lawyers, the most liberal and intelligent of the people of New-York, I find a willingness to give the colored man a chance that is a bright contrast to the general experience of my people in their dealings with the whites, and quite a number of our young men who have secured good educations have been given situations of responsibility, with good salaries, in their offices—not as mere messengers or watchmen, but as clerks and bookkeepers. But they are a very small fraction of the colored men who are equally competent and who can find no such positions."

April 28, 1889

THE AFRO-AMERICAN LEAGUE.

It appears that a newspaper in Detroit, which is a special organ of our colored fellow-citizens, is urging the formation of an "Afro-American League," the object of which is to benefit the colored people in some way or ways not very clearly defined. One of the promoters declares that the business of the organization will be "the protection and promotion of all interests pertaining to the dignity and welfare of the Afro-American." This is large Afro-American language, but it is not precisely explicit, and it does not help us much toward ascertaining what it is that the league is meant to do.

Educated men of color, of whom the number is comparatively very small, complain that they are subjected in this country to certain inconveniences and slights which they do not encounter in Europe. This is unquestionably true, and if it can be made untrue or less true by the formation of a league every humane and fair-minded person will be glad. Nevertheless, it is hard to see how the league is to soften the obdurate hearts of the keepers of those hotels and theatres in which men and women of color do not find themselves welcome. The Civil Rights acts that were intended to enable Afro-Americans to make good their footing in places where they were not wanted have been decided to be unconstitutional. Before that decision was reached they had become a dead letter, as all laws designed to affect sentiments and social relations must necessarily do. The prejudice against association with negroes, except in capacities of recognized inferiority, is one of which every manly man is ashamed and one which he tries to conquer so as to treat every colored man he meets exactly as he would treat a white man whose deportment was the same. Nevertheless, it exists, and it is very nearly if not quite as strong at the North as at the South. In the days of slavery the Southern politicians used to maintain that it was a natural instinct, and if they were pious, as they mostly were, to explain it by reference to the Scriptural curse of HAM. That it is not a natural instinct is sufficiently proved by the fact that it does not exist in Europe, where a negro in a hotel or a theatre excites more curiosity by reason of his rarity, but no resentment. The difference is that African slavery has never existed in Europe, and it is slavery that is responsible for the prejudice against negroes, as negroes, in the United States. The fact that the race has been recently in servitude ought perhaps only to excite sympathy, but as a matter of fact it seems to excite aversion and contempt. The innkeepers and managers who turn away negroes under false pretenses are merely pursuing their own interests by reflecting the wishes and feelings of the great majority of their customers.

African slavery has been abolished but a quarter of a century, and it cannot be expected that the prejudice that has proceeded from it will perish at once. There were in the early history of this country many white slaves, men who were bound to labor for a term of years in payment of their passage money. They were objects of contempt, doubtless, to their contemporaries, but they did not carry the evidence of their degraded condition in their faces as the negroes do, and their children had as fair a chance in life as the children of their neighbors. The misfortune of the negro is that he bears about with him the unmistakable evidence of his past subjection. Time will no doubt remove the prejudice of color so far as it is groundless—that is to say, so far as it is not strengthened by the behavior and manners of the race against which the prejudice exists.

This is the business of the colored people themselves. If they can do it any better by forming a league, by all means let them do so. One of the founders of the league considers that "the race must demonstrate its ability to grasp the questions in all their curious ramifications, bury personal aggrandizements, and success will follow." In fact, "success will follow" in the degree in which colored men show themselves worthy and useful citizens, and no other device will be of any avail. It may be suggested that a league of the kind proposed is a very unlikely means of "burying personal aggrandizements." There will be a tendency on the part of smart Afro-Americans to make use of it in order to urge their own claims to public office, and this tendency will require a great deal of resisting, so much, indeed, that the elevation of the colored race seems more likely to be effected without than with the aid of the Afro-American League.

October 20, 1889

FORTUNE STIRRED THEM UP

EXCITEMENT AT THE COLORED MASS MEETING.

SHOUTS, CHEERS, AND HISSES—SOUTHERN LYNCHINGS CONDEMNED—AN ADDRESS TO THE PUBLIC AND AN APPEAL TO THE PRESIDENT.

The burning at the stake of a negro at Texarkana not long ago, and sundry lynchings which have lately aroused the negroes in the South, brought about in Cooper Union last night what was probably the largest meeting of colored persons ever held in New-York City.

The crush was remarkable. The great majority of the persons in the hall were well dressed and orderly, and their faces indicated mingled anger and sympathy with what the speakers many times characterized as outrages upon their race. The effect of the eloquence which some of these speakers displayed, however, was in a measure dampened by the tremendous uproar which was caused in the audience by the unnecessary dragging of politics into the expressions of protest.

The colored ex-Congressman John R. Lynch of Mississippi follow d T. Thomas Fortune, the colored editor of this city, in an impassioned address upon the alleged uprightness of the new Mississippi Constitution.

"Any negro in the South, or in New-York City, for that matter," he said, "who shows any inclination to concur in the political beliefs of the framers of this Constitution, is a party to the crimes against negroes which have lately arrested the attention and aroused the indignation of the civilized world."

Mr. Fortune, whose Republicanism is said to be not above reproach, immediately began to blaze with wrath at this allusion, which he took to himself. Stepping to the front of the platform and waving his arms wildly he exclaimed:

"I am as good a Republican as any—"

He got no further. The meeting, which up to that point had been orderly, and even dignified, became a pandemonium. A volley of hisses like the sound of many locomotives blowing off steam proceeded from the audience. Then followed groans and shouts. The whole assemblage, already roused to a highly nervous state by the addresses, became hysterical. For ten minutes men and women screamed, shouted, and hissed.

Fortune had friends there, but they were in a minority. They persisted in shouting out words of encouragement to the editor, who stood immovable among the crowd on the platform. For a time it looked as if the preventing of a fight between the two factions was impossible. The appeals of the strong-voiced Chairman, Dr. W. B. Derrick, were all vain. Finally the band struck up "The Star-Spangled Banner." Even this did not in any degree lessen the uproar, which only ceased when half a dozen strong men on the platform seized Editor Fortune and forcibly pushed him to the rear.

Then the crowd, the burden of whose cry since the beginning of the din had been "Throw him out!" "Shut him off!" "Get off the platform!" became quiet, and Mr. Lynch went on with his speech.

Dr. W. B. Derricks opened the meeting with an address, in which he characterized the Southern lynchings as "the most appalling occurrences in American history."

"This meeting evidences the fact," said the speaker, "that the negro is not asleep, in the woodpile or anywhere else. He is awake, and has been awakened by piteous voices from Pine Bluff, from Texarkana, from Memphis, from Louisiana, and now by a voice in New-York that speaks not from the grave, but in resounding protest against the slaughterers who have brutally used the knife and the pistol and applied the torch in the South."

Lawyer T. McCants Stewart then read a long "Address to the friends of liberty and humanity everywhere," which had been prepared by the organizers of the meeting. Among other things this address says:

"We recommend that the race in the South maintain their trust in God, but we also recommend that they unite for mutual protection; that they seek to bring to their support and into public expression the opinion of that part of the white people in the South who are disgusted and who feel compromised by the lawless elements among them.

"We urge Afro-Americans in the South to keep always within the law, to bring, as far as practicable, actions for damages against cities and against counties in which these lawless acts are allowed to occur by public officers, who are sworn to execute the law and to protect life and property.

"We urge organization; we urge agitation; we urge the prosecution of every peaceful remedy; but we also advise our brethren to protect to the extent of their ability their defenseless fellows charged with crimes against the lynchers and midnight marauders, who are always brutal outlaws, and we advise our brethren to let it be known that endurance has a limit, and that patience under same conditions may cease to be a virtue."

The address of Mr. Fortune, whose unpopularity with the crowd only began after his interruption of ex-Congressman Lynch, created tremendous enthusiasm on account of its very decided tone. At the conclusion of a sentence reciting a long list of "outrages upon American civilization," Mr. Fortune asked: "What are you going to do about it?"

"Fight! fight!" was the word that was hoarsely shouted from a thousand throats, and it was many minutes before the Chairman could restore order.

Editor Fortune had to cut his speech short, so fiery did the expressions from the audience become, and to quell the disturbance the band struck up "My Country, 'Tis of Thee." This had a soothing effect and the whole audience joined in the chorus.

The address of ex-Congressman Lynch was an attack upon what he termed the "Jim Crow" railroad laws in the South, which forced the negroes to occupy separate cars from the whites. Mr. Lynch also urged a stronger enforcement of the civil rights laws in the Northern States. His arraignment of the new Mississippi Constitution caused the uproar, which came dangerously near being a general fight.

Gen. James R. O'Beirne, Assistant Superintendent of Immigration, followed in a speech in which he likened the negroes of this country, struggling for justice, to the people of Ireland struggling for freedom. His suggestion that the negroes appeal to President Harrison to send a special message to Congress on the subject of negro outrages in the South met with tremendous applause.

The suggestion was immediately adopted, and Dr. Derrick announced that delegations of colored men from New-York, New-Jersey, Connecticut, and Pennsylvania would soon go to Washington for the purpose of presenting the "Address to the American People" to Mr. Harrison, and asking for his aid in causing special legislation in the matter.

Col. Elliott F. Shepard was expected to deliver an address, but he did not appear. The meeting closed with the reading of a set of resolutions regarding the lynching.

Among those who occupied seats on the platform were the Rev. Dr. A. N. Monroe, the Rev. Dr. J. P. Sampson, the Rev. Dr. Rufus L. Perry, the Rev. Dr. W. H. H. Butler, the Rev. Theodore Gould, D. M. Webster, J. H. Simms, J. H. Mitchell of Richmond, Va.; L. Christy of Indianapolis, W. H. Barnett of Chicago, Calvin Chase of Washington, Isaac B. Johnson, Ida B. Wells, Memphis, Tenn.; J. C. Cooper, Indianapolis; C. Perry, Philadelphia; J. C. Daucy, Salisbury, N. C.; Prof. Charles R. Reasons, Prof. Richard Robinson, Pierre Barquet, Col. Lee, and Thomas Brown.

April 5, 1892

MUTUAL BENEFIT CLUBS

They Are Many and Popular Among the Southern Negroes.

ARE A GREAT BLESSING TO THE RACE

Organized Chiefly for the Purpose of Caring for the Sick and Burying the Dead—Treasurers Grow Rich.

MACON, Ga., June 8.—There are no people in the world who believe so strongly in organization for self-protection and mutual benefit as the negroes of the South, and their nature affords fine study for labor organizations and insurance companies, especially mutual benefit associations—in fact, there are no insurance companies conducted so successfully and satisfactorily as the thousands of negro societies of the South that are conducted solely by negroes on the mutual benefit plan. These societies are formed in every city in the South, and as a rule are purely local affairs; but some of them extend over several States, with auxiliary branches in hundreds of cities and towns. Ninety-nine out of a hundred of these societies work a blessing to the negro, but every one of them proves a great inconvenience and a hardship to the Southern white people, especially housewives, in one sense, and a blessing in another.

Take Macon, for instance, with its 12,000 negro inhabitants, and it is not too broad an assertion to make that more than half of this number belong to one or more societies. This is especially true with the women, ninety-nine out of every one hundred of whom belong to a society, and some of them to half a dozen. For this reason there are from one to three half days in each week when the Southern housewife is without a servant, as there are few of them who would not give up the most lucrative position it is possible for a negro servant to obtain to attend her "s'ciety" meeting. They will not work for a family that will not give them time off to attend these meetings, and as the Southern people are fast finding this out, they are reaching the point where, when they employ a servant, they have an understanding as to when and how often she is to attend society meetings.

The objects of the negro societies are too manifold to be fully explained, but they are principally for the purpose of caring for the sick and burying the dead. Nearly all of them are duly chartered, and can sue and be sued. Monthly assessments are levied on the members, and it is seldom a member is expelled for non-payment of dues or assessments. Some of the societies in Macon have from 600 to 800 members, and one society, The United Brothers and Sisters, has over 1,000. The dues are, in most cases, 50 cents per month, although some of them go as high as $1 per month, and some as low as 25 cents. The children's societies, and there are a number of them, levy an assessment of 10 and 15 cents per month.

Every society has its own uniform, and some of them are decidedly fantastic. Some have both male and female members, while others confine themselves to one sex. Each of them has some preacher or other prominent negro at its head as President, and he also acts as Secretary and Treasurer, or, in other words, he is a whole insurance company within himself, as he collects and pays out all moneys and keeps a record of the society's business. This is not done by a system of bookkeeping, as there are few of them who can write. He simply sees that every member pays his or her monthly assessment, and when a member is sick he pays him or her the weekly benefit, generally $2 per week, and in case of death he pays from $25 to $50, the amount being governed by the charter. If he does not pay it, a lawsuit follows, but, to their credit be it said, they prove faithful to their trusts in most instances. Few of them are required to give a bond. Viewed from a business standpoint, they would be foolish to violate the trust reposed in them, for so long as they conduct themselves properly they are in control of the funds, to use them as they desire, without accounting to any one. In fact, all the members ask is that they receive their weekly benefits when sick and that each member is buried decently. It makes no difference to them if the Treasurer receives $5,000 per year and pays out only $2,000, for, in fact, they rarely know how much he does receive.

It is surprising to learn how many negroes get rich out of these societies. Most of them, however, are careful not to invest their money in real estate, as the other members would know of their wealth, and that would arouse jealousy and they would soon be asked to resign. One of them said recently that his income was something over $6,000 a year from his societies. This, however, is an exceptional case, as the negro in question is President, Secretary, and Treasurer of thirteen distinct societies, one of which has branches in Alabama, Mississippi, Texas, and Arkansas. He is also the pastor of a half dozen churches in Middle Georgia. These churches are not all of the same denomination, Methodist, Baptist, and Primitive or "Hardshell" Baptist being represented among them. He is very popular with a majority of the negroes, but has a large number of enemies who are constantly getting him before the courts, first on one charge and then another, but he always comes out with flying colors. The women worship him, and on one occasion, when he was put in jail on a charge of stealing and swindling, hundreds of them quit work and congregated about the jail, while others raised $500 in a few hours for the purpose of furnishing his bond.

Up to a very few years ago nearly every negro who died was given a pauper's burial by the city or county, few of them being able to bury their dead, but for the past two years there have been only three negro paupers buried by the county, and none by the city. This is solely due to the societies. Formerly, when a negro became ill he was carried to the county poorhouse, but now they are cared for at home with the weekly benefits, and when the sick have no family a sick committee is appointed to wait on them. As some of them belong to as many as a half dozen societies, they receive a great deal more when sick than when well. This also accounts for the large and expensive funerals some negroes have, as each society has to pay the full amount of its stated death benefit.

From the pomp and splendor of some negro funerals a person not acquainted with the facts would think the deceased a person of great wealth, but in all probability inquiry will reveal the fact that she was a washerwoman whose sole dependence was $2 per week. When a member dies the society turns out and attends the burial in full uniform and regalia, there being a fine for non-attendance at a funeral. This is done principally to insure a good attendance at the funeral and to keep up the interest in the society.

Where the negroes get the names for their societies is a mystery, and why they select such odd and inappropriate, not to say nonsensical, names is a still greater mystery. Imagine such names as "The Sons and Daughters of the Ring Tailed Doves," and "The Mourning Doves," "The New Jerusalem Sisters," "The Golden Star Brothers," "The Nightingales," "The Honey Bees," "The Sons and Daughters of Butterflies," "The Hamburg Sisters," "The Boston Brothers," "The Fairy Queens," "The North Star Sisters and Brothers," and many similar names, all of which are bona fide societies here in Macon. All of the societies are semi-religious, and have their regular places and times of meeting. Few of them are secret, but no one is allowed at the meetings except members.

The white people have never antagonized the societies, except to forbid their servants taking so much time to attend the meetings and funerals, but as they have found that it is useless to try to stop this they give them encouragement and endeavor to make the best of it, as they are awakening to the fact that the societies are doing a great work, both in caring for the negro and in elevating him.

June 9, 1895

NEGROES TO BUILD A COTTON MILL.

Interesting Experiment to be Made by the Race in North Carolina.

From The Charleston (S. C.) News and Courier.

A dispatch from Concord, N. C., announces that "negro capitalists" of that place are about to build a cotton mill "for the express purpose of manning it with negro labor, and thus trying to settle the disputed question whether the negro will make a good mill operative. The capitalists, it is estimated, are rather small ones, but one of them, N. C. Coleman, who is the leader of the venture, is worth between $25,000 and $50,000, which he has made at industrial pursuits since the war, and he speaks confidently regarding his scheme.

If it is found to be necessary, he says, he and his associates will be glad to have their white friends help them out with subscriptions, but if possible they "would like to have all the stock taken by negroes in order to make the mill essentially a negro enterprise," and he thinks "there is little doubt that the mill will be built and put in operation this year."

This is a very different experiment, of course, from that of employing colored operatives in mills owned and run by white capitalists, and it must be considered accordingly. We think the promoters of the experiment have erred in determining to try it in a town where there are mills operated by white labor, as they would doubtless have more sympathy and support and would have less reason to fear friction of any kind in an unoccupied field. This is a question of their own decision, however, and, as they appear to haye decided it already to their own satisfaction, it only remains to await the issue of their venture. There is no good reason that we know of why they should not make it, and be encouraged to make it, if they can afford to risk their money in it.

All kinds of pursuits that can be conducted by members of their own race are now open to the colored people; there is no reason why an exception should be made of cotton manufacturing or any other kind of manufacturing. If they can make a success of spinning and weaving cotton goods they are entitled to any measure of success, from the lowest to the highest, that they can achieve, and all right-thinking men will approve and encourage their self-helpful efforts. The colored people have comparatively few opportunities for skilled and profitable employment; it is to their interest and the interest of the white people among whom they live that they should have more, and all that they can make and use to advantage.

It may be added that the proposed experiment at Concord will not be altogether as novel in its way as its promoters appear to regard it. We are informed that a small mill was operated by colored laborers on a large cotton plantation ner Bainbridge, Ga., before and during the war. It was under white management, of course. The special interest of the Concord mill will be due to its management as well as its operation by colored men. The success of their enterprise will hardly disturb the conditions of cotton manufacturing in the South, in any event, but it may have a very important bearing on the future of the industry and of the negro race in Africa.

July 28, 1896

A STUDY OF NEGRO LIFE.

Investigations into the Social and Economic Conditions of a Typical Negro Community in Virginia.

WASHINGTON, Jan. 29.—Under instructions from the Government Department of Labor a study of a typical negro community, that at Farmville, the county seat of Prince Edward County, Virginia, was made during the months of July and August, 1897, by W. E. B. Du Bois, Ph. D., who lived in the community during that time and tabulated such statistics as he could obtain to show the social conditions. Farmville has a population of 3,684, of which 2,438 are negroes and 1,246 whites. It is the geographical centre of a typical "black" Virginia county, which had in 1890 a population of 14,694, of which 9,924 were negroes. The same proportion holds good at the present time. These negroes held 17,555 acres of land in 1895, as against 202,962 held by whites, the respective values being $132,189 and $1,064,180. It is therefore as representative a county for social science purposes as Farmville is a town.

Farmville is the trading centre for six counties. It has an opera house, normal school for white girls, armory, court house and jail, bank, depot, churches, sixteen tobacco warehouses, and factories and substantial dwellings. On "market day," which is Saturday, the town is swollen to twice its size by the influx of country people, mostly negroes. So Mr. Du Bois concludes it to be as good a spot as could be found to study the negro as he actually is under such social and economic conditions as exist to-day.

The investigator found a considerable excess of females in the population, due to the emigration of males to the North for employment, which is a habit among both sexes between the twenties and thirties in age, these frequently leaving their children behind them, so that there is an excess of old people and children. As to their social condition Mr. Du Bois found of 351 males over fifteen years of age 147, or 41.9 per cent., were single; 178, or 50.7 per cent., were married, and 14, or 4 per cent., were widowed. The remaining 12, or 3.4 per cent., were in no case regularly divorced, but were permanently separated from their wives. Of the 392 women, 126, or 32.1 per cent., were single; 178, or 45.4 per cent., were married; 76, or 19.4 per cent., were widowed, and 12, or 3.1 per cent., were permanently separated.

Of the population in 1897 15 per cent. were of illegitimate birth, which is a decided change for the better in twenty years, especially when it is considered that marriages are now contracted at a much later period in life than during the slavery days. The standard of morals is constantly improving. The social life of the community has developed marvelously, says the report, due to the better influence of church life—which has been much elevated and has passed beyond the "excitable and screaming stage of hysterical revivals" among the better classes—and the sécret and beneficial organizations.

Negroes have divided themselves into three social classes in the little own, the upper being the land owners and people in trade and of rather austere habits and strict morals, the middle class of laborers, servants, and farm hands, and the lowest strata, which includes criminals, male and female, who are ostracised by the rest of the community. Mr. Du Bois found two cases of miscegenation, which are undisturbed despite the law. He could get no statistics as to percentage, except by observation, and that gave him these results: "Of 705 negroes met face to face, 333 were apparently of unmixed negro blood; 219 were brown in color and showed traces of white blood, and 153 were yellow or lighter, and showed considerable infusion of white blood. According to this, one-third to one-half the negroes of the town are of mixed blood, and verifying this by observations on the street and in assemblies, this seemed a fair conclusion." He also concludes that the concubinage which brought about these results is rapidly decreasing.

The negroes are nearly all anxious to own lands and houses. They have established a building and loan association made up of men of their own race and are as rapidly as possible abandoned the old one-room cabin for frame houses with two or more distinct rooms. The school system is indifferent, yet the percentage of illiteracy has greatly decreased and he found that of 908 people above ten years of age, 42.5 per cent. could both read and write, 17.5 per cent. could read, but not write, and 40 per cent. were wholly illiterate, the excess of illiteracy being now among the men instead of women, as formerly.

As to the occupation of the negroes above ten years of age, according to the popular classification of pursuits, he found in professional occupations, 22; in domestic, 287; in commercial, 45; in agricultural, 15; in industrial, 282; not engaged in gainful occupations, 259, and not reported, 14. Using a different classification, he found those working on their own account, 36; laboring class, 350; house service, 92; day service, 149; at home, unoccupied, and dependent, 259; professional and clerical, 24, and not reported, 14.

There was no colored physician or lawyer in the town, but there were two preachers of clean life and high repute, both graduates of theological seminaries; several young women earning from $100 to $250 a year as teachers during the six months' school period which is the rule in such communities, one undertaker, seven grocers, two blacksmiths, one wheelwright, a hotel and bakery conducted by a Hampton graduate and her husband, and the only steam laundry in the county, conducted by two colored men, who also own a laundry in Richmond, fifty-seven miles distant, and both paying well.

The entire brickmaking business of Farmville and vicinity is in the hands of a colored man—a freedman, who bought his own and his family's freedom, purchased his master's estate, and eventually hired his master to work for him. He owns a thousand acres or more of land in Cumberland County and considerable Farmville property. In his brickyard he hires about fifteen hands, mostly boys from sixteen to twenty years of age, and runs five or six months a year, making from 200,000 to 300,000 bricks. The richest negro in the county is a barber, who is worth over $10,000. Among the skilled trades, negroes are found as painters, shoemakers, cabinetmakers, coopers, blacksmiths, wheelwrights, brick masons, plasterers, carpenters, bakers, butchers, and whipmakers. There are fourteen carpenters, three painters, and three masons who live in the town, besides several who live in the country and work in town. There are apparently more negroes with trades than white men, but there is a dearth of young negro apprentices, so that colored contractors often have to hire white mechanics.

Farming is abandoned for the industrial chances of the town, large numbers being engaged in tobacco stripping, though many of the men work on the farms in the Spring and Autumn, to eke out their incomes, which average about $250 a year, as nearly as can be ascertained. Negroes look on domestic service as a relic of slavery, and only enter it as a necessity or a temporary makeshift.

Speaking of the distinctly social life of the community, as exemplified in their group or social assemblages, Mr. Du Bois says: "Among this class of people the investigator failed to notice a single instance of any action not indicating a thoroughly good moral tone. There was no drinking, no lewdness, no questionable conversation, nor was there any one in any of the assemblies against whose character there was any well-founded accusation. The circle was, to be sure, rather small, and there was a scarcity of young men."

Of the better group of inhabitants, socially, he says: "It is pervaded by a peculiar hopefulness on the part of the people themselves. No one of them doubts in the least that one day black people will have all rights they are now striving for, and that the negro will be recognized among the earth's great peoples. Perhaps this simple faith is, of all the products of emancipation, the one of the greatest social and economic value." His final conclusion is: "After an impartial study of Farmville conditions the industrious and property-accumulating class of the negro citizens best represents, on the whole, the general tendencies of the group. At the same time, the mass of sloth and immorality is still large and threatening."

January 30, 1898

NEGRO LIFE IN THE SOUTH

A Study of the Residents of the Georgia "Black Belt."

MUCH DEPRAVITY IS FOUND

Whisky, Tobacco, and Snuff Used to Excess—The People in the Towns Better than in the Country.

A recent bulletin of the National Department of Labor contains a social study of "The Negro in the Black Belt," by W. E. Burghardt Du Bois, Ph. D. The section selected for this study, Dekalb and Newton Counties, Ga., is not generally, however, considered in "the Black Belt," which is usually indicated, in Government and other maps, as passing through Georgia some little distance to the south of this country. It is noteworthy that the information upon which the study is made was gathered by the members of the senior class of the Atlanta University, a negro institution.

Eleven negro families were selected as typical of a section of country about Doraville, a small village seventeen miles from Atlanta. "In general," says Dr. Du Bois, "these negroes are a degraded set. Except in two families, whisky, tobacco, and snuff are used to excess, even when there is a scarcity of bread. In other respects also the low moral condition of these people is manifest, and in the main there is no attempt at social distinctions among them."

In the eleven families there were 131 individuals, an average of nearly twelve to the family. One family had twetny-one members. The report continues:

"The fecundity of this population is astonishing. Here is one family with 19 children—14 girls and 5 boys, ranging in age from 6 to 25 years. Another family has 1 set of triplets, 2 sets of twins, and 4 single children. The girls of the present generation, however, are not marrying as early as their mothers did. Once in a while a girl of 12 or 13 runs off and marries, but this does not often happen. Probably the families of the next generation will be smaller.

"Four of the 11 heads of families can read and write. Of their children a majority, possibly two-thirds, can read and write a little. Five of the families own their homes. The farms vary from 1 to 11 acres in extent, and are worth from $100 to $400. Two of these farms are heavily mortgaged. Six families rent farms on shares, paying one-half the crop. They clear from $5 to $10 in cash at the end of a year's work. They usually own a mule or two and sometimes a cow.

"Nearly all the workers are farmhands, women and girls as well as men being employed in the fields. Children as young as six are given light tasks, such as dropping seed and bringing water. The families rise early, often before daylight, working until breakfast time and returning again after the meal. One of the men is a stonecutter. He earns $1.50 a day owns a neat little home, and lives comfortably. Most of the houses are rudely constructed of logs or boards, with one large and one small room. There is usually no glass in the openings which serve as windows. They are closed by wooden shutters. The large room always contains several beds and home-made furniture, consisting of tables, chairs, and chests. A few homes had three rooms, and one or two families had sewing machines, which, however, were not yet paid for."

So much for negro life in the country. The next study is of negro life in a village, Lithonia, the rock quarry of the South. The place is twenty-four miles east of Atlanta, and has a population of between 800 and 1,200. "Nearly all the workingmen of the town," says the report, "are employed in the rock quarries, which furnish the chief business of the village. The negro stonecutters here used to earn from $10 to $14 a week, but now they receive from $5 to $8.50 a week." Less than a dozen houses, it is said, are owned by the negroes. There are two negro schools.

In this community sixteen negro families were studied. These averaged only six to the family. Six families owned homes.

The "study" neglects to show the results of a sharp and instructive competition in this town between the negro laborers and the stonecutters that were brought over from Scotland. These Scotch cutters were expert, and could by a single blow of a stone hammer do more toward breaking up a big block of granite into paving blocks than a negro could accomplish in a dozen much harder strokes. The negroes could not readily acquire the knack of doing this work. The result was that they could not earn and did not get the good wages—ranging from $3 to $4 a day—paid to the Scotchmen.

As to the moral and intellectual condition of the negroes in Lithonia, Dr. Du Bois says:

"The morals of the colored people in the town are decidedly low. They dress and live better than the country negroes, however, and send their children more regularly to school. There are three churches—two Baptist and one Methodist—whose pastors are fairly intelligent."

In Covington, Newton County, Georgia, fifty families were studied out of about 300. They averaged only three and three-quarters persons to the family. The men usually earn from $10 to $12 and the women from $4 to $6 a month. There are four negro preachers, who earn about $400 a year, with house rent. There is a great deal of idleness and crime. The mass, however, is said to be composed of "hard-working people with small wages."

Somewhat better conditions were found to exist in Marietta and Athens, Ga., which are larger and more prosperous towns, and where there are a number of industries in which negroes are employed at fairly good wages for that section. "The average negro family can live," says the study, "on from $2 to $4 a week," and the negro hands can earn from 50 to 75 cents a day.

The report sums up the results of this local study of the negro as follows:

"The communities fall easily into three classes: A country district of 131 persons and 11 families; a small village of 101 persons and 16 families; town and city groups of 688 persons and 168 families. In the first class is had a glimpse of the deepest of the negro problems, that of the country negro, where the mass of the race still lives in ignorance, poverty, and immorality, beyond the reach of schools and other agencies of civilization for the larger part of the time. Small wonder that the negro is rushing to the city in an aimless attempt to change, at least, if not to better, his condition. Perhaps, on the whole, this is best; certainly it is if this inflow can be balanced by a counter migration of the more intelligent and thrifty negroes to the abandoned farms and plantations. In the second class we catch a glimpse of the small village life with one indutsry, more material prosperity, but traces of shiftlessness and thrift, immorality and a better family life, curiously intermingled. In both these classes the sketches furnished are, unfortunately, meagre. In the third class we have a wider field of observation—four thriving Southern towns—but here, again, there is a limitation. We have studied that part of the population which has succeeded best in the struggle of town life, and have seen little of the crime, squalor, and idleness of some of the rest of the negro population. Nevertheless, these 168 families have a peculiar interest. They represent, so far as they go, a solution of the negro problem, in that they are law-abiding, property holding people, marrying with forethought, careful of their homes, working hard in new lines of economic endeavor, and educating their children."

July 17, 1899

Needs of the Negro.

Passages from the Writings and Speeches of Booker T. Washington.

I think a part of his [the negro's] mission is going to be to teach white men a lesson of patience, forbearance, and forgiveness. I think he is going to show people of this country what it is possible for a race to achieve when starting under adverse conditions. Again, I believe he is destined to preach a lesson of supreme trust in God and loyalty to his country, even when his country has not been at all times loyal to him. I think my people will excel in the missionary spirit. It will take the form of reaching down after the less fortunate both at home and abroad.

It takes 100 per cent. of Caucasian blood to make a white American. The minute it is proven that a man possesses one one-hundredth part of negro blood in his veins, it makes him a black man. He falls to our side; we claim him. The 99 per cent. of white blood counts for nothing when weighed against 1 per cent. of negro blood.

The prejudice against the negro is not on account of color, but because of the badge of slavery—the slavery we used to be in and the industrial slavery we are now in.

It is all very well to bewail our wrongs. I feel them as keenly as any one else. But, I think, we have had quite enough talk about them, and that the thing to do now is to try to get our rights.

Education is the sole and only hope of the negro race in America. Transportation, colonization, and other schemes of misguided enthusiasts, are impracticable and futile.

No schoolhouse has been opened for us that has not been filled.

We are a new race, as it were, and the time, attention, and activity of any race are taken up during the first 50 or 100 years in getting a start.

We are led into saying that there is no difference between us and other people. We must admit that there is a difference produced by the unequal opportunities. To argue otherwise is to discredit the effects of slavery.

I only ask an equal chance in the world for the negro.

The negro's present great opportunity in the South is in the matter of business, and success in the South is going to constitute the foundation for success and relief along other lines.

If in the providence of God the negro got any good out of slavery, he got the habit of work.

Immediately after freedom we made serious mistakes. We began at the top. We made these mistakes, not because we were black people, but because we were ignorant and inexperienced.

Do not think life consists of dress and show. Remember that every one's life is measured by the power that that individual has to make the world better—this is all life is.

Any colored man with a reasonable education, common sense, and business ability, can take $1,000 in cash and go into any Southern community and in five years be worth $5,000. He does not meet with the stern, relentless competition that he encounters when he butts up against a Northern Yankee.

The Georgia Legislature has before it a bill, recently introduced, proposing to greatly reduce the amount of money annually appropriated for the education of the black youth of that State, on the ground that it cannot afford to spend so much money for negro education. I would reverse the proposition. I would say, with all the earnestness of my soul, that the State of Georgia is not able to let the 800,000 negroes within her borders grow up in ignorance. It will cost Georgia more not to educate them than to educate them.

In spite of all talk of exodus, the negro's home is permanently in the South; for, coming to the bread and meat side of the question, the white man needs the negro, and the negro needs the white.

The North should help the South educate the negro, if it would finish the work begun by Abraham Lincoln.

One man in the West, riding behind two fine horses, sitting upon a machine that laps off two furrows at a time, and drops and covers the corn at the same time, does as much work as four Southern corn planters of the present method of planting corn. So long as this is true, so long will the South buy corn from the West.

The prime condition of slavery was to keep closed every avenue to knowledge. The negro had no estate, no family life. His sole inheritance was his body.

One of the saddest sights I ever saw in the South was a colored girl, recently returned from college, sitting in a rented one-room log cabin attempting day by day to extract some music from a second-hand piano, when all about her indicated want of thrift and cleanliness.

Learn all you can, but learn to do something, or your learning will be useless.

The fact that a man goes into the world conscious that he has within himself the power to create a wagon or a house gives him a certain moral backbone and independence in the world that he could not possess without it.

Not much religion can exist in a one-room log cabin or on an empty stomach.

The progress along material lines is marked, yet the greatest lesson that we have learned during the last two decades is that the race must begin at the bottom, not at the top, that its foundation must be in truth and not in practice. We have learned that our salvation does not lie in the direction of mere political agitation or in hating the Southern white man, but that we are to find a safe and permanent place in American life by first emphasizing the cardinal virtues of home, industry, education, and peace with our next-door neighbor, whether he is white or black.

There are a million and a half black men in the South who have never worn a necktie, but send them to school and educate them and they will want neckties, cuffs, and, instead of the bare floors in their little log cabins, they will want carpet in neat frame houses.

There are several reasons why the South should give special attention to the matter of industrial education. Slavery taught both the white man and the negro to dread labor—to look upon it as something to be escaped, something fit only for poor people and slaves.

There is a custom that prevents a black man in some parts of our country from sleeping in a hotel, or eating in a restaurant, or riding in a first-class car. The average black man has the opportunity only to be denied this privilege about twice a year, but, thank God! there is no law or custom that prevents him from occupying the most convenient, comfortable, and attractive residence, and sleeping in the most luxurious bed, and dining at the best table in the country for 365 days in the year.

We might as well settle down to the uncompromising fact that our people will grow in proportion as we teach them that the way to have the most of Jesus and in a permanent form is to mix with their religion some land, cotton, and corn, a house with two or three rooms, and a little bank account. With these interwoven with our religion, there will be a foundation for growth upon which we can build for all time.

To right his wrongs, the Russian appeals to dynamite, Americans to rebellion, the Irish to agitation, the Indian to his tomahawk, but the negro, the most patient, the most unresentful and law-abiding, depends for the righting of his wrongs upon his songs, his midnight prayers, and his inherent faith in the justice of his cause.

I would call Lincoln the emancipator of America—the liberator of the white man North, the white man South, the one who in unshackling the chain from the negro has turned loose the enslaved forces of nature, and has knit all sections of our country together by the indissoluble bonds of commerce.

Another element which shows itself in the present stage of the civilization of the negro is his lack of ability to form a purpose and stick to it through a course of years, if need be, years that involve discouragement as well as encouragement, till the end is accomplished. The same, I think, would be true of any race with the negro's history.

Just so sure as the rays of the sun dispel the frosts of Winter, so sure will brains, property, and character conquer prejudice.

The longer I live and the more I study the question, the more I am convinced that it is not so much the problem of what you will do with the negro, as what the negro will do with you and your civilization.

No race can prosper until it learns that there is as much dignity in tilling a field as in writing a poem.

It will be needless to pass a law to compel men to come in contact with a negro who is educated and has $50,000 to lend.

Close to the spot where Jefferson Davis took the oath of office, swearing to sustain African slavery, a negro has erected a three-story drug store. When people go and see the thousands of dollars invested there, it is not in reason that they should try to drive that man away.

The education that the American negroes most need for the next 50 or 100 years should be mostly, but not exclusively, along scientific and industrial lines. When I say scientific, I mean science so applied that it will enable the black boy who comes from a plantation where ten bushels of corn were being raised to return to the farm and raise fifty bushels on the same acre.

One of the most encouraging things in connection with the lifting up of the negro race in this country is the fact that he knows that he is down, and wants to get up—he knows that he is ignorant, and wants to get light.

Just as sure as right in all ages and among all races has conquered wrong, so sure will the time come, and at no distant day, when the negro in the South shall be triumphant over the last lingering vestige of prejudice.

My first acquaintance with our hero was this: Night after night, before the dawn of day, on an old slave plantation in Virginia, I recall the form of my sainted mother, bending over a bundle of rags that enveloped my body, on a dirt floor, breathing a fervent prayer to Heaven that "Massa Lincoln" might succeed, and that one day she and I might be free.

Our greatest danger is that in the great leap from slavery to freedom we may overlook the fact that the masses of us are to live by the production of our hands, and fail to keep in mind that we shall prosper in proportion as we learn to dignify and glorify common labor, and put brains and skill into the common occupations of life.

The wisest among my race understand that the agitation of the question of social equality is the extremest folly, and that progress in the enjoyment of all the privileges that will come to us must be the result of severe and constant struggle rather than of artificial forcing. No race that has anything to contribute to the markets of the world is long in any degree ostracized.

At the end of 250 years of enforced labor, the negro finds himself without warning, and with no preparation, competing with the world for a market for his labor.

Sooner or later this country is going to realize that it has at its very doors the best labor that the world has seen.

August 20, 1898

DESTINY OF THE NEGRO.

Booker T. Washington Tells the League for Political Education His Idea of It.

Booker T. Washington, Principal of the Tuskegee (Ala.) Normal and Industrial Institute, addressed the League for Political Education yesterday on the subject of "Race Problems." He said that the solution of the race problem might be summed up in the statement that in proportion as the negroes were improved along educational, industrial, and moral lines in that same proportion would the evils affecting the race be eliminated.

"You cannot, however," said the speaker, "bring about these necessary reforms in a day, or by any short cut. It is a matter of time and of patience. Bring proper influence to bear upon the negro, make his environments congenial and uplifting, work on him not so much directly as unconsciously, and the matter will in time solve itself. For let me say that we are not going to be gotten rid of as a race. We are going to stay here in order to help you, our Government, and our country."

The speaker then referred to the movement some years ago which sought to solve the race problem by transporting negroes to Liberia. "But that won't solve it," the speaker added. "Those people who thought that sending 600 black men to Liberia would be the beginning of a satisfactory ending of the race problem forgot that on the very morning that those negroes were transported 600 or more black babies were born. Nor will the problem be solved by putting the blacks in any territory in the West. For, first, you would have to put up a wall about the territory to keep the negroes in, and then you would have to put up another wall to keep the white men out.

"No, there is a better and a higher solution of the difficulty than that. The negroes ought to be taught to do as good work as the white man in agriculture and in all industrial lines. The conditions with which they are surrounded should be studied carefully. They should be taught the dignity, the beauty, and the civilizing power of labor, and that there is no disgrace in any kind of work. Above all, they should be educated wherever and as fully as it is possible, and then their ignorance would not be made a pretext for keeping them down.

"Let me remind you that there are nearly ten millions of us, and that wherever our life touches yours we either help or hurt you. There can be no separation of our destinies. Let us then settle the problem in a sensible and logical manner. Let us, in short, make the negro a man so that he can do as much work as a white man and meet him with full confidence in a business way. Make him a useful and an important member of the community in which he lives and then, believe me, he will work out his own salvation."

April 16, 1899

THE FUTURE OF THE NEGRO.

Mr. Booker T. Washington deserves the gratitude of his own race, for he has done much for it in a way that no one else has been at once able and willing to do. But he deserves the cordial approval and sympathy of the whites as well, and of the whites of the South in particular, for his influence is in the direction of a settlement of the race difficulties in the only manner that we can hope will be practical and permanent. The principle on which Mr. Washington seeks to work and to get his people to work is that the trouble for the blacks will tend to disappear in the proportion that they are honest, industrious, and intelligent. The goal he sets before them is to be able to do any task that they undertake as well as white men can do it, and as well as it can be done. In the ratio that they approach this goal he believes that they will win the fair treatment to which they are entitled.

Neither Mr. Washington nor those that are working with him in the same field have any illusions as to the difficulty or the complexity of the problem on which they are engaged. They know only too well that whatever success they may attain will come very slowly, but it is precisely for this reason that they are pursuing a plan and a principle that looks to future results as well as to the present. The main object of their labors is industrial education. This, of course, includes and starts with the fundamental portions of ordinary schooling, and this is carried as far as circumstances will permit. But the essential aim is to produce not scholars, but self-supporting men and women, as thoroughly equipped and trained as possible for the various callings in which a living is to be gained. Among these callings those connected with the farm and the household are regarded as most important, because the greater number of the colored people are in agricultural States, and life on the farm offers generally better opportunity for independent life and comparative freedom from the worst forms of prejudice with which they have to contend. In the schools established mainly through the efforts of Mr. Washington the training for agricultural work, however, is directed to making not merely farm hands, but farmers, versed in the essential knowledge of crops and the conditions of raising them, in the care of stock, and in the mechanical work of the farm. In the same spirit it is sought to make the girls fitted for the life of the farm as well as competent housewives.

As Mr. Washington constantly urges on the people of the North and of the South in his addresses, this work is for the children and the children's children, and thus for the race as an element in the industrial, social, and political life of the Nation. At best the life of the negro in our land has many cruel possibilities. His race has been more heartlessly oppressed than any other in modern times. In the consequences of that oppression the whole Nation has been forced to share, and they are not yet exhausted. Serious confusion and disturbance, with some danger, still attend the settlement of what we call the race problem. The work of Tuskegee and of Hampton is an intelligent and valuable contribution to that settlement, and no one can doubt that in the proportion of their success the difficulties will lessen. They surely deserve the cordial support of all who can understand their influence and the need of it.

April 17, 1899

THE NEGRO QUESTION.

Essays and Sketches Touching Upon It by a Colored Writer.*

T is generally conceded that Booker T. Washington represents the best hope of the negro in America, and it is certain that of all the leaders of his people he has done the most for his fellows with the least friction with the whites who are most nearly concerned, those of the South. Here is another negro "educator," to use a current term, not brought up like Washington among the negroes of the South and to the manner of the Southern negro born, but one educated in New England—one who never saw a negro camp-meeting till he was grown to manhood and went among the people of his color as a teacher. Naturally he does not see everything as Booker Washington does; probably he does not understand his own people in their natural state as does the other; certainly he cannot understand the Southern white's point of view as the principal of Tuskegee does. Yet it is equally certain that "The Souls of Black Folks" throws much light upon the complexities of the negro problem, for it shows that the key note of at least some negro aspiration is still the abolition of the social color-line. For it is the Jim Crow car, and the fact that he may not smoke a cigar and drink a cup of tea with the white man in the South, that most galls William E. Burghardt Du Bois of the Atlanta College for Negroes. That this social color line must in time vanish like the mists of the morning is the firm belief of the writer, as the opposite is the equally firm belief of the Southern white man; but in the meantime he admits the "hard fact" that the color line is, and for a long time must be.

The book is of curious warp and woof, and the poetical form of the title is the index to much of its content and phraseology. To a Southerner who knows the negro race as it exists in the South, it is plain that this negro of Northern education is, after all, as he says, "bone of the bone and flesh of the flesh" of the African race. Sentimental, poetical, picturesque, the acquired logic and the evident attempt to be critically fair-minded is strangely tangled with these racial characteristics and the racial rhetoric. After an eloquent appeal for a fair hearing in what he calls his "Forethought," he goes in some detail into the vexed history of the Freedman's Bureau and the work it did for good and ill; for he admits the ill as he insists upon the good. A review of such a work from the negro point of view, even the Northern negro's point of view, must have its value to any unprejudiced student—still more, perhaps, for the prejudiced who is yet willing to be a student. It is impossible here to give even a general idea of the impression that will be gained from reading the text, but the underlying idea seems to be that it was impossible for the negro to get justice in the Southern courts just after the war, and "almost equally" impossible for the white man to get justice in the extra judicial proceeding of the Freedman's Bureau officials which largely superseded the courts for a time. Much is remembered of these proceedings by older Southerners—much picturesque and sentimental fiction, with an ample basis of truth, has been written about them by Mr. Thomas Nelson Page and others. Here we have the other side.

When all is said, the writer of "The Souls of Black Folk" is sure that the outside interference of which the Freedman's Bureau was the chief instrument was necessary for the negro's protection from supposed attempts of his former masters to legislate him back into another form of slavery, yet he admits that "it failed to begin the establishment of good-will between ex-masters and freedmen." It is proper to place beside this, of course, the consensus of fair Southern opinion that the interference in question and the instrumentalities it employed were the cause of the establishment of an ill-will previously non-existent. Here is a point where Booker T. Washington, as a Southern negro, has the advantage of his present critic in this that he knows by inherited tradition what the actual antebellum feeling between the races was. Du Bois assumes hostility.

While the whole book is interesting, especially to a Southerner, and while the self-restraint and temperateness of the manner of stating even things which the Southerner regards as impossibilities, deserve much praise and disarm harsh criticism, the part of the book which is more immediately concerned with an arraignment of the present plans of Booker T. Washington is for the present the most important.

In this matter the writer, speaking, as he says, for many educated negroes, makes two chief objections-first, that Washington is the leader of his race not by the suffrage of that race, but rather by virtue of the support of the whites, and, second, that, by yielding to the modern commercial spirit and confining the effort for uplifting the individual to practical education and the acquisition of property and decent ways, he is after all cutting off the negro from those higher aspirations which only, Du Bois says, make a people great. For instance, it is said that Booker Washington distinctly asks that black people give up, at least for the present, three things:

First, political power;
Second, insistance on civil rights;
Third, higher education for negro youth, and concentrate all their energies on industrial education, the accumulation of wealth, and the conciliation of the South. This policy has been courageously and insistently advocated for over fifteen years, and has been triumphant for perhaps ten years. As a result of this tender of the palm branch what has been the return? In these years there have occurred:

1. The disfranchisement of the negro.
2. The legal creation of a distinct status of civil inferiority for the negro.
3. The steady withdrawal of aid from institutions for the higher training of the negro.

These movements are not, to be sure, direct results of Washington's teachings, but his propaganda has, without a shadow of doubt, helped their speedier accomplishment.

The writer admits the great value of Booker Washington's work. However, he does not believe so much in the gospel of the lamb, and does think that a bolder attitude, one of standing firmly upon rights guaranteed by the war amendments, and alluded to in complimentary fashion in the Declaration of Independence, is both more becoming to a race such as he conceives the negro race to be, and more likely to advance that race. "We feel in conscience bound," he says, "to ask three things: 1. The right to vote; 2. Civic equality; 3. The education of youth according to ability" and he is especially insistent on the higher education of the negro—going into some statistics to show what the negro can do in that way. The value of these arguments and the force of the statistics can best be judged after the book is read.

Many passages of the book will be very interesting to the student of the negro character who regards the race ethnologically and not politically, not as a dark cloud threatening the future of the United States, but as a peculiar people, and one, after all, but little understood by the best of its friends or the worst of its enemies outside of what the author of "The Souls of Black Folk" is fond of calling the "Awful Veil." Throughout it should be recalled that it is the thought of a negro of Northern education who has lived long among his brethren of the South yet who cannot fully feel the meaning of some things which these brethren know by instinct—and which the Southern-bred white knows by a similar instinct; certain things which are by both accepted as facts—not theories—fundamental attitudes of race to race which are the product of conditions extending over centuries, as are the somewhat parallel attitudes of the gentry to the peasantry in other countries.

*THE SOULS OF BLACK FOLK. Essays and Sketches, By W. E. Burghardt Du Bois, Chicago: A. C. McClurg & Co.

April 25, 1903

TAKES ISSUE WITH BISHOP TURNER.

To the Editor of The New York Times:

I have read in this morning's TIMES what is reported to be an interview with Bishop Turner of the African Methodist Church, in Atlanta, Ga., and it seems to me plausible that, at his age, he should be seeking notoriety which in the end will do him more harm than good. His conviction as he terms it, not supported by any reasonable exposition of fact by which we may be guided to believe the negro emigration justifiable, is time and energy spent on a question that has been lost by the many able solutions presented for its elimination.

The question I would like to ask the venerable Bishop to answer is this: Why should he wish the negroes in the United States to emigrate to Africa, and what benefit will they derive and gain by emigration? Will it solve any of the many contentions?

I am under the impression that there is much field for improving the existing conditions of the negro in the United States; if things are worked and carried out properly by those charged with looking after the moral and practical side of his life, it would be a great satisfaction. There is nothing in Africa for the negro to better his condition, compared with the opportunities that exist in the United States. Let the Bishop give us some reasonable views why Africa is better than the place of his birth and the surroundings of his childhood. This is the fatherland of the negro born in the United States. If he cannot prosper here, then he will nowhere else under the sun.

"Bishop Turner saying that "I will give the world another Rome" is a dream and deception. I am sure he cannot do it, (ex nihilo nihil fit.) If he can, let him start the "Rome" in the United States with honest people. The Government may give some tract of land for the noble aspirations.

I believe that if we were to try to better things, try to be a little more optimistic in our views, these matters can be remedied; we may succeed in time in partly overcoming the existing conditions without sacrificing genuine affections for doubtful ones.

You have explicitly said in your editorial of to-day that "the black is here to stay," and, I think, till the laws of evolution eradicate him completely from this planet.

ARTHUR A. SCHOMBURG.

New York, Aug. 29, 1901.

September 3, 1901

SOUTHERN DEMOCRATS BERATE THE PRESIDENT

Condemn Him for Sitting at Table with a Negro.

Say It Will Do the Republican Party No Good in Their Section of the Country.

Special to The New York Times.

CHICAGO, Oct. 18.—Dispatches from many points in the South contain opinions of the press and politicians on President Roosevelt's action in entertaining Booker T. Washington at dinner. Democrats condemn the President and Republicans say that his action will redound to the credit of the party. The opinions expressed are as follows:

NEW ORLEANS.—The Picayune says: The President assumes to do officially that which he would not dream of in the way of violating accepted social usages.

THE TIMES DEMOCRAT.—White men of the South, how do you like it? White women of the South, how do you like it? When Mr. Roosevelt sits down to dinner with a negro, he declares that the negro is the social equal of the white man.

THE DAILY STATES.—In the face of the facts it cannot but be apparent that the President's action was little less than a studied insult to the South, adopted at the outset of his Administration for the purpose of showing his contempt for the sentiments and prejudices of this section, and of forcing upon the country social customs which are utterly repugnant to the entire South. In addition to all this, he is revivifying a most dangerous problem, one that has brought untold evil upon the whole country in the past, but which it was hoped, and believed, had been removed by the firmness and wisdom of the South.

THE DAILY ITEM, the only Republican paper in the city, says in the course of a long apology for the action of the President: The Item does not believe that the courtesy shown Booker T. Washington by the President is an attempt to break down the barriers that society has erected between the races. Society is abundantly able to take care of itself, but politics demand the use of many and various means for obtaining information and arriving at beneficent results.

MONTGOMERY, ALA.—Editor HOOD: Roosevelt has destroyed the threatened Republican boom in the South.

Ex-Gov. OATES: No respectable white man in Alabama would ask Washington to dinner or go to dinner with him.

NASHVILLE, TENN.—THE NASHVILLE BANNER: It was a mistake which will be deplored by a strong public sentiment.

CHARLOTTE, N. C.—Editor SMITH of THE NEWS: Roosevelt is the only consistent Republican, so far as the race question is concerned, who has been in public office since the war.

Ex-Mayor McCALL: The President's attempt to make the negro equal to the white man socially, is an insult to Southern people.

RALEIGH, N. C.—JOSEPHUS DANIELS, National Democratic Committeeman: It is not a precedent that will encourage Southern men to join hands with Mr. Roosevelt.

Dr. PRIEST, Republican: It will have the effect of inspiring the Republican Party all over the South.

LOUISVILLE.—C. F. GRAINGER, Democratic candidate for Mayor: I can't believe the President did that.

ATLANTA, Ga.—Gov. CANDLER: No self-respecting Southern man can ally himself with the President after what has occurred.

AUGUSTA, GA.—Mayor PHINISEY: It will hurt his influence, not only in the South, but will be condemned in the North.

State Senator SULLIVAN: It is a serious blunder

P. M. MULHERIN: No President can force social equality.

Lieut. Gov. TILLMAN of South Carolina, in Augusta to-day: Social equality with the negro means decadence and damnation.

LITTLE ROCK, Ark.—President FLETCHER of the First National Bank: I think the President has made a serious and grave mistake.

APPROVAL IN BOSTON.

BOSTON, Oct. 18.—Judging from a score or more of interviews with representative Bostonians, which are published here this evening, President Roosevelt's course in entertaining Booker T. Washington is approved in this city.

WILLIAM LLOYD GARRISON said: It was a fine object lesson and most encouraging. It was the act of a gentleman, an act of unconscious, natural simplicity.

Major HENRY L. HIGGINSON said: I have invited Bocker T. Washington to my house. He has been my guest at my table. When he comes to Boston I shall be glad to do it again.

Col. THOMAS WENTWORTH HIGGINSON: I heartily approve of President Roosevelt's course.

THE REV. DR. GEORGE A. GORDON: Every good citizen of the country admires President Roosevelt and every good citizen admires his guest.

PRESIDENT ELIOT OF HARVARD: Harvard dined Booker T. Washington at her table last comencement. Harvard conferred an honorary degree on him. This ought to show what Harvard thinks about the matter.

PROFESSOR CHARLES E. NORTON: I uphold the President in the bold stand that he has taken.

PROFESSOR N. S. SHALER: If I were in Roosevelt's place I would do the same thing myself.

MOORFIELD STOREY: The President is just right. I applaud him in that position every time.

Col. N. P. HOLLOWELL: I think it is altogether admirable on the part of the President.

HENRY B. BLACKWELL: I think the action of President Roosevelt in entertaining Booker T. Washington at the Executive Mansion was eminently wise, timely, and proper.

October 19, 1901

THE BLACK NORTH

A SOCIAL STUDY.

First of a Series by W. E. Burghardt DuBois, Atlanta University

NEW YORK CITY

THE negro problem is not the sole property of the South. To be sure, it is there most complicated and pressing. Yet north of Mason and Dixon's line there live to-day three-quarters of a million men of negro lineage. Nearly 400,000 of these live in New England and the Middle Atlantic States, and it is this population that I wish especially to study in a series of papers.

The growth of this body of negroes has been rapid since the war. There were 150,000 in 1860, 225,000 in 1880, and about 385,000 to-day. It is usually assumed that this group of persons has not formed to any extent a "problem" in the North, that during a century of freedom they have had an assured social status and the same chance for rise and development as the native white American, or at least as the foreign immigrant.

This is not true. It can be safely asserted that since early Colonial times the North has had a distinct race problem. Every one of these States had slaves, and at the beginning of Washington's Administration there were 40,000 black slaves and 17,000 black freemen in this section. The economic failure of slavery as an investment here gave the better conscience of Puritan and Quaker a chance to be heard, and processes of gradual emancipation were begun early in the nineteenth century.

Some of the slaves were sold South and eagerly welcomed there. Most of them staid in the North and became a free negro population.

They were not, however, really free. Socially they were ostracized. Strict laws were enacted against intermarriage. They were granted rights of suffrage with some limitations, but these limitations were either increased or the right summarily denied afterward.

North as well as South the negroes have emerged from slavery into a serfdom of poverty and restricted rights. Their history since has been the history of the gradual but by no means complete breaking down of remaining barriers.

To-day there are many contrasts between Northern and Southern negroes. Three-fourths of the Southern negroes live in the country districts. Nine-tenths of the Northern negroes live in cities and towns. The Southern negroes were in nearly all cases born South and of slave parentage.

About a third of the Northern negroes were born North, partly of free negro parentage, while the rest are Southern immigrants. Thus in the North there is a sharper division of the negroes into classes and a greater difference in attainment and training than one finds in the South.

From the beginning the Northern slaves lived in towns more generally than the Southern slaves, being used largely as house servants and artisans. As town life increased, the urban negro population increased. Here and there little villages of free negroes were to be found in the country districts of the North tilling the soil, but the competition of the great West soon sent them to town along with their white brothers, and now only here and there is there a negro family left in the country districts and villages of New England and of the Middle States.

From the earliest settlement of Manhattan, when the Dutch West India Company was pledging itself to furnish the new settlers with plenty of negroes, down to 1900, when the greater city contained 60,000 black folk, New York has had a negro problem. This problem has greatly changed from time to time. Two centuries ago it was a question of obtaining "hands" to labor. Then came questions of curbing barbarians and baptizing heathen. Long before the nineteenth century citizens were puzzled about the education of negroes, and then came negro riots and negro crime and the baffling windings of the color line.

At the beginning of the eighteenth century there were 1,500 negroes in New York City. They were house servants and laborers, and often were hired out by their masters, taking their stand for this purpose at the foot of Wall Street. By the middle of the century the population had doubled, and by the beginning of the nineteenth century it was about 9,000, five-sixths of whom were free by the act of gradual emancipation.

In 1840 the population was over 16,000, but it fell off to 12,500 in 1860 on account of the competition of foreign workmen and race riots. Since the war it has increased rapidly to 20,000 in 1880 and to 36,000 on Manhattan Island in 1900. The annexed districts raise this total to 60,666 for the whole city.

The distribution of this population presents many curious features. Conceive a large rectangle through which Seventh Avenue runs lengthwise. Let this be bounded on the south by a line near Sixteenth Street and on the north by Sixty-four or Sixty-fifth Street. On the east let the boundary be a wavering line between Fourth and Seventh Avenues and on the west the river.

In this quadrangle live over 20,000 negroes, a third of the total population. Ten thousand others live around the north end of the Park and further north, while 18,000 live in Brooklyn. The remaining 10,000 are scattered here and there in other parts of the city.

The migration of the black population to its present abode in New York has followed the growth of the city. Early in the eighteenth century the negroes lived and congregated in the hovels along the wharves and of course in the families of the masters. The centre of black population then moved slowly north, principally on the east side, until it reached Mulberry Street, about 1820. Crossing Broadway, a generation later the negroes clustered about Sullivan and Thompson Streets until after the war, when they moved northward along Seventh Avenue.

From 1870 to 1890 the population was more and more crowded and congested in the negro districts between Twenty-sixth and Sixty-third Streets. Since then there has been considerable dispersion to Brooklyn and the Harlem districts, although the old centres are still full.

The migration to Brooklyn began about 1820 and received its great impetus from the refugees at the time of the draft riots. In 1870 there were 5,000 negroes in Brooklyn. Since then the population has increased very rapidly, and it has consisted largely of the better class of negroes in search of homes and seeking to escape the contamination of the Tenderloin.

In 1890 the Brooklyn negroes had settled chiefly in the Eleventh, Twentieth, and Seventh Wards. Since then they have increased in those wards and have moved to the east in the Twenty-third, Twenty-fourth, and Twenty-fifth Wards and in the vicinity of Coney Island.

Let us now examine any peculiarities in the colored population of Greater New York. The first noticeable fact is the excess of women. In Philadelphia the women exceed the men six to five. In New York the excess is still larger—five to four—and this means that here even more than in Philadelphia the demand for negro housemaids is unbalanced by a corresponding demand for negro men.

This disproportion acts disastrously today on the women and the men. The excess of young people from eighteen to thirty years of age points again to large and rapid immigration. The Wilmington riot alone sent North thousands of emigrants, and as the black masses of the South awaken or as they are disturbed by violence this migration will continue and perhaps increase.

The North, therefore, and especially great cities like New York, has much more than an academic interest in the Southern negro problem. Unless the race conflict there is so adjusted as to leave the negroes a contented, industrious people, they are going to migrate here and there. And into the large cities will pour in increasing numbers the competent and the incompetent, the industrious and the lazy, the law abiding and the criminal.

Moreover, the conditions under which these new immigrants are now received are of such a nature that very frequently the good are made bad and the bad made professional criminals. One has but to read Dunbar's "Sport of the Gods" to get an idea of the temptations that surround the young immigrant. In the most thickly settled negro portion of the Nineteenth Assembly District, where 5,000 negroes live, the parents of half of the heads of families were country bred. Among these families the strain of city life is immediately seen when we find that 24 per cent. of the mothers are widows—a percentage only exceeded by the Irish, and far above the Americans, (16.3.)

In these figures lie untold tales of struggle, self-denial, despair, and crime. In the country districts of the South, as in all rural regions, early marriage and large families are the rule. These young immigrants to New York cannot afford to marry early. Two-thirds of the young men twenty to twenty-four years of age are unmarried, and five-eighths of the young women.

When they do marry it is a hard struggle to earn a living. As a race the negroes are not lazy. The canvass of the Federation of Churches in typical New York tenement districts has shown that while nearly 90 per cent. of the black men were wage earners, only 92 per cent. of the Americans and 90 per cent. of the Germans were at work.

At the same time the work of the negroes was least remunerative, they receiving a third less per week than the other nationalities. Nor can the disabilities of the negroes be laid altogether at the door of ignorance. Probably they are even less acquainted with city life and organized industry than most of the foreign laborers. In illiteracy, however, negroes and foreigners are about equal—five-sixths being able to read and write.

The crucial question, then, is: What does the black immigrant find to do? Some persons deem the answer to this question unnecessary to a real understanding of the negro. They say either that the case of the negro is that of the replacing of a poor workman by better ones in the natural competition of trade or that a mass of people like the American negroes ought to furnish employment for themselves without asking others for work.

There is just enough truth in such superficial statements to make them peculiarly misleading and unfair. Before the civil war the negro was certainly as efficient a workman as the raw immigrant from Ireland or Germany. But whereas the Irishmen found economic opportunity wide and daily growing wider, the negro found public opinion determined to "keep him in his place."

As early as 1824 Lafayette, on his second visit to New York, remarked "with astonishment the aggravation of the prejudice against the blacks," and stated that in the Revolutionary War "the black and white soldiers messed together without hesitation." In 1836 a well-to-do negro was refused a license as drayman in New York City, and mob violence was frequent against black men who pushed forward beyond their customary sphere.

Nor could the negro resent this by his vote. The Constitution of 1777 had given him full rights of suffrage, but in 1821 the ballot, so far as blacks were concerned, was restricted to holders of $250 worth of realty—a restriction which lasted until the war, in spite of efforts to change it, and which restricted black laborers but left white laborers with full rights of suffrage.

So, too, the draft riots of 1863 were far more than passing ebullitions of wrath and violence, but were used as a means of excluding negroes all over the city from lines of work in which they had long been employed. The relief committee pleaded in vain to have various positions restored to negroes. In numerous cases the exclusion was permanent and remains so to this day.

Thus the candid observer easily sees that the negro's economic position in New York has not been determined simply by efficiency in open competition, but that race prejudice has played a large and decisive part. Probably in free competition exslaves would have suffered some disadvantages in entering mechanical industries. When race feeling was added to this they were almost totally excluded.

Again, it is impossible for a group of men to maintain and employ itself while in open competition with a larger and stronger group. Only by co-operation with the industrial organization of the Nation can negroes earn a living. And this co-operation is difficult to effect. One can easily trace the struggle in a city like New York. Seventy-four per cent. of the working negro population are common laborers and servants.

From this dead level they have striven long to rise. In this striving they have made many mistakes, have had some failures and some successes. They voluntarily withdrew from bootblacking, barbering, table waiting, and menial service whenever they thought they saw a chance to climb higher, and their places were quickly filled by foreign whites.

Some of the negroes succeeded in their efforts to rise, some did not. Thus every obstacle placed in the way of their progress meant increased competition at the bottom. Twenty-six per cent. of the negroes have risen to a degree and gained a firmer economic foothold. Twelve per cent. of these have gone but a step higher; these are the porters, packers, messengers, draymen, and the like—a select class of laborers, often well paid and more independent than the old class of upper house servants before the war, to which they in some respects correspond.

Some of this class occupy responsible positions, others have some capital invested, and nearly all have good homes.

Ten per cent. of the colored people are skilled laborers—cigarmakers, barbers, tailors and dressmakers, builders, stationary engineers, &c. Five and one-half per cent. are in business enterprises of various sorts. The negroes have something over a million and a half dollars invested in small business enterprises, chiefly real estate, the catering business, undertaking, drug stores, hotels and restaurants, express teaming, &c. In the sixty-nine leading establishments $800,000 is invested—$13,000 in sums from $500 to $1,000 and $269,000 in sums from $1,000 to $25,000.

Forty-four of the sixty-nine businesses were established since 1885, and seventeen others since the war. Co-operative holders of real estate—i. e., hall associations, building and loan associations, and one large church, which has considerable sums in productive real estate—have over half a million dollars invested. Five leading caterers have $30,000, seven undertakers have $32,000, two saloons have over $50,000, and four small machine shops have $27,500 invested.

These are the most promising enterprises in which New York negroes have embarked. Serious obstacles are encountered. Great ingenuity is often required in finding gaps in business service where the man of small capital may use his skill or experience.

One negro has organized the cleaning of houses to a remarkable extent and has an establishment representing at least $20,000 of invested capital, some ten or twelve employes, and a large circle of clients.

Again, it is very difficult for negroes to get experience and training in modern business methods. Young colored men can seldom get positions above menial grade, and the training of the older men unfits them for competitive business. Then always the uncertain but ever present factor of racial prejudice is present to hinder or at least make more difficult the advance of the colored merchant or business man, in new or unaccustomed lines.

In clerical and professional work there are about ten negro lawyers in New York, twenty physicans, and at least ninety in the civil service as clerks, mail carriers, public school teachers, and the like. The competitive civil service has proved a great boon to young aspiring negroes, and they are being attracted to it in increasing numbers. Already in the public schools there are one Principal, two special teachers, and about thirty-five classroom teachers of negro blood. So far no complaint of the work and very little objection to their presence have been heard.

In some such way as this black New York seeks to earn its daily bread, and it remains for us to ask of the homes and the public institutions just what kind of success these efforts are having.

November 17, 1901

THE BLACK NORTH

A SOCIAL STUDY.

Second of a Series by W. E. Burghardt DuBois, Atlanta University

NEW YORK CITY

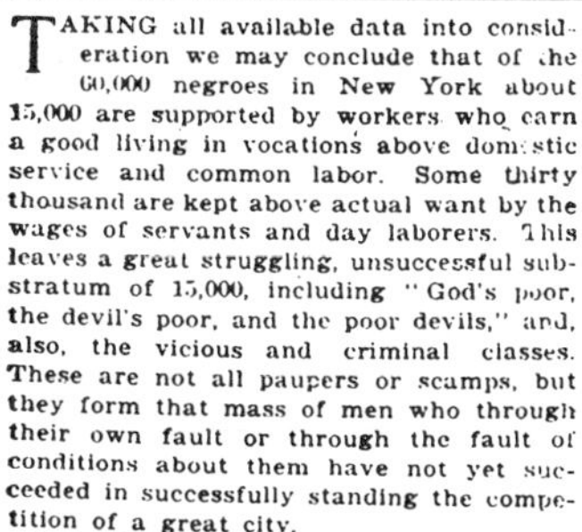

TAKING all available data into consideration we may conclude that of the 60,000 negroes in New York about 15,000 are supported by workers who earn a good living in vocations above domestic service and common labor. Some thirty thousand are kept above actual want by the wages of servants and day laborers. This leaves a great struggling, unsuccessful substratum of 15,000, including "God's poor, the devil's poor, and the poor devils," and, also, the vicious and criminal classes. These are not all paupers or scamps, but they form that mass of men who through their own fault or through the fault of conditions about them have not yet succeeded in successfully standing the competition of a great city.

Such figures are of course largely conjectural, but they appear near the truth. So large a substratum of unsuccessful persons in a community is abnormal and dangerous. And yet it is certain that nothing could be more disgraceful than for New York to condemn 45,000 hard working and successful people, who have struggled up in spite of slavery, riot, and discrimination, on account of 15,000 who have not yet succeeded and whom New Yorkers have helped to fail.

In no better way can one see the effects of color prejudice on the mass of the negroes than by studying their homes. The work of the Federation of Churches in the Eleventh and Thirteenth Assembly Districts, where over 6,000 negroes live, found 19 per cent. living in one and two room tenements, 37 per cent. in three rooms, and 44 per cent. in four or more rooms. Had the rooms been of good size and the rents fair this would be a good showing; but 400 of the rooms had no access to the outer air and 655 had but one window. Moreover, for these accommodations the negroes pay from $1 to $2 a month more than the whites for similar tenements—an excess rent charge which must amount to a quarter of a million dollars annually throughout the city. One-fourth of these people paid under $10 a month rent; two-thirds paid from $10 to $20.

We may say, then, that in the Tenderloin district, where the newer negro immigrants must needs go for a home, the average family occupies three small rooms, for which it pays $10 to $15 a month. If the family desires a home further from the vice and dirt of New York's most dangerous slum, it must go either to Brooklyn or, far from work, up town, or be prepared to pay exorbitant rents in the vicinity of Fifty-third Street.

More than likely the new-comer knows nothing of the peculiar dangers of this district, but takes it as part of the new and strange city life to which he has migrated. Finding work scarce and rent high, he turns for relief to narrow quarters and the lodging system. In the more crowded colored districts 40 per cent. of the families take lodgers and in only 50 per cent. of the cases are the lodgers in any way related to the families. Unknown strangers are thus admitted to the very heart of homes in order that the rent may be paid. And these homes are already weak from the hereditary influence of slavery and its attendant ills.

The very first movement of philanthropy in solving some of the negro problems of New York would be the separation of the decent and vicious elements, which the lodging system and high rent bring in such fatal proximity. Thus the movement of the City and Suburban Homes Company to build a model negro tenement on Sixty-second Street is an act of far-seeing wisdom. To-day it is the intricate and close connection of misfortune and vice among the lower classes that baffles intelligent reform.

A great mass of people, bringing with them a host of unhealthful habits, living largely in tenements, with wretched sanitary appliances, and in poor repair—such a mass must necessarily have a higher death rate than the average among the whites. Before the war this excess was very great, and even this year the colored death rate is 28 per thousand, against 20 for the whites. Since 1870 the death rate of negroes has been:

1870, 36 per thousand.
1880, 37 per thousand.
1890, 38 per thousand.
1900, 28 per thousand.

The decrease in 1900 is due to the inclusion of the healthier negro districts of the greater city, as, e. g., Brooklyn, and the immigration of young people. In itself a death rate of 28 is not high; the death rate of the whole city was 29 in 1870. Nevertheless, the disparity between whites and blacks shows plainly that the difference is due primarily to conditions of life and is remediable.

The most sinister index of social degradation and struggle is crime. Unfortunately, it is extremely difficult to-day to measure negro crime. If we seek to measure it in the South we are confronted by the fact that different and peculiar standards of justice exist for black and white. If we take a city like New York we find that continual migration and concentration of negro population here make it unfair to attribute to the city or to the permanent negro population the crime of the new-comers. Then, again, it has been less than a generation since, even in this city, negroes stood on a different footing before the courts from whites, and received severer treatment. In interpreting figures from the past, therefore, we must allow something at least for this.

There was complaint of negro misdemeanors back in the seventeenth century, as, for example, in 1682, when the city was suffering "great inconvenyencys" from the "frequent meetings and gatherings of negroes," and the City Council passed ordinances against such disorder and gambling. There was continual fear of negro uprisings, and when, after the establishment of Neau's Negro School, in 1704, a family of seven were murdered by their slaves, a great outcry was raised against negro education.

In 1712 and 1741 there were negro conspiracies—the first a fierce dash for freedom, the second a combination of negro thieves, white women of evil repute and their aiders and abettors. The city on both these occasions was vastly scared, and took fearful vengeance on those whom they thought guilty, burning and hanging twenty-nine blacks in 1741.

No very exact data of negro crime are available until about seventy-five years ago. In 1827, 25 per cent. of the convicts in New York State were negroes, although the negroes formed but 1 per cent. of the population. Twenty years later the negroes, forming the same proportion of the population, furnished 257 of the 1,637 convicts, or more than 15 per cent. In 1870 the proportion had fallen to 6 per cent.

Since then we may use the arrests in New York City as a crude indication of negro crime. These indicate that from 1870 to 1885 the negroes formed about 2 per cent. of the arrests, the best record they have had in the city. From 1885 to 1895 the proportion rose to 2½ per cent., and since then it has risen to 3½ per cent. A part of this rise is accounted for by the increase in the proportion of negro to white population, which was 1 1-3 per cent. in 1870 and 1⅔ per cent. in 1900. The larger portion of the increase in arrests is undoubtedly due to migration—the sudden contact of new-comers with unknown city life. From a mere record of arrests one can get no very good idea of crime, and yet it is safe to conclude from the fact that in the State in 1890 every 10,000 negroes furnished 100 prisoners that there is much serious crime among negroes. And, indeed, what else should we expect?

What else is this but the logical result of bad homes, poor health, restricted opportunities for work, and general social oppression? That the present situation is abnormal all admit. That the negro under normal conditions is law-abiding and good-natured cannot be disputed. We have but to change conditions, then, to reduce negro crime.

We have so far a picture of the negro from without—his numbers, his dwelling place, his work, his health, and crime. Let us now, if possible, place ourselves within the negro group and by studying that inner life look with him out upon the surrounding world. When a white person comes once vividly to realize the disabilities under which a negro labors, the public contempt and thinly veiled private dislike, "the spurns that patient merit of the unworthy takes"—when once one sees this, and then from personal knowledge knows that sensitive human hearts are enduring this, the question comes, How can they stand it? The answer is clear and peculiar: They do not stand it; they withdraw themselves as far as possible from it into a world of their own. They live and move in a community of their own kith and kin and shrink quickly and permanently from those rough edges where contact with the larger life of the city wounds and humiliates them.

To see what this means in practice, let us follow the life of an average New York negro. He is first born to a colored father and mother. The mulattoes we see on the streets are almost invariably the descendants of one, two, or three generations of mulattoes, the infusion of white blood coming often far back in the eighteenth century. In only 3 per cent. of the New York marriages of colored people is one of the parties white. The child's neighbors, as he grows up, are colored, for he lives in a colored district. In the public school he comes into intimate touch with white children, but as they grow up public opinion forces them to discard their colored acquaintances, and they soon forget even the nod of recognition. The young man's friends and associates are therefore all negroes. When he goes to work he works alongside colored men in most cases; his social circle, his clubs and organizations throughout the city are all confined to his own race, and his contact with the whites is practically confined to economic relationships, the streets, and street cars, with occasionally some intercourse at public amusements.

The centre of negro life in New York is still the church, although its all-inclusive influence here is less than in a Southern city. There are thirty or forty churches, large and small, but seven or eight chief ones. They have strongly marked individuality, and stand in many cases for distinct social circles. The older families of well-to-do free negroes who count an unspotted family life for two centuries gather at St. Philip's Episcopal Church, on Twenty-fifth Street. This church is an offshoot of Trinity and the lineal descendant of Neau's Negro School early in the eighteenth century. The mass of middle-class negroes whose fathers were New Yorkers worship at Mother Zion, Tenth and Bleecker Streets. This church is far from the present centre of negro population, but it is a historic spot, where the first organized protest of black folk against color discrimination in New York churches took place. Up on Fifty-third Street, at Olivet, one finds a great Baptist church, with the newer immigrants from Georgia and Virginia, and so through the city.

Next to the churches come the secret and beneficial societies. The Colored Masons date from 1826; the Odd Fellows own a four-story hall on Twenty-ninth Street, where ninety-six separate societies meet and pay an annual rental of $5,000. Then there are old societies like the African dating back to 1808, and new ones like the Southern Beneficial with very large memberships. There is a successful building association, a hospital, an orphan asylum, and a home for the aged, all entirely conducted by negroes, and mainly supported by them. Public entertainments are continually provided by the various churches and by associations such as the Railway Porters' Union, the West Indian Benevolent Association, the Lincoln Literary, &c.

Here, then, is a world of itself, closed in from the outer world and almost unknown to it, with churches, clubs, hotels, saloons, and charities; with its own social distinctions, amusements, and ambitions. Its members are rarely rich, according to the standards of to-day. Probably less than ten negroes in New York own over $50,000 worth of property each, and the total property held may be roughly estimated as between three and four millions. Many homes have been bought in Brooklyn and the suburbs in the last ten years, so that there is a comfortable class of laborers.

The morality and education of this black world is naturally below that of the white world. That is the core of the negro problem. Nevertheless, it would be wrong to suppose here a mass of ungraded ignorance and lewdness. The social gradations toward the top are sharp and distinct, and the intelligence and good conduct of the better classes would pass muster in a New England village. As we descend the social distinctions are less rigid, and toward the bottom the great difficulty is to distinguish between the bad and the careless, the idle and the criminal, the unfortunate and the impostors.

November 24, 1901

THE BLACK NORTH

A SOCIAL STUDY.

Third of a Series by W. E. Burghardt DuBois, Atlanta University

PHILADELPHIA

FROM early times there was a steady northward flow of free negroes and fugitive slaves, and these immigrants invariably sought the cities. Thus we find the negro population concentrating and growing in the City of Philadelphia. At the time of the first census there were 2,500 negroes there, and a century later there were 40,000. To-day there are 62,613 negroes in Philadelphia—a population larger than that of the whole city when it was the capital of the Nation.

Practically then a study of the negro in Pennsylvania means a study of the metropolis of the State. We have here without doubt an interesting group of men. To-day if one visits Philadelphia one plainly sees that this city of 1,293,970 souls has a considerable number of black folk.

They live in the central part of Philadelphia—the historic Seventh Ward, while the main white residence districts have stretched northward toward Germantown and westward across the Schuylkill. The Seventh Ward was itself a residence section fifty years ago, and then the negroes were strictly confined to a ghetto bordering the Delaware River.

Gradually as they grew in numbers they moved up into the city. Finally in the last decade they have scattered more among the white population, moving northward in smaller numbers toward Germantown and southward toward the bay.

The Seventh Ward, however, is still their chief dwelling place, its 10,000 negroes being more than a third of its total population. It is an interesting sight of a Sunday morning to walk down Lombard Street, Philadelphia, and watch this mass of people. Here the chief churches and halls are situated. Here is the general promenade of all classes and conditions.

Here one can see the Northern negro in a peculiar way. They differ decidedly from the crowds of a Southern city in dress and carriage, in their demeanor toward the whites, and their general air of self-reliance. They are well dressed and clean, and they have about them a certain air of prosperity. The first question one naturally asks is, How do they earn a living amid the competition of a large Northern city?

The answer is of so much importance that a glance into the past will best help us to understand it. When, early in the last century, the negroes began to come to Philadelphia they easily found there an assured economic footing. By 1810 they formed nearly 16 per cent. of the population, and did practically all the house-work and common labor and a good part of the work in the mechanical trades. White mechanics had protested against this, back in the eighteenth century, but it was to the interest of the masters then to hire out slave mechanics on good terms.

After emancipation, the negro mechanics lost this protection, and a hard economic struggle in the trades took place. The negroes were fairly successful in holding their own until outside forces began to tell. Philadelphia, feeling the influence of the industrial revolution after the war of 1812, began to change from a provincial capital to a great manufacturing centre. White mechanics and laborers hurried to the city and far outnumbered the blacks.

Even then the negro mechanics might have held their own had it not been for the peculiar limitations which prejudice placed about them. Their better class could not rise above the mass. Public opinion placed the successful, law-abiding city negro along with the untrained and lazy immigrant. And when these immigrants began to sink into poverty and crime, public opinion turned against the negro, and he rapidly lost ground as an industrial factor.

About 1840 new obstacles appeared in the propaganda of the Abolitionists and the increased influx of foreign immigrants. The wrath of the pro-slavery party spent itself on the free negroes, and the new foreign workmen were not slow to use race prejudice as a means of shutting negroes almost entirely out of the new industries which were now arising on all sides. For ten years and more the negroes were repeatedly the object of mob violence, the right of suffrage was taken away, and the growth of the black population seriously checked.

To take away any considerable part of the customary livelihood of a mass of people means that their ingenuity will be stimulated to find new avenues of labor, if they are really energetic and resourceful. And the Philadelphia negroes were energetic. They had established their own churches and beneficial societies, they had accumulated some property, they had sent their children to school, and they were not willing to be simply and always servants and common laborers.

Led by a group of excellent business men they developed the catering business of Philadelphia to such an extent as to make it famous all over the land. This was a natural resource. They had among them numbers of well-trained servants. By serving several families instead of one, investing capital in table furniture, and adding taste and skill, there was evolved the negro caterer. He invented the business, conducted it without a rival for two generations, and only lost his pre-eminence when industrial changes after the war substituted the stock company for the individual undertaker of limited capital.

Moreover, after the war this group of negroes was again overwhelmed and almost submerged beneath a mass of immigrants from the freed masses of the South. How have they emerged from this crisis, and what do they do for a living to-day? The negroes of Philadelphia to-day are occupied about as follows:

	Per Cent.
Professions, (physicians, lawyers, clergymen, &c.)	1.5
Conducting business, (merchants, caterers, expressmen,) with their own capital	4.0
Artisans, (carpenters, masons, upholsterers, &c.)	8.5
Clerks and responsible workers, (cooks, stewards, messengers, managers, foremen, policemen, actors, &c.)	3.0
Laborers, select, (teamsters, janitors, stevedores, elevator men, hostlers, &c.)	12.0
Laborers, ordinary	32.0
Servants	39.0
Total	100.0

In other words, of the 63,000 negroes in Philadelphia, 37,500 actually work in gainful occupations. Of these at least 26,500 are servants and ordinary laborers, while 4,500 others are laborers of a little higher grade. Another 4,500 are clerks and artisans, while 2,000 are business and professional men.

The servants and laborers are composed mainly of the recent Southern born immigrants to the city. They find little else open to them, and only a few are fitted for other work. Some of them are artisans and they find some work in the building trades, and in a few large establishments, notably the Midvale Steel Works. For the most part, however, they are entirely shut out from mechanical labor by the trades unions, which in nearly all cases frankly or covertly debar the negro. The cigarmakers' union is almost the only exception.

The descendants of the free negroes and other Northern trained colored people, together with some of the best of the Southerners, have gone into business and professional lines. The opening in the large business firms for colored boys are very few. The business ventures by negroes are mostly small shops. The physicians are the most successful professional men.

A careful consideration of the ages of the members of any community and of the relative number of men and women always teaches the student something of the social conditions there. When we find a great many more men than women, as in the Western States, we know this is due to migration and unusual opportunities for work. So, too, an excess of children and old people in any place would indicate the migration of those in the prime of life to find work elsewhere.

Among the negroes of Philadelphia and of most large cities there is a marked excess of females and a preponderance of young people between the ages of twenty and thirty years. Sixty years ago the negro women of Philadelphia outnumbered the men more than three to two. To-day the proportion is a little less than six to five.

This is the result of open chances for young women to work as servants and restricted chances for young men. The age distribution shows that the young people of the South are hurrying Northward in search of large opportunities. Such a migration, however, has its dangers, for all experience shows that the ages of twenty and thirty are peculiarly a period of temptation to excess and crime.

Moreover, the negro home life is, on account of slavery, already weak, sending forth children poorly equipped to meet the allurements of a great city. The excess of young women further complicates the situation, so that under this peculiar moral and economic stress it is well to ask just how this population is standing the strain.

There is without doubt a great deal of crime among negroes. When any race passes through a vast and sudden change like that of emancipation from slavery the result is always that numbers unable to adjust themselves to the new circumstances easily sink into debauchery and crime. We should expect then to find the greatest excess of crime a generation after emancipation and then to see it gradually decrease to normal conditions.

This is partially exemplified in Philadelphia. The act of gradual emancipation did not begin to have full effect until about 1810. From 1829 to 1834 the negroes, who were less than 9 per cent. of the total population, committed 29 per cent. of the serious crime. This outbreak of crime among young negroes was met, not by efforts at regeneration and opportunities for betterment, but by repression, denunciation, and restricted openings. The result was that from 1835 to 1839 the negro committed over 40 per cent. of the serious crime, although they formed but 7½ per cent. of the population.

As the better class of negroes, however, continued diligently at work, conditions gradually grew better. By 1850 the negroes were charged with but 16 per cent. of the serious crime, and by 1874 less than 4 per cent. of the arrests in Philadelphia were arrests of colored persons. In other words, negro crime had become about normal.

Then came a change. The Centennial Exhibition was the beginning of a new immigration of negroes. They came from the South, and represented young people only a few years removed from slavery.

It must be expected that the percentage of criminals and social failures among such a class would be large. It was large. From 1876, and coincident with the influx of Southern immigrants, the percentage of crime for which negroes are responsible in Philadelphia has steadily risen from 4 per cent. to 9 per cent. To-day the negroes, forming one-twentieth of the population, commit about one-eleventh of the crime, judging by the crude measure of arrests.

While this is not nearly as bad as the record in the past, it is nevertheless a serious problem. It is not fair, of course, to charge Philadelphia negroes with this amount of crime without discrimination. We must remember that only a fifth of those committing the more serious crimes are Philadelphia-born, and three-fifths of them are immigrants from the South and the product of its peculiar social conditions.

The North has a direct interest in the race problem in the South, and cannot expect permanent improvement in the criminal rate of its negro population as long as Southern conditions breed crime and send it North. At the same time the city itself is partially responsible.

Receiving, as it does, a population easily tempted to crime, it ought not to make yielding to temptation easier than honest labor. And yet, by political protection to criminals and indiscriminate charity, it encourages the worthless and at the same time, by shutting negroes out of most avenues of honest employment, the city discourages labor and thrift.

The fight for a livelihood and the temptation to crime are both a severe strain to physical health. It was a generation ago confidently asserted that Northern cities would never have a large negro population because of the cold climate and the stress of competition. The continued growth of this population afterward was laid solely to migration. This is only partially true.

The death rate of negroes in Philadelphia is higher than that of whites for obvious reasons. The majority live in the most unsanitary sections of the city, and in the worst houses in those sections; high rents lead to crowding, and ignorance of city life to unhealthful habits. As a consequence, the death rate of the negroes exceeds that of the whites just as the death rate of an unsanitary region exceeds that of a clean, orderly city.

In the decade 1820-30, the Negro death rate was over 47½ in every thousand; in 1830-40 it was 32½ in every thousand, against 24 for the whites. If, however, this difference in death rate is due solely to difference in conditions of life, we should expect an improvement in the death rate to-day as compared with the past, and a smaller death rate in the residence districts of the better class of negroes.

The death rate of the negroes has fallen from 47.6 per thousand in 1820-30, to 28.02 in 1891-96. Moreover, while the death rate in the slums of the Fifth Ward is 48½, in the better home life of the Thirtieth Ward it is practically the same as that of the whites, 21 per thousand. While then the high death rate of the negro is a misfortune to him and a menace to his neighbors, the evil is one which will easily yield to the influence of cleaner streets, better houses, and better homes.

Looking now at the social conditions of this mass of 60,000 souls, we are first struck by the fact that a rapid differentiation into social classes is going on. The common measure of this would be that of accumulated property.

There are no rich negroes according to the standards of modern wealth. Indeed, the income of a people who find the problem of breadwinning so difficult would preclude this. These incomes are something like this:

EARNINGS PER WEEK PER FAMILY.

	P.C.		P.C.
$5 or less	18	$10 to $15	26
$5 to $10	48	$15 or over	8

The negroes are earning to-day as a mass probably more than formerly. From these small incomes they have accumulated property as follows:

1821	$281,162	1855	$2,685,693
1838	322,532	1898	5,000,000
1848	531,809		

Only one negro is reputed to be worth over $100,000. Four have estates worth from $50,000 to $100,000; eleven, $25,000 to $50,000, and thirty-seven from $10,000 to $25,000. One obstacle to saving is the high rent which negroes are compelled to pay for houses. Over $1,250,000 are annually spent for rent by this race, and the rents paid are from 10 to 30 per cent. higher than whites pay.

The average monthly rent per family is $10.50. Twenty-two per cent. of the families pay less than $5 a month; the mass, or 57 per cent., pay between $10 and $20. The result of high rents is crowding, so that through a system of sub-letting fully a third of the families in one of the most populous wards live in one-room apartments.

If in addition to economic differences we bring in considerations of education and morals we may divide the colored people of Philadelphia, roughly, into four classes. The lowest class are the slum elements—criminals, gamblers, and loafers who form the "submerged Tenth." They live in the alleys of the older ward, centre about "clubs" and near saloons, and form, perhaps, 6 per cent. or more of the city negro population. They are a dangerous class both to their own people and to the whites, are responsible for much serious crime, and tempt the hardworking immigrants from the South into excess and immorality.

Above these come the poor and unfortunate. They are the class of negroes who for various reasons have not succeeded in the sharp competition of city life. They include unfortunates who want work and cannot find it, good-natured but unreliable workmen who cannot keep work, hand workers who spend regularly more than they earn, and in general people poor but not criminal nor grossly immoral. Thirty per cent. of the negroes would probably fall in this class.

Above these would be the bulk of the laborers—hard-working, good-natured people, not as pushing and resourceful as some, but honest and faithful, of fair and rapidly improving morals, and with some education. Fifty-two per cent. of the negroes fall into this class. Their chief difficulty is in finding paying employment outside of menial service.

Finally about 12 per cent. of the negroes form an aristocracy of wealth and education. They correspond to the better middle-class population of modern cities, and have, usually, good common-school training, with here and there a high school or college man. They occupy comfortable homes, are educating their children, and own property.

They, too, have difficulty in finding careers for their children, and they are socially in an anomalous position. The world classes them with the mass of their race and even in a city like Philadelphia makes but little allowance for their culture or means. On the other hand, not being to any considerable extent themselves employers of colored labor, or bound to them by ties of industrial interest, they cannot easily assume leadership over their own people. Indeed, a natural instinct of self-defense and self-preservation drives them away from the lower masses of their people.

They feel that they can only maintain their position and advance further by drawing social lines against the incompetent and criminal of their own race. Thus they face a peculiar paradox, and stand between black and white, the representatives of all that is best in the one and at the same time suffering vicariously at the hands of the other for all that is worst among their own people.

THE BLACK NORTH

A SOCIAL STUDY.

Fourth of a Series by W. E. Burghardt DuBois, Atlanta University

BOSTON

SLAVERY in Massachusetts began with the undoing of the Pequods in 1638 and the sailing of the slaver Desire about the same time. At the beginning of the eighteenth century there were 400 negroes in Boston. A century later there were 1,200.

At the beginning of the twentieth century there were 12,000 negroes in Boston. The growth of this population was slow and steady until 1850, when it increased more quickly. Since 1880 it has grown rapidly.

The negroes from the first were house servants almost exclusively, and the comparatively small demand for these in a thoroughly democratic community early put a stop to further importation. The presence of those already there, however, was as puzzling as in other Colonies.

James Otis declared slavery inconsistent with the principles of the Revolution, and it certainly seemed so after the songs of Phillis Wheatly and the bold fighting of black men at Bunker Hill. Nevertheless, even early emancipation did not secure equal rights for negroes.

After the Revolution negroes complained that in the streets of Boston they were "shamefully abused," and that to such a degree that "we may truly be said to carry our lives in our hands." They were for some time taxed without having the right of suffrage. Separate schools were maintained until 1855. Intermarriage of the races was forbidden until 1843, and as late as the civil war often "large audiences have been thrown almost into spasms by the presence of one colored man."

The 12,000 negroes of Boston to-day are largely immigrants since the war, as the excess of adults shows. Nevertheless, this urban negro population is peculiar in being the only one among the larger Northern cities to have a normal distribution of the sexes—5,904 males and 5,687 females in 1900.

This immediately leads us to expect better family life and social conditions than we have yet studied. Thanks to Massachusetts schools, the illiteracy of this group is about that of England and France, (13½ per cent.)

The oldest colored settlement was on the north side of Beacon Hill, in the West End. For a century and more this has been the historic centre of the blocks. Here was the African meeting house, where the anti-slavery movement was launched, here was the first negro school, and here still are the chief churches and halls. The bulk of the population, however, has moved.

By 1860 this district had become crowded and congested, and by 1870 the influx of low characters made the problem of home life among the better class of negroes as difficult as it is in the New York "Tenderloin" of to-day.

Even as late as 1880 it was very difficult for a negro to rent or buy a house outside this district. Gradually, however, the newer tenement houses of the South End, beyond the Common and out Tremont Street, were hired and purchased. To-day some 4,000 or 5,000 negroes live there, leaving perhaps 2,500 still in the vicinity of Joy Street, and many are scattered in the little settlements between.

It is, however, in the purchase of pretty suburban homes scattered here and there throughout the towns surrounding Boston that the negroes have been peculiarly successful in solving the problem of living. Since the marvelous development of Boston's street car service thousands of the better class of negroes have bought homes in Cambridge, Roxbury, Dorchester, and Chelsea, and other places.

These settlements are not usually in colonies, but the black families are scattered here and there among the whites, and the opposition which black neighbors at first aroused has largely died away, save, I believe, in the aristocratic suburb of Brookline.

It was always thought pretty certain that Boston would have no considerable negro population on account of the high death rate. Early in the eighteenth century the negro deaths for ten years averaged 87 per 1,000, and during ten years in the middle of the century it was 73. By the middle of the nineteenth century the rate had, however, fallen to 40, and since the war it has steadily decreased. The following table will illustrate this:

	Death Rate Per 1,000. Whites.	Negroes.
1875	23	31
1880	23	33
1885	24	42
1890	23	32
1895	23	30
1900	20	25

Economic conditions in Boston also show considerable difference from those in the other cities. The different environment, the less eager rush for wealth, and the New England ideals of home and society have plainly influenced the negroes. Of those who work for a living only 60 per cent., a little over half, are servants and ordinary laborers—less than half of the men, and three-fourths of the working women. This is an unusually small proportion for a negro population. At the same time, it renders the negro a relatively unimportant part of the common labor force of Boston.

In Boston as elsewhere the typical negro employments are disappearing. The negro waiter is going from the hotels. The negro barber is disappearing. The negro bootblack has almost gone.

Ordinary observers have supposed that this displacement leaves no work for negroes. This is true in the sense that there are few employments now which are his by right and common consent. As long as there were such employments they fell under the stigma of race prejudice and they failed to attract the best talent of the negroes—hence negro waiters who were poor, careless barbers, and the like. To-day as these employments are open to all the negro is pushing as a competitor into higher walks.

In New York and Philadelphia the negro is too largely handicapped by race prejudice to make much headway, but he has made some. In Boston the atmosphere has been more liberal, although by no means unbiased, and he has had correspondingly better success. The select laborers, as janitors, porters, messengers, draymen, &c., are an unusually trustworthy and respectable class in Boston. They form perhaps 15 per cent. of the workers, and they are connecting links between menial labor and skilled labor or business.

In the skilled trades there are about 15 per cent. of the colored workers, chiefly barbers, dressmakers, railway employes, tailors, carpenters, masons, painters, &c. In the mechanical industries which fill East Boston and South Boston there are few negroes, although large concerns like the American Tool and Machine Company have a few colored apprentices.

There are quite a number of negroes in the building trades, and they work side by side with whites on some jobs. Often they work for colored contractors. The negro tailors are well represented and very skillful.

In the various lines of business enterprise will be found about 7 per cent. of the colored people. This includes merchants, peddlers, clerks, salesmen, agents, &c. Among the leading merchants are:

Five merchant tailors, with a trade of $20,000 to $50,000 a year.

Three undertakers, with a trade of $5,000 to $10,000 a year.

Five caterers, with a trade of $5,000 to $25,000 a year.

Four real estate dealers, with a trade of $5,000 to $10,000 a year.

Besides these there are two tobacconists, a florist, a butcher, and a bookdealer, who do considerable business. The tailors and real estate men are most successful. The chief of the former conducts a large and well-known establishment on Washington Street. The real estate men have made money in supplying the demand for better homes, especially in the suburbs.

The caterers are also to be noticed. They are not here, as in Philadelphia, the last of a noted guild, but are instances of individual push. One conducted a fashionable suburban hotel until the crash of 1892. After losing this, he pluckily started again and now has another suburban inn and a large city restaurant. He is the inventor of a bread machine which the Bread Trust hopes to use.

Much more frequently than in the other two cities, Boston negroes have gained positions in large mercantile establishments. Yet here the black boy's chance of promotion is nothing like the white boy's. The son of a prominent negro, who had been graduated from a college of the first rank and had some capital of his own, wished a place in a mercantile establishment to learn business methods. He was promised several openings, but at last all frankly told him that their employes objected, and they could not take him.

He nevertheless went into business for himself, and is to-day very successful. Usually the positions gained are by reason of long service as a laborer. A porter in a shoe store has recently been made head of the stock department, but it took him fifteen years to gain this promotion.

A large bicycle firm has put a negro in charge of its repairing department. One of the largest furniture stores has a black floor walker. A wholesale clothing house has a colored salesman. An East Boston shipbuilder has a negro draughtsman. Notman, the photographer, has a black assistant.

And there are instances of negroes employed as chemists, druggists, architects, and engravers. Some half-dozen stenographers hold good positions—one in the general office of the Fitchburg Railroad. In a few cases business men have taken bright bell boys and servants out of hotels and given them a chance in their stores. The pity is that ability, when hidden by color, is so seldom sought out and put to use!

A little less than 3 per cent. of the Boston negroes are in the professions and Government service. There are twelve negro lawyers, and nowhere else in the country, save in Chicago, are colored lawyers so successful. One is a Master in Chancery, and at least seven have a practice of $5,000 or more annually. Their clients are largely white foreigners. One negro held the position of Judge in the Charleston Court some years ago, but is now dead.

Among the medical men in Boston are nine general practitioners, four dentists, and one veterinary surgeon. Five of these make large incomes. One of the dentists was formerly a demonstrator in Harvard University and is one of the finest dentists in Boston, having many of the best families of the city among his patrons.

There are always a considerable number of negro students in Boston—six or eight in Harvard College, a number in the professional school, in Boston University, the Institute of Technology, and the public high schools.

The chief political positions held in Boston by negroes are a Deputy Collector of Taxes, a Deputy Sealer of Weights and Measures, a Deputy Sheriff, a Postmaster of a city sub-station, and a Sergeant of Police. Through the civil service competitive examinations considerable numbers have secured appointments in the postal service and the city civil service.

In Boston and the immediate suburbs there is one Principal of a public school with white teachers and pupils, and there are five colored teachers in various other schools. In the governing bodies of the city and vicinity are two Aldermen and three Common Councilmen.

There have been several colored members of the Legislature and one negro in the Governor's Council. One of the Aldermen mentioned was formerly the famous black class orator at Harvard in 1890. He has made an excellent record as a public servant.

It is noticeable that only 62 per cent. of the Boston negroes are in gainful occupations, a smaller proportion than in the other cities, showing a larger number of children in school and a larger number of mothers and daughters making and keeping homes. There are no very wealthy negroes in Boston, but a large number of persons owning homes worth from $2,000 to $10,000.

There are in Boston a half dozen estates of $25,000 or more, ten or fifteen from $10,000 to $25,000, belonging to negroes. Their total wealth is probably between two and three millions of dollars.

It would be wrong to suppose that beneath the fair conditions described there was not the usual substratum of crime and idleness. Seventy years ago the negroes of Massachusetts furnished 14 per cent. of the convicts. Just before the war they still furnished 11 per cent., although forming less than 1 per cent. of the population.

To-day, forming 2 per cent. of the population, they furnish 2¼ per cent. of the prisoners, 3½ per cent. of the penitentiary convicts, and 1½ per cent. of the paupers. While not ideal, this record is very encouraging.

On the whole Boston negroes are more hopeful than those in New York and Philadelphia. A prominent negro author said recently in The Boston Globe, in answer to the question "Do negroes expect to attain perfect equality with the whites?":

> Contrast the changes which have taken place here during the last half century in relation to this question. What have we to-day in place of all that inequality and wrong?
>
> Complete equality before the law, in the public schools, at the polls, and in public conveyances, and substantial equality in hotels, restaurants, and places of amusement, while on that self-same common from which colored boys were once driven by what seemed at the time a relentless race prejudice stands one of the noblest monuments of genius in America, erected to commemorate the heroic services of the Union of a regiment of black troops in the War of the Rebellion.

December 8, 1901

THE BLACK NORTH

A SOCIAL STUDY.

Sixth of a Series by W. E. Burghardt DuBois, Atlanta University

SOME CONCLUSIONS.

WE have followed in some detail the history and condition of the colored people in the three chief cities of the Northeast, and there are some manifest conclusions which may here be gathered up and considered.

And first, in regard to the inner life of the negroes, it is apparent that we have been dealing with two classes of people, the descendants of the Northern free negroes and the freed immigrants from the South. This distinction is not always clear, as these two elements have mingled often so as to obliterate nearly all differences.

Yet it has everywhere been manifest in the long run that while a part of the negroes were native-born and trained in the culture of the city, the others were immigrants largely ignorant and unused to city life. There were, of course, manifold exceptions, but this was the rule. Thus the history of the negro in Northern cities is the history of the rise of a small group growing by accretions from without, but at the same time periodically overwhelmed by them and compelled to start again when once the new material had been assimilated.

Philadelphia is perhaps the best example of this. Four times the freed men started forward and four times they were overwhelmed and dragged back by a mass of immigrants. The first two times the newcomers were gradually incorporated and the group started with renewed energy.

Before the negroes, however, had recovered from the invasion of 1876 the new stream of 1890 started and has not yet stopped, although the native-born Philadelphia negroes constitute but little over a

quarter of the colored population. In New York the native-born have been perhaps even more completely overwhelmed. In Boston alone have they held their own sufficiently to retain considerable influence in the leadership of the group.

To realize what this cleft in the black world means we must remember that, as has been said, to all intents and purposes, the negroes form a world among themselves. They are so organized as to come in contact with the outer world as seldom as possible.

The average negro to-day knows the white world only from afar. His family and relatives are colored, his neighbors with few exceptions are colored, and his acquaintances are chiefly colored. He works with colored people, if not for them; he calls on colored people, attends meetings and joins societies of colored people, and goes to a colored church.

He reads the daily paper just as the whites read foreign news, chiefly for its facts relating to his interests; but for intimate local and social notes he reads The Age or The Tribune or The Courant—colored sheets. The chances are to-day that he is served by a colored physician, consults now and then a colored lawyer, and perhaps buys some of his supplies of colored merchants.

Thus the white world becomes to him only partially real, and then only at the points where he actually comes in contact with it—on the street car, in taking his employer's directions, and in a few of his amusements. This contact is least in New York and broadest in Boston, where it extends to restaurants, theatres, and churches.

Now there arises from such facts as these a peculiarly baffling question: Ought the black man to be satisfied with and encourage this arrangement, or should he be dissatisfied? In either case what ought he to do?

From the earliest times the attitude of the free negroes has been opposed to any organization or segregation of negroes as such. Men like Fortune, McCune, Smith, and Remond insisted that they were American citizens, not negroes, and should act accordingly. On the other hand, the Southern immigrants had of necessity been used to herding together. When they came North the clan spirit prevailed, partly from instinct, chiefly because they felt their company was not desired and they dreaded refusals and rebuffs.

The free negroes deeply resented this action; they declared it was voluntarily drawing the color line; that it showed cowardice, and that wherever the negro withdrew his pretensions to being treated as a man among men, he lost ground and made himself a pariah. Notwithstanding all this opposition the new immigrants organized, slowly but surely, the best and only defense of the ostracized against prejudice. They built negro churches, organized negro societies, settled in negro neighborhoods, and hired out to work in gangs.

They made a negro world and then in turn taunted the free negroes with wishing to escape from themselves, and being ashamed of their race and lineage. Here stood the paradox, and here it still stands to puzzle the best negro thought. How can negroes organize for social and economic purposes and not by that very organizing draw and invite the drawing of the color line? Every inpouring of freed Southern immigrants into the North has naturally forced their ideas of clan life on the community of blacks, estranged the older free negro element, and deprived the whole group of its best natural leadership.

The Northern negro needs to-day intelligent, far-seeing leadership. His problem differs from the problems in the South, because his history, condition, and environment are different. And such leadership demands leisure for thought and education—the emerging again of a dominant intelligent class such as the free negroes formerly were. This would rapidly happen could the negro find work.

The problem of work, the problem of poverty, is to-day the central, baffling problem of the Northern negro. It is useless and wrong to tell the negro to stay South where he can find work. A certain sort of soul, a certain kind of spirit, finds the narrow repression and provincialism of the South simply unbearable. It sends the aspiring white man North, it sends also the negro. It is a natural movement and should not be repressed.

And yet the surest way to pervert the movement and ruin good immigrants and encourage criminals is the policy of refusing negroes remunerative work. That this is the case at present all evidence proves; the evidence of strain in domestic relations as shown by late marriage, deserted wives, children out of school, and unhealthful homes. Such strain falling as it does on the weakest spot in the negro's social organization—the home—is a partial explanation of idleness and crime, while the encouragement of professional criminals and gamblers in our large cities furnishes whatever additional explanation is needed.

Turning now to the white population of these cities, it is probably true that the larger part of them are aware of no particular desire to hinder the progress of negroes. The question is one of mere academic interest to them, belonging principally, they imagine, to the South, and in regard to which they have inherited most of their opinions from their fathers. In practice they seldom meet negroes, know little of them, and are quite indifferent to them.

Then there is a large class of people who dislike negroes, and still others with positive views as to the natural inferiority of this branch of the human family. Further than this there are numbers of people who have a direct pecuniary interest in color discrimination—skilled workmen, who can thus further limit the supply for the labor market; clerks, salesmen, and managers, who fear for the dignity and respectability of their positions, and merchants and dealers, who dare not offend the public prejudice.

It is plain that with negative indifference, positive prejudice, and some pecuniary interests against the negro, and with simply a general humanitarian feeling on the part of a few for him, it is easy to make his opportunities narrow and his life burdensome. There can be no reasonable doubt but that the Northern negro receives less wages for his work and pays more rent for worse houses than white workmen, and that it is not altogether a matter of fitness that confines his work chiefly to common labor and menial service.

That the question of fitness plays a considerable part in determining the condition of the negro is without doubt true. His training hitherto has but ill-fitted him for the stern competition of world-city life. He knows but little of modern skilled labor, he has not been trained in thrift, he is not as a rule neat and tidy, and he is apt to be slow and unreliable.

True as this is of the mass everybody knows personally of individual exceptions to the rule. Even the most prejudiced admit this. If this be so then the only thing that the best interests of the black race and of the Nation demand is that these exceptions be treated as their deserts entitle them—that the energetic, resourceful black man, the thrifty, neat black woman, the negro carpenter, tanner, and steel worker, the boy of promise and the man of education be given that chance to make a living and enjoy life which America offers freely to every race and people except those whom she has most cruelly wronged.

Nor is this merely a matter of sentiment. To keep down the black men who are fit to rise costs the city something to-day and will cost more to-morrow. The black population of the North is growing. Despite the phenomenal increase in the white North the black North has silently kept abreast or been but little behind. It has doubtless increased more rapidly than the native whites and is still growing.

There can be no doubt of the drift of the black South northward. It is said that the Wilmington riot alone sent a thousand negroes to Philadelphia. Every failure of the South, every oppressive act, every unlawful excess shifts the black problem northward.

Many who see the cloud coming have thought that passive if not active discouragement of negroes might keep them away. The census figures do not bear them out in this. But what the census figures, taken in conjunction with the statistics of crime and the history of municipal misgovernment, do prove is this: That the exclusion of honest negro workmen from earning a living in the North means direct encouragement to the Northward migration of negro criminals and loafers.

The crime of negroes in New York is not natural or normal. It is the crime of a class of professional negro criminals, gamblers, and loafers, encouraged and protected by political corruption and race prejudice. The only sort of negro that is generously encouraged in Philadelphia is the criminal and the pauper.

The black man who wants charity or protection in crime in the Quaker City can easily get it. But the black man who wants work will have to tramp the pavements many a day. Thus crime is encouraged, politics corrupted, energy and honesty discredited, and a reception prepared for simple-minded negro immigrants such as Dunbar has so darkly painted in his "Sport of the Gods."

What is the remedy? First, the negroes must try to make the deserving and fit among them as numerous as possible. So long as the majority are mediocre workmen, and a considerable minority lazy and unreliable, those who seek to attack the race will have ample ammunition. Whatever, therefore, may be best as to negro organization in many lines, certainly all must unite in keeping the blacks from succumbing to the present temptations of city life. Three lines of effort here seem advisable: First, systematic search for work; second, better homes; third, political reform.

There must be no idleness. Work, even if poorly paid and menial, is better than no work. Rebuffs and refusals, though brutal and repeated, should not discourage negroes from continually and systematically seeking a chance to do their best.

Homes in respectable districts and healthful places should be had. No respectable negro family should linger a week in the Tenderloin of New York or the Fifth Ward of Philadelphia, or in the worst parts of the West End in Boston.

Concerted, organized effort can bring relief here, even if it costs something in comfort and rent. The home training of children should be more strict even than that of whites. Social distinctions should be observed. A rising race must be aristocratic; the good cannot consort with the bad—nor even the best with the less good.

Negroes have an interest in honest government. They should not allow a few minor offices to keep them from allying themselves with the reform movements in city government. The police riot of New York is but one clear proof of this.

Finally the white people of these three great cities should—but is it necessary here in the twentieth century to point out so plain a duty to fair-minded Americans?

December 15, 1901

DEMANDS OF THE NEGROES.

Afro-American Council Issues an Address Asking Equality and Highly Praising the President.

WASHINGTON, Jan. 26.—The Executive Committee of the National Afro-American Council in session here to-day passed a resolution urging confirmation by the Senate of the nomination of Dr. W. D. Crum to be Collector of Customs at Charleston, S. C., and adopted an address to the country on the race question in general. The address is signed by Alexander Walters, Chairman of the Executive Committee; Cyrus Field Adams, Secretary, and William A. Pledger, Acting President National Afro-American Council. The address in part is as follows:

It is evident to the thoughtful among us that we are passing through one of the most critical periods of our existence in this country. Questions that immediately concern the liberty and well-being of one-eighth of the population of the United States and scarcely to less degree the whole population of the country are pressing for treatment as never before.

A systematic effort has been inaugurated on the part of the South which has for its object the withdrawal of the franchise from the Afro-Americans of that section and their reduction to a position of absolute subserviency in all the relations of life.

It has been openly declared by some of the most prominent leaders of the South that it was the intention of the framers of the new Constitutions to disfranchise as many Afro-Americans as possible and leave every Caucasian in full possession of the suffrage. The effect has been that not only has the Afro-American been disfranchised, but also that a very large number of Caucasians who previous to the adoption of these Constitutions participated in elections have ceased to register and vote.

We contend for our constitutional rights on the ground that the right of suffrage has been conferred upon its citizens by the Federal Government.

We heartily commend the Afro-Americans of Virginia, Alabama, Louisiana, and other States, who are seeking redress through the courts of the land, and we pledge them our moral and financial support.

We denounce the mob murders now so prevalent in this country. We call the attention of the country to a condition of service on many farms in a number of the Southern States resembling very much the old peonage system, and ask for legislation looking to the remedying of the evil.

We submit our protest against the unfair practices in the transportation of passengers in Southern States discriminating unjustly against Afro-Americans, requiring of them the highest rates for travel and providing in return the poorest accommodations in carriage, and we invoke the exercise of the powers of the Inter-State Commerce Commission, by that tribunal, to prevent discriminations in rates and accommodations against inter-State passengers.

We appeal to Congress for favorable action upon one of the several measures now pending therein for the appointment of a commission to inquire into the condition of the Afro-Americans of the country.

The address closes with an indorsement of the deliverances of President Roosevelt with reference to the fitness of the negro for appointment to office, and the attitude of the Federal Government toward all of its citizens regardless of race or color. He has, the address continues, imparted new life and vigor to the time-honored principles and traditions of human rights, and has given hope and inspiration to a people struggling heroically beneath the burden of hate and proscription.

"We commend Theodore Roosevelt," the address concludes, "to the affection and confidence of our people regardless of party affiliation."

January 27, 1903

Negroes Organize a Bank.

Special to The New York Times.

MOUND BAYOU, Miss., Jan. 14.—The Bank of Mound Bayou, the only town in the county absolutely owned and controlled by colored people, was organized to-day, with a capital of $10,000.

The officers are: John W. Francis, President; B. H. Creswell, Vice President; Charles Banks, Cashier. The bank will open its doors Feb. 10.

January 15, 1904

NEGRO'S COTTON MILL FAILS.

Only One of Its Kind Sold at Auction to the Dukes.

CHARLOTTE, N. C., June 28.—The Coleman Cotton Mill, at Concord, was sold at public auction to-day under an execution of two mortgages held by the Dukes of Durham, N. C. The property was bid in for the mortgage and at $10,000. The concern owes $20,000. The Coleman Mill was the first in North Carolina to run with colored help under a negro owner.

The organizer of the mill was Warden Coleman, a well-known negro of Concord. He had considerable means, and it is said that the failure of the venture cost him most of his property. Coleman died some months ago.

June 29, 1904

DR. W. E. B. DU BOIS, the well-known negro man of letters, has resigned the Chair of Sociology at Atlanta University and come to New York as Director of Research and Publicity for the National Association for the Advancement of Colored People. Dr. Du Bois's "Souls of Black Folk" has just gone into a twelfth edition.

November 26, 1910

THE DARKER RACES

Something About The Crisis, the Negro Magazine of New York

WE have received for review a number of copies of The Crisis, a magazine which announces itself as "a record of the darker races." It undertakes to give every month a resume of the situation in regard to the race problem in this country and is the organ of the National Association for the Advancement of Colored People, at 20 Vesey Street, in this city.

It carries out its mission very well, as might be expected from the names of the editorial board. The editor in chief is Dr. W. E. B. Du Bois, the well-known negro writer and sociologist and the author of that remarkable book. "The Souls of Black Folk." Assisting him are Messrs. Oswald Garrison Villard, the journalist and historian of John Brown; Charles Edward Russell, Kelly Miller, the essayist, and others. The purpose of the magazine is declared to be the collection and publication of the exact facts in the effort to secure justice for the black man. An editorial sets forth that:

> There is to-day a tendency among colored people and their earnest friends to tell the half truth concerning the situation of the colored people and to condemn those who seek to tell the whole truth.

Several instances are mentioned of help and good-will on the part of the white man, but while The Crisis welcomes the publication of these facts, it goes on to say that they are not the whole truth, and that the "widespread injustice" toward the black man should not be passed over. The Crisis continues:

> There are friends of black folk in this land. There is continual advance in human sympathy. There is an awakening in the White South on the race problem. All that is true. It is also true that the Negro-American to-day faces the crisis of his career; race prejudice is rampant, and is successfully overcoming humanitarianism in many lines, and the determination of the dominant South to beat the black man to his knees, to make him a docile, ignorant beast of burden, was never stronger than to-day. This is the truth. Let us tell the truth, unpleasant though it be, and through the truth seek freedom. There is no other way.

The little magazine contains many interesting contributions, with illustrations, as well as a digest of the news and of the current newspaper discussion of the problem. Mr. Russell has an article relating his experiences with the South which are not reassuring, while a Southern woman in the latest issue gives her views, which are distinctly reassuring. But which she frankly says are not those of the majority. There is evidently an honest effort made to be just, though many instances of oppression are recorded, and there is a standing column of the number of black men lynched without a trial since 1885. It amounts to 2,458, an average of nearly two for each week.

Among the members of the association's committee, which is made up of colored and white alike, are Miss Jane Addams, Prof. John Dewey, Miss Sophronisba Breckinridge of Kentucky, Prof. E. R. A. Seligman, Mrs. Florence Kelley, Jacob Schiff, the Hon. Thomas M. Osborne, Miss Lillian D. Wald, and Judge Stafford of Washington, and many well known persons interested in social problems. The Crisis is printed very creditably by a colored man, Robert N. Wood.

August 6, 1911

WANTS NEGROES TO PROTEST

Washington Urges Them to Fight Discrimination by Railroads.

Dr. Booker T. Washington, Principal of the Tuskegee Normal Industrial Institute has asked the negroes of this country to set aside Sunday, June 7, and Monday, June 8, as special days on which to protest to railroads against showing discrimination in the accommodations provided for colored passengers.

He has asked that the churches, secret societies, business leagues, women's clubs, and other agencies organize and send on those days representatives to the officials of the lines that have provided poorer accommodations for negro than for white travelers.

April 18, 1914

How Negroes Are Helping to Win the War

WHAT of the American negro in this war? Earlier reports about him were disquieting. His loyalty at home was under severe test, his fortitude under trench strain in doubt. Men returning from abroad were wont to say at first that the negro, whose bravery had been demonstrated beyond peradventure in former conflicts, wasn't standing up under the more trying conditions there.

How far mistaken were those reports the news has shown. The high exploits of negro soldiers on the battlefield thrilled all America. But not so much has been known about the situation on this side of the water. German agents have sought to create widespread disaffection among the colored folk here, and apparently they used much the same tactics as in Ireland.

But, while the answer in Ireland was commonly understood to be Sinn Feinism, the answer here was a heightened patriotism and a more zealous support of the war.

"This is not the time to discuss race problems," said Emmett J. Scott, Special Assistant to the Secretary of War, in Washington the other day. "Our first duty is to fight, and to continue to fight until this war is won. Then we can adjust the problems that remain in the life of the colored men. This is the doctrine we are preaching to the negroes of the country, and every day indications reach this office that it is meeting with a larger and larger response."

Mr. Scott, who is Secretary of Tuskegee Institute, himself a colored man, has been lent to the War Department, to aid in smoothing out any problems that might arise in connection with the negroes and the war. All questions of this kind, whether the complaint is from some negro organization alleging an unjust discrimination or from a colored mother asserting that her son has been treated unfairly in the draft, is referred to Mr. Scott, who has a staff of assistants under him. A single mail sometimes contains a hundred letters, from all parts of the country. It is the best place in the country to measure the state of negro public opinion as to the war. There have been many complaints, but through them all, Mr. Scott says, there runs a note of sound loyalty, that whether or not the grievance can be straightened out now, as most of them readily are, the complainant will do his part in the war and trust to obtaining at some later time what he thinks is due him.

Mr. Scott, acting by direction of the War Department, recently called a conference in Washington of thirty-one representatives of the negro press, including publications with more than 1,000,000 circulation, at which every man was encouraged to voice any complaint or objection that might be apparent in his section of the country. The talks gave proof of unwavering loyalty. Means were discussed of mobilizing the resources of the 12,000,000 negroes in the United States, and resolutions were adopted expressing the earnest and resolute temper of the meeting. Among those who spoke were the following high officials:

Newton D. Baker, Secretary of War; George Creel, Chairman of the Committee on Public Information; Franklin D. Roosevelt, Assistant Secretary of the Navy; Edward N. Hurley, Chairman of the United States Shipping Board; Major Joel E. Spingarn, attached to the General Staff, United States Army; Captain Arthur S. Spingarn of the Medical Reserve Corps, National Army; General Paul Vignal, Military Attaché of the French Embassy; Major Edouard Requin and Major L. P. de Montal of the French High Commission.

Emmett J. Scott, Special Negro Assistant to Secretary Baker.

"Our purpose here," said Mr. Scott, "is to obtain the maximum of efficiency and the minimum of friction. I think the absence of this so far is remarkable, especially with reference to white and negro soldiers in the same camps. It is at variance with some of the predictions that were made. Both races have been represented at some twelve camps, and there have been no misunderstandings that were not soon adjusted. One precaution that was taken in breaking in the new colored troops was to obtain staff officers for the command of the colored troops who had previous experience with negro soldiers in the regular army and who understand them.

"Few persons realize what the negroes as a whole are doing in this war. We have just been getting together some information on this, for distribution mainly among the colored people, to stimulate to action any of those who are not contributing a full share; also, we will show what opportunities our Government has provided for the negro, as an incentive to a responsive part.

"As to what the negroes are doing as soldiers: Besides the volunteer enlistment in large numbers since the declaration of the war, the colored man has cheerfully responded to the call of arms under the draft, the percentage of those who asked for exemptions being low. Under the first draft there were 737,628 registrants, or close to 8 per cent. of the total registration of the country. Of the negro registrants, close to 100,000 have been called into camp for active military service. There have been commissioned in the United States Army as Captains, First Lieutenants, and Second Lieutenants about 1,000 colored men, including about 250 colored medical officers in the Medical and Dental Reserve Corps. The 92d Division and the 93d (Provisional) Division, each finally to consist of approximately 30,000 negro soldiers, have been organized, under the command of Major Gen. C. C. Ballou and Brig. Gen. Roy C. Hoffman, respectively.

"The company units of these arms of service will, in large measure, be commanded by colored line officers. About 650 commissioned officers were graduated from the first training camp for colored officers at Des Moines, and these officers, according to reports, have for the most part made good and are in command of troops of their race at several camps. There are thirty-four colored Chaplains in the various branches of the army.

"As I have indicated, colored men have been afforded opportunities to enlist or qualify for such branches of service as the field artillery. There are now three such regiments, the 349th, the 350th, and the 351st Field Artillery. At Camp Meade a training camp has been established to prepare officers of the colored race for these regiments; it is said that the men undergoing this training are showing marked adaptability for the work.

"The personnel officer of the Coast Artillery has thrown open the doors of his branch to colored men not yet called into active service, and has announced that there is a need for negroes with college training. At Camp Sherman, at Chillicothe, Ohio, there is a colored regiment of engineers, many of the officers of which are colored.

"The Government has recently made provision for special training of young men of the colored race in technical and mechanical work for army service. At standard colored schools, such as Howard University, at Washington, D. C.; Hampton Institute, Hampton, Va., and Tuskegee, Tuskegee, Ala., selected students will be taught, during the Summer and Fall, such branches as radio engineering, auto-mechanics, woodworking, carpentry, general and electrical engineering.

"One hundred and fifty colored men are engaged in Y. M. C. A. work in the various camps where negro soldiers are stationed; some of these workers are in France with the colored regiments under General Pershing. Wherever they are they are working out, in addition to their religious duties, systems whereby the illiteracy so prevalent among troops from certain parts of the country may be reduced to a minimum.

"Much is being done in an organized way in many quarters to speed up the labor of colored men and women, to increase their technical skill in rendering service in war work, and to cut down illiteracy. The War Council of the Y. M. C. A., it is stated, is devoting $200,000 of its budget to work among negro women.

"In the purchase of Liberty bonds, War Savings Stamps, and in subscribing to war philanthropies the negro has made an excellent showing, when it is remembered that few are wealthy.

"As to War Savings Stamps, I will only refer to what the negroes in one city did, Washington, and that only in part. A campaign of education among the colored people in the District of Columbia resulted in the sale of $52,000 in stamps. This does not include $800 a week bought by the children in the schools and the amounts bought in the Federal departments by colored employes, nor the individual purchases of colored persons."

July 7, 1918

ASK FOR GERMAN COLONIES.

Negroes Want Land Captured in Africa Turned Over to Natives.

That the captured German colonies in Africa be turned over to the natives and that educated negroes be placed in leadership there, is one of the requests that the negroes of New York as represented by the Universal Negro Improvement Association and African Communities League, will make to this Government and to the Allies. These requests were contained in a resolution adopted at a meeting of 5,000 negroes in the Palace Casino in 135th Street, near Madison Avenue, yesterday.

The meeting was presided over by Marcus Garvey, head of the League and representatives of the race made speeches. The resolution set forth that it would only be through the granting to the negro his rights and the rights of all weaker peoples at the Peace Conference that future wars would be obviated. It was also asked that negroes be permitted to travel and to reside in any part of the world; that they be permitted the same educational facilities as Europeans; that all segregative and proscriptive ordinances against negroes be repealed and that they be given political, industrial and social equality.

November 11, 1918

FOR ACTION ON RACE RIOT PERIL

Radical Propaganda Among Negroes Growing, and Increase of Mob Violence Set Out in Senate Brief for Federal Inquiry

EVEN though recurring race riots have made the public aware that the negro problem has entered upon a new and dangerous phase, only those in touch with the inner forces that are playing on ignorance, prejudice, and passion, realize how great this menace is. Bloodshed on a scale amounting to local insurrection at least will be threatened in more than one section where large white and black populations face each other unless some program of conciliation is adopted to forestall influences that are now working to drive a wedge of bitterness and hatred between the two races.

So far this problem, in some respects the most grave now facing the country, has been allowed to drift. The States have done nothing. The Federal Government has done nothing. The only move made at Washington is the introduction by Senator Charles Curtis of Kansas of a resolution calling for the appointment of a subcommittee of the Senate Committee on Judiciary to investigate recent riots and lynchings and to report what remedies should be employed to prevent their recurrence. Senator Curtis said in Washington the other day that information in his possession made it clear that there should be no delay in grappling the problem and that he would press for action. A brief containing new information as to the extent of race clashes which he will lay before the committee accompanies this article. It shows that since the beginning of the year there have been since Jan. 1, 1919, thirty-eight race riots and clashes in cities and other communities in various parts of the country. Senator Curtis is uncertain whether Congress has the authority to pass a law against riots and lynchings; this may be a question for action by the States, but he is certain that, after an investigation has laid bare the causes of the growing antagonism between white and black, recommendations can be made that will show the urgent need of a policy of organized conciliation, backed by the better elements in each race, in every community wherever whites and blacks confront each other in considerable numbers.

The War's Responsibility.

Out of the war has come a new negro problem—that, observers agree, is the first fact to be recognized in taking up the question. Before the war negro leaders, still under the influence of Booker Washington, were in the main for a policy of conciliation. For all the scattered injustice and oppression that the negro still suffered the majority of the negro leaders still held in clear prospective the great benefits granted the negro race in this country, the fact that their freedom had been won by the sacrifice of an immense number of white men's lives, that in no other country in the world where a large colored population lived in contact with the white race did the principle of the laws confer equal recognition to the black man. In a word, there was still active among the negro leaders a sense of appreciation tracing back to the civil war period. Whenever friction threatened, leaders of this type, believing that by forbearance and thrift on the part of the black man a fair and harmonious adjustment of the two races would be attained, steadily argued conciliatory methods.

Some of these leaders remain, but they are growing fewer. The assertion is made in explanation that these moderate leaders have been without the support of white leaders. Under heavy attack of radicals and militants, charged with being at heart the betrayers of the negro race, they have been unable, according to this attempted elucidation of the situation, to point to any organized co-operation on the part of the whites to see, for example, that police officers and courts dealt justly with the negro, to remove unjust treatment of the negro wherever found. The other side proclaims that there is no evidence that the great majority of the white men in this country, who as the result of a civil war had bestowed on the black man opportunities far in advance of those he had in any other part of the white man's world, were still the friend of the negro. They ask proof that forbearance, not militancy, is the course to follow.

Reds Inflaming Blacks.

Every week the militant leaders gain more headway. They may be divided into general classes. One consists of radicals and revolutionaries. They are spreading Bolshevist propaganda. It is reported that they are winning many recruits among the colored race. When the ignorance that exists among negroes in many sections of the country is taken into consideration the danger of inflaming them by revolutionary doctrine may apprehended. It is held that there is no element in this country so susceptible to organized propaganda of this kind as the less informed class of negroes.

The other class of militant leaders confine their agitation to a fight against all forms of color discrimination. They are for a program of uncompromising protest, "to fight and to continue to fight for citizenship rights and full democratic privileges." The former leadership of Booker Washington is derided. A negro paper of wide circulation said in a recent editorial: "* * * the late Booker T. Washington was selected by a group of Southern and Northern philanthropists and business men who sympathized with them to teach the negro to know his place. * * * Under the Frederick Douglass propaganda we gained freedom and citizenship. Under the Booker Washington propaganda we lost our citizenship in the Southland and saw the spread of mob violence all over the country. The thing to do is to get back to the teachings of Frederick Douglass. * * *"

W. E. B. Du Bois, a foremost leader in this class of militants, says in the leading editorial in the current issue of his magazine, The Crisis:

"We have cast off on the voyage which will lead to freedom or to death. For three centuries we have suffered and cowered. No race ever gave passive submission to evil longer, more piteous trial. Today we raise the terrible weapon of self-defense. When the murderer comes, he shall no longer strike us in the back. When the armed lynchers gather, we too must gather armed. When the mob moves, we propose to meet it with bricks and clubs and guns."

There is no doubt that owing to recent experiences many negroes have provided themselves with arms, and that unless Governmental efforts, based on some carefully considered policy, are made to stop the riots and race clashes and to remove their causes, that outbreaks of far greater extent than any of those that have yet occurred may take place. The one approach to a betterment of conditions is asserted to be through those negro leaders who are opposed to militant methods, but it is pointed out, while they preach co-operation, they insist also that the only solution is "full justice, manhood rights and full opportunities for the negro American."

Industrial Clashes.

New industrial contacts between white and negro workers aggravate the problem. Three weeks before the riot last week in Omaha investigators from Washington reported that a clash was imminent owing to ill-feeling between white and black workers in the stockyards. It is estimated that during the war period 500,000 negro workers migrated from the South to the North. In whatever Northern city they have settled in numbers there is the menace of racial clash, and consequently the immediate need of some agency of conciliation, in which both whites and negroes shall be represented, as a medium for clearing those misunderstandings that spring from rampant prejudice. An illustration of changes in Northern industrial centres is provided by the case of Detroit. In 1914 there were probably not 1,000 negroes in that city. At present it is estimated there are between 12,000 and 15,000 engaged in the automobile industry there. In the steel plants of Pittsburgh the number of negro workers has increased 100 per cent in some of the plants. In New York City many negro girls are now at work in the cheaper branches of the garment trade. This is one of the many industries in the North in which they have won or are seeking to win a place.

On this phase of the problem, Dr. George E. Haynes, a leading negro educator, and now Director of Negro Economics of the Department of Labor, recommends:

"To be concrete, the first step in this direction seems to be for local and national officials to call into conference and counsel the liberal-minded citizens of both races and with them to map out some plan to guarantee greater protection, justice, and opportunity to negroes that will gain the support of law-abiding citizens of both races. Co-operative local committees on matters involving race relations, both under private and governmental auspices, especially during the war, have demonstrated that far-reaching practical results can be secured by such efforts."

The brief on which the projected investigation of the race problem by the Senate Committee on Judiciary will be based follows, the more detailed information being summarized at several points. It is headed: "Why Congress Should Investigate Race Riots and Lynchings," and is divided into five heads:

I. The Facts—1919.

A. Race riots:

Washington, D. C.: "Nation's Capital at Mercy of the Mob"—headline on Page 1 of Washington Post Tuesday, July 22, 1919. Rioting in the main streets of national capital was unchecked during four nights from Saturday, July 19, until Wednesday, July 23. Six persons were killed outright, 50 severely wounded, and a hundred or more less seriously wounded.

Chicago, Ill.: At least 36 persons were killed outright, by official report, in race rioting which lasted from Sunday, July 27, to Friday, Aug. 1. According to unofficial reports, the number killed was much larger. Houses were wrecked and burned, mobs roamed the streets, and it was necessary to put seven regiments of State militia under arms.

Knoxville, Tenn.: On Aug. 30 a mob of white persons stormed the Knox County Jail, firing on officers of the law, liberating 16 white prisoners, of whom several were convicted murderers; looting the house of the Sheriff, stealing stocks of confiscated whisky. The mob then wrecked and looted shops and invaded the colored district. At least seven persons were killed and twenty or more injured.

Longview, Texas.: Four or more men were killed outright in a riot on the night of July 10, when a mob of white men invaded the negro residence district, shooting and burning houses.
Norfolk, Va.: Receptions of the home-coming negro troops had to be suspended because of riots July 21, in which six persons were shot, necessitating the calling out of the marines and sailors to assist the police.
Philadelphia, Penn.: A riot call was sent to all West Philadelphia stations July 7; eight arrests were made and one man was taken to a hospital in consequence of a race riot at a carnival.
Charleston, S. C.: One or more men were killed and scores were shot or beaten in a race riot led by United States sailors May 10; city placed under martial law.
Bisbee, Ariz.: Clashes occurred on July 3 between local police and members of the 10th United States Cavalry, (colored.) Five persons were shot.

There were in addition race clashes in the following cities:
Tuscaloosa, Ala., July 9.
Hobson City, Ala., July 26.
New London, Conn., June 13.
Sylvester, Ga., May 10; one reported killed.
Putnam County, Ga., May 29.
Mullen, Ga., April 15; seven reported killed.
Blakely, Ga., Feb. 8; four reported killed.
Dublin, Ga., July 6; two reported killed.
Ocmulgee, Ga., Aug. 29; one reported killed.
Bloomington, Ill., July 31.
New Orleans, La., July 23.
Annapolis, Md., June 27.
Baltimore, Md., July 11.
Monticello, Miss., May 31.
Macon, Miss., June 27.
Hattiesburg, Miss., Aug. 4.
New York City, N. Y., Aug. 21.
Syracuse, N. Y., July 31.
Coatesville, Penn., July 8.
Philadelphia, Penn., July 31.
Scranton, Penn., July 5.
Darby, Penn., July 23.
Newberry, S. C., July 28.
Bedford County, Tenn., Jan. 22.
Memphis, Tenn., March 14; one killed.
Memphis, Tenn., June 13.
Port Arthur, Texas, July 15.
Texarkana, Texas, Aug. 6.
Morgan County, W. Va., April 10.

B. Lynchings:
Forty-three negroes, four white men lynched from Jan. 1 to Sept. 14.
Eight negroes burned at the stake, one of the burnings extensively announced beforehand in newspapers of Louisiana and Mississippi. Copies of these papers are filed as exhibits with the brief. Of the number sixteen were hanged. Others were shot. One was cut to pieces.

1889-1918.

Two thousand four hundred and seventy-two colored men, 50 colored women, 691 white men and 11 white women lynched. Less than 24 per cent. of these lynchings were ascribed to be on account of attacks on women.

1918.

Five negro women, 58 negro men and 4 white men lynched. No member of any mob was convicted. In only two cases were trials held.

II. The Failure of the States.

The States have proven themselves unable or unwilling to stop lynchings, as the figures show. Even attempts to prosecute are so rare as to be exceptional. Before the burning at the stake of John Hartfield at Ellisville, Miss., June 26, 1919 Governor Bilbo of Mississippi said:

"I am utterly powerless. The State has no troops, and if the civil authorities at Ellisville are helpless, the State's are equally so. Furthermore, excitement is at such a high pitch throughout South Mississippi that any armed attempt to interfere with the mob would doubtless result in the death of hundreds of persons. The negro has confessed, says he is ready to die, and nobody can keep the inevitable from happening."

The Houston Post, Texas, in a widely quoted editorial, said:

"The Post believes * * * that the half-century old lynching problem is about to pass from the jurisdiction of State authority into the domain of Federal action. Surely, in the light of half a century of lynchings, in which the victims have been numbered by the thousands, the failure of the States must be confessed.

III. A National Problem.

Lynching and mob violence have become a national problem. President Wilson was aroused by the danger of mob violence to make a statement July 26, 1918, in which he called the subject one which "vitally affects the honor of the nation, and the very character and integrity of our institutions. * * * I say plainly that **every American who takes part in the action of a mob or gives any sort of countenance is no true son of this great democracy, but is its betrayer, and does more to discredit her by that single disloyalty to her standards of law and right than the words of her statesmen or sacrifices of her heroic soldiers in the trenches can do to make a suffering people believe in her, their savior."**

The extension of lynching to Northern States with white men as victims shows it is idle to suppose murder can be confined to one section of the country or to one race.

IV. Consequences of Lynching.

1. **Race riots: Persistence of unpunished lynchings of negroes fosters lawlessness among white men imbued with the mob spirit, and creates a spirit of bitterness among negroes. In such a state of public mind a trivial incident can precipitate a riot.**
2. Industrial: Property values and productivity are lessened and business is disturbed in districts from which people are forced to migrate to escape mob violence.
3. Psychological: Brutalization of men, women, and children who take part in and witness hangings, burnings at stake, and the horrors of lynchings. Dr. A. A. Brill, neurologist, assistant Professor of Psychiatry at the Post Graduate Medical School, says:

"The torture which is an accompaniment of modern lynchings shows that it is an act of perversion only found in those suffering from extreme forms of sexual perversion. Of course, not all lynchings are conducted in that way, but it is not uncommon to read accounts telling that the victim was tortured with hot irons, that his eyes were burned out, and that other monstrous cruelties were inflicted upon him. Such bestiality can be recognized only as a form of perversion. Lynching is a distinct menace to the community. It allows primitive brutality to assert itself and thus destroys the strongest fabric of civilization. Any one taking part in or witnessing a lynching cannot remain a civilized person."

4. Political: The position of the United States before the world is impaired by its failure to accord protection and trial by law to its own citizens within its own borders.

V. The Danger.

1. Disregard of law and legal process will inevitably lead to more and more frequent clashes and bloody encounters between white men and negroes and a condition of potential race war in many cities of the United States.
2. Unchecked mob violence creates hatred and intolerance, making impossible free and dispassionate discussion not only of race problems, but questions on which races and sections differ.

October 5, 1919

HARDING SAYS NEGRO MUST HAVE EQUALITY IN POLITICAL LIFE

Special to The New York Times.

BIRMINGHAM, Ala., Oct. 26.—Following ovations accorded to him by crowds conservatively estimated to have numbered more than 100,000 persons, President Harding, speaking today before a great audience of whites and colored people in Capitol Park, declared that the negro is entitled to full economic and political rights as an American citizen. He added that this does not mean "social equality." The white man and the negro also should stand, he asserted, uncompromisingly against "every suggestion of social equality." Racial amalgamation, he added, can never come in America.

Perhaps not one person among ten in the thousand that jammed Capitol Park could hear the President, but those who were up in front and in a position to understand heard the Presidential message with the closest attention. Some approved of his utterances. Others, it appeared, did not, although there was not one word uttered from the audience that would substantiate the last observation. Parts of the speech appealed to the negroes in the audience and they gave vent to loud and lusty cheers to evidence their approval. On the other hand only once or twice was there any applause from the white section and in both instances it was scattered.

The race problem, the President declared, was no longer a sectional question applicable only to the Southern States, but a national question which must be met as such. In recent years, he pointed out, great numbers of negroes had left the South to seek homes in the North and West and as a result of this migration, he said, the "race question" had been brought closer to the people of the North and West and "I believe," he added, "it has served to modify somewhat the views of those sections on this question."

After warning the negroes that social equality was a dream that could not be realized, the President in words that held no doubt told them that in his opinion the time had come when they should vote not as Republicans but as they thought. He wanted, he said, to see the tradition of a solid black Republican vote broken and the time come when negroes would vote the Democratic ticket when they considered Democratic candidates and policies best for the country and when only for those same reasons would they vote for Republican candidates.

During the first part of the President's speech he paid his tribute to Birmingham, the "Magic City," as he styled it, of the South. He also warmed the hearts of the old Confederates and the sons and daughters of the "Lost Cause" veterans when he expressed the earnest hope that some day the history of the "Aladdin-like industrial wonder" of the civil war South will be written and the world will realize how the agricultural and aristocratic South of ante-bellum days met a great crisis and gave what he described as one of "the greatest demonstrations in all history of the possibilities of adaptation, organization and industrial development under stress of great necessity."

It was following his glowing tribute to the Confederate South that the President brought forward the race problem which, he told his audience, he was going to discuss frankly and honestly "whether you like it or not."

"If the civil war marked the beginnings of industrialism in a South which had previously been almost entirely agricultural, the World War," he said, "brought us to full recognition that the race question is national rather than sectional."

"While there are no authentic statistics, it is common knowledge that the World War was marked by a great negro migration from the Southern into the Northern States, where the negroes were attracted by the demand for labor and the higher wages offered in the North and West. The movement had been slowly under way for decades. But in the World War, because of conditions already described, it was greatly accelerated and has subsequently continued at only a slightly reduced rate."

Those in the crowd near the speakers' stand, white as well as black, were all attention. The negroes were particularly attentive, and the smiles on their faces indicated that they anticipated that their status in the South was about to be championed by the President of the United States. When the President, a few minutes later, referred to the war record of the black soldier they cheered at the tops of their voices, and when, still later, the President said that the black man should be permitted to vote when he was fit to vote and the white man deprived of his vote when he was unfit for the suffrage, the black element in the audience again shouted their approval of the sentiment in unmistakable fashion. The whites were silent.

On the stand with the President were Governor Thomas E. Kilby, Dr. N. A. Barrett, President of the Commission Government of Birmingham; E. W. Bowie, President of the Birmingham Semi-Centennial Celebration Association, and a majority of the members of the Alabama Legislature, who came on special trains from Montgomery to join in the welcome to the President and Mrs. Harding, Senator and Mrs. Underwood, Secretary of War Weeks, Secretary of the Interior Fall, and the sixty-seven girls who have been voted the "most beautiful" in the sixty-seven counties that comprise the Commonwealth of Alabama.

Dr. Barrett, in the name of Birmingham, welcomed the President, while Governor Kilby extended the welcome for the State. Birmingham welcomed the President, said Dr. Barrett, because he was a man of "great heart" and with vision big enough to take in as President "every mother's son of us," a statement that brought forth a vigorous nod of approval from President Harding.

The President was presented to the audience by Mr. Bowie. When the applause seemed to be gathering momentum and another ovation was threatened the President waved for silence. It was very hot and there was no shade over the speakers' stands.

The President said:

"I entered the Senate when you commissioned Senator Underwood to that body, and somehow, I never knew just why, we began with a 'paired' agreement to protect each other's votes. That arrangement held until I retired from the Senate, and we rarely, if ever, had to ask each other for instructions. There was a confident, respectful and cordial friendship from the beginning, and it was never embarrassed. Perhaps I need not tell you that my high opinion and affectionate regard still abide. Not so very long ago it became my duty to choose four outstanding Americans to represent our Republic in a conference with the statesmen of the leading nations of the world. It was not a personal regard alone, but that feeling combined with a high estimate of his statesmanship and his lofty devotion to country impelled me to name him as one of four to speak for America in a conference pregnant with incalculable possibilities.

"Politically and economically there need be no occasion for great and permanent differentiation, provided on both sides there shall be recognition of the absolute divergence in things social and racial," said the President. "I would say let the black man vote when he is fit to vote; prohibit the white man voting when he is unfit to vote. I wish that both the tradition of a solidly Democratic South and the tradition of a solidly Republican black race might be broken up. I would insist upon equal educational opportunity for both.

"Men of both races may well stand uncompromisingly against every suggestion of social equality. This is not a question of social equality, but a question of recognizing a fundamental, eternal, inescapable difference.

"Racial amalgamation there cannot be. Partnership of the races in developing the highest aims of all humanity there must be if humanity is to achieve the ends which we have set for it. The black man should seek to be, and he should be encouraged to be, the best possible black man and not the best possible imitation of a white man.

"The World War brought us to full recognition that the race problem is national rather than merely sectional. There are no authentic statistics, but it is common knowledge that the World War was marked by a great migration of colored people to the North and West. They were attracted by the demand for labor and the higher wages offered. It has brought the question of race closer to North and West, and, I believe, it has served to modify somewhat the views of those sections on this question. It has made the South realize its industrial dependence on the labor of the black man and made the North realize the difficulties of the community in which two greatly differing races are brought to live side by side. I should say that it has been responsible for a larger charity on both sides, a beginning of better understanding; and in the light of that better understanding perhaps we shall be able to consider this problem together as a problem of all sections and of both races, in whose solution the best intelligence of both must be enlisted.

"Indeed, we will be wise to recognize it as wider yet. Whoever will take the time to read and ponder Mr. Lothrop Stoddard's book on 'The Rising Tide of Color,' or, say, the thoughtful review of some recent literature on this question which Mr. F. D. Lugard presented in a recent Edinburgh Review, must realize that our race problem here in the United States is only a phase of a race issue that the whole world confronts. Surely we shall gain nothing by blinking the facts, by refusing to give thought to them. That is not the American way of approaching such issues.

"Mr. Lugard, in his recent essay, after surveying the world's problem of races, concludes thus:

"Here, then, is the true conception of the interrelation of color—complete uniformity in ideals, absolute equality in the paths of knowledge and culture, equal opportunity for those who strive, equal admiration for those who achieve; in matters social and racial a separate path, each pursuing his own inherited traditions, preserving his own race purity, and race pride; equality in things spiritual; agreed divergence in the physical and material.

"Here, it has seemed to me, is suggestion of the true way out. Politically and economically there need be no occasion for great and permanent differentiation, for limitations of the individual's opportunity, provided that on both sides there shall be recognition of the absolute divergence in things social and racial. When I suggest the possibility of economic equality between the races, I mean it in precisely the same way and to the same extent that I would mean it if I spoke of equality of economic opportunity as between members of the same race. In each case I would mean equality proportioned to the honest capacities and deserts of the individual.

"Men of both races may well stand uncompromisingly against every suggestion of social equality. Indeed, it would be helpful to have that word 'equality' eliminated from this consideration; to have it accepted on both sides that this is not a question of social equality, but a question of recognizing a fundamental, eternal and inescapable difference. We shall have made real progress when we develop an attitude in the public and community thought of both races which recognizes this difference.

"Take the political aspect. I would say let the black man vote when he is fit to vote; prohibit the white man voting when he is unfit to vote. Especially would I appeal to the self-respect of the colored race. I would inculcate in it the wish to improve itself as a distinct race, with a heredity, a set of traditions, an array of aspirations all its own. Out of such racial ambitions and pride will come natural segregations, without narrowing and rights, such as are proceeding in both rural and urban communities now in Southern States, satisfying natural inclinations and adding notably to happiness and contentment.

"On the other hand I would insist upon equal educational opportunity for both. This does not mean that both would become equally educated within a generation or two generations or ten generations. Even men of the same race do not accomplish such an equality as that. There must be such education among the colored people as will enable them to develop their own leaders, capable of understanding and sympathizing with such a differentiation between the races as I have suggested—leaders who will inspire the race with proper ideals of race pride, of national pride, of an honorable destiny; and important participation in the universal effort for advancement of humanity as a whole. Racial amalgamation there cannot be. Partnership of the races in developing the highest aims of all humanity there must be if humanity, not only here but everywhere, is to achieve the ends which we have set for it.

"I can say to you people of the South, both white and black, that the time has passed when you are entitled to assume that this problem of races is peculiarly and particularly your problem. More and more it is becoming a problem of the North; more and more it is the problem of Africa, of South America, of the Pacific, of the South Seas, of the world. It is the problem of democracy everywhere, if we mean the things we say about democracy as the ideal political state.

"The one thing we must sedulously avoid is the development of group and class organizations in this country. There has been time when we heard too much about the labor vote, the business vote, the Irish vote, the Scandinavian vote, the Italian vote, and so on. But the demagogues who would array class against class and group against group have fortunately found little to reward their efforts. That is because, despite the demagogues, the idea of our oneness as Americans has risen superior to every appeal to mere class and group. And so I would wish it might be in this matter of our national problem of races. I would accept that a black man cannot be a white man, and that he does not need and should not aspire to be as much like a white man as possible in order to accomplish the best that is possible for him. He should seek to be, and he should be encouraged to be, the best possible black man, and not the best possible imitation of a white man.

"It is a matter of the keenest national concern that the South shall not be encouraged to make its colored population a vast reservoir of ignorance, to be drained away by the processes of migration into all other sections. That is what has been going on in recent years at a rate so accentuated that it has caused this question of races to be, as I have already said, no longer one of a particular section. Just as I do not wish the South to be politically entirely one party; just as I believe that is bad for the South, and for the rest of the country as well, so I do not want the colored people to be entirely of one party. I wish that both the tradition of a solidly Democratic South and the tradition of a solidly Republican black race might be broken up. Neither political sectionalism nor any system of rigid groupings of the people will in the long run prosper our country.

"With such convictions one must urge the people of the South to take advantage of their superior understanding of this problem and to assume an attitude toward it that will deserve the confidence of the colored people. Likewise, I plead with my own political party to lay aside every program that looks to lining up the black man as a mere political adjunct. Let there be an end of prejudice and of demagogy in this line. Let the South understand the menace which lies in forcing upon the black race an attitude of political solidarity.

"Every consideration, it seems to me, brings us back at last to the question of education. When I speak of education as a part of this race question, I do not want the States or the nation to attempt to educate people, whether white or black, into something they are not fitted to be. I have no sympathy with the half-baked altruism that would overstock us with doctors and lawyers, of whatever color, and leave us in need of people fit and willing to do the manual work of a workaday world. But I would like to see an education that would fit every man not only to do his particular work as well as possible but to rise to a higher plane if he would deserve it. For that sort of education I have no fears, whether it be given to a black man or a white man. From that sort of education, I believe, black men, white men, the whole nation, would draw immeasurable benefit.

"It is probable that as a nation we have come to the end of the period of very rapid increase in our population. Restricted immigration will reduce the rate of increase, and force us back upon our older population to find people to do the simpler, physically harder manual tasks. This will require some difficult readjustments.

"In anticipation of such a condition the South may well recognize that the North and West are likely to continue their drafts upon its colored population, and that if the South wishes to keep its fields producing and its industry still expanding it will have to compete for the services of the colored man. If it will realize its need for him and deal quite fairly with him, the South will be able to keep him in such numbers as your activities make desirable.

"Is it not possible, then, that in the long era of readjustment upon which we are entering, for the nation to lay aside old prejudices and old antagonisms, and in the broad, clear light of nationalism enter upon a constructive policy in dealing with these intricate issues? Just as we shall prove ourselves capable of doing this we shall insure the industrial progress, the agricultural security, the social and political safety of our whole country, regardless of race or sections, and along the lines of ideals superior to every consideration of groups or class, of race or color or section or prejudices."

When the speech was ended, Governor Kilby was one of the first to shake the President's hand. He was followed by scores of other prominent citizens. If any in the great throng resented what the President had said none indicated it by their remarks. As a matter of fact, THE TIMES correspondent has not met a Birmingham citizen who has expressed disapproval of the President's views.

There are many who do not agree with him as to political equality, but what he said about the impossibility of "social equality" more than offset anything he said on other lines. The warmth and enthusiasm of the demonstration that followed his every appearance during the afternoon and night proved this.

October 27, 1921

NEGROES ENDORSE SPEECH.

Against Social Equality, Says Leader in Congratulating Harding.

Marcus Garvey, President General of the Universal Negro Improvement Association, sent the following telegram to President Harding yesterday congratulating him on his speech on the negro question as delivered at Birmingham, Ala.:

"Hon. Warren G. Harding, President of the United States, Birmingham, Ala.

"Please accept heartfelt thanks of four hundred million negroes of the world for the splendid interpretation you have given of the race problem in your today's speech at Birmingham. The negroes of the world at this time, when the world is gone wild in its injustice to weaker peoples, greet you as a wise and great statesman, and feel that with principles such as you stand for humanity will lose its prejudice and the brotherhood of man will be established. All true negroes are against social equality, believing that all races should develop on their own social lines. Only a few selfish members of the negro race believe in the social amalgamation of black and white. The new negro will join hands with those who are desirous of keeping the two opposite races socially pure and work together for the industrial, educational and political liberation of all peoples.

"The negro peoples of the world expect the South of the United States of America to give the negro a fair chance, and your message of today shall be conveyed to the four hundred millions of our race around the world. Long live America! Long live President Harding in his manly advocacy of human justice! I have the honor to be your obedient servant, MARCUS GARVEY,

"President General, Universal Negro Improvement Association and Provisional President of Africa."

October 27, 1921

ASKS NEGRO VOTERS TO CUT PARTY LINES

James Weldon Johnson Says They Must Break Republican Domination.

KEEP POLITICIANS GUESSING

Can Get Nothing as Long as Bosses Know How They'll Vote, Says Their Leader in "Crisis."

"Keep the politicians guessing," is the advice to negroes of James Weldon Johnson, Secretary of the National Association for the Advancement of Colored People, in an article which will appear in the October issue of The Crisis.

Mr. Johnson calls on negroes to work out by that method their emancipation from the Republican Party, which is trying to hold the negro "and do as little for him as possible."

He points to the negroes who reside in Harlem as a group which has succeeded in working out its own political salvation from Republican bondage.

"The politicians can never foretell to a dead certainty how much of that vote will be Republican, Democratic, Socialist or for the third party," he writes. "Colored Harlem is now represented in the Legislature of New York State by a negro Democrat and in the Aldermanic Board of New York City by a negro Democrat. Has this made the Republican Party less solicitous about negro votes? On the contrary, the hard-boiled Republican machine has just designated for that district a colored man as candidate for Congress."

A "gentleman's agreement" against the negro has been concluded by the leaders of the two older parties, Mr. Johnson writes. This agreement, he declares, is made possible by the disfranchisement of 4,500,000 negroes in the South through the "white primary" and with the connivance of the Republican politicians in Southern States, who are determined not to build up a strong organization, as they would be in a position to do with the aid of the negroes, because if they did there would be too many to divide the Federal patronage dispensed by a Republican Administration at Washington.

Seek Only aPtronage, He Says.

"The Republican Party in the South is not a political party; it is an office-holding oligarchy," Mr. Johnson writes. "The bosses are not interested in building up a party; they are interested solely in taking a hand-picked delegation every four years to the National Convention, and landing on the band wagon. If a Republican President is elected these bosses have all the Federal jobs in the whole empire of the South to parcel out among themselves and their friends. These are fat pickings and are exclusively reserved to relatively a very few persons, for the mass of white Southerners are barred by being Democrats. It is here we have the reason for the rise and growth of Lily-White-ism: the white men in the game simply wanted all the jobs.

"Indeed not only do the Republican bosses in the South neglect to build up a strong party—they could make a fair beginning with nearly five million negro voters to draw from—but a strong Republican Party is precisely what they do not want; such a party would develop too much competition for the Federal jobs. These bosses, without protest, allow the white South to control the local situation and reduce the negro to a political zero in exchange for full control of Federal patronage. This arrangement suits the white South. It is not considered too great a price to pay for the elimination of the negro."

"Where negroes can vote, in the North, the race is still in the Fourth of July stage of politics. The gentlemen's agreement between the Republican and Democratic Parties is possible because practically every negro vote is labeled, sealed, delivered and packed away long before the election. How can the negro expect any worth-while consideration for his vote as long as the politicians are always reasonably sure as to how it will be cast? The Republicans feel sure of it, and the Democrats don't expect it."

Asks for Action This Fall.

The way out for the negro, Mr. Johnson counsels, is for the negro to smash the gentlemen's agreement. It can be smashed in two ways, he concludes:

"In the gradual assertion of political independence on the part of the negro, or it can be done at one blow. The negro can serve notice that he is no longer a part of the agreement by voting in the coming elections in each State against Republicans who have betrayed him, who are in league with the Ku Klux Klan, who are found to be hypocrites and liars on the question of the negro's essential rights and by letting them know he has done it. I am in favor of doing the job at once."

September 21, 1924

NEGRO CITY IN HARLEM IS A RACE CAPITAL

Community of 175,000, Now Largely in Negroes' Hands, Has Become a Cultural Centre

THE following article is based on material contained in the current issue of the Survey Graphic, devoted to Harlem and the negro culture it represents.

By MARY ROSS.

BLACK fingers whipping furiously over the white keys, beating out cascades of jazz; black bodies swaying rhythmically as their owners blow or beat or pluck the grotesque instruments of the band; on the dancing floor throngs, black and white, gliding, halting, swinging back madly in time to the music—this is the Harlem of the cabarets, jazz capital of the world.

Downtown specialists who have wearied of the tricks of Broadway come northward to this new centre of pleasure. In some of its fifteen cabarets black and white eat together, dance together in the rich abandon of the race which evolved that first jazz classic, the Memphis Blues; which refined the cakewalk into the fantastic fling of the "Charleston." But this only one Harlem. Jazz is only the foam on the surface. Beneath it lie depths of which New York, rumbling under it on the subway or past it in elevated or bus, little dreams.

Fifteen years ago Harlem looked much as it does today on the surface—a prosperous district north of Central Park, with brownstone houses bordering the broad streets, with "new-law" tenements, costly churches, magnificent avenues. Originally a Dutch village, it had become in turn predominantly Irish, then Jewish and Italian. About 1904 a few negroes began to trickle in east of Lenox Avenue. The district had been overbuilt, and a negro real estate operator, Philip A. Payton, offered to fill the vacant houses with self-respecting negro tenants. As the move of the more enterprising negroes northward from the old negro district in West Fifty-third Street became more and more pronounced, there was first organized opposition by the whites, then a panic which left rows of houses vacant. If one negro family moved into a block every one else moved out.

With the war a black tide of laborers rolled in from the South and the West Indies to the metropolitan district, where work was waiting at wages beyond their rosiest dreams. The white resistance in Harlem broke. And then a strange thing happened. According to all popular legends, those negroes who were earning big money for the first time in their lives should have spent it all on silk shirts and white buckskin shoes. But they did no such thing. They bought real estate.

Realty Ownership Shifts.

There was a feverish taking up of those handsome houses and shops and churches of Harlem, which in the war slump could be had at prices far below their actual value. During the height of that fever it was no unusual thing to see a negro cook or washerwoman or workman go into a real estate office and plunk down $1,000, or $2,000, or $5,000 on the price of a house. Fifteen years ago there were only a few negroes in all Manhattan who owned real estate. Now it is estimated conservatively by John E. Nail, a negro real estate operator in Harlem, that the property owned and controlled by negroes in that district alone exceeds $60,000,000 in value. Negro Harlem is practically owned by members of that race.

The solid current of opportunity beneath the beat of the jazz has lured the most enterprising negro men and women from all corners of the world, till Harlem now is a city of 175,000 negroes, larger than Worcester, Mass., or Dallas, Texas. They have poured in from the South. Probably there are no less than 25,000 negro Virginians in New York City, more than 20,000 North and South Carolinians, and 10,000 Georgians. They have come from every island of the Caribbean Sea. More than 25,000 West Indians have arrived in the past four years. A few even have come over the old slave trail from Africa, voluntary immigrants this time. Within the seventy or eighty blocks of Harlem for the first time in their lives negroes formerly living under the Spanish, Dutch, French, Arabian, Danish, Portuguese and British flags, as well as those native to Africa, meet and move together in one of the largest negro communities ever known.

The West Indians are the business men of the group. Coming from countries where there have been fewer discriminations against their choice of occupation than in the United States, they have not been content to follow the unskilled trades and personal service jobs. They have been compared to the Jews in their tendency to set up in independent retail business—grocery stores, tailor shops, jewelry stores, fruit markets. They are numerous in the fields of real estate and insurance. They are saving, ambitious, aggressive. It is due to the enterprise and energy of women from the West Indies that the needle trades have been opened to negro women.

For, all in all, negro workers in New York City are shunted chiefly into "blind alley" jobs. By the last census count half the negro men at work in New York City were porters, waiters, messengers, elevator tenders, chauffeurs and janitors. More than half the negro women were laundresses and servants. It is hard for negroes to get into the trades requiring long periods of apprenticeship, and some of the trade unions discriminate against them. Yet of the 5,387 negro longshoremen at work in 1920 (the largest number of negro men in any one occupation and more than a tenth of all negro workmen in the city) about 5,000 were organized.

In certain of the skilled trades, too, negroes are becoming more common. Between 1910 and 1920 the number of negro carpenters was trebled: electricians, machinists and musicians more than doubled in number; the number of shoemakers jumped from 14 to 581; stationary firemen from 249 to 1,076, real estate agents from 89 to 247, and 462 negro mechanics, 2,685 workers in the textile industries and more than 6,000 clothing workers appeared where there had been practically none ten years earlier.

The most enterprising, ambitious and adventurous of the race come to New York—to Harlem–as a golden dream of opportunity. In the aggregate, they find it. Individually they are likely to get a job as a porter and rent a room or a flat that absorbs the greater part of their wages. Rents are high everywhere in New York. In Harlem they are exorbitant. A Judge who handles rent cases in the Municipal Court in Harlem says that negroes pay twice as much as white tenants for the same apartments. Result: the family takes in lodgers until the flat is crowded to twice its capacity, or they go without fun and food and clothing to stretch their money over the rent. Sometimes one room will be let out one the double shift principle—one tenant by day (a night worker, ofcourse), another by night. Black owners as well as white have taken advantage of the pressure on Harlem to turn easy pennies.

As a result, at least in part, of these conditions, comes much of the "running wild" of Harlem. It is the old story of the tired business man–dull by day, reckless by night. It is behind the development of the cabarets, of the street gambling on "numbers" (the day's total of bank exchanges and bank balances announced by the Clearing House each day in the newspapers), in which even the poorest wager their nickels and dimes: the brass band parades which liven Lenox Avenue. The city has many pitfalls for the negroes who hurdle a half century of economic development at one leap, coming from their little quiet country towns or tropical islands to a city on edge with competition, ready to exploit their color, their credulity, their simple faith in dreams or "yarbs." The medicine men of Harlem–many of them cannot be dignified as doctors or druggists—and their preying on the simple immigrants would make a story in themselves.

Harlem Jazz.
Winold Reiss, in the Survey Graphic.

To one institution the negro has held tightly under the new conditions of city life. That is his church. The last Sunday of September, 1924, was a dramatic day in Harlem. Members of the Salem Methodist Episcopal Church, a congregation of negroes, had assembled as usual in rooms where they had worshiped for fourteen years—a converted apartment house in which partitions had been torn down to allow for the expansion from a mission to a church. At a designated hour they rose and marched quietly up Seventh Avenue to the Metropolitan Methodist Episcopal Church, where a white pastor and his white congregation waited to turn over the buildings–church, parish house and parsonage–which the negro group had purchased. There were representatives, negro and white, from the common denominational board to witness this historic event. After addresses of welcome and hymns, the outgoing Board of Trustees turned over the keys of the church to the incoming board, and here was another imposing negro church in Harlem.

One of the earliest of the large negro churches in this part of the city is St. Philip's Protestant Episcopal Church, erected fifteen years ago on 133rd Street under the supervision of a negro architect, after the congregation had sold its buildings in the Pennsylvania zone for a large sum. The Abyssinian Baptist Church sold its building in Fortieth Street and built a church and community house on 138th Street at a cost of about $325,000. "Mother Zion" built in 136th Street twelve years ago, and has recently added a new structure, running through to 137th Street. St. Mark's Protestant Episcopal Church is building in 138th Street at a cost of half a million. The Grace Congregational Church of Harlem has taken over the building of a Swedish congregation west of Eighth Avenue. The imposing Lutheran Church at Edgecombe Avenue and 140th Street has been bought and occupied by the Calvary Independent Methodist Church, and recently it was announced that the Mount Olivet Baptist Church had purchased for $450,000 the white limestone Adventist Temple in 120th Street. The seating capacity of the churches and missions of Harlem is estimated at 24,000 by George E. Haynes, Secretary of the Commission on Church and Race Relations of the Federated Churches of Christ in America. Around these churches centres much of the social and cultural life of Harlem.

An Intellectural Center.

Far beyond its physical setting of a negro capital in brick and stone, Harlem has become an intellectual capital, a Paris of the race. The negro race has furnished the stage such actors as Paul Robeson, Gilpin, Bert Williams; its musicians include Roland Hayes, Harry Burleigh, Nathaniel Dett; with writers and poets such as DuBois, Paul Laurence Dunbar, Claude McKay, Countee Cullen, Walter White, James Weldon Johnson. Probably no form of race segregation has bitten deeper into the consciousness of the leaders of the race than the intellectual isolation to which its more talented members often were condemned because they were barred from other racial groups with like interests. They cannot enter a downtown restaurant without fear of insult or ejection. The first centre in New York of fashionable and intellectual negro life was the Marshall, a hotel which flourished seven or eight years ago in the West Fifty-third Street district. Artistic achievement is no new attainment of the negro race. The famous Barnes Foundation at Merion, Pa., has subtle and delicate sculptures made in Africa centuries before the first colonists landed on American shores. But Harlem is the setting for a new negro art, built on race consciousness, influenced by the America of which it is one element.

While consciousness and pride of race are characteristics of the younger generation of negro intellectuals, who ask neither partronage nor philanthropy but the chance to work out their own racial salvation, there is another group who take the easiest way out. In a city such as New York, where there are many swarthy races, it is comparatively easy for the lightest of the negro group to deny their negro blood. This is known as "passing." Walter F. White, novelist and Assistant Secretary of the National Association for the Advancement of Colored People, is authority for the statement that in New York there is a well-known surgeon who grew tired of the proscribed life of a negro in a Southern city and moved North and forgot his negro blood. At least one man high in the field of journalism, a certain famous singer, several persons prominent on the stage and many others pass for white despite their mixed race, he says. A very few of those who might escape the racial yoke do so, though the total probably is much larger than is popularly supposed.

To many more, equally gifted, Harlem is the proud capital of their race—something entirely new on the face of the earth–a modern negro metropolis, a city of movement, gayety and color, singing, dancing, boisterous laughter and loud talk; of luxurious homes and shops that would do credit to Fifth Avenue; of churches and cabarets and overcrowded tenements; the meeting place of the negroes of the world, where the habits and customs of half a dozen different decades are suddenly cast into one of the liveliest corners of the twenty-fifth year of the twentieth century. Harlem is not a ghetto or a quarter or a slum. It is an American city, in tones of black, brown and yellow. The negroes have bought it; can they hold it?

Yes, answers James Weldon Johnson, who speaks for his race as journalist, editor, poet, Executive Secretary of the National Association for the Advancement of Colored People. When the negroes leave Harlem, Mr. Johnson predicts, it will be because the property they have bought has become so valuable that they no longer can afford to hold it; they will sell it at an even greater profit than the transactions of the past few years have registered, with the power to choose another favorable home. But another move northward seems remote.

In most of the other Northern cities into which negroes have migrated they are employed in gangs in great industries, such as the steel mills of Pittsburgh or the automobile plants of Detroit. In a period of economic depression they are the first to go, often losing all they have gained in prosperity. But negro labor in New York is scattered and diversified. There is little danger here that thousands of negroes will be thrown out of employment at once.

Stability of the Colony.

Their city has therefore stability; it can dig in, anchored to its own homes, churches, shops and theatres, with its own working classes and business and professional groups. It is not a section shut off, but a zone through which run the main thoroughfares of the city; its people speedily become New Yorkers. With the passing of the first indignation of the white residents, many of those who did not escape at the time have settled down to live side by side with the negroes, in some cases occupying apartments in the same buildings. White merchants still do business there. As a result of these conditions, and especially of the fact there are large numbers of negroes on the police force in that district, there is perhaps less racial friction than is observable in any other part of the country where there are large groups of negroes. In Harlem there are 75,000 more negroes than in any Southern city, yet there is little irritation or disorder, Harlem is not typical, but it may be prophetic.

March 1, 1925

URGES NEGRO TO ORGANIZE.

De Priest Makes an Address at Knoxville.

KNOXVILLE, Tenn., June 24 (.T).—Oscar De Priest, negro Representative from Illinois, tonight told an audience of 1,500 negroes that the negro would never gain his political and civil rights under the Constitution until he organized politically. Mr. De Priest received an ovation from a throng of negroes who met him at the train. Following his talk, he was honor guest at a dinner.

Discussing the question of social equality, Mr. De Priest said in his address:

"All that I want for the negro is white equality before the law.

"The American white people are not to blame for the economic, social, political or civil status of the negro. His salvation is laid at his own doorstep."

Mr. De Priest told his audience that he was "not in Congress to put over any special legislation in behalf of the negro."

Mayor James A. Fowler, former solicitor general for the United States, introduced Mr. De Priest.

June 25, 1929

NEGRO SEEN AS GAINER IN THE RECOVERY DEAL

What has the American Negro got out of the New Deal? What further may he expect to receive as the various recovery measures are more fully developed? These questions are dealt with in the following article by the adviser on the economic status of Negroes in the Department of the Interior.

By CLARK FOREMAN.

MUCH has been said and written about the effect of the recovery program on Negroes. There has been little realization, however, of the deeper implications for minority groups of the integration of our national economy, which is proceeding apace in Washington.

It is said that in many cases the Negroes have suffered rather than benefited from the NRA and other recovery programs. Doubtless in some cases this is true. What is by no means always true is that this suffering is the result of race prejudice or discrimination.

If we examine the occupational distribution of the Negro population in 1930 we find that of the total 5,503,535 of those over 10 years of age who were gainfully employed, 1,807,521, or 32.9 per cent, were farm tenants and farm workers; 1,387,527, or 25.2 per cent, were domestics, and 1,012,830, or 18.4 per cent, were common laborers. This was the situation at the peak of the laissez-faire period, and it is from this point, intensified after three years of depression, that the recovery program had to begin.

The Negroes made their first effective entry into the industrial life of the North in the years of the World War when immigration was restricted. The number engaged in manufacturing and mining grew from 692,409 in 1910 to 960,039 in 1920—an increase of 35.8 per cent. During the same period the total of Negroes gainfully employed decreased from 5,192,535 in 1910 to 4,824,151 in 1920. After 1920, when the immigration bars were lowered, many Negroes nevertheless held their jobs and successfully demonstrated their economic usefulness.

A Lower Market Price.

Because of prejudice and the related restricting influences on the stability of Negro labor, employers have been able to buy it for less than the usual market price for white labor. This fact has had two definite effects on the situation. First, it has furnished Negroes with an entering wedge whereby they may penetrate the structure so jealously protected by the vested interests of the labor world; second, the very importance of this opportunity has led the employers often to use Negroes as strikebreakers and under-cutters of the established wage scale.

Negroes seem to have held more than their share of the marginal and underpaid jobs in those industries which were most ruthless in their efforts to reduce wages and extend hours. The "chiseling" industries have had the advantage of unusual opportunities for exploitation and have very generally availed themselves of it. With a planned and ordered control of our economic life such industries must be eliminated or changed.

The first efforts of the NRA have severely attacked the exploitative methods and many industries have been forced to better their wage scales. With a codification of hours and wages the particular advantage derived from the employment of Negroes disappears, provided the code is enforced. Thus in many cases Negroes have lost their jobs to whites. The pressure on industries to employ white workers instead of colored exists in most localities, but was fairly well resisted when additional profits could be made by employing Negroes for less.

A Wage Differential.

In those industries which cannot so easily displace their Negro labor for white there has come a cry for a wage differential for Negroes. Some of these industries say, perhaps truthfully, that they cannot make profits on the wage scale set by the NRA. The question then arises as to whether such industries are desirable.

There are, on the other hand, many indications that Negroes are profiting by the code regulations for the larger industries. For example, a Negro investigator (Ralph N. Davis of Tuskegee Institute) wrote from Birmingham recently:

"The attitude of Negro labor to the large industries in most instances, as reflected in the discussions of the investigator with workmen and their reports to local Negro leaders, indicated that they felt that there was no discrimination in the application of the code to Negro labor."

Few of the larger industries have displaced Negroes for white labor and may have added to the number of Negro employes at the code scale of hours and pay.

Thus we see that the effort to regulate industry has caused some Negroes employed by unsound or marginal industries to lose their jobs. It has at the same time greatly helped the Negroes employed by the larger industries. Moreover, with the development of a planned economic life and the prosperity of industry on the basis of fair wages, there is every reason to believe that Negroes will gain directly and indirectly from the more wholesome situation.

A Broader Program.

As the NRA program is extended to domestic servants and farm laborers, the great mass of the Negro population will be more affected. But already the work of the NRA, the PWA and the CWA has produced results that every student of Negro life must welcome, despite the fact that in many cases the effect has not been so complete as was hoped for.

The whole question of the future of the Negro population is intimately involved in the administration's attempt to regulate and plan industry. Although undoubtedly there will be displacement and misfortune in the transitional period, the Negroes along with all minority groups seem certain to benefit from a greater centralized control.

The new policy of our government is particularly important for Negroes. The efforts of the administration to improve housing conditions and to raise educational standards will be especially beneficial to the colored population, which heretofore has been greatly in need of this help. The assistance which health authorities are receiving will enable them to bring about much-needed improvements in general health conditions, and again the Negroes who have been neglected will profit most.

It is only natural that all efforts to raise the purchasing power and economic status of the population above a certain minimum will be welcomed by Negroes who have been kept in the lowest economic brackets.

Unskilled Labor.

There are certain dangers connected with the organization of our national life. With organization there is always a tendency toward crystallization. People unfamiliar with the development of our Negro population are likely to think of Negroes only in terms of unskilled or domestic labor.

According to the 1930 census, 2,400,357 Negroes were so classified. Such figures tend to overshadow the excellent progress of Negroes in the professions and in business. In the seventy years that have passed since emancipation from slavery and enforced illiteracy, Negroes have become prominent in our literary, art and scientific life. Some have also been remarkably successful in amassing fortunes in the real estate and life insurance and other businesses.

Officials who are called on to plan, too often think that the situation of race prejudice can be handled by giving the unskilled job to Negroes and the skilled and white-collar jobs to white people. The discouraging and unfair effect of such settlements is obviously felt keenly by those Negroes who have struggled against many difficulties to prepare themselves for higher work.

Discrimination has occurred in local communities in the administration of several of the government's big programs. Such irregularities, when reported, have often been corrected, but it is not held that even now the various programs work fairly. In too many cases local prejudice thwarts the best plans of the central authority.

Nevertheless, the government is well aware of its obligations to the Negro population and has undertaken them with an energy comparable to its activities in other fields of reconstruction. Every one interested in achieving a fair opportunity for advancement on the part of the colored population must welcome the social theories of the New Deal and the clear intention of the administration to apply these theories without reference to race.

April 15, 1934

Southern Social Relationships

CASTE AND CLASS IN A SOUTHERN TOWN. By John Dollard. 502 pp. New Haven: Yale University Press. $3.50.

By WILLIAM SHANKS MEACHAM

THERE is a classic story about a Governor of one of the States of the deep South who replied to an invitation to attend a conference on interracial relations with the statement that the race problem in his State had been solved. The story is genuinely illustrative of the method through which some political dignitaries solve the problem by rationalization. Nowhere in the South is it really solved, and this method results in precisely the kind of fixation that retards its solution and social progress in Southern regions. Dr. John Dollard's study of the caste and class system of a small Southern town of the Far South is one more book to prove it. Here a social psychologist gets beneath the intellectual surface of the fixed status and discovers some of the reasons for the relative social immobility of the town in the Black Belt. If its outlook has changed little since the Civil War, it has been partly because it has been too much preoccupied in maintaining an elaborate formula of taboos and restrictions at the lines of racial relations to offer hospitality to new ideas.

Life in Southerntown, where the Negro must live across the railroad tracks, is disfranchised, cannot move in any direction outside his own group and finds his economic opportunities extremely limited even within his own caste, is at marked variance with the dominant pattern of American civilization or that of democracy anywhere in the world. Both races are at least subconsciously aware of this. The conflict in mores is met by the white group with a series of defensive beliefs (such as a belief that what is being done is for the good of the colored man), and the Negro is forced into accomodation attitudes. In order that he may be kept "in his place," the middle-class white masters of Southerntown (the few members of the aristocracy have arrived at that bourne of social security where struggle is no longer necessary) maintain at all times an aggressive pressure against the Negro group. Dr. Dollard, who has an excellent command of the techniques of psychoanalytical investigation, gives us the interior psychological view of the working of this system. Through interviews with numerous representative clients, and by systematic observation, he learned how constant was the interplay of emotions in the psychological atmosphere above the racial caste lines. And it may be said that the net result of all the penetrating research that went into this book is to show how wide and deep is the fallacy that the two races in the South can have any distinctly separate destiny.

Dr. Dollard shows how caste distinctions pattern religion and politics, as well as social life, and play upon the economic forces at work in Southerntown. But of course this special social system within American democracy does not function smoothly. It is characterized by a number of forms of aggression. To justify swift action when a colored man oversteps any of the bounds laid down for him, the fear of Negro violence is always exaggerated. The actual violence that is the result of the caste system occurs chiefly in the Negro group, and Dr. Dollard's conclusion that this is the result of a hostility that would be properly aimed at the white group is undoubtedly accurate. Does the white jury of the South unconsciously recognize this fact in its leniency toward Negroes charged with offenses against Negroes? "It is clear," Dr. Dollard writes, "that the differential application of the law amounts to a condoning of Negro violence and gives immunity to Negroes to commit small or large crimes so long as they are on Negroes. . . . One cannot help wondering if it does not serve the ends of the white caste to have a high level of violence in the Negro group, since disunity in the Negro caste tends to make it less resistant to white domination."

There could be no doubt that Dr. Dollard went to the right research site to study the caste system in the South, where it is least adulterated. Nevertheless, it should be noted that there is a middletown of interracial relations in a South of greater social nobility. Southerntown (never located on the map except as a part of the deep South) is both small and wholly rural in outlook, without industry. In the industrial middletown of the South, white labor organizers are today asking Negro workers to join labor unions. This tendency toward the recognition of the identity of economic interests between white and Negro workers is of transcending significance in considering the thesis of caste and class in the South. Collective bargaining was formerly the exclusive right of white workers in the South, yet in Richmond during the past year Negro workers have been out on strike, in the course of exercising this right, and the fact caused no noticeable excitement. While Dr. Dollard's conclusions are essentially valid for the entire South, there is a flexibity of the caste line under industrial pressure, not suggested by anything in the picture of his particular Southern town.

Two excellent appendices add to the value of this book. One by Dr. Dollard presents material on the life history of middle-class Negroes, and Dr. Leonard W. Doob contributes a study of another frustrated Southern group —the poor-white sharecroppers.

December 5, 1937

Harlem Compact Gives Negroes Third of Jobs in Stores There

Uptown Chamber and Racial Group Pledged to End Old Source of Unrest—Mayor Hails 'Common Sense and Justice'

An agreement intended to end racial unrest in Harlem by guaranteeing to Negroes at least one-third of all white-collar jobs in Harlem retail establishments was announced yesterday by the Uptown Chamber of Commerce, acting for hundreds of independent and chain stores, and the Greater New York Coordinating Committee for Employment, representing more than 200 Negro organizations.

Mayor La Guardia hailed the compact as "a tribute to common sense and justice." Its signers on both sides proclaimed it a "historic" step toward improved relations between the races in a district long regarded as a danger zone.

A joint statement by the merchants' group and the coordinating committee emphasized that white employes of Harlem stores would not be forced out of their jobs through the agreement. In establishments where fewer than one-third of the sales persons, clerks and executives are Negroes, members of that race will replace white clerks as the latter quit, are transferred to other branches or are discharged for cause.

The Rev. Dr. William Lloyd Imes, acting chairman of the coordinating committee in the absence of its chairman, Dr. A. Clayton Powell Jr., estimated that stores subscribing to the agreement employed 10,000 persons in the 125th Street area.

The retailers promised not to limit opportunities for advancement for Negroes, to exert pressure on non-cooperating labor unions to persuade them to admit Negroes, to avoid retaliation against Negroes employed in stores outside Harlem and to eschew discriminatory wage scales or disproportionate lay-offs of Negro workers in periods of staff reduction.

The Negro groups, for their part, agreed to abstain from picketing, boycotts and other mass demonstrations against stores, even those not participating in the agreement, until charges of discrimination against them had been sustained by a joint arbitration committee. An arbitration board of ten, five from each side, will settle all disputes and both sides will be bound by its findings.

Cooperating stores will be identified by a special sign and the Negro organizations have promised that they will seek to create more jobs in those stores by promoting Negro patronage. A central employment bureau operated jointly by the New York Urban League, the Harlem Young Men's Christian Association and the Harlem Young Women's Christian Association is to be used by employers in obtaining Negro workers under the agreement.

The agreement was signed after four months of negotiation by Colonel Leopold Philipp, president, and Matthew J. Eder, executive secretary, for the Uptown Chamber of Commerce, and by Dr. Powell and Arnold P. Johnson, executive secretary, for the coordinating committee.

Colonel Philipp and Dr. Imes predicted yesterday that the agreement would abolish prejudice against Negro workers in Harlem and bring a new and more peaceful era to the community's Negro population of 300,000.

Leaders Give Views

"The settlement reached today is historic," they declared. "It is the first agreement of its kind ever negotiated and leaders of both races in Harlem hail it as a constructive step toward more harmonious relations between the two races.

"It will contribute immeasurably to the advancement of Negroes and will help quiet unrest in Harlem because it is proof that white business leaders have a sympathetic interest in the economic problems of the colored race.

"We believe the formula developed for Harlem may also be used to pave the way for a peaceful settlement of the racial employment problem in other large cities. Given equal opportunities for employment, Negroes will improve their economic status and become better citizens and better business men.

"As more and more colored workers are transferred from relief rolls to the security of decent paying jobs in private industry, we shall see a vast improvement in the social, living and economic conditions of Harlem."

Fear of racial uprisings last April, when the people of Harlem threatened reprisals against stores accused of discriminating against Negro workers, brought leaders of both races together in the hope of an amicable settlement.

Among those who praised the agreement were Newbold Morris, President of the City Council; Borough President Stanley M. Isaacs, Representative Joseph A. Gavagan, Mrs. Elinore M. Herrick, Regional Director of the National Labor Relations Board; State Senator Duncan T. O'Brien and Norman Thomas.

The list of chain stores subscribing to the compact, the joint statement said, includes F. W. Woolworth & Co., Ludwig Baumann, S. H. Kress & Co., McCrory's, W. T. Grant & Co., Davega, Busch Kredit Jewelers, A. S. Beck & Co., and the Wise Shoe Company.

August 8, 1938

NEGROES' WAR AID TOLD IN BROCHURE

OWI Gives Official Recognition to Their Role in Civilian Life and the Services

Special to THE NEW YORK TIMES.

WASHINGTON, Jan. 16—The achievements of Negro Americans in many fields are recorded in a booklet issued by the Office of War Information today, in official recognition of their important contribution to the war effort and in the armed forces. The OWI plans to distribute 2,000,000 copies of the brochure.

Under the title, "Negroes and the War," the publication has seventy-two pages and is done in rotogravure with 141 reproductions of protographs. A six-page preface is written by Chandler Owen, Chicago publicist. The booklet tells:

What Negroes are doing in agriculture, industry, and in the armed services; what Negroes have to gain by an American victory, and what they would lose should the Axis win.

In addition, the brochure outlines the progress of Negroes in recent years in education, economically, and in the arts and sciences.

"There were 1,643 students in Negro colleges in 1916," the booklet states. "By 1941 the number had grown to 40,000. There are approximately 100 universities and colleges devoted exclusively to Negro education."

In a section on "The Young Generation on the Land," it says:

"In the seventeen Southern States during the year 1915 only 58 per cent of the Negro children between 6 and 14 were enrolled in school. By the school year 1939-40 some 85.9 per cent of the children between 5 and 17—a much wider range—were regularly in attendance. There were 2,174,260 in elementary school and 254,580 in high school. The number of youngsters in high school has more than doubled in ten years."

January 17, 1943

NEW BOARD SET UP TO END HIRING BIAS

President Appoints Mgr. Haas Head of Committee on Fair Employment Practice

Special to THE NEW YORK TIMES.

WASHINGTON, May 28—President Roosevelt issued an executive order today setting up a new Committee on Fair Employment Practice, with additional powers to prevent discrimination in war industry employment, and appointed Mgr. Francis J. Haas of Catholic University as chairman.

The action was designed to end widespread complaints among some groups, especially Negroes, that effective governmental action against discrimination had become impossible when the old committee was made a part of the War Manpower Commission, headed by Paul V. McNutt. Malcolm S. MacLean, president of Hampton Institute, resigned as committee chairman after Mr. McNutt ordered postponement of hearings into charges that railroads refused to employ Negroes for certain jobs.

The President has not yet appointed the new committee to serve with Mgr. Haas, but it is expected that one of its first official acts will be to reschedule the railroad hearing. Informed sources said that the committee probably would have six members, equally representative of industry and labor, in addition to the chairman.

The power given the new committee to carry out the President's policy forbidding discrimination in war industries because of race, creed, color or national origin consists of the authority to require not only war contractors but subcontractors to include a clause in their contract with the government forbidding such discrimination. The old committee could request inclusion of such a clause only in prime contracts.

THE PRESIDENT'S ORDER

By The Associated Press.

WASHINGTON, May 28—Following is the text of President Roosevelt's executive order establishing a new Committee on Fair Employment Practices:

In order to establish a new Committee on Fair Employment Practice, to promote the fullest utilization of all available manpower, and to eliminate discriminatory employment practices, Executive Order No. 8802 of June 25, 1941, as amended by Executive Order No. 8823 of July 18, 1941, is hereby further amended to read as follows:

"Whereas the successful prosecution of the war demands the maximum employment of all available workers, regardless of race, creed, color or national origin; and

"Whereas it is the policy of the United States to encourage full participation in the war effort by all persons in the United States regardless of race, creed, color or national origin, in the firm belief that the democratic way of life within the nation can be defended successfully only with the help and support of all groups within its borders; and

"Whereas there is evidence that available and needed workers have been barred from employment in industries engaged in war production solely by reason of their race, creed, color, or national origin, to the detriment of the prosecution of the war, the workers' morale, and national unity:

"Now, therefore, by virtue of the authority vested in me by the Constitution and statutes, and as President of the United States and Commander in Chief of the Army and Navy, I do hereby reaffirm the policy of the United States that there shall be no discrimination in the employment of any person in war industries or in government by reason of race, creed, color or national origin, and I do hereby declare that it is the duty of all employers, including the several Federal departments and agencies, and all labor organizations, in furtherance of this policy and of this order, to eliminate discrimination in regard to hire, tenure, terms or conditions of employment, or union membership because of race, creed, color or national origin.

Ruling on Contracts

"It is hereby ordered as follows:

"1. All contracting agencies of the Government of the United States shall include in all contracts hereafter negotiated or renegotiated by them a provision obligating the contractor not to discriminate against any employe or applicant for employment because of race, creed, color or national origin and requiring him to include a similar provision in all sub-contracts.

"2. All departments and agencies of the Government of the United States concerned with vocational and training programs for war production shall take all measures appropriate to assure that such programs are administered without discrimination because of race, creed, color or national origin.

"3. There is hereby established in the Office for Emergency Management of the executive office of the President a Committee on Fair Employment practice, hereinafter referred to as the committee, which shall consist of a chairman and not more than six other members to be appointed by the President. The chairman shall receive such salary as shall be fixed by the President, not exceeding $10,000 per year. The other members of the committee shall receive necessary traveling expenses and, unless their compensation is otherwise prescribed by the President, a per diem allowance not exceeding $25 per day and subsistence expenses on such days as they are actually engaged in the performance of duties pursuant to this order.

"4. The committee shall formulate policies to achieve purposes of this order and shall make recommendations to the various Federal departments and agencies and to the President which it deems necessary and proper to make effective the provisions of this order. The committee shall also recommend to the chairman of the War Manpower Commission appropriate measures for bringing about the full utilization and training of manpower in and for war production without discrimination because of race, creed, color or national origin.

"5. The committee shall receive and investigate complaints of discrimination forbidden by this order. It may conduct hearings, make findings of fact, and take appropriate steps to obtain elimination of such discrimination.

Status of Old Committee

"6. Upon the appointment of the committee and the designation of its chairman, the Fair Employment Practice Committee established by Executive Order No. 8802 of June 25, 1941, hereinafter referred to as the old committee, shall cease to exist. All records and property of the old committee and such unexpended balances of allocations or other funds available for its use as the Director of the Bureau of the Budget shall determine shall be transferred to the committee. The committee shall assume jurisdiction over all complaints and matters pending before the old committee and shall conduct such investigations and hearings as may be necessary in the performance of its duties under this order.

"7. Within the limits of the funds which may be made available for that purpose, the chairman shall appoint and fix the compensation of such personnel and make provision for such supplies, facilities and services as may be necessary to carry out this order. The committee may utilize the services and facilities of other Federal departments and agencies and such voluntary and uncompensated services as may from time to time be needed. The committee may accept the services of State and local authorities and officials, and may perform the functions and duties and exercise the powers conferred upon it by this order through such officials and agencies and in such manner as it may determine.

"8. The committee shall have the power to promulgate such rules and regulations as may be appropriate or necessary to carry out the provisions of this order.

"9. The provisions of any other pertinent executive order inconsistent with this order are hereby superseded."

FRANKLIN D. ROOSEVELT.

The White House, May 27, 1943.

May 29, 1943

REPORT OUTLINES REMEDY FOR RIOTS

Urban League Inquiry Traces Troubles in Detroit to Neglect of Negroes

The National Urban League, following an investigation of the race riots in Detroit, issued a report yesterday which stated that back of the rioting were very poor and inadequate housing for a rapidly growing Negro population, discrimination in employment, and a feeling among the rank and file of Negroes that they have been "entirely neglected by the white community and by much of the Negro leadership."

The report outlined immediate steps to curb influences that tend toward lawlessness and racial conflict and a long-term program "of social betterment for Negroes, and education of both Negroes and whites in the fundamentals of good citizenship."

The report said the rioting started late in the evening as thousands of Detroit citizens returned from a Sunday at Belle Isle Park. "At least 75 per cent of the vast crowd at the island were Negro workers in Detroit war industries and their families," it said. Fighting broke out on the bridge between the island and the mainland and soon wild rumors spread throughout the city, which gave rise to rioting in many places that continued until Tuesday, when Federal and State troops arrived. Thirty-one, including twenty-four Negroes, died, 700 were wounded and 1,400 jailed.

Points to Rapid Influx

Discussing underlying causes, the report said "pressure of population, due to the influx of new workers, black and white, may be considered the first contributing cause." In 1910 there were 5,741 Negroes in the city, in 1920 there were 40,838 and in 1930 there were 120,066, with the 1940 census reporting 149,120, a number that has increased materially in the last two years, the report said.

Declaring there was a feeling that the police had shown racial discrimination in arresting 1,200 Negroes and only 200 whites, the report urged that the full strength of the police department and courts should be marshaled to preserve order and respect for the laws and that "public officials lacking in courage and willingness to discharge their responsibilities should be removed." It recommended an increase in Negro police officers beyond the present forty.

Better Housing Is Urged

Troops should be retained in the city at least ninety days, the report said. It recommended that "efforts to increase housing facilities for Negro and white war workers should be vigorously pushed" and asserted that "a study of the operations of the city's division of sanitation will reveal that the concentrated Negro residence areas are neglected in matters of garbage and rubbish removal."

The long-term program proposed that education in race and human relations should be developed in the city, as "hatreds and animosities which have been known to exist for so long cannot be eradicated in a short time." The report called upon all social agencies to devote serious study to the problems and praised the United Automobile Workers of America and other labor groups for promoting interracial understanding.

July 3, 1943

WARNS OF REDS' INROADS

Porters' Head Says They Are Trying to Organize Negroes

A. Philip Randolph, head of the Brotherhood of Sleeping Car Porters, joined yesterday with Walter White, executive secretary of the National Association for the Advancement of Colored People, in warning that Communists are trying to organize a strong left-wing element among Negroes.

He pointed to the recent election of Benjamin Davis to the New York City Council as an illustration of how Communists are backing Negroes in politics. Mr. Randolpn also asserted that Communists were creating good-will for themselves among Negroes in other fields as well.

To counter this trend, he suggested that the other political parties assume a realistic approach to the Negro problem and make a sincere effort to combat existing prejudices.

December 22, 1943

Democracy—the Negro's Hope

AN AMERICAN DILEMMA: THE NEGRO PROBLEM AND MODERN DEMOCRACY. By Gunnar Myrdal with the assistance of Richard Sterner and Arnold Rose. 2 volumes; 1,483 pp. New York: Harper & Brothers. $7.50.

By FRANCES GAITHER

AS a young social scientist on his first visit to America back in the Twenties, Gunnar Myrdal was challenged on all sides to say what was "wrong with this country." Like Bryce before him, he saw in our eagerness to be criticized, however embarrassing to a visitor, a healthful and hopeful sign in a young nation. More lately, called from Sweden by the Carnegie Corporation to delve into the "least clean corner of the national household," the American Negro problem, Dr. Myrdal has set himself to be as thorough as possible.

In the mirror he holds up we appear as the most idealistic of all the Western nations, ready to a man to recite the doctrines which make up what the author names and honors as the American creed. But we do not always by any means practice the high truths we preach. Nearly any American will, for instance, declare himself against discrimination in principle—even the convinced segregationist has his "separate but equal" slogan—but when the color question presents itself to the average man personally, in his own neighborhood, his own factory, he often behaves in direct opposition to his declared beliefs. Thus, the conflict is here conceived not simply as between one group and another but within the soul of every American. "An American Dilemma" is, therefore, a study of our split personality.

For this woeful gap between our rational creed and our irrational customs, Dr. Myrdal holds the white man responsible. If within this most democratic of nations there are two castes with an impassable gulf between, it is because the white man says there must be. His feelings, his wishes determine the rules. He alone has power to alter them, if ever they are altered. Quoting James Weldon Johnson ("The main difficulty does not lie so much in the actual condition of the blacks as it does in the mental attitudes of the whites"), Dr. Myrdal examines these attitudes one by one, to discover, if he can, what part of their content stands the test of reality—and what stems from myth.

Incidentally, one of the interesting services performed by this book is the running commentary it offers on American social science to date, from Sumner with his fixed conviction that "stateways cannot change folkways" through those more recent adventures in field research which have produced such reports as "Deep South" by three of the Warner group and John Dollard's excellent "Caste and Class in a Southern Town." Dr. Myrdal believes that the theory of folkways and mores, proper "for studying primitive cultures and isolated, stationary folk-communities under the spell of magic and sacred tradition" must be applied with extreme caution to "a modern Western society . . . characterized by . . . a virtually universal expectation of change and a firm belief in progress."

He finds that the degree to which caste control is felt to be necessary varies in emotional intensity, not only from region to region (and within each region from class to class) but even within the breast of every man from the most to the least prejudiced. The anti-amalgamation maxim is among white men generally "held holy," a "consecrated taboo," of all others the most fiercely defended, so fiercely that even Dr. Myrdal grants this may be "a remnant of something really in the mores." Second in emotional intensity he places all those ideas clustering about the phrase "social equality" which, in the South particularly, have given rise to an elaborate caste etiquette that even the most liberal white man does not often or publicly dare infringe. Other white requirements Dr. Myrdal classes in descending order: segregation in public facilities, disfranchisement, inequality before the law and finally discrimination in all that relates to earning a living.

Just here a point is made which, to this reader at least, is the most hopeful in implication of the whole study: The Negro's desires are graded in exactly the reverse order—that is, his strongest wants, his direst need, lie in the field where white men find it easiest to reason and so to grant concessions; and he wants least, or usually not at all, that which touches the white man's central taboo. If this is true, may not the area of real agreement be steadily widened? And if every rise in the Negro's real status is accompanied (as Dr. Myrdal believes it is) by a corresponding fall in prejudice, surely every step the white man takes in objective fairness will bring him nearer, too, to inner unity.

Not all of Dr. Myrdal's conclusions are so hopeful. In the South—where once the human ties at least were strong—he sees the distance between the two races widening, the emotional tensions growing. He fears clashes. But basically, one feels, he counts on the American Creed to make us whole at last.

It is practically impossible in a brief review to give any adequate idea of the scope and impact of this overwhelming book, to more than hint at what a challenge to social self-analysis it is. Its very [illegible]ze defies summarization—4[illegible] [illegible]napters followed by over [illegible] fine-printed pages of appendices, bibliography and footnotes, where, by the way, lies some of the best material. Every aspect of the tangled problem has the light directed on it in turn: birth control, slum housing, sharecroppers, suffrage restrictions, institutions in the Negro world and so on. Only the hardiest appetite can digest it all. Perhaps only a student of social science will hope to. But it is a book which nobody who tries to face the Negro problem with any honesty can afford to miss.

Inevitably one wishes that Dr. Myrdal could have relied relatively less on other sociologists and more on first-hand experience—could have attended more luncheons like the one in the Deep South (p. 32) at which he so shrewdly knew how to make men lower their social guard and show him their true feelings; or that at least, since he sees our caste society as the direct descendant of slavery, he had not so largely dispensed with the historical approach. His thinking would be warmed and enriched by the feeling comment of those travelers, by no means socially unobservant, who moved among us in other times—such as Dr. Myrdal's distinguished countrywoman, Frederika Bremer, to name but one. And perhaps more space is given to arguing theories of how far racial differences may be real or imagined, innate or environmentally acquired, than is strictly necessary or relevant—since few readers of this book, surely, can fail to agree to the premise that "it is highly improbable that such differences . . . could justify a differential treatment in education, suffrage and entrance to various sections of the labor market."

"Negro Boy," by Joe Jones.

April 2, 1944

SEGREGATION RULE TESTED

Mixed Group Makes Bus Trip in South—Twelve Arrests

Following a fourteen-day bus trip through the Upper South by a mixed group of whites and Negroes, the Fellowship of Reconciliation issued yesterday a statement asking bus passengers to ignore the race-segregation pattern in many Southern communities.

Basing procedure on a United States Supreme Court decision in 1946, which was said to outlaw racial segregation for interstate bus passengers, the mixed group violated the race-segregation pattern in various places. Most bus passengers were declared to be apathetic about segregation.

Members of the group were arrested on twelve occasions in Virginia and North Carolina. Two members were convicted and sentenced to thirty days in jail in Asheville, N. C.

April 28, 1947

TRUMAN ORDERS END OF BIAS IN FORCES AND FEDERAL JOBS; ADDRESSES CONGRESS TODAY

PRESSES FOR RIGHTS

President Acts Despite Split in His Party Over the Chief Issue

LITTLE 'FEPC' IS CREATED

'Merit, Fitness' Set as U. S. Employment Guides—Military Equality Is Demanded'

By ANTHONY LEVIERO

Special to THE NEW YORK TIMES.

WASHINGTON, July 26—President Truman ordered today the end of discrimination in the armed forces "as rapidly as possible" and instituted a fair employment practices policy throughout the civil branch of the Federal Government.

On the eve of his appearance before Congress, the President issued two executive orders to carry out his sweeping aims. He said that men in uniform should have "equality of treatment and opportunity" without regard to race, color, religion or national origin.

Similarly, he decreed that "merit and fitness" should be the only application for a Government job, and that the head of each department "shall be personally responsible for an effective program to insure that fair employment policies are fully observed in all personnel actions within his department."

The two orders were expected to have a thunderbolt effect on the already highly charged political situation in the Deep South, a situation which is expected to be aggravated further tomorrow when Mr. Truman makes his omnibus call on Congress for action. The message, in one of its eleven major elements, is expected to go down the line for his ten-point civil rights program, which last February started the deep fissures in the Democratic party.

Enforcement Machinery Set Up

The Presidential orders, which require no Congressional sanction, specified in detail the machinery that would be employed to monitor both anti-discrimination programs.

In the National Military Establishment, Mr. Truman created an advisory panel, called the President's Committee on Equality of Treatment and Opportunity in the Armed Services. It will consist of seven members, none of whom was named today.

It was said, however, that one man who probably would be recommended for membership is Dr. Frank Graham, president of the University of North Carolina. It was believed he would be acceptable to North and South, Negro and white.

The civilian employe order directed that a Fair Employment Board be formed from among members and employes of the Civil Service Commission. This, too, is to be a seven-member board, as yet unnamed.

The committee of the armed forces received the mission of determining how present practices might be altered to carry out the Presidential order. In stipulating rapid application of the policy, Mr. Truman said that it should be done with due regard to "the time required to effectuate any necessary changes without impairing efficency or morale."

Hearings, Appeals Provided

In the civil departments, the head of each was directed to designate an official as "Fair Employment Officer," who was charged with full operating responsibility for the non-discrimination program. Provision was made for hearings of complaints, appeals and disciplinary action.

As the top agency in this program, the Fair Employment Board in the Civil Service Commission received a six-point program providing for review of decisions, drafting of regulations, advice on problems to all departments, publication of information relating to the program, coordination of the policy in the departments and reports to the Commission and to the President.

While the orders did not so state, they implied that they were designed to deal with the Negro problem. Both orders have their roots in the President's civil rights program, and further, in the report of his Committee on Civil Rights, which was issued late last year.

In recent months, Mr. Truman has been caught between two fires on the civil rights issue. The more extreme Southern Democrats, arguing that the program infringed on states rights, have named their own Presidential and Vice Presidential candidates. Their feelings, it was predicted, would become even more exacerbated by the executive orders.

On the other hand, the President has been under pressure by organizations with opposite views. For instance, A. Philip Randolph, president of the Brotherhood of Sleeping Car Porters, AFL, recently visited Mr. Truman with a disturbing report.

He told the President that there was a widespread feeling among Negroes that they would not obey the draft unless discrimination was ended in the armed forces. Mr. Truman asked him if they, too, were not Americans who should loyally serve their country.

More recently, Representative Leo Isacson, American Labor, of New York, demanded that Mr. Truman start his civil rights program in the Government, where he did not need the authority of Congress. Mr. Isacson is prominent in Henry Wallace's Progressive party, which put proposals similar to the President's in its platform.

A federal official interested in promoting the rights of Negroes said tonight that while the orders were a step in the right direction, they called for the end of "discrimination," but made no mention of ending "segregation." Proponents of the Negro cause declare that segregation is prima facie evidence of discrimination.

This official recalled that integration of Negroes in the Army went down to companies who served alongside of companies of white troops. He believed that the President's order on the armed forces, which was not specific on the degree of integregation or mixing to be attained, might push the line down to platoon.

He did not believe that it would go down to the squad, the smallest unit, or twelve men, as that would involve white men and Negroes eating and sleeping in the same quarters. The official expressed the opinion that high officers would not for a long time carry intègration lower than the platoon.

In the absence of a deadline for action in the military order and its omission of the question of segregation, the official saw "a pretty good political approach," which should not too greatly affront Southern Democrats.

July 27, 1948

'INTEGRATION' HELD NEW AIM FOR NEGRO

'Separate but Equal' Theory Is Dropped, Walter White Says —Group Convenes Today

ST. LOUIS, June 22 (AP)—The National Association for the Advancement of Colored People has abandoned its "separate but equal" theory in favor of a goal of total integration in its fight to abolish racial discrimination.

Walter White of New York, executive secretary of the association, made the statement here today at a press conference preceding the organization's annual convention, which opens tomorrow.

Mr. White said "the myth of white supremacy is crumbling but has not been crushed."

Unless America can show its good faith toward the Negro, Mr. White said, the country stands to lose support from the dark-skinned countries of the world.

"These people are skeptical because of what they have heard about the treatment of Negroes," he said.

Mr. White said there were 2,000 Negroes enrolled in universities in nine Southern states and litigation over admission of Negroes was pending in five other Southern states. He said the association would try to increase materially the number of voters in the seventeen Northern and Border states.

June 23, 1953

TUSKEGEE OMITS 'LYNCHING LETTER'

Southern Institute Proposes a New Annual Report on Racial Relations

Special to THE NEW YORK TIMES.

TUSKEGEE, Ala., Dec. 30—Tuskegee Institute today announced its abandonment of its annual Lynching Letter.

"Lynching * * * as a barometer for measuring the status of race relations in the United States, particularly the South, seems no longer to be a valid index to such relationships," Tuskegee's Department of Records and Research observed in a letter released by the institute's president, Dr. L. H. Foster.

For the second straight year, Tuskegee reported that no lynchings had been recorded in the United States. During the years 1949-53 there had been only six lynchings, the report added.

As a substitute for the annual Lynching Letter, issued for the last forty-one years, Tuskegee propeses to publish a new annual report upon which the applicable year's race relations will be based.

"This can and should be as objective and as factual as were the lynching reports," Dr. Foster's letter stated.

The new standard for race relations will be based on employment and other economic conditions, politics, education and health, as well as possible other fields "significant for the present times," the Tuskegee report said.

Will Note Future Lynchings

Dr. Foster added, however, that Tuskegee's Department of Records and Research would continue to record any future lynchings. The report noted three instances in 1953 in which lynchings had been prevented. It said:

"On Jan. 17, near Mobile, Ala., Henry Lee Brown, 17-year-old Negro, escaped from sheriff's deputies while being taken to Kilby Prison near Montgomery. Convinced that he was about to be lynched when his handcuffs were removed and the deputies stopped their cars, the prisoner jumped from the car in which he rode and ran into the swamps, while shots were fired at him. He later gave himself up, was tried and acquitted of the charge of slaying a white woman.

"In March, near Willcox, Ariz., Arthur Thomas, 29-year-old Negro migrant cotton worker, accused of murdering a white woman storekeeper, was saved from lynching by an officer of the law.

"On May 24, at New York City, Edward Cartagena, 42-year-old Puerto Rican, was rescued from a mob by two umpires of a baseball game and the mounted police. Before several hundred persons watching the game in Central Park, Cartagena stabbed his estranged common-law wife to death. Players and spectators tried to kill him."

New Standard Planned

The report added:

"Lynching as traditionally defined and as a barometer for measuring the status of race relations in the United States, particularly in the South, seems no longer to be a valid index to such relationships. This is due to significant changes in the status of the Negro and to the development of other extra-legal means of control, such as bombings, incendiarism, threats and intimidation, etc.

"We believe that a new standard of measuring race relations is needed. This can and should be as objective and as factual as were the lynching reports. This tandard, we think, can best be established in such areas as employment and other economic conditions; in political participation; in education; in law and legislation; in health and perhaps in other fields.

"We propose then, in future annual releases, to issue a statement with information as significant for the present times as was the annual Lynching Letter in the past."

December 31, 1953

HIGH COURT BANS SCHOOL SEGREGATION; 9-TO-0 DECISION GRANTS TIME TO COMPLY

1896 RULING UPSET

By LUTHER A. HUSTON

Special to The New York Times.

WASHINGTON, May 17—The Supreme Court unanimously outlawed today racial segregation in public schools.

Chief Justice Earl Warren read two opinions that put the stamp of unconstitutionality on school systems in twenty-one states and the District of Columbia where segregation is permissive or mandatory.

The court, taking cognizance of the problems involved in the integration of the school systems concerned, put over until the next term, beginning in October, the formulation of decrees to effectuate its 9-to-0 decision.

The opinions set aside the "separate but equal" doctrine laid down by the Supreme Court in 1896.

"In the field of public education," Chief Justice Warren said, "the doctrine of 'separate but equal' has no place. Separate educational facilities are inherently unequal."

He stated the question and supplied the answer as follows:

"We come then to the question presented: Does segregation of children in public schools solely on the basis of race, even though physical facilities and other 'tangible' factors may be equal, deprive the children of the minority group of equal educational opportunities? We believe that it does."

States Stressed Rights

The court's opinion does not apply to private schools. It is directed entirely at public schools. It does not affect the "separate but equal doctrine" as applied on railroads and other public carriers entirely within states that have such restrictions.

The principal ruling of the court was in four cases involving state laws. The states' right to operate separated schools had been argued before the court on two occasions by representatives of South Carolina, Virginia, Kansas and Delaware.

In these cases, consolidated in one opinion, the high court held that school segregation deprived Negroes of "the equal protection of the laws guaranteed by the Fourteenth Amendment."

The other opinion involved the District of Columbia. Here schools have been segregated since Civil War days under laws passed by Congress.

"In view of our decision that the Constitution prohibits the states from maintaining racially segregated public schools," the Chief Justice said, "it would be unthinkable that the same Constitution would impose a lesser duty on the Federal Government.

"We hold that racial segregation in the public schools of the District of Columbia is a denial of the due process of law guaranteed by the Fifth Amendment to the Constitution."

The Fourteenth Amendment provides that no state shall "deny to any person within its jurisdiction the equal protection of the laws." The Fifth Amendment says that no person shall be "deprived of life, liberty or property without due process of law."

The seventeen states having mandatory segregation are Alabama, Arkansas, Delaware, Florida, Mississippi, Missouri, North Carolina, Oklahoma, Georgia, Kentucky, Louisiana, Maryland, South Carolina, Tennessee, Texas, Virginia and West Virginia.

Kansas, New Mexico, Arizona and Wyoming have permissive statutes, although Wyoming never has exercised it.

South Carolina and Georgia have announced plans to abolish public schools if segregation were banned.

Although the decision with regard to the constitutionality of school segregation was unequivocal, the court set the cases down for reargument in the fall on questions that previously were argued last December. These deal with the power of the court to permit an effective gradual readjustment to school systems not based on color distinctions.

Other questions include whether the court itself should formulate detailed decrees and what issues should be dealt with. Also, whether the cases should be remanded to the lower courts to frame decrees, and what general directions the Supreme Court should give the lesser tribunals if this were done.

Associated Press Wirephoto

LEADERS IN SEGREGATION FIGHT: Lawyers who led battle before U. S. Supreme Court for abolition of segregation in public schools congratulate one another as they leave court after announcement of decision. Left to right: George E. C. Hayes, Thurgood Marshall and James M. Nabrit.

Cases Argued Twice

The cases first came to the high court in 1952 on appeal from rulings of lower Federal courts, handed down in 1951 and 1952. Arguments were heard on Dec. 9-10, 1952.

Unable to reach a decision, the Supreme Court ordered rearguments in the present term and heard the cases for the second time on Dec. 7-8 last year.

Since then, each decision day has seen the courtroom packed with spectators awaiting the ruling. That was true today, though none except the justices themselves knew it was coming down. Reporters were told before the court convened that it "looked like a quiet day."

Three minor opinions had been announced, and those in the press room had begun to believe the prophesy when Banning E. Whittington, the court's press information officer, started putting on his coat.

"Reading of the segregation decisions is about to begin in the court room," he said. "You will get the opinions up there."

The courtroom is one floor up, reached by a long flight of marble steps. Mr. Whittington led a fast moving exodus. In the courtroom, Chief Justice Warren had just begun reading.

Each of the Associate Justices listened intently. They obviously were aware that no court since the Dred Scott decision of March 6, 1857, had ruled on so vital an issue in the field of racial relations.

Dred Scott was a slave who sued for his freedom on the ground that he had lived in a territory where slavery was forbidden. The territory was the northern part of the Louisiana Purchase, from which slavery was excluded under the terms of the Missouri Compromise.

The Supreme Court ruled that Dred Scott was not a citizen who had a right to sue in the Federal courts, and that Congress had no constitutional power to pass the Missouri Compromise.

Thurgood Marshall, the lawyer who led the fight for racial equality in the public schools, predicted that there would be no disorder and no organized resistance to the Supreme Court's dictum.

He said that the people of the South, the region most heavily affected, were law-abiding and would not "resist the Supreme Court."

Association Calls Meetings

Mr. Marshall said that the state presidents of the National Association for the Advancement of the Colored People would meet next week-end in Atlanta to discuss further procedures.

The Supreme Court adopted two of the major premises advanced by the Negroes in briefs and arguments presented in support of their cases.

Their main thesis was that segregation, of itself, was unconstitutional. The Fourteenth Amendment, which was adopted July 28, 1868, was intended to wipe out the last vestige of inequality between the races, the Negro side argued.

Against this, lawyers representing the states argued that since there was no specific constitutional prohibition against segregation in the schools, it was a matter for the states, under their police powers, to decide.

The Supreme Court rejected the "states rights" doctrine, however, and found all laws ordering or permitting segregation in the schools to be in conflict with the Federal Constitution.

The Negroes also asserted that segregation had a psychological effect on pupils of the Negro race and was detrimental to the educational system as a whole. The court agreed.

"Today, education is perhaps the most important function of state and local governments," Chief Justice Warren wrote. "Compulsory school attendance laws and the great expenditures for education both demonstrate our recognition of the importance of education in our democratic society. It is the very foundation of good citizenship.

"In these days it is doubtful that any child may reasonably be expected to succeed in life if he is denied the opportunity of an education. Such an opportunity, where the state has undertaken to provide it, must be made available to all on equal terms."

As to the psychological factor, the high court adopted the language of a Kansas court in which the lower bench held:

"Segregation with the sanction of the law, therefore, has a tendency to retard the educational and mental development of Negro children and to deprive them of some of the benefits they would receive in a racially integrated school system."

1896 Doctrine Demolished

The "separate but equal" doctrine, demolished by the Supreme Court today, involved transportation, not education. It was the case of Plessy vs. Ferguson, decided in 1896. The court then held that segregation was not unconstitutional if equal facilities were provided for each race.

Since that ruling six cases have been before the Supreme Court, applying the doctrine to public education. In several cases, the court has ordered the admission to colleges and universities of Negro students on the ground that equal facilities were not available in segregated institutions.

Today, however, the court held the doctrine inapplicable under any circumstances to public education.

This means that the court may extend its ruling from primary and secondary schools to include state-supported colleges and universities. Two cases involving Negroes who wish to enter white colleges in Texas and Florida are pending before the court.

The question of "due process," also a clause in the Fourteenth Amendment, had been raised in connection with the state cases as well as the District of Columbia.

The High Court held, however, that since it had ruled in the state cases that segregation was unconstitutional under the "equal protection" clause, it was unnecessary to discuss "whether such segregation also violates the due process clause of the Fourteenth Amendment."

However, the "due process" clause of the Fifth Amendment was the core of the ruling in the District of Columbia case. "Equal protection" and "due process," the court noted, were not always interchangeable phrases.

Liberty Held Deprived

"Liberty under law extends to the full range of conduct which an individual is free to pursue, and it cannot be restricted except for a proper governmental objective," Chief Justice Warren asserted.

"Segregation in public education is not reasonably related to any proper governmental objective, and thus it imposes on Negro children of the District of Columbia a burden that constitutes an arbitrary deprivation of their liberty in violation of the due process clause."

Two principal surprises attended the announcement of the decision. One was its unanimity. There had been reports that the court was sharply divided and might not be able to get an agreement this term. Very few major rulings of the court have been unanimous.

The second was the appearance with his colleagues of Justice Robert H. Jackson. He suffered a mild heart attack on March 30. He left the hospital last week-end and had not been expected to return to the bench this term, which will end on June 7.

Perhaps to empnasize the unanimity of the court, perhaps from a desire to be present when the history-making verdict was announced, Justice Jackson was in his accustomed seat when the court convened.

May 18, 1954

NEGRO RIGHTS: KEY DATES

JAN. 1, 1863—President Lincoln puts into effect his Proclamation, declaring all persons held as slaves in rebel states to be free.

DEC. 18, 1865—Thirteenth Amendment ratified, abolishing slavery in the United States and its territories.

JUNE 16, 1866—Guarantee of "equal protection of the laws" for all citizens is written into the Constitution as the Fourteenth Amendment.

MARCH 30, 1870—Fifteenth Amendment ratified, declaring that the right to vote should not be denied to anyone on grounds of "race, color or previous condition of servitude."

MAY 31, 1870—Enforcement Act, designed to bring Federal Government pressure to bear against any effort to circumvent the Fourteenth and Fifteenth Amendments, passed by Congress. (Six years later the act was nullified by the Supreme Court as exceeding the Federal Government's proper role.)

MARCH 1, 1875—Congress, in an act subsequently nullified by the courts, passes Civil Rights Law guaranteeing all persons, regardless of race, the use of "inns, public conveyances on land or water, theatres, and other places of amusement."

MAY 18, 1896—Supreme Court, in Plessy v. Ferguson, establishes principle of "separate but equal" facilities for Negroes, in a decision upholding a Louisiana railway segregation law as not violating "equal protection of Fourteenth Amendment.

JUNE 21, 1915—Supreme Court declares unconstitutional the so-called "grandfather clause," a voting-qualification device employed by Southern states to restrict Negro suffrage. The clause exempted persons who had voted or whose progenitors had voted prior to 1867 from the fulfillment of educational tests and property qualifications required of other voters.

MARCH 7. 1927—Texas law barring Negroes from voting in Democratic primary elections is unanimously overruled by Supreme Court. (Five years later, Texas Democrats' attempt to bar Negroes from primaries by party resolution is declared illegal by the Supreme Court.)

NOV. 9, 1935—C. I. O. adopts constitution barring discrimination in union affairs on grounds of race or color.

DEC. 12, 1938—Supreme Court rules a state must admit a Negro to its law school or establish comparable separate facilities.

NOV. 25, 1940—Conviction of a Southern Negro is overturned by Supreme Court on grounds that Negroes had been barred from jury service at his trial.

JUNE 25, 1941—President Roosevelt establishes Federal Fair Employment Practices Commission, with instructions to seek the elimination of discriminatory practices in industry.

DEC. 18, 1944—Supreme Court rules that Railway Brotherhoods cannot act as bargaining agents under Railway Act of 1934 unless they grant equality of membership rights to Negroes.

MARCH 12, 1945—New York passes first state Fair Employment Practices law forbidding "discrimination because of race, creed, color or national origin" in employment, and sets up a State Commission against discrimination.

JUNE 3, 1946—Jim Crow practices on interstate buses barred by the Supreme Court.

OCT. 29, 1947—President Truman's Committee on Civil Rights calls for an end to all forms of segregation.

MAY 3, 1948—Supreme Court rules that "racial covenants," private real estate agreements involving residential race restrictions, cannot be enforced by the courts.

JULY 26, 1948—President Truman issues executive order establishing as a policy for the armed forces "equality of treatment and opportunity for all persons * * * without regard to race, color or national origin."

JAN. 7, 1949—Alabama constitutional provision giving local registrars wide powers to deny franchise nullified by a Federal court.

JUNE 5, 1950—Segregation in railroad dining cars is banned by the Supreme Court.

JUNE 5, 1950—Supreme Court, in a decision acknowledging that inequality is inherent in segregation, directs University of Oklahoma to stop segregating a Negro student in classrooms, library and other facilities.

MAY 17, 1954—Supreme Court, in a ruling based on the principle that separate facilities are inherently unequal, outlaws segregation in public schools as "a denial of the equal protection of the laws" guaranteed by the Fourteenth Amendment.

May 23, 1954

Effects of Social Change

Some Violence to Be Expected With Desegregation, It Is Felt

The writer of the following letter is the author of "Race Prejudice and Discrimination" and "The Negro's Morale." He is Professor of Sociology at the University of Minnesota.

TO THE EDITOR OF THE NEW YORK TIMES:

The violence which accompanied the desegregation of some of the public schools in Baltimore and Washington may have shocked many good Americans who favor social changes except when they involve physical violence. Yet some degree of violence is to be expected when rapid social change occurs in basic social structures. Social scientists would not be fulfilling their duty to the public if they did not announce that sporadic violence is to be expected in the coming desegregation in various areas of social life.

There already has been a considerable reduction of discrimination and segregation in the South, and the surprising thing to the social scientist is that there has been practically no violence until now. The school decision is the keystone in the current process of social change, and the manner in which desegregation is handled now will determine whether the United States will remain a racially divided nation with classes of citizenship or will achieve internal unity under the principle of equality of opportunity.

Signs of Readiness

From every sign available to us it appears that the Border States are ready for desegregation now, and the Deep South will be ready in two or three years. It is quite likely that some violence will continue to accompany these changes.

Some individual whites are anxious and afraid, for they have had built up in their minds—through more than a century of myth-making—a terrifying image of the Negro as a social equal. This is a great opportunity for the unscrupulous demagogue who wants either to earn an easy living by exploiting people's fears or to achieve personal leadership without going through the usual channels of advancement and recognition.

Thus the United States has to decide now whether it will tolerate sporadic outbreaks of violence for a few years until whites become accustomed to seeing Negroes in positions of social equality. The alternative is to continue segregation indefinitely, at the cost of maintaining a continuing and increasingly costly social conflict, with the prospect of ultimate disaffection of the minority elements of the population and of the future domination of many communities by instigators to bigotry.

Question of Intermarriage

A frank word should be said about the bogy of "intermarriage." Choice of marital partners will continue to be an individual matter. Thus far there is no evidence of a trend toward intermarriage despite the considerable amount of desegregation that has already occurred. However, it is quite possible when some among the different races get to know each other as individual human beings that they will want to marry each other. That is, in twenty or thirty years from now we may expect a slow upturn in the intermarriage rate, and it will be only because people want it and expect it, as they do not now.

From considerable numbers of careful researches we can assure the population that no biological harm will come from this probable future trend and that the prospects are for social advantages.

The country is thus facing a few internal growing pains resulting from the great mixture of peoples who make up the American population. The pains will be eased if people will examine their anxieties calmly and think of what they really want the future of this country to be.

ARNOLD M. ROSE.
Minneapolis, Oct. 8, 1954.

October 12, 1954

BUSES BOYCOTTED OVER RACE ISSUE

Montgomery, Ala., Negroes Protest Woman's Arrest for Defying Segregation

MONTGOMERY, Ala., Dec. 5 (AP)—A court test of segregated transportation loomed today following the arrest of a Negro who refused to move to the colored section of a city bus.

While thousands of other Negroes boycotted Montgomery city lines in protest, Mrs. Rosa Parks was fined $14 in Police Court today for having disregarded last Thursday a driver's order to move to the rear of a bus. Negro passengers ride in the rear of buses here, white passengers in front under a municipal segregation ordinance.

An emotional crowd of Negroes, estimated by the police at 5,000, roared approval tonight at a meeting to continue the boycott.

Spokesmen said the boycott would continue until people who rode buses were no longer "intimidated, embarrassed and coerced." They said a "delegation of citizens" was ready to help city and bus line officials develop a program that would be "satisfactory and equitable."

Released Under Bond

Mrs. Parks appealed her fine and was released under $100 bond signed by an attorney, Fred Gray, and a former state president of the National Association for the Advancement of Colored People, E. D. Nixon.

Mr. Gray and Charles Langford, another Negro lawyer representing the 42-year-old department store seamstress, refused to say whether they planned to attack the constitutionality of segregation laws affecting public transportation.

The Supreme Court in Washington already has before it a test case against segregation on buses operating in Columbia, S. C. The United States Court of Appeals in Richmond, Va., has ruled in this case that segregation must be ended. If the Supreme Court sustains the decision, the effect will be to outlaw segregation in all states and cities.

Mrs. Parks was charged first with violating a city ordinance that gives bus drivers police powers to enforce racial segregation. But at the request of City Attorney Eugene Loe, the warrant was amended to a charge of violation of a similar state law. The state statute authorizes bus companies to provide and enforce separate facilities for whites and Negroes. Violation is punishable by a maximum fine of $500.

Other Negroes by the thousands, meanwhile, found other means of transportation or stayed home today in an organized boycott of City Lines Buses, operated by a subsidiary of National City Lines at Chicago.

The manager, J. H. Bagley, estimated that "80 or maybe 90 per cent" of the Negroes who normally used the buses had joined the boycott. He said "several thousand" Negroes rode the buses on a normal day.

December 6, 1955

Eisenhower Bids Negroes Be Patient About Rights

By FELIX BELAIR Jr.
Special to The New York Times.

WASHINGTON, May 12—President Eisenhower told a Negro audience today that "there are no revolutionary cures" for racial discrimination in the United States. In an address to 350 editors and business and community leaders, the President said that human problems depend for their solution on "patience and forbearance" and on more and better education rather than "simply on the letter of the law."

General Eisenhower spoke extemporaneously at a luncheon session of the National Newspaper Publishers Association.

The President was given a special citation by the group for "the prestige and power he has used in behalf of civil rights." The award was for the President's action in using troops to enforce school integration at Little Rock, Ark., as well as other support for civil rights goals.

The President's address also touched on his Pentagon reorganization plan, as well as the reciprocal trade and foreign aid programs. But his central theme was "the special problem of civil rights."

Audience Applauds

It is absolutely essential, the President said, that all citizens consider themselves equal under the law. Every American, regardless of color or creed must have respect for the law to be true to his constitutional heritage and to know that "he is equal before the law," General Eisenhower continued.

Prolonged applause greeted the President when he said:

"Every American must have respect for the courts. He must have respect for others. He must make perfectly certain he can, in every single kind of circumstance, respect himself."

The Grand Ball Room of the Presidential Arms, a downtown dining club, was filled before the President's arrival. Round on round of applause greeted his repeated references to civil rights and racial discrimination.

The President's appeal for patience and forbearance received some applause. However, William O. Walker, president of the publishers' association, when questioned later, said that this was "the kind of advice we have been getting" since the 1954 school integration decision of the Supreme Court. He said that today Negroes were asking the question:

"How long is deliberate speed?"

"Deliberate speed" was a phrase used by the court in ordering integration.

One of the most prolonged waves of applause followed the President's observation that "you may be Negroes—but you are Americans."

The President invited attention at the outset of his remarks to his custom of avoiding the use of "adjectives in describing the American group I am talking to."

"Though some are farmers or some are Chambers of Commerce people, or some like you may be Negroes, we are Americans and we have American problems," President Eisenhower said.

Indifference Opposed

No citizen can be indifferent to racial or any other type of discrimination, he said, then added:

"To my mind, every American, whatever his religion, his color, his race or anything else, should have exactly the same concern for these matters as does any individual who may have felt embarrassment or resentment because those rights have not been properly observed."

After referring to laws to meet the problems of racial discrimination, the President said:

"But I do believe that as long as they are human problems—because they are buried in the human heart rather than ones merely to be solved by a sense of logic and right—we must have patience and forbearance. We must depend on more and better education than simply on the letter of the law. We must make sure that enforcement will not in itself create injustice."

The President emphasized that he was not decrying laws.

"But I say that laws themselves will never solve problems that have their roots in the human heart and in human emotions," he said.

He said that he still believed, as he did in 1952, that the final answer is in the "communities where we live."

Every American who opposes inequality, every American who helps in even the smallest way to make equality of opportunity a living fact, is doing the business of America," he said.

Before President Eisenhower addressed the group, Mr. Walker, its keynote speaker observed that while the Negro has progressed in the last eighty years, his advancement has not kept pace with the rest of the country.

He said that the "Negro is not taller in America than he can reach in Mississippi."

He continued:

"Our freedom must not be measured in Ohio but in South Carolina. The level of justice is not what it is in Illinois, but what the gauge shows in Georgia."

Today's national conference here was the first such since Negro leaders met in 1906 and paved the way for the National Association for the Advancement of Colored People.

May 13, 1958

N.A.A.C.P. LEADER URGES 'VIOLENCE'

North Carolina Aide Makes Statement—Association Quickly Suspends Him

MONROE, N. C., May 6 (UPI) —A Negro leader here, charging that members of his race cannot obtain justice in the courts, threatened last night to "meet violence with violence."

The leader is Robert Williams, president of the Union County (N. C.) Chapter of the National Association for the Advancement of Colored People. He said that Negroes "must even be willing to kill if necessary" to protect themselves against white people.

In New York, Roy Wilkins, executive secretary of the N. A. A. C. P., suspended Mr. Williams from his position in the association, pending "consideration of your status." Other Negro leaders assailed Mr. Williams' statement.

Mr. Wilkins said the N. A. A. C. P. repudiated any "pro-lynching statement by one of our officers regardless of provocation." In a telegram to Mr. Williams, he said the association's board of directors would consider the case in New York next Monday.

Mr. Williams said "we cannot take these people who do us injustice to the court, and it becomes necessary to punish them ourselves."

He issued the statement after a jury had acquitted a white man charged with attempting to rape a Negro woman here.

"In the future," Mr. Williams said, "we are going to have to try and convict these people on the spot. This opens the way to real violence."

Mr. Williams' statement went on:

"We cannot rely on the law. We can get no justice under the present system.

"If we feel that an injustice **is done, we must right then and there on the spot be prepared to inflict punishment on these people."**

Mr. Williams, 34 years old, said "I was speaking for myself and not the N.A.A.C.P., and the N.A.A.C.P. or no organization is going to tell me what to do or say." He said he intended to remain in his post here.

May 7, 1959

Negro Sitdowns Stir Fear Of Wider Unrest in South

By CLAUDE SITTON

Special to The New York Times.

CHARLOTTE, N. C., Feb. 14—Negro student demonstrations against segregated eating facilities have raised grave questions in the South over the future of the region's race relations. A sounding of opinion in the affected areas showed that much more might be involved than the matter of the Negro's right to sit at a lunch counter for a coffee break.

The demonstrations were generally dismissed at first as another college fad of the "panty-raid" variety. This opinion lost adherents, however, as the movement spread from North Carolina to Virginia, Florida, South Carolina and Tennessee and involved fifteen cities.

Some whites wrote off the episodes as the work of "outside agitators." But even they conceded that the seeds of dissent had fallen in fertile soil.

Backed by Negro Leaders

Appeals from white leaders to leaders in the Negro community to halt the demonstrations bore little fruit. Instead of the hoped-for statements of disapproval, many Negro professionals expressed support for the demonstrators.

A handful of white students joined the protests. And several state organizations endorsed it. Among them were the North Carolina Council on Human Relations, an inter-racial group, and the Unitarian Fellowship for Social Justice, which currently has an all-white membership.

Students of race relations in the area contended that the movement reflected growing dissatisfaction over the slow pace of desegregation in schools and other public facilities.

It demonstrated, they said, a determination to wipe out the last vestiges of segregation.

Moreover, these persons saw a shift of leadership to younger, more militant Negroes. This, they said, is likely to bring increasing use of passive resistance. The technique was conceived by Mohandas K. Gandhi of India and popularized among Southern Negroes by the Rev. Dr. Martin Luther King Jr. He led the bus boycott in Montgomery, Ala. He now heads the Southern Christian Leadership Conference, a Negro minister's group, which seeks to end discrimination.

Wide Support Indicated

Negro leaders said that this assessment was correct. They disputed the argument heard among some whites that there was no broad support for the demonstrations outside such organizations as the National Association for the Advancement of Colored People.

There was general agreement on all sides that a sustained attempt to achieve desegregation now, particularly in the Deep South, might breed racial conflict that the region's expanding economy could ill afford.

The spark that touched off the protests was provided by four freshmen at North Carolina Agricultural and Technical College in Greensboro. Even Negroes class Greensboro as one of the most progressive cities in the South in terms of race relations.

On Sunday night, Jan. 31, one of the students sat thinking about discrimination.

"Segregation makes me feel that I'm unwanted," McNeil A. Joseph said later in an interview. "I don't want my children exposed to it."

The 17-year-old student from Wilmington, N. C., said that he approached three of his classmates the next morning and found them enthusiastic over a proposal that they demand service at the lunch counter of a downtown variety store.

About 4:45 P. M. they entered the F. W. Woolworth Company store on North Elm Street in the heart of Greensboro. Mr. Joseph said he bought a tube of toothpaste and the others made similar purchases. Then they sat down at the lunch counter.

Rebuked by a Negro

A Negro woman kitchen helper walked up, according to the students, and told them, "You know you're not supposed to be in here." She later called them "ignorant" and a "disgrace" to their race.

The students then asked a white waitress for coffee.

"I'm sorry but we don't serve colored here," they quoted her.

"I beg your pardon," said Franklin McCain, 18, of Washington, "you just served me at a counter two feet away. Why is it that you serve me at one counter and deny me at another. Why not stop serving me at all the counters."

The four students sat, coffeeless, until the store closed at 5:30 P. M. Then, hearing that they might be prosecuted, they went to the executive committee of the Greensboro N. A. A. C. P. to ask advice.

"This was our first knowledge of the demonstration," said Dr. George C. Simkins, who is president of the organization. He said that he had then written to the New York headquarters of the Congress of Racial Equality, which is known as CORE. He requested assistance for the demonstrators, who numbered in the hundreds during the following days.

Dr. Simkins, a dentist, explained that he had heard of a successful attempt, led by CORE, to desegregate a Baltimore restaurant and had read one of the organization's pamphlets.

CORE's field secretary, Gordon R. Carey, arrived from New York on Feb. 7. He said that he had assisted Negro students in some North Carolina cities after they had initiated the protests.

The Greensboro demonstrations and the others that it triggered were spontaneous, according to Mr. Carey. All of the Negroes questioned agreed on this.

The movement's chief targets were two national variety chains, S. H. Kress & Co. and the F. W. Woolworth Company Other chains were affected. In some cities the students demonstrated at local stores.

The protests generally followed similar patterns. Young men and women and, in one case, high school boys and girls, walked into the stores and requested food service. Met with refusals in all cases, they remained at the lunch counters in silent protest.

The reaction of store managers in those instances was to close down the lunch counters and, when trouble developed or bomb threats were received, the entire store.

Hastily painted signs, posted on the counters, read: "Temporarily Closed," "Closed for Repairs," "Closed in the Interest of Public Safety," "No Trespassing," and "We Reserve The Right to Service the Public as We See Fit."

After a number of establishments had shut down in High Point, N. C., the S. H. Kress & Co. store remained open, its lunch counter desegregated. The secret? No stools.

Asked how long the store had been serving all comers on a stand-up basis, the manager replied:

"I don't know. I just got transferred from Mississippi."

The demonstrations attracted crowds of whites. At first the hecklers were youths with duck-tailed haircuts. Some carried small Confederate battle flags. Later they were joined by older men in faded khakis and overalls.

The Negro youths were challenged to step outside and fight. Some of the remarks to the girls were jesting in nature, such as, "How about a date when we integrate?" Other remarks were not.

Negro Knocked Down

In a few cases the Negroes were elbowed, jostled and shoved. Itching powder was sprinkled on them and they were spattered with eggs.

At Rock Hill, S. C., a Negro youth was knocked from a stool by a white beside whom he sat. A bottle of ammonia was hurled through the door of a drug store there. The fumes brought tears to the eyes of the demonstrators.

The only arrests reported involved forty-three of the demonstrators. They were seized on a sidewalk outside a Woolworth store at a Raleigh shopping center. Charged with trespassing, they posted $50 bonds and were released.

The management of the shopping center contended that the sidewalk was private property.

In most cases, the demonstrators sat or stood at store counters talking in low voices, studying or staring impassively at their tormenters. There was little joking or smiling. Now and then a girl giggled nervously. Some carried bibles.

Those at Rock Hill were described by the local newspaper, The Evening Herald, as "orderly, polite, well-dressed and quiet."

'Complicated Hospitality'

Questions to their leaders about the reasons for the demonstrations drew such replies as:

"We feel if we can spend our money on other goods we should be able to eat in the same establishments," "All I want is to come in and place my order and be served and leave a tip if I feel like it," and "This is definitely our purpose: integrated seating facilities with no isolated spots, no certain seats, but to sit wherever there is a vacancy."

Some newspapers noted the embarrassing position in which the variety chains found themselves. The News and Observer of Raleigh remarked editorially that in these stores the Negro was a guest, who was cordially invited to the house but definitely not to the table. "And to say the least, this was complicated hospitality."

The newspaper said that to serve the Negroes might offend Southern whites while to do otherwise might result in the loss of the Negro trade.

"This business," it went on, "is causing headaches in New York and irritations in North Carolina. And somehow it revolves around the old saying that you can't have your chocolate cake and eat it too."

The Greensboro Daily News advocated that the lunch counters be closed or else opened on a desegregated basis.

North Carolina's Attorney General, Malcolm B. Seawell, asserted that the students were causing "irreparable harm" to relations between whites and Negroes.

Mayor William G. Enloe of Raleigh termed it "regrettable that some of our young Negro students would risk endangering these relations by seeking to change a long-standing custom in a manner that is all but destined to fail."

Some North Carolinians found it incomprehensible that the demonstrations were taking place in their state. They pointed to the progress made here toward desegregation of public facilities. A number of the larger cities in the Piedmont region, among them Greensboro voluntarily accepted token desegregation of their schools after the Supreme Court's 1954 decisions.

But across the state there were indications that the Negro had weighed token desegregation and found it wanting.

When commenting on the subject, the Rev. F. L. Shuttlesworth of Birmingham, Ala., drew a chorus of "amens" from a packed N.A.A.C.P. meeting in a Greensboro church, "We don't want token freedom," he declared. "We want full freedom. What would a token dollar be worth?"

Warming to the subject, he shouted:

"You educated us. You taught us to look up, white man. And we're looking up!"

Praising the demonstrators, he urged his listeners to be ready "to go to jail with Jesus" if necessary to "remove the dead albatross of segregation that makes America stink in the eyes of the world."

John H. Wheeler, a Negro lawyer who heads a Durham bank, said that the only difference among Negroes concerned the "when" and "how" of the attack on segregation.

He contended that the question was whether the South would grant the minority race full citizenship status or commit economic suicide by refusing to do so.

The Durham Committee on Negro Affairs, which includes persons from many economic levels, pointed out in a statement that white officials had asked Negro leaders to stop the student demonstrations.

"It is our opinion," the statement said, "that instead of expressing disapproval, we have an obligation to support any peaceful movement which seeks to remove from the customs of our beloved Southland those unfair practices based upon race and color which have for so long a time been recognized as a stigma on our way of life and stumbling block to social and economic progress of the region."

It then asserted:

"It is reasonable to expect that our state officials will recognize their responsibility for helping North Carolina live up to its reputation of being the enlightened, liberal and progressive state, which our industry hunters have been representing it to be."

The outlook for not only this state but also for the entire region is for increasing Negro resistance to segregation, according to Harold C. Fleming, executive director of the Southern Regional Council. The council is an interracial group of Southern leaders with headquarters in Atlanta. Its stated aim is the improvement of race relations.

"The lunch-counter 'sit-in'," Mr. Fleming commented, "demonstrates something that the white community has been reluctant to face: the mounting determination of Negroes to be rid of all segregated barriers.

"Those who hoped that token legal adjustments to school desegregation would dispose of the racial issue are on notice to the contrary. We may expect more, not less, protests of this kind against enforced segregation in public facilities and services of all types."

February 15, 1960

Bi-Racial Buses Attacked, Riders Beaten in Alabama

By The Associated Press.

ANNISTON, Ala., May 15—A group of white persons ambushed today two buses carrying Negroes and whites who are seeking to knock down bus station racial barriers. A little later, sixty miles to the west, one of the buses ran into another angry crowd of white men at a Birmingham bus station.

The integrated group took a brief but bloody beating, and fled. No serious injuries were reported.

Both buses were carrying members of the Congress of Racial Equality on a swing through the Deep South, testing segregated facilities in bus stations. They call themselves "Freedom Riders."

State Investigator Ell M. Cowling, acting on a tip, was aboard a Greyhound bus attacked near Anniston. He barred the bus door with his body when the crowd of white men tried to board the bus. Two highway patrolmen fired their pistols into the air to quiet the crowd of about 200 that had followed the bus from Anniston.

Somebody threw a fire bomb through a bus window. Twelve persons were hospitalized, most of them for smoke inhalation. Ten of them later were released.

The bus, stalled about six miles out of Anniston by a flat tire, was destroyed by the blaze.

Those who were not hospitalized were taken back to Anniston and placed on another bus. They completed their ride to Birmingham and arrived at Birmingham's Greyhound station without incident.

The C. O. R. E. members left Washington ten days ago with six white and seven Negro Freedom Riders. The number has fluctuated at various stops. On the Anniston bus the C. O. R. E. group included five Negroes and four white persons. The number on the Trailways bus that reached Birmingham was not known.

They split into two groups in Atlanta and took separate buses into Alabama. The trouble started when the Greyhound, carrying the nine Freedom Riders and five other passengers, reached this city of about 30,000.

The Trailways bus that later ran into difficulty at Birmingham had its first trouble at Anniston, also.

Dr. Walter Bergman, 61 years old, a former Michigan State University professor and a member of the C. O. R. E. group, said a fight broke out on the bus.

The Trailways station was closed but Mr. Bergman said he got off, got sandwiches nearby and was getting back on the bus when a policeman came up.

"The driver said he wasn't going to move until the Negroes moved to the back of the bus," Mr. Bergman said.

"At that time, about ten white men attacked Charles Person, a student at Morehouse University, Atlanta.

"And then James Peck stepped forward, then they turned on us. Peck was beaten about the face and got a deep cut on his scalp.

"They beat me and were kicking me. And then they threw the Negroes and others over me. There was no other violence until we got to Birmingham."

Mr. Bergman said three policemen stood outside the bus at Anniston while this took place.

Mr. Bergman explained the group as a coordinated organization of nonviolent action groups that believed in racial equality and in achieving it through nonviolent means.

The courts have outlawed enforced segregation among interstate bus passengers, but the two bus stations in Birmingham still maintain white and Negro waiting rooms.

A spokesman at Greyhound said its waiting room signs read "white Intrastate passengers" and "Negro Intrastate Passengers." A neon sign at the Trailways station says "Negro Waiting Room."

The bus continued to Birmingham, where Mr. Peck was admitted to a hospital in fair condition. The hospital said later that Mr. Peck, of New York, would be discharged tonight.

When the bus arrived at Birmingham it met more trouble.

Several white men attacked the group inside the Birmingham station, beating a Negro youth and a white man who was apparently accompanying the group.

The white men obviously were waiting for the bus, and covered telephone booths and exits.

The fighting broke out in several areas around the station in downtown Birmingham. No police were in evidence until several minutes after the outbreak.

A white man left the bus and shook hands with one of the Negro passengers at the Birmingham station.

The Negroes hesitated momentarily, then walked into a passageway leading to the waiting rooms.

To the left was a sign that designated the Negro waiting room. The group hesitated, then a young Negro walked ahead.

He walked about ten feet into the white waiting room. The other Negroes followed, a few feet behind.

Several white men stopped the young Negro, and one of them told him:

"The Negro waiting room is back that way."

They turned him around forcefully, and pushed him. The Negro group turned back into the passageway, but were met by another group of white men entering from the opposite end of the hall.

The Negro man tried to walk through, but a husky white man knocked him against the wall.

Several white men entered the passageway from the white waiting room. They stood behind the young Negro.

"Hit him," one of them said.

A man slammed his fist into the young Negro's face. He fell to the floor, bleeding from the nose.

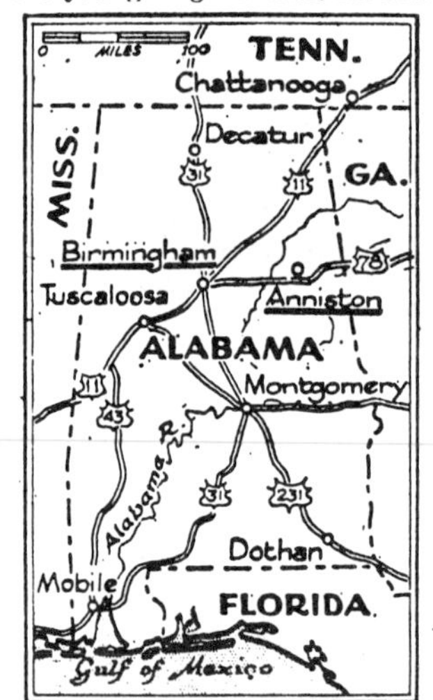

The New York Times May 15, 1961

ALABAMA OUTBREAKS: Bus riders attacked at Birmingham and Anniston.

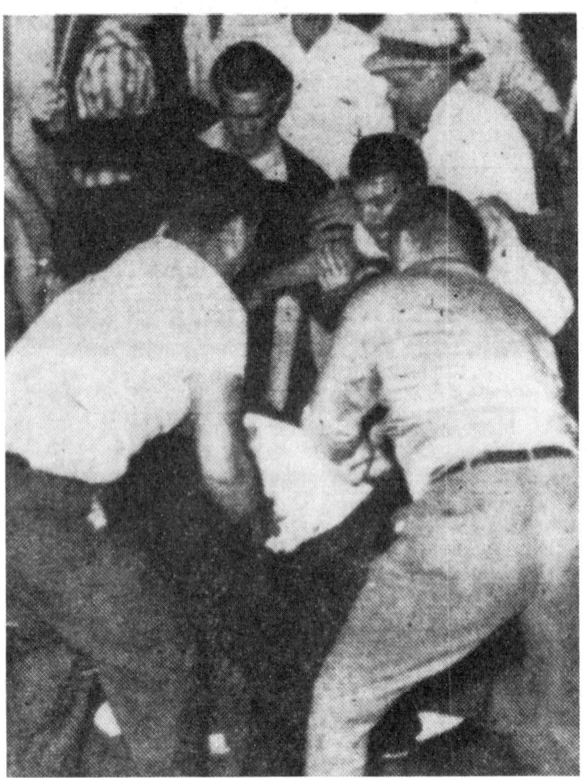

United Press International Telephoto

VIOLENCE IN BIRMINGHAM: James Peck of New York is beaten by whites at a Trailways bus station. Mr. Peck was taking part in a bus ride with Negro passengers to test desegregation laws at interstate stations in South.

As he got up, the white man hit him again. This time, the Negro fell backward into the arms of the white men. They pushed him up again. The white man struck him again.

A white man who had been riding the bus with the Negroes attempted to interfere. He was beaten in the face and fell on his back, blood streaming from his nose.

The injured Negro youth and the white man staggered outside to the bus parking area, with the white attackers behind them.

It was all over in about three minutes.

There was no way to determine the number of attackers. Some apparently had been waiting in the bus station before the incident.

An aged Negro woman who was on the bus cried:

"It started on the bus. It started on the bus."

Two newsmen were attacked by the crowd outside the bus station. One, Tom Langston of The Birmingham Post-Herald, was hurt painfully, but was not hospitalized.

Clancy Lake, a radio newsmen, was attacked as he sat in his closed automobile broadcasting an account of the violence. He escaped serious injury.

In an alley behind the bus station, a group of photographers had leaped into their automobile, but were blocked by another crowd of white men who seized several cameras. They smashed some and made off with the others.

Those hospitalized at Anniston were Genevieve Hughes, 28 years old, a white woman of Chevy Chase, Md., and Edward Blackenheim, 28, a white man of Tucson, Ariz., members of C. O. R. E.

They were reported in good condition.

Other members of C. O. R. E. who were treated and released were listed at the hospital as:

Henry Thomas, 19, a white man, of Washington; Albert Bigelow, 55, white, of Cos Cob, Conn.; James McDonald, a Negro, New York City, and Mae Frances Moultrie, a Negro, of Sumter, S. C.

Also treated were two reporters accompanying the Freedom Riders. They were Moses J. Newson, 34, a Negro, of The Baltimore Afro-American, and Charlotte Devree, 50, a white woman, of New York City.

These passengers aboard the bus also were treated and released:

Roberta Holmes, a Negro, of Birmingham; Roy J. Powers, 39, white, Clinchport, Va., and Larry A. Harper, 22, a Negro, of Margaret, Ala., and Mr. Cowling, an Alabama state investigator.

Thomas J. Jenkins of Birmingham, an official of the Federal Bureau of Investigation, said his agents were making an inquiry into the bus-burning to determine whether there had been any violation of Federal law.

He declined to comment further.

May 15, 1961

Assertive Spirit Stirs Negroes, Puts Vigor in Civil Rights Drive

Racial Pride Increases Throughout U. S. —'Black Nationalism' Is Viewed as Powerful Force for Change

By M. S. HANDLER

A new assertive mood, characterized by some Negro leaders as "Black Nationalism," is spreading throughout the United States.

This spirit is usually identified by the general public with the Black Muslims, the extremist group dedicated to the establishment of a Negro nation carved out of America.

But, in the opinion of many responsible Negro observers, this new mood also lies behind movements with much greater Negro support—the voter registration drive in Greenwood, Miss., and the assault by the Rev. Dr. Martin Luther King Jr. on segregation in Birmingham, Ala., to cite two examples.

For the term Black Nationalism is now used by Negroes to embrace a multitude of feelings, many of them contradictory, but all having one concept in common—identification of America's Negro population as a group with a common heritage of suffering and achievement, engaged in mass action to compel the white majority to recognize and implement their legal rights.

A Way of Life

A Negro social philosopher, Dr. C. Eric Lincoln of Clark College in Atlanta, explains that Black Nationalism developed because "the Negro is required to be Negro before, and sometimes to the exclusion of, anything else."

But Dr. Lincoln says that this new mood goes beyond earlier appeals for equality.

"Black Nationalism is more than courage and rebellion," he says, "it is a way of life," rejecting white symbols and culture and emphasizing pride in being black.

As yet Black Nationalism has not developed into an organized political force, although the potentiality worries both Republicans and Democrats.

Democrat Takes Note

Louis Martin, a Negro, who is deputy chairman of the Democratic National Committee, notes the rising demands of his race to improve its lot and says Negro aspirations can be summed up in one phrase:

"White man, move over."

Negro leaders differ in their assessment of this new mood, but they agree that it is a powerful new force in American public life.

It is threatening the old order of Negro leaders, the business and professional class and ministers. It is questioning the coleadership of liberal whites in the struggle against discrimination.

Moreover, it is helping destroy the stereotypes of the shiftless Negro as well as the Uncle Toms.

Heartaches For Some

It is bringing new heartaches as Negro parents find they no longer understand their children, who are filled with new racial pride. It has also brought to the surface widespread hatred for the white man.

One highly placed Negro leader, who has dedicated his life to integration within the present framework of society, sadly remarked:

"There is a bit of hatred in every Negro's heart. The Negro would not be human if he did not resent the oppression to which he has been subjected. But this feeling would be stifled if the Negro were to receive the equal status he is entitled to as a citizen of this country."

Black Nationalism as a mood is to be found in varying degrees in all segments of the Negro population, but only the Black Muslims expound a complete renunciation of coexistence with the whites.

Malcom X, the dynamic leader of the Black Muslims in New York, contends that the Negroes "have already seceded" emotionally and intellectually from American society.

To this, Roy Wilkins, the executive secretary of the National Association for the Advancement of Colored People, replies:

"Hogwash. Ninety-nine per cent of the Negroes want in."

In an interview published this week in U. S. News & World Report, Mr. Wilkins described the dominant mood of the Negroes today as "very great impatience" with the slow progress toward the "status of full citizenship." He said the Black Muslims are an outgrowth of this sentiment. But he pointed out that the Muslims, "idea of black for black has always been around in one form or another."

However, even the most bitter opponent of the Black Muslims agrees that they have contributed to the growth in Negro racial pride, group identification, self-respect and the conscious worth of Negro values.

The Black Muslims have never published any figures about their membership, but estimates have ranged from as low as 25,000 to as high as 250,000, with the number often put as 70,000. None of these estimates, however, is based on verifiable data. There are more than 18,000,000 Negroes in the country.

Followers of Islam

The Muslims assert that they are faithful followers of Islam, a claim that raises some doubts among their opponents. The Muslims, note, however, that Elijah Muhammed, head of the movement, has made a pilgrimage to Mecca, a journey reserved for Islam's faithful.

The Black Muslims reject Christianity as a white man's religion imposed on the Negroes to insure their prolonged servitude and inferior status. Christianity promises rewards in the hereafter, and according to the Black Muslims, this belief is designed to condition Negroes to accept their inferior status on this earth. But all the old-line Negro leaders insist that the Christian churches are the most powerful force in the Negro community, without which no gains can be made.

The Black Muslims are organized in a vertical chain of command that makes for authoritarian control vested in Elijah Muhammed, who resides in Chicago. A highly trained praetorian guard, called Fruit of Islam, assures almost military discipline and unquestioning execution of orders.

This élite was recruited in part in penitentiaries among prisoners who were later rehabilitated.

Malcolm X, who is of impressive bearing and is endowed with a shrewd mind, today overshadows Elijah Muhammed.

Rising Popularity

Other Negro leaders are very conscious of the growing popularity of Malcolm X. While they reject his contention that alienation of the Negroes from the American social order is "an accomplished fact," they concede that he is able.

They also grudgingly agree that the puritanical code of the movement—its prohibition of smoking, drinking, gambling, prostitution and narcotics, is destroying the old symbols affixed to Negroes.

The obscure origins of the Black Muslims used to generate derision. But one visit to the Black Muslim restaurant on Lenox Avenue is enough to silence those who make fun of the movement. Waiters dressed in immaculate white jackets and black trousers move about quietly. Mirrors glisten and tables, chairs and floor are almost antiseptically clean.

The old-line leaders also note that Black Muslim ideology, which preaches racial pride, is sweeping Negro communities.

For example, a distinguished Negro who is committed to a "social contract" between the white and Negro populations tells the story of his daughter, the only Negro pupil in an upper middle class school in Westchester County. Though friendly and helpful at school, white pupils refused to attend her Christmas party. She was deeply hurt but said nothing. Several months later, without the knowledge of her parents she wrote a moving essay for school on the meaning of being black.

Warns of Dangers

James Farmer, executive secretary of the Congress of Racial Equality warns of the perils of Black Nationalism. CORE is the organization identified with the sit-ins, freedom rides and voters' registration.

"There is the possibility that Black Nationalism could develop into hate of everything that is different," he says. "How it turns out will depend on how successful the Negroes will be in bringing down the barriers to desegregation."

Whitney Moore Young Jr., executive director of the National Urban League, which has contributed greatly to equal job opportunity, assessed the situation even more seriously.

"In previous generations the Negro believed what is white is right, he said. "This was the standard to which the Negro aspired. This has changed. The Negro is taking a second look and is rejecting many of the white man's values and codes of behavior."

Mr. Young attributed the radical change in the Negro's attitude to the fact that "he no longer feels innately inferior to the white man."

This point was brought home to a white man who drove a few blocks out of his way to take a Negro home. As the Negro got out of the car, he turned and said:

"Thanks for the lift. It was mighty black of you."

Negroes are Divided

There seems to be a sharp cleavage between Negro intellectuals and Negro organization officials in their assessment of the Negro's viewpoint. The intellectuals are closer to James Baldwin, the Negro author, sharing his anger, despair and rage.

In his essay, "Letter From A Region In My Mind," Mr. Baldwin, said:

"The treatment accorded the Negro during the Second World War marks, for me, a turning point in the Negro's relation to America. To put it briefly, and somewhat too simply, a certain hope died, a certain respect for white Americans faded. One began to pity them, or to hate them."

Many New York City Negro intellectuals are categorical in their belief that the Negro fundamentally hates the white man.

They insist that this hatred is spreading as the entrenched white men in the North resort to more refined and resourceful techniques of confining the Negroes to black ghettos, denying them equal work opportunities and hospital facilities and arranging de facto school segregation and token desegregation of the major institutions of higher learning.

Dr. Kenneth Clark, Negro professor of psychology at the City College of New York, says "the Negro does not feel himself a part of this white dominated country, and there is no reason why he should concern himself with this country's international relations."

"There is a colossal indifference to the United States

Nancy Sirkis

BLACK MUSLIM HEAD: Malcolm X, leader of extremist group in New York.

Government's foreign problems," he said. "If I were to talk to a Negro crowd about foreign affairs people would say to me, 'man, what are you talking about?' "

Mr. Farmer says this lack of interest is due more to Negroes' preoccupation with their own problems than to a conscious withdrawal from the problems of the United States.

Thus students at the all-Negro Atlanta University said they were not interested in and did not favor the American Government's Cuban policy until last fall when war seemed near.

Asked why they changed their minds, one replied:

"Well we wanted to find out why we might have to get killed."

Widening Gulf

The evidence seems to indicate that the widening gulf between the dominant white majority and the Negro minority in the United States is paralleled by a growing gulf separating the educated Negroes from the Negroes masses.

For example, Atlanta University students from rural districts said they had lost contact with their own families after a brief time at the university. They said they found it difficult to communicate when they returned home during the holidays.

Such alienation is found in all societies, but it apparently is more acute for the Negro because of his struggle for civil rights. Professor Samuel duBois Cook, Negro sociologist of Atlanta University, explains the effect of education on Negroes:

"In a sense the Negro's very progress tends to magnify his frustrations, insecurities, anxieties, tensions and pathetic dilemmas. For his unsatisfied desires expand, his longings intensify, his expectations are greater, his level of aspiration soars, and his imagination and vision are more lucid, inspired and vivid. Accordingly, his disappointments are more bitter, roadblocks more intolerable, barriers more baffling and gnawing, suffering more poignant. Success aggravates failure. Failure erodes success."

Southern students publicly accept Dr. King's approach of non-violence as the guideline for all assaults on segregation.

Nevertheless, they seem attracted to the extremist views of Malcolm X, although reluctant to discuss them.

Leslie Dunbar, of the Southern Regional Council, one of the white men most intimately associated with Negro communities in the South, says it is difficult, if not impossible, to ascertain the true feelings and aspirations of the Negroes. He doubts whether the Negro leaders themselves are fully cognizant of what the Negro masses want because, he says, the Negro almost instinctively hides his feelings, even from members of his own race.

Malcolm X goes even further:

"Because of his special history the Negro has had to become a two-faced individual. I could take you to a round of meetings in one evening and you would find the same Negroes responding enthusiastically to the most divergent point of view on the same theme. And this is when he is among his own people."

But Malcolm X believes he knows where the Negro masses stand:

"The black masses in this country are not divided, it is **the leadership that is divided.**

"The established Negro leaders have lulled the white man into believing everything is under control. There is a difference of night and day when a white man and a Negro say 'our country.' The white man means it. The Negro behaves like a house dog and wags his tail."

Malcolm X maintains that the Negroes should dispense with white liberal leadership and guidance in the civil rights fight. This view, which is shared in part by moderate Negroes, is based on two assumptions.

The first is that the Negroes have developed enough leaders capable of taking charge of their own struggle. The second is that the white liberal cannot be trusted because he is identified with the white man's power structure and when the chips are down will support the white man's interests.

The consensus among Negro leaders is that the key to the problem of race relations is the refusal and inability of the white majority to understand the prevailing mood of the Negro minority. There is no escape from blackness for the Negro, they said, and this goes for all Negroes regardless of their station in life.

Thus a Negro girl with Caucasian features, a graduate cum laude from Vassar, recalled that whenever her white college friends held a party they included her but always invited a Negro man as her escort. She knows this was done out of kindness, but to the Negro

girl it meant there was no escape for her from her race. Now she is Negro-conscious and is working actively on racial problems.

Although Malcolm X regards the Negro Christian Ministry as an evil influence, he found it politically expedient to share the same platform with Representative Adam Clayton Powell, who is pastor of the Abyssian Baptist Church, one of the largest Protestant congregations in the United States.

Malcolm X had no compunction on this score because Mr. Powell shares his opposition to whites in policy-making positions in Negro civil rights organizations.

Negro leaders believe there are other clues indicating a tactical shift by the Black Muslims to exploit the old-line Negro organizations.

Muhammed Speaks, the Black Muslim newspaper, has begun to report Mr. Powell's speeches with approval. It also published a lengthy interview with A. Philip Randolph, president of the Sleeping Car Porters Union.

There is no doubt in Mr. Farmer's mind that Malcolm X is beginning to emerge as a political power. Interviewed last Sunday on the WINS program, "News Conference," Mr. Farmer suggested that Mr. Powell might consider the Black Muslims a political threat to him and that this fear might explain his friendly public attitude toward Malcolm X.

Regardless of tactical shifts Malcolm X remains firm on one point:

"You cannot integrate the Negroes and the whites without bloodshed. It can't be done. The only peaceful way is for the Negroes and whites to separate."

April 23, 1963

PRESIDENT SEES GAIN FOR NEGRO IN ORDERLY WASHINGTON RALLY; 200,000 MARCH FOR CIVIL RIGHTS

ACTION ASKED NOW

10 Leaders of Protest Urge Laws to End Racial Inequity

By E. W. KENWORTHY
Special to The New York Times

WASHINGTON, Aug. 28 More than 200,000 Americans, most of them black but many of them white, demonstrated here today for a full and speedy program of civil rights and equal job opportunities.

It was the greatest assembly for a redress of grievances that this capital has ever seen.

One hundred years and 240 days after Abraham Lincoln enjoined the emancipated slaves to "abstain from all violence" and "labor faithfully for reasonable wages," this vast throng proclaimed in march and song and through the speeches of their leaders that they were still waiting for the freedom and the jobs.

Children Clap and Sing

There was no violence to mar the demonstration. In fact, at times there was an air of hootenanny about it as groups of schoolchildren clapped hands and swung into the familiar freedom songs.

But if the crowd was good-natured, the underlying tone was one of dead seriousness. The emphasis was on "freedom" and "now." At the same time the leaders emphasized, paradoxically but realistically, that the struggle was just beginning.

On Capitol Hill, opinion was divided about the impact of the demonstration in stimulating Congressional action on civil rights legislation. But at the White House, President Kennedy declared that the cause of 20,000,000 Negroes had been

Associated Press

VIEW FROM THE LINCOLN MEMORIAL: The scene during the march looking tow: :d the Washington Monument

advanced by the march.

The march leaders went from the shadows of the Lincoln Memorial to the White House to meet with the President for 75 minutes. Afterward, Mr. Kennedy issued a 400-word statement praising the marchers for the "deep fervor and the quiet dignity" that had characterized the demonstration.

Says Nation Can Be Proud

The nation, the President said, "can properly be proud of the demonstration that has occurred here today."

The main target of the demonstration was Congress, where committees are now considering the Administration's civil rights bill.

At the Lincoln Memorial this afternoon, some speakers, knowing little of the ways of Congress, assumed that the passage of a strengthened civil rights bill had been assured by the moving events of the day.

But from statements by Congressional leaders, after they had met with the march committee this morning, this did not seem certain at all. These statements came before the demonstration.

Senator Mike Mansfield of Montana, the Senate Democratic leader, said he could not say whether the mass protest would speed the legislation, which faces a filibuster by Southerners.

Senator Everett McKinley Dirksen of Illinois, the Republican leader, said he thought the demonstration would be neither an advantage nor a disadvantage to the prospects for the civil rights bill.

The human tide that swept over the Mall between the shrines of Washington and Lincoln fell back faster than it came on. As soon as the ceremony broke up this afternoon, the exodus began. With astounding speed, the last buses and trains cleared the city by mid-evening.

At 9 P.M. the city was as calm as the waters of the Reflecting Pool between the two memorials.

At the Lincoln Memorial early in the afternoon, in the midst of a songfest before the addresses, Josephine Baker, the singer, who had flown from her home in Paris, said to the thousands stretching down both sides of the Reflecting Pool:

"You are on the eve of a complete victory. You can't go wrong. The world is behind you."

Miss Baker said, as if she saw a dream coming true before her eyes, that "this is the happiest day of my life."

But of all the 10 leaders of the march on Washington who followed her, only the Rev. Dr. Martin Luther King Jr., president of the Southern Christian Leadership Conference, saw that dream so hopefully.

The other leaders, except for the three clergymen among the 10, concentrated on the struggle ahead and spoke in tough, even harsh, language.

But paradoxically it was Dr. King—who had suffered perhaps most of all—who ignited the crowd with words that might have been written by the sad, brooding man enshrined within.

As he arose, a great roar welled up from the crowd. When he started to speak, a hush fell.

"Even though we face the difficulties of today and tomorrow, I still have a dream," he said.

"It is a dream chiefly rooted in the American dream," he went on.

"I have a dream that one day this nation will rise up and live out the true meaning of its creed: "We hold these truths to be self-evident, that all men are created equal.'

Dream of Brotherhood

"I have a dream . . ." The vast throng listening intently to him roared.

". . . that one day on the red hills of Georgia, the sons of former slaves and the sons of former slave-owners will be able to sit together at the table of brotherhood.

"I have a dream . . ." The crowd roared.

Associated Press

LEADERS MEET WITH KENNEDY: From left Whitney M. Young Jr., of National Urban League; the Rev. Dr. Martin Luther King Jr., Southern Christian Leadership Conference; John Lewis, partly hidden, Student Nonviolent Coordinating Committee; Rabbi Joachim Prinz, American Jewish Congress; the Rev. Dr. Eugene Carson Blake, United Presbyterian Church in U.S.A.; A. Philip Randolph, Negro American Labor Council; the President; Walter P. Reuther, the United Automobile Workers; Vice President Johnson, almost hidden, and Roy Wilkins, N.A.A.C.P. Mr. Kennedy and Mr. Johnson met with leaders of the civil rights march at the White House after the ceremonies at the Lincoln Memorial.

". . . that one day even the State of Mississippi, a state sweltering with the heat of injustice, sweltering with the heat of oppression, will be transformed into an oasis of freedom and justice.

"I have a dream . . ." The crowd roared.

". . . that my four little children will one day live in a nation where they will not be judged by the color of their skin but by the content of their character.

"I have a dream . . . " The crowd roared.

". . . that one day every valley shall be exalted, every hill and mountain shall be made low, the rough places will be made plain, and the crooked places will be made straight, and the glory of the Lord shall be revealed and all flesh shall see it together."

As Dr. King concluded with a quotation from a Negro hymn—"Free at last, free at last, thank God almighty"—the crowd, recognizing that he was finishing, roared once again and waved their signs and pennants.

But the civil rights leaders, who knew the strength of the forces arrayed against them from past battles, knew also that a hard struggle lay ahead. The tone of their speeches was frequently militant.

Roy Wilkins, executive secretary of the National Association for the Advancement of Colored People, made plain that he and his colleagues thought the President's civil rights bill did not go nearly far enough. He said:

"The President's proposals represent so moderate an approach that if any one is weakened or eliminated, the remainder will be little more than sugar water. Indeed, the package needs strengthening."

Harshest of all the speakers was John Lewis, chairman of the Student Nonviolent Coordinating Committee.

"My friends," he said, "let us not forget that we are involved in a serious social revolution. But by and large American politics is dominated by politicians who build their career on immoral compromising and ally themselves with open forums of political, economic and social exploitation."

He concluded:

"They're talking about slowdown and stop. We will not stop.

"If we do not get meaningful legislation out of this Congress, the time will come when we will not confine our marching to Washington. We will march through the South, through the streets of Jackson, through the streets of Danville, through the streets of Cambridge, through the streets of Birmingham.

"But we will march with the spirit of love and the spirit of dignity that we have shown here today."

In the original text of the speech, distributed last night, Mr. Lewis had said:

"We will not wait for the President, the Justice Department, nor the Congress, but we will take matters into our own hands and create a source of power, outside of any national structure, that could and would assure us a victory."

He also said in the original text that "we will march through the South, through the heart of Dixie, the way Sherman did."

It was understood that at least the last of these statements was changed as the result of a protest by the Most Rev. Patrick J. O'Boyle, Roman Catholic Archbishop of Washington, who refused to give the invocation if the offending words were spoken by Mr. Lewis.

The great day really began the night before. As a half moon rose over the lagoon by the Jefferson Memorial and the tall, lighted shaft of the Washington Monument gleamed in the reflecting pool, a file of Negroes from out of town began climbing the steps of the Lincoln Memorial.

There, while the carpenters nailed the last planks on the television platforms for the next day and the TV technicians called through the loudspeakers, "Final audio, one, two, three, four," a middle-aged Negro couple, the man's arm around the shoulders of his plump wife, stood and read with their lips:

"If we shall suppose that American slavery is one of the offenses which in the providence of God must needs come but which having continued through His appointed time, He now wills to remove. . . ."

The day dawned clear and cool. At 7 A. M. the town had a Sunday appearance, except for the shuttle buses drawn up in front of Union Station, waiting.

By 10 A. M. there were 40,000 on the slopes around the Washington Monument. An hour later the police estimated the crowd at 90,000. And still they poured in.

Because some things went wrong at the monument, everything was right. Most of the stage and screen celebrities from New York and Hollywood who were scheduled to begin entertaining the crowd at 10 did not arrive at the airport until 11:15.

As a result the whole affair at the monument grounds began to take on the spontaneity of a church picnic. Even before the entertainment was to begin, groups of high school students were singing with wonderful improvisations and hand-clapping all over the monument slope.

Civil rights demonstrators who had been released from jail in Danville, Va., were singing:

Move on, move on,
Till all the world is free.

And members of Local 144 of the Hotel and Allied Service Employes Union from New York City, an integrated local since 1950, were stomping:

Oh, freedom, we shall not,
we shall not be moved,
Just like a tree that's
planted by the water.

Then the pros took over, starting with the folk singers. The crowd joined in with them.

Joan Baez started things rolling with "the song"—"We Shall Overcome."

Oh deep in my heart I do
believe
We shall overcome some
day.

And Peter, Paul and Mary sang "How many times must a man look up before he can see the sky."

And Odetta's great, full-throated voice carried almost to Capitol Hill: "If they ask you who you are, tell them you're a child of God."

Jackie Robinson told the crowd that "we cannot be turned back," and Norman Thomas, the venerable Socialist, said: "I'm glad I lived long enough to see this day."

The march to the Lincoln Memorial was supposed to start at 11:30, behind the leaders. But at 11:20 it set off spontaneously down Constitution Avenue behind the Kenilworth Knights, a local drum and bugle corps dazzling in yellow silk blazers, green trousers and green berets.

Apparently forgotten was the intention to make the march to the Lincoln Memorial a solemn tribute to Medgar W. Evers, N.A.A.C.P. official murdered in Jackson, Miss., last June 12, and others who had died for the cause of civil rights.

The leaders were lost, and they never did get to the head of the parade.

The leaders included also Walter P. Reuther, head of the United Automobile Workers; A. Philip Randolph, head of the American Negro Labor Council; the Rev. Dr. Eugene Carson Blake, vice chairman of the Commission on Religion and Race of the National Council of Churches; Mathew Ahmann, executive director of the National Catholic Conference for Interracial Justice; Rabbi Joachim Prinz, president of the American Jewish Congress; Whitney M. Young Jr., executive director of the National Urban League, and James Farmer, president of the Congress of Racial Equality.

All spoke at the memorial except Mr. Farmer, who is in jail in Louisiana following his arrest as a result of a civil rights demonstration. His speech was read by Floyd B. McKissick, CORE national chairman.

At the close of the ceremonies at the Lincoln Memorial, Bayard Rustin, the organizer of the march, asked Mr. Randolph, who conceived it, to lead the vast throng in a pledge.

Repeating after Mr. Randolph, the marchers pledged "complete personal commitment to the struggle for jobs and freedom for Americans" and "to carry the message of the march to my friends and neighbors back home and arouse them to an equal commitment and an equal effort."

August 29, 1963

Cites Jews' Progress

By M. S. HANDLER

The experience of the Jews in achieving an important place in American society within a relatively short time can, according to Malcolm X, provide valuable lessons to the 20 million American Negroes who are struggling for equal status.

The black nationalist leader expressed this view at a news conference Thursday, after his return from a pilgrimage to Mecca and a tour of many newly independent African nations.

Before journeying to Mecca, Malcolm withdrew from Elijah Muhammad's Black Muslim organization to form a new movement open to Negroes of all religions.

At his news conference, which was held at the Theresa Hotel in Harlem, Malcolm said:

"The American Jews have raised their own status in this country through their philosophical, cultural, and psychological migration to Israel.

"In the same way, the American Negroes can raise their own status by becoming deeply involved philosophically, culturally and psychologically with the new African nations."

In interviews before his journey to Mecca, Malcolm formulated his ideas in more concrete and elaborate terms. He said:

"The Jews have strengthened their own group consciousness and their own individual consciousness as Jews through their strong emotional attachment to the state of Israel. This close identification with Israel has intensified the individual Jew's personal identification with the great Jewish historical tradition, and he knows who he is as a man.

"This knowledge of one's self has enabled the Jew to become a highly effective man in this society and explains the psychological foundations of his tremendous success.

"This is an important lesson for the American Negroes who have no sense of cultural or historical identity because they don't know who they are. Their historic connections with Africa and African culture were destroyed by the slave owners.

"The result is that the American Negroes, ignorant of their African past, and therefore lacking in any justifiable pride in this past, are in a sense zombies because they don't know who they are."

May 24, 1964

PRESIDENT SIGNS CIVIL RIGHTS BILL; BIDS ALL BACK IT

By E. W. KENWORTHY
Special to The New York Times

WASHINGTON, July 2 — President Johnson signed the Civil Rights Act of 1964 tonight.

It is the most far-reaching civil rights law since Reconstruction days. The President announced steps to implement it and called on all Americans to help "eliminate the last vestiges of injustice in America."

"Let us close the springs of racial poison," he said in a short television address.

The President signed the bill in the East Room of the White House before television cameras shortly before 7 o'clock. That was about five hours after the House of Representatives had completed Congressional action on the sweeping bill.

Among other things, it prohibits discrimination in places of public accommodation, publicly owned facilities, employment and union membership and Federally aided programs.

Adopts Senate's Changes

The House approved, by a vote of 289 to 126, the changes that the Senate had made. All provisions of the measure became effective with President Johnson's signature except the one prohibiting discrimination in employment and union membership. This one goes into effect a year from now.

In announcing his implementation program, the President said he was, as had been previously indicated by White House sources, appointing former Gov. LeRoy Collins of Florida as director of the new Community Relations Service. Mr. Collins is now president of the National Association of Broadcasters.

Among other implementation steps, the President said he would name an advisory commission to help Mr. Collins resolve disputes arising under the bill. He will also ask Congress, Mr. Johnson said, for a supplemental appropriation to finance initial operations under the new law.

Surrounded by the leaders of both parties in both houses, who had labored to frame and pass the bill, President Johnson began his address to the nation by recalling that 188 years ago this week, "a small band of valiant men began a struggle for freedom" with the writing of the Declaration of Independence.

That struggle, he said, was a "turning point in history," and the ideals proclaimed in the Declaration of Independence still shape the struggles "of men who hunger for freedom."

Nevertheless, he declared, though Americans believe all men are created equal and have inalienable rights, many in America are denied equal treatment and do not enjoy those rights, or the blessings of liberty, "not because of their own failures, but because of the color of their skins."

The reasons, the President said, can be understood "without rancor or hatred" because they are deeply embedded in history, tradition and the nature of man.

"But it cannot continue," the President said with great earnestness.

Treatment Forbidden

The Constitution, the principles of freedom and morality all forbid such unequal treatment, he declared.

"And the law I will sign tonight forbids it," he said.

The President then sought to set fears at rest and to correct misapprehensions by stating that the new law would not restrict anyone's freedom "so long as he respects the rights of others," and would not give special treatment to any citizen.

"It does say that those who are equal before God shall now be equal in the polling booths, in the classrooms, in the factories, and in hotels, restaurants, movie theaters, and other places that provide service to the public."

Turning to the future, the President besought the cooperation of state and local officials, religious leaders, businessmen, and "every working man and housewife" in bringing "justice and hope to all our people—and peace to our land."

"My fellow citizens," he closed in solemn adjuration, "we have come to a time of testing. We must not fail.

"Let us close the springs of racial poison. Let us pray for wise and understanding hearts. Let us lay aside irrelevant differences and make our nation whole. Let us hasten that day when our unmeasured strength and our unbounded spirit will be free to do the great works ordained for this nation by the just and wise God who is the father of us all."

Discrimination Banned

The new law — the most sweeping civil rights legislation ever enacted in this country — goes beyond the proscribing of various forms of discrimination.

It gives the Attorney General authority to initiate suits to end discrimination in jobs and public accommodations when he finds such discrimination is part of a practice or pattern. It also gives him new powers to speed school desegregation and enforce the Negro's right to vote.

While the bill provides for the final enforcement of its sanctions by Federal court orders, it also gives state and local agencies primary jurisdiction to deal with complaints for a limited time.

The new Community Relations Service that Mr. Collins will head will be situated in the Commerce Department. It will assist communities in meeting problems and resolving disputes arising out of the desegregation of public accommodations.

The vote today was close to that of last Feb. 10 when the House passed its original bill by 290 to 130.

The party line-up today was 153 Democrats and 136 Republicans for the Senate substitute, and 91 Democrats and 35 Republicans against it.

Last February, 152 Democrats and 138 Republicans voted for the House version, and 96 Democrats and 34 Republicans against it.

Six members—five Republicans and one Democrat—switched from the positions they took last February.

Charlotte T. Reid of Illinois, Bob Wilson of California and Earl Wilson of Indiana, who voted for the bill before, opposed it today. All are Republicans.

On the other hand, two Republicans, Edward Hutchinson of Michigan and John J. Rhodes of Arizona, and one Democrat, Charles L. Weltner

Associated Press Wirephoto

RIGHTS BILL BECOMES LAW: President Johnson signing the bill yesterday in East Room of White House. Standing are, from left: Senator Everett McKinley Dirksen, Illinois Republican; Representative Clarence J. Brown, Ohio Republican; Senator Hubert H. Humphrey, Minnesota Democrat; Representatives Charles A. Halleck, Indiana Republican; William M. McCulloch, Ohio Republican, and Emanuel Celler, Brooklyn Democrat.

Associated Press Wirephoto

A view of the Civil Rights Act of 1964 as President Johnson completed his signature. The other signatures are those of Representative John W. McCormack, Speaker of the House, and Carl Hayden, President pro tem of Senate.

United Press International Telepnoto

Mr. Johnson gives one of pens used to sign bill to the Rev. Dr. Martin Luther King Jr., president of Southern Christian Leadership Conference. Senator Jacob K. Javits, New York Republican, who backed bill, looks into camera.

of Georgia, changed their nays to ayes.

Goldwater Backer Changes

Particular interest was aroused by the switches of Mr. Rhodes and Mr. Weltner.

Mr. Rhodes not only comes from the same state as Senator Barry Goldwater, the leading contender for Republican Presidential nomination, but he is also a prominent supporter of Mr. Goldwater.

When he voted aye today, Mr. Rhodes parted company with the Senator, who voted against closure on June 10 and against the Senate bill on June 19.

Mr. Weltner was the only Southern opponent of the bill to change position. He comes from Atlanta where, under the leadership of Mayor Ivan Allen Jr., much progress has been made on desegregation of public accommodations and beginnings made on school desegregation.

On June 17, 54 House Republicans announced their support for Mr. Goldwater, contending that his nomination would "result in substantial increases in Republican membership in both houses of Congress."

Twenty-eight of these Republicans voted for the bill today. Of the 25 who opposed it, nine are Southerners and three come from border states. The one who did not vote, James B. Utt of California, was paired against the bill.

It seemed obvious from this division that a 2-to-1 majority of the Northern Republicans who came out for Mr. Goldwater either did not agree with his position on the civil rights bill or thought a vote against it would not be to their advantage.

Representative Adam Clayton Powell Jr., Democrat of Manhattan, who has contended the bill does not go far enough, was not present for the vote but was paired for it.

Mr. Powell's office said he was attending the International Labor Conference in Geneva, Switzerland, on appointment by Speaker John W. McCormack.

Representative William E. Miller of New York, the Republican National Chairman, was also not present but paired for the measure.

The Senate substitute bill, which the House Rules Committee cleared for action two days ago, was the first order of business when the House convened at noon.

Under House rules, only one hour of debate was allowed. The outcome was certain. Nevertheless, the galleries were packed by those eager to witness the culmination of the long battle for a bill that civil rights advocates regard as a Magna Carta in the struggle to secure equal treatment and opportunity for the Negro.

They were not disappointed. The air of a great occasion hung over the chamber from the moment that Chaplain Bernard Braskamp began his prayer with the quotation from Leviticus engraved on the Liberty Bell:

"Proclaim liberty throughout all the land unto all the inhabitants thereof."

But the high point of the drama came when 36-year-old Representative Weltner of Georgia rose and explained, in a subdued voice and somewhat stumbling delivery, why he was going to switch his vote.

In February, he said, he voted against the bill because he questioned its means while approving its purpose.

"Now, after the most thorough and sifting examination in legislative history," he went on, "this measure returns for final consideration. It returns with the overwhelming approval of both houses of Congress.

"Manifestly, the issue is already decided, and approval is assured. By the time my name is called, votes sufficient for passage will have been recorded.

"What, then, is the proper course? Is it to vote 'no,' with tradition, safety—and futility?

"I believe a greater cause can be served. Change, swift and certain, is upon us, and we in the South face some difficult decisions.

"We can offer resistance and defiance, with their harvest of strife and tumult. We can suffer continued demonstrations, with their wake of violence and disorder.

"Or, we can acknowledge this measure as the law of the land. We can accept the verdict of the nation.

"Already, the responsible elements of my community are counseling this latter course. And, most assuredly, moderation, tranquility, and orderly processes combine as a cause greater than conformity.

"Mr. Speaker, I shall cast my lot with the leadership of my community [Atlanta]. I shall cast my vote with the greater cause they serve. I will add my voice to those who seek reasoned and conciliatory adjustment to a new reality.

Sees Unfinished Task

"And finally, I would urge that we at home now move on to the unfinished task of building a new South. We must not remain forever bound to another lost cause."

Applause burst from the civil rights advocates on both sides of the aisle as Mr. Weltner finished. His fellow Southerners sat stunned.

Time was controlled in the debate by Ray J. Madden of Indiana, ranking Northern Democrat on the Rules Committee, and Clarence J. Brown of Ohio the ranking Republican. Both of them allotted time so that opponents of the bill had at least half of the hour.

Howard W. Smith of Virginia, chairman of the Rules Committee, led off for the opponents. He said the history of the bill was one of "heedless trampling upon the rights of citizens from the time the first bill was introduced" in June, 1963.

"But the bell has tolled," he said. "In a few minutes you will vote this monstrous instrument of oppression upon all of the American people."

"You have sowed the wind. Now an oppressed people are to reap the whirlwind. King, Martin Luther, not satisfied with what will then be the law of the land, has announced his purpose, with the backing of the Executive department, to begin a series of demonstrations inevitably to be accompanied by bloodshed, violence, strife and bitterness."

"With all due disrespect to the Supreme Court of the United States," he concluded, "May I still be permitted to utter the pious and prayerful words 'God save the United States of America.'"

Louis G. Wyman, Republican of New Hampshire, who is a supporter of Senator Goldwater's candidacy, said he would have no fear "if we had a Supreme Court worthy of the name," because then the unconstitutional aspects of the bill "would soon be struck down."

"Of this we could be confident," he said. "Unfortunately, it is otherwise, and has been virtually ever since the incumbency of the present Chief Justice (Earl Warren)."

The Southerners gave him a rousing ovation.

William M. McCulloch of Ohio, ranking Republican on the Judiciary Committee, who was a principal architect of the House bill, wound up for the Republican advocates.

The Senate bill, like the original House measure, is "comprehensive in scope yet moderate in application," he said.

July 3, 1964

POWELL DEMANDS POWER FOR NEGRO

By AUSTIN C. WEHRWEIN
Special to The New York Times

CHICAGO, May 28—Representative Adam Clayton Powell, Manhattan Democrat, issued a call tonight for a "black revolution" in America.

It was not a call for an armed revolt but a rhetorical challenge for seizure of "audacious power."

Mr. Powell explained this as meaning that Negroes should seek to be "mayors, United States Senators, presidents of companies and members of stock exchanges."

His remarks were made in a speech at the annual dinner of the Ebenezer Missionary Baptist Church, held in McCormick Place, Chicago's convention hall.

'Audacious Power' Cited

Mr. Powell, who is the minister of the Abyssinian Baptist Church in New York's Harlem, said:

"As the Negro revolt was our Sunday of protests, so the black revolution must become our week of production."

"This can only be done by black people seeking power — audacious power," he continued.

The Negro Representative also sent out a call for Negro control of civil rights organizations, a move of considerable significance in the eyes of observers who regard these organizations as a meeting ground that has helped to control passions in the current racial crisis.

Mr. Powell made detailed references to Chicago politics.

There has been speculation that he wishes to extend his power beyond Harlem into the balliwick of Representative William L. Dawson, the aging Southside Negro Democratic political leader here.

Representative Powell recently made a scathing attack on the Office of Economic Opportunity, centering his attention on Chicago where he said the program was under the control of the political machine of Mayor Richard J. Daley, whose political organization embraces that of Mr. Dawson.

Mr. Dawson in the past has been called America's most powerful Negro leader.

Mr. Powell also said that black organizations should be black led and black financed.

"Jews control Jewish organizations," he said. "There are no Italians or Irish on the board of directors of B'nai B'rith. Poles control Polish-American organizations.

"But the moment a black man seeks to dominate his own organization he is labeled 'racist' and frightened black Uncle Toms quickly shun him and cuddle up to Mr. Charlie to prove their sniveling loyalty to the doctrine that 'white must be right.'"

Negroes call the white man Mr. Charlie.

Mr. Powell said that Negroes must refuse to accept anything less than "proportianate percentage of the political spoils."

"Black people must support and push black candidates for political office," he said. "This is a lesson you Chicago Negroes might well learn."

Holds News Conference

The Representative said that Negroes must reject the "white community's ceremonial Negro leaders." He said black leaders, such as ministers, politicians and businessmen "must come back to the Negroes who made them in the first place."

The New Yorker also declared that no Negro over 21 should be permitted to participate in a demonstration or picket line, or be any part of any civil rights effort if he is not a registered voter.

Mr. Powell said at a news conference that he wanted all Negro organizations, from the separatist Black Muslims to conservative integration groups, united for integration.

After that is achieved, he said, the various organizations would be free to go their separate paths.

John Ali, national secretary and public relations man for the Black Muslims, whose headquarters are here, was present at the news conference. Mr. Ali told a reporter, "He's on the right track."

Mr. Powell, who is chairman of the House Labor and Education Committee, also said that he would block passage of the repeal of Section 14(b) of the Taft-Hartley Act—the section that permits states to pass "right-to-work" laws—until an antisegregation clause was written into it.

A Position Paper

Mr. Powell delivered what he called a "Black Position Paper for America's 20 million Negroes," an allusion to diplomatic white papers, to an audience of 900 and was applauded 35 times. There was a sprinkling of whites.

Frequently he used heavy sarcasm and "inside" jokes that drew murmurs of agreement.

In what could be a bid for national leadership of the Negro movement, he urged, as he had at his earlier press conference, that all Negro groups, starting with the Black Muslims, "come together now and fight for desegregation."

Mr. Powell said that demonstrations should be nonviolent and violence should be curbed "even when it erupts recklessly in anger among our teenagers." However, he said, "Black people must continue to defy the laws of man when such laws conflict with the laws of God."

He declared that the white power structure must be "cracked" but said he did not want a black power structure but rather a structure of whites and blacks working together.

May 29, 1965

JOHNSON SIGNS VOTING RIGHTS BILL, ORDERS IMMEDIATE ENFORCEMENT; 4 SUITS WILL CHALLENGE POLL TAX

Room in Which Lincoln Freed Some Slaves in 1861 Is Used

By E. W. KENWORTHY
Special to The New York Times

WASHINGTON, Aug. 6 — President Johnson signed today the Voting Rights Act of 1965 and announced steps to bring about its quick and vigorous enforcement.

Tomorrow, he said, the Justice Department will officially certify the states where discrimination exists under the definition of the act.

On Monday morning, the President continued, the Justice Department will designate the counties in those states "where past experience" had shown that Federal action was needed to register Negroes.

On Tuesday, he said, Federal examiners will begin registering Negroes "in 10 to 15 counties."

Poll Tax Challenged

The President also announced that tomorrow at 1 P.M., Attorney General Nicholas deB. Katzenbach would file suit challenging the poll tax of the state of Mississippi on the ground that it is used to abridge the right of Negroes to vote in violation of the 15th Amendment to the Constitution.

Next Tuesday, the President went on, additional suits will be filed against the three other states — Alabama, Texas and Virginia—that require payment of a poll tax as a prerequisite to voting in state and local elections.

The signing took place in the President's Room of the Capitol, just off the Senate chamber. There, 104 years ago today, President Lincoln signed a bill freeing slaves impressed into the service of the Confederacy.

Will Act Quickly

Gathered in the small, ornate room, dominated by a great gilt chandelier, were Vice President Humphrey, the Cabinet, Congressional leaders, members of the Senate and House Judiciary Committees who had refined

United Press International Telephoto

IN THE ROTUNDA: President Johnson speaks at Capitol before signing Voting Rights Act. To left are daughter Luci Baines, Vice President Humphrey and members of Congress. "Surrender of Cornwallis" by John Trumbull is on wall.

and strengthened the bill, and Negro and white leaders of the civil rights movement.

Earlier, in the Rotunda of the Capitol, the President said that Congress had acted swiftly in passing the bill and that he intended to act "with equal dispatch" in enforcing it.

In his speech, which was broadcast and televised nationally, the President compressed into three sentences the history, the meaning and the hope of the occasion. He said:

"Today is a triumph for freedom as huge as any victory won on any battlefield.

"Today we strike away the last major shackle of those fierce and ancient bonds.

"Today the Negro story and the American story fuse and blend."

The Voting Rights Act, the President declared, "flows from a clear and simple wrong," and its only purpose "is to right that wrong."

It is nearly five months since the President sent his draft bill to Congress with the adjuration to work "long hours and nights if necessary to pass this bill."

Today the President said that the product of the labors of the Justice Department and Congress was long and complicated.

"But," he added, "the heart of the act is plain."

Wherever states and counties are using literacy tests or other devices to deny the right to vote, the President said, those tests and devices "will be struck down."

And wherever state and local officials persist in discriminating against Negro applicants, "Federal examiners will be sent in to register all eligible voters," he continued.

The states and counties that will be automatically covered by the act are those that have literacy tests and had less than 50 per cent of their voting-age population registered or voting in the Presidential election of 1964.

This formula covers the states of Alabama, Mississippi, Louisiana, Georgia, South Carolina and Virginia, and about 34 counties in North Carolina.

2 Million Unregistered

There are more than 2 million unregistered Negroes of voting age in the areas covered by the law. Justice Department officials believe a good portion of these may be registered before next year's general election.

The influx of Negro voters is expected to temper the traditionally conservative tone of Deep South politics because Negroes usually favor welfare legislation and expanded public services.

In areas of the South where Negroes have been registered in large numbers — Atlanta and Memphis, for instance — the effect has been not only to liberalize the political platforms but also to end the traditional contest in racial demagoguery by candidates for office.

The President drove to the Capitol shortly before noon and went first to the office of House Speaker John W. McCormack where the Vice President, Congressional leaders, Cabinet members and prominent civil rights leaders awaited him.

Among the Negro leaders were Roy Wilkins, head of the National Association for the Advancement of Colored People; James Farmer, national director of the Congress of Racial Equality; John Lewis, chairman of the Student Nonviolent Coordinating Committee, and the the Rev. Dr. Martin Luther King Jr., president of the Southern Christian Leadership Conference.

Led Drive in Selma

Dr. King conceived and led, beginning last January, the voting rights demonstrations in Selma, Ala. It was the club-wielding response of Sheriff James G. Clark Jr., and his deputies — what the President today called "the outrage of Selma" — that spurred the Administration to speed a voting rights bill to Congress.

The President and his invited guests then walked to the Rotunda under the dome of the Capitol.

His lectern had been placed with attention to history and symbol. On his left was the model of Gutzon Borglum's head of Lincoln on Mount Rushmore in South Dakota. On his right was the standing figure of Lincoln by Vinnie Ream. Behind him was the Trumbull painting of Cornwallis's surrender to George Washington at Yorktown.

Behind the President, as he spoke, were the Cabinet, Congressional leaders and his daughter Luci. Before him were gathered members of Congress, the diplomatic corps and other invited guests.

Associated Press Wirephoto

AS JOHNSON SIGNED VOTER BILL: Attending ceremony at Capitol are, front row, from left: Vice President Humphrey, House Speaker John W. McCormack, Representative Emanuel Celler, Brooklyn Democrat; Luci Baines Johnson, and Senator Everett McKinley Dirksen, Illinois Republican. Behind Mr. Humphrey is Representative Carl Albert, Oklahoma Democrat. Behind Mr. Celler is Senator Carl Hayden of Oklahoma, President Pro Tem of Senate.

In his speech the President had a special message for those hard-core areas of resistance to Negro suffrage. He said:

"Under this act, if any county in the nation does not want Federal intervention it need only open its polling places to all people."

But he sounded a note of caution when he said:

"This act is not only a victory for Negro leadership; this act is a great challenge to that leadership. It is a challenge which cannot be met simply by protests and demonstrations. It means that dedicated leaders must work around the clock to teach people their rights and their responsibilities and to lead them to exercise those rights and to fulfill those responsibilities and those duties to their country."

Continuing in this vein he told Southern Negroes and their leaders:

"Only the individual Negro can walk through those doors [of the polling place], use that right and transform the vote into an instrument of justice and fulfillment.

"Let me say now to every Negro in this country: You must register. You must vote. And you must learn, so your choice advances your interest and the interest of the nation."

And he exhorted Negro leaders to remember that they could not meet their challenge simply by demonstrations.

Finally the President had a word both for the North and the South.

While, he said to the North, "there is no room for injustice in the American mansion," there is room for "understanding toward those who see old ways crumbling."

And to the South, he said:

"It must come. It is right that it should come. And when it has, you'll find a burden has been lifted from your shoulders too."

The burden would be lifted from the South, and it would from the nation, because the central fact of American civilization is "freedom, and justice, and the dignity of man," Mr. Johnson declared.

As long as some are oppressed, the President said, belief in these things is blunted and the strength of the nation's high purpose is sapped.

"It is not just a question of guilt, although there is that," he said. "It is that men cannot live with a lie and not be stained by it."

Turning to the future, the President said that the struggle for true equality must move toward a different battlefield, so that the Negro would not merely be secure in his legal rights but be able "to enter the mainstream of American life."

After the speech the President walked to the Senate side of the Capitol and made his way to the President's Room. More than 100 persons crowded in to watch the signing with multiple pens.

The green baize-covered walnut table—known as the "Lincoln table"—that occupies the center of the room had been pushed aside and replaced with a small, simple desk, also covered in green baize.

There is some dispute whether Lincoln used this desk or the "Lincoln table" when he signed the bill freeing slaves forced into the service of the Confederate Army.

But the desk had a personal meaning for the President because it was this desk that he used as majority leader, and it was from this desk that he guided through the Senate the 1957 and 1960 civil rights acts.

Vice President Humphrey was given the first pen, and the Senate Republican leader, Everett McKinley Dirksen, the second.

In this way the President paid tribute to the two men who had played the leading roles in Senate passage of the omnibus Civil Rights Act of 1964. Mr. Dirksen worked closely with Attorney General Katzenbach on drafting the voting rights bill as he had on the civil rights bill.

Among those whom the President invited to witness the signing were Vivian Malone, the first Negro to enter the University of Alabama at Tuscaloosa, and Mrs. Rosa Parks, who started the Montgomery, Ala., bus boycott in 1956. That boycott was the beginning of the Negro protest movement and led to the 1957 Civil Rights Act, the first civil rights legislation since Reconstruction.

August 7, 1965

'The Wonder Is There Have Been So Few Riots'

By KENNETH B. CLARK

IT is one measure of the depth and insidiousness of American racism that the nation ignores the rage of the rejected—until it explodes in Watts or Harlem. The wonder is that there have been so few riots, that Negroes generally are law-abiding in a world where the law itself has seemed an enemy.

To call for reason and moderation, to charge rioters with blocking the momentum of the civil-rights movement, to punish rioters by threatening withdrawal of white support for civil rights may indeed ease the fears of whites and restore confidence that a firm stern hand is enforcing order.

But the rejected Negro in the ghetto is deaf to such moral appeals. They only reinforce his despair that whites do not consider equal rights for Negroes to be their due as human beings and American citizens but as rewards to be given for good behavior, to be withheld for misbehavior. The difficulty which the average American of goodwill—white or Negro—has in seeing this as a form of racist condescension is another disturbing symptom of the complexities of racism in the United States.

It is not possible for even the most responsible Negro leaders to control the Negro masses once pent-up anger and total despair are unleashed by a thoughtless or brutal act. The prisoners of the ghetto riot without reason, without organization and without leadership, as this is generally understood. The rioting is in itself a repudiation of leadership. It is the expression of the anarchy of the profoundly alienated.

In a deeper sense such anarchy could even be a subconscious or conscious invitation to self-destruction. At the height of the Harlem riots of 1964, young Negroes could be heard to say, "If I don't get killed tonight, I'll come back tomorrow." There is evidence these outbreaks are suicidal, reflecting the ultimate in self-negation, self-rejection and hopelessness.

It was the Negro ghetto in Los Angeles which Negroes looted and burned, not the white community. When white firemen tried to enter the ghetto, they were barred by Negro snipers. Many looters did not take the trouble to avoid injury, and many were badly cut in the looting orgy. So one cannot help but wonder whether a desire for self-destruction was not a subconscious factor. Of the 36 people killed in the Los Angeles riot, 33 were Negroes, killed in the campaign to restore law and order. The fact of their deaths—the senseless deaths of human beings—has been obscured by our respectable middle-class preoccupation with the wanton destruction of property, the vandalism and the looting.

Appeals to reason are understandable; they reflect the sense of responsibility of Governmental and civil-rights leaders. But they certainly do not take into account the fact that one cannot expect individuals who have been systematically excluded from the privileges of middle-class life to view themselves as middle-class or to behave in terms of middle-class values. Those who despair in the ghetto follow their own laws—generally the laws of unreason. And though these laws are not in themselves moral, they have moral consequences and moral causes.

THE inmates of the ghetto have no realistic stake in respecting property because in a basic sense they do not possess it. They are possessed by it. Property is, rather, an instrument for perpetuation of their own exploitation. Stores in the ghetto—which they rarely own—overcharge for inferior goods. They may obtain the symbols of America's vaunted high standard of living—radios, TV's, washing machines, refrigerators—but usually only through usurious carrying costs, one more symbol of the pattern of material exploitation. They do not respect property because property is almost invariably used to degrade them.

James Bryant Conant and others have warned America it is no longer possible to confine hundreds of thousands of undereducated, underemployed, purposeless young people and adults in an affluent America without storing up social dynamite. The dark ghettoes now represent a nuclear stock pile which can annihilate the very foundations of America. And if, as a minority, desperate Negroes are not able to "win over" the majority, they can nevertheless effectively undermine what they cannot win.

A small minority of Negroes can do this. Such warnings are generally ignored during the interludes of apparent quiescence and tend to be violently rejected, particularly when they come from whites, at the time of a Negro revolt.

When Senator Robert Kennedy incisively observed, after Watts, "There is no point in telling Negroes to observe the law. . . . It has almost always been used against [them]," it was described by an individual who took the trouble to write a letter to The New York Times as an irresponsible incitement to violence. The bedeviling fact remains, however, that as long as institutionalized forms of American racism persist, violent eruptions will continue to occur in the Negro ghettoes. As Senator Kennedy warned: "All these places—Harlem, Watts, South Side—are riots waiting to happen."

When they do happen, the oversimplified term "police brutality" will be heard, but the relationship between police and residents of the ghetto is more complicated than that. Unquestionably, police brutality occurs. In the panic probably stemming from deep and complex forms of racism, inexperienced policemen have injured or killed Negroes or Puerto Ricans or other members of a powerless minority. And it is certainly true that a common denominator of most, if not all, the riots of the past two summers has been some incident involving the police, an incident which the larger society views as trivial but which prisoners of the ghetto interpret as cruel and humiliating.

In spite of the exacerbating frequency of police racism, however, the more pertinent cause of the ghetto's contempt for police is the role they are believed to play in crime and corruption within the ghetto—accepting bribes for winking at illegal ac-

KENNETH B. CLARK is a professor of psychology at City College of New York and the author of "Dark Ghetto," published this past spring.

tivities which thrive in the ghetto. The police, rightly or wrongly, are viewed not only as significant agents in exploiting ghetto residents but also as symbols of the pathology which encompasses the ghetto. They are seen matter-of-factly as adversaries as well as burdens. The more privileged society may decide that respect for law and order is essential for its own survival, but in the dark ghettoes, survival often depends on disrespect for the law as Negroes experience it.

Thus the problem will not be solved merely by reducing the frequency of police brutality or by increasing the number of Negro policemen. It will require major reorganization and reeducation of the police and a major reorganization of the ghetto itself. To say as Police Chief William Parker did of the Los Angeles Negroes, "We are on top and they are on the bottom," is to prove to Negroes that their deep fears and hatred of established law and order are justified.

While the riots cannot be understood by attempts to excuse them, neither can they be understood by deploring them—especially by deploring them according to a double standard of social morality. For while the lawlessness of white segregationists and rebellious Negroes are expressions of deep frustrations and chronic racism, the lawlessness of Negroes is usually considered a reflection on all Negroes and countered by the full force of police and other governmental authority, but the lawlessness of whites is seen as the primitive reactions of a small group of unstable individuals and is frequently ignored by the police—when they are not themselves accessories. Moreover, rarely do the leaders of a white community in which white violence occurs publicly condemn even the known perpetrators, while almost invariably national and local Negro leaders are required to condemn the mob violence of Negroes.

As long as these double standards of social morality prevail, they reflect the forms of accepted racism which are the embers of potential violence on the part of both Negroes and whites. And it should be obvious also, although it does not appear to be, that the violence of the Negro is the violence of the oppressed while the violence of white segregationists seeks to maintain oppression.

It is significant that the recent eruptions in Negro communities have not occurred in areas dominated by more flagrant forms of racism, by the Klan and the other institutions of Southern bigotry. They have occurred precisely in those communities where whites have prided themselves on their liberal approach to matters of race and in those states having strong laws prescribing equal opportunity, fair employment and allegedly open housing. (Some observers see a relationship between the defeat of the open housing referendum in California and the Los Angeles outbreak, but it would seem misleading to attempt to account for the riot by any single factor.)

It is revealing to hear the stunned reaction of some top political officials in Los Angeles and California who are unable to understand that such a thing could happen in Los Angeles. Here, they said, whites and blacks got along fine together; here, as reporters constantly pointed out, ghetto streets are lined with palm trees, some with private homes surrounded by tended lawns.

Americans are accustomed to judging the state of people's minds by the most visible aspects—the presence of a TV antenna indicates affluence and a neat lawn a middle-class home. The fact is a ghetto takes on the physical appearance of the particular city—in New York, rat-infested tenements and dirty streets; in Los Angeles, small homes with palm trees—but in many a small home live numerous families and in every house live segregated, desperate people with no jobs or servile jobs, little education, broken families, delinquency, drug addiction, and a burning rage at a society that excludes them from the things it values most.

It is probably not by chance that the Federal Civil Rights Act of 1964 and the Voting Rights Act of 1965 were followed by violence in the North. This was important legislation, but it was more relevant to the predicament of the Southern Negro than to Negroes in Northern ghettoes.

It may well be that the channeling of energies of Negroes in Southern communities toward eliminating the more vicious and obvious signs of racism precludes temporarily the dissipation of energy in random violence. The Northern Negro is clearly not suffering from a lack of laws. But he is suffering—rejected, segregated, discriminated against in employment, in housing, his children subjugated in *de facto* segregated and inferior schools in spite of a plethora of laws that imply the contrary.

He has been told of great progress in civil rights during the past 10 years and proof of this progress is offered in terms of Supreme Court decisions and civil rights legislation and firm Presidential commitment. But he sees no positive changes in his day-today life. The very verbalizations of progress contribute to his frustration and rage. He is suffering from a pervasive, insensitive and at times self-righteous form of American racism that does not understand the depth of his need.

Not the civil-rights leaders who urge him to demonstrate, but the whites who urge him not to "in the light of present progress" contribute to the anger which explodes in sudden fury. He is told by liberal whites *they* contribute to civil rights causes, *they* marched to Washington and journeyed to Selma and Montgomery to demonstrate their commitment to racial justice and equality.

But Negroes see only the continuing decay of their homes, many of them owned by liberal whites. He sees he does not own any of the means of production, distribution and sale of goods he must purchase to live. He sees his children subjected to criminally inefficient education in public schools they are required to attend, and which are often administered and staffed by liberal whites. He sees liberal labor unions which either exclude him, accept him in token numbers or, even when they do accept him en masse, exclude him from leadership or policy-making roles.

And he sees that persistent protest in the face of racism which dominates his life and shackles him within the ghetto may be interpreted by his white friends as a sign of his insatiability, his irrationality and, above all, of his ingratitude. And because this interpretation comes from his friends and allies it is much harder to take psychologically than the clear-cut bigotry of open segregationists.

It is precisely at this point in the development of race relations that the complexities, depth and intensity of American racism reveal themselves with excruciating clarity. At this point regional differences disappear. The greatest danger is an intensification of racism leading to the polarization of America into white and black. "What do they *want?*" the white man asks. "Don't they know they hurt their own cause?" "Get Whitey," cries the Negro. "Burn, baby, burn." At this point concerned whites and Negroes are required to face the extent of personal damage which racism has inflicted on both.

It will require from whites more than financial contributions to civil-rights agencies, more than mere verbal and intellectual support for the cause of justice. It will require compassion, willingness to accept hostility and increased resolve to go about the common business, the transformation and strengthening of our society toward the point where race and color are no longer relevant in discussing the opportunities, rights and responsibilities of Americans.

Negroes, too, are confronted with difficult challenges in the present stage of the civil-rights struggle. The bitterness and rage which formed the basis for the protests against flagrant racial injustices must somehow be channeled into constructive, nondramatic programs required to translate court decisions, legislation and growing political power into actual changes in the living conditions of the masses of Negroes. Some ways must be found whereby Negro leadership and Negro organizations can redirect the volatile emotions of Negroes away from wasteful, sporadic outbursts and toward self-help and constructive social action. The need for candid communication between middle-class Negroes and the Negro masses is as imperative as the need for painful honesty and cooperation between Negroes and whites.

These demands upon whites and Negroes will not be easy to meet since it is difficult, if not impossible, for anyone growing up in America to escape some form of racist contamination. And a most disturbing fact is the tendency of racism to perpetuate itself, to resist even the most stark imperatives for change. This is the contemporary crisis in race relations which Americans must somehow find the strength to face and solve. Otherwise we will remain the victims of capricious and destructive racial animosities and riots.

The key danger is the possibility that America has permitted the cancer of racism to spread so far that present available remedies can be only palliative. One must, however, continue to believe and act on the assumption that the disease is remediable.

It is important that all three branches of the Federal Government have committed themselves to using their power to improve the status of Negroes. These commitments must be enforced despite overt or subtle attempts to resist and evade them.

But this resistance must be seen not only in the bigotry of segregationists. It must be recognized in the moralizing of Northern whites who do not consciously feel themselves afflicted with the disease of racism, even as they assert that Negro rioting justifies ending their involvement in the civil-rights cause. It must be recognized in the insistence that Negroes pull themselves up by their own bootstraps, demonstrating to the liberal and white communities they have earned the right to be treated as equal American citizens. These are satisfying, self-righteous arguments but they cannot disguise the profound realities of an unacknowledged racism.

If it is possible to talk of any value emerging from the riots it would be this: They are signals of distress, an SOS from the ghetto. They also provide the basis for therapeutically ventilating deeply repressed feelings of whites and Negroes—their underlying fear and the primitive sense of race.

In the religiously oriented, nonviolent civil rights movement in the South, courteous, neatly dressed Negroes carrying books fitted into the middle-class white image more adequately than the vulgar whites who harassed them. The middle-class white, therefore, identified with the oppressed, not the oppressor. But empathy given as a reward for respectable behavior has little value. Understanding can only be tested when one's own interests are deeply threatened, one's sensibilities violated.

These feelings of hostility must be exposed to cold reality as the prelude to realistic programs for change. If under the warmth of apparent support for civil rights lies a deeply repressed prejudice, no realistic social change can be effective.

It would be unrealistic, of course, to expect the masses of whites and Negroes who have grown up in an essentially racist society suddenly to love one another. Fortunately, love is not a prerequisite for the social reorganization now demanded. Love has not been necessary to create workable living arrangements among other ethnic groups in our society. It is no more relevant to ask Negroes and whites to love each other than to ask Italians and Irish to do the same as a prerequisite for social peace and justice.

Nevertheless, real changes in the predicament of previously rejected Negroes—changes compatible with a stable and decent society—must be made, and soon.

The Negro must be included within the economy at all levels of employment, thereby providing the basis for a sound family life and an opportunity to have an actual stake in American business and property.

The social organization of our educational system must be transformed so Negroes can be taught in schools which do not reinforce their feeling of inferiority. The reorganization, improvement and integration of our public schools is also necessary in order to re-educate white children and prepare them to live in the present and future world of racial diversity.

The conditions under which Negroes live must be improved —bad housing, infant mortality and disease, delinquency, drug addiction must be drastically reduced.

Until these minimum goals are achieved, Americans must accept the fact that we cannot expect to maintain racial ghettoes without paying a high price. If it is possible for Americans to carry out realistic programs to change the lives of human beings now confined within their ghettoes, the ghetto will be destroyed rationally, by plan, and not by random self-destructive forces. Only then will American society not remain at the mercy of primitive, frightening, irrational attempts by prisoners in the ghetto to destroy their own prison.

September 5, 1965

Rights March Disunity

By GENE ROBERTS
Special to The New York Times

JACKSON, Miss., June 27—The civil rights march through Mississippi has made it clear that a new philosophy of "black consciousness" is sweeping the civil rights movement. When the march began, the philosophy was only an idea in the mind of Stokely Carmichael and other members of the Student Nonviolent Coordinating Committee. But when the three - week - old march ended yesterday, the new philosophy had given rise to a distinct movement within a movement. It had Mr. Carmichael as its leader and the late Malcolm X as its prophet. It also had a battle cry, "Black Power," and a slogan directed at whites, "Move on Over, or We'll Move on Over YOU."

News Analysis

Just how many adherents the black consciousness movement has is a matter of heated debate, but there appear to be hundreds in Mississippi, with hundreds more fascinated by it.

Its principal appeal is for Negroes between the ages of 15 and 30, but it also provides an emotional release for older Negroes.

As the rights march drew to an end yesterday a truckload of "black power" advocates passed a 61-year-old Negro domestic sipping ice water under a Mimosa tree.

"Say black power," those in the truck yelled to the woman under the tree.

"Black power," she yelled back and then giggled and shook her glass of ice cubes at the audacity of it all.

Appeal to the Young

For all of its appeal to younger Negroes, however, the movement is potentially the most disruptive force yet in the rights movement. The Rev. Dr. Martin Luther King Jr. and his aides in the Southern Christian Leadership Conference are alarmed that the "black power" chant might alienate white supporters from the rights drive and make it more difficult to get new legislation through Congress and financial and moral help from Northern white liberals.

All this has strained an already tenuous working relationship among civil rights organizations, but Dr. King has stopped short of disassociating himself from the Student Committee, a development that would strip him of much of the youthful support he needs to mount successful demonstrations.

Meanwhile, Dr. King is in the position of helping to popularize the black consciousness movement and the "black power" chant.

Reporters and cameramen drawn to a demonstration by the magic of Dr. King's name stay to write about and photograph Mr. Carmichael as he demonstrates and talks alongside the older civil rights leader. This is an advantage that Malcolm X never had as he exalted the Negro and attacked the "duplicity" and "hypocracy" of the white man.

Should Dr. King decide to step up his attacks on the new philosophy, his task would be difficult because the philosophy almost defies definition. It has been called black nationalism, but that term has been used in the past to describe paramilitary movements, back to Africa movements and "separate Negro state" movements. And none of these is what most Negroes are calling for when they chant "Black power."

The new philosophy has also been called racist and separatist, but the Student Committee, unlike the Black Muslims, is prepared to work with whites, providing the working relationship is built upon terms laid down by Negroes.

Although the movement is amorphous and could grow more strident at any moment, the term "black consciousness" appears to suit it best for the present.

An Angry Movement

It is an angry movement, designed to shake the Negro into political action by telling him repeatedly that he is as good as the white man, that white society has failed him, that it is demeaning to ask the white man for anything, including school desegregation, and that whatever the Negro wants he should take by drowning "white paternalism" in a sea of black votes.

All this is enough to make the new philosophy an "umbrella" philosophy, broad enough to give shelter to paramilitary organizations, such as the Deacons for Defense and Justice and to Negroes, who dislike white people because they are white.

It is also emotional enough to lead Mr. Carmichael, in the heat of a speech, to talk of "tearing down" courthouses, and Willie Ricks, a Student Committee field secretary, to declare that "white blood will flow" if black blood does.

After the speeches are over and the emotionalism is gone, Mr. Ricks and Mr. Carmichael spend hours explaining that they are not antiwhite but are simply trying to make the Negro feel pride in his race.

If white liberals do not understand this, Mr. Carmichael says, then this proves that they are not willing to grant the Negro full equality.

But there are Negroes, as well as whites, who fail to follow the Student Committee's dialectic, and thus, the civil rights movement appears to be entering its stormiest period.

June 28, 1966

Wilkins Says Black Power Leads Only to Black Death

By M. S. HANDLER
Special to The New York Times

LOS ANGELES, July 5—Roy Wilkins, executive director of the National Association for the Advancement of Colored People, denounced the concept of black power tonight as black racism that could only lead Negroes to a "black death."

[The Rev. Dr. Martin Luther King Jr. said Tuesday that he was opposed to the term "black power" because it "connotates black supremacy and an antiwhite feeling that does not or should not prevail."]

In his keynote address at the opening of the annual convention of the N.A.A.C.P., Mr. Wilkins seemed to be challenging the Congress of Racial Equality and the Student Nonviolent Coordinating Committee.

Besides denouncing their concept of black power, he rejected any modification of the doctrine of nonviolence as a prelude to "lynchings" and "counterviolence."

If nonviolence is reinterpreted to mean "instant retaliation in cases adjudged by aggrieved persons to have been grossly unjust," Mr. Wilkins said, "this policy could produce in extreme situations lynchings, or, in better-sounding phraseology, private vigilante behavior."

"Moreover, in attempting to substitute for derelict law enforcement machinery, the policy entails the risk of a broader, more indiscriminate crackdown by law officers under the ready-made excuse of restoring law and order," he declared.

However, Mr. Wilkins said, the idea of black power opens a more serious division in the civil rights movement than does a new concept of nonviolence.

He said:

"No matter how endlessly they try to explain it, the term 'black power' means antiwhite power. In a racially pluralistic society, the concept, the formation and the exercise of an ethnically tagged power means opposition to other ethnic powers.

"In the black-white relationship, it means that every other ethnic power is the rival and the antagonist of 'black power.' It has to mean 'going it alone.' It has to mean separatism."

Separatism, he warned, offers "a disadvantaged minority little except the chance to shrivel and die."

Throughout the speech, Mr. Wilkins received warm applause.

Mr. Wilkins said that until recently there had been some differences in methods and emphasis among civil rights groups, but none in ultimate goals. The aim, he said, was always the inclusion of the Negro into society.

Now, he said, there is "a strident and threatening challenge to a strategy widely employed by civil rights groups, namely nonviolence."

"One organization, which has been meeting in Baltimore, has passed a resolution declaring for defense of themselves by Negro citizens if they are attacked," he said.

Mr. Wilkins argued that the N.A.A.C.P had always defended in court persons who had defended themselves and their homes with firearms.

Members of the American Nazi party caused two disturbances at the convention, which was held at the First Methodist Church, near Wilshire Boulevard.

In the eevning, a Nazi in civilian clothes dashed across the front of the hall, shouting and cursing Jews and Negroes. He rushed up on the platform, but was seized and dragged off. The more than a thousand delegates sat quietly during the incident.

The Nazi appeared while Bishop Stephen Gill Spotsswood, chairman of the N.A.A.C.P. board, was eulogizing Arthur B. Spingarn, who had retired from the presidency of the N.A.A.C.P. last year.

This afternoon a carload of Nazis passed the Statler Hilton Hotel, where most of the delegates are staying, and screamed denunciations of Jews and Negroes. They left after delegates ignored them.

The convention opened without representatives of the N.A.A.C.P. Legal Defense and Educational Fund, Inc., the chief legal arm of the civil rights movement.

It is the first time that the fund, which has argued almost all the major civil rights cases in the Supreme Court and the lower Federal courts since the 1954 school desegregation decision, has not been represented at a convention of the N.A.A.C.P.

Although convention sources said that there had been no open break between the two organizations, it was understood that friction had developed as a result of competition for financial aid and a feeling among some members that the legal defense fund should be directed by a Negro.

In past years, it has been customary for the director of the fund, Jack Greenberg, a white lawyer, to hold luncheons and discussion sessions parallel to the functions of the N.A.A.C.P. convention.

This year, however, Mr. Greenberg received a letter from Dr. John A. Morsell, the assistant executive director of the N.A.A.C.P., which, in effect, suggested that he and his associates not attend the convention.

One reason put forward by Dr. Morsell was that the N.A.A.C.P. did not wish to have the activities of the fund divert national attention from the deliberations of the convention.

Today, Dr. Morsell said that the meetings of the fund might also have detracted from tonight's keynote address by Roy Wilkins, executive director of the N.A.A.C.P.

Dr. Morsell asserted that the absence of Mr. Greenberg and his associates did not conceal any ideological or personal difficulties between the N.A.A.C.P. and the fund.

The fund, an incorporated, separate tax-exempt entity, raised $1.75-million last year to prepare and fight major civil rights legal actions. This year it is seeking to raise $2-million.

It has received financial support from the white community and from foundations, and has been able to enlist some of the best legal minds in the United States to help prepare its briefs.

The N.A.A.C.P. also has a legal section that is becoming progressively more involved in civil rights action. Its cases deal mostly with specific grievances that do not necessarily raise constitutional questions. Last year the legal section of N.A.A.C.P. raised $350,000.

Some delegates said privately that they believed the invitation to Mr. Greenberg and his associates to ignore the convention might be traced to the competition in fund-raising between the fund and the N.A.A.C.P. legal office.

Another reason offered was that there was a growing feeling among Negro lawyers that the fund, since it is the chief legal arm of the civil rights movement, should be directed by a Negro.

This attitude was said to reflect the general trend of replacing whites in leadership positions in the rights movement with Negroes. Nonetheless, the president of the N.A.A.C.P., Kivie Kaplan, is white.

July 6, 1966

Bloc Voting for Power

To the Editor:

"Black power" and rejection of nonviolence are essentially distinct issues. Yet both involve the question of whether American Negroes are prepared to accept American political and moral ideology.

Black power is nothing more or less than bloc power. There is nothing sinister, improper, or even unusual for a group of Americans with common interests to seek political influence through peaceful action as a bloc in voting or in publicizing a cause. Throughout our history, bloc power has been exercised by minority religious blocs (such as Jews, Catholics and Mormons), economic blocs (such as labor unions, farmers, manufacturers and doctors), and ethnic blocs (such as Irish, Italian and Polish Americans).

Negro Americans have a community of interests, and a need for united action, that far transcend the scope of common interests and needs of other blocs in our society. Moreover, it is entirely proper, and a psychological necessity, for Negroes themselves to control Negro political and social action groups. Catholics lead and control the National Catholic Welfare Conference, and doctors lead and control the American Medical Association. It would be surprising to find it otherwise.

Doctrine of Self-Defense

The doctrine of nonviolence in the face of physical attack is foreign to American law and morality. Its rejection by Negroes is long overdue. The vast majority of Americans have consistently recognized the moral imperative of self-defense, both in international relations and in private action.

Just as our view of international law sanctions our counterattack after Pearl Harbor, our civil law recognizes the right of the individual to counter physical attack by sufficient force to protect himself from harm. This right is all the more important when, as with beatings and murders of Negroes, society has failed effectively to substitute public sanctions for self-defense.

MONROE H. FREEDMAN
Professor of Law
The George Washington University
Washington, July 12, 1966

July 24, 1966

PANEL ON CIVIL DISORDERS CALLS FOR DRASTIC ACTION TO AVOID 2-SOCIETY NATION

By JOHN HERBERS
Special to The New York Times

WASHINGTON, Feb. 29—The President's National Advisory Commission on Civil Disorders gave this warning to Americans tonight: "Our nation is moving toward two societies, one black, one white—separate and unequal."

Unless drastic and costly remedies are begun at once, the commission said, there will be a "continuing polarization of the American community and, ultimately, the destruction of basic democratic values."

The commission said "white racism" was chiefly to blame for the explosive conditions that sparked riots in American cities during the last few summers. But it also warned that a policy of separatism now advocated by many black militants "can only relegate Negroes to a permanently inferior economic state."

As for the civil disorders that ravaged American cities last summer, the commission said they "were not caused by, nor were they the consequences of, any organized plan or 'conspiracy.'"

Broad Proposals

The panel made sweeping recommendations at Federal and local levels in law enforcement, welfare, employment, education and the news media. It made no attempt to put a price tag on these recommendations, but they go far beyond social programs that are now in trouble in Congress because of a tight budget. They would cost many billions of dollars.

"The vital needs of the nation must be met," the commission said. "Hard choices must be made, and, if necessary, new taxes enacted."

The 11-member commission, headed by Gov. Otto Kerner of Illinois, was appointed by President Johnson last July 27 to find the causes of urban riots and recommend solutions.

Its report amounts to a stinging indictment of the white society for its isolation and neglect of the Negro minority. Its pages are filled with findings to bear this out.

"Segregation and poverty have created in the racial ghetto a destructive environment totally unknown to most white Americans," the commission said. "What white Americans have never fully understood—but what the Negro can never forget—is that the white society is deeply implicated in the ghetto. White institutions created it, white institutions maintain it, and white society condones it."

The report was considered remarkable in that it was chiefly the work of white, middle-class Americans, several of them politicians with white constituencies. Most of the commission members are known as moderates.

Some, however, said they were shocked by the conditions they had found in Negro slums during their seven months of work. Some believed at the outset that because of the extent of the rioting there was bound to be some conspiracy involved, some plan for rioting that had been carried out.

But the most exhaustive investigations could find no evidence of this, the report indicated, even though it was clearly established that black militants had created a climate for rioting in their calls for violence. What the commission found over and over was evidence of white prejudice or ignorance that had led to Negroes being crowded into the inner city under a "destructive environment."

New Attitudes Urged

"Reaction to last summer's disorders had quickened the movement and deepened the division," the commission said. "Discrimination and segregation have long permeated much of American life; they now threaten the future of every American."

But the movement can be reversed, the commission said.

"The alternative is not blind repression of capitulation to lawlessness," it said. "It is the realization of common opportunities for all within a single society. This alternative will require a commitment to national action — compassionate, massive and sustained, backed by the resources of the most powerful and richest nation on this earth. From every American it will require new attitudes, new understanding, and, above all, new will."

Although the report was signed by all members, there was a close division on what approach to take. The very strong report, with very broad recommendations, was said to be chiefly due to the consistent pressures of Mayor Lindsay of New York, the vice chairman, Senator Fred Harris, Democrat of Oklahoma, and several others.

They were said to be aligned on most votes with Governor Kerner, Roy Wilkins, national director of the National Association for the Advancement of Colored People, Senator Edward W. Brooke, Republican of Massachusetts, and Herbert Jenkins, Atlanta police chief.

Drive for Jobs Asked

Other members of the commission were Representative James C. Corman, Democrat of California; Representative William M. McCulloch, Republican of Ohio; I. W. Abel, president of the United Steelworkers of America; Charles B. Thornton, president of Litton Industries, and Mrs. Katherine G. Peden, former Commerce Commissioner of Kentucky.

The following were among the commission's scores of recommendations for bringing about equality and integration:

¶A revamping of the welfare system, with the Federal Government assuming a much higher percentage of the cost—up to 90 per cent—and with changes in administration that would help to hold families together.

¶Immediate action to create 2 million new jobs, 1 million by the state, local and Federal governments and 1 million by private industry.

¶Federal subsidy of on-the-job training for hard-core unemployed "by contract or by tax credits."

¶Long-range approach to a "guaranteed minimum income" for all Americans through a "basic allowance" to individuals and families.

¶Bringing 6 million new and existing dwellings within reach of low and moderate income families in the next five years, starting with 600,000 next year.

¶Decentralizing city governments to make them more responsive to the needs of the poor and placing police officers, with a higher percentage of Negroes, in the slums to act as advocates of the people as well as keepers of the peace.

¶Creation of a privately organized and financed Institute of Urban Communications to train and educate journalists in urban affairs and to bring more Negroes into journalism.

News Media Assailed

Turning to news coverage of last summer's racial disorders, the commission charged that some newspapers and radio and television stations so badly portrayed the scale and character of the violence that "the over-all effect was . . . an exaggeration of both mood and event."

The commission credited news media "on the whole" with trying to give a balanced report of the summer unrest. But, the commission added, "important segments . . . failed to report adequately on the causes and consequences of civil disorders and on the underlying problems of race relations."

"They have not communicated to the majority of their audience — which is white — a sense of the degradation, misery and hopelessness of life in the ghetto," the commission said.

The commission declared that reportorial improvement must come from within the media and not through governmental restrictions. They recommended that newspapers and radio and television stations take the following steps:

¶Expand and intensify coverage of the Negro community and racial problems so that the resulting newspapers and radio and television programs "recognize the existence and activities of Negroes as a group within the community."

¶Recruit more Negro reporters and editors and employ more newsmen familiar with urban and racial affairs.

¶Adopt self-regulating reporting guidelines for racial news and improve coordination with police information officers.

¶Organize an Institute of Urban Communications that will study urban affairs, train newsmen in covering them, recruit more Negro reporters and editors and develop better police-press relations.

The report has deep political implications. The commission is advocating that the nation go much further than the President has recommended in seeking new social legislation. It comes at a time, too, when the nation is deeply involved in the war in Vietnam and there have been reports that the President might send in additional troops, further increasing the cost of the war.

Congress has been reluctant to increase domestic programs while the war is draining so much of the nation's resources —$2-billion a month.

But the thrust of the commission's report is that the nation cannot afford to continue on its present domestic course, even if new sacrifices are needed.

"Large-scale and continuing violence could result," the group said, "followed by white retaliation, and, ultimately, the separation of the two communities in a garrison state."

"Even if violence does not occur," it said, "the consequences are unacceptable."

The commission added that the second choice—"enrichment of the slums and abandonment of integration"—is also unacceptable. "It is another way of choosing a permanently divided country" in which Negroes would continue to have a "permanently inferior economic status," it declared.

The commission also warned that "there is a grave danger that some communities may resort to the indiscriminate and excessive use of force." It went on:

"The commission condemns moves to equip police departments with mass destruction weapons, such as automatic rifles, machineguns and tanks. Weapons which are designed to destroy, not to control, have no place in densely populated urban communities."

Of the black militants who preach separation of their race as a means of advancing, the commission had this to say:

"They have retreated from a direct confrontation with American society on the issue of integration and, by preaching separatism, unconsciously function as an accommodation to white racism."

But the commission said "white racism is essentially responsible for the explosive mixture which has been accumulating in our cities since the end of World War II." It listed the following as the "most bitter fruits" of this racism:

¶Pervasive discrimination and segregation in employment, education and housing have resulted in the exclusion of great numbers of Negroes from economic progress that whites have enjoyed.

¶Migration of Negroes into the cities and movement of whites to the suburbs has created a "growing crisis of deteriorating facilities and services and unmet human needs."

¶In the slums, segregation and poverty converge on the young to destroy opportunity and enforce failure.

The commission devoted one section of its report to other minorities who have worked their way out of poverty and segregation and are now asking why Negroes do not do the same.

"Today, whites tend to exaggerate how well and quickly they escaped from poverty," it said. "The fact is that immigrants who came from rural backgrounds, as many Negroes do, are only now, after three generations, finally beginning to move into the middle class.

"By contrast, Negroes began concentrating in the city less than two generations ago, and under much less favorable conditions. Although some Negroes have escaped poverty, few have been able to escape the urban ghetto."

March 1, 1968

PRESIDENT SIGNS CIVIL RIGHTS BILL; PLEADS FOR CALM

Acts a Day After Final Vote on Measure That Stresses Open Housing in Nation

Special to The New York Times

WASHINGTON, April 11—With another plea against violence and for the legal redress of injustice, President Johnson signed today the Civil Rights Act of 1968.

Its major provision is intended to end racial discrimination in the sale and rental of 80 per cent of the nation's homes and apartments.

Mr. Johnson read swiftly but with feeling through a brief speech that invoked the memories of the slaying of the Rev. Dr. Martin Luther King Jr. a week ago and the rioting that ensued in many cities.

But the President also displayed the pride that comes, he said, with the signing of the "promises of a century." Few thought when he proposed it at a White House meeting two years ago that fair housing would "in our time" become the law of the land, he said, "and now at long last this afternoon its day has come."

'Roots of Injustice'

"We all know that the roots of injustice run deep," the President asserted, "but violence cannot redress a solitary wrong or remedy a single unfairness.

"Of course all America is outraged at the assassination of an outstanding Negro leader who was at that meeting that afternoon in the White House in 1966.

"And America is also outraged at the looting and the burning that defiles our democracy. And we just must put our shoulders together and put a stop to both. The time is here. Action must be now."

The action he wants, the President said to leaders of Congress who received a pen symbolizing attendance at the signing, is the enactment of all his domestic programs and appropriations.

Associated Press

RIGHTS BILL BECOMES LAW: Around President Johnson are, from the left: Senator Hugh Scott, Representative William M. McCulloch, Senators Edward W. Brooke and Jacob K. Javits; House Speaker John W. McCormack; Representative Emanuel Celler; Senator Walter F. Mondale and Supreme Court Justice Thurgood Marshall.

These include programs for the accelerated construction of low-cost housing, improved vocational training for the poor, Federal aid to law enforcement, urban development, education and health and for a tax increase to pay the costs.

There is much to do, the President said, and after his program is enacted there will be still more to do.

As he indicated to moderate Negro leaders whom he assembled at the White House last Friday after the slaying, it would be quite an achievement to enact what he had already suggested.

But Mr. Johnson had also scheduled an address to a special joint session of Congress. On reflection and a closer look at the crowded legislative agenda, he postponed the address, apparently indefinitely.

Mr. Johnson seems to believe that new proposals for faster action for the poor will only arouse false hopes in the country while meeting resistance in Congress.

Although some of his aides have wanted him to make an inspiring address to the nation as the rioting subsides, his principal concern, as usual, is said to be with effectiveness on Capitol Hill.

The House passed the civil rights bill yesterday in the same form in which it was rescued from a Senate filibuster last month by a single vote. As the President scheduled the signing ceremony this noon, Federal troops were still on duty in the Capital to help patrol the tenser sections of the Negro slums.

One of the first invitations to the East Room ceremony was telephoned to Mrs. King in Atlanta, but she was out and did not return the call for several hours.

The Rev. Ralph D. Abernathy, Dr. King's successor as head of the Southern Christian Leadership Conference, had still not been reached when the ceremony began at 5 P.M.

Hence, only a few of the better known civil rights leaders were among the 300 guests who were mostly members of Congress and the executive branch, along with staff officers who had worked on the measure.

Present were Justice Thurgood Marshall, the first Negro on the Supreme Court; Sen. Edward W. Brooke, Republican of Massachusetts, the first Negro Senator since Reconstruction; Walter E. Washington, Chief Commissioner of the District of Columbia, and Clarence Mitchell Jr., local director of the National Association for the Advancement of Colored People.

Mrs. Johnson, wearing a bright green dress, was also present for the speech, which was carried live over television, and then she led her husband by the hand to a receiving line. He was introduced to every guest.

The White House said that an effort had been made to reach every group that had participated in the drafting of and the lobbying for the bill.

The act immediately bars discrimination in federally owned housing and in multi-unit dwellings insured with Federal funds.

On Dec. 31, it will cover all multi-unit dwellings and homes in real estate developments except those occupied by the owners with four or fewer units, such as boarding houses.

Effective Jan. 1, 1970, it will extend to all single-family homes that are sold or rented through brokers.

The act provides for Federal conciliation efforts, allows civil suits for damages and permits the Attorney General to seek injunctions against discernible patterns of discrimination.

Other provisions provide heavy penalties for persons convicted of threatening or injuring persons exercising their civil rights; makes it a Federal crime to travel or broadcast from one state to another with intent to incite a riot; makes it a Federal crime to manufacture, sell or demonstrate firearms, firebombs or other explosives for use in riots, and extends broad rights to American Indians in their dealings with their tribal authorities and with local, state and Federal courts.

April 12, 1968

'Benign Neglect' on Race Is Proposed by Moynihan

By PETER KIHSS

Daniel Patrick Moynihan, counselor to President Nixon, has reported in a memorandum to the President that Negroes have made "extraordinary progress" and has suggested that "the time may have come when the issue of race could benefit from a period of 'benign neglect.'"

Mr. Moynihan urged the Administration to avoid building up "extremists of either race" and to ignore "provocations" from the Black Panthers.

The memorandum described "a virulent form of antiwhite feeling" among "black lower classes" and even "portions of the large and prospering black middle class," and urged more recognition for a working and "silent black majority."

"Greater attention to Indians, Mexican-Americans and Puerto Ricans would be useful," Mr. Moynihan's report said. It asserted that Negro problems had been "too much talked about" and "too much taken over to hysterics, paranoids and boodlers on all sides."

The 1,650-word "Memorandum for the President" became known in New York yesterday, and its existence was confirmed by Mr. Moynihan at the White House later, when he also expressed hope that its views would be considered as a whole.

Ronald L. Ziegler, the White House press secretary, declined yesterday to report any Presidential reaction to Mr. Moynihan's memorandum and added that Administration action could not be related to any particular report.

"The President receives a number of memos every day on a number of subjects from key advisers," Mr. Ziegler said, "and he encourages them to provide him with their thoughts. He likes to receive different points of view."

The phrase "benign neglect," Mr. Moynihan said in a telephone interview, came from an 1839 report on Canada by the British Earl of Durham. The Durham report, he said, described Canada as having grown more competent and capable of governing herself "through many years of benign neglect" by Britain, and recommended full self-government.

"What I was saying," Mr. Moynihan continued, "was that the more we discuss the issue of race as an issue, the more people get polarized, the more crazy racists on the left and maybe crazy racists on the right shout and yell and make things seem so much worse than they are, when in fact the nineteen-sixties have been a period of enormous progress.

"If we get the Maddoxes [Gov. Lester G. Maddox of Georgia] and the Cleavers [Eldridge Cleaver, a leader of the Black Panthers] to shut up, or pay less attention to them, and really try to solidify the gains of the sixties, making absolutely certain they are not lost in the readjustment after the Vietnam war, we might look up at the end of the nineteen-seventies and say, 'This has kind of worked.'"

The Moynihan memorandum became known only two days after Leon E. Panetta departed as director of civil rights in the Department of Health, Education and Welfare, contending that "political pressures" were hampering school desegregation.

On Thursday, Robert E. Hampton, chairman of the Civil Service Commission, was quoted as saying that the Nixon Administration had stopped putting pressure on Federal agencies to employ members of minority groups as discrimination in reverse.

The Moynihan memorandum was dated Jan. 16. A covering note indicated that copies were sent Feb. 10 to Vice President Agnew; Attorney General John N. Mitchell; Postmaster General Winton M. Blount; Secretary of Labor George P. Shultz; Robert H. Finch, Secretary of Health, Education and Welfare; Donald Rumsfeld, director of the Office of Economic Opportunity; Peter M. Flanigan, an assistant to the President; Leonard Garment, special consultant to the President; and Edward L. Morgan, deputy counsel to the President.

Mr. Moynihan offered four recommendations, involving a meeting of top officials to discuss the issues, the move for "benign neglect" in public while continuing progress, research on preventing crime and recognition of such Negro leaders as Dr. Jerome H. Holland, a former Cornell football star who was recently named Ambassador to Sweden.

Mr. Moynihan said in his memorandum that he did not know of any advance in understanding crime since the start of the Nixon Administration, and commented that "lawyers are not professionally well-equipped to do much to prevent crime." This and his suggestion that the Black Panthers should be ignored were construed in some quarters as soft criticism of Attorney General Mitchell.

Mr. Moynihan is a Democrat who served in the Kennedy and Johnson Administrations as Assistant Secretary of Labor. In 1965, he wrote a controversial report describing the disintegration of Negro family life as a major cause for Negro social and poverty problems.

His new report asserted that the Nixon Administration had made perhaps more effort to help Negroes than any other administration had, but was still "a long way from solving" its relationships with Negroes. In part, the report said, this was because of "political ineptness in some departments."

Mr. Moynihan, who is 42 years old, was executive secretary of the Cabinet-level Council on Urban Affairs from the start of the Nixon Administration until last Nov. 4.

Since then, he has been a counselor, with Cabinet rank, a post designed to give him an opportunity to develop long-term strategies for urban and other domestic problems and to free him from day-to-day administrative chores.

Last year, Mr. Moynihan was credited with developing the Nixon welfare reform proposal, including help for the working poor and Federal aid to guarantee $1,600-a-year incomes for families of four, to replace the present aid-to-dependent children program.

March 1, 1970

Blacks 'Within the System' Gloomy Despite Progress

By THOMAS A. JOHNSON

"There has been some progress, yes, but there is more resistance to Negro aspirations today than at any time in recent history," said Gloster B. Current, director of field administration for the N.A.A.C.P.'s 1,700 branches across the country.

This judgment by Mr. Current, who is considered a racial moderate on the black American scale of activism, reflects the strong concerns of scores of other black Americans interviewed in the last three weeks.

All of those questioned are working somewhere "within the system," subject to periods of frustration and doubt but not yet ready to break away and work from outside.

An immediate concern of all of them, one shared by many city officials during the traditionally volatile summer months, was the inadequacy of funds to provide summer jobs for urban youths. But all voiced much longer-range worries.

Most saw traditional and entrenched American racism, detailed two years ago in the report by the President's Commission on Civil Disorders, as a continuing deterrent to black progress.

And many were disturbed by new expressions of antiblack bias from working class whites as black people have attempted to compete more and more with whites for jobs.

Generally, they admit to being encouraged by the national black successes, like the recent election of a black Mayor in Newark, but this is often more than counterbalanced in their minds by police killings of blacks, as in Augusta, Ga., and Jackson, Miss.

They are despondent about rising unemployment and are convinced that the Nixon Administration's reported "Southern strategy" works to the detriment of black Americans because, they say, it seeks to expand the Republican party in and around the white South and among white, conservative "middle Americans" elsewhere.

"We are finding across the nation," said Mr. Current, "that there is an emergence of expressions of anti-Negro prejudices from the white ethnic minorities who were mostly silent before. The fact that the Nixon Administration will apparently not contradict them has given much encouragement to suburban bigots."

Aaron E. Henry, a druggist from Clarksdale, Miss., who heads the Mississippi branch of the National Association for the Advancement of Colored People, was one who also saw Southern white hostilities toward blacks as taking on "the more polite and smiling subtleties of the North."

Mr. Henry suggested that "black frustrations are more kinetically explosive today, as the result of white resistance, because more and more blacks are becoming aware that these repressions of freedom need not continue and they are putting forth efforts to stop them."

And the rising black frustrations, across the nation, can still be seen in the areas that have for many decades been focal points in the black struggle—police conduct, unemployment, housing (new urban renewal) and education.

Forgotten Promises

Still another area of black frustration, defined by a high Los Angeles civil servant, is that of "official promises—either stated promises to prevent riots or implied promises like a war on poverty. The frustration is imminent when none of the promises is kept."

One of the more volatile black-white relationships, according to those interviewed, continues to be the day-to-day relationship between the basically white police department and the urban, black, poor citizen. Most of the urban disorders of recent years started with police-citizen confrontations.

"Generally, white policemen feel they are better than blacks and they take the attitude that they're in a war where its 'either them or us,'" said Renault Robinson, a Chicago patrolman who heads the Afro-American Patrolmen's League.

The league was formed, said Mr. Robinson, to counter "within the system" both the physical and verbal brutalization of blacks and the poor by Chicago law officers.

While all those interviewed felt serious racial antagonisms existed in most police forces, several praised New York City's Commissioner Howard B. Leary for promoting better understanding "from the top."

An urban housing specialist in Buffalo, Joseph L. Easley, traced black frustrations in his city to "the many promises that never materialize." He pointed to an urban renewal program under which hundreds of acres of a black community razed more than a decade ago, still lie abandoned. Black former residents now pack an area called "the fruit belt" where, he said, an even worse slum condition has grown.

Another frustration, he said, "comes from the promise of some 700 jobs by businessmen for the summer and delivering less than 200."

Most blacks interviewed expected the impact of rising unemployment—plus the immediate lack of the usual "riot preventative" summer jobs—to add greatly to black frustrations in coming weeks.

"This is very serious for many blacks," according to the N.A.A.C.P. executive director, Roy Wilkins, who reasoned that the "five per cent unemployment rate generally means at least twice or three times that amount for blacks and even higher in some geographic areas—much higher among black youths."

The national black frustrations are not necessarily the exclusive concern of the black poor.

A black Wall Street executive said recently: "I want very much to believe in this country and it's making me schizoid. I feel good—ecstatic—when white people vote for Ken Gibson in Newark, but then I realize that so many others don't give a damn when they can live with a 5 per cent national unemployment figure, knowing what it does to black people.

"I find I bounce back and forth, happy about a new training program for blacks and then angry as hell about an Augusta or Jackson or a Fred Hampton killing. I believe in America—I like my job—I like 'making it' in a 'white' world. But sometimes I have to get away by myself—sometimes I can't stand to be around white people."

July 6, 1970

BLACK ASSEMBLY VOTED AT PARLEY

Agenda at Convention Asks Permanent Group to Build Up Political Power

By THOMAS A. JOHNSON

Special to The New York Times

GARY, Ind., March 12—Delegates to the first National Black Political Convention adopted today a far-ranging political agenda that would establish an "independent black political assembly" to seek to strengthen the over-all effectiveness of some 7.5 million black voters. Such an assembly would also perform follow-up duties recommended by this convention.

The proposed National Black Assembly would tie black voters together across philosophical and political differences and bring blacks together in a political convention every four years prior to the Democratic and Republican conventions. The next convention would be in Philadelphia in 1976.

The proposed agenda was drafted last weekend by a number of black professionals from many disciplines who met at Howard University in Washington. The planners tried to put together a document that would be acceptable to a majority of black Americans.

The document does not call specifically for a black political party. It does not call for stronger black influence on the Democratic and Republican parties, although that is implied. In addition, it does not consider the endorsement of candidates for nomination for President.

Each of these considerations had loomed as convention-splitting controversies during the last three days as the 2,776 delegates and 4,000 alternates and observers from 43 states were meeting in the modern, sprawling West Side High School here

The convention passed a resolution that condemned busing to achieve school integration. The resolution called for blacks to control the schools in their neighborhoods. The delegates also opposed the merger of black and white colleges in the South.

Roy Innis, executive director of the Congress of Racial Equality, told newsmen that the South Carolina delegate proposing these steps, as well as many people on the floor pushing for their support, were all CORE members. He said that black people "were tired of being guinea pigs for social engineers and New York liberals."

Mr. Innis said: "Busing is obsolete and dangerous to black people—we are ready for a change, ready to control our own destiny."

However, some delegates walked out to show their disapproval of the resolution. Some said it expressed a form of self-segregation.

Later the delegates voted not to endorse a political candidate.

After the convention ended tonight, Mayor Richard Hatcher of Gary said that a steering committee, formed to implement the convention's resolutions, would refine the national black political agenda even further and present it for publication on May 19, Malcolm X's birthday. He said that the goal was to have it accepted by as wide a group of black Americans as possible.

Early today, more than 20 nationally known political figures and social activists gathered in special sessions to refine the basic political agenda even further. While they worked, crowds gathered in the school's gymnasium for nonstop music of the Band from Chicago's People United to Save Humanity before the business sessions began at noon.

Imamu Amiri Baraka, also known as LeRoi Jones, the poet-playwright, was the chairman of the often tumultuous general session. He called often for unity, for speeding up the proceedings and for keeping controversies within the state caucuses and not on the floor during the general meetings.

Mr. Baraka, a nationalist leader and head of the Pan-Africanist Congress of African People, was a major force in recent months in setting up the three-day meeting.

Mr. Baraka induced the audience to accept a proposal to create an "ongoing structure naming 50 state chairmen plus a chairman from the District of Columbia as a steering committee" to see that the plans growing out of this convention are implemented.

The proposed preamble to the political agenda was condemned last week by the National Association for the Advancement of Colored People because, in draft form, it called for an independent black political movement. The N.A.A.C.P. insisted that such a preamble promoted racial separatism.

The remainder of the original document does not deal with traditional political activity but rather seeks to set general guidelines aimed at across-the-board advancement for black people.

In politics, the document calls

The New York Times/Gary Settle

CLENCHED-FIST SALUTE was given at the start of the opening ceremonies of the first National Black Political Convention in Gary, Ind. Representative Charles C. Diggs. Jr., Democrat of Michigan, is at the lectern at left.

for "a minimum of 66 Congressional Representatives and 15 Senators" and a "proportionate black employment and control at every 'evel of the Federal Government structure," among other things.

The agenda makes other recommendations in areas of economic, human and rural development. It suggests that black Americans exert an influence on American foreign policy and especially with regard to the nation's relations with Africa and the Caribbean. It is critical of United States relationships with South Africa, Rhodesia and Portugal.

It calls for the setting up and controlling of black communications media and for quality education "for every black youth in the country" and for massive Federal funds to accomplish this.

Basil Patterson, former New York State Senator, was one of the political figures supporting the document because it does not take up the possibilities of political parties or endorsements.

Representative Shirley Chisholm, Democrat of Brooklyn, did not attend the convention. Her supporters said she had sent a telegram saying that she was ill in Florida.

March 13, 1972

Civil Rights Unity Gone In Redirected Movement

By PAUL DELANEY

Ten years ago, Hannah D. Atkins, a State Representative in Oklahoma, and J. L. Chestnut, a Selma, Ala., lawyer, frequently entertained whites in their homes.

Today, neither bothers, unless it is absolutely necessary.

Ten years ago, there was a spirit of togetherness among blacks and whites and it was the guiding principle of the civil rights movement.

Today, blacks snicker at the refrain, "black and white together," from the theme song of the movement, "We Shall Overcome." If anybody still sings the theme, it is possibly a nostalgic white liberal.

Neither Mrs. Atkins nor Mr. Chestnut is regarded as a wild-eyed radical or as antiwhite.

But because both may fairly be regarded as solid black bourgeois, their attitudes raise one of the central questions worth asking 10 years after the Rev. Dr. Martin Luther King Jr.'s dramatic speech during the march on Washington on Aug. 28, 1963: What has become of the civil rights movement of the sixties?

On the face of it, the answer is clear: the "movement"—at least to those who knew it as mass protest, spirited rhetoric, and a deep and widespread feeling of interracial fellowship — exists no more. The quest for human equality goes on, but its targets have changed and its passions have been redirected.

Ten years ago, the movement's major goal was to win for blacks basic legal rights that had long been routinely granted to whites. Today, the residual energies of that movement are focused on politics and economics, on the formidable task of translating into reality the right to wield influence at the ballot box and in the marketplace.

Meeting in Denver

On Monday, for example, black Democratic and Republican leaders met in Denver to begin charting new political strategies and, specifically, to devise ways of increasing minority participation in national conventions and arresting the gradual decline in black voters in national elections. The meeting was arranged by three of the most prominent members of the new breed of black politicians—State Senator George Brown of Colorado, State Representative Julian Bond of Georgia, and Mayor Richard G. Hatcher of Gary, Ind.

Ten years ago, there were only three blacks in Congress. Today, there are 16 Representatives and one Senator. Ten years ago, there were only a few state and locally elected blacks; today, there are over 2,600, according to the Joint Center for Political Studies.

Ten years ago, the gross national product of the black community was estimated at less than $20-billion; today, it is said to be more than $50-billion. To be sure, the gross national product as a whole has also grown rapidly; but unlike 1963, there is, in just about every major city and many towns and rural areas, one or more fledgling organizatian devoted exclusively to black economic development.

Ten years ago, four major civil rights organizations dominated the black community and commanded the headlines: the National Assocation for the Advancement of Colored People, the Student Non-Violent Coordinating Committee, the Congress of Racial Equality, and Dr. King's Southern Christian Leadership Conference.

Of these, only the N.A.A.C.P. retains its original identity as, increasingly, blacks have directed their commitment to smaller groups with more narrowly defined purposes; the National Welfare Rights Organization, the National Association of Community Developers, the National Sharecroppers Fund, the National Tenants Organization, the Movement for Economic Justice, the Urban Coalition, and others—literally, an organization for every cause

Key Leadership

The National Urban League, which was not regarded as a civil rights organization similar to S.N.C.C. or S.C.L.C., nevertheless provided the movement with key leadership, most prominently that of its late director, Whitney M. Young, Jr. Even today it remains a vital force in the black quest for more and better job opportunities.

Few of the gains made in the last decade would have been possible without the movement and, in particular, its legislative offspring: the Public Accommodations Act of 1964, the Voting Rights Act of 1965, and sections of the Elementary and Secondary Education Act of 1965, all of which owed much to the defiant tactics and unyielding convictions of those who marched not just in Washington but in Albany, Ga., Birmingham, Ala., Nashville, Tenn., Selma, Ala., St. Augustine, Fla., and Bogalusa, La.

Moreover, the movement gave blacks a strong sense of self-awareness that allowed them to be more comfortable with the fact of their blackness and gave them a keener appreciation of their common cause.

As Lonnie King, Jr., who led demonstrations in Atlanta and now owns a consulting company there, put it not long ago: "The movement awakened a very sleepy population to the fact that it can aspire to the highest."

And to others — including Mary King, a former S.N.C.C. publicist who now lives in Washington — the movement was a phenomenon that would never be duplicated. "It was incredible that a bunch of college kids were able to make the impact we made," she says. "It was one of the great events of history."

Beyond that, the movement gave hope to other groups who freely borrowed the tactics, symbols and slogans of civil rights advocates. These included women, white ethnics, Spanish-Americans, Indians, college students and homosexuals.

Slogans Are Copied

The slogan "black power" was copied and became "senior power" and "gay power," among others. It was not uncommon for Polish-Americans and Italian-Americans to raise a clenched fist, the black power salute. Eventually, President Johnson was to utter, "we shall overcome," and President Nixon was to adopt the Black Panther slogan, "power to the people."

The movement represented a drive for economic opportunity as well as legal parity, and here again there were tangible results. "It helped raise the lower economic class a little," comments Charles Black, a leader in the Atlanta student movement who is now a consultant in that city. His friend, Lonnie King, adds: "A very substantial number of blacks are better off now."

But both men agree that black economic progress has been too little and too slow, leaving millions untouched, and both share with Hannah Atkins, the Oklahoma State Representative, a nagging sense that "progress has been kind of mixed; we seem to take one step forward and two steps backwards."

This continuing and pervasive sense of economic injustice—the feeling that American institutions were fundamentally unresponsive to the requirements of the black masses—angered blacks and did much to damage the movement and discredit the feeling of black-white togetherness that helped shape it.

Newly Won Rights

"While the movement raised the aspirations of blacks it left many of them frustrated when they could not enjoy the benefits of those newly won rights," said Lonnie King. "The riots of the mid-to-late sixties were a result of that frustration. Most blacks couldn't buy a steak at the desegregated restaurant because they didn't have the money, and they couldn't buy a $30,000 house on a $2 an hour salary."

Charles Hamilton, a black historian who teaches at Columbia, makes much the same point in somewhat different terms, asserting that nearly all struggles tend to be governed by a small elite and tend to yield most of their gains to a relatively small and highly visible group.

"It is more difficult," he said in an interview, "for society to accommodate to the demands or interests of the masses [unless it is prepared to] question some of the fundamental bases of the system."

But economic dissappointments alone did not kill the movement. It suffered, too, from increasingly abrasive relations between whites and blacks who had once been friends and allies, from the violence that characterized black protests as time passed, from a fragmentation of leadership, from the increasing complexity and difficulty of the problems that the movement confronted as it shifted its concerns to the industrial centers and urban ghettoes of the North.

The feeling that white liberals ran out on black problems before the problems were solved is pervasive in the black community, and is considered a prime reason that even middle-class blacks are hostile.

"There is now a rejection of whites by blacks, great animosity, even repulsion," said Mrs. Atkins, the Oklahoma legislator.

"Blacks are repulsed because they feel betrayed by whites, feel that white liberals turned and ran when the crunch came. We feel that liberals were part of schemes to get around integration laws, in housing and schools.

"Ten years ago, we entertained whites a lot in our home —my husband was head of the local Urban League. We went to dances with whites, but we don't any more. We go to cocktail parties now out of obligation. But I just don't feel like entertaining whites in my home."

Mr. Chestnut, the movement lawyer in Selma, says he now gets more protests from his two daughters over entertaining whites in his home. His daughters were 8 and 11 years old a decade ago.

"They marched when Dr. King pulled the students out of school and they expected a lot out of integration," Mr. Chestnut recalls. "But then they saw that whites didn't mean it. They had bitter experiences with school integration, too, where whites abandoned the public schools rather than go to school with blacks. All of this left a bitter taste in my daughters and many of their friends."

White America too, was puzzled as liberals found themselves increasingly shut out from the movement to which they had given their energies, increasingly disillusioned—or frightened—by the anger of a newly militant black community.

Quite Another Thing

For Northern whites, it was one thing to redress grievances in the South but quite another when the issue was brought home to New York, where bridge traffic was stopped; to Cleveland, were bulldozers were blocked at school construction sites; and to Chicago, where Dr. King tried—unsuccessfully—to rally white opposition to segregated housing in suburban Cicero.

Thus in 1965, whites in Detroit were able to eulogize Viola Liuzzo, the white mother of five who was shot to death during the Selma movement. But two years later, white Detroiters seemed baffled when their black neighbors rioted; still later, when a Federal judge ruled that the city must undertake areawide busing to correct racial segregation in its schools, whites reacted with the undisguised hostility that has marked busing disputes in a number of Northern cities.

"The movement was happening in the South so Northern liberals could feel self-righteous and feel guilty without really hurting themselves," said the Rev. Jesse L. Jackson, president of Operation PUSH (People United to Save Humanity), a Chicago based group devoted to black economic progress.

"Hope was raised in the South and fulfillment came," he went on. "Hope was raised in the North but there was no fulfillment."

And as the focus of the movement shifted northward so, too, did black people themselves, forming a steady procession of humanity that reinforced old problems and hastened the search for new organizations to solve them.

Between 1960 and 1970, for example, 1,380,000 blacks left the South and journeyed to the already over-crowded ghettos and marginal neighborhoods of the urban North — a loss of nearly one-eighth of the South's entire black population. Of these, nearly half settled in the industrial Northeast, the rest in the big cities of central states and on the West Coast.

Deep in the hot and crowded slums, the movement took on a more anti-white and problack fervor. The Black Panthers were formed and found easy recruitment among restless urban youth. Intellectuals turned once again to African roots for solace and inspiration, just as blacks had done at the turn of the century.

Meanwhile, men like George Wiley—who drowned earlier this month—turned from the established institutions and began attacking specific urban ills. First Mr. Wiley organized welfare recipients into the militant National Welfare Rights Organization; then he formed the Movement for Economic Justice, which he hoped would attract allies from other minorities.

The toll on the older organizations was heavy. The N.A.A.C.P. returned to its legal fight to end discrimination through the courts; the Southern Christian Leadership Conference, deprived of its leader by an assassin's bullet, is today broke and disorganized.

The Congress of Racial Equality became a black nationalist organization pushing a philosophy without a program, gaining attention only because of its charismatic leader, Roy Innis. And the Student Non-Violent Coordinating Committee—the young commandos of the early nineteen-sixties flirted briefly with a militant, all-black nationalism and then went out of business altogether.

The history of the Student Non-Violent Coordinating Committee—a history at once brief and incandescent—tells much about the rise and fall of the movement, its joys and sorrows.

Beginning on Feb. 1, 1960—when four students at North Carolina A. & T. College in Greensboro decided to sit in at the white-only lunch counter at the local Woolworth's — and sweeping through the next four years, the young men and women of S.N.C.C. brought the movement to nearly every black campus in the nation, helped organize demonstrations at the rate of 40 or 50 a week, and lured a seemingly unending stream of students into the

South to fight the laws and customs of Jim Crow. Its philosophy was one of love and nonviolence; its tactics were those of confrontation.

From the beginning, the organization of youngsters daringly, almost arrogantly, flaunted integration before the eyes of whites, who after getting over being flabbergasted retaliated with a wave of bombings, beatings, jailings and killings that shook the nation.

The student committee dispatched integrated teams of students with the word to poor, scared, disbelieving blacks that they had the numbers, and most certainly the right, to change the system if they would register to vote.

"We began in Southwest Georgia and Mississippi because we knew they would be tough to crack and because there was an abundance of organizations working urban areas and almost no national groups in rural areas," said Charlie Cobb, a former S.N.C.C. leader who was beaten and arrested many times.

"In many places, we were the first visible organization to come in, and that was good for local people," he said. "S.N.C.C. made voter discrimination a national issue in the North as well as the South. We destroyed the myth that blacks didn't want to vote, as white Southerners claimed in denying blacks the right to vote."

The young students of S.N.C.C. carried their enthusiastic idealism with them wherever they went, and as it turned out their unyielding zeal was their downfall. The end began, perhaps, with the march on Washington, a day they had envisioned as one of civil disobedience intended to stop the Federal Government until it recognized and dealt with the problems in Mississippi.

But older, more experienced leaders and organizations, in concert with the Administration of President Kennedy, made the march the biggest in history for civil rights, with the goals of "jobs and freedom."

The S.N.C.C. youths were bitter, and took their bitterness back South, with an increasing distrust of all others.

"The march was the crack in he movement, that laid the seed for the destruction of the movement," remarked Ed Brown, a S.N.C.C. leader and the brother of H. Rap Brown, a former chairman of the organization.

The distrust of white liberals and other civil rights organizations was further confirmed the next year, 1964, when S.N.C.C. led a challenge of the regular party delegates to the Democratic National Convention. That challenge failed and S.N.C.C. felt betrayed by politicians and older black leaders.

"We were politically naive," commented Charlie Cobb. "We really believed that since our cause was just and moral that we would win."

The student committee plunged into a leadership crisis from which it never recovered. And sticking to its principles, it began alienating its supporters, first by kicking out whites, then by becoming the first civil rights organization to oppose the Vietnam War and finally by its opposition to Israel in the Six Day's War of 1967.

The organization, Charlie Cobb said, "sort of just dwindled away after losing its campus base, its Southern base and its on-going programs. The organization became one of personalities rather than programs under Rap and Stokely Carmichael."

Today, many of the leaders of the defunct student committee still consider themselves activists, but chiefly from inside the system that they fought so hard to change. Most are connected in some way with the cause of blacks, other minorities and the poor as a whole.

A partial roll might include: John Lewis, director of the Voter Education Project, which conducts registration campaigns and voter education among minorities. Ten years ago as chairman of S.N.C.C., his stinging speech before the throng that gathered at the Lincoln Memorial was toned down after Archbishop Patrict O'Boyle of Washington objected to some passages.

Julian Bond is not only a member of the Georgia General Assembly but also is president of the Southern Elections Fund, which raises money to aid mostly black political candidates in the South. A decade ago, he was director of public relations for the student committee.

Robert Zellner, who led more demonstrations than any other white S.N.C.C. leader and who, like so many others of the militant youthful organization, bears the scars of numerous beatings, heeded the advice of his black colleagues that whites go into their own community to bring about change. He is organizing white pulpwood workers in Louisiana and Mississippi.

Bob Moses, once regarded as the spiritual leader of the student committee and author of many of its programs is teaching school in Tanzania.

Marion Barry, the first chairman of S.N.C.C., is president of the board of education of Washington.

Rap Brown, also a former chairman, was recently sent to prison in connection with a robbery attempt and shootout with New York policemen. A fiery speaker, Brown was accused of inciting riots in several cities, including Cambridge, Md., and Dayton, Ohio, where disturbances followed his appearances.

A New Era

Stokely Carmichael, perhaps the best known former head of S.N.C.C., who helped usher in a new era of black awareness with his shouts of "black power," has organized a black political party with the aim of fostering stronger ties between American blacks and black Africans.

The roll would also include some leaders of other organizations who are now dead: Whitney Young, head of the National Urban League and considered a bridge between the conservatism of the N.A.A.C.P. and the militancy of S.N.C.C. and C.O.R.E.; Medgar Evers, head of the N.A.A.C.P. in Mississippi, Dr Wiley, the organizer of welfare recipients, and of course Dr. King.

A. Philip Randolph, former president of the International Brotherhood of Sleeping Car Porters and dean of Civil Rights Leaders, is retired, ailing and writing his memoirs in his New York apartment.

Bayard Rustin, former aide to Mr. Randolph, is now in the labor movement. James Farmer, who went on to become Assistant Secretary of Health, Education and Welfare was head of C.O.R.E., in the Nixon Administration, is now speaking and writing. Floyd McKissick, who succeeded Mr. Farmer at C.O.R.E., is now building an all-black community in North Carolina.

And Jesse Jackson, who was a student at North Carolina A.&T. College when the first sit-ins occurred and eventually became an aide to Dr. King, is now a leading spokesman for political, social and economic equality.

Seeds Bear Fruit

"The seeds sown in 1963 have borne fruit," Mr. Jackson remarked in a recent interview. "We have more ethnic consciousness now. We are less likely to be put back into slavery. We are better equipped intellectually, emotionally and technically to protect ourselves. We finally had to come to that independence to get to the next step."

"This time," he added, "we are moving from the civil rights struggle where the object was citizenship, to the fight for equality and parity."

In retrospect, many blacks agree that the struggle—though different in its tactics, its rhetoric and its style of leadership—is still on course. As Charles Black, the former student leader who is now a consultant in Atlanta, put it:

"We are older and maybe a little wiser, and certainly a little more conservative than 10 years ago. We are in the system now and we should be there. But there should be younger, more militant people pushing and prodding us, as we did then."

August 29, 1973

Poorer and Poorer

By Frances Fox Piven and Richard A. Cloward

Ghetto protests against injustices and hardships in the nineteen-sixties were a factor in bringing gains to the poor, mainly in civil-rights legislation and welfare liberalization, which benefited the white poor as well as the black. If those injustices and hardships were sufficient to stir the ghettos to revolt then, there is cause for disquiet now.

Conditions are rapidly worsening again. Unemployment is more than 5 per cent officially, and from two to three times as high in the ghettos. Minimum wages lag far behind the level necessary to support families. Inflation, particularly in food and utility costs, which consume a disproportionate share of the low-income budget, has had disastrous effects. Some poor families are eating dog food.

Finally, housing deterioration is accelerating, partly because of the inflated costs of maintenance, and partly because Federal programs to expand or improve low-cost housing have been terminated or cut back by the Nixon Administration.

Meanwhile, the welfare system, which might have eased some of these hardships by higher grant levels or eased eligibility rules, has instead been made more restrictive. Grant levels are being kept down, and in some places are being reduced. Getting on the rolls and staying on the rolls is now much more difficult. The result is that the rolls are actually falling for the first time in fifteen years.

The grievances are real and worsening but the people who suffer them do so silently. In part, they have become intimidated. Led by the Nixon Administration, the political climate has shifted sharply to the right, and the black poor have been singled out for attack by the most influential voices in the land. In 1965, President Johnson ended a nationally-televised speech with the civil right refrain, "We Shall Overcome"; in 1973, President Nixon proclaimed in his inaugural address that Americans ought not to ask what government can do for them but what they can do for themselves. In the ensuing assault on blacks, Federal programs that aided the ghettos were slashed and Federal procedures for antidiscrimination enforcement were moderated. The antagonism of Federal leaders encouraged similar assaults by state and local authorities. When people rose up in protest in the nineteen-sixties, they had reason for hope. Now they have reason for fear.

But even if the masses of black poor were restive, the agitators and organizers who led them are gone from the streets. One response to the turmoil of the nineteen-sixties was the opening of American institutions to those blacks who were poised to enter the middle class. Veteran black activists, and many younger people who might have become activists, can now be found in the colleges and universities, in the ranks of government, and in industry and commerce. Their concerns are individual advancement instead of group struggle. No one can fault them; the black's turn had come. Like the leaders of earlier ethnic group struggles, they merely took advantage of the opportunities that political turmoil had produced. And now, in an increasingly hostile climate, this new black middle class is concerned with preserving its precarious stakes; its ranks are hardly likely to produce leaders of mass insurgency.

■

Furthermore, the new black middle class is exerting an ideological influence that reduces the likelihood of mass protest among the urban poor. To take advantage of greater opportunities for electoral office, black leaders now prefer conventional electoral political activity to protest.

Conventional electoral power, it is said, is far superior to the vicissitudes of protest and will yield the black masses larger and more permanent concessions from society. It is a wholly ahistorical argument. For decades, the poor of America were an electoral majority, but they rarely exerted much influence on Government policies. What then can a weak and isolated minority expect? But it is neither history nor the needs of the black poor that informs this argument. A rising class of black leaders requires the rhetoric of conventional politics to justify its advance. Meanwhile, the ghettos remain quiescent.

Frances Fox Piven and Richard A. Cloward are authors of "Regulating the Poor."

July 16, 1974

They Have Still to Overcome

By William V. Shannon

WASHINGTON—A major American problem is how large numbers of low-income and often poorly educated, poorly trained blacks can move out of the shallows of poverty and into the mainstream of society. This country has made long strides toward racial justice in the last 25 years but no one believes that the "American Dilemma" has been fully resolved.

Yet paradoxically the issue of social justice for blacks has scarcely surfaced in this national election. Even more curiously, blacks themselves seem sunk in political apathy and are expected to turn out on Election Day in relatively small numbers.

The most interesting development has instead been an intellectual event. Prof. Herbert G. Gutman has published excerpts from his forthcoming book, "The Black Family in Slavery and Freedom, 1750-1925," in which he argues against the view "that migration [to the North] and urbanization, per se, caused widespread family dissolution among poor blacks."

Unfortunately, the Gutman study ends in 1925. He merely notes that "far more family disorganization followed the migration of the Southern black poor to Northern cities between 1940 and 1970 than before 1930."

Prof. Nathan Glazer of Harvard points out that "something quite serious" must have happened because the stable black families described by Professor Gutman have not survived into the present. Last year, 35 percent of black families were headed by females. Only 56 percent of black children were living with both parents.

My own view is that there are at least three unrelated causes for the family deterioration measured by these statistics. First, in normal times, immigrants are a self-selected group. Only the strongest and most venturesome dare to emigrate whether from the old country in Europe or a farm in Mississippi. Most people stick with the familiar no matter how miserable it is.

But in times of catastrophe, everyone flees—weak and strong, young and old. The mechanization and economic upheaval in Southern agriculture after 1940 was the equivalent of the major crop failures in 19th-century Europe that drove people off the land. These later black immigrants included some persons less strongly motivated than those who preceded them. They came in such large numbers and so swiftly that they overwhelmed the neighborhood institutions and the self-help networks slowly built up by the earlier generations of black immigrants, and suffered worse personal disorganization.

Secondly, the later black immigrants arrived when the labor market had changed to their disadvantage. New York and Chicago had far less need of unskilled manual labor in 1970 than they had in 1910. The result was a higher rate of unemployment and a harder time for family men trying to make the adjustment from Southern farm to Northern city.

Finally, the last 30 years are the period of severe drug addiction. This curse was unknown to blacks living in the rural South. As Prohibition with its flood of illegal—and often lethal—whiskey was a crime-inducing, pathological force in the lives of many Americans in the 1920's, so drug addiction has been an independent demoralizing force in recent decades, particularly among younger blacks of marrying age.

Why are the statistics on disrupted family life significant? Obviously, the psychological deficiencies and distortions inherent in being raised in a single-parent family can be overcome but only with some effort and at some cost, financial or psychic or both. Ideally, a child should have both parents available as models and guides.

Government policies such as family allowances and the guaranteed minimum income should be developed to help parents stay together and cope with their common problems. There ought to be a greater Federal investment in cheap rental housing for families with children. Economic programs should seek full employment and particularly stress industries and projects that are labor-intensive.

But beyond such help as broad Federal policy can provide, blacks can best help themselves by creating

strong neighborhoods in the Northern cities. They need and want schools that maintain discipline and teach their children to read, honest police that arrest drug-pushers, and sanitation departments that actually pick up the garbage and keep the streets clean.

To gain these elementary yet essential objectives and to build political coalitions with predominantly white neighborhoods, black voters have to mobilize their political strength. Blacks habitually have a low turnout because they are younger and poorer than the general population, and the young and poor of every race have a lower rate of political participation. Blacks, moreover, never acquired the voting habit during their generations of disenfranchisement in the rural South. But like other groups that have preceded them on the urban frontier, the blacks must learn the habits of politics and the arts of coalition-building if they are to make the final move from a client group to sharers of power.

October 9, 1976

Southern Blacks Shift Goals From Rights to Economics

By ROY REED
Special to The New York Times

NEW ORLEANS, Jan. 23—A group of tough, publicity-shy black Mississippians has built a $10-million business that runs four manufacturing plants and a Delta plantation. It operates without fanfare and shuns do-gooders because, the president says, "we're not in the sympathy market."

Southern black leaders no longer find much profit in white sympathy. The civil rights movement of the late Rev. Dr. Martin Luther King Jr., which was built as much on white conscience as on black expectations, is almost unrecognizable as his fellow Georgian, Jimmy Carter, moves into the White House.

Activism as it was known during the 1960's has practically disappeared. In most Southern towns, one cannot find a trace of civil rights activity.

The target has shifted. The wall of white resistance that was once so visible is virtually gone.

Some organized white supremacy, a kind of museum sample of Ku Klux Klan bullyragging, still exists in the narrow coastal crescent between Pensacola, Fla., and Mobile, Ala. It is as if the forces of virulent racism have retreated southward to the sea and stand now at the water's edge, fighting for a last strip of land.

But elsewhere in the South, except for an occasional eruption like a firecracker going off in January, overt racial hostility has almost disappeared. What has taken its place is in some ways harder to handle.

"I perceive us to be in a war of attitudes," Michael Figures, a young black lawyer in Mobile, said recently. Many agree with him.

That attitudinal war is being fought in two main areas, politics and economics. Activists who demanded "freedom now" during the 1960's talk now of "the long hard years ahead." Those who once would have marched in the street are going into business, working in consumer organizations, battling utility companies, arguing tedious lawsuits, bulldozin gerrymandered political districts, running for office and poring endlessly over bureaucratic structures to discover ways of siphoning power to the powerless.

The changes that have taken place in the South are well-known. Blacks are voting freely and effectively, barely a decade after they were dying for the right. Schools are desegregated more widely in the South than in the North. The racial barriers are down in a vast majority of Southern institutions, from restaurants to libraries. A black middle class is growing with the Southern economy.

Black gains in education have been spectacular in some areas. In 1960, only 15 percent of Southern black men 25 years old and older had finished high school. That percentage had almost exactly doubled by 1975. In the same years, the number who had graduated from college increased from 60,000 to 125,000. Gains for Southern black women have been only a little less striking.

The changed way is taken for granted by the young of both races. A black mother in Atlanta went shopping with her two young daughters the other day. As they stopped for a drink of water in a department store, the mother mused out loud about the time when the store had had two fountains, labeled "colored" and "white." Her daughters looked at her in disbelief and said, "Oh, mother." She was stunned, and when she got home she sat them down for a history lesson.

Complacency Growing

Many blacks who are old enough to know better have become as complacent as those daughters.

One reason is the growing black middle class. In the mid-1960's, 9 percent of Southern black families earned $10,000 a year or more; in 1974 the figure was 31 percent

Aside from complaints about prices and inflation, the main complaints from the black middle class concern job mobility. Many are still frozen out of the top positions.

On the surface, it appears that Southern black men are closing the income gap between them and Southern white men. Figures from the Census Bureau show that black men in the South improved their annual income 115 percent between 1967 and 1975. In the same period, white men's income increased 83 percent.

Dollar Gap Widened

But in dollars, the gap between the two groups widened. The black men's average income went from $3,703 to $7,978; the white men's from $6,827 to $12,536.

The fairly visible gains of middle-class blacks in the South have tended to obscure the fact that millions of black Southerners still live in poverty.

Improvements have been made. About 68.5 percent of all Southern blacks, or 7.5 million people, lived below the Federal Government's poverty income level in 1959. That had dropped to 36.6 percent, or 4.7 million people, by 1975.

For all the gains, Southern welfare rolls are still loaded with unemployed blacks. Malnutrition is still a chronic problem in black households of the Deep South, as it is in white households of the Appalachians.

'Still Trapped'

One leader who still worries about poverty is John Lewis, the director of the voter education project. "Those people are still trapped," he said recently in his Atlanta office. He is getting ready to run for Congress. If he wins, he said, one of his main aims will be to make life better for the rural poor.

The big farms of the Deep South have been mechanized, so masses of black people who once worked the land are now on welfare. The industrialization that has revived much of the rest of the South has bypassed the rural areas with large black populations. Corporate leaders are said to regard black former agricultural workers as undependable industrial employees.

Government programs, for all their publicity, have brought few steady jobs to the black belt. The task of solving rural black unemployment has been largely left to self-help organizations, most of which, relying on white sympathy and foundation money from the north, have failed.

The Delta Foundation of Greenville, Miss., is one that has not failed. It has used foundation money along with any other money it can get, including government aid and private bank loans. But it has been primarily a hard-nosed business operation from the start, and that apparently has made a difference.

"Our first objective is profitability," Charles Bannerman, the young black man who serves as president and board chairman, said on the telephone from Greenville the other day. "We're not a romantic quilting bee. We're not in the sympathy market. Those things don't work."

Beginning with $20,000 of their own money, a small group of blacks from the Mississippi Delta borrowed $350,000 from a Greenville bank in 1970 and opened a blue jeans manufacturing plant. They called the company Fine Vines, which is ghetto slang for good clothes.

Fine Vines advertised for 50 plant workers. It got 750 applications.

The company still is not large enough to wipe out unemployment in the Delta. But it now owns plants in Sardis and Canton, Miss., and Memphis. The other plants make attic stairs, exhaust fans, metal stampings and electronic switches. It also has a 4,000-acre plantation growing soybeans, cotton and rice.

Management Is White

The parent company, Delta Enterprises, now employs more than 350 persons. More than 90 percent are black. All the plant managers are white.

"There are no black managers," Mr. Bannerman said. "I'm looking for people with experience. I'm trying to create that experience as we go along. We have a development program for training managers. But I'm not going to sacrifice the company."

He said the firm's manufacturing divisions would sell about $7 million worth of goods this year. The profits go back into the business.

"We've brought in close to $20 million to this area. We've paid out over $6 million in salaries, which is about equal to what the state of Mississippi has paid in welfare in this area. We've got people making $4 and $5 an hour for the first time in their lives," Mr. Bannerman said.

"There's a lot of potential in the Mississippi Delta. Politically, we're moving ahead. But there's still a lot of institutional racism that wasn't solved by the Voting Rights Act.

"The town I live in has 50,000 people. It's 55 percent black. It has one black elected official. This county is 62 percent black and it has one black elected official. Education is still poor here. There's no compulsory attendance law."

"And farming—farming is one of the most important things here, and blacks are getting out of it. The number of black families in farming goes down every year."

An Exception

Delta Enterprises is exceptional. Most black Southerners must rely for jobs, not on the occasional flourishing black-owned business, but on white-dominated businesses and institutions. Because of that, black progress in employment is slow and strewn with the obstacles of custom and subtle racism.

Michael Figures, who has represented many clients in employment discrimination cases, says blacks are still the last hired and first fired by the big national companies that harvest the timber and other resources of South Alabama.

Peter J. Petkas, executive director of the Southern Regional Council in Atlanta, says the Council's monitoring indicates that government at all levels continues to discriminate against blacks in hiring. There are exceptions in cities like Atlanta, Austin and Houston, where blacks have gained significant political power.

Mr. Lewis, who has seen black voter registration in the South more than double in the last 10 years, says black political participation has made "a fantastic difference."

21 in Georgia Legislature

Georgia, for example, has 21 black legislators in the House of its state legislature. Some are veterans and sit on powerful committees. Black members have fought successfully for such measures as a tenants rights law, and have been influential, if not yet successful, against the death penalty.

Mr. Lewis and others agree that the branch of government with the worst employment record in the Southern states is the Federal Government. The South has no black Federal judges or United States attorneys. Only a handful of blacks hold other important Federal jobs.

The states, in contrast, have growing lists of black employees and officials. Joseph W. Hatchett made history in Florida last year by becoming the first black in modern times to win a statewide election there. He solidly defeated a white opponent for a seat on the State Supreme Court after having been appointed to the court by Gov. Reuben Askew 13 months earlier.

Mr. Hatchett was elected with the support of large numbers of white voters. Representative Andrew Young, just appointed to be Ambassador to the United Nations, made the same breakthrough with white voters in Atlanta.

There are other examples scattered around the South—all apparently supporting the contention of President Carter and others that the civil rights movement liberated whites as well as blacks in the South.

The last concerted resistance to black progress in the South is found in that shrinking coastal crescent from Pensacola to Mobile.

There the Ku Klux Klan still parades in the streets, burns crosses and holds rallies. In Pensacola the Klan sponsors a telephone recording warning against the activities of blacks and Jews.

Blacks still complain regularly of police brutality, not only in Pensacola but also as far aong the coast as New Orleans.

In Mobile, a group of white policemen picked up a black robbery suspect one night last year. With intentions known only to themselves, they got a rope out of a car, tied it around the suspect's [illegible]k and threw the other end over a tree [illegible] Mobile blacks call it "the hanging," though the man was not physically harmed.

Several of the policemen were charged with assault. An all-white jury acquitted two of them, and the white prosecutor decided that trying the others would be "an exercise in futility."

Black Leaders Arrested

Black leaders in the coastal crescent have been subjected to what some say is a campaign of selective law enforcement. The leaders of one civil rights group after another have been arrested over the last few years. Several have been sent to prison on such charges as tax evasion and violation of the drug laws.

One black leader in Pensacola was recently charged with rape. His accuser dropped the charge, and he complained that he had been framed. Nevertheless, his influence reportedly was destroyed.

Mr. Figures, who graduated from the University of Alabama Law School in 1972 and is now one of the most influential black leaders in Mobile, said, "It would be almost impossible to prove, but there appears to be a pattern [of selective law enforcement]. I know I walk a very thin line because of the position I'm in. I'm very careful in my personal affairs."

Mr. Figures and others believe that white resistance has lasted longer on the Gulf Coast because the area was overlooked during the 1960's by civil rights activists. The public, overt racism that was allowed to grow undisturbed is only now being rooted out, they say.

Most of the South went through the rooting-out stage five to 10 years ago. Those responsible for it hope that it will not be necessary again.

Winifred Green of Atlanta, the white director of the Southeastern Public Education Program of the American Friends Service Committee, said recently that if the Federal presence—Justice Department lawyers, Federal court decisions, various Federal agents pressing for integration—had been withdrawn from the South 10 years ago, "we could have resegregated in 24 hours."

She said she doubted that re-segregation could happen now. A lot of white people are satisfied with the new way, she said.

And as for the black people who once were locked out of the mainstream—"in two years, there will be thousands of black children graduating from Southern high schools who have never been to segregated schools. There's no way they're going back—no way."

January 24, 1977

BLACK AMERICANS: IMAGE AND IDENTITY

WHAT IS A COLORED PERSON?

The Superintendent of Public Instruction in Indiana is endeavoring to decide what constitutes a "colored person." A Floyd County teacher refused admittance to a couple of children whom he considered colored. The parents appealed to the School Trustee, and he sustained the teacher. Another appeal was made to the County Superintendent, who refers the subject to head-quarters for a solution. The evidence shows that the children's great grandmother was an octoroon. The colored element is, therefore, only one sixty-fourth of the whole. Considerable importance attaches to this ruling, as it affects every school in the State, directly or indirectly, in city and country.

December 8, 1873

AN ARRANGEMENT IN BLACK AND RED.

An interesting specimen of humanity—the smallest prisoner ever taken into the Jefferson Market Police Court—was before Justice Murray in that court yesterday. A colored woman named Lizzie Curtis, of No. 104 Greene-street, had the specimen in charge, and told the Justice that she wanted to complain of "William Henry" as a boy who, although only 8 years old, was so utterly incorrigible a little scamp that she was compelled to ask to have him sent to a reformatory institution. His father, she said, was white, and his mother colored. Both were dead, and she had taken the boy to bring up. The Justice leaned over the bench and gazed at the prisoner in astonishment. The boy, standing erect in his ragged little shirt and trousers, was not as high as an ordinary walking-stick, had unmistakably black skin, but bright red hair. The red-headed colored youngster had nothing to say for himself, and the Justice sent him to one of the institutions.

July 12, 1879

THE CAKE WALK.

The proud Caucasian is apt to sneer at the institution of the cake walk, although in that institution there are expressed, in a manner most interesting to the philosopher and most touching to the philanthropist, the æsthetic yearnings of the African race. It has been remarked, possibly by EMERSON, and repeated by many soulful persons of the female sex, that there is a better thing than writing poems, and that is to make one's life a poem. Now the cake walker does at least attempt, for the space during which he walks for a cake, to convert himself into a work of art. One would expect that the representatives of what boasts itself to be a superior race would cheer him on in this effort. As a matter of fact they do nothing of the kind. Cake walks, as they have been known heretofore, are attended by whites exclusively for the purpose of guying the cake walker and, if possible, breaking him up in his stride. Indeed, it is often the design of the white spectators of the baser sort to break up the whole cake walk, and this intention is frustrated only by the superior numbers of the African race and their promptness with the lethal razor. The cake walk is commonly given in comparatively secluded quarters and intended for the race to which the cake walkers belong. When the largest place of entertainment in New-York was secured for the cake walk of last night, and the utmost possible publicity given to the enterprise, it was not an unreasonable suspicion that the intention was not to hold out the cake walkers as models for the reverent imitation of the spectators, but to expose them to the derision of an unsympathetic concourse of whites.

This is all very wrong. The African race is comic mainly for the reason that it is imitative, but the cake walk is an institution evolved from the African intellect. It is a progress in which style is of more account than speed, and hence the precise opposite of the go-as-you-please pedestrianism, which is the latest production of the Caucasian mind in this kind. Upon the whole, the cake walk must be pronounced to be far in advance of the go-as-you-please race in point of civilization, inasmuch as grace is a more civilized quality than brute strength or brute endurance. There is, indeed, a moral element in endurance, but it is a moral element in which a bulldog surpasses a man. Of the two it is safe to say that the cake walk would have been more in favor at the Olympian games than the go-as-you-please. The Assyrian sculptures and the Egyptian mural decorations prove that in those ancient days there was a sense of the value of "grace, style, and execution," which are the qualities in which the contestants of last night competed. Nay, it has been justly observed that the frieze of the Parthenon itself is but the representation of a Pan-Athenaic cake walk.

Instead of jeering at the cake walkers, therefore, we ought to do them homage for keeping alive the regard for graceful locomotion to which we pay so little heed. It is, indeed, unlikely that the modern sculptor would find inspiration in a cake walk for a rivalry with the carved processions of antiquity. His search for grace and style might be impeded by his painful consciousness of a too prognathous jaw or a too protrusive heel. The more credit is due to the performers who, more or less in spite of nature,

> "Go past as marshaled to the strut
> Of ranks in gypsum quaintly cut,"

and exhibit a dignified indifference to the guyings of spectators in the boxes, who would themselves cut a very indifferent figure on the floor.

February 18, 1892

A WISE WOMAN OF COLOR

A TYPICAL SERVANT OF THE TIMES FAST PASSING AWAY.

Her Honesty and Dignity—Her Many Maxims—When the Glove Fitted—Her Small Hands and Feet—Opinions in Regard to the Education of Her Race—Criticisms on Servants of To-day-Fearful of Their Future in the South.

Mammy may be sixty-five, as black as the ace of spades, and with the whitest of souls. She is as broad as she is long, but with the hand and foot of a Princess of the blood and if Mammy has a weakness, it is in regard to these physical distinctions.

This is what she says: "I done sot no partikler vartue on a small han." Dere's good wukkin' people, better Christians nor I be, what has foots that stretch' cross two cornhills, and dey han's-sakes!-is jes like bedspreads—but the good Lord He make 'em so—but it seem to me dey gets 'emselves tangle up wid dese yere members. 'De very fuss time you give me a pa'r of your old gloves, Miss Clare,' says I, 'if dey fits me, I am gwine to take up wid dese people.' Sure enuf, dem gloves was neither too tight nor too loose—fit me jess snug. 'Den,' says I, 'dese here folks is quality,' which you was sure enuf. I never see, howsumever, dat dese hands of mine don't do no less wurruck nor odders as had regular b'ar paws. Dese yere big-handed women just interferes wid demselves, like a hoss dat ain't got no amble. I done say dat it is allers the case, but dar was Belindy, a fair likely cook. You remember her sweet potater pone, Miss Clare, and sakes! what han's she had. When she try and hide 'em under her apron, 'twan't no use. Dere dey was Dat Belindy lose her grip when she try to make an egg sauce. 'Twan't no use. Dem han's of hers spile de sauce ebery time."

The beauty of Mammy's hands is not exaggerated. She always tells how a cast was made of her right hand. A distinguished sculptor saw that plaster hand and expatiated over its singular beauty. He was all at sea about its origin, though certain it was not European. When its source was explained to him he said: "In the Viennese anthropological collection there are the casts of the hands and feet of a particular tribe of Africans, and they excel everything else in purity of outline and distinction of form."

It is rather Mammy's quaint philosophy, with manner of expression, which is worthy of mention. For tact and good sense, for sterling honesty, she is beyond compare, and with that there is a certain austerity. Thrown in by the vicissitudes of fortune with white servants in the North, she quietly asserted her position. Maybe rebuffed at first, her innate dignity soon made her take a leading position. Soon in the kitchen she was absolute queen. There was no dish she could not prepare. It was an inheritance from a Martinique grandmother. Once Gen. Joe Johnston partook of her prawns with Madeira sauce, and sent her his compliments on the perfection of the plate, and she said: "I hears you folk say dat Gineral Joe Johnston is a grandfather; dat may be so, but he jus' tech my heart and de whole Souf when he compliments my prawn."

Here are specimens of Mammy's proverbial philosophy, and in her language she follows Mrs. Malaprop: "Done you nebber open ice cream shop when de snow fall, an' done you sell sheep fleece in Summah." "Done you kiss around promiscus like, some folks keep der mouf cram full needles like pincushi'n." "Fust drink you got him. Second drink he got you." "When one dog yelps dat's nuffen, but when de pack howl, der fox is aroun." "Done you follow hazzard, for you be shu to come to carrion." "Blacksnake hut [hurt] you much, but black talk hut you most." "Ugly ooman a deal uglier in fine clothes."

"You live nex' to meetin' house, what for dat make you good." "Little Billy sut on de mountain. Little Billy no bigger for dat."

Some one was talking of the waste in this world, unnecessary luxuries, when Mammy remarked: "Dey talk heap about the supperfluse. 'Pore people is breakfas' fluse and dinner fluse mose of de time. Befo' de wah I might have seen a fluse time, but nebber since." Did Mammy mix up superfluous with the "flush times"?

"People says I'se fault findin. Mebby I is. Polyphemee, dat was my niece, she dig taters, and she bring me a small bushel, after she wurruck an hour, and the sweet potater patch was close by. I fault-finded her. She turn back and come back after an hour wid a few more taters in her apron. Then I fault-finded her wid a broom handle and she brung me in fifteen minute a heapin' bushel. Fault findin' wid a broomstick put me a bushel ahead and de chile, she larn her lesson.

"Done you tell me nuffin' about sottin' a teaf to kotch a teaf. Dat ain't Christianlike. You'se a transgressin' wid de transgressors. Dar war my Chulius. He drink Miss Mary's cup of new milk what I sot by her door ebery mornin', and Chulius swallow 'em right down, and Miss Mary, pore sick thing, complainin' she got no milk. Then I sot Chulius's own brudder Hemanuel, to watch that milk, and I give Hemanuel a hunk of gingerbread for to do the bizness. It didn't wurruck. Fust thing I knows Hemanuel he was drinkin' the milk with Chulius, and dey was goin' snacks wid de gingerbread. Mebbe sot a teaf to kotch a teaf is scriptur, but it don't wurruck on dis here wicked yearth."

Mammy had listened with marked attention to a political discussion. Impressed with what she had heard, she wanted further information.

"Miss Clare," she asked, "what's dis yere big word about contamminigation of sutthen like that? Dat sounds Biblelike, 'case I remember Brudder Wilson, who once run a turkentine mill in Norf Karlinet, he say: 'You can't tech tar widout being contamminigated.' Now you tell Mars William not to have nuffin' to do wid dese yere politics, or he's like to be contamminigationed, and dis yere family hasn't come Norf for to lose their souls."

Mammy never by chance gets the words of a maxim exactly right, although she at once appreciates the meaning of it. Somebody said: "Familiarity breeds contempt," and this was her explanation of this old saw and how it worked:

"Miss Clare, you remember dem Yankee Hawkinses and dat shop dey opened? Dey was from the Norf, and dey was hungry for trade, and went for customers, ebery time and all de time. Dey was familious wid all sorts of folks–de niggers and white trash. Hannah Kershaw, she war a buyin' all de time Klone water and pink silk stockings. When you gwine pay for that sweet scent and dem shin kivvers?' says I to Hannah, for Hannah was a good-for-nothing ooman, 'Dey is mighty nice people,' says Hannah, 'familious like—and dey don't turn up dere hoses on niggers as you does, for you is always down on your own color,' says Hannah.

"You knows the end of that Hawkins ship? Not a nigger paid a cent. Nor as for that did the white trash. Hawkins didn't get a pound of cotton —and he busted. Dat was familious bring contempt." Mammy is sensitive to a degree, and an unkind woed on the prt of those who do not know her special qualities is highly irritating to her. The other day she was disconsolate. Some white person had doubted a statement of hers, and she said: "Miss Clare, you know my temperature." As it was an excessively hot day, we could not judge Mammy thermometrically. Then Mammy went on: "Dis yere way some folks has of paying no 'tention to anudder's temperature is de cause of half de contentions in dis yearth, specially wid oomen." Then we "caught on." Mammy meant "temperament."

It never would do, if Mammy were to be put in charge of the education of her race. Maybe she has been too thoroughly indoctrinated in the ideas of the past. Born in slavery times, she was fortunate in having had for owners people who, acknowledging her sterling qualities, regarded her as a friend more than as a servant, for from her childhood to her old age her life has been one of devotion. The old colored woman's peculiarities of pronunciation are but scarcely imitated in what follows.

"The colored people Souf," Mammy says —for she has made frequent visits there of date, in the State where she was born and raised—"is nuffin' like as good—as decent—as moral—as dey was when dey was slaves. Dat's all bad enuff but dey don't see de main chance. Dey is lazy. Dey won't wurruk. Dey has no bess to tell um, 'Do dis or do dat.' Dey gets married, and don't keer for their chillums—don't feed 'em—don't put no clothes on 'em. Dey can some on 'em read and write, but readin' and writin' don't give 'em no meat no bread. Dey has no 'pinions of der own, or ef dey has dey follows the ijeas of some white man dat only keers to get on top—using the niggers like a ladder—and when he gets away up he kicks the prop from under him, and down goes de nigger on his head. Jess my heart is sore—when I sees old families of honest colored people, good simple souls, in slavery time, jest disappeared from off de face of the yerth. Not one of 'em left. Yet I thanks God for liberty—but maybe I was lucky, for I hain't felt no difference in freedom. I had to do less when I was a slave than I did when I was free. It was the good trainin' I had Souf—that was everything. Decent folk is the same, Norf or Souf—and I ain't no smarter nor the next one—but bein' brung up with ladies and gentlemen, I always tried to get service with such.

Dere is some good colored folks in the Souf yet, but dey is fast runnin' to seed. It's the wanderin' way they has, that is a killin' of 'em out. Ninety-nine times in a hundred when a colored man moves from Alabamy to Georgy or from Georgy to Texas, he just signs his death warrant. He may have been badly off in one place, but he's wuss off in the next.

"I hain't been twenty-five years in New-York, consortin' with my own people, without knowing something. Southern colored folks, say from Virginny, that comes to New-York, particularly the oomen, it just good or nothing. They was born of the new stock just after the war, and don't know anything about home service. Can't help it, Miss, they don't know their place, and don't want to be taught. They never had no bringing up. A mistress in Richmond knows just what they are, and what is to be expected of them. But a Northern lady don't. Educashun! Why don't they learn how to cook or sew? It's about one colored ooman in a thousand that knows how to hem a kitchen towel. As to fixin' decent food for a table, save fryin' in grease, or cooking greens, that's all they knows or want to know. I don't know, Miss, but it is this here poor Southern stock comin' to the Norf that isn't doin' no good to the colored people in the big cities. It's a crowdin' out the better class of workers. It's creatin' a kind of prejudish against colored oomen as house servants. Why shouldn't there be more of 'em in service? Just because they are worthless. That is a hard word, but they's not worth the money they ask. I've known some real capable colored oomen, neat, handy, fitten for the kitchen or the bedroom, but this was the trouble. They couldn't make up their minds to stay in the one place, but wanted variety, and I never seed a oomen change places three times in one year that didn't in time get lost. Done you talk to me about the Washington colored persons. They is mighty poor people to have in a house. There are ever so many decent colored people in Baltimore as is brung up thar, and you can get a good servant thar. Now these here remrks of mine is mostly to be taken as to oomen. There are lots of Southern colored men in New-York, occupying tip-top positions, janitors and sich, and what is more, holding them. I been trying to account for that. They is the colored people as has been tried in the fire, and has not melted; dese isn't dross niggers. I tell you, Miss—and I know–there never was such a good man in every way as the old-fashioned body servant in the Souf. I don' mean alone the waiting and attendance, but he was the most honest of men, and his master's interests was his own. But the last body servant died ten years ago, and the whole race has gone to hebbenly rest. See here, Miss Clare. There's an old grave-yard in Mississippi–and I have seen it–and I can't read, but I have had the writin' on the tombstone read to me, and I remember not the last names, but the fust ones. The master's name was Henry and the servant's name was Andrew, and the time when the two men died was in 1853, and the twin tombstones is jined with a bit of stone, as if to bind them together and this is the readin': 'Here lies Mr. William So-and-So, * * * and his very best friend, Andrew. For seventy-three years, boy and man, Andrew followed his master, and was his faithful friend and adviser: nor shall death now divide the white and the colored man.' "

August 23, 1896

SCORES NEGRO EDUCATION

Mississippi's Governor Calls It the Black Man's Curse.

Says the Negro Is Deteriorating Morally Every Day — Wants Fifteenth Amendment Repealed.

JACKSON, Miss., Jan. 19.—In his inaugural address, delivered to-day before a joint session of the Mississippi Legislature, Gov. Vardaman declared that the growing tendency of the negro to commit criminal assault on white women was nothing more nor less than the manifestation of the racial desire for social equality. In strong terms he declared that education was the curse of the negro race, and urged an amendment to the State Constitution that would place the distribution of the common school fund solely within the power of the Legislature. Continuing, Gov. Vardaman said of the negroes:

"As a race they are deteriorating morally every day. Time has demonstrated that they are more criminal as freemen than as slaves; that they are increasing in criminality with frightful rapidity, being one-third more criminal in 1890 than in 1880.

"The startling facts revealed by the census show that those who can read and write are more criminal than the illiterates, which is true of no other element of our population. I am advised that the minimum illiteracy among the negroes is found in New England, where it is 21.7 per cent. The maximum was found in the black belt—Louisiana, Mississippi, and South Carolina—where it is 65.7 per cent. And yet the negro in New England is four and one-half times more criminal, hundred for hundred, than he is in the black belt. In the South, Mississippi particularly, I know he is growing worse every year.

"You can scarcely pick up a newspaper whose pages are not blackened with the account of an unmentionable crime committed by a negro brute, and this crime, I want to impress upon you, is but the manifestation of the negro's aspiration for social equality, encouraged largely by the character of free education in vogue, which the State is levying tribute upon the white people to maintain.

"The better class of negroes are not responsible for this terrible condition, nor for the criminal tendency of their race. Nor do I wish to be understood as censuring them for it. I am not censuring anybody, nor am I inspired by ill-will for the negro, but I am simply calling attention to a most unfortunate and unendurable condition of affairs. What shall be done about it?

"My own idea is that the character of the education for the negro ought to be changed. If, after years of earnest effort and the expenditure of fabulous sums of money to educate his head, we have only succeeded in making a criminal out of him and imperiling his usefulness and efficiency as a laborer, wisdom would suggest that we make another experiment and see if we cannot improve him by educating his hand and his heart. There must be a moral substratum upon which to build or you cannot make a desirable citizen."

The Government also declares that the people of the Nation should demand the repeal of the Fifteenth Amendment.

January 20, 1904

TOPICS OF THE TIMES

"Negro" and Its Equivalents. Discussing with equally good temper and sense the proper appellation of the race it chiefly addresses, The New York Age expresses a preference for "negroes" over "colored people," on the ground that the former is "a frank, manly name." It declares, however, that "colored people" is "certainly" free from any implication of inferiority. In so doing, we think, The Age forgets that, while the colored people themselves may use the phrase without any admission of inferiority, it is difficult for other to do so without at least a subconscious assertion of it, since human nature is constructed that any recognition by one group of men of the fact that another differs from itself amounts at least to an intimation that the difference is for the worse. So "colored people" does convey an implication of inferiority, and its popularity among negroes is therefore somewhat remarkable for that reason as well as for the reason that it links them closely with races which they have a right to regard as inferior. "Negro," on the other hand, despite its etymological significance, is at present purely a racial designation; it brings to mind merely racial and geographic distinctions, the distinctions are perfectly definite, and the same for all whohear or use the word, and there is no denying any of them The Age says that its sole objection to "negro" is its inaccuracy, which it establishes in this peculiar way: " 'Negro' is the race name of certain uncivilized black tribes in Africa; can it therefore be applied with entire truthfulness to us, modified as we are by 200 years' contact with American civilization, and in the case of 2,000,000 of us, by the infusion of a greater or less proportion of Caucasian blood?" Herein The Age is itself inaccurate, it seems to us, since universal usage is against it, and universal usage is decisive as to the meaning of words. All things considered, The Age pronounces in favor of "Afro-American," which, in our opinion, is thoroughly bad, for reasons that have been discussed a thousand times in the case of other "hyphenated Americans," and need not now be repeated. We are really puzzled, however, when The Age asks us why, giving a capital letter to practically all other race names, we deny it to "negro." Possibly it is because practically all the other race names come from words which themselves begin with capitals, while "negro" harks back to plain, ordinary adjective. That's is pretty poor reason, but the best we can offer at the moment—in addition to inherited habit.

April 8, 1905

A LACK OF RACE PRIDE.

In the columns of our contemporary The New York Age, which professes to be devoted to the best interests of the negro race, appears a large advertisement inviting its readers to buy seven cosmetic preparations, by the use of which colored men may obtain "better situations in banks, clubs, and business houses," and colored women may "occupy higher positions socially and commercially, marry better, get along better." One of the preparations purports to be a cream that "makes dark skin lighter colored," and "not with artificial white, but naturally; makes the skin itself lighter colored every time it is applied." A pomade "uncurls kinks in hair and keeps it straight," and so on. The advertisement implies that negroes should be ashamed of their own features and should by all means mask them into some resemblance to the Caucasian race.

We do not take this occasion to inquire whether the company advertising the cosmetics is acting in good faith, and if they are as represented. We presume The Age's business ethics would prompt it immediately to reject a spurious advertisement. But, whether genuine or not, its admission to the columns of this organ of negro uplift affords a revelation of racial psychology that is both curious and saddening. The exhortation to stand proudly upon nature's endowments—to be a man, or a mouse, or a long-tailed rat—is not needed by most races. While the negro disesteems himself and seeks to be something else will he be respected as he is?

August 8, 1909

Play of Negro Life by Negro Actors.

Mrs. Emelie Hapgood, who entered the ranks of the professional play producers recently when she presented Chesterton's "Magic" at the Maxine Elliott Theatre, has placed in rehearsal a play of negro life. The play will be acted by negro players. The play is by Ridgley Torrence, whose one-act drama "Granny Maumee" was presented by the Stage Society several years ago.

March 3, 1917

"NEGRO" WITH A CAPITAL "N."

The tendency in typography is generally toward a lessened use of capital letters. Yet reverence for things held sacred by many, a regard for the fundamental law of the land, a respect for the offices of men in high authority, and certain popular and social traditions have resisted this tendency. Races have their capitalized distinction, as have nationalities, sects and cults, tribes and clans. It therefore seems reasonable that a people who had once a proud designation, such as Ethiopians, reaching back into the dawn of history, having come up out of the slavery to which men of English speech subjected them, should now have such recognition as the lifting of the name from the lower case into the upper can give them. Major ROBERT R. MOTON of Tuskegee, the foremost representative of the race in America, has written to THE TIMES that his people universally wish to see the word "Negro" capitalized. It is a little thing mechanically to grant, but it is not a small thing in its implications. Every use of the capital "N" becomes a tribute to millions who have risen from a low estate into "the brotherhood of the races."

THE NEW YORK TIMES now joins many of the leading Southern newspapers as well as most of the Northern in according this recognition. In our "style book" "Negro" is now added to the list of words to be capitalized. It is not merely a typographical change; it is an act in recognition of racial self-respect for those who have been for generations in "the lower case."

March 7, 1930

HARLEM DISORDERS MARK LOUIS DEFEAT

Negroes Set Upon Whites, Beat One Seriously—Man Is Shot —Windows Broken.

A series of disorders staged by hoodlums occurred in Harlem late last night following the defeat of Joe Louis, Detroit Negro, by Max Schmeling of Germany. In one of them a 50-year-old man, white, was knocked down and kicked by thirty men, Negroes, and was taken to Harlem Hospital.

The victim, Samuel Kulin, a WPA worker of 246 Amberg Street, Brooklyn, who had been attending a meeting of his local union in Harlem, was set upon at 119th Street and Fifth Avenue. He suffered a cut over the right eye and bruises of the body. His assailants escaped.

An unidentified white man, walking at 113th Street and Fifth Avenue, was surrounded and beaten by forty men, Negroes. He managed to run away and refused medical aid, the police said.

At 133d Street and Lenox Avenue fifteen mounted policemen dispersed after a half hour of manoeuvring a large crowd of men and youths, Negroes, who had been shouting and milling about. Radio squads were sent to 116th Street and Amsterdam Avenue, where Negro boys had been throwing stones through the windows of motor cars returning from the fight and to 155th Street and Bradhurst Avenue, where the windows of several buses from the Yankee Stadium, scene of the fight, had been stoned by men and boys.

While twenty-five police were investigating a shooting in a restaurant at 401 Lenox Avenue, at 130th Street, a group of Negroes threw a stone through the plate glass window. None was hurt. In the shooting, William Moss, 26, a Negro, of 100½ West 130th Street, was wounded in the right side. He was taken to Harlem Hospital. He had been eating when one of several men who had been arguing about the fight at a near-by table fired several shots, one of which struck Moss.

June 20, 1936

ROBESON HITS HOLLYWOOD

SAN FRANCISCO, Sept. 22 (AP) —Paul Robeson said today he was through with Hollywood until movie magnates found some other way to portray the Negro besides the usual "plantation hallelujah shouters."

In an interview the Negro baritone said he was particularly despondent over his recent return to Hollywood to play a sharecropper sequence in "Tales of Manhattan."

"I thought I could change the picture as we went along," Robeson said, "and I did make some headway. But in the end it turned out to be the same old thing—the Negro solving his problem by singing his way to glory. This is very offensive to my people. It makes the Negro child-like and innocent and is in the old plantation tradition. But Hollywood says you can't make the Negro in any other role because it won't be box office in the South. The South wants its Negroes in the old style."

September 23, 1942

'MUGGING' HELD TERM TO SLANDER NEGROES

Young Communists Here Told Harlem Has No Crime Wave

Recent emphasis on "muggings" by Negroes was characterized yesterday as "exaggerated crime wave slander" and "an attack upon the war effort" by Benjamin Davis Jr., executive secretary of the Harlem division of the Communist party, in an address at a convention of the Young Communist League in the Central Opera House in Sixty-Seventh Street, near Third Avenue. Eight hundred young men and women from all parts of the State attended the session.

"Actually," Mr. Davis said, "mugging is a new handle of slander and libel to be used against the Negro people to exaggerate and to create the impression that the Negro people are a criminal element and that the white population should regard them as such.

"There is no crime wave in Harlem. There is a social problem which has existed there for a long time. There is demoralization among sections of Negro youth, which is understandable, because although we are fighting a people's war for liberty, freedom, self-determination and the right of all peoples to stand up like men and women, to have full citizenship in our country, Negro youth has not been granted the full citizenship which it is more and more demanding."

Michael Saunders, executive secretary of the organization, announced that 2,500 members were in the armed services.

March 28, 1943

RACE IN THE NEWS

When race hate explodes into violence, as it did recently in Georgia and has done so many times, North and South, the original causes do not lie on the surface, and are never found in police records. Always the outbreak is prepared for by thousands of unrecorded, unremembered acts and words of discrimination. The vast majority of Southerners and Northerners never think in terms of violence. Yet when they are unfair toward the Negro in little ways they help build up the bad moral climate which does encourage violence among the ignorant, the weak and the vicious.

Each of us contributes, in his daily acts, toward good-will or ill-will in our communities. The press, we believe, has a special and heavy responsibility, not merely editorially—what American newspaper would defend the Georgia lynching?—but in its treatment of news. Some newspapers treat news about Negro citizens in separate sections or columns, extending Jim Crowism to the printed page. In others discrimination is less flagrant but Negroes are often identified, whereas members of other races are not.

This may seem a small thing. The Negroes do not think so. This newspaper has considered its obligations to its readers in the light of these facts and principles. This consideration has led us to adhere to the rule that the race of a person suspected or accused of crime shall not be published unless there is a legitimate purpose to be served thereby. By this we mean that it is correct to refer to race when the accused is still at large and race seems one mark of identification. It is also correct when, as in the case of race riots or racial antagonisms, it becomes essential to the understanding of the news. And even though, inconsistently, we would hold that in view of difficulties under which the Negro labors it is fair to identify Negro artists or others whose achievements are a matter of pride to all of us.

We are sure that this policy meets the approval of fair-minded readers. News that encourages racial discrimination may sometimes be of interest, but responsible journalism has a higher law than a passing interest.

August 11, 1946

To Beam Negro Story to Italy

OKLAHOMA CITY, April 13 (AP) —The story of a Negro couple who made a fortune in the junk business and built a $431,000 hospital for their race will be beamed to Italy by radio on Saturday. The State Department's "Voice of America" broadcast will cite the example of free enterprise to counteract Italian Communist propaganda that Negroes in America have no opportunities. On the same day that Italy's crucial election is held, April 18, the hospital, built by Mr. and Mrs. W. J. Edwards will be dedicated here. The 105-bed institution, patterned after the Mayo Clinic in Rochester, Minn., will be ready for its first patients April 22.

April 14, 1948

Picture Tests for Mental Patients

At last week's Boston meeting of the American Psychological Association, the Thompson Revision of the Murray Thematic Apperception Test was discussed. The revision is designed to leap the hurdle that Negroes face when they try to identify themselves ("empathize" is the technical word) with white persons, especially in areas where there are insuperable racial barriers. Charles E. Thompson, graduate psychology student at Tulane University, New Orleans, developed the revised method.

The Murray test takes its name from Dr. Henry A. Murray, Harvard psychologist and psychoanalyst. It consists of a set of pictures, ten each for men, women and children. Each picture shows one or more persons in some kind of situation or action. Some of the pictures are either drawn especially for the test, or taken from photographs or magazine illustrations. All are exhibited, one at a time, in the process of mental treatment or diagnosis.

"What story does this picture tell?" the patient is asked. There is a diagnostic purpose behind the question: to lead the patient back—by association, perhaps—to the possible source of his worry and obsession. The kind of story he tells will reveal facets of his personality.

Patient's Reaction

The psychiatrist is interested in the patient's reaction to the situations depicted in the pictures. The aroused reaction may be cheerful, or it may be depressed. The fantasies are revealing. Through these tests the psychiatrist can also observe how the patient reasons.

For years it was not realized that these pictures, which portray white persons exclusively, might not evoke the same responses in all races and cultural groups. Thompson, a 34-year-old former Army psychologist from Duncan, Okla., last year hit on the idea of modifying the Murray test for use with Negroes.

He noticed that a Negro patient in a Veteran's Administration hospital was giving particularly flat, unimaginative responses to the Murray pictures. On a hunch, Thompson asked the man to imagine that the persons in the pictures were Negroes. Immediately the responses were more colorful and imaginative. The man identified himself with the figures in the pictures and acquired a new insight into the situations portrayed.

Group Differences

Sociologists know that among different racial or cultural groups in the same society there is an "in-group, out-group" feeling; that is, a Negro in a mixed community, or any person belonging to a class or group whose economic, racial, religious or cultural standards and customs are different from those of neighboring groups, will meet members of his own, or in-group, on common ground.

But the member of an alien or rival out-group he tends to see as a stereotype, a mere representative of the group. The in-group person cannot understand that the outsider has his own feelings and emotions and personality characteristics.

Thompson's patient had run into this cultural wall, but the minute he imagined the people in the pictures as Negroes—as members of his own in-group—he developed more complex stories.

After several impromptu experiments, Thompson decided to modify the Murray plates by changing the white figures to Negroes, but to retain the expression on the faces and the situations as originally pictured. He found that Negroes almost invariably gave longer, more detailed responses to the modified pictures than to the Murray plates and hence responses of greater diagnostic importance. W. K.

September 12, 1948

STUDY SHOWS LACK OF AID TO NEGROES

Parent education and guidance in Negro-white relations among Negroes are insufficient and fail to help the Negro child develop a feeling of personal adequacy, according to Dr. Regina M. Goff. Her study, "Problems and Emotional Difficulties of Negro Children," was published yesterday by Teachers College, Columbia University.

In interviews with 150 youngsters and their parents, Dr. Goff investigated the problems, fears, annoyances, frustrations and other emotional difficulties Negro children experience because they are Negroes.

The influence of these experiences on their feelings and their behavior; the kinds of guidance given by their parents and its effectiveness, along with the attitudes of parents t. ward certain social conditions were also looked into.

Study Is Widespread

Seventy-five girls and seventy-five boys, ranging in age from 10 to 12 years, made up the group. Ninety of the youngsters lived in New York, sixty in St. Louis. Their parents had different kinds of jobs and different cultural backgrounds.

Colored youngsters "perceive clues which enable them to attach significance to themselves as different and early learn to limit their expectations of freedom of movement and gratification of desires," Dr. Goff reports.

Of the 150 children, 95 per cent had experienced situations they interpreted as problems in inter-racial relations. Of these 77 per cent had been ridiculed; 56 per cent had felt indirect disparagement through observation of stereotyped movie and radio characters; 41 per cent through aggressive behavior of white children; 33 per cent through discriminatory situations; 18 per cent through rude, ill-mannered behavior and 11 per cent through physical ill-treatment by white adults.

Most Prefer to Withdraw

The youngsters' first impulse in these situations, according to the report, was to strike back, but the most prevalent response was withdrawal.

According to the parents, the churches contribute little to racial understanding. Inter-racial activities are practically nonexistent in low-income groups and are looked upon with suspicion. Upper-income parents reported more belief in the sincerity of these groups, but there was no evidence that their activities had any influence on parental guidance.

More information on these problems for Negro parents, more education among the white population of the results of unthinking behavior, are recommended along with an "enriched program of training for the Negro child which places more emphasis on building his self-esteem, self-respect and self-confidence."

February 11, 1950

DOLLS TO COMBAT BIAS

Negro and Other Races Will Be Represented in New Toys

A line of fine quality Negro dolls will be used to combat racial prejudice among small children, it was announced yesterday by David Rosenstein, sociologist, and president of the Ideal Toy Corporation.

His company will make dolls of other races based on anthropological research. They will be manufactured under the suprevision of the originator of the idea, Miss Sara Lee Creech of Belle Glade, Fla., leader in interracial activities. She conceived the idea when she observed that Negro children in her town had only white dolls to play with.

September 11, 1951

Theatrical Types Portrayed

Objection Voiced to Presentation of Minorities in Unfavorable Light

TO THE EDITOR OF THE NEW YORK TIMES:

In Topics of The Times you mention the foolish objections of lawyers and hotel maids to recent individual theatrical portrayals, and add your own objection to other "relatively unimportant" sensitivities that "threaten to deprive us of much that has been pleasant in American humor."

May I take the liberty of suggesting that you have, in part, begged the question? First, by way of context, let me note that I oppose any kind of picketing against points of view or portrayals. But this is not necessarily to agree with you in deploring the passing of certain portrayals, notably Moran and Mack. (Incidentally, this type of portrayal has, unfortunately, not passed, but remains all too prevalent even now.)

Your attempted parallel between this kind of thing and the portroyal of a slattern maid misses in one important respect: There does not exist in America a pressing problem of community attitudes and practices against hotel maids; there does exist such a problem regarding Negroes and other minority groups. Very few people, if any, project from seeing one sloppy maid to thinking of maids as sloppy en masse. We know, however, that many people do project to a group generalization from seeing the portrayal of a drawling, happy-go-lucky, stupid, shiftless Negro. So the presentation of this stereotyped "stage Negro" works a direct adverse effect on the community welfare.

The question begging I mentioned lies in stating as a premise that such portrayals are "pleasant in American humor." Are they, indeed, pleasant? If so, to whom? They are not pleasant to me, and to countless other Americans. They contribute directly, I have observed, to the maintenance of attitudes and practices which are anti-libertarian and are, accordingly, downright unpleasant.

As I have said, I object strenuously to efforts to censor such portrayals. But just as strenuously I advocate "improved" portrayals—whereby, for instance, writers would act on their responsibility (when describing the current scene) to portray Negroes, and other minorities, as every-day members of the community—lawyers, merchants, policemen, housewives, students, etc.—not merely menials, criminals and joke butts. In such a changed context the portrayal of foolish, depraved, bandanna-wearing Negroes would become what you suggest now (and properly suggest as regards maids and lawyers): mere individual portrayals, having no detrimental community effect.

PAUL L. KLEIN.
Cleveland, Ohio, March 20, 1953.

March 25, 1953

AFRICAN ORIGINS CITED

28 Million Whites in U. S. Said to Have Negro Ancestors

COLUMBUS, Ohio, June 14 (AP)—An Ohio State University professor of sociology and anthropology says more than 28,000,000 white persons in the United States are descendants of persons of African origin.

The majority of these are classifed as white, Robert P. Stuckert reports in the current issue of the Ohio Journal of Science.

Mr. Stuckert is assistant professor in the university's Department of Sociology and Anthropology. He said he had drawn these conclusions from a genetic study of African and non-Afrcan population in this country from 1790 through 1950.

"The data indicate that approximately 21 per cent of the persons classified as white in 1950 have an African element in their inherited biological background," Professor Stuckert write. He said he had found in addition, that during the last four census years, from nearly 65 to 73 per cent of United States Negroes had some degree of non-African ancestry.

June 15, 1958

NEGRO CURB CITED BY PSYCHIATRISTS

Study Finds Social Factors Slowing the Intellectual Growth of Non-Whites

By EMMA HARRISON

Special to The New York Times.

CHICAGO, Feb. 27—Dramatic evidence of the impartment of intellectual functioning in the Negro child by socio-cultural factors was presented here today.

Negro and white children who scored nearly the same in early intellectual and developmental potentials show greater variance at the age of 3, when environmental factors enter into the picture, two psychiatrists said.

The differences occur largely in adaptive and language behavior, areas most susceptible to socio-cultural influences, while motor behavior, which is largely the result of basic neurological endowment, remains the same.

Drs. Hilda Knoblock and Benjamin Pasamanick, a husband-wife team conducting a study on causes and effects of prematurity, made the report to the annual meeting of the American Orthopsychiatric Association here.

Studied 1,000 Cases

The researchers, whose studies of 1,000 Baltimore children have turned up much vital statistical material on results of both organic and environmental defects on the infant, reported on a study of 300 of these youngsters, measured for capacities as infants and later as 3-year-olds. The group comprised about half white and half Negro youngsters.

The youngsters were used as "controls" on normal births in the larger study of prematurity being conducted by the Pasamanicks, who are now in Columbus, Ohio. The measurements were made by means of the Gesell Developmental Test, which operates very like intelligence-quotient tests, relating achievement and ability of infants to their chronological age.

Behavior Change Noted

A comparison of the full-term, controlled youngsters at the age of 3 showed a "marked racial divergence in adaptive and language behavior, while the motor and personal-social behavior are essentially unchanged."

The general developmental quotient rose for the white children, while it fell for the Negro youngsters; language ability rose in the whites and fell in the non-whites.

Failure in these areas is quite understandable among the non-white groups, usually of the lower socio-economic background, Dr. Pasamanick said. There is less motivation to learning in most of these non-white homes in the study, because there is more of a deterrent to it.

There is more sickness, more working mothers, lower nutrition and less, if any, stimulus to intellectual achievement. The Negro children in the study just did not have the motivation to achievement in all phases of their development, intellectual and physical, he observed.

They observed that in infancy, the factors producing differences in intellectual potential were largely organic. Other of their studies have maintained that even organic differences are the result of certain environmental factors, such as malnutrition, time of conception, etc.

Social Differences Stressed

Thus, the study is putting statistical measurements to the environmental factors in intelligence that social scientists have long discussed.

"It is perhaps fitting," they say, "to comment here that evolution in man no longer appears to be on an organic structural level. The major changes are social and cultural and small differences in intellectual potential due to organic factors cannot be detected within this larger framework."

They noted that while the studies of the intelligence indicated need for a hard look at educational and social welfare activities, the emphasis was still on genetic factors. They asked for a reversal of emphasis, with increased socio-cultural research into problems related to intelligence, and organic research into such mental disturbances as schizophrenia.

It is true, they observed, that there are organic factors determining mental behavior, but intellectual functioning can also be controlled by environmental conditions.

February 28, 1960

Foreign Affairs

The American Negro and Free Africa

By C. L. SULZBERGER

CHICAGO.

The savage and chaotic Congo problem is only one aspect, if a deeply disturbing aspect, of the burgeoning independence movement in Africa which is now having a profound effect upon the 18,000,000 Negroes of the United States. These Americans are worried and distressed by what is occurring in Belgium's former colony. But, on the whole, they take pride in and draw emotional satisfaction from the fact that the continent of their ancestors is at last becoming free.

This pride and emotional satisfaction produce no attachment in any way comparable to the Zionist movement among American Jews with its particular relationship to tiny Israel. Nevertheless, one perceptible result has been the immense acceleration of our Negroes' demands for complete equal rights. Another has been the growth of a new self-confidence and racial pride.

Negro delegates at the recent Los Angeles convention cited the former attitude as at least partially responsible for the Democratic party's insistence, despite Southern opposition, on passing the strongest civil rights plank in its history. They claim that even die-hard advocates of segregation no longer defend segregation as such, but merely argue against too rapid a change.

Even those American Jews who have not actively supported Zionism tend to admire the success of their co-religionists in Israel. The interest of American Negroes in new African states and new African leaders is equally sympathetic if less precise and, although embarrassed by the Congo disaster, delights in the progress already registered in Ghana and in the renown of such African statesmen as Kwame Nkrumah and Tom Mboya.

While a few American Jews became Israeli citizens, virtually no American Negroes contemplate accepting the nationality of any African state. There is, after all, no similar connection, religious, philosophical or otherwise. Negro families here have no way of knowing from which African land their ancestors were kidnapped by slave traders. Thus, there is no "Ghanaism" or "Nigerianism" in our Negro community.

Stimulus of Freedom

However, American Negroes identify themselves more and more with independent Africa. Also they report that African leaders visiting this country feel at ease with them and are simultaneously repelled by the social inequality of our Negroes and impressed by their opportunities and high living standards.

American Negro leaders admit that in the past their community tried to forget its racial history. Now, they say, under the stimulus of African freedom, a new sense of vanity has been awakened. One odd result is that the market for such humiliating products as hair-straighteners and skin-whiteners is shrinking. Negro business men are pooling assets to invest in African enterprises and specialists are seeking temporary jobs.

Intellectuals see a link between the spread of the Negro sit-in movement, which aims at breaking down discriminatory barriers, and the growth of African independence. The latter has tended to encourage the Gandhi-like technique originated by the Rev. Martin Luther King.

For the first time prosperous young people, strongly supported by their parents, have moved to the fore and Negro leaders relate this to concern with African freedom. They say that earlier efforts were guided by labor organizers and poorer elements and sought primarily to end economic discrimination.

Now the middle class, influenced strongly by such men as Mboya, is more active and concentrates on political and social advances. Hitherto, this class played a passive role. Its doctors, teachers and lawyers had a ready-made clientele with a vested interest in segregation. But this interest is being discarded.

Thus, once again, our country's internal evolution is affected profoundly by external events. From the seventeenth through the nineteenth century revolution and persecution in Europe were mirrored in our changing social structure.

Africa's contribution has hitherto been of an ugly nature: the initial supply of unwilling immigrants. But today the expression of African influence is changing, and changing drastically. As Africa's own Negroes take the path of liberty they encourage ours to complete the process of their political and social emancipation.

July 20, 1960

A Protest Of His Own

NOBODY KNOWS MY NAME: More Notes of a Native Son. By James Baldwin. 241 pp. New York: The Dial Press. $4.50.

By IRVING HOWE

TWELVE years ago a young Negro writer named James Baldwin printed an impassioned essay, "Everybody's Protest Novel," in which he attacked the kind of fiction, from "Uncle Tom's Cabin" to "Native Son," that had been written in America about the sufferings of Negroes. The "protest novel," said Baldwin, began with sympathy for the Negro but soon had a way of enclosing him in the tones of hatred and violence he had experienced all his life; and so choked up was it with indignation, it failed to treat the Negro as a particular human being. "The failure of the protest novel * * * lies in its insistence that it is [man's] categorization alone which is real and which cannot be transcended."

To transcend the sterile categories of "Negro-ness," whether those enforced by the white world or those erected defensively by Negroes, became Baldwin's central concern as a writer. He wanted, as he says in "Nobody Knows My Name," his brilliant new collection of essays, "to prevent myself from becoming *merely* a Negro; or, even, merely a Negro writer." He knew how "the world tends to trap and immobilize you in the role you play," and he knew also that for the Negro writer, if he is to be a writer at all, it hardly matters whether this trap is compounded of hatred or uneasy kindness.

AVOIDING the psychic imprisonment of a fixed role, however, is more easily said than done. It was one thing for Baldwin to rebel against the social rebelliousness of Richard Wright, the older Negro novelist who had served him as a literary hero, and quite another to establish his personal identity when there was no escaping that darkness of skin which in our society forms a brand of humiliation. Freedom cannot always be willed into existence; and that is why, as Baldwin went on to write two accomplished novels and a book of still more accomplished essays, he was forced to improvise a protest of his own: nonpolitical in character, spoken more in the voice of anguish than revolt, and concerned less with the melodrama of discrimination than the moral consequences of living under an irremovable stigma.

Jacket photograph by Roy Hyrkin for "Nobody Knows My Name."
James Baldwin.

This highly personal protest Baldwin has released through a masterly use of the informal essay. Writing with both strength and delicacy, he has made the essay into a form that brings together vivid reporting, personal recollection and speculative thought. One of his best pieces, for example, begins as an account of his return to the streets of Harlem where he was raised; moves toward a description of why Negroes living in housing projects resent the liberal authoritarianism with which these are often managed; rushes to some sharp observations about the residents of Harlem who "know they are living there because white people do not think they are good enough to live anywhere else"; and comes to a reflective climax with an outburst of eloquent speech: "*Negroes want to be treated like men* * * *. People who have mastered Kant, Hegel, Shakespeare, Marx, Freud, and the Bible find this statement utterly impenetrable * * *. A kind of panic paralyzes their features, as though they found themselves on the edge of a steep place."

THERE are other essays in "Nobody Knows My Name" composed with equal skill: a saddening account of Baldwin's first visit South, a report on an international conference of Negro intellectuals debating whether they share a common culture, a chilling polemic against William Faulkner's views on segregation. And especially noteworthy are three essays on Richard Wright, which range in tone from disturbed affection to disturbing malice and reflect Baldwin's struggle to achieve some personal equilibrium as writer and Negro by discovering his true feelings toward the older man.

That Baldwin has reached such an equilibrium it would be foolish to suppose, and he himself would surely be the first to deny it. One great merit of his essays is their honesty in reflecting his own doubts and aggressions, and in recording his torturous efforts to find some peace in the relations between James Baldwin the lonely writer and James Baldwin the man who suffers as a Negro. This honesty, I would suggest, has driven him to abandon some of the more sanguine assumptions of "Everybody's Protest Novel"—it is, alas, not so simple to shed the categories imposed by society—and to come closer to Richard Wright's anger than he might care to admit. For if he began by attacking Wright for writing as a Negro rather than an individual artist, the pressures of experience have forced Baldwin to do his best work as an individual artist precisely when writing as a Negro.

I have only one complaint to register against "Nobody Knows My Name." Partly because his work relies so heavily on a continuous scrutiny of his own responses, Baldwin succumbs at times to what Thorstein Veblen might have called the pose of conspicuous sincerity. In the essays on Wright and especially in a piece on Norman Mailer, the effort to expose the whole of his feelings slips occasionally into a mere attitude, and the confessional stance reveals some vanities of its own.

These are small blemishes on a splendid book. James Baldwin is a skillful writer, a man of fine intelligence and a true companion in the desire to make life human. To take a cue from his title, we had better learn his name.

July 2, 1961

Mr. Howe, chairman of the Department of English at Brandeis University, is author of "Politics and the Novel" and critical studies of Faulkner and Sherwood Anderson.

Advertising: Negro Market Plays Big Role

By PETER BART

Of late, American Negroes have been displaying a new militancy in their continuing quest for racial equality. Repercussions of this new posture have been felt in many spheres of American life—politics, education and labor, for example. They also have been felt in the field of marketing.

Leading advertisers are thus becoming aware of two basic facts of life regarding the Negro community. The first is that the Negro represents an increasingly important market, both because of population growth and because of rising income. Second, the Negro consumer poses a special sort of marketing problem—a problem that grows more and more complex as the new militancy takes firmer hold of the Negro population.

Indications that the Negro is playing an increasingly important role in this country's economic life are readily apparent. The Negro population now numbers nearly 20,000,000, an increase of more than 25 per cent in the last decade. This compares with a 16 per cent rise in the white population. The wage of the average Negro worker is now nearly 60 per cent of the average white worker, compared with 41 per cent in 1939. The total gross income of American Negroes now is put at more than $19,000,-000,000 a year.

Spending Increases

In addition to making more money, the Negro also is spending more money. In fact, Negro families tend to spend more than white families in the same income group. This spending follows some highly unpredictable patterns to be sure. For example, the average Negro man during his lifetime buys about 80 per cent more shoes than the average white man. Moreover, Negroes, although, accounting for only 11 per cent of the adult male population, purchase 20 per cent of all the men's dress shoes sold.

Because of these unpredictable patterns of consumption, the Negro poses a puzzle to many white marketing men. School integration, the sit-in demonstrations, the emergence of the new African states—all these forces have created a greater sense of racial pride and awareness, which in turn has affected the Negroe's behavior in the marketplace.

According to specialists in Negro marketing, the heightened pride and awareness have manifested themselves at two levels. At the first level, the Negro consumer has developed an increasing responsiveness to advertisements that are beamed directly at him. The use of Negro models in an advertisement, for example, might determine his decision to purchase one brand in preference to another that uses only white models.

At the second level, the Negro has suddenly shown a willingness to demonstrate his growing economic power through use of the boycott. Apparently without coordination or instigation by any national organization, local boycotts have sprung up in Cincinnati, Fort Wayne, Philadelphia, Baltimore and other cities against a variety of companies. The aim of the boycotts usually is to induce the companies in question to extend greater employment opportunities to Negroes.

An Effective Weapon

"The boycott has suddenly started to spread across the country and has become a highly effective weapon," says Harvey Russell, manager of special markets for the Pepsi-Cola Company. "The difficult thing is that once the boycott is called the harm is already done —even if a company alters its employment practices and hires Negroes, the local Negro community still retains its reluctance to use that company's products."

How does a company go about serving the Negro consumer? How can the Negro be reached with an advertising message?

One obvious way, of course is through the Negro media. Advertising revenues of the Negro media have risen sharply in recent year. The number of Negro radio stations has grown from three to 700 over the last twenty years. And today more than half of the 100 top advertisers make use of the leading Negro national magazine, Ebony.

In addition, companies have grown increasingly skillful in tailoring their sales messages to a Negro audience. Benjamin H. Wright, Ebony's sales production manager, notes that about 90 per cent of the models used in Ebony advertisements are Negroes, compared with about 50 per cent five years ago.

Besides the Negro media, Negro consumers also can be reached through conventional media. But here advertisers run into problems. "Most Negroes, when they see white models acting out a television commercial, say, 'that's for them, that's not for me,' " Mr. Wright says.

One answer, of course, is to put Negro models into the advertisements, but this in turn creates problems. Even in the North, whites are unaccustomed to seeing whites and Negroes appearing together in ads.

Still, many Negro leaders continue to campaign for the wider use of Negro models. Dick Gregory, the young Negro comedian, has started something of a personal crusade in this area. And some companies have written him to the effect that they have tried to make use of Negro models in ads. For example, a Boston manufacturer of stereophonic record players, Lang& Taylor, Inc., under Mr. Gregory's prodding, used Negro model's recently in an ad in The Boston Herald and received a favorable reaction.

Though the problem of white vs. Negro models is far from being resolved, advertisers have turned their attention to other techniques for reaching the Negro consumer that are less charged with controversy. The Negro, advertisers have found, is more responsive to certain appeals than is the typical white consumer. One such appeal is that of status. A sociologist recently asked a group of Negroes what whisky they served and recorded their responses. Later in the interview he asked them what brand of whisky they thought was served at most plush, high-society functions and, in a vast majority of cases, the answers corresponded with the earlier expressed brand preference.

Two Basic Approaches

In a recent series of articles in the Harvard Business Review, Henry Allen Bullock, Professor of Sociology at Texas Southern University, noted that advertisers basically had made two approaches to the Negro market. Either they have given it special treatment or they have given it no treatment at all. Professor Bullock advocates a third course—namely, developing an "integrated" advertising approach. Such an approach, he says, is feasible because "the motivational forces guiding the behavior of black and white consumers, though different in detail, have common denominators which advertisers can manipulate in favor of the sale of a product or a service."

September 24, 1961

SCHOLAR DEFENDS NEGROES' DIALECT

It Should Be Judged on Its Own Terms, She Says

By PAUL L. MONTGOMERY

A specialist in linguistics said yesterday that looking at the American Negro dialect as merely a deviation from standard English was "blind ethnocentrism."

The scholar, Dr. Beryl Loftman Bailey of Hunter College, said it was "meaningless" and "unenlightened" for philologists to apply English grammatical categories to the dialect — as is traditionally done — because such categories do not take into account the dialect's historical roots.

She said that the Negro dialect, characterized by such phrases as "I glad he gone," "Chester say" or "I two people and this one ain't me," had undergone development like pidgin English and the Creole tongues.

Like them, she concluded, it should be treated on its own terms rather than on terms foreign to its development.

Dr. Bailey's paper, "Toward a new perspective in Negro English dialectology," was one of 20 presented at the 10th annual National Conference on Linguistics. The two-day gathering at the Biltmore Hotel, sponsored by the Linguistic Circle of New York, ended yesterday.

Similar to Creoles

Dr. Bailey said most scholars accepted the thesis that the origin of the grammatical structure of Negro speech was similar to that of the Creoles.

In linguistics, any trade language used by an economically, socially or politically subject group, class or race is described as a Creole. It is characterized by an extreme simplification of the language of colonization from which it derives.

Although this is recognized, she said, philologists persist in criticizing the dialect for such things as its "confusion of persons," "frequent use of present forms in the past," and "a tendency to omit all forms of 'to be.' "

While this is superficially true, she said, structural analysis of the dialect might show it is "so ordered as to make it possible to ignore certain categories which are basic to English."

Dr. Bailey, a small woman whose black straw hat barely peeped over the speaker's rostrum, used the dialect phrase "He a cool killer" as an illustration of one of her points. In Jamaican Creole, she said, the phrase would be "Him a one cool killer," and in English "He is a cool killer."

To Each Its Rules

Each form, she said, follows its own structural rules, but to describe one form in terms of the rules of the others—as has been done in Negro dialectology—would result in "partial descriptions that serve to obfuscate the true state of affairs."

In another paper, Dr. Marshall D. Berger of City College outlined a method by which the place of origin of a person can be determined by certain telltale pronunciations. These include whether the speaker distinguishes among "merry," "marry" and "Mary," "cod" and "cawed," "don" and "dawn" and whether he pronounces the first syllable of "horrible" to rhyme with "bore" or "bar."

Dr. Berger said that he had had success with his method, particularly in areas that he knew well, and that he was able to detect social and economic differences, as well as geographical ones, among speakers.

The scholar said in an interview that the method had more then entertainment value. He said it could be useful for persons like announcers who wanted to "modify their speech intelligently" and to sociologists because the degree to which a person reflects the speech patterns of his town or parents is indicative of his social or psychological make-up.

March 15, 1965

REPORT FOCUSES ON NEGRO FAMILY

Aid to Replace Matriarchy Asked by Johnson Panel

By JOHN HERBERS
Special to The New York Times

WASHINGTON, Aug. 26—A confidential report submitted to President Johnson last spring by a committee on civil rights has become one of the most widely discussed and quoted papers in Washington.

Still unpublished in its entirety and still officially confidential, the report has come in for new attention since the Los Angeles riots, for it pinpoints the causes of discontent in the Negro ghettos and says the new crisis in race relations is much more severe than is generally believed.

Entitled "The Negro Family—the Case for National Action," the 78-page report constitutes a devastating indictment of what white Americans have done to Negro Americans in 300 years of slavery, injustice and estrangement—the result of which is a "tangle of pathology" that will require a unified national effort to correct.

MADE RACIAL STUDY: Daniel P. Moynihan played key role in study of Negro family life while he was with the Labor Department.

A leading role in compiling and writing the report was played by Daniel P. Moynihan, then Assistant Secretary of Labor for policy planning and research. Mr. Moynihan later resigned to become candidate for President of the New York City Council on the Democratic ticket of Paul R. Screvane.

The essence of the report, as published in The New York Times on July 19, is that deterioration of the Negro family has resulted in a deterioration in the fabric of Negro society. As a result, Negroes as a group are not able to compete on even terms in the United States.

As a synthesis of the published works of a number of social scientists, reinforced by government statistics, the report contains little new information. It points out, however, that "probably no single fact of Negro American life is so little understood by whites" as the breakdown of the Negro family.

President Johnson has accepted the thesis of the report and is expected to make it the basis for White House conferences on civil rights next fall. In a speech at Howard University last June, the President said white Americans must accept the responsibility for the breakdown of the Negro family structure.

The report was prepared by a committee appointed by the President to help chart the Government's course in race relations. The members came from several agencies of the Government.

Essentially what has happened, the report said, is that white Americans by means of slavery, humiliation and unemployment have so degraded the Negro male that most lower class Negro families are headed by females.

"The very essence of the male animal from the bantam rooster to the four-star general is to strut," the authors said. "But, historically, the instincts of the American Negro male have been suppressed. Indeed, in the 19th century America, a particular type of exaggerated male boastfulness became almost a national style. Not for the Negro male. The 'sassy nigger' was lynched."

"In essence," the report continued, "the Negro community has been forced into a matriarchal structure which, because it is so out of line with the rest of American society, seriously retards the progress of the group as a whole, and imposes a crushing burden on the Negro male and, in consequence, on a great many Negro women as well."

Disintegration of the family, the report said, has been speeded by poverty, isolation and displacement of Negroes from southern farms into urban ghettoes of the north. The result has been a high rate of crime, delinquency, school dropouts and escape from reality.

August 27, 1965

WHITE HOUSE STUDY OF NEGRO ASSAILED

James Farmer, national director of the Congress of Racial Equality, sharply criticized yesterday a report on the Negro family prepared by a White House study group.

Writing in The New York Amsterdam News, a Harlem weekly, Mr. Farmer said the report had given racists "a respectable new weapon" while "insulting the intelligence of black men and women everywhere."

The thrust of the report, written by Daniel P. Moynihan, former Assistant Secretary of Labor, was that three centuries of injustices had created for the Negro a matriarchical society, the result of which has been "a tangle of pathology capable of perpetuating itself without assistance from the white world."

Mr. Farmer said that the Negro was "sick unto death of being analyzed, mesmerized, bought, sold and slobbered over while the same evils that are the ingredients of our oppression go attended."

"It has been the fatal error of American society for 300 years to ultimately blame the roots of poverty and violence in the Negro community upon Negroes themselves," he added.

December 17, 1965

The Negro Today Is Like the Immigrant Yesterday

By IRVING KRISTOL

LET us suppose that, a century ago, Harvard had been host to a conference on "the crisis in our cities." Let us suppose further that there were sociologists in those times (sociologists such as we know them today, I mean) and city planners and professors of social work and directors of institutes of mental health and foundation executives—and that these assembled scholars were asked to compose a description of the u.ban conditions in the United States. They would have been at no loss for words; and their description would have gone something like this:

"Our cities are suffering a twofold crisis. First, there is the critical problem arising from the sheer pressure of numbers upon the amenities of civilized life. Our air becomes ever more foul from the activities—both at work and at play—of this large number; our water is shockingly polluted; our schools are overcrowded; our recreational facilities vandalized; transportation itself, within the city, requires ever more heroic efforts.

"As if this were not enough, there is this second phenomenon to observe: our cities are being inundated *by people who are themselves problems.* These are immigrants—Irish, mainly —who are more often than not illiterate and who are peculiarly unable to cope with the complexities of urban life. Their family life is disorderly; alcoholism is rampant among them; they have a fearfully high rate of crime and delinquency; not only do they live in slums, but they create slums wherever they live; they are bankrupting the resources of both public and private charities; they are converting our cities into vast cesspools of shame, horror and despair; they are——" And so on and so forth.

Any American of the nineteen-sixties could complete this bleak catalogue without overly exerting himself: it is the identical catalogue that any such conference today would come up with.

NOW, it is important to realize that the scholars of a hundred years ago would have been telling the truth. Because we surmounted the particular crisis they endured, we would be inclined to think their concern bordered on hysteria, and that they unduly exaggerated the evils around them. They did nothing of the sort—I say "did" because, while this conference is hypothetical, the urban crisis was real enough at that time. We sneer gently at the agitation of years ago as representing a lack of imagination on the part of "The Brahmins"—the "old Americans"—and as testifying to a fear of historical change combined with an overrefined distaste for plebeian realities. We think of Henry James lamenting his "sense of dispossession"; and we do not think too flatteringly of him for doing so. But the question might be asked: are we not all Brahmins now?

I am not saying that the problems of American cities today are identical with, or even perfectly parallel with, those of yesteryear. Such identities do not exist in history; and all historical parallels are, in the nature of things, less than perfect. But I do think it important that we keep American urban history always in mind, lest we be carried away by a hysteria all our own.

Just how close we are to such a hysteria may be discovered by directing to ourselves the question: *how much of a disaster would it be if some of our major cities were to become preponderantly Negro?* I rather doubt we would answer this question candidly, but I am sure we would find the prospect disturbing—just as disturbing, probably, as the 19th-century "proper" Bostonian found the fact of his city becoming preponderantly Irish. *His* disaster happened to him; our disaster is still only imminent. I do believe that there is a sense in which we can properly speak of such transformations as "disasters"; but I also believe that it takes an impoverished historical imagination to see them *only* as disasters.

NO one acquainted with the historical record can fairly doubt that American cities such as Boston and New York were much nicer places to live in before the immigrant mobs from Western Europe descended upon them in the eighteen-thirties and forties. Our conventional history textbooks—sensitive to the feelings (and to the political power) of yesterday's immigrants, who are by now very important people—tend to pass over this point in silence. They concentrate, instead, on the sufferings and privations of the immigrants, the ways in which they were discriminated against by older settlers, the fortitude they displayed in adverse circumstances, and the heroism of so many of them in coping with this adversity.

That is all true enough, and fair enough. Still, though it may not be advisable for our textbooks to make the point, it would be helpful to all of us if we could somehow do justice to the feelings of the older urban settlers. While chauvinism and xenophobia and gross self-interest certainly affected their attitudes, it is also true that their complaints and indignation had a quite objective basis. The fact is that American cities in the early decades of the 19th century seemed to have relatively few of those "urban problems" which were a traditional feature of the older cities of Europe, and which we now tend to regard as inherent in the urban experience itself.

There is no difficulty in explaining why this should have been the case. It had to do with no peculiar American virtues or unique American "genius." The reason there was no "urban crisis" is that the kinds of people who create an urban crisis simply didn't live in those cities. There was no urban "proletariat" to speak of — the comparatively high standard of living, the existence of free land, the constant creation of "new towns" out West, the general shortage of labor, the traditional mobility of the average American, and the religion of "self-reliance" that most Americans subscribed to: all this made it impossible for the condition of the urban working classes in New York or Boston or Philadelphia to resemble that of the working classes of London or Manchester or Glasgow.

Even more important: there was no dispossessed *rural* proletariat whom the cities had to absorb—the rise of commerce and industry in this country, as contrasted with their rise in Europe, was not connected with the displacement of masses of people from country to city. American cities, in those early decades of the 19th century, grew larger, wealthier, and more populous; *but on the whole they performed no assimilatory role*—unlike the older cities of Europe, or the American cities subsequently.

ALL this changed, of course, with the arrival of European immigrants —heterogeneous in religion, language and customs, with few skills and no money who settled in the larger seaboard cities. Instead of assimilating individually to American life, they challenged the city to assimilate them en masse. Despite the "melting pot" myth that later developed, it was a challenge that the American city did not meet with either grace or efficiency. The main reactions were resentment and anxiety and anger. Public order, public health, public education and public life were all thrown into disarray—and who can blame the older citizens for disliking these consequences? The transformation of the American city was a very real and very personal disaster for most of them. It destroyed their accustomed amenities, disrupted their

IRVING KRISTOL is senior editor and a vice president of Basic Books, Inc., of New York and co-editor of the quarterly The Public Interest.
This article will be included in a volume of essays, "A Nation of Cities," edited by Robert A. Goldwin and published by Rand McNally & Co.

neighborhoods, and quite ruthlessly interrupted their pursuit of happiness. Many of them began to move into what was then suburbia.

Well, that urban crisis was overcome, if ever so slowly. Not many people now think this would be a better country had those immigrants never come: their contributions to American life have been too notable, their indispensability for our American civilization is too obvious. And today it is the children and grandchildren of those immigrants who, faced with the mass movement of a rural Negro proletariat into "their" cities, echo all the old American laments and complaints.

They, too, have good reason—their discomfort and distress are not at all imaginary, despite what many liberal sociologists seem to think. (It is one thing to say abstractly that the great American city is, and has been for more than a century now, a social mechanism for the assimilation of "foreign" elements into American society. It is quite another thing for a concrete individual to try to live out a decent life in this "social mechanism.") But it does help to see this discomfort and this distress in historical perspective. And in such a perspective, the key question—often implicitly answered, less often explicitly asked—is: will we be able, decades from now, to look back upon our present "urban crisis" as but another, perhaps the final, stage in the "assimilation" of a new "immigrant" group, or is this crisis an unprecedented event that requires unprecedented and drastic social action?

* * *

THAT the American Negro is different from previous "immigrant" groups is clear enough. (I use quotation marks because there is patently something ironical in referring to Negroes as immigrants, when most of them are technically very old Americans indeed. Nevertheless, I think it is accurate enough, if one has in mind movement, not to America, but to the city; and I shall henceforth refer to them as immigrants, simply.) The color of his skin provokes all sorts of ancient racial fears and prejudices; and he lacks a point of "national origin" that could provide him with an authentic subculture of his own one on which he could rely for psychological and economic assistance in face of the adversary posture of American civilization toward him.

The very special problems of the American Negro have been the subject of a literature so copious and so insistent that it is surely unnecessary to do more than refer to it. What I should like to emphasize, instead, is the danger we are in of *reducing* the Negro to his problematic qualities.

Underlying practically all of the controversies about the American city today there lies the question: can the Negro be expected to follow the path of previous immigrant groups or is his a special, "pathological" case? This word "pathological" turns up with such surprising frequency in sociological literature today—on slums, on poverty, on education—that one might suspect a racist slur, were it not for the fact that those who use it most freely clearly intend to incite the authorities to corrective action by presenting the Negro's condition in the most dramatic terms.

Indeed, there has developed an entire rhetoric of liberal and melioristic slander that makes rational discussion of "the Negro problem" exceedingly difficult. Anyone who dares to suggest that the Negro population of the United States is not in an extreme psychiatric and sociological condition must be prepared for accusations of imperceptiveness, hardheartedness, and even soullessness. And when it is a Negro who occasionally demurs from this description, he runs the risk of being contemptuously dismissed as an "Uncle Tom."

From my own experience as a book publisher, I think I can say confidently that if a Negro writer today submits a manuscript in which dope addiction, brutality and bestiality feature prominently, he has an excellent chance of seeing it published, and of having it respectfully reviewed as a "candid" account of the way Negroes live now; whereas, if a Negro writer were to describe with compassion the trials and anxieties of a *middle-class* Negro family, no one would be interested in the slightest — middle-class families are all alike, and no one wants to read about Negroes who could just as well be white.

It is worth lingering on this last point for a moment—precisely because this unrelieved emphasis on the hellishness of the Negro condition reminds us, paradoxically, of the literature of previous immigrant groups. No one can doubt that, of all immigrants, it is the Jews who have been most successful in exploiting the possibilities that America offered them. Yet if one examines the literature that American Jews created about themselves, in the years 1880-1930, one discovers that it was a literature of heartbreak and misery. All during this period, we now know, the Jews were improving their condition and equipping themselves for full participation in American life.

There is not much trace of this process in the Jewish novels and stories of the period. This is not to be taken as a deficiency of the literature: it is never literature's job to tell the whole sociological truth—the literary imagination is "creative" exactly because it transcends mere social description and analysis. But it does serve to remind us that such a book as Claude Brown's "Manchild in the Promised Land," powerful and affecting as it is, cannot be taken to represent a definitive statement of the facts of Negro life in America—any more than one could take Michael Gold's "Jews Without Money," written in 1929, as a definitive statement of the facts of Jewish life in America.

EVERY year, tens of thousands of Negroes are moving out of poverty, and thousands more are moving into the middle class, both in terms of income and status of employment. Moreover, the rate of such movement is noticeably accelerating every year. More than half of Negro families in the North have incomes greater than $5,000 a year; and, over the nation, the proportion of Negroes living in dilapidated housing has been cut in half during the past decade.

These people exist and their numbers are increasing — just as the number of poor is decreasing: approximately one half-million Negroes per year are moving above the poverty line. It is similarly worth noting that there are now something like a half-million Negroes in *each* of the following occupational categories: **(a)** professional and technical, **(b)** clerical, and (c) skilled workers and foremen. The total is about equal to that of Negro blue-collar factory operatives, and will soon be substantially larger.

But all this receives little attention from our writers and sociologists, both of whom are concerned with the more dramatic, and less innocently bourgeois, phenomena of Negro life. This is to be expected of the writers; it is less expected of sociologists, and the antibourgeois inclination of so much of current American sociology would itself seem to be an appropriate subject for sociological exploration.

"The density of habitation in the urban Negro slum today is less than it used to be—and is *considerably* less than it was for the Irish, Italian and Jewish immigrants when *they* lived in slums."

The fact, incidentally, that so much of our indignant attention is centered on the northern urban Negro—who is, by any statistical yardstick, far better off than his Southern rural or small-town counterpart—is in itself reassuring, since it follows a familiar historical pattern. In England and France, in the 19th century, the movement of the rural proletariat to the cities was accompanied by an increase in their standard of living, and a vast literature devoted to their urban miseries. These miseries were genuine enough —and it might even be conjectured that there is something qualitatively worse about urban poverty than about rural poverty, even where the latter is quantitatively greater.

But, in the absence of a corresponding literature about the life of the rural poor, one can too easily see, in this process of urbanization, nothing but mass degradation, instead of a movement toward individual improvement—which, in retrospect, one can perceive it was. Obviously, it is absurd to expect the average Negro immi-

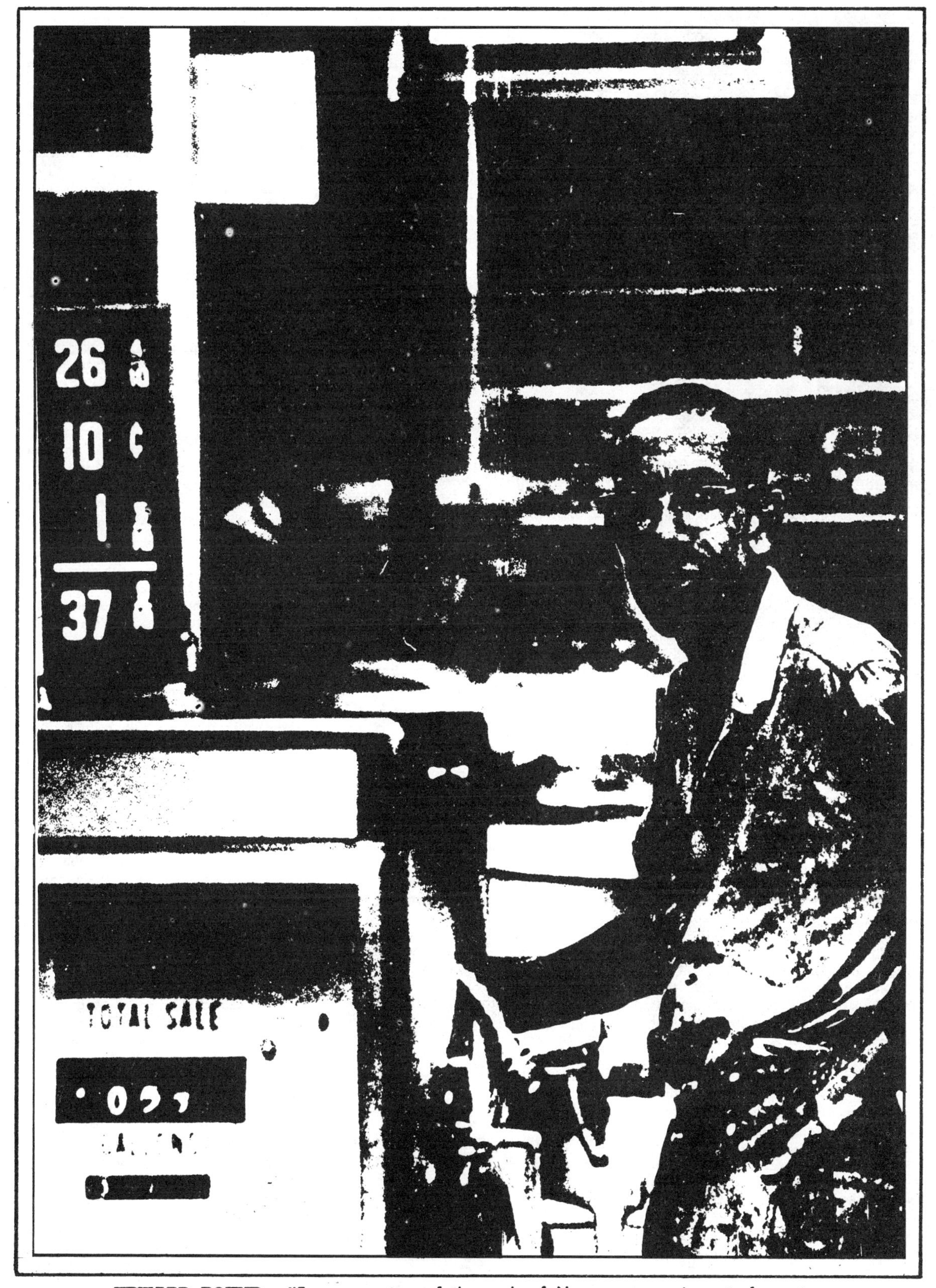

UPWARD BOUND — "Every year, tens of thousands of Negroes are moving out of poverty, following the path of previous immigrant groups." (above) an employe of a Negro-owned station.

grant to the American city to have such a historical perspective on himself—he would have to cease being human and become some kind of sociological monster to contemplate his situation in this detached and impersonal way. But one does wish that those who are professionally concerned with our Negro urban problem, while not losing their capacity for indignation or their passion for reform, could avail themselves of such a longer view. After all, that presumably is what their professional training was for.

One could also wish that these same scholars were less convinced a priori of the uniqueness of the Negro's prob-

SPIRIT—A self-appraisal on a Harlem wall. "We have convincing data that American Negroes, despite their hardships, face the future with confidence and hope."

lem and more willing to think in terms of American precedents. A casual survey of the experiences of the first two generations of Irish immigration can be instructive in this respect. There is hardly a single item in the catalogue of Negro "disorganization," personal and social, that was not first applied—and that was not first applicable—to the Irish. "The Dangerous Classes of New York" was the title of a book published in 1872; it referred primarily to the Irish. Ten years later, Theodore Roosevelt confided to his diary that the average Irishman was "a low, venal, corrupt and unintelligent brute."

Alcoholism wreaked far greater havoc among the immigrant Irish than all drugs and stimulants do today among the Negroes. The "matrifocal family"—with the male head intermittently or permanently absent—was not at all uncommon among the Irish. Most of the Irish slums were far filthier than they need have been—and, if we are to believe contemporary reports, were not less filthy than the worst Negro slum today — because the inhabitants were unfamiliar with, and indifferent to, that individual and communal self-discipline which is indispensable to the preservation of civilized amenities in an urban setting. (We easily forget that our extensive public services rely, to a degree not usually recognized, upon rather sophisticated individual cooperation: for garbage to be collected efficiently, it must first be neatly deposited in garbage cans.)

It is one of the ironies of this matter that some of the very improvements in the life of the urban Negro are taken to represent the more problematic aspects of his condition. Everyone, for instance, is terribly concerned about the spread of Negro slums, both in the central city and, for some time now, in our suburbs. Why? The answer has been provided by Raymond Vernon in his excellent little book "The Myth and Reality of Our Urban Problems":

"As long as the slum was contained in a small congested mass within the old center of the city, most of the middle-income and upper-income inhabitants of the urban area could live out their lives without being acutely aware of its existence. As the slum dweller has taken to less dense living, however, the manifestations of his existence have not been quite so easy to suppress."

It will doubtless come as a surprise to most people that the density of habitation in the urban Negro slum today is less than it used to be—and is *considerably* less than it was for the Irish, Italian and Jewish immigrants when *they* lived in slums. Nevertheless, this is indisputably the case. (There is one slum area, on the Lower East Side of New York, where the Negro population is *one-third* of what the population was when this same area was a white slum 50 years ago.)

The population of the slum ghettos in our central cities is steadily decreasing, as the Negroes use their improved incomes to acquire more dwelling space per capita. To be sure, this means that poor Negroes now spread out more, that their slums and poor neighborhoods are more extensive, that they impinge far more powerfully upon the white neighborhoods than used to be the case. The whites, in turn, become highly agitated, as they discover that the problems of the slums and its inhabitants are now becoming their problems too, and can no longer be blandly ignored. Things *look* worse, and are *felt to be* worse, precisely because they have been getting better. **This is one of the most banal of all sociological phenomena—but even sociology itself is being constantly caught off guard when confronted by it.**

Another such ironic instance is the by-now famous "cycle of dependency." This ghost now insistently haunts all discussions of Negro poverty, and is invoked by all the authorities of the land, from the highest to the lowest. We are constantly being told, and are provided with figures to prove it, that not only do poor Negroes tend to beget poor Negroes—this tendency is true for whites as well, and does not astonish—but that poor Negroes on public welfare tend to beget poor Negroes who also end up as recipients of public welfare. Public policy, we are told, must—at whatever cost—aim at breaking this "vicious cycle" of dependency, if the Negro is ever to be truly integrated into American society and American life.

What we are *not* told, and what few seem even to realize, is that this "vicious cycle" is itself largely created by public policy—and that, indeed, so far from this being a vicious cycle, it is a function of humanitarianism. The point is really quite simple and, once made, exceedingly obvious: *the more money we spend on public welfare, and the easier we make it for people to qualify for public welfare, the more people we can expect to find on welfare.* Moreover, since—as we have noted—the children of poor people are always and everywhere more likely to end up as poor than the children of rich people (not *as* poor as their parents, of course, but poor by new and more elevated standards of poverty), it follows that dependency on welfare may easily flow from one generation to the next.

I have discovered, to my cost, that one must be very cautious in making this point, and I should therefore like to emphasize that I am *not* calling for a reduction in welfare expenditures, or for more restrictive qualifications for welfare. On the contrary, I believe that in most parts of the country such expenditures are niggardly and the qualifications idiotically rigid. I also believe that, everywhere in our affluent society, the poor and the distressed get too little money with too much fuss. We can afford to be more generous and —as a matter of both equity and morality—should be more generous. But what I do *not* think makes any sense is for us simultaneously to give more money to more poor people—and then to get terribly excited when they take it!

Does the "cycle of dependency" come down to much more than this? I doubt it. If anything like our present welfare system had been in existence 50 years ago, or 100 years ago, this same "cycle of dependency" would have been a striking feature of the Italian and Irish immigrant communities. In those days, however, the ideology of "self-reliance" was far more powerful than the ideology of "social welfare." In effect, American society coerced poor people into working at any kind of jobs that were available, at any rate of pay, in whatever disagreeable conditions. American society did this in easy conscience because Americans then thought it was "good" for poor people to experience the discipline of work, no matter how nasty.

We have changed our views on this matter, and I should say for the better. But if we hadn't changed our views, we would not be witnessing, or worrying about, the "cycle of dependency." All of those women who now, as heads of households, get welfare and Aid-to-Dependent Children grants, would be forced to go into domestic labor—of which there is a great and growing shortage — or to take in washing and mending at home, as they used to. Our more humane welfare policies have liberated them from this necessity. It is this liberation that is the true meaning of the "cycle of dependency.'"

It is, to be sure, a frightening thought that generation after generation of a whole segment of society will be cut off from the mainstream of American life, by virtue of their status as welfare clients. But just how valid is such a projection? Robert Hunter, in his book, "Poverty," written in 1904, was already concerned about the "procreative" power of poverty, over the generations. His concern turned out to be baseless—will ours be less so?

After all, the basic premise of our welfare ideology is that people's moral fiber, their yearning for self-improvement and economic advancement,' is not sapped by a more generous system of social welfare. We assume that most of those who receive welfare would prefer not to, and that once they are qualified to join the labor force, and once the labor force is ready to receive them, they will happily remove themselves from the welfare rolls. If this assumption is false, the very essence of the welfare state will be called into question. I happen to believe that the assumption is not false. And that is why I regard the vision of self-perpetuating and self-generating dependency as spectral rather than sociological.

• • •

SINGALONG—At a Brooklyn summer school where songs are used to familiarize underprivileged children with new words. "Urban Negroes have it harder than yesterday's immigrants, but they do get more help."

TO say that the problems of Negro migration into our large cities have relevant precedents in American history is not to assert that these immigrants do not face unique and peculiar dilemmas of their own. The fact that their "ethnicity" is racial rather than cultural, and the corollary fact that racial prejudice seems more deeply rooted than cultural prejudice, are certainly not to be minimized. It is not likely, for instance, that an increase in social acceptance—holding out the prospect of a substantial degree of intermarriage—will keep pace with an improvement in the Negro's social and economic status.

Even here, however, one cannot be sure: the whole world gets a little less bemused by skin color every day, as the new nations of Asia and Africa shed the colonial stigma; and many young people are militantly color-blind. In any case, unless one really believes in inherent and significant race differences, which I do not, then this question of social acceptance does not appear to be so terribly important for the visible future.

What is important is that, if anti-Negro prejudice is more powerful than, say, anti-Irish prejudice ever was, it is also true that public policy today is far, far more powerfully anti-discriminatory than it ever was. The Negro migrants to the city start under a more onerous handicap than their predecessors—but they are receiving much more assistance than their predecessors. It is impossible to strike any kind of precise equation out of these opposed elements; but my own feeling is that they are not too far from balancing each other.

Another problem, one which is receiving a considerable amount of controversial attention, is that of the Negro family: specifically, the fact that the Negro father so often—at least relative to the white population—refuses to assume a permanent position as head of household, while the Negro mother so often (again relative to the white population) will have a large number of children by different and transient "husbands." The Moynihan Report dramatically focused on this issue. Anyone who has taken the trouble to read Daniel Moynihan's study of the United States' Negro family cannot fail to be impressed by the truth of his claim that this unstable family situation makes it particularly difficult for the urban Negro both to cope with the disadvantages of his condition and to exploit the possibilities for advancement that do exist. Nevertheless, this problem, too, must be kept in perspective. Without going too deeply into a subject whose ramifications are endless—involving, as they do, the entire sociology of family life and the whole history of the Negro race—the following points can be made:

(1) We tend to compare the Negro urban family today with the white suburban family of today, rather than with the white urban family of yesteryear. Family life among the raw urban proletariat of 19th-century America, as in 19th-century Britain, showed many of the "pathological" features we now associate with the Negro family. Statistics on broken homes and illegitimacy are impossible to come by for the earlier times. But anyone familiar with the urban literature of that period cannot but be impressed by the commonplace phenomenon of Mrs. Jones or Mrs. O'Hara raising her brood while Mr. Jones and Mr. O'Hara have vanished from the scene.

(2) Having said this, one must also say that it does seem to be the case that the Negro family—not only in the United States, but in Canada, the West Indies and Latin America as well—is a less stable and less permanent unit than the "bourgeois" white family, as the latter has developed over the past four centuries. But it is not at all certain that this instability will persist indefinitely: there is no more passionately "bourgeois" a group than the middle-class Negro family in the United States today. And even if it should persist, to one degree or another, there is no reason I can see why American society cannot quite easily cope with it. Working mothers are not exactly a rare occurrence these days; and a comprehensive network of crèches and nursery schools should be able to provide the children with a decent home-away-from-home during the working hours.

(3) The evidence clearly suggests that the major problem of the Negro family is, quite simply, that there are too many children. Three-quarters of the poor children in the United States today are in families of five or more children. If the average Negro family size were no larger than the white's, Negro per capita income would soar, a large section of the Negro proletariat would automatically move above the poverty line, and the question of the fatherless Negro family would become less significant.

A mother who has one or even two young dependents can manage, given a decent program of social assistance; the larger the family, the closer her situation approaches the impossible. The availability of birth-control information, and growing familiarity with birth-control techniques, should eventually work their effects. ("Eventually" is the operative word here: our experience with poor and highly reproductive people all over the world demonstrates that the deliberate control of family size is not something that can be achieved in a single generation.) In the more immediate future, a program of family allowances could certainly be helpful.

But all this is, if not beside the point, then all around the point. We can legitimately worry about the Negro's capacity to achieve full inclusion in American society only after our society has seriously tried to include him. And we have consistently shirked this task. The real tragedy of the American Negro today is not that he is poor, or black, but that he is a late-comer—he confronts a settled and highly organized society whose assimilatory powers have markedly declined over the past decades. The fact that the urban Negro is poor is less important than the fact that he is poor in an affluent society. This has both subjective and objective consequences.

DEVELOPMENT—A new middle-income apartment house in Harlem.

Subjectively, it means that the poor urban Negro feels himself at odds with the entire society, in a way that was not true 50 years ago. He was then much worse off than he is today—but he was then also surrounded by lots of poor whites who were obviously not much better off. Misery loves company, as we know; and when misery has its company, it is far more tolerable. The lonely misery of the poor Negro in our society today takes a tremendous psychological toll, and it can so exacerbate his sensibilities as to hinder him from taking advantage of the modest opportunities to improve his status that do present themselves.

When Bayard Rustin writes—and I am quoting literally—that "to want a Cadillac is not un-American; to push a cart in the garment center is," he is writing absolute nonsense. Pushing a cart in the garment center is the traditional point of departure for pushing one's way into the garment industry—and, at the very least, it has always been thought to offer advantages over hanging around a street corner and perhaps "pushing" less innocent items than ladies' wear. But one can easily understand how, in view of the isolation of the Negro and his poverty, such nonsense can be persuasively demoralizing to those Negroes who listen.

But I myself suspect that fewer Negroes listen to "their" spokesmen (or should it be their "spokesmen?") than is commonly assumed. We have convincing opinion-poll data to the effect that the overwhelming majority of American Negroes look to the future with confidence and hope, and that they feel they have been making real progress since the end of World War II. Even on specific, highly controversial issues there is a marked divergence between what Negro leaders say and what the average Negro thinks.

School integration, for instance, is one such issue. Despite the fact that this matter has been so urgently pressed by the civil-rights movements, all public-opinion polls show that a clear majority of Negroes—in the city's ghettos and out—think the quality of education in their neighborhood schools is a more important issue than the racial integration of these schools.

One could presumably infer from this that the mass of the Negroes are lagging in their "social consciousness.". I think the more accurate inference is that the mass of Negroes are more rational in their thinking, and less affected by demagogic slogans, than their spokesmen. After all, it is a fact—and one which becomes more certain with every passing year—that, simply as a consequence of demographic forces now at work, the great majority of school children in our central cities are going to be Negro, and that there just aren't going to be enough white school children around to integrate. To be sure, one could bus white children in from the suburbs, or Negro children out to the suburbs. But this is political fantasy; and, interestingly enough, the majority of Negro parents who have been polled on this question are indifferent to or opposed to both kinds of busing.

In short, then, while the lonely poverty of the Negro in our affluent society renders his situation more difficult, it does not place him beyond the reach of intelligent social policy. There may be—there doubtless are—some unhappy few who, not seeing a Cadillac in their future, will decide this future is beyond redeeming. One cannot help but sympathize with their resentment and their resignation—it is unquestionably better to have a Cadillac than not to have one. But one is also glad they constitute, as they clearly do, a tiny minority. And even if and when they picket, or riot, they are still a tiny minority.

But will social policy achieve its potentialities? Here we run up against an "objective" consequence of the affluent society that is infinitely depressing. For this society is constantly in the process of so organizing itself as to exclude those who, like the Negroes, are poor, uneducated, and without previous ownership of a monopoly in any craft or trade. I am not referring to the process of "automation" which—as the report of the President's commission on this subject pointed out—has not yet had any observable effect on the labor market. I do mean the process of "pseudo-professionalization," which has received very little critical attention but whose consequences are notably pernicious.

We are all familiar with the way in which, over these past years, plumbers have become "sanitary engineers" and elevator operators have become "transportation specialists." One does not begrudge these men their fancy titles—we are all allowed our harmless vanities. Only it's not really so harmless as it seems. For this change in nomenclature is the superficial expression of a basic restrictionist attitude that is remorselessly permeating the whole society. The majority of the population that has secured its position in this society seems determined to make it ever more difficult for others to gain new entry.

A superb example is provided by the New York City civil service, which now establishes formal educational requirements for jobs—firemen, maintenance men, mechanics' helpers, etc.—where none were required 10 years ago. The jobs themselves have not changed; even the official descriptions of tasks, duties, and skills have not been revised; but the barriers have gone up.

THIS sort of thing is happening all over, with the enthusiastic cooperation of unions and management. There is no economic sense to it, or economic justification. Sociologically, of course, it is easy to understand—it is a reflection of that protectionist-guild state of mind that comes with a society's advancing years: everyone and everything seeks to "establish" itself. One would be content to go along with this trend, for the peace of mind it seems to provide—were it not for the fact that, in present circumstances, it threatens to *disestablish*, more or less permanently, a not insignificant proportion of America's Negroes.

Our social policy is *not* to provide suitable jobs for the Negroes; it is to provide suitable Negroes for the jobs. We have undertaken to re-educate, rehabilitate, retrain, readjust, recompense and re-just-about-everything-else the American Negro. Faced with the choice between modifying our occupational structure and transforming the people who seek a place in it, we have chosen the latter.

A nice instance of this choice can be found in the present activities of the U. S. Post Office. It is generally agreed that our postal service is in a sorry state, with only one mail delivery a day to noncommercial residences. To revert to a twice-daily delivery in our cities — one doesn't have to be even middle-aged to remember when this was the practice—would demand something like 60,000 new employes. These, in turn, would require little else than a minimum of literacy to be able to do the job—and if even this minimum were found to be an obstacle, a brief apprenticeship could be used to teach them what they needed to know.

This proposal had a short life in Washington. It was rejected in favor of (a) spending money trying to automate the postal system, and (b) spending money on the Job Corps, on elaborate programs of vocational training, and so on. Meanwhile, over the last three years the nonwhite proportion of postal workers has declined by a couple of percentage points.

This episode evoked no comment or protest from Negro spokesmen, civil rights leaders or warriors against poverty. They, too, seem more interested in ultimate Cadillacs than in actual mailbags. If asked about the matter, they would have doubtless dismissed it—as one did to me—with a few clipped words about the pointlessness of placing people in "dead-end" jobs. To which one can only reply that most people, even in our affluent society, end up in "dead-end" jobs of one kind or another; and that, for most *poor* people, "social mobility" is something that happens to their children, not to themselves. It goes without saying that it is preferable to have the unemployed or underemployed become engineers instead of postmen. But to try to enforce such a preference, as a matter of social policy, is utopian to the point of silliness.

BEHIND all this is a more fundamental choice: to put the emphasis on the elimination of relative *inequality* between Negro and white, rather than on the mere improvement, in absolute terms, of the Negro's condition. That the elimination of such relative inequality is a worthwhile goal, needs no saying; and all of those programs directed toward this end are highly meritorious. But the goal is a distant one, depending as it so largely does on equality of educational experience. And progress toward this goal is not likely to be at a steady and uniform pace; it will proceed through sharp and intermittent spurts, separated by long periods during which nothing seems to be happening. (The experience of the Japanese-Americans is illustrative of this point.)

In contrast, *poverty* can be abolished within the next decade—if we concentrate on the task. Right now, one of every four Negroes in their early twenties has not gone beyond the eighth grade; over half have not completed high school. These people exist; the formative years of their lives are passed beyond recall; it is cruel and demogogic to offer them an impossible "second chance"—while blithely refusing to offer them a realistic first chance.

We go around in a circle which, while one can hardly call it vicious, is nevertheless decidedly odd. We begin by prissily categorizing the Negroes as "pathological," we end by proclaiming vast and dubious programs for their instant conversion to middle-class values and upper-middle class status. Within this circle, the majority of our urban and suburban Negroes — these latter, incidentally, rapidly growing in numbers — are making substantial progress in their own way, at their own tempo, and largely by virtue of their own efforts.

In comparison with previous waves of immigration to the great cities, they are "making out" not badly at all. They need, and are entitled to, assistance from the white society that has made them — almost our oldest settlers — into new immigrants. But the first step toward effective help would seem to be a change in white attitudes. Until now, we have spent an enormous amount of energy and money trying to assimilate Negroes into "our" cities. Is it not time we tried helping them to assimilate into "their" cities? September 11, 1966

NEGROES AND IMMIGRANTS

TO THE EDITOR:

After reading Mr. Kristol's article, I am left disturbedly cold and convinced that the author is well-meaning but naive.

There is something very cruelly ironic about being compared with Caucasian immigrants of years past. After all, as the author pointed out, we Negroes are among America's earliest settlers. How utterly unbelievable it is to hear an alien or a newly naturalized citizen say, "Keep the niggers out of *our* neighborhoods" in English so belabored as to be almost unintelligible.

Middle-class Negroes are so affected, I believe, by such occurrences that it is all but impossible to be *like* the white middle-class family. The blanket restrictiveness against the Negro is so pervasive and harsh as to be felt daily and stingingly, no matter what his socio-economic standing. Persons who are Negro cope with this social reality in diverse ways, of course.

Public policy and socially oppressive forces may *seem* to be in some balance, but I would venture to say that the lag in implementation of this public policy upsets this possible balance.

Perhaps the author is really writing for Caucasians and, from a Caucasian's vantage point, this may account for the distance I felt from the writer —that he was talking about the "out-groups" as a member of the "in-group."

On the other hand, the distance felt as a member of the "out-group" may be simply an expression of the reality. At any rate, no comfort was gained from Mr. Kristol's attempt to explain "the Negro today" through an historical understanding or appreciation of yesterday's immigrant.

ALMA C. NORMENT, A.C.S.W.
Clinical social worker.
New York.

September 25, 1966

Some Negroes Accuse Styron Of Distorting Nat Turner's Life

By JOHN LEO

William Styron's "The Confessions of Nat Turner," the No. 1 best-selling novel, is stirring bitter controversy among Negroes.

"It's the worst thing that's happened to Nat Turner since he was hanged," William Strickland, a Harlem writer, said in an interview. "It's a racist book designed to titillate the fantasies of white America."

John Morsell, assistant executive director of the National Association for the Advancement of Colored People, said his organization planned a "presentation" to Wolpert Pictures, Ltd., which bought screen rights to the novel for $600,000, to make certain that "offensive aspects are purged from the movie."

The Random House book, which Mr. Styron calls "a meditation on history," is a fictional re-creation of an 1831 slave rebellion in Southampton County, Virginia. Fifty-five whites were killed. The leader of the short-lived rebellion, Nat Turner, was captured and hanged.

The novel has been hailed as a literary triumph by most critics and is considered a leading contender for the National Book Award and the Pulitzer Prize.

But several Negro leaders and intellectuals argue that the Virginia-born Mr. Styron distorted the historical record and promoted racial stereotypes.

"Styron's Nat Turner, the house nigger, is certainly not the emotional or psychological prototype of the rebellious slave," said Michael Thelwell, a Jamaican graduate student at the University of Massachusetts in Amherst. "He is the spiritual ancestor of the contemporary middle-class Negro, that is to say, the Negro type with whom whites, and obviously Mr. Styron, feel more comfortable."

A collection of essays by Negroes critical of the book is being edited for Beacon Press by John Henrik Clark, an editor of Freedomways, the radical intellectual quarterly. "Coming at this time," said Mr. Clark, "it's not accidental that a white Southerner should write about Nat Turner, altering his character to express and justify a lot of current white Southern anger."

Other Negro intellectuals disagree. "I thought it was a great book," Dr. John Hope Franklin, chairman of the history department at the University of Chicago, said. Dr. Franklin, reviewing it in The Chicago Sun-Times, wrote: "In his meditation, Mr. Styron makes many salient comments and observations that reveal his profound understanding of the institution of slavery."

"The book showed tremendous insight into the Negro psyche," said Dr. Rembert Stokes, president of Wilberforce (Ohio) University, which awarded Mr. Styron an honorary doctorate in humanities last November. "It expresses what a number of contemporary Negroes, both militant and non-militant, feel about whites."

Reached by phone at his home in Roxbury, Conn., Mr. Styron said he was generally pleased by acceptance of the book by Negroes but "shocked" by some attacks on it.

"What I regret" the 42-year-old writer said, "is that some younger people take the criticism of Herbert Aptheker, for whom neither I nor anyone else in the field of history has any respect as valid criticism of my book."

Mr. Aptheker, a member of the National Committee of the Communist Party, a historian and the author of a book on Nat Turner, has published several attacks on the Styron novel.

"First," said Mr. Aptheker in a telephone interview, "Styron presents the Turner rebellion as unique. This fits the image of the American Negro as docile and passive. Actually there were 250 uprisings, plots and conspiracies, including several that cost more lives and lasted longer."

"Second, Styron has the Turner revolt being put down by blacks armed by their owners. This is false and inconceivable. He also has Turner broken at the end of the book, when the record shows he went to his death saying, "Was not Christ crucified?'"

Mr. Styron replied: "Mr. Aptheker is grinding his ideological ax. His evidence doesn'ttv convince me or any other responsible historian. The only effective sustained revolt was Nat Turner's."

Sex Life at Issue

Another dispute that has struck a sensitive nerve among some Negroes was Mr. Styron's decision to make Nat Turner a lifelong celibate, whose sole object of erotic interest was Margaret Whitehead, a teenage white girl.

Many Negroes feel that this portrayal of Turner promotes sexual fears of Negroes among whites.

"All this was unnecessary. Nat Turner was married," said Howard Meyer, a white New York lawyer and biographer of Thomas Wentworth Higginson, colonel of the black regiment that fought for the North in the Civil War.

"In an 1861 essay in Atlantic Monthly," Mr. Meyer said, "Higginson clearly states that Turner had a young wife, and that she was a slave belonging to a different owner." Mr. Meyer's complaint has been picked up by the Negro press and promoted as a major issue.

In reply, Mr. Styron said: "I'm aware of the Higginson essay, but I really can't accept a word-of-mouth reference put down 30 years after the fact. I made Turner a celibate because he seemed to have the single-minded, ascetic character you often find in revolutionary figures."

Mr. Styron said his fictional account of Nat Turner's unfulfilled sexual love for Margaret Whitehead grew out of his meditation on the historical account that she was the sole person killed by Turner.

"As a novelist, that got to me," he said. "Why did he kill only her? Hate? If you're a certain kind of man, you only kill the thing you love, and I think Nat Turner was that kind of man."

"I was fully aware that this might cause a considerable stir among those who wished to misconstrue it, and I wasn't wrong. The attack on me has gone so far that recently I was accused of having the girl instigate the revolt."

February 1, 1968

MILITANTS OBJECT TO 'NEGRO' USAGE

'Black' or 'Afro-American' Replacing Barred Word

BY JOHN LEO

One by-product of the black power movement is an assault on the word "Negro."

"Negro is a slave word," said H. Rap Brown of the Student Nonviolent Coordinating Committee. He, Stokely Carmichael and other black power spokesmen insist on the word "black" and often refuse to talk to reporters who speak of "Negroes."

Prominent leaders such as Dr. Martin Luther King of the Southern Christian Leadership Conference and Roy Wilkins of the National Association for the Advancement of Colored People have been heckled and denounced for using the word "Negro."

The New York Amsterdam News, the largest Newspaper published in Harlem, has barred the word from its columns and generally uses "Afro-American." Letters from readers are running 9-to-1 in favor of the change, according to Richard Edwards, assistant managing editor.

"There seems to be violent objection to the term 'Negro' among young people, who link the word with Uncle Tom," he said.

A Cultural Reference

Keith Baird, the coordinator of the Afro-American History and Cultural Center of the New York City Board of Education, said: "This is not a minor semantic dispute. It engages the emotions and intellect of a vast number of people, from Southern campuses to the corner of 125th Street and Seventh Avenue in Harlem."

Mr. Baird, a Jamaican, prefers "Afro-American," which, he says, "Clearly relates us to land, history and culture, whereas 'black' doesn't have that kind of cultural reference."

Separate surveys by Ebony and Jet, sister publications in the black press, show that "Afro-American" is the term most preferred by readers, with "black" second and "Negro" a weak third. Recently, many small civic and campus organizations have put "Afro-American" or "African-American" in their titles. The Negro Teachers Association of New York City, for instance, is now the African-American Teachers Association.

A shorter form—"Afram"— is popular among some younger people, partly in response to the objection that "Afro-American" is too much of a mouthful for everyday use.

Hoyt W. Fuller of Negro Digest says "Afro-American" is being promoted by internationally minded men as a semantic reminder that the fate of the black struggles today in America and in Africa are closely linked.

Usage Catching On

In recent years, the greatest pressure for the world "black" has come from the Nation of Islam (Black Muslims) and the late Malcolm X, who referred contemptuously to "so-called Negroes" who refuse to struggle for their rights.

This usage has caught on widely. Ossie Davis, the playwright-actor, wrote recently: "A black man means not to accept the system as Negroes do, but to fight hell out of the system as Malcolm did."

"Mr. Fuller said: "There is definitely a generation gap in usage. Those who have adjusted to things as they are use 'Negro'; those that haven't, use 'black and Afro-American.' "

He said Negro Digest was under "great pressure" to change its name.

Mixed usage has become common. Dr. Ralph J. Bunche, Under Secretary for Special Political Affairs in the United Nations Secretariat, says he now used "black" as often as "Negro," though he attributes little significance to the dispute. C. Eric Lincoln, the sociologist, now at Union Theological Seminary, said:

"I use 'black' when talking to young people, and 'Negro' when addressing those past 40 who are comfortable with that term. In writing, I find myself shifting from 'Negro' to 'Black American.' "

Mayor Lindsay rarely uses the word 'Negro' any more, using 'black' instead.

Because "black" has become associated with militancy, few prominent Negroes seem willing to attack the new usage publicly. However, John Morsell of the N.A.A.C.P. said:

"I don't think it's worth such a storm to replace a well-established word. 'Negro' is a precise and useful word. And, after all, 'black' is just as much a reminder of the slave period as Negro."

"Negro" is the Spanish and Portuguese word for black and was used from the earliest days of their slave trading in Africa. "Black" as a noun, apparently a shortening of "blackamoor," dates from 1625, according to the Oxford English Dictionary.

Dr. William E. B. Dubois, one of the founders of the N.A.A.C.P., used the terms 'black,' 'Negro' and 'Afro-American' interchangeably.

Current clamor for a more "meaningful" label began with the parade to independence, in the late nineteen-fifties, of black African nations. With the end of the myth of dark, uncivilized Africa, Negroes became more willing to identify with that continent.

Satirized by Feiffer

At the same time, the development of the civil rights movement in the United States contributed to a pride in blackness.

Jules Feiffer, the writer-cartoonist, satirized the problack movement with a cartoon of a militant who said: "As a matter of racial pride, we want to be called 'blacks,' which has replaced the term 'Afro-American,' which replaced 'Negroes,' which replaced 'colored people,' which replaced 'darkies,' which replaced 'blacks.' "

"Feiffer misses the point," said Preston Wilcox, a white professor at the Columbia School of Social Work. "Black' is now a symbol of pride because it's being imposed by blacks themselves. They are defining themselves now."

Ivanhoe Donaldson, a member of the S.NN.C.C. executive committee, agrees: " 'Negro' is an enforcement from the white establishment," he says. "Those who would liberate themselves must first define themselves."

February 26, 1968

Schools Turn to Negro Role in U.S.

By J. ANTHONY LUKAS

Who was the real McCoy?

According to most historians, he was Kid McCoy, a famed prize ring and barroom battler of the eighteen-nineties. One day, the story goes, he was taunted by a saloon heckler who said if he were the real McCoy he should put up his dukes and prove it. McCoy did just that. When the heckler came to, his first words were: "That's the real McCoy, all right."

But the educator William Loren Katz has another version. According to Mr. Katz, the real McCoy was Elijah McCoy, who gained more than 75 patents in the late 19th century for various mechanical devices. His best known invention was the drip cup, which fed oil to moving parts of heavy machinery. The tiny cup was so highly valued by machinists, Mr. Katz says, that they insisted on "the real McCoy."

Elijah McCoy was born in Canada, the son of runaway American slaves. But he figured in few American history courses until Mr. Katz, a pioneer in the teaching of Negro history, devoted nearly a full page to him in a new textbook.

Mr. Katz does not maintain a barroom dogmatism about McCoy. He concedes that the Kid's backers may have a point.

The New York Times (by Don Charles)

At the New York Public Library, James Kyle, 32, of Stamford, Conn., studies Negro history for a course he is taking.

But he argues that what has been wrong with the teaching of American history for so long is that few students have had a chance even to hear about Elijah McCoy.

This kind of omission, comparatively petty though it may be, is the kind that advocates of "black history" are seeking to correct. Demands for the teaching of 'black studies"—including literature, art and other aspects of the black experience—have swelled on campuses across the country during the last year.

These demands will have their first major results this fall when many colleges and universities introduce courses in black studies, often using new books that have been rushed into print by publishers.

The new focus on black studies raises anew the question: who is the real American?

Is he, as most American history textbooks have seen him, a bland, homogenized product of the melting pot, with all his ethnic peculiarities and angularities boiled out? Or is he a very particular person, formed by his own racial, ethnic and religious background and proud of it, yet living together with many other proud and particular persons in a richly variegated, pluralistic society?

Advocates of the teaching of black history are not demanding that it replace white history.

"I don't say we should rewrite all history to stress the Negro's role," Mr. Katz, who is now editing a series of books on black history, said in a recent interview.

Occasionally the Extreme

"Let the kids read several books, some with the old white approach and some with the pluralistic approach," he went on. "If this should lead to an argument in class—great, marvelous! Can you imagine anything better than kids—many of them so-called 'nonreaders' —actually getting into an argument over something on a printed page?"

The call for black studies is the product of Negroes' new pride in their race a new awareness of their blackness, which contrast sharply with the assimilationist, integrationist impulse of only a decade ago.

At its peak, this often abrasive pride in race can lead to a kind of voluntary self-segregation, such as among those Cornell students who are demanding courses in black studies open only to black students — a demand the university is unlikely to accept.

But this is not typical. When black students, administrators and faculty have sat down together in good faith, they have usually had no great difficulty finding common ground on curriculum matters at least. For many educators have now reached the conclusion that the traditional American curriculum does not properly prepare either black or white students to live in a pluralistic society.

STAGE STAR: Ira Aldridge, an American Negro actor, depicted as Othello in a playbill. Mr. Aldridge (1807-1867) played leading roles in European theaters for 40 years.

New Texts Multiracial

There is now substantial agreement among students and professional educators that measures to correct this imbalance should fall into two broad areas: First, introduction of materials on the black experience in all areas of the curriculum in which they are appropriate; and second, the development of specific courses, such as Negro history, African history, black literature or the sociology of the ghetto.

How this is to be done depends on the level of education and the composition of the student body. In the elementary schools, it is generally agreed, new material is needed throughout the curriculum. Examples in all subjects — whether reading, mathematics, or basic sciences — are commonly drawn from daily life, but for years they have been drawn primarily from white, suburban, middle-class life.

In many schools, the "Dick and Jane" brands readers ("watch Dick run; watch Jane run, too") have given way to texts, such as the "Bank Street Readers," which reflect a more multiracial society.

However, Dr. Kenneth B. Clark, the prominent Negro psychologist, says many of these new books have "simply colored some of the children brown."

A study by Dr. Clark's Metropolitan Applied Research Center, in cooperation with Harper & Row, found one elementary reader that had "a few brown children and adults and a few urban phenomena such as tall buildings, escalators, buses or department stores." But it also found that the central family in the book "takes an airplane ride to visit grandparents and lives in a split-level house with a neatly tended lawn, with morning milk delivery. This does not reflect the urban experience of white or Negro students."

Moreover, the study noted that "readers showing happy families with mother, father, two children living in comparative affluence are at best nonstimulating to children of the poor from broken homes; at worst, their insensitivity may present a barrier to reading."

High School Courses

At the secondary-school level, likewise, most educators feel that the prime need is for material integrated with the regular curriculum in virtually all subjects. However, in some high schools, particularly those with heavy Negro student bodies, there may be a demand for courses on Negro or African history.

Educators differ on the advisability of giving way to this demand at the high-school level. Some argue that this is just another form of segregation and that the history of blacks and whites in America is so intertwined that they should not be separated for high-school students.

Others, however, say that if there is a demand from the students themselves it should not be rejected. John Crawley, executive director of the Urban League in New Jersey's suburban Bergen County, cites a letter he recently received from a 13-year-old white girl as evidence that there is a thirst for knowledge about these subjects that is not being met by the current curriculum.

Two Approaches in College

"My class is presently studying United States history," the girl wrote. "I feel that it is very important for us to study American minority groups. As my teacher is somewhat reluctant to get into this, I have had to try and put pressure on her. After insisting that at least a short amount of time be spent on the American Indian, I was left no alternative but to do research and instruct the class myself."

The girl then asked Mr. Crawley whether he would come to the school and lecture on the Negro.

At college and graduate school, most educators agree that the two approaches are both required. They maintain that the physiology major should learn that Dr. Charles Drew, a Negro, was largely responsible for the development of blood plasma; a drama major should be taught about Ira Aldridge, the American Negro actor who was a star of European stages for 40 years in the 19th-century; an English major should read the poetry of Langston Hughes.

But at the same time, many universities and colleges are now setting up courses that will enable students to delve more deeply into the black experience.

"Certainly, 'The Negro Writer in America' is as valid a subject for serious study in this country today as 'The English Restoration Drama,'" one Ivy League professor said recently.

Reliability a Question

At Yale, a joint student-faculty committee has recommended the establishment of a major in Afro-American studies, and other universities seem on the verge of the same step.

At a recent seminar at Yale, Gerald McWhorter, a professor at Fisk University, warned that few universities or colleges are adequately prepared to establish departments, much less courses, in Afro-American studies. He proposed a foundation-financed study of the black experience to prepare such a curriculum.

Many educators concede that Professor McWhorter may have a point about postgraduate studies, in which considerable sophistication and specialization is required. There are special problems in teaching and research in Afro-American Studies, most of them deriving from the paucity of written materials.

For example, there are few reliable documents written by Africans about Africa before the slave traders arrived, or by slaves about slave life. Benjamin Quarles, a Negro historian, has cautioned against what he calls the "heroic fugitive" school of American literature——the narratives nominally written by runaway slaves, but often "ghostwritten by abolitionists and hence representing a white reformer's idea of how it felt to be a slave."

However, this shortage of original source materials is less critical in college and high schools. There, secondary sources often will suffice, and these are rapidly becoming available.

Dr. Austin J. McCaffrey, executive director of the American Textbook Institute, said the country is now going through its second textbook revolution in 10 years.

"The first, which began with Sputnik and the fears that raised about American teaching of science, set off a vast and thoroughgoing revision of science textbooks. The second, which may have begun with the Supreme Court's 1954 school-desegregation decision and has certainly accelerated since, is producing a whole host of new or revised textbooks which give us a better picture of the Negro's role in our country."

N.A.A.C.P. Notes Gain

June Shagaloff, the education specialist of the National Association for the Advancement of Colored People, largely agrees.

"The textbook picture is getting better—particularly in general elementary school books and in specialized books on aspects of Negro history," she said. "It is still far from good, particularly in general American history texts, but for any teacher who recognizes the need, there are certainly books available."

Miss Shagaloff and others pinpoint teachers themselves as the greatest single obstacle to the teaching of the Negro experience.

In most school systems, teachers have wide discretion in what and how they teach. New Jersey recently required the introduction of material on Negro history into the compulsory two-year American history course in high schools, but the implementation of this directive is largely left up to local principals and teachers. New York has no such legislative requirement.

When asked about what is being done in this field, state, county and city public education officials in the New York metropolitan area said it was virtually impossible to say because, as one put it, "you'd have to interview every individual teacher in every classroom."

But Miss Shagaloff fears that if it left to teachers alone little will be done.

Fischer Admits Lacks

Dr. John H. Fischer, president of Teachers College at Columbia University, conceded in an interview that it was still possible for a student to come out of his college without taking a single course in the area of Negro history or urban and racial problems. However, he said, the college recognized the need and was moving to fill it.

Dr. Fischer warned against any "concentrated attempt to rewrite history that would smack of what the Russians do from time to time. You won't find any effort to set up an indoctrination program here. What you will find is a readiness to admit errors of commission and omission."

What are these errors that the advocates of teaching Negro history now demand be corrected?

'DEADWOOD DICK,' the cowboy, is a nearly legendary figure in American frontier history, but few schoolboys learn that he was a Negro. His real name was Nat Love.

Part is the failure to teach about the thousands of Negroes, like Elijah McCoy, who have made major contributions to American life.

George Washington Carver, the Negro educator at Tuskegee Institute, won fame for his development of peanuts for industrial uses, but how many graduates of an American high school know that Benjamin Banneker, a Negro mathematician, inventor and gazetteer, was appointed by George Washington to the commission that helped lay out the nation's capital?

How many know that Crispus Attucks, a runaway slave, was the first man killed in the Boston Massacre and therefore probably the first martyr of the American Revolution? How many know that Estevanico, a Moorish slave, participated in some of the major explorations of 16th-century America and is credited by some with the discovery of a large part of the area now called Arizona and New Mexico?

Who Was Matzeliger?

How many have ever heard of Jan Matzeliger, who invented the lasting machine that revolutionized the American shoe industry; of Dr. Revlon Harris, the first American to make an analysis of the German V-2 rocket; of Dr. Percy Julian, who helped develop drugs that are now in widespread use by those afflicted by arthritis; of Matthew Henson, who played a major role in Adm. Robert E. Peary's expedition to the North Pole and who was the first man to stand atop the world?

How many know that Negroes are credited with having invented the potato chip (Hyram B. Thomas), ice cream (Augustus Jackson), the golf tee (George F. Grant), the mop holder (Thomas W. Stewart) or the player piano (J. H. and S. L. Dickinson)?

Or, for that matter, how many know that Deadwood Dick, the nearly legendary cowboy, or James Beckwourth, the Indian fighter for whom California's "Beckwourth Pass" is named, were Negroes?

But those interested in the teaching of Negro history stress that the "name game" is only a small part of what needs to be done.

For the omission of most of the names from American history courses is only one aspect of what these historians see as a persistent myopia about the role of the Negro in America.

One of these myths is that all Negroes were slaves until after the Civil War. Yet Professor Quarles in his book, "The Negro in the Making of America," notes that there were 59,557 nonslave Negroes in the United States as early as 1790 (7.9 per cent of the total black population) and that by 1860 there were 488,070 (11 per cent).

Many of these were originally indentured servants who completed their terms of service. Others were former slaves who obtained their liberty through military service. Others ran away or purchased their freedom. The result was a fairly significant number of free Negroes who performed roles far from that of the traditional cotton-picking Negro on the plantation. They were artisans, land-owning ranchers, cowboys, tailors, shoemakers, barbers.

However, most did remain slaves until the war, and in the treatment of their condition the new historians see an equally giant error.

Mr. Katz points to a paragraph from the 1940 printing of "The Growth of the American Republic," a widely used textbook by Samuel Eliot Morison and Henry Steele Commager, as a vivid example of the distortions he is trying to correct. It reads:

"As for Sambo, whose wrongs moved the abolitionists to wrath and tears, there is some reason to believe that he suffered less than any other class in the South from its 'peculiar institution.' The majority of the slaves were adequately fed, well cared for, and apparently happy . . . Although brought to America by force, the incurably optimistic Negro soon became attached to the country, and devoted to his 'white folks.'"

Among other things, this account overlooks the series of violent protests by Negro slaves against their condition. Herbert Aptheker has found records of some 250 slave conspiracies, some dating back to Colonial times. The best known, the one led by Nat Turner in Virginia, has been fully documented in William Styron's Pulitzer-Prize-winning novel, "The Confessions of Nat Turner." But there were scores of other bloody rebellions that make one wonder about just how happy "Sambo" was.

Another area in which the historians are trying to correct the record is the obstacles put up to even free Negroes in their efforts to better themselves.

"Over and over these days you hear the Irish-American or the Italian-American say 'if we could do it why can't they?'" Mr. Katz says.

"So few of them realize the degree of organized, statutory

repression to which Negroes have been subjected, but which the Irish and Italians never faced. Few realize that in many parts of America for a long time it was illegal to teach a slave—or even a free Negro—to read and write.

"Nobody pulled books out of the Jews' or the Italians' hands and said 'you can't read.' But for a long time the white man in America consciously sought to maintain the Negro's ignorance and illiteracy. After all, as late as 1909, President Taft told a black audience in Charlotte, N. C.: "Your race is adapted to be a race of farmers, first, last and for all times."

Cautions Are Added

Many of the new historians recognize the danger of letting historical revisionism run away with them and turning the Negro into a heroic figure who could do no wrong—a people that have produced only Marian Andersons, Jackie Robinson and Sidney Poitiers. Most are on guard against this kind of antiseptic history and try to show where the Negro stumbled too, where he gave in to the weaknesses of all human kind.

Mr. Katz, in his book "Eyewitness: the Negro in American History" tells not only about McCoy and Banneker, but about John (Mushmouth) Johnson and Dan Jackson, two of the toughest Negro racketeers in American history.

The study by the Metropolitan Advanced Research Center and Harper & Row calls for books that would more often inject controversy into history courses.

"Possible topics," it suggests, "might include dramatic accounts of such events as the 1920 Palmer raids, the Haymarket affair, the 1943 Detroit race riot, the rise and defeat of the Tweed Ring, the history of Al Capone and prohibition gangsters, the cleanup of the Barbary Coast, the evacuation of Japanese-Americans to detention camps in World War II."

Historians interviewed said the effect of this kind of vital history on ghetto youth can often be miraculous. Mr. Katz tells of a 17-year-old boy he once had in class who was described by previous teachers as a "nonreader" and a troublemaker."

"One day I got up and said 'there were 22 Negro Congressmen from the South after the Civil War.' Before I could even turn around this kid yells 'there were not.' For nearly an hour this kid and the others grilled me about it. 'How could Negroes have played such a role? How did it get back to the way it is?' Finally I said, 'Would you believe it if I showed it to you in the words of a white Mississippi historian?' They said they would.

"But the next day when I walked in the classroom door, the kid comes over and shows me a book by Carter Woodsen and says, 'Hey, you're right. Here's B. K. Bruce, the Senator from Mississippi.' From then on the kid woke up. He did me a 10-page paper on the Negro in World War II, and another one on James Baldwin and Gordon Parks. This kid, this nonreader, was turned on."

July 8, 1968

FORGOTTEN MEN: Three of the thousands of Negroes whose contributions to American life have been largely ignored by most American history textbooks. From the left are: Dr. Charles Drew, who was largely responsible for the development of blood plasma; Matthew Henson, who played a major role in Admiral Peary's expeditions to the North Pole, and Elijah McCoy, 19th-century inventor who held more than 75 patents for mechanical devices.

NEGRO 'PARANOIA' ASSAYED IN BOOK

White Racism Said to Push Blacks to the Brink

By JOHN LEO

White racism forces the American Negro to lead a life of "cultural paranoia" and often pushes him over the brink into true paranoid schizophrenia, a black psychiatrist said here yesterday.

Paranoia, a withdrawal from reality with delusions of percution, is by far the most common form of mental illness among black Americans, according to Dr. William H. Grier, assistant professor of psychiatry at the University of California Medical Center, San Francisco.

"That's because a black person has to develop a suspiciousness and defensive posture just to survive in America," he said in an interview. "He has to develop a 'healthy,' adaptive 'cultural paranoia,' which pushes him close to the line of mental illness."

Dr. Grier is the author of "Black Rage," a psychological portrait of the American Negro, published today by Basic Books. Dr. Price M. Cobbs, another Negro psychiatrist at the San Francisco medical center, is co-author.

The book argues that the rage of black men is beginning to break through a complex set of psychic defenses, erected in the time of slavery and little changed since.

Suffering Is Masked

Beneath "the cool style" and "the postal-worker syndrome" of ingratiating deference and passivity, the authors say, the Negro has been spending enormous amounts of psychic energy to mask suffering and rage.

"As a sapling bent low stores energy for a violent backswing," the authors write, "blacks bent double by oppression have stored energy which will be released in the form of rage — black rage, apocalyptic and final."

They also make these arguments:

¶It is the role of the Negro mother to suppress assertiveness in her sons, so they can survive in white society. As a result, Negro men develop considerable hostility toward Negro women as the inhibiting agents of an oppressive system.

¶Negro family structure is weak because it cannot fulfill its primary function: protection of its members. "Nowhere in the United States can the black family extend an umbrella of protection over its members in the way that a white family can," they say.

¶The black woman is prone to depressive, self-deprecatory attitudes. By white beauty standands, she is unable to develop a healthy narcissism, or self-love. She tends to see the sexual act as a degrading submission, which further lowers her self-esteem.

¶After early promise, many talented Negroes fail to do well in their careers because accomplishment is often perceived as a major move beyond the family, and thus as a form of abandonment of loved ones.

¶Many black men weep frequently — without warning and without feeling. It occurs while the black man is passively witnessing another man's triumph, and "the tears are for what he might have achieved if he had not been held back ... by some inner command not to excel, not to achieve, not to becoming outstanding, not to draw attention to himself."

"Under slavery," the authors write, "the black man was a psychologically emasculated and totally dependent human being. Times and conditions have changed, but black men continue to exhibit the inhibitions and psychopathology that had their genesis in the slave experience."

The "Black Norm," the authors write, is a set of defensive character traits that the American Negro must acquire.

They list these traits as cultural paranoia (every white man and every social system is the enemy until proven otherwise), cultural depression (sadness and intimacy with misery) and cultural antisocialism (an "accurate reading of one's environment" in which laws are never quite respected because they are designed to protect white men, not Negroes.

Essential Characteristics

"To regard the Black Norm as pathological," the psychiatrists write, "and attempt to remove such traits by treatment would be akin to analyzing away a hunter's cunning or a banker's prudence. This is a body of characteristics essential to life for black men in America and woe be unto that therapist who does not recognize it."

Rage is rising rapidly in the black community, they write, and whites must "get off the backs" of Negroes if they wish to avoid a conflagration.

"Today it is the young men who are fighting the battles, and, for now, their elders, though they have given their approval, have not yet joined in. The issue finally rests with the black masses. When the servile men and women stand up, we had all better duck."

July 25, 1968

AFRICA SURPRISES BLACK AMERICANS

Visitors Discover They Are Regarded as Foreigners

ILE DE GOREE, Senegal, Sept. 5 (AP)—A young black American visiting the remains of the appalling slave dungeon here reached down, scooped up a fistful of sand and told the curator: "I've got to take this back with me."

Now it is no longer only rich white Americans on safari who come to Africa. In Senegal, in Nigeria, in Kenya, black Americans are exploring Africa as a Boston O'Reilly visits Ireland.

They come singly and in Government - funded groups. Some study, others just wander around.

What they find depends largely on what they expected, but every one of hundreds interviewed encountered surprises.

Africa, they discover, is neither jungles full of white hunters nor model Governments run by black nationalists whom all white men call sir.

They Are Still Foreigners

Black Africans might offer them more understanding, but to blacks steeped in French, British and Belgian culture, black Americans are still foreigners. True African unity, much less unity with distant cousins, is still a dream, Africans say.

"I know Africa is the motherland, but I'm of a different tribe," said Ted Joans, a jazz poet of the nineteen-fifties who was passing through Togo compiling a book called "A Black Man Guides You Through Africa."

"A black man at face value gets one or two more seconds of tolerance from Africans," observed A. J. Franklin, a young black sociologist from Brooklyn who taught for two years at the University of Lagos.

Even the militant extremists underscore the American's need to adjust thinking on automatic black brotherhood. "An American just can't fall in here and pick things up," one said.

A doctoral candidate visiting Nigeria for the first time looked around and said: "I figure that 10 per cent of the people are real warm and enthusiastic, and maybe 5 to 6 per cent are mistrustful and against me, and the rest just treat me as any other foreigner."

Few Came to Stay

American blacks have been looking earnestly at Africa at least since the late Malcolm X toured West Africa in 1964 as a returning Black Muslim pilgrim.

Before then men like Marcus Garvey and W. E. B. DuBois turned to Africa for cultural identity, but with limited following. A scattered few like James W. Flemister who left Cincinnati for Liberia, came to settle.

Mr. Joans remembers how he tried to run an African restaurant in Harlem in the early nineteen-fifties.

"People were laughing at me when I wore African clothes, when I talked about Africa," he recalled. "I couldn't get a black woman to wear African dresses. I had to use white girls in African clothes as waitresses."

Now Africa is big business for Americans. Last year a team of American designers and artisans toured West Africa looking for inspiration for fashions, jewelry and cloth. They found it.

Deeper Man Clothing

The black identity goes deeper than clothing. A young filmmaker came here on a Government grant and told a white reporter: "I'm home, man, this is it! These are my brothers." Weeks later, when Nigeria currency restrictions prevented his changing money back into dollars, his attitude had changed markedly.

For Mr. Joans the problem is preserving African art and customs before they are lost and exposing American blacks to them.

A gentle but imposing man hung with amulets, he talked over beer at a former Peace Corps hostel in Togo that a black American couple had turned into a hamburger restaurant and inn.

On the Veranda was Mrs. Eulalia Barrow Bobo, sister of Joe Louis, the heavyweight champion, who was an early symbol of black pride. She was visiting fom her home in Beverly Hills.

Like Those in the U.S.

"Here," she said, "the educated blacks are the same as those in the States—they are materialistic. The uneducated blacks here are the same as in the States, because they don't know anything."

Mrs. Bobo, a strong advocate of the Bahai religion, makes no case for black unity but she has been encouraged by tracing her roots to the Mandingos of West Africa and finding that "they are a big, tall, handsome people —they are doctors, lawyers educated men." It was once considered degrading to be descended from Africans, she added, but no longer.

Among older Africans tribal feelings remain strong. An American black is not a brother; neither is a black from a neighboring tribe.

One of the strongest unifying factors is American soul music. The Afro hairdo, a new phenomenon to Africa, is catching on, and as black studies mushroom in America, awareness of America grows here.

September 6, 1970

Whites Report Rise in Contacts With Blacks Over Last Decade

By PAUL DELANEY
Special to The New York Times

CHICAGO, Aug., 17—Whites say their contacts with blacks slowly but steadily increased between 1964 and 1974.

A series of surveys over that period by the Institute for Social Research, which is located at the University of Michigan in Ann Arbor, documented the increasing mixing of the races, with a concomitant change in attitude about blacks on the part of whites from negative to positive. As a result the authors say, there appears to be growing acceptance of blacks by whites.

The surveys found diminishing numbers of whites who said their environment was all white—their friends, their neighborhoods, the schools nearest them, the people at work and the places they shop.

The surveys were conducted in 1964, 1968, 1970, 1972 and 1974. The sampling consisted of between 1,500 and 2,000 persons, a tenth of them black, all over the country. Thus, for the five surveys, up to 10,000 persons were interviewed, according to Dr. Angus Campbell, director of the institute.

Dr. Campbell, and Shirley Hatchett, a research assistant, put together the report on racial trends.

"The material pretty clearly tells us that white people have a strong sense of feeling of more change taking place now in their contact with blacks in all phases of life than in the past," Dr. Campbell said in a telephone interview.

The surveys found that in 1964, 81 per cent of the whites said all of their friends were white. Last year the percentage was 53.

In 1964, 80 per cent of the whites interviewed said that their neighborhood was all white. The figure was 61 per cent last year.

In 1964, 53 per cent said their coworkers were white; last year 39 per cent said so.

A decade ago, 39 per cent reported that the people they came into contact with while shopping were all white; in 1974, the figure was 15 per cent.

The surveys also showed the following:

¶Perceived contact with blacks is clearly associated with education. Whites with little schooling tended to have the least contact with the blacks, while college graduates had the most. Whites in metropolitan centers had more contact with blacks than those living elsewhere, and, with younger whites and those with more education, became more favorable in their attitude toward blacks as the decade passed—although the differences between metropolitan and nonmetropolitan residents had narrowed considerably by 1974.

¶The proportion of whites believing in "strict segregation" declined from one-fourth to one-tenth during the decade.

¶The proportion believing the Federal Government should protect the rights of blacks to equal accomodation rose from 56 per cent to 75 per cent.

¶The proportion feeling that blacks should have the right to move into any neighborhood they can afford rose from from 65 per cent to 87 per cent.

The report said that an improved attitude toward blacks had been noted throughout the population. However, it added: "The South, which had been the most negative in 1964, was still the most negative region in 1974, although the changes in these attitudes were greater in the South than in any of the other regions and as a result the regional differences were smaller at the end of the decade than they had been at the beginning."

The authors said they had found two areas in which what they saw as negative attitudes prevailed in the nineteen-seventies. Those areas were desegregation of jobs and schools, and the Federal role in desegregation efforts.

Slightly less than a majority of whites in 1964 said that the Federal Government should "see to it that black people get fair treatment in jobs." The proportion remained almost the same a decade later.

Also in the 1964 survey a little fewer than half the whites interviewed agreed that the Federal Government should "see to it that white and black children go to the same schools."

By 1970, the percentage had climbed some to a small majority, the report said. But since 1970, white support has dropped sharply to slightly better than a third, and stands at the lowest point of the 10-year period, the report said.

Neverthless, the findings on schools were significant, especially for the South where the data tended to confirm reports that more schools had been desegregated there than elsewhere. In 1964, 59 per cent of whites interviewed nationwide said the grade school nearest them was all white, while 43 per cent said the high school was all white. Last year, the percentages were 26 and 16 per cent respectively.

Great Change in South

But in the South, the statistics showed that in 1964, 78 per cent said the grade school nearest them was all white, and 61 per cent said the high school nearest them was all white. In 1974, those figures were down to 16 and 10 per cent, respectively.

As a comparison, in the Northeast in 1964, 48 per cent said the grade school nearest them was all white. Last year, 38 per cent said it was all white. A decade ago, 38 per cent said the high school was all white. The figure was 21 per cent last year.

While noting the importance of the breaking down of negative racial attitudes, Dr. Campbell and Miss Hatchett expressed concern about some of the implications of their findings. Both agreed that there was little correlation between expressed attitudes and action.

Further, Dr. Campbell said he agreed with the contention of some blacks that whites feel satisfied with the racial progress and have become less enthusiastic about civil rights.

He said surveys that showed racial progress, along with the fact that whites were seeing "black faces on television and seeing blacks move into high positions such as Cabinet members and on the Supreme Court," gave some whites the feeling that racial injustice no longer existed.

August 18, 1975

Just How Unstable Is the Black Family?

The view of the black community held by many a social scientist is by now familiar: a disaster area plagued by social disorganization, family disintegration and a host of conditions that breed emotional pathologies. It is a view based, in large measure, upon the statistics published periodically by the Bureau of the Census and the Bureau of Labor Statistics.

Last week, a new set of figures on American blacks was released—a special study by the Federal Government entitled "The Social Economic Status of Negroes in the United States, 1970." As expected, it painted a somber portrait. But by a coincidence most unexpected, the issuance of the study coincided exactly with the release of another report, this one set forth by the National Urban League at its annual convention in Detroit. And though both studies were based on the same statistics, conclusions drawn by many social scientists and those drawn by the league were worlds apart.

Thus, for example, the Government study shows that 28.9 per cent of black families are headed by females, an increase from 22.4 per cent in 1960. The familiar sociological analysis: A significant indication of continuing social deterioration and family instability. The view of the league: The assumption of instability in "matriarchal" households ignores the extended-family adaptation common in the black community — the strong kinship bonds between aunts, uncles and grandparents and the family's children. Some black sociologists go further; they argue that, in fact, roughly 70 per cent of these families actually do have a father present.

Another example: The Government study showed that, in order to obtain and maintain a median family income comparable to that of whites, both the black husband and his wife must —and often do—work. Conventional wisdom holds that this is a negative fact, since it is claimed that such families tend to be less stable than those in which the father is the sole breadwinner. But many black social scientists deny the claim, citing the prevalence of the extended-family adaptation—and they see the fact not as an indication of family deterioration but as proof of an attitude of cooperativeness and a strong work orientation in these families.

Sometimes the different approaches to statistics partake of a sparring match. Thus, one social scientist may point to figures that clearly show that, in 1970, black persons 14 to 19 years of age were more likely to be high school drop-outs

The Negro in the Sixties—A Statistical Portrait

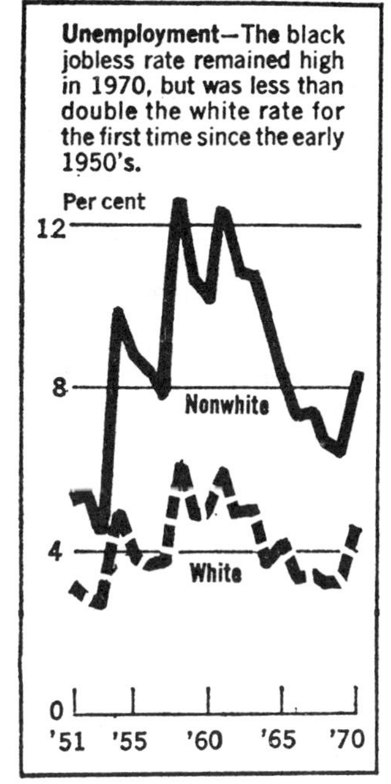

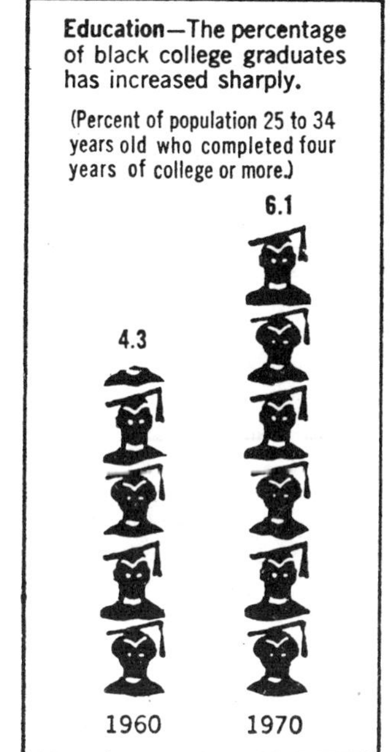

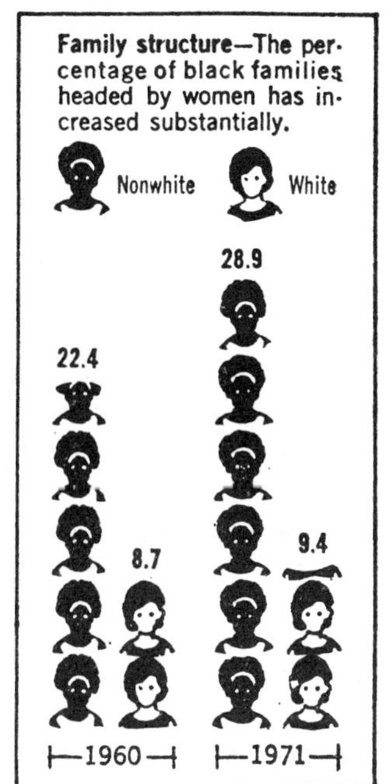

than were white persons in that age group. But another sociologist will counter with data showing that, since 1964, the number of blacks going to college has more than doubled, from 234,000 to 522,000. And he might add that these new students are mostly the first generation of their families to attend college, indicating a strong and increasing orientation toward achievement.

On the face of it, these arguments among social scientists over statistics that, by and large, both sides accept may seem to be nothing more than an exercise in academic semantics. But for the Urban League leadership, and for many black sociologists, the issue has far greater importance.

The manner in which these figures are interpreted, they feel, serves to delineate and identify the black community—in the eyes of whites and blacks alike. Statistics, heedlessly broadcast, are dangerous. And the customary negative interpretations reinforce negative generalizations, ignoring the actual and potential strengths of the black community.

—CHARLES V. HAMILTON

Mr. Hamilton is a professor of political science at Columbia University.

August 1, 1971

The Black Family in Slavery And Freedom, 1750-1925

By Herbert G. Gutman.
Illustrated. 664 pp. New York: Pantheon Books. $15.95.

By RICHARD SENNETT

Over a decade ago Daniel Patrick Moynihan issued a controversial report on black family life in America. He argued that poor blacks were caught in a "tangle of pathology" today because the black family had been destroyed by slavery. Slavery, he said, had disrupted the mother-father pair, and set in its place the female-headed household; today's black children are thus deprived of the complete family life that would give them the psychological strength to make it in a hostile world. Moynihan was immediately assailed as a racist. His critics said he was reviving all the old clichés about blacks as demoralized and unable to help themselves—in the guise of social-science jargon about "pathology." Whatever the justice of these charges, Moynihan should not have been seen as the sole culprit, for his report had popularized a view of the black family put forward 30 years before by the black sociologist E. Franklin Frazier, and it can be traced even further back to such sympathetic 19th-century observers of the plight of the slaves as Tocqueville.

Soon after Moynihan's report appeared, the scholarly community began hearing about research being

Richard Sennett is the author of "The Uses of Disorder," "Families Against the City," and co-author of "The Hidden Injuries of Class." His new book "The Fall of Public Man" will be published in January.

conducted by an American historian, Herbert G. Gutman, that attacked the core of the "tangle of pathology" thesis—the idea that slavery destroyed the black family. Gutman, it was said, was finding evidence that slaves had made determined efforts to keep their families together and succeeded more often, than not. Gutman was also thought to be uncovering evidence that the vast majority of poor black families in the 80 years after Emancipation took the two-parent nuclear form. If these findings were substantiated, the argument that the black family has a history of pathological divisions could be discarded simply on the grounds it isn't true.

But from year to year Gutman delayed publishing his results in book form. In the past year, 10 years of his work has suddenly appeared in a rush: "Work, Culture and Society in Industrializing America," a book of collected essays, the major one being about black coal workers; "Slavery and the Numbers Game," an intense, Swiftian attack on an econometric view of the slave family; and now, most important, "The Black Family in Slavery and Freedom, 1750-1925," the weaving together of his work on two centuries of black family history. But even now his readers do not have it all. This last book is so fat that Gutman has had to move all his material on black families during the critical years 1861-1868 to another volume, yet to appear. With what has appeared, however, Gutman has established himself as a leading social historian. "The Black Family in Slavery and Freedom" is an intellectual event, but it is not the last word on the subject.

Gutman's work may be seen as one part of a two-pronged attack on the broken black family thesis. While Gutman has been working to show that this broken family in no way typifies the black family of the past, a group of sociologists have been looking carefully at the implicit equation of a mother-headed household with a "pathological" family situation among the poor. The most reliable evidence shows that single-parent households appear among the poor not in simple conjunction with race or even with the most meager income, but rather in conjunction with unemployment: poor males are forced to move around to find work more than poor females, who can usually get jobs as domestics. (In fairness to Moynihan, it should be said his report laid great stress on this.)

What are these father-deprived families like? Joyce Ladner and Carol Stack have shown them to be like defensive networks: kids are passed from mothers to uncles and aunts to neighbors and back to mothers and/or fathers as the economic situation of various parts of the family web rises and falls. Life under these conditions is no picnic, but it is oriented to taking care of the children and to preserving a sense of the family as a community. No reliable data correlate single parenting as such to crime-rates or rates of emotional disorder among the very poor—although such correlations can be drawn statistically among the middle class (and even here, it's debatable). In any event, if blacks and other poor people unstably employed were organized into the nice, isolated, nuclear homes Moynihan envisioned, they probably would have perished.

Gutman's effort on his side of the argument has been immensely ambitious. Not only has he undertaken to present the black family throughout its history in America, but he has taken this subject as the occasion to write a new kind of social history. It is one in which quantitative records, from the U. S. Census, slave plantation journal books and the like are intermixed with qualitative materials, such as letters slaves wrote each other, testimony given to Government commissions, or observations of foreign travelers. Gutman is a master at combining this numerical and intimate material, and he has put this mastery to the service of a particular cause: to write history from an avowedly left-wing perspective that is never mechanical in its explanations and that reveals the complexities of everyday life. Thus, in his book of essays "Work, Culture and Society" we see him changing what he means by "class" and "work" in articles written over the years, as he shifts scenes and times.

Because of his technical mastery of social history, Gutman entered the debate about Stanley Engerman and Robert Fogel's econometric study, "Time on the Cross," as few others were equipped to do. He was not intimidated by the welter of statistics that Fogel and Engerman proffered in support of the thesis that if slavery was profitable it must have been minimally benign (you wouldn't abuse your human property any more than you would beat your tractor). Gutman simply proceeded to decimate their data step by step and then showed that, even if the thesis of "Time on the Cross" were true, it was trivial because their definition of slavery was a kind of economic science-fiction with little resemblance to the actual slave economy. Gutman's attack, a compact book called "Slavery and the Numbers Games," will endure long after "Time on the Cross" itself is forgotten—as a classic exposé of what can now be regarded as little more than an intellectual hoax.

The strategy of Gutman's own book on the black family is to array an immense amount of general material on slave families in the ante-bellum South around the study of six plantations he has studied in detail. He begins with the Good Hope plantation in South Carolina, which was an extraordinarily long-lived cotton plantation owned by the forebears of John Foster and Allen Dulles, then treats five plantations with different economic bases and life spans in Virginia, North Carolina, Alabama and Louisiana. Gutman wants to show the aspects of slave family life that cut across these very different settings. The most striking is the tenacity with which slaves held to the family bond either through conserving the African rituals they had known before bondage, or through making up new rituals.

Gutman shows, for instance, how slaves strove to maintain and pass down kin names, how individuals forced together created appellations of "aunt," "uncle" and "cousin" for each other, and how these preserved or created tissues of family identity formed a basis for defining obligations between people in the larger slave community. Gutman's analysis of the passing on of last names is, by the way, an example of how an insignificant detail in the hands of a first-class historian can reveal a great deal, in this case the transmission of a sense of family even when families were broken up and the individual slaves sold to different masters. Gutman's ferreting out of these kin ties is all the more impressive since by the 19th century most slaves had been given only the most common Anglo-Saxon names.

In the last third of the book, Gutman begins drawing together the materials he has assembled on the antebellum slave family and to pursue such interesting special problems as what happened to the family life of runaways and how did slaves react as family units to the outbreak of the Civil War. In a way "The Black Family in Slavery and Freedom" is a misleading title, since the book offers far less about black families after the Civil War than before it. Only 25 of its nearly 500 pages of text present material on the period from the 1870's to 1925, although the U.S. Census data discussed in these few pages show graphically how slaves, once freed, formed themselves into two-parent families.

For all its virtues, "The Black Family in Slavery and Freedom" is not the definitive work on the subject, because this book is not an interpretation of black family history so much as the presentation of an immense amount of data about these families set against a false stereotype of their history. Granted, Gutman has performed an immense service in burying the idea that slavery destroyed the black family, but the material he has assembled is much richer and demands much more analysis than this interpretative framework. Gutman does assert that the blacks' sense of kinship has been their main weapon in establishing themselves in the midst of a harsh culture, but he gives the reader no sense of how, say, the slave family of 1800 differed from the slave family on the eve of the Civil War. I do not mean that he fails to assemble the data to make such an analysis, but that he fails to make it. He is so intent on disproving the mythology of the broken black family, on showing what the black family was not, that he fails to show what it was.

For instance, Gutman has gathered a great deal of material on the sexual abuse of slave women by their white masters and also documented the abuse of black women by Northern white troops during the Civil War. These two experiences are never clearly compared, nor are the effects of white sexual abuse on the conduct of black family life drawn to the reader's attention. Instead, example after example piles up and the narrative becomes shapeless. While the author ruthlessly exposes the errors of his historical colleagues, he writes as though, in the history of the black family, apart from these misconceptions, the evidence speaks for itself. And it doesn't.

In his zeal to rout historiographic error, Gutman is often unfair to his colleagues. Gutman is at his worst when he cuts Eugene Genovese's "Roll, Jordon, Roll" down to smaller than its true size: Genovese is portrayed, falsely, as someone who believes black culture was little more than a response to the conditions the white masters set up. Making Genovese this sort of whipping boy may help Gutman underline his own position that the black family transcended those limitations, but simplifying his colleague's work can only, in the end, demean his own.

Without romanticizing their suffering or distorting the peculiar conditions of their lives, I think one ought to learn from the experiences of black families something about the family bond itself. In what ways are the ties of kinship the ultimate defense against tyranny? Gutman shows us families who, without these ties, would have been subject to all the demoralization imputed to them by popular mythology — but what was it about the kinship bond that helped people resist this collapse? If we could answer this question, if a writer like Gutman could show us why family ties work this way, then we might have a concrete understanding of why no person is ever wholly the creature of another person's will. Commitment to one's kin is, under the most adverse conditions, the essence of asserting one's freedom. This is the lesson of Gutman's book, a lesson he doesn't explain.

Moreover, the experiences of these black families call into question the current fashionable belief that the family per se is a form of slavery. This belief is based on the often-unspoken conviction that making a commitment to another human being limits one's own freedom to act, as though everything vouchsafed to another is subtracted from oneself. How rarified and mean-spirited this fashionable notion of "liberation" from the family is is made clear by the families of black slaves, post-Civil War sharecroppers, and modern urban workers. The capacity to commit oneself to another person enlarges and enriches one's life; it is this desire to be committed, to be tied down, which led the black slaves to track down their namesakes before the Civil War, to invent kin-relations with neighbors when no natural blood relatives were nearby, or to be willing to support children passed from mother to cousin to neighbor during crises of unemployment in modern times.

In sum, Gutman has closed one chapter in the history of the black family and begun another that is still to be finished. His three books announce the appearance of a formidably equipped researcher who is no narrow specialist. These gifts should now be put in the service of a coherent interpretive scheme, one which will do full justice to the importance of his subject. ■

October 17, 1976

My Furthest-Back Person-'The African'

By Alex Haley

My Grandma Cynthia Murray Palmer lived in Henning, Tenn. (pop. 500), about 50 miles north of Memphis. Each summer as I grew up there, we would be visited by several women relatives who were mostly around Grandma's age, such as my Great Aunt Liz Murray who taught in Oklahoma, and Great Aunt Till Merriwether from Jackson, Tenn., or their considerably younger niece, Cousin Georgia Anderson from Kansas City, Kan., and some others. Always after the supper dishes had been washed, they would go out to take seats and talk in the rocking chairs on the front porch, and I would scrunch down, listening, behind Grandma's squeaky chair, with the dusk deepening into night and the lightning bugs flicking on and off above the now shadowy honeysuckles. Most often they talked about our family—the story had been passed down for generations—until the whistling blur of lights of the southbound Panama Limited train *whooshing* through Henning at 9:05 P.M. signaled our bedtime.

So much of their talking of people, places and events I didn't understand: For instance, what was an "Ol' Massa," an "Ol' Missus" or a "plantation"? But early I gathered that white folks had done lots of bad things to our folks, though I couldn't figure out why. I guessed that all that they talked about had happened a long time ago, as now or then Grandma or another, speaking of someone in the past, would excitedly thrust a finger toward me, exclaiming, "Wasn't big as *this* young 'un!" And it would astound me that anyone as old and grey-haired as they could relate to my age. But in time my head began both a recording and picturing of

Alex Haley, is a magazine writer and author of "The Autobiography of Malcolm X." He is currently completing a new book, "Roots," tracing his family history. Columbia Pictures plans a film of the book, to be published by Doubleday next year.

the more graphic scenes they would describe, just as I also visualized David killing Goliath with his slingshot, Old Pharaoh's army drowning, Noah and his ark, Jesus feeding that big multitude with nothing but five loaves and two fishes, and other wonders that I heard in my Sunday school lessons at our New Hope Methodist Church.

The furthest-back person Grandma and the others talked of—always in tones of awe, I noticed—they would call "The African." They said that some ship brought him to a place that they pronounced "'Naplis." They said that then some "Mas' John Waller" bought him for his plantation in "Spotsylvania County, Va." This African kept on escaping, the fourth time trying to kill the "hateful po' cracker" slave-catcher, who gave him the punishment choice of castration or of losing one foot. This African took a foot being chopped off with an ax against a tree stump, they said, and he was about to die. But his life was saved by "Mas' John's" brother—"Mas' William Waller," a doctor, who was so furious about what had happened that he bought the African for himself and gave him the name "Toby."

Crippling about, working in "Mas' William's" house and yard, the African in time met and mated with "the big house cook named Bell," and there was born a girl named Kizzy. As she grew up her African daddy often showed her different kinds of things, telling her what they were in his native tongue. Pointing at a banjo, for example, the African uttered, *"ko"*; or pointing at a river near the plantation, he would say, *"Kamby Bolong."* Many of his strange words started with a *"k"* sound, and the little, growing Kizzy learned gradually that they identified different things.

When addressed by other slaves as "Toby," the master's name for him, the African said angrily that his name was *"Kin-tay."* And as he gradually learned English, he told young Kizzy some things about himself—for instance, that he was not far from his village, chopping wood to make himself a drum, when four men had surprised, overwhelmed, and kidnaped him.

So Kizzy's head held much about her African daddy when at age 16 she was sold away onto a much smaller plantation in North Carolina. Her new "Mas' Tom Lea" fathered her first child, a boy she named George. And Kizzy told her boy all about his African grandfather. George grew up to be such a gamecock fighter that he was called "Chicken George," and people would come from all over and "bet big money" on his cockfights. He mated with Matilda, another of Lea's slaves; they had seven children, and he told them the stories and strange sounds of their African great-grandfather. And one of those children, Tom, became a blacksmith who was bought away by a "Mas' Murray" for his tobacco plantation in Alamance County, N. C.

Tom mated there with Irene, a weaver on the plantation. She also bore seven children, and Tom now told them all about their African great-great-grandfather, the faithfully passed-down knowledge of his sounds and stories having become by now the family's prideful treasure.

The youngest of that second set of seven children was a girl, Cynthia, who became my maternal Grandma (which today I can only see as fated). Anyway, all of this is how I was growing up in Henning at Grandma's, listening from behind her rocking chair as she and the other visiting old women talked of that African (never then comprehended as *my* great-great-great-great-grandfather) who said his name was *"Kin-tay,"* and said *"ko"* for banjo, *"Kamby Bolong"* for river, and a jumble of other *"k"*-beginning sounds that Grandma privately muttered, most often while making beds or cooking, and who also said that near his village he was kidnaped while chopping wood to make himself a drum.

The story had become nearly as fixed in my head as in Grandma's by the time Dad and Mama moved me and my two younger brothers, George and Julius, away from Henning to be with them at the small black agricultural and mechanical college in Normal, Ala., where Dad taught.

To compress my next 25 years: When I was 17 Dad let me enlist as a mess boy in the U.S. Coast Guard. I became a ship's cook out in the South Pacific during World War II, and at night down by my bunk I began trying to write sea adventure stories, mailing them off to magazines and collecting rejection slips for eight years before some editors began purchasing and publishing occasional stories. By 1949 the Coast Guard had made me its first "journalist"; finally with 20 years' service, I retired at the age of 37, determined to make a full time career of writing. I wrote mostly magazine articles; my first book was "The Autobiography of Malcolm X."

Then one Saturday in 1965 I happened to be walking past the National Archives building in Washington. Across the interim years I had thought of Grandma's old stories—otherwise I can't think what diverted me up the Archives' steps. And when a main reading room desk attendant asked if he could help me, I wouldn't have dreamed of admitting to him some curiosity hanging on from boyhood about my slave forebears. I kind of bumbled that I was interested in census records of Alamance County, North Carolina, just after the Civil War.

The microfilm rolls were delivered, and I turned them through the machine with a building sense of intrigue, viewing in different census takers' penmanship an endless parade of names. After about a dozen microfilmed rolls, I was beginning to tire, when in utter astonishment I looked upon the names of Grandma's parents: Tom Murray, Irene Murray . . . older sisters of Grandma's as well—every one of them a name that I'd heard countless times on her front porch.

It wasn't that I hadn't believed Grandma. You just *didn't* not believe my Grandma. It was simply so uncanny actually seeing those names in print and in official U.S. Government records.

During the next several months I was back in Washington whenever possible, in the Archives, the Library of Congress, the Daughters of the American Revolution Library. (Whenever black attendants understood the idea of my search, documents I requested reached me with miraculous speed.) In one source or another during 1966 I was able to document at least the highlights of the cherished family story. I would have given anything to have told Grandma, but, sadly, in 1949 she had gone. So I went and told the only survivor of those Henning front-porch storytellers: Cousin Georgia Anderson, now in her 80's in Kansas City, Kan. Wrinkled, bent, not well herself, she was so overjoyed, repeating to me the old stories and sounds; they were like Henning echoes: "Yeah, boy, that African say his name was *'Kin-tay'*; he say the banjo was *'ko,'* an' the river *'Kamby-Bolong,'* an' he was off choppin' some wood to make his drum when they grabbed 'im!" Cousin Georgia grew so excited we had to stop her, calm her down, 'You go' head, boy! Your grandma an' all of 'em—they up there watching what you do!"

That week I flew to London on a magazine assignment. Since by now I was steeped in the old, in the past, scarcely a tour guide missed me—I was awed at so many historical places and treasures I'd heard of and read of. I came upon the Rosetta stone in the British Museum, marveling anew at how Jean Champollion, the French archaeologist, had miraculously deciphered its ancient demotic and hieroglyphic texts . . .

The thrill of that just kept hanging around in my head. I was on a jet returning to New York when a thought hit me. Those strange, unknown-tongue sounds, always part of our family's old story . . . they were obviously bits of our original African *"Kin-tay's"* native tongue. What specific tongue? Could I somehow find out?

Back in New York, I began making visits to the United Nations Headquarters lobby; it wasn't hard to spot Africans. I'd stop any I could, asking if my bits of phonetic sounds held any meaning for them. A couple of dozen Africans quickly looked at me, listened, and took off—understandably dubious about some Tennesseean's accent alleging "African" sounds.

My research assistant, George Sims (we grew up together in Henning), brought me some names of ranking scholars of African linguistics. One was particularly intriguing: a Belgian- and English-educated Dr. Jan Vansina; he had spent his early career living in West African villages, studying and tape-recording countless oral histories that were narrated by certain very old African men; he had written a standard textbook, "The Oral Tradition."

So I flew to the University of Wisconsin to see Dr. Vansina. In his living room I told him every bit of the family story in the fullest detail that I could remember it. Then, intensely, he queried me about the story's relay across the generations, about the gibberish of *"k"* sounds Grandma had fiercely muttered to herself while doing her housework, with my brothers and me giggling beyond her hearing at what we had dubbed "Grandma's noises."

Dr. Vansina, his manner very serious, finally said, "These sounds your family has kept sound very probably of the tongue called 'Mandinka.'"

I'd never heard of any "Mandinka." Grandma just told of the African saying *"ko"* for banjo, or *"Kamby Bolong"* for a Virginia river.

Among Mandinka stringed instruments, Dr. Vansina said, one of the oldest was the *"kora."*

"Bolong," he said, was clearly Mandinka for "river." Preceded by *"Kamby,"* it very likely meant "Gambia River."

Dr. Vansina telephoned an eminent Africanist colleague, Dr. Philip Curtin. He said that the phonetic *"Kin-tay"* was correctly spelled *"Kinte,"* a very old clan that had originated in Old Mali. The Kinte men traditionally were blacksmiths, and the women were potters and weavers.

I knew I must get to the Gambia River.

The first native Gambian I could locate in the U.S. was named Ebou Manga, then a junior attending Hamilton College in upstate Clinton, N. Y. He and I flew to Dakar, Senegal, then took a smaller plane to Yundum Airport, and rode in a van to Gambia's capital, Bathurst. Ebou and his father assembled eight Gambia government officials. I told them Grandma's stories, every detail I could remember, as they listened intently, then reacted. "*'Kamby Bolong'* of course is Gambia River!" I heard. "But more clue is your forefather's saying his name was *'Kinte.'*" Then they told me something I would never even have fantasized—that in places in the back country lived very old men, commonly called *griots*, who could tell centuries of the histories of certain very old family clans. As for *Kintes*, they pointed out to me on a map some family villages, Kinte-Kundah, and Kinte-Kundah Janneh-Ya, for instance.

The Gambian officials said they would try to help me. I returned to New York dazed. It is embarrassing to me now, but despite Grandma's stories, I'd never been concerned much with Africa, and I had the routine images of African people living mostly in exotic jungles. But a compulsion now laid hold of me to learn all I could, and I began devouring books about Africa, especially about the slave trade. Then one Thursday's mail contained a letter from one of the Gambian officials, inviting me to return there.

Monday I was back in Bathurst. It galvanized me when the officials said that a *griot* had been located who told the *Kinte* clan history—his name was Kebba Kanga Fofana. To reach him, I discovered, required a modified safari: renting a launch to get upriver, two land vehicles to carry supplies by a roundabout land route, and employing finally 14 people, including three interpreters and four musicians, since a *griot* would not speak the revered clan histories without background music.

The boat Baddibu vibrated upriver, with me acutely tense: Were these Africans maybe viewing me as but another of the pith-helmets? After about two hours, we put in at James Island, for me to see the ruins of the once British-operated James Fort. Here two centuries of slave ships had loaded thousands of cargoes of Gambian tribespeople. The crumbling stones, the deeply oxidized swivel cannon, even some remnant links of chain seemed all but impossible to believe. Then we continued upriver to the left-bank village of Albreda, and there put ashore to continue on foot to Juffure, village of the *griot*. Once more we stopped, for me to see *toubob kolong*, "the white man's well," now almost filled in, in a swampy area with abundant, tall, saw-toothed grass. It was dug two centuries ago to "17 men's height deep" to insure survival drinking water for long-driven, famishing coffles of slaves.

Walking on, I kept wishing that Grandma could hear how her stories had led me to the *"Kamby Bolong."* (Our surviving storyteller Cousin Georgia died in a Kansas City hospital during this same morning, I would learn later.) Finally, Juffure village's playing children, sighting us, flashed an alert. The 70-odd people came rushing from their circular, thatch-roofed, mud-walled huts, with goats bounding up and about, and parrots squawking from up in the palms. I sensed him in advance somehow, the small man amid them, wearing a pillbox cap and an off-white robe—the *griot*. Then the interpreters went to him, as the villagers thronged around me.

And it hit me like a gale wind: every one of them, the whole crowd, was *jet black*. An enormous sense of guilt swept me—a sense of being some kind of hybrid . . . a sense of being impure among the pure. It was an awful sensation.

The old *griot* stepped away from my interpreters and the crowd quickly swarmed around him—all of them buzzing. An interpreter named A. B. C. Salla came to me; he whispered: "Why they stare at you so, they have never seen here a black American." And that hit me: I was symbolizing for them twenty-five millions of us they had never seen. What did they think of me—of us?

Then abruptly the old *griot* was briskly walking toward me. His eyes boring into mine, he spoke in Mandinka, as if instinctively I should understand—and A. B. C. Salla translated:

"Yes . . . we have been told by the forefathers . . . that many of us from this place are in exile . . . in that place called America . . . and in other places."

I suppose I physically wavered, and they thought it was the heat; rustling whispers went through the crowd, and a man brought me a low stool. Now the whispering hushed—the musicians had softly begun playing *kora* and *balafon*, and a canvas sling lawn seat was taken by the *griot*, Kebba Kanga Fofana, aged 73 "rains" (one rainy season each year). He seemed to gather himself into a physical rigidity, and he began speaking the *Kinte* clan's ancestral oral history; it came rolling from his mouth across the next hours . . . 17th- and 18th-century *Kinte* lineage details, predominantly what men took wives; the children they "begot," in the order of their births; those children's mates and children.

Events frequently were dated by some proximate singular physical occurrence. It was as if some ancient scroll were printed indelibly within the *griot's* brain. Each few sentences or so, he would pause for an interpreter's translation to me. I distill here the essence:

The *Kinte* clan began in Old Mali, the men generally blacksmiths ". . . who conquered fire," and the women potters and weavers. One large branch of the clan moved to Mauretania from where one son of the clan, Kairaba Kunta Kinte, a Moslem Marabout holy man, entered Gambia. He lived first in the village of Pakali N'Ding; he moved next to Jiffarong village; ". . . and then he came here, into our own village of Juffure."

In Juffure, Kairaba Kunta Kinte took his first wife, ". . . a Mandinka maiden, whose name was Sireng. By her, he begot two sons, whose names were Janneh and Saloum. Then he got a second wife, Yaisa. By her, he begot a son, Omoro."

The three sons became men in Juffure. Janneh and Saloum went off and found a new village, Kinte-Kundah Janneh-Ya. "And then Omoro, the youngest son, when he had 30 rains, took as a wife a maiden, Binta Kebba.

"And by her, he begot four sons—Kunta, Lamin, Suwadu, and Madi . . ."

Sometimes, a "begotten," after his naming, would be accompanied by some later-occurring detail, perhaps as ". . . in time of big water (flood), he slew a water buffalo." Having named those four sons, now the *griot* stated such a detail.

"About the time the king's soldiers came, the eldest of these four sons, Kunta, when he had about 16 rains, went away from this village, to chop wood to make a drum . . . and he was never seen again . . ."

Goose-pimples the size of lemons seemed to pop all over me. In my knapsack were my cumulative notebooks, the first of them including how in my boyhood, my Grandma, Cousin Georgia and the others told of the African *"Kin-tay"* who always said he was kidnapped near his village—while chopping wood to make a drum . . .

I showed the interpreter, he

showed and told the *griot*, who excitedly told the people; they grew very agitated. Abruptly then they formed a human ring, encircling me, dancing and chanting. Perhaps a dozen of the women carrying their infant babies rushed in toward me, thrusting the infants into my arms —conveying, I would later learn, "the laying on of hands . . . through this flesh which is us, we are you, and you are us." The men hurried me into their mosque, their Arabic praying later being translated outside: "Thanks be to Allah for returning the long lost from among us." Direct descendants of Kunta Kinte's blood brothers were hastened, some of them from nearby villages, for a family portrait to be taken with me, surrounded by actual ancestral sixth cousins. More symbolic acts filled the remaining day.

When they would let me leave, for some reason I wanted to go away over the African land. Dazed, silent in the bumping Land Rover, I heard the cutting staccato of talking drums. Then when we sighted the next village, its people came thronging to meet us. They were all—little naked ones to wizened elders—waving, beaming, amid a cacophony of crying out; and then my ears identified their words: *"Meester Kinte! Meester Kinte!"*

Let me tell you something: I am a man. But I remember the sob surging up from my feet, flinging up my hands before my face and bawling as I had not done since I was a baby . . . the jet-black Africans were jostling, staring . . . I didn't care, with the feelings surging. If you really knew the odyssey of us millions of black Americans, if you really knew how we came in the seeds of our forefathers, captured, driven, beaten, inspected, bought, branded, chained in foul ships, if you really knew, you needed weeping . . .

Back home, I knew that what I must write, really, was our black saga, where any individual's past is the essence of the millions'. Now flat broke, I went to some editors I knew, describing the Gambian miracle, and my desire to pursue the research; Doubleday contracted to publish, and Reader's Digest to condense the projected book; then I had advances to travel further.

What ship brought Kinte to Grandma's "'Naplis" (Annapolis, Md., obviously)? The old *griot's* time reference to "king's soldiers" sent me flying to London. Feverish searching at last identified, in British Parliament records, "Colonel O'Hare's Forces," dispatched in mid-1767 to protect the then British-held James Fort whose ruins I'd visited. So Kunta Kinte was down in some ship probably sailing later that summer from the Gambia River to Annapolis.

Now I feel it was fated that I had taught myself to write in the U. S. Coast Guard. For the sea dramas I had concentrated on had given me years of experience searching among yellowing old U. S. maritime records. So now in English 18th Century marine records I finally tracked ships reporting themselves in and out to the Commandant of the Gambia River's James Fort. And then early one afternoon I found that a Lord Ligonier under a Captain Thomas Davies had sailed on the Sabbath of July 5, 1767. Her cargo: 3,265 elephants' teeth, 3,700 pounds of beeswax, 800 pounds of cotton, 32 ounces of Gambian gold, and 140 slaves; her destination: "Annapolis."

That night I recrossed the Atlantic. In the Library of Congress the Lord Ligonier's arrival was one brief line in "Shipping In The Port Of Annapolis—1748-1775." I located the author, Vaughan W. Brown, in his Baltimore brokerage office. He drove to Historic Annapolis, the city's historical society, and found me further documentation of her arrival on Sept. 29, 1767. (Exactly two centuries later, Sept. 29, 1967, standing, staring seaward from an Annapolis pier, again I knew tears). More help came in the Maryland Hall of Records. Archivist Phebe Jacobsen found the Lord Ligonier's arriving customs declaration listing, "98 Negroes"—so in her 86-day crossing, 42 Gambians had died, one among the survivors being 16-year-old Kunta Kinte. Then the microfilmed Oct. 1, 1767, Maryland Gazette contained, on page two, an announcement to prospective buyers from the ship's agents, Daniel of St. Thos. Jenifer and John Ridout (the Governor's secretary): "from the River GAMBIA, in AFRICA . . . a cargo of choice, healthy SLAVES . . ."

Spin-offs of a family search

I have told of the most dramatic successes of my search, but it has also led to other finds. In New York one morning, W. Colston Leigh, who arranges my lectures, asked if I'd fly on short notice to Simpson College in Indianola, Ia., to take the place of an ailing client. Barely making it, I spoke of my long search—then the Academic Dean, Waller Wiser, told me that from things I'd said, he knew he was of the seventh generation from the Wallers of Spotsylvania County, Va., who had bought my forefather. The college since has given me the honorary Doctorate of Letters; Waller presented it. Every time that he and I get the chance to talk now, we speculate upon why we met after 200 years.

I also explored the lineage of my Dad's mother. She was small, very fair blonde, with blue eyes and such quickness of movement that she had always been called "Cricket." Grandma Haley often told of her girlhood on a plantation in Alabama, where her mother was "Easter," a slave, and her father was a Civil War colonel. Well, just curious, I identified the colonel in microfilmed census rolls in the National Archives, and also his father. Then I found that the father had come to America in 1799, from his native County Monaghan, Ireland. That rocked me: I'd never felt any Irish within me. But sheer curiosity kept bugging me until I flew to Ireland, tracing the Colonel's lineage finally back as far as 1707 to a little town of Carrickmacross, where once again I got rocked: They were most hospitable—until they learned I'm Protestant.

Here at home, deep discussions occur these days with my two younger brothers, both now Washingtonians, George as Chief Counsel of the U.S. Urban Mass Transportation Administration, and Julius as a Navy Department architect. We are giving to our ancestral Juffure village a new mosque—as a personal symbol. But Kunta Kinte really was but one among those many millions of enslaved African forebears, whose seeds are the black generations of North and South America and the West Indies. Just recently, we have incorporated the nonprofit Kinte Foundation in the District of Columbia. It exists to create a black genealogical library.

For my research has revealed that an unsuspected wealth of records exists to identify slaves and early free blacks. In antebellum documents in innumerable courthouses and libraries, in countless homes' old trunks and chests and attics, in no end of other places, there are plantation slave lists; records of slaves sold, bought, hired out and run away. There are slave traveling passes, manumission (freedom) papers; there are wills bequeathing slaves, lists of black churchgoers, marriage records and old personal diaries, journals, notebooks and letters which contain the names of 19th-century blacks. All of these the library will gather.

If my efforts have led to dramatic finds, though, some have ended in failure. Is there anybody who can help me to document my great-great grandpa "Chicken George" Lea's sailing to England around the eighteen-fifties? Grandma and Cousin Georgia always told of a "rich Englishman" hiring away the great slave gamecocker from Tom Lea. They said this Englishman took "Chicken George" on a ship "to fight his roosters against anybody's" in English, then French gamecock pits. Then returning to North Carolina, and being given his freedom by his father-master shortly before the Civil War, "Chicken George" participated in the gory battle of Fort Pillow.

I've dug into accounts of cockfighting in North Carolina, England and France; what little I have been able to find finally came from cockfighting historian Robin Walker in Glasgow, Scotland. If you know of any promising sources, please write me c/o Kinte Foundation, 716 National Press Building, Washington, D. C. 20004.—A. H.

July 16, 1972

80 Million Saw 'Roots' Sunday, Setting Record

By THOMAS A. JOHNSON

Eighty million Americans, the largest television audience in the history of the medium, watched the final episode of "Roots," the dramatization of a black family's life during the American slavry era, the National Nielsen Ratings reported yesterday.

Last Sunday's broadcast on the American Broadcasting Company's network was watched on 51.1 percent of all the television sets in homes across the country, the Nielsen service said. And the network reported that the show went into 36,380,000 homes and exceeded by 2.4 million the audience reached by the first portion of the broadcast of "Gone With The Wind," the previous record holder.

The program was based on a book of the same name by Alex Haley, whose 12 years of research traced his family back through old records in the United States and England to a remote farming village in Gambia, West Africa. The televised production with vivid depictions of slave conditions, has had a wide impact on black Americans and white, evoking shock, disbelief, tears and anger.

"Since the program started," said Dr. James Turner, director of the Africana Center at Cornell University, we have been getting calls from teachers groups, businessmen and housewives to sit in our class on 'Afro American Life and Culture in America, Roots.'"

The network reported that a total of 130 million viewers (representing 85 percent of all television homes in the United States) saw all or part of the series. The eight episodes averaged a 44.9 percent rating and for the week ending last Sunday, the top seven rated shows were all "Roots" episodes.

One significant impact of the series has been an early groundswell of interest in the history of early America and the black man's involvement, a random telephone survey of university campuses has revealed.

Dr. Melvin Drimmer, chairman of the history department at Cleveland State University, reported that "more than 200 students, mostly white, showed up for a hastily called seminar on 'Roots,' on Monday."

"The notice went out late Friday," he said. "The impact here has been phenomenal."

Alma W. Robinson, director of black studies for Stanford University, said she hoped to get copies of the total "Roots" presentation and "build a number of courses around it."

And the provost for the University of California at Santa Cruz, Dr. Herman Blake, said he was encouraged that "Roots" looks like "it will help us in trying to move blacks, other minorities and poor whites to the place where they recognize more fully the legitimacy of their needs and the hopes to achieve their goals."

The upsurge in the interest in history, the educators all said, would probably be a passing phase for the layman but, they pointed out, the surge should assist the serious scholars in their own work

Dr. Turner, of Corneil, said serious scholarship and analysis between black and white scholars "should definitely follow" the increased interest. He said also that the total impact of "Roots" would strengthen "the blck community's attempts to influence American foreign policy toward Africa, feel more relaxed and proud of their African connections and encourage even more black travel to Africa."

Three years ago, James Dyer, a program officer for the Carnegie Foundation, authorized a grant of $459,000 after hearing Mr. Haley lecture to encourage groups interested in historical projects ealing tiwh slave conditions and also the tracing of ancestry.

One part of the grant went to the Kente Foundation, a program run by Mr. Haley and his brothers, George, the general counsel for the United States Information Agency, and Julius, an architect with the Navy Department.

The aim of the fund is to assist other people wanting to trace their own family histories, George Haley said that his brother Alex had prepared a manuscript detailing how he had made a successful search of old records.

February 2, 1977

THE PUERTO RICANS

PUERTO RICANS SEEK BETTER JOBS HERE

Columbia Report Says Migrants Want Improvement, Not 'Any Work' or Relief From City

1,113 FAMILIES IN SURVEY

Population Put at Only 160,000 —Study Differs With Murtagh Findings on Aid, Travel

Desire for economic betterment brings the typical Puerto Rican migrant to New York rather than the quest for just a job or support from the city's relief program, according to a nine-month study of migration conducted by Columbia University's Bureau of Applied Social Research for the Puerto Rican Government.

The principal findings were summarized in the form of a letter to Jesus T. Pinero, Governor of Puerto Rico, which was made public yesterday at a press conference in Columbia's School of Journalism.

The sixteen-page summary of the 175-page study was presented to Fernando Sierra Beridecia, Commissioner of Labor of Puerto Rico, by Dr. Robert K. Merton, Professor of Sociology and acting director of the Bureau of Applied Social Research, and Clarence Senior, associate director of the migration study. Dr. C. Wright Mills, director of labor research for the bureau, who directed the study, was unable to attend the conference.

Manuel Cabranes, head of a Puerto Rican Government office established here to aid migrants, attended the conference. A comparable office is operated by the Labor Department in Puerto Rico.

Migrants Put at 160,000

The study estimates the number of migrants ranges between 160,000 and 200,000, "with the weight of evidence favoring 160,000." It was pointed out that this figure was considerably smaller than some recent estimates, which have set the Puerto Rican migrant population as high as 500,000 or 600,000. An accurate count of the Puerto Ricans here is held impossible because of the absence of a complete census enumeration.

At the conference Mr. Senior suggested that such estimates might be partly derived from a failure to distinguish between Spanish-speaking persons and Puerto Ricans. The latter can be recognized by their pronunciation.

In obtaining data for the $35,000 project, which was begun last October, interviews were held with one person in each of 1,113 Puerto Rican households in Harlem and the lower East Bronx, the two principal migrant residential districts, making available data on 5,000 persons.

Interviews of at least one and a half hours were conducted by Spanish-speaking persons, most of whom were college educated Puerto Ricans. Recruited from various sources and including two candidates for Ph.D. degrees in Spanish from Columbia, the interviewers received trial interviewing assignments before they were hired permanently. A weeding out process of interviewers followed the period of initial instructions for and checks upon their work.

Throughout the study, three checkers examined a quota of each interviewer's reports to ascertain "the internal consistency" of answers given to 174 questions. The questionnaire employed was structured to have answers to the same question—phrased variously in different parts of the questionnaire—check against one another.

43% Employed Last Month

The study noted that 43 per cent of the 5,000 were part of the city's labor force in April and May. In contrast, 35 per cent of the total Puerto Rican population was working or seeking work.

Information on the relief situation among the 1,113 interviewed shows that of 98 per cent who had heard of the existence of city relief, only 8 per cent knew of it before leaving Puerto Rico. Ninety per cent of the interviewed learned of relief "only after they came to New York City," the study says.

"Only eleven people among all the 1,113 cases we interviewed," the study reports, "were aware that there is no fixed period of residence required before one is eligible for relief in New York. More than half of the sample [54 per cent] were under the impression that there is a length of residence requirement in the city. The remainder had no information whatever on the subject.

"Of the 5,000 people in the 1,113 households where we made interviews, about 3,000 were adults [20 years old and over]. Six per cent of the total number of these adults were relief cases—161 persons."

Twenty-seven per cent of the 161 arrived after the war; 15 per cent during the war and 58 per cent before the war.

Women constitute 77 per cent of the 161 relief cases. Of them, 55 per cent were widowed, divorced or separated and of this group, 80 per cent have dependent children.

"It should be noted that these women with dependent children constitute one of the groups included in the Federal Social Security program," the study says. "That is, funds are provided jointly by the state and Federal Government, not by the city, which simply administers them."

Data on the number of persons on relief who have made trips back and forth to Puerto Rico were available for only 132 cases. The study says that only four cases (3 per cent) of those on relief who were studied had ever made a trip to the island and returned here. Seven per cent of those not on relief have made such trips.

"Fifty-four of these relief cases entered New York since 1940," the study says. "Of these, not one person had ever made a visit back to the island. Those four relief cases who did make trips had all migrated to New York before 1940. None of them was on relief at the time he made the journey."

Contrasts to City Findings

The bureau's figures on travel to Puerto Rico contrasted with the findings of John M. Murtagh, Commissioner of Investigation, in an interim report to Mayor O'Dwyer on June 3. Terming that part of his report evidence of the "existence rather than the extent" of misuse of public funds, Mr. Murtagh said that in the week of Dec. 17 to Dec. 24, 1947, twenty-three persons then on relief made airplane flights to Puerto Rico.

At the press conference Dr. Merton pointed out that nine out of ten migrants learned about New York from letters, friends and relatives. The role of newspapers, radio and motion pictures was negligible, he said, in informing migrants about the city.

In this respect the report says:

"The Puerto Rican family ties are generally as strong as those in the United States a generation ago. Any member of either the immediate or extended family who is in distress can usually count on support from other members. This is an obligation backed up by the mores of the society.

"In this context, it is understandable that 28 per cent of our total sample sends remittances to the island regularly. Among those on relief, of course, the proportion drops considerably; only 6 per cent send money back to the island. Thus 94 per cent of those on relief are in this respect not able to meet any social obligations they might have to other family members as demanded by their folkways. What this means in terms of stresses and strains within the families on relief is something on which one can only speculate.

"No relief client sends more than $5 monthly. The amount of money sent monthly to Puerto Rico by all of our cases currently on relief totals no more than $40."

In his report to the Mayor, Mr. Murtagh said that in one Manhattan postoffice on Dec. 16, 17 and 18, 1947, after the issuance of relief checks, relief clients purchased 129 money orders, totaling $1,406.50. Of the 129 orders, 118 were sent out of the country, he reported.

Defining the Pureto Rican as a person born on the island or of Puerto Rican parents, the study indicates that New York Puerto Ricans represent a selected group that is not typical of the total island population, the migrant differing in regard to literacy, rural or urban origin and occupation and skills.

The average migrant has had 5.5 years in school. Although there are no comparable figures for the island population, "there is strong indication that the New York group has had more education than the island group." The sample of migrants used in the study revealed that 7 per cent were illiterate, compared with 32 per cent in Puerto Rico in 1940. Illiteracy is negligible among New York Puerto Ricans up to 45 years of age.

Seventy-nine per cent of the 1,113 persons interviewed were born in the urban centers of the island. Just before coming to New York, 70 per cent had lived in either Puerto Rico's two largest cities—San Juan or Ponce. Only 36 per cent of the island's population in 1936 was composed of urban dwellers. The migrants, therefore, represent over twice as high a proportion of city dwellers as there are in the island's population.

Migrants More Skilled

In further distinction from the total population of Puerto Rico, the migrants include a high proportion of persons who had achieved skill or training for employment before leaving. Last year 39 per cent of Puerto Rico's labor force was in agriculture compared to 5 per cent among the migrants who had done such work. In manufacturing, 47 per cent of the New York Puerto Ricans had had experience prior to migration, while only 24 per cent of the island's labor was so employed.

A comparison of skills between migrants and the island's population follows:

Skilled workers: migrants, 18 per cent; island, 5 per cent.

Semi-skilled workers: migrants, 37 per cent; island, 20 per cent.

Unskilled workers: migrants, 25 per cent; island, 50 per cent.

Six occupational categories for the New York Puerto Ricans were used in the study and showed the following percentage groupings:

Professional—2 per cent.
Small business—4 per cent.
White collar—6 per cent.
Skilled worker—16 per cent.
Semi-skilled worker—51 per cent.
Unskilled worker—21 per cent.

The flow of migration for the study's 5,000 persons shows that 43 per cent came here before the war, 22 per cent arrived during the war and 35 per cent came after the war. The median age for the study's sample is 24.2 years. As a group they are considerably younger than the total population of the city.

A slightly higher proportion of women than men was found among the migrants, the percentages being 54 for women and 46 for men. The sex ratio in Puerto Rico as a whole is 50-50 while the urban ratio on the island is 53-47.

In respect to color, 63 per cent of the 1,113 interviewed were white; 5 per cent were Negro; 15 per cent mulatto. A mixed group known in Puerto Rico as "grifo" and "Indio" constituted 17 per cent. In the Bronx, 77 per cent of those interviewed were white, while in Harlem 57 per cent were white.

Regarding housing of the migrants, there is an average of 4.4 persons in each household. The typical Harlem apartment has 2.7 rooms, while the comparable Bronx apartment has about 3.1 rooms. For Manhattan, each room has 1.3 persons with rooms in the Bronx having 1.1 persons.

Mr. Senior said that the interviewers had been impressed by "the cleanliness and neatness" of the migrants' households, which are only slightly more crowded than the typical New Yorker's. Clean and orderly households were found even in buildings where the entrance might be littered with garbage, he declared.

Economic motivation was found to be the main reason for migrants coming here, according to the study. Other reasons included a migrant's desire to join his family, to attend schools, use hospitals and other facilities of the city.

85% Quit Jobs to Come Here

Respecting pre-migration employment, the study said:

"When only those migrants are considered who were in the labor force [working or seeking work] in Puerto Rico—that is, excluding school children, housewives, those who were too old to work, or who had already retired—we found that 85 per cent of the migrants in the labor force had quit jobs to make the trip; 15 per cent had been unemployed at the time they left Puerto Rico. The majority [71 per cent] had worked the entire twenty-four months of the two years just before they left for New York. In other words, they were not in search of jobs as such, but of better jobs.

"Ninety-one per cent of those in the labor force arrived in New York City without prior arrangements for a job. There has been some speculation lately that many Puerto Rican migrants have been brought to New York by labor contractors. We found, however, that only 3 per cent of those now in the New York labor force had come on contract; another 6 per cent had assured themselves of jobs before they left, mostly through friends and relatives."

The Puerto Rican migrant who comes to New York increases his earning power immediately, the report notes. The average weekly cash income of those migrants who came during the post-war period was $14.60 on his last job in Puerto Rico and $28.05 on his first job here.

Speaking for the Bureau of Applied Social Research, Dr. Merton said he was very pleased with the study and hoped that it could be of real use. "We are delighted with the possibility," he said, "that the Puerto Rican Government may be able to use the study. So many surveys die after the day they are completed."

Mr. Berdecia praised the project "as an objective study which goes beyond my personal expectations about the facts in connection with the so-called Puerto Rican situation." The Puerto Rico Government, he said, makes no effort to encourage or discourage migration, but is anxious to equip the prospective migrant for a successful adjustment to New York.

June 16, 1948

MIGRANTS HELPED BY PHILADELPHIA

Expansion of Puerto Rican Populace Sets Problems for City's Agencies

By WILLIAM G. WEART
Special to The New York Times.

PHILADELPHIA, Aug. 15—The Puerto Rican populace of Philadelphia, concentrated in a crowded old section, has nearly tripled in five years.

In 1954, about 7,300 Puerto Ricans lived here. Today the number has grown to 20,000, the third largest in the continental United States. New York has more than 600,000 and Chicago 50,000.

The Commission on Human Relations, an agency of the city government, cited these figures this week in a report prepared to assist health and welfare agencies in planning for the needs of Puerto Ricans.

It is trying to help them adjust to new urban ways. It said the job was difficult because of the language barrier and the conflict between their island culture and that in the city."

Speech Causes 'Alarm'

Aside from misuse of English, the report noted, Puerto Ricans speak a brand of Spanish that "alarms many purists from the old country and from parts of South America."

The ordinary migrants are often accused of speaking neither tongue well, of being "illiterate in two languages," it continued.

A special committee of educators has just completed a study of 1093 Puerto Rican children in elementary schools here. The committee found some of the children had been here five years but still had only a meager knowledge of English.

The language barrier not only hinders the children's progress in school, the committee said, but also makes it difficult to draw conclusions as to how much talent they have. The committee recommended, among other things, the establishment of a special testing service, orientation classes and the gathering of suitable instruction materials.

The Puerto Ricans are concentrated in a rundown section of North Philadelphia, bounded by Fifth, Twenty-fourth, and Spring Garden Streets and Columbia Avenue.

The commission attributed the five-year population increase to a movement to the mainland in search of jobs and better standards of living.

Unlike European migrants, the report noted, Puerto Rican migrants do not come for political or religious reasons, but entirely for economic reasons. Thus, it said, the arrivals represent families in economic distress.

However, the commission emphasized, Puerto Ricans "have difficulty in finding well-paying jobs and, for the most part, live in inadequate housing accommodations."

"The men generally work for small non union contractors, who pay wages below union levels," it said. "Nevertheless, these jobs are preferred over other types of available work which pay even less. Few Puerto Ricans work for large construction companies because they are unable to pay the union entrance fee of about $100.

"Few Puerto Ricans have found their way into the heavy industry of the city or into skilled trades, despite their marked dexterity and industriousness."

The study also disclosed that because of their language difficulties and cultural background, Puerto Ricans had not become socially acceptable to the general residents.

In a series of interviews, the commission found that 75 per cent of their neighbors considered Puerto Ricans "different."

"The word 'different' usually had an unfavorable connotation," it added.

"About half of the neighbors considered the Puerto Ricans to be hard-working, to be living in overcrowded conditions and less clean and noisier than other people in the neighborhood," it went on.

"The net result indicated that Puerto Ricans and their neighbors have little to do with each other socially."

In developing its program for the growing Spanish-speaking community, the commission is concentrating on bridging the language barrier and communicating information about city services available to Puerto Ricans. It hopes to create more understanding between Puerto Ricans and the rest of the population.

Several other agencies have established special programs to help Puerto Ricans overcome their difficulties. The Department of Public Health is operating a "parasitosis" clinic in the Puerto Rican community, as well as dental, chest and baby clinics.

The report drew these other conclusions:

"The major problems of the Puerto Rican in Philadelphia, in the order of their importance to him, are employment, housing, health and social acceptance.

"A second important factor is the concept of 'self,' which is of great importance to the Puerto Rican. This feeling is best characterized as 'dignidad,' a feeling of pride or self-esteem. The Puerto Rican tends to withdraw from any situation that he fears will expose him to pity, or censure or, especially, to contempt. This feeling will narrowly limit the places from which he will seek support or help."

August 16, 1959

SCAPEGOATS SEEN IN PUERTO RICANS

They Take Place of Negroes in Being Blamed for Social Ills Here, Parley Is Told

By HOMER BIGART

Special to The New York Times.

SAN JUAN, P. R., Oct. 15—The migrant Puerto Rican has replaced the Negro as the scapegoat for all the social ills of New York City, a Fordham University sociologist said today.

The Rev. Joseph P. Fitzpatrick, an associate professor of sociology, blamed the press, radio and television for fostering "the stereotype of the Puerto Rican as a criminal."

Father Fitzpatrick told the conference of the National Association of Intergroup Relations officials that the repeated identification of Puerto Ricans with crime news in August had had the effect of making it "now polite to speak openly of a Puerto Rican as a criminal."

Another speaker, Ricardo Meana, executive director of the Human Relations Commission of Grand Rapids, Mich., said the handling of New York crime news by wire services had created a similar stereotype in the Midwest.

For the first time, he said, the Grand Rapids police showed in their thinking a tendency to separate Puerto Ricans from the rest of the Spanish-speaking community. Previously, he explained, Midwest communities usually failed to identify Puerto Ricans as such.

Father Fitzpatrick said Puerto Rican migrants could make a "great contribution" to American social life by continuing their practice of marrying without regard to color.

A recent study of Puerto Rican marriages in six Catholic parishes of New York showed, he said, that 25 per cent of the marriages involved persons of "noticeably different color."

But there is still the danger, he added, that pressures of color discrimination might break the Puerto Rican community into two camps in the United States. Those who could pass as whites would then be assimilated by the white community and would break contacts with the darker Puerto Ricans, who would be absorbed by the Negro community, he said.

October 16, 1959

CHICAGO GOOD CITY TO PUERTO RICANS

Many Find Work and Living Better Than in New York

By DONALD JANSON

Special to The New York Times.

CHICAGO, June 2—For Luis Maymi, New York City was not the answer. Chicago was.

Like an increasing number of his fellow Puerto Ricans, Mr. Maymi found New York wanting in job possibilities and living conditions. After working in a foundry in Brooklyn and living in the Bronx, he returned home.

But his ambition to offer his daughter, an only child, a better education than the island might provide spurred him to try again. He found a good job in Chicago and sent for his wife and daughter. Since then he has seen his daughter achieve success in the field of human relations after undergraduate and graduate work at the Universities of Illinois and DePaul.

In the last decade the number of Puerto Ricans seeking economic betterment in the Chicago area has mounted steadily. Many came from New York, which still has three-fourths of the Puerto Rican population of the United States. Still more arrived on twice-daily flights from San Juan. Until the recession cut the influx to 4,000 in 1960, the net annual increase in recent years had been about 6,000.

As a result, the Chicago area is second to New York in Puerto Rican population. Small Puerto Rican communities have sprung up in Chicago, in its industrial suburbs and as far away as Gary, Ind., to the southeast and Milwaukee to the north. The migration has not yet spread to other Midwestern cities in significant numbers.

Totals Are Still Small

By comparison with the 720,000 Puerto Ricans living in New York City, the estimated total of 50,000 in Chicago and the few thousand more in near-by cities still is tiny. But since 1957 only 60 per cent of the Puerto Ricans migrating to the mainland have settled in New York, compared with the 95 per cent in the upsurge of immigration that followed World War II.

Of those going elsewhere than New York, the largest number has picked the Chicago area, where diversified industry and a great variety of jobs have been the principal magnets. A recent study of one group of Puerto Ricans here shows that they were able, on the average, more than to quadruple their annual income, raising it from $780 to $3,828.

While hotels and restaurants hire a great many Puerto Ricans, in the Chicago area the major employer is industry, from the steel mills of Gary to the foundries of Milwaukee.

The Midwest office of the Migration Division of the Commonwealth's Department of Labor here said that in normal times a Puerto Rican newcomer could be placed in a job in the Chicago area in a week or two.

The recession has thrown this timetable out of line, but since the Puerto Rican immigration ebbs and swells with economic conditions in the United States, at present there are few new arrivals to place.

Often Live With Relatives

While job hunting, the newcomer to the Midwest usually moves in with friends or relatives, as in New York. The dilapidated housing available to the migrant, sometimes at outrageous rentals for the value received, is a shock. But adjustment to new surroundings and neighbors here usually is rapid.

Once employed and adapted, and after experiencing the shafts of discrimination and suspicion visited on foreign-language newcomers, Puerto Ricans here find themselves accepted as tenants throughout the city. Although their means have limited most of them so far to rundown neighborhoods, they do not concentrate in such areas to the extent that they do in New York.

"There is no such thing as a Spanish Harlem in Chicago," said the Rev. Leo T. Mahon of the Cardinal's Committee for the Spanish Speaking in Chicago. "There is not one single block in Chicago that is all Puerto Rican."

He conceded that entrenched prejudices prevented Puerto Ricans from renting apartments in some sections of Chicago, but he described housing here as "a paradise compared with New York."

"Here it is possible for the migrant to put down two to three thousand dollars and buy a little home in the old Polish area on the northwest side," Father Mahon said.

A few Puerto Ricans have bought homes in the suburbs, but most of them live near the city's core.

Before the Puerto Rican is able to find an adequate apartment in the near-slum areas most easily available to him, he usually has experienced overcrowding, inflated rents, poor plumbing and sometimes life among cockroaches.

Seldom Make Complaints

Because of the language barrier and fear of losing even what he has for a home, the new arrival seldom presses charges of violations of the building code.

"It probably will be ten years before the bad housing conditions in Chicago can be cleaned up," said John J. L. Hobgood of the city's Commission on Human Relations.

Considerable redevelopment activity is under way, however. Five per cent of Chicago's Puerto Ricans now live in public housing projects. A year's residence in the city is the eligibility requirement.

Besides hampering job hunting and house hunting, the language barrier has handicapped the arriving Puerto Ricans in Chicago in other ways. The most flagrant example of abuse of the barrier here has been by "easy credit" merchants.

Last year William Rodriguez, 24-year-old father of four, swallowed poison and died after his indebtedness had mounted to $850. His $60 weekly paychecks had been garnisheed three times. He had misunderstood terms offered by high-pressure salesmen.

On one occasion his wife signed a "receipt" for a quilt a salesman left with her "for the neighbor next door." There were no neighbors next door. The "receipt" was a time-payment contract.

New Legislation Pushed

Mr. Rodriguez' death spurred action for remedial legislation, now pending in Springfield. Passage of reform measures, including exemption of 85 per cent of wages from garnishment, seems to be near.

A frontal attack on the language barrier was started in Chicago this year — several years too late, according to critics. The Commonwealth's Migration Office, the Board of Education and other agencies are offering courses in Spanish and in the Puerto Rican culture to welfare workers, policemen, nurses, teachers and others who work with Puerto Ricans.

Miss Carmen Maymi, daughter of proud parents and since her graduation from DePaul an interpreter and community organizer for the Commonwealth's office here, opened the first course last February. It is continuing, and others have followed. They concentrate on the words, phrases and customs the "pupils" may need the next day or any day.

"The policemen practice their Spanish on the streets," Miss Maymi said. "The Puerto Rican people are thrilled."

In recent years Puerto Ricans in Chicago complained of police brutality. The straight-from-the-island custom of congregating noisily on the streets, which lacked plazas, was "loitering" to some policemen. The language barrier did not help.

Handicaps Are Easing

The new classes amplify the effectiveness of the Board of Education's efforts to teach English to the newcomers, efforts that were hampered at first by a dearth of bilingual teachers.

Puerto Rican children generally have learned English, sometimes in supplemental classes, while keeping up with classmates in regular courses.

There is no segregation of minority groups in schools here. Everyone is assigned to the nearest school. Relations of the Puerto Rican students with both native whites and Negroes have been smooth, just as have relations among adults of these groups.

Nor has there been any notable Puerto Rican involvement in juvenile delinquency or crime here.

"The Puerto Ricans stay out of trouble," said Mr. Hobgood, Spanish-speaking human relations officer for a referral office of the Mayor's committee on new residents in the heart of a Puerto Rican neighborhood on the West Side.

"They are very well behaved. But we have few teen-agers so far and some people fear for the second generation when it comes along."

The Mayor's committee was created in 1957 to help ease the adjustment of migrants. It functions as a catalyst among the many public and private agencies set up to aid newcomers in many ways. It is staffed by professional workers from the Commission on Human Relations. The preamble of the order creating it said:

"Chicago is historically a city of opportunity. Waves of immigrants have come to it seeking political freedom and economic opportunity, learning its language, adapting culture and habit to the demands of a strange industrial complex.

"The present migration differs in that most of these new residents are already United States citizens and a good many of them know the language. These new residents, however, have a rural background and are unfamiliar with the priorities of industry, the demands of city living, the resources a city makes available to them."

Clarence Senior of New York, a leading consultant in migration studies, said recently that Chicago deserved first place as a city that had tried to educate newcomers in ways of living in a metropolis. He praised the work done by the Mayor's committee.

An Easy Place to Live

Miss Maymi, who completed high school here before going away to college, agreed.

"Chicago is an easy place for Puerto Ricans to live," she said. "Agencies are well organized to help them. No public facilities are denied them. There is no general prejudice against Puerto Ricans."

Puerto Ricans have adjusted to the new environment without strife. Moreover, although comprising little more than 1 per cent of Chicago's population, they have begun to make contributions to the community.

The Puerto Rican Institute of Culture was formed and has brought music and drama to Chicago from the home island. Chicago schools have graduated Puerto Rican doctors, lawyers, policemen and entrepreneurs who have established small busisses of several kinds.

"Of all the foreign-language groups that ever migrated here," Father Mahon said, "the Puerto Ricans integrate fastest."

June 4, 1961

PUERTO RICAN BID ON RIGHTS IS MADE

By LAYHMOND ROBINSON

Puerto Rican leaders are establishing a citywide civil rights organization to spearhead their people's fight against discrimination and segregation.

The organization will be the first of its kind representing the city's 650,000 Puerto Rican residents. Its aim is to provide a powerful, central voice for the group in economic, social, educational and political affairs.

The formation of the organization will be announced in the next several weeks.

As outlined by Joseph Monserrat, one of its principal organizers, the new group will draw under its wing 200 to 300 existing civic, civil rights, fraternal, educational and social organizations scattered through New York's Puerto Rican community.

"We are trying to develop the kind of civil rights organization that does not exist anywhere —one that is representative of the entire group," Mr. Monserrat said.

"We have many organizations doing different things," he went on, "but they don't speak for the whole community."

Mr. Monserrat is chief of the migration division of the Puerto Rico Labor Department here. As such he serves as a troubleshooter on problems affecting the Puerto Rican community.

Action by Majority

The new organization will not replace any Puerto Rican organization now in operation, Mr. Monserrat emphasized. Instead, it will serve as a clearing house and principal spokesman for them when they wish it to act as such by majority vote.

"What we are trying to do," said Mr. Monserrat, "is to pull these diverse groups into a citywide organization. We want a mechanism that will enable us to say that when this organization speaks it is a cross-section of the Puerto Rican population."

He noted that New York's Puerto Rican population, which makes up about two-thirds of all Puerto Ricans on the United States mainland, has sometimes been accused of shying away from a militant civil rights position. Much of this criticism has come from the Negro civil rights groups.

"If the only way people are going to listen is for us to start demonstrating then we'll have to do it," Mr. Monserrat declared.

He contended, however, that Puerto Ricans had been "extremely militant" in many areas, particularly in campaigns to get more of their people on the voting rolls.

"I don't believe any other group has done as much as we have in so short a time," he declared. "In 1953 we had only 30,000 Puerto Ricans on the voting rolls in the city. Through the most diligent effort, we now have somewhere between 150,000 and 200,000."

Special Problems

While some of the Puerto Rican's problems of adjustment and acceptance in New York are similar to those of Negroes, Puerto Rican leaders say their people have special language, cultural and other considerations that make creation of their own civil rights organization essential.

For example, many Puerto Rican leaders contend that one of the major barriers to voting by their people is the state law that requires facility in the English language to qualify for the ballot.

They estimate that half of the adult Puerto Ricans in the city are disqualified because they cannot pass the English literacy test, but could, if the law were changed, pass it in Spanish.

Permitting Puerto Ricans to take the test in either English or Spanish will be a major objective of the new organization. Other objectives will be to represent the Puerto Rican community in cases involving charges of police brutality and in matters affecting school and housing integration.

Representatives will be selected from existing groups. One of the problems in establishing the new group, which has been under discussion for several months, has been the determination of how the votes shall be distributed among member organizations.

Organizational meetings have been held at the migration division of the Puerto Rican Labor Department, at 322 West 45th Street.

January 3, 1964

Latins Have a Label For Their Uncle Toms

Puerto Ricans have a name for members of their community who are counterparts of the individuals Negroes call "Uncle Toms."

Gilberto Gerena-Valentin, a member of the steering committee of the National Association for Puerto Rican Civil Rights, said his people called such persons lagartos, or lizards.

Mr. Gerena-Valentin brought up the label yesterday on a WCBS radio interview when told that a Negro clergyman had said "the Negro should improve his own image before complaining about civil rights" and had added that school boycotts were "taking the bread out of children's mouths."

February 24, 1964

POVERTY CALLED CHIEF RIOT CAUSE

By HENRY RAYMONT

Poverty—not racial antagonism—is the leading cause of urban violence, an authority in Puerto Rican affairs asserted yesterday.

Joseph Monserrat, director of the migration division of the Department of Labor of the Commonwealth of Puerto Rico, said that events of recent months had proved that the "major detonators" of unrest in disadvantaged neighborhoods were unemployment and inadequate housing.

In an interview, he cautioned against the insistence of many Federal and municipal officials and of social scientists that race and cultural conflicts were the chief cause of tensions.

"The real problem is that everybody wants to simplify everything by blaming urban tensions on minorities and racial frictions," Mr. Monserrat said.

Very often outbursts of violence which some police officials consider to be racial flare-ups really originate from hostility against slum owners and thoughtless employers, without any racial overtones, he maintains.

"It should be more properly called a revolt and rejection of poverty and inhumanity to which the element of race is incidental," he said.

Mr. Monserrat, a 41-year-old sociologist with a husky frame and prematurely grey hair, has been waging a vigorous campaign for better understanding of the economic and cultural pressures that confront New York's 730,000 Puerto Ricans. He has headed the migration office at 322 West 45th Street for almost a decade.

Many city officials and civil rights leaders regard him as one of the most knowledgeable spokesmen for the Puerto Rican community. His views are often solicited by city labor and welfare agencies and by university and public affairs study groups.

In the early 1950's when migration from Puerto Rico was at a peak, Mr. Monserrat worked closely with Gov. Luis Muñoz Marin and Dr. Clarence Senior, a noted sociologist, in trying to bring qualified personnel into the New York City government to help the newcomers.

Among other things, they brought hundreds of New York teachers, police and welfare officials to the island to equip them with a better understanding of the cultural background of Puerto Ricans.

His experience in recent years has convinced Mr. Monserrat that one of the great problems confronting America's urban industrial centers "with their anonymity, their mechanized ways" was how to absorb groups of people without destroying human values.

Although he attributes some of the recent outbreaks in East New York and Chicago to job and housing inequities, Mr. Monserrat has devoted much effort to explaining the problems confronting Puerto Ricans in cultural terms.

At a recent panel discussion by civil rights leaders held at the Harlem Y.M.C.A., Mr. Monserrat bitterly scored allegations that a possible factor of violence was rivalry between the Negro and Puerto Rican communities.

"One of the problems which does create racial tension," he angrily told the panel moderator, "is making of myths sound somehow a fact and a reality."

Amid applause and cheers from the audience, made up mostly of Negroes, he went on:

"We Puerto Ricans, as a race, are the only really integrated group in the city. We are black, and we are white, and we are mixed. Our problem in this society is to be able to maintain the rights and the liberties that we've enjoyed in Puerto Rico, here."

In closing, Mr. Monserrat told a story of a fight between a young Negro and a Puerto Rican boy while he wasrunning a Settlement House in East Harlem.

When one of the social workers went over to separate the youngsters, one of them turned around and asked, "Don't you teach us that we ought to live together?" The social worker looked perplexed and said "Yes."

"And don't you teach us that we ought to work together?" the boy asked. Again the social worker replied affirmatively.

Then, Mr. Monserrat recalled, the youngster finally asked: "Well, if we fight together, is that a race riot?"

August 21, 1966

NEW BOOK SCORED BY PUERTO RICANS

By PAUL HOFMANN

Puerto Rican civic leaders here clashed in a debate yesterday with the author of a new book that starkly portrays a slum family in San Juan and New York.

"It's not a pretty picture of life in the slums," conceded the writer, Oscar Lewis. "It may trouble a great many people." He said that even social workers active in urban pockets of poverty did not really know what was actually going on in them.

The Puerto Rican critics of the book, "La Vida," published by Random House, did not question the authenticity of its taped interviews with slum dwellers. However, they expressed fear that the family presented by Dr. Lewis would be taken as typical Puerto Ricans by readers in the United States, and thus add to prejudice.

"I used to be called a spick," remarked José Montserrat, who heads the migration division here of the Commonwealth government's labor department. Now, he said, he isn't. Instead, "I can be called culturally deprived," Mr. Montserrat said.

Those who were thus describing Puerto Ricans with a "scientific euphemism," he added, overlooked the fact that New York "has no second-generation Puerto Ricans" yet. Thus they have not had the chance to prove that their people have the capacity to emerge from poverty as the Irish, Jewish and Italian immigrants had before them, Mr. Montserrat said.

Mr. Montserrat and other participants in yesterday's debate at a luncheon in the auditorium of the Time and Life Building also protested that "La Vida" gave the impression that Puerto Ricans were preoccupied with sex.

"Do you think Puerto Ricans are more indecent than other people?" Samuel Quiñones Jr., son of the president of the Puerto Rican Senate, asked.

"La Vida" is Spanish for "the life," but also may be understood as "prostitution." Dr. Lewis' book reveals that several women in the family on which it focuses were at one time or another prostitutes in Puerto Rico or New York. The book also shows what the author calls the brittleness of family life in the subculture of poverty—the five major characters have had a total of 20 marriages, seven of which were consensual and only three legal.

Nevertheless, in an introduction, Dr. Lewis, an anthropologist, says he is impressed by the strength of the family he examined, "their fortitude, vitality, resilience and ability to cope with problems" that would paralyze many middle-class people.

Dr. Lewis retorted that Puerto Rican authorities were subordinating urban renewal to the interests of tourism and real estate operator

The Puerto Rican leaders charged the author with "biased science." They said he completely ignored industrialization and other social-economic processes that had brought many islanders from rural to urban areas. They also said he glossed over the fact that "Puerto Ricans didn't make the New York slums—they found them."

"There was nothing scientific about publishing verbatim transcripts of interviews with individuals belonging "to the worst case history of the Welfare Department of Puerto Rico," Mr. Montserrat asserted.

"You got it all confused," rebutted Dr. Lewis. The author insisted that he had not meant to equate poverty with delinquency, and that he hoped his work would promote better understanding between the middle class and the poor along with a realization of how inadequate current antipoverty programs were.

November 16, 1966

City's 2d-Generation Puerto Ricans Rising From Poverty

By PAUL HOFMANN

The second-generation Puerto Ricans are entering New York's scene in force, and their dramatic growth in number is accompanied by significant social advances.

There are at least 250,000 of them, one-third of the city's total Puerto Rican population of 765,000. There will be many more in the years to come, for the average Puerto Rican family remains prolific.

The Puerto Rican population of New York has increased 10 times since World War II, and more second-generation Puerto Ricans than ever before have reached adulthood and joined the labor force since the 1960 census.

Dearth of Statistics

There will be much guesswork about the second generation of the city's youngest large minority group until the next census in 1970, and since statistics are scanty at present any assessments are controversial. But interviews with more than 40 second generation Puerto Ricans and experts have brought out the following trends:

¶A lingering identity crisis that perplexes them on whether their values are urban American, or traditionally Latin.

¶Their accelerating rise from poverty to middle-class conditions.

¶Their ambivalence toward Negroes—sometimes seen as allies, sometimes as rivals.

¶Their new militancy under a young, slum-bred leadership.

However, Puerto Rican militants are not extremists and do not openly advocate violence. They have kept out of the recent wave of racial outbreaks following the assassination of the Rev. Dr. Martin Luther King Jr.

"We've learned a lot from the black-power boys," a young Puerto Rican slum leader said. "But we're different. We are more polite, even if we do some saber-rattling."

Every Friday night in a basement at 158 East 103d Street a group of second-generation Puerto Ricans grapples with the problems of growing up, and living, in New York. The group, the Citizens Committee for Unity, one of scores of new Puerto Rican grassroots organizations, meets to discuss how the neighborhood can be improved and its residents aided.

Why do they attend the meetings?

"This block is my home; I know nothing else," said George Vega, a 14-year-old student and native New Yorker.

As he spoke, his father, Roberto Vega, an aide at Public School 83, 219 East 109th Street, and vice president of the committee, was discussing neighborhood problems. The president is José Quiñones. Maria Martínez, a secretary, is a regular at the Friday night meetings, as are Delia Martínez, Rafael De Gracia and Cesar Rodríguez, young people who all live in the block, and scores of others who were born in the neighborhood or have been living there ever since they were small children.

Changing Image of Home

Unlike earlier waves of immigrants, the Irish, the Jews and the Italians, the Puerto Ricans are not likely to become fully assimilated in their second generation. The chances are that large segments even of their third generation will remain a bilingual and bicultural group.

At the same time, most of the second-generation Puerto Ricans do not share with many of the city's half million first-generation Puerto Ricans the dream of going home some day; to most of the second generation home isn't a dusty plaza on a tropical island, but a New York block where smells of fried fish and hamburgers mingle with the Latin beat drifting from the corner record store —like 103d Street between Third and Lexington Avenues.

A factor in the difficulty to assimilate completely is the air fare—as low as $45 one way—between the mainland and Puerto Rico. This contributes to what has been called the "commuter status" of New York's Puerto Ricans. A million and a half persons traveled between Puerto Rico and the American mainland, most of them to and from New York, last year. The ease of "commuting" makes assimilation here less urgent for all Puerto Ricans than it was for earlier immigrants.

For a fraction of Puerto Ricans, including their second generation, there is another handicap to assimilation: color. Puerto Ricans are statistically always lumped together regardless of the fact that they range from white to very dark.

However, the vast majority of Puerto Ricans on their island—and in New York—are light skinned, and this can help in the long run to bridge the cultures and bring assimilation.

Second-generation Puerto Ricans and many thousands of others who came here as children are known as Mike and Manny rather than Miguel and Manuel. Some speak genuine Brooklynese, though more often they retain a soft accent.

"I am proud of my accent, I don't want to drop it," says Miss Carmen Olivieri, the leader of a slum human-relations group who was born in New York and educated in Puerto Rico and here.

Puerto Rican children switch easily from one idiom to another as they do their homework, talk to parents and friends, and act as interpreters in family dealings with welfare workers and other officials. When they visit the old island, they often discover that they are treated, and feel, like strangers.

"When I first saw Puerto Rico, it was a culture shock," says Bob Ortiz, a 26-year-old radio announcer. "It was so very rural. To me, it's not an enchanted island." He spent six days there, visiting relatives, and has no desire to return.

He says that although he ardently hated the welfare caseworkers who used to come to see his parents here, he feels now pretty much assimilated, doesn't read Spanish-language newspapers, and prefers dating girls who aren't Puerto Ricans.

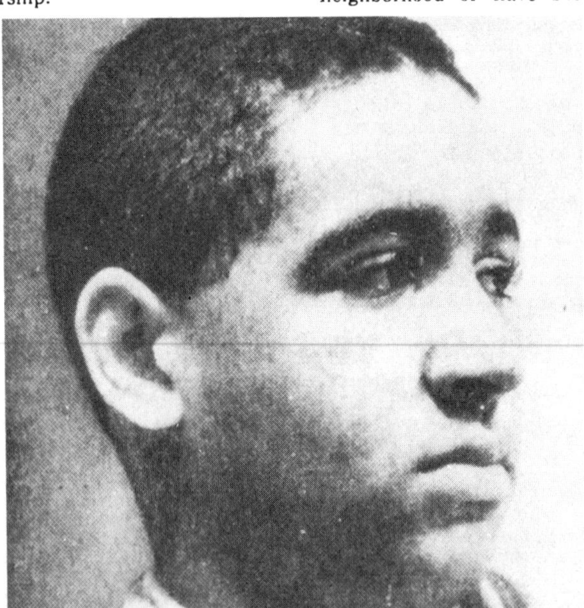

Cesar Rodriguez, 14, is one of the youngest in the group

Rafael De Gracia attends organization's weekly meeting

Yet, the college-educated Mr. Ortiz remarks, "most Puerto Ricans feel they have no place to go." Gilberto Gerena-Valentín, a veteran Puerto Rican civil rights leader and chief of the Hispanic section of the city's Human Rights Commission, puts it this way: "Second-generation Puerto Ricans find themselves in a very awkward position — they don't know where they belong."

Consequently, they cling to part of the old culture while they reach out for the new. Current research tends to support the view that Spanish is not doomed in New York, as Yiddish and Italian seem to be. A Yeshiva University psychologist, Dr Joshua A. Fishman, predicts that "Puerto Ricans won't disappear into the American melting pot as early immigrants did."

Notes Political Power

Recording speech patterns of Puerto Rican college students in a current project, Dr. Fishman and his team of researchers noted that the youths were communicating in different "languages"—informal Spanish, a mixture of street English and Spanish, informal English and college English—with their families neighborhood friends and intellectual peers.

To a leading authority on Puerto Ricans, the Rev. Joseph Fitzpatrick of Fordham University, the identity crisis of the latest large group of newcomers to the city will be solved by what may be called Puerto Rican power.

They will integrate from a position of strength," Professor Fitzpatrick says in an article written for the International Migration Review, a journal published in Staten Island. "But the strength will rest not on the continuation of a traditional culture in the form of an immigrant community, but on the solidarity which results from organizing their efforts for the pursuit of group interests in the political arena."

The example of the black power movement is a central factor in the Puerto Rican experience here, according to Professor Fitzpatrick. "The aggressiveness and success of the black citizens in antipoverty programs has resulted in realization among Puerto Ricans that they must do likewise," the Jesuit sociologist writes.

Professor Fitzpatrick points to "great, open hostility" between Puerto Ricans and Negroes over control of antipoverty programs. ("We are at each other's throats fighting for the crumbs of antipoverty funds," Mr. Genera-Valentín remarked.

Photographs for The New York Times by LARRY MORRIS

George Vega observing weekly citizens committee meeting recently. His father, Roberto, is the vice president.

The Jesuit notes in his article that the Puerto Ricans came to New York just when automation started eliminating many jobs that were traditional channels of immigrant advancement, and when government was providing a wide range of new public services, from housing to welfare.

Weekly Meetings

Jack Agueros, a 33-year-old college graduate who may well epitomize the second-generation Puerto Rican militant, said the other day: "We have to share everything. We share poverty with the Negroes, religion with the Irish and Italians. We don't even have our slums for us alone."

Mr. Agueros has just been named a deputy comissioner of the city's Community Development Agency, an arm of the Human Resources Administration. Last month another Puerto Rican community leader, Manuel Diaz Jr., was appointed as a deputy commissioner in the same superagency. The two men worked together earlier in the Puerto Rican Community Development project, a major antipoverty agency. Mr. Agueros, Mr. Diaz and a dozen or so other Puerto Ricans have for some time met every Wednesday night, each time in another place.

The group calls itself the Citywide Puerto Rican Action Movement. It is radically different from the several hundred Puerto Rican associations here, whose activities range from dominoes to civil rights. It couples militancy with underground techniques, organizes demonstrations, and plans political strategy.

"It's a kind of informal think tank," Mr. Agueros explained. He said that among the new second-generation leaders "there's practically not one guy who hasn't been in a street gang."

The city called on the new breed of Puerto Rican leaders last July when East Harlem, which Puerto Ricans call El Barrio (the neighborhood), boiled up into street violence. Mayor Lindsay met at Gracie Mansion with a group led by Arnold Segarra, a 24-year-old community organizer who developed youth programs for East Harlem and is now working in the Mayor's office. He, too, is a member of the Wednesday night circle.

To Mr. Gerena Valentin, who is 53 years old, and doesn't belong to it, the new generation's leadership may expect to become "very powerful." He says: "They're real professionals. They have been associated with black people in the community, they believe in social change."

April 23, 1968

PUERTO RICANS REMAIN IN BLOCS

Study Also Finds Negroes Enlarge Their Smaller Areas

By WILL LISSNER

The notion that Puerto Ricans are rapidly dispersing throughout the city is a myth, according to a population analyst who has studied the growth of Negro and Puerto Rican communities in New York City since 1900.

The demographer, Prof. Nathan Kantrowitz of Kent State University, who made the study for the American Geographical Society, reports that "on the contrary, Puerto Rican settlements appear almost without exception as foothills to Negro mountains."

Contemporary Negro settlements in New York City "are only enlargements of long-standing ones," Professor Kantrowitz found in mapping Negro and Puerto Rican population movements into the city.

While many Negroes are newcomers to the city, their settlements, even in suburban places such as Queens, have as much a history as the settlements of white groups of similar ethnic backgrounds, he reports.

New York's Negro community, he said in a report just published by the society, has spread from its major center in Manhattan's central Harlem to coalesce and enlarge its smaller centers throughout the city.

3 Puerto Rican Areas Cited

The growth of the Puerto Rican community since World War II has followed the same pattern, he said. Typical of the process are three separate Puerto Rican areas in Brooklyn. These, he finds, are actually extensions, on three sides, of Negro Bedford-Stuyvesant, north of Prospect Park.

Mapping the distribution of the Negro and Puerto Rican residents of the city at different times since the turn of the century shows "that New York's numerous Puerto Rican settlements are established on the fringes of Negro areas," Professor Kantrowitz reports.

"This pattern of overlapped edges" between the Negro and the Puerto Rican settlements, he declares, "persists even where the Puerto Ricans are widely dispersed," as in the outlying areas of Queens.

This leads him to suspect that nonwhite Puerto Ricans, who are believed to be less than 10 per cent of the total, "may link Negro and Puerto Rican settlements."

Limited data show, Professor Kantrowitz believes, that central Harlem, for the Negroes, and the El Barrio neighborhood in East Harlem for the Puerto Ricans, are not communities in themselves. He contends that the poorer sections of communities that spread throughout the city, with the more affluent members living in the more outlying parts of the residential boroughs.

Migration Noted After War

As New York's population grew from 3.4 million in 1900 to 7.8 million in 1960, its Negro population swelled from 1.8 to 13.6 per cent. Only after World War II did large scale migration of Puerto Ricans begin.

Not until after World War II did any appreciable decline in the concentration of Negro populations in the city take place, Professor Kantrowitz reported. After the depression of the nineteen-thirties, Puerto Rican dispersion paralleled the Negro dispersion, spreading first into the Bronx and Brooklyn and then into Queens, he added.

December 21, 1969

Hospitalization Rate of Puerto Ricans for Mental Disorders Said to Be High

An "abnormally high" rate of hospitalization of Puerto Ricans for mental disorders in New York State was reported yesterday, along with a warning that "the high rates may be the result of intercultural misunderstanding."

First-time admissions for schizophrenia to the state's mental hospitals — excluding those for the criminal insane — showed a rate for Puerto Ricans in proportion to population that was almost triple that of the general population: 102.5 for each 100,000 Puerto Ricans in contrast to 34.5 for the state as a whole, as of 1967.

The Rev. Dr. Joseph P. Fitzpatrick, Fordham University professor of sociology and a long-time student of Puerto Rican affairs, and Dr. Robert E. Gould, Bellevue Hospital chief of psychiatric adolescent services, offered the data but they suggested that some such Puerto Rican cases might be only a "temporary disturbance due to a convergence of problems of the poor."

Give Hospital's Data

If so, they said that the remedy might be "neither hospitalization nor clinical treatment, but an effective attack on the complexity of problems in the poor person's life."

They cited a special Governeur Hospital unit in behavioral sciences as having reported 891 cases, largely Puerto Rican, under care in the first half of 1967, but only four of these needing hospitalization.

Father Fitzpatrick and Dr. Gould made their report yesterday to the American Orthopsychiatric Association, which is holding its 47th annual meeting in San Francisco. They also released a report based on a task force study for the Joint Commission on Mental Health for Children in Chevy Chase, Md.

The State Mental Hygiene Department's data, they said, showed 6,317 first-admissions for schizophrenia from the general population in the fiscal year ended June 30, 1967. These were 3,223 men and 3,094 women, for a rate of 36.6 men and 32.6 women for each 100,000 population, based on Census Bureau estimates.

Puerto Rican Figures

Within this total, there were 769 persons, or 12.2 per cent, of Puerto Rican birth or parentage. These were 445 men and 324 women, for rates of 122.0 for men and 84.2 for women for each 100,000 people, based on population estimates of the Puerto Rican Migration Division.

For all mental disorders, Father Fitzpatrick and Dr. Gould reported 255.8 Puerto Ricans for each 100,000 of the population admitted to the state mental hospitals. They held this to be a considerable rise since an indicated annual rate of 157.7 for Puerto Ricans in New York City reported in a 1956 study by Benjamin Malzberg in The Journal of Nervous and Mental Disease.

They also noted a 1965 report by Lloyd H. Rogler and August B. Hollingshead that found schizophrenia to be the most common mental illness among 1,500 Puerto Rican patients in the public mental hospital in Puerto Rico. This study had estimated that there will 10,000 more psychotics living at home. The island's population when the study began in 1958 was 2.3 million.

Cultural, migration and poverty problems have been among explanations offered for such cases, Father Fitzpatrick and Dr.Gould said. They cited:

¶Male anxieties about masculimty or "machismo" and female "martyr complex or excessive fear resulting from cloistering."

¶Mother-son relationships, leading to a son's difficulty in overcoming dependence on women or establishing a love relationship with a wife, and to a woman's depression over "an idealized female role."

¶Patterns that may "generate hostility against authority figures."

¶A "keen sense of personal dignity" that "may build up a potential for violence."

¶An "overwhelming burden" of poverty.

¶Stress from migration, including uprooting, adjustment to a new way of life and intergenerational conflict.

Describe a Syndrome

The two analysts described a "specific form of hysteria," known as "the Puerto Rican syndrome" or "the ataque," as a "tendency to resort to a hyperkinetic seizure at a time of acute tension or anxiety."

They noted that Dr. Beatrice Berle, in a 1959 study of 80 Puerto Rican families, had indicated that the ataque was "a culturally expected reaction to situations of serious stress, and that Puerto Ricans manage it as an ordinary occurrence."

She further indicated, they said, that "for someone not familiar with the cultural background, it could be interpreted as a symptom of more serious mental disturbance."

Father Fitzpatrick and Dr. Gould observed that "those in control of the mental health facilities determine the definition, the methods and condition of treatment."

They expressed concern over "the acute shortage of fluent Spanish-speaking personnel and auxiliary staff" in New York City mental health clinics and facilities. "Lack of communication is a continual hazard to the Puerto Rican," they said.

March 26, 1970

LACK OF UNITY LAID TO PUERTO RICANS

A Fordham University sociologist says New York City's Puerto Ricans have found it difficult to achieve "community solidarity," but suggests they may work out adjustment "in very new ways" differing from those of past immigrants.

The sociologist, the Rev. Joseph P. Fitzpatrick, says he foresees strength growing from new "militancy about their interests in antipoverty programs, education, public welfare and housing" and then organizing "in the political arena."

The latest of Father Fitzpatrick's many studies of Puerto Rican affairs is being published by Prentice-Hall as a 192-page book, "Puerto Rican Americans: The Meaning of Migration to the Mainland."

Despite the Puerto Ricans' concentration in some geographic areas, Father Fitzpatrick says the disruptions of city renewal programs have caused frequent moving. Coupled with the integration policy of public housing, he says this has hampered growth of strong neighborhoods such as those of past immigrants.

"The proximity of the island and the ease of return," he says, "seem to prompt the Puerto Ricans to find in the island the sense of strength, support and identity that former immigrants found in the clusters of their own kind in the immigrant communities of American cities.

"There is a great deal of truth in the comment that this is not a Puerto Rican migration but a porcess of Puerto Rican commuting."

Generation Gap Noted

Many aspects of the Puerto Ricans' cultural background, Father Fitzpatrick says, "are not well adapted to modern styles of confrontation, protest and power." The school, teacher and judge in Puerto Rico, he says, have been regarded as extensions of the home and family, not targets for demonstrations.

The "militant action which may give the youth a sense of identity and unity behind a common cause may alienate many other people in the Puerto Rican community" of the older generation, he adds.

Nevertheless, Father Fitzpatrick reports "impressive improvement" in occupation levels among second-generation Puerto Ricans here.

Father Fitzpatrick says "migration and assimilation are processes which regularly involve unrest, conflict and hostility."

The Puerto Rican migration, he says, is the first great airborne arrival, the first by people from a different cultural background but still citizens, the first from a tradition of "widespread intermingling and intermarriage" of different colors.

The newcomers, he says, find automation eliminating "by the hundreds of thousands" the jobs that once were channels for immigrant advance.

The great majority of Puerto Ricans, he goes on, are Roman Catholics, but integrated parishes—rather than Puerto Rican ones—eliminate the identifying community role of former ethnic parishes.

New York City's one million Puerto Ricans, Father Fitzpatrick says, are mainly "poor working people, the backbone of the labor force for hotels, restaurants, hospitals, the garment industry, small factories and shops, without whom the economy of the city would collapse."

Only 6 Puerto Rican Priests

He says Puerto Ricans constitute about one-fourth of the entire public school population and represent half the membership of the New York Archdiocese. The city has only six Puerto Rican priests, however. Twenty-five per cent of eligible Puerto Rican voters are registered, but none holds an elective office in the city government.

Although only about 15 per cent of the city's population, Puerto Ricans constitute perhaps 40 per cent of welfare recipients in the Aid to Dependent Children category, Father Fitzpatrick says. By relating 1968 estimates, he accordingly suggests 35 per cent of 800,000 Puerto Ricans then here might have been on welfare.

"All this adds up to a struggling, suffering, poor, but vital segment of the community," he says. He pictures it as highly dispersed, without "the level of sophistication in organizational activity" of older groups, and "troubled by internal divisions and controversies."

"The community is continually losing experienced persons to the island in return migration, and is replenishing its poorest ranks with newcomers from the island," Father Fitzpatrick says.

In cultural background, he reports "a deep sense of family obligation," a tradition of superior authority of the man, an acceptance of consensual marriage, a recognition of children born out of wedlock without stigma, an emphasis on the importance of the person, thinking "in terms of transcendent qualities such as justice, loyalty or love" rather than materialism, and a fatalism that "softens the sense of personal guilt for failure."

September 12, 1971

Police Change Designation

The Police Department announced yesterday that in the future it would refer to Puerto Ricans and Spanish-speaking peoples as Hispanic. The department will no longer use the term Puerto Rican when referring to people of Spanish origin.

March 30, 1972

U.S. CALLED UNFAIR TO PUERTO RICANS

Special to The New York Times

BOSTON, Feb. 26—A report to the United States Commission on Civil Rights contends that antipoverty and Model Cities programs are not equitably delivering service to the Puerto Rican communities in Boston and Springfield, Mass.

It warns that black and Puerto Rican minorities could become adversaries because of what is called shortsighted actions by agencies administering the programs.

The report, issued Thursday, was the first of four state advisory committee reports to be published in a study by the Civil Rights Commission of problems facing Puerto Ricans in Eastern cities.

"Most of the agencies' programs are directed to the black community; the Puerto Rican communities' needs are seldom considered," the committee said.

Dividing the Poor

The report said the Spanish-speaking community was the fastest-growing minority population in the Boston Model Cities area and would achieve majority status within five to eight years if present trends continued.

"It was a sad day in the history of American civil rights when the Populist movement at the turn of the century was destroyed by racial bigotry," the report said. "The attempt to divide poor whites from poor blacks by appealing to man's prejudices was eminently successful in that era.

"It would be equally sad today if the black and Puerto Rican minorities became adversaries rather than allies because of the shortsighted actions of antipoverty and Model Cities."

The report said the Federal Government had been guilty "to some degree" by failing to develop projects "designed by and for the Spanish-speaking community and by failing to pressure local programs into making such decisions."

Blacks Exhorted

However, the report went on, blaming the Government tends to shift the spotlight to Washington and away from the black and Puerto Rican poor in the neighborhoods.

"If equality is to become a reality," the report said, "minorities are the first who should espouse it. The black directors and staffers should examine their positions and then open the benefits of these programs and the power to make policy to their Puerto Rican neighbors.

"The results may well be that all concerned will have to share an already small pie. But the alternative of no pie or a racially divided pie would, in the long run, be no boon for either blacks or Puerto Ricans."

Washington Blamed

Gabriel Guerra-Mondragon, staff member of the United States Civil Rights Commission and director of its Puerto Rican project, said the Massachusetts report, and those yet to be published by advisory groups in New York, Connecticut and New Jersey, would be used by the commission in preparing its report on Puerto Rican problems.

He said the commission report would be presented to President Nixon and Congress in the fall.

Mr. Guerra-Mondragon said the major responsibility for the asserted lack of response to the Puerto Ricans' problems rested with the Federal Government.

"It doesn't lie at the local level," he said in apparent disagreement with the Massachusetts advisory committee.

February 27, 1972

A New Study Details Puerto Rican Role

The New York Times/Fred R. Conrad

Arturo Rivera teaching his students Puerto Rican history at the Carlo Tapia School in Williamsburg

By DAVID GORDON

The generation gap seems to be as strong among Puerto Ricans in Brooklyn as among other ethnic groups, a new study indicates. Older Puerto Ricans, who toiled in factories, on the waterfront, in sweatshops on Pacific Street and Atlantic Avenue, and in hotels and restaurants, say the younger Puerto Ricans do not have the same feeling for hard work as did the early arrivals.

This is one of the impressions drawn from the first history of the Puerto Ricans in Brooklyn, which has just been completed by Prof. John D. Vazquez of New York City Community College. Another impression in the study is that, despite the alleged attitude toward work, considerable progress is being made by the Puerto Ricans.

From the taped history, now containing more than 70 personal interviews, there emerges a story of a struggling immigrant group that has managed to adapt to another culture and language without losing its own identity.

There are now 271,769 persons of Puerto Rican birth or parentage living in Brooklyn, as against a citywide total of 1,300,000.

"The retired individuals think the young people do not like to work as hard," Professor Vazquez said. "They also think that social services have spoiled some of the young people and the newest Puerto Rican arrivals in the United States."

However, the retired Puerto Ricans think advances have been made. "My impression is that the early Puerto Ricans set a very positive pace for the Puerto Ricans who came after them," he pointed out.

"They set a direction and an attitude, which, to some extent, opened doors for Puerto Ricans who sought work. Remember, the earlier arrivals came predominantly from rural, agricultural backgrounds."

Most Came After 1923

Although some went to Brooklyn after United States citizenship was granted to Puerto Ricans on March 17, 1917, most of the early arrivals came after 1923, Professor Vazquez said.

Pier 35 in Red Hook, near the former site of the Hamilton Avenue ferry, was the first "Puerto Rican bridgehead" in Brooklyn. The pioneer Puerto Ricans arrived on ships there and got their first look at Brooklyn.

Professor Vazquez said the Puerto Ricans gathered in downtown Brooklyn, living in houses on Johnson, Adams, Columbia, High and Sands Streets.

Others moved to DeGraw, Sackett and Union Streets, and some, eventually, went to the Bay Ridge area between 39th and 60th Streets, hugging the waterfront districts between Second and Third Avenues.

"The men came first and were unskilled," Professor Vazquez said. "They looked for factory jobs. After World War II, they started moving into the garment industry.

"The early arrivals, prior to World War II, got jobs in such places as a battery factory on Pearl Street and a candy factory on York Street. A few later managed to open small bodegas, or grocery stores, which have been social institutions where people gather, and play games like dominos."

In 1928, he said, there was a social club in East Harlem called Hijos de Salinas (Sons of Salinas). Salinas is a town in the southern part of Puerto Rico. Political social clubs were also formed in Brooklyn at that time, aided by the late James Kelly, borough historian. These were the Aqueybana, De Hostos and Betances clubs. Most of the Puerto Ricans in the nineteen-twenties and thirties were Democrats, Professor Vazquez added.

Professor Vazquez singled out Carlos Tapia and Luis Felipe Weber as Brooklyn folklore figures in the history of the Puerto Ricans. Tapia, a powerful, 250-pound man, standing 6 feet 3 inches, worked on the Brooklyn docks and later owned a restaurant.

"He helped to clothe, shelter and feed many early arrivals," he said. Public School 120 in Williamsburg was named after Tapia, who died in 1945.

Weber, who came from Mayaguez, was another legendary paternal figure. With Tapia, he distributed food baskets regularly during Christmas and Thanksgiving, and frequently staked needy Puerto Ricans with hard cash.

Professor Vazquez interviewed another leading figure in the Puerto Rican history in Brooklyn, Sister Carmelita, who, he said, was the first Puerto Rican nun in New York City.

"Sister Carmelita served as the liaison between the early arrivals and the church, and helped them get oriented to their new surroundings," Professor Vazquez said. "She took mothers with their children to Cumberland Hospital to get preventive shots, found homes for them, and acted as interpreter. Sister Carmelita came from St. Peter's Church, Warren and Hicks Streets."

The problems of the Puerto Ricans are still severe. A regional report in 1971 called the Puerto Ricans "the most deprived of all workers living in the city's major poverty neighborhoods." It added that 50 per cent of Puerto Rican families have less than $5,000 annually, and that one out of every three Puerto Rican families was headed by a woman.

The oral history was started in 1972 when the Long Island Historical Society, which is supervising the project, obtained a grant from the State Council on the Arts. James Hurley, executive director of the society, said that a second grant was obtained in January, 1974, and Professor Vazquez was then asked to continue the history.

June 29, 1975

PUERTO RICAN PLIGHT IN U.S. IS DEPLORED

Rights Panel Says Percentage in Poverty Rose From 29 to 33 in Five Years of Study

By DAVID VIDAL

Special to The New York Times

WASHINGTON, Oct.13—After 30 years of significant migration, the poor socioeconomic position of the 1.7 million Puerto Ricans on the United States mainland remains largely unchanged and in some respects is even worsening, and the prospects for future improvement are "uncertain," according to the United States Commission on Civil Rights.

In the first national study of Puerto Rican immigrants undertaken by a Federal agency, the five-member commission found that "a dismayingly high percentage of Puerto Ricans are still trapped in poverty." It says that this percentage rose from 29 percent in 1970 to nearly 33 percent in 1974.

Part of the reason for this persisting pattern is government "insensitivity," the commission said. It recommended that the Federal Government "should officially recognize that Puerto Ricans are a minority group whose problems require specific forms of aid."

The report, entitled "Puerto Ricans in the Continental United States: An Uncertain Future," was published in a 170-page English version and a 190-page Spanish version.

Five Years of Hearings

The report, released here today, was compiled from diverse official data, from testimony at hearings from 1971 to this year throughout the nation, and from interviews with hundreds of persons.

"Official insensitivity, coupled with private and public acts of discrimination, has assured that Puerto Ricans often are the last in line for benefits and opportunities made available by the social and civil rights legislation of the last decade," the report says.

To help counter this trend, which the commission said was influenced by government "ignorance of Puerto Ricans," the group, appointed by the President, said that a new advisory group on Puerto Rican problems should be created by the director of the President's Domestic Council.

The commission's chairman, Arthur S. Flemming, a former Secretary of Health, Education and Welfare, said at a news conference:

"A decade or so ago, it was quite common to hear that Puerto Ricans would, in a matter of years, make it in American society in the same fashion that some immigrant families from Europe have climbed the economic, political, and social ladder in the United States. Unfortunately, these optimistic predictions have not been realized for most Puerto Ricans."

Comparison With Others

The report said that as of March, 1975 while 11.6 percent of all American families were below the low-income level, thi was the case for 32.6 percent of mainland Puerto Ricans. That compared with 24 percent for Mexican-Americans and 14.3 percent for Cuban-Americans.

Thus, it said, the incidence of poverty and unemployment among Puerto Ricans "is more severe than that of virtually any ethnic group in the United States."

Also, while about 5 percent of all United States families depended on some public assistance or welfare according to the 1970 census, this was the case for 24.5 percent of Puerto Rican families.

The report tempered the gloomy statistics by stating that "three-fourths of the Puerto Rican families on the mainland are wholly self-sufficient and receive not 1 cent of welfare or other Federal aids" and adding that "the purpose of stating these facts is that in the face of hostility, prejudice and Government neglect, many Puerto Ricans have successfully made the transition from their native land to the United States."

In 1970, the report said, 57 percent of the mainland Puerto Ricans lived in New York City, and the commission said that "the future health of the city is inextricably bound to the development" of the community. Today, it added, the migration of Puerto Ricans has extended itself to every single state and to major cities that include Chicago, Philadelphia, Cleveland, Newark, Hartford and Boston.

The commission recommended that the Federal Government widen job opportunities for Puerto Ricans by increasing funds for the Comprehensive Employment Training Act and setting up affirmative action programs in United States Employment Service offices in target cities.

It also urged the Government to reduce educational disadvantage for Puerto Ricans by increasing funds to aid needy children and to training bilingual teachers.

War on Poverty of Little Help

The report noted that the War on Poverty program of the Johnson Administration in the 1960's was of small benefit to Puerto Ricans.

To help devise policies to correct discrimination against Puerto Ricans, the report urged that the director of Domestic Council create an advisory panel with representatives from Puerto Rico and from states and cities in the United States that have large Puerto Rican populations.

Louis Nunez, a deputy staff director of the commission, said that one of the purposes of the report "is to show that this is a national and not just a regional problem."

October 14, 1976

MEXICAN-AMERICANS

TEXANS GIVE HEED TO MEXICAN RAPS

Bias Against People From South of the Border Is Charged in Complaints

GOOD-NEIGHBOR EFFORTS

By WALTER C. HORNADAY

DALLAS, Texas, Feb. 12—The controversy that Mexico has raised with Texas over discrimination against Mexicans and American citizens of Mexican blood has caused Texans to hasten to cultivate more friendly relations with its neighbor to the south.

Texans and Latin Americans, a term which in some sections of the State seems preferred by Texas citizens of Mexican descent, have long lived in peace and harmony for the most part in every section of the State, particularly in the southwestern border areas.

The economic role of the Mexicans has been largely that of providing agricultural labor, although many gain their livelihood from other pursuits. In the sections of the State where the Mexican population is heavy, they are engaged in every line of trade.

The social status of the Mexican, or Latin American if you prefer, is that of any other person in the lower economic scale, accustomed to a lower standard of living, satisfied with poorer living conditions. Racial lines have never been drawn in Texas as in some other sections. B t the Mexicans have their own communities in the larger cities. A voluntary segregation exists. Children of Latin-American descent are found largely in certain schools; for instance, because the school is adjacent to the section where they live. Jim Crow laws affect the Negro, not the Mexican.

Political Role

Mexicans form blocs of considerable political importance in the southwestern border sections, and in cities where they are numerous, such as in San Antonio. Their vote, even where used in the past by political bosses, is of local and sectional and not of State-wide importance. They pay their poll taxes where they have been aroused by politically interested leaders. Mexicans and citizens of Mexican descent form almost one-sixth of the State's population, and are not far behind the Negroes in numbers.

The Mexican press has pointed ou that Mexico is helping agriculture in Texas and other border States by furnishing badly needed farm labor, and that thousands of men of Mexican descent are serving in the American armed forces. These facts are listed as making it imperative that Mexicans receive fair treatment.

Texans have been told that the State plays an important role in maintaining the country's Good-Neighbor policy for the Latin-American nations, and that Nazi propaganda seizes on incidents to create rifts in the Western Hemisphere solidarity.

The belief seems to be taking hold in Mexico that Texans are making sincere efforts to end discrimination and that good-will does exist toward Mexicans. Signs are the praise given Governor Stevenson by Mexican officials and by American Ambassador Messersmith for his efforts in fighting discriminatory practices.

February 13, 1944

MEXICANS IN U.S. BATTLE ANXIETIES

Special to The New York Times

MEXICO CITY, Aug. 1 — The emotional problems of a million Mexican-Americans trying to adjust themselves to living in the United States are growing. The process of adaption is becoming more difficult rather than less so, it has been found.

The problem, manifest in the heavily populated Mexican immigrant colonies along the North-American border, was investigated at a symposium last month at the University of Arizona in Tucson. Its report has now been made public.

The meeting, under the chairmanship of Dr. Arnold Meadow of the University of Arizona's department of psychology, brought together psychiatrists, psychologists, social workers and anthropologists from Mexico and North America.

Anxiety and tension are increasing among certain categories of transplanted Mexicans in the United States, the conference was told, and increased clinical facilities are becoming necessary for them. Young Mexicans of school age, both male and female, generally do fairly well in adapting until they leave school and their families. After that, it was found, anxiety sets in as a result of increased competition with the Anglo-American elements for jobs and social acceptance.

Mexican-Americans male schizophrenic patients outnumber the Mexican-American female patients 2 to 1 in the Arizona State Hospital, it was reported by Dr. Meadow. A lesser number of Mexican-American females received high schizophrenic scores in tests than did Mexican-American males, Dr. Meadow said. The Mexican-Americans have not been affected by alcoholism to the degree that Anglo-Americans or Negroes have in the same areas and those of like size, he said.

The study was sponsored by the Mexican National Institute of Health in conjunction with the Arizona State Hospital and the Mental Health Association of Tucson.

The working class of the Mexican-American population, although not adapting readily to the North-American way of life, is exposed to less tension than the rising middle class of Mexican-American, the conference was told. The former relies more heavily on the strong Mexican family system and on folkways. The rising middle class, on the other hand, deserts old family ways and attempts to integrate into the Anglo-American culture.

In many instances, conference speakers said, these advances are rejected outright by the Anglo-American culture. In other cases, the alien impersonality of the Anglo-American society is interpreted by the Mexican-American as rejection. In either case the results can be traumatic.

Dr. Ramon Parres, an American-trained Mexican psychiatrist practicing here, maintained that the problem of integration and adaption would never be completed until the concept "Mexican-American" was abandoned. Mexican immigrants, who by choice or by Anglo-American rejection, cling to their Mexicanization would never be integrated into North-American society, he argued.

Dr. Parres also objected to the line of discussion suggesting differences in the methods of treating disturbed Mexican-American and Anglo-Americans. The mental thearapy, he argued, should be the same for each group. Other conferees argued that Mexican immigrants reacted favorably to therapy that Anglo-Americans refused to accept.

A significant factor in the increase of emotional disturbance among Mexican immigrants has been the "culturelessness" among Mexicans in transition between their own culture to that of Anglo-Americans. Dr. Hans Leder of Arizona University noted a high incidence of "pathology-inducing stresses" in this area. Among the patients, he informed the symposium, there was a high incidence of divorce, abandonment and alcoholism. Among the children, delinquent behavior and increased school dropouts resulted.

Arizona, owing largely to its Mexican-American population, has one of the highest dropout rates in the United States. The three key factors in the dropout rate were culture conflicts, prejudice and low socioeconomic status. Most American school programs are based on the culture of the American middle class, it was noted, while most Mexican immigrants to North America come from the lower classes.

August 2, 1964

NEGRO GAINS VEX COAST MEXICANS

By PETER BART

Special to The New York Times

LOS ANGELES, Oct. 16—The stepped-up public and private aid being channeled into the Negro community as a result of the Watts riots has stirred sharp resentment among a much larger minority group in Los Angeles — the Mexican-Americans.

Anti-Negro feeling is running so strong among Mexican-Americans that some residents and community leaders express fear of renewed rioting. "The thing could blow up any time," says Ann Ramirix, a mother of four who lives in a combined Negro-Mexican neighborhood.

Officials estimate that there are as many as 2.5 million Mexican-Americans in California, or nearly twice the number of Negroes. About half live in the Los Angeles area.

Though the Mexican-Americans' median income is slightly higher than that of the Negroes, their general educational level is far lower. Surveys indicate that only about 8 per cent of the Mexican-American community has attained a high school education. Unemployment is believed to hover around 10 per cent and median income around $2,800.

Mexican-Americans believe their problems have been completely shunted aside as a result of the Watts riots, in which 35 persons were killed, 947 injured and 3,873 arrested.

This fear prompted a conference last weekend of some 90 Mexican-American community leaders. They described the meeting as a "quiet riot."

"Our riot isn't one of force and uncontrolled emotions, but a riot of ideas designed to relieve the growing frustration and despair in our community," Ray Gonzales, the conference chairman, said.

This sense of despair was given abundant documentation at the conference. Ralph C. Guzman, leader of a $500,000 foundation-supported research project that is investigating the problems of Mexican-Americans in the United States, reported that a "terrible bitterness" was building up because of a widespread belief that Mexican-Americans were being dismissed from many jobs to make way for Negroes.

This bitterness, he said, was revealed in surveys that showed many Mexican-Americans feel it has "now become fashionable to hire colored people" and reject Mexican-American applicants.

Fernando Del Rio, a consultant with the Youth Training and Employment Project, financed by Federal antipoverty funds, reported that Mexican-Americans would no longer patronize Federal or state employment offices in downtown Los Angeles because they complained that Negroes were being given first chance at all jobs.

"The truth of the matter is that the typical poor Mexican-American is now convinced that the 'Anglo' power structure is kicking him out," Mr. Del Rio said.

When questioned by a visiting "Anglo," many Mexican-Americans expressed resentment similar to that of the Negroes. One 28-year-old Mexican carpenter who lives near Watts said:

"When I was a teen-age kid in high school the police would stop me almost every day on my way home from school to ask me where I was going or what I was doing. They would hound you to be sure you felt you were unwanted in this city."

Mrs. Ramirix said:

"The schools don't seem to understand the problems of our children. The children speak Spanish at home and then are expected to be able to keep up with the English spoken by the teachers. There should be more bilingual teachers who understand that our children come from two cultures—the Anglo and the Mexican."

This dual heritage is often cited as a deep-seated cause of the Mexican-Americans' problems in adjusting to conditions in California. "The average Mexican-American doesn't know whether he wants a taco or a hamburger," said Marcos de Leon, a community coordinator for the Los Angeles school system. "We belong to two cultures and are influenced by two value systems."

With all their problems, the Mexican-Americans have been relatively passive in their dealings with the rest of the community. Though a series of bloody "zoot suit" riots flared in the Mexican sections some 20 years ago there has been no violence since that time. Nevertheless, crime rates and narcotics addiction run high in the Mexican-American sections.

The largest concentration of Mexican-Americans is in eastern Los Angeles, though there are also many living in the predominantly Negro South Los Angeles district.

There are many pleasant streets in East Los Angeles, but there are also many bleak sections of dilapidated, crumbling homes that represent greater squalor than that in the worst parts of Watts.

The Watts riots could have two potential effects on the Mexican-Americans, community officials believe. The searing animosity of Mexican-Americans toward Negroes could erupt into new rioting, or the Mexican-Americans, noting the impact of the Watts violence, could be spurred into organizing their own civic committees.

This sort of leadership has been lacking in the past, according to Mr. Gonzales.

Mexican-Americans who succeed eductionally and financially abandon the low-class "barrios" (ghettos) and disappear into the "mainstream of middle-class Anglo society," he said, and thus there is no one left to guide the less fortunate.

Mr. Gonzales hopes that conferences such as the one he led last weekend can help develop new leadership.

October 17, 1965

A New Mexican-American Militancy

By HOMER BIGART
Special to The New York Times

LOS ANGELES—Five million Mexican-Americans, the nation's second largest minority, are stirring with a new militancy. The ethnic stereotype that the Chicanos are too drowsy, too docile to carry a sustained fight against poverty and discrimination is bending under fresh assault.

The Chicano revolt against the Anglo establishment is still in the planning stage, however. No national leader has arisen. La Causa, as the struggle for ethnic identity is called, has only a fragmented leadership of regional "spokesmen." No one really seems to want a chief, for as one young militant explained: "It's too easy to co-opt, buy off or assassinate a single leader."

The Mexican-Americans are a distinctive minority, separated from the dominant culture by a great gulf of poverty and differences in language and culture.

California, with two million, and Texas, with a million and a half, have the most Chicanos. New York probably has fewer than 10,000 and they are completely submerged by the massive Puerto Rican presence.

Some Mexican-Americans, notably in New Mexico, claim descent from Spanish explorers. Others say they were derived from the ancient Aztecs, and stress their Indianness. But the vast majority describe themselves as mestizos, people of mixed Spanish and Indian blood.

They all have a common complaint: they say the Anglos treat Chicanos as a conquered people by suppressing their Spanish language in the schools and discriminating against them in jobs, housing and income.

Consigned in the main to menial jobs, they earn a little more money than the Negro, but because their families are larger, the per capita income is generally lower: $1,380 for Mexican - Americans, against $1,437 for nonwhites in the Los Angeles area.

The worst-off Chicanos are the farm workers. Testifying last December before the Civil Rights Commission in San Antonio, the local Roman Catholic Archbishop, the Most Rev. Robert E. Lucey observed that migrant farm workers lived "in the awful reality of serfdom."

Nelson Tiffany for The New York Times
Mexican-American at work in California's Imperial Valley

Like other ethnic groups, the Chicanos are drawn to cities. The crowded urban barrios are usually adjacent to the Negro ghettos, and the rising ferment among Mexican-Americans has been stimulated in part by the Negro civil rights movement.

There are varying degrees of Chicano militancy:

In the Spanish-speaking ghetto of East Los Angeles, barrio toughs boast of grenades and other explosives cached for the day of revolt against the gringo.

In Denver, Rodolfo (Corky) Gonzales plans a massive nation-wide school walkout by Chicano students on Sept. 16, Mexico's Independence Day. Corky, a former prize fighter, claims total victory in last month's strike at a high school in the west side barrio, a strike marred by violence in which, Corky says, a dozen police cars were disabled.

Quixotic Courthouse Raider

In New Mexico, Reies Lopez Tijerina, the quixotic former evangelist who raided a courthouse two years ago to make a "citizen's arrest" of a district attorney, takes a visitor on a tour of a "pueblo libre," a proposed free city-state in the wilderness where Chicanos will control their own destiny.

Unfortunately, 90 per cent of the pueblo is national forest. This does not bother Tijerina's followers. They claim the land under Spanish royal grants made prior to American sovereignty. They have chopped down the boundary markers and other signs of gringo occupation.

They have even held a mock trial for a couple of forest rangers who fell into their hands. Tijerina himself is under a two-year Federal sentence for aiding and abetting an assault on a ranger. His conviction is under appeal.

Tijerina, who has been alternately snoozing and crunching sunflower seeds in the back seat while his lawyer, Bill Higgs, takes the wheel, suddenly comes to life. At a high pass where the road cowers under skyscraper rocks, the leader shouts: "Here's our port of entry for the Free City of Abiquiu."

Straight ahead, gleaming in the sun, is the Abiquiu Reservoir of the Chama River and on either side, sloping gently to the mountains, are wide stretches of grazing land. The black tower of Flint Rock Mesa looks down on a bowl completely empty of cattle and men.

"To me, this is holy ground," cries Tijerina with some of his old Pentecostal fervor. "Here we will build a city dedicated to justice. This is our Israel! And just like the Jews we are willing to die for our Israel, yes sir."

A Diverse People

Mexican-Americans are as diverse as any other people. Cesar Chavez, the gentle, introspective, sad-eyed director of the California grape strike, is totally unlike either the fiery Tijerina or the somberly wrathful Corky Gonzales.

Mr. Chavez has been called the spiritual leader of the Chicano moderates. His tiny bedroom at Delano, Calif., where he spends most of his time (he is afflicted with muscular spasms) is adorned with photos of his heroes—Gandhi and Martin Luther King, both apostles of nonviolence—and of his political mentor, the late Senator Robert F. Kennedy.

His belief in nonviolence seems unshakeable. He told a visitor: "Those of us who have seen violence never want to see it again. I know how it tears people apart. And in the end we lose.

"I'm not saying we should lay down and die. I think I'm as radical as anyone. But I think we can force meaningful change without the short cut of violence."

The strength of the militants is impossible to gauge. Tijerina contends he has 35,000 members in his Alianza; Corky Gonzales says he can muster 2,500 for a demonstration in Denver. Barrio militants in Los Angeles say they have "gone underground" and refuse to discuss strength.

"Our people are still frightened, but they are moving," commented Mr. Chavez, who said he had no wish to become a national leader. "I'm at most a leader of our union, and that union is very small," he said.

Three years ago, the Mexican-American community had no staff-funded organization except Mr. Chavez's organizing committee. Today there are several, including the Mexican-American Legal Defense and Educational Fund (which resembles the N.A.A.C.P. Legal Defense and Educational Fund, Inc.) and the Southwest Conference of La Raza (The People), both of which are supported by the Ford Foundation.

The grape strike is now in its fourth year. The main issue is no longer money. Most of the table grape growers against whom the strike is directed have raised wages. The main issue now is recognition of the United Farm Workers Organizing Committee, and Mr. Chavez says he expects a long tough fight before that is achieved.

This week Mr. Chavez extended the strike to the Coachella Valley of Southern California. The strikers expect even more trouble in organizing the workers there than in the San Joaquin Valley, for the Coachella vineyards are only 90 miles from the border and a plentiful supply of strike breakers can be recruited from the hordes of "green carders" who pour across the frontier each day in search of work.

These green carders, so-called from the color of identification cards, are aliens who are allowed to commute to jobs in this country. They are a constant source of cheap labor, undermining wage scales in the border region and frustrating union attempts to organize not only the farms but also the new industries that are settling in dozens of frontier towns from Brownsville, Tex., to San Diego.

Chicanos are demanding a tightening of the immigration laws. They would curb the commuting by requiring the green carders to reside in the United States. Then, confronted by higher living costs on this side of the border, the Mexicans would no longer be willing to work at depressed wages and might be more receptive to joining a union, the Chicanos believe.

The grievances of the Mexican-Americans, most of whom live in California, Texas, New Mexico, Arizona and Colorado, with sizable colonies in the Middle West (founded in the last century by construction gangs for the Santa Fe Railroad) sound familiar: job discrimination, miserable housing, social isolation, lack of political power (the result of gerrymandering the urban barrios) and exposure to a school system completely insensitive to Mexican-American history and cultures.

In only one respect is the Mexican-American better off than the Negro. Provided he is not too swarthy and provided he has money, the Chicano can escape from the barrio and move into Anglo middle-class districts.

He is worse off in other respects. Of all the minorities, only the American Indian makes less money than the Chicano. A linguistic and cultural gap separates the Mexican-American from the Anglo. Proud of his ancient Spanish-Indian heritage, the Chicano is less eager for assimilation than the Negro.

Most Speak Little English

Most Chicano children speak only a few words of English when they enter school. It can be a traumatic experience, especially in districts where Chicano pupils are spanked if they are overheard using Spanish in the halls and on the playground.

Recalling his first encounter with the strange and threatening atmosphere of an Anglo public school, Arnulfo Guerra, now a successful lawyer in Starr County, Tex., said that when a Chicano wanted to go to the toilet he had to wave his hand and try to say: "May I be excused?" Mr. Guerra said with a laugh that for a long time he believed that "bisquez" (be excused) was the Anglo word for toilet.

Children caught speaking Spanish were sometimes humiliated, he said, by having to stand with their nose pressed against the blackboard inside a circle of chalk. If overheard on the playground, they were made to kneel and ask forgiveness.

Besides being confronted with a foreign language, the Chicano pupil finds that the attitudes, social relationships and objects depicted in his lessons are entirely outside his home experience. He is constantly admonished that if he wants to be an American, he must not only speak American but think American as well.

Their school dropout rate (34 per cent for Chicano children enrolled in Grades 7-12 in Texas) is the highest for any minority group.

In San Antonio, which has the second largest Mexican-American colony (about 350,000; Los Angeles is first with about one million), a hearing conducted last December by United States Civil Rights Commissioner J. Richard Avena disclosed subtle forms of discrimination.

School officials admitted, according to Mr. Avena, that junior high school counselors tended to steer Chicanos into predominantly Mexican-American vocational high schools. This betrayed the counselors' ethnic stereotype of the Chicano as an individual inherently equipped only for vocational training and unsuited for the Anglo college preparatory schools, he said.

The school system is a prime target of Chicano wrath. "Cultural rape" is a term frequently used by Mexican-Americans to describe what they call the system's attempt to make little Anglos out of their children.

School strikes and boycotts in the Southwest are becoming an almost daily occurrence. In Texas, Chicano pressure has obliged the school districts of San Antonio, Austin, El Paso and Edcouch-Elsa (adjacent towns in the lower Rio Grande Valley) to stop the punishment of children using Spanish in schools or playgrounds.

In Denver a few weeks ago, Corky Gonzales made the school board suspend a teacher accused of "racist" remarks.

The teacher denied having called a Chicano "stupid," denied having said: "If you eat Mexican food you'll look like a Mexican," and his denials were supported by some students who said he had been quoted out of context.

However, the school board seemed intimidated by the disorders that attended the walkout. Stones and bottles were thrown at police cars; a 26-year-old Mexican-American was struck by a charge of birdshot fired by a policeman; 16 others were injured, and more than 40 persons, including Corky, were arrested.

Concessions Granted

The board made a number of concessions: more emphasis on Mexican history and literature in west side barrio schools, a re-evaluation of the counselling programs (Corky charged that some counsellors were urging Chicano youths to join the armed forces) and Mexican food in the cafeteria.

A grand jury returned no indictments on the Denver outbreak, although it found that "the inflammatory statements of Rodolfo (Corky) Gonzales at Lincoln Park bordered upon violations of the anarchy and sedition laws of the state." It exonerated the patrolman for shooting the demonstrator and praised the police for "remarkable self-restraint in the face of vile abuse and obscene taunts."

Chicanos Confronted By Gringos and Anglos

Special to The New York Times

AUSTIN, Tex., April 19 — Three words that have come into common use in the confrontation in the Southwest are "Chicano," "gringo" and "Anglo."

Chicano is never used in Mexico, but in the Southwest it is considered a diminutive of Mexicano, or little Mexican.

Gringo is similar to a Spanish word that has come to mean gibberish, so it is said that a gringo is someone who speaks a language that the people of Mexico do not understand.

Anglo is used to distinguish a white person who is not of Mexican ancestry.

The precise derivations of the words remains unclear, despite their common use in the area.

Corky Gonzales, 40 years old, father of eight children, was one of the top 10 featherweights from 1947 to 1955. A former Democratic district captain in the barrio, he gave up politics because, he said, "I was being used." Then he founded a militant organization, "Crusade for Justice."

On a recent warm April day, a visitor to Corky's headquarters, a former Baptist church in the decaying Capitol Hill district of Denver, was led upstairs to a barnlike room where four or five hairy, unkempt youths were watching the funeral of Dwight D. Eisenhower on television. They were offensive and rude.

"C'mon, stick him in the ground and get it over with," one of them said, and the others laughed.

Accompanied by Guard

Corky, when he arrived with a bodyguard, went directly to his office, a musty cluttered room that had been the minister's study. He was no longer a featherweight, but he still looked trim and tough. He had grown a bushy black mustache, and he wore a pendant symbol of his movement—a three dimensional head representing Spanish father, Indian mother and mestizo offspring, mounted on an Aztec calendar plaque.

"How can there be justice," he demanded bitterly, "if we don't have our people on the jury system and the draft boards?"

Denver Chicanos had lost faith in the political system, he said, because every Mexican-American who achieved office in the country was "absorbed into the Anglo establishment and castrated by it."

Chicano schoolchildren were being perverted, he said, by "middle class aspirations," and the middle class was "dying and corrupt." He was against competitive society: "Success today in this country is learning how to cut throats."

Corky said he believed the best way to unify Mexican-Americans was through nationalism.

To foster Chicano nationalism Corky held a five-day conference in Denver at the end of March. About 1,000 youths from five southwestern states showed up, and they represented an ideological spectrum that included the New Left, Communists and liberals.

Coalition in Dispute

The convention nearly broke up on the issue of coalition with Negroes. Some barrio youths, resentful of Negro dominance in the civil rights movement, insisted on maintaining racial separateness.

Corky, who had quarreled with the black leadership of the Poor People's March on Washington a year ago, preached a modified ethnic nationalism, and he prevailed. Coalition with the blacks might be feasible later, he said, but meanwhile the Chicano must first achieve enough self-reliance to "do his thing alone."

As a first step toward liberating the Chicanos, Corky told the youths to go home and prepare a nationwide walkout of Mexican-American students on Sept. 16.

Down in Albuquerque, meanwhile, Corky's main rival for leaderhsip of the Chicano youth, Tijerina, was plotting his own demonstration. It would be held on June 5, the second anniversary of his shootout at the Rio Arriba County courthouse, an event as significant to Mexican-Americans, Tijerina believes, as the Boston Tea Party was to the American colonists.

Two years ago Tijerina and his band raided the courthouse in the northern New Mexico hamlet of Tierra Amarilla to "arrest" the district attorney for "violation of our civil rights."

Acquitted by Jury

He said that the district attorney, the sheriff, the state police and the forest rangers were all conspiring to deprive the Mexican-Americans of ancestral land, insisted that the Federal Government had welched on a promise, contained in the protocol to the Treaty of Guadelupe-Hidalgo (which ended the Mexican-American war in 1848) to honor some old Spanish and Mexican land grants.

A jury acquitted Tijerina of kidnaping and other charges growing out of his bloodless coup.

Tijerina's headquarters are in a blue and white two-story adobe building on a quiet Albuquerque street — quiet except when terrorists are trying to bomb the place. Tijerina, a hawk-faced man vibrant with nervous energy, said he suspected the Minutemen, a right-wing Anglo organization, of perpetuating three explosions, the last of which wrecked a dozen automobiles in the headquarters parking lot.

The leader of the Alliance of Free City-States has taken a few precautions. His apartment above the ground-floor meeting hall is protected by a steel door, by 18-inch concrete walls and by a triple-layered steel and cement floor.

Inside this fortress Tijerina discussed the future. The June 5 anniversary would be peaceful, he said, unless the gringo interfered. Some new Chicano families would be settled in the free city-state of San Joaquin and there would be a barbeque.

"Are you in rebellion?" he was asked.

"I don't know," he replied thoughtfully. "It's a matter of interpretation. The Government has raped our culture. So I think the Government is in revolt against the Constitution. It's our constitutional obligation to go on the cultural warpath to save our honor and identity. We demand that the Government cease the illegal occupation of our pueblos."

Tijerina said he had signed a treaty of mutual respect with the Hopi Indians, pledging mutual support against any aggressor.

Another plan for territorial revision was being advanced in Texas by Dr. Hector P. Garcia, founder of the American GI Forum, an organization of moderate Mexican-Americans.

Dr. Garcia proposed that South Texas, which has a large Chicano concentration, be made a separate state. This would give the Mexian-Americans a chance to send one or two Senators and several Congressmen to Washington, he said, thereby easing the frustrations of political impotence.

The new Chicano militancy, with its cry of "Brown Power," can be heard even in Texas, where Mexican-Americans have long complained of brutal suppression by the Texas Rangers and by the state and local police.

Last month more than 2,000 Chicanos paraded through the border town of Del Rio, ostensibly to protest Gov. Preston Smith's decision to shut down the local projects of VISTA, the domestic Peace Corps, but also to cry out against discrimination.

Normally such demonstrations are small and sedate, the Chicanos parading behind a priest carrying the banner of the Virgin of Guadelupe.

But this time the priest and the Virgin were forced to yield the front of the line to militants of the Mexican-American youth Organization (MAYO), and they tacked a manifesto on the courthouse door warning that violence might erupt if demands for equality were not met.

One of the founders of MAYO, José Angel Gutierrez, 22, said the organization's goals were the formation of political units independent of the Republican and Democratic parties ("only Mcxians can really represent Mexican interests") gaining control of schools, and the building of economic power through the weapon of boycott.

But the cause has had serious setbacks in the Rio Grande Valley. Attempts to organize farm labor have failed completely. Unemployment is high. And a powerful friend of the Chicanos, the Rev. Ed Krueger, was recently dismissed by the Texas Conference of Churches as its field representative in the lower valley.

Mr. Krueger said he had been under pressure from conference officials to "work with the establishment instead of with the poor," and that his superiors were also displeased because he refused to withdraw a suit against the Texas Rangers, a suit alleging that the Rangers manhandled Mr. Krueger and his wife when they tried to photograph a farm strike in Starr County two years ago.

The dismissal of Mr. Krueger was investigated by a panel headed by Dr. Alfonso Rodriguez, in charge of the Hispanic-American ministry of the National Council of Churches. The panel reported "tragic conditions of alienation, polarization, conflict and tension" in the valley, adding that the tension had been aggravated by Mr. Krueger's dismissal.

Farther west, El Paso and Phoenix show scant signs of Chicano militancy, despite their teeming barrios. In El Paso, where thousands of Mexican-Americans still live in squalid, rat-infested, barrack-like "presidios," some of which have only one outhouse for 20 families, about the only recent demonstrations have been peaceful "prayer-ins" on the lawn of a slumlord's agent.

In Phoenix a Roman Catholic priest, the Rev. Miguel Barragan, field representative of the Southwest Conference of La Raza, said it was difficult to involve the older Chicanos because they were prejudiced against political solutions, recalling the turmoil in Mexico. And the newer migrants feared police harassment and loss of jobs.

Yet the priest warned:

"If there are no immediate changes in the Southwest, no visible improvement in the political and economic status of the Mexican-American, then I definitely foresee that our youths will resort to violence to demand the dignity and respect they deserve as human beings and as American citizens.

"I see the barrios already full of hate and self-destruction. I see an educational system doing psychological damage to the Mexican-American, creating a self-identity crisis by refusing to recognize his rich cultural heritage and by suppressing his language.

"And therefore, to me, burning a building and rioting is less violent than what is happening to our youth under a school system that classes as 'retarded and inferior' those with a language difficulty."

In California Mexican-American demands for larger enrollments of Chicanos at the Berkeley and Los Angeles campuses of the University of California were receiving sympathetic attention. And Berkeley was planning a Department of Ethnic Studies in which Mexican history and culture would be taught.

But in East Los Angeles and Boyle Heights, these concessions were taken as insignificant crumbs.

Basically, people are tired of talking," said a youth in the Boyle Heights barrio. "A confrontation is inevitable. It's not unusual to see people going around with grenades and TNT. The tension is here; the weapons are here. The new underground organizations of ex-cons, addicts and dropouts make the Brown Berets look like Boys Scouts."

Across town, on the U.C.L.A. campus, a neutral observer gave a pessimistic but somewhat milder assessment. Prof. Leo Grebler, a German-born economist who directed a four-year study of Mexican-Americans for the Ford Foundation, a study soon to be published, recalled how Gunnar Myrdal in his classic study of the Negro in the United States had been over-optimistic about the nation's ability to cope with the racial crisis.

Professor Grebler said that he and his coauthors, Prof. Joan W. Moore, a sociologist, of the University of California, Riverside, and Dr. Ralph Guzman, a professor of political science at California State College in Los Angeles, were making no such error in their projections about the Mexican-Americans.

The study will conclude that the Anglo Establishment must quickly remove obstacles to the socio-economic development of the Mexican-Americans and broaden its understanding of this minority.

Some idea of the ignorance and apathy displayed toward Mexican-Americans by the dominant institutions was reported in Washington by Vicente T. Ximenez, who resigned recently as chairman of the Federal Interagency Committee on Mexican-American Affairs.

Mr. Ximenez, who is the first Mexican-American member of the Equal Employment Opportunities Commission, said he had invited 50 of the nation's largest foundations to send representatives to a conference on Mexican-American problems on March 22. Only two of the 50 accepted, he said, and the conference was canceled.

"The negative replies didn't bother me so much as the reasons they gave," Mr. Ximenez said. He quoted a letter from the Guggenheim Foundation that noted: "Our program is confined to awarding fellowships through annual competition to advanced workers in science, scholarship and the arts."

"I can't understand," Mr. Ximenez said rather dryly, "why we Mexican-American can't qualify for advanced work in science. scholarship and the arts."

A tour of the Southwest revealed many Chicanos were aware of the strategic position they might hold in an increasingly divided white and black nation.

Tijerina has said: "We are reaching the point where the black and white color gap will demand a brown, middle-color peacemaker."

And a would-be Chicano power-broker predicted in Los Angeles that a coalition of Mexican-Americans and blacks would sweep Councilman Thomas Bradley, a Negro, to victory over Mayor Sam Yorty in next month's Los Angeles mayoralty election.

"With this strong coalition," he said, "we can control every political job in the city by 1972."

April 20, 1969

MEXICAN-AMERICANS ASSAIL COMMERCIALS

Special to The New York Times

WASHINGTON, Dec. 9—A group of Mexican-Americans, charging that they have been victims of negative stereotyping on public airwaves, said today they intended to file a complaint with the Federal Communications Commission.

The announcement, made by the National Mexican American Antidefamation Committee, complained of several specific television commercials.

But the brief to be filed with the F.C.C. will be directed "against the negative stereotype of the Mexican-American presented by the Frito-Lay Company in its Frito Bandito commercials," according to Joseph L.Gibson, general counsel of the committee.

Nick Reyes, executive director of the committee, charged that the Frito Bandito commercial was the "most blatant offender" in stereotyping Mexican-Americans.

December 10, 1969

The Urban Chicano's Struggle: 'Escape If You Can' to Equality

By STEVEN V. ROBERTS
Special to The New York Times

SAN JOSE, Calif., Nov. 13—Helen Garcia worked at the Salvation Army sorting old clothes. She had to leave every morning at six because it took three buses and three hours to reach the job from her home here on the East Side of San Jose. With five children, Mrs. Garcia had to work, but she did not like the travel time, and a friend got her a job as a grocery clerk. She had been working only a few days when a young man, apparently crazed by drugs, entered the store last week waving a gun. Everyone thought he was kidding until he fired a shot into an ice machine. Then he shot Mrs. Garcia. She died 15 minutes later.

The Talk of Sal Si Puedes

Helen Garcia was one more victim of the barrio, the Mexican-American section of San Jose. The area has been called Sal Si Puedes, which means "escape if you can."

Cesar Chavez, the leader of the Farm Workers' Union, who lived here for many years, once explained the name: "That's what that barrio was called, because it was every man for himself, and not too many could get out of it, except to prison.'

Mainly because Mr. Chavez's efforts on behalf of field workers have received so much attention, there is a popular impression that most Mexican-Americans are rural harvest hands. In fact, at least 80 per cent of the more than one million Chicanos in California live in urban barrios like this one, and here is where their problems — and potential power — are most significant. Nationally, the percentage of Chicanos living in urban areas is almost as high.

At first glance, Sal Si Puedes looks like a rather pleasant working-class suburb, with small frame and stucco houses, cars in the driveways, well-kept yards and sprays of flowers.

But it is like the impression made by a middle-aged woman with heavy make-up in a dim light. When you look harder you see the cracks and wrinkles — the broken windows, the sagging roofs, the overcrowding, the shacks hidden in tiny alleys that would be unfit for chickens, let alone human beings.

Still, many residents have a certain affection for their community. "I didn't know I lived in a ghetto. I thought I lived in a pretty nice place," said Mrs. Armida Rivera, an active member of Our Lady of Guadalupe Church.

"Remember when Robert Kennedy came to San Jose?" she went on. "That's the church he went to, our little church. I have a paper to prove it."

The tides of community life flow through the tiny grocery store where Mrs. Garcia died. One evening recently a steady procession of people came in to buy milk and bread (mainly with the food stamps the Federal Government issues to poor families) or beer, cigarettes and candy.

A Chicano man and an Anglo woman, carrying a black baby, picked out some bread and soda and paid with pennies.

"You know how it is when you run out of money," the girl said sheepishly.

When the pennies were counted she had a few extra —enough to buy the boy a yellow lollipop.

The black population on the East Side is growing steadily, and while there is some tension and competition between the two groups, it was not evident in the store.

A young black man came in and Lydia Torres, the clerk, knowing his craving for them, automatically brought him a jar of green olives. "It's just something I'm stuck with," the man laughed. "I picked it up in the service. This jar will be gone in an hour."

The store is owned by Ernie Abeytia, a local political leader, who performs little services for his customers. He cashes checks for people with no identification and gives customers a break when they're a few cents short on the bill.

The store is filled with the intoxicating aroma of fresh tamales, shells of ground corn stuffed with meat and sold hot from a big blue pot on the counter. The store sells at least a dozen kinds of chili and chili powder, and Mr. Abeytia has learned to stock black-eyed peas during the New Year holiday for the recent arrivals from the South.

On Sunday morning the store features hot home-made menudo, a Mexican soup made with beef tripe and hominy and flavored with cows' feet.

"Menudo is very traditional throughout the Southwest on Sunday morning," Mr. Abeytia said. "It's great for hangovers—it's a real belly liner."

Although Chicano traditions like menudo endure, the old image among Anglos of the lazy, feckless Mexican is changing rapidly in San Jose.

Mr. Abeytia ran for the State Legislature this year and lost by only 145 votes, even though many Chicanos are prevented from voting by complicated citizenship and language requirements.

After years of complaints about police brutality, the Chicanos recently decided to organize a unit called the Community Alert Patrol to monitor police actions.

A radio operator listens to police radio communications and dispatches an observer car to the scene of any trouble. The observers, usually responsible older men with no arrest records, do not interfere with the police. They just take photographs and make tape recordings.

Reports of brutality have dropped sharply in the two months the patrol has been operating. "But the cops," said one member of the group, "never know when they're being watched."

A police group, the Peace Officers Association of San Jose, charged that the patrol was getting aid from the Mexican-American Community Service Association, a group that in turn receives money from San Jose's United Fund campaign. The peace officers threatened to boycott the United Fund if the Community Service Association got any more money. Within days the fund suspended the association.

The incident made Chicano leaders more aware than ever of how little real power they have in the political system. Yet, despite the rhetoric of some young militants, most Chicanos remain committed to that system.

"A lot of people like me don't go along with this militant deal. They get so wound up they want to force issues," said one community leader who refused to be identified. "There's all this talk about revolution, but that's not what the majority of Chicanos want. They're fighting the system just to get ahead, to get some breaks, to get a job. We just want our equal rights like everyone else."

November 17, 1970

Panel Says 5 States Deny Rights of Chicano Pupils

By PAUL DELANEY

Special to The New York Times

WASHINGTON, Feb. 4—Five Southwestern states were accused today by the United States Civil Rights Commission of failing to educate Mexican-American school children.

The commission, in its sixth and final report on the education of the nation's second largest minority group, accused Arizona, California, Colorado, New Mexico and Texas of not providing Mexican-American children with an education equal to that of white pupils.

"How well are the schools of the Southwest serving Mexican-American students? Are they providing equal educational opportunities for them?" the commission asked in the 269-page report. It added:

"These are the fundamental questions the commission has addressed in its four-year study of Mexican-American education. On the basis of the five reports already issued, the unavoidable conclusion is that the schools are failing."

Governments Blamed

The commission blamed Federal, state and local governments, as well as institutions of higher learning in the five states, for the alleged failure.

The Federal Government, in particular the Office of Civil Rights of the Department of Health, Education and Welfare, has failed to enforce firmly constitutional and legislative guarantees of equal education, the commission said. It contended, for example, that the department had not cut the funds of school districts found to be discriminating against Spanish-American children.

While states and higher education institutions play key roles in the education of Mexican-American pupils, the commission said direct responsibility belonged to the local school districts and the schools themselves, and that changes should thus be instituted on the local level.

"The commission, however, believes that the problems of unequal educational opportunity are of such magnitude and so widespread that it would be unwise to rely entirely on the good faith efforts of individual school districts to bring about the kind of uniform and comprehensive educational reform needed."

Among the 51 recommendations made by the independent agency were: State governments should prohibit at-large election of school board members in favor of single-member districts; the Federal and state governments should impose sanctions, including cutting funds, of districts violating the rights of Mexican-Americans; states should requre bilingual education; states, along with the Department of Health, Education and Welfare, should set numerical goals and timetables "for securing equitable Chicano representation," and all levels of government should reorder their budget priorities to provide the needed funds to implement the recommendations.

The commission said the clearest indication of the failure of the schools in the Southwest was that, of every 10 Mexican-American children who started the first grade, "only six graduate from high school." The commission added:

"By contrast, nearly nine of every 10 white students graduate. Moreover, by graduation, the Mexican students are more than twice as likely to repeat a grade and seven times more likely to be overage for their grade than white students."

In its series of studies, the commission also said it found that the language and culture of Mexican-American students are ignored and even suppressed by the schools," that Mexican-American schools were under-financed in comparison with white schools, and that teachers failed to involve Mexican-American pupils as active participants in the schools.

The commission added that it found serious deficiencies in the curriculums of Mexican-American schools; that Mexican American students were victims of ability grouping and were over-represented in low-ability groups and under-represented in high-ability groups; over-represented in mentally retarded classes, and had not been given adequate counseling.

Lack of Influence

The commission contended that few Mexican-Americans were in positions of influence in teacher-preparation programs and were in small percentages on classroom-teaching staffs. The report added:

"Although ethnic data on teacher trainees are not systematically maintained, the under-representation of Chicanos, both as public schoolteachers and college students in the Southwest, strongly suggest that Chicanos are severely under-represented as teacher trainees."

The commission said that the Office of Civil Rights of the Department of Health, Education and Welfare had developed the mechanism to determine the denial of equal educational services, but added that "weaknesses remain in enforcement and implementation of the law." It said this was caused by inadequate staffing of the office and failure of the school districts to receive technical aid.

Several times in the past, the commission has criticized H.E.W., as well as the Nixon Administration, for its alleged failure to enforce adequately civil rights guarantees.

February 5, 1974

Depression Is a Reality To Chicanos

Special to The New York Times

ALBUQUERQUE, N. M. — Anyone with grass outside his adobe home in South Valley, near this city, is affluent by Chicano standards. It means he can afford to use water on his little plot of land.

Among the approximately 40,000 Chicanos in this area, usually living on unemployment insurance, welfare, food stamps, or short-lived jobs, there are few patches of grass anymore. The pale land is as barren as the snow-capped mountains in the distance. The land is so hard that even a chill north wind does not raise much dust.

There are no sidewalks in the South Valley. No street lights. Most roads are unpaved and each house has a cesspool. There is no sewage system here. Gaunt dogs roam the flat land.

"People are surviving here, not living," says Larry Lopez, and official of a community center.

What most Americans, in this era of spreading unemployment and inflation, have come to know as a recession, has been normal for these Chicanos for many years. They are now in a full depression. Unemployment here is in excess of 12 per cent.

"I do not think I will be able to buy food stanps," says Roberto Lujan. "It is too much."

Lean, gray and work-bent, he sits on an old sofa in his home, his wife, Aurora, standing behind him, her hands on the back of the sofa. On the wall, in a gilt frame, is a picture of Christ among sheep.

"My family goes to church," Mr. Lujan says. "Sometimes we have to push them. But they go."

He lives on $295 a month in Social Security, $66 a month welfare and $15 a month from a son who has a job. Ten children live in the house.

Competition for Jobs

In the basement of the Laborers International Union, some unemployed men in work clothes sit on metal folding chairs, telling how nonunion contractors are taking advantage of the sharp competition for jobs to pay nonunion scale considerably below the average rate of $5 an hour.

Some are paying only $2 an hour. And in some food stands, they say, the rates are as low as $1.50 an hour. Unemployment insurance of $69 a month is what one man says he gets to support his wife and four children.

"I got to finagle around to stay alive," says another man. "We all do. My wife works, but I don't tell them when I go for food stamps. I will have to pay more for food stamps. The less you tell them you got, the more you can get on the food stamps. You almost have to lie to them to stay alive."

One man, about 60, says he is finding it difficult to get work because he's considered too old. A young man says the excuse he gets is that they would rather have jobs for older men.

"Last year," says the young man, who is married, "I saved $900. Now my savings are down to $300. I was saving to buy a house."

The house is a dream now. The down payment would be $3,000 on a $30,000 home. Another man tells how he sold his car, a necessity to the Chicanos in their search for jobs. He could no longer afford gasoline. Many have had to sell their cars for this reason. They come to the union office in groups, sharing a single old car or truck.

"I can't feed my dog any more," says one. "I beg bones from the supermarket."

Children Sell Tamales

They talk of wives who are baking cakes to sell house to house, of children who go forth with buckets of tamales to sell in the streets.

With reluctance, even with anguish, they talk of the illegal immigrants from Mexico, brought in to work for small wages on construction, ranches, restaurants. They try not to be too critical of the immigrants, who are taking their jobs.

"The employers are being protected so they can exploit the workers," says one of the men.

"We are being forced to fight over crumbs," says another.

Eventually they get around to politics. The Chicanos are Democrats, their commitment strong since the New Deal legislation of the Depression. But now, while they still speak with anger against the Republicans in Washington, they no longer have much faith in the Democrats either.

At the Southwest Valley Neighborhood Center, a one-story building, Mr. Lopez told of an incident that morning.

"I found these two people who had nothing left in the house but leftover gravy. That's what they were living on. I finally talked them into going for food stamps. Many people here are ashamed."

March 25, 1975

ATTEMPTS TO UNITE HISPANICS

The Goal Among the Spanish-Speaking: 'Unidos'

WASHINGTON — In the current era of interest-group politics, militant organizations have arisen to press a host of causes: black power or flower power; the liberation of women or homosexuals; opposition to foreign war or domestic pollution; equality for the man of red skin or blue collar.

Last week, it was the turn of Spanish-speaking groups. Individual nationalities, like Puerto Ricans, Mexican-Americans and Cuban migrants, have long been organized. What was striking about the crowd packed into a suburban Washington motel was that for the first time, these groups were trying to join forces.

The goal was alluring indeed—a coalition of at least 9 million, ranging from New York to Florida to Texas to California, with a roster of common complaints: law enforcement brutality, unilingual education, job discrimination and poor housing.

The allure notwithstanding, the conference soon turned into such an uproar that one official dubbed it the "raucous caucus." And by the end, there was considerable doubt whether the aim of coalition had in fact been advanced.

Many conferees left elated at the progress made toward the unity betokened by the ubiquitous badges that proclaimed, in red on orange, "Unidos." Task forces quickly set to work organizing a national executive committee and raising funds for a full-time Washington office to press for legislation and to monitor law enforcement agencies.

But others—including the four Spanish-speaking members of Congress who organized it—left the conference discouraged, even angry. The conference proved, Representative Herman Badillo, Bronx Democrat, said, that unity "is an idea whose time has not yet come."

For ultimately, the conference endorsed two proposals that Mr. Badillo and others scorn as distracting or romantic—Puerto Rican independence and a separate Spanish-speaking political party.

Moderates from community groups and Hispanic organizations made up a majority and could easily have defeated both proposals, pressed by young militants. But, Mr. Badillo said acidly, "All of a sudden, when a floor fight looked likely, the moderates had to catch a one o'clock plane. If a group of 1,500 melts to 150, what can you do? Unless the moderates stop being afraid of offending somebody, coalition will have to wait—another generation maybe."

A more fundamental obstacle to unity was illuminated earlier this month when the Census Bureau issued its first general report on families of Spanish-speaking background. Its central finding was that such families typically earn a good deal more than black families even though they are less educated.

Census analysts were the first to warn against putting too fine a Pangloss on this finding. More detailed tables showed great variations, indeed, among different Spanish-speaking groups.

In the Cuban-American population of 626,000, the typical family earns 36 per cent of the typical white family income of $10,236. But among five million Mexican-Americans, the figure was 70 per cent. And among the 1.4 million mainland Puerto Ricans, it was 58 per cent, possibly a little *under* the figure for blacks.

In the view of a number of conferees, such divergences mean that, at best, only a limited alliance is possible—of poor Chicanos (Mexican-Americans) and poor Boricuas (Puerto Ricans). The unity conference may have promoted that kind of coalition. But its consummation remains a long way off and even that result would fall considerably short of "unidos."

—JACK ROSENTHAL

October 31, 1971

To Spanish-Speaking In Chicago, Numbers Don't Mean Strength

By WILLIAM E. FARRELL

CHICAGO, May 10—The Mexican and Puerto Rican populations are Chicago's most rapidly growing ethnic groups, and each year this city, like many other large urban centers across the nation, shows more signs of their presence.

That presence is evident in the proliferation of quick-stop tacos parlors, in the bilingual signs on the city buses and in advertisements such as the placard in a barber shop window promoting a "baile primavera," a spring dance.

The soaring growth rate, however, is not yet reflected in the arena where it counts most—the carefully crafted Chicago political machine of Mayor Richard J. Daley.

The 1970 Federal census placed the number of Spanish-speaking Chicagoans at about 250,000, or a little more than 7 per cent of the city's total population of 3.4 million.

That number is regarded as a sizable undercount. Experts on the city's Spanish-speaking population, such as Prof. Pastora San Juan Cafferty of the University of Chicago, estimated that the city's "Latinos," as they are called here, number about 600,000.

Mrs. Cafferty pointed out in an interview that Chicago had the largest Mexican population in the United States except for Los Angeles and the largest Puerto Rican population in the country outside of New York.

She and others said that while those outside the Spanish-speaking community regarded it as monolithic, this was not so.

Chicago's growing Spanish-speaking numbers reflect their increase in other large American cities such as New York, Los Angeles, Houston, Miami and San Antonio. According to the 1970 Federal census, the Spanish-speaking population in the country was 9.6 million, or 4.7 per cent of the total population.

There is little mixing here of the city's Mexican and Puerto Rican populations. Their residential enclaves tend to be separate.

A Major Difference

"Our cultures are different," said Arturo Velasquez, a Mexican-American who is one of the few Spanish-speaking Chicagoans with political entrée to City Hall. "We speak the same language, but our heritages are different."

A main difference is that Puerto Ricans are automatically United States citizens, while Mexicans who emigrate to Chicago must be naturalized.

That separation, which some feel will wane over the years, is at least part of the reason why one looks in vain for a Spanish surname on the list of Chicago's 50 aldermen. Another is that the city's aldermanic ward lines are drawn so that despite large concentrations of Mexicans and Puerto Ricans in some areas, they do not yet have sufficient numbers in any one ward to win a seat.

That will all change by 1979 in the opinion of Judge David Cerda of Cook County Circuit Court.

"We're growing, and the trend is to cooperate more among ourselves," Judge Cerda said. "At first we didn't know each other, but there's growing confidence."

"Up until recently we've been ignored," Judge Cerda, one of the area's few elected Spanish-speaking officials, said. "Benign neglect. But in the last 10 years we've been getting people employed in city, county and Federal offices. We are, however, nowhere near getting our representation in these organizations."

The Classic Pattern

Mr. Velasquez, who operates a jukebox rental company in the city's "back of the yards" section, views the progress of the city's Spanish-speaking population in the classic immigrant pattern. The pattern applies particularly to Chicago, which has absorbed many ethnic groups over the years.

Mr. Velasquez, who came to Chicago at the age of 8 from the Jalisco section of Mexico, has been working in the city for 50 years and is regarded as a key operative for Mr. Daley in the Spanish-speaking community. The walls of his makeshift office are dotted with pictures of him and the Mayor at political fetes.

"We weren't able to fight him so we had to stick with him to get what we wanted," Mr. Velasquez said. "Out of that, we got the judge, a commissioner and a trustee of the University of Illinois.

"We still need a state representative and we're working on that right now.

"This generation is a different generation. My generation—our level of education was grammar school, if any. The second generation started going to high school. This generation is starting in college."

All four of his children have college degrees and two hold official posts. His daughter Carmen was recently appointed to the Chicago Board of Education and his son Arthur was recently elected as a trustee of the University if Illinois.

Educators such as Miss Velasquez are pushing for meaningful bilingual school programs, something that has been met with indifference or hostility by some outside the Spanish community.

Another source of irritation in the Spanish-speaking residents has been what they feel is the tendency of the outside news media, when it pays attention to them at all, to focus on illegal immigrants.

There are illegal immigrants in Chicago. But some, such as Mrs. Cafferty, see an overemphasis on it to the exclusion of other characteristics of the Spanish-speaking community.

May 11, 1975

HISPANICS CREATE ACTION COALITION

Special to The New York Times

CHICAGO, June 29—A national organization for political and economic advancement of Hispanic Americans was announced last night by Mayor Maurice A. Ferré of Miami.

The new organization, called the American Coalition for Hispanic Action, "will give us common identity and diminish divisions that have for decades hampered our advancement," Ferré said.

"We have been on the bottom of the totem pole for too long," said the 40-year-old Mayor. "The only weapon that 15 million Spanish-speaking Americans have is political pressure and militancy. The time is now."

Mr. Ferré, the only Hispanic Mayor of a major American city, made his comments in a speech to about 500 Latin businessmen and politicians at a banquet of the Cuban American Chamber of Commerce of Chicago.

Earlier in the day the Mayor, accompanied by scores of Chicago's Latin leaders, walked through Hispanic sections of the city.

Political Role Urged

In shops, restaurants and street encounters, he stressed the need for unity and political involvement. He sounded the same theme in private meetings with Spanish-speaking leaders of the city yesterday and today.

Mr. Ferré who had been making plans for the new national organization for 18 months, chose Chicago for his announcement because he believes it has the best conditions to be a major nucleus of the new group.

With a Spanish-speaking population estimated at more than 600,000, Chicago has the nation's second largest Mexican population after Los Angeles, the second largest Puerto Rican population after New York and the third largest Cuban population after Miami and New York, as well as a sizable group of Ecuadorians, Colombians, Nicaraguans and others.

Even though the Democratic political machine of Mayor Richard J. Daley can count on most of the Hispanic vote here, there are few elected Spanish-speaking officials and policemen in Chicago.

"Political parties have taken us for granted for too long, too," Mr. Ferré said at a working breakfast today with a group of politicians. "Don't forget that John F. Kennedy and Richard Nixon won by a few hundred thousand votes. That's a few votes per precinct. If we organize and register to vote we, the Spanish-speaking Americans, could even swing national elections in the future."

Few Officeholders

Despite its numerical strength, Mr. Ferré said, the nation's Hispanic community has been able to elect only two Governors, one United States Senator, fewer than a dozen Representatives and one Mayor, all Democrats except for one Representative.

According to Government statistics, persons of Spanish origin are also economically and socially behind whites and in many areas, blacks.

Hispanic schoolchildren, statistics show, have the highest dropout rate in the nation. Many of them speak little or no English.

Bilingual education, made mandatory by a Supreme Court decision, has reportedly made limited progress in major cities.

"There is no organization to demand the rights of a New York Puerto Rican boy who drops out of school because he doesn't understand English," Mr. Ferré said. "There is no Hispanic urban league, no N.A.A.C.P. or B'nai B'rith. This is where we have to come in.

"We have little money, our people are divided, with the fragmentation often fostered by the white majority. But we are confident that we shall prevail. There is more that unites us than divide us. And we have a strong common bond which is the Spanish language." Mr. Ferré and his father, Jose, have an international cement and real estate concern that is said to be worth more than $100-million. His uncle, Luis, was Governor of Puerto Rico.

Active for years in national Democratic campaigns, Mr. Ferré is on friendly terms with most national political figures. He was recently appointed by President Ford to the advisory board of the Vietnamese Refugee Committee.

June 30, 1975

Latins Announce Creation of National Hispanic Caucus Affiliated With Democratic Party

Special to The New York Times

WASHINGTON, Nov. 2—The creation of a National Hispanic Caucus, a new organization affiliated with the Democratic Party, was announced here today.

The caucus, designed to have a degree of independence within the party, was created at a weekend conference of Spanish-surnamed elected officials that ended today.

Co-chairmen for the conference were three Governors, Jerry Apodaca of New Mexico, Raul H. Castro of Arizona and Rafael Hernandez Colón of Puerto Rico; Senator Joseph M. Montoya of New Mexico; Representatives Herman Badillo of Manhattan, Eligio de la Garza and Henry B. Gonzalez of Texas, and Edward R. Roybal of California, and Jaime Benitez, resident commissioner of Puerto Rico in Washington.

More than 150 persons of Mexican, Puerto Rican, Cuban and South American heritage, mostly elected officials from 14 states and Puerto Rico, participated in the conference.

A steering committee with members from all state delegations was given 90 days to prepare proposals for the structure of the new organization and its program.

At the meetings, various leaders demanded a larger voice in the drafting of the 1976 Democratic platform. They complained, often bitterly, that the 1972 McGovern platform failed to recognize the needs and aspirations of 15 million Hispanic-Americans in the United States.

The conference, according to observers, contrasted with a similar effort in 1971 when radicals who took over the floor denounced United States foreign policy and quarrelled with one another.

"It's different now," said Mr. Badillo. "Now we are definitely on the move. We have not only agreed to unite, but we have also agreed that the best way to exercise 'Latin Power' is through the electoral process."

Experts have said that the political power of citizens of Hispanic background could be decisive in some states. The estimate that only about two million of the six million citizens of Spanish-American ancestry eligible to vote are registered.

The caucus is expected to open a nationwide registration drive, with a goal of increasing Hispanic voters by one million before the 1976 election.

It has been shown in selected districts and states that when Latin voters feel an election is of importance to them, they vote in proportionally larger numbers than other citizens.

With over 85 percent of the Hispanic vote going Democratic, the Democratic party sees a special stake in the success of the new organization.

"We have no other way but to unite and demand our rights," said Governor Castro in a ringing speech last night. "Somos del mismo barro — we are of the same clay."

November 3, 1975

RED POWER

A NEW PATTERN OF LIFE FOR THE INDIAN

By FRANK ERNEST HILL

MORE than a mile above the sea level, on a plateau of the American Southwest, two hundred and fifty men are building a new capitol. It is not the capitol of a State. Its stone walls rise in shapes that are strange to most Americans; its name—Nee Alneeng—falls with a strange accent. Nee Alneeng belongs to a world far from Manhattan and Main Street. It is an Indian world, and the capitol belongs to the Navajo, now the largest of the North American tribes.

This little centre is symbolic of a new way of life among the Navajo: in fact, a new way of life for the 340,000 Indians of the United States. A year ago the Wheeler-Howard Act gave to the tribes the right to decide whether they would accept important privileges in education, self-determination and self-government. A popular vote was asked; the essential question was: "Do you want to help save yourselves?" So far 134 reservations containing 128,468 Indians have voted to come under the act, while fifty-four reservations with 85,179 Indians have excluded themselves.

Thus the Wheeler-Howard Act embodies an Indian policy far different from that pursued in the past. The Federal Government could have conferred self-government upon the American Indian without asking him if he wanted it. To understand why he was asked, one must take a brief but discriminating glance at American history as it has affected the Red man.

* * *

THE record may be thought of as falling into three stages. The first dates from the earliest white settlements in the Southwest and in Virginia and marks the beginning of a protracted struggle between European and Indian cultures. The struggle ended with the sporadic Western wars of the 1880s—in the inevitable defeat of the Indian. The last of the aboriginals entered United States Government reservations, and a second stage began: the government's effort to control and protect the Indian and adapt him to white American ways.

For more than fifty years this persisted. The possibility of a normadic hunter's life for the Indian was gone; as a substitute, the government sought to educate him and make him a stock raiser or farmer. It is clear now that in many ways the system failed to protect him from cruel exploitation and yet prevented him from acting for himself. It led him to lean passively on the rather precarious bounty the government extended. Presently the Indian had suffered the loss of much of his allotted land, much of his separate culture, and had developed a deep inferiority complex with an accompanying resentment. Disease and bitter poverty menaced him. The days of his vitality seemed numbered.

The third stage may be said to have begun with a growing conviction among many thoughtful Americans that Indian life had latent strength and important cultural values and that the Indian if given the right opportunities could do what the government had failed to do: he could arrange a place for himself and his customs in this modern America. The appointment of John Collier as Commissioner of Indian Affairs in April, 1933, brought into power a leader of this trend of opinion.

* * *

MR. COLLIER, slight, almost scholarly in appearance, at his desk in Washington describes what the administration is trying to do for the Indian and why he believes the new policy to be enlightened.

"In the past," he says, "the government tried to encourage economic independence and initiative by the allotment system, giving each Indian a portion of land and the right to dispose of it. As a result, of the 138,000,000 acres which Indians possessed in 1887 they have lost all but 47,000,000 acres, and the lost area includes the land that was the most valuable. Further, the government sought to give the Indian the schooling of the whites, teaching him to despise his old customs and habits as barbaric. Through this experiment the Indian lost much of his understanding of his own culture and received no usable substitute. In many areas such efforts to change the Indian have broken him economically and spiritually.

"We have proposed in opposition to such a policy to recognize and respect the Indian as he is. We think he must be so accepted before he can be assisted to become something else, if that is desirable. It is objected that we are proposing to make a 'blanket Indian' of him again. That is nonsense. But if he happens to be a blanket Indian we think he should not be ashamed of it. We believe further that while he needs protection and assistance in important ways, these aids should be extended with the idea of enabling him to help himself. We are sure that he can and will do this. But he must have the opportunity to do it in his own way. This is what we have been trying to extend to him. It is an opportunity he has not had since he entered the reservations, where he has 'been discouraged from thinking and acting for himself.

"It is all an educative process. Perhaps the most drastic innovation of the last two years has been our effort not only to encourage the Indians to think about their own problems but even to induce them to. Our design is to plow up the Indian soul, to make the Indian again the master of his own mind. If this fails, everything fails; if it succeeds, we believe the Indian will do the rest."

THE people whom the commissioner is trying to reanimate, and to incite to this crusade for self-survival, are in one sense heterogeneous. There is no typical Indian but rather a hundred different types. These are scattered. The 220 tribes that comprise the race are to be found here and there in twenty-two States. They are of many different stocks physically, and they speak dozens of different languages.

Their cultures vary, and so does the degree to which they have adopted the white man's ways. The five civilized tribes, now in Oklahoma, were farming when De Soto discovered the Mississippi. So were the Pueblo Indians, who were also skilled weavers and master potters. On the other hand, the roving tribes of the Northwestern plains did little cultivating, and, though skillful in crafts, were esthetically far less developed. Similar differences persist today. Some Indians are competent farmers and stock raisers; others are less happy and successful in the settled life.

Some speak no English, are inexpert with tools and live in crude shelters; others have acquired modern houses and automobiles and serve as teachers, doctors, lawyers and storekeepers. Some tribes find a personal "planned economy" difficult; others, like the Hopi, are thrifty and far-seeing. Unquestionably Indians generally are willing to use much of the white man's equipment and means to knowledge, but often are backward because their economic grip on life is a precarious one. Many of the tribes hold grants of land that is inferior or insufficient in extent, yet manage well with their facilities, and are deft as artisans and mechanics, sometimes eager for better tools, machinery and methods.

Underneath all their differences lie identical, unifying instincts, habits, aptitudes and spiritual feelings. Fine qualities are to be observed in almost any Indian group: artistic cleverness, tenacity, courage, dignity and a decent pride. Under the parochial control of the past, with its effort to make the Indian a white man, these qualities have shown but little. They have come out best where the Indian, as in the Southwest, has lived his own life.

* * *

IN attempting to "plow up the Indian soul" and put these qualities into action, Mr. Collier has not depended on the Wheeler-Howard Act alone. This law is important; it may justly come to be regarded as an Indian Magna Carta. It repeals the Allotment Act of 1887 and so makes the further loss of Indian lands impossible. It provides for the purchase of additional badly needed land for the tribes up to a valuation of $2,000,000 a year.

It creates a revolving credit fund of $10,000,000 against which the Indians can borrow (if they accept the new law) when they have governmentally approved farm or industrial projects. This is wholly novel: the government had never previously recognized, in Mr. Collier's phrase, "the cold fact that capital in some form is needed to transform even a piece of raw land into a productive farm." There is a fund for scholarships also, and preference is given to Indians who seek positions in the Indian Service. Finally, there is the right of every tribe accepting the act by majority vote to adopt a constitution and take over most of the powers now exercised by the Federal Government.

All these privileges are important. Those providing economic and political sinews are especially significant because of the independence and self-reliance which they may develop. Yet the Indian Office regards the Wheeler-Howard Act as a step only. "It is merely a beginning," Mr. Collier points out, "in a process of liberating and rejuvenating a subjugated and exploited race living in the midst of an aggressive civilization far ahead, materially speaking, of its own. Even that beginning is oppressively difficult."

This difficulty has been recognized by the creation by the Indian Office of an organization unit of field agents and special men who will cooperate with tribal councils, business committees and special tribal commissions in framing the Constitution now permitted. The organization unit will advise the Indians, seeking to make the governments they set up both effective and legal. Definite educative work will be done to give the Indians an understanding of their new civic powers.

* * *

THE possibilities in economic and political development here are dynamic. However, they follow a spirit and practice fostered since the Spring of 1933. This called for a much greater use of Indians both as officials in the Indian Service and as routine workers outside the permanent staff. The results have been notable.

In the case of the permanent staff, changes come slowly for all positions are subject to civil service rules. However, while in 1932 the Indian Office used 6,172 employes, of which 1,296 were Indians, its reduced force of 5,322 today contains 2,037 Indians. The Indians have derived other benefits by being utilized on ECW and CWA projects. Last year these workers swelled the total of government Indian employes to 19,616. This figure takes no account of the quota of 14,000 Indians in the CCC camps.

The work that Indians have done in the last two years in building roads, dams, bridges, trails and improving forest lands has been impressive. More than half the supervisory force consisted of Indians. Mr. Collier regards as important the demonstration they have given of skill, initiative and responsibility.

Beyond its successful effort to give the Indian a fair trial as a worker, the Indian Service has undertaken several specific projects of considerable importance to him. The most comprehensive of these has been going forward in the Navajo country. It touches all phases of Navajo existence: the preservation of the soil, its better use for farming and grazing, the character of the stock used, self-government, health, education, and, indirectly, art and spiritual life.

* * *

THE Navajo nation, the largest of all Indian tribes, was confronted two years ago, and still is, with an economic crisis. On its great reservation in Arizona and New Mexico, with an area equal to Maryland, Massachusetts and New Hampshire combined, the tribe had developed sheep raising. In 1870 a population of 10,000 Indians was existing on its arid plateaux. From their sheep they got mutton, their chief food. From them also they took the wiry wool for the best known of native loom products—the Navajo rug. They raised a little corn. They hammered silver ornaments from Mexican silver dollars—creating the best known of all American Indian metal work. These activities sustained the tribe.

But meanwhile the Navajo increased from 10,000 to almost 50,000, and the sheep, under government encouragement, increased with the population. Carefully used, the range might have supported 1,000,000 head. But in 1933 there were 1,300,000. Furthermore, the land had long been overgrazed; experts reported its actual capacity had sunk to 550,000 head. Cropped too intensively, grass and bushes were losing their strength and were pulled up by hungry animals. Then the wind churned the uncovered soil into drifting hillocks. Rain, which falls seldom on the Navajo reservation but then usually in torrents, ran off the denuded land, carrying soil with it.

• • •

IN order to live the Navajo must have his sheep. Having his sheep, he seemed doomed to economic ruin. Into this situation stepped the Indian Office. It had Emergency Conservation funds for work in the Navajo region. The office said:

"Reduce the number of your sheep. We will study how to control the destruction of the soil. We will employ your young men on government projects. We will show you how to use what water there is for irrigation. Gradually you will be able to increase your herds again. We will develop better stock for you, consuming no more but producing two-fold. In the end the land will give more than it has ever given."

In separate meetings and in their tribal councils the Navajo debated. What if the work gave out before the range was restored and the herds built back? This question is still in their minds. They have reduced their stock to 900,000 head; now they hesitate to reduce it further.

Meanwhile they have cooperated in the establishment of work projects and demonstration projects in various areas. Some are under the farm agents of the Indian Office; the greatest number are under the control of the new Soil Erosion Service. About 200 square miles of Navajo territory are now being managed as demonstration areas by this agency alone, sixty-seven of them about Mexican Springs, N. M.

In addition to the work with the land there are health and education and governmental projects in process. Schools are being built for the first time in the Navajo country. Navajo teachers will constitute the greater part of the teaching force, a new experiment. The staff of the first ten schools will consist of fifty Indians and five whites. Some instruction will be given in Navajo. The "Longhairs," the older men of the tribe, will be asked to teach the children tribal tradition, folklore and conduct of life. In health work Navajo girls are being trained as nurses to carry the fight against tuberculosis and trachoma into the remoter districts. Finally, at Nee Alneeng, twenty-five miles from Gallup, N. M., 250 Navajo workmen have been raising the walls of a new capitol which will make a centre for Navajo political life.

H. Armstrong Roberts.

The Redman as a Shepherd.

All activities are going forward with the agreement and participation of the Navajo, and their cooperation means a training in modern methods of work, in management, in government.

• • •

THE activities of the Indian Office have nowhere been so intensive as in this many-sided development and conservation of Navajo resources. But they have been country-wide. The school program has sought everywhere to bring the Indian children into a closer relation to their homes by increasing the number of day schools and reducing the number of boarding schools. Many new schools for day use are rising—in California, in Montana, in Minnesota, in North Dakota. Economic and soil erosion work on a large scale is being pushed by the Indian Service and the Soil Erosion Service on the Rio Grande watershed, and the Indian is sharing in it.

All this is a part, with the Wheeler-Howard Act, of the new policy of setting the Indian to save himself. On the whole the response has been a revelation as to his capacity as a worker and his eagerness to lead. He has shown independence of spirit—often to the point of rebellion. The Navajo, by a narrow margin, have rejected the Wheeler-Howard Act because of unbased allegations that it would unduly curtail their herds.

But Mr. Collier, regretting such actions, prefers rebellion to dry rot. "The Indians may be confused and thrown back for a time," he says, "but it is a part of their life and education. They will win through in the end, in their own way."

If they win through it will, in the Commissioner's opinion, mean a victory for both Indian and white man. Economically independent, the Indian will cease to be a financial burden to the nation. And spiritually and culturally he will bring something valuable too.

The new policy has already started a renaissance in Indian arts. Young Indians are painting murals on the walls of school houses and government buildings. They are studying the ancient pottery of their tribes in museums, and devising new designs and textures in their workshops. The young people are flocking to the ceremonial dances, which for a time they had avoided. This cultural revival goes hand in hand with an interest in self-government and economic independence. In Mr. Collier's opinion, it is equally valuable.

"The Indian," he says, "can use white technologies and remain an Indian. Modernity and white Americanism are not identical. If the Indian life is a good life, then we should be proud and glad to have this different and native culture going on by the side of ours. Anything less than to let Indian culture live on would be a crime against the earth itself. The destruction of a pueblo is a barbarous thing. America is coming to understand this, and to know that in helping the Indian to save himself we are helping to save something that is precious to us as well as to him.

July 14, 1935

NEW U.S. ATTITUDE STIRS INDIAN HOPES

Population of First Americans Is Finally on the Increase Under a Helpful Policy

BUT THEY HAVE LOST MUCH

By FRANK L. KLUCKHOHN

WASHINGTON, Dec. 9—Often when Americans have criticized the treatment accorded minorities in other countries, foreign critics have referred bitingly to this country's treatment of its Indian minority, but at long last those criticisms have lost much of their validity.

In great part as a result of a shift in Washington's attitude and policy toward the Indians, which started in 1924 and has received perhaps its greatest impetus during the Roosevelt Administration, the number of original Americans has increased to 351,878, as compared with 270,000 in 1900. In 1928 the Indian birth rate finally leaped ahead of the death rate, and in the last fiscal year the Indian population increased by 9,381, the birth rate exceeding that of the whites.

Behind this vital change, which offers tangible human proof that the government attitude has changed from one of suppression and elimination of the Red Man to one of aid and assistance, there is a tale of improved health conditions, with Indian babies born in hospitals instead of tepees; of better food and a cessation of wars, but most of all, probably, of a new will to live induced by the restoration of hope.

Concentrated in Five Zones

Although there are Indians in almost every State of the Union, the great concentrations are in five regions of the country. About one-third are in Oklahoma. About a quarter live in the Southwest generally, the great stronghold of full-blooded Indians. The Sioux, Blackfeet,. Cheyennes, Shoshones and Arapahoes dwell in the Dakotas, Montana and Wyoming. The Chippewas are situated in the Great Lakes region of Minnesota, Wisconsin and Michigan, while, on the Pacific Coast, Indians are scattered through Oregon, Washington and California.

The form of life differs greatly as between the Indians in these far separated regions. In Oklahoma, where one-tenth of the people claim Indian blood, most of them do not live on reservations. In the Southwest generally almost all are living upon reservations, devoting themselves principally to sheep-herding. The Northwestern Indians live in a territory where white land is checkerboarded with that of the Indians and the Indian has an existence much like that of his white neighbor.

But almost everywhere, in more or less degree, the Indian has been taken advantage of. In 1887 the Indians owned 139,000,000 acres of land, much of it good. In 1933 they owned only 52,000,000 acres, much of it bad.

The Story of the Sioux

Typical in some ways is the story of the Sioux Indians on, or near, the Rosebud Reservation in South Dakota. Theirs is a grazing country and, after the buffalo were destroyed, they were without means of support. For over a decade after 1890, the government kept them on rations and did not allow them to acquire any sizable number of cattle. Eventually, when they did start cattle grazing, the World War came, white men leased their land to grow wheat, and they lived on this bcunty. It soon disappeared.

These Sioux found that the leasing of two-thirds of their land to whites had broken up their communal lands, and had made it impossible for them to maintain their integrity and customs.

Under the Indian Reorganization Act of 19[illegible]4, these Indians were again put in a position to handle their own affairs. Under a Federal charter, they elect representatives to a tribal council, have a constitution and are authorized as an incorporated tribe to do business with the Federal and State governments.

The government lent $50,000 to this tribe of 8,891 and the tribe, in turn, had made loans at low interest rates to cooperatives, largely engaged in cattle raising, and to individuals. Another move to make them a responsible and self-respecting community has been a Federal attempt properly to correlate their land holdings through consolidation of properties, exchanges of land with white ranchers and purchase or lease of vital areas.

Schools have been established to teach soil conservation, proper treatment of the range and livestock handling so that proper use can be made of what land these Sioux possess. With variations, because of different local conditions and tribal customs, much the same basic work has been done in other parts of the country.

Economic Difficulties

The economic problem, officials here say, is the biggest one facing the American Indian. On the Navajo reservation, for instance, the birth rate is increasing, but it has been impossible, because of the opposition of white interests, to get Congressional authorization for extending the reservation.

The government has attempted to answer this problem by teaching the Indians to make the best possible use of their land and of their resources.

Critics of the government's Indian policy make three principal charges. They hold that in plans for self-government by the Indians and in advancing credit Washington has proceeded in such a manner as to tie up the Indians in red-tape.

Another group maintains that in moving so fast to change the status of Indians, the government 's stampeding and overstimulating the Red M[illegible] to a damaging extent. A third set charges that the government is spending too much money on the program.

December 10, 1939

Nation Is Still Palefaces', but Its Defense Is Given Back 'Reluctantly' to the Indians

The six proud nations of the Iroquois Indian Confederacy may never have been conquered by the United States, but their members are citizens of this country whether they like it or not, the United States Circuit Court of Appeals here ruled yesterday. The court, in an opinion written by Judge Jerome Frank, admitted that it had reached this conclusion reluctantly but inevitably.

The question had been brought before the white man's tribunal by descendants of those Indians who for more than a century had made and kept treaties with the young American government. Adopting, for the sake of the dispute, the rules developed in the Federal court, the Indians had sued out a writ of habeas corpus, demanding that their kinsman, Warren Eldreth Green, be released from serving under the Selective Service Act of last year.

Through their attorney, they contended that the men of the Confederacy were not citizens, because they belonged to an independent, unconquered nation and the United States Congress had no right to make laws affecting them, particularly when they did not consent to the laws.

Congress made Indians citizens in 1924, and last year made all citizens subject to the draft. Thus the Iroquois youths were subject to the military training requirements, unless it could be shown that they were not among those who received the gift of American citizenship seventeen years ago. The elders of the six tribes set out to do just that, and they raised funds for their fight by holding dances and other modern parties.

The court, however, after considering the various aspects of the matter, had this to say:

"We find ourselves compelled to decide against Green, although because of the historic relations of the United States to the Indians we reach that conclusion most reluctantly. Assuming, arguendo, the validity of his argument as to the treaty status of the Indian tribe to which he belongs, and that the statutes of 1924 and 1940 are at variance with that status, yet those statutes are not on that account unconstitutional."

The law is clear, the court held, that where domestic statutes conflict with treaties, the domestic courts are bound by the domestic statutes.

November 25, 1941

AMERICAN INDIANS FIGHT AXIS

Thousands Volunteer for All Branches of the Service and Many Are in Industry

By RICHARD L. NEUBERGER

COEUR D'ALENE, Idaho, Aug. 29—Although the Nazis have specifically addressed their propaganda to the Indians of America ever since 1933, the descendants of this continent's original inhabitants are active in all phases of the war effort. Indians are buying War Bonds, working in aircraft factories and shipyards, and enlisting in large numbers in the Army, Navy, Coast Guard and Marine Corps.

Of the 60,000 Indian males in the United States and possessions between the ages of 21 and 44, approximately 8,800 are serving in the armed forces of the nation. Army officials maintain that if the entire population was enlisting in the same proportion as Indians there would be no need for selective service. At some reservations half the male inhabitants have volunteered for Army duty.

This is the first war in history in which the American Indians have been required to participate as an obligation of citizenship, for not until 1924 did Congress make all Indians in this country full-fledged citizens. In recent years many radio broadcasts from Berlin predicted there would be an Indian uprising in the United States if the first Americans were asked to fight against Germany.

Financial Aid Given

Indian purchases of War Bonds have been extremely high. Numerous reservations have sent money to President Roosevelt from their tribal funds to help finance the war. In Oregon a big tribe, the Klamaths, has built its own school to train Indians for defense work. Several tribes have postponed land-claim suits against the government for the duration of the war. The late Maj. Gen. Clarence L. Tinker, lost in action during the Battle of Midway, was a member of the Osage Tribe. Private Sampson One Skunk, a Sioux Indian from Cherry Creek, S. D., is one of Uncle Sam's Rangers, who took part in the recent raid on Dieppe.

"An Inspiration"

Senator D. Worth Clark of Idaho has called the record of participation in the war effort by America's Indians "an inspiration to patriotic Americans everywhere."

An all-Indian platoon in the Marine Corps has made a notable record for marksmanship at San Diego. In Arizona Indians still scout the uplands and frontier. When the Navajos registered for Selective Service they brought along their rifles, thinking they were to go to war then and there. Blackfeet in Montana said they disapproved of the draft because every one should fight, rather than deciding the matter by lot.

Statistics compiled by the Office of Indian Affairs indicate that practically all Indians of eligible age have registered for Selective Service—the roll is more than 99 per cent complete. At Fort Defiance in Arizona, Indians stood in the snow for hours to sign their draft cards in the old headquarters of Christopher (Kit) Carson, famous scout of the frontier. A number of Western newspapers have urged that Indian Army officers and men be included in the delegations and missions which the United States sends to colonial countries and to nations with large aboriginal populations.

August 30, 1942

INDIANS RETARDED, LA FARGE CHARGES

Federal Paternalism Prevents Realization of Full Status as Citizens, He Asserts

Government paternalism is holding back the American Indians from full status as citizens by robbing them of "the most fundamental of all rights—the right to make their own mistakes," Oliver La Farge, president of the Association on American Indian Affairs, told its annual meeting yesterday.

The Bureau of Indian Affairs, he said, insists on "protecting" Indians far beyond the minimum that is necessary for their own good. Mr. La Farge said the public "must demand from the bureau increasing delegation of responsibility and authority to the tribes themselves."

Mr. La Farge cited restrictions on the Indians' rights to spend their own money and choose their own legal counsel. The latter, he said, sometimes seems intended to "prevent them from hiring attorneys that will be troublesome to the Bureau of Indian Affairs."

Backs Pending Legislation

Under present law, he noted, Congress appropriates their money for them, "frequently much against their will." He said that, as a start, the association would work for pending legislation that would give the tribes authority to spend the interest on their funds and —subject only to the approval of the Secretary of the Interior—their capital as well.

As for the civil rights of Indians, Mr. La Farge said, "it is painfully obvious that they are freely and habitually violated in almost all states with a sizable Indian population." He asserted that "chronic police brutality" had almost reached the point of a "perpetual open season on Indians."

"In large areas of the United States," he said, "Indians are treated as sub-human, as second-class citizens, because their skins are a trifle darker.

"Even their freedom to come and go as they please is pretty theoretic. There is no such freedom for people so shackled by ignorance, so circumscribed by paternalism—benevolent or not."

Finds Discrimination Rising

Miss Mârie Sandoz, author of several books on Plains Indians, agreed that segregation of Indians and discrimination against them had increased "alarmingly" in the last decade or more.

"Veterans who made fine records in World War II returned to find signs in public places, "No Indians Allowed," Miss Sandoz said. "But they found few jobs to match their new skills and little job-training or education to develop their skills and leadership.

"I am writing to twenty-three Indian G. I.'s in Korea," she said. "All of them are learning things. All of them are excited, talking about what they're going to do when they come home. But I'm not so sure. I saw what happened before."

Dillon Myer, Commissioner of Indian Affairs, attended the meeting, held at the Museum of Natural History. He answered informational questions from the floor but made no statements on bureau policy.

April 19, 1951

Striking at Indians

Directive Viewed as Aimed at Destruction of Trusteeship

The writer of the following letter is president of the Institute of Ethnic Affairs.

TO THE EDITOR OF THE NEW YORK TIMES:

The situation in American Indian affairs has entered a crisis phase. The facts should be known before, not after, the coming election. A directive from Commissioner Dillon S. Myer to all Indian Bureau officials, dated Aug. 5, occasions the crisis and also describes it.

The thirty-four-page directive of Commissioner Myer was sent to Indian Bureau personnel but not to the press, nor to the Indians or the Indian welfare groups.

In brief, the Myer directive is aimed at the destruction of federal trusteeship toward Indians, and the obliteration of that complex of achievements for and with Indians which commenced with President Hoover's Administration and became enlarged under President Roosevelt, and was continued until 1950. The document is extraordinary because it contravenes, sweepingly, statutory directions of Congress and a host of bilateral treaties, agreements, contracts and commitments which the United States and the Indians have entered into across more than one hundred years and up the years until 1950.

Directive's Purpose

The subject of Commissioner Myer's covert, if not secret, order is "Withdrawal Programming." Indian Bureau personnel are commanded to make "withdrawal," i. e., the wrecking of federal trusteeship, into the overriding objective. "We must proceed," the directive states, "even though Indian co-operation may be lacking in certain cases."

"Withdrawal" means the termination of United States educational, health and welfare services to Indians; the unilateral voiding of the Government's trusteeship toward the Indians' lands and other properties; the individualizing of much or most of the Indian corporate estate, and the casting of Indian landholdings under the local land taxes of states and counties.

The "basic methods to the development of withdrawal" already have been laid down in the California Indian "withdrawal" bill which Commissioner Myer pressed for enactment in the last Congress. This California bill was the entering wedge for substantially identical bills affecting all or nearly all the Indians.

This bill empowered the Commissioner (nominally the Secretary of the Interior) to dispose, without Indian consent, of tribal lands, funds, water rights, irrigation systems and all other tribal properties; to terminate tax exemptions without application or consent of the Indians concerned; to sell or give away, in his discretion, without Indian consent, federal Indian schools, hospitals, etc., built and until now maintained for the benefit of Indians; and to prescribe, virtually without limit, rules and regulations governing Indian property and local organizations; and these rules and regulations were made exempt from review or correction in any court of law. The bill did not become law.

Commissioner Myer's Aug. 5 directive is silent as to the position of Secretary of the Interior Oscar Chapman, and of the President. Nor has the Administration either publicly endorsed or repudiated the Myer directive.

Federal Commitments

One line in the thirty-four-page Myer document mentions "treaty rights pertinent to withdrawal." Nowhere are mentioned the Congressional agreements and executive orders which have been enacted in lieu of the unwithdrawn treaties since 1872, nor the hundreds of federal commitments in the Indian Reorganization Act of 1934 and the federal constitutions and charters bilaterally established pursuant to that act. The personnel is commanded, in effect, to proceed on the assumption that the Government's contractual obligations have been or can be, through direction or indirection, voided by Commissioner Myer. That assumption is, plainly speaking, totalitarian and unconstitutional.

The oldest and also the most living, and the most profoundly buttressed, of United States trusteeship commitments is the American Indian commitment. The "pilot projects" of Point Four, now our world-wide enterprise, were the Indian projects of the Hoover and Roosevelt Administrations; and they were hugely successful. The concentrated unilateral assault against the trusteeship obligation, now revealed in Commissioner Myer's document, concerns every citizen, and must be disturbing to every supporter of the United Nations Charter.

JOHN COLLIER.

New York, Oct. 10, 1952.

October 19, 1952

'POINT 4' IS URGED FOR U. S. INDIANS

A Point Four program to aid American Indians was proposed yesterday by the grandson of a Sioux medicine man.

The Rev. Dr. Vine Deloria also recommended a cloak of humility for white men searching for a solution to the "Indian problem." The minister is assistant secretary in charge of Indian work in the Division of Domestic Missions of the Protestant Episcopal Church.

Preaching at the Church of the Epiphany, York Avenue and Seventy-fourth Street, Dr. Deloria ascribed the plight of American Indians to their long exile on reservations.

"There they were isolated from the rest of the world," he said, "kept ignorant of the advances in world thought, living a Stone Age existence as the world was approaching the atomic era."

Although the Indian was permitted to leave the reservation in 1924, when he received United States citizenship, Dr. Deloria asserted that the damage already had been done.

"Before the white man came," he said, "the Indian was like an eagle, flying all over this country. The white man cooped him into a cage—the reservation—and pulled out his Indian cultural feathers."

According to Dr. Deloria, the first white missionaries entered the Indian country as superiors. They scoffed at the Indian customs and equated Christianity with Americanism.

The Indians today, he said, need experts to show them modern ways of raising stock, cultivating land, all within the framework of Indian traditions. A desperate need, he went on, is more priests, nurses and social workers.

Dr. Deloria called upon the congregation to urge Congress to repeal or amend Public Law 280. This law authorizes any state to extend civil and criminal jurisdiction over Indian lands.

Many Indians believe they will not receive justice from state courts, Mr. Deloria said. They prefer Federal supervision over their affairs, he asserted.

March 11, 1957.

TV WESTERNS DRAW FIRE OF INDIANS

By LEONARD BUDER

TELEVISION Westerns have come under new attack from the old frontier foe of the cowboys—the Indians. This time, however, the Indians are up in arms over video portrayals that depict them as ruthless savages intent only on plunder and atrocity. These blood-curdling TV "war whoopers," they feel, are giving viewers, particularly impressionable children, a distorted and harmful picture of frontier life and the historical role of the Indians.

According to one Indian spokesman, Harry J. W. Belvin, principal chief of the Choctaw Nation, "there is no excuse for TV producers to ignore the harm that may be done to the children of America by repetitious distortion of historical facts pertaining to the way of life of any race or creed, including the American Indian. Many TV programs show Indians as bloodthirsty marauders and murderers."

Contemporary Indian leaders make it clear that they do not want television to rewrite frontier history, but merely to show what really happened—including the fact that the Indians were fighting to defend their homes against what they considered to be invasions by the white man.

Organization

The new Indian uprising, which is being supported by many Americans who are not Indians, was initiated recently at a meeting in Fort Gibson, Okla., of delegates from eleven Indian tribes. The delegates appointed a committee, headed by Creek Tribal Chief Turner Bear, to submit a petition to President Eisenhower protesting false portrayals on the television screen.

The tribal action has been endorsed by the Public Education Committee of the Association on American Indian Affairs, which has headquarters at 475 Riverside Drive, New York. The association is concerned that, "by constant exposure to distorted frontier history, our younger generation will be taught the Indian is not worthy of his goodwill."

The A.A.I.A. committee is hoping to change the present state of TV Western relations through a program of public information. The unit, which is under the chairmanship of Arthur Ochs Sulzberger, assistant to the general manager of The New York Times, also has offered consultation services to television and motion-picture producers to insure accuracy in the dramatization of frontier life.

"Accurate portrayal of frontier history and Indian wars," LaVerne Madigan, A. A. I. A. executive director asserts, "does not require that the white man be presented as a ruthless invader; he was that—and yet he was more, because he built a democracy when he could have built a tyranny.

"Accurate portrayal, however, does require that the American Indian be presented as a brave defender of his homeland and a way of life as good and free and reverent as the life dreamed of by the immigrants who swarmed to these shores."

Exceptions

Not all TV Westerns have incurred the Indians' ire. Some programs are considered by the association to be "outstanding portrayals of frontier life."

One of these is the "Lone Ranger." In this series, Samuel Birnkrant, public education assistant of the association, notes, the masked rider's faithful Indian companion, Tonto, is accorded "co-hero status."

Another favorable series is "The Plainsman," in which "Sam Buckhart, pure-blooded Indian and Harvard graduate, is a United States marshal who wages a brave fight on the side of law and order and morality."

"'Gunsmoke' and 'Have Gun, Will Travel,'" Mr. Birnkrant adds, "also portray the Indians as human beings, rather than bloodthirsty killers."

However, on the other side of the ledger, Mr. Birnkrant cites the following examples of what Indians object to in television presentations:

"'Wagon Train': In many episodes, Indians are shown as drunken, cowardly outlaws. Indians are usually attacking wagon trains. Curiously, the Indians hardly, if ever, score a hit on the white men, whereas they are mowed down with ease. The resultant portrait indicates that the Indians are poor, inept fighters.

"'Riverboat' often depicts Indians as bloodthirsty, inhuman fiends. * * *

"'Overland Trail' portrays our original Americans as unbelievably stupid savages, believing in the most ridiculous witchcraft. Example: In one episode, two whites face twenty Indians on the opposite ridge. 'The Indians will be scared of us if we jump up and down and throw sand at each other,' one of the white men says. They performed this bit of mumbo-jumbo and lo! the simple-minded Indians flee."

"Perhaps it is true that a script needs villains," Mr. Birnkrant adds, "but must it always be the Indian?"

June 26, 1960

INDIAN TRIBESMEN DECRY U. S. POLICY

By DONALD JANSON
Special to The New York Times.

CHICAGO, June 14—One word that inflames tribesmen now attending the American Indian Chicago Conference is "termination."

"Termination" is the term used to describe the Government's recent policy of ending Federal trusteeship over the assets of tribes that are considered capable of handling their own money, property, services and affairs.

The policy is apparently being held in abeyance while a Kennedy Administration task force investigates it. Seven tribes or bands of Indians have been "terminated" since the policy was enunciated in 1953.

The largest was the Menominee tribe, which had been under Federal guardianship for more than a century. Termination became final for the Menominees last April 30. The 234,000-acre Menominee Reservation established by negotiations with Chief Oshkosh in 1854 is now Menominee County in Wisconsin.

Now Most Pay Taxes

The tribe's assets, principally timberland and a sawmill, have been turned over to the Indians to be operated by a tribal corporation. For the first time, each Menominee's interest in common property is represented by alienable securities. For the first time he must pay taxes and doctor bills and adhere to state laws limiting his right to hunt and fish, even in the county.

Legislation is pending in Washington to provide aid to ease the transition. Mrs. Irene Mack, a member of the pre-termination advisory council, the administrative body of the tribe, said that today the tribe's situation was "a chaotic mess."

"It's steadily getting worse," she said. "Taxation is not understood and is scaring the people to death. How are we going to get money to pay taxes? We'll have to hope it rains down from heaven."

Many members of the tribe who are attending the conference did not like the way the undertaking in free enterprise and political democracy was going. They were fearful that their tribal identity and culture were in danger and that economically they would fare badly without Federal protection.

Hope for Logging Jobs

Edwin and Wilbur Wilber, young Menominees from Keshena, their county's only town, have been jobless since termination.

Logging operations and the sawmill represent the chief source of income for most of the Indians in the county. The tribe's lumber business is operating in the red now because of a decline in demand. Employment is half of normal.

The county's only hospital was closed earlier this year. Funds for bringing it up to state standards are lacking, Monroe Weso said. He said that the reservation's two doctors had left the county.

Hospitals in adjacent counties prefer to exclude Indians, Mrs. Mack said.

"We have not been able to continue paying mission school expenses," said Mrs. Weso, a mother of seven who said that she had to leave her husband and family on the reservation to come to Chicago for a job.

"No other tribe should accept termination under the conditions under which the Menominees were forced to accept it," Mr. Weso said.

A tribal council in 1954 accepted the principal of termination when informed by a Senate subcommittee that the Menominees could not expect to get a per capita disbursement of $1,500 from tribal funds unless they also approved termination to demonstrate that they considered themselves capable of administering their own affairs.

"There would have been no termination if we hadn't been bribed," Mrs. Wilber said.

Government attempts to force or speed the assimilation of Indians into the broader American society were roundly assailed at the conference last night.

"American Indians can be integrated into the total American society without giving up the inherent right of human beings to be different," Edward P. Dozier said in a keynote address to the opening session of the week-long meeting. About 1,000 Indians representing more than 100 tribes are attending the sessions at the University of Chicago.

One target of Mr. Dozier, a Professor of Anthropology at the University of Arizona and a member of the Santa Clara Pueblo group of Indians, was the termination policy.

June 15, 1961

Infant Deaths Among Indians Found 3 Times U. S. Figure

By DONALD JANSON
Special to The New York Times.

CHEROKEE, N. C., Sept. 8—Nearly 5 per cent of all babies of American Indians die before they are a year old.

This is more than three times the national infant mortality rate.

Indian babies fare as well as others during their first few weeks—at the hospital and shortly after returning home.

The high death rate is in the period between 1 month and 1 year of age. Twenty-one per cent of all Indian deaths occur in this period.

In large part, the United States Public Health Service blames insanitary living conditions on many of the 200 reservations in the country.

The service is conducting a campaign to stamp out this blight. Legislation passed in 1955 is providing money for modern plumbing and village sewer systems, where practical.

Sixty-seven projects are planned for the current fiscal year.

Forrest J. Gerard of Montana's Blackfeet Tribe, tribal relations officer for the service, said in an interview today that dysentery and diarrhea, two of the biggest killers of Indian babies, would be "considerably reduced" by the campaign to bring sanitation to the reservation.

He is attending the nineteenth annual convention of the National Congress of American Indians, an all-Indian group representing some seventy-five tribes and more than half of the nation's 600,000 Indians.

For the first time in the centuries that they have lived in them, modern plumbing has come to the adobe pueblos of the Zuñis, one of the first tribes to benefit from the Public Health Service program in New Mexico.

Solving the challenge presented by the irregular, piled-up, apartment-like dwellings, engineers installed a maze of piping and underlaid the entire community with a sewer system.

The Zuñis provided the work force and agreed to destroy their outdoor toilets and move their horse corrals out of town.

Mr. Gerard said that acceptance of the idea of modern sanitation facilities "has been very good" throughout the reservations. Tribes that can afford it are sharing in the cost. The Navahos of Arizona, for example, are putting up money as well as labor to equip villages that never before had running water.

September 9, 1962

U.S. INDIANS SEEK TO REMAIN APART

Special to The New York Times

BISMARCK, N. D., Sept. 14—Indian tribes convening here this week made it clear that that their aims did not match those of Negroes struggling for integration.

Quite the opposite is true. Indians want to remain a group apart, to preserve their ancient culture.

This was borne out by speeches of Indian leaders to the 20th annual convention of the National Congress of American Indians.

It was also borne out by resolutions denouncing efforts to assimilate reservation Indians into the white man's society by such means as state assumption of civil and criminal jurisdiction now largely reserved to Federal and tribal courts.

Treaties guarantee the Indians the right to live their own lives on land set aside for them, and they will resist all efforts to "violate" the treaties, delegates declared.

Rights and Customs

As citizens the Indian wants all the constitutional rights of whites, Robert Burnette of the Rosebud Sioux tribe said in an interview. But the Indian also insists, Mr. Burnette said, on the additional, treaty-secured right to a large degree of self-government and to freedom to retain tribal customs and values that are not a part of Caucasian culture.

These include such characteristics as non-competitiveness and seemingly spendthrift generosity, as well as freedom to hunt and fish and to work or not.

Instead of acceding to Congress's expressed policy of fostering assimilation, Indians here left no doubt that they want to retain tax-free, Federally aided reservation life for all of their number who prefer it.

At the same time, Mr. Burnette said, Indians will resist any efforts to subvert their civil rights off the reservation.

Demonstrations Opposed

Mr. Burnette said that prejudice in towns near some reservations had led to violations of Indians' rights to service in public accommodations, to equal treatment at the hands of local law enforcement agencies, and to equal opportunities in jobs and housing.

But unlike Negroes battling to overcome the same sort of discrimination, he said, Indians will not engage in protest demonstrations.

"This kind of belligerence does not square with our feeling of patriotism," said the 37-year-old former marine. "We want to give no aid to Communist propaganda about dissension."

Instead, he said, Indians will go to the Federal courts and to the conscience of the American public for relief when they encounter illegal discrimination.

Mr. Burnette has been executive director of the National Congress of American Indians for several years. As such, he represents the 90 affiliated tribes in dealing with Washington. Two-thirds of the nation's 600,000 Indians belong to these tribes. Mr. Burnette's organization is the country's only national all-Indian group.

Indians at the week-long gathering agreed that they suffered much less from racial discrimination than Negroes did.

The major problem, according to Walter Wetzel, chairman of the Blackfeet in Montana, is economic underdevelopment on the reservations.

Indians on reservations live in varying degrees of poverty, despite national affluence, because of lack of jobs on and near their homes. This is partly attributable to the poor, though improving, education in schools provided by the Federal Bureau of Indian Affairs.

Mr. Burnette said 9,000 Indian children on reservations were out of school now because there were no seats for them. Many others drop out before finishing high school because of lack of motivation.

Many schools are overcrowded and dilapidated, the bureau concedes. It also rates 90 per cent of Indian housing as substandard.

September 15, 1963

Coast Navajos Win Right to Use Peyote In Religious Rites

Special to The New York Times

SAN FRANCISCO, Aug. 24—The Supreme Court of California ruled today that the use of peyote, a hallucinatory drug, in religious ceremonies does not violate state narcotics laws.

The court set aside the convictions of Jack Woody and two other Navajo Indians who were arrested in 1962 in a hogan near Needles, Calif.

They had been watched by authorities while they sat in a circle sipping peyote tea. Peyote, which contains mescaline, grows in buttons on top of cacti throughout the Rio Grande Valley.

The defendants, who were given suspended sentences, were members of the native American Church, which is incorporated under California law.

Indian tribes, the court noted, are known to practice the religion of peyotism in Arizona, Montana, Oklahoma, Saskatchewan and Wisconsin as well as California.

"To forbid the use of peyote is to remove the theological heart of peyotism," the court said.

The faith's adherents believe that the partaking of peyote brings one into direct contact with God. They also address prayers to it and consider it to be a protector.

The court compared Indian servicemen who carry the buttons in beaded bags with Roman Catholics who wear medallions.

August 25, 1964

Eskimos and Indians Reject Negro Bid to Join Civil Rights Drive in Alaska

By HOMER BIGART
Special to The New York Times

BARROW, Alaska—The Eskimos and Indians of Alaska, now busy forming native associations that they hope will generate enough political pressure to lift them from their squalor, have rejected overtures from Alaskan Negroes to make common cause in a struggle for equal rights.

Surprisingly, there is a large Negro minority in Alaska. Anchorage has the biggest community, more than 4,000, and Fairbanks is next with more than 1,500.

Elsewhere in the state there are only a few hundred Negroes, living mainly in Juneau, Kodiak and the Kenai Peninsula.

Most Alaskan Negroes came from the West Coast in the late 1940's and early 1950's, lured by the building boom and the prospect of earning good money as construction workers. Reconstruction after the earthquake havoc in the Anchorage area two years ago brought another boom, but now building has tapered off. There are many Negroes unemployed and some of them, ruefully, have offered a new definition for the Alaskan sourdough, or prospector—"sour on the country but not enough dough to leave it."

Here in Barrow, the current focus of Eskimo political ferment, there are no Negroes, at least no permanent Negro residents. However, educated natives in this northernmost community have followed with sharp interest the civic rights agitation in the "lower 48."

Although Barrow Eskimos say they have no intention of resorting to marches, sit-ins and other dramatic manifestations of protest—which, after all, would be pointless in a community where discrimination in public facilities vanished 20 years ago—they are convinced that economic security for their people can only be attained by the creation of a political power bloc.

There was even talk last winter of forming a native political party that would unite Alaska's 27,000 Eskimos who live on the Arctic slope and the Bering coast, the 17,000 Indians of the interior and southeastern Alaska, and the 6,000 Aleuts, a people of mixed Eskimo and Russian blood, who live on the Aleutian Islands and the Kenai and Alaskan Peninsulas.

Vast Tracts Sought

This plan was shelved when the native leaders came to realize they could exercise more leverage if they remained a powerful minority within the Democratic party.

Although eager to bring about a confederation of Eskimo and Indian groups, the leaders reject an alliance with Negroes because "our problems are different."

Their most pressing concern, they explain, is obtaining title to vast tracts of Arctic and Subarctic land claimed "by reason of occupancy from time immemorial" and guaranteed them, they contend, by the Treaty of Cessation in 1867.

"Some" discrimination is encountered in housing and employment when natives go to the cities, but leaders are afraid that the land issue would be submerged if Eskimos were to involve themselves with Negroes in the civil rights movement.

"The Negro wants complete integration, we Eskimos tend to segregate ourselves by choice," said Hugh Nicholls, a carpenter who is executive director of the Arctic Slope Native Association.

Exploitation Is Cited

"Let me be blunt," said Sam Taalak, president of the association. "I've seen exploitation of Alaska. All the money they dug out of the land went out of the state and didn't help us a darn bit. We want to preserve this land for our children. I tell you this is a harsh country. We run out of game animals and we will hit the relief rolls pretty hard."

"I don't dislike white people," Willie A. Hensley, a leader of the newly formed Northwest Alaska Native Association, told a visitor in Kotzebue. "But we have been exploited—maybe unconsciously—but it has happened."

"Our people are more aware of native rights than ever before," said State Senator Eban Hopson, a Barrow Eskimo. "We are after flush toilets and running water. Compared to what the Government is putting into Vietnam, it isn't much."

Of 300 native houses in Barrow, five have running water, Mr. Hopson said. Barrow is the most prosperous Eskimo settlement in Alaska.

In Fairbanks, Howard Rock, an Eskimo who is the editor of The Tundra Times, which says it has a native circulation of 2,500, said he saw "no direct relation between Negro and Indian problems."

"To natives, discrimination is not really a big problem," Mr. Rock declared. "We are more concerned with education sanitation and political action. Personally, I have a lot of sympathy for these Negro people, but I don't know how to help them."

An Athabascan Indian leader, Ralph W. Perdue, who is a jeweler in Fairbanks, said that "Negroes want to join forces with us, but we don't want to inherit their problems."

"I have a lot of colored friends and I respect them," he continued, "but the general feeling of those who have worked with colored people is this: They have a tendency to be lazy, and if you force them to work harder they yell 'discrimination.'"

However, a Negro leader in Fairbanks, the Rev. Larmon Stennis, head of the local branch of the National Association for the Advancement of Colored People, said that "all native leaders I talked with said our problems are identical."

Mr. Stennis and Willard L. Bowman, a Negro from Anchorage who is head of the Alaska Human Rights Commission, said that natives were victims of discrimination, but to a lesser extent than Negroes. Discrimination is encountered mainly in the fields of employment and housing, they said.

Negroes encounter "basically the same discrimination in Alaska as in any other state," Mr. Bowman said.

"The department stores want a Jackie Robinson or a Lena Horne," he asserted, "and when they hire one or two Negroes they think they've done the job."

Down in southeastern Alaska, where the politically sophisticated Tlinget and Haida Indians founded the first Indian association, the Alaska Native Brotherhood, 53 years ago to fight discrimination, John Hope of Ketchikan, a member of the brotherhood's executive committee, said that vestiges of discrimination persisted.

"All of us know we can't join the Moose or the Elks or the Improved Order of Redmen," Mr. Hope said. "They deny it, but they can't point to one native who's a member."

"Ask a non-Indian about discrimination and he'll say it doesn't amount to much," said Stanley J. McCutcheon, an Anchorage lawyer who is a former Speaker of the Alaskan House. "But if you ask an Indian, it amounts to a hell of a lot."

Mr. McCutcheon was the lawyer who helped the Tyonek Indians win title to oil lands that the tribe has just leased for $12.5-million, thinks the Alaskan native has been shabbily treated.

"When they talk about the underprivileged in Vietnam and let our own people go to hell, it doesn't make sense," he declared. "America ought to be ashamed of herself."

August 1, 1966

A Novelist Fights for His People's Lore

By HENRY RAYMONT

N. Scott Momaday, the Pulitzer Prize-winning novelist, is fighting to keep the poetry and folklore of the American Indian from disappearing with "the vanishing American."

Mr. Momaday, a Kiowa Indian who won a Pulitzer this year for his novel "House Made of Dawn," has just moved to Berkeley, where he will start a program in American Indian literature at the University of California.

The program, to be incorporated into the department of comparative literature next spring, is believed to be the first to be wholly dedicated to the preservation of the legends and folk tales of the Navahos, Kiowas and other Indian tribes.

'The Emerging American'

"It is a wonderful idea, and I hope other universities and colleges will soon take a new interest in the legacy of the American Indian," the 35-year-old author said in a telephone conversation from Berkeley yesterday. "If such a trend develops, I am sure that the notion of 'the vanishing American' may be transformed into 'the emerging American.'"

"The American Indian has a great tradition of poetry and mythology," he continued, "but so appallingly little has been preserved in writing that it will soon be extinguished unless the universities make a systematic effort to preserve it."

Mr. Momaday brought the same message to New York publishers and literary critics during a recent series of talks here, using his newly won fame to draw attention to the plight of the American Indian heritage.

"I am convinced that American Indian culture has greater sensibility for language than almost any other culture," he said puffing a blackened briar pipe at a luncheon given by Harper & Row, his publisher.

Possession of a Few

"The Kiowas, for example, tell remarkable stories full of beautiful and intricate plots with great concern for human attitudes and their relation to nature. But they are based on an oral tradition—word by mouth—which has been the private possession of a very few.

"An appalling amount has been lost, but a great deal still remains and has to be preserved. This will require a determined effort by scholars and publishers to systematically record these tales before the tradition is completely lost as the young leave the reservations and become absorbed by the technological world."

Mr. Momaday, stocky and quiet-spoken, sees in the revival of Indian folklore a rare opportunity for a "linguistic antidote" to what he calls the harshness and materialism that has crept into the American idiom.

"Our children are being bombarded with television commercials, radio and the slang propagated by junk mail," he said. "As a result, the sensibility for language has been dulled. Yet because of its isolation, because it is not written, the Indian language is not so vulnerable. It has retained a sense of poetry and an affinity to nature that could be a boon to the young."

A Painter's Son

Mr. Momaday, the son of a Kiowa painter, grew up on Southwestern Indian reservations and attended Indian schools before he began writing at the University of New Mexico and Stanford. He has divided his time between researching Indian folklore and doing scholarly work on 19th-century American literature.

The emphasis on the oral tradition, he maintains, has long impeded the emergence of the American Indian as a writer, very much as the Old Testament prohibition against reproducing the image of God caused Jews to be latecomers in the visual arts.

In his poetry he seeks to retain the Indian perspective ("looking closely at the landscape, severed from the world of technology") as illustrated in "Headwaters," a poem in "The Way to Rainey Mountain," a collection of Kiowa folk tales published recently by the University of New Mexico Press with his father's illustrations. It reads:

Noon in the intermountain plain:
There is scant telling of the marsh—
A log, hollow and weather-stained,
An insect at the mouth, and moss—
Yet waters rise against the roots,
Stand brimming to the stalks.
What moves?
What moves on this archaic force
Was wild and welling at the source.

July 26, 1969

This Country Was a Lot Better Off When the Indians Were Running It

By VINE DELORIA Jr.

ON Nov. 9, 1969, a contingent of American Indians, led by Adam Nordwall, a Chippewa from Minnesota, and Richard Oakes, a Mohawk from New York, landed on Alcatraz Island in San Francisco Bay and claimed the 13-acre rock "by right of discovery." The island had been abandoned six and a half years ago, and although there had been various suggestions concerning its disposal nothing had been done to make use of the land. Since there are Federal treaties giving some tribes the right to abandoned Federal property within a tribe's original

VINE DELORIA Jr. is the author of "Custer Died for Your Sins: An Indian Manifesto," published last year.

territory, the Indians of the Bay area felt that they could lay claim to the island.

For nearly a year the United Bay Area Council of American Indians, a confederation of urban Indian organizations, had been talking about submitting a bid for the island to use it as a West Coast Indian cultural center and vocational training headquarters. Then, on Nov. 1, the San Francisco American Indian Center burned down. The center had served an estimated 30,000 Indians in the immediate area and was the focus of activities of the urban Indian community. It became a matter of urgency after that and, as Adam Nordwall said, "it was GO." Another landing, on Nov. 20, by nearly 100 Indians in a swift midnight raid secured the island.

The new inhabitants have made "the Rock" a focal point symbolic of Indian people. Under extreme difficulty they have worked to begin repairing sanitary facilities and buildings. The population has been largely transient, many people have stopped by, looked the situation over for a few days, then gone home, unwilling to put in the tedious work necessary to make the island support a viable community.

THE Alcatraz news stories are somewhat shocking to non-Indians. It is difficult for most Americans to comprehend that there still exists a living community of nearly one million Indians in this country. For many people, Indians have become a species of movie actor periodically dispatched to the Happy Hunting Grounds by John Wayne on the "Late, Late Show." Yet there are some 315 Indian tribal groups in 26 states still functioning as quasi-sovereign nations under treaty status; they range from the mammoth Navajo tribe of some 132,000 with 16 million acres of land to tiny Mission Creek of California with 15 people and a tiny parcel of property. There are over half a million Indians in the cities alone, with the largest concentrations in San Francisco, Los Angeles, Minneapolis and Chicago.

The take-over of Alcatraz is to many Indian people a demonstration of pride in being Indian and a dignified, yet humorous, protest against current conditions existing on the reservations and in the cities. It is this special pride and dignity, the determination to judge life according to one's own values, and the unconquerable conviction that the tribes will not die that has always characterized Indian people as I have known them.

I WAS born in Martin, a border town on the Pine Ridge Indian Reservation in South Dakota, in the midst of the Depression. My father was an Indian missionary who served 18 chapels on the eastern half of the reservation. In 1934, when I was 1, the Indian Reorganization Act was passed, allowing Indian tribes full rights of self-government for the first time since the late eighteen-sixties. Ever since those days, when the Sioux had agreed to forsake the life of the hunter for that of the farmer, they had been systematically deprived of any voice in decisions affecting their lives and property. Tribal ceremonies and religious practices were forbidden. The reservation was fully controlled by men in Washington, most of whom had never visited a reservation and felt no urge to do so.

The first years on the reservations were extremely hard for the Sioux. Kept confined behind fences they were almost wholly dependent upon Government rations for their food supply. Many died of hunger and malnutrition. Game was scarce and few were allowed to have weapons for fear of another Indian war. In some years there was practically no food available. Other years rations were withheld until the men agreed to farm the tiny pieces of land each family had been given. In desperation many families were forced to eat stray dogs and cats to keep alive.

By World War I, however, many of the Sioux families had developed prosperous ranches. Then the Government stepped in, sold the Indians' cattle for wartime needs, and after the war leased the grazing land to whites, creating wealthy white ranchers and destitute Indian landlords.

With the passage of the Indian Reorganization Act, native ceremonies and practices were given full recognition by Federal authorities. My earliest memories are of trips along dusty roads to Kyle, a small settlement in the heart of the reservation, to attend the dances. Ancient men, veterans of battles even then considered footnotes to the settlement of the West, brought their costumes out of hiding and walked about the grounds gathering the honors they had earned half a century before. They danced as if the intervening 50 years had been a lost weekend from which they had fully recovered. I remember best Dewey Beard, then in his late 80's and a survivor of the Little Big Horn. Even at that late date Dewey was hesitant to speak of the battle for fear of reprisal. There was no doubt, as one watched the people's expressions, that the Sioux had survived their greatest ordeal and were ready to face whatever the future might bring.

In those days the reservation was isolated and unsettled. Dirt roads held the few mail routes together. One could easily get lost in the wild back country as roads turned into cowpaths without so much as a backward glance. Remote settlements such as Buzzard Basin and Cuny Table were nearly inaccessible. In the spring every bridge on the reservation would be washed out with the first rain and would remain out until late summer. But few people cared. Most of the reservation people, traveling by team and wagon, merely forded the creeks and continued their journey, almost contemptuous of the need for roads and bridges.

THE most memorable event of my early childhood was visiting Wounded Knee where 200 Sioux, including women and children, were slaughtered in 1890 by troopers of the Seventh Cavalry in what is believed to have been a delayed act of vengeance for Custer's defeat. The people were simply lined up and shot down much as was allegedly done, according to newspaper reports, at Songmy. The wounded were left to die in a three-day Dakota blizzard, and when the soldiers returned to the scene after the storm some were still alive and were saved. The massacre was vividly etched in the minds of many of the older reservation people, but it was difficult to find anyone who wanted to talk about it.

Many times, over the years, my father would point out survivors of the massacre, and people on the reservation always went out of their way to help them. For a long time there was a bill in Congress to pay indemnities to the survivors, but the War Department always insisted that it had been a "battle" to stamp out the Ghost Dance religion among the Sioux. This does not, however, explain bayoneted Indian women and children found miles from the scene of the incident.

STRANGELY enough, the Depression was good for Indian reservations, particularly for the people at Pine Ridge. Since their lands had been leased to non-Indians by the Bureau of Indian Affairs, they had only a small rent check and the contempt of those who leased their lands to show for their ownership. But the Federal programs devised to solve the national economic crisis were also made available to Indian people, and there was work available for the first time in the history of the reservations.

The. Civilian Conservation Corps set up a camp on the reservation and many Indians were hired under the program. In the canyons north of Allen, S. D., a beautiful buffalo pasture was built by the C.C.C., and the whole area was transformed into a recreation wonderland. Indians would come from miles around to see the buffalo and leave with a strange look in their eyes. Many times I stood silently watching while old men talked to the buffalo about the old days. They would conclude by singing a song before respectfully departing, their eyes filled with tears and their minds occupied with the memories of other times and places. It was difficult to determine who was the captive—the buffalo fenced in or the Indian fenced out.

While the rest of America suffered from the temporary deprivation of its luxuries, Indian people had a period of prosperity, as it were. Paychecks were regular. Small cattle herds were started, cars were purchased, new clothes and necessities became available. To a people who had struggled along on $50 cash income per year, the C.C.C. was the greatest program ever to come along. The Sioux had climbed from absolute deprivation to mere poverty, and this was the best time the reservation ever had.

World War II ended this temporary prosperity. The C.C.C. camps were closed; reservation programs were cut to the bone and social services became virtually nonexistent; "Victory gardens" were suddenly the style, and people began to be aware that a great war was being waged overseas.

The war dispersed the reservation people as nothing ever had. Every day, it seemed, we would be bidding farewell to families as they headed west to work in the defense plants on the Coast.

A great number of Sioux people went west and many of the Sioux on Alcatraz today are their children and relatives. There may now be as many Sioux in California as there are on the reservations in South Dakota because of the great wartime migration.

THOSE who stayed on the reservation had the war brought directly to their doorstep when they were notified that their sons had to go across the seas and fight. Busloads of Sioux boys left the reservation for parts unknown. In many cases even the trip to nearby Martin was a new experience for them, let alone training in Texas, California or Colorado. There were always going-away ceremonies conducted by the older people who admonished the boys to uphold the old tribal traditions and not to fear death. It was not death they feared but living with an unknown people in a distant place.

I was always disappointed with the Government's way of handling Indian servicemen. Indians were simply lost in the shuffle of 3 million men in uniform. Many boys came home on furlough and feared to return. They were not cowards in any sense of the word but the loneliness and boredom of stateside duty was crushing their spirits. They spent months without seeing another Indian. If the Government had recruited all-Indian outfits it would have easily solved this problem and also had the best fighting units in the world at its disposal. I often wonder what an all-Sioux or Apache company, painted and singing its songs, would have done to the morale of élite German panzer units.

ON "THE ROCK": Two young Indians in a corridor of the abandoned Alcatraz prison. The island has become "a focal point symbolic of Indian people" since the new tenants moved in.

After the war Indian veterans straggled back to the reservations and tried to pick up their lives. It was very difficult for them to resume a life of poverty after having seen the affluent outside world. Some spent a few days with the old folks and then left again for the big cities. Over the years they have emerged as leaders of the urban Indian movement. Many of their children are the nationalists of today who are adamant about keeping the reservations they have visited only on vacations. Other veterans stayed on the reservations and entered tribal politics.

THE reservations radically changed after the war. During the Depression there were about five telephones in Martin. If there was a call for you, the man at the hardware store had to come down to your house and get you to answer it. A couple of years after the war a complete dial system was installed that extended to most of the smaller communities on the reservation. Families that had been hundreds of miles from any form of communication were now only minutes away from a telephone.

Roads were built connecting the major communities of the Pine Ridge country. No longer did it take hours to go from one place to another. With these kinds of roads everyone had to have a car. The team and wagon vanished, except for those families who lived at various "camps" in inaccessible canyons pretty much as their ancestors had. (Today, even they have adopted the automobile for traveling long distances in search of work.)

I left the reservation in 1951 when my family moved to Iowa. I went back only once for an extended stay, in the summer of 1955, while on a furlough, and after that I visited only occasionally during summer vacations. In the meantime, I attended college, served a hitch in the Marines, and went to the seminary. After I graduated from the seminary, I took a job with the United Scholarship Service, a private organization devoted to the college and secondary-school education of American Indian and Mexican students. I had spent my last two years of high school in an Eastern preparatory school and so was probably the only Indian my age who knew what an independent Eastern school was like. As the program developed, we soon had some 30 students placed in Eastern schools.

I insisted that all the students who entered the program be able to qualify for scholarships as students and not simply as Indians. I was pretty sure we could beat the white man at his own educational game, which seemed to me the only way to gain his respect. I was soon to find that this was a dangerous attitude to have. The very people who were supporting the program—non-Indians in the national church establishments—accused me of trying to form a colonialist "élite" by insisting that only kids with strong test scores and academic patterns be sent east to school. They wanted to continue the ancient pattern of soft-hearted paternalism toward Indians. I didn't feel we should cry our way into the schools; that sympathy would destroy the students we were trying to help.

IN 1964, while attending the annual convention of the National Congress of American Indians, I was elected its executive director. I learned more about life in the N.C.A.I. in three years than I had in the previous 30. Every conceivable problem that could occur in an Indian society was suddenly thrust at me from 315 different directions. I discovered that I was one of the people who were supposed to solve the problems. The only trouble was that Indian people locally and on the national level were being played off one against the other by clever whites who had either ego or income at stake. While there were many feasible solutions, few could be tried without whites with vested interests working night and day to destroy the unity we were seeking on a national basis.

In the mid-nineteen-sixties, the whole generation that had grown up after World War II and had left the reservations during the fifties to get an education was returning to Indian life as "educated Indians." But we soon knew better. Tribal societies had existed for centuries without going outside themselves for education and information. Yet many of us thought that we would be able to improve the traditional tribal methods. We were wrong.

For three years we ran around the conference circuit attending numerous meetings called to "solve" the Indian problems. We listened to and spoke with anthropologists, historians, sociologists, psychologists, economists, educators and missionaries. We worked with many Government agencies and with every conceivable doctrine, idea and program ever created. At the end of this happy round of consultations the reservation people were still plodding along on their own time schedule, doing the things they considered important. They continued to solve their problems their way in spite of the advice given them by "Indian experts."

By 1967 there was a radical change in thinking on the part of many of us. Conferences were proving unproductive. Where non-Indians had been pushed out to make room for Indian people, they had wormed their way back into power and again controlled the major programs serving Indians. The poverty programs, reservation and university technical assistance groups were dominated by whites who had pushed Indian administrators aside.

Reservation people, meanwhile, were making steady progress in spite of the numerous setbacks suffered by the national Indian community. So, in large part, younger Indian leaders who had been playing the national conference field began working at the local level to build community movements from the ground up. By consolidating local organizations into power groups they felt that they would be in a better position to influence national thinking.

Robert Hunter, director of the Nevada Intertribal Council, had already begun to build a strong state organization of tribes and communities. In South Dakota, Gerald One Feather, Frank LaPointe and Ray Briggs formed the American Indian Leadership Conference, which quickly welded the educated young Sioux in that state into a strong regional organization active in nearly every phase of Sioux life. Gerald is now running for the prestigious post of chairman of the Oglala Sioux, the largest Sioux tribe, numbering some 15,000 members. Ernie Stevens, an Oneida from Wisconsin, and Lee Cook, a Chippewa from Minnesota, developed a strong program for economic and community development in Arizona. Just recently Ernie has moved into the post of director of the California Intertribal Council, a statewide organization representing some 130,000 California Indians in cities and on the scattered reservations of that state.

BY the fall of 1967, it was apparent that the national Indian scene was collapsing in favor of strong regional organizations, although the major national organizations such as the National Congress of American Indians and the National Indian Youth Council continued to grow. There was yet another factor emerging on the Indian scene: the old-timers of the Depression days had educated a group of younger Indians in the old ways and these people were now becoming a major force in Indian life. Led by Thomas Banyaca of the Hopi, Mad Bear Anderson of the Tuscaroras, Clifton Hill of the Creeks, and Rolling Thunder of the Shoshones, the traditional Indians were forcing the whole Indian community to rethink its understanding of Indian life.

The message of the traditionalists is simple. They demand a return to basic Indian philosophy, establishment of ancient methods of government by open council instead of elected officials, a revival of Indian religions and replacement of white laws with Indian customs; in short, a complete return to the ways of the old people. In an age dominated by tribalizing communications media, their message makes a great deal of sense.

But in some areas their thinking is opposed to that of the National Congress of American Indians, which represents officially elected tribal governments organized under the Indian Reorganization Act as Federal corporations. The contemporary problem is therefore one of defining the meaning of "tribe." Is it a traditionally organized band of Indians following customs with medicine men and chiefs dominating the policies of the tribe, or is it a modern corporate structure attempting to compromise at least in part with modern white culture?

The problem has been complicated by private foundations' and Government agencies' funding of Indian programs. In general this process, although it has brought a great amount of money into Indian country, has been one of cooptation. Government agencies must justify their appropriation requests every year and can only take chances on spectacular programs that will serve as showcases of progress. They are not willing to invest the capital funds necessary to build viable self-supporting communities on the reservations because these programs do not have an immediate publicity potential. Thus, the Government agencies are forever committed to conducting conferences to discover that one "key" to Indian life that will give them the edge over their rival agencies in the annual appropriations derby.

Churches and foundations have merely purchased an Indian leader or program that conforms with their ideas of what Indian people should be doing. The large foundations have bought up the well-dressed, handsome "new image" Indian who is comfortable in the big cities but virtually helpless at an Indian meeting. Churches have given money to Indians who have been willing to copy black militant activist tactics, and the more violent and insulting the Indian can be, the more the churches seem to love it. They are wallowing in self-guilt and piety over the lot of the poor, yet funding demagogues of their own choosing to speak for the poor.

I did not run for re-election as executive director of the N.C.A.I. in the fall of 1967, but entered law school at the University of Colorado instead. It was apparent to me that the Indian revolution was well under way and that someone had better get a legal education so that we could have our own legal program for defense of Indian treaty rights. Thanks to a Ford Foundation program, nearly 50 Indians are now in law school, assuring the Indian community of legal talent in the years ahead. Within four years I foresee another radical shift in Indian leadership patterns as the growing local movements are affected by the new Indian lawyers.

THERE is an increasing scent of victory in the air in Indian country these days. The mood is comparable to the old days of the Depression when the men began to dance once again. As the Indian movement gathers momentum and individual Indians cast their lot with the tribe, it will become apparent that not only will Indians survive the electronic world of Marshall McLuhan, they will thrive in it. At the present time everyone is watching how mainstream America will handle the issues of pollution, poverty, crime and racism when it does not fundamentally understand the issues. Knowing the importance of tribal survival, Indian people are speaking more and more of sovereignty, of the great political technique of the open council, and of the need for gaining the community's consensus on all programs before putting them into effect.

One can watch this same issue emerge in white society as the "Woodstock Nation," the "Blackstone Nation" and the block organizations are developed. This is a full tribalizing process involving a nontribal people, and it is apparent that some people are frightened by it. But it is the kind of social phenomenon upon which Indians feast.

In 1965 I had a long conversation with an old Papago. I was trying to get the tribe to pay its dues to the National Congress of American Indians and I had asked him to speak to the tribal council for me. He said that he would but that the Papagos didn't really need the N.C.A.I. They were like, he told me, the old mountain in the distance. The Spanish had come and dominated them for 300 years and then left. The Mexicans had come and ruled them for a century, but they also left. "The Americans," he said, "have been here only about 80 years. They, too, will vanish, but the Papagos and the mountain will always be here."

This attitude and understanding of life is what American society is searching for.

I WISH the Government would give Alcatraz to the Indians now occupying it. They want to create five centers on the island. One center would be for a North American studies program; another would be a spiritual and medical center where Indian religions and medicines would be used and studied. A third center would concentrate on ecological studies based on an Indian view of nature—that man should live *with* the land and not simply *on* it. A job-training center and a museum would also be founded on the island. Certain of these programs would obviously require Federal assistance.

Some people may object to this approach, yet Health, Education and Welfare gave out $10-million last year to non-Indians to study Indians. Not one single dollar went to an Indian scholar or researcher to present the point of view of Indian people. And the studies done by non-Indians added nothing to what was already known about Indians.

Indian people have managed to maintain a viable and cohesive social order in spite of everything the non-Indian society has thrown at them in an effort to break the tribal structure. At the same time, non-Indian society has created a monstrosity of a culture where people starve while the granaries are filled and the sun can never break through the smog.

By making Alcatraz an experimental Indian center operated and planned by Indian people, we would be given a chance to see what we could do toward developing answers to modern social problems. Ancient tribalism can be incorporated with modern technology in an urban setting. Perhaps we would not succeed in the effort, but the Government is spending billions every year and still the situation is rapidly growing worse. It just seems to a lot of Indians that this continent was a lot better off when we were running it. ■

March 8, 1970

PRESIDENT URGES WIDER INDIAN ROLE IN AID FOR TRIBES

Proposes to Congress That They Get Control Over Their Own Destiny

'INJUSTICE' CONDEMNED

By JAMES M. NAUGHTON
Special to The New York Times

WASHINGTON, July 8 — President Nixon denounced today "centuries of injustice" to the American Indians and proposed a comprehensive program to give them dignity and control over their destiny.

The President said that he would send to Congress legislation to enable tribal groups to assume operational control of many existing Indian aid programs.

Mr. Nixon also formally renounced a 17-year-old national policy under which the Government has sought to cut Indians loose from their dependence on the national administration.

But, at the same time, he charted a course that would avoid continuation of Federal paternalism toward Indians, assuring them instead of assistance while granting them authority to decide how it should be used.

Record Is Recalled

In a gesture that has symbolic significance to all Indians, the President endorsed a pending House resolution that would return 48,000 acres of sacred land in New Mexico to the Taos Pueblo tribe.

Mr. Nixon, in a message to Congress, deplored a history of white "aggression, broken agreements, intermittent remorse and prolonged failure" in treatment of the Indians.

He said that "as a matter of justice and as a matter of enlightened social policy" it was time for an era in which the 462,000 Indians living on reservations and perhaps as many more in urban slums might determine how best to use the Federal help made available to them.

Toward that end, the President said that he would send legislation to Congress giving Indian groups authority to assume control over federally administered programs — such as schools, housing, medical services, economic development, and public works — on which the Government spends some $400-million annually.

Mr. Nixon called for establishment of tribal boards of education to take the place of federally administered school systems and he urged Congress to amend the existing law so that funds for Indian education could go directly to Indian groups as well as to public school systems.

He proposed the Indian Financing Act of 1970, which would raise from the current $27-million to $77-million the amount in a revolving fund for Indian economic development projects and would provide $200-million for privately financed economic projects on Indian lands.

The legislation would also permit tribes to lease parts of their lands for as long as 99 years, as an incentive to private developers who are limited now by Federal law to leases ranging between five and 10 years.

The President announced plans to add $10-million to the current $113-million in appropriations for Indian health programs and announced plans to set up or expand urban centers to aid Indians in Los Angeles, Denver, Phoenix, Omaha, Minneapolis and Fairbanks.

The White House program called for creation of a new post of Assistant Secretary for Indian and Territorial Affairs in the Department of the Interior.

'A Complete Reversal'

And, touching on an issue that has been a sore point with Indian tribal leaders, Mr. Nixon conceded that it amounted to "an inherent conflict of interest" for the Interior and Justice Departments to serve at the same time as counsel to the Government and to Indians pressing claims against it. He proposed an independent agency to be called the Indian Trust Counsel Authority, to represent Indian groups in future legal disputes.

Mr. Nixon's program was hailed by Alvin Josephy, vice president of the American Heritage Publishing Company in New York, an authority on Indian affairs who urged many of the same reforms in a report to the President in February, 1969.

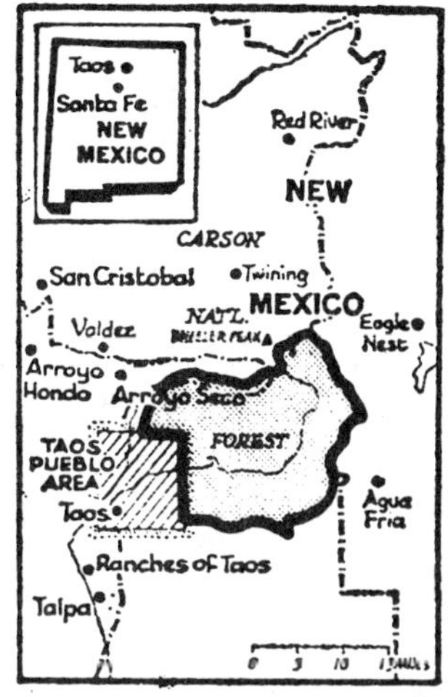

The New York Times July 9, 1970

President endorsed bill that would give Taos Pueblos, in diagonally lined area, the 46,000 acres in shaded area.

"This represents a complete reversal in the Federal attitude toward Indian affairs," Mr. Josephy said in a telephone interview. He called it the first positive action in the long history of Federal relations with Indians.

Many of the programs Mr. Nixon called for today had also been urged by Indian members of the National Council on Indian Opportunity, who met with Vice President Agnew in January to press for "more services, more self-determination and relief from the hovering specter of termination."

Since 1953, when Congress declared that termination of the Government's trustee relationship with Indians should be a long term goal, Indian groups have grown increasingly wary about the implications of the policy.

In rejecting the termination policy, and prodding Congress to do the same, the President said that it was not possible to end the commitments to Indians that had been written over the centuries in treaties and other agreements and that "carry immense moral and legal force."

Although Mr. Nixon's proposals granted to Indians many of the things they have been seeking, it did not provide all the benefits the Indian leaders want.

There was no guarantee, for example, that Indian demands for massive increases in Federal education and health assistance would be met. Nor was there any reference in the President's message to the insistent demands by all Indians that Eskimos, Indians and Aleuts in Alaska be granted title to ancestral lands and compensation for property taken from them.

Nonetheless, Mr. Nixon made it clear that "no Government policy toward Indians can be fully effective unless there is a relationship of trust and confidence" involved. To help restore such trust, he allied himself with the 64-year-old legal effort by the Taos Pueblo tribe

The New Mexico tribe has sought the return of the 48,000 acres surrounding Blue Lake, the focal point of a sacred area they used for six centuries before the Government appropriated it in 1906 for a national forest.

The House of Representatives has twice passed legislation returning the land to the Taos Pueblos, but it has been blocked by Senator Clinton P. Anderson, Democrat of New Mexico, the powerful second-ranking Democrat on the Interior Committee.

The Interior Committee will consider the Blue Lake legislation tomorrow. The President's endorsement of the House bill —"as an important symbol of this Government's responsiveness to the just grievances of the American Indians" — may help to dislodge it.

July 9, 1970

Senate Votes to Give Land to Taos Indians

By **WARREN WEAVER Jr.**
Special to The New York Times

WASHINGTON, Dec. 2—The Senate voted overwhelmingly today to give the Taos Pueblo Indians title to 48,000 acres of Carson National Forest in New Mexico, land they have long claimed as part of their cultural and religious heritage.

It includes Blue Lake, an area sacred to the Taos Pueblos.

The vote was 70 to 12 in favor of legislation backed by the Nixon Administration and a bipartisan bloc of Senators. It followed the defeat, 56 to 21, of an alternative proposal that would have given the tribe "exclusive use" but not ownership of the property.

The measure now goes to President Nixon, whose approval is assured. Senator Robert P. Griffin, Republican of Michigan, said that it was the first of a series of Administration-backed efforts to open a new era for the nation's Indians.

The chief spokesman for the measure was Senator Fred R. Harris, Democrat of Oklahoma, whose wife, LaDonna, a member of the Comanche tribe and head of Americans for Indian Opportunity, sat in the gallery during the three hours of debate.

Mr. Harris argued that the Taos Indians had a unique claim to the New Mexico property because the land and Blue Lake, which it encompasses—had been an intimate part of the tribe's religious life since the thirteenth century.

If other tribes seek to obtain public land for their use in the future, the Oklahoma Senator said, the Senate can judge each case on its merits rather than consider that any binding precedent has been set.

Moments after the final Senate vote, the cacique, or spiritual and religious leader of the Taos Indians, 90-year-old Juan de Jesus Romero, told a news conference in his native tiwa dialect, through an interpreter, that the tribe was "grateful and thankful."

The Governor, or chief, of the tribe, Quirino Romero, brandished a ribbon-decked, silver-headed staff that President Nixon had given him to symbolize recognition of the Pueblo people. And the Governor welcomed the Indian victory "over the power here in Washington."

Under the bill approved by the Senate, title to the New Mexico tract will go to the Taos Indians, to be held in trust by the Secretary of the Interior. The land will be preserved as a wilderness area, and the Indians will be permitted to use it for religious ceremonials and for hunting, fishing and maintaining live stock.

The defeated alternate, drafted by the Senate Interior Committee, would have continued the property under the jurisdiction of the Secretary of Agriculture with the Forest Service directly in charge and the Indians given its "exclusive use."

Senator Henry A. Jackson, Democrat of Washington and chairman of the Interior Committee, argued that giving the Taos Indians title to the land would set a precedent and encourage other tribes with claims against the Government to seek property rather than financial reimbursement.

Senator Jackson predicted that various tribes would "invade our national parks, our national monuments and our national forests" if the bill were passed.

One of the Republicans defending the Indians' right to the land was Senator Barry Goldwater of Arizona, who described how intimately land and growing things could be connected with the religious observances of certain tribes.

"Who got here first is the real question," Mr. Goldwater said. "And yet we try to tell them what they can do with land that they've occupied for 2,000 or 3,000 years."

December 3, 1970

1970 Census Finds Indian No Longer the Vanishing American

By JACK ROSENTHAL
Special to The New York Times

WASHINGTON, Oct. 19—The vanishing American is not vanishing. According to a 1970 census computation published today, the American Indian population is growing four times as fast as the population as a whole.

The principal explanation, experts say, lies in dramatic public health gains. These have, in 15 years, added four years to Indian life expectancy and cut the infant mortality rate in half.

Hence, the Indian population has more than doubled since 1950, when there were 343,410 Indians. In 1960, there were 523,591 and in 1970, 792,730.

The growth rate among Indians in both decades was slightly more than 5 per cent a year, against a general national growth rate of 1.3 per cent year in the nineteen-sixties and 1.8 per cent in the fifties.

Indians are not the fastest growing racial component of the population. That distinction belongs to Filipinos, whose population rose 95 per cent in the nineteen-sixties, to 343,060, according to the census report.

Analysts say the reason for this increase was the Immigration Reform Act of 1965, which revised the prior "national origins" quota system under which immigrants from Western Europe had been favored.

Other results were that the Chinese population grew 83.3 per cent, to a total of 435,062, and the Japanese population increased 27.4 per cent, to 591,290. The relatively lower growth rate in the latter category, according to analysts, reflects Japan's economic health.

The new census computation is limited to race and relies on each person's description of his own race. Ethnic breakdowns, like persons of Irish, Italian, or Spanish-speaking origin, are to be reported later.

Also to be reported later are details about "other" races — including Koreans, Haitians, Aleuts, Eskimos, Malayans and Polynesians.

Persons of races other than white or black live predominantly in the 13 western states, the census report disclosed.

Those states contain 17 per cent of the white population and 8 per cent of the black population. But for other races, the figures are far higher: Japanese, 81 per cent; Filipino, 73 per cent; Chinese, 56 per cent; American Indians, 50 per cent; all other races, 51 per cent.

After the West, the region with the highest Indian population is the South, with 201,222, followed by the Midwest, 151,287, and the Northeast, 49,466.

Some of the Indian population gains may be illusory, in the view of Mose E. Parris, tribal affairs director of the Federal Indian Health Service.

"Indians are more likely to describe themselves as Indians now, rather than white," he said in an interview. "There's increased Indian consciousness, increased pride in their heritage."

But the importance of improved health as an explanation for the increases is overriding, he and other officials believe.

One measure of these improvements is increased life expectancy among Indians. In 1950, it was 60 years for Indians, compared with 68 for whites, according to Mozart I. Spector, the Indian Health Service's statistical chief. By 1967, the Indian figure had improved four years, while the white figure had gained only two.

Traditionally Indians have had a higher birth rate than whites, but they have also had a strikingly higher infant death rate. This is still substantially higher than among whites, Mr. Spector said, but has been greatly reduced.

In 1950, the Indian mortality rate was 63 infant deaths per 1,000 live births. By 1968, the rate had dropped to 31. The white figure in 1968 was 22.

In the same period, mortality rates for all Indians declined sharply in numerous categories. For example, he said, "The death rate from tuberculosis, a traditional scourge of the Indians, declined 75 per cent."

October 20, 1971

Census Table of Racial Trends

Special to The New York Times

WASHINGTON, Oct. 19—The following table shows the number of persons who stated their race for the United States census as white, Negro, American Indian, Japanese, Chinese, Filipino or a category grouped as "other" in 1960 and 1970; the percentage growth for each group and the percentage of each group that lived in the West in 1970.

Group	1960	1970	Growth Percentage	Percentage of Group Living in West*
White	158,831,732	177,748,975	11.9%	17%
Negro	18,871,831	22,580,289	19.7	8
American Indian	523,591	792,730	51.4	50
Japanese	464,332	591,290	27.4	81
Chinese	237,292	435,062	83.3	56
Filipino	176,310	343,060	94.9	73
Other #	218,087	720,520	230.0	51
Total	179,323,175	203,211,926	13.3	17

*The Census Bureau lists as Western states Hawaii, Alaska, Washington, Oregon, California, Nevada, Utah, Arizona, Colorado, Wyoming, Idaho, Montana and New Mexico.

#Includes Koreans, Hawaiians, Aleuts, Eskimos, Malayans and Polynesians.

An Indian Affair

American Indian Students Concerned About Nicknames, Mascots in Sports

By MARTY RALBOVSKY

Three years ago, a group of students at Dartmouth College, assembled by Howard Bad Hand, Dwayne Birdbear, Travis Kingsley and Rick Buckanaga, demanded that the athletic department stop entertaining fans at home football games by employing undergraduates, dressed as Indians, to do imitations of war dances along the sidelines and at midfield during half-time.

The students said the practice was demeaning to the American Indian and that they were insulted because the tribal customs of their ancestors "were being used to feed the fantasies of the insensitive." They also suggested that Dartmouth consider changing its nickname to something other than "The Indians." School officials, after several meetings on the subject, decided to retain the nickname, but to abolish the Indian mascots.

"The mascots had been jeered and laughed at for years and we just decided to put an end to that kind of nonsense," said Bill Yellowtail, a student in the American Indian studies program at Dartmouth and a member of the group.

"The old grads, especially, used to get a big kick out of them every time they'd come back to see a game; they'd point them out to their kids or to their grandchildren, just like they'd point out a monkey at a zoo, 'Look, look, there it is, the Indian.' To this day, a lot of the old grads jump on us for what we did; they say we destroyed one of the school's oldest and most enjoyable sports traditions.

"But we feel we did our part in eliminating another false illusion. Too many people in this country still think of Indians as savages doing war dances and wearing feathered headdresses and having two-word vocabularies: 'How and Ugh.' People in sports are as responsible as anybody for perpetuating these illusions, with their Indian mascots and their Indian half-time shows and their Indian nicknames. I've often wondered to myself if the people who owned these teams ever stopped to think what goes through the mind of a 10-year-old Indian kid on a reservation in North Dakota when he picks up a sports page and reads a headline, 'Redskins Scalp Chiefs'?"

Marquette Mascot Goes

Last year at Marquette University, a similar incident occurred: Indian students petitioned the governing student organization, the Associated Students of Marquette, asking that the school's Indian mascot, nicknamed "Willie Wampum," be abolished because it was portraying the American Indian in a demeaning manner.

The student senate subsequently passed a resolution calling for the mascot's abolition and, last April, the school announced that the mascot, like Dean Meminger's basketball jersey, would be permanently retired.

The Association for American Indian Affairs here is hardly ecstatic over the number of sports teams in the country bearing nicknames alluding to the American Indian. Six professional teams and 97 colleges, according to the Blue Book of College Athletics, have Indian nicknames, ranging from Redskins (Washington, National Football League) to Choctaws (Mississippi College) to Scalpers (Huron, S.D., College).

Jeffrey Newman, assistant director of the association, said: "If we had the money, we would file suits against every college and professional team in the country with an Indian nickname we found offensive. I think the first team we would go after would be the Atlanta Braves; if any team in the country is exploiting the American Indian for its own purposes it is this one. It is outrageous, I feel, to have a man dressed as an Indian, sitting in an alleged tepee outside the outfield fence, doing a silly dance every time some player hits a home run. Even the name they gave him, Chief something or other, is discriminatory. Would they hire a black man to sit in a tar-paper shack out there and come out picking cotton every time a player hit a home run? No, they wouldn't dare."

'No Criticism of Act'

The man portraying the Indian in Atlanta, Levi Walker, said he was an American Indian by birth and a showman by profession. He said he had never been criticized for his act, but that he was aware of a faction among Indians "that was trying to do away with the feather-leather-skin-and-beads image." Walker also said he did not feel he was exploiting the American Indian; that his war dances were authentic "in a sense," and that his most fervent followers were young people—"I get more fan mail from kids than most of the players on the team."

"The danger in these shows and mascots," said Newman, "is this: They keep alive false myths about the American Indian . . . you know, being bloodthirsty and a warmonger and, always, the aggressor. Young people, like the white, middle-class, suburban kid who may never meet an Indian in his entire life, are particularly vulnerable to them. Subconsciously he's developing an inaccurate image of what Indian people are, and were. The same kid wouldn't think of calling a team the 'Blackskins' or the 'Yellowskins,' but he has Redskins' pennants on his bedroom walls.

"Somebody once asked me if Indian children look up to and identify with athletes in America? I said I didn't think so for two reasons: First, there are very few of their own kind in sports, a Sonny Sixkiller, a Jim Plunkett, a Johnny Bench, maybe, and that's about it, and, second, how can you expect an Indian kid to identify with sports people when he can't identify with his own?"

November 14, 1971

American Indian Activists Winning Bureau Reform

By HOMER BIGART

Special to The New York Times

WASHINGTON, Jan. 8—President Nixon's doctrine of self-determination for the American Indian is finally moving toward implementation after a long and bitter power struggle within the Bureau of Indian Affairs.

That struggle has apparently ended in victory for the Commissioner of Indian Affairs, Louis R. Bruce, and for the "fearless fourteen," a group of young Indian activists brought in by Mr. Bruce to galvanize a complacent and predominantly white bureaucracy.

The decline and fall of the bureaucrats entrenched in an agency responsible for the well-being of the Indians since 1824 began 18 months ago when Mr. Nixon sent a message to Congress saying that Indians, not "outsiders" should run programs directly serving Indians.

The bureaucrats did not give up without a fight. Early attempts at reform were frustrated so long by the stalling tactics of senior officials that some impatient Indians began calling the new Commissioner "Bruce the Goose" and "the Indian from Greenwich Village."

At first glance Mr. Bruce did not seem a likely innovator of radical change. Although his father was a Mohawk and his mother an Oglala Sioux, Mr. Bruce in his dark conservative suits and fraternity pin looked more like a Madison Avenue executive than an Indian. In fact, he was vice president of a New York advertising firm and a gentleman farmer who lived in Greenwich Village and spent weekends on a 400-acre dairy farm near Cooperstown, N. Y.

But he had kept up with Indian problems. He was a mainstay of the National Congress of American Indians, which he helped establish as the leading Indian lobby in Washington. And he was one of the few Republican Indians extant. So, Mr. Nixon appointed him commissioner in 1969.

Mr. Bruce's early modest attempts to reinvigorate the bureau were supported by the then Secretary of the Interior, Walter J. Hickel. But after Mr. Hickel's dismissal by President Nixon on grounds of "mutual lack of confidence," the new Secretary, Rogers C. B. Morton, looked coldly on reform and gave ear to the embattled bureaucrats and their allies, a group of conservative tribal chiefs who complained that Mr. Bruce was packing the bureau with young "red power" militants.

Diluted His Power

Last July Secretary Morton tried to dilute the power of Mr. Bruce by appointing as Deputy Indian Commissioner an old-line official, John O. Crow, a Cherokee from Oklahoma. Mr. Crow was given sole power to redelegate authority, including that reserved for Mr. Bruce.

Mr. Crow moved swiftly against the activists. He replaced Leon F. Cook as acting director of economic development and ordered him transferred to Denver to participate in a study of Western water resources. Mr. Cook, a Red Lake Chippewa and one of the original "fearless fourteen," resigned in disgust, charging a conspiracy by the Interior and Justice Departments to "destroy the Indian community." In November he was elected president of the National Congress of American Indians.

A similar attempt by Mr. Crow to transfer William H. Veeder, a non-Indian lawyer and an outspoken and controversial advocate of Indian water rights, stirred the tribes to wrath.

Tribal leaders saw these actions as part of a general retreat from President Nixon's explicit endorsement of Indian self-determination. Thirteen months had passed since the President outlined to Congress his "historic step forward in Indian policy."

In that time, the Indians noted, there was "much thunder but little rain," and while they were pleased with the restoration of the sacred Blue Lake to the Taos Pueblo Indians—they conceded that the bill returning the 48,000-acre tract would never have passed the Senate without the intervention of the White House—action was slow on other proposals.

In September, a group of tribal chiefs headed by Peter McDonald, chairman of the Navajo Tribal Council, descended on Washington with a request that the Indian Affairs Bureau be removed from the Department of the Interior and placed in "receivership" in the White House.

Simultaneously a band of young militants invaded the bureau and attempted a citizens' arrest of Mr. Crow. Mr. Crow was not in. The Indians barricaded themselves inside an office and were forcibly ejected by the police.

'Shake It Up Good'

These demonstrations brought a quick reaction from Mr. Nixon. There was a series of high-level conferences between Interior officials and the White House. "Shake it up, shake it up good," was the President's advice to Secretary Morton on the agency's bureaucracy.

Meanwhile, Commissioner Bruce had been restored to full power. By November he was able to announce that his team of activists had taken over key jobs at top echelon. Mr. Veeder's transfer was rescinded. Ernest Stevens, an Oneida from Wisconsin, and Alexander McNabb, a Micmac from Maine, have emerged as Mr. Bruce's chief lieutenants and innovators.

The President had called for a "new and coherent strategy." Commissioner Bruce was convinced from the start that the old strategy of relocating Indians in the cities had failed miserably. So, instead of shipping young reservation Indians to job training centers in distant cities, he proposed that the $40-million employment assistance program be centered on the reservation for the development of a local labor force. "Relocation is dead," said one of his aides.

There are no jobs on the reservations now, so Mr. Bruce's next objective was to start developing "truly viable Indian economic systems."

Developing an Indian economy did not mean bringing in white industries that were only looking for cheap Indian labor, he said. It meant the creation of Indian-owned services, so that a dollar earned on the reservation would stay there instead of being sucked out by white cities like Gallup, N.M., he explained.

There would be Indian carpenters, Indian plumbers, Indian electricians, even Indian bankers, he said.

"I want to see Indians buying cars from an Indian dealer and having them serviced there," he told an Indian manpower conference in Tulsa, Okla., Dec. 8. "I want to see Indians buying food at an Indian food market where the food has been brought in Indian trucks from Indian food distributors. I want to see houses built by Indian construction companies on designs by Indian architects."

For 15 Enterprises

Meanwhile, the tribes were being encouraged to come up with comprehensive economic development plans. The Standing Rock (North Dakota) Sioux, for example, have submitted a plan for 15 enterprises, including a bank, a motel, a shopping center and a cattle ranching operation.

The Navajos started operating their own telephone system a year ago. The Pine Ridge (South Dakota) Sioux, angry at the sight of white contractors building houses for Indians on a reservation where unemployment exceeded 70 per cent, are setting up their own construction company.

In the past, most contracts for reservation projects had to be let to white contractors because of the lack of skilled Indian technicians. Today the bureau is developing Indian action teams to provide training on reservations in the technical skills needed to conduct, operate and manage their housing, roads and public works facilities.

A typical action team consists of seven instructors: the team leader who teaches basic engineering, drafting, surveying and so forth, an equipment operator, a mechanic, a steel worker, a builder, a plumber and a electrician.

But the full implementation of the President's self-determination doctrine must await Congressional action on a number of proposals that were submitted by the White House 17 months ago.

Only a few of these bills have passed. Congress gave Blue Lake back to the Taos Pueblo. And the Senate finally voted in December to repeal the so-called "termination" policy which the President had called "morally and legally unacceptable." (Termination was designed to free Indians from Federal paternalism, but the policy brought extreme hardship to many Indian communities).

Congress went home for the holidays without voting on a bill to establish an Indian Trust Counsel Authority. This three-man directorate, to be appointed by the President, would provide independent legal representation for the Indians' land and water rights.

Also waiting action is a bill guaranteeing the right of Indians to contract for the control of services now operated by the Bureau of Indian Affairs—an important measure, according to Mr. Nixon, because "it would directly channel more money in to Indian communities, since Indians themselves would be administering programs and drawing salaries which now often go to non-Indian administrators."

There are some Indians who complain that the White House has been less than vigorous in pushing these legislative proposals.

Most Indian leaders seem to agree, however, that the Nixon message of July 8, 1970, marked a historic break with the past. All they want now, they said, is less thunder, more rain.

January 9, 1972

Militancy of Urban Indians Spurs Hope for Change

By HOMER BIGART

Special to The New York Times

CHICAGO — There are some faint stirrings of hope among the more than 300,000 "urban Indians" who have left their reservations and are living in a wretched stew of poverty and squalor in the darkest slums of the nation's cities.

Groups of young militant Indians have organized to fight for better housing and an end of discrimination in jobs and schooling. They have tried to shock the nation into an awareness of their plight by seizing unused Federal land in a number of cities and squatting there in conspicuous misery. This new generation of activists is shaking the apathy not only of the city-dwelling Indians, but also of the 477,000 who remain on reservations.

Indians are a mini-minority so thinly dispersed that they are a political force in only a half-dozen Western states — Arizona, New Mexico, South Dakota, Oklahoma, North Dakota, Montana—and in Alaska. But 60 cities harbor more than 1,000 Indians each, with the largest concentrations in Los Angeles, Chicago, Minneapolis, Milwaukee, Phoenix, Albuquerque, Tulsa, Oklahoma City, Seattle and New York.

In none of these cities do Indians constitute more than 3 per cent of the population. And usually they are the most unstable element, shifting from slum to slum, shuttling back and forth between city and reservation. They are hard to count. It is believed that the census estimates are wildly inaccurate and on the low side.

Despite their political impotence, the Indians are beginning to draw sustained attention from the White House. In his message to Congress on national Indian policy on July 8, 1970, President Nixon recognized that the urban Indians were ineligible for free Federal services such as health and education that were available to reservation Indians. He pledged that this "most deprived and least-understood segment" would be helped by expanded Federal support to seven urban Indian centers.

Last year, with approval of the Nixon Administration, a group of young, tough-minded Indian activists was installed in the Bureau of Indian Affairs in Washington. That Federal agency has long been accused of smothering Indian initiative by treating the natives as children incapable of decision. In the last few months it has gone through the throes of an Indian take-over.

The bureau has no responsibility over urban Indians; its funds must be spent for specific programs "on or near" reservations. Yet the mere presence of the activists in the bureau has heartened the city militants and even stimulated a modest wave of reservation unrest.

The Milwaukee Story

Consider Milwaukee. In that city, where 3,300 Indians, mostly Oneidas, Chippewas and Menonomies, constitute less than 1 per cent of the population, members of the American-Indian Movement (A.I.M.) seized an abandoned Coast Guard station last August on the Lake Michigan shorefront — and they still hold it.

They justified the seizure by citing an old Indian treaty that they said held that abandoned Federal land should revert to Indian ownership.

Federal authorities made no effort to evict the Indians, who promptly proceeded to set up a school in the old station and a halfway house for alcoholics in an adjacent building.

Then, in November, the Bureau of Indian Affairs, in a startling break with tradition, moved to acquire title to the property as a center for a broad range of services for American Indians.

The bureau sent one of its education officers to Milwaukee to discuss aid to the makeshift Indian school. The General Services Administration in Washington, caretaker of the property, has reached no decision.

The fact that the Coast Guard site was under Lake Michigan when the 1868 treaty was signed — the station was built on landfill in the nineteen-twenties—seemed a legal quibble to the Indians.

"We feel it's ours anyhow," said Herbert Powless, chairman of A.I.M.'s Milwaukee chapter, as he escorted a visitor through the station. Mr. Powless, a burly former marine, said he kept several Indian guards on the post at all times to resist any possible attempts at eviction.

The Coast Guard station is in an upper-class, higher-rent neighborhood adjoining a handsome lakefront park and the Milwaukee Yacht Club. The neighbors have been friendly.

The Indians have converted the two upper floors above the boathouse into small cell-like classrooms. The place swarmed with Indian children, toddlers to teen-agers, and from the kitchen came the smell of Indian corn bread. Heaps of second-hand textbooks lay about, awaiting the scrutiny of Indian censors eager to take out all mention of "savages" or any tendentious accounts of early massacres.

Indian Vantage Point

The director, Mrs. Dorothy LePage Ogrodowski, a Menominee who said she had been certified as a teacher in elementary education, explained: "We teach all the traditional courses, social studies, math, science and language development — but all from the Indian vantage point.

"I think it's the answer to Indian education — a school controlled by Indians for Indian children. We try to develop a sense of pride in Indianness."

The benign attitude of Federal authorities toward the Milwaukee project contrasts sharply with the official reaction to other confrontations.

In San Francisco Bay, Indians who seized Alcatraz were forcibly evicted after a 19-month occupation.

In Minneapolis, birthplace of the militant A.I.M., Indians were chased out of an abandoned naval air station.

The same thing has happened in Chicago, where Michael Chosa, leader of a group called Chicago Indian Village, has been repeatedly frustrated in attempts to take over idle Government property.

Mr. Chosa, the most militant of Chicago's estimated 15,000 Indians, visited Washington in December in another effort to persuade the Nixon Administration to yield some empty buildings at the Argonne National Laboratory, on the southwest outskirts of Chicago.

Returning to Chicago, he said he had received a flat "no" from the General Services Administration, with the official explanation that radioactivity from the laboratory made the site unsafe for residents.

Mr. Chosa, a 36-year-old Chippewa, said 50 homeless Indians had lived in tepees and tents outside the laboratory grounds and some of them became sick from exposure.

The Indians involved are from Chicago's "Uptown", a mixed North Side slum of Indians from the West and hillbillies from the South. There may be worse streets in New York than Uptown's West Wilson, where a few respectable frame houses still resist the creeping slum of rotting tenements, tawdry saloons and employment bureaus that hire Indians at $10 to $12 a day, when work is available. But Mr. Chosa says living conditions there are the worst in the country.

Three years ago, after working on the West Coast as a farm labor organizer and pool hustler, Mike Chosa returned to Chicago. He began planning a campaign to shame the city into doing something for the Indians.

When Mrs. Carol Warrington, a Menominee, and her six children were evicted from a tenement in May, 1970, Mr. Chosa set up a large tepee and six tents behind Wrigley Field to shelter the Warringtons and other Indians. Two portable toilets and $300 were donated by DePaul University students, and there were other gestures, but the city was cold to demands for an Indian housing complex.

Last June, Mr. Chosa and 40 Indians seized the abandoned Belmont Harbor Nike site, demanding that it be set aside for 200 housing units and a school for 500 Indian children. Evicted forcibly by the Chicago police after two weeks' occupancy, the Indians marched to one of the city's wealthiest churches—Fourth Presbyterian, on Chicago's Gold Coast — and requested asylum.

"They let us into the basement and locked the door," said Mr. Chosa. "Next morning they gave us four hours to get out."

"We were told by the police we'd be arrested if we didn't get out of the city" he said. "So we moved to Big Bend Lake in Des Plaines and lived in tents. Three weeks later we got a tip that the police were planning to evict us. I sent a scouting crew out and we found that the Argonne Nike site was abandoned. We sneaked out of Big Bend and went to Argonne."

Indian residency on these allegedly radioactive premises lasted three weeks. Mr. Chosa said the Indians had left voluntarily, thinking they had a firm agreement with the Chicago Housing Authority for 132 permanent dwelling units in Uptown.

Temporary Quarters

It was late August. The Indians were offered temporary quarters at a Methodist summer camp near Naperville. Until they were evicted on Dec. 11, the Indians lived in wooden cabins uninsulated against the cold.

Reluctantly, the Methodists went to court and obtained an eviction order. Camp Seager was not a fit place in winter, one official said.

Frustrated in the attempt to locate at Argonne and later at Fort Sheridan, Mr. Chosa and a group of followers, now reduced to about 20 adults and children, are encamped at Camp Logan, near Zion, Ill., north of Chicago.

Senator Adlai E. Stevenson Jr., Democrat of Illinois, said he had learned that the Bureau of Indian Affairs had been "pressured" not to support an application for the Argonne site because of the General Services Administration's opposition to it.

In Minneapolis, a cradle of Indian activism, the Minnesota Council of Churches established an Indian housing committee, which finally succeeded last year in obtaining the first federally funded urban housing development primarily for Indians. The $5.4-million project will contain 216 units.

"But it's just a drop in the bucket," said E. G. (Hap) Holstein, the church council's coordinator for Indian work. "There are 8,000 Indians here, mainly Chippewa and Sioux, and they are in a worse position than ever in housing, jobs and health. Unemployment has gone up to 60 per cent."

Most Minneapolis Indians live in the East Franklin Avenue slum on the near South Side. A recent report by the Minnesota League of Women Voters noted that "uncollected garbage, mice, cockroaches, exposed wiring and debris piled in the yards of old houses plague Indian tenants, but the tenants do not generally complain even where civil rights violations are evident."

The militants have been trying to change this. Before 1965 there were no Indian complaints about housing, according to Dennis Banks, national director of the American Indian movement. Then 60 complaints were filed with the Minneapolis Civil Rights Department, resulting in 12 convictions, he said.

"Now, there has been a very slight improvement," he said. "Landlords are trying to live up to minimal standards."

To combat what it called police harassment, A.I.M. set up patrols with walkie-talkies and borrowed cars to observe the police and report incidents of alleged brutality.

The Minneapolis Indians have another new safeguard—a free legal rights center staffed by three lawyers and financed by the city's big law firms.

Generally, Indian leaders in Minneapolis were not optimistic. Some said there was a lack of communication with the new Mayor, Charles Stenvig, a former detective in the police department, who is strong for law and order Mr. Banks saw increasing Indian misery in the recent factory layoffs.

"There'll be frustration and anger," he said. "You'll see Indians forced to the brink of violence. Someday Americans will wake up to the fact that they are spending too much time and money on Vietnam, a national disgrace, and that there are still a lot of problems at home."

February 10, 1972

A Long List of Grievances

WASHINGTON—Few observers of Indian life were surprised last week when the "Trail of Broken Treaties," a cross-country caravan of Indian activists, led straight to this city. Tribal grievances have been blossoming here much longer than the cherry trees—since 1792, in fact, when the Seneca Chief Red Jacket journeyed to Washington to lecture the Senate on the importance of keeping its promises.

What startled some, however, was the fury of the protest, its tone of noisy desperation. "We have now declared war on the United States of America—seek your stations," Vernon Bellecourt, one of the leaders, told protesters who had barricaded themselves inside the Bureau of Indian Affairs (B.I.A.) Building.

The week-long drama began on Nov. 2 when a band of 500 Indians, including women and two dozen children, took over the building. Subsisting on food and other supplies brought into the building by sympathizers, the Indians set about fashioning makeshift tomahawks and clubs from legs of tables and chairs and vowed to resist any effort to oust them by force.

Last Wednesday they walked out voluntarily, having reached agreement with White House aides that a study of their problems would be made. They also carried with them a number of documents from Government files that, they declared, contained "highly incriminating" evidence of exploitation of Indians by present and past members of Congress. Interior Department officials said they had also carried out "priceless Indian art and artifacts" and that the building had been heavily damaged.

The protest was organized by the American Indian Movement (A.I.M.), a young activist group with a taste for high drama. The caravan set out weeks ago from Seattle, where much of the new militancy is centered, and picked up members along the way.

They carried a list of 20 demands dealing with such matters as health, education, housing, water rights and treaty provisions.

Official Washington response to the demonstration: The protesters, said Secretary of the Interior Rogers C. B. Morton, were a "splinter group of militants . . . not supported by a majority of reservation Indians."

A veteran Indian spokesman offered a different view: "We may disagree with some of their tactics," he said, "but there probably isn't a single Indian organization anywhere that would disagree with those 20 points. A lot of Indians out there are watching the protest and saying 'right on!'"

It could fairly be said that the Indians were engaged in an old and honorable enterprise: They were taking over territory which they claimed had been theirs all along. Three years ago, and for similar reasons, they had occupied Alcatraz, and more recently they had squatted atop the head of Teddy Roosevelt on Mount Rushmore.

But Indian anger has long focused on the Bureau of Indian Affairs in

An Indian activist—a member of the band that took over the Bureau of Indian Affairs building to dramatize Indian grievances. "The new mood stems from more than broken treaties."

Chie Nishio/Nancy Palmer

THE PLIGHT OF THE INDIAN

	American Indian	U.S.
Suicides (1970)	32.0 per 100,000	16.0 per 100,000
Life expectancy (1970)	47 years	70.8 years
Unemployment rate (1972)	45% estimated	5.8%
Median family income (1971)	$4,000	$9,867
Infant mortality (1970)	30.9 per 1,000 live births	21.8 per 1,000 live births
Per cent entering college (1971)	18%	50%

Washington. One of the leaders of last week's demonstration told Bureau officials: "You don't know anything about Indian affairs. You should turn the Bureau over to us."

Actually, the Bureau has been gradually "going Indian" ever since John Collier reorganized it during New Deal days. Many of its top officials now, including Commissioner Louis R. Bruce, are Indians. Yet the Bureau's essential outlook, most Indians believe, remains white and paternalistic. On nearly all reservations, where three-fourths of the nation's one million Indians reside, its word is still law, despite recent gains in tribal autonomy.

All of which points up the Indian's peculiar position in America. Though a fully qualified citizen with full voting rights, he retains strong allegiances to his tribe; and when he calls for self-determination, as the protesters did, he is thinking not so much of individual rights as he is of tribal sovereignty.

Like a spider in a web, he and his tribe live in the center of an intricate network of treaties and agreements with the rest of America, many of them old and all but forgotten, some only now being brought to light. Worst, wherever he turns he finds these treaties in tatters: state governments seize his land; developers drain his lakes; tourists and business ignore his fishing rights. In the end he declares war on the B.I.A.—not because it is the only source of his miseries (it isn't) but because it is there.

The new mood stems from more than broken treaties. Much of it is the natural consequence of what Indians perceive to be their second-class status. "We're tired of being kicked around," said a Sioux Indian leader. Thanks to nearly a decade of Federal anti-poverty education, the Indian now knows that he and his brothers are rated "the poorest of the poor." The accompanying table gives some indication of the Indian's problems.

Ironically, much of the Indian's suffering occurs within a pervasive "welfare" setting: On every reservation the Public Health Service provides free medical service; the Department of Agriculture supplies commodity foods; and the B.I.A. offers a variety of welfare-oriented services. But jobs are scarce and morale remains low. Like the Southern blacks, many Indians have been drifting off the land and into the city in search of jobs and housing. There they suffer from alienation, fewer social services and, in many cases, open racial hostility.

They have therefore begun to adopt the same tactics of pride and protest which they saw blacks use successfully in the nineteen-sixties. Young Indian activists have been heard to call their elders "Uncle Tomahawks," or to label an Indian bureaucrat, who may have neglected his origins, "an apple"—red on the outside, white on the inside.

Increasingly now, Indians are seeking redress of their grievances in the courts. And much of the friendly white money is being channeled into this area. The Ford Foundation, for example, recently gave $1.2-million to the Native American Rights Fund, a legal services group, "in the belief that the law can be made to . . . help fulfill the rights of Indian people."

The question raised by the "Trail of Broken Treaties" protest is whether Indians—by reputation a patient race—will be content to wait while the wheels of justice slowly turn; or will they, as they did last week, rush to the barricades?

—RICHARD J. MARGOLIS

Mr. Margolis is a freelance who frequently writes about Indian affairs.

November 12, 1972

Armed Indians Seize Wounded Knee, Hold Hostages

By United Press International

PINE RIDGE, S. D., Feb. 28—Militant Indians held at least 10 persons hostage in the historic settlement of Wounded Knee today, exchanging gunfire with Federal officers and firing on cars and low-flying planes that came within rifle range.

Federal marshals, F.B.I. agents and Bureau of Indian Affairs policemen surrounded the Oglala Sioux hamlet near where the Indians' forefathers fought their last tragic battle with the United States cavalry 83 years ago. Two armored personnel carriers were brought in.

A trading post and a church at the Pine Ridge Sioux Reservation settlement were occupied last night by from 200 to 300 members of the American Indian Movement.

The embattled Indians relayed demands to Washington that the Senate Foreign Relations Committee hold hearings on treaties made with the Indians, that the Senate start a "full-scale investigation" of Government treatment of the Indians, and that another inquiry be started into "all Sioux reservations in South Dakota."

The protesters vowed they would stay in Wounded Knee until they got answers from the Federal Government, but pledged that no harm "by Indians" would come to the hostages.

Federal officials refused to disclose any plans they might have to rout the militants and free the hostages.

Joseph H. Trimbach, special agent of the Federal Bureau of Investigation in charge of the Dakotas and Minnesota, said "We know of 10 hostages." Carter Camp, a spokesman for the Indians, said, "We have 10 or 12."

Marshals exchanged fire with the Indians early today. No one was reported injured, but earlier one Indian man was wounded in another outburst of gunfire.

At least 17 persons were arrested when they attempted to flee through the cordon of nearly 100 Federal officers surrounding the area. United States Attorney William Clayton said in Sioux Falls that the Indians would be charged with burglary, larceny and attempt to commit burglary.

Low-flying planes carrying photographers were fired on, and local airports were warned to keep planes away from the area.

Cars that came within or near rifle range were fired on. The rear window of a car

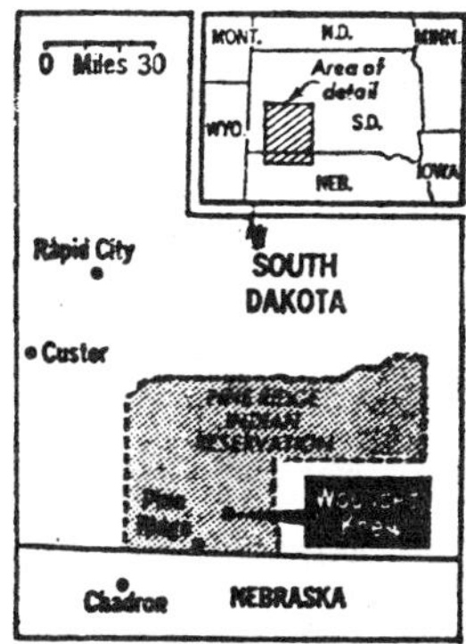

The New York Times/March 1, 1973

carrying an Indian man, his wife and baby was riddled with bullets when it passed the trading post and the driver refused to heed the militants' demands that he stop.

Chief Dick Wilson, of the Oglalas whom AIM members sought unsuccessfully to oust from his post last week, said the militants were armed with hunting rifles and ammunition seized in the takeover of the trading post and might have at least two machine guns. They could hold out "for weeks," he said.

Wounded Knee is about 14 miles northeast of Pine Ridge near where United States cavalry fired on an encampmen of Sioux men, women and children in 1890. A historical marker at the site says 146 Indians were killed. Some historians place the figure at nearly 300.

Mr. Camp, an AIM coordinator said in Oklahoma and Kansas, said:

"We will occupy this town until the Government sees fit to deal with the Indian people, particularly the Oglala Sioux tribe in South Dakota. We want a true Indian nation, not one made up of Bureau of Indian Affairs puppets."

An Assistant United States Attorney said early today a decision on what action to take "will be made in Washington and not here."

March 1, 1973

Indians Sign Accord to End 68-Day Seizure of Town

By ANDREW H. MALCOLM

Special to The New York Times

PINE RIDGE, S. D., May 6—Federal and Indian negotiators announced here today an agreement to end the 68-day occupation of the village of Wounded Knee by militant Indians and their supporters.

Unlike a previous agreement that collapsed early last month, the accord announced today includes detailed, written procedures for the disarmament and evacuation of Wounded Knee.

The new agreement, signed last night when Federal negotiators were summoned from their supper table, stipulates that Government armored personnel carriers will begin pulling back from roadblocks and bunkers Wednesday at 7 A.M. (9 A.M. New York time).

Some to Be Arrested

At the same time 150 persons estimated to be inside the historic hamlet, the scene of the last clash of the Indian Wars, will begin surrendering all weapons to Government intermediaries.

The occupiers will then evacuate the village. Officials say that 35 to 40 will be arrested, and Government agents will occupy the bullet-pocked settlement, seeking other weapons and possible holdouts.

Representatives of both sides seemed cautiously optimistic tonight that this agreement would end the occupation and siege that began Feb. 27 and in which two Indians were killed, one United States marshal was paralyzed by injuries and dozens of other persons were wounded.

"This has been a historic and tragic event," said Ramon Roubideaux, an Indian lawyer for the American Indian Movement whose members led the armed take-over to protest the Government's Indian policies and the present elected tribal government headed by Richard Wilson.

Bail Funds Sought

"What we have here," said Richard R. Hellstern, the chief Federal negotiator, "is an agreement that can be implemented, a complete package that we have been striving for over such a long time."

Mr. Hellstern, who is Deputy Assistant Attorney General, said that the Government had agreed, somewhat reluctantly, to a 65-hour delay in implementation of the disarmament agreement at the request of Indian lawyers, who are seeking bail funds.

Kent Frizzell, solicitor for the Interior Department, who is a chief negotiator, praised the Federal agents surrounding the village for their patience during the often-stalled, time-consuming talks.

But Mr. Hellstern commented that the Government had learned much during the 68-day seizure and that such patience "will not necessarily be the pattern followed in any future confrontation."

The agreement followed countless hours of sometimes hostile negotiations over the last 10 weeks. Federal negotiators said recently that they were operating in "an atmosphere of distrust" over earlier agreements that they felt were not adhered to by the Indian militants. "I sure hope this one works," a weary Justice Department attorney remarked.

Note From White House

The signed accord was spurred by a letter from Leonard Garment, counsel for President Nixon. At Mr. Frizzell's request it was delivered yesterday to a group of Oglala Sioux elders who support the occupiers of Wounded Knee against Mr. Wilson in a bitter tribal division.

The letter, a reaffirmation of an earlier guarantee, promised that at least five White House representatives would arrive at the remote 2,500-square-mile Pine Ridge Indian Reservation within two weeks to discuss Indian grievances with the headmen. These grievances include broken treaties and compensation sought for lands once ceded to the tribes but no longer their property.

The meeting was made contingent on a complete end to the armed confrontation at Wounded Knee by next Friday. Assuming that the disarmament goes as planned, Mr. Frizzell said today that the meeting would take place here at the home of Frank Fools Crow, probably on May 17. Later meetings may also be held.

With the signed agreement in hand, Federal officials approved the burial at Wounded Knee of Lawrence D. Lamont, a 31-year-old tribe member who died in a gunfight with Federal marshals nine days ago.

This afternoon, under a blue sky, Mr. Lamont's body and three dozen mourners, including his mother, Mrs. Agnes Lamont, were allowed through Federal roadblocks surrounding the hamlet. He was buried in the rain-soaked soil near the mass grave of 153 other Sioux Indians, who were slain by the United States Army in 1890.

The agreement was announced at 10:30 A.M. at a news conference here. It followed an afternoon and night of fast-moving events.

Yesterday, at a sacred mesa more than 40 miles north of here, Mr. Garment's letter was delivered to Mr. Fools Crow and other elders by Hank Adams, an Indian adviser.

At 4:30 P.M. the elders entered the besieged village with the letter and a detailed written statement of the latest Federal proposals for disarmament, or "dispossession of weapons," as the Washington officials here call it.

At 7 P.M. Mr. Roubideaux, seeking clarification of the proposals, met with Mr. Hellstern and Mr. Frizzell in a parked car at the intersection of Route 18 and Big Foot Trail.

At 11 P.M. Mr. Roubideaux telephoned news of the agreement to the two Federal officials at the Food Bowl in Rushville, Neb. Over the week that restaurant had become the semi-official office of Government officials and reporters, probably because it is the only dinner restaurant for more than 50 miles.

The agreement was signed by Mr. Hellstern, Mr. Frizzell, Wayne B. Colburn, who is the Chief United States Marshal, and 11 Indians, including Frank Kills Enemy, Isaac Brave Eagle, Eugene White Hawk and Roger Iron Cloud.

Dennis J. Banks, a top leader of the American Indian Movement, did not sign the document because, he said, it was "outside the protection of the United States Constitution." But he said that he would adhere to the disarmament plan because "A.I.M.'s job is done here." Mr. Roubideaux, the group's lawyer, said there was general agreement among the occupiers that nothing further would be accomplished by continuing the seizure.

"There will be more Wounded Knees," Mr. Roubideaux warned, "unless it is realized that current Indian government is a failure that must be corrected."

The main provisions of the latest agreement are:

¶The Government was to receive an inventory of all weapons in Wounded Knee late today.

¶All perimeters and bunkers will be maintained. But at 7 A.M. Wednesday all Government armored personnel carriers will withdraw and one Indian chief or elder will enter each Government bunker as an observer.

¶Simultaneously, the occupiers of Wounded Knee will evacuate all fortifications and assemble at the chapel where all weapons, ammunition and explosives will be given to 19 members of the Community Relations Service, a mediating arm of the Justice Department.

¶They, in turn, will give the weapons to Federal officers near a roadblock. Legal weapons will be returned to their owners within 24 hours.

¶The occupiers will be processed and those arrested taken to Rapid City, 120 miles northwest of here, for speedy bond hearings.

¶Officers will then escort Wounded Knee's residents back to their homes and search the area. Government officers also will remove all roadblocks and bunkers surrounding the village. A force of perhaps 40 marshals will remain here for two to four weeks, according to Chief Marshal Colburn, who said that he would permit no vigilante activity to sabotage the agreement.

In addition to the meeting with White House representatives, the Government agreed to undertake "an intensive investigation" of the operation and finances of this reservation and to protect all legal rights of individual Indians against unlawful abuses by tribal governing authorities.

Mr. Wilson, the present tribal chairman, was handed a copy of the agreement at today's news conference. He left during the announcement and could not be reached for comment.

May 7, 1973

Census Statistics Indicate Indians Are the Poorest Minority Group

By PAUL DELANEY

Special to The New York Times

WASHINGTON, July 16—the Bureau of the Census offered statistical evidence today that American Indians are the poorest minority in the country.

A bureau report showed that Indians lagged behind the rest of the nation in just about every socio-economic barometer, based on the 1970 census.

The report, composed merely of statistics, coincided with a study by the United States Commission on Civil Rights, published in May, that concluded that Indians were worse off than any other minority. The commission's study was of Indians in New Mexico and Arizona and attributed the problems to the Federal Government.

Today's census report said that the median income of Indian families was $5,832 in 1969, against the national median of $9,590. Further, nearly 40 per cent of the Indian population lived below the Federal poverty level in 1969, against 13.7 per cent of the total population.

The $5,832 median income for Indians in 1969 was less than the $6,191 for all minorities, including the $5,999 for blacks.

The census report noted that in 30 metropolitan areas with at least 2,500 Indians, median family income ranged from as low as $3,389 in Tucson, Ariz., to more than $10,000 in Washington, D.C., and Detroit. However, the range was lower on reservations, from $2,500 on the Papago Reservation in Arizona to $6,115 on the Laguna Reservation in New Mexico.

'Assimilate or Starve'

While Indians have been urged to leave the reservations and go to the cities, a Civil Rights Commission official charged last year that such an attitude amounted to an ultimatum to Indians to "assimilate or starve."

Joe Muskrat, a regional director of the commission, writing in the October, 1972, issue of The Civil Rights Digest, monthly publication of the agency, said:

"Assimilate or starve! This has been the choice offered the American Indian by the dominant society, a choice based on the fundamental misunderstanding of Indians, their needs and aspirations.

"When it has not been genocidal, the traditional approach to the American Indian has been to seek his assimilation into the larger society, an attitude based on a feeling of cultural superiority."

The Census Bureau report today indicated that the only bright spot for Indians was in education. The report showed that 95 per cent of Indian children from 7 to 13 years old and more than half the Indians from 3 to 34 were attending school in 1970, and the number attending college doubled between 1960 and 1970.

Indians in Washington, D.C., ranked above the national averages in both median number of school years completed, 12.6, and in the percentage of high school graduates, 66. On the other hand, the median school years completed on the Navajo Reservation, the largest, was 4.1, and only 17 per cent of persons there 25 and over had finished high school.

Nationally, one-third of all Indians over 25 had completed high school, compared with less than one fifth in 1960, and median schooling with 9.8 years, the same as for blacks. The national median was 12.1 years, and 52.3 per cent of the total population had finished high school.

The report said that the Indian population in 1970 was 792,730, compared with 523,591 a decade earlier. Half lived on reservations.

The census survey showed further that 55 per cent of the Indians over 16 years old worked in urban areas, with 70 per cent of the males employed in four broad categories—craftsmen and foremen, operatives, laborers and service workers. Only 9 per cent of Indians worked in technical and professional jobs.

In another report issued today, the Civil Service Commission noted a slight increase in the employment of minorities by the Federal Government last year.

Blacks gained 2,950 additional Federal jobs between November, 1971, and November, 1972, and constituted 389,762, or 15.3 per cent, of the work force. There were 77,577 Spanish-surnamed employes, up 1,860 over 1971, and 20,440 Indians, an increase of 1,182 jobs.

July 17, 1973

Legacy of Wounded Knee: Hatred, Violence and Fear

By GRACE LICHTENSTEIN

Special to The New York Times

PINE RIDGE, S.D. — The week of April 20-26 has been proclaimed as Law and Order Week on the Pine Ridge Oglala Sioux Reservation.

The designation—announced by the local Roman Catholic Church in conjunction with tribal and Federal officials—is ironic, because, for the last several months, law and order have been virtually nonexistent here.

Since Jan. 1, according to Federal Bureau of Investigation statistics, six people have been killed on this reservation, the site of the original Wounded Knee massacre in 1890 and of the take-over by militant Indians of that landmark in 1973. There have been 67 assaults, including tomahawk and hammer bludgeonings.

On a per-capita basis, the Pine Ridge homicide rate is six times greater than that of the city of Chicago, the assault rate four times greater. Residents are afraid to leave their houses, even in daylight.

Two years after the militant American Indian Movement took over Wounded Knee for 71 days to dramatize a new Indian consciousness, the Pine Ridge Reservation is worse off in almost every way than it was then.

It is a caldron of violence, intimidation, alleged economic corruption and virulent political animosity between Russell C. Means, a leader of the 1973 take-over, and Richard Wilson, the tribal council president—a power struggle that has erupted into gunfire at times.

A Federal grand jury convened to investigate the new troubles has returned assault indictments against four men. An investigating team from the Federal Bureau of Indian Affairs has called for more police and an overhaul of the tribal court system.

The superintendent of the reservation and the police chief were recently replaced. Mr. Means was jailed on charges involving another murder just off the reservation. He has since been released on bail put up by Marlon Brando, the actor, who is a supporter of the Indian Movement.

The Indians' Legacy

Behind the recent breakdown of law is the legacy of both Wounded Knee episodes. It is a vastly complicated situation involving the warrior heritage of the Oglala Sioux, the scars left by United States colonization of the Indians and the long-standing feud between the poor, traditionalist, mostly full-blooded reservation members and the more affluent, more powerful mixed-bloods.

Everyone connected with Pine Ridge—Mr. Means and his A.I.M. supporters, Mr. Wilson and his supporters (known locally as "goons"), the ordinary Oglala who just want peace and Bureau of Indian Affairs officials—agrees that the 1973 Wounded Knee uprising brought about nothing to improve the overwhelming economic and social ills here, which make this reservation look to an outsider like a rural South Bronx.

Instead, as R. Dennis Ickes, director of the Justice Department's Office of Indian Rights, said: "Wounded Knee '73 created a climate of hate. The damage was not so much to buildings as to the relationships among the people. It set brother against brother, family against family." Those hates are now being fought out almost daily.

Members of the American Indian Movement occupied the tiny town of Wounded Knee in the middle of the reservation in February, 1973, as part of a battle to overthrow the elected tribal government headed by Mr. Wilson, whom they regarded as the white man's puppet, a dictator and a corrupt leader. Mr. Wilson, they said, awarded jobs and money to his relatives and friends, most of whom were mixed-bloods.

Division Within Tribe

The tribal division pitted the more full-blooded Sioux (and some mixed-bloods) who believe in a return to traditional Indian values against mixed-bloods who hold most of the jobs, accept the Federal system of controlling Indian reservations and are the recipients of the meager benefits of that system.

Then, as now, the mixed-bloods, a majority of whom live in poverty-ridden, dusty Pine Ridge village, were far better off than those with a higher percentage of Indian blood. (There are few totally full-bloods left.)

Most of the full-bloods live in the "districts," the outlying

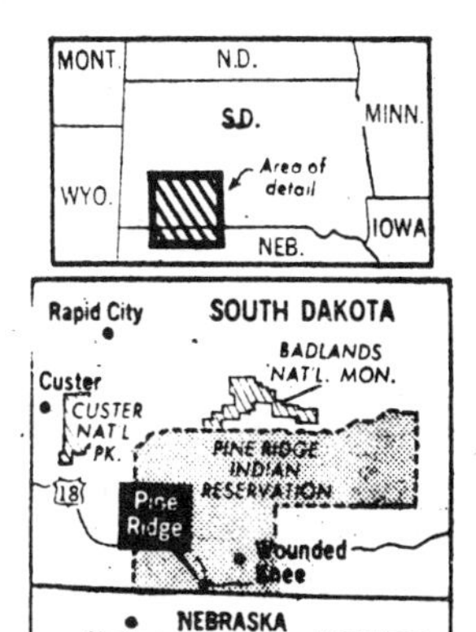

The New York Times/April 22, 1975

regions of the huge reservation that covers about 2.7 million acres of southeastern South Dakota.

A June, 1974, study done for the Bureau of Indian Affairs, for example, reported that 81 per cent of the mixed-bloods had electricity in their homes compared with 40 per cent of the full-bloods. More importantly, the study declared that the bureau's tiny police force on the reservation was ineffective, that there was "political interference" with tribal judges and that the police were "hesitant" to arrest politically connected residents. Those with half or more Indian blood accounted for 91 per cent of the arrests in January, 1973.

Charges Dismissed

After the American Indian Movement ended its occupation at Wounded Knee (where, on Dec. 29, 1890, some 153 Sioux men, women and children were slaughtered by United States Army troops) Mr. Means challenged Mr. Wilson's leadership as tribal council president. In an election that the Civil Rights Commission later declared was permeated by fraud, Mr. Means lost by 200 votes. A new election is set for December.

Meanwhile, Mr. Wilson continued as president while Mr. Means stood trial in Federal court for his part in the 1973 take-over. Last September, charges against Mr. Means were dismissed by the judge in St. Paul.

The most recent troubles began soon afterward when Mr. Means returned to the reservation to hold a victory celebration. Mr. Wilson vowed not to let it take place. But Al Trimble, an Oglala Sioux who had been appointed as the new Bureau of Indian Affairs' superintendent of Pine Ridge, refused to allow the bureau's police to interfere.

Since then, the antagonism between Mr. Means and Mr. Wilson has grown more vicious. The president refers openly to Mr. Means and his supporters as "hooligans," "lawbreakers" and worse. Mr. Means calls Mr. Wilson the head of a "fascist regime" working in collusion with the Federal Government to suppress the American Indian Movement and systematically kill its members.

The antagonism erupted into warfare this February Mr. Means and some friends accompanied a companion, Vine Richard Marshall, to the tribal court in Pine Ridge village on Feb. 26, where he was to stand trial for a minor offense.

Accounts of what followed vary, but the consensus is that the members of the Indian Movement verbally harassed the judge, demanding a jury trial. (The tribal court is so poorly funded it has not had a jury trial in years.) When they left the courtroom, a group of Wilson supporters followed their car and shooting broke out on the highway.

Plane Damaged

Later the same day a group of the movement's lawyers who had flown to the reservation in a small plane returned from an auto trip to find their aircraft riddled with shotgun pellets.

According to a participant, Eda Gordon of the Wounded Knee Legal Offense/Defense Committee, the lawyers climbed into a car to drive out of the reservation. They were blocked, she said, by 30 Wilson supporters who surrounded the car and began to break in the windshield.

Mr. Wilson drove up and Roger Finzel, a lawyer for the movement, demanded that the attackers consult Mr. Wilson, contending, "There must be some mistake." According to Miss Gordon, Mr. Wilson replied, "Stomp 'em!" The representatives of the Indian Movement were stabbed and beaten, Miss Gordon said. Mr. Wilson said this week that his lawyers had advised him not to comment on the incident.

From that day, the violence increased. On Feb. 28, a woman was stabbed in the throat. On March 1, Martin Montileaux, a reservation resident, was shot to death in a bar in Scenic, a dirty little town just north of the reservation boundary. The next day, Mr. Means and Mr. Marshall were arrested and later charged by the state with the murder.

Further Violent Incidents

On March 4, shots were fired into the house of Matthew King, supporter of the movement. On March 9, William Josh Steele, a Wilson supporter, was killed in Manderson, one of the outlying reservation towns. On March 20, Stacy Cottier, another supporter of the movement, was killed in Manderson. On March 26, Burgess Red Cloud was clubbed in the forehead with a hammer. On the same day, Jeanette Marie Bissonette, a 35-year-old mother of six and the sister-in-law of a movement supporter killed in 1973, was slain in her car by a sniper.

On April 14, Louis Moves Camp, a movement supporter who testified against Mr. Means in St. Paul, was critically wounded by a rifle shot in Wanblee, another reservation town. The same night, a parent-child center in the town of Porcupine was burned down.

No one here knows how much of the violence is politically motivated, how much can be attributed to family feuds or how much is merely the result of drunken fights that have plagued the reservation for years. The four men indicted by the grand jury are American Indian Movement supporters, although informed Government sources say new indictments will include Wilson people as well.

'Constant Fear'

For the Oglalas on the reservation, the violence is routine and terrifying. "We pray all the time," whispered Agnes Yellow Boy, the sister-in-law of Jeanette Bissonette, as she sat in her traditional, low-slung log cabin on the outskirts of Pine Ridge village. "We are afraid of guns. They used them on my brother and my sister-in-law. It doesn't look like anyone is doing anything about it."

Residents of the districts who have turned up at meetings with the Bureau of Indian Affairs investigating team no longer tell what side they are on, if any. They simply speak of the "constant fear" as Stella Brown Eyes did recently. "Someone comes up to you and says, 'You're an A.I.M.,' or 'You're a goon,' and boom, you go down. Too many guns," she said.

Then there are others, such as Charlie Red Cloud, who talk about the betrayal of the white man, the land claimed and lost by the tribe, the domination of the mixed-bloods, the deep-felt desire of traditionalists to do away with tribal council rule and a return to the older system of individual tribal groupings.

Now 90 years old, Mr. Red Cloud, a grandson of the great Sioux Chief, stood in a threadbare black suit in the Holy Rosary Mission library speaking to the investigating team in the Sioux language. "Money is sent here but as far as I'm concerned it disappears," he declared, adding: "In 1870 Chief Red Cloud went to Washington and signed a treaty. That treaty has been broken."

The reverence for their traditions and the knowledge of the glory of their warrior ancestors stand in sad counterpoint to the pitiful conditions of the Oglala Sioux today. Although the Government spends $25-million on the reservation, there is little evidence of it.

Unemployment reaches 70 per cent in the winter. Some cluster housing projects have been built in the towns but anti-Wilson people say most of the homes are doled out to his friends. The landscape of Pine Ridge is littered with abandoned cars that lie on their backs like dead metal cockroaches alongside empty roads. Few Indians have money for cattle herds to put on the land they own. The best land is leased to whites.

Most of the reservation consists of acres upon acre of empty mustard and beige prairie, punctuated by weirdly beautiful eroded buttes. Only Pine Ridge village, at the southern tip near the Nebraska border, has the basic amenities of a real town—gas station, brick government buildings, a few stores.

The other towns spotted across Pine Ridge are hardly more than trading posts and patches of cheap suburban-style cluster houses. At Wounded Knee itself, a new grave, that of L. D. Lamont, who was killed in the 1973 take-over, lies near the monument to the Indians who died in the 1890 massacre.

Mr. Lamont's grave is covered with plastic flowers. There are no local florists from whom mourners can buy real ones.

Two Leadership Choices

In terms of leadership, the reservation has only two choices: Mr. Wilson or Mr. Means. Some neutralists felt that Al Trimble might have been a moderating force. Although he had been brought in as superintendent with Mr. Wilson's blessings, he had in his 17-month tenure, begun to speak up for more services to full-bloods in outlying districts.

Movement supporters such as Marvin Ghost Bear, a tribal council member, thought he was on his way to "doing right by the tribe." But Mr. Trimble ran afoul of Mr. Wilson, especially in the control of the bureau's police force. Mr. Wilson was instrumental in his ouster. Mr. Trimble, now temporarily assigned to New Mexico by the Bureau of Indian Affairs, said on a recent visit to Pine Ridge that he did not think the answer to the reservation's problems was simply more policemen.

"The real problem is the lack of participation by the people, and they are intimidated by Wilson," he remarked. As for the tribal council president, a burly man who wears dark glasses, the issue is power. "There is no room on this reservation for A.I.M. and me," he said.

Indians' Benefits Cited

Asked about charges that his so-called "goons" have threatened and shot supporters of the movement, he replied, "If they beat anyone up, it was in self-defense." Asked about his relatives on the tribal payroll, he said, "There's nothing in tribal law against nepotism." He added that he had brought talented relatives, including his brother James, a Ph.D. who in charge of the planning center, back to Pine Ridge.

Mr. Wilson thinks the American Indians enjoy the best of two worlds—the protection of the Federal Government and the right to run their own reservation. At the opposite pole is Mr. Means, who spoke by telephone from the Rapid City jail. The Government uses Mr. Wilson "in its strategy that began with the Puritans—divide and conquer," he said.

Acknowledging that material conditions at Pine Ridge had worsened since Wounded Knee '73, Mr. Means declared that the 71-day occupation had accomplished one major victory: It gave the Sioux a new feeling of pride. "The children are growing their hair long, wearing sacred eagle feathers," he said. "I count that an immeasurable plus. First, one has to have self-pride, then you have to have political change and then follows economic change."

Despite the current violence, despite the six trials he now faces, despite the seeming hopelessness of the tribe's economic plight, Mr. Means insists those changes will come—sometime. "I don't see a really immediate change in the lifestyle on Pine Ridge," he said. "But neither do the Indian people on Pine Ridge expect it. Our concept of time, which makes up part of our reason for being Indian, is that we have no concept of time."

April 22, 1975

Indian Sovereignty Is Reborn in U.S.

By BEN A. FRANKLIN

It was in 1532 that Francisco de Victoria, a theologian who helped lay the foundations of modern international law, persuaded the king of Spain that land claims in the New World would not be legally and morally perfected without the consent of the resident sovereigns, the Indians.

As the history of conquest and containment of the native Americans demonstrates, for most of the four and a half centuries that followed, "consent" retained its 16th-century definition: It was to be given by purchase, or if not, taken by force. Now, in a growing number of cases, the courts of the United States are redefining consent by retroactively introducing the notion of what sovereign rights really means. As a result of one of the judgments, Indian chiefs and lawyers are seeing President Carter on Tuesday.

The case in question is one of the more recent and more publicized, that of the Passamaquoddy and Penobscot tribes, whose demand for 5 to 10 million acres, or as much as half of the state of Maine, or billions of dollars in back rent and damage has been declared to be on valid ground. What is to be discussed is the remedy. In 1971 in the largest settlement to date, Alaskan Indians, Eskimos and Aleuts were given $1 billion and 40 million acres of Alaska as belated compensation for land that had been taken

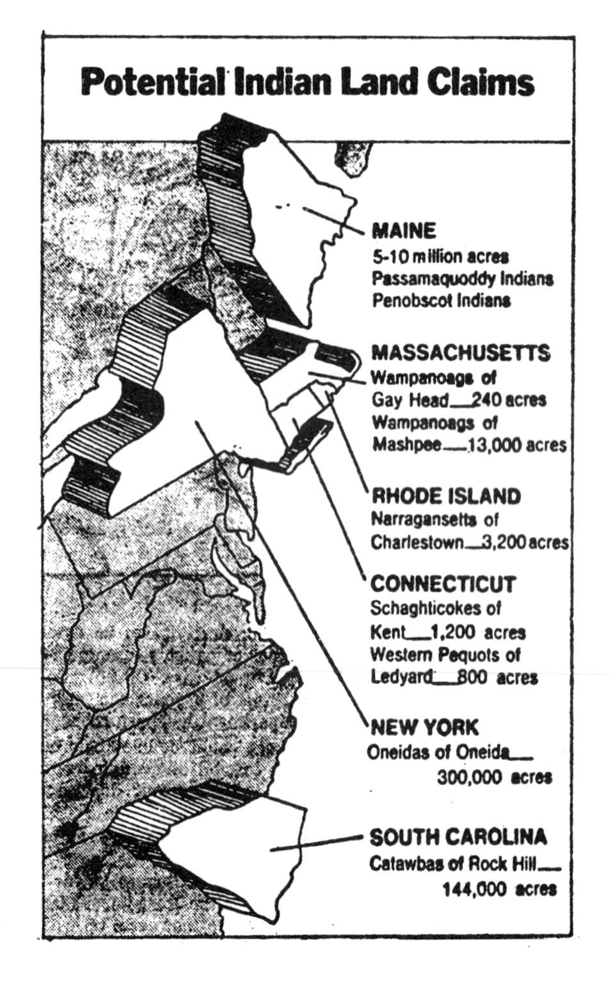

randomly and in officially sanctioned chunks.

The more renowned cases assert land claims, and have lately been in the Northeast. In Massachusetts, for example, the Wampanoags are claiming 13,000 acres of the Cape Cod resort town of Mashpee. The Narragansett, Schaghticoke and Western Pequot tribes are asserting their rights to the settlements of Charlestown, R.I., and Kent and Ledyard, Conn., respectively. The Oneida tribe contends that it owns and has always owned 300,000 acres of upstate New York between Syracuse and Utica. And in South Carolina, the small Catawba tribe wants back perhaps all of the 144,000-acre English land grant from which its members may have been unlawfully dispossessed.

The unexpected success of the Maine case and others is the result of the rediscovery by tribal lawyers, many of whom were trained as young practitioners in the Johnson poverty programs, of the Indian Non-Intercourse Act of 1790. Its intention was to protect permanently "a simple, uninformed" Indian population from the greed and fraud of whites. The act says flatly that any and all land transactions between Indians and non-Indians—say, the canny settler citizens of Maine—must have the ratification of Congress, or else be "null and void." In the 1794 treaty with the Passamaquoddy, four members of the tribe signed over almost all the tribe's land in Maine in exchange for nothing, without the requisite Congressional review.

Whether the Eastern land cases are finally settled by negotiation or by litigation, or with cash rather than acreage, the process may finally bring many of the scattered, outcast remnants of the Eastern tribes under the umbrella of Federal law and protection. They never were included in the Federal trusteeship granted the Western reservation Indians after the 13 original states formed their union.

But it is in the Western states where the impact of reborn Indian sovereignty may be the greatest. One quarter—perhaps one-third—of all the 400,000 reservation Indians (120,000 Navajo alone) are in Arizona and New Mexico, and what appears to be coming on reservations in the West is a return of Indian political independence that in the long run may be more dramatic than the partial recapture or payment for land.

The Western reservation Indians have been the fiduciaries of the United States since the 1830's when Chief Justice John Marshall wrote a majority opinion holding them to be not sovereign in the sense of a foreign state, but "domestic dependent nations," with the relationship to the Federal Government of "a ward to his guardian." The wards have not always been treated well and their descendants have been with increasing frequency turning to the courts for redress.

Slowly, at first, in the 1960's, and now faster and faster, cases on nearly every aspect of Indian life have been brought—and many have been won. The Western victories have yielded money for stolen Indian land but also, more importantly, peripheral pieces of sovereignty: the right to collect taxes or to be exempt from white men's taxes; the right to arrest and try white transgressors in tribal courts and those courts' jurisdiction over Indian civil actions; the right to hunt or fish on (or sometimes off) reservations without state game permits; the right to sell liquor at an Indian-owned reservation resort without a state license.

The courts' language has often been as encouraging as the rulings. Upholding the right of the Suquamish tribe to arrest a white man who came on its Washington reservation and got into a fight, the United States Circuit Court of Appeals for the Ninth Circuit, in a decision that stands without further appeal, said: "Surely the power to preserve order on the reservation, when necessary by punishing those who violate the tribal law, is a sine qua non of the sovereignity that the Suquamish people originally possessed."

The idea of tribal sovereignty has also begun to obtain outside the courts as well. In the early 1970's, both the North Cheyenne and the Crow, on neighboring reservations in Montana that cover some of the West's rich deposits of strip-minable coal, complained to the Interior Department that the mining leases negotiated for them by the Bureau of Indian Affairs with major energy corporations were substandard not only in regard to payments, but also to the minimal control over environmental impact. The contracts were administratively canceled. That would have been unthinkable a few years ago. If they are renegotiated, tribal spokesmen say they will insist on the right to establish environmental and land reclamation standards many states now have.

Still in that unthinkable stage is a reconsideration of Indian rights to the West's shrinking supply of water. Indian lawyers have found parallels in the Maine land case. In 1908, in Winters v. United States, the United States Supreme Court ruled that Indian reservations hold inalienable, aboriginal and "paramount" rights to all the water flowing through them. That includes the Colorado, the San Juan and the Rio Grande, and the decision stands. The "Winters doctrine" has rarely been asserted, but in the past decade it has been expanded. But Indian water lawsuits could put Arizona and New Mexico in the same state of siege as Maine.

Ultimately whether Indians can regain a form of sovereign status will depend not on the courts but on Congress, and what the Ninth Circuit considered a philosophical sine qua non has not often been considered essential by the legislators in the past. A two-year study by the American Indian Policy Review Commission, scheduled to be published in May, reportedly will dwell less on restitution by legislation than on living up to what is left of promises and treaty obligations through the ongoing case-by-case consideration of Indian rights by the courts. What Congress may then have to do is to resist white political pressures to undo favorable court decisions as they come.

Ben A. Franklin is a correspondent for The New York Times.

March 27, 1977

SATURDAY
OCTOBER 8
1:30 PM
COLLEGE DAY
大學日
BASMENT WORKSHOP
199 LAFAYETTE ST.
925-3258
BY: BASEMENT WORKSHOP, YALE & PRINCETON MINORITY RECRUITMENT
PORT ARTHUR
HONG YING
RICE SHOPPE
合飯店
RESTAURANT
15
WO-HOP
HONG KONG
HOUSE
WO-HOP
QUON HING COMP
CHEVROLET

CHAPTER 4

Problems of a Multi-Ethnic Society

Competition for a limited amount of places in American universities has recently engendered conflict between ethnic groups. Here two of the nation's oldest and most prestigious schools actively recruit members of one group.

Courtesy Gene Brown

SURVEY DISPUTES IMMIGRANT MENACE

Protestant, Catholic, Jewish Leaders Join in Plea for Policy of Hospitality

FIGURES ON INFLUX QUOTED

Editorial in Magazine Says U. S. Built Power by Ability to Assimilate Aliens

A widespread impression that European unrest has resulted in a flood of immigrants to the United States is disputed by facts presented in a special issue of the magazine, Social Work Today, published yesterday.

The publication contains results of a study of the refugee situation in this country by leading Protestant, Catholic and Jewish groups. It is edited by an advisory editorial committee consisting of the Rev. Dr. John P. Boland, chairman of the State Labor Relations Board; Dr. Eduard C. Lindeman of the New York School of Social Work, and Harry L. Lurie, director of the Council of Jewish Federations and Welfare Funds.

Urging the maintenance of the traditional American policy of hospitality to the immigrant, the magazine showed that from 1931 to 1939, 457,675 immigrants entered the United States. This was the smallest number coming to this country since the decade 1831 to 1840, when nearly 600,000 arrived here.

Immigration by Years

The total immigration to the United States from 1931 to 1939 by years, according to the magazine, follows: 1931—97,139; 1932—35,576; 1933—23,068; 1934—29,470; 1935—34,956; 1936—36,329; 1937—50,244; 1938—67,895, and 1939—82,998. All figures are those of the Immigration and Naturalization Service of the United States Department of Labor.

An editorial signed by the three members of the advisory editorial committee pointed out that the North American Continent in three centuries had "peopled itself with those seeking a new life free from national, racial, religious, economic and political repression."

"By accepting man as man," the editorial added, "the United States has built herself into nationhood. As an earnest of man's future, she has found not weakness, but power unexcelled, in her ability to receive, to assimilate, to respect—to become.

"Today all this is challenged by a period of international and national unrest. The last quarter century has witnessed an unprecedented constriction of our policy of the open door. During this time, in which we have all but shut the gates, national, racial and religious groups have been uprooted in many parts of Europe.

Propaganda Effective

"Here at home, powerful propaganda has been alarmingly effective in 'popularizing' the idea that immigrants represent a destructive, subversive influence in our common life. In the last decade a depression (which none can claim arose out of overpopulation) has been viewed, all too widely, as an additional reason for the exclusion of immigrants from our shores.

"In these opening months of what may become a major European war, and perhaps a devastating world war, it seems nearly inevitable that the disinheritance of national, ethnic and religious groups will become intensified. With increased emphasis, modern history is going to ask the United States: 'Have you the continuing ability to convert Europe's problems into your cultural and economic enrichment!' * * *

"It is our belief that when history offers a larger perspective still, it will not indicate that the America of these times was the loser because of any limited willingness she may be showing further to enrich herself. Rather the question will be asked: 'Why, in this particular period, was she so hesitant upon honoring a tradition upon which her greatness was built?' "

Agency Activities Outlined

The activities of the four principal agencies assisting refugees in the United States from various European countries are described in the same issue of the magazine by the heads of these groups.

They are the American Friends Service Committee, of which Clarence E. Pickett is executive secretary; the Committee for Catholic Refugees from Germany, the Rev. Joseph D. Ostermann, executive director; the National Refugee Service, Inc., William Haber, executive director, and the American Committee for Christian Refugees, Evelyn W. Hersey, executive director.

Special articles dealing with immigrants also appear in the same issue by Edward Corsi, former United States Commissioner of Immigration and Naturalization; Marian Schibsby, associate director of the Foreign Language Information Service; Cecilia Razovsky, director of the Migration Department of the National Refugee Service, Ind.; Emma S. Schreiber, director of the Service for Foreign Born, National Council of Jewish Women, and Erika Mann, daughter of Thomas Mann.

Also Maxwell S. Stewart, editor of Public Affairs Pamphlets; Bertha Josselyn Foss, executive secretary, National Emergency Conference; Professor Madge M. McKinney of Hunter College, Robert G. Spivack, secretary, International Student Service, and others.

December 10, 1939

EUROPEANS PRESS FOR ADMISSION HERE

Applications From the Disturbed Nations Doubled in One Year

GERMANS FAR IN THE LEAD

Special to THE NEW YORK TIMES.

WASHINGTON, Feb. 21—Because of the disturbed conditions in Europe applications for admission to the United States on immigration quotas more than doubled during the year ended June 30, 1939, the State Department reported today.

On that date, 657,353 persons were registered at United States Consulates abroad, awaiting their turn to enter America, as compared with 317,606 on June 30, 1938. Almost half the number, or 309,782, were chargeable to the quota for Germany, enough to fill that nation's total immigration to the United States for eleven years, under present admission regulations.

Moreover, in countries adjacent to Germany and since then incorporated in the Reich, the application ran to high figures. In Czecho-Slovakia 51,271 were registered, exhausting the quota almost eighteen years ahead. Pre-war Poland had 115,222 registrants, filling the total for seventeen years.

Despite the large numbers hoping to depart, actual visas issued under quota regulations were 62 per cent below the permissible total of 153,774 for all countries, according to the State Department figures. Non-quota visas numbered 23,813 and quota visas 58,853, a total of 82,666 during 1939 as opposed to 74,948 in 1938.

Forty-six per cent of those receiving the quota visas during 1939 came from the Reich, including former Austrian territory.

Quotas were filled during the year for Albania, Australia, Bulgaria, China, Czecho-Slovakia, Danzig, Germany, Greece, Hungary, Latvia, Lithuania, Palestine, Poland, Rumania, Syria, Turkey, Yugoslavia and the Philippines.

Of the 82,666 immigration visas issued in the fiscal year 1939, "new" immigrants received the 58,853 quota visas and 20,493 of the non-quota visas, or a total of 79,346. The remaining 3,320 non-quota visas were issued to students, whose admission into the United States is on a temporary basis, and to aliens previously lawfully admitted into the United States for permanent residence who were returning from temporary absences.

Of the 79,346 "new" immigrants, 15,627, or approximately 20 per cent, consisted of fathers, mothers and husbands of American citizens, and wives and unmarried, minor children of lawful alien residents of the United States.

February 22, 1940

ALL-EUROPE PURGE OF JEWS REPORTED

Hitler Said to Have Ordered Continent Cleared Before End of the War

By Wireless to THE NEW YORK TIMES.

STOCKHOLM, Sweden, Oct. 7—Well-informed circles here said today that a decree had been issued in Berlin ordering the removal of all Jews from Europe before the end of the war. The sources said that the order was issued by Adolf Hitler himself.

This report coincided with an article in Stockholm's Social Demokraten saying that informed circles in Berlin believe the Swedish protest to Berlin in behalf of Danish Jews has caused a delay in their transportation to Poland. The newspaper adds that arrangements are being made by prominent Nazi circles to have the Jews interned in Denmark.

The power behind the Nazi persecution of Danish Jews is the so-called "Jew Dictator," Storm Trooper Eighmann, the newspaper is informed. Eighmann, who was born in Palestine of German emigrants and brought up there, is known for his sadistic hatred of Jews. He engineered all the extermination action against Jews in Germany and the occupied territories. The intervention of Eighmann in Denmark is defended by Germans in the following manner: There is no Danish Government and therefore General Hanneken has demanded that the Gestapo aid in the execution of military orders.

In official quarters in Germany there is much indignation over the Swedish action and this country is "accused of meddling in matters that are none of her business," since Sweden is not the protecting power of Denmark. The German Nazis even cite international law, saying there is no paragraph therein justifying the Swedish move because Denmark is occupied and under "German protection."

October 8, 1943

No Room In the Inn

WHILE SIX MILLION DIED. A Chronicle of American Apathy. By Arthur D. Morse. 420 pp. New York: Random House. $6.95.

By E. W. KENWORTHY

IN early January, 1944, Randolph Paul, a staid tax lawyer who was then the Treasury Department's general counsel, handed to Secretary Henry Morgenthau Jr. an 18-page document entitled "Report to the Secretary on the Acquiescence of This Government in the Murder of the Jews."

The report, the substance of which is published for the first time by Arthur D. Morse in his history, "While Six Million Died," narrated in detail eight months of inaction and outright opposition by the State Department on a proposal to license the remittance to Switzerland of some $600,000 for the rescue and support of Jews still remaining in Rumania and occupied France. The money was to be reimbursed by wealthy Rumanian Jews who had escaped.

The report, prepared by Josiah E. DuBois Jr., Paul's assistant, and John Pehle, head of the Treasury's Foreign Funds Control Division, had been ordered by Secretary Morgenthau for a meeting with President Roosevelt on January 16. Its conclusion stated: "[State Department officials] have not only failed to use the Government machinery at their disposal to rescue Jews from Hitler, but have even gone so far as to use this Governmental machinery to prevent the rescue of these Jews.

"They have not only failed to cooperate with private organizations in the efforts of these organizations to work out individual programs of their own, but have taken steps designed to prevent these programs from being put into effect. . . .

"While the State Department has been thus 'exploring' the whole refugee problem, without distinguishing between those who are in imminent danger of death and those who are not, hundreds of thousands of Jews have been allowed to perish."

Mr. Morgenthau changed the title to "Personal Report to the President," and tempered the language slightly. But the report which the President read in his presence on January 16 still said: ". . . there is a growing number of responsible people and organizations today who have ceased to view our failure as the product of simple incompetence on the part of those officials in the State Department charged with handling this problem. They see plain anti-Semitism motivating the actions of these State Department officials and, rightly or wrongly, it will require little more in the way of proof for this suspicion to explode into a nasty scandal. . . . The matter of rescuing the Jews from extermination is a trust too great to remain in the hands of men who are indifferent, callous and perhaps even hostile."

President Roosevelt, who had repeatedly expressed his horror at the wholesale murder of Europe's Jews by the Hitler government and its satellites but who had also given repeated instructions to officials dealing with refugee matters that it was not "desirable or practicable to recommend any change in the quota provisions of our immigration laws," saw the looming political danger pointed out by Morgenthau. Six days later he created the War Refugee Board under Pehle.

It was late in the day, yet the Board, by zeal and imagination, succeeded in saving thousands of Jews and others from the gas chambers, and there is good reason to believe Ira A. Hirschmann, the Board's representative in Turkey, that "individual thousands of refugees could have been saved" had the Board been created a year or two earlier.

Mr. Morse, who resigned from his position as executive producer of CBS Reports to write this book, has provided a documentation of Secretary Morgenthau's charges—a documentation provided from the Government's archives, much of which has not been previously disclosed. And on the whole, the record he sets down supports his own conclusion: "One might describe the American response to Nazi racism as an almost coordinated series of inactions . . . it was one thing to avoid interference in Germany's domestic policies, quite another to deny asylum to its victims."

Mr. Morse subtitles his book "A Chronicle of American Apathy." Yet one wonders whether the apathy would have been so great had the public been privy to the attitude of State Department officials revealed in cables and memos dug up by Mr. Morse.

Item: In August, 1942, Gerhart Riegner, Geneva representative of the World Jewish Congress, sent a cable through the United States consulate in Geneva to Rabbi Stephen S. Wise, head of the American Jewish Congress, stating that he had learned from a trusted informant, a leading German industrialist, that Hitler had many months before ordered the "total solution" of the Jewish question in Europe by the extinction of all Jews. A covering memo to the State Department from Vice Consul Howard Elting Jr. endorsed Riegner as "a serious and balanced individual."

In the Department's Division of European Affairs, Elbridge Durbrow recommended that the cable not be forwarded to Rabbi Wise "in view of the . . . fantastic nature of the allegations and the impossibility of our being of any assistance if such action [by Hitler] were taken." (Rabbi Wise later got the same information from Riegner through British diplomatic channels.)

Item: On February 10, 1943, the State Department sent an order to the Bern legation ordering the minister not to accept any more reports for "transmission to private persons." Thus did the Department close its channels to Riegner's intelligence, from which it might have learned much.

THE holocaust has ended. The six million lie in nameless graves. But what of the future? Is genocide now unthinkable, or are potential victims somewhere in the world going about their business, devoted to their children, aspiring to a better life, unaware of a gathering threat?—"While Six Million Died."

Item: An undated memo of the Division of European Affairs on the origins of the Evian Conference of July, 1938, which created the ineffective Intergovernmental Committee on Refugees, stated that Secretary of State Cordell Hull and other officials had decided the conference was necessary, in view of the public outcry against Nazi atrocities, if the Department was to "get out in front and attempt to guide the pressure, primarily with a view toward forestalling attempts to have the immigration laws liberalized."

Item: On April 19, 1943—again largely in response to public pressure—the Anglo-American Conference on Refugees opened in Bermuda. The next day Assistant Secretary of State Breckinridge Long recorded in his diary: "One danger in it all is that their activities [the telegrams to the President by Jewish and other groups] may lend color to the charges of Hitler that we are fighting this war on account of, and at the instigation and direction of, our Jewish citizens. In Turkey . . . and in Spain . . . and in Palestine's hinterland and in North Africa the Moslem population will be easy believers in such charges. It might easily be a definite detriment to our war effort."

It is a serious question whether the United States could have saved large numbers of Jews from Hitler's ovens and firing squads, and Mr. Morse does not make due allowance for all the practical difficulties. But the burden of his indictment of the President and the State Department is that almost no effort was made until 1944, and that most proposals were met with indifference or, even worse, active opposition.

Nowhere was this opposition so open and forceful as in the refusal to enlarge immigration quotas or to provide for the transfer to Germany and its satellites of unused quotas of other countries. And even the quotas of Germany and the occupied countries were not fully used because of the severity with which consulates—on State Department orders—administered the laws. From 1933 until 1943, there were altogether 1,244,858 unfilled places, and of these 341,567 were allotted to citizens of countries dominated or occupied by Germany.

In her "This I Remember," Mrs. Roosevelt, who urged her husband to relax the quota system, wrote: "While I often felt strongly on various subjects, Franklin frequently refrained from supporting causes in which he believed, because of political realities."

There were political realities—the campaigns of the American Legion, the Veterans of Foreign Wars, much of organized labor, and various "patriotic societies"—against admitting Jews outside quota limits, and it is true that politics is the art of the possible. Nevertheless a strong and determined leader can do much to expand the possible and make it acceptable. This Roosevelt failed to do, and it is a grievous mark against him. Indeed, altogether this is a grievous book that should rend the heart and smite the conscience of a nation that began as a refuge for the despairing, the homeless and the persecuted.

MR. KENWORTHY covers the State Department for The Washington Bureau of The Times.

EXCLUSION REPEAL IS LAW

President Signs Bill to Admit Chinese on Quota Basis

WASHINGTON, Dec. 17 (AP)—The sixty-year-old Chinese exclusion laws were taken off the statute books today when President Roosevelt signed a bill to repeal them and rectify what he has termed an "historic mistake."

The repealing act makes all Chinese residents of this country eligible for nationalization and lowers immigration barriers to permit 105 Chinese to enter the United States annually, on a quota basis.

Mr. Roosevelt asked Congress in October to repeal the exclusion laws and thereby "silence distorted Japanese propaganda."

December 18, 1943

PRESIDENT ORDERS SPEEDY ADMISSION OF MORE REFUGEES

Directive States Quota Limits Set by Immigration Laws Will Be Observed

VISAS FOR 3,900 MONTHLY

Instructions Are Given That Eligible Members of Oswego Group May Stay Here

By ANTHONY LEVIERO
Special to THE NEW YORK TIMES.

WASHINGTON, Dec. 22—President Truman issued a directive tonight designed to expedite the admission to America of displaced persons and refugees from Europe, up to the limit permitted by the immigration laws, and he specified that the greatest help be given to orphaned children. Thus this country would set an example to the world in relieving human misery, Mr. Truman declared.

To most of the 1,000 refugees who have been living in the war relocation camp at Oswego, N. Y., for eighteen months, and facing a return to a war-torn and hungry Europe, the President offered a Christmas gift of American freedom.

He said it had been determined "that if these persons were now applying for admission to the United States most of them would be admissible under the immigration laws," and added:

"In the circumstances, it would be inhumane and wasteful to require these people to go all the way back to Europe merely for the purpose of applying for immigration visas there and returning to the United States. I am therefore directing the Secretary of State and the Attorney General to adjust the immigration status of the members of this group who wish to remain here, in strict accordance with the existing laws and regulations."

Directive to Six Officials

The President issued a directive to carry out the plan to the Secretary of State, the Secretary of War, the Attorney General, the War Shipping Administrator, the Surgeon General of the Public Health Service, and the Director General of the United Nations Relief and Rehabilitation Administration. He ordered them to send representatives to Europe, establish consular and other facilities near the displaced person and refugee assembly areas in the American zones of occupation, and speed the most destitute to a refuge here.

"Visas should be distributed fairly among persons of all faiths, creeds and nationalities," Mr. Truman stated in his directive. "I desire that special attention be devoted to orphaned children to whom it is hoped the majority of visas will be issued."

In anticipation of possible criticism in Congress, which has bills pending to prohibit further immigration or severely restrict it, President Truman stressed the point in his announcement and in his directive that nothing would be done beyond the scope of quotas of the immigration laws.

He also specified that every person admitted must be supported by a welfare organization and even the cost of travel and visa fees must be guaranteed so that their transportation "will not cost the American taxpayers a single dollar."

No Time "to Narrow Our Gates"

Referring to the restrictive measures, Mr. Truman said:

"I hope that such legislation will not be passed. This period of unspeakable human distress is not the time for us to close or to narrow our gates. I wish to emphasize, however, that any effort to bring relief to these displaced persons and refugees must and will be strictly within the limits of the present quotas as imposed by law."

To a demobilization-conscious public Mr. Truman said:

"I wish to emphasize above all that nothing in this directive will deprive a single American soldier or his wife or children of a berth on a vessel homeward bound, or delay their return."

The President's plan represents an effort to assist a comparatively small number. Only 3,900 may receive visas monthly in central and eastern Europe and the Balkans, 10 per cent of the annual immigration quota for the areas designated. It may offset to some extent the bitter criticism aimed at the agreement announced last month by Mr. Truman of the formation of the Anglo-American Committee of Inquiry on Palestine, which opens its investigation Jan. 7 and will not report for 120 days. The President had previously asked Great Britain to admit 100,000 of Europe's homeless Jews to Palestine and the British proposed the joint study.

Unused Quotas Not Cumulative

In his statements today Mr. Truman said that two-thirds of the annual immigration quota was allotted to Germany and that very few persons reached this country from Europe during the war. He stressed that the quotas were not cumulative and added that he did not intend to ask Congress to change this rule on unused quotas.

Explaining that consular offices in all parts of the world were put out of operation by the war, the President stated that it was impossible to restaff them overnight.

"The decision has been made, therefore, to concentrate our immediate efforts in the American zones of occupation in Europe," he continued. "This is not intended, however, entirely to exclude issuance of visas in other parts of the world.

"In our zones in Europe there are citizens of every major European country. Visas issued to displaced persons and refugees will be charged, according to law, to the countries of their origin."

Promising strict adherence to law prohibiting admission of persons who might become public charges, Mr. Truman added that responsible welfare organizations already engaged in the refugee field would guarantee each person. He said further that he was informed that no person admitted under sponsorship of these organizations had ever become a burden to his community. He also said that many of the immigrants would have close family ties here and would receive assistance until they could provide for themselves.

December 23, 1945

LEGION WILL FIGHT ADMISSION OF DP'S

The American Legion will oppose all plans to admit displaced persons to this country lest refugees deprive veterans of housing and jobs and "invite a flood tide of immigrants," Col. Paul H. Griffith, national commander of the Legion, declared here last night.

Colonel Griffith, the elected spokesman for an estimated 3,500,000 veterans of World Wars I and II, said that the addition of refugees to the country's population at this time would "increase the distress" of more than a million veterans who are still unemployed. The competition, he said, would extend to consumer goods. There was also the possibility, Colonel Griffith asserted, that the refugees would not be competent to earn a living and would become public wards.

"There are probably 4,000,000 of our own people tonight who are half-housed, ill-housed or actually unhoused," he said. "A shocking proportion of these is made up of homeless veterans."

Colonel Griffith outlined the Legion's position during a radio broadcast at a "Town Meeting" in Town Hall, 123 West Forty-third Street. Dr. James G. McDonald, a member of the Joint Anglo-American Committee of Inquiry on Palestine, the other main speaker, declared that true Americanism calls for non-partisan and enthusiastic support of the displaced persons plan set forth recently by President Truman.

The Legion head said he had been informed that there are about 20,000,000 persons in Europe who wish to come to America. The admission of displaced persons, he said, would lead to a drive to relax immigration laws to permit their relatives and others to enter the country. "There will be no end to it," he added.

He cited the 1,237,116 veterans who were drawing unemployment compensation during the week of Oct. 5. Hundreds of thousands of other veterans, he said, had exhausted their compensation and were still unemployed.

"These veterans are entitled to first consideration to homes and to jobs," he asserted. "The American Legion will not budge from its stand that Americans must come first in these respects."

Colonel Griffith questioned the assertion of Dr. McDonald that the refugees could be absorbed into our economy, that the 270,000 refugees who had entered the United States since 1933 "had demonstrated that our country could readily absorb the proposed additional immigrants."

"If they are valuable workers and producers," Colonel Griffith replied, "no country needs them as much as their own."

Points to Our Early Times

Dr. McDonald described President Truman's displaced persons' proposal as a program which has continued in various forms since the beginning of the country's history.

"Little Americans," he asserted, "in every stage of our history have argued that the country was 'mature' and that the door should be closed to further immigration."

Refugees who entered the United States during the period of Nazi persecution not only have become adjusted, Dr. McDonald said, but in significant instances have established businesses in non-competitive fields that have given employment to Americans.

A recent Government estimate, he said showed that every adult industrial worker coming to this country represents a $5,000 addition to the national wealth.

Colonel Griffith disclosed earlier in the day that the Legion intended to back legislation at the next session of Congress which would define a veteran's seniority in regaining his old job.

November 1, 1946

NAZIS SENT TO U. S. AS TECHNICIANS

By DELBERT CLARK

Special to THE NEW YORK TIMES.

BERLIN, Jan. 3—Infiltration of outstanding Nazis into the United States with the tentative promise of citizenship for themselves and their families was revealed today as one result of "Project Paperclip," which involves the transportation of technical experts to the United States.

Among the so-called experts now said to be in the United States as part of "Project Paperclip" is Herbert Axster, a Berlin attorney. There appeared some puzzlement at first as to how he qualified as a technical expert but later it developed he had a temporary wartime job in a purely administrative capacity in connection with buzzbomb manufacture.

Herr Axster, while apparently never a member of the Nazi party, is known to have been very active during the war in responsible positions with the high commands of both the Wehrmacht and the Luftwaffe.

Frau Axster, who at latest reports was in Bavaria with her children waiting to join her husband in the United States, was less discreet. She was an officer of the National Socialist Party Frauenschaft.

Under denazification laws this automatically makes her a major offender. She is barred from activity in any public or private enterprise or profession except ordinary labor.

Another interesting expert said to be in the United States is Herr Paten, former owner of factories engaged in the manufacture of auditory devices for the Luftwaffe. A well-known specialized industrialist, Herr Paten left his wife behind but she is said now to be en route to the United States, if not already there.

Expediency in rounding up genuine technical experts and research scientists and making temporary use of their services is well recognized. However, there have been many misgivings here, even among some Germans, at the liberal provisions of "Project Paperclip" in virtually promising United States citizenship to ardent Nazis or pro-Nazis and families, in return for services that could be required in any event.

Fear that the War and State Departments, in what is deemed a misguided deal, are setting up trained leadership for a potential fifth column or at least granting the citizenship prize to our worst enemies has been freely expressed.

Germans Criticize Project

The Berlin Tribune, new organ of the organized trade unions, noted with approval the formal protest of Albert Einstein and other United States leaders, quoting liberally from it under the headline "Bacillus Carriers of Nazism."

Under "Project Paperclip," instituted in Washington, the military government was ordered to round up available experts in certain categories, send them to the United States as soon as possible, maintain them in camps while awaiting transportation and permit their families to follow them. The promise of citizenship was held out if they "prove themselves worthy."

Efforts to obtain from military officials an elaboration of the phrase "prove themselves worthy" met with ambiguous replies. At a recent press conference Gen. Joseph T. McNarney, theatre commander, said in response to a question from this correspondent that the technicians were screened politically, as well as for skills.

In any event the revelation that Class I and II Nazi are being sent to the United States is taken by many military government officials as further evidence that there is an unhappy lack of comprehension of the magnitude of the responsibility of the United States in stamping out nazism.

While the Germans themselves are ordered to try party members and others under laws promulgated by the military government some of the most intellectual of such offenders are either employed by the military government, permitted to occupy elective or appointive posts of great responsibility or sent to the United States.

Most of these critics within the United States organization here absolve Lieut. Gen. Lucius D. Clay, deputy military governor of knowledge of such matters. They insist, however, that many of his subordinates do not know what makes a Nazi or do not care so long as the German has good table manners, speaks good English and is efficient in the job assigned to him.

January 4, 1947

PRESIDENT SCORES DP BILL, BUT SIGNS

By ANTHONY LEVIERO

Special to THE NEW YORK TIMES.

WASHINGTON, June 25—The displaced persons bill to admit 205,000 refugees into this country became law this afternoon, but in signing it President Truman denounced the measure as a mockery.

In a long statement filled with scathing denunciations of the compromise measure, passed in the hectic closing hours of the session, the Chief Executive declared he would have vetoed the bill if Congress had not adjourned.

Mr. Truman said the bill was anti-Semitic and he also asserted that it would exclude some Catholics who had fled into the American zone of Germany as anti-Communist refugees.

In explanation of why he had signed the bill so reluctantly, Mr. Truman said he did not wish to penalize its beneficiaries and that he did so with the hope that Congress would rectify its "injustices" as soon as possible.

Whether the bill was "worse than no bill at all" was a hairline question which Mr. Truman said he had decided in favor of the 200,000 persons who would be admitted in the next two years, along with 3,000 orphans and 2,000 recent refugees from the Communist coup in Czechoslovakia.

Against these three points of merit, the only ones he conceded in the bill, Mr. Truman listed "numerous" defects which he said had resulted from a compromise that combined "the worst features of both the Senate and House bills."

"If the Congress were still in session," Mr. Truman said, "I would return this bill without my approval and urge that a fairer, more humane bill be passed. In its present form this bill is flagrantly discriminatory. It mocks the American tradition of fair play.

"Unfortunately, it was not passed until the last day of the session. If I refused to sign this bill now, there would be no legislation on behalf of displaced persons until the next session of the Congress."

With his blistering statement, Mr. Truman raised his score of criticism of the Republican-dominated Congress another notch. It was believed that if he called a special session of Congress, an action which it was understood he was deliberating, he would ask for a revision of the bill, along with action on his discarded housing program and possibly other problems.

Rosenman Is a Caller

Judge Samuel Rosenman of New York, an adviser to President Roosevelt, was a White House caller this morning, but it was denied that he had anything to do with the preparation of the message. It is reported, however, that he was taking an increasingly important part in Administration affairs as the Presidential campaign approached.

In his recent cross-country speaking tour, President Truman made a plea for displaced persons in a major speech at Chicago on June 4. He said there that displaced persons were "heroes of democracy" and that some legislation then being considered by Congress was unfairly discriminatory.

Today Mr. Truman added that "the bill discriminates in callous fashion against displaced persons of the Jewish faith," and he said this was "a brutal fact," which its sponsors could not obscure by the maze of technicalities in the bill.

The major "device" for discriminating against Jews, he said, was the provision limiting immigration to displaced persons who entered Germany, Austria or Italy on or before Dec. 22, 1945.

Mr. Truman pointed out that most Jewish refugees who had entered those countries by that date have already left, and that most of the Jewish displaced persons now in those areas had arrived after Dec. 22, 1945. "and hence are denied a chance to come to the United States under this bill."

Organizations which have been urging Mr. Truman to veto the bill as unfair have pointed out that 100,000 to 150,000 Jews fled Poland after the anti-semitic pogroms in Kielce in July, 1946.

President Truman said the "device" of the deadline of Dec. 22, 1945, "definitely excluded" 90 per cent of the remaining Jewish displaced persons.

In objecting to the deadline as one which also discriminated against Catholic refugees from communism, Mr. Truman characterized the "device" as follows:

"It is inexplicable, except upon the abhorrent ground of intolerance, that this date should have been chosen instead of April 21, 1947, the date on which General Clay [Gen Lucius D. Clay, American Military Governor] closed the displaced persons camps to further admissions."

Mr. Truman's statement echoed previous criticism heard in Congress that the bill was framed to favor Balts and persons of "German ethnic origin" who were largely of the Protestant faith.

The new law stipulates that not less than 40 per cent of all visas issued under it must be made available to DP's "whose place of origin or country of nationality has been de facto annexed by a foreign power." This in effect means persons from the Baltic countries of Lithuania, Latvia and Estonia and Poland east of the Curzon line.

Senator Chapman Revercomb, Republican, of West Virginia, a sponsor and supporter of the bill in its final form, denied in debate that the Baltic-East Poland priority was discriminatory.

Critics of the ethnic German feature objected that it would permit the entry of persons suspected of close association with the Nazis, but Senator Revercomb said this was covered by a provision denying visas to any one who had participated in a movement hostile to this country.

June 26, 1948

TRUMAN SIGNS BILL EASING D. P. ENTRIES; 415,744 GET REFUGE

End of the 'Discriminations Inherent' in Old Measure Hailed by President

PRAISE FOR BOTH PARTIES

By ANTHONY LEVIERO
Special to THE NEW YORK TIMES.

WASHINGTON, June 16—The American franchise for liberty and a home was held out today to additional men, women and children as President Truman signed the Displaced Persons Bill. Refuge to virtually every kind of victim of Europe's wars and persecutions is offered under the new law, which expands to 415,744 the total of those admitted or to be admitted under the special post-war legislation.

The bill signed today rubbed out the provisions that have been denounced as un-American in the Displaced Persons Law of 1948.

It opened the gates wide for Jews stigmatized by Hitler, to Catholics who have crept out from under Russia's iron curtains, to Italians who fled Fascism, to the Volksdeutsch who were turned out of the Sudetenland, Yugoslavia, Hungary, Rumania, Poland and the Baltic countries early in the war era, to orphans and to those in a few other categories of human misery.

"Corrects Discriminations"

President Truman hailed the measure as one which "corrects the discriminations inherent in the previous act."

"The countrymen of these displaced persons have brought to us in the past the best of their labor, their hatred of tyranny and their love of freedom," President Truman said. "They have helped our country grow in strength and moral leadership. I have every confidence that the new Americans who will come to our country under the provisions of the present bill will also make a substantial contribution to our national well-being."

He expressed gratification for the support given by Republicans, in combination with Democrats, to produce a generous bill. He characterized this bipartisanship as "a splendid example of the way in which joint action can strengthen and unify our country."

He issued an Executive Order required by the bill designating the Displaced Persons Commission to investigate the character and eligibility of displaced persons and persons of German ethnic origin who apply for entry. This is a security provision designed to comb out subversive persons who might seek to come to this country with evil intent.

Like all bills that have a major position in President Truman's program, the displaced persons bill received ceremonial treatment. The President used twelve pens, and after the signing distributed them to persons gathered about his desk, leaders in the efforts to solve the D. P. problem and in conducting the bill through Congress.

Students Receive Asylum

The bill provides for the issuance of a grand total of 400,744 visas, including 172,230 which had been issued up to May 31, and for granting permanent entry to 15,000 students and other persons already here on temporary permits who cannot return to their homes without danger of death or persecution. The over-all total is 415,744.

The new law primarily offers a home to 341,000 displaced persons. the largest category affected. Officials familiar with the intent and provisions of the law believe that it will lead to the effectual emptying of the displaced persons camps of Europe.

They estimate that 300,000 displaced persons are in these camps now. In addition to those who have already received its visas the United States would take about 120,000 of the 300,000. The others would go to Australia, Canada and South America.

Officials expect that about 20,000 persons will be left as a tragic remnant unwanted or ineligible to emigrate for various reasons. These will somehow have to be assimilated in Europe.

Division of Eligibles

The eligible entrants under the new law may be divided into four main groups as follows:

1. The basic group of 341,000 refugees, compared with 205,000 provided for in the 1948 law.
2. The 54,744 so-called expellees or persons of German ethnic countries known as the Volksdeutsch. This compares with 27,000 permitted to enter under the old law. Under the old law persons in this group had to pay for their own transportation and many could not leave the camps. The new law gives them travel on the same basis as the basic D. P. group.
3. Orphans, 5,000 of them from eighteen Western European countries, including Germany and Austria, and from Italy, Greece and Turkey.
4. The 15,000 persons here on temporary visas who may qualify for permanent admission.

In the first and largest group 32,000 may be subdivided thus:

1. The 18,000 Polish veterans of the Polish exiled army formed in Great Britain.
2. About 10,000 persons displaced by the recent guerrilla war in northern Greece, including 2,500 persons who can be admitted on the request of their relatives here and 7,500 destitute.
3. Four thousand persons now in temporary refuge on the island of Samar, in the southern Philippines. They fled Germany eastward early in the war and had reached Shanghai via Siberia. They fled Shanghai to escape the Russians, and include Jews as well as White Russians.
4. Another group of 5,000 orphans, mostly under care of the International Relief Organization.
5. A total of 500 persons who have or may flee from Russia and her satellites and who for some special qualification are admitted to the United States in the national interest upon the certification of the Secretary of State and the Secretary of Defense.
6. Two thousand persons of the Venezia Giulia, or the Fiume-Trieste region, who found themselves stateless in the transfer of territory between Italy and Yugoslavia.

New Deadline Provided

A fundamental feature of the old law was wiped out in the new one. President Truman denounced the provision of the old law as flagrantly discriminatory against Jews and Catholics. This was the deadline that a D. P. had to be in camps in Germany, Italy or Austria before Dec. 22, 1945, to qualify to come here.

The new law makes the deadline Jan. 1, 1949, thus providing for those who fled persecutions in Poland and Rumania in 1946 and 1947 and Catholics who have been fleeing Communist pressures since 1947.

Also wiped out were the provisions that 40 per cent of the D. P.s should be from the Baltic countries, which was regarded as anti-Jewish and partly anti-Catholic by some critics, and the requirement that 30 per cent should be farmers. The quota of farmers in the old law was never filled, but other D. P.'s could not take their places.

The old law would have expired on June 30, this month, but the new law permits general issuance of visas until June 30, 1951, and visas for orphans and the 54,744 expellees until July 1, 1952.

Among those who witnessed the signing ceremony was Representative Emanuel Celler of Brooklyn, the author of the bill. Conspicuously absent was Senator Pat McCarran, Democrat of Nevada, chairman of the Senate Judiciary Committee, which handled the bill in the upper chamber. He had been an unrelenting opponent of the new bill. Instead, present from the Senate was Senator Harley M. Kilgore, Democrat of West Virginia, who fought Senator McCarran to bring out the bill.

June 17, 1950

CONGRESS ENACTS IMMIGRATION BILL OVER TRUMAN VETO

Senate, 57-26, Follows House on Overriding President—Law Effective in 6 Months

POLITICAL LINES BROKEN

Debate Is Bitter on Measure to Overhaul and Codify Alien and Naturalization Laws

By C. P. TRUSSELL

Special to THE NEW YORK TIMES.

WASHINGTON, June 27—The Senate today joined the House of Representatives in overriding President Truman's veto of the McCarran-Walter immigration bill, and thus wrote the measure into law. The Senate's vote was 57 to 26.

The bill, designed to codify and overhaul immigration and naturalization statutes enacted piecemeal through many generations, will become effective six months from today. It represents the first complete redrafting of such laws since 1798.

Mr. Truman had called the bill infamous. He said the measure, the result of more than three years of Congressional committee work, smacked of thought control. He held its bad points far outweighed the good and would injure the country's international position.

The fifty-seven votes to override were one more than the two-thirds majority required to kill a veto. A switch of two votes would have sustained the President's action. Yesterday the House gave seventeen votes more than a two-thirds majority to the motion to override. The vote was 278 to 113.

Party Lines Shattered

As in the House, party lines were shattered in the Senate. Voting to override were twenty-five Democrats, mostly from the South, but including Senator Ernest W. McFarland of Arizona, the majority leader. Thirty-two Republicans voted to override.

Eighteen Democrats voted to sustain the veto. Eight Republicans joined them.

Five of the Senate's aspirants for Presidential nominations were not present. They were Senator Robert A. Taft, Republican of Ohio, and Senators Estes Kefauver of Tennessee, Richard B. Russell of Georgia, Brien McMahon of Connecticut and Robert S. Kerr of Oklahoma, Democrats.

Senators Kefauver and McMahon were announced as favoring upholding the veto.

Debate before the vote was limited but bitter.

Senator Pat McCarran, Democrat of Nevada, principal sponsor of the bill and chairman of the Judiciary Committee, accused the President of making "unfounded and untrue" statements in his veto message. He also declared that "a person or persons" in the President's office had dealt in "chicanery" in futile attempts to break down State Department endorsement of his bill.

McCarran Makes Plea

The President, Mr. McCarran held, had vetoed the bill although recommendations that he sign it had been made by all other Federal agencies that would be concerned with its administration and enforcement, including the Federal Bureau of Investigation, the Central Intelligence Agency and the Immigration and Naturalization Service. The Senator said the principal issue was internal security at a critical time.

"In God's name, in the name of the American people, in the name of America's future," he declared, "let us override this veto."

The opposition, led by Senators Hubert H. Humphrey of Minnesota, Herbert H. Lehman of New York, William Benton of Connecticut, Blair Moody of Michigan, Paul H. Douglas of Illinois and John O. Pastore of Rhode Island, all Democrats, appeared stunned by the result of the voting, for it had expected the veto to be sustained.

Senator Lehman contended that the bill would make immigration "a myth" by "reducing it to a trickle," in violation of the American spirit of welcoming worthy foreigners.

Senator Humphrey said the measure would bar anyone from Poland and "slam the door" on most immigrants from other Baltic countries. Senator Moody argued that overriding the veto would be "a blow for Stalin," by discriminating against entry of persons seeking to escape from Iron Curtain countries.

The measure retains the quota system of immigration based on national origins. Mr. Truman criticized that system bitterly in his veto message.

Under the system, immigration quotas are assigned to countries in a ratio comparable to that of the various groups of foreign origin in this country's population in 1920. Since the heaviest immigration before 1920 came from Great Britain, Ireland, Germany and other Northern European lands, opponents of the measure argued it would discriminate against immigrants from Southern and Eastern Europe.

Senator Pastore said today the bill was "born in bigotry, founded in hate, and sought to reaffirm the discriminations devised in the present national origin quota system." Senator Benton warned that "more than one seat in this chamber is going to change hands, based on this vote alone."

Today's overriding of a Presidential veto was the first since Oct. 20, 1951. The issue then was contribution by the Government of $1,600 toward the purchase of an automobile for every veteran of World War II or of the Korean fighting who had been blinded or had lost a limb. It was the first enactment over veto on major legislation since 1947, when Congress enacted the Taft-Hartley labor law.

Applause broke out in the House today when it was announced that the Senate had concurred in overriding the veto of the McCarran-Walter bill. The co-author of the measure was Representative Francis E. Walter, Democrat of Pennsylvania.

June 28, 1952

IMMIGRATION SET BRISK PACE IN '56

U. S. Reports Total Exceeded 350,000, Most Since '24—Hungarians in Spotlight

By LUTHER A. HUSTON

Special to The New York Times.

WASHINGTON, Jan. 2—The head of the Immigration and Naturalization Service reported today that immigration into the United States in 1956 exceeded 350,000.

This was the highest total of any year since 1924. In 1925 the Immigration Quota Act of 1924, sharply restricting immigration, became effective. The result was a drop from 706,896 in 1924 to 290,725 in the following year.

The highlight of the immigration year, according to Commissioner J. M. Swing, was the Hungarian refugee program. The Refugee Relief Act had but a few weeks to run when thousands of Hungarians were forced to flee their homeland in the wake of the revolt against the Soviet yoke. Only about 6,500 visas were available to Hungarians.

More than 82,000 immigrants were admitted in 1956 under the Refugee Act, and 6,500 were Hungarians who came in during the last five weeks of the year. In addition, 15,000 Hungarians for whom no visas were available were admitted as parolees.

Besides these refugees, some 90,000 quota immigrants, principally from European countries, were admitted last year. More than a third of the 350,000 total came from nonquota countries in the Western Hemisphere, with Mexico the heaviest contributor.

Families United

Another 35,000 were admitted under the principle of family unification, which makes wives, husbands and children of citizens admissible as nonquota immigrants.

Mr. Swing said that smuggling activities increased during the year. Stepped-up operations at the land borders, he said, resulted in the arrest of approximately 600 smugglers.

Five major attempts to bring stowaways into the United States from Italy and South America were detected. Criminal prosecutions followed against thirty-seven stowaways, five alien crewmen and seventeen residents of the United States.

As a result of improved control of the Mexican border, the commissioner reported, deportations dropped to 4,600. Aliens illegally in the United States who were required to depart without formal deportation proceedings, however, numbered 66,000.

Included among those deported were twenty-four who were expelled as subversives. Deportation proceedings were instituted against 1,073 persons on criminal, immoral or narcotics grounds and 527 in these categories were expelled.

During the year 145,000 aliens became citizens.

January 3, 1957

A New Life in the U. S. Is Problem for Family

By MARTIN TOLCHIN

THE GOLDEN DOOR, a beacon to 1,000,000 immigrants a year prior to World War I, continues to beckon the homeless and tempest-tossed refugees.

Although most of the 270,000 refugees last year came from such friendly lands as Canada, Mexico and West Germany, tens of thousands arrive here in search of political freedom and economic opportunity.

Into this category fall the 180,000 Cuban exiles who have fled to the United States since Fidel Castro's rise to power, and, a few years back, the 40,000 Hungarian refugees who fled after the uprising. A steady stream of Poles and Yugoslavs have also found asylum here in recent years.

Not all of these refugees are the tired and the poor. Typically, political emigrés represent a cross-section of their native countries, according to the International Rescue Committee, a nonsectarian agency that helps resettle and rehabilitate the victims of totalitarianism. The refugees come from all social classes and all walks of life, the committee notes.

Helps to Find Jobs

The International Rescue Committee directs its efforts at the bread-and-butter concerns of the refugees. It helps them find jobs and homes, teaches them English, helps them bring relatives and friends to the United States, provides counseling, medical aid, clothing and cash grants. The committee's offices in New York are at 460 Park Avenue South.

What happens to a family that is compelled to pull up stakes was described in a report by the committee on Cuban refugees.

"Sudden mass flights always disrupt family life," the report observed. "People exposed to the danger of police repression frequently must leave alone. Yet once they reach safety, to become reunited with their families is their major preoccupation.

"Often it was not the heads of families who preceded their wives and children into exile. Thousands of Cuban families sent their children to the United States because they did not want to expose them to totalitarian indoctrination in Cuban schools."

Solidarity Is Asset

In many respects, the Cuban immigration bore the earmarks of the large-scale pre-World War I immigrations.

"Their great asset is solidarity," Charles Sternberg, director of the committee's resettlement department, said. "They stick together."

Relatives and former neighbors temporarily adopt children and other newcomers. This is reminiscent of an earlier day when the Eastern European Jews were shown the ropes by "landsmen," and Italian immigrants found temporary havens with "paisanos."

"One of the reasons why their homes are so crowded," a caseworker said, "is that they're so kind to each other. They take each other in. I had a case of 22 people living in four rooms."

In the Hungarian immigration, this loyalty was not as manifest. But compensation came in a groundswell of American support for the Hungarian "freedom - fighters." Renting agents went out of their way to offer apartments to the Hungarians, and companies made job openings available to them.

Nevertheless, language difficulties led to job difficulties for both Hungarians and Cubans. A Hungarian refugee in his mid-20's who had been a college teacher found employment as a file clerk, in which position he is still employed. Asked what he found hardest to adjust to in the new world, he replied:

"To lose your profession and your position, and to become nothing."

Similarly, an upper middle-class Cuban has finally found employment emptying garbage in a factory. Many Cubans have been forced to take jobs as menials in restaurants and hotels.

For those who, like their predecessors of 50 and 75 years ago, expected the city's streets to be paved with gold, the disillusionment could be devastating.

The Hollywood films that glamorized American life were held partly responsible for the "automobile complex" that afflicted many refugees, including the Hungarian youth who came to the committee with a provisional sales slip from a car dealer. He could not understand the agency's refusal to give him $50 for a down payment.

"But I started work this morning," he protested.

Because it is easier for women to find employment in low-wage industries, there are many families in which both husband and wife work, or in which the wife works while the husband cares for the children.

"This is terrible for a man in any culture," said a woman who has worked extensively with Cuban refugees, "but for a man from a Spanish culture, it is intolerable."

Like the children of the large-scale immigrations, Cuban and Hungarian children are eager to become Americanized, and sometimes become ashamed of their parents who cling to old customs and the native tongue.

Names Americanized

A Hungarian girl in the first grade told her teacher that her mother was coming to open school week, adding:

"My mommy has a funny way of talking, but you'll understand her."

Typically, the children Americanize their names. Janos and Juan become John. Rozsika and Rosita become Rose. Studies of third-generation Americans indicate that their children's children will begin to revert to the old names, and take pride in their heritage as they become secure in their new life.

Many Cuban parents regard the United States as a temporary haven, and intend to return to Cuba. Moreover, most Cuban exiles have entered this country under a visa waiver in which they are ineligible to become citizens, according to the United States Immigration and Naturalization Service. Thus, their basic outlook is at odds with the attitude of their children, who can hardly wait to become Americans.

"A person who is waiting for the day when he will be able to return home does not meet the exigencies of a new existence in the same way as a person who has made a break with the past," the International Rescue Committee observes. "The first remains a refugee. The second has become an immigrant."

October 8, 1962

IMMIGRATION LAW GOES INTO EFFECT

by CABELL PHILLIPS

Special to The New York Times

WASHINGTON, Dec. 1—The new immigration law abolishing the national-origins quota system became effective today, but its full effects will not be felt for about two and a half years.

The principal change that occurred today creates an international pool of unused quota numbers accumulated in under-subscribed countries like Britain and makes them available on a first-come, first-served basis to certain "preference" applicants in other nations.

The pool for the current fiscal year is approximately 56,000. Its principal beneficiaries will be close relatives—living in such low-quota countries as Italy, Greece and Nationalist China—of persons who have already become citizens or resident aliens of the United States.

Quick Action Ordered

The visa division of the Department of State has ordered consular offices around the world to process as many of these preference cases as possible in the next two weeks so that families can be reunited for Christmas.

The old national-origins quota system will remain substantially in effect until June 30, 1968, but the transfer of unused quota numbers to the world pool will mitigate some of its more objectionable features.

Until today, these unused quota numbers were not transferable from one country to another.

After June 30, 1968, an annual over-all ceiling of 170,000 immigrants will apply on a basis of strict equality to nations outside the Western Hemisphere. There will be no country by country quotas, and the national of an Asian or African country will receive the same consideration as a citizen of France or Germany. No country, however, will be allowed more than 20,000 immigrant visas in a single year.

Ceiling for Hemisphere

On the same date, an over-all ceiling of 120,000 annually will apply to the Western Hemisphere, but without the 20,000 country by country limitation.

This will be the first time that such a restriction has been placed on immigration from Canada and the independent nations of Latin America and the Caribbean. This feature of the bill was strongly opposed by the Administration on the ground that it would cause resentment on the part of some Latin-American governments.

A spokesman for the State Department said today, however, that there had been no substantial protest from any of those governments since the bill became law.

The chief complaint against the national-origins quota system, adopted in 1924, was that it favored the nations of Northern Europe and discriminated against those of Southern and Eastern Europe and of most other regions of the world.

Most nations of Asia and Africa, for example, have had token annual quotas of only 100 under this system.

Another aspect of the bill that became effective today requires all immigrants from the Western Hemisphere who plan to seek work here, and most of those from other parts of the world who do not have a preference rating as close relatives of United States citizens, to obtain a certificate from the Secretary of Labor before they can be granted visas.

The certificate must affirm that whatever job the immigrant intends to take in the United States, he will not displace an American worker.

December 2, 1965

Filipinos: A Fast-Growing U.S. Minority

By EARL CALDWELL
Special to The New York Times

SAN FRANCISCO, March 2—On the sidewalk outside a theater in the Mission District here, a casually dressed young man studied a poster that billed a movie in a language called Tagalog.

Frowning slightly, he turned to a stranger and asked: "What's Tagalog?"

Tagalog is the national languages of the Republic of the Philippines and it is appearing more and more often these days, not only on movie billboards here in the Mission, but also in shop windows, stores and other places in the urban United States.

All of this reflects a sharp increase in the immigration of Filipinos—they now rank second, slightly behind Mexicans, in total yearly numbers—since a new immigration law was passed in 1965.

That law abolished a "national origins" quota system designed to preserve the ethnic balance of the United States population as reflected in the census of 1920. In an attempt to put immigration on a first-come, first-served basis, the new law allocated 170,000 visas a year to Europe, Asia and Africa and 120,000 a year to Canada and Latin America.

The most dramatic result of the change was the increase in the immigration of Filipinos—from 2,545 in fiscal 1965 to 25,417 in fiscal 1970, according to government statistics.

Five years ago, the Philippine Islands ranked 22d in the number of immigrants to the United States, with less than 6 per cent of the number admitted from the leading nation, Canada.

Most of the Filipinos immigrating to the United States tend to settle on the West Coast, the area closest to their homeland. But with their increased numbers, and in seeking jobs, they are moving across the country and settling in cities as far east as Chicago and New York.

Impact in the West

The impact of the immigration of Filipinos to San Francisco is clearly evident.

They make up the city's fastest growing minority group. In less than five years their numbers have doubled and now exceed 20,000.

And San Francisco is not unique.

In the Los Angeles area, the Filipino population has jumped from about 20,000 in 1965 to about 45,000. In San Diego in the last five years the population of Filipinos has tripled. And in Portland, Ore., where there were only about 1,000 Filipinos in 1965, there are now nearly 3,000.

In the predominantly Mexican-American Mission District here in San Francisco, theaters that cater to Filipinos by showing films in the national language of their homeland are not uncommon.

There is also at least one Roman Catholic Church where mass on Sunday is read in Tagalog. As on Ninth Avenue in New York, there are clusters of grocery stores that specialize in supplying foods from the Philippines. And across the city there are a number of restaurants that feature the cuisine of the Philippines.

The Filipinos coming here now are neither poor, nor illiterate nor unaccustomed to American ways, in contrast to former immigrants from the Philippines, who were mostly farm workers and household servants whose only hope was to exist.

Today a majority of the Filipino immigrants are doctors, lawyers, engineers, teachers, nurses and other professionally trained persons.

Yet the life that many of them accept here is one that most native Americans would shun.

Lawyers work as file clerks, teachers as secretaries, dentists as aides, engineers as mechanics and, in some instances, as common laborers.

However, while they differ from their predecessors in goals and skill, their prime reason for immigrating is still economic. Not only are jobs scarce in the Philippines, but average salaries provide little more than subsistence.

"Salaries are very high here," said Mrs. Mercedes Nicanor, a Filipino who is public relations manager for the Philippine Communities Executive Council of New York and New Jersey. "Here we've found a green pasture."

In addition to the poor economy, many of the immigrants cite what they consider to be a bad political situation back in the islands as a reason for moving to the United States. Some feel that a revolution may be in the wind and thus prefer the stability that life in this country represents to them.

Since English is taught in the public schools on the islands, most of the Filipino immigrants find the adjustment to life here somewhat less difficult than other immigrants.

Many of the immigrants are married and bring their families here. However, unlike the Chinese, Japanese and other groups that have immigrated from the Eastern Hemisphere, the Filipinos do not tend to live together in ethnic neighborhoods. They do tend to meet socially, sometimes through ethnic clubs and more often through religious activities, since they are predominantly Roman Catholic.

But since they are also predominantly professionals, or relatively highly skilled workers, they do not follow the location patterns previously associated with immigrants from the Philippines. Once, Filipinos lived primarily in rural areas; now, the Filipino population in such areas is decreasing. The new arrivals are settling in and working in urban areas.

Surplus of Professionals

One reason for the preponderance of professionals in the new wave of immigrants is economic: It costs about $1,000 just to obtain a visa. Even so, there are waiting lists, but professionals reportedly are given first preference for admission to the United States.

Thus the immigrants are relatively well educated, a factor that many observers say contributes to the ease with which the Filipinos have been assimilated here.

"They adapt very well," a Catholic priest in San Francisco said. "If they have to work as janitors they can do it—and they will do it very well."

Jose Arcega, who is 41 and who was a lawyer in the Philippines, offers another explanation. He said there was a surplus of professionals in the Philippines and that some were unemployed.

He also said that many immigrants could make more money here working outside their professions than they could there even if they were employed in the jobs for which they were trained.

Mr. Arcega himself is employed here as a file clerk. But that does not bother him. What does, he said, is the feeling that "we're out of the current, we're just watching things."

Financially, Mr. Arcega believes that he is better off in America. "But socially," he says, "I think I'd prefer to stay back in the Philippines. There people just love to talk to you. Here they are just too busy. There is no time to talk."

There are differing opinions as to whether the immigrants from the Philippines are victims of discrimination in the United States.

Issue of Prejudice

Alex Intal said he saw no prejudice. He is 47 years old and has been here less than a year. He taught in the public schools in the Philippines for 23 years. He retired and moved here with his wife and four children.

He has not been able to get a teaching license in California. He works as a clothier. But he is confident that once he obtains his license and a position opens, he will be teaching again.

Frederico Austria, who is 43 and lives in Portland, is not so certain about prejudice. He was a practicing attorney, dean of liberal arts at Palaris College and executive secretary to the Mayor of San Carlos before he left the Philippines in 1963.

He intended to enroll in a law school to get a master's degree but needed a job to earn tuition. "I tried to get a job at law offices as a researcher or legal clerk," he said. "I answered 25 ads in the newspaper, but every time I went out to the interview, they said the job had been filled. I really don't know."

Carlos Rivera of the Civil Rights Division of the State Bureau of Labor in Oregon says flatly: "They (Filipinos) have the same problems as Cubans. While they may be professional and meet the qualifications, they meet this pattern of employer bias."

Mr. Rivera added that, "given the economic situation, when you talk about hiring a minority group member, you are talking about not hiring a white, and this isn't going to happen."

March 5, 1971

Catholic Diocese of Brooklyn to Help New Immigrants Adjust to City

By GEORGE DUGAN

A diocesan migration office was established last week by the Most Rev. Francis J. Mugavero of Brooklyn to serve the growing numbers of new Roman Catholic immigrants settling in Brooklyn and Queens.

Headed by the Rev. Anthony J. Bevilacqua, the office established by Bishop Mugavero will help Spanish-speaking Puerto Ricans and Dominicans, and Italians, Haitians, Germans, Poles and Croats adjust to their new surroundings. The office will be at the headquarters of the Roman Catholic Diocese of Brooklyn, at 75 Greene Street, in Brooklyn.

"The American church was until recently an immigrant church and as such it had a special structure, attitude, behavior and pastoral theology," Father Bevilacqua said. "While much of the rest of the country is out of that period, the diocese of Brooklyn must be regarded once again as largely a church of migrants.

"It is not exaggerating to say that the future of the diocese lies in these newcomers."

Father Bevilacqua said in an interview Friday, that there were about 800,000 recently arrived Roman Catholics now living in the diocese, which comprises both Brooklyn and Queens. As of last year an estimated total of 1.5 million Catholics resided in the diocese.

About 600,000 of the new arrivals, he said, are Spanish-speaking Puerto Ricans and Dominicans concentrated in the Brooklyn neighborhoods of East New York, Brownsville, Williamsburg, South Brooklyn, Park Slope and parts of Bay Ridge, and in the northern Queens communities of Astoria, Woodside and Jackson Heights.

On the basis of these estimates, Father Bevilacqua said that it must be concluded that for more than 50 per cent of the Catholics who are permanent residents of the diocese, English is not the native language.

In addition, he continued, the church also must be concerned with helping other immigrants who do not settle permanently in the diocese, such as workers, students, Government officials, business people, seamen, airline personnel and tourists.

For example, he said, 400,000 foreign seamen come to the Port of New York each year and 100,000 non-immigrant Italians arrive in this country each year as temporary workers or tourists.

According to Father Bevilacqua, the image of the church in Brooklyn and Queens has been to a large extent that of an English-speaking, middle-class, educated community. "This is unrealistic in a diocese where 800,000 Catholics are immigrants," he said.

The new office will analyze the needs of newcomers, encourage the training of priests in the languages of the immigrants and act as an intermediary between Bishop Mugavero and the immigrants.

March 14, 1971

Eased Laws Alter U.S. Ethnic Profile

By BILL KOVACH

Special to The New York Times

FALL RIVER, Mass., June 13—There is a new Fall River in the making as a result of shifting immigration patterns that are adding new features to the profile of the American population.

Lured by earlier settlements from the Old World, Portuguese immigrants are flocking to southeastern Massachusetts in numbers exceeding 4,000 a year. Since the immigration laws were liberalized in 1965, the Portuguese have overwhelmed older groups from Northern European countries and now form the predominant ethnic group of the region.

Fall River is only one among many cities in the United States undergoing a radical shift in ethnic make-up because of the liberalized immigration laws.

Los Angeles and other West Coast cities have developed entirely new communities of Filipinos; The Boston Globe carries a weekly column in Spanish for thousands of new Spanish-speaking residents; Greek and Middle Eastern restaurants are opening in a number of towns across the country.

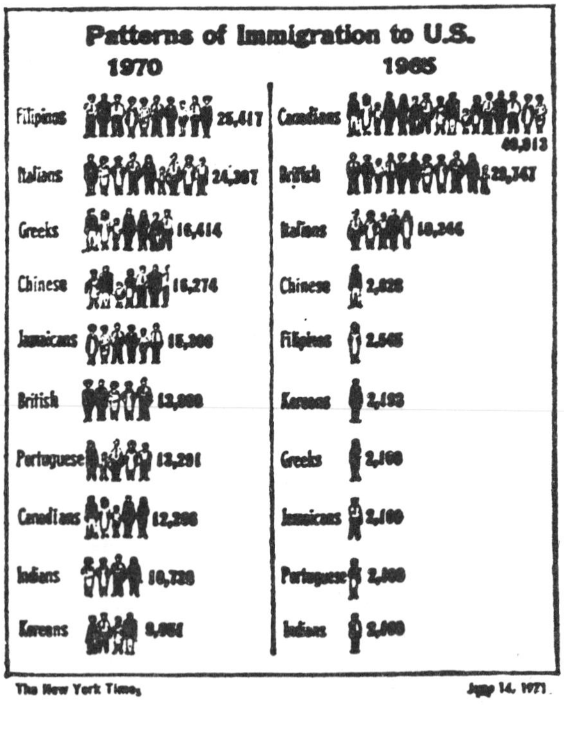

Indeed, the Immigration Act of 1965, which allowed for the first time large-scale immigration from countries outside Northern Europe, is changing America fast.

The Census Bureau recently reported that nonwhite immigration reached significant proportions for the first time from 1960 to 1970—accounting for nearly 14 per cent of all immigrants—for a total of half a million people. Immigration accounted for about 20 per cent of the total population growth of the United States—a total of 3.9 million people—in the decade.

Fall River, which is 15 miles southeast of Providence, R. I., typifies the new immigration patterns.

A few years ago, to serve the needs of the changing population of southeastern Massachusetts, the Roman Catholic Church entered into an agreement with the family of Luciano J. Pereira, a young native of St. Michael Island in the Portuguese-owned Azores. The church, it was agreed, would educate young Luciano (public education was provided by the Government only through the third grade) if he would agree to serve the needs of the church.

An assistant at St. Michael Church here, Father Pereira administers to the growing community of Portuguese immigrants.

When Father Pereira came to perform his service to the Portuguese-American community, he was dealing with a small band of people. The Azores had served as a station of cheap labor for the New Bedford and Fall River ships. Some settled here and each generation lured others to work in cordage factories and later in the textile factories that saved the region from economic ruin in the face of steamship competition.

Twentieth-century immigration from the Azores and the Cape Verdean Islands—as well as the mainland of Portugal—was miniscule, however, under the National Origins Act of 1924, which favored Northern Europeans almost to the exclusion of the Southern and Eastern Europeans and Asians.

In 1965, the last year that the quota system begun by the 1924 law was in operation, 42 Portuguese immigrants came to Fall River. In 1970, there were 700.

The Immigration Act of 1965, which became fully operative in 1968, abolished the old National Origins quota system, which tried to keep the same ethnic balance in immigration that was reflected in the population census of 1920.

Quotas by Nations

Quotas were allocated by nations before 1965; Britain, Germany and Ireland accounted for 70 per cent of all immigrant visas issued for Europe, Asia and Africa. Southern and Eastern Europeans, Orientals and Africans were admitted only in tightly controlled dribbles.

By 1965, a Democratic party

Patterns of Immigration to U.S.

1970		1965	
Filipinos	25,417	Canadians	40,013
Italians	24,307	British	28,747
Greeks	16,414	Italians	10,346
Chinese	16,274	Chinese	2,628
Jamaicans	15,300	Filipinos	2,545
British	13,000	Koreans	2,193
Portuguese	13,291	Greeks	2,100
Canadians	12,398	Jamaicans	2,100
Indians	10,728	Portuguese	2,000
Koreans	9,314	Indians	2,000

The New York Times June 14, 1971

campaign for a more equitable system bore fruit. The act passed that year put all potential immigrants on a first-come, first-served basis. Europe, Asia and Africa were granted 170,000 immigrant visas each year; Canada and Latin America were allocated 120,000. No national quotas were imposed and the only limit set was a ceiling of 20,000 from any one country in a year.

At the same time, the law gave a clear priority to relatives of United States residents seeking immigration visas.

The result was like opening a flood gate for those countries whose entry had been restricted. Italians, Chinese, Filipinos and others formerly curbed streamed in.

The results are graphically shown in comparisons of the country of origin of immigrants in 1965 and 1970. The Philippines and Italy replaced Canada and Britain by 1970 in the top three countries sending new citizens to America.

Ireland, France, the Netherlands, Russia, Sweden and Norway dropped out of the list of the top 25 nations by 1970 to be replaced by countries such as Jamica, Portugal and India.

Indian and Portuguese communities are springing up in cities that once had Irish and German colonies; Yugoslavs are filling restaurant jobs once dominated by Frenchmen; Spanish is the first language with more and more residents of New York, Boston and Chicago.

Problems, Too

The sudden shift in the character and size of immigration to the United States has created a number of problems, including the following:

¶Chinatown in San Francisco is bursting at the seams because of immigration that reached a peak of 10,000 in 1969. The new immigrants do not fit into the traditional life style set by the long-time residents.

¶Chicago's schools are filling with Spanish-speaking and Chinese-speaking youngsters attempting to cope with little in the way of special programs in language training.

¶Hawaii is receiving unskilled Filipinos and Orientals who compete with local labor and force wage rates down.

Such problems are magnified in places like Fall River, a city of 50,000 population.

Clustering around churches like St. Michael and St. George, named for the home islands and serving communities from those islands, the Portuguese have spread across Fall River, dominating the north and east arms of the city.

Names like Silva, Cabral and Escobar predominate. Restaurants feature povo (octopus), couves (kale soup) and de bulho (a mixture of tripe and liver).

Inside neat frame houses that reflect the pride in ownership that drove most of the immigrants to invest their first American money in homes, scenes of rediscovery are common. Scenes like the one at 139 Merchant Street, where 72-year-old José M. Silva, tears streaming from his eyes, was reunited with his sisters for the first time in 50 years.

A scene so frequent as to be characteristic of Fall River these days is that of a middle-aged or elderly Portuguese couple, carefully dressed in a dark suit or dress, being led through stores and offices by a young, dark-haired 12-year-old. The youngster, testing his still-imperfect English is necessary as an interpreter for the newly arrived relatives.

In supermarkets and employment offices, the young one transmits information back and forth as his elders begin to cope with the job of becoming Americanized.

John R. Correiro, who in 1967 organized the English as a Second Language Center here to help immigrant children move more smoothly into the school system, has studied the problem in detail.

"Primarily," he said, "it is a problem of rapid adjustment with little or no help. Most come directly from farms, many never wore shoes. Can you imagine the shock? One day you leave a peasant culture by airplane and the next day you're in the middle of a complex, dizzy city."

There is also the problem of an increase in illegal immigration brought about by the rising hope fostered by the new law. As more immigrants come to the United States from restricted countries, they lure more friends and relatives — many of whom come without proper clearance.

At a recent hearing held in New Bedford by Senator Edward W. Brooke, Republican of Massachusetts, one lawyer admitted to inducing more than 200 Portuguese to remain in the country illegally with the hope of forcing the Government to keep them.

Barbara M. Watson, administrator of the Bureau of Security and Consular Affairs for the State Department, which is in charge of immigrant visas, takes a philosophical view of the problem. "As long as there are any restrictions on immigration into the United States," she said, "there is going to be the problem of illegal entry."

June 14, 1971

Illegal Aliens Pose Ever-Deepening Crisis

By PAUL L. MONTGOMERY

They come on the commercial jet from Bogotá, nervous and silent among the tourists, and they come stuffed three together in the trunk of a Mexicali smuggler's sedan. They come hunted ragged through the scorched Texas brush country, alert for rattlesnakes, and they come pushing their luggage before them in the air-conditioned reception center at the Kennedy International Airport.

While there is no way of accurately determining their numbers, a consensus of informed estimates is that there are between one and two million illegal aliens working and living now in the United States. They continue to come at the rate of at least 2,000 a day.

They come from the arid backlands of Mexico and the teeming, hopeless towns of Haiti, from Trinidad and Lima, Athens and Hong Kong, or any one of 50 countries around the world. They come, some of them, to begin a new life in America against all odds, but most seek only temporary surcease from poverty for themselves and their families.

The goal of almost all the aliens is honest work, usually at minimum pay, but their petty offenses in entering the country sometimes lead to deeper entanglements. Although they come innocently, they are prey for document forgers, smugglers, drug traffickers, unscrupulous lawyers and landlords, and employers whose profits depend on a ready supply of cheap labor.

The influx of illegal aliens has had clearly deleterious effects on the economy, the balance of payments, the labor market, immigration policy and social services such as welfare.

At least four Congressional committees are preparing investigations on various aspects of the problem.

The illegal aliens exist at least in part because of an immigration system, revised in 1965, that appears unworkable. In many cases, the law seems to invite circumvention.

The undermanned Immigration and Naturalization Service appears overwhelmed by the deluge. While it captures a substantial number of illegal aliens—420,126 in the fiscal year 1971—hundreds of thousands and more find it easy to get away.

In Washington, officials limit themselves to the minimum expression of concern about the problem, but immigration officers in the field continually use words like "fantastic" and "unbelievable" to describe the situation.

There is a spreading underground of illegal aliens in every major city. They can be found in any job, but most gravitate to work requiring few skills. Some still pursue the traditional callings of farm worker or live-in maid, but most now seek factory or service jobs.

Raid at Packaging Plant

In Los Angeles last week, Federal agents found 36 illegal immigrants among 300 workers in a food packaging plant owned by Mrs. Romana A. Banuelos, President Nixon's nominee for Treasurer of the United States.

Most of the aliens work in private industry, but they are by no means limited to it. Last week, a raid at the United States Military Academy at West Point uncovered 18 illegal aliens working there as busboys and in similar jobs.

Last month in New Jersey, immigration officials were finding an average of 25 illegal aliens per factory before they ran out of money to continue the sweep. There are few large hotels, restaurants or hospitals on the East Coast that do not employ illegal aliens as dishwashers, busboys or orderlies.

Some of the aliens find the decent income and stability they came to the United States for. Many more lead lives of little joy—their families in disarray, fearful of capture, exploited and cheated, corroded with that peculiar loneliness of a person without a country. Those wait only for that magic moment when they have earned enough, or have put up with enough, to go home.

Encarnación A. is 25, deeply religious, very pretty, has a husband and two children in Ecuador. Her husband is ill and can't work. She came to New York on a tourist visa in the spring of 1970 and, through *friends, found a job in New Jersey in a garment shop. "I had to leave," she says softly. "The boss there was all over me." At her second job the same thing happened. She won't talk directly about the situation at the next place, where she still works as a sewing-machine operator in a factory making bathrobes. "I have to work, my family needs the money," is all she will say. She makes about $95 a week with piecework, and sends $150 a month to Ecuador. Her landlord has raised her rent three times in the last eight months and lately has started hanging around her apartment. She keeps telling herself she should go home, but somehow each month she is still here.*

Bulk Are Mexicans

The bulk of the illegal entrants are Mexicans who sneak across the 3,000-mile border between Brownsville, Tex., and San Ysidro, Calif. Of the 420,000 illegal aliens caught in the last fiscal year, 320,000 were Mexicans.

The entrants, commonly called "espaldas mojadas" in Spanish and "wetbacks" or "wets" in English, are pursuing a calling that has been endemic in northern Mexico for 30 years.

The first phase of the influx reached its peak in 1954, when nearly one million Mexicans were caught trying to cross the border. The traffic was reduced to a dribble within two years by an immigration crackdown in the Eisenhower Administration and the institution of the bracero program, under which Mexican contract workers were allowed to enter for seasonal farm work.

Congress ended the bracero program in 1965 to benefit American labor. The immigration laws were revised in the same year to make it generally more difficult for Western Hemisphere residents to immigrate to the United States legally.

The consequence was resumption of the illegal traffic across the border, a traffic that has increased at least 25 per cent each year since and shows no sign of abating.

Set Out for Cities

At the same time, mechanization on the farms and shortage of urban labor willing to work at the minimum wage drew the new wave of braceros to the cities.

Where before the illegal workers generally stayed within 50 miles of the border, now they set out immediately for cities like Houston, Los Angeles, Denver, Detroit and Chicago.

Last month, during the harvest in the Northeast, officials were regularly picking up carloads of illegal Mexicans in New York and New Jersey who had crossed the Texas border a few days before.

There are occasional cases of Mexicans who have crossed the border illegally years before and started families and new lives in the United States, only to be caught and deported. Most of the traffic, however, is transient, made up of young men and young women trying to make enough money to get their families through the Mexican winter.

Manuel Najara Borjas, 18, is one of seven children of a farm laborer in Santa Martá del Oro, Durango state, Mexico. His family lives in a three-room shack, he says, with "hardly enough to eat." Until September, 1970, he worked on a farm near home for $1.25 a day. Then his father told him to go to the United States, find a job, and send back all the money he could. The youth jumped a fence in the early morning near El Paso and caught a plane to Chicago the same day with money his father had given him. He soon found work as a dishwasher at the Sheraton-Chicago Hotel, earning $14.75 a day plus two meals. He lived frugally with two other illegals in a $175-a-month apartment. By last month, when he was caught, he had sent back $700 to his father. "If I can, I'll come back," he said, just before he was put aboard a bus for deportation.

Entry by Braceros

Some of the braceros still use the traditional method of entry—in Texas, swimming the Rio Grande or being ferried across in boats made of car hoods welded together, and in California and Arizona crossing the desert at unpopulated places. Then they trek north to get around the checkpoints the Border Patrol maintains on roads leading from the border, cooking over mesquite fires, walking in the cool parts of the day. When they are past the 30-mile Border Patrol zone, they emerge and look for work.

The crossing through the brush or desert is perilous. There are dozens of drownings a month in the Rio Grande, and several cases a month of rattlesnake bite. Almost every week, the Border Patrol finds a body or two in California's Imperial Valley desert, desiccated by the heat.

Because of the danger and the distances to be traveled, the smuggling of braceros has been on the rise. Last year, the Border Patrol encountered 349 such cases; of the operators, about 2,000 were American citizens or legal permanent residents and 1,400 were themselves ill[illegible]l aliens.

In the [illegible] smuggling operation, the braceros cross on their own and assemble at a border town. The smugglers, working at night, then transport them in trucks or campers through the checkpoints and on to their destinations. The usual fee for a run from the border to Chicago or San Francisco is $250 a head. The size of the groups can range from 15 to 70.

Chipping In for Car

Another common method is for the braceros to chip in and buy an old car to take them past the checkpoints and on to the work. The driver gets the car as his payment.

For many of the braceros, the smuggling operation is the first taste of the rough treatment and exploitation they will encounter throughout their time in the United States.

In July, near Hollister, Calif., the bodies of two Mexicans were found in a roadside ditch. Jesus Ayala Dominguez, 35, of Michoacan state, was dead of carbon monoxide poisoning. His companion, Delores Blanco, 25, was unconscious. They had been kept locked in the trunk of a car for two days by a smuggler who was driving them from San Ysidro to the Salinas Valley for $250 each. When the smuggler saw their condition, he dumped them out. The two had crossed the border with 19 other clients of the same smuggling operation.

On the Canadian border, there is also a substantial illegal traffic, including many Canadians—estimates run as high as 100,000 at any one time —who work illegally in the United States. The border-crossers also include aliens being smuggled to the United States, and seamen who have jumped ship in Canada.

A substantial part of the traffic involves people who are already living illegally in the United States—usually on an expired tourist or student visa —who want to go home for a visit without passing through United States Immigration.

New Wave After 1965

After the Mexicans and Canadians, the other large section of the illegal alien population consists of people from elsewhere in the Western Hemisphere who have entered since the revision of the immigration laws in 1965.

Under the old law, there was no numerical restriction on immigration from independent countries in the Western Hemisphere, but travel distances, literacy and financial requirements and lack of family ties limited immigration in effect to citizens of Mexico and Canada.

The 1965 law established annual quotas of 120,000 for the Western Hemisphere and 170,000 for the rest of the world. While the law removed grave inequities for Asians and Southern Europeans, it appears to have had the opposite effect in the Western Hemisphere.

The new law came at a time of growing interest in Latin America and the Caribbean to working in the United States. The result was a concerted effort to avoid the law.

The usual method of entry for illegal aliens from Latin America or the West Indies is to get a tourist or student visa and to use it to get past immigration.

Such visas allow the holders to stay in the United States for a short period, as much as three months for tourists, or the term of study for students. Aside from exceptional circumstances, the visa holders are prohibited from working; they must sign a statement before getting a visa affirming that they do not intend to work.

Non-immigrant Visas

In 1965 in the Western Hemisphere excluding Mexico and Canada, the State Department issued 205,358 non-immigrant visas. The comparable figure for 1970 was 488,914. It is probable that illegal aliens account for most if not all of the increase.

Adelaida S., 45, has two children in Cartagena, Colombia, where she worked as a seamstress. She is separated from her husband. In August, she came to Miami on a tourist visa, intending to look for work to support her family. A week after her arrival she answered a newspaper ad and was hired on the spot as a $65-a-week housekeeper by a Coral Gables businessman with two school-age children. When her three-month visa is up, she will apply for a three-month extension, which will probably be granted. After six months, when further extension is unlikely, she will return to Colombia for a week (the air fare is about $100) and then come back to Miami to start the process over again. She figures she can save $500 in each six-month period. Her employer is pleased with her. "I hope she will stay with us," he said. "In the past few years many of my friends have become apprehensive about employing American black domestic help."

Most of the aliens do not go through the ritual of renewing their visas once they are in the United States. The chances of their being apprehended are slim, since the Immigration and Naturalization Service has stopped the practice of tracking down holders of expired visas.

In explanation, James F. Greene, associate commissioner for operations, says that the process had resulted in finding only 12,380 illegal aliens out of 673,589 investigated.

Illegal aliens form what amounts to a subculture in American cities. It is likely, for example, that the majority of the West Indian and Haitian residents of Bedford-Stuyvesant in Brooklyn, and of the Latin Americans in Elmhurst, Queens, are not lawful immigrants. The illegal alien population of Miami is estimated at 100,000, made up mostly of Dominicans and Haitians.

Because of fear of being found out, the aliens cannot participate in many aspects of American life. As a substitute, they have developed a kind of outlaw culture of their own in which information about jobs, immigration regulations, helpful lawyers and travel agents and other matters is exchanged.

A continuing concern for some is trying to find an American citizen who will marry them, either permanently or for convenience, so they can acquire citizenship and stop their furtive life.

Joan F. entered from Guyana in October, 1967, at the age of 17 as a tourist, and has been working since then in a pocketbook factory. She has been living with a 25-year-old fellow illegal, Brinsley T., in Bedford-Stuyvesant. They have a child, who is a United States citizen since he was born here. Joan F. and her companion are waiting out a complicated game. Brinsley T. has married a 54-year-old legal permanent resident, a welfare recipient, but does not live with her. The couple's hope is that Brinsley T. can get legal permanent residence through his marriage, divorce his wife of convenience, then confer legal permanent residence on Joan F. by marrying her.

In the illegal alien enclaves, something of the people's home culture is kept alive by parties, national celbrations, soccer games and the like.

"In our hearts, we have never left our countries," one man said.

There is also an edge to life in the enclaves that is not present elsewhere, since disputes and broken romances often end with one party denouncing the other to the immigration authorities.

While exploitation at work and in housing, and unpleasantness generally appear to be the rule, some illegal aliens have been able to escape the grind and achieve stable, fairly prosperous lives.

Jaime Charleswell, 34, of San Pedro de Macoris, the Dominican Republic, came to the United States as a tourist in September, 1969, and has stayed. He has five natural children of two wives in the Dominican Republic cared for by relatives. He processes orders for a shoe company in Long Island City. By last year he had saved enough to bring one of his common-law wives to the Bronx and marry her. This year he bought a two-family house on East 141st Street for $1,000 down and an $18,000 F.H.A.-guaranteed mortgage. He has converted one-half of the house for roomers. He has applied for an immigrant visa, but probably will have to wait at least another year before it is granted.

Aside from Western Hemisphere residents, the traffic of illegal aliens is small, although sizable groups of Portuguese, Greeks, Italians, Chinese and Filipinos show up in the statistics.

Chinese Immigration

Illegal Chinese immigration has been prevalent since the Chinese Exclusion Act of 1884; in its present form, it consists largely of seamen who jump ship.

Yim Shing Chau, 33, and Tsang Chung Chiu, 38, both of Hong Kong, entered the United States in June, 1970, as seamen on the motor vessel Deganya. Both have wives and children in the East—Mr. Yim in mainland China and Mr. Tsang in Hong Kong. Both of the men jumped ship in Newark and came to Chinatown by bus, where they were put up by an association of people from their home city of Swatow. After some odd jobs in laundries and restaurants, they went to work at a restaurant in East Hampton, L. I. in August, 1970. Mr. Yim was making $125 for a 78-hour week as a dishwasher and Mr. Tsang was making $140 for 66 hours a week as a waiter. Both lived on the premises. When they were caught last month, Mr. Tsang had $3,015 in the bank and Mr. Yim about $3,000.

The size of the illegal alien population in the United States is a subject of debate. Mr. Greene, like many officials, says the figure is impossible to determine or even estimate, but that some of the larger estimates, such as five million, are excessive.

The size of the traffic across the Mexican border is the hardest to approach, since the braceros leave no documentary traces if they are not caught. Experienced Border Patrol officers, speaking privately, concede that a figure of one million for the number of illegal Mexicans in the United States at any one time is a reasonable estimate.

The number of people who have entered the United States on tourist or other non-immigrant visas and then stayed is also elusive, since a careful record of entrance to the country is kept but no reliable record of departure appears in statistics.

Mr. Greene says that since 1965 his office has received documents on 1,785,127 people who have entered the country on non-immigrant visas and have apparently failed to depart. However, he says that 57 per cent of that figure is the result of filing errors, and that most of the rest would be cleared if the cases were investigated.

Mr. Greene's figures, which would result in an estimate of perhaps 40,000 for the number of visa holders illegally in the United States, conflict with much of the available evidence.

In Haiti, for example, 10,000 tourist visas were issued in 1970; more than half of the "tourists" have failed to return from the United States. Officials in Haiti and the Dominican Republic estimate that 300,000 of their citizens have gone to the United States as non-immigrants and failed to return.

Immigration officers who have had experience in the field say their feeling is that there are hundreds of thousands of non-Mexican illegal aliens in America.

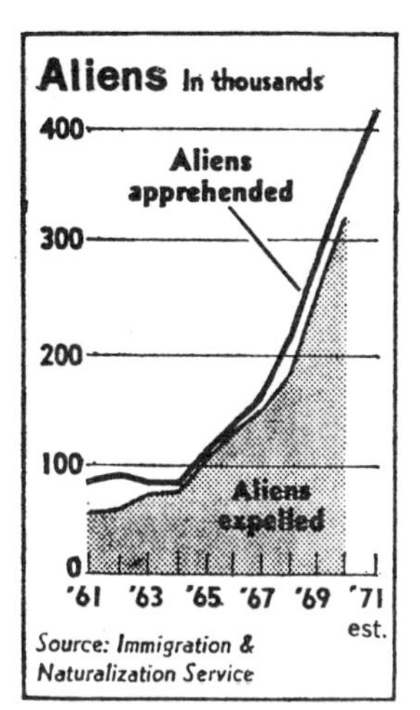

The New York Times/Oct. 17, 1971

"In this office alone, there are 100,000 live leads to illegal aliens that we don't have the manpower to follow up," says Edward Kavazanjian, an investigator in the New York office who has frequently tried to call attention to the problem.

The influx of illegal aliens has many social and economic ramifications. Although employers argue that the aliens do necessary jobs that Americans would not take, it is apparent that in some areas—farm work in California, for example—the braceros compete directly with American labor, and are a significant factor in unemployment.

The presence of a tractable, exploitable pool of cheap alien labor also enables employers to continue paying substandard wages and retards reforms in working conditions. Unions, for example, find it very difficult to organize in factories where illegal aliens are employed, since the aliens want to avoid all quasiofficial connections.

Deception and Ignorance

Most of the aliens pay Social Security and income taxes through withholding, but it is common either through deception or ignorance to claim more dependents than are allowed (under the law, only dependents living in the United States can be deducted). The money they send home also contributes to the balance of payments deficit; Mr. Kavazanjian has said the loss could be as much as $5-billion a year.

Although most officials would say that statements about the number of illegal aliens receiving welfare payments and the like are greatly exaggerated, the problem has emerged recently in California and elsewhere.

There is also the question of the social cost to the aliens themselves. While most appear to be able to make the money they came to the United States for, the price they pay is often large.

They must put up with low wages, bad working conditions and the knowledge that anyone at any time could turn them in. In many cases, their status is little better than that of an indentured laborer; in some cases, conditions approach that of slavery.

While the courts have ruled that illegal aliens have the same constitutional rights as everyone else, they cannot realistically enjoy them because to go to the courts would probably mean deportation.

The corruption of the system also makes them extremely vulnerable to loss of everything they have worked for. For example, consulates in the Western Hemisphere have become quite strict in issuing tourist visas because of their obvious abuse in the past.

In consequence, an applicant in, say, Santo Domingo who has been turned down will often be approached by someone who says that, for $600 or $1,000, they can secure a visa.

The would-be immigrant gets his passport back with what looks like a visa in it, but when he tries to enter the United States the obvious fraud is detected and he is put back on the airplane for home. This can mean a loss of as much as $1,500, which in Latin America can be 10 years' savings.

Illegal aliens are also prey for unscrupulous lawyers and travel agents who, for a fee, promise to regularize the aliens' status. The present conditions are such that such an agent could accept the fee, do absolutely nothing, and be fairly sure that the alien will not be caught.

A Denial of Talents

Manoel Vital of the Azores was on the immigration quota list for 26 years without receiving a visa. Last year he tired of waiting and, at the age of 63, not able to read or write, came on a tourist visa and stayed illegally. He found work as a painter in New Bedford, Mass. A Boston lawyer who has since been indicted for larceny allegedly told him that for $800 he would get Mr. Vital American citizenship. Mr. Vital paid. At a hearing this spring where Mr. Vital testified, Senator Edward Brooke asked how many people attending had had the same experience. Fifty-seven stood up.

Aside from the dollars sent home, it seems clear that the illegal traffic is of no social benefit to the home countries. Despite the menial jobs the aliens have here, many have at least a grade school education, are ambitious and experienced in business. What many regard as the lax American immigration system thus, in effect, deprives developing countries of needed talents.

The obvious increase in illegal immigration has heightened calls for reform by the Congress. The system of labor certification for legal immigration, which most observers of immigration call unworkable, has come in for particular attack.

There is also widespread agreement that the Immigration and Naturalization Service needs more money and manpower. To handle the entire illegal alien problem in the United States, there are 1,704 Border Patrolmen and 750 immigration inspectors.

Immigration officials also say their job is complicated by Social Security regulations, which provide that a card can be issued to anyone who works and makes payments. As a result, illegal aliens have no difficulty obtaining cards. Employers tell the immigration authorities that, since their workers have Social Security cards, they presume they are entitled to work.

Possible Reforms

Another subject being discussed for reform is the length of time allowed on tourist visas. Many officials feel that the term should be shortened to that needed by a bona fide vacationer or visitor instead of the present three or six months.

Almost all immigration officials are agreed that the most meaningful countermeasure to the illegal traffic would be a law making it a Federal crime to employ illegal aliens. A similar bill has been introduced several times in the California Legislature, but has been defeated by pressure from the big farm operators.

There is evidence that some members of Congress are coming to a realization of the gravity of the illegal influx. This is what Senator Walter F. Mondale, Democrat of Minnesota, had to say before the Migratory Labor Subcommittee:

"We will never end the curse of a stream of migrant workers as long as this hemorrhage exists along the Mexican border. I don't think there are any meaningful restrictions there today. In all fairness to all of the Federal agencies, I don't think any of them give a damn. It is getting worse, not getting better, and the damage caused by that one phenomenon exceeds by 1,000 per cent anything we are doing to try to correct the problem from the other end."

October 17, 1971

Latinization of the Miami Area Is Showing No Signs of Abating

Special to The New York Times

MIAMI, April 17 — The Latinization of this metropolitan area of 1.3 million does not show any signs of abating even though the seven-year-old Cuban airlift ended earlier this month.

More than 261,000 refugees came to the United States on the airlift, adding to the 200,000 who arrived here beween 1959 and 1962, when commercial flights between Cuba and the United States were suspended.

Last week, a special flight with more than 60 new exiles aboard arrived here from Spain, where more than 25,000 refugees are awaiting immigrant visas to join their families in the United States.

While few Cuban refugees living in Mexico have been illegally crossing the Rio Grande lately, some still manage to do so. And from whatever place the Cubans enter the United States, sooner or later most of them end up here. The New York-New Jersey area has the second largest concentration of exiles.

400,000 Cubans

According to the most recent estimates, Greater Miami has more than 450,000 Spanish-speaking residents, of whom about 400,000 are Cubans — either naturalized American citizens, permanent residents or parolees, as refugees are called by the Federal immigration authorities.

But citizens from virtually every other Latin-American country have also found Miami a congenial place to live.

Last month, for example, when the newly created Brazilian Cultural Society held its first Carioca Carnival Ball here, about 1,000 Brazilians turned out to make it the liveliest and the most picturesque event in years.

Besides the 450,000 Spanish-speaking residents of Greater Miami, at any given time there are perhaps another 150,000 Latins here: visitors who stay up to six months and sometimes work illegally, or other visitors who fail to return to their native countries and become illegal immigrants.

Bilingual Area

Asserting that the Latin population of the area is growing more rapidly than other ethnic groups, Mayor Jack Orr of Dade County has introduced a resolution declaring the county a bilingual and bi-cultural area.

Yesterday, the Dade County Commission unanimously passed the Mayor's resolution, making Spanish the second official language here. The commission also created a department of bilingual and bicultural affairs to implement the resolution.

In Tallahassee, a state Senate Education Committee recently passed a bill requiring public school districts to teach students in their native language. The Cuban population here would be the main beneficiary of the program.

So far, the Latin influence here has been felt mostly in the economic area, with several important sectors, among them the services and the garment and construction industries, being dominated by Latins.

Church Is Powerful

The Latin influence is also felt here in many other ways. The Roman Catholic Church, once a minority group, has become the most numerous and powerful religious organization, with Miami as the seat of an archdiocese.

Hundreds of Latin restaurants have sprung up here. Their patrons, once almost entirely Spanish-speaking, are now mixed. Café Cubano, a thick, strong brew made with Italian coffeemakers, is consumed on many street corners by a growing number of Americans. Practically all floor shows at Miami Beach hotels, whose personnel is almost entirely Cuban, are staged by Latin producers.

Not all Latin influence is positive, however. The so-called Latin Mafia has been accused of converting Miami into the top cocaine-importing city in the country.

Latins have also preserved a native interest in the occult, palmistry and African beliefs and Caribbean voodoo, exemplified by the existence of many "botanicas" — herb and charm shops — which offer all kinds of remedies to superstitious buyers.

Little Political Impact

Although a recent University of Miami study envisions Latin mayors in Miami and Hialeah in the nineteen-eighties, so far the impact of the Spanish-speaking people on Miami's political life has been largely negligible.

Observers attribute this principally to the lack of qualified Latin candidates, not only for elective offices but also for appointive consultative city and county boards.

Thus, while the Latin population here keeps growing, its political clout remains limited. "Latins are too busy making money to think about politics," one local official said. "It will probably take another five to 10 years before they realize that political power is just as important as economic power."

April 18, 1973

LAST VIETNAMESE RESETTLING IN U.S.

By RICHARD D. LYONS

Special to The New York Times

FORT CHAFFEE, Ark., Dec 20—On this weatherbeaten military base today, 68 Vietnamese men, women and children, the last group of 140,000 refugees who fled Indochina to seek new lives in an Occidental world, boarded buses and headed for cities from New York to Los Angeles.

The departure of this final group ended the large resettlement efforts of the United States Government and a score of welfare agencies that started with the collapse of the Cambodian and South Vietnamese Governments last April.

The wave of refugees has now fanned out over 50 states, Guam American Samoa, Puerto Rico and 24 other countries.

Some of the refugees have quickly adapted to their new environments, resuming careers as doctors, teachers and helicopter pilots. Others, numbed by the displacement, refuse to learn English, are unable to find jobs, feel lost in an alien culture, and long for their old homes.

The American officials who have directed the program insist, however, as one put it, that "These refugees are as successful, if not more successful, than any immigrant group ever to arrive in the United States."

Two hours after the last refugees left this old farming camp, the last of four in the United States that had housed the refugees, Fort Chaffee was officially closed in a ceremony attended by Gov. David Pryor, Senator Dale Bumpers, Democrat of Arkansas, and Jack Freeze, the Mayor of Fort Smith, the nearest city.

Aside from a few tattered signs and some grahiti in Vietnamese, virtually the only remaining evidence of the of humanity that was once here is a triangular rocky cairn by the fort's front gate.

Attached to the memorial are three bronze plaques each inscribed in a different language—English, Vietnamese or Cambodian. The plaque in English states:

"In Search Of New Lives 50,796 Refugees From Indochina Passed Through Fort Chaffee May 2-December 20, 1975."

Dr. Liem Comg Vu, a pathologist who arrived at the camp on May 2 and now practices medicine in Fort Smith, expressed on behalf of himself and his fellow refugees "gratitude to the American people for welcoming us to their homeland."

Donald MacDonald, the civil administrator of the camp, termed the resettlement program "highly successful." He noted that the United States absorbed 42,000 Hungarian refugees in the 1950's and 600,000 Cuban refugees in the 1960's and that they had "quickly landed on their feet as those from Indochina will also."

Few See Last Ceremony

The austere closing ceremony was attended by only a few dozen persons, a marked contrast to the frenetic start of the refugee evacuation effort when mobs battled to board airlift planes while shuttles of helicopters snatched people off roofs in Saigon and some took to makeshift rafts to reach American ships in the South China Sea.

For those with marketable skills and serviceable English the transition has been relatively easy. Such persons include Huynh Van Hue, 26 years old, and his wife, half-sister and nephew, who left today for Southold, L. I., where they will live and next year open a restaurant.

They are being sponsored by a group of private investors and the resettlement has been arranged by the United States Catholic Conference Other religious groups, such as the Lutheran Immigration and Refugee Service and the Hebrew Immigration Aid Service, have also helped in the resettlement.

But the effort to get started in the new world has been frustratingly difficult, for other refugees.

One such case involves Phan Kim Khank, a 47-year-old university graduate from Saigon who sat in an old wooden barracks here yesterday and recounted in fluent French and passable English the troubles of being a displaced person.

Mr. Khank, a slight, personable man with flashing eyes, spent seven months here with his wife and infant daughter while seeking a new home. He spoke without rancor about records being lost and the illness of his doe-eyed daughter, Nguyen Ngoe Long, now a year old.

Needs More English Study

Mr. Khank, who has degrees in French and philosophy and had been an editor at a government radio station in Saigon, said he would "like to work in a college library, then get into newspaper work perhaps."

"But first I must continue my studies in English," he said.

Through the Church World Service he and his family left the camp by bus for Lynchburg, Va., where an apartment, but not a job, has been found. Members of the Westminster Presbyterian Church in Lynchburg which is sponsoring his resettlement, said they were seeking a post for Mr. Khank as a French teacher. Failing that, they will try to find him a factory job.

Mr. Khank's case may be typical of thousands of refugees who, according to a Federal analysis of early refugee resettlements, find work, but at low pay levels and at a lower level of skill than their customary one.

The survey made for the Department of Health, Education and Welfare, deduced that 65 percent of the adult refugees were employed, but that their incomes only averaged $2,500 a year, meaning many were working part-time.

"The employment level is higher than we had expected, but the earnings level is lower than expected, which means that the refugees are not being employed up to their abilities," said Robert Keeley at an interview in Washington earlier this week.

Mr. Keeley is deputy director of the President's Interagency Task Force on Indochina Refugees, which was established in April to coordinate the relief work of a dozen Federal agencies. It will be dissolved at the end of this month.

The survey made for H.E.W. found that 9 percent of the refugees were accepting some form of Government welfare, but Mr. Keeley said he believed that the figure was probably closer to between 15 and 20 percent.

He noted that the educational level of the refugees was close to that of American citizens and that three-quarters of the heads of households had at least a high school education, while about a quarter of them were college graduates. Only slightly more than one-third of the heads of households were able to speak English fluently.

Mr. Keeley offered an admittedly subjective psychological profile of the refugee population, which he divided into three categories.

"The most grateful group are those who originally had lived in North Vietnam and chose to settle in the South at the time the country was divided in the middle-1950's," he said. "This is their second displacement and many arrived here in large families, some even with their own priests since many are Roman Catholics."

"The other extreme is the group that chose to be repatriated," he continued. "They have a tremendous attachment to their country, but they are non-ideological." Some of these, Mr Keeley said, still think they will return to Vietnam, but not until they find out if it would be safe for them to do so.

He said the last group, was composed of fatalists "who accept what is done for them and will work hard and hope for a better life."

While some are still to be counted in the final tally, 129,775 refugees sought American aid and have been resettled in the United States. California has received the largest group, 27,357, while 9,364 are in Texas, and 7,439 in Pennsylvania. New York was chosen by 3,915 refugees, while 1,529 settled in New Jersey and 1,222 in Connecticut.

In addition, 6,629 refugees chose to settle in other countries, the majority in Canada and France. The four resettlement camps also processed 1,087 persons who were found to be either United States citizens or had the status of permanent resident aliens and 127 persons who were not citizens of the United States, South Vietnam or Cambodia. The camps also reported 810 births, and 77 deaths.

In all, 1,546 Vietnamese and Cambodians have thus far returned to their home countries, while 115 Cambodians, now living in Philadelphia, insist they wish to return home. Their repatriation has yet to be determined.

Most From South Vietnam

Of the total number of refugees, about 135,000 are from South Vietnam, and 5,000 from Cambodia.

Thousands of other refugees are in Thailand. Some of these, including more than 3,000 Laotians, are seeking to enter the United States. Some may be allowed to enter this country through immigration quotas or acts of Congress.

Congress authorized $405 million for the resettlement effort, and $100 million in funds of the Agency for International Development were spent to transport the refugees to the United States. It is expected that $30 million will be left over from the resettlement fund and that this will be turned over to the Department of Health, Education and Welfare, which will use it for the special social services of the refugees.

Part of the Federal funds were turned over to the private charitable agencies to help buy food, clothing and housing for the refugees. The amounts averaged $500 a person.

There has been widespread speculation that some refugees arrived here with large stocks of gold and American currency. While reports that the total gold hoard was $150 million could not be verified, one person is known to have cashed in $350,000 in gold.

In the pattern of former immigrant groups, some of the refugees are Anglicizing their names. This is especially true of their children, who are sometimes given the names of camps such as Fort Chaffee here and Camp Pendleton, Calif.

This led Mr. Keeley to muse that sometime in the future a refugee, now a child named Nguyen Pendleton Chin, may wind up a Jim Pendleton, and play quarterback for Stanford.

December 21, 1975

Ford Signs Immigration Bill Aiding Residents of Western Hemisphere

By PHILIP SHABECOFF
Special to The New York Times

WASHINGTON, Oct. 23—Just before midnight last Wednesday night, President Ford signed without fanfare a bill effecting the most sweeping changes in the nation's immigration laws since 1965.

The bill, which will give immigration from the Western Hemisphere the same status as that from the Eastern Hemisphere, had been supported by many Hispanic-Americans.

But the new laws will sharply reduce the number of Mexicans who receive permanent visas each year, and the bill was strongly opposed by some leaders of the Mexican-American community, where Mr. Ford may now have some new political problems just as Election Day approaches.

A spokesman for the Immigration and Naturalization Service said that as a result of the change in the law there may be a sharp increase in the number of Mexicans seeking to enter the country illegally.

However, Mr. Ford indicated in a written statement issued when he signed the legislation that he was aware of the impact on the Mexican-American community. He said that he would introduce legislation next year to raise the immigration quota for Mexico.

Some Differences Erased

The new law retains the same overall annual immigration ceilings established in 1965: 170,000 for the Eastern Hemisphere and 120,000 for the Western Hemisphere. But the other differences in status between the hemispheres have been erased.

Under the old law, immigration from the Western Hemisphere was on a first-come, first-serve basis. There was no quota for any individual country.

For the Eastern Hemisphere there was a ceiling of 20,000 permanent visas for any one country per year. There was also a system of "preferences" related to such goals as reuniting families and attracting professionals and skilled workers.

These rules are retained in the new law and are now applied to the Western Hemisphere as well.

Representative Joshua Eilberg, Democrat of Pennsylvania, who was a chief sponsor of the law, said: "For over 10 years, our immigration laws have severely disadvantaged our neighboring countries in the Western Hemisphere and have created hardship and inconvenience to many persons in this hemisphere seeking to join their families in the United States. This unjust situation is enacted into law."

Mexican-Americans are strongly opposed to the new law, however. Legal Mexican immigration to this country has been averaging 40,000 people a year in recent years. Under the new law, that number would be cut in half.

A spokesman for the Immigration Service said that the pressures caused by the reduced quota for Mexico would probably produce an increase in the number of Mexicans seeking to enter the United States illegally.

The service has estimated that there are a million aliens now in this country illegally, and that 500,000 illegal aliens have been apprehended each year, most of them from Mexico. Spokesmen for the Mexican-American community were sharply critical of President Ford for signing the immigration bill. They did not appear to be aware of his plans to introduce new legislation next year.

Al Perez associate counsel of the Mexican-American Legal Defense and Educational Fund, said that Chicanos were "very opposed" to the new law because its impact on Mexicans, he said, is discriminatory.

Mr. Perez said that "Mexico is right across the Rio Grande from the United States, and Mexicans have more compelling reasons than others for coming into this country."

He said that the Mexican-American community wanted more immigration from Mexico because that would mean more political power for Chicanos in the United States.

Meanwhile, the Immigration Service spokesman, Verne Jervis, said that among other things the new law was likely to increase the immigration of Latin Americans into New York City. He said that the Hispanic population in New York was largely from the Caribbean and Central America rather than from Mexico, and that that was the case in other large American cities.

October 24, 1976

CORRECTION

Representative Joshua Eilberg, Democrat of Pennsylvania, was misquoted by The Times last Sunday in a statement intended to convey praise of a new immigration law. The correct statement, about discriminatory immigration policies for Western Hemisphere nations, said that "this unjust situation is now eliminated and I am very pleased that these overdue reforms have been enacted into law."

October 31, 1976

THE SECOND, THIRD AND FOURTH GENERATION

HIS SON TOO AMERICAN.

Quong You Chong Has Lad Arrested for Acquiring Our Prejudices.

Quong You Chong, a Chinese merchant at 16 Pell Street, brought his son, Feng, 12 years old, into the Elizabeth Street Police Station shortly before midnight last night, and made a charge of incorrigibility against him. He told Lieutenant Duane he brought the boy from China three years ago, and sent him to school, intending to make a doctor of him. He said the lad had associated too much with American boys lately, and had become a Boy Scout, and had shown a dislike for the Chinese.

Feng said he was too thoroughly an American to live longer with his father; the smell of fish in his father's store was more than he could stand, and he had run away and had been working in a restaurant in Long Island City. He said the neat suit of clothes he wore was purchased with money he had earned himself.

August 23, 1915

TOPICS OF THE TIMES.

Our Second Foreign-Born Generation.

In The Interpreter, the bulletin of the Foreign Language Information Service, is the translated report of a letter written to an Italian newspaper in New York by a young girl, "Concetta," was troubled, and, like many young American girls, she wrote to the paper about her difficulties.

She did not describe her situation in relation to that of the thousands of other girls like herself, unhappy in the homes of their foreign-born parents, but confined herself to a personal complaint. Nevertheless, she spoke for those other thousands, and the Italian paper replied to her in general and thoroughly unsentimental terms.

The friction in "Concetta's" home is caused by the reluctance and inability of her parents to accept the conditions of their adopted land. They want their daughter to grow up in the tradition of their own youth. She must associate only with Italians, though she has met young people of other nationalities in school and at work and likes them. She must turn over her entire salary to her father, who saves it as a dowry which she knows quite well her prospective husband, when he comes along, will not expect. She must live in the way that her mother enjoyed as a girl.

The Italian newspaper, like many of the foreign-language papers in such a situation, frankly advised the parents to make reasonable concessions to the girl. They must try not to distrust the freer manners of a new day and country, but, by yielding, keep the confidence of their child while they learn the new customs with her.

May 30, 1927

PUPILS NOW TEACH PARENTS IN CITY'S NEW SCHOOL PLAN

Experiment in Education of the Foreign-Born Adult Tried by Extension Division

A NOVEL experiment in educating the foreign-born adult has been started by the Board of Education of New York. The effort is being made to enlist the cooperation and aid of school children in teaching their parents the rudiments of the English language and the fundamentals of American institutions.

The experiment is directed by the Division of Extension Activities and operates through the medium of the elementary schools. Sets of forty lesson sheets are sent out to the ten schools used for the experiment. Teachers in charge of this work talk to pupils whose parents have not mastered the language of the country and who are not enrolled in evening schools.

The children are instructed in the use of the lesson sheets. They are told to stress pronunciation above spelling in such a way as to lead them to recognize early individual printed words. In sentence work, the young teachers are told, the aim is to develop grasp of the meaning of the sentence as a whole, rather than construction or style. There follow instruction in question drills and writing exercises, including "fill-ing in" drills, the adult students supplying the missing words in sentences.

Thus tutored, the pupil teachers receive the first lesson sheet, no time limit being set for the satisfactory completion of any one lesson. At home, in the leisure hours of the evening, parent and child go over the lesson.

The first lesson begins with easy language exercises such as "What is your name?" "Where do you live?" "How old are you?" There is a writing drill attached, and as the lessons proceed more difficult combinations of words, phrases and clauses are introduced. There are date lessons, number lessons and drills giving directions in city transportation, how to get to Times Square, or from the square home.

More advanced lesson sheets offer drills in the form of civic, educational and social paragraph essays. In fact, the entire course is conceived from the point of view of an elementary civics course, apart from the fundamental purpose of teaching the language. When a lesson is completed the pupil reports on its success to the teacher in charge of the experiment and the next lesson sheet is issued.

Adult Schools Neglected.

One of the reasons for inaugurating the experiment is the failure of many adults to take advantage of the public evening classes offered by the Board of Education. Some adults do not attend such groups because of the frequent embarrassment of crowded classrooms and unsympathetic, tired teachers. Results of the experiment will be studied carefully with a view of determining the advisability of making the new method a permanent addition to the public school system.

However, the tests have already been subjected to unfavorable criticism. The underlying theory of "child teaching parent" has been seriously challenged on the ground that the procedure might tend to create tension in home relations. Some of the pupil-teachers have been surprised at the slowness of the process of learning. A teacher reported that several pupils had complained that parents forget previous lessons "just like children."

March 3, 1929

QUOTA SCATTERS RACIAL GROUPS

Alien-Born Becoming Naturalized — Business Based on Nationality Appeal to New Immigrants Is Affected

By EMERY DERI.

ABOVE the racial groups of foreign-born in the United States who cling to allegiances, languages and tribal differences brought from abroad, dark clouds are gathering—harbingers of an approaching change.

Since the 3 per cent. quota law began to be applied to immigration the little cafés of the east side, each a meeting place for one or another of the various racial groups of this cosmopolitan city, have gloomily discussed the signs of decline in their respective colonies and the grim aspects of the future. After four years of restricted immigration the large foreign-born communities with their different newspapers, special business enterprises and financial institutions are facing a crisis.

The stream of new immigrants that kept the ranks of the foreign-born population filled with fresh accessions has dried up; and the rising generation never sticks to the colonies formed by its parents. There are no new audiences for the foreign-language theatres, and no new readers for the more than 2,500 foreign-language newspapers published within the United States.

Aside from the death rate, which cuts wide swaths into the masses of our foreign population, the population of the "foreign quarters" are being lessened by the ever increasing number of foreign-born citizens who are leaving the different linguistic ghettos for the open spaces of American life, where the horizons are wider and chances for success manifold. Business men and professional men, eager to save themselves from the approaching break-up which is threatening the solidarity of the quarters where the alien-born grouped themselves are making a new start among Americans by essaying to become integral parts of the actual American life.

Schools Americanize Children.

It is a well known fact that the members of every racial colony represent only the first generation of immigrants. Every attempt to preserve the second generation, the so-called junior class, for an alien-born community, has invariably failed even within such racial groups as the Jews, where religious ties are interwoven with racial links, or in the case of the Germans, where love and admiration for German culture and literature aided cohesion. Every girl or boy who attends American schools becomes lost to the racial community of the parents. The second generation does not want to be different from other Americans, and looks down upon the limited possibilities of the racial community's ghetto life.

But a considerable part of the foreign-born immigrants also can be regarded as lost to the "foreign quarters." Though there are no statistics available regarding the Americanization of immigrants—the adoption of American citizenship does not necessarily mean the immigrants' Americanization—it is observed that those immigrants who come to our shores under the age of 30 become Americanized so rapidly that after the lapse of a few years they depart from the sharply drawn limits of their racial group.

To the "lost" class may be added those immigrants who came here after the war and who represent a new type. Before the war and the practice of selective immigration the bulk of immigrants consisted of peasants and laborers almost without education; but these newer immigrants belonged in their homeland to that intellectual middle class which lost its fortune as a consequence of the war. These new immigrants arrive in this country with a considerable knowledge of English; they refuse to take an active part in the provincial and paltry community life of their racial groups and try to become Americanized as quickly as possible.

Racial Business Drops.

Thus in quarters that still use a foreign language the communities are composed of older immigrants, around or above 40 years of age, who did not try to learn the language of the country, and of a thin upper stratum of business and professional men who are linked to the community by business interests. In reality, these foreign-language communities are cemented together by this upper stratum, for it has special concern in the longest possible preservation of the community.

The aim of the business men of the communities using alien speech is to retard the influence of Americanization, to keep as a privileged domain their business with their compatriots and preserve it against competition from outside. Other than business motives are often instrumental in keeping the community together. However, speaking broadly, one may say that foreign-born communities were held together by business services arranged especially for them. Therefore, the decline of businesses that depended on a particular racial clientele is the most significant sign of the decay of the "foreign communities."

As the first racial community commerce created by the ever-increasing immigration was steamship ticket agencies and the banking and money-forwarding business this has been the first to feel the shock of restricted immigration. Banking for foreign-born groups in the United States has declined considerably. A great number of banks, with an exclusively foreign-born patronage, and a still greater number of money-forwarding and steamship agencies have either closed their doors during the last four years or turned to other lines of trade that offer possibilities of expansion in the native field.

Many Yugoslav money-forwarding and shipping agencies have combined their former business with that of selling imported goods; others have ventured into enterprises not solely based on the patronage of their fellow-countrymen. Large Italian banks, eager to insure their future in due time, took interest in American banking by buying up the majority of shares of American banking firms.

Hungarian banks in New York City have transformed themselves, too. A few years ago there were four Hungarian banks and many Hungarian money-forwarding and steamship agencies in New York; today practically only one banking firm depends exclusively on a Hungarian patronage. The others have become American banks or have sold their business to American banking institutions which continue them as branch banks. The small money-forwarding agencies simply disappeared as a result of greatly reduced immigration.

Alien-Born Adopt America.

The decrease of banks serving only this or that racial group throws light upon the changes in the business. While before and immediately after the war almost every immigrant sent part of his income to the old country for the amortization of the cost of a piece of land or of a house. That stream of American dollars to the native lands of the immigrants has dried up. Those newcomers who up to recently had intended to go back and invest their savings in a tract of land or a business of their own changed their resolution and decided to stay here. After the war many of them returned to their native land, lured by low exchange rates, but hurried back to the United States, after observing the desperate economic situation on the other side, and resolved never to leave again "the blessed soil of America."

Their attitude toward America and everything American changed simultaneously with that decision. Previously, they had had no interest in America; they had lived within the spiritual walls of their racial communities. Their interest in their adopted country came with the resolution to "stay here for good." Presently they drifted away from the closed world to their racial groups and augmented the number of those "lost to the alien community."

Immigrant Societies Diminish.

Another sign of the dispersal of racial colonies in the United States is the diminishing of the number of immigrants' societies and the decrease in the membership of the surviving associations. Analysis of various racial groups discloses that nearly all immigrants' societies are waging an uphill struggle against the indifference of their members or against the reluctance of others of their motherland to join their racial associations. The number of Italian societies has decreased considerably within the last few years. Other racial groups are complaining about the membership losses of their established associations.

Mutual benefit societies which acted as substitutes for insurance companies within the foreign communities are badly off. There are no new members to pay dues and with a large number of old members dying every year the time is approaching when the societies will be unable to pay premiums to the families of the dead. Some of these benefit associations, having seen ahead the rock of bankruptcy, arranged with strong American insurance companies to take over their business.

Changed State of Mind.

The withdrawals from racial organizations and groups can be traced to two motives. One is that the speedier process of Americanization now authorized has caused wholesale desertions from the closed "nationality islands," and the other is that the interest of the foreign-born elements has turned from their own racial communities to the country generally. This shows that the decline of the foreign-born communities is explained by the changed state of mind of the immigrant, rather than by an actual decrease of the foreign-born element.

This discounting of the coming depopulation of the "nationality islands" or racial centres is perceptible in the nervousness prevailing among the foreign-language newspapers. Their circulation has not fallen off much as a whole since the quota law went into operation. Some have lost, but others have managed to hold their own. The publishers of the foreign-language newspapers, however, are not deceived by the fact that so far they have suffered little. They can see that the present stock of their readers is bound to diminish through the working of the death rate, which is almost twenty per thousand among the foreign-born who are over 25 years old. They know that the rate of mortality cannot be cheated.

Their only hope is that laws are not made for eternity and that Congress may be induced to change the ratios in the present Immigration act. For the last ten years the most important source of income of the foreign-language press has been the advertisements of their own people. The rato for advertising is not based on the fact that this or the other newspaper all but monopolizes the market among the racial group it represents, but is regulated by the paper's total circulation. A decrease in circulation would diminish advertisements. That, in turn, would mean nothing less for racial organs than a lessened reason for being.

September 5, 1926

JEWISH SURVEY REVEALS MANY GRAVE PROBLEMS

DURING the last ten years there has been a vast movement of people within the five boroughs of New York. Hundreds of thousands have shifted their homes from one section to another, new communities have been developed, and new social and economic problems have as a consequence arisen. These facts are revealed in the report of the Jewish Communal Survey of Greater New York, which has just been made public by the Bureau of Jewish Social Research.

The survey was promoted by the Jewish Citizens' Committee, composed of men and women prominent in social, educational and financial circles, with the object of obtaining an actual picture of the existing conditions in New York's Jewish community. It was felt that with the material made available by the survey all philanthropic and social activities could be reviewed in the light of the changing conditions, and, where necessary, reorganized. The facts to be provided by the statistical study, the committee believed, would make it possible to outline a policy designed to meet future needs as well as those of the present generation.

Changes by Boroughs.

Ten years ago, as now, the Jewish inhabitants of New York formed 30 per cent. of the total population, their number increasing proportionately with the growth of population of the city. There are today 1,750,000 members of the Jewish faith in Greater New York. Of the 1,503,000 who lived here in 1916, 696,000 were located in Manhattan; in the Bronx

THE JEWISH COMMUNITY OF 1916

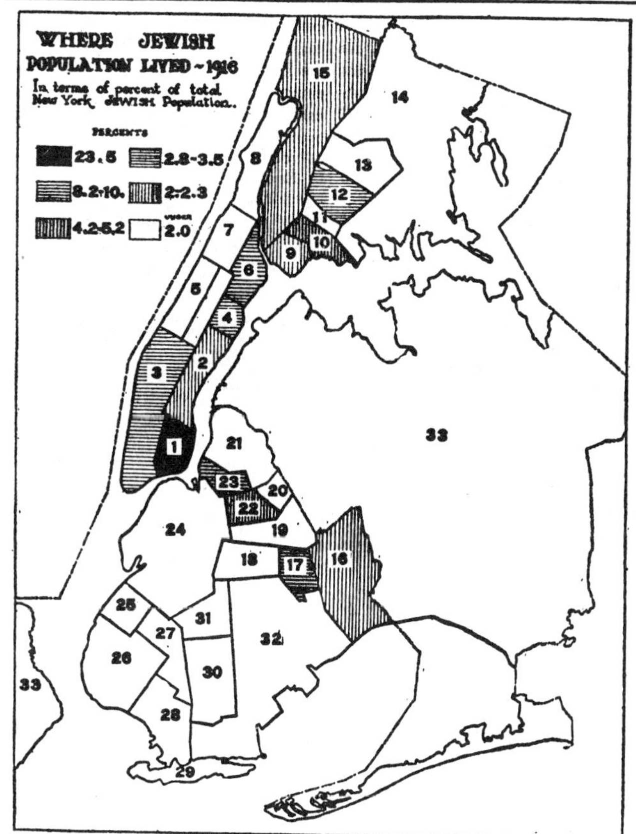

Key to Map: 1—Lower East Side; 2—Central East Side; 3—Lower West Side; 4—Yorkville; 5—West End; 6—Harlem; 7—West Harlem; 8—Washington Heights; 9—South Bronx; 10—Lower Central Bronx; 11—Upper Central Bronx; 12—Tremont; 13—Fordham; 14—North Bronx; 15—Grand Concourse; 16—New Lots; 17—Brownsville; 18—Eastern Parkway; 19—Bushwick; 20—Ridgewood; 21—Greenpoint; 22—Willoughby; 23—Williamsburgh; 24—South Brooklyn; 25—Bay Ridge; 26—Fort Hamilton; 27—Borough Park; 28—Bath Beach; 29—Coney Island; 30—Flatbush; 31—North Flatbush; 32—East Flatbush; 33—Queens and Staten Island.

THE JEWISH COMMUNITY OF 1925

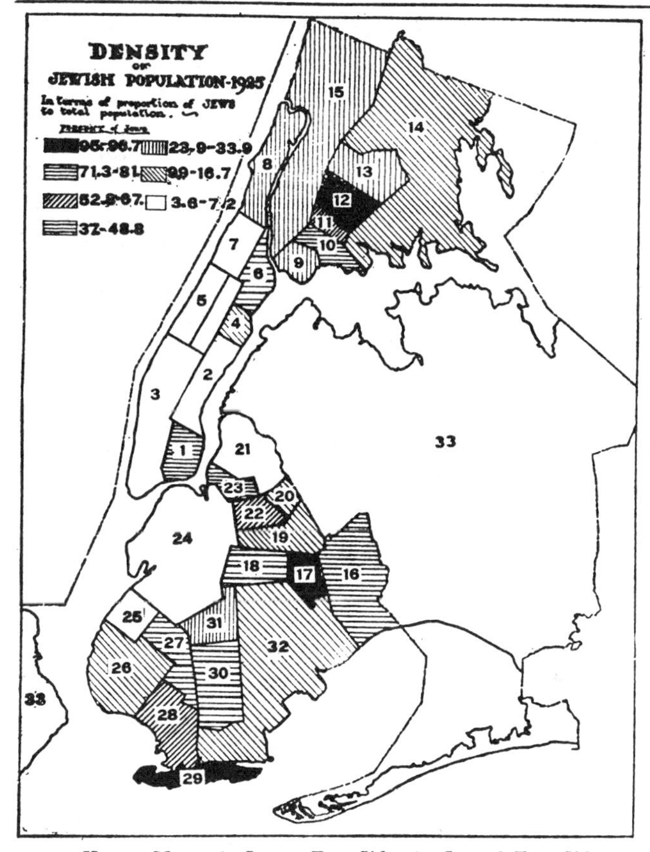

Key to Map: 1—Lower East Side; 2—Central East Side; 3—Lower West Side; 4—Yorkville; 5—West End; 6—Harlem; 7—West Harlem; 8—Washington Heights; 9—South Bronx; 10—Lower Central Bronx; 11—Upper Central Bronx; 12—Tremont; 13—Fordham; 14—North Bronx; 15—Grand Concourse; 16—New Lots; 17—Brownsville; 18—Eastern Parkway; 19—Bushwick; 20—Ridgewood; 21—Greenpoint; 22—Willoughby; 23—Williamsburgh; 24—South Brooklyn; 25—Bay Ridge; 26—Fort Hamilton; 27—Borough Park; 28—Bath Beach; 29—Coney Island; 30—Flatbush; 31—North Flatbush; 32—East Flatbush; 33—Queens and Staten Island.

there were 211,000; in Brooklyn, 568,000; in Queens, 23,000, and in Richmond, 5,000.

In the post-war years considerable shifts of population have taken place. Two hundred thousand Jews have quitted Manhattan, leaving only 500,000 in that borough, which since the earliest days of the migration from Europe had been their main centre in New York. Of the former 279,000 have been added to the Bronx, making a total of 390,000, an increase of 84.7 per cent. in ten years; in Brooklyn the number has risen to 800,000—40.7 per cent above the 1916 figure. Queens more than doubled its Jewish population in the same period, having 56,000 in 1926. In Richmond, it appears, there was a decrease of 1,500, 3,500 remaining.

Perhaps the most striking feature of the survey is the revelation that with growing economic strength the Jews have rapidly abandoned the congested sections of older New York to move into newly opened areas, where health conditions are better. The current of progress has swept them from the tenements of the lower east side to Harlem, Yorkville, the Bronx, Brooklyn and Queens.

The decentralization has been marked in Manhattan where the radiation has been chiefly from the lower east side and Harlem. Here figures speak eloquently. In 1916 the lower east side had 23.5 per cent. of all New York's Jews; by 1925 this figure had dropped to 15.2 per cent. In Harlem the decrease was from 9.9 per cent. of the total Jewish population of the entire city in 1916, to 6.6 in 1925.

The movement in the Bronx has been northward and westward, into the Tremont, Fordham and Grand Concourse sections. Tremont has surpassed Harlem in the density of its Jewish population. In Brooklyn the shift has been south and west into Borough Park, Eastern Parkway, Bath Beach and Coney Island.

As wave followed wave of settlers across the East River, land values began to rise. The first settlers lived in one-family houses; later arrivals were obliged to move into two-family houses, and now the newcomers are confronted with the inevitable apartment house. The sequence of events is more or less true of the conditions met in other sections.

While numerically fewer than in Brooklyn, the Jews make up 44.7 per cent. of the Bronx's population, the Brooklyn figure being 36.3 per cent. of the total. In Manhattan the figure is 25.7, while for Queens and Richmond the ratios are 8 and 2.5 per cent., respectively.

Going into details the survey points out that some sections of the boroughs are almost solidly Jewish, among these being Coney Island, Tremont and Brownsville, where the ratio is more than 95 per cent.

Causes of the Change.

This shift of the population is attributed largely to economic adjustment. Jews are to be found in every industry, the largest number being in the garment and fur trades. But gradually they have been leaving the machines and work benches to go into business for themselves. Working men have became proprietors. The survey attributes to such social and financial betterment the desire to move into districts in keeping with improved conditions, where the families of the successful will find greater opportunities.

Before scientific methods in dealing with social conditions were put into practice migrations frequently went unexplained. The present survey has attempted to explain and, further, to answer such questions as: How do these people live? When do they die and what are the causes?—

questions asked and answered by them with the expectation of helping the people concerned to make a successful fight against the ills by which they are beset.

According to the report the "crude" Jewish death rate in the greater city was 7.91 per 1,000 in 1925—a lower figure than that for either New York State, which was 13.0 in 1923, or New York City, which in the same year was 11.7. To reach its conclusion the survey set up a unit of living population whose death rates it could study for the purpose of arriving at rates by age and sex. Having obtained these specific death rates and figures on all the deaths in New York City, what was held to be an accurate estimate for the entire Jewish population was reached.

Some Vital Statistics.

Among the facts that became apparent by a study of statistics relating to the Jews of New York City as compared to the rest of the white population of the United States were that the Jewish population is a younger group and that it is not increasing as rapidly as the general population. It was determined that the Jewish death rate after the forty-fifth year has been passed is greater than the general rate, and that the birth rate among the Jews is lower.

To ascertain the birth rate a comparative method was used, as in the case of the death rate. Brooklyn was chosen as the working unit. Eighteen children were born per 1,000 of Jewish population, the census figures for the same year (1925) indicating that the rate for the entire population of Brooklyn was 22.95 per 1,000.

The survey, devoting attention to the question of the diseases which have made the greatest inroads on the Jews, disclosed that up to the age of 45 they are freer from tuberculosis than the other inhabitants of the United States—the comparative figures being 36.5 per 100,000, as against 86.3 for the total population. The account is not so bright when it turns to heart disease. Here it is found that, in spite of its younger age, the New York group death rate in 1925 was 191.3 per 100,000, while in the country at large the rate is 180.7 per 100,000.

A review of all the information gathered on the subject of diseases leads to the conclusion that deaths from tuberculosis, nephritis and cerebral hemorrhage are considerably lower, proportionately, than in the non-Jewish white population; on the other hand, there are more deaths from cancer, diabetes and diseases of the heart.

Other Problems.

"It is evident, therefore, that as the New York Jewish population grows older and into middle age," states the report, "it must be confronted by these degenerative diseases. Unless effective measures are taken to fight them, the conclusion is inevitable that the Jewish population will be hard hit. With smaller replacements from Europe, with a smaller birth rate, and with higher death rates after forty-five years, it is likely that the Jews of New York will not display the same net population increase as will other groups. The tentative conclusion may be drawn that the future will see the New York group dwindling, proportionately."

The problem of the scientific care of such a large population has become yearly more important. Those intimately concerned with the social, educational and spiritual welfare of the Jews have been conscious of the changing conditions, but have been unable to cope with them because of their lack of knowledge of the actual situation. For this reason representatives of all the boroughs got together about two years ago and decided to form a Citizens' Committee with Judge Otto A. Rosalsky at its head. After due consideration the committee invited the Bureau of Jewish Social Research to conduct the survey under the direction of Samuel A. Goldsmith. The movement was supported by all of the important Jewish bodies of the community, both orthodox and liberal.

The investigators are of the conviction that the trend of Jewish population from Manhattan to Brooklyn and the Bronx is a permanent one. There are in Greater New York two federations for the support of the Jewish philanthropic societies, operating on the community-chest principle — one in New York, the other in Brooklyn. The drift toward Brooklyn and the Bronx began to put too heavy a strain on the resources of those boroughs, the more so since the bulk of the wealth of the entire community lies in Manhattan.

What is to be the future distribution of responsibility and where is the cost of bearing it to rest? "Shall there be one federation, to include all Jewish charitable institutions in the greater city?" is the question raised by those responsible for the survey.

What is to be the outcome of the studies none as yet can foresee, according to Mr. Goldsmith, who has been supervising the survey for the last two years. It is the organization's belief that in future a broader basis will be sought and that all social welfare will be handled from the point of view of the greater city instead of being treated as a local or parochial problem.

Manhattan, it is pointed out, is more fully equipped with hospitals and social centres, and yet the majority of Jews are in other boroughs. Where institutions can minister to the needs of those residing in other parts of the city, their policies will in the future be so directed that city-wide service will be made possible. Again to quote Mr. Goldsmith: "A bridge of community consciousness will have to be built between Brooklyn and New York which will bring those living across the river to the already existing centres."

In the general readjustment which is expected to follow the publication of the survey, it is hoped that Jewish institutions organized on a neighborhood or local basis, finding that the population they have served is dwindling, will take stock of the services they will be able to render those remaining and of their duties to the general field in which they are engaged.

"These institutions," continues Mr. Goldsmith, "will need to study the possibility of securing a greater return for both the capital and the maintenance funds invested in their work by transferring their activities to localities where there may be increasing Jewish settlements or much larger and stronger Jewish communities than are now surrounding them."

As the statistical information was being gathered observations were made of the social, economic and recreational life in each borough section. The material gathered will be embodied in another report, which will also show whether or not a saturation point in building of Jewish institutions has been reached.

Auxiliary committees covering all branches of social service have been appointed. Every department of the organization that deals with charity among the Jews will be brought under consideration. Committee heads and their assistants will discuss the advisability of institutional care of children and try to ascertain whether educational groups are doing follow-up work among the school children who have moved from their old surroundings. The recreations of boys and girls in the outlying districts of New York. are to be compared to those in the heart of the city and an attempt made to learn if delinquency is increasing or decreasing.

As the investigation progresses step by step it will touch also on the life of the aged, both as to institutional and home care.

"The survey as it stands today," says Mr. Goldsmith, "is comparable to the steel structure of a skyscraper. It is fundamental; without it we could not advance; but only time can tell what its ultimate appearance will be. We hope for great results. Those Jews of New York who have generously contributed to the support of their fellow-beings have a right to know how well and how wisely their money is being handled and what changes if any should be made in its administration."

On the Executive Committee of the survey are Dr. Lee K. Frankel, Chairman; B. C. Vladeck, Albert Goldman and Colonel Herbert H. Lehman. Chairmen of the various sections include Felix M. Warburg, recreation; Judge William N. Cohen, child care; Fred M. Stein, health; Mrs. Sidney C. Borg, family welfare; Ludwig Vogelstein, Jewish education, and Judge Otto A. Rosalsky, community organization.

March 18, 1928

Topics of The Times

Some Newer Americans

The superintendent of the National Tube Company's mill at McKeesport, Pa., said in our story yesterday that 55 per cent of the 8,000 employes are native born. But also present, as they say in the society columns, were 765 Slavs, 443 Hungarians, 395 Croatians and 359 Poles, or a total of nearly 3,000 foreign born. In addition, many of the native born at McKeesport are the sons of Slavs, Hungarians, Croats, Poles and others. In the 1930 census there were in McKeesport 53,000 white people, of whom more than 32,000 were born abroad or the children of foreign parents.

Thick of the Fray

Next after the blast furnaces and the rolling mills the best place where to look for Slavic, Hungarian, Croatian and Polish names is the college football teams. Not so many gridiron seasons ago Carnegie Tech, playing Duquesne, pitted TABOVICH, MIHON and TEREBUS against CHAPALA, RADO, MALKOVICH, KAKASIC, CUTORNA and ZANIENSKI. It was the same day when Fordham sent into the firing line LUBIMOWICZ, SABO, PALA and SARAUSKY. On the same day Catholic University summoned her stout sons to battle and there sprang forward KARPOVICH, ANTHONOWAGE, YANCHULIS, LAJOUSKY, DRANGANIS and GLADECK. It was about the same time that Columbia University hurled RICHAVICH, COVIELLO, CIAMPA, WURZ, HUDASKY and BROMINSKI into the fray against Syracuse, whose banner was held aloft by MINSAVAGE, STEENMA, JONTOS, VAVRA, WAMSACH and ALBANESE. Their presence on the college elevens testified to their being all good Americans. Some, indeed, were of All-American caliber.

How They Came

The connection between "foreign" names in the steel mills and foreign names on the football teams is not altogether whimsical, not to say frivolous. Capital and labor are girding for a unionization contest in the big industries where foreign-stock labor is heavily represented. In the course of the struggle silly things are sure to be said on both sides concerning native Americans and foreign-born Americans and good Americans. So it is well to have the facts readily at hand.

One of the silly things often uttered on this subject comes from the Labor-Radical side. In the revaluation of America's past, which has been actively under way for some years, the big employers of labor have not escaped. This country contains so many Slavs, Hungarians, Poles, Croatians, Czechs and others because they were brought over here in droves by rapacious capitalists who wanted cheap factory fodder. Over here they were sweated and exploited and otherwise mistreated and degraded until their usefulness was over. They were then thrown on the scrap-heap like so much human slag in the Pittsburgh area.

Not Exactly Kidnapped

Now the steel employers certainly did want cheap labor and probably did something to recruit it abroad. And the history of the twelve-hour day in the steel mills is well known. The silly part begins when exaggeration begins. American employers are depicted almost as luring huge numbers of Slav peasants from their rich and happy homes under their Magyar landlords, from the soft embraces of the Czarist police, to become the victims of American greed.

The thing becomes silly at the other end when the ultimate fate of these exploited European workers is described as cinders on the Pittsburgh refuse heaps. But it could not be that, for the high schools are filled with the sturdy sons of these exploited European peasants; and the football scouts come down from the colleges and hire these husky young men bearing Slav and Hungarian and Italian names, and the latter proceed to play rings around the Cabots and the Saltonstalls.

In other words, the steel mills have molded more than tinplate and girders. They have molded new Americans. There has been much individual pain, no doubt, and a vast amount of toil and sweat, but there has been no mass tragedy. On the contrary, a nation-building and man-building process has been working itself out.

No Lethal Effects

But that, too, is the answer to silly outcries from the companies' side. In 1890 people worked eighty-five hours a week in the blast furnaces. Ten years ago it was down to sixty hours, and America was not destroyed in the process. In fact, workers in this country probably grow more discontented as they grow more American. Some day the foreign-born of mill workers may get to be almost as restless as American farmers.

In 1922 a committee of the Federated American Engineering Societies published a detailed study, "The Twelve-Hour Shift in Industry." The volume had a brief foreword containing this sentence: "The old order of the twelve-hour day must give way to a better and wiser form of organization of the productive forces of the nation, so that proper family life and citizenship may be enjoyed suitably by all of our people. WARREN G. HARDING.

"The White House, Washington, Nov. 9, 1922."

July 10, 1936

FAITHFUL TO U. S., ITALIANS HERE SAY

Young men with Italian parents are as ready as any other Americans to shoulder a Springfield if the United States goes to war. This and a sentiment for universal military service were expressed by many yesterday in a sampling of opinion among youths playing handball on the streets of the upper East Side.

In a group found in a Mulberry Street bicycle shop two or three "stuck up" for Mussolini but all the young men were unanimous in avowing their willingness to fight for the defense of the United States. A few older persons, naturalized citizens or of Italian descent, were cynical and sad. They were not keen about American involvement, but they said their sons belonged to Uncle Sam.

The curbstone interview began with the question: "Do you think Mussolini was right in declaring war?"

"The quicker we go over the better," said Frank Praino of 208 East 114th Street. "I'm unemployed anyway." He is 24 years old.

A block further north, late in the afternoon, a group stopped a handball game to give their opinions. Two of the players, it turned out, can already handle a Springfield. They learned in the C. M. T. C. They were Vincent Romano, 26, a clerk, of 206 East 115th Street, and his brother, Cosmo, 19, who will be graduated from Textile High School this month.

"We Don't Like," Youth Says

"We expected it and we don't like it," was Vincent's response to the question.

He added:

"If Congress passes that bill for universal military training it'll be the best thing that can happen."

"What would you do if the United States gets into the war?" they were asked.

"What could we do but serve our country," was the answer. The brothers said they had taken the basic course in the Citizens Military Training Camps.

The neighborhood iceman walked over. He is Joseph Rella of 232 East 115th Street and is 44 years old.

"The first thing I'm an American citizen," said Rella. "I hope we stay out, but if they stick their nose in North or South America we'll fight to the last drop of blood."

Peter Camarano, a truck driver, of 208 East 115th Street, proved to be an isolationist. He said neither Italy nor the United States got anything out of the last war, in which he served more than eighteen months in the Aviation Signal Corps, A. E. F.

John Malangone, barber, of 200 East 115th Street, said the United States should mind its defenses. He added he had three sons ready to fight for their native America.

Mulberry Street Gives Ideas

Down in Mulberry Street, Pasquale Cuomo of No. 235, a pupil in Metropolitan Vocational High School, did not like "the idea" of Mussolini plunging. He said Europe should be left to battle out its own destiny. It turned out that Pasquale was only 15 years old, but big enough to fool any recruiting officer. He said he would now be in the Marine Corps Reserve if his mother had given him permission to join.

Michael Chico, 19, of 421 Kenmare Street, favored Mussolini's action and said Italy had always been "pushed back by England and France." He said he was willing to fight for the defense of America but not to go "over there."

Anthony Napolitano, 19, of 196 Broome Street, said he would have to know "what I'm fighting for, who I'm fighting for and why I'm fighting."

Dominick Pelosi, 22, of Kenmare Street appeared just then and said, "I think he's (Mussolini) a regular phoney for going in." There were a few others who agreed with him but one young fellow pronounced, with a laugh, the Communist slogan, "The Yanks are not coming."

The Italian Consulate, as on Monday, was glum and silent on Italy's plunge. The Mazzini Society condemned Italy's Fascist Government for its war on the Allies and for refusing President Roosevelt's good offices. It called on all Italians and those of Italian origin to support the United States in any action against Fascists and Nazis.

An observer in the "Little Italy" in East New York, Brooklyn, commented on sentiment there as follows: "The older men were bragging like hell before Italy got in. Now they're all quiet. Their sons are good Americans."

June 12, 1940

JEWS ARE FOUND DOUBTING VALUES

Changes in Population Said to Alter Their Concepts

BY IRVING SPIEGEL

Special to The New York Times

BOSTON, April 18—An authority on American Jewish life cited today sociological reasons that "American Jews have become unsure about the meaning of Jewishness in America for themselves and for their children."

Sanford Solender, executive vice president of the National Jewish Welfare Board, in an interview, said that the radical changes in the population patterns of American Jews in the last 50 years "has substantially altered Jewish attitudes, mores and regard for traditional Jewish moral and ethical concepts." This, in turn, has established a whole new pattern of individual, group and community needs among American Jews, he said.

The board, the national association of all Jewish community centers and Young Men's and Young Women's Hebrew Associations, is holding its biennial convention at the Statler Hilton Hotel.

Mr. Solender, former President of the National Conference of Social Welfare, said that American Jews, in achieving security and a large measure of acceptance, "are now beset with doubts and uncertainties as to whether the price of equality is assimilation."

Ambivalence Arises

As American Jews "have become acculturated with an American society," Mr. Solender said, "they encounter an ambivalence between emulation of non-Jewish neighbors in all aspects of social and cultural life and the desire for Jewish group survival."

Mr, Solender said that "concern over interreligious dating and intermarriage on the one hand and eagerness to take full advantage of the open American society on the other create an emotional tug-of-war."

Mr. Solender said that reconciling the urge for full integration in American society and the need for Jewish identification was the basic problem confronting every form of Jewish community activity in the United States.

However, Mr. Solender held that few American-born Jews sought to hide or escape their Jewishness, and that their ties to Judaism, "however shallow, superficial or sentimental, remain strong."

But in affirming their Jewish identity "they have little knowledge or understanding of its content or meaning, all too many are Jewish by habit and by ignorance," he contended.

Boredom Found

Mr. Solender declared that American Jews could not be put off with generalities "and they find participation in Jewish affairs boring and unrewarding."

"Because they regard their Jewishness, whatever its quality," he said, "as a dimension of their Americanism, the task is to strengthen this dimension."

American Jews, Mr. Solender said, need help in formulating their own beliefs as Jews, "to find their own Jewish roots and nourish them through greater knowledge."

Speaking at the dinner session, Dr. Abram L. Sachar, president of Brandeis University, said that American Jews "as Americans have a freedom and security that has never been enjoyed in such measure by our ancestors in any land."

Dr. Sachar received the board's 1964 Frank L. Weil Award for a distinguished contribution to the advancement of American Jewish Culture.

Similar awards were presented to Arthur Kling of Louisville, Ky. for a contribution to Jewish community center work, and to Walter D. Heller of San Francisco, for contribution to the welfare of Jewish military personnel.

April 19, 1964

Success Story, Japanese-American Style

By WILLIAM PETERSEN

ASKED which of the country's ethnic minorities has been subjected to the most discrimination and the worst injustices, very few persons would even think of answering: "The Japanese Americans." Yet, if the question refers to persons alive today, that may well be the correct reply. Like the Negroes, the Japanese have been the object of color prejudice. Like the Jews, they have been feared and hated as hyperefficient competitors. And, more than any other group, they have been seen as the agents of an overseas enemy. Conservatives, liberals and radicals, local sheriffs, the Federal Government and the Supreme Court have cooperated in denying them their elementary rights—most notoriously in their World War II evacuation to internment camps.

WILLIAM PETERSEN is a professor of sociology at the University of California at Berkeley. His latest book is "The Politics of Population."

Generally this kind of treatment, as we all know these days, creates what might be termed "problem minorities." Each of a number of interrelated factors—poor health, poor education, low income, high crime rate, unstable family pattern, and so on and on — reinforces all of the others, and together they make up the reality of slum life. And by the "principle of cumulation," as Gunnar Myrdal termed it in "An American Dilemma," this social reality reinforces our prejudices and is reinforced by them. When whites defined Negroes as inherently less intelligent, for example, and therefore furnished them with inferior schools, the products of these schools often validated the original stereotype.

Once the cumulative degradation has gone far enough, it is notoriously difficult to reverse the trend. When new opportunities, even equal opportunities, are opened up, the minority's reaction to them is likely to be negative—either self-defeating apathy or a hatred as all-consuming as to be self-destructive. For all the well-meaning programs and countless scholarly studies now focused on the Negro, we barely know how to repair the damage that the slave traders started.

The history of Japanese Americans, however, challenges every such generalization about ethnic minorities, and for this reason alone deserves far more attention than it has been given. Barely more than 20 years after the end of wartime camps, this is a minority

PAST—In World War II, 117,116 Japanese Americans — citizens and noncitizens alike — were taken from their homes and held in internment camps. Above, an internee's sketch of a kindergarten class.

that has risen above even prejudiced criticism. By any criterion of good citizenship that we choose, the Japanese Americans are better than any other group in our society, including native-born whites. They have established this remarkable record, however, by their own almost totally unaided effort. Every attempt to hamper their progress resulted only in enhancing their determination to succeed. Even in a country whose patron saint is the Horatio Alger hero, there is no parallel to this success story.

FROM only 148 in 1880 to almost 140,000 in 1930 the number of Japanese in the United States grew steadily and then remained almost constant for two decades. Then in 1960, with the more than 200,000 Japanese in Hawaii added to the national population, the total reached not quite 475,000. In other words, in prewar years Japanese Americans constituted slightly more than 0.1 per cent of the national population. Even in California, where then as now most of the mainland Japanese lived, they made up only 2.1 per cent of the state's population in 1920.

Against the perspective of these miniscule percentages. It is difficult to recapture the paranoiac flavor of the vast mass of anti-Japanese agitation in the first decades of this century. Prejudice recognized no boundaries of social class; the labor-dominated Asiatic Exclusion League lived in strange fellowship with the large California landowners. The rest of the nation gradually adopted what was termed "the California position" in opposing "the Yellow Peril" until finally Asians were totally excluded by the immigration laws of the nineteen-twenties.

Until the exclusion law was enacted, Japanese businesses were picketed. In San Francisco, Japanese were assaulted on the streets and, if they tried to protect themselves, were arrested for disturbing the peace. Since marriage across racial lines was prohibited in most Western states, many Japanese lived for years with no normal family life (there were almost 25 males to one female in 1900, still seven to one in 1910, two to one in 1920). Until 1952 no Japanese could be naturalized, and as noncitizens they were denied access to any urban professions that required a license and to the ownership of agricultural land.

But no degradation affected this people as might have been expected. Denied citizenship, the Japanese were exceptionally law-abiding alien residents. Often unable to marry for many years, they developed a family life both strong and flexible enough to help their children cross a wide cultural gap. Denied access to many urban jobs, both white-collar and manual, they undertook menial tasks with such perseverance that they achieved a modest success. Denied ownership of the land, they acquired control through one or another subterfuge and, by intensive cultivation of their small plots, helped convert the California desert into a fabulous agricultural land.

THEN, on Feb. 9, 1942, a bit more than two months after war was declared, President Roosevelt issued Executive Order 9066, giving military commanders authority to exclude any or all persons from designated military areas. The following day, Lieut. Gen. John L. DeWitt, head of the Western Defense Command, defined the relevant area as major portions of Washington, Oregon, Idaho, Montana, California, Nevada and Utah.

In this whole vast area all alien Japanese and native-born citizens of any degree of Japanese descent — 117,116 persons in all—were subjected in rapid succession to a curfew, assembly in temporary camps within the zone and evacuation from the zone to "relocation centers." Men, women and children of all ages were uprooted, a total of 24,712 families. Nearly two-thirds were citizens, because they had been born in this country; the remainder were aliens, barred from citizenship.

"Some lost everything they had; many lost most of what they had," said the official report of the War Relocation Authority. The total property left behind by evacuees, according to the preliminary W.R.A. estimate, was worth $200-million. After the war, the Government repaid perhaps as much as 30 or 40 cents on the dollar. The last claim was settled only in November, 1965, after two out of the three original plaintiffs had died.

What conceivable reason could there have been for this forced transfer of an entire population to concentration camps, where they lived surrounded by barbed wire and watched by armed guards? The official explanation was that "the evacuation was impelled by military necessity," for fear of a fifth column. As General DeWitt said: "A Jap's a Jap. It makes no difference whether he is an American citizen or not. . . . They are a dangerous element, whether loyal or not."

The cases of injustice are too numerous to count. One of the more flagrant was that of the so-called renunciants. After years of harassment, a number of Japanese Americans requested repatriation to Japan, and they were all segregated in the camp at Tule Lake, Calif. On July 1, 1944, Congress passed a special law by which Japanese Americans might renounce their American citizenship, and the camp authorities permitted tough Japanese nationalists seeking converts to proselytize and terrorize the other inmates. Partly as a consequence, 5,371 American-born citizens signed applications renouncing their citizenship. Many of them were minors who were pressured by their distraught and disillusioned parents; their applications were illegally accepted by the Attorney General. A small number of the renunciants were removed to Japan and chose to acquire Japanese citizenship. A few cases are still pending, more than 20 years after the event. For the large majority, the renunciation was voided by the U.S. District Court in San Francisco after five years of litigation.

WHO are the Japanese Americans; what manner of people were subjected to these injustices? Seen from the outside, they strike the white observer as a solidly unitary group, but even a casual acquaintanceship reveals deep fissures along every dimension.

The division between generations, important for every immigrant group, was crucial in their case. That the issei, the generation born in Japan, were blocked from citizenship and many of the occupational routes into American life meant that their relations were especially difficult with the nisei, their native-born sons and daughters. Between these first and second generations there was often a whole generation missing, for many of the issei married so late in life that in age they might have been their children's grandparents. This was the combination that faced General DeWitt's forces—men well along in years, with no political power and few ties to the general community, and a multitude of school children and youths, of whom the oldest had barely reached 30.

The kibei, American-born Japanese who had spent some time as teen-agers being educated in Japan, were featured in racist writings as an especially ominous group. For some, it is true, the sojourn in the land of their fathers fashioned their parents' sentimental nostalgia into committed nationalism. In many instances, however, the effect of sending a provincial boy alone into Tokyo's tumultuous student life was the contrary. Back in the United States, many kibei taught in the Army language schools or worked for the O.S.S. and other intelligence services.

Camp life was given a special poignancy by the Defense Department's changing policy concerning nisei. Until June, 1942, Japanese Americans were eligible for military service on the same basis as other young men. Then, with the evacuation completed and the label of disloyal thus given official sanction, all nisei were put in class IV-C—enemy aliens. The Japanese American Citizens League (J.A.C.L.), the group's main political voice, fought for the right of the American citizens it represented to volunteer, and by the end of the year won its point.

Not Guilty

Hawaii's Japanese in World War II were spared the injustices inflicted upon mainland Japanese Americans. Immediately after Pearl Harbor, there were, inevitably, allegations that they had sabotaged the defenses there. These rumors were investigated in full four times over—by Honolulu's chief of police, by the Secretary of War, by the director of the F.B.I. and, finally and most completely, by John A. Burns, now Governor of Hawaii and then a Honolulu police lieutenant in charge of counter-espionage and a liaison officer with military intelligence—and proved completely false. But the lesson of this quadruple vindication was lost in the morass of mainland racial prejudice.—W.P.

Most of the volunteers went into a segregated unit, the 442d Infantry Combat Team, which absorbed the more famous 100th Battalion. In the bloody battles of Italy, this battalion alone collected more than 1,000 Purple Hearts, 11 Distinguished Service Crosses, 44 Silver Stars, 31 Bronze Stars and three Legion of Merit ribbons. It was one of the most decorated units in all three services.

With this extraordinary record building up, the Secretary of War announced another change of policy: the nisei in camps became subject to the draft. As District Judge Louis Goodman declared, it was "shocking to the conscience that an American citizen be confined on the ground of disloyalty, and then, while so under duress and restraint, be compelled to serve in the armed forces, or be prosecuted for not yielding to such compulsion." He released 26 nisei tried in his court for refusing to report for induction.

The Government's varying policy posed dilemmas for every young man it affected. Faced with unreasoning prejudice and gross discrimination, some nisei reacted as one would expect. Thus, several hundred young men who had served in the armed forces from 1940 to 1942 and then had been discharged because of their race were among the renunciants at Tule Lake. But most accepted as their lot the overwhelming odds against them and bet their lives, determined to win even in a crooked game.

In John Okada's novel "No-No Boy," written by a veteran of the Pacific war about a nisei who refused to accept the draft, the issue is sharply drawn. The hero's mother, who had raised him to be a Japanese nationalist, turns out to be paranoid. Back in Seattle from the prison where he served his time (he was not tried in Judge Goodman's court), the hero struggles to find his way to the America that rejected him and that he had rejected. A nisei friend who has returned from the war with a wound that eventually kills him is pictured as relatively well-off. In short, in contrast to the works of James Baldwin, this is a novel of revolt against revolt.

THE key to success in the United States, for Japanese or anyone else, is education. Among persons aged 14 years or over in 1960, the median years of schooling completed by the Japanese were 12.2, compared with 11.1 years by Chinese, 11.0 by whites, 9.2 by Filipinos, 8.6 by Negroes and 8.4 by Indians. In the nineteen-thirties, when even members of favored ethnic groups often could find no jobs, the nisei went to school and avidly prepared for that one chance in a thousand. One high school boy used to read his texts, underlining important passages, then read and underline again, then read and underline a third time. "I'm not smart," he would explain, "so if I am to go to college, I have to work three times as hard."

From their files, one can derive a composite picture of the nisei who have gone through the Berkeley placement center of the University of California over the past 10 years or so. Their marks were good to excellent but, apart from outstanding individuals, this was not a group that would succeed solely because of extraordinary academic worth. The extracurricular activities they listed were prosaic — the Nisei Student Club, various fraternities, field sports, only occasionally anything even as slightly off the beaten track as jazz music.

Their dependence on the broader Japanese community was suggested in a number of ways: Students had personal references from nisei professors in totally unrelated fields, and the part-time jobs they held (almost all had to work their way through college) were typically in plant nurseries, retail stores and other traditionally Japanese business establishments.

Their degrees were almost never in liberal arts but in business administration, optometry, engineering, or some other middle-level profession. They obviously saw their education as a means of acquiring a salable skill that could be used either in the general commercial world or, if that remained closed to Japanese, in a small personal enterprise. Asked to designate the beginning salary they wanted, the applicants generally gave either precisely the one they got in their first professional job or something under that.

To sum up, these nisei were squares. If they had any doubt about the transcendental values of American middle-class life, it did not reduce their determination to achieve at least that level of security and comfort. Their education was conducted like a military campaign against a hostile world; with intelligent planning and tenacity, they fought for certain limited positions and won them.

The victory is still limited: Japanese are now employed in most fields but not at the highest levels. In 1960, Japanese males had a much higher occupational level than whites —56 per cent in white-collar jobs as compared with 42.1 per cent of whites, 26.1 per cent classified as professionals or technicians as compared with 12.5 per cent of whites, and so on. Yet the 1959 median income of Japanese males was only $4,306, a little less than the $4,338 earned by white males.

FOR all types of social pathology about which there are usable data, the incidence is lower for Japanese than for any other ethnic group in the American population. It is true that the statistics are not very satisfactory, but they are generally good enough for gross comparisons. The most annoying limitation is that data are often reported only for the meaninglessly generalized category of "nonwhites."

In 1964, according to the F.B.I.'s "Uniform Crime Reports," three Japanese in the whole country were arrested for murder and three for manslaughter. Two were arrested for rape and 20 for assault. The low incidence holds also for crimes against property: 20 arrests for robbery, 192 for breaking and entering, 83 for auto theft, 251 for larceny.

So far as one can tell from the few available studies, the Japanese have been exceptional in this respect since their arrival in this country. Like most immigrant groups, nisei generally have lived in neighborhoods characterized by overcrowding, poverty, dilapidated housing, and other "causes" of crime. In such a slum environment, even though surrounded by ethnic groups with high crime rates, they have been exceptionally law-abiding.

Prof. Harry Kitano of U.C.L.A., has collated the probation records of the Japanese in Los Angeles County. Adult crime rates rose there from 1920 to a peak in 1940 and then declined sharply to 1960; but throughout those 40 years the rate was consistently under that for non-Japanese. In Los Angeles today, while the general crime rate is rising, for Japanese adults it is continuing to fall.

According to California life tables for 1959-61, Japanese Americans in the state had a life expectation of 74.5 years (males) and 81.2 years (females). This is six to seven years longer than that of California whites, a relatively favored group by national standards. So far as I know, this is the first time that any population anywhere has attained an average longevity of more than 80 years.

FOR the sansei — the third generation, the children of nisei—the camp experience is either a half-forgotten childhood memory or something not quite believable that happened to their parents. They have grown up, most of them, in relatively comfortable circumstances, with the American element of their composite subculture becoming more and more dominant. As these young people adapt to the general patterns, will they also— as many of their parents fear —take over more of the faults of American society? The delinquency rate among Japanese youth today is both higher than it used to be and is rising—though it still remains lower than that of any other group.

Frank Chuman, a Los Angeles lawyer, has been the counsel for close to 200 young Japanese offenders charged with everything from petty theft to murder. Some were organized into gangs of 10 to 15 members, of whom a few were sometimes Negroes or Mexicans. Nothing obvious in their background accounts for their delinquency. Typically, they lived at home with solid middle-class families in pleasant neighborhoods; their brothers and sisters were not in trouble. Yori Wada, a nisei member of the California Youth Authority, believes that some of these young people are in revolt against the narrow confines of the nisei subculture while being unable to accept white society. In one extreme instance, a sansei charged with assault with the intent to commit murder was a member of the Black Muslims, seeking an identity among those extremist Negro nationalists.

In Sacramento, a number of sansei teen-agers were arrested for shoplifting—something new in the Japanese community but, according to the police, "nothing to be alarmed at." The parents disagreed. Last spring, the head of the local J.A.C.L. called a conference, at which a larger meeting was organized. Between 400 and 500 persons—a majority of the Japanese adults in the Sacramento area—came to hear the advice of such professionals as a psychiatrist and a probation officer. A permanent council was established, chaired jointly by a minister and an optometrist, to arrange for whatever services might seem appropriate when parents were themselves unable (or unwilling) to control their offspring. According to several prominent Sacramento nisei, the publicity alone was salutary, for it brought parents back to a sense of their responsibility. In the Japanese communities of San Francisco and San Jose, there were similar responses to a smaller number of delinquent acts.

Apart from the anomalous delinquents, what is happening to typical Japanese Americans of the rising generation? A dozen members of the Japanese student - club on the Berkeley campus submitted to several hours of my questioning, and later I was one of the judges in a contest for the club queen.

I found little that is newsworthy about these young people. On a campus where to be a bohemian slob is a mark of distinction, they wash themselves and dress with unostentatious neatness. They are mostly good students, no longer concentrated in the utilitarian subjects their fathers studied but often majoring in liberal arts. Most can speak a little Japanese, but very few can read more than a few words. Some are opposed to intermarriage, some not; but all accept the American principle that it is love between the partners that makes for a good family. Conscious of their minority status, they are seeking a means both of preserving elements of the Japanese culture and of reconciling it fully with the American one; but their effort lacks the poignant tragedy of the earlier counterpart.

Only four sansei were among the 779 arrested in the Berkeley student riots, and they are as atypical as the Sacramento delinquents. One, the daughter of a man who 20 years ago was an officer of a Communist front, is no more a symbol of generational revolt than the more publicized Bettina Aptheker.

It was my impression that these few extremists constitute a special moral problem for many of the sansei students. Brazenly to break the law invites retribution against the whole community, and thus is doubly wrong. But such acts, however one judges them on other grounds, also symbolize an escape from the persistent concern over "the Japanese image." Under the easygoing middle-class life, in short, there lurks still a wariness born of their parents' experience as well as a hope that they really will be able to make it in a sense that as yet has not been possible.

THE history of the United States, it is sometimes forgotten, is the history of the diverse groups that make up our population, and thus of their frequent discord and usual eventual cooperation. Each new nationality that arrived from Europe was typically met with such hostility as, for example, the anti-German riots in the Middle West a century ago, the American Protective Association to fight the Irish, the national-quota laws to keep out Italians, Poles and Jews. Yet, in one generation or two, each white minority took advantage of the public schools, the free labor market and America's political democracy; it climbed out of the slums, took on better-paying occupations and acquired social respect and dignity.

This is not true (or, at best, less true) of such "nonwhites" as Negroes, Indians, Mexicans, Chinese and Filipinos. The reason usually given for the difference is that color prejudice is so great in this country that a person who carries this visible stigma has little or no possibility of rising. There is obviously a good deal of truth in the theory, and the Japanese case is of general interest precisely because it constitutes the outstanding exception.

What made the Japanese Americans different? What gave them the strength to thrive on adversity? To say that it was their "national character" or "the Japanese subculture" or some paraphrase of these terms is merely to give a label to our ignorance. But it is true that we must look for the persistent pattern these terms imply, rather than for isolated factors.

The issei who came to America were catapulted out of a homeland undergoing rapid change — Meiji Japan, which remains the one country of Asia to have achieved modernization. We can learn from such a work as Robert Bellah's "Tokugawa Religion" that diligence in work, combined with simple frugality, had an almost religious imperative, similar to what has been called "the Protestant ethic" in Western culture. And as such researchers as Prof. George DeVos at Berkeley have shown, today the Japanese in Japan and Japanese Americans respond similarly to psychological tests of "achievement orientation," and both are in sharp contrast to lower-class Americans, whether white or Negro.

The two vehicles that transmitted such values from one generation to the next, the family and religion, have been so intimately linked as to reinforce each other. By Japanese tradition, the wishes of any individual counted for far less than the good reputation of his family name, which was worshiped through his ancestors. Most nisei attended Japanese-language schools either one hour each weekday or all Saturday morning, and of all the *shushin*, or maxims, that they memorized there, none was more important than: "Honor your obligations to parents and avoid bringing them shame." Some rural parents enforced such commandments by what was called the *moxa* treatment—a bit of incense burned on the child's skin. Later, group ridicule and ostracism, in which the peers of a naughty child or a rebellious teen-ager joined, became the usual, very effective control.

This respect for authority is strongly reinforced in the Japanese-American churches, whether Buddhist or Christian. The underlying similarity among the various denominations is suggested by the fact that parents who object strongly to the marriage of their offspring to persons of other races (including, and sometimes even especially, to Chinese) are more or less indifferent to interreligious marriages within the Japanese groups. Buddhist churches have adapted to the American scene by introducing Sunday schools, Boy Scouts, a promotional effort around the theme "Our Family Attends Church Regularly," and similar practices quite alien to the old-country tradition.

On the other hand, as I was told not only by Buddhists but also by nisei Christian ministers, Japanese Americans of whatever faith are distinguished by their greater attachment to family, their greater respect for parental and other authority. Underlying the complex religious life, that is to say, there seems to be an adaptation to American institutional forms with a considerable persistence of Buddhist moral values.

It is too easy, however, to explain after the fact what has happened to Japanese Americans. After all, the subordination of the individual to the group and the dominance of the husband-father typified the family life of most immigrants from Southern or Eastern Europe.

Indeed, sociologists have fashioned a plausible theory to explain why the rate of delinquency was usually high among these nationalities' second generation, the counterpart of the nisei. The American-born child speaks English without an accent, the thesis goes, and is probably preparing for a better job and thus a higher status than his father's. His father, therefore, finds it difficult to retain his authority, and as the young man comes to view him with contempt or shame, he generalizes this perception into a rejection of all authority.

Not only would the theory seem to hold for Japanese Americans but, in some respects, their particular life circumstances aggravated the typical tensions. The extreme differences between American and Japanese cultures separated the generations more than in any population derived from Europe. As one issei mother remarked to the anthropologist John Embree: "I feel like a chicken that has hatched duck's eggs."

Each artificial restriction on the issei—that they could not become citizens, could not own land, could not represent the camp population to the administrators — meant that the nisei had to assume adult roles early in life, while yet remaining subject to parental control that by American standards was extremely onerous. This kind of contrast between responsibility and lack of authority is always galling; by the best theories that sociologists have developed we might have expected not merely a high delinquency rate among nisei but the highest. The best theories, in other words, do not apply.

One difficulty, I believe, is that we have accepted too readily the common-sense notion that the minority whose subculture most closely approximates the general American culture is the most likely to adjust successfully. Acculturation is a bridge, and by this view the shorter the span the easier it is to cross it. But like most metaphors drawn from the physical world, this one affords only a partial truth about social reality.

The minority most thoroughly imbedded in American culture, with the least meaningful ties to an overseas fatherland, is the American Negro. As those Negro intellectuals who have visited Africa have discovered, their links to "negritude" are usually too artificial to survive a close association with this—to them, as to other Americans—strange and fascinating continent. But a Negro who knows no other homeland, who is as thoroughly American as any Daughter of the American Revolution, has no refuge when the United States rejects him. Placed at the bottom of this country's scale, he finds it difficult to salvage his ego by measuring his worth in another currency.

The Japanese, on the contrary, could climb over the highest barriers our racists were able to fashion in part because of their meaningful links with an alien culture. Pride in their heritage and shame for any reduction in its only partly legendary glory—these were sufficient to carry the group through its travail. And I do not believe that their effectiveness will lessen during our lifetime, in spite of the sansei's exploratory ventures into new corners of the wider American world. The group's cohesion is maintained by its well-grounded distrust of any but that small group of whites—a few church organizations, some professors, and particularly the A.C.L.U. in California—that dared go against the conservative-liberal-radical coalition that built, or defended, America's concentration camps.

The Chinese in California, I am told, read the newspapers these days with a particular apprehension. They wonder whether it could happen here—again.

January 9, 1966

Nine Tailors, Etc.

In 1942, the U.S. State Department arranged for the deportation from Peru to the United States of nine Japanese tailors, to be followed by more than 2,100 other Japanese residents of Peru and other South American nations. Many were native-born Peruvian citizens, and all had been declared politically innocuous after a full investigation by the Department of Justice.

The United States wanted them in order to exchange them for American citizens interned in countries occupied by Japan. Actually, none was ever so used.

They were shifted from one camp to another. Then, after the war, the Federal Government began proceedings to deport them to Japan for having entered the country without proper papers (although under escort of U.S. military police). Their plight came to the attention of Wayne Collins, the San Francisco lawyer of the American Civil Liberties Union and, for years, a defender of the rights of Japanese Americans. Collins telephoned a lawyer in the Justice Department in Washington. When he heard who was calling, this Government attorney audibly exclaimed to a colleague: "Oh, oh, Collins has found them!"

In a seemingly endless series of legal moves, for which there were no precedents, Collins won their right not to go to Japan but back to Peru — and then not to go to Peru (which for years refused to permit their re-entry) but to stay in the United States. In 1954, after 12 years in and out of camps, and in constant litigation, the South American Japanese were permitted to apply for permanent residence in the United States, and many became citizens.

—W.P.

20,000 Greek-Americans Having Words Here

By PETER MILLONES

It has been said that when two Greeks get together they open a restaurant; when three get together they start a revolution. Twenty thousand are in New York now, and they are having a convention.

The infectious strains of Greek music and the smell of simmering shish kebab have accompanied the Greeks and Greek-Americans attending this week the 46th annual convention of the Order of AHEPA (pronounced "uh-HEH-puh"), the most influential of the Greek fraternal organizations.

But while the cultural ties to Greece remain strong, the Order of AHEPA, organized to ease the assimilation of Greek immigrants into American society, is now so American it holds debutante balls and its members drink more Scotch than ouzo.

Gone are the struggles to gain acceptance, struggles that produced, for example, a sign in a Santa Rosa, Calif., restaurant in the nineteen-twenties that said: "John's Restaurant — Pure American — No Rats — No Greeks."

The Political Outlook

Instead, those gathering at the convention headquarters at the Waldorf-Astoria Hotel tonight will point with pride to Gov. Spiro T. Agnew, the first American of Greek descent to be nominated for the Vice Presidency, who will address the meeting.

Ever since Governor Agnew was chosen as the Republican Vice Presidential candidate, comments have been made similar to those of the Republican from Iowa who said sarcastically:

"Boy, this is just great. We'll get every Greek vote in Iowa—all eight or nine of them."

Some Greek-Americans think he may have been overoptimistic.

They note that the estimated total of 450,000 first- and second-generation Greek-Americans in the nation have a tradition of squabbling over politics and anything else that will produce a good rousing fight. (Thus the saying when three get together they start a revolution.)

Right now, for example, there is widespread disagreement among Greek-Americans over the evils or virtues of the military regime in Greece.

Agnew's Not Typical

On a less significant matter, members of AHEPA—the American Hellenic Educational Progressive Association—have in the past argued over whether to have their official business conducted in Greek or English. Younger men favor English because it would mean that the older men, with less facility in English, would likely speak less.

Another factor that is working against any bloc vote is that Governor Agnew is not a typical first-generation Greek-American.

For one thing, his name has been shortened from Anagnostopoulos, so that it is not readily identifiable as Greek. For another, he is now an Episcopalian, not Greek Orthodox. And for a third, he does not speak Greek.

But he is typical in another respect. His father was an immigrant who owned a restaurant.

Although Americans of Greek descent are now in virtually all the professions and in many different businesses, Greek-Americans have probaby had their greatest impact on America's digestive tracts.

Many of the immigrants in the late 19th and early 20th centuries drifted into some form of small restaurant ownership, selling everything from hot dogs to the traditionally Greek lamb and rice.

According to Theodore Saloutos, author of "The Greeks in the United States," no particular aptitude for cooking led the Greeks into restaurant ownership. Rather, he suggests, it may have been their common desire to be "their own boss."

Customers Hostile

At times, the combination of a language barrier for the Greeks and a customer attitude against foreigners caused difficulties. The Mayor of Roanoke, Va., in a letter to the Governor in 1907, noted that some patrons had refused to pay their food bills to show their contempt for the Greeks.

He added:

"Very few of these Greeks can speak English, and if a customer gets a 15-cent lunch, and there is a misunderstanding as to the price, there is hardly a Greek in the place that will not, upon the least provocation, grab a butcher knife or some other weapon and make for the complaining customer. I am somewhat inclined to believe that they are too presumptuous."

Mr. Saloutos came to a different conclusion. "Wherever one turned," he noted, "the admonition was to work hard, save, invest, succeed, become independent."

Many of the immigrants, who settled primarily in big cities, such as New York, Chicago, Detroit, Boston, and San Francisco, also became florists, furriers, shoemakers and candy and pastry makers. They were so successful as lobster fishermen that in Rhode Island in 1909 legislation was introduced to ban Greeks from fishing. It was defeated.

Few Greek-Americans have become prominent in the public life of the nation. Governor Agnew was the first Greek-American Governor and there are three Greek-American Representatives.

But if Greek-Americans are making slow inroads, non-Greeks appear—from the success of New York Greek-owned restaurants—to be enthusiastically learning Greek dancing and acquiring a taste for egg lemon soup.

August 21, 1968

New Power Struggles Roil Chinatown's Future

By MURRAY SCHUMACH

The Establishment of Chinatown — its family associations, tongs and trade groups, all interlocked in the conglomerate Chinese Consolidated Benevolent Association —is bending to strong winds stirred up by thousands of Chinese immigrants, hundreds of college-educated Chinese-Americans, dozens of government-sponsored agencies.

"The subsurface currents of Chinatown and the internal politics are pulling in many directions," says a knowledgeable Chinese-speaking Occidental who is persona grata with almost all faction. "The situation is so complex it's almost impossible to predict what will emerge from this ferment."

To tourists, browsing in gift stores, sampling tasty foods, overwhelmed by staccato Cantonese dialects, these developments are as unknown as the mah-jongg and card games in backrooms or upstairs. These trends, more important than occasional clashes among teenage gangs, include:

¶The growth of major family associations, which had been declining in recent years, because of the influx of nearly 1,000 immigrants a month.

¶Signs of a power struggle between the Chinese Benevolent Association and social agencies in which young Chinese-Americans are active.

¶Efforts by influential Chinese-American professionals and businessmen to bridge the gap between old and young, between private association and public agency, between new immigrant and older resident. Their drive is for greater unity for more political power.

¶Growing interest among young Americans of Chinese parentage in Chinatown and its problems.

One symptom of Chinatown's quiet revolution is a new attitude by the Chinese Benevolent Association. It used to look upon government aid as somewhat leprous. Now it is getting public help in teaching English and is formulating ambitious plans for government-assisted housing in the tenement-packed streets.

Financial Assistance

Or there is the establishment by the Lee Family Association, one of the largest, of a credit union, which already has almost $1-million in assets, for its members. And nearly every important group in Chinatown is participating in the Chintaown Advisory Council set up under the tactful hands of Manhattan Borough President Percy E. Sutton.

At the heart of the changes is the growing realization that the Establishment groups have been moving too slowly in dealing with serious community problems and that great progress cannot be made without government help.

"Some groups, unfortunately, are beginning to feel that confrontation is the only solution," says Irving S. K. Chin, a product of Yale ('53) and the Harvard Law School, who lives in Chinatown and has been active in pushing for a full-scale crash program in teaching English to immigrants.

"I do not agree with the advocates of confrontation as the only way," says Mr. Chin, a member of the big Chin Family Association, and one of the key persons in Mr. Sutton's council. "Black Panther methods will not work with the Chinese. The Benevolent Association is not going to kowtow to some young people who shout 'non-negotiable demands.'"

At the hub of all plans in Chinatown is the Benevolent Association. Its power is conceded not only by family associations and merchants, but also by such social agencies as the Chinese Youth Council and the Chinatown Planning Council, which tend to attract young Chinese-Americans from local colleges.

The friction between the Benevolent Association and some young insurgents is epitomized by the disagreement about the association's decision to close its gym about three years ago after it was vandalized.

About six months ago a Sports Committee, sponsored mainly by young Chinese-Americans, was formed to force the association to reopen the gym. The association has said it is willing to do so if the Sports Committee will assume responsibility for supervision and guarantee to make good any damage. Young Chinese-Americans are willing to supervise, but not obligate themselves to pay for any vandalism.

One young man in strong sympathy with the Sports Committee's objective says:

"Nobody would be crazy enough to sign a paper to make good any damage after the gym is opened."

The vandalism was generally attributed to a small group of recent teen-age immigrants. It had been used regularly for years before that without any trouble.

The extent of the Benevolent Association's hold on the people becomes particularly apparent on Sundays, when Chinese-Americans come from all over the metropolitan area to shop, gossip with friends, go off to Chinese movies and Chinese restaurants, gather in family association buildings.

Families Meet

But from 10 A.M. to 5 P.M. the Benevolent Association's glass doors, at 62 Mott Street, rarely close. From Long Island, New Jersey, Westchester and Connecticut, as well as many parts of the city, come children and adolescents born here of Chinese parents to study Chinese language and culture. And when classes break, parents and grandparents are often in the lobby and on the sidewalk to meet the students.

"We are the over-all organization for Chinatown," says Y. K. Chu, secretary of the association, with confidence. "There are students who try to serve the community. It will take them time to acquaint themselves with the community's problems."

Nevertheless, the Benevolent Association, prodded by some concerned leaders in Chinatown's 59 family associations, realizes the young people have taken the initiative in attacking such problems as helping the old get Social Security and welfare payments and improved housing and English instruction.

During the summer, for instance, the college-educated Chinese-Americans descend on Chinatown from their suburban homes by the score to take children from congested Chinatown on picnics, to movies, to beaches; to organize them in arts and crafts groups and to tutor them in English.

Their weakness, however, is that they have almost no rapport with the immigrant high school dopouts who become the core of the young street gangs. The language problem is the main reason for dropouts. Most college students do not speak Chinese. And even when they do, the dropouts tend to resent the success of the college youths.

According to Dr. C. T. Wu, associate professor of geography at Hunter College, dropouts are talking about returning to Hong Kong.

The attitude of the Benevolent Association toward the dropouts is philosophical. "There are always a few bad ones," says Mr. Chu. "We are more interested in setting up a scholarship fund for the good ones."

One possibility—and this is suggested almost in whispers in Chinatown—is that these gangs may become useful material for the Hip Sing and On Leong Tongs. These tongs have long been peaceful since their bloody warfare early in this century. They were formed in the last century on the West Coast to defend their countrymen against attacks and lynchings by whites.

The word tong means club, but with military connotations. One widely believed rumor is that the tongs may use these gangs to quash tiny Maoist factions that try to foment violence in the area.

No one doubts the power of the tongs. Their obvious strength is that they include influential businessmen. But there are also reports—with no proof—that there is organized gambling in Chinatown, controlled by the tongs. No one familiar with the intrigues of Chinatown questions that the tong membership in the Benevolent Association is one reason the association, continues to wield so much power.

Wide Membership

Another reason is that the 60 family associations belong to the over-all group. Most of these family organizations, which, like the tongs began on the West Coast, are little more than mail drops. But others, such as the Lee, Chin, Gee, Wong, are sizable, though no figures are available.

"The family associations used to have more power many years ago," says Andrew P. Lee, an affluent lawyer and insurance man. "They were a sort of judicial body, where they would settle disputes between members of the Lee association. There is not much of that any more. But as a social organization the family groups still have a great deal of influence."

However, with the immigration growth, the family associations can help find jobs, offer a place to meet friends, talk Chinese, make the first step toward adjustment in a strange land. Their membership is growing, but not among the young.

"We keep hoping," says Mr. Lee, "that the Benevolent Association will bring all groups together. The young seem to feel that the Benevolent Association is too much concerned with the old people and moves too slowly."

Mr. Chin, who has judged dance competitions at outings of the Chin association, notes that at many colleges the Chinese-Americans have formed social groups that organize ski weekends, trips to dude ranches and Halloween parties.

"It is dreadfully important to start a realistic program to teach English," says Mr. Chin. "Then there would be very few high school dropouts. Adults would be able to find jobs away from Chinatown and live away from Chinatown. We have to get all Chinese groups together and bring real pressure for government funds and foundation-backing for English programs.

"We've lost a great deal by not being united. This is what I think the young and the old are both beginning to realize."

November 19, 1970

The Last of the American Irish Fade Away

By ANDREW M. GREELEY

CHICAGO.

NOT so long ago I was wandering through the halls of a progressive Catholic women's college with a young woman I know and noticed a sign announcing that the Irish Club was to meet that afternoon.

DR. ANDREW M. GREELEY, a Roman Catholic priest, is head of the Center for the Study of American Pluralism of the National Opinion Research Center, University of Chicago. His book, "Why Can't They Be Like Us?", about ethnicity, will be published next month.

"Are you a member of the club, Peggy?" I asked.

"A member? I'm the president!"

"And what do you do at your meetings?"

"Why, we plan the St. Patrick's Night Ball."

"Peg, have you ever heard of the Easter Rising?" She had not.

"And, Peg, what about the Sinn Fein?" She thought it might be Chinese.

"And, what's the Irish Republican Army?" As a loyal Cook County Democrat, she wanted no part of it.

Finally, "Peg, who is Eamon de Valera?"

"Oh, I know! He's the Jewish man who's Mayor of Dublin."

And so the last of the American Irish fade away into the mist.

The American Irish are in a most unusual position. For the first time in the history of American society, it is legitimate to be an ethnic. Black pride has become acceptable, even obligatory, for American blacks. The Chicanos and the American Indians are right behind them proclaiming pride in their traditions. If black is beautiful, and if red is beautiful, the children and grandchildren of the immigrants are asking why they should not think the same way. Polish is beautiful, Hungarian is beautiful, Italian is beautiful, Greek is beautiful; in Chicago, even Luxembourger is beautiful. One may not only be publicly proud of one's heritage on certain days of the year; one can be proud every day of the year. Nay, increasingly, one *must* be proud. He who does not have any racial or ethnic tradition to fall back upon is forced to stand by in shamed silence. If you are neither black nor ethnic, how can you be part of the black-ethnic dialogue? One confidently expects the dawn of the day when some New England prophet will announce that Yankee is beautiful.

But in the midst of all this resurgence of ethnic pride, the state of the Irish is distressful (to use their word), indeed. At long last, it would be legitimate for them to act as if they were Irish.

Only they've forgotten how.

The legitimation of ethnicity came too late for the American Irish. They are the only one of the European immigrant groups to have over-acculturated. They stopped being Irish the day before it became all right to be Irish. The WASP's won the battle to convert the Irish into WASP's just before the announcement came that permanent peace had been made with ethnic diversity. Daniel Patrick Moynihan makes the lonely hegira from Washington to Cambridge with the melancholy thought that he and Richard J. Daley are the last of the American Irish.

Some will argue that the Irish have never been more powerful. All kinds of citizens whose Celtic origins are problematic at best will sport green emblems on the 17th of March (even though the Roman Church threw St. Patrick off the calendar—for which God forgive the Curia—together with St. Christopher). Richard M. Nixon goes to Ireland in search of his roots. Gaelic names have become part of the common pool of names that American mothers impose upon their children: Brian, Kevin, Maureen, Sheila; Eileen stopped being exclusive Irish property a generation ago; now Sean, Seamus, Moira and Deirdre get attached to Polish, Jewish, black and even Oriental faces. Only Liam and Phionna have not yet been expropriated (and when I hear of a Liam Levy, I'll know it's the end). The Clancy Brothers and the Irish Rovers will set hippie feet stomping in packed concert halls. Tens of thousands of Saxons, Africans, Teutons, Semites, Slavs and Latins will march down Fifth Avenue and proclaim the glories of the Emerald Isle. And there's even Bernadette Devlin.

So much of the nation seems to have become Irish. Only it's too late because there's nothing left. Characteristically, the Irish quit just before they won.

The political star fades. If there is another Catholic President in the near future, he is likely to be Polish. Mr. Moynihan and his colleague, Mr. Nathan Glazer, have documented the fall of the Irish from political power in New York City in their book, "Beyond the Melting Pot." Unless one counts James L. Buckley (and west of the Alleghenies we'd rather not), the Irish have vanished from office and even from election slates in New York. Mayor Lee is gone from New Haven; Boston has a Yankee mayor, and few doubt that Richard Daley will be the last of the breed.

The voice of the storyteller is mute. The O'Connors, Flannery and Edwin, are dead. J. F. Powers stands in his storefront window in St. Cloud and dreams of a church which exists no more. Fitzgerald has become the object of scholarship and Jimmy Farrell dreams still of 57th and Indiana, but no one is interested. There are no entertainers to replace Cagney or O'Brien or Crosby, and Notre Dame has an Armenian coach and a Methodist quarterback.

Even in the church the Irish power wanes. Boston is Portuguese, Philadelphia is Polish, Brooklyn is Italian, and the U.S. Catholic Conference in Washington is presided over by another Italian. Seventeen per cent of American Catholics are Irish as are 35 per cent of the priests and 50 per cent of the bishops; but the young Irish are no longer flocking to the seminaries, the Irish Catholic intellectuals vie with one another to see who can reject his own past with more enthusiasm, and the clergy and people drift apart as the former get caught up in a painful identity crisis and the latter permit the respect of old to be transmuted into contempt. The hierarchy, apparently lacking the political skills of its civil counterparts, can no longer keep its priests in line; and Irish priests and bishops have little time to ecumene with their separated Protestant brothers (formerly heretics) because they are so busy fighting with each other.

The Irish young buried their political hopes with Robert Kennedy. The flags of Camelot are furled, never more to dance merrily in the brisk spring breezes. Young men slide their permanently unfinished novels into their brief cases as they board the commuter train, and their matrons leave half-written poems in Buick glove compartments when they drive to the country club.

The Irish sun was shining brightly in the heavens at the beginning of the nineteen-sixties; Mr. Moynihan could write with pardonable pride of the President of the United States and the Prime Minister of the United Kingdom standing on a yacht in front of a life preserver marked "Honey Fitz—U.S.A." The "Honey Fitz" is gone and in a few years no one will even remember who he was; all we will have left from the glory days of the early sixties is memories—and tragic ones at that.

The Irish have finally proved to the WASP's that they could become respectable. But they paid a price: they are no longer Irish. What does it take to be Irish? A distinguished American sociologist of Florentine extraction once observed to me, "The Irish don't have a cultural heritage; they don't have a unique cuisine or an art or a family structure. They're just lower-middle-class WASP's with a political style and a religious faith of their own."

But a religious faith and a political style are not all that bad as a heritage (I even believe that I made a fairly nasty comment to my colleague that I was unaware that the Italians had either of these). In fact, they have done rather well for a millennium and a half; the only trouble is that the American Irish are in the process of losing both.

The 19th-century myth about the Irish was that they were dreamers, mystics, brawlers, political pragmatists and far too committed to their otherworldly faith to become good Americans. Some of the 20th-century Irish self-critics have elaborated on this theme to argue that Irish Catholics were too otherworldly to make it as economic successes in American society, too Catholic to become good intellectuals and too pragmatic to be good Democratic liberals.

But the implicit standard for these judgments (and in Mr. Moynihan's case the standard is quite explicit) is set by the achievements of New York Jews. The Irish are not as "successful," not as "liberal" and not as "intellectual" as the New York Jews (though they recently displayed that they are still somewhat better at winning elections). However, they are more "successful," more "liberal" and more "intellectual" than just about anyone else in sight—a[illegible] data to support this statement are overwhelming. In fact, on various measures of the achievement syndrome called the "Protestant ethic" the Irish Catholics score higher than anyone else—save the Jews.

AND that's just the trouble. The Irish are not mystics, dreamers, brawlers, storytellers and political pragmatists. They have given up all these splendid traits to become accepted as full-fledged Americans. And for the Irish and for the rest of American society this sacrifice has been a tragedy. Of WASP's we had a sufficient supply; but of Celts not nearly enough. And now we have almost none.*

Where indeed are the mystics to be found, the dreamers who look into the mists and see visions of splendid new worlds—with or without the aid of a mug of the "creature" which 19th-century cartoons invariably placed in Paddy's hand? Surely not in the Roman Church of the Irish-American variety. The bishop and the parish priest have become executives of the standard American corporation variety — though somewhat pre-Harvard Business School in their style. The intellectuals as manifested, for example, in such journals as Commonweal are no more interested in being mystics or dreamers than are their role models on The New York Review of Books.

The young Irish clergy are trying their best to act like Methodists and in the process have finally yielded to the WASP's on those twin bugaboos of 19th-century Protestantism: the parochial school and clerical celibacy. And a mystic could hardly be ordained a Catholic priest in the United States today. He would do very poorly on his group-adjustment scores. When he ought to be in his T-group, he would very likely be in church praying or out on the bogs dreaming. What in the world would American Catholicism do with a priest that dreams and prays? Of what relevance are saints?

The only dreaming that goes on among the Irish laity is that which is aided by that very non-Celtic concoction, the martini, to which they turn for surcease from their worries about their career achievements. A little bit of the "creature" never did anyone any harm; but the Irish upper middle-class—male and female alike—drinks as if alcohol were following the route of TV cigarette advertisements. Even though the alcoholism problem was solved long ago in the mother country (by the characteristic Irish strategy of repression), the American Irish are 25 times more likely to be alcoholics than is the typical American —though, for whatever consolation it will bring them, much less prone to this affliction than the Mormons.

Wine, stout, even Irish whisky, may be powerful assets for seeing visions, but the only sort of mystical experience to be produced by the martini is a hangover. (And what can one say of that terrible Russian subversion, the vodka martini?)

THE most serious political failure of the Irish results not from an excess of pragmatism but from not being pragmatic enough. Even if they didn't have a revolutionary tradition of their own, even if the Blackstone Rangers do not call to mind the Ragan Colts and the Black Panthers do not look like the Molly Maguires (though in truth the Panthers are much more moderate), even if Bernadette Devlin is not the Julian Bond of Ulster, the Irish, on the most pragmatic of political grounds, should have identified with the aspirations of the blacks and Spanish-speaking in American society. This failure of the most elementary political intelligence allowed the alliance of Jewish and Yankee "liberals" to sweep the Irish from political power in New York—and then, of course, put Mr. Buckley in the United States Senate.

This mistake, being duplicated in some fashion or other in many American cities, has been avoided in Chicago for the time being. But after Daley, what? The problem is not so much that the Mayor will be succeeded by someone who is not concerned about the problems of blacks but that a new generation of Irish will appear on the scene that is not seriously concerned about politics at all. As one young Irish ideologue put it—having successfully learned to imitate the style of American liberalism—"All Daley is interested in is winning elections." The alternative to winning elections is losing them; and since it has been clearly demonstrated that WASP and Jewish Democrats would sooner lose elections than win, the Irish were the only winners left in the party. After a thousand years of losing in the Old Country, the American Irish were once quite hungry to win. But now they are not hungry and can permit themselves the luxury of self-righteous defeat. So the young Irish amateur politicians (nothing is more "immoral" than a "professional" politician; he belongs to a "machine") proclaim Adam Walinsky and Allard Lowenstein their heroes and switch allegiance from McCarthy to McGovern as they sally forth in search of more windmills to joust with.

The storytellers have all vanished. Poor, weary John O'Hara spent his last years apparently proud of how far he had come from his Irish origins. Thomas Flemming, in his recent "The Sandbox Tree," delights in the escape of his characters from their heritage (one of his characters finds redemption by marrying a Unitarian, surely an extraordinary fate); Elizabeth Cullinan, in "House of Gold," meticulously describes the destructiveness of an Irish mother, but after a while we stop caring. And if James T. Farrell had grown up at 93d and Hoyne instead of 57th and Indiana, he would never have set a word on paper. What are the rest of the potential storytellers doing? Why, they're pushing their careers, investing in the stock market, vacationing in Mexico, playing golf at the "club," drinking martinis and watching professional football on Sunday afternoons.

A Jewish colleague once observed to me, "The big difference between the Jews and the Irish is that we know we have a self-hatred problem and take it into account. You don't even know you have the problem."

So the American Jewish novelist turns his self-hatred to literary and financial profit, and the American Irish novelist doesn't write at all. Studs Lonigan was a 1925 Irish Portnoy, but one looks in vain for an updated model of the Irish American who suffers defeat "in an atmosphere of spiritual poverty."

HOW did it all happen? Whatever happened to the Irish? They were seduced by the possibility of becoming respectable. A millennium of oppression by a foreign invader would work havoc with even the strongest of national ego strengths—and there is no particular evidence in Irish history to indicate that emotional security was ever one of their notable assets. British tyranny did to the Irish something analogous—though obviously not so severe — to what white tyranny did to American blacks. The Clancy Brothers lyric which proclaims

When we at last are civilized,
Won't Mother England be
surprised!

is more than just a joke. It manifests the mixture of envy, hatred, admiration and need to imitate which every subject people displays toward its masters. The Irish would be dreamers, mystics, prophets, pragmatists and occasionally even saints when there was no other alternative available to them.

But in this land of the free, the Mick was told that there was no reason why he couldn't succeed, no reason why he couldn't be respectable, no reason why he couldn't even be President of the United States—so long as he went to Harvard. Respectability was his for the asking, so long as he was willing to settle down to the serious business of becoming just like everyone else. The temptation proved irresistible. You could sit by the bog and dream, so long as respectability was not a real possibility. But in the United States, the Irishman finally found an opportunity to be accepted—though not fully, of course—by the "real world" of the Anglo-Saxon. Aided and abetted by his mother and his priest he set

***Besides their ethnic identity, the Irish have also lost: [1] their ability to laugh at themselves, and [2] their ability to spot blarney when it crosses their path. I lament both these losses. God forgive them for losing their wit. As for me, I not only kissed the fabled stone, I very much fear that I swallowed it — as should be clear to any authentic Celt who chances to read this article.**

out to prove definitively that he was just as good as anyone else, if not a little bit better. Unfortunately, he succeeded.

But respectability has always meant something very lower middle class for an Irishman, even if he owns a steel mill. The Irishman's concept of respectability is shaped by the narrow parochialism of the "lace curtain"—probably because most of his aristocrats long since went over to the enemy. The sad truth is that the Irish new rich do not really know how to spend their money and their idea of "class" and "style" rarely exceeds flying to Dallas to watch Notre Dame play in the Cotton Bowl.

In Chicago it is not merely that they do not go to Orchestra Hall, but they do not know that it exists; and the comment of one suburban community about the son of a prominent citizen is typical, "What did he want to go to Harvard for? He had already been accepted by Notre Dame." Respectability does not mean aristocratic responsibility for culture and the arts; it means that you act in such a way that no one will dare to say, "Who do you think you are?" Such an attitude spares most Irish communities Mercedes 600's—or even 300's—but it also means that a young person who wants to be a poet or an artist is packed off to the local psychiatrist.

THE primary agent of this seduction by respectability is the Irish mother. The Jewish mother may kill her children with kindness (if the novelists are to be accepted as accurate reporters). The Irish mother rather manipulates her children by starving them for affection. The Jewish mother may, according to what her sons write about her, say: "Eat your chicken soup: it's good for you." But the Irish mother announces: "There is not enough chicken soup to go around; and if you don't love Mother enough, you'll go to bed hungry."

The lack of respectability was an even heavier cross for her than for her husband—who could always escape to the local pub. There is no need for women's liberation in Ireland. The mothers have been running the country for centuries—together with the clergy, of course, who, in their turn, are dominated by their mothers and housekeepers. When a time finally came when there seemed to be a real chance for respectability, the Irish matriarch wasted no time in setting her male off in hot pursuit. If his enthusiasm seemed to flag, she set him in motion once again with her most powerful weapon, "What will people say?"*

And so the WASP's won; seduced by the bright glitter of respectability and egged on by their mothers, the Irish have become just like everyone else; and the parades on St. Patrick's Day are monuments to lost possibilities of which few people in the parade are even aware.

***I might note in passing that it is unfortunate that the mores—Irish and American alike—do not permit the women to get involved in politics. It has been my experience that Irish women make their husbands look like rank political amateurs. Indeed, the three most astute politicians I have ever met were young women in their late teens or early 20's. You'd not catch them pursuing lost causes.**

YET, one is permitted to sit by the side of the peat bog, stare into the mist and think of all the things that might have been.

What if American Catholicism had been presided over by more Irishmen like John England, who democratized his diocese in 1825, or John Ireland, who preached and practiced ecumenism in 1890—long before even the word was invented?

What if there had been mystics or poets or storytellers who could have raised long ago for all Americans the question of whether technological and economic success were really enough?

What if there had been a few Irish political leaders who could have realized how much the Irish had in common with more recent migrants to the city and allied themselves with the poor and the oppressed, not so much for losing elections as for winning them?

What if one white immigrant group had been able—without the aid of sensitivity training—to keep alive some of the wild passion of its tribal days, a passion which suburban culture abhors but desperately needs? What, in other words, would have happened if the Irish had hung on to not only their souls but also their "Soul"?

What if John Kennedy had not been shot?

But, then, we Irish have always been good at sitting by the side of the bog, staring into the rainy mist and dreaming lofty dreams of what might have been.

And yet...on the South Side of Chicago a toddler named Liam does battle with a Burmese cat while his red-haired mother sits at a kitchen table and scrawls poetry on a yellow notepad.

I shall not give to you, my son, a heritage of splintered dreams
that slushed down the sink with stale beer and squeezed out tears of pain
for all the years that might have been if I had lived
instead of killing dead my heart and ours, bit by bit,
with breaking rage and chunks of sorrow,
passion which guilt turned sour and then, misunderstanding
took my soul and crashed it in the night.
I shall not go desperate dying into life.
The enemy that dare to take the sea's surge from our eyes
I shall defy and drag to hell and back and shake the skull of suicide
which says the gift will be ungiven and the hope denied.
No, for we shall sing, my son, and eat the sweets of victory.
Lie peaceful down your head
This hunger will not be quieted, nor ever fully fed
Not before we hold the stars and until then,
we'll go a-brawling and wooing life.
There're fights to be fought and battles won
But never in the name of life direct our own undoing
Nor allow while we breathe that Life should be undone!

Maybe, to lift one final line from Mr. Moynihan, there are still some of us around.

March 14, 1971

THE AMERICAN IRISH ARE ALIVE AND WELL . . .

To the Editor:

The Rev. Andrew M. Greeley's article, "The Last of the American Irish Fade Away" (March 14), is misleading in several regards.

First, he persists in making the disastrous distinction between Irish Americans of the Catholic and Protestant faiths. As a priest, he naturally overemphasizes the impact of the clergy upon the Irish American Catholic. Not only is he incorrect in saying that the Irish young buried their hopes with Robert Kennedy (there is still Edward, as well as Eugene McCarthy, George McGovern, who by the way is of Irish extraction but does not fit Father Greeley's notion of being Irish, since he is not Catholic), but the author grossly underestimates the voting power of Irish Americans—they do, to a very large extent, vote on Election Day, and politicians today, as well as yesterday, have always recognized this fact.

Father Greeley's comments concerning the Irish mother border on sheer nonsense. He holds that Irish mothers starve their children for affection; what he should have said, if there was any depth to his understanding of the Irish American family, is that the opposite holds—Irish mothers, and Irish fathers, shower their sons and daughters with affection. This affection, however, is more often of the type described by comedian Bill Cosby—"Make your bed, get home early"—rather than of the "I love you, I love you" variety.

As far as the author's contention that the Irish family structure is a matriarchal one, nothing could be further from the truth. The Irishman has always been the unquestioned head of his family. In fact, one of the major reasons why Bernadette Devlin has received such a cool reception among the Irish of this country as well as the Irish of the fatherland, aside from her political views, may be indeed due to the fact that she is a civil-rights leader who is also a woman — and Irishmen today, as throughout the centuries, have always recoiled from the idea of being led by women, in anything.

What Father Greeley has failed to do is give an accurate picture of the American Irish. For millions of Irish Americans, the rich heritage of Ireland is as treasured as ever, and no WASP will ever be able to take that from them. Irish Americans are achieving great successes in politics, the arts, sports, etc., and they will continue to do so. John Tunney, Senator from California, actress Ali McGraw and baseball star Pete

Rose are just some examples.

Perhaps Father Greeley suffers from the self-hatred problem he describes in the article. Fortunately, millions of other Irish Americans do not suffer from this problem—instead of self-hatred, they substitute pride in being Irish—and they feel, rightfully so, that they don't have to answer to any man because of it.

JAMES CLERKIN.

New York.

●

TO THE EDITOR:

Oh, come now, Father Greeley, because the Irish have gone from shanty to lace curtain, they have blown their heritage?

You say, "The Irish have finally proved to the WASP's that they could become respectable." Must one believe that the John Lindsays of the world have a patent on respectability, that being Irish and being respectable is mixed marriage at its worst?

There are many of us who did not go to Harvard; who buy our suits at Brooks Brothers; who enjoy the Philharmonic, opera and ballet; who are creative, sensitive, talkative, argumentative, politically astute, and who dream and brawl with relish.

The Irish did not move the pigs out of the parlor to please the WASP's; they worked hard enough to afford something other than home-grown work. They gave up tugging their forelocks in the presence of their overlords—although, in Massachusetts, many of them still do it in the presence of a Kennedy.

They have done many things to better themselves; not to please the WASP's, a group that itself may fade away—by boring itself to death.

EDWARD GALLIGAN.

New York.

●

TO THE EDITOR:

You have escaped my wrath and ire as you have castigated the Army that I admire, but you go too far when you deprecate my dear Mother, née Nora McCullough.

For shame, for Father Greeley to even think that anyone could love her children more than my dear Mother! With such wild ideas as his, he will accomplish what the strangers could not in 300 years of fire, famine and sword—he will drive us all to the Church of England!

ARTHUR J. JOHNSON.

Scarsdale, N. Y.

●

TO THE EDITOR:

Apropos of the American Irish "fading away" (not unlike old soldiers), I'm reminded of the remark someone once made, maybe a WASP at the time of Harrigan and Hart: "The Irish are a funny people without a sense of humor."

GEORGE SALVATORE.

Bellmore, N. Y.

April 4, 1971

CENSUS EXAMINES 7 ETHNIC GROUPS

Families of Russian Origin Report Highest Income

Families headed by persons of Russian origin report the highest incomes—a median of $11,554 a year—among the nation's nearly 20 million families that consider themselves to be in seven major ethnic groups.

By contrast, families whose head is of Spanish origin report a median income of $5,641, according to a Census Bureau study.

The new information comes from the first census study based on self-identification by an individual as to his origin or descent. Previous reports have inferred ethnic origin, using indicators such as place of birth, country of origin, mother tongue and surname.

About 75 million persons among the nation's 200 million identify themselves by one of the major ethnic categories. The study reported these as 20 million German, 19.1 million English, 13.3 million Irish, 9.2 million Spanish, 7.2 million Italian, 4 million Polish and 2.2 million Russian.

Italy in the Lead

This self-identification by ethnic group compared with only 11 million reported as foreign-born. The countries of birth were headed by Italy, with 1,353,000, followed by Britain, 1,006,000; Germany, 1,004,000; Mexico, 938,000; Poland, 550,000; Cuba, 504,000; Russia, 412,000; Ireland, 277,000; Austria, 236,000, and Sweden, 166,000.

"Some respondents having a diverse ethnic background or having several generations of residence in the United States may have reported the ethnic association they felt most strongly," the Census Bureau said.

The study was based on a sampling supplement to the November, 1969, current population survey. The income differences reflected such factors as age. Persons of Spanish origin were youngest, with a median age of 19.9 years, while those of Russian origin were oldest, at 45.8.

The median family income—that is, half the families had more and half had less—was reported for 1968, and showed the following:

Ethnic Origin	Family Heads	Income
Russian	598,000	$11,554
Polish	1,149,000	8,849
Italian	1,924,000	8,808
German	5,674,000	8,607
English	4,997,000	8,324
Irish	3,639,000	8,127
Spanish	1,927,000	5,641

About 29 per cent of the Spanish-headed families had incomes under $4,000, while 32.8 per cent of the Russian-headed families had incomes of $15,000 and more.

Occupation Categories

Among males 16 years and older in the ethnic groups, four-fifths were in the civilian labor force, with an over-all unemployment rate of 2.8 per cent. About 44 per cent of females 16 and older were in the work force, with over-all unemployment of 4.8 per cent.

In terms of occupation categories, the ethnic group with the largest number of males in professional and technical jobs was Russian, with 27,100. Farmers were led by Germans, 6,700; managers and proprietors, Russian, 29,300; clerical workers, Italian, 9,100; sales workers, Russian, 12,800; craftsmen and foremen, Polish, 24,400.

Men of Spanish origin had the largest number of jobs classified as operatives, 28,600; service workers, 10,500; farm laborers and foremen, 4,800, and other laborers, 11,800.

May 23, 1971

Twenty Million Italian-Americans Can't Be Wrong

By RICHARD GAMBINO

THE persistent myth of a monolithic Italian-American subculture called the Mafia presents a set of dilemmas for Italian-Americans. To the average person in this country, the Sicilian-American norm is seen as identical with Mafia in a self-locking stereotype. Italian-Americans are offered a bigoted choice of two identities somewhat paralleling the two imposed on blacks. Blacks could be child-like, laughing, Uncle Tom figures or sullen, incorrigible, violent, knife-wielding criminals. Similarly, the nativistic American mentality, born of ignorance and nurtured by malice, insists that Italian-Americans be either/or creatures. They must be either spaghetti-twirling, opera-bellowing buffoons in undershirts (as in the TV commercial with its famous line, "That'sa some spicy meatball"), or swarthy, sinister hoods in garish suits, shirts and ties. The incredible exploitation of "The Godfather" testifies to the power of the Mafia myth today.

RICHARD GAMBINO is assistant professor of educational philosophy at Queens College, the City University of New York.

To understand the special identity problems of Italian-Americans, we must begin with a very popular Sicilian proverb quoted by Leonard Covello in his pioneering work, "The Social Background of the Italo-American School Child": *"Che lascia la via vecchia per la nuova, so quel che perde e non sa quel che trova"*—"Whoever forsakes the old way for the new knows what he is losing, but not what he will find."

In Sicily, *la via vecchia* was family life. The Sicilian immigrants to America were mostly *contadini* (peasants) to whom there was one and only one social reality, the peculiar mores of family life. *La famiglia* and the personality it nurtured was a very different thing from the American nuclear family with the personalities that are its typical products. The *famiglia* was composed of all of one's "blood" relatives, including those relatives Americans would consider very distant cousins, aunts and uncles, an extended clan with a genealogy traced through paternity. The only system to which the *contadino* paid attention was *l'ordine della famiglia*, the unwritten but all-demanding and complex series of rules governing one's relations within, and responsibilities to, his own family and his posture toward those outside the family. All other social institutions were seen within a spectrum of attitudes ranging from indifference to scorn and contempt.

One had absolute responsibilities to family superiors and absolute rights to be demanded from subordinates in the hierarchy. All ambiguous situations were arbitrated by the *capo di famiglia* (head of the family), a position held within each household by the father, until it was given to—or taken away by—one of the sons, and in the larger clan, by a male "elder." The *contadino* showed calculated respect to members of other families which were powerful, and haughtiness or indifference toward families less powerful than his own. He despised as a *scomunicato* (pariah) anyone in any family who broke the *ordine della famiglia* or otherwise violated the *onore* (honor, solidarity, tradition, "face") of the family.

Thus, Sicily survived a harsh history of invasion, conquest and colonization by a procession of tribes and nations.* What enabled Sicilians to endure was a system of rules based solely on a phrase I heard uttered many times by my grandparents and their contemporaries in Brooklyn's "Little Italy": *sangu di me sangu*, Sicilian dialect for "blood of my blood." (As is typical of Sicilian women, my grandmother's favorite and most earnest term of endearment when addressing her children and grandchildren, and when speaking of them to others, was *sangu mio*—literally, "my blood.")

It was a norm simple and demanding, protective and isolating, humanistic and cynical. The unique family pattern of Sicily constituted the real sovereignty of that island, regardless of which government nominally ruled it.

As all of us are confronted with the conflicts of our loyalty to a sovereign state *vs.* our cosmopolitan aspirations, so the Italian-American has found himself in the dilemma of reconciling the psychological sovereignty of his people with the aspirations and demands of being American. Most Italian-Americans are derived from areas of Italy south of Rome, about 25 per cent of them from Sicily. In his book "The Italians," Luigi Barzini reminds us that "Goethe was right when he wrote: 'Without seeing Sicily one cannot get a clear idea of what Italy is.' Sicily is the schoolroom model of Italy for beginners, with every Italian quality and defect magnified, exasperated and brightly colored."

This background illustrates the confused situation of Italian-Americans. It would not be an exaggeration to say that to Middle America the Chinese character is more scrutable than that of the Italian-American. Although the problems of Italian-Americans are less desperate than those of groups whose fate in this country has been determined by color, they are no less complicated. And they are rising to a critical point.

The solutions have been too long delayed, and we have thus lagged behind other large groups of European "ethnics," notably Jews and Irish-Americans, in social and economic terms. We live in a time of ethnic consciousness, when each group asserts its presence and insists on determining its character and destiny. It remains to be seen how the 20 million to 23 million Italian-Americans, many of them third- and fourth-generation citizens, will determine theirs—and how much upset it will bring.

To the immigrant generation of Italians, the task was clear: Hold to the psychological sovereignty of the old ways and thereby seal out the threats of the new "conqueror," the American society that surrounded them. This ingrained disposition was strongly reinforced by the hatred and insult with which the Italian immigrant was assaulted by American bigots who regarded him as racially inferior—a "dago," a "wop," a "guinea." Although one might assume that such indignity has altogether disappeared, we need only recall well-known tests used to identify these prejudices. In one study, American college students were shown photographs of members of the opposite sex with what purported to be the name of each person on the photograph. The students were asked to evaluate the attractiveness of the person in the photograph. Then, some time afterward, names were changed and the procedure repeated with the same students. The result was that those people who were regarded as "handsome" or "pretty" when they had names like "Smith" were found not attractive when their names were changed to Italian ones. (The same result was found using names commonly thought of as Jewish.)

As a blond, blue-eyed American who was never spotted as Italian, I was sometimes exposed to gibes and jokes such as the one I overheard at the University of Illinois in 1961. Question: "What sound does a pizza make when you throw it against the wall?" Answer: "Wop!" The joke was greeted with much laughter, not by red-necks, but by a roomful of graduate students, some of whom became belligerent rather than embarrassed when as a Sicilian-American I expressed irritation at the joke.

Discriminatory attitudes toward Italians are also indicated in the restrictionist immigration policies aimed at them. Not only did these continue until 1968, but they were made more stringent in recent times than in the period following the flood of immigrants before World War I. Whereas the Emergency Quota Law of 1921 imposed an annual limit of 42,000 Italian immigrants, the Immigration and Nationality Act of 1952 dropped the Italian quota to its lowest point in history, less than 6,000 per year. The law was passed by Congress over the veto of President Truman.

The Italian immigrants admitted in the earlier years responded to an alien, hostile society by clustering together in crowded Italian "ghettos,"* euphemistically called "Little Italies," and exhausted their energies in America's sweatshops. So desperate was the poverty of Southern Italy from which they had escaped (my father came to this country as a boy with rags wrapped around his feet because his family could not afford shoes) that their lives in the promised land were regarded as progress. Oppression and economic exploitation were woven into the fabric of life in

*Including the Phoenicians, Carthaginians, Greeks, Romans (Sicily was the first province of ancient Rome), Vandals, Goths, Byzantines, Saracens, Arabs, Normans, the French, the Spanish, the Savoians and various northern Italian powers, the Sardinians and the Allies of World War II.

*The very word is Italian, deriving from the experience of the Jews in Venice. According to a publication of the Italian Tourist Agency, the word came from the Venetian *geto* or *gheto*, meaning "to cast," which indicates a foundry. In 1516, Jews in Venice were segregated into the *Ghetto Nuovo*, a neighborhood at the edge of the town where a "new foundry" existed.

Southern Italy. Most of Sicily's foreign governors had two things in common: the subjugation of Sicilians, and systematic exploitation of them and their land. The first was done by force, the second by perennial systems of absentee landlordships.

The complicated customs and institutions developed by the Sicilians were marvelously effective in neutralizing the influence of the various alien masters. The people of the island survived and developed their own identity not so much by overtly opposing the oppressor, a suicidal approach given the small size, exposed location and limited resources of the island, but by *sealing out* the influence of the strangers.

The sealing medium was not military or even physical. It was at once an "antisocial" mentality and a supremely social psychology, for it forms the very stuff of Sicilian society. It constitutes the foundation and hidden steel beams of a society that historically has been denied the luxury of more accessible (and vulnerable) foundations or superstructure. This is the reason for the Sicilians' pride. A system of social attitudes, values and customs that is impenetrable to the *sfruttamento* (exploitation) of any *stranieri* (strangers), no matter how powerful their weapons or clever their devices. But like all defenses, their life-style has exacted costs from the people of Sicily—the vexing social and economic problems that Italians lump together as the *Problema del Mezzogiorno* or *Questione Meridionale* meaning "the Southern Problem."

THIS year, I saw evidence of the continuing conflict between Northern and Southern Italy. Three companions and I hired a car in Rome to drive south to the area of Naples. Our driver and guide was a native Roman in his 30's, a graduate of Boston University. His sophisticated and friendly manner changed as we drove south. He spoke rather rudely to everyone in the South, except policemen. At a restaurant in Naples, he found fault with everything and expressed it loudly to the staff and other patrons. Yet, in my observation, the restaurant was as good as any I had seen in Rome, Florence or Venice. Our guide explained to me that Southerners are uncultured, brutish creatures, even as they treated me in a warm, civilized way. He called them "not really Italian," a phrase that evoked an incident in my childhood.

I attended P.S. 142 in Brooklyn for eight years, a school about 90 per cent Italian-American in my recollection. One day, when I was about 9 years old, I was told by one of my teachers (a non-Italian, like almost all the staff), "You're not *real* Italians." When I went home that day, I asked my immigrant grandfather what the teacher meant. He sat there, looked me level in the eye, shook his head slowly and said simply, *"i americani!"* (To Italian immigrants, all other people in this country were "Americans," whatever their ethnic background). The prejudice of the Northerners, experienced by my grandfather in the old land, had been transplanted to his new home and visited upon his grandchild. For this reason, among others, the mores the immigrants brought with them from the old land gave them psychological stability, order and security, and were held to tenaciously. But in the United States, the price was isolation from the ways of the larger society.

The immigrants' children, the "second generation," faced a challenge more difficult to overcome. They could not maintain the same degree of isolation. Indeed, they had to cope with American institutions, first schools, then a variety of economic, military and cultural environments. In so doing, what was a successful social strategy for their parents became a crisis of conflict for them. Circumstances split their personalities into conflicting halves. Despite parental attempts to shelter them from American culture, they *attended* the schools, *learned* the language and confronted the culture.

It was a rending confrontation. The parents of the typical second-generation child ridiculed American institutions and sought to nurture in him *la via vecchia*. The father nurtured in his children (sons especially) a sense of mistrust and cynicism regarding the outside world. And the mother bound her children (not only daughters) to the home by making any aspirations to go beyond it seem somehow disloyal and shameful. Thus, outward mobility was impeded. Boys were pulled out of school and sent to work at the minimum legal age, or lower, and girls were virtually imprisoned in the house. Education, *the* means of social and economic mobility in the United States, was largely blocked to the second generation, because schools were regarded not only as alien but as immoral by the immigrant parents. When members of this generation did go to school the intrinsic differences between American and Southern-Italian ways was sharpened even further for them. The school, the employer and the media taught them, implicitly and often perhaps inadvertently, that Italian ways were inferior, while the immigrant community constantly sought to reinforce them.

IMMIGRANTS used "American" as a word of reproach to their children. For example, take another incident from my childhood: Every Wednesday afternoon, I left P.S. 142 early and went to the local parish church for religious instruction under New York State's Released Time Program. Once I asked one of my religious teachers, an Italian-born nun, a politely phrased but skeptical question about the existence of hell. She flew into a rage, slapped my face and called me a *piccolo americano*, a "little American." Thus, the process of acculturation for second-generation children was an agonizing affair in which they had not only to "adjust" to two worlds, but to compromise between their irreconcilable demands. This was achieved by a sane path of least resistance.

Most of the second generation accepted the old heritage of devotion to family, and sought minimal involvement with the institutions of America. This meant going to school but remaining alienated from it. One then left school at the minimum legal age and got a job that was "secure" but made no troubling demands on one's personality, or the family life in which it was imbedded. This explains why so many second- and even third-generation Italians fill civil service, blue-collar and low-echelon white-collar jobs—many of those in the last category being employed in more static, low-growth industries, for example utilities like Con Edison, where my father worked for 40 years.

So we see in New York City that the Fire Department, Police Department and Sanitation Department are filled with Italians. The top positions in these services and in their counterpart private corporations remain conspicuously and disproportionately free of Italian-Americans.

Another part of the second-generation compromise was the rejection of Italian ways which were not felt vital to the family code. They resisted learning the Italian culture and language well, and were ill-equipped to teach it to the third generation.

Small numbers of the second generation carried the dual rebellion to one extreme or the other. Some became highly "Americanized," giving their time, energy and loyalty to schools and companies and becoming estranged from the clan. The price they paid for siding with American culture in the culture-family conflict was an amorphous but strong sense of guilt and a chronic identity crisis not quite compensated for by the places won in middle-class society. At the other extreme, some rejected American culture totally in favor of lifelong immersion in the old ways, which through time and circumstance virtually dissipated in their lifetimes, leaving underdeveloped and forlorn people.

THE tortured compromise of the second-generation Italian-American thus left him permanently in lower middle-class America. He remains in the minds of Americans a stereotype born of their half-understanding of him and constantly reinforced by the media. Oliver Wendell Holmes said a page of history is worth a volume of logic. There are, with very few exceptions, no serious studies of the history of Italian-Americans. It is easy to see why this has left accounts of their past, their present and their future expressed almost exclusively in the dubious logic of stereotypes.

The second-generation Italian-American is seen as a "good employe," i.e., steady, reliable but having little "initiative" or "dynamism." He is a good "family man," loyal to his wife and a loving father vaguely yearning for his children to do "better" in their lifetimes, but not equipped to guide or push them up the social ladder. He maintains his membership in the church, but participates in it little beyond ceremonial observances; while he often sends his children to parochial schools, this represents more his social parochialism than enthusiasm for the American Catholic Church, which has very few Italian-Americans at the top of the hierarchy and has never had an Italian-American cardinal. (Since the Irish immigrants who controlled the Church in its earlier years often discriminated against them, the Italian-Americans tended to view it as an alien "Irish institution.")

He is a loyal citizen of America, but conceives of his political rôle as protecting that portion of the status quo which he has so painfully carved out by his great compromise. Thus, his political expressions are reactive rather than active. He tends to feel threatened by social and political change, and he is labeled "conservative" or "reactionary" by the larger society. His political reactions are usually to *ad hoc* situations and individual candidates. A bloc of Italian-American voters has not been identified. And with few exceptions, Italian-Americans have not achieved visible positions of major political power. There was not an Italian-American in a Cabinet post until 1962 and none in the Senate until 1950. And even on state and local levels, this ethnic group remains under-represented, although it constitutes 11 or 12 per cent of the national population.

CLEARLY the most prominent — and pernicious — element in the Italian-American stereotype is that of the Mafia. The Italian immigrant carried with him a memory of the Mafia in Sicily as a kind of ultimate family. Formed in feudal times, the organization began with bands of men who were secretly organized on *ad hoc* bases to protect families (and their customs) from foreign oppression. The term Mafia is thought to be an acronym of the 13th-century Sicilian battle cry, *"Morte ai francesi gl'italiani anelano!"* ("Death to the French in Italy"), part of a violent rebellion against French oppression that began in Palermo on Easter Sunday, 1282, and spread to all of Sicily. Almost all of the French on the island were in fact killed. With characteristic sarcasm, Sicilians dubbed their bloody insurrection "Sicilian Vespers."

The early Mafia bands defended the people by appropriating two old Sicilian institutions: a) they adopted as their own code the traditional family code, and b) they adopted the *vendetta*, i.e., revenge against the enemy by deadly violence and terrorism. These *vendette* were not just political wars. They were struggles of honor, demanding total allegiance and sacrifice. By the late 19th century, the Mafia had become an institutional force in the life of Western Sicily. But by this time it had lost its role of defender of the Sicilian morality and people. The code of family honor had become hopelessly corrupt. With its members bribed by alien rulers and growing in greed, the Mafia became, over the years, a federation of gangs controlling Western Sicily largely for its own profit and power. To the Italian immigrant in America, the Mafia was both feared in its present form and respected in its archaic ideal as the supreme representative and protector of the family morality. *Rispetto* (respect in the sense of awe of this ideal) was expressed in colloquialisms he brought from Sicily. The immigrant used the word *Mafioso* as an adjective synonymous with "good" and "admirable." For example, he would speak of a *Mafioso* horse, meaning the animal was strong and spirited. Similarly, he would say of a fine specimen of a man, *"Che Mafioso!"*

EVERY group living in poverty in this country—or any country—has spawned crime. The poor Italian ghettos spawned criminals whose inherited Southern-Italian morality led them naturally to band together into groups that combined both elements of that morality and their own criminal bent. This amalgam is what the so-called Mafia was, and perhaps still is, in America. *The common characteristic of the at most 5,000 to 6,000 Italian-American gangsters claimed by governmental agencies, and the 23 million other Italian-Americans, is the family morality, a noncriminal morality. The distinguishing characteristic that separates the relatively few gangsters from the millions of law-abiding Italian-Americans is that the gangsters have turned that morality to serve criminal ends.*

THE responses to this by Italian-Americans of the second generation have fallen into several categories:

(1) Some have adopted as the lesser of two evils the "Uncle Giovanni" image. Interestingly, Sicilians have a contemptuous word for this kind of vulgar fool, *cafone*.

(2) Some have sought to get some public-relations mileage out of what has today become a kind of "Mafia chic" by half-jokingly adopting the Mafia myth. For example, I arrived at a plush restaurant at the Plaza Hotel recently and gave my name to the maitre d' in charge of reservations. He looked at me and said hello in Italian, and we exchanged a few words in the language. Then, searching for something else in our common background, he said, jokingly, referring to my name: "I expected somebody from 'The Godfather.' "

At least the *Mafioso* is taken seriously as an individual of some importance, an improvement over being considered a buffoon, or being ignored. Moreover, the myth of an extrasocietal, almost omnipotent power has great appeal to people in a complex society who are exasperated by feelings of confusion, impotence and defeat.

(3) Some have patiently repeated, "We are not criminals. . . . The percentage of racketeers among us is small." In the scream of the public media-exploitation of Mafia chic, such voices are, year after year, all but lost. Through the power of the media, Mafia chic has even crossed the Atlantic to Italy itself. In bookstores in Italy, I saw copies of a translation of "The Godfather," called *"Il Padrino,"* a very popular book there. And even in Florence I was asked, as I have been for years in the United States, whether I am related to the reputed boss of the American Mafia, Carlo Gambino. (I am not.)

(4) Some recently have begun denying that the Mafia exists in this country at all This hyperdefensive reaction is in reality an expression of rage, a reaction to years of abuse. The fact that some racketeers attempt to exploit the outrage for their own ends is beside the point. There is opportunism in every social expression.

Most recently, the media

have gone beyond their role in transforming a difficult social problem of organized crime into a romantic myth of fantastic proportions. Just as many media people found that the creation and exploitation of the fantasy was lucrative, so many are discovering that ridiculing overstatements of denial of the fantasy is also lucrative. (At the height of the Italian Civil Rights League's campaign to deny that the Mafia exists, New York magazine ran a parody of a fictional gangster named Salvatore Gambino who kept insisting there is no such thing as the Mafia.)

But the Mafia issue must be kept in perspective. The myth has become an example of an interesting expression I heard used in Italy, *americanata*, meaning something spectacular, wild, exaggerated. We are in danger of focusing only on this side show, which merely diverts attention from the larger issues. So far, none of the responses of Italian-Americans to the old Mafia myth and the new Mafia chic have been effective in improving either their image or their real, complex problems.

WE come at last to the compound dilemma of the third- and fourth-generation Italian-Americans, who are now mostly young adults and children with parents who are well into their middle age or older. (The number of those in the third and fourth generations is estimated as at least 10 million, compared to more than 5 million in the second generation and no more than 5 million living members of the first generation.) The difference between the problems of the second and those of the third generation is great—more a quantum jump than a continuity.

Perhaps a glimpse at my own life will serve as an illustration. I was raised simultaneously by my immigrant grandparents and by my parents, who were second generation, notwithstanding my father's boyhood in Italy. So I am both second and third generation at one time. I learned Italian and English from birth, but have lost the ability to speak Italian fluently. In this, my third-generation character has won out, although I remain of two generations, and thus perhaps have an advantage of double perspective.

My grandfather had a little garden in the back yard of the building in which we all lived in Brooklyn. In two senses, it was a distinctly Sicilian garden. First, it was the symbolic fulfillment of every *contadino's* dream to own his own land. Second, what was grown in the garden was a far cry from the typical American garden. In our garden were plum tomatoes, squash, white grapes on an overhead vine, a prolific peach tree, and a fig tree! As a child, I helped my grandfather tend that fig tree. Because of the inhospitable climate of New York, every autumn the tree had to be carefully wrapped in hundreds of layers of newspaper. These in turn were covered with waterproof linoleum and tarpaulin. The tree was topped with an inverted, galvanized bucket for final protection.

“'Get an education, but don't change,' exhort the Italian-American college girl's parents. 'Grow, but remain within the image of the houseplant Sicilian girl.'”

But the figs it produced were well worth the trouble. Picked and washed by my own hand, they were as delicious as anything I have eaten since. And perhaps the difference between second-and third-generation Italian-Americans is that members of the younger group have not tasted those figs. What they inherit from their Italian background has become so diluted as to be not only devalued but quite unintelligible to them. It has been abstracted, removing the possibility of their accepting it or rebelling against it in any satisfying way.

I WAS struck by this recently when one of my students came to my office to talk with me. Her problems are typical of those I have heard from Italian-American college students. Her parents are second-generation Americans. Her father is a fireman and her mother a housewife. Both want her to "get an education" and "do better." Yet both constantly express fears that education will "harm her morals." She is told by her father to be proud of her Italian background, but her consciousness of being Italian is limited to the fact that her last name ends in a vowel. Although she loves her parents and believes they love her, she has no insight into their thoughts, feelings or values. She is confused by the conflicting signals given to her by them: "Get an education, but don't change"; "go out into the larger world but don't become part of it"; "grow, but remain within the image of the 'houseplant' Sicilian girl"; "go to church, although we are lacking in religious enthusiasm." In short, maintain that difficult balance of conflicts which is the second-generation's life-style.

Her dilemma becomes more widespread as unprecedented numbers of Italian-Americans enter colleges today. The nineteen-seventies are bound to see a sharp increase in the number of Italian-American college graduates. (In New York City only 5 per cent of native-born Italian-Americans graduated from college during the nineteen-sixties. One reason the figure is bound to jump during this decade is that large numbers of Italian-Americans, along with other lower middle-class whites, have taken advantage of the open-admissions policy of the City University of New York.) Moreover, my impression is that Italian-American college students are more inclined to move into areas of study other than the "utilitarian" ones preferred by their predecessors, which for boys were chiefly engineering, music for entertainment, and occasionally pharmacology or medicine, and for girls, were teaching and nursing

When the third-generation person leaves school and his parents' home, he finds himself in a peculiar situation. A member of one of the largest minority groups in the country, he feels isolated, with no affiliation with or affinity for other Italian-Americans. This young person often wants and needs to go beyond the minimum security his parents sought in the world; in a word, he is more ambitious. But he has not been given family or cultural guidance upon which this ambition can be defined and pursued. Ironically, this descendant of immigrants despised by the old WASP establishment embodies one of the latter's cherished myths. He sees himself as purely American, a blank slate

upon which his individual experiences in American culture will inscribe what is to be his personality and his destiny

But it is a myth that is untenable psychologically and sociologically. Although he usually is diligent and highly responsible, the other elements needed for a powerful personality are paralyzed by his pervasive identity crisis. His ability for sustained action with autonomy, initiative, self-confidence and assertiveness is undermined by his yearning for ego integrity. In addition, the third generation's view of itself as a group of atomistic individuals leaves them politically unorganized, isolated, diffident and thus powerless in a society of power blocs.

THE dilemma of the young Italian-American is a lonely, quiet crisis, so it has escaped public attention. But it is a major ethnic group crisis. As it grows, it will be more readily recognized as such, and not merely as the personal problem of individuals. If this is to be realized sooner rather than later, then these young people must learn whence they came and why they are as they are. A "page of history" will expose the logic of their problems and thus make them potentially solvable. *How* they will solve them is unpredictable.

They may opt for one of the several models that have served other ethnic groups. For example, they may choose to cultivate their Italian culture, pursue personal careers and fuse the two into an energetic and confident relationship—which has been characteristic of the Jewish-Americans. They may also turn toward the church, revive it and build upon its power base a political organization and morale, as Irish-Americans did. Or, they may feel it necessary to form strictly nationalistic power blocs, as some black-Americans are doing. On the other hand, they may forge their own models of individual and group identity out of an imaginative use of their unique inheritance.

Peter L. and Brigitte Berger, in an intriguing article called "The Blueing of America," write, "As the newly greened sons of the affluent deny the power of work, blue-collar class youth quietly prepare to assume power within the technocracy." No group of young people surpasses Italian-American youth in its sense of the power of work, which although derived from the old notion of labor for the family, is now, in the third generation, independent of it. It now remains for young Italian-Americans to root their sense of work in broader identity—and to take their proportionate share of what America has to offer. The dream denied their immigrant grandparents, sacrificed by their culture-conflicted parents, can be realized by them. It can be achieved by adding consciousness, knowledge and imagination to the legacy of courage, work and fortitude inherited from past generations. ■

April 30, 1972

Baltic Americans Are Saddened by Pact

By **JAMES T. WOOTEN**
Special to The New York Times

CHICAGO, Aug. 1—Standing near the busy cash register in his bustling grocery here today, his English nearly smothered by his accent, Jurgis Janusaitis had reached 1949 in the story of his life.

"So then, I'm coming to America with eight dollars in pocket," he remembered.

"But, ho boy, look at him now!" a friend interrupted. "Ho boy, now Jurgis almost Lithuanian millionaire."

The 62-year-old merchant was not amused.

"Yes, yes, I have done good," he solemnly conceded. "I have much, it is true. I have house, I have store, I have cars, I have money—but I do not have my country, my nice little country."

In a variety of ways and in dozens of United States cities, those same sentiments were expressed today by thousands of Baltic immigrants who believed that President Ford's signature on a multination agreement in Helsinki formalized forever the incorporation of their homelands—Latvia, Estonia and Lithuania—in the Soviet Union.

Like Mr. Janusaitis, many were saddened, but others were moved to protest, and here and in such places as Detroit, Cleveland, Dayton, Washington and New York there were marches, vigils and angry public statements against the pact.

"They sign anyway," Mr. Janusaitis sighed, shrugging his shoulders and lifting his palms—and behind him, the constant ringing of his cash register seemed to symbolize that life for Baltic Americans was moving inexorably on.

Up to now, it has gone fairly well. Census estimates place their number at close to a million, the bulk of them Lithuanians, and their various ethnic organizations are proud to point to an almost complete absence of poverty wherever they have chosen to live.

Among this country's well-known Lithuanians are Charles Bronson, the actor, and Johnny Unitas and Dick Butkus, former professional football stars.

Accustomed to hard work and unimpeded by color or exotic creed, they have found assimilation relatively painless and capitalism remarkably comfortable, especially on this city's South Side, where the largest concentration of Lithuanians in the United States resides.

In the general area of Marquette Park, at least 25,000 live in immaculately kept houses on narrow, tree-lined streets that all seem to lead to a Roman Catholic church or a Lithuanian bar.

It is middle, middle class, with new and late-model cars parked at the curbs, color television sets in the living rooms and pictures on the walls of children gone off to college or the suburbs.

The whole area has been officially designated by the Chicago City Council as "Lithuanian Square," and although about 75,000 of their countrymen or their descendants live elsewhere in the city, the "square" is still the refuge, the old citadel to which many of them frequently return.

In addition to the churches and the bars, there are Lithuanian schools, Lithuanian hospitals, Lithuanian sanitariums, Lithuanian bakeries, Lithuanian delicatessens and, like Parama, Mr. Janusaitis's market, Lithuanian groceries.

There, the shelves are stocked not only with the staples common to most American stores but with the stuff of Lithuanian life as well: stiprus lifuviski krienai (a potent horse rdish), saltiena (a strong sausage), Andrulis (a zesty tea), midu-olis, a honey-liqueur long a Lithuanian tradition and vodka by the gallons, usually Skaidrioji.

"In spite of the assimilation, there is even a surprising degree of ethnic curiosity among our young people," Josephine J. Dauzvardis, who has the title of Lithuanian consul general in Chicago, said in her office today.

Consequently, the native opera company, the cultural museum, a journalists' association, the daily newspaper, the chamber of commerce, the medical association and a variety of other Lithuanian groups and institutions receive energetic support from the community.

On Saturday, there are numerous high school and grammar schools teaching the language, the folklore, the music and the customs of the old country and on Sundays, the Lithuanian boy scouts and girl scouts continue the study of their ancestors' native land.

The Lithuanian Foundation, created little more than a year ago, has already raised more than $1-million, the proceeds of which are spent on cultural activities ranging from a Lithuanian debutantes' ball to the celebration of independence day on Feb. 16 and of St. Casimir's Day on March 4.

Mrs. Dauzvardis, whose husband was consul general here until his death in 1971, is one of several dozen people throughout the Western world still considered in the U.S. to be serving in the foreign service of the Republic of Lithuanian, a state that, for all practical purposes, has not existed since the Russians invaded Lithuania in June, 1940, and Lithuania became a Soviet repubic. More than 34,000 Lithuanians, most of them young people, were sent into Siberian exile.

Until then, most immigration to this country had been from the ranks of the poor and less educated—those responding to the promise of a better economic life.

After World War II, however, many doctors, lawyers, teachers, clergy, engineers and other professionals began making their way out of the country—and working in America as janitors, pressers and the like.

"But they persevered," Irene Blinstrubas, the office manager for the Lithuanian American Council, said today. "Having lost so much in their leaving, they worked incredibly hard to regain it here."

Some of them have not regained everything, some of them have, some of them are still trying and some of them have simply settled for the life of the middle class.

"You see many of them now moving away from Marquette Park out to the suburbs," said the Rev. Joseph Prunskis, a Roman Catholic priest who came here in 1940, "but you always know the Lithuanians out there by the white birch trees they've planted in their front yards."

August 2, 1975

Hamtramck Strives to Retain Its Polish Character

By WILLIAM K. STEVENS

Special to The New York Times

HAMTRAMCK, Mich., Oct. 5—They once called it "the Wild West of the Middle West," this 2.09-square-mile island of Slavic zest, an autonomous city surrounded by Detroit, a tight little village where for more than half a century the Polish-Americans have loved, hated, worked hard in their yards and on the assembly line, and lived life with gusto.

Starting in 1910 they came from the farms of Poland, and sometimes from the Ukraine, to work at "Dodge Main," the Chrysler Corporation's huge assembly plant that still dominates the southern end of town. By the late nineteen-twenties, they had so filled Hamtramck that it was said to be the most densely populated city in the country.

The boom days are long gone. The second and third generations have largely moved to the suburbs, and Hamtramck has only half as many inhabitants (26,000) as at its peak about 1930. This has brought the city to a crucial stage in its history, and therein lies one chapter in the story of social change that is overtaking industrial America.

Hamtramck's Polish flavor remains. At Jaworski's market, wonderfully wrinkled grandmothers in babushkas still shop for herring, fresh okra, parsley, dill and leek; for black radishes, barley and kasha (buckwheat); for peppers and sorrel and cabbage; for kielbasa and kiszka, the pungent Polish sausages; for the seeds with which to plant the backyard gardens that are the heirs to the fields of old Poland.

A Run on Leeches

Until a few weeks ago, Jezewski's pharmacy sold leeches, the old-world remedy for a black eye. Every Saturday morning, after the Friday night bar fights, there would be a run on the leeches.

Hamtramck has long been seen from outside as a volatile, rough-and-ready, sometimes troubled town—a vice den in the twenties and thirties, when two of its mayors were sent to prison on anticorruption charges and once, it is said, the City Council president pulled a gun on the Mayor in a council meeting. Once the city had two police chiefs, each backed by a rival faction sharing the chief's desk as friends. In 1971, the city government went bankrupt, or nearly so, depending on whose analysis one accepts.

Detroiters have at times reacted gleefully to Hamtramck's troubles—sometimes, it is said here, so that attention would be diverted from Detroit's difficulties. It is unfair, Hamtramckans say. What happened here happened in some form everywhere in the country, they assert.

Now there is a new setback for Hamtramck, and it is resented by the fiercely proud community.

The United States Court of Appeals for the Sixth Circuit upheld last week a lower court decision finding Hamtramck guilty of "Negro removal" in its urban renewal policies.

The courts held that the city had demolished sub-standard housing in areas inhabited primarily by blacks, but were not providing new housing within Hamtramck for displaced residents. As a result, the court said the black popoulation of the city had fallen from 14 per cent to 8.5 per cent. This, it was held, had been the city's intention and it was ordered that the city provide housing for a portion of those displaced.

'We're Decent People'

The citizens and their acting Mayor, William V. Kozerski, object to charges of racism.

"It hurts," he says, "because we're not what they claim we are. We're decent people."

Hamtramck's record in race relatons, in fact, has in some ways been outstanding compared with many other communities.

While one could hardly contend that prejudice and discrimination are unknown here, blacks have lived here as long as the Poles, the public schools have been integrated for decades, and opinion surveys have shown blacks to be happy here. Flaxen-haired Slavic and black youths can be seen intermingling on the street.

Ricardo Stovall, a 22-year-old black auto worker who has lived here all his life, said the other day that he had grown up side by side with Polish friends, that they went to parties together, and that it had always been like that. Hamtramck, he said, has been "one big, happy family."

Other blacks say the same, as do many whites, even those who speak in the most virulent terms about "Detroit blacks." One reason may be the smallness of the town, its old-world village quality, its sense of community.

Community identity focuses on the Roman Catholic churches, the numerous Polish-American clubs and its sports teams. Some years ago, Hamtramck won the Little League World Series. Its teams, thoroughly integrated, remain strong. Jane Bartkowicz Krot, the professional tennis player, is a local heroine.

The shops, banks and utility companies that line Joseph Campau Street, the main thoroughfare, are within walking distance of the immaculately kept one-and-two-family homes on postage-stamp plots, typically with wrought-iron railings and posts that fill the town.

"You feel better here," said 79-year-old John Wozniak, a retired auto worker who immigrated from Poland in 1913. "I know every stone in this sidewalk. I could go to Florida, but everything would be strange. You feel like home here."

Most Hamtramck residents are older people. This poses major questions as to the city's future and offers a clue as to what the urban renewal decision means to that future.

The people wonder whether Hamtramck's Polish character can be preserved after Mr. Wozniak's generation is gone. They worry that it will die from lack of tending by the second and third-generation Poles who now are thoroughly Americanized, as their parents and grandparents wanted them to be, and live in the suburbs.

The city fathers say they want fervently to lure back the young people, who often return to the old neighborhood on weekends to visit and shop. City officials had planned to use the cleared urban-renewal land for, among other things, middle-income homes for returnees from the suburbs.

The courts, however, have ruled that this cannot be done at the expense of those already living there. Some observers, however, express hope that the good condition and low price of housing ($50 a month for a decent apartment is common and a $14,000 house is considered expensive) will bring back many young people who are being priced out of housing markets elsewhere by inflation. There is some evidence that this is already happening.

Some see in Hamtramck a potential as a regional center for the preservation and perpetuation of Polish culture. The Detroit Workingman's Co-Op, a Polish-Ukrainian restaurant that once was a haven for union organizers in the auto industry, might be one element of such a center.

It draws from throughout the metropolitan area customers who like its low-priced specialties—golombkis (stuffed cabbage), nalesniki (crepes suzette filled with apple, strawberry, cheese or prune) and pierogi (dumplings).

Culture Center Seen

Another element, should it get off the ground, would be a block-long center for Polish theater, dance and other cultural activities that is being planned by a group of young Polish-Americans who are attempting to keep alive "true" Polish culture, not just the peasant aspects of it that came to America with the immigrants.

Copernicus, they say, is more important than kielbasa and beer. It is largely because of this group that Polish history, literature and language is undergoing a revival in Hamtramck's parochial schools.

Some even see the city as a setting in which effective integration of races and racial viewpoints can be nurtured.

The situation is in flux. Yugoslavians and Albanians—the latter-day immigrants — are moving here in increasing numbers. Establishing their coffee houses as community centers, where men gather to play cards and drink thick Turkish coffee that might keep one awake for days, they are starting a new ethnic nucleus.

How it will all turn out is anyone's guess. Some think that, in the end, Hamtramck will become all black, as most of the surrounding areas of Detroit have. The older Poles of Hamtramck talk about that prospect with the utmost sense of balance.

"What's the difference?" asks John Wozniak. "America, America for everybody."

October 6, 1974

Ethnic Study Portrays Polish-American Workers

By JOHN NOBLE WILFORD
Special to The New York Times

BOSTON, Feb. 23—A study of a working-class Polish-American community in Detroit, reported at a scientific meeting here yesterday, has produced a portrait of the hard work, sacrifice and low self-esteem of blue-collar workers among one of the nation's largest ethnic groups.

Dr. Paul Wrobel, an anthropologist at the Merrill-Palmer Institute in Detroit, described the results of his research in a paper entitled, "Polish-American Men: As Workers, as Husbands and as Fathers." He spoke to the American Association for the Advancement of Science, which ends its week-long annual meeting tomorrow.

To conduct his study, Dr. Wrobel, a third-generation Polish-American himself and the son of a foundry worker, lived and taught school for 16 months in the Detroit community. He observed 20 families in depth and interviewed several hundred other men and women.

Nearly all of them were second-generation Polish-Americans who belonged to the same Roman Catholic parish, had a median income of $11,413, had a median education of 10 years of schooling and worked in factories. He declined to identify the community.

"Generally speaking," Dr. Wrobel reported, "the men in this community view themselves as unintelligent factory workers unworthy of respect and incapable of accomplishing anything worthwhile except supporting a family through hard work and the ability to sacrifice."

Their dissatisfaction with themselves as blue-collar workers, Dr. Wrobel noted, is "evident from their lack of interest in discussing their jobs."

"I just work in a factory," one man said to Dr. Wrobel. "Nothing special. Same old thing day after day now for 20 years. But it pays the bills. So I can't complain."

The women in the community viewed a man's attitude toward work as the most important factor in choosing a future husband.

One woman said, "You don't have to know that much about a man, just make up your mind that you can put up with anything."

Speaking of the difficulty of getting along with their husbands "on a day-to-day level," another woman said:

"But I make it. By God I make it. I go to church, I pray. I go to communion regularly. And it all helps, I guess. After all, we are in this for life."

Goals for Children

As fathers, the major goal of the Polish-American men interviewed seemed to be to raise children whose lives would be significantly different — and presumably better—from their own.

One man told Dr. Wrobel: "You're asking me how I would feel if my sons followed in my footsteps? Are you kidding? That's the last thing I would want to happen. They're gonna stay in school and study so they can get into college and get a good job—like working in an office, teaching school. Yeah, my kids are going to wear suits and ties to work. And they're not gonna come home all smelly and dirty like me."

Dr. Wrobel observed that sacrifices for the children sometimes led to a sense of betrayal. If the children failed to achieve a better life, the anthropologist said, the parents felt that all their work had gone for nought. If their children were successful, the parents sometimes resented the social and economic gap that it placed between them and their children.

"Our son thinks he's too good for us now that he's gone to college, got himself a high-paying job, and bought a big home in the suburbs," a Polish-American mother complained.

In the conclusion to his report, Dr. Wrobel said:

"While there's nothing wrong with hard work, there is something very wrong with a society that says who you are is based on what you do for a living; there is something very wrong with a society which uses the color of a man's collar as a measure of his intelligence. In America a man is considered unintelligent if he operates a drill press in a factory. And he is considered stolid and dull if he is a Polish-American. The men in the community I studied are fully aware of what society says about them. It is tragic that so many believe it."

Dr. Wrobel's study was the basis for his Ph.D. dissertation in anthropology at Catholic University in Washington. The Merrill-Palmer Institute is a private organization specializing in social and family research.

February 24, 1976

Amid Bounty, Longing

By Leo Hamalian

My father, like most Armenian survivors of the Turkish genocide, was a man who never wanted to leave home. Until he was forced to flee, he loved the place where he had been born and brought up. It was a milieu alien to the American mentality, and as a result my father never really adapted to the customs of this country. As I look back upon his memory, more in sadness than in the anger I used to feel flaring so often in his presence, I think that his life was about the damage done to the human spirit by exile.

From the time that he set foot in the New World in 1911, an early victim of the Turkish pogroms against Christians, to the day of his lonely death in 1939, neither the chimera of the American Dream nor the bounty of material rewards could numb the pain of a refugee who found himself uprooted in a strange land where he was forced to flourish or founder. He did all those things that transformed other transplanted Armenians into lovers of this land.

Yet I remember him as a ghost who gestures, talks, but utters no words, not even the smallest murmur, of that interior grief that I now realize he had held within like a stone for 25 years.

He must have left recognizing the grim shadow that the future threw before him. Of a large family of prosperous peasants in the Lake Van area, only he and his sister got out of Turkey before the Turks got them, he to America, she to Egypt, where she lived out her days as a stateless person. But not even a futile reality is easily replaced nor the wounds of seperation quickly healed.

He tried to be a good American as he understood the idea. He became a photoengraver, took his family to picnics in Hudson Park, argued politics while he played backgammon with his cronies, attended church on 34th street, and perhaps hoped that the Big Dream would materialize, as it did for so many other Armenians. Instead, I suspect, it only emphasized his sense of loss. His emotional attachment to the place that had treated him like dirt was so massive, so monumental that he was almost blind to the bounty he had reaped in his new homeland.

Why did he resist resurrection when other Armenians were rising out of the ashes of the Turkish tragedy? I am not sure I know, but I think that the stone of sorrow in his guts may simply have stayed stone. Nothing softened it, and this stonified sorrow showed itself in excessive sternness with his children. The more American we became, the more infuriated he became. We couldn't tell whether his anger was directed against America or against us. We felt that we had somehow misbehaved by becoming what we had to become in face of the heavy claims made upon our malleable natures.

I think my father believed that he could regain, magically, some part of his past, even alleviate the pain of his exile if he could keep his children Armenian. Thus, he would triumph over the Turk, who had sought to destroy his Armenian identity. So we spoke only Armenian at home, ate only Armenian food, and saw mainly Armenian friends. In those days, the nativist elements used the public schools to disparage the cultural origins of foreigners; I must confess that I was an innocent but willing collabo-

rator. I had no notion that my childish gestures of rebellion might have been torture to my father.

Now I think I know what was eating like acid at my father. Did he deserve the bounty and safety that the New World offered for the earning? Were those signs of success in reality the fruits of his failure as a man? Should he have stayed behind with his parents? Should he have left his sister? Should he have had the courage to confront his enemies, no matter what the cost to him?

I think my father felt guilty that he had escaped the fate of his family. Though he knew that he had avoided terror and even death, in one part of himself he became persuaded that he had betrayed his family by not sharing their destiny, that he had—this will sound irrational to all those but the survivors of concentration camps—survived at their expense.

Thus far his insight took him, but no further. The act of sorting out and comprehending these ambivalent feelings proved too much for this uneducated though intelligent immigrant. And indeed why should he have been proud that he had had to run away, even to save his life? This frame of mind was made doubly difficult to endure by obtuse neighbors and America-firsters. He was in America. He was safe. He was prospering. His children had opportunities. What more did he want? Let the dead bury the dead. But my stubborn father could not bring himself to congratulate himself for what he considered to be an act of betrayal.

Fortunately, our society no longer puts pressure on immigrants to forget their former associations, or to deny anything dear left behind. We deplore the bitterness of a destiny that displaces people from their homes, that uproots and deracinates, that creates a league of dislocated persons. Such people are no longer debarred from the ranks of "good and true" Americans by virtue of their tragic sense of life. We can be thankful that we have developed this dimension of spiritual tolerance. I prize it and my father, were he alive, would have prized it.

Leo Hamalian is professor of English at the City College of New York.

December 1, 1976

ETHNIC POLITICS

How Long Will Protestants Endure?

There is a steady and insensible change going on in this State in the seat of political power, which involves most important consequences, and which our readers ought to carefully weigh. The population of this City and the surrounding counties, owing to immigration and the prolific power of a laboring class, is increasing at an enormous rate. The interior and agricultural counties are growing in a much less rapid degree. It is true that in the last decade our Metropolis has greatly fallen off in growth, compared to the previous; but this has arisen mainly from the fact that the middle classes are transferring themselves to the adjacent counties. The increase of Westchester and Kings is still immense, as compared with that of St. Lawrence and Oneida. It is also the lowest laboring classes which increase the most rapidly, as is the experience everywhere in the civilized world. Our political power follows population, and the result is that the governing power of this portion of the State, and in consequence the whole State is fast centering itself in the ranks of the lowest and most ignorant class of the whole community—the Irish Catholic laborers and tenement-house population of New-York and its vicinity, led by shrewd native demagogues. Each year gives this class a greater numerical value. They work together as a compact battalion under able and audacious leaders. They control in the City administration enormous sums of money. Where they are deficient in votes, they can create them. The timid or the ambitious Americans who have belonged to the same party organization, have not nerve or principle enough to separate themselves from these useful associates, whom socially they despise.

Thus it happens that this accumulation of ignorant voters in one corner of the State controls more and more every year the interior counties. Were it left to itself, it could do little, as even the plundering of the City treasury would soon be checked by the honest yeomanry of the rural districts. But this mass of voters here is in affiliation with a large party in the country, the majority of whom are directly opposed to them in all their ideas and habits. Party links, and the hopes of emolument and office, bind the two opposing wings together, and the Democratic Party of New-York State is simply the tool of the Irish Catholic laborers and their demagogues in this City.

The course the latter have marked out for themselves is simple and clear. They had first to get absolute possession of the government and income of this wealthy capital. This they have done. Next, they aimed at founding the Roman Catholic Church, so that it could not be easily shaken. This they have nearly accomplished by State and City grants of land and moneys. The amounts which the various Romanist churches have received, either from the Common Council or the Legislature, during the past few years, would be incredible, were we not so hardened to such appropriations.

Their next blow was aimed at the free schools, in carrying through the appropriation for "sectarian schools." Though but partially effective, this blow will be repeated either this or some succeeding year, with more complete success. Already some of our Ward schools are supplied entirely with Catholic teachers, and everything is "expurgated" from the books taught which might seem to smack too much of liberty of conscience and of thought. Many of our citizens will undoubtedly see the day—unless some great revolution breaks forth—when the Board of Education of this City will be as thoroughly Roman Catholic as Tammany is now.

The next blow—perhaps the most insulting of all—has been aimed, during this session, in the bill for "hereditary religion." By this it is assumed that every Protestant charity dealing with the *enfants perdus* of our streets, is engaged in spreading a false religion, and therefore must annually expurgate (or criminate) itself before the Legislature, under a penalty of one hundred dollars for each offense! As a correspondent suggests, the natural amendment to this act would be a provision requiring every Protestant householder to make an annual statement, under oath, that he had never invited his Roman Catholic servants to family devotions, or "otherwise interfered with their religious belief."

These incredible insults to the courage of our Protestant bodies would never be given by these demagogues if our own leaders had not shown themselves in the whole question such utter cowards. When the most eminent public men of the country are afraid to speak a word for one of the grandest events in the history of liberty, because the priests will denounce them before the ignorant Roman rabble, what can be expected but that such tools of the priests as Senator NORTON and his associates, will propose such insulting acts as this in the Senate of the New-York Legislature? If our Protestant bodies do not arise and show some manhood, they will deserve to be thus trampled on and insulted by the delegates of the Catholic masses in these counties. And they may be certain that the treatment they have thus far received from the Tammany Ring is mild and considerate to what is in store for them.

February 2, 1871

THE CHINAMEN ORGANIZING.

WONG CHIN FOO AROUSING THEM TO A SENSE OF THEIR DUTY AS CITIZENS.

A novel meeting was held last evening at No. 32 Pell-street, when the naturalized Chinamen of New-York and vicinity met and organized themselves into a political association. There were present about 50 Chinamen, half a dozen American sight seers, and two Malayan Chinamen. Li Quong, who is President of the Chinese Cigarmakers' Union, was made temporary Chairman of the meeting, and Wong Chin Foo Secretary. After a few preliminary remarks by the Chairman in reference to their first efforts in America to take an active part in politics and the necessity of unity in all matters concerning their dignity here as citizens, the Secret ry, Wong, was introduced amid cries of "Hear! hear!" He spoke substantially as follows:

FELLOW COUNTRYMEN: I congratulate you heartily on this sudden movement on your part to obtain representation and recognition in American politics. We are a small drop in the mighty ocean of American politics, but small and insignificant as we are, I feel certain that had we only attended to this duty as citizens of this country we might have prevented the passage of the shameful Anti-Chinese bill by a Republican Congress, led by that arch-Republican politician, Mr. Blaine. [Applause and hisses.] For you must remember that the politician who lords it over you to-day is an arrant coward, and trims his sails to every breeze that blows. When you don't vote and don't wish to vote he denounces you as a reptile; the moment you appear at the ballot box you are a man and a brother and are treated (if you consort with such people) to cigars, whiskies, and beers. Why can't we make our marks in politics as well as any of our brother races? Why can't we become good and substantial citizens like those from England, Ireland, Germany, and other European and Asiatic and even African countries? We had our diplomats, nay, philosophers and sages, when those about us were half-naked savages, and they are yet but in their infancy in point of true civilization compared with the peaceful, industrious, and conscientious sons of the Middle Kingdom. I do not understand American politics enough to give you a scientific lecture just at the present moment, but I will say this of the three great political parties now contending for power. The Republicans have been in power for more than 20 years, and they have during that time rallied into their ranks some of the ablest men in American politics. They were supposed to represent the true principles of a republican form of government, of which this country has been so pre-eminently distinguished before the eyes of all civilized nations. Under this impression countless thousands of well-meaning and respectable people were led to support them. It is needless for me to say how much in this regard the people at large have been disappointed. It was under this administration that our own race was ostracized. It was under this administration that we have been prohibited from becoming citizens of the United States, a privilege that is granted to all other races of men, even to the blackest people under the sun. It was under this administration that our fellow-countrymen are for 10 years forbidden upon a free and Republican shore. These are but a few instances of their bad faith to those who supported them to preserve the honor and dignity of the glorious Constitution of the United States. The Democrats are robbers and thieves, who will plunder and rob the United States Treasury worse than any of our late bank Presidents have done. When they have once got into power, then look out. Every citizen will have to go about then armed to the teeth for self-protection. New York City is an example of what they will do. As to the third and last party—the Greenbackers—well! They are too young yet! They have no character to amount to anything. Ben Butler will ride them to the devil if they don't get there themselves.

Ah Chong, of No. 34 Pell-street, asked as to the platform the association would support, to which the Chairman answered: "That will be decided upon at our next meeting, some time in August."

July 30, 1884

A SUMMONS FROM ROME

DR. M'GLYNN TO ANSWER FOR POLITICAL ACTIVITY.

HIS COURSE DURING THE RECENT CAMPAIGN DISAPPROVED BY HIS CHURCH SUPERIORS.

A statement published that the Rev. Edward McGlynn, D. D., is "to be deposed from his priestly office in the Catholic Church," on account of his advocacy of the Henry George movement, was, to say the least, extremely premature. No one has the right to say that Dr. McGlynn is to be "deposed," for the sufficient reason that he has not yet been found guilty of anything which warrants deposition. He must be tried by the proper tribunal at Rome before he can be convicted.

A reporter of THE TIMES called at the Archbishop's palace yesterday and obtained the facts in the case. It seems that as far back as last September Archbishop Corrigan received a letter from the Propaganda at Rome complaining of the course which Dr. McGlynn was pursuing. Another letter came, and at last a letter arrived which the Holy Father himself had ordered to be written. In it was the statement that the doctrines Dr. McGlynn was espousing were contrary to the teachings of the Catholic Church.

In consequence of these letters, and particularly of the last, the Archbishop communicated with Dr. McGlynn and Henry George. He read one or more of the letters to the latter and all of them to the former. At length he prohibited Dr. McGlynn from taking any further part in Mr. George's canvass for the Mayoralty. Nevertheless, Dr. McGlynn persisted, and at the great George mass meeting in Chickering Hall on the evening of Oct. 1 he spoke for an hour in his most eloquent and impressive manner, eulogizing Mr. George, calling him the greatest man in this country, and saying that he was fit not only for the office of Mayor, but for President. During the rest of the campaign Dr. McGlynn continued to support Mr. George, and on election day he made the rounds of the polling places, sitting in an open barouche with Mr. George, Mr. Powderly, and the Rev. J. H. Kramer.

It was owing more than anything else to the letters which the Archbishop received from Rome in regard to Dr. McGlynn and Dr. McGlynn's actions thereupon that the Archbishop introduced in his pastoral letter his now celebrated paragraphs upon the right to hold private property.

Nothing further was done in the premises until last Friday, when the Archbishop received a cable dispatch from Cardinal Simeoni, the Prefect of the Propaganda, summoning Dr. McGlynn to Rome and ordering him to set out at once. With a heavy heart the Archbishop performed his duty and gave the summons and order to Dr. McGlynn.

It is not known at the palace what Dr. McGlynn will do about the matter, nor was any information to be obtained yesterday of Dr. McGlynn himself. A reporter, who called upon him at the pastoral residence of St. Stephen's Church, was admitted after a great deal of hesitancy, and after some delay was informed that Dr. McGlynn would not see him, but would answer any question which the reporter would put in writing and send to him. The reporter wrote: "What truth, if any, is there in the rumor that you are to be deposed?"

The servant came back with the answer: "The doctor has no information to give on the subject."

"Will you ask Dr. McGlynn if he means that he has no information to give me, or that he possesses no information on the subject?"

The servant delivered the message and came back with this answer: "The doctor has no information to give on the subject, because he possesses no such information."

After learning at the Archbishop's palace that Dr. McGlynn had been fully informed by the Archbishop last Friday that he had been summoned to Rome, the reporter called again at Dr. McGlynn's residence. He was unable to see him or to communicate with him in any way.

Said a prominent priest yesterday: "Dr. McGlynn will have to change his opinions and show sorrow for his actions or take the consequences. He has repeatedly disobeyed the orders of his ecclesiastical superiors, and they have borne and forborne with him because he is personally liked, is a man of ability, and, if rightly guided, is a force for good. Dr. McGlynn went into active political work with his eyes open. He knew that the recorded opinion of the heads of the church was against it. As early as 1855 Archbishop Hughes, in an open letter to the *Freeman's Journal*, used this plain language:

"When I took charge of this diocese I prescribed for its numerous clergy, as a rule of conduct, to abstain wholly from interference in politics. I did not deny them the right to vote as other citizens, merely in consequence of their being clergymen—a right, I believe, they have seldom, if at all, exercised. I myself have not exercised it. I have ever considered that the most appropriate position for a clergyman, whether Catholic or Protestant, to occupy in the midst of political struggles is one of absolute neutrality, and complete abstinence from all partisanship. There are few congregations in which the members are not divided in their political opinions, and a Catholic clergyman who would take sides on such an occasion would be sure to impair the usefulness of his own ministry."

The news makes a great impression on Catholics and Protestants alike. Dr. McGlynn has been a prominent figure in many important public movements, and no Catholic clergyman is so well known to the people of other faiths. Several of his acquaintances have an impression that he will not obey the summons to Rome, but will leave the Catholic Church and become one of the leaders in the labor movement. It is a week ago that he received the summons and he has not gone yet, nor has he made any sign. If he should go to Rome, as the United States is still regarded as a mission by the Catholic Church, Dr. McGlynn will be required to appear before the Sacred Congregation of the Propaganda and answer charges of unpriestly conduct. Should he not satisfy the court of his innocence he will have to undergo such penalty as may be imposed. It is thought by some that he is delaying his departure for the reason that he has communicated to the Propaganda in writing, or has, by cable, asked for and obtained an extension of time.

December 10, 1886

CITIZENS' UNION AFTER ALIEN VOTES

Opens an Italian Naturalization Bureau, Breaking Tammany's Monopoly.

IT WILL ASK NO FAVORS

But It Is Not Expected That the New Citizens It Aids Will Turn to the Tiger—3,000 of Them.

The Citizens' Union is making a new attempt to beat Tammany at its own game by opening an Italian naturalization bureau, and before Aug. 10 an effort will be made to round up at its offices, 252 Fourth Avenue, as many as possible of the possible Italian voters, of which there are said to be about 3,000.

Heretofore Tammany has been looking after these prospective voters, taking them around to the Naturalization Bureau in the Federal Building and expecting favors at election time in return. It is the Citizens' Union's intention to put an end to this system, and for that purpose the new bureau has been opened. The Union will, however, ask no favors of the Italians whom it aids to become citizens.

William Schieffelin, Jr., the Chairman of the Union, has turned over the work to the Italian Workers' Committee and placed Dr. Albert Pecorini in charge. He is a scholar of considerable experience in social work both here and in Europe, who has done considerable investigating and research for both his own country and the United States. He obtained his doctor's degree from Columbia and speaks several languages besides Italian. Therefore the work of the bureau will not be confined to Italians alone, but will include every other race that can offer candidates for citizenship.

"It is called an Italian Naturalization Bureau," said Robert S. Binkerd, Secretary of the Union, yesterday, "because, as a matter of fact, the proportion of possible citizens is greater among the Italians than among the other races who come to this country in large numbers. The Jews and the Irish become citizens almost as soon as they can by law, but the Italians do not unless they are urged, and there must be about 3,000 of them now who can be made into good citizens and voters.

"The Tammany district leaders have been in the habit in the past of rounding up these votes shortly before election time and expecting them to vote the Tammany ticket. For our part we ask no favors of these new citizens, and consider them under no obligation to us.

"The Italian Committee of the Union has records which will make it possible to bring many of these people to us at once, but it will probably not be possible to make a thorough canvass of the possible voters before Aug. 10, after which it will be too late to have any effect on the Fall election."

As fast as the agents of the committee working in the various Settlements get lists of their men they will marshal them at the appointed places and conduct them to the Federal Building. It is said that, although the list of possible Italian voters is getting longer every day, yet unless the men are directed in every move what to do, they do not become voters, for they neither know where to go nor how to comply with the law.

The Citizens' Union believes that aliens who honestly desire to become citizens appreciate heartily all that has been done for them, and will be in a mood sympathetic with the aims and purposes of the organization. At all events, it is the opinion that Tammany will not henceforward enjoy a monopoly of turning out new citizens.

July 24, 1909

DR. LIPSKY FIGURES ON THE JEWISH VOTE

Shows by Analysis That This Vote Favored Straus Generally in Last Election.

EAST SIDERS ARE UNITED

Inclined to Support a Jew Against a Gentile on Even Grounds—Doubt in Sulzer's Case.

The question whether or not there is among the Jews of this city a "Jewish vote" that can be depended upon for political purposes is ingeniously analyzed in a current issue of The American Hebrew by Dr. Abram Lipsky in connection with the sudden and interesting outburst of popularity of William Sulzer on the east side among people with whom as a class he seemed to feel he had earned gratitude. Dr. Lipsky answers the question in the affirmative, but with modifications.

Dr. Lipsky uses the so-called "Cohen method" of analysis, which consists of counting the frequency of a selected group of Jewish names (in this case Cohens) in the registration lists and then multiplying it by a coefficient derived from a similar examination of a catalogue of some 30,000 purely Jewish names. It was found that there were in New York city 113,000 registered Jewish voters out of a total registration of 617,809. They were distributed among the boroughs as follows: 68,000 in Manhattan, 12,000 in the Bronx, 30,000 in Brooklyn, 2,000 in Queens, and 300 in Richmond.

Tracing the distribution of Jewish voters among the Assembly districts, Dr. Lipsky came upon the striking, although not altogether unexpected phenomenon that the districts with odd numbers showed comparatively few and the districts with even numbers comparatively many Jewish voters as evidenced by the percentage of Cohens.

"It should be remembered that the odd numbered districts are on the West Side and the even numbered on the East Side," explains the writer.

The following is the table showing this percentage of Cohens in the various districts:

MANHATTAN.

Assembly District.	Percentage of Cohens.
1	.5
2	5.9
3	2.6
4	6.5
5	.4
6	7.1
7	.5
8	12.4
9	.5
10	4.5
11	.4
12	1.8
13	.3
14	.7
15	1.6
16	1.3
17	1.8
18	1.8
19	2.0
20	2.5
21	2.1
22	2.0
23	2.2
24	2.7
25	.8
26	8.3
27	.6
28	3.5
29	3.6
30	1.7
31	7.4

BRONX.

Assembly District.	Percentage of Cohens.
30	.9
32	2.2
33	2.0
34	2.5
35	1.0

BROOKLYN.

Assembly District.	Percentage of Cohens.
5	1.2
19	1.3
20	.5
6	4.0
21	5.3
7	6.6
22	7.0
23	4.5

Turning from the Jewish voters to the so-called "Jewish vote," Dr. Lipsky says:

"The election of 1912 was in the nature of a crucial experiment to test the existence of the Jewish vote. On the Progressive ticket the candidate for President ran on the strength, mainly of his personal popularity, and with him the candidate for Governor was nominated, in the opinion of most men, to draw the Jewish vote, if it existed. We ask now, was there a Jewish vote for Mr. Straus? A comparison of the election returns with the table presented above gives a definite answer to the question."

In a table Dr. Lipsky gives the percentage of the total vote of each assembly district received by Mr. Roosevelt and by Mr. Straus, respectively. All the assembly districts of Manhattan and the Bronx are given, and two most significant groups of districts in Brooklyn. He observes:

"In the first eleven districts Roosevelt's percentage is invariably greater in the odd districts, while Straus's is greater in the even, this alternation corresponding perfectly with the excess and deficiency of Jewish voters.

"At the Twelfth district the sharp line dividing East Side Jews from West Side Gentiles ceases. But wherever the figures in the first table indicate a sharp rise in the number of Jewish voters in the district, the second table shows a corresponding excess of Straus's percentage over Roosevelt's. In the Twenty-sixth, for example, the number of Cohens, &c., jumps to 8.3 per cent., and we find that Straus's per cent. of the gubernatorial vote was 58.2 per cent., while Roosevelt's percentage of the Presidential vote was 42.7 per cent., a difference of 15.5 per cent. In the Thirty-first there is a similar correspondence, the per cent. of Cohens, &c., being 7.4 and Straus's margin over Roosevelt 16.4 per cent. In the Thirtieth the percentage of Cohens being small, Roosevelt runs ahead of Straus; in the Twenty-eighth and Twenty-ninth, the percentage of Cohens being fairly high, Straus has a good lead over Roosevelt.

"The same correspodence is found in Brooklyn. Two groups of contiguous districts are given in our tables including districts that show a markedly high percentage of Cohens. In the Fifth, the percentage of Cohens is low and Roosevelt runs ahead of Straus. In the Sixth the percentage of Cohens is high and Straus soars above Roosevelt. In the Seventh there are few Jews and Roosevelt is again ahead. The same correspondence may be observed in the group of districts from the Nineteenth to the Twenty-third, the Twenty-first and Twenty-third being especially noteworthy. The fact that here and there, as, for instance, in the Seventeenth district, Manhattan, a low percentage of Jews is accompanied by an excess of votes for Straus need not invalidate the conclusion drawn from the striking coincidences that we have observed.

November 3, 1913

LUSTGARTEN ASSAILS G. O. P.

Loyal American League Head Charges Catering to Hyphen Vote.

William Lustgarten, President of the Loyal American League, issued a statement last night accusing the managers of the Republican Party of catering to the so-called hyphen vote. The attack made by the President against the small, mischievous element among the foreign-born Americans, he said, must not be construed as an attack upon the genius, the institutions, and traditions of the people who have come to America from other lands. He said:

The unanimity with which foreign language newspapers and professional foreign-born politicians are supporting Mr. Hughes is possible of but one interpretation—that is, the managers of the Republican Party have given assurances to the hired agents of European feudalism that their nefarious and traitorous conduct will be tolerated under a Republican National Administration.

The subsidized newspapers printed in foreign languages and the conscienceless and catering politicians who court the hyphenated vote have misled many honest foreign-born Americans whose hyphen was not at all prominent before the war began. That America's adopted children are as loyal to her as her native sons cannot and must not be questioned; nevertheless, there has been a small element among the foreign-born, who have carried on in these United States the propaganda of their foreign sympathies, not as Americans participating in open public debate, but secretly in alliance with the agents of foreign monarchies. No one can doubt that this element has through subsidized organizations been guilty of acts designed and executed to embarrass the President in the troubled hours of this European war.

No mere pronouncement on "undiluted Americanism" can purge Mr. Hughes and the Republican Party of the charge that they are catering to these royalists parading under the cloak of American citizenship. He must repudiate the support of these foreign language newspapers, the fake truth societies and leagues, and the representatives of autocratic feudalism.

Mr. Lustgarten added that the President was entitled to and deserving of the support of all loyal-minded Americans regardless of their origin, particularly in all matters of foreign policy, "as it is incumbent upon foreign citizens to support the President as the mouthpiece of the American people."

June 20, 1916

APPEAL TO JEWS RESENTED.

Leading Men Protest Against Dragging Religion Into Politics.

Recent efforts by political supporters of President Wilson to line up the Jewish vote for his re-election have aroused twenty-six prominent Jews of the city to prepare and issue a protest against such mixing of religion or race and politics. The protest, which was made public yesterday, reads:

We the undersigned earnestly protest against drawing religion into politics.

Wide publicity has recently been given to a direct appeal calling upon the "Jews of America" to form a "Ten Thousand Club" and to contribute $1 each to a fund in aid of the campaign for the re-election of President Wilson. The appeal purports to recite in detail various official acts declared to have been favorable to the Jewish people. The sponsors of this appeal seem to have had some misgivings as to propriety of their course, as they later attempted to explain that the address "was prepared for publication in the Yiddish press and for the guidance of their readers."

Within the past few days a letter has been circulated which indicates a continuance of the effort to disseminate campaign literature containing a like appeal to the Jewish voters.

It is not our purpose to discuss the candidates or the principles and achievements of any political party. We differ in our political affiliations; but we are agreed in condemning any appeal for votes whether to Jews or to the members of any other race or creed, as such. We regard such methods as an insult to the intelligence of the voters who are sought to be influenced by them, and as tending to degrade them politically.

We desire to emphasize the fact that the American Jews regard their citizenship as a sacred possession and resent as a reflection upon their manhood the intimation that they can be influenced in the exercise of the right of suffrage by any considerations which do not apply equally to all of their fellow-citizens.

Leo Arnstein.	Louis Marshall.
Julius Ballin.	Eugene Meyer, Jr.
George Blumenthal.	Leopold Plaut.
Joseph H. Cohen.	E. R. A. Seligman.
W. N. Cohen.	I. N. Seligman.
H. L. Einstein.	Louis Stern.
Henry Goldman.	Oscar S. Straus.
D. Guggenheim.	Benjamin Tuska.
L. J. Horowitz.	Israel Unterberg.
Lee Kohns.	William I. Walter.
Arthur Lehman.	Felix M. Warburg.
David Leventritt.	Charles Wimpfheimer.
Adolph Lewisohn.	Henry F. Wolff.

It was understood from several of the signers of this protest that a substantial number of Wilson advocates were among them, one or two estimates placing the number as high as half of those protesting.

October 16, 1916

POLITICAL EFFECT OF THE DYER BILL

Delay in Enacting Anti-Lynching Law Diverted Thousands of Negro Votes.

By ERNEST HARVIER.

An important contributing cause to the sensationally large majority given John F. Hylan for Mayor last November was the defection of the negro voters from the Republican Party in what is known as "New York's Black Belt," in and around 125th and 145th Streets, Madison and Eighth Avenues. There was a like defection in the colored districts of Brooklyn and Queens. To the number of many thousands and for the first time they voted the Democratic ticket. This was not a sporadic outcome of independence, but a milestone in a political fight which began two years ago and may affect decisively the elections of this year.

Negro suffrage became general in the United States in 1870. Between 1870 and 1920, a period of fifty years, the colored voters of the North voted solidly, stolidly and, to some extent, blindly for the candidates of the Republican Party, or rather against the candidates named by the Democrats. In many Northern or Border States, notably Ohio, Indiana, New Jersey, Maryland, Kentucky, New York, Illinois, Delaware and West Virginia, their loyal and undeviating support assured Republican success in many doubtful elections.

They secured few appointments; their influence on Republican policies was small and in the Southern States they were practically disfranchised. The recurrence of lynchings in Southern States, accompanied by many hideous barbarities and, in some cases, by flagrant injustice, led negro leaders to formulate their one and first demand upon the party which they had so long supported. This was the adoption of a law which would make the taking of human life by mob violence a Federal offense whenever the local authorities were supine or accessory.

The Republicans at the National Convention in Chicago which nominated Mr. Harding agreed to this demand and adopted in their platform a plank which the colored leaders accepted as a pledge to be carried out in good faith, if a Republican Congress were chosen in November.

The new Administration went into office on March 4, 1921, but for nearly a year the pledge, if it was a pledge, was ignored. Not until Jan. 26, 1922, did the House of Representatives, in response to insistent demand, pass the Dyer Anti-Lynching bill. Congressman Dyer, the author of the bill, is a St. Louis Republican who represents a constituency in which negro voters are numerous. The action of the Republicans in passing the bill was regarded by their negro followers as belated and the delay over its adoption as an evidence of bad faith. Hence their defection last year from the Republicans in New York City and in many other cities throughout the country and this year in Chicago, Kansas City and Philadelphia.

The Dyer Bill.

Rigid conciseness and clarity do not always mark bills introduced into Congress. The Dyer bill is no exception. What it provides is this:

The phrase "mob or riotous assemblage" used in the act means an "assemblage composed of three or more persons acting in concert for the purpose of depriving any person of his life without authority of law, as a punishment for, or to prevent, the commission of some actual or supposed public offense." Such is the language of the statute.

A mob of three persons would not be very formidable, and lynchings in the South do not usually occur to "prevent the commission of some actual or supposed public offense." They follow, rather, some offense—or there would be no reason for lynching.

However, this is what is provided: "That if any State or county fails, neglects or refuses to secure and maintain protection to the life of any person within its jurisdiction against a mob or riotous assemblage, such State or county shall by reason of such failure, neglect or refusal be deemed to have denied to such person the equal protection of the laws."

Any State or municipal officer charged with the duty, or who possesses the power or authority as such officer, to protect the life of any person that may be put to death by any mob or riotous assemblage, or who has any such person in his charge as a prisoner, who fails, neglects, refuses to make all reasonable efforts to prevent such person from being put to death, or any State or municipal officer charged with the duty of apprehending or prosecuting any person participating in such mob or riotous assemblage who fails, neglects or refuses to make reasonable effort to perform his duty in apprehending all persons so participating, "shall be guilty of a felony and upon conviction shall be punished by imprisonment not exceeding five years or by a fine not exceeding $5,000 or by both."

United States courts are given jurisdiction to try such cases after a failure of the local authorities to apprehend the guilty parties within thirty days of the lynching. Further, the Dyer law provides that any county in which a person is put to death by a mob shall forfeit $10,000, which sum may be recovered by an action in the name of the United States for the use of the family, if any, of the person so put to death; if he had no family, then to his dependent parents, if any; otherwise for the use of the United States. Such action is to be prosecuted by the District Attorney of the United States of the district where the lynching occurred.

If such forfeiture is not paid upon recovery of a judgment, the Court has jurisdiction to enforce payment by execution upon any property of the county, or may compel the levy and collection of a tax.

In the event that the person put to death shall have been transported from one county to another during the time intervening between his capture and putting to death, the county in which he is seized and the county in which he is put to death shall jointly and severally be liable to pay the forfeiture. In construing the law, the District of Columbia is deemed a county, as are each of the parishes of Louisiana.

Whether the Dyer bill would mitigate the evil of lynching by mobs either in the South or in other parts of the country is a matter of controversy, but that the colored voters of the country regard the failure of a Republican Congress to pass it as a breach of faith and a repudiation of the pledge given in the Republican National platform of 1920 there can be no doubt.

From Cold Harbor to San Juan Hill and since colored soldiers in the uniform of the United States have acquitted themselves creditably and with courage. Many of their leaders have come to look upon the action of the Republican Party as two-faced and recalling in this the lines of Kipling's:

"O, it's Tommy this, an' Tommy that, an' 'Tommy, go away,'
But it's 'Thank you, Mister Atkins,' when the band begins to play."

July 9, 1922

ALIEN CONSCIOUSNESS

Older Americans Helping to Extend It From Generation to Generation by Their Attitude Toward New-Comers

To the Editor of The New York Times:

From time to time the so-called "alien bloc" has been attacked in the public press as menace to American institutions, and few people have endeavored to justify the existence of such an institution. While no attempt will be made here to prove that the "alien bloc" is indispensable or desirable, there are some fundamental reasons for its existence. * * *

Aliens are here and expect to remain and can remain legally. What are we going to do to make good citizens out of them? What are we going to do to Americanize them? Surely we cannot do either by oppressing them in one way or another, either materially or sentimentally, because oppression in one way or another has made necessary the organization of "alien bloc" in the past and will make their organization all the more imperative in the future. Oppression and prejudice have made imperative the entrance into politics of the various purely racial organizations, like the Hibernians, Sons of Italy, the Alliance Francaise and other organizations of Germans, Poles, Slovaks, Greeks, Jews, &c.

The continual criticism and petty persecution of foreigners has made the foreign element all the more conscious of the fact that they are not Americans. And this consciousness, while it originally extended only to the first generation, has become so strong that intelligent professional and business men of foreign origin—even of the second or third generation in this country—are obliged to go to the defense of their race because they bear a foreign name and feel the criticism, although they may not ever have been to the land of their ancestors. There is a much graver menace to American institutions in the latter class, because they are intelligent and know how to make advantageous use of any political influence which they might be able to gain. Then there have been instances of their being susceptible to suggestions on the part of foreign diplomats or propagandists. While no attempt will be made here to justify this method surely there is a reason for its use.

All of these deplorable racial discriminations should cease. They are contrary to the principle which made possible the foundation of this Republic. Today, on one hand, we are talking about the League of Nations being the only means of insuring universal peace, while, on the other hand, we are continually spoiling the good work by encouraging prejudice at home against all foreigners instead of helping them to become good citizens. If foreign organizations are able to keep a strong hold on their nationals it is because we are in many ways negligent in not calling aliens' attention to the advantage of becoming at least 75 per cent. Americans. We cannot expect the Italian American of any station in life to forget that he comes from the land of Caesar, of Dante, of Garibaldi, but we can help him to become proud of the fact that he is or can become, a citizen of the land of Washington, Jefferson, Lincoln and Roosevelt. LUIGI CRISCUOLO.

New York City, July 8, 1924.

July 13, 1924

FOREIGN-BORN VOTERS NEED CARE

Tact Is Necessary in Instructing These New but Earnest Americans

To the Editor of The New York Times:

The job of supervising the millions of voters represented in the racial groups of the country is not one usually sought for. There are few who understand the sentiments, emotions and temperaments of these fine upstanding Americans, many of whom have just received their final papers, and all too frequently a deplorable error is made by placing someone in charge who has not the slightest idea of the complexities of the work. Improper leadership in either of the major political parties brings untold griefs and annoyances to the national Chairman and the complaints and criticisms are manifold.

To properly estimate the political feelings of a man or woman born in lands other than our own, more than a passing knowledge of the political, economic, ethnologic, social and psychologic conditions in the mother countries is necessary.

Director's Work Cut Out.

The director of this work must classify his groups throughout the various States into their racial, political and religious affiliations. He must organize workers and speakers among the foreign born and see that each faction has proper representation. It must be considered that an ethnologic entity may be divided into several distinct language and subracial groups together with branches of religious sects.

In 1920 there were 6,274,193 naturalized American voters. It is safe to assume that approximately 2,000,000 have taken out their final papers since then. And it is well to remember that most of these are within the confines of New York City.

The greatest percentage of naturalized citizens voting in the elections is to be found in North Dakota (preponderantly Norwegian) with 31 per cent. of the total number of voters. Minnesota, where the foreign-born population is composed largely of Norwegians, Swedes and Germans, has a naturalized vote of 27 per cent.

There are good reasons for the assumption that the number of foreign-born voters who avail themselves of their privilege to vote is larger proportionately than the native born. The foreign-born voter has had to go through certain difficulties in order to obtain citizenship, and for that reason is very likely to value his rights more highly.

Principal Campaign Issues.

The principal issues appearing to me in the campaign of interest to the foreign born is, first, the Johnson Immigration law; second, foreign debts; third, the Ku Klux Klan, still a menace in some parts of the country; fourth, radicalism and dissatisfaction among certain of these peoples that Senator La Follette usually attempts to capitalize; fifth, the wet and dry issue, and sixth, the tariff.

There are about 1,200 foreign language newspapers in the United States. Probably 100 are solvent financially with good editorial writers and well-trained reporters. Campaign material must be sent many of these publications; a work of some proportions when it is realized that all matters of a political slant must be checked several times in the English texts and again after translated.

I have merely touched the high spots of the manifestations in the work with the foreign-born voter. Sight must not be lost of the fact that he is, in almost every case, an ardent and enthusiastic American. He is a power politically, and a few hundred votes cast by one of these groups in a congested election district may easily reverse the vote in that district.

One reads of crime being committed by our foreign born and one knows of certain radicals among them, but the final analysis will show that over 87½ per cent. of those holding citizenship papers are true to the ideals of this country and have a proper respect for its laws. The in-telligentsia among certain native-born men and women cause the ulcers of discontent to make headway, but don't blame the foreign born for too much until you know him.

Let us recall that eight men of foreign birth signed the Federal Constitution. They were Elbridge Gerry, who was born in England; Francis Lewis, Wales; Robert Morris, England; James Smith, Ireland; Mathew Thornton, Ireland; George Taylor, Ireland; James Wilson, Scotland, and John Witherspoon, Scotland.

W. A. SCULLY.

Washington, D. C., May 28, 1928.

June 3, 1928

END OF BLOC VOTE SEEN IN WISCONSIN

Lack of Support for La Follette Is Held to Explode Theory of German-American Front

By JAMES RESTON
Special to THE NEW YORK TIMES.

MILWAUKEE, Aug. 16—Woodrow Wilson, referring to the foreign-born groups in this country, observed once that they were called "hyphenated-Americans" because only half of them had come across. His view seemed to be that many of them had left their hearts in the old country and that they tended to vote here for what they believed to be the best interests of the country from which they came.

This old political idea dies hard. Even "Young Bob" La Follette seems to have followed it to the end. A study of this week's election returns in Wisconsin, however, does not sustain the myth that there is a solid isolationist German-American vote. Only one-half of the first generation Germans may have come across, but the second, third and fourth generations are voting their pocketbooks, not their origins.

German-Born Counted On

There are roughly 90,000 German-born voters in this State. They and their families are located mainly in Milwaukee, Dodge, Jefferson, Waukesha, Washington and Outagamie Counties. In Milwaukee most of them are in the Fifth Congressional District, the north side of the city. A generation ago, "Old Bob" La Follette appealing to the isolationist and radical tendencies of these first generation Germans, could count on their solid support and this support was the basis of his political dynasty.

In this week's election, however, every one of these counties went against "Young Bob" LaFollette in his campaign for renomination to the Senate. Forgetting Outgamie County, because that was the home county of his young soldier-opponent, Joseph R. McCarthy, the other strong German counties not only gave McCarthy a margin in the Republican primary, but what is equally important, many of these Germans followed the labor vote into the Democratic primary.

In the Fifth district of Milwaukee County, where the Germans are strongest for example, every predominantly German ward gave a slight majority vote to McCarthy and piled up enough votes between McCarthy and Howard J. McMurray in the Democratic primary to wipe out the early lead LaFollette had taken in the rural areas of the State.

The same tendency is even more marked among the Poles, who are the second largest foreign-born group in the Milwaukee region. The Poles voted against La Follette all right, but they did so mainly by voting for the unopposed Democratic Senatorial candidate, Mr. McMurray. He ran a full-page advertisement in the newspapers here calling La Follette an isolationist and a conservative who had "betrayed the progressives, betrayed President Roosevelt" and gone over to the conservative party.

The Polish wards, more than the German wards, voted in the Democratic primary, and when they did vote in the Republican primary they voted against "Young Bob." Moreover, in the Fourth District Democratic primary, Congressman Thaddeus F. B. Wasielewski, who had been an outspoken supporter of the conservative London Polish Government, was defeated by a 27-year-old former Army corporal, Edmund V. Bobrowicz. This defeat is undoubtedly attributable more to the charge that he was anti-labor than that he was appealing to the interests of Poland, but the outcome nevertheless supports the thesis that the so-called "foreign groups" are following the political trend of their economic group rather than the interests of the old fatherland.

In view of this trend, the victory of the Republican candidate for the Senate does not seem so assured here as it did yesterday in Madison.

The issue in Wisconsin this fall, for the first time in two generations, will not be whether the Senatorial candidates are isolationist or internationalist—both are running on an internationalist platform. The issue this time will be the same as it will be in most other areas of the country, the so-called "conservative-liberal" issue on domestic affairs.

August 17, 1946

Topics of The Times

Negro Voters' Shift?

One estimate of the Negro vote in the North speaks of it as being potentially decisive in a "reasonably close election" in eighteen States with no less than 281 electoral votes. This figure seems high in itself and is to be further weighed in light of the fact that for more than twenty-five years we have had no close Presidential elections.

Nevertheless, the basic fact remains that there were times when national elections were close and the Negro vote may have been a decisive factor in such contests within a single generation after the Civil War. Today it is a question of serious moment whether the Negro voters who went over in large numbers from their traditional Republican allegiance to the Franklin Roosevelt camp are drifting back or not. They can make themselves felt if we ever get back to the close elections of the Grover Cleveland era.

Few Weigh More

This factor of the Northern Negro vote somehow manages to get itself overlooked in most discussions of Negro disfranchisement in the deep South. We strongly suspect that in Ilya Ehrenburg's report on America to his Soviet countrymen it appears that the Negro is disfranchised everywhere in the United States.

Against this stark picture one might recall the bold thesis suggested by one ingenious native writer. He advanced the paradox that as a result of anti-Negro discrimination in the South the political role of the Negro in the country as a whole has been greatly enhanced. He argues that if the Negro were allowed to vote freely all through the South, along with freedom from all other forms of discrimination, then it is conceivable that the electoral vote of a Southern State here or there might be decided by the Negro vote; conceivable, though not very likely.

But a heavy Negro migration to the North, as the result largely of a denial of political and civic rights at home, has brought it about that three million Negroes in the North actually exercise a more decisive influence in our national elections than if thirteen or fourteen million Negro people in the South had been allowed to vote freely.

October 26, 1946

How Big Is the Bloc Vote?

By SEYMOUR MARTIN LIPSET

TO speak and write of bloc voting on a religious, racial or ethnic basis has been deprecated alike by politicians and by the leaders of the groups involved. But why, in this day and age, some people object to the idea that bloc voting exists while others, recognizing that there are such blocs, consider their behavior undemocratic or un-American, is difficult to understand.

SEYMOUR MARTIN LIPSET, director of the Institute of International Studies at the University of California, wrote "The First New Nation: The United States in Historical and Comparative Perspective."

Although bloc voting may seem to violate the democratic *mythos* about rational man, most politicians have always known what all great observers of American life, from Tocqueville, Bryce and Ostrogorski to Brogan, have noted—that Americans differ in their voting behavior not only according to whether they are poorer or richer, rural or urban, Northerners or Southerners but also because they are of different religious, racial and ethnic backgrounds.

What is meant by the term "bloc"? Basically, a bloc is a group of people who have a "consciousness of kind," who react to others having the same identifying characteristics with the feeling, "He is one of mine." This does not mean that a bloc is monolithic, that group members vote only for others in their group. What it does imply, however, is that membership in a bloc, particularly of a religious or ethnic character, is a major influence on political behavior.

THUS we have the deep study of bloc trends this year. Public opinion polls suggest that over 90 per cent of Jews and Negroes will support President Johnson. It also appears from such surveys that about three-quarters of the Catholic vote will go to the national Democratic ticket, despite the nomination of a Catholic as the Republican Vice-Presidential candidate.

Why should Catholics be much more likely to vote for Johnson and Humphrey than for Goldwater and Miller? The answer lies not only in the candidates themselves, but also in the long-term association between Catholics and Democrats.

From Jefferson's day to Lyndon Johnson's, the principal vehicle for Catholic expression and recognition in politics has been the Democratic party. In 1960, at the end of the Eisenhower era but before the Kennedy nomination, over 80 per cent of the Catholic members of Congress were Democrats. One-quarter of all Democrats in Congress were Catholics, as contrasted with but 7 per cent among Republicans. If we compare, as political scientist Peter Odegard did, appointments to the Federal judiciary under Republican Presidents Harding, Coolidge and Hoover with those made by Democrats Roosevelt and Truman, we find that one of every four judges appointed by the Democrats was a Catholic, as contrasted with but one of every 25 designated by the Republicans.

CLEARLY, the Democratic party has played a considerable role in helping to raise the status of the Irish and other Catholic groups. From the early 19th century onward, most immigrant Catholic ethnic groups found the local Democratic parties to be of considerable assistance in obtaining jobs and power. The Federalists, Whigs and Republicans were the parties of the social, economic and religious Establishment, composed largely of persons of Anglo-Saxon background. With rare exceptions, these parties showed little interest or ability in helping the lowly immigrants who settled in the cities. And as these parties of the privileged saw the immigrants providing massive infusions of new Democratic votes, they often openly campaigned for restrictions on the political rights of the foreign-born, thus welding them and their offspring even closer to the Democrats.

The predominant identification of Catholics with one party in the 19th century did not matter too much, since they were not that large a segment of the electorate. Today, however, Catholics constitute over one-quarter of the eligible voters, and their continuing commitment to the Democratic party makes the job of electing a Republican extremely difficult.

But Catholic attachment to the Democratic party, long after the end of massive immigration and the decline of Catholic slum ghettos, cannot be explained by historic loyalties alone. Continuity would seem, in part, to be related to the congruence between the positions fostered by the Democratic party in recent decades and certain values endemic in Catholic teachings.

AS early as 1917, Catholic bishops in the United States called for a more heavily graduated income tax, social security, unemployment insurance and minimum-wage legislation. Whether such pronouncements affect the laity is difficult to demonstrate, but voting studies have shown that Catholics are not only more Democratic, they are also much more likely to favor trade unions and welfare measures than are socially and economically comparable Protestants. Even Republican Catholics have been found to be, on the average, more favorable to welfare-state or New Deal measures than their Protestant co-partisans.

JEWS constitute but 3 to 4 per cent of the eligible electorate, but since they are concentrated in a few large states, such as New York, Illinois, Pennsylvania and California, they are in a position to influence considerably the politics of key states with many electoral votes. And their strong liberal propensities give to the Democrats a sizable number of needed middle-class votes and financial contributions.

Much has been written seeking to account for the reformist propensities of Jews, even among many of the well-to-do. To a considerable extent this behavior would seem to reflect the fact that Jews, even when wealthy, have suffered social discrimination at the hands of non-Jewish privileged classes. Although anti-Semitic attitudes and behavior are now at a low ebb, one finds that many of the élite city clubs, as well as many suburban country clubs, do not admit Jews to membership. These clubs tend to be overwhelmingly Republican in their membership. Such discrimination places even wealthy Jews among the out-groups, and helps them to continue to identify with others who are low in the stratification system.

THERE is also the fact that Jewish values, as developed under the pressure of living as a persecuted minority for two millenia, have emphasized the responsibility of the privileged to help others. The pattern of philanthropic giving, which shows much higher contributions by Jews to both Jewish and non-Jewish causes than is usually found among non-Jews in the same economic bracket, reflects the great strength of community responsibility felt by Jews. And such values seem also to stimulate support for the party which favors the welfare state, which presses for community help to the underprivileged.

To these general factors must be added the specific conditions in 1964 that have made the Jews more predisposed to vote Democratic than ever before: the civil-rights issue and the visible presence of extreme rightists in the Goldwater movement.

For most Jews, the civil-rights cause is as much theirs as it is the Negroes'. He who hates the Negro, the Catholic or the foreign-born is probably also anti-Semitic. The Nazi experience is obviously still the most crucial historical event in the minds of Jews; extreme rightists, whether anti-Semitic or not, are seen as threats to the democratic process and consequently a danger to Jews. Hence the phenomenon in this election year that seemingly most Jewish conservatives and Republicans will vote for Johnson and Humphrey.

NEGROES, who make up about 11 per cent of the population, represent somewhat less than that percentage of the electorate. Their overwhelming backing for the Democratic ticket in 1964 is clearly linked to party differences on the civil-rights issue, with most Negro Republican leaders the country over supporting Johnson and Humphrey because of Senator Goldwater's attitude on the Civil Rights Bill and the conscious effort to recruit segregationist support in the South.

Although in the recent decades we have become accustomed to counting Negroes as a bulwark of the Democratic party, this voting pattern is relatively new. From the Civil War to 1932, studies of the vote of Negro residential districts show that they voted Republican for the most part. The G.O.P. was the party of Lincoln, the party that freed the slaves, and insofar as racial issues entered into elections before the New Deal, the party that supported the Negro's cause.

The shift from the Republican to the Democratic camp occurred largely between 1932 and 1936. Federally financed relief and public-works programs established during the first Roosevelt Administration gave money and employment to a large section of the Northern Negro population, and gave such help in nondiscriminatory fashion. Labor unions, closely identified with the Democrats, organized large numbers of Negro workers for the first time in history in the late thirties. And gradually the Democratic party took over the role of advocate of the Negro's social and economic rights.

WHAT about the voting patterns of the white Protestant majority—now a declining majority in relation to the total electorate? Clearly, if they had ever voted as a predominant bloc, the party they backed would have had a permanent majority. But various factors have divided the Protestants. In addition to class and ethnic cleavages, the most visible factor, of course, has been sectional; the South, until recently, was one of the most Protestant and Democratic parts of the country.

Class and sectional factors apart, white Anglo-Saxon Protestants — the so-called WASPs—would seem to have been developing tendencies toward bloclike links to the Republicans. Various historians have documented trends in this direction in the 19th century, when the WASPs were still a majority nationally.

Thus, a study of British immigration after the Civil War shows that many British immigrants joined the Republican party because cities were often controlled by an Irish-Catholic-dominated Democratic party, and Democratic officeholders were wont to give strong support to the cause of Irish independence. More recently, students of urban New England politics, communities in which those of old-stock Anglo-Saxon Protestant background find themselves in a small minority, suggest that they have been behaving much like a self-conscious minority ethnic group, usually backing Republicans but showing a disposition, like other deeply self-conscious minorities, to break away and support Democratic candidates from their own ethnic group.

Many observers have called attention to the processes through which Protestantism, particularly in its ascetic forms, has contributed to individualism, self-reliance, feelings of personal responsibility for success and failure, and interpretation of social evils in terms of individual moral failings. Catholicism, on the other hand, has tended to stress community responsibility and does not stress as much the responsibility of the individual for the moral consequences of action.

A SURVEY of the values of Protestants and Catholics by a sociologist, Gerhard Lenski, reports that when comparing members of these two broad religious communities, holding class background and length of family immigrant history constant, Protestants are more likely to have positive feelings toward hard work, to be less active in trade unions, to expect to agree with businessmen rather than with unions on various issues, and to be critical of installment buying. Data from Gallup surveys indicate that as recently as 1959, one-third of all Protestants as compared with 9 per cent of Catholics favored a national law prohibiting the sale of liquor.

These varying attitudes of Catholics and Protestants, especially Protestants adhering to the historic sectarian groups, may incline the Protestants to oppose the welfare state and the Democrats, much as Catholic and Jewish values may press *their* adherents to accept liberal political principles.

Foreign policy issues, particularly those revolving around entry into the two World Wars and relations with the Soviet Union, have had visible effects on the voting patterns of different ethnic groups. **Harding's great victory in 1920 was in large part a result of his picking up support from normally Democratic ethnic groups who had opposed entry into World War I or resented the treatment of their ancestral mother country in the Versailles Treaty. Similarly, in the late 1930's and the 1940's, national groups were pressed in different political directions by their attitudes toward Germany, Italy, and the Soviet Union.**

PAST elections show how parties may pick up needed support from groups normally predisposed against them by fostering policies or nominating candidates who appeal to the values or interests of the given group. In New York, for example, Governor Dewey, running for re-election, was able to carry Harlem by a small majority after signing the first state fair-employment law enacted in the country. But though New York Negroes rewarded Dewey in the state election, they voted overwhelmingly against him for President in 1948.

Senator Jacob Javits of New York, a liberal Republican with a very favorable record on civil rights and welfare issues, has been able to make strong inroads into Jewish and Negro Democratic strongholds whenever he has run for election. His strength cannot be credited primarily to the fact that he himself is Jewish. Other Jews running as Republicans against non-Jewish Democrats have frequently been defeated by large majorities in Jewish districts. A major example of this occurred in the New York City mayoralty election in 1945, when Jonah Goldstein, running as a Republican, was trounced in Jewish and Catholic areas alike by William O'Dwyer.

Today, all signs point to a large segment of New York Jews, perhaps a good majority, voting for Senator Keating

against Robert Kennedy. Interviews with Jewish and other liberally disposed voters indicate that they are moved by Senator Keating's seeming liberalism in his pre- and post-convention opposition to the candidacy of Senator Goldwater. Voting blocs can be split when a party normally opposed by the bloc moves in the direction of the values of the bloc; they are no more immutable than patterns of class voting.

Whether or not one deplores bloc voting, the fact remains that it will continue to affect American politics. No party can afford to ignore such groups; yet the Goldwater Republicans would seem to be doing so.

NEGROES, Catholics and Jews constitute about 40 per cent of the American population today. Collectively, they will probably give Lyndon Johnson a larger majority than any other candidate has ever achieved among them. At the moment it would appear that of this 40 per cent, over three-quarters — constituting more than 30 per cent of the entire electorate—will back Johnson and Humphrey. Less than one-quarter of these three blocs, or less than 10 per cent of the total voting population, favor Goldwater and Miller.

If this estimate is correct, then the Democrats need but one out of every three Protestant voters to have 50 per cent of the voters. Any votes above that figure will add to their majority. In fact, all available evidence suggests that the Democratic Presidential ticket may secure *a majority of white Protestant votes*, coming disproportionately from trade-unionists, intellectuals, government employes and white-collar workers in large industry, in that order.

THERE is strong evidence from opinion surveys, both academic and non-academic alike, that the Democrats were gradually gaining in Protestant backing long before the Goldwater nomination. At the same time, they have not been losing their religious, ethnic and racial bloc support. The welfare program of the Democrats has an appeal to many Protestants, particularly workers and farmers. And the growth of education and general cosmopolitanism has greatly increased the number of white Protestants who join the minorities in seeing the Negro's claim for equality as a moral one which must be acknowledged by government action.

THE growing weakness of the Republicans among Protestants was concealed in 1960 by the considerable antipathy which many of them had to backing a Catholic president. The real meaning of the 1960 vote may be seen by comparing the votes of religious groups in the Congressional elections of 1958 and the Presidential contest of 1960.

In the former year, without Kennedy on the ticket or in the White House, Democratic Congressional candidates outside the South took over 70 per cent of the Catholic vote, as much or more among Jews and Negroes, and apparently slightly less than half of the white Protestant electorate. Two years later, with a Catholic heading up the ticket, over 80 per cent of the Catholics voted Democratic, but only about one-third of the Protestants. Seemingly, approximately one out of seven Protestants who favored Democrats in the 1958 Congressional elections voted against Kennedy in 1960. But in 1964, as I have already suggested, Lyndon Johnson should equal or surpass Kennedy's vote among the Catholic-Jewish-Negro segment of the population while also gaining a majority among white Protestants.

All this suggests, of course, the same result as the evidence adduced so far by pollsters and political analysts about 1964 suggests—an overwhelming majority for Johnson and Humphrey. If the analysis is correct, it also points up the long-term problem faced by the Republican party if it is not to go the way of the Federalists and the Whigs: Either it must modify its program and image so as to appeal to groups which now decisively reject it, or else it must reconcile itself to a permanent minority status.

October 25, 1964

Republicans Seek To Win Ethnic Vote

By BILL KOVACH

Special to The New York Times

CHICAGO, Oct. 20—The Republican party, after 40 years of standing wistfully by, this year is pursuing the white ethnic vote with a vigor that is making ethnic politics a major matter in many of the nation's large industrial centers.

In districts like those on the Northwest Side of Chicago, there is for the first time a sprinkling of names ending in "ski" and "wicz" on tickets traditionally dominated by Anglo-Saxon names. And in the Republican party headquarters there is a special organization to define and cater to separate units of hyphenated Americans.

"There is a feeling of having been exploited by the Democrats," explained a member of the Republican Nationalities Committee. "They appeal to the ethnic identity during election year, but the rest of the time they pretend there is no such thing as ethnicity. We think we can capitalize on the frustration this sort of politics has developed in the ethnic communities."

This drive is stirring political counterthrusts in Chicago, Pittsburgh, Cleveland and other Northern cities as the Democratic party strives to retain the dominant position it has held among ethnic voters since the days of Franklin Delano Roosevelt.

Daley Aides Comb Suburbs

Here in Chicago, for instance, Mayor Richard J. Daley has ordered precinct captains out into the suburbs to find the ethnic voters who have fled the city and try to reverse the slow Republican growth that threatens to surround and smother the Daley organization's power in Cook County.

Because of a deep-seated belief in the "melting pot" theory of America, few statistics have been compiled on ethnic Americans, and figures are hard to come by. Most authorities seem to agree that the identifiable ethnic population includes about 40 million people, concentrated mostly in the 58 major industrial cities of the Midwest and North.

Throughout the country politicians have been forced by reaction to a decade of civil rights agitation to take another look at the white urban community, and they have found that the "melting pot" has not done its job. Angry, frustrated and fearful collections of voters who had been welded to the Democratic party in the nineteen thirties are looking for new answers to new questions.

In Pittsburgh, this reawakening of ethnic identity has taken the form of a renewal of the Pan Slavic Alliance, which seeks Government-sponsored ethnic cultural and research centers and is making itself felt as a political pressure group.

Who the "Ethnics' Are

In Cleveland, an Italian-American president of the City Council is attempting to rebuild the Italian community of the city and an ethnic task force is demanding representation on policy-making boards and commissions.

"Ethnic American" is a term almost as old as the country but one that needs to be redefined each generation or so. Those so defined in the current political struggle are largely eastern and southern Europeans — Italians, Poles, Hungarians and the welter of Slavic nationalities from the Balkan states.

They are considered by most political strategists as the largest identifiable group now ripe for political exploitation.

There was a time when a politician could win an ethnic presence on Election Day by appearing at the fraternal hall on feast day to be photographed with leading citizens, by praising the value of the voters nationality, by donating to a local cause, and by maintaining a permanent representative who could "speak in their language" in the community. But the social dislocation of the nineteen-sixties ripped away the cocoon that the ethnic communities had wrapped around themselves and they now fear loss of their identity in a still-alien world.

Calling a Pole a Pole

Today the ethnic voter is waiting to see who will talk to him about law and order (without accusing him of a racism he thinks is best reflected by those who left the city) and who will cater to his ethnic pride with programs like cultural research and American history text books that call a Pole a Pole.

Curiously, politicking among ethnic Americans is a secretive thing — as though to admit it would be to deny the American dream of a melting pot.

The appearance of a spate of ethnic-Americans on the Republican ticket here, for example, is strenuously argued as "just a happenstance."

According to Peter Piotroicz, deputy chairman of the Republican state committee in Illinois and himself a candidate for a local office and a moving force in the ethnic-oriented thrust of the G.O.P. "You don't need a Serb to attract a Serb anymore — the 'skis' are not necessarily on the ticket because of their names."

Yet at the national and local level the Republicans have spent two years developing an intensive appeal to white ethnic Americans.

Volpe on Mission to Chicago

Secretary of Transportation John A. Volpe was sent on "a goodwill mission" to Chicago early this month. His sole purpose in visiting Chicago, in the words of The Italian American News, a leading ethnic newspaper, "was to have members of the sponsoring organization [including all major Italian-American organizations in the city] to meet first-hand the leading ranking U.S. official of Italian heritage."

Then there is Vice President Agnew, whose heritage and rhetoric speak to the white ethnic Americans' pride and frustrations, and Gov. Richard B. Ogilvie who regularly finds space in the ethnic press to catalogue the ethnic Americans he has appointed to posts in his administration.

Joseph Sable, Republican Mayor of Duquesne, a suburb of Pittsburgh, who is himself a leader in the Pan Slavic Alliance in that area, expects his party to achieve some success, but only some.

"In practice the community has always been Democratic," Mr. Sable says. "Little by little they are beginning to break away or see an advantage to the other side. Mr. Agnew's speeches have had some impact, but as a group I would guess they are still 75 to 80 per cent Democratic."

The Democrats, who have practiced the art since early Irish immigrations showed the way to power through patronage politics, still seem to maintain their basic hold on the ethnic vote.

"To function in the ethnic community" explains a Cleveland Democrat whose job is to keep the ethnic vote in line, "you have to understand how it works. Our most effective weapon is word of mouth. I go to a store or restaurant where the old folks hang out and I plant my message. These people spend their entire day visiting and talking. For each one I tell, 40 will know by the end of the day."

Nevertheless, professionals in the shadowy world of ethnic politics agree that these old formulas are dissolving.

October 21, 1970

Why Blacks Don't Vote

By Lisle C. Carter Jr.

ATLANTA—The Voting Rights Act of 1965 is up for renewal this year, and it is essential that it be re-enacted The law, one of the most important pieces of legislation in United States history, has contributed enormously to the transformation of the South. Not only has it enfranchised large numbers of black citizens and led to the election of hundreds of them to public office, it has also helped release the region's dynamic growth. Much more remains to be done.

The impact of the act outside the South has been marginal. Existing proposals for changes in the act would tend to dilute the effectiveness of the legislation without doing very much for the non-South. The issue in the non-South is not so much voting rights as voting outcomes—that is, what benefits are derived, or are perceived to be derived, from the exercise of voting rights?

A question often asked is, Why don't blacks outside the South vote in closer proportion to whites? There are obviously specific answers for specific situations. But generally one could add this:

The black population is younger than the white population, and young people are less likely to vote; the black population is poorer than the white population, and poor people are less likely to vote. Moreover, eligible whites have not been voting in very large proportions, and blacks who have less to get or keep should not be expected to do nearly as well.

But there is a different explanation that is much more fundamental.

The movement of blacks to the Northern cities has often been compared to the immigration of the Irish, Italians and Polish. There is probably much less to that comparison than appears to the intellectual eye. West-Indian blacks may provide a closer analogy, but native-born blacks are bound to have been shaped, in some measure, by the illusion that they were United States citizens when they arrived in Chicago, Detroit or New York.

Nevertheless, European immigrants and Southern blacks did share a lack of empathy for the abstract notions of public interest so dear to WASP reformers. This shared difference of perspective was rooted in certain cultural similarities and was reflected in the immigrants' approach to electoral politics.

The social institution that mattered to these groups was the family—the immediate family and relatives, friends, one's church and fraternal organization—one's own kind. It was on the family that black and white immigrants relied and whose welfare they were concerned about protecting.

Government as an entity functioning for the general good had little reality. Policemen, social workers, teachers, judges and other local officials had meaning as individuals able to affect the family welfare for good or bad.

Electoral politics, through the vehicle of the local political organization, in turn became an extension of the family, because that kind of politics enabled immigrants to influence the behavior of these functionaries and ultimately to join their ranks and share their power.

Although blacks have benefited to a degree from this process, it is the great difference in benefits awarded blacks and white immigrants that has undermined the confidence and the participation of blacks in electoral politics. Race accounts for those differences. So long as there were white immigrants with needs, they went to the head of the lines.

By the time demographic changes made expansion of black political power all but inevitable, the whites who had preceded them had discovered the virtues of civil service, zoning laws, and special tax districts, not as instruments of the public good but as weapons of family protection.

Accordingly, the payoff from the accession of blacks to electoral office has become meager indeed, and so has incentive for blacks to take part in the elective process.

This is not intended as a brief for the maintenance or return of the patronage system, though it is impossible not to emphasize the role that the "patronage" function of the anti-poverty program played in providing incentive and access for minorities to take part in the political system.

Nor do I want to probe old wounds between blacks and other ethnic groups. I want, rather, to make a point to those who are critical of low black voter participation, particularly to those who would count 90 per cent of that vote in their pockets.

The public interest defined as the greatest happiness for the greatest number, which has dominated our economic and political thought since the 19th century, has hardened into an increasingly impervious "majoritarianism" and thus cannot be perceived as the way to happiness by those who see themselves consistently counted out of the greatest number. Only when political leaders have the courage and vision to define the public interest as social justice in their response to the basic needs of the decent life will blacks and millions of other Americans be won to the polls.

Lisle C. Carter Jr., chancellor of the Atlanta University Center, was Assistant Secretary of Health, Education and Welfare in the Johnson Administration.

March 28, 1975

The 'Ethnics' Vote in the States That Really Count

By R. W. APPLE Jr.

Driven from their Eastern European homelands by economic privation early in the 20th century and by Communist oppression 50 years later, Czechs and Poles, Hungarians and Bohemians, Slovaks and Lithuanians came by the hundreds of thousands to America. They settled in and near the great industrial cities of the North—the northwest side of Chicago, the blue-collar enclave of Hamtramck within the city of Detroit, the "cosmo" wards of Cleveland—and in small clusters elsewhere. Set incongruously in the cornfields of Nebraska, for example, is the almost entirely Czech-American settlement of Wilmer, and that state's junior Senator, Roman L. Hruska, is of Bohemian descent.

Illinois and Ohio, and the other northern industrial states in which the Eastern Europeans are concentrated matter a great deal to President Ford this year. He must carry most of them to beat Jimmy Carter, and it is now a very live possibility that through his gaffe in the debate last week he has killed his small but realistic hope of winning those states by winning many of the Eastern European-Americans away from the Democrats.

Like most other immigrants, they turned early to the Democratic Party, whose urban leaders in return for their political allegiance supplied them with some of the necessities of life in harsh and unfamiliar circumstances. After a time the sons of the pioneers began to win high office as Democrats, among them Senator Edmund S. Muskie, Gov. Frank Lausche of Ohio and Representatives Clement Zablocki of Wisconsin and Charles Vanik of Ohio and Daniel Rostenkowski of Illinois. Their constituents were New Dealers all, passionately anti-Communist, devoutly Catholic, builders and conservators of neatly kept "ethnically pure" neighborhoods. Once wed to the Democratic Party, the ethnics expected it to be marriage for life.

Except lately. No sooner had they begun to gain a tenuous hold on prosperity than their values were threatened on every side. To these true believers, the Roman Catholic Church seemed less a rock than before in the aftermath of Vatican II, with its bold encyclicals and ecumenicism. The neighborhoods seemed less a shelter than before, in view of growing black assertiveness (and occasional disorderliness). The Democratic Party seemed less an anchor than before, as it swung from the politics of anti-Communism to the politics of accommodation. The Vietnam war, so reviled by the party in the later years, always seemed a just cause to Hamtramck, where perhaps one family in five had one relative fighting in Southeast Asia and another living in Warsaw or Cracow.

Some of them turned to Richard M. Nixon in 1968 and 1972, and some supported Republicans in local elections, too. There was Congressman Edward Derwinski, a crewcut conservative who seemed to represent the goals of the new suburban constituency outside Chicago that elected him, and there was Mayor Ralph P. Perk of Cleveland, a city where the tension between blacks and Eastern European whites ran particularly high. By the 1970's, these particular "unmeltable ethnics"—in the phrase of one of their spokesmen, Michael Novak—were the most conservative ethnic bloc except the Germans. Or at least so the public-opinion polls suggested.

There was good reason, therefore, for President Ford to think he might win the support of enough of them in November to tip some important states his way. Jimmy Carter, the Democratic nominee, could not have been more foreign to their experience, and he was less than a roaring success during the primaries in south Milwaukee and in the coal fields of Pennsylvania, with their concentrations of Eastern Europeans. Moreover, he was in trouble with the church over his stand on abortion; it was no coincidence that one of the most aggressive prolife demonstrations Mr. Carter encountered early this fall came in northeastern Pennsylvania. Finally, he had used the kind of language in a Playboy interview that no self-respecting mother in Parma, Ohio, a heavily Polish-American suburb of Cleveland, would want her children to hear.

But Gerald Ford may well have fatally damaged himself with his comments in his second debate with Mr. Carter on Wednesday night, the comments in which he insisted that Eastern Europe was not now and never would be, as long as he was President, under Soviet domination.

Andrew Greeley, a Catholic scholar with a special interest in ethnic groups, commented: "The Poles hadn't made up their minds, but they have now and there's nothing Ford can do to change it."

Already made nervous by the implication in the Helsinki accords that the United States was accepting a Soviet sphere of influence east of the Elbe, leaders of the ethnic groups exploded in anger at the President's statement and his subsequent attempts to explain himself. However impractical it may be, the Eastern-European pressure groups cling to the notion that their homelands must be liberated, and they want their government to cling to it, too. At a minimum, they want their President to deplore the situation in Poland and Czechoslovakia.

Although they are not nearly so numerous as the Irish or the Germans in America—those of Eastern European stock number perhaps 7 percent of the population, with the Polish Americans by far the most numerous—their anger could count for something on Nov. 2 because of their strategic location. They are concentrated in the very states, from New York in the East to Wisconsin in the West, in a broad arc around the Great Lakes, where Mr. Ford's strategists say he must win or perish. They could be pivotal in Illinois and Wisconsin, where the race is close. In Ohio, President Ford's managers have been telling visiting reporters for weeks that he would carry the state, which has been carried by every Republican elected President since the Depression, because Mr. Carter's unusual strength in fundamentalist rural areas downstate would be more than offset by defections to the President in the "cosmo" wards. This weekend, they were not so sure.

R. W. Apple Jr. is national political correspondent of The New York Times.

October 10, 1976

PREJUDICE AND DISCRIMINATION

HELP WANTED.

FEMALES.

WANTED—A FIRST-RATE PROTESTANT cook for small family, where a kitchenmaid is kept; English, Scotch, or Swedish woman preferred. Address, with references, A. B. C., Box 385 Times Up-town Office, 1,269 Broadway.

WANTED—AN ENGLISH PROTESTANT housemaid; one who has been trained in England; best references required. Address J., Box 386 Times Up-town Office, 1,269 Broadway.

WANTED—A THOROUGH WAITRESS; Protestant; best city references required; must be competent to fill a man's place. Address W., Box 387 Times Up-town Office, 1,269 Broadway.

WANTED—AN EXCELLENT PROTESTANT laundress; Scotch or Swedish preferred; best city references required. Address K., Box 388 Times Up-town Office, 1,269 Broadway.

WANTED—FIRST-CLASS LAUNDRESS AND assist with chamberwork; best references. Apply at 77 West 52d-st.

WANTED—GERMAN, SWEDE, OR ENGLISH girl; must be good cook and laundress. Call at 209 West 21st-st.

WANTED—A GERMAN WOMAN AS LAUNdress and chambermaid; good references required. Apply at 28 East 58th-st., before 2 o'clock.

WANTED—FIRST-CLASS SHIRT IRONER. Apply at 982 9th-av., between 62d and 63d sts.; ring first bell, left side.

February 8, 1887

CHIEF HENNESSY AVENGED

ELEVEN OF HIS ITALIAN ASSASSINS LYNCHED BY A MOB.

NEW-ORLEANS, March 14.—In every paper in the city this morning appeared the following call:

"All good citizens are invited to attend a mass meeting on Saturday, March 14, at 10 o'clock A. M., at Clay Statue, to take steps to remedy the failure of justice in the Hennessy case. Come prepared for action."

John C. Wickliffe,
B. F. Glover,
J. G. Pepper,
C. E. Rogers,
F. E. Hawes,
Raymond Hayes,
L. E. Cenas,
John M. Parker, Jr.,
Harris R. Lewis,
Septeme Villere,
Dickson Bruns,
William H. Deeves,
Richard S. Venables,
Samuel B. Merwin,
Omer Villere,
H. L. Fovrot,
T. D. Mather,
James P. Mulvey,
Emile Dupré,
W. P. Curtiss,
William M. Railey,
Lee McMillan,
C. E. Jones,
J. F. Queeny,
D. R. Calder,
Thomas Henry,
James Lea McLean,
Felix Couturie,
T. D. Wharton,
Frank B. Hayne,
J. G. Flower,
James Clarke,
Thomas H. Kelley,
H. B. Ogden,
Ulric Atkinson,
A. Baldwin, Jr.,
A. E. Blackmar,
John V. Moore,
William T. Pierson,
C. L. Stegal,
W. S. Parkerson,
Charles J. Roulett,
T. S. Barton,
C. J. Forstall,
J. Moore Wilson,
F. Henry,
Hugh W. Brown,
C. Harrison Parker,
Edgar H. Farrar,
J. C. Aby,
R. C. D. Hahse,
C. A. Walscher,
W. Gosby,
Charles M. Barnwell,
H. R. Labouisse,
Walter D. Denegre,
George Denegre,
R. H. Hornbeck,
S. P. Walmsby,
E. H. Pierson,
James D. Houston,
E. T. Leche.

The verdict of the jury in the Hennessy case had startled and angered everybody. The statements of the jury bore out the suspicion that the members had been purchased. Consequently at 10 o'clock there was a large crowd at Clay Statue on Canal Street. Mr. Parkerson appeared with a number of gentlemen "prepared for action." There was a crowd of young and old men, black and white, but mostly of the best element. Speeches were made by Messrs. Parkerson, Denegre, and Wickliffe.

Shortly before 10 o'clock Mr. Parkerson appeared at the base of the monument. The crowd quickly swarmed around him. He requested them to fall into line. At their head he marched around the monument three times, and ascending the pedestal he turned and made the following address:

PEOPLE OF NEW-ORLEANS: Once before I stood before you for public duty. I now appear before you again actuated by no desire for fame or prominence. Affairs have reached such a crisis that men living in an organized and civilized community, finding their laws fruitless and ineffective, are forced to protect themselves. When courts fail, the people must act. What protection or assurance of protection is there left us when the very head of our Police Department, our Chief of Police, is assassinated in our very midst by the Mafia Society and his assassins are again turned loose on the community? Will every man here follow me and see the murder of Hennessy avenged? Are there men enough here to set aside the verdict of that infamous jury, every one of whom is a perjurer and a scoundrel?

There is another viper in our midst and that is Dominick C. O'Malley. This community must get rid of the man who has had the audacity to enter a libel suit against one of our daily papers that boldly came out and denounced him to the public in his true colors. I now, right here, publicly, openly, and fearlessly, denounce him as a suborner and procurer of witnesses and a briber of juries. Men and citizens of New-Orleans, follow me! I will be your leader.

Mr. Parkerson was enthusiastically cheered throughout his entire speech, and at its close the cries and cheering of the multitude were deafening. Mr. W. Denegre followed and his opening words were drowned by the cries of the crowd, "We have had enough of words! Now for action!" Quiet was restored and Mr. Denegre proceeded:

"To-day is the 14th of March. On the 14th of September you assembled on this very spot for a purpose similar to that which has convoked you here this morning. When our late lamented Chief of Police, David C. Hennessy, had been so cruelly stricken down by red-handed assassins, an indignation meeting was held at Lafayette Square. It was there decided that we peacefully and quietly await the action of the law. This we have done. The law has proved a farce and mockery. It now reverts to us to take upon ourselves the right to protect ourselves. Are we to tolerate organized assassination? Not one of those jurors told the truth. While perhaps not all of the twelve accepted a bribe, some of them did. They were bribed, and bribed by whom? By that scoundrel D. C. O'Malley, than whom a more infamous monster never lived. The Committee of Fifty have already notified him to leave town without avail. More forcible action is now called for. Let every one here now follow us with the intention of doing his full duty."

Mr. J. C. Wyckliffe followed in the same denunciatory manner, saying among other things that self-preservation is the first law of nature, and that the time had now come for the citizens of New-Orleans to protect themselves. "If such action as the acquittal of these assassins is to be further tolerated, if nothing is done to forcibly portray the disapproval of the public of this infamous verdict, not one man can expect to carry his life safe in the face of the organized assassination that so powerfully exists in our midst as to openly set law and order at defiance. We met in Lafayette Square to talk. We now meet at the foot of Henry Clay's statue to act. Let us therefore act, fellow-citizens. Fall in under the leadership of W. S. Parkerson. James D. Houston will be your First Lieutenant, and I, J. C. Wickliffe, will be your Second Lieutenant."

Arms had been provided at Royal and Bienville for about fifty men, and the members of the committee who had called the assemblage went there, secured pistols and shotguns, and then the crowd marched on the Parish Prison.

The starting of the crowd had an electric effect on the city. Soon the streets were alive with people running from all directions and joining the main body, which moved sullenly down Rampart Street to the jail near Congo Square. Doors and windows were thrown open, and men, women, and children crowded on the galleries to encourage those who were taking part and to witness the scene.

When the main crowd from Canal Street reached the prison a dense throng had already collected there, all eager to take a hand in whatever might happen. When the vanguard of armed citizens reached the jail, which is many squares from Canal Street, that grim old building was surrounded on all sides.

Sheriff Villere, when he heard that a movement was on foot to take the prisoners, armed his deputies and then started on a hunt for Mayor Shakespeare. The Italian Consul and Attorney General Rogers joined in the pursuit but, his Honor does not reach his office until noon, and he was not to be found at any of his regular haunts. The Governor had not heard of the uprising and had no time to act and the police force was too small to offer much resistance to the army of avengers. Superintendent Gaster had ordered an extra detail of officers to be sent to the jail and the small crowd kept the sidewalks around the old building clear until the great multitude, swelling all the time like a mighty roaring stream, surged around the door and crowded the little band of bluecoats away.

Capt. Lem Davis was on guard at the main entrance with a scant force of deputies. They were swept away like chaff before the wind, and in an instant the little ante-room leading into the prison was jammed with eager, excited men.

Meanwhile the prisoners were stricken with terror, for they could hear distinctly the shouts of people without, madly demanding their blood. Innocent and guilty alike were frightened out of their senses, and those who were charged with crimes other than complicity in the murder of the Chief also shared in the general demoralization. Some of the braver among the representatives of the Mafia wanted to die fighting for their lives, and they pleaded for weapons with which to defend themselves, and when they could not find these they sought hiding places. The deputies, thinking to deceive the crowd by a ruse, transferred the nineteen men to the female department, and there the miserable Sicilians trembled in terror until the moment when the doors would yield to the angry throng on the outside.

Capt. Davis refused the request to open the prison, and the crowd began the work of battering in the doors. Around on Orleans Street there was a heavy wooden door, which had been closely barred in anticipation of the coming of the avenging mass. This the crowd selected as their best chance of getting in. Neighboring houses readily supplied axes and battering rams and willing hands went to work to force an opening. This did not prove a difficult task to the determined throng. Soon there was a crash, the door gave way, and in an instant armed citizens were pouring through the small opening, while a mighty shout went up from 10,000 throats. There was more resistance for the intruders, however, but it too was soon overcome with the huge billet of wood which a stout man carried. Then the turnkey was overpowered and the keys were taken from him.

By that time the excitement was intense, none the less so when a patrol wagon drove up with a detachment of policemen, who were driven away under a fire of mud and stones. When the leaders inside the prison got possession of the keys the inside gate was promptly unlocked, and the deputies in the lobby promptly got out of harm's way. The avengers pressed into the yard of the white prisoners. The door of the first cell was open and a group of trembling prisoners stood inside. They were not the men who were wanted, and the crowd very quickly, though with remarkable coolness, burst into the yard. Peering through the bars of the condemned cell was a terror-stricken face which some one mistook for Scoffedo. A volley was fired at the man and he dropped, but none of the shots struck him, and it was subsequently found that he was not one of the assassins. The inmates of the jail were ready to direct the way to where the Italians were.

"Go to the fem[illegible] department," some one yelled, and thither [illegible]e men, with their Winchesters, ran. But t[illegible] door was locked. In a moment the key was produced. Then the leader called for some one who knew the right men, and a volunteer responded and the door was thrown open. The gallery was deserted, but an old woman, speaking as fast as she could, said the men were up stairs. A party of seven or eight quickly ascended the staircase, and as they reached the landing the assassins fled down at the other end. Half a dozen followed them. Scarcely a word was spoken. It was the time for action. When the pursued and their pursuers reached the stone court yard the former darted toward the Orleans side of the gallery and crouched down beside the cells. Being unarmed they were absolutely defenseless. In fear and trembling they screamed for mercy. But the avengers were merciless, and a deadly rain of bullets poured into the crouching figures.

Gerachi, the closest man, was struck in the back of the head, and his body pitched forward and lay immovable on the stone pavement.

Romero fell to his knees, with his face in his hands, and in that position was shot to death.

Monastero and James Caruso fell together under the fire of half a dozen guns, the leaden pellets entering their bodies and heads, and the blood gushing from the wounds.

The executioners did their work well, and beneath the continuing fire Cometex and Trahinia, two of the men who had not been tried, but who were charged jointly with the other accused, fell together. Their bodies were literally riddled with buckshot, and they were dead almost before the fusillade was over.

When the group of assassins was discovered on the gallery, Macheca, Scoffedi, and old man Marchesi separated from the other six and ran up stairs. Thither half a dozen men followed them, and as the terror-stricken assassins ran into cells they were slain. Jo Macheca, who was charged with being the arch-conspirator, was a short, fat man, and was summarily dealt with. He had his back turned when a shot struck him immediately behind the ear, and his death was instantaneous. There was no blood from the wound, and when the body was found the ear was swollen so as to hide the wound, which the Coroner had great difficulty in locating.

Scoffedi, one of the most villainous of the assassins, dropped like a log when a bullet hit him in the eye. Old man Marchesi was the only one who was not killed outright. He was struck in the top of the head while he stood beside Macheca, and though he was mortally wounded, he lingered all the evening before dying.

Pollize, the crazy man, was locked up in a cell up stairs. The doors were flung open and one of the avengers, taking aim, shot him through the body. He was not killed outright and in order to satisfy the people on the outside, who were crazy to know what was going on within, he was dragged down the stairs and through the doorway by which the crowd had entered. A rope was provided and tied around his neck and the people pulled him up to the crossbars. Not satisfied that he was dead, a score of men took aim and poured a volley of shot into him, and for several hours the body was left dangling in the air.

Bagnetto was caught in the first rush up stairs and the first volley of bullets pierced his brain. He was pulled out by a number of stalwart men through the main entrance to the prison and from the limb of a tree his body was suspended, although life was already gone.

Just as soon as the bloody work was done Mr. Parkerson addressed the crowd, and asked them to disperse. This they consented to do with a ringing shout, but first they made a rush for Parkerson, and lifting him bodily, supported him on their shoulders while they marched up the street. The avengers came back in a body to the Clay statue and then departed. Immense crowds rushed from all directions to the neighborhood of the tragedy, while the streets in front of the newspaper offices were blocked with people anxious to see the latest bulletins.

W. S. Parkerson, who was the Captain of the mob, was the political leader of the Democracy of 1888, and is one of the leading lawyers of the city. James D. Houston, who was announced as First Lieutenant, is also a prominent political leader. John C. Wickliffe, the Second Lieutenant, is one of the editors of the *New Delta*. He is a Kentuckian and a West Pointer. The mob's work was done quickly and without unnecessary violence. No one was injured but the men against whom there was proof of complicity in the assassination of the late Chief of Police, and men who are known to be active agents of the Mafia. The shotguns and ropes of the mob have executed eleven men. Public sentiment condemned the men indicted for the crime; public action has put them to death. The city is unanimous in upholding the action of the mob.

Coroner Lemonnier spent the day at the prison. The first inquest was on the body of Joe Macheca, the recognized leader of the party, who was stretched out on the gallery of the row of condemned cells on the third floor. He looked perfectly natural, the fatal wound being just below and behind the left ear. He was the only one upon whom much money was found, he having over a hundred dollars in his pocket. Scaffedi, the man in the oilcloth coat, who stood at the corner of Basin and Girod and fired the last shot at the Chief, was also shot through the brain. Marichesi, the old bald-headed man, identified as the party who jumped into the street and fired two volleys into the tottering body of Hennessy, also had a bullet in the brain. He evidently grasped at one of the guns with his left hand, and several of his fingers were shot away. Although unconscious, he did not die until evening, being the only one to escape immediate death.

As soon as the approach of the mob was known, the Italians were released from the rooms where they were confined. They scattered in different directions. Macheca and those just mentioned were terror-stricken, and ran from one corridor to the other without getting out of the way. The crowd found them on the gallery of the condemned cells, and they were shot from the yard and from the gallery entrance. The statement that Macheca had a weapon and faced the mob is denied by the prison officials. Politz and Bagnetto were also caught near by and carried outside the prison, where they were hanged to the limbs of the nearest trees on the neutral ground. With a gun to his head, Bagnetto was asked who killed the Chief, and said he did not know. They were his last words. The police cut the bodies down and carried them into the Fourth Precinct Police Station.

A part of the Italians, as soon as they were told to hide, got over into the women's side of the prison. The majority got into a cell on the lower floor, from which they were driven out and shot down in a body.

Frank Romero, Rocco Geracci, Caruso, Charles Trahina, and Monasterio, the shoemaker who lived in the shanty where the assassins gathered, and Loreto Camites were laid out in a row. Most of them had been shot through the brain, and made a horrible sight as they lay weltering in blood and brains. Natali and Sunzeri, two of the men not on trial, and for whom an alibi was claimed, were caught hidden in a doghouse, but one of the leaders claimed protection for them, as they had not yet been tried, and they were turned over to the prison officers. The same course was pursued with Charlie Pietzo, an Italian grocery keeper, at whose place the Mafia are said to have met, and where the guns were gathered before the shooting. He took refuge in the wash house, from which he was pulled out. Incardona, ordered acquitted by the court, was not touched, and John Caruso slipped into a cell with prisoners for minor offences, and the latter were locked up for safe keeping. The boy Marchesi, who is said to have given the signal of the Chief's approach was caught, but the crowd refused to wreak vengeance on so young a victim. When the boy heard that his father was shot, he tried to tear out his eyes.

Two men who concealed themselves effectually and have not yet been seen are Charles Matranga, who was supposed to be a fellow-chieftain with Macheca in the murderous plot, but against whom no evidence was obtained, and Charles Patorno, a brother of the ex-Alderman, against whom there was also no evidence. The crowd did not want them, anyhow.

The jurors in the Hennessy case are also reaping a bitter harvest to-day. Walker Livandais, a clerk in the Southern Pacific Railroad, was discharged, as his fellow-clerks refused to work with him. J. M. Seligman, the foreman, was partner with his brother in the jewelry business. The brother dissolved the partnership to-day. The clubs and Exchanges of which he was a member expelled him, and he sought to leave town this afternoon. A mob captured him on the way to the station, but he was rescued by the police, and has been concealed by his friends.

The *New Delta*, published by Col. Parker and edited by Wickliffe, foreshadowed the result in a double-leaded editorial, of which the following are extracts:

"The time to call a halt has arrived. The individual resigns to organized justice his natural right to protect himself. Organized justice has failed to discharge its functions. Human law is unable to hold at bay or even to punish the midnight murderer. There is nothing left save a resort to that law of God—self-preservation.

"Citizens of New-Orleans, you are brought face to face with the question whether your city shall be ruled by orderly government or by organized assassination. You are to-day to decide whether you are to be governed by laws made by yourselves, or by the edicts of the Sicilian Mafia. Before the setting of the sun to-night, it will be determined whether you are to enjoy the security of orderly citizens of a free republic, or to carry your lives at the mercy of a band of organized assassins.

"Your Chief of Police, the executive head of law itself, has fallen before the vengeance of a band of criminals. For four months you have patiently waited for the law to act. A perjured jury has brought the law to naught, and to-day the officers who conducted the prosecution, the citizens who supported it, and the witnesses who testified for it, are living at the mercy of the men who took the life of your officer. At any moment any one of you may become the object of the Mafia's vengeance, and the target for the murderous musket of the hired assassin. Will you hold your life at the mercy of these law-proof murderers, or will you protect it by the only means this foresworn jury has left you? Go to the Clay statue this morning at 10 o'clock and answer the question for yourselves."

As soon as the purpose of the mass meeting became known to Gov. Nicholls he ordered the State militia to be called out to protect the prisoners, but before Gen. Meyer could find his Adjutant General to promulgate the order, the work had been accomplished.

The Stock Exchange held a meeting this morning and expelled from membership therein Mr Jacob Seligman, the foreman of the jury that brought in the verdict of acquittal. He was charged with having manipulated the jury so that no one should be convicted, and is said to have been interested in large wagers that not one of the accused would be condemned. The *Picayune* accumulated evidence of Seligman's unreliability, and laid it before District Attorney Luzenberg, but the latter refused to entertain it, not realizing the extent of the depravity of the defense in its methods.

O'Malley is a private detective here, the partner of Lionel Adams, ex-District Attorney, and he had charge of the "fixing" of the witnesses and jury. He has a bad criminal record, has been in jail in Cleveland, Ohio, and has been indicted on several occasions for various offenses. His special forte is subornation of perjury and bribery. He and Hennessy were deadly enemies, and it is generally supposed that he was back of Macheca in the conspiracy for the assassination. He has always been the agent of the Sicilians whenever they appeared in the Criminal Court, and has been frequently seen in close consultation with the leaders of the Mafia. He is thoroughly fearless and daring, full of insolence and bravado. When the mob meeting was assembling he walked right through it, jokingly talking to acquaintances. He heard Mr. Parkerson's denunciation of his acts, and straightway proceeded to a gun store and bought twenty-five shotgun cartridges loaded with buckshot. Since that time he has been in hiding. He will either leave town or be killed.

Joe Macheca, who was shown by the testimony in the trial to have been the head of the conspiracy, was a wealthy merchant, the founder of the house of Macheca Brothers, though not of late connected with it, and the pioneer of the steamship fruit trade with Central America, he owning the first steamer to make the venture twenty years ago, where a score of steamships are kept busy to-day. He was worth some hundreds of thousands of dollars, and was a pleasant-mannered, popular gentleman. He has always taken an active interest in Democratic politics. In the Seymour and Blair campaign of 1868 he organized and commanded a company of Sicilians 150 strong, known as the Innocents. Their uniform was a white cape, bearing a Maltese cross on the left shoulder. They wore side arms, and when they marched shot every negro that came in sight. They left a trail of a dozen dead negroes behind them every time. Gen. James B. Steedman, managing the campaign here at the time, finally forbade them making further parades, and they disbanded.

On the 14th of September, 1874, just seventeen years and a half ago, when the White League turned out and dispossessed the Kellogg Government, Macheca commanded the Italian company. One of the incidents of the battle between the White League and the metropolitan police is worthy of recollection. Gen. A. S. Badger, commanding the police, was shot in half a dozen places and fell in the street. The mob rushed for him to administer the *coup de grace*. Capt. Macheca took in the situation at a glance, threw his men around the fallen foe, drove back the howling mob, and, lifting the wounded man on a stretcher, detailed a squad of his men to escort him to the Charity Hospital. Macheca thus saved his life. Badger is now a member of the committee of fifty appointed by the Mayor to ferret out the Italian assassins.

THE STORY OF THE MURDER.

HOW CHIEF HENNESSY WAS ASSASSINATED AT HIS OWN DOORSTEP.

David C. Hennessy, Chief of the police force of New-Orleans, walked home from his office on the night of Wednesday, Oct. 15, 1890. A friend and fellow-officer accompanied him to within one block of his house, which he reached just a little before midnight. The street was then apparently deserted, but no sooner had Chief Hennessy mounted his doorstep than a volley of bullets struck him and he fell upon the sidewalk mortally wounded.

The number of shots fired simultaneously indicated that there were several assassins, and as they skulked away under cover of the darkness the dying Chief's friend, from whom he had parted but a moment before, came hurriedly upon the scene. Many citizens hearing the shots also hastened to the spot. The Chief, it was ascertained, had received three bullets in the stomach besides one just under the heart and one in the left forearm. He was removed to a hospital, where he died at 9 o'clock the following morning.

Although the murdered man did not recognize his assailants, circumstances soon pointed to a bloodthirsty gang of Sicilians as the assassins. It was learned that just before Hennessy reached his house an Italian lad ran ahead of him and gave a peculiar whistle, which was undoubtedly the signal to the concealed murderers. The dead Chief's associates on the police force recalled the fact that some years ago Hennessy had incurred the bitter hatred of a certain element of Italians by his vigorous efforts to break up a bloody vendetta of long standing and to bring the guilty men to justice. He arrested a famous bandit, Esposito, who was in hiding in New-Orleans and sent him back to Italy, where he was convicted and punished for his crimes. Threats of vengeance were anonymously sent to Chief Hennessy many times, but, disregarding them, he persisted in his warfare against the lawless portion of the large Italian community in New-Orleans. He learned many of the secrets of the Sicilian societies and "murder circles," and thereby became a marked man in the eyes of the stealthy and dreaded Mafia.

Public excitement rose to an intense pitch in New-Orleans, when the realization became general that the well-liked Chief of Police had fallen a victim to the Sicilian assassins. For years the people of New-Orleans had read and shuddered over the accounts of atrocious murders committed by the Sicilians among themselves, but never before had the secret vengeance of the Mafia been wreaked upon an American. Chief Hennessy's body was removed to the home of his mother, where it was visited by thousands of people. His murder was the only theme of conversation, and public indignation quickly reached a point where only the most earnest counsels of conservative citizens prevented a wholesale lynching of Italians. A public meeting was called by Mayor Shakespeare, and stirring resolutions were adopted denouncing the murder of Hennessy and avowing a determination on the part of the people to exterminate the Sicilian secret assassination societies. A committee of fifty citizens of the highest standing was appointed by the Mayor to thoroughly investigate the whole subject.

The New-Orleans police force set to work energetically, and within a few hours after Chief Hennessy's death half a dozen arrests had been made. Thanks to the persistent investigations that had been made by the dead chief the Police Department was in possession of a very complete history of the Sicilian vendetta in New-Orleans, and of the names of the leaders of the various factions. One of the first persons arrested was Antonio Scoffedi, who was suspected of having fired one of the fatal shots at Chief Hennessy. He was confined in the Parish Prison. A young paper carrier named Thomas Duffy obtained permission to see the prisoner, and deliberately shot him in the neck. Duffy regarded Chief Hennessy as his best friend. It was at first thought that Scoffedi would die, but the surgeons pulled him through, and he fell a victim to the lynchers yesterday. The police made fifteen arrests in two days, and all of the prisoners who stood trial were among the number.

The case which is supposed to have incited the killing of Chief Hennessy was narrated at the time as follows: The Provenzanos enjoyed a monopoly in discharging fruit vessels at New-Orleans. Matranga, a noted leader of bandits and the proprietor of a negro gambling and dance house, finding the police too severe on him concluded to change his business, and by persuasions, threats, and other methods he succeeded in ousting the Provenzanos. A deadly enmity sprang up between the two factions. One night as a gang of Matranga's men were returning from their work they were fired upon and several, including Tony Matranga, were wounded. Chief Hennessy arrested the Sicilians who were accused by the Matrangas, and they were convicted. A new trial was secured after persistent efforts, and it was to have begun a few days after the date of Hennessy's assassination. During the investigations occasioned by the application for a new trial, the Chief, it appears, became convinced that the witnesses on the Matranga side had perjured themselves, and that the principal witness for the defense had been assassinated by one of the Matranga gang. He also obtained from Sicily the record of the Matrangas, which would have been very damaging to their case. The police were assured by many Italians that it was to prevent Chief Hennessy from telling what he knew about the Matrangas that he was murdered.

Matranga was described as having been the head of the dreaded Mafia, or Stoppaghera, Society in New-Orleans. There were about 20 leaders and 300 ignorant Sicilians in the society. Members of the Provenzanos faction declared that there were Mafia societies in San Francisco, St. Louis, Chicago, and New-York. The purpose of the society is to further its own ends by any means, assassination being the favorite method. Whenever a member is told to put a certain man "out of the way" he is bound to obey under penalty of death.

THE NEWS IN NEW-YORK.

THE ITALIAN COLONY EXCITED OVER THE LYNCHING.

Italians in this city were greatly excited when the news of the lynching of their fellow-countrymen in New-Orleans was received yesterday. At 5 o'clock they began to gather at the Italian banks and places of resort to hear the latest news. In forcible language they denounced the lynching. Early in the evening arrangements were made for a mass meeting of the various Italian societies and citizens to be held to-night, when the action of the mob in New-Orleans will be protested against. The meeting will probably be held in Webster Hall.

Il Progresso Italo-Americano will publish an extra this morning in regard to the affair. Charles Barsotti, the editor of the paper, and his associates were discussing the matter at the office, 40 Duane Street, yesterday afternoon. They said that two or three of the men reported killed were naturalized citizens, but for all of them the Italian Government claimed protection. What action the Italian Government would take, if any, of course could not be said, but it was believed that an explanation would be asked of the Federal Government why persons proclaimed innocent by an American jury should be submitted to such violence, and that a thorough prosecution of those engaged in the jail breaking would be demanded. "It is horrible," added Mr. Barsotti, "that a mob should break down the gates of the jail and kill these men in any event, but the more because they had been fairly tried, and a jury made up of citizens had by their verdict decided that there was not sufficient evidence to convict them. It was well enough to call them murderers before the trial, but the jury had decided that they were not, or at least that it could not be proved against them. One was only seventeen years old.

Signor Barsotti said that a little over $500 had been raised in this city and sent down to New-Orleans with a view that the men might be ably represented at the trial and given a fair one. The money was raised by subscriptions sent in to *Il Progresso Italo-Americano*. This was the only money subscribed in New-York. A similar subscription, however, had been opened by an Italian journal in New-Orleans. It was believed by the Italian editors yesterday that the impression that pervaded Italy that America was a paradise of freedom would now receive a severe shock.

THE ACTION OF THE MOB INDORSED.

NEW-ORLEANS, March 14.—A meeting of the Cotton Exchange was called to order at 1:30 o'clock P. M. by President Chaffee, who stated that he had been called upon by a large committee of members with the request that he convene the institution in general meeting for the purpose of adopting a suitable resolution indorsing the action of the citizens of New-Orleans in the deplorable event of the morning. Mr. Chaffee said that inasmuch as all were familiar with the events, it was not necessary to dilate upon them. They knew the facts, and they knew the necessity of the situation. He then caused the following preamble and resolution to be read by the Secretary:

Whereas, The deplorable administration of criminal justice in this city, and the frightful extent to which the bribery of juries has been carried, rendered it necessary for the citizens of New-Orleans to vindicate outraged justice; be it

Resolved, That, while we deplore at all times the resort to violence, we consider the action taken by the citizens this morning to be proper and justifiable.

On motion of Mr. Lapeyre, seconded by Mr. Emmett, the preamble and resolution were unanimously adopted. Resolutions of similar purport were also adopted by the Board of Trade, (Produce Exchange,) the Sugar Exchange, and the Stock Exchange.

Mr. Pasquale Corte, the Italian Consul in this city, states that he called on the Mayor of New-Orleans and Governor of Louisiana and asked for protection for the Italian subjects among the prisoners. None was given by either official. He immediately communicated with the Italian Minister in Washington, and also the home Government at Rome. Four of the eleven were Italian subjects: Monasterio, Marchesi, Cornilez, and Traina. The others were of American birth or naturalized.

March 15, 1891

WHY THE ITALIAN SHOOTS.

Explanation of His Frequent Recourse to Gun and Knife.

To the Editor of The New York Times:

"What is the matter with the Italians in this country, anyway?" asks "An American" in THE TIMES of to-day.

The Italians are the best workingmen in this country; they are sober, economical, honest, and very resisting to hard work. They are not be found among the burglars, the assaulters of women, the kidnappers, the poisoners, the highwaymen, and the worst species of criminals.

They usually mind their own business, but when American or Irish loafers insult them, hoot them, jeer them, assault them, stone them as these ruffians are wont to do, then they react and use any weapon is at their reach. Patience is a great virtue, but has its limits, and we Italians are men and not slaves.

What is remarkable is that they never are assaulted by equal numbers. Not one against one, but cowardly ten against one or a hundred against a few. That the police or Magistrates like Mr. Flammer side by the drunken loafers or the wicked ragamuffins is only to be explained by the hatred they bear to anything that is Italian

BRUTO V. GIANNINI.

New York, June 9, 1904.

June 10, 1904

The Italian as a Citizen.

To the Editor of The New York Times:

In response to the special plea by an Italian for his country in THE TIMES, I am prompted to ask, Are not most of the attempts and acts of assassination the work of Italians? Are they not to-day the mercenaries whose votes are for hire to the corrupt politician on election day, and therefore the most dangerous element in our local politics?

No other nationality can be charged with such damnable societies as the "Mafia" and the "Black Hand," nor has there ever appeared in print a condemnation of these criminal organizations by an Italian. The record of the police and criminal courts will show whether they are law abiding, more or less, than other foreigners. Ninety-five per cent. will tell you that when they earn money enough to live in comfort in Italy they will return there.

And finally does not your Italian correspondent justify and by implication advise the criminal practice of carrying concealed weapons, contrary to the laws of all civilized people? These conclusions are the result of many years' observation, though I am not willing to condemn these people if it can be shown that I am wrong.

FOR LAW AND ORDER.

New York, June 10, 1904.

June 23, 1904

WRONG ABOUT JEWS, BINGHAM ADMITS

Statement That They Supplied Half New York Criminals Was an Error.

HIS STATISTICS WERE BAD

Didn't Collect Them Himself for His North American Review Article—Jewish Citizens Satisfied.

One of the first things that Police Commissioner Bingham did on returning from his vacation yesterday was to admit that the statement in his recent article in The North American Review that about half the crime of New York City was committed by Jews was based upon incorrect figures, which he had not himself gathered. The Commissioner declined to say who had furnished the incorrect statistics, but admitted frankly that examination had disclosed that they were incorrect. The incident seems now to be about closed.

Mayor McClellan said yesterday that the Commissioner's acknowledgment of error was a straightforward, manly statement which did credit to him. Prominent Jews of the city expressed themselves in the same terms about the Commissioner, and hoped that the talking on the subject was over.

Commissioner Bingham himself issued a formal statement. This he evidently considered his final word on that subject, for when asked what were the true figures as to the races who commit the crime in this city he said:

"No more about that. I won't say anything further on that subject now."

Mr. Bingham's Statement.

Here is his statement:

My attention has been called to a serious complaint, made by Jewish citizens, concerning a passage in an article recently published in The North American Review, entitled "Foreign Criminals in New York," in which I said that, under existing conditions, "it is not astonishing that with a million Hebrews, mostly Russian, in the city (one-fourth of the population) perhaps half of the criminals should be of that race," and in which comment was made on the percentage of Jewish boys in the House of Refuge.

My purpose in writing the article was not to publish statistics nor to enter upon a scientific inquiry into the race or religion of those charged with criminality, but solely to make a plea for a secret service fund in order that criminality might be more effectively dealt with than is now possible.

To indicate the necessity for additional weapons to cope with crime, it was pointed out that crimes of various kinds are committed by those of our population who are of foreign origin, which cannot be adequately reached by the police force as now constituted. It was only incidentally that the remarks were made which have been challenged.

The figures used in the article were not compiled by myself, but were furnished me by others and were, unfortunately, assumed to be correct. It now appears, however, that these figures were unreliable. Hence it becomes my duty frankly to say so and repudiate them.

The idea which I sought to impress was that the number of foreigners with whom the police come in contact is very large and that a special knowledge of racial customs and manners is essential to the attainment of the best results by the police in the investigations of crimes committed by and against those of foreign origin.

The percentage given of Jewish boys in the House of Refuge is, it appears, also misleading. This proportion of boys, it is now pointed out, should be considered not solely in relation to the total number of boys in the House of Refuge, but with reference to the total number of boys in all similar institutions where boys of this age and of other races and faiths are sent. The proportion of Jewish boys considered in relation to the inmates of other like institutions would be, of course, radically different.

In view of all this and of the fact that many estimable citizens feel hurt by what I wrote without the slightest malice, prejudice, or unfriendliness, for I have none, I withdraw the statements challenged, frankly and without reserve. I shall look forward with interest to the result of the research of the committee of representative citizens which is now engaged in making an accurate and exhaustive study of this whole question and shall be glad to meet them.

Manly Disclaimer, Mr. Marshall Says.

Louis Marshall of 37 Wall Street said after reading this:

Commissioner Bingham is entitled to credit for the manly and courageous manner in which he has acknowledged his error, and has retracted what he said in his article in The North American Review with regard to criminality among the Jews. I am convinced that in writing his article he was not actuated by the slightest feeling of hostility toward the Jews, but that he was merely seeking to support his thesis by what he considered to be proper arguments, but which, unfortunately, were in part based on incorrect premises. His frank recognition that he had unwittingly wronged the Jewish people will be accepted by them in the same frank and manly spirit. The incident should be considered closed.

Personal investigation has convinced me that his estimate was grossly inaccurate, from any point of view, and that his figures, which were not in any way based on statistics or official data, must have been in part influenced by the mental habit which prevails among the police of deeming every man convicted as soon as he is arrested, and of considering every man a criminal no matter on what charge he is summoned before a court. Such an estimate would include a large body of offenses which are not crimes and a number of individuals who are in no sense criminals. This Commissioner Bingham now recognizes, and every good citizen will gain added respect for him, because he, of his own accord, has made public amends for his error.

Francis J. Oppenheimer, the writer, said last night, speaking for the Federated Jews, that all was now harmony between Commissioner Bingham and those who raised the protest against his racial crime statistics. He had seen the Commissioner yesterday, Mr. Oppenheimer said, and he and his associates were convinced that the Commissioner was sincere in his retraction, as he verbally praised the progressive upright Jew.

September 17, 1908

INVEIGHS AGAINST 'NEUTRALS.'

Chicago Paper Charges Germans There Are Treated Unfairly.

Special to The New York Times.

CHICAGO, Ill., May 10.—The Chicagoer Presse, afternoon edition of the Illionis Staats-Zeitung, the leading German publication of Chicago, editorially complains today that Americans in Chicago are discriminating against Chicago German-Americans, and as a consequence their business languishes. The Chicagoer Presse suggests that inasmuch as in many parts of the city German dealers have been blacklisted, it would be in order for German-Americans to patronize only German-American merchants and read only German papers. The editorial in part follows:

"America is neutral. The Government at least says it is, and terms everyone a lout who doubts this and dares undertake to prove the* contrary. * * Here in Chicago grocers and butchers especially are affected by this continually increasing 'neutral' feeling, for neutral Americans no longer patronize their German-American fellow citizens. Everybody has the privilege of geting his necessary supplies from whoever he chooses to purchase them from. But if we have no right to harm others because of their creed or race, neutral Americans certainly are not privileged to blacklist their fellow-citizens on account of their descent. This equals a sort of conspiracy which would be severely punished by law if America was less neutral.

"It is nevertheless a fact that in many districts in Chicago German dealers have been blacklisted. But what if the Germans in Chicago would resort to the same retaliatory measures adopted by Germany against England? What if the local Germans, finding their economic existence endangered, should follow the example set and declare that they will buy from Grmans only, will employ German workmen only, read no other than German papers and patronize such firms only which advertise in German papers?"

May 11, 1915

Bohemian Parentage a Bar.

To the Editor of The New York Times:

Daily advertisements are appearing in the newspapers calling for girl stenographers needed in the Government service. The writer, American born but of Bohemian parentage, at present in a comfortable position, wishing to do her bit, applied in answer to the advertisement. After filling out the application blank at 55 Wall Street she was told by the officer in charge that the fact that her parents were born in Bohemia will count against her. Is this red tape or ignorance of the Government officials? Notwithstanding the fact that the Bohemians (Czechs) are heart and soul with the cause of the Allies, is not every American except the Indians a descendant of some foreign-born parents? If the above-mentioned practice be systematically applied, there would be something to count against every American.

The way some of the Government agencies are managed is a damper to one's enthusiasm to help the United States win the war.

STENOGRAPHER.

New York, March 14, 1918.

March 16, 1918

GERMAN BECOMING DEAD TONGUE HERE

Schools All Over America Banishing Study of the Tongue from Courses —A Survey of Their Attitude

By JOHN WALKER HARRINGTON.

EMPTY benches confront teachers of the German language all over the United States. Instruction in the Teutonic tongue has fallen off in American schools by at least 50 per cent. and possibly 65, and by next September it will be near the vanishing point in the elementary classes.

Action hostile to teaching the speech of the enemy has been taken either by State or local authorities in thirty-six out of the forty-eight American Commonwealths. In Delaware, Florida, Idaho, New Mexico, and Wyoming this amounts to prohibition direct from the officials at the capitals. Iowa and North Dakota, through their State Boards of Education, recommend the dropping of German from all schools, while Oklahoma bans it from elementary classes and gives the local educators their choice regarding advanced grades.

In many of the large centres of population the municipal Boards of Education have either banished German or so cut down schedules that it seems only a question of a year or so before its influence will have disappeared. In twenty States of the Union communities have been vigorously exercising their option and have been steadily eliminating the language.

Before proceeding with the details of this survey it should be said that the facts were gathered from many sources. They come from investigations made by the National Education Association's Commission on the Present Emergency in Education, from a canvass made by the National Security League of this city, from the American Universities Association, and from independent inquiries made in other directions. The close of the school year last month was marked by much feeling against the continuance of German language teaching, and there was an additional reflex of that sentiment in the annual meeting of the National Education Association held a few days ago in Pittsburgh. On that occasion many attacks were made on the cultural value of Teuton language and ideals, and a resolution was adopted urging that in future the pupils of American schools be taught only in English.

Much information as to the attitude of our educational authorities throughout the United States was gained from their correspondence with the New York City Board of Education, which was among the first to proceed against German instruction. Acting on the recommendation of Dr. William L. Ettinger, City Superintendent of Schools, it ordered that no classes in German should begin next September in the high schools. Pupils in those schools who are trying to make certain credits for entering college may continue their German study. Under these conditions the language will have disappeared from the high schools in two years. It will not be taught in the elementary grades.

"There is every reason to believe," says the argument presented by the Board of Superintendents, "that the high school pupils will themselves abolish the teaching of German by refusing to elect it. This is shown by the fact there are at present in twenty-four high schools only 12,954 pupils studying German, whereas 23,898 were in German classes in February, 1917. There are only 1,097 at present enrolled in first term German classes, when in February, 1917, there were 5,859 in those same classes.

"Nevertheless, a hands-off policy is not meeting this issue squarely. New York should lead the way in the abolishing of the teaching of German as a means, though a slight means, of winning the war by making a dent in Pan Germanism, by shaking possibly the morale of the German people, as they come to realize that the great city of New York is unwilling to endure any longer their language, and desires to break off more completely the possibility of intimate relations with them through that medium."

The falling off in the study of German in the high schools of the metropolis, which was 50 per cent. before the action of the board, has been taken up by the Spanish classes. A few days ago a conference of New York public school teachers of Spanish was held, at which there were 140 present.

In the State of New York an order has been issued by the Department of Education that all schools must teach in English, and an investigation is being made of all foreign language schools. In Syracuse the classes in German were one-third of their usual size at the close of the academic year.

New Jersey through its State Board of Education began last April to stimulate the introduction of Spanish as a substitute for German in her schools whenever possible. Calvin N. Kendall, the State Superintendent of Public Instruction, on June 1 of this year sent a letter to all local boards of education throughout the Commonwealth recommending that they exclude textbooks, magazines, or newspaper publications from the schools which would tend, directly or indirectly, to establish German propaganda.

The cutting of German out of the schools of such a large industrial city as Newark had a marked effect. East Orange claims the distinction of being the first city in the United States whose school board passed a resolution, as it did in May, 1917, eliding German from the course of study in her schools. Paterson forbade the singing of any German songs in her schoolhouses; Glassborough got rid of German in her high school and two days later substituted French. Vineland dispensed with German in all classes.

The State Board of Education of Connecticut adopted a resolution that all books used in its schools, except in the high schools, should be in the English language. It is also provided that the use of any book, leaflet, periodical, or newspaper printed in any foreign language be prohibited in any class on and after July 18, 1918, except by authorization of an agent of the board.

Massachusetts adopted a law providing that every public high school of not less than 150 pupils, offering a business course, should give instruction in commercial Spanish. Boston is reluctant to exile German from its classic portals. Beginning with next Autumn, Worcester will reduce the schools where German is taught from five to two in the preparatory division of the grammar department. In smaller cities German is being dropped because the students are no longer electing it.

New Hampshire a few weeks ago passed a law that no textbooks should be used in her schools which favored any political party or sect. The statute was a thinly disguised attack on the German propaganda. Vermont is making an investigation of German influences on her pupils. Maine has never had much of a Teutonic tide, and there are only twenty-nine of her schools which teach German. In Rhode Island no definite action has yet been taken regarding the instruction in the language, although a committee was appointed to examine the courses used in the schools of Providence for evidences of German propaganda.

Even in the Keystone State, famed as the home of the "Pennsylvania Dutch," the trend against the learning of German is marked. The State Superintendent of Public Instruction has only gone so far as to direct the elimination of unpatriotic literature, but German is being dropped in many schools by the direction of the local boards. The Common Council of Philadelphia demanded that the language be barred from the public schools. In Pittsburgh the German texts were not only taken from the students, but tons of the volumes were burned as though they were under the ban of heresy. Harrisburg, which is in the centre of a population largely German by descent, drove the study of the language from many public schools, and in its high schools discontinued the German electives, although students already registered were permitted to continue. Johnstown dispensed with German last May, as did also Lebanon. Erie took it from all grammar grades and voted to keep it out of high schools next Fall.

The speech of Kultur has fallen out of educational favor along the Atlantic seaboard. The State Board of Education of Delaware forbade instruction in German in all the schools under its control after the academic year just closed. Although Maryland has not acted, the authorities of Baltimore directed that no German should be taught in the elementary schools of the City of Monuments.

Education in the District of Columbia is supervised by Congress and there are movements on foot destined to remove German from all the schedules in the national capital. In the five white high schools of Washington the attendance in the German classes last year decreased 50 per cent., as compared with the same terms in 1917. In the Central High School there were eleven classes in German in the term recently closed and thirty-five in the same period last year.

Opposition to the teaching of German rears an almost solid front in the South.

"German will be discontinued and French taught instead in the ensuing year in the schools," writes Frank Evans, Superintendent of the schools of Spartanburg, S. C.

"I doubt whether German is being taught anywhere in Mississippi," reports W. F. Bond, the Superintendent of Public Instruction in that Commonwealth "Practically all the schools of the State make it elective, but both pupils and parents have eliminated it by declining to elect it."

As so very few pupils wished German, the New Orleans Parish School Board voted to discontinue German instruction with the close of the academic year.

Florida discarded all German books from the State-adopted texts. Texas has taken no action. There is a strong sentiment against German instruction in Georgia, and such cities as Athens have dropped it from their high schools.

The State Superintendent of Alabama is ridding the classrooms of that region German propaganda, and it is the intention of the State Board, he writes, to submit a bill to the Legislature providing for the teaching of patriotism in the public schools.

In Tennessee, German was excluded from the entire county school system. Chattanooga will not have it in her public schools next year, but will emphasize French and Spanish. Virginia has as yet taken no action, but in West Virginia there is a strong anti-Teutonic movement. In Wheeling the lessons in German have been banished from the elementary classes and in some of the advanced grades.

"Many of our high schools," to quote the State Superintendent of Public Instruction in Kentucky, "have cut out German entirely, and others have censored their textbooks.

German immigration, especially in the period from 1848 until just before the civil war, swept into the Middle States and those of the Northwest, and so across to the prairies toward the Southwest. Some of the settlers became sturdy American citizens, others kept alive the traditions of Germany, founded clubs, societies, and schools for the perpetuation of speech and song. Here were communities as Teutonic as Berlin. The German became a factor in politics, and in return for the German vote so called, he was able to obtain special schools for his language. In many of the Central and Western Commonwealths laws were passed providing that on the petition of a certain number of inhabitants of a district schools might be started which would give instruction in a foreign language. There were a few Polish and Scandinavian schools, but usually this provision in the laws resulted in the establishment of German centres. Where the instruction was not given outright in German, the Teutons insisted that there should be classes for imparting their language. Many teachers of the language of their Fatherland established themselves in the public school systems of the central part of the United States and formed powerful associations which were always ready to battle and to lobby for the perpetuation of their calling. Efforts were made a quarter of a century ago to break the German grip in these regions, but as long as there were parts of these Western cities under Pan Germanic spells the task was difficult.

One of the surprises of the year was the abolishing of German in the schools of Cincinnati, for in that Ohio city the Germans were so numerous that when one spoke of going "over the Rhine," as the canal was called, he meant that he was disappearing into a realm where all English was left behind. At one time pupils had to go to the City Hall to make excuses for not studying German.

Cleveland, where year after year the German party had foisted their language on the schools, not only put the German textbooks out of the schools, but provided cans in the principal streets, where pupils and the public might throw all the volumes they wished to have destroyed. At Columbus German was dropped from the entire system, including the elementary, intermediate, and high school grades.

"German," writes Horace Ellis, State Superintendent of Public Instruction of Indiana, "is being practically excluded from the schools next year."

Indianapolis has eliminated the study of the language from all her elementary classes.

Illinois permits probably wider latitude to the local school boards than any State in the Union, and for this reason it is difficult to learn the attitude of her educators. In Chicago there has been much discussion and many conferences. The high schools at Peoria have discontinued German, and at Dundee the board voted the language out of the schools and disposed of all the textbooks.

Michigan's Superintendent of Public Instruction says that many of the high schools will discontinue all German instruction next year.

"Our students of German have been reduced 64 per cent.," writes W. W. Warner, at the head of the public schools of Saginaw. Houghton, Marine City, Ypsilanti, and Adrian are among the Michigan cities which are barring the language.

Milwaukee, famed for its Germans, has risen against the language to such an extent that in the high schools those electing to study German were fifty less than they were a year ago, and many have chosen French and Spanish from the modern language lists. Wisconsin leaves much of her education to local option, but strong adjurations to patriotism and unity of spirit have been issued by the State board.

"In Nebraska," said Richard L. Metcalfe of the State Council for Defense, "there are eighteen districts where the public school has been driven out by German schools. In those German schools nothing but the German language is spoken or taught. Out of 379 teachers investigated 350 were Germans. In three counties where those German schools predominated the German national hymn was generally sung. In 100 of these schools the American national anthem, up to thirty days ago, had never been sung."

Disclosures such as those made by Mr. Metcalfe led to the movement for the repeal of the Market law, passed by the Nebraska Legislature a few years ago at the request of the German-American Alliance, which required a district to provide instruction in a foreign language when twelve citizens petitioned for it.

The Deputy State Superintendent writes that the teaching of German has been generally discontinued in the Nebraska high schools.

On the recommendation of the State Department of Public Instruction all public schools in Iowa decided last April to discontinue the training in German. This includes the high schools as well. German was dropped also from nearly every high school in the State of Arkansas. The State Superintendent of Public Instruction in North Dakota, a Commonwealth in which there are many citizens of Teutonic descent, recommended that umlaut drills be omitted for those in the elementary and high schools as rapidly as possible after July 1, 1918. The sister State of South Dakota has as yet taken no definite action.

Despite the fact that Missouri has a strong Teuton element, the State Superintendent of Schools declares that only one high school in the State, outside of the cities, is maintaining a German course. The urban high schools are confining their instruction of this kind to such students as are preparing for technical work. The common schools of St. Louis abolished German.

R. H. Wilson, State Superintendent of Public Instruction of Oklahoma, says in an official circular to local boards: "So far as I have been able to learn there will not be a high school in the State teaching German when school opens next September. The language has been removed from the elementary classes, and next Fall the State Board of Education will require every teacher who draws public money to subscribe to an oath to uphold the Constitution of the United States and that of the Commonwealth of Oklahoma."

Idaho has prohibited German entirely in her schools up to the ninth grade, and the advisability of such instruction in the high schools was left to the communities. Only a few of these schools of the higher grade include it in their courses. The local authorities in Wyoming dropped the German classes, while the City of Laramie burned all its German textbooks. The State Superintendent of Nevada writes that the teaching of German has been largely discontinued in his jurisdiction. The University of Colorado abolished its German courses after adopting scathing resolutions on the subject. The schools of that State are following its lead. Montana and Utah, at this writing, are making an investigation of their school books in quest of Prussian propaganda.

All courses in German were discarded from the high schools of California by a resolution of the State Board of Education. Los Angeles abandoned German the first of January, with the exception of instruction to a few advanced pupils who had already studied it and needed it for credits. No German will be taught in her elementary schools, high schools, and junior colleges hereafter. By order of the State Board of Education the study of the German language was discontinued in New Mexico. Arizona has not yet reached a decision. In Spokane schools have dropped it.

How greatly the demand for German language teachers is declining is shown in the falling off in the registration of normal courses for them at the Summer school of Teachers College, Columbia University, which opened last Monday. The program provided for demonstration classes in the pedagogy of German both in primary and secondary schools. There were four of these, in charge of educators of high ability and ordinarily there would have been from fifty to sixty students in each division. As there were no applicants, the courses will not be given. For the special class on German phonetics, which is of high value to instructors in English and language students in general, there were only four applications, and it was decided not to organize it.

In higher educational circles strong pressure is being brought upon the colleges and universities to withdraw credits for German for entrance to the freshman classes. The College of the City of New York has already abolished the study of German in the high school and junior high school departments, and has so lessened the credits for German in the collegiate courses as to discourage the study of the language. Members of the Faculties of Yale and Harvard Universities recommended to the National Educational Association the substitution of Spanish and French for German as entrance subjects as well as for college studies.

Many leading college educators, however, are disposed to uphold the study of German for advanced students, as they think it will be of value to them in scientific and technical work. Some of them believe, as does Dr. Nicholas Murray Butler, President of Columbia, that access would be cut off to the idealism of Goethe and Schiller if this Teuton tongue were removed from the curricula of our higher institutions of learning. This is the view, in a measure, which is taken by Philander P. Claxton, the United States Commissioner of Education.

Dr. L. D. Coffman, Dean of Education of the University of Minnesota, in his recent address on "Competent Teachers for American Children," delivered before the association, quoted from the official quarterly of the League for Germanism in Foreign Lands the following significant passage:

"Work rendered in the interest of the German school is a noble service rendered to the German nation; for the most effective means for perpetuating Germanism in foreign countries is the school. Within its sacred walls the strange land is transformed for children and teachers and parents into a fatherland."

"Not until recently," to quote Dr. Coffman further, "were we aware that there were many un-American schools and many un-American teachers. Our ignorance of the situation was appalling and our stupidity colossal. Not until we entered the world conflict did we pause, take stock, and discover the sinister influence of German Kultur in the schools of the country. Now we find that there has been an organized program the Germanizing of America."

July 14, 1918

KUKLUX KLAN AGAIN IN THE SOUTH

By LITTELL McCLUNG.

THE first "Invisible Empire," which was brought into being by General Forrest after the civil war to offset the evils of carpetbag rule, has been succeeded by a second Invisible Empire. The Kuklux Klan—silent, daring, and terrible—is once more organized in many localities of the South. In hundreds of communities its warnings are being posted.

Not many days since, at Mobile, Ala., the unseen hand of the Kuklux Klan stretched forth. Great shipbuilding yards are already turning out ships there for the Government, and other docks are in process of construction. A double-barreled strike of workers had been arranged—the stevedores at the docks and the washerwomen of the city were to be called out simultaneously by a self-appointed but locally powerful leader.

There was no desire on the part of the workers to quit work. Both stevedores and washerwomen were making good wages and hundreds of them had put savings into Liberty bonds and War Savings Stamps. They had contributed to the Red Cross and the Y. M. C. A. But the strike was to be called, nevertheless.

Then a rumor spread, from what source is not known, that the man planning the strike was in personal danger. Ostensibly to protect him from what threatened, he was arrested, put into a patrol wagon and was being taken to the police station one night when, as the patrol wagon was rounding a dark corner, it was stopped by a squad of automobiles. Each car was covered with white cloth and each bore the insignia of the Invisible Empire—"the fiery cross of old Scotland." Men leaped out of the cars, each completely disguised under the ghostly garb of the Kuklux Klan. The driver of the patrol wagon saw rifles pointed at him, gave up his prisoner, and drove back to the police station.

Was the arrest, with the patrol wagon, a part of the Klan's plan for capturing its victim and spiriting him away? Who knows? No strike was called. Just what became of the strike agitator has not been told.

Virtually the same incident, with the patrol wagon left out, has been repeated at Birmingham, Ala. A strike leader appeared in the neighborhood of the great mills in the suburbs of the town. His appearance was the first intimation that any one, even the workers in the mills, had that a strike might take place. Despite the fact that there was no desire on the part of the workers to quit work, this leader laid his plans for a walkout of considerable proportions.

But suddenly the arm of the Invisible Empire was put forth in the darkness. There was no strike. The man has not been heard of since.

The Invisible Empire has caused the information to be circulated that it is, first of all, on the lookout for alien enemies, for the disloyal, and for the fellow who is seeking to begin a strike.

Presumably the anonymous leaders of the Klan make a distinction between the loyal union laborer and the laborer who is infected with the I. W. W. spirit, even though he may not be a member of that organization.

When there is no trouble brewing in labor circles, or among disturbers suspected of being alien enemy sympathizers, the Klan goes after the idlers and slackers. For these the preliminary action is a public warning. It is believed that this warning will be heeded. If not, individual idlers and slackers will be picked out by the Klan and special warnings will be given them. Should these fail in any case, the Klan is prepared to take more drastic action.

In Montgomery the Klan has warned immoral women that they must keep away from the soldiers in Camp Sheridan.

The Invisible Empire apparently has public sentiment behind it, as in the seventies. Its organization is much the same as in the seventies. Wherever it is organized it is made up of some of the best men in the community. Neither strangers nor half-strangers are taken in. The shrewdest detective might find it impossible to penetrate into the inner circles of the Empire.

The disguise, as in the old days, is completely effective. The long, white robe hides a man's body as well as his face. Even the horses or automobiles are similarly covered. General Forrest and his associates knew full well the fear of the unseen and the unknown. The founders of the second Invisible Empire are depending again on this fear to make drastic action unnecessary except in special instances. But, in addition to the terror of the unseen in the darkness, the Klan is ready to make every warning good, and it is known to be composed of men who are sworn to do their part fearlessly, whatever consequences may impend.

While many communities in the South are much larger than in the days of the first Empire, there are few so large as to make its operations a failure. It might be impossible for the Klan to do its work in the larger cities of the North, but in Southern cities it has not been thwarted. It is notably active in Mobile, which has now nearly 100,000 population, and Birmingham, with 200,000.

Each community in the South knows whether it is as yet organized into the Invisible Empire. We know of one community that two weeks ago had no Klan. Now there is a Klan in the community and everybody knows there is. But no one, except the Klansmen, knows who its members are, and probably no outsider will know until the war is over.

September 1, 1918

FORD EXPLAINS ATTACKS.

Caused by Statements Made to Him by Jews on Peace Trip.

Special to The New York Times.

FLORENCE, Ala., Dec. 4.—Henry Ford today told reporters the fundamental reason why for the last two years he has attacked the Jew in his weekly magazine, The Dearborn Independent. He said that the course of "instruction on the Jew which he intends to give the United States will continue for five years."

"It was the Jews themselves that convinced me of the direct relation between the international Jew and war, in fact, they went out of their way to convince me," he said.

"You remember the effort we made to attract the attention of the world to the purpose of ending the war through the medium of the so-called peace ship in 1915. On that ship were two very prominent Jews. We had not been to sea 200 miles before these two Jews began telling me about the power of the Jewish race, how they controlled the world through their control of gold and that the Jew, and no one but the Jew, could stop the war.

"I was reluctant to believe this and said so—so they went into detail to tell me the means by which the Jew controlled the war, how they had the money, how they had cornered all the basic materials needed to fight the war and all that, and they talked so long and so well that they convinced me. They said, and they believed, that the Jews had started the war; that they would continue it as long as they wished and that until the Jew stopped the war it could not be stopped. We were in mid-ocean and I was so disgusted that I would have liked to have turned the ship back.

"When I got back to the United States I still had in mind what the Jews had told me. In Europe, I had looked about quite a bit and I could see that a lot of the things the Jews had told me were so. Once at home, I set about investigating a bit, and the more I investigated the more I found to substantiate what the Jews had told me. I determined that the situation should be made clear to the people of the United States through publicity. But do you think I could get a newspaper to print it? Not on your life. It seemed that there was no newspaper in the United States that dared print the truth.

"Then a funny thing happened just at this juncture. An old chap in Dearborn came to my office and wanted to sell the local paper, The Dearborn Independent, a weekly newspaper. The thought came to me like a flash. Surely some place in the United States there should be a publisher strong and courageous enough to tell the people the truth about war. If no one else will I'll turn publisher myself. And I did."

"How long will your paper continue to deal with the Jewish question?" he was asked.

"We've got a five years' course in sight, and we are going to tell the people, among other things, some American history that they don't teach in the schools. We will show indisputably that one of the great factors behind the Civil War, that brought it on and made peaceable settlemen of the issues impossible, was the Jew. And that isn't the whole story either. There will be more than that."

Mr. Ford and Mr. Edison spent Sunday morning looking over the site of dam No. 3 at Muscle Shoals, which is still to be started, and which, when built, will create a great reservoir for control of the back waters above the power plant. The afternoon was spent at a Southern barbecue at the home of E. A. O'Neal, head of the Alabama Farm Bureau.

December 5, 1921

ASCRIBE TO KLAN ATTACK ON SMITH

Legislators Get Series of Letters Assailing Jews and Catholics.

BOAST OF FIRING CHURCHES

Missives Call Governor Jesuit Pawn and Laud Ku Klux as Protestant Champion.

ALL ARE ANONYMOUS

Communications Pour In After Introduction of Bill Requiring Divulging of Klan Membership.

Special to The New York Times.

ALBANY, Feb. 28.—During the last forty-eight hours legislators have been deluged with letters, bearing no signature but generally attributed by the recipients to the Ku Klux Klan, in which the burning of Catholic churches and institutions is boasted of. The Klan set up as a champion of American principles and the Protestant faith and a general appeal made to the Legislature "to guard against Jesuit scheme for the mastery of America."

Governor Smith, "groomed as a democratic presidential candidate in 1924" is assailed as a pawn in the "Jesuit" game. The Governor, when a copy of the communication was shown to him this evening, read it through with knitted brow but declined to comment.

Three letters, all anonymous and the last of a series mailed in Brooklyn, have been received by the legislators since the introduction last week by Senator Walker of a bill which would require the Klan and other secret organizations to file lists of their membership with the Secretary of State. The first of the communications was received by Senators and Assemblymen when they returned to the Capitol on Monday night. The last, received this morning, contains approximately 2,000 words.

Until this latest missive was read, the lawmakers had paid no attention to the communications. Tonight, however, Assemblyman Jesse of New York City made it known that he would call the attention of the Legislature to the letters and demand an investigation with a view to revealing their source.

One of the communications was in the form of a circular, headed "The Assassination of Abraham Lincoln the Foul Work of the Roman Catholic Church."

This contained what purported to be quotations from utterances of Lincoln attributing the Civil War to Jesuits. Like the other communications, this was mimeographed.

The attack on Governor Smith was contained in the missive which reached the capital today. This began with quotations from a book bearing the title "Fifty Years in the Church of Rome," the author being the Rev. Charles Chiniquy, described as a former Roman Catholic Priest of Canada who turned Protestant "and became the firm friend of Lincoln." In his book the former priest describes what he calls a "papal plot to destroy Protestant America."

"'Al' Smith, Governor of New York, a Tammany Hall Roman Catholic politician, is being groomed as Democratic Presidential candidate in 1924 and his propaganda for this purpose is already seen in the daily press," said the attack.

"The Jesuit schemes for mastering America, by starving out what they call 'the godless public schools' in favor of parochial schools, where papal idolatry and criminal ignorance are bred and by subverting our political institutions to the domination of Roman priest-craft, seemed until lately to be on the crest of the wave.

"But, lo! at the blackest hour, what is this strange light which gleams in the gloom like a flaming torch, causing a hundred sceptred 'Princes of the Holy Roman Empire' in America to tremble on their cushioned thrones, sweating in an agony of fear? Whence comes this flaming torch of truth that ignites Roman Catholic churches, academies and nunneries, fills the land with patriotic anti-Catholic literature and inspires millions of aroused Americans to organize for action to destroy this triple-headed beast which threatens America's life.

"A wave of Protestant, Anglo-Saxon Americanism is beginning now to gather force. It shall mount and mount, higher and yet higher, increasing its strength into irresistible might, until at last it breaks in a torrential flood that shall sweep every disloyal and Christ-hating Jew from the face of our beloved land, washing it clean and pure as a fit habitation for Christ and His followers.

"Under the guidance of the King of Kings, Christ, who is always watching over His people, the Ku Klux and its allies are 'fighting fire with fire.' The Roman Catholic and the Jew are notoriously lawless. The Klansman is law abiding, and he is executing the law of God, who says to evil and evil's mouthpieces, 'Thou shalt surely die.' "

March 1, 1923

SEES A BAR TO JEWS IN MEDICAL STUDY

President McConaughy Warns 12 at Wesleyan That Graduate Schools Discourage Them.

ANTI-SEMITISM IS DENIED

Students of Other Races Also Are Told of Difficulties Put in Their Way, He Says.

Special to THE NEW YORK TIMES.

MIDDLETOWN, Conn., Nov. 30.—The text of a letter sent to twelve Jewish students in the pre-medical course at Wesleyan University advising them that it would be difficult for them to enter medical schools was made public here today by Dr. James L. McConaughy, president of the university.

The letter, written by Dr. McConaughy and Dr. Edward Christian Schneider, Professor of Biology, pointed to an estimate that 17 per cent of the freshmen students in medical schools were Jews. Citing reports that the Jews in this country constituted 5 per cent of the population and that over 50 per cent of the applicants for medical schools last year were of Jewish ancestry, the letter pointed out that "it is difficult for Wesleyan to place her graduates of the Jewish race in medical schools."

Commenting on the letter, Dr. McConaughy said it was not intended to convey the impression that medical schools were anti-Semitic, or that Wesleyan was seeking to discourage Jews from studying medicine. The letter was written, he said, merely to present the difficulties of entering medical schools. Students of other races have been informed by their teachers of difficulties they may find in seeking a medical career, he said.

TEXT OF THE LETTER.

The text of the letter was as follows:

"We are sending you this memorandum in order that there may be no misunderstanding regarding the rather difficult situation which you may face if you are planning to seek admission to a medical school.

"It is important to recognize that the opportunity to study medicine is definitely limited by the medical schools of this and other lands. In our country each year between 6,200 and 6,300 medical freshmen are chosen from a field of approximately 14,000 applicants.

"Furthermore, foreign study is restricted in that almost impossible barriers have been established in many countries abroad and by legislation in most of our States. Any young man wishing to study abroad must first get permission from the State in which he wishes to practice. Most States are refusing to accept the foreign diploma.

"While the racial question does enter somewhat into the selection of students, it does not enter as much as some claim. The Association of Medical Colleges reports that 17 per cent of the freshmen students in medical schools are Jews. It is pointed out that in this country, out of a population of more than 120,000,000, there are probably between 5,000,000 and 6,000,000 Jews. They, in round numbers, make up not more than 5 per cent of the entire population.

"It has further been reported that over 50 per cent of the applicants for entrance to the medical schools in 1933 were of Jewish ancestry.

"Very Little Room Left."

"The above facts explain why it is difficult for Wesleyan to place her graduates of the Jewish race in medical schools. It should be apparent that in selecting its freshmen each medical school will feel some degree of responsibility for the graduates of the institution with which it is associated, and it therefore is impelled to accept the promising applicants within its own borders.

"It is now quite generally admitted that after that selection has been made very little room is left for Jewish candidates from other institutions.

"We have no desire to discourage you in your hope for a medical career, but feel that it is only fair that you should know the circumstances. We have been disturbed at the difficulties which some of our students, even after a very good Wesleyan record, have encountered in the last few years in securing admission to medical schools."

Dr. McConaughy said there were about thirty Jewish students among 650 young men at Wesleyan. When the twelve told their teachers of their definite intention to study medicine, he said he felt that they should be informed of the difficulties they might face.

December 1, 1934

PLEDGES LOYALTY TO U. S.

Japanese Group Here Disavows Concern With Tokyo Policy

The Committee for Democratic Treatment for Japanese Residents in Eastern States, 152 West Forty-second Street, issued a statement yesterday pledging its membership to support the government of the United States and disavowing concern with Japanese policy. The statement, copies of which were sent to President Roosevelt and to Congress, was signed by Tom Kume, chairman, and other officers of the organization.

The statement said:

"We shall fully support the national defense and encourage our sons to do their share in the protection of this country and the democratic principles on which it is founded. We shall oppose fifth columnists of any nation working against the interest of this country. We shall cooperate with any American organizations in the promotion and extension of democratic principles in this country. We shall oppose any discrimination against minorities on the basis of race, color, religious beliefs or national origin. We shall help the Japanese residents to participate fully in American community life."

August 14, 1941

Chinese to Wear Lapel Buttons

LOS ANGELES, Dec 8 (AP)—T. K. Chang, Chinese Consul, said today that Chinese in America are designing a label button which will show the nationality of the wearer. "This is being done," he said, "to save the Chinese from embarrassment and mistaken identity."

December 9, 1941

EVACUATION AREA SET FOR JAPANESE IN PACIFIC STATES

By LAWRENCE E. DAVIES

Special to THE NEW YORK TIMES.

SAN FRANCISCO, March 3—Practically a quarter of a million square miles of the Western coastal region, an area almost as large as the entire Japanese Empire, was set apart today by Lieut. Gen. John L. De Witt, Western Defense Commander, as a military zone from which all Japanese, American citizens as well as aliens, must move.

Starting at the Canadian border, the 2,000-mile-long "evacuation line" will force Japanese to vacate about two-thirds of Washington, two-fifths of Oregon, far more than half of California to the west of the Sierras and the southern two-fifths of Arizona.

General De Witt's proclamation, issued under authority of a Presidential order, was the blow which aliens and Japanese-Americans had been expecting for more than two weeks. The proclamation did not actually order an evacuation and the general said that "immediate compulsory mass evacuation" of all Japanese and other aliens from the coast was "impracticable." But he left no doubt that every person of Japanese lineage must get out of the region defined under orders to be issued "eventually."

Cites Gain to Early Movers

"Those Japanese and other aliens who move into the interior out of this area now," he added, "will gain considerable advantage and in all probability will not again be disturbed."

There was no official announcement as to how much time would be allowed for voluntary evacuation, but some estimates put the "deadline" at around April 15. Japanese-American leaders begged for further information as to resettlement plans, so that they might give proper advice to an estimated 100,000 persons of Japanese ancestry who will be affected by the later moving orders.

Officials in charge of resettlement were understood to be planning a registration and reception center for the Japanese in California's Owens Valley, now owned by the city of Los Angeles. Already materials are being assembled there and buildings are being fabricated. The first unit, it is said, will house between 5,000 and 10,000 and eventually perhaps 50,000 persons can be taken care of in that Southern California area, east of the Sierras.

General De Witt said the Federal Government was "fully aware" of the problems involved, particularly with respect to "property, resettlement and relocation" of the groups to be affected.

"Since the issuance of the Executive order," he went on, "all aspects of the various problems have been subjected to careful study by appropriate agencies of the Federal Government. Plans are being developed to minimize economic dislocation and the sacrifice of property rights. Military necessity is the most vital consideration, but the fullest attention is being given the effect upon individual and property rights."

The man in whom the President reposed full authority to set up military zones and exclude citizens and aliens alike as a precaution against fifth column activities declared that the evacuation would be carried out as "a continuing process."

Affected Groups Are Classified

To simplify procedure General De Witt set up these five classes:

Class 1—All who are suspected of espionage, sabotage, fifth column or other subversive activity.

Class 2—Japanese aliens.

Class 3—American-born persons of Japanese lineage.

Class 4—German aliens.

Class 5—Italian aliens.

"Persons in Classes 2 and 3," General De Witt explained, "will be required by future orders to leave critical points within the military areas first. These areas will be defined and announced shortly. After exclusion has been completed around the most strategic areas, a gradual program of exclusion from the remainder of Military Area No. 1 will be developed."

Next on the list for evacuation after the Japanese aliens and American-born Japanese, the General indicated, would be the German and I an aliens, but persons 70 years old or more would not be required to move except when individually suspected. The families, including parents, wives, children, sisters and brothers, of Germans and Italians in the armed forces would not be moved unless for some specific reason.

General De Witt pointed out that persons in Class 1 were being seized daily by agents of the Federal Bureau of Investigation and other intelligence services.

His "Proclamation No. 1," after setting up Military Area No. 1 along the 2,000-mile front in Washington, Oregon, California and Arizona, defined Military Area No. 2 as the remaining parts of those States. The coastal area extends on an average about 100 miles inland, but in Washington and in parts of California the width is much greater.

Inland States May Get Some

General De Witt's language indicated that persons who moved from Area 1 to Area 2 would not be further disturbed. This presumably will be good news to the Governors of a group of inland States who had protested against any influx of Japanese from the coastal areas; still it does not necessarily mean that California, Oregon and Washington will retain all, for study is being given to the idea of letting some go to the interior for farm jobs.

The general already has indicated that he will not pay much attention to other States' protests if they are against the interests of security and "military necessity."

Military Area No. 1, from which all Japanese and some German and Italian aliens will be moved, includes most of the leading cities of the four States, although Spokane, Wash., lies outside its boundaries.

Eastern Border of Area

Its eastern border begins, on the north, where United States Highway 97 intersects the international boundary line between this country and Canada. Thence it runs south along Highway 97 to a projection of United States Highway 10-A near the junction of the Columbia River with the Wenatchee River. It roughly follows the Columbia to a point two miles south of Maryhill, Wash., where Queen Marie of Rumania dedicated a museum some fifteen years ago, and then extends south along United States Highway 97 through Oregon into California to a point where this road, projected, intersects United States Highway 99.

The line then proceeds southward to the vicinity of Red Bluff, running near Auburn, Mariposa, Raymond, Coarse Gold, Fresno, Visalia, Exeter and Ducor. Near Isabella, in Southern California, it turns eastward, eventually following United States Highway 66 across the Colorado River to a point near Topock, Ariz. The line runs through or near Yucca, Signal, Congress Junction, Phoenix and Florence Junction to the Arizona-New Mexico line.

This, in a general way, shows the extent of Military Area No. 1. It is divided by the Army, however, into two zones. The western boundary of Military Zone A-1 runs roughly three miles off the coast from all continental parts of the United States, plus the Farallon and Santa Barbara Islands, and its eastern boundary is from thirty-five to eighty miles inland. Zone B includes all of Military Area No. 1 lying outside Zone A-1.

"Prohibited" Areas Designated

In addition, General De Witt's first proclamation designated "prohibited" regions within Military Area No. 2, which he labeled "Zones A-2 to A-99." He did not explain his reasons for dividing Military Area No. 1, apparently leaving the explanation to future proclamations.

The zones in Military Area No. 2, however, are all in the vicinity of defense plants, communication centers, dam sites, power houses and the like, many of which already were "prohibited" areas under an order of Attorney General Francis Biddle.

Thus, General De Witt's proclamation appeared to leave the eastern parts of Oregon, Washington and California and Northern Arizona open for resettlement of persons required to leave Military Area No. 1, with such resettlement forbidden, however, in the military zones established in that area.

"Such persons or classes or persons as the situation may require," he said, "will by subsequent proclamation be excluded from all of Military Area No. 1 and also from such of those zones herein described as Zones A-2 to A-99, inclusive, as are within Military Area No. 2.

"Certain persons or classes of persons who are by subsequent proclamation excluded from the zones last above mentioned may be permitted, under certain regulations and restrictions to be hereinafter prescribed, to enter upon or remain within Zone B.

"The designation of Military Area No. 2 as such does not contemplate any prohibition or regulation or restriction except with respect to the zones established therein."

Notice of Moving Required

General De Witt warned that any Japanese, German or Italian or any person of Japanese ancestry now resident in Military Area No. 1, who changed his place of habitual residence was required to obtain and executive "a change of residence notice" at any postoffice within the four States.

"Such notice must be executed not more than five days or less than one day prior to affecting any such change of residence," he stated; "nothing contained herein shall be construed to affect the existing regulations of the United States Attorney General which require aliens of enemy nationalities to obtain travel permits from United States Attorneys and to notify the Federal Bureau of Investigation and the Commission of Immigration of any change in permanent address."

General De Witt said further that the designation of prohibited and restricted areas by Attorney General Biddle on Dec. 7 and 8 and regulations prescribed by him were "hereby adopted and continued in full force and effect."

As the proclamation defining new military areas was issued at the Presidio, FBI agents and police seized sixty-nine crates of skyrockets and colored flares in the home of George Makamura, an alien Japanese who lived a few yards from the seashore at Santa Cruz.

"Arsenal" Fills Jail Storage

The arsenal of fireworks, described as of "the most powerful type," more than filled all storage space in the Santa Cruz jail. The crates were the size of large office desks. The raiders saw in the flares a potentially dangerous signalling system.

Illegal cameras were found in possession of Joe Moreno, also of Santa Cruz, who had failed to register as an alien. Police in Oakland seized Carlo Lenandrino, 35, an alien Italian itinerant, who, they declared, "cussed the United States" when arrested.

Japanese families, leaving some restricted areas in advance of formal orders from the Army, were reported wandering about California districts uncertain where they should go.

This subject of destination was the theme of a statement given out by Mike Masaoka, national secretary and field executive of the Japanese-American Citizens League, with a membership of 20,000. He said:

"We trust that our government will treat us as civilian citizens who are voluntarily cooperating in national defense and not as military wards.

"The Japanese-American Citizens League is interested in a positive constructive program of resettlement for the evacuees so that they may continue to contribute on the production lines to the inevitable victory of the democratic forces. With this purpose in mind, we are instructing the sixty-five chapters of our organization in 300 communities to call meetings immediately in their localities to discuss methods by which they can correlate their energies and cooperate extensively in the evacuation process."

The statement urged that an opportunity be given to American-born Japanese "to collaborate in the work and planning of resettlement."

The Committee on National Security and Fair Play, headed by Dr. Henry F. Grady, a former Assistant Secretary of State, said in a statement that there appeared to be "only three methods of caring for the evacuees: either allow them to settle where they can work freely and produce; or set up supervised work projects; or support them in whole or in part at public expense."

Some officials are understood to be hopeful that Owens Valley can be turned into a highly productive agricultural area by the Japanese and that vocational schools and workshops can be established, making it adaptable to rehabilitation of United States soldiers after the war.

March 4, 1942

Kansas Bars Japanese

TOPEKA, Kan., April 1 (AP)—Governor Ratner today ordered Kansas highway patrolmen to turn back any Japanese trying to enter the State. "We don't want them here," he said.

April 2, 1942

GERMANS, ITALIANS ARE AIDED ON COAST

General De Witt Abolishes All Prohibited and Restricted Areas Set Up in Winter

THOUSANDS CAN RETURN

Ending of Stop-Gap Measure Means More Agricultural Help for California

Special to THE NEW YORK TIMES.

SAN FRANCISCO, June 28—Several thousand German and Italian aliens who were required to move out of prohibited or restricted areas set up in California by Attorney General Biddle in late January and early February will be able to move back into or work in those districts under a new proclamation just issued by Lieut. Gen. John L. DeWitt, Western defense commander.

Guiseppo DiMaggio, 69-year-old father of Joe DiMaggio, the baseball player, may return to Fisherman's Wharf to keep an eye on Joe's restaurant and reminisce with his old cronies, many of whom likewise had been barred from that picturesque district. Hundreds of Italians who had to leave homes they had occupied for many years in industrial Pittsburgh may now move back.

General DeWitt abolished all "prohibited and restricted areas within the Western defense command" as established by the Attorney General, thus making available a substantial number of farm-hands and fruit pickers for work in California's fields and orchards situated within those areas.

Italians Help Truck Farming

Italian aliens had been an important factor in truck farming in the Half Moon Bay district near San Francisco. They had been barred also from considerable farm land in Monterey County as well as in Southern California and elsewhere in the State.

Although General DeWitt's proclamation thus should help the State's economic situation, this consideration was not a motivating factor in his decision, according to observers.

A spokesman explained the situation this way. The Attorney General's orders, which cleared the prohibited areas of German and Italian aliens by Feb. 24, were issued before a long-range pattern for dealing with the enemy alien problem had been worked out. They were thus looked upon as a stop-gap measure to protect the coast from possible sabotage and fifth column activity until there was opportunity to develop a broad program.

When General DeWitt received, under Presidential order, the right to deal with the Japanese, German and Italian situation as he deemed best consonant with requirements of "military necessity," he retained all the prohibited and restricted areas as designated by Mr. Biddle.

Some Places Still Proscribed

But, it is understood, retention of these areas would not follow the present pattern of dealing with enemy aliens. There are still certain places into which an Italian or German may not go, and all these places are guarded by physical barriers and sentries. They are along waterfronts and near military and naval installations.

The new order has the effect of restoring almost normal family life to thousands of Germans and Italians, but compliance with curfew, residence and travel restrictions is still required. German and Italian aliens in California must, in spite of the relaxation of the earlier orders, remain in their homes between 8 P. M. and 6 A. M., may not travel more than five miles from their homes, and must obtain permission from a federal attorney before changing their residences.

June 29, 1942

Supreme Court Upholds Return Of Loyal Japanese to West Coast

By LEWIS WOOD

Special to THE NEW YORK TIMES.

WASHINGTON, Dec. 18 — The constitutionality of the wartime regulations under which American citizens of Japanese ancestry were evacuated from Pacific Coast areas in 1942 was upheld by a vote of 6 to 3 in the Supreme Court today, but in another decision the court ruled unanimously that Japanese-Americans of unquestioned loyalty to the United States could not be detained in war relocation centers.

The Supreme Court rulings came only twenty-four hours after the Army announcement that exclusion of Japanese-Americans from the West Coast would be ended Jan. 2. They came also at about the time Secretary Ickes declared in a statement that he did not foresee a "hasty mass movement" of evacuees back to the West Coast.

Justice Hugo L. Black wrote the majority opinion on the evacuation question, which involved the 1942 order of Maj. Gen. John L. De Witt as applied to Fred Toyosaburo Korematsu.

Upholding the order as "of the time it was made and when (Korematsu) violated it," he deplored compulsory exclusion, but said that Korematsu was excluded because we were at war with Japan, adding:

"When under conditions of modern warfare our shores are threatened by hostile forces, the power to protect must be commensurate with the threatened danger.

"We are unable to conclude that it was beyond the war powers of Congress and of the Executive to exclude those of Japanese ancestry from the West Coast area at the time they did."

Justices Owen J. Roberts, Frank Murphy and Robert H. Jackson all entered dissents on the ground that the majority finding violated the Constitution.

The unanimous opinion regarding confinement in war relocation centers was written by Justice William O. Douglas in the case of 22-year-old Mitsuye Endo, held at Topaz, Utah. Without going into constitutional issues, Mr. Douglas held that, as Miss Endo's detention was not related to espionage or sabotage, she must be released. He noted that she was a concededly loyal citizen and stated:

"Loyalty is a matter of the heart and mind, not of race, creed or color."

'Hardships Are Part of War'

As in the case of Miss Endo, no question has been raised concerning the loyalty of Korematsu, native American. Accused of remaining in a military zone after his exclusion was ordered, he was convicted and sentenced to serve five years, but was placed on probation.

In upholding the exclusion order, Justice Black said that the court was "not unmindful of the hardships" which it imposed on a large group of American citizens.

"But," he stated, "hardships are part of war, and war is an aggregation of hardships. All citizens alike, both in and out of uniform, feel the impact of war in greater or lesser measure."

Pressing public necessity may sometimes justify the existence of restrictions on a racial group, but "racial antagonism never can," Justice Black remarked.

He pointed out that in the Hirabayashi case of several months ago the Supreme Court had held that a curfew applied properly to the program controlling Japanese-Americans.

"We upheld the curfew order as an exercise of the power of the Government to take steps necessary to prevent espionage and sabotage in an area threatened by Japanese attack," he said.

Steps to Prevent Sabotage

"In the light of the principles we announced in the Hirabayashi case, we are unable to conclude that it was beyond the war power of Congress and the Executive to exclude those of Japanese ancestry from the West Coast area at the time they did. True, exclusion from the area in which one's home is located is a far greater deprivation than constant confinement to the home from 8 P. M. to 6 A. M., the curfew hours.

"Nothing short of apprehension by the proper military authorities of the gravest imminent danger to the public safety can constitutionally justify either. But exclusion from a threatened area, no less than curfew, has a definite and close relationship to the prevention of espionage and sabotage."

Justice Black declared that it was wrong to "cast this case into outlines of racial prejudice," for, he contended, Korematsu was not excluded "because of hostility to him or his race," but because of the war circumstances. He pointed out that while many persons of Japanese origin were loyal to the United States, 5,000 refused to swear allegiance and several thousands sought repatriation to Japan.

In one of the dissents, Justice Jackson said the majority seemed to be "distorting the Constitution to approve all that the military may deem expedient." He asserted that he could not determine on the evidence whether General De Witt's orders were reasonable cautions.

Challenges Constitutionality

"But even if they were permissible military procedures," he commented, "I deny that it follows that they are constitutional. If, as the court holds, it does follow, then we may as well say that any military order will be constitutional and have done with it."

Justice Roberts declared that the "indisputable facts exhibit a clear violation of constitutional rights." He held also that Korematsu was faced by two orders, one, not to remain in the zone unless in an assembly center; two, not to leave the zone.

Justice Black held for the majority, however, that only the exclusion order was the issue.

Justice Murphy charged in his dissent that the exclusion of all Japanese, alien and non-alien, from the West Coast, "goes over the 'very brink of constitutional power' and falls into the ugly abyss of racism."

"No reasonable relations to an immediate, imminent and impending public danger is evident," he said, "to support this racial restriction which is one of the most sweeping and complete deprivation of constitutional rights in the history of this nation in the absence of martial law."

Justice Felix Frankfurter agreed with the majority, but said independently:

"To find that the Constitution does not forbid the military measures now complained of does not carry with it approval of that which Congress and the Executive did. That is their business, not mine."

State Civil Service Employe

The detention program was challenged by Miss Endo, a Civil Service employe of the California State Government at Sacramento, who appealed from a denial of a writ of habeas corpus. She contended that the resettlement of evacuated Japanese-Americans had been depriving her of her rights as an American citizen.

"We are of the view that Mitsuye Endo should be given her liberty," said Justice Douglas. "In reaching that conclusion we do not come to the underlying constitutional issues which have been argued. For we conclude that, whatever power the War Relocation Authority* may have to detain other classes of citizens, it has no authority to subject citizens who are concededly loyal to its leave procedure."

Justice Murphy, in a concurrence, asserted that the whole program of detention was unauthorized and only "another example of the unconstitutional resort to racism inherent in the entire evacuation program."

This "racial discrimination," he argued, was "utterly foreign to American ideals and traditions."

Also concurring, Justice Roberts said that, just as in the Korematsu case, the court sought to avoid constitutional issues.

Plans Outlined by Ickes

Secretary Ickes, in his statement, said that, under the Army order of yesterday, the War Relocation Authority would "intensify its efforts" to relocate loyal West Coast evacuees in other parts of the country, but would also assist those who "prefer to exercise their legal and moral right of return" to the West Coast.

He stated also that the WRA would work toward early liquidation of its centers. None would be closed in less than six months, he added, but it was hoped to close all inside of a year.

He called on State and local officials, especially on the West Coast, and public and private agencies, including church and welfare groups, and the American Legion and other veterans organizations "to aid these people and by so doing to show their devotion to the American principles of charity, justice and democracy."

December 19, 1944

URGE NISEI TO STAY AWAY

But Hood River Legionnaires Pledge Aid to Bar Violence

HOOD RIVER, Ore., Dec. 23 (AP)—The Hood River American Legion Post published quarter-page advertisements in the local newspapers today urging Japanese not to return to Hood River County.

The post, which recently scratched Japanese-American service men's names from its honor roll, has been backing a campaign to prevent Nisei from resettling in this apple valley.

"Public records show that there are about twenty-four or thirty families out of some 600 Japanese who have not already sold their property in Hood River County," said the advertisement. "We strongly urge these to dispose of their holdings. If you desire assistance from this post in disposing of your land we pledge ourselves to see that you get a square deal.

"If you do return we also pledge that to the best of our ability we will uphold law and order and will countenance no violence."

December 24, 1944

SNUBBED JAPANESE DIES A PACIFIC HERO

UNITED STATES ARMY PACIFIC HEADQUARTERS, Feb. 15 (AP)—Frank T. Hachiya, 25, of Portland, Ore., one of sixteen Japanese-Americans whose names have been stricken from the county memorial roll by the Hood River, Ore., American Legion Post, died after performing a dangerous volunteer mission, the Army reported today.

Hachiya, attached to the Seventh Division, was fatally wounded on Leyte Dec. 30. He died Jan. 3 after most of the men in his regiment had volunteered to give him blood transfusions.

Lieut. Howard Moss, Hachiya's commanding officer, said he had volunteered to cross a valley under Japanese fire to scout an enemy position. Information on enemy disposition was essential. At the bottom of the valley Hachiya worked ahead of his protecting patrol.

"A Jap sniper let Frank have it at close range," Moss related. "Frank emptied his gun into the sniper. Shot through the abdomen, Frank walked back up the hill. Medics gave him plasma and started him to a hospital. He was operated on immediately, but the bullet had gone through his liver and he died."

Hachiya attended the University of Oregon. He enlisted shortly after the Pearl Harbor attack. He had served through the Kwajalein and Eniwetok invasions.

His father, Junkichi Hachiya, is in a War Relocation Authority camp.

February 16, 1945

RANKIN SHAKES FIST AT CELLER IN HOUSE

WASHINGTON, Feb. 7 (AP)—A debate in the House today on religious rights and beliefs ended with Representatives Emanuel Celler of New York and John E. Rankin of Mississippi, both Democrats, with the latter shaking his fist at the former as they stood face to face.

The controversy started when Mr. Celler said the American Dental Association had urged that religious tests be required for entrance into dental colleges. Such a requirement, he said, would be "un-American."

Mr. Rankin, jumping to his feet, shouted:

"I am getting tired of the gentleman from New York raising the Jewish question in the House and then jumping on every man who says anything about it. Why attack the American Dental Association? That organization has done what it had the right to do. I wonder if the gentleman knew that 90 per cent of the doctors who get on the Civil Service roll are Jews, and 60 per cent of the ones we are compelled to accept in our veterans' hospitals are Jews.

"Remember that the white Gentiles of this country also have some rights."

Mr. Celler contended that he had not raised the Jewish question and that the Southerner's statements were "false, unfair and outrageous."

February 8, 1945

Anti-Bias Ad Demand High

The Advertising Council reported yesterday that it will issue soon a new set of ads in support of its "United America" campaign against racial and religious prejudice. Response on the first set, released nearly four weeks ago, has been unexpectedly high, surpassing that on many council campaigns, said Ted Royal, account executive on the project. Within the past twenty-four days orders for newspaper mats have averaged 2,500 from every state except four. These four, he noted, are not Southern states. Requests for reprints have reached more than 10,000. The council also plans to issue some special ads for magazines, and a kit of cartoons by a dozen noted artists.

March 23, 1948

Students Win Anti-Bias Battle; Fraternity Board Reverses Stand

By MORRIS KAPLAN
Special to THE NEW YORK TIMES.

WASHINGTON, Nov. 26 — The National Interfraternity Conference struck a blow today against discrimination in college Greek-letter societies by recommending the elimination of restrictive membership provisions. This action was a reversal of its previous stand that the controversial issue was "not proper" for consideration by the conference.

A statement of its resolutions committee was approved by a standing vote of 36 to 3. Nineteen of the fifty-eight member fraternities abstained from voting.

Delegates refused to adopt a more strongly worded resolution offered by Alexander Goodman of Baltimore, executive secretary of the Phi Alpha Fraternity. This sought to "repeal and abolish" any by-law or constitutional provision that discriminates "against any college student because of his religion, race, color or creed."

The delegates adopted without discussion, however, the precedent-making statement that could alter membership patterns in some 2,700 chapters of fraternities throughout the country. The action culminated a victorious struggle by undergraduate fraternity leaders from New England and the Middle West to bring the issue out into the open.

With no vote in the conference of graduate fraternity officers, the students met informally yesterday and demanded the repeal of discriminatory clauses. They presented their views through the resolution offered by Mr. Goodman, thus prevailing on delegates to revive the issue after it had been excluded from the program.

Text of Adopted Resolution

The resolution as adopted read as follows:

"Resolved that it is the sense of this conference that (1) it recognizes that many member fraternities have had and now have restrictive provisions. (2) It recognizes that the question is of concern to many interested parties. (3) It calls these facts to the attention of all member fraternities, appreciating that membership is an individual fraternity responsibility.

"(4) It recommends that member fraternities that do have selective membership provisions consider this question in the light of prevailing conditions and take such steps as they may elect to eliminate such selectivity provisions."

The statement emphasized also that the question was one which had interested many college officials, likewise fraternity men, both undergraduate and alumni.

The conference tabled a resolution that labeled the fraternity as "an instrument of student self-government," and holding that qualifications for membership should be determined by the fraternities themselves "free from dictation on the part of the colleges and universities."

The number of fraternities affiliated with the conference rose to fifty-eight with the reinstatement of Tau Kappa Epsilon, suspended last year because it activated a chapter at Tri-State College at Angola, Ind., not accredited by the conference.

Poor Scholarship Stirs Action

Delegates expressed concern over the reported failure of fraternity members to achieve good scholarship. Ralph W. Wilson, scholarship counselor, reported that only 803 out of 2,027 chapters were equal to or above the all-men's average in their respective colleges in the 1947-48 academic year. With 1,224 below that average, he said, "some of the worst" have yet to be heard from.

Delegates resolved to seek to stimulate and encourage good scholarship as the major objective of college life to the end that it be made synonymous with good fraternity membership.

The conference recommended also that its members participate in measures to combat "subversive influences and activities" on college campuses. This was urged in an address yesterday by Attorney General J. Howard McGrath. Members also were urged to endorse the giving of assistance to local agencies working in the field of "eliminating juvenile delinquency."

The conference recognized too the desirability of "increased public relations efforts," appreciating that the public attitude depended primarily on the performance and accomplishments of fraternities.

The conference closed its forty-first annual session with the election of William J. Barnes, a patent lawyer of 20 Exchange Place, New York, as chairman succeeding Judge Frank H. Myers of the Municipal Court, Washington. Mr. Barnes is a graduate of Stevens Institute of Technology in 1924 and a member of Theta Xi fraternity.

Other officers elected were these:

Vice chairman, A. Ray Warnock, retired Dean of Men at Pennsylvania State College, a graduate of the University of Illinois, Beta Theta Pi.

Secretary, Charles E. Pledger Jr. of Washington, George Washington University and Theta Delta Chi.

Treasurer, Clarence E. Yeager of Attleboro, Mass., University of Kentucky and Pi Kappa Alpha.

Educational adviser, Joseph A. Park, Dean of Men at Ohio State University and Alpha Tau Omega.

November 27, 1949

GROUP BIAS TERMED A FAILURE'S ESCAPE

Survey Shows How Blame for Shortcomings Is Shifted to a 'Weaker' Unit

By LUCY FREEMAN

The function of anti-Semitism, or any other form of group prejudice, is self-defense against one's own failures by projecting them on a weaker group, according to the conclusion of a comprehensive study of prejudice sponsored by the American Jewish Committee in cooperation with leading universities and colleges over the country.

Dr. John Slawson, executive vice president of the committee, explained in an interview that this finding represented the one common denominator in the five separate studies that are part of the project.

Based on research by anthropologists, historians, psychologists and other social scientists, these studies represent a pioneer attempt in this country to investigate the nature of racial, religious and ethnic prejudices, and to analyze their effect upon those who harbor them, as well as upon the community as a whole, Dr. Slawson explained.

Projection of Responsibility

The studies all show that "the prejudiced person does not accept responsibility for his own failures but projects them on to the group that is different in race, religion or color from his own," Dr. Slawson said.

"In other words," he continued, "all the findings point to the fact that prejudice and its expression, discrimination, are due to a sense of weakness that needs bolstering up by producing psychologically in one's mind a group inferior to one's own."

The weakness in a person that uses a "scapegoat" for its relief need not be economic—it can be emotional or social, he said. Conformity with the patterns of group prejudice held by the "elite" groups in the population becomes a great incentive to pursue this escape mechanism of prejudice, be it religious or racial, he declared.

The studies also point up the finding that a person who is hostile to one religious or racial group tends to be hostile to a number of others, and frequently to all others except his own.

Dr. Slawson said that the most practical approach to the problem of prejudice and discrimination was to establish the pattern that such attitudes and acts are "contrary to good taste and the moral code of one's group."

Five Separate Studies

The first study, "Prophets of Deceit," describes the techniques of the American agitator; the second, "Rehearsal for Destruction," traces the historical development of political anti-Semitism in imperial Germany from 1870 to 1914; and the third, "The Authoritarian Personality," is an intensive psychological study of the anti-Semitic and prejudiced personality on the basis of tests and interviews with 2,000 persons.

The fourth, "Anti-Semitism and Emotional Disorder," shows the connection between anti-Semitism and emotional disturbances in the prejudiced person, and the final volume, "Dynamics of Prejudice," shows the relationship between the veteran's social and economic adjustment and his prejudices toward other groups in the American population.

The series, which has been several years in preparation, is expected to constitute a reservoir of knowledge yielding better methods for the protection of democracy against the inroads of bigotry and totalitarianism, Dr. Slawson declared.

The studies were initiated by him in May, 1944, and represent the first part of a large program to be undertaken by the American Jewish Committee's Department of Scientific Research in cooperation with universities and colleges.

The first two volumes of the five-volume series, to be called "Studies in Prejudice," have just been published by Harper & Brothers. The other three will be issued in the next few months.

December 27, 1949

CLASS PREJUDICE IS FOUND POTENT

By EMMA HARRISON
Special to The New York Times.

WASHINGTON, Aug. 28—Social class distinction was described here today as a "sleeper" more powerful than racial prejudice in producing stereotyped thinking about people.

The theory was put to a test with white and Negro school children in a segregated school system immediately following racial disturbances. Class prejudice still oustripped racial biases in producing glittering generalities about types of people, a psychologist told the American Psychological Association's sixty-sixth annual meeting.

Dr. Rachel T. Weddington, research associate at the Merrill Palmer School in Detroit, made the study during a period of race tension in Gary, Ind., in 1948. He said that both Negro and white pupils had ascribed more favorable traits to middle class persons of both colors than to those of lower social status.

Dr. Weddington conducted her tests by showing pictures of persons to 374 elementary school children about equally divided as to color, class status and sex. The paired pictures presented the youngsters, all between the ages of seven and ten, with combinations of color and class comparisons.

Middle Class Bias Shows

Designations such as "honesty" were almost invariably ascribed to the person in the picture more readily identifiable as "middle class" than to one obviously "lower class." This was true regardless of the color of the subjects and of the children making the identifications.

August 29, 1958

EDUCATION FOUND NO CURE FOR BIAS

By IRVING SPIEGEL

A sociologist and educator said here yesterday that formal education alone would not eliminate deep-rooted prejudices against minority groups.

Dr. Charles Herbert Stember, Professor and Chairman of the Department of Sociology at University College, Rutgers University, held that the better educated people were more likely to reject social contacts of more than a casual nature with minority group members.

However, the better educated, he said, are less likely to harbor sterotyped conceptions about minority groups.

His study was made public by the American Jewish Committee at its annual meeting, held at the Roosevelt Hotel.

Dr. John Slawson, executive vice president of the agency, said that although the study did not attempt to predict how "social gains can be achieved through education," it raised "fundamental questions which must be answered before education can make its maximum contribution to better intergroup relations."

Dr. Stember found that the impact of education was limited, maintaining that its chief effects were "to counteract the notion that members of minority groups are strange creatures with exotic ways" and "to diminish fear of casual personal contact."

Noting that there seemed to be no appreciable change in personal prejudice during a student's college years, the study found education most effective "among persons of lower socioeconomic status."

Dr. Stember suggested that the better educated were more susceptible to the prevailing political and social climate, saying that "when anti-Semitism in American society has risen or fallen the changes have been sharpest among educated people."

His study maintained that members of the families in the lower economic brackets tended to be less prejudiced than those of higher status, and that the highest level of prejudice at the end of the college experience could be found among those who majored in business.

April 29, 1961

TEACHERS ADVISED ON RACIAL TERMS

By FRED M. HECHINGER

The school system asked the city's teachers yesterday to remember to speak positively in talking with parents. In a "compilation of observations of speech patterns that lead to difficulty in meeting with parents," it urged teachers to avoid words and phrases that "inadveterntly suggest prejudice" of all kinds.

Among the phrases the pamphlet lists are references to minority groups like "you people" or "your kind."

The pseudo-complimentary terms it asks teachers to avoid are "Negroes naturally have rhythm," "thrifty as a Scotsman" or "smart as a Jew." It cites as offensive stereotypes such "symbols" as "watermelon for Negroes; pushcarts for Jews; laundries for Chinese; rice and beans for Puerto Ricans; spaghetti and meatballs for Italians."

Basis for Discussion

The report is suggested "as a basis for stimulating discussion by school staffs at conferences and as material for inservice courses in human relations for teachers, supervisors and other school employes."

It was prepared by a committee under the chairmanship of Dr. Frederick H. Williams, director of the school system's human relations unit.

It warns against the use of terms that remind anyone "of his group membership and of the status of his group in our social structure."

From a warning against camouflages for prejudice—"some of my best friends * * *," etc.—the report branches out into a discussion of what it considers misuse of professional terms.

It urges teachers to avoid or use "with extreme caution" such labels as "underprivileged children, culturally deprived, slum areas, low socio-economic," as well as "fear of walking in neighborhood" or "complete apathy of parents." Somewhat mysteriously, the same list includes the term "dedicated teacher."

Parents' Faith Cited

Negative words, the report says, "emphasize disadvantages" and destroy the parents' faith in "the upward mobility of their children."

In its glossary of positive words, the report suggests that "deprived, depressed or disadvantaged" children be described as "having a likelihood for good intellectual development, or children with untapped potential, or children with latent ability."

In a more elaborate list of substitutions, "low-income or underprivileged children" become "children unable to secure much beyond the necessities of today's world because of the modest finances of the family."

"Culturally deprived children," a term originally introduced by educators as a substitute for "poor," becomes, in the new glossary, the term "children whose experiences, generally speaking, have been limited to their immediate environment."

October 4, 1961

DISTORTED 'IMAGE' DEPLORED BY JEWS

Anti-Defamation Unit Says False Views Persist

By IRVING SPIEGEL
Special to The New York Times

WASHINGTON, Feb. 4—The Anti-Defamation League of B'nai B'rith asserted today that despite a decrease in discrimination, distorted views concerning Jews "remain deeply embedded in the minds of too many Americans."

This disclosure came from a national sampling of attitudes toward Jews that is part of a $500,000 research program into anti-Semitism in the United States. The program is being conducted by the Survey Research Center of the University under a grant from the league.

Presenting the report at the league's 52d annual meeting at the Shoreham Hotel, Dore Schary, the agency's national chairman, deplored the fact that "these mythical or unclear images of Jews persist."

'Ancient Canards'

The report indicated that between 29 to 59 per cent of the adult American population still believed "the ancient canards that Jews engage in sharp business practices, control international finance, are excessively clannish, and have more money than anyone else." About the same range of percentages is "not sure" whether such statements are true or false.

At the same time, the report disclosed that "62 per cent of those queried said they liked Jews more after getting to know them better." Mr. Schary said that there had been "a subtle, steady decline" in anti-Jewish discrimination "especially where there has been increased contact between Jews and non-Jews."

The University of California report, based on a nationwide sampling, showed that "stereotypic thinking" about the physical appearance of Jews remained strong. The report said that 43 per cent of those questioned "thought they could tell whether a person was a Jew just by the way he looks."

The study cited the following figures as "encouraging and an improvement over times past" based on answers to direct questions:

¶76 per cent said Jews were warm and friendly.

¶74 per cent said Jews "are becoming more and more like other Americans."

¶83 per cent said it would make no difference to them if they had Jewish neighbors.

¶87 per cent said companies "should hire the best people available whether they are Jewish or not."

Mr. Schary said that "homogenization" was not the answer to anti-Jewish prejudice and discrimination.

"The ideal society," he said, "is not one in which all Americans blend into the scenery but rather one in which legitimate differences are as much respected as similarities."

February 5, 1965

German-Americans Provoked By Portrayal of Germans on TV

By PAUL HOFMANN

German-American community leaders started "exploratory" discussions last week of a proposed effort to combat disparagement of their ethnic group.

Television is the main target of complaints by German-Americans. They say that programs presented by various networks portray Germans either as spies, "mad scientists," sadists and other sinister characters, or as bumbling fools.

"We object to anti-German programs, not to anti-Nazi ones," said Erwin Single, who has been investigating prospects for a German-American counterpart to the anti-defamation organizations of other ethnic communities. Mr. Single is editor of The New Yorker Staats-Zeitung und Herold, the 132-year-old German-language daily.

Like the Others

"Other groups have their 'anti' organizations, the Italians have even two," Mr. Single said in an interview Friday. "Maybe we ought to have one."

The two Italian groups he referred to are the National Italian-American League to Combat Defamation, Inc., and a younger body, formerly known as the American Italian Anti-Defamation League, Inc., now called simply A.I.D. The latter group is in search of a new name as a result of a court order in October to drop its original name.

The injunction was obtained by the Anti-Defamation League of B'nai B'rith, the Jewish service organization, which contended that the words "anti-defamation league" have been closely identified with its own activities during the 53 years of its existence.

The creation of a German-American Anti - Defamation League was announced in Chicago recently. It was reportedly prompted by a television documentary on present-day Germany that German-American and West German spokesmen denounced as biased.

Mr. Single said that German-Americans in Canada had recently started letter-writing campaigns to protest what they considered anti-German slurs by television and other media. The Staats-Zeitung also has received many complaints from readers about television shows and publications they considered damaging to the German image, its editor said.

"Effective fight against anti-German television films and documentaries as well as improvement of German prestige in America," was a major item on the agenda of a meeting of the German-American Committee here Friday night. The committee, which includes representatives of major German-American organizations in the metropolitan area, met at the Liederkranz Club, 6 East 87th Street.

Reporters were not admitted to the meeting, and no immediate information on any decisions was made available.

"No interviews," Gustave I. Jahr, secretary of the committee, said yesterday.

Asked whether any conclusion had been reached on the proposals for an antidefamation effort, Mr. Jahr, a lawyer, said: "I am not at liberty to talk. Nothing has been formulated, or put on paper. This matter must be digested."

The committee's agenda also included a project for the creation of a German House in the city as a cultural center for German-Americans in the metropolitan area, estimated at 350,000. Work for a similar community center has been started in Detroit recently.

February 25, 1968

Poles and Italians Threaten Legal Action on Ethnic Jokes

By PAUL HOFMANN

Polish-American and Italian-American groups are planning a joint drive to fight alleged ethnic slurs.

A recent spate of "Polish jokes" and "Italian jokes" has depicted members of the two minorities in an unflattering or even unsavory way, adding to other forms of ethnic disparagement, spokesmen for the two communities said yesterday.

They said they had considered legal action against television networks as prime offenders.

In a related move, an official of Americans of Italian Descent, Inc., announced yesterday that the 31,000-member organization contemplated a formal protest to the Italian Government against a proposed film on Christopher Columbus.

The Italian-American spokesman, Joseph Jordan, quoted information from Hollywood and Italy as having indicated that the Genoese who discovered America would be portrayed as a liar, a bungler, a lecher and a poor navigator in the proposed film.

"This would tarnish the Italian heritage and offend America, too," Mr. Jordan, program coordinator of the Italian-American organization, said in an interview. "We plan to protest with the Italian Embassy requesting that the Italian Government stop the project."

According to Mr. Jordan, reports about the proposed film said it was to be shot in Italy with Marcello Mastroianni, the star of "La Dolce Vita" and "Divorce—Italian Style," as Columbus. Edward Dmytryk of Hollywood was said to have been mentioned as director of the picture.

Representatives of the Italian-American group and the Polish-American Guardian Society, a six-year-old organization that says it has 3,000 members and is endorsed by many Polish-American community groups, are scheduled to meet here tomorrow to discuss collaboration.

The conference was arranged in connection with a membership meeting of Americans of Italian Descent at the Waldorf-Astoria Hotel at 8:15 P.M. tomorrow.

From the Polish-American group's Chicago headquarters, its president, Leonard Jarzab, said in an interview yesterday that ethnic jokes were causing prejudice "that harms our children when they are looking for jobs."

Italians are depicted in jokes and television programs as gangsters or vicious characters, Mr. Jarzab asserted, while Polish people are represented as being vulgar and ignorant.

Jokes Are Cited

Mr. Jarzab said that his organization objected to such gags in recent television shows as one having a Polish travel bureau accept reservations on the Titanic and the Lusitania, and another one with a Polish swimming champion emerging from the water with an inner tube.

The Polish-American leader said he had proposed to the Italian-American group to retain a nationally known lawyer to take court action to enforce the Television Code by purging programs of ethnic insults. The code is a voluntary agreement setting standards for the television industry; most television stations adhere to it.

Asked for his explanation for the asserted slurs against his community, Mr. Jarzab ruled out religious motives—most Poles, like most Italians, are Roman Catholics—and suggested roots in World War II.

"The Polish question was a very sensitive issue in efforts to disunite war allies," he remarked. "We think defamation of Poles may have started at that time."

Mr. Jarzab said that "Polish people in America have been submissive to ethnic insults, whereas Jewish and Negro groups were combative."

The Polish-American spokesman, who described himself as a sales representative for a large company, expressed confidence that the proposed Polish-Italian alliance to fight defamation would eventually benefit other minority groups as well.

At the headquarters of Americans of Italian Descent, 400 Madison Ave., Mr. Jordan, who formerly operated a talent agency and now works full time for the community organization, said yesterday that a meeting of its board of directors last week approved in principle the proposals for an anti-defamation front made by the Polish-American group.

June 18, 1968

Observer: Here Come the Wasps

By RUSSELL BAKER

Wasp-Americans are as angry as wet hornets, according to John Doe. "We are sick and tired of ethnic slurs that make Wasps the object of ridicule and contempt, and we intend to fight back," Doe said the other day at the first meeting of Sting, a new organization formed to boycott or sue anybody who speaks comically or disparagingly of a Wasp.

"Wasp," of course, is an acronym for "White Anglo-Saxon Protestant." Why, Doe was asked, is it necessary to include "white"? Is this not drawing the color line against purple and green Anglo-Saxon Protestants? Not at all. Without the "white" they would have to be called Asps, a term of derogation that would be intensely degrading, Doe pointed out.

The Job to Be Done

John Doe is Sting's first president. Its vice presidents are John Smith and Bill Jones. Smith's task is to purge American television, literature and humor of all references that tarnish or belittle Wasp-Americans. Jones's job is to promote public awareness of the great contributions Wasps have made to American life.

Sting's creation, according to John Doe, is a natural result of the rising number of other ethnic and racial organizations dedicated to stamping out slurs upon Negroes, Indians, Jews, Catholics, and Americans of Irish, Italian, Polish and sundry other national backgrounds.

Disparaged With Impunity

"Things have reached a stage where Wasps are the only ethnic group that can be disparaged with impunity any more," Doe asserted. "Just the other day Polish and Italian groups announced a joint fight to stamp out Polish and Italian jokes. Jewish, Negro and Catholic jokes have been forbidden for years. If the Chinese ban Confucius-say jokes, there won't be anybody but Wasps left to tell jokes about."

Sting will begin its campaign against Wasp jokes by instituting a number of suits against nightclub comics who tell traveling salesman jokes. These jokes, Doe said, typically paint a daughter's farmer-father as a buffoon, and, since most farmers are Wasps, tend to tarnish Wasps as an ethnic group.

One thing that particularly infuriates Doe, Smith and Jones is television's tendency to portray Wasps as gangsters and idiots. "When television wants to show us a boob, it's always a Wasp like Gomer Pyle," Smith said.

Since Italian-Americans objected to the high incidence of Italian gangsters on "The Untouchables," TV gangsters have increasingly become people named Smith and Jones.

"Whenever the police want a warrant for someone they can't identify, why do they automatically give him the Wasp name of John Doe?" asked Doe. "Because they know if they called him Giovanni Verdi, Running Cloud or Sean O'Hoolihan, they'd soon have a defamation suit on their hands. We have to win the same respect for Wasps."

One of John Smith's first projects will be to get "Huckleberry Finn" banned in any school systems where it has not already been banned by Negro pressure groups. By depicting the Negro slave Jim as superior in wisdom and humanity to the Wasp thieves and cutthroats along the Mississippi, the book creates an unflattering and unsavory portrait of the Wasp character, Smith said.

"There has been an unhealthy and degrading emphasis on the Waspishness of our country's robber barons, thieves, imbeciles and Southern politicians," Bill Jones said. "All too little attention is paid to the contributions that Wasps like Mickey Mantle have made to baseball."

Jones is concerned, too, about the degradation that may result for Wasps from the contemporary tendency to equate the European conquest of America with genocide. Last fall, American Indians demanded rebuttal time to counter a TV series depicting George Custer, a Wasp, as a hero of the Old West.

Since the American conquest was effected in the time when Wasps ran the country, Bill Jones fears the present trend to turn cowboys and settlers into villains may end by making modern Wasps the object of contempt and persecution.

In the Old Days

"We certainly do not want to hurt our Indian friends by going back to the old days when they could be publicly defamed at will," Jones said. "On the other hand we do not want to have Wasps defamed simply because there is no other group that the entertainment industry can safely make villains of."

If Sting is successful, there will be nobody at all left for the entertainment industry to turn into villains, and nobody left to make jokes about, to write books about, to play the gangster, the thief, the pirate, the buffoon, or anything else relevant to life.

"And what," John Doe was asked, "will America do then?"

"Listen to Lawrence Welk," he said.

June 20, 1968

AGNEW EXPLAINS 'POLACK' AND 'JAP'

Says He Meant No Offense When He Used the Terms

Special to The New York Times

KAHULUI, Hawaii, Sept. 23 — Gov. Spiro T. Agnew said today he meant no offense when he used the word "Polack" in referring to Polish-Americans and when he referred to an American newsman of Japanese descent who was traveling with him as "the fat Jap."

Apparently upset by wire-service reports on his use of the terms, Mr. Agnew threw aside his prepared remarks at a Hawaiian Republican picnic and spoke extemporaneously.

"Those who read their papers this morning," he said, "will find this Vice-Presidential candidate is being accused of an insensitivity.

"This is a ridiculous charge to make against the son of a Greek immigrant, accused of an insensitivity to the national pride and heritage of other people."

He said he wanted to apologize to anyone who read into his remarks a racial slur.

Mr. Agnew used the word Polack about 10 days ago at a news conference in Chicago when, after being asked if he intended to make an appeal to minority groups, he replied:

"When I look out at a crowd, I don't see there a Negro, there an Italian, there a Polack."

Incident on Plane

Three days ago, while on the way from Las Vegas, Nev., to Los Angeles, Governor Agnew came back into the section of the plane where the newsmen were. He noticed Gene Oishi, a correspondent of The Baltimore Sun.

Mr. Oishi was asleep. Governor Agnew looked at him and said to the other newsmen, "What's the matter with the fat Jap?"

Today Governor Agnew called Mr. Oishi "a reporter who I consider a friend of mine —because I never jest with my enemies."

"How important it is that we don't lose our sense of humor," Mr. Agnew said. He pointed out that what he called "racial appellations" were often used in good humor.

With a quaver of emotion Mr. Agnew explained his use of the word Polack.

"I confess Ignorance," he said, "because my Polish friends have never apprised me of the fact that when they call each other by that appellation it is not in the friendliest context."

For his clarification of his racial attitudes, Mr. Agnew was applauded by an ethnically mixed crowd of 300, of whom many were of Japanese stock.

September 24, 1968

Thousands of Italians Rally to Protest Ethnic Slurs

By PAUL L. MONTGOMERY

Tens of thousands of Italian-Americans filled Columbus Circle yesterday afternoon for a communal outpouring of pride in the land of their forebears and outrage at the practice of equating Italians with criminals.

In a vast, waving field of American flags and Italian tricolors, the crowd sang and danced and cheered the two-and-one-half-hour stream of speakers. Their greatest applause was reserved for Joseph Colombo, who is carried on Justice Department lists as the leader of one of the six Mafia "families" in the city.

The first Italian-American Unity Day rally drew men, women and children from all over the metropolitan area. Many stores and restaurants in Italian neighborhoods were closed in honor of the day. Activity in the Port of New York was scanty as only 10 per cent of the longshoremen usually shaping up on Mondays reported for work.

The rally — a combination protest and celebration — attracted dozens of politicians, including Richard L. Ottinger, the Democratic nominee for the Senate; Paul O'Dwyer, who lost to him in last week's primary; former Controller Mario A. Procaccino, Deputy Mayor Richard R. Aurelio and Representative Adam Clayton Powell.

In the midst of the festive atmosphere and ethnic rhetoric, however, there was an edge of pent-up anger — what Italians call being "affocato." It came out after the rally, when 12,000 from the gathering marched through Central Park to the headquarters of the Federal Bureau of Investigation at Third Avenue and 69th Street.

The F.B.I. office has been picketed nightly since April 30 by supporters of the Italian-American Civil Rights League, which organized the rally. The group accused the bureau of discriminating against Italian-Americans in its surveillance of underworld activities. The surging crowd of marchers forced the police to fall back twice around the F.B.I. office as barriers were trampled underfoot and tempers flared.

Two policemen on crowd-control duty were stabbed in a scuffle on Third Avenue near 70th Street during the confrontation. Although there were thousands of marchers in the area, the police could find no one who said he had seen the incident.

The two patrolmen, one stabbed in the back and one in the chest, were in fair to satisfactory condition last night at Lenox Hill Hospital.

Mr. Colombo's eldest son, Anthony, one of the organizers of the rally, mounted a police sound truck and asked the crowd to disperse.

"We have achieved a greatness here today," he shouted to the milling crowd. "We are one now. We must keep this peaceful and nonviolent. As a favor to me, be a super-human-being now—turn around and go home."

There was some booing from the gathering, but in a few minutes they were straggling down the side streets toward home. A rally marshal with a bullhorn urged them on gently with: "Have a glass of wine now and a plate of spaghetti in your house."

The picketing of the F.B.I., which gave rise to yesterday's rally, began on the day that the elder Mr. Colombo's second son, Joseph Jr., was arrested on a charge of conspiring to melt silver coins into more valuable ingots. The father and Anthony, joined by perhaps 30 friends on the first night, organized the early protests.

The elder Mr. Colombo, who lives in Brooklyn, spent 30 days in jail in 1966 for refusal to answer questions of a rackets grand jury. It is the only time he has been in prison, although he has a record of 12 other arrests on charges of minor offenses.

Anthony, who is a furniture dealer, and Joseph Jr., who has a custom tailor shop, both have families. Through the picketing and formation of the Civil Rights League, they have been vociferous in their complaints of F.B.I. surveillance of their homes and alleged harassment of their private lives.

When the plans for the rally were announced early this month, they apparently evoked genuine and deeply felt sympathy among many Italian-Americans in the metropolitan area.

"This thing just snowballed," said Natale Marcone, a retired union official who is president of the Civil Rights League.

Based on a computation of the area filled by the crowd in Columbus Circle yesterday, and assigning two square feet to each person, the gathering would have numbered 31,200. The police estimate was 40,000. Mr. Marcone's was 620,000.

Columbus Circle was filled with red, white and green streamers and Italian and American flags formed of carnations. Hawkers moved through the crowd in the sun, selling buttons that said "I'm Proud to be Italian" and "Kiss Me, I'm Italian" and "Italian Power" for 50 cents each.

Anthony Colombo taking part in unity day program.

Supporters of the protest carried such signs as "Italians Are Beautiful" and "25% of World War II Veterans were Italian-Americans — Be Proud" and "There Are No Italo-Americans on the Board of Education." One man in a gorilla costume held a placard that said: "F.B.I. is using gorilla tactics."

Throughout the rally, which began at noon, the eight-piece Bob Chevy Orchestra played "Neapolitan Tarantella," "Volare," "You've Got to Change Your Evil Ways" and other tunes.

Although the crowd was overwhelmingly Italian-American, it was not exclusively so. Steven Teitel, who is 13 years old, circulated with an Italian ice in one hand and a placard reading "Jews of Flatlands Support Americans of Italian Descent."

Thomas D. Carpenter, a bearded film worker, said he was attending out of curiousity and because he believed that "any generalization leads to discrimination."

Representative Mario Biaggi, Democrat of the Bronx, noted that of 22 million Italian-Americans in the United States, only 5,000 were involved in organized crime.

"That's better than the 99 and 44/100ths purity of Ivory Soap," he said, continuing:

"Without a doubt, the F.B.I. and its director, J. Edgar Hoover, deserve the respect of us all. Let us not fall into the trap of employing for our own use that which we condemn. Because of the misconduct of a few, let us not use a wide black brush on the F.B.I.—and let not the F.B.I. or any other law-enforcement agency use the same brush on us."

Former Controller Procaccino was loudly cheered by the crowd.

"Don't let anyone imagine that when looters, arsonists and bombers are being coddled and treated like heroes, we are going to stand by and permit the smearing and harassment of innocent people whose sole crime is that they are related, friends or neighbors, or just happen to be Italian-American," he said.

Deputy Mayor Aurelio, who represented Mayor Lindsay, was roundly booed at the mention of the Mayor's name. His brief address was inaudible to the crowd. He was asked later if he had found the auspices of the meeting prejudicial.

"The only auspices that I know of is a group of Italian-Americans protesting against discrimination in the Italian-American community," he replied.

Representative Powell was also heavily booed when he tried to speak, but his cries of, "Right on, right on" pierced the uproar. "This nation is for all, not just the Wasps," he told the crowd.

"Anyone who doesn't understand what America owes to Italy doesn't understand America," said Representative Allard K. Lowenstein, Democrat of Nassau County, to loud cheers.

Anthony LaRosa, a 12-year-old from Brooklyn, his head barely peering above the massed microphones, stilled the crowd with his address.

"I am a young Italian-American boy who doesn't want to grow up labeled," he said. "I want to grow up with my constitutional rights, not to be harassed and discriminated against."

John F. Malone, assistant director in charge of the New York division of the F.B.I., said late in the day that he would have no comment on the rally.

Two men were arrested near the F.B.I. building. The police identified them as Anthony Auciello, 32, of 1333 42d Street, Brooklyn, who was charged with harassment, and Vittorio Nuciforo, 28, of 526 Sixth Avenue, Brooklyn, who was charged with obstructing Government administration, disorderly conduct and attempted escape.

Neither was being questioned in the stabbing of the policemen, the police said.

A number of merchants, most of them not Italian, with shops in Italian areas complained anonymously of intimidation to close their businesses for the Italian-American unity day.

The manager of a clothing store on Kings Highway in Brooklyn said two men came into his store last week and told him "it would be wise" if he closed. Yesterday, he said, he opened his store as usual and about noon received an anonymous call that he would be "in a whole lot of trouble" if he didn't close.

"I closed," the merchant said. "Everybody here did."

June 30, 1970

'Mafia' Loses Its Place In Federal Vocabulary

WASHINGTON, July 23 (UPI)—With President Nixon's concurrence, Attorney General John N. Mitchell has told the Justice Department and the Federal Bureau of Investigation to stop using the terms "Mafia" and "Cosa Nostra" because they offend "decent Italian-Americans."

In a confidential memorandum to all division and agency heads, including the director of the F.B.I., J. Edgar Hoover, Mr. Mitchell said:

"It has become increasingly clear that a good many Americans of Italian descent are offended. They feel that the use of these Italian terms reflects adversely on Italian-Americans generally, and there is no doubt that their concern is genuine and sincere."

A department spokesman said that the order resulted from complains to Mr. Mitchell and was not inspired by recent protest demonstrations against the F.B.I. by Italian-Americans in New York City.

July 24, 1970

TIMES IS PICKETED BY PROTEST GROUP

Italian-Americans Complain on Use of Crime 'Labels'

A group of Italian-Americans picketed The New York Times's West Side Plant in the Lincoln Center area of Manhattan for five and a half hours yesterday morning, protesting the use of Italian "labels" for a section of organized crime.

The pickets, estimated at 150 by the police and 300 to 500 by spokesmen for the demonstrators, ranged their line to cover the plant's trucking bays on West End Avenue at 65th Street. This stopped 19 trucks from leaving and a tractor-trailer truck from entering because drivers feared endangering the pickets.

As 50 policemen prepared to clear the pickets from the truck bays, Barry Slotnick of Slotnick & Narral, attorney for the Italian American Civil Rights League, persuaded the pickets to stop blocking the driveway and they dispersed.

Meet Times Executives

Mr. Slotnick and officers of the league, headed by Natale Marcone, president, a retired furniture dealer, met for two hours with executives of The Times to present their complaints.

They told the Times executives that their chief complaint was against the use of the terms "Mafia" and "Cosa Nostra" to describe a section of organized crime. They insisted that the words not be used by The Times.

The Times does not use the term Cosa Nostra. "Mafia" has been used to describe a specific segment of organized crime in the United States.

Last week Attorney General John N. Mitchell ordered Justice Department officials to stop using the terms "Mafia" and "Cosa Nostra" because of the objections of Italian-Americans.

Complains About Story

The committee also complained about a story reporting that law enforcement agencies were convinced that organized crime had initiated the protests against discrimination by the league through Joseph Colombo, identified by the Federal Bureau of Investigation as the head of a crime syndicate. Mr. Columbo was one of the pickets and his son Anthony attended the meeting at the Times. The committee said the story was untrue.

Another complaint was the The Times did not pay sufficient attention to discriminations suffered by Italian-Americans.

A. M. Rosenthal, managing editor, told the group that The Times was sensitive to all aspects of the problem of discrimination in America and would review its performance with reference to Italian-Americans to see if there were deficiencies.

He said that other points raised would be discussed among editors and reporters but that under no circumstances would The Times respond to pressure tactics.

July 31, 1970

Orientals Find Bias Is Down Sharply in U.S.

When J. Chuan Chu came to the United States as a student at the end of World War II from his home in North China, he had trouble finding a place to live. Having an Oriental face, he discovered, was a liability.

But Mr. Chu, with an engineering degree from the University of Pennsylvania, has now risen to become a vice president of Honeywell Information Systems. He lives today in the wealthy Boston suburb of Wellesley, near his concern's headquarters.

"If you have ability and can adapt to the American way of speaking, dressing, and doing things," Mr. Chu said recently, "then it doesn't matter any more if you are Chinese."

His story reflects a quiet, little noted American success story—the almost total disappearance of discrimination against the 400,000 Chinese and 500,000 Japanese Americans since the end of World War II and their assimilation into the mainstream of American life.

Some Chinese have been left behind in the nation's depressed Chinatown ghettos, unable to speak English or too old, too poorly educated, or too fixed in their traditional ways to be assimilated.

And some younger Asian-Americans have become increasingly sensitive, like blacks and Indians, to what they consider white Americans' patronizing attitude toward them. They resent the tourists in Chinatown who politely ask if they can speak English. They indignantly reject the old Oriental stereotype of the slant-eyed, pig-tailed Chinaman, eating chop-suey and mumbling "ah-so." And they insist that many whites, behind a facade of believing in equality, are still prejudiced.

But the great majority of Chinese and Japanese Americans, whose humble parents

Joyce Dopkeen for The New York Times

J. Chuan Chu, a Chinese immigrant who became an official of Honeywell information, at his office in Waltham, Mass.

had to iron the laundry and garden the lawns of white Americans, no longer find any artificial barriers to becoming doctors, lawyers, architects, and professors.

Some have **achieved national reputations, a feat unimaginable 20 years ago: I.M. Pei and Minour Yamasaki as architects, Gerald Tsai as head of the Manhattan Fund; Tsung Dao Lee and Chen Ning Yang as Nobel prize winners in physics; S.I. Hayakawa as President of San Francisco State College, and Daniel Inouye as Senator from Hawaii.**

With one Senator and two Representatives, the Japanese may well be the nation's most over-represented minority.)

In interviews with dozens of Chinese and Japanese Americans, from executives like Mr. Chu to militant students in Chinatown, very few complaints of discrimination in jobs, housing, or education could be found.

Many people below 30 could not recall a single personal instance of discrimination. The Los Angeles Housing Opportunities Center, which helps people who feel they have been discriminated against, reports that it has had only one complaint from an Oriental in the last three years.

Some prejudice still exists. Mr. Chu, for example, was told last year that he would not be allowed to join the Wellesley Country Club.

No Oriental Members

A spokesman for the club, in a telephone interview, denied that the club discriminates against people of any race. He said that he could not recall Mr. Chu's case. However, he also said that the club at present had no members of Oriental ancestry.

Most of the problems confronting Asian Americans today are more subtle.

Dr. Ai-li Chin, a Chinese sociologist at the Massachusetts Institute of Technology who is doing a study of the Chinese experience in the United States, feels that "discrimination against Orientals has definitely diminished."

"In any statement you make about prejudice you must be very careful," she feels. "But for most Chinese the problem is not so much physical barriers, as it is for blacks, as it is the question of identity. Who are you as a Chinese in the United States?"

This is particularly true, Dr. Chin believes, for the younger people who have grown up in the United States, still having an Oriental face but not speaking their parents' language.

Reject Old Stereotype

"Ironically, at the same time as prejudice has diminished, some of these younger people have now begun to become concerned about white Americans' attitudes toward them," Dr. Chin, a diminutive, soft-spoken woman points out. "They refuse to accept, as their parents did, the old humiliating Oriental stereotype."

Under the influence of the Black Panthers and Young Lords, some young Chinese have organized radical groups with names like the Boxers and the Red Guards to try to stop tourists from visiting New York's and San Francisco's Chinatown. "This is our community, it is not a zoo," reads one sign in the Boxers' dingy store-front headquarters in New York.

Even the old stereotype, however, has undergone a metamorphosis. The pig-tailed coolie has been replaced in the imagination of many Americans by the earnest, bespectacled young scholar.

"My teachers have always helped me because they had such a good image of Chinese students," recalled Elaine Yuehy, a junior at Hunter College whose father used to run a laundry. " 'Good little Chinese kid,' they said, 'so bright and so well-behaved and hard-working.' " Her comment was echoed by many of those interviewed.

For many years, from the time the first Chinese came to California in the eighteen-fifties, prejudice and discrimination against Orientals were very real indeed. Many were barred from obtaining citizenship. In California, where the great majority settled, none born outside the United States were allowed to own land.

1854 Report Is Quoted

A report written by the San Francisco Board of Aldermen in 1854 typified the 19th century American view. "The Chinese live in a manner similar to our savage Indians," the report said. "Their women are the most degraded prostitutes and the sole enjoyment of the male population is gambling." The aldermen recommended "the immediate expulsion of the whole Chinese race from the city."

The ultimate indignity came during World War II when thousands of Japanese on the West Coast were interned in concentration camps and their property sold for 10 cents on the dollar.

Many Japanese Americans feel that white Americans' guilt over the internment has helped lead to their increased acceptance since the end of the war.

Another factor, cited by Dr. Chin, "is the general growth in awareness of racial problems since the end of World War II." Dr. Chin points to the beneficial effect of the Negro civil rights movement, with its fair housing and fair employment laws.

The old California law barring foreign-born Orientals from owning land was ruled unconstitutional by the California State Supreme Court in 1952 after a test case brought by a Japanese lawyer. And the McCarran Act of 1952 made foreign-born Orientals eligible for citizenship.

Respect for Achievements

Dr. Chin also feels that the respect Chinese and Japanese have gained by their recent achievements in the sciences and the professions has helped ease discrimination.

Where before the war San Francisco's Chinese were all crowded into Chinatown and Los Angeles' Japanese were confined to a small area in west Los Angeles, today they have spread out over all parts of their cities.

According to Jeffrey Matsui of the Japanese American Citizens League, Los Angeles' 150,000 Japanese have been able to move into exclusive communities like Bel Air, Beverly Hills, and Pacific Palisades. Two judges and the county coroner are Japanese.

Mr. Matsui reported that although Japanese on the

West Coast still have some difficulty finding executive or sales jobs, which require large amounts of personal contact, they have no trouble getting jobs as secretaries, accountants, doctors, or engineers. Before the war they were largely restricted to working as gardeners or small farmers.

The situation is much the same across the country. Chicago's 14,000 Japanese live in every section of the city and its suburbs. "There used to be discrimination here," said the Rev. Shinei Shigefuji, minister of the Midwest Buddhist Church, "but I don't think there's much now."

Like Whites in South

In the South, Orientals are now considered whites rather than Negroes. Young Chinese and Japanese growing up in such widely scattered areas as Richmond, Durham, N. C., and Atlanta remember attending white schools and going in the white entrance to movie theaters.

In Augusta, Ga., where a group of Chinese laborers was brought in 1910 to dig the Augusta Canal, their descendants have gone to college and become pharmacists, doctors, and insurance salesmen: In its small way, a typical American immigrant success story.

But for the minority of Chinese who have not been assimilated, Chinatown, behind the glitter of its red-tile store fronts and pagoda-like roofs, is a slum.

In San Francisco's Chinatown, with 50,000 people the nation's largest, 45 per cent of the families live below the Federal poverty level of a $3,700 annual income; 15 per cent are unemployed; 60 per cent share a bathroom with another family or have none at all.

Half of the people cannot speak or read English. In the heart of Chinatown, nearly half have never been to school. Employment for many means long hours sewing dresses in backroom sweatshops at wages below the legal minimum.

Still Controlled by Elders

Control in Chinatown rests as it always has with the mysterious family and district associations, ruled by elders who see no need to change their old patterns.

Some younger Chinese radicals, like those in the Boxers and Red Guards, blame white racism for the situation.

But discrimination appears to be much less of a factor than old age, preference for Chinese tradition, and difficulty with English. Japanese-Americans, who have not clung so closely to their traditional ways, have not encountered the Chinatown problem.

"We resent ignorant Americans coming in here and staring at us as if we were animals," explained Richard Lee, a 20-year-old resident of New York's Chinatown. "But we're not really discriminated against. If you can speak English you can get through school and get a job like any American."

"We still have problems," said Mr. Lee, who wears blue jeans and lets his straight black hair grow long over his ears, "but they are more internal problems, problems of what you are, than of discrimination."

Looking around his tiny, cramped room at a picture of Raquel Welch, he said:

"Our parents suppressed this identity problem for themselves, but we think about it a lot. Chinese have always considered that they are Chinese no matter where they are. Now we have to figure out what this means for us."

December 13, 1970

Chairman of Joint Chiefs Regrets Remarks on Jews

By JOHN W. FINNEY
Special to The New York Times

WASHINGTON, Nov. 13—Gen. George S. Brown, chairman of the Joint Chiefs of Staff, expressed regret today for comments that he made a month ago in which he suggested that Israel had undue political influence in Congress and that Jews controlled the newspapers and banks in this country.

In statements issued through the Defense Department, General Brown said that his comments were "unfortunate," "ill-considered," "unfounded," and "all too casual" and did not represent his convictions.

The White House, in what amounted to a reprimand to the nation's top military officer, said that President Ford felt "very strongly" that General Brown's comments were "ill-advised and poorly handled." Ron Nessen, the Presidential press secretary, said that Mr. Ford also wanted it emphasized that the comments of General Brown "in no way represent his views or the views of any senior officials of his Administration, military or civilian."

United Press International
Gen. George S. Brown

Despite General Brown's apologies, which stopped short of a direct retraction, it appeared that he was enmeshed in a major personal controversy as a result of comments that he made Oct. 10 at Duke University but that went unreported nationally until a Duke law student provided a tape recording of his statements to The Washington Post.

In response to a question after a speech on international law before Duke law students, General Brown said that if there was another Arab oil embargo, the American people might "get tough-minded enough to set down the Jewish influence in this country and break that lobby."

General Brown went on to say, however, that the Jewish influence "is so strong, you wouldn't believe, now."

"They own, you know, the banks in the country, the newspapers," he said. "Just look at where the Jewish money is."

At least for the moment, there appeared to be no inclination within the Administration to dismiss General Brown. However, because of the protests made by Jewish groups, it was apparent that his political standing, particularly on Capitol Hill, was seriously injured, and perhaps his future as Chairman of the Joint Chiefs was in jeopardy.

He took over that post last July after serving as Air Force Chief of Staff.

Mr. Nessen, while emphasizing the President's disapproval of the general's statements, said that he had "not heard of any plans for him not to remain in office."

Pentagon Statement

At the Pentagon, a spokesman said that Defense Secretary James R. Schlesinger "continues to have confidence" in General Brown and "realizes this was a very unfortunate misexpression of the general's opinions."

Various Jewish groups, however, demanded General Brown's resignation. The Jewish War Veterans said that if President Ford did not dismiss him, it would work through Congress to secure his dismissal.

The Anti-Defamation League of B'nai B'rith told Mr. Ford that General Brown's remarks were "not only false but contemptible, and have an illiterate odor of prejudice and malice" and that the general should be dismissed immediately.

Rabbi Arthur Hertzberg, president of the American Jewish Congress, sent a telegram to Mr. Ford saying that the general's remarks demonstrated "a degree of ignorance and susceptibility to classic anti-Semitic propaganda that cast grave doubts on his ability to serve in his presently critically important position."

Investigation Asked

The telegram asked the President to investigate General Brown's statements and "determine whether it is fitting for him to serve in his present ca-

Elmer R. Winter, president of the American Jewish Committee, welcomed General Brown's action in repudiating some of his remarks but expressed concern that "he has not yet withdrawn his reckless canard about 'Jewish control.'"

Senator William Proxmire, Democrat of Wisconsin; Representative Bella S. Abzug, Democrat of Manhattan—and Representative Elizabeth Holtzman, Democrat of Brooklyn, all issued statements charging General Brown with anti-Semitic statements and demanding his dismissal.

In a transcript provided by the Pentagon today, the general said at Duke:

"Now, in answer to the question would we use force in the Middle East. I don't know—I hope not. We have no plans to. It is conceivable, I guess. It would be almost as bad as the seven days in May. You can conjure up a situation where there is another oil embargo, and people in this country are not only inconvenienced and uncomfortable, but suffer.

"They get tough-minded enough to set down the Jewish influence in this country and break that lobby. It is so strong, you wouldn't believe, now.

"We have the Israelis coming to us for equipment. We say we can't probably get the Congress to support a program like this. And they say don't worry about the Congress. We will take care of the Congress. This is somebody from another country, but they can do it. They own, you know, the banks in this country, the newspapers. Just look at where the Jewish money is."

General Issues Statement

This afternoon, the general issued a statement in which he said that his "remarks might mistakenly lead to the wholly erroneous inference that American citizens and groups do not enjoy in this nation the privilege of expressing their views forcefully."

"What are called pressures lies at the very heart of democracy," he said. "We in Defense know that. We experience pressures from contractors, pressures from those opposed to defense expenditures, pressures from foreign governments.

"Moreover, my improper comments could be read to suggest that the American Jewish community and Israel are somehow the same. Americans of Jewish background have an understandable interest in the future of Israel—parallel to similar sentiments among other Americans, all of whom at one time or another trace their descent to other lands.

"I do in fact appreciate the great support and the deep interest in the nature of our security problems and our defenses that the American Jewish community has steadily demonstrated, and I want to reemphasize that my unfounded and all-too-casual remarks on that particular occasion are wholly unrepresentative of my continuing respect and appreciation for the role played by Jewish citizens, which I have reiterated to the Jewish War Veterans."

November 14, 1974

FORD REBUKES BUTZ FOR SLUR ON BLACKS

Agriculture Secretary Apologizes —Brooke, Javits Ask Resignation

By DIANE HENRY

Special to The New York Times

WASHINGTON, Oct. 1—In a rare public upbraiding of a Cabinet official, President Ford summoned Agriculture Secretary Earl L. Butz to the White House today and gave him a "severe reprimand" for making racist remarks that were "highly offensive," a White House spokesman said.

The Presidential reprimand came after New Times magazine attributed to Mr. Butz quotations in which black people were referred to as "coloreds" and described as wanting only three things. The things were listed, in order, in obscene, derogatory and scatological terms.

Upon learning of the remarks, Senator Edward Brooke, Republican of Massachusetts and the only black member of the Senate, issued a statement calling for Mr. Butz's resignation.

Senator Jacob K. Javits, Republican of New York, also called for the Secretary to step down.

Butz Calls Brooke

Mr. Butz telephoned Senator Brooke to apologize at about 6 P.M. this evening, according to Mr. Brooke's press assistant.

Mr. Brooke's reply was: "I just don't feel that a man who would make such a statement is fit to serve in the Cabinet of the President of the United States."

The origin of the quotes was an article in the most recent issue of Rolling Stone magazine, but the author, John W. Dean 3d, the Watergate figure and former assistant to the President in the Administration of Richard M. Nixon, did not name Mr. Butz. Instead he attributed the quotes to an unnamed Cabinet official.

Mr. Butz reportedly made the remarks to Mr. Dean during an airplane flight to California following the Republican National Convention in Kansas City in August.

The New Times article said that Pat Boone, the actor, who was also on the flight, had confirmed the quotes by Mr. Butz.

"The President told the Secretary the remarks were highly offensive to him and to the American people," according to a statement issued by the White House. It continued:

"The President was informed of Secretary Butz's comments late Thursday evening. Secretary Butz was summoned to a meeting with the President in the Oval Office this morning. The President informed the Secretary that such language and attitudes were not acceptable from a member of his Administration."

William Roberts, a Presidential spokesman, said he could not recall any prior incident in which another member of Mr. Ford's Cabinet has been so sternly rebuked for the behavior.

However, this is the second time Mr. Butz has been reprimanded by the President. In November 1974 Mr. Ford ordered Mr. Butz to make a public apology for a comment that seemed critical of Pope Paul VI. In a meeting with reporters prior to that rebuke, Mr. Butz, in a mock Italian accent, had told the story of an "Italian lady" who, speaking of the Pope's birth-control policy, said: "He no plays da game, he no maka da rules."

Following much public criticism at the time, Mr. Butz was chastised by the President in a 15-minute private meeting, and Mr. Butz made a public apology.

'Unfortunate Choice of Language'

Tonight an Agriculture Department spokesman, Claude Gifford, read the following statement:

"Secretary Butz issued an apology for an unfortunate choice of language used in a recent conversation and reported in the press. He regretted any offenses which may have been given any person or any group and issued a complete apology. He said that although he was merely repeating a comment made decades ago by a ward politician in a large midwestern city, even that was no excuse for the incident."

Richard B. Cheney, the President's chief of staff, said that Mr. Butz had informed him of the publication of the remarks and that he had, in turn, informed the President last night.

When asked if he had made any attempt to determine who the unnamed Cabinet official was that had made the derogatory remarks in the Rolling Stone article, Mr. Cheney said that he did not "normally read that publication" and that no one had brought the article to his attention in the two weeks it has been on the newstands.

'Vulgar and Offensive'

Senator Brooke called Mr. Butz's remarks "vulgar and offensive" and added: "No man who harbors such thought is fit to serve in the Cabinet of the President of the United States.

Senator Charles McC. Mathias Jr., Republican of Maryland, said the language attributed to Mr. Butz was "coarse and crude and certainly is not befitting any member of the Cabinet of the President of the United States—I hope the President will feel this way."

Mr. Javits said that Mr. Ford should ask for Secretary Butz's resignation. Representative John B. Anderson of Illinois, one of the top Republican leaders in the House and a long-time friend of the President, also called on the Secretary to resign.

Senator Brooke said that when Mr. Butz had telephoned him this evening to apologize, the Secretary had said he "intended no disrespect."

According to Senator Brooke, Mr. Butz had said he was asked, "Why aren't many blacks in the Republican party, and I said we're getting more all the time." Mr. Butz then told Senator Brooke that his comments after that were "not an original thought," according to the Senator.

Senator Brook said that he had asked Mr. Butz who had made the derogatory statement and that Mr. Butz replied, "An old-time ward politician." But the Senator said that Mr. Butz did not know the politician's name.

Mr. Brooke said he had told Mr. Butz: "I know you're not fit to serve in the President's Cabinet."

October 2, 1976

Dixon Writes an Apology to Nader For 'Derogatory' Ethnic Remark

By LINDA CHARLTON
Special to The New York Times

WASHINGTON, Feb. 2—Federal Trade Commissioner Paul Rand Dixon, who publicly described Ralph Nader, the consumer advocate, as a "dirty Arab," has written to Mr. Nader to apologize for what he called "a derogatory reference to your ethnic background."

In the letter received by Mr. Nader today, Mr. Dixon alluded to the disagreements between the two men, which date back at least to the issuance of a Nader-sponsored report in 1969 that was highly critical of the Federal Trade Commission in general and Mr. Dixon in particular.

As for the "derogatory reference" made at an appearance before a grocery industry group last month. Mr. Dixon wrote, "I deeply regret having made this remark and I apologize for it."

Mr. Nader, in a telephone interview, said that he viewed the letter as only a partial apology. "He didn't apologize for his more conventional obscenities," Mr. Nader said. He was refering to the fact that on the same occasion, Mr. Dixon also described Mr. Nader with a four-word vulgar phrase.

But Mr. Nader went on to say that neither the "ethnic slur" nor the other insult was the issue that concerned him. "The issue is his [Mr. Dixon's] fitness to hold office," Mr. Nader said. He said that Mr. Dixon's record in his 16 years at the commission and his "deep-rooted prejudice" should disqualify him from continuing as a member of the commission.

Mr. Dixon's term in office expires in 1981. He can be removed by the President only for cause—a serious dereliction of duty approaching criminal activity.

Among the continuing protests against the Dixon remarks made public today was a statement from Leonard Woodcock, president of the United Automobile Workers. He said that Mr. Dixon had "crossed the line of human decency" and added: "His ugly remark is no more easily erased by simply an apology than the infamous Earl Butz joke. It is about time we begin to demand decent respect for fellow human beings from high government officials."

The joke by Mr. Butz about blacks led to his resignation under fire as Secretary of Agriculture in the Ford Administration.

The American Jewish Committee, in a letter to chairman of the Federal Trade Commission, Calvin Collier, wrote: "There is no place in our governmental structure for people with such views. Mr. Dixon should resign." The letter also said that "descent to racial slurs" took controversy into the "sewer of racial prejudices."

The Anti-Defamation League of B'nai B'rith sent a telegram to Mr. Dixon saying, "We believe it is incumbent upon you to make a public apology not only to Mr. Nader but to all Americans."

February 3, 1977

PRIDE IN ETHNIC ROOTS

White Ethnic Groups in Nation Are Encouraging Heritage Programs in a Trend Toward Self-Awareness

By GENE I. MAEROFF

America's white ethnic groups, apparently stirred by the rise of black consciousness and a heightened sense of their own group identities, have begun perssing for programs to help young people of European extraction explore their heritage and the immigrant experience of their forebears.

The mounting interest at schools and colleges in white ethnic studies, complete with slogans such as "Italian Power" and "Polish Is Beautiful," is evidenced in New York City and across the country.

"There is growing self-awareness among European-American groups that were so long raised to believe in the idea of the melting pot," says Dr. Richard Kolm of Catholic University in Washington. A sociologist, he also is the founder and chairman of the National Ethnic Studies Assembly.

315 Courses

A survey financed by the United States Office of Education and two other Federal agencies found last year that 135 colleges and universities were giving 315 courses in the area of white ethnic studies.

Invariably, the offerings have been added to curriculums that already included programs focusing on blacks, American Indians, Chicanos and Puerto Ricans.

Contrary to the philosophy of assimilation, ethnic pride and ethnic indentification are being stressed in the new programs. Some educators contend that a better understanding of one's ancestry make a person more tolerant of others.

The attention to white ethnic studies seems related to an awakening of interest generally in ethnicity. An outpouring of books such as Michael Novak's "The Rise of the Unmeltable Ethnics" and a proliferation of conferences on the subject have occurred.

Federal funds are being made available to support ethnic studies under the Ethnic Heritage Studies Act that is to provide $2.5-million this year for projects involving white groups as well as the black and brown minorities.

"The action marks the first time in American history that Congress has recognized that the melting pot was not the reigning theory of American society but that ethnic pluralism was much more the reality," says Irving M. Levine, director of the National Project on Ethnic America for the American Jewish Committee.

Among the schools adopting white ethnic studies is the City University of New York. Courses in Jewish and Italian-American studies are under way. One of the university's units, City College, has a program in Slavic-American heritage, which is taught under the Puerto Rican Studies Department for want of its own home base.

Polish-American Studies

At Wayne State University in Detroit, Polish-American studies began last fall and a course entitled "The Italian-American Experience" was started this month by Dr. Richard Raspa. He grew up in the heavily Italian neighborhood on Philadelphia't south side and got a doctorate in English at the University of Notre Dame.

Similar courses are being considered at Wayne State for students of Irish, Armenian, German and Greek extraction.

Bruce Spector, a student at the University of Florida, which has an enrollment of about 100 in its Jewish studies program, said he was glad to see such courses "because when you go to the public schools and study history you get the Pilgrims, the Civil War and then a little Roman and Greek history and if you're in an ethnic group you realize they're skipping important things."

Now, however, even the elementary and secondary schools are exploring white ethnic studies.

In New York City, Dr. Seymour P. Lachman, the president of the Board of Education, is encouraging the movement and has asked for a report on the progress of ethnic studies in the city's schools.

One Manhattan public school, P.S. 75 on the west side, serving many cultural backgrounds, recently received a $45,000 grant from the Rockefeller Brothers Fund to develop a multi-ethnic project that will include pupils of European ancestry.

Illnois is requiring a statewide program of ethnic studies by 1975, to 'include a study of the role and contributions of American Negroes and other ethnic groups including but not restricted to Polish, Lithuanian, German, Hungarian, Irish, Bohemian, Russian, Albanian, Italian, Czech, French and Scots."

The feeling is spreading that ethnicity should be a source of pride and not an embarrassment, as it was to the grandparents of so many of today's students.

"Ethnicity is no longer considered something to get rid of as quickly as possible," says Dr. Richard Gambino, who formed the Italian-American studies program at Queens College.

Dr. Gambino's book, "Blood of My Blood: The Dilemma of Italian-Americans," will be published in April. In addition, textbook publishers are increasingly taking notice of white ethnics in their presentations.

In Detroit, which has done more than most cities with ethnic studies, a visitor the other day in a sixth-grade classroom at the Casimir Pulaski Public School observed the pupils studying Italian-American heritage.

Anna Castiglione brought some loaves of Italian bread that her mother had baked; Steve Vieto brought a huge green, white and red Italian flag that he had made at home from a bedsheet, and Joey D'Angelo brought a jug of chianti that was examined but not consumed.

As the lesson progressed, the teacher asked the children to stand and declare their ancestry.

Pupils of Polish, Italian and German background predominated. Finally, only one girl had not spoken.

In a barely audible voice, she said "Jewish" and sank into her seat, apparently feeling alone and isolated.

Some critics have mentioned the possibility of such incidents and have suggested that ethnic studies could exacerbate polarization. They have expressed concern that an emphasis on differences, improperly handled, could undermine whatever unity exists in America.

Writing in Commentary magazine in 1972, Dr. Harold R. Isaacs, a political scientist at the Massachusetts Institute of Technology, pointed to possible negative results in an article entitled "The New Pluralists." He wrote of white ethnic studies:

"Sounds reasonable, doesn't it, just like the black studies programs? But the same questions hold. Just who will represent the 'community?' Who will decide what is to go into each program?"

He asked: "Which Hungarians, for example, will decide what version of contemporary Hungarian history is to be taught to young Hungarian-Americans? And what kind of pressure will this system apply, in the public domain, on the child whose family has other ideas about its own or any other ethnic background?"

The Rev. Andrew M. Greeley, who has written widely on ethnicity, concedes that such programs "could become chauvinistic and hostile to other groups," but he believes that beneficial effects are more likely.

"I don't think that pretending diversity is not there will make it go away," says Father Greeley, who directs the Center for the Study of American Pluralism in Chicago. "Some think that talking about differences will make things worse, but I think it will make it better because he who understands his own traditions is more likely to be sympathetic to someone else."

Despite the current activity, the future of white ethnic studies is unclear. Even the black studies movement, which is older, stronger and more pervasive, is experiencing a slackening of interest.

But proponents of white ethnic studies say that, at the very least, the movement, black and white, offers promise of getting teachers, textbook writers and students to think about the roles and contributions of ethnic groups.

Michael Novak sugests that ethnic studies could eventually be integrated into social studies, history and economics rather than be pursued as separate programs.

Visits to ethnic studies programs in classrooms have disclosed a number of instances in which understanding was promoted. However, there were exceptions. One girl at Chadsey High School in Detroit, when asked why she was taking Polish studies, snapped: "I ain't no Polak. I'm Irish and I'm only in here because it's easy and there's nothing else for me to take."

January 28, 1974

Pure as the Driven Slush

By Richard Sennett

A myth is an idea people need to believe in, whether or not it is true. All societies, from the primitive to the overcivilized, are held together by such ideas. One of the most powerful of these ideas is the myth of decline, the conviction that the present is inferior to the past. Today, American society is finding a new set of symbols to express a peculiar version of the myth of decline—symbols derived from the experience of ethnic groups.

The communal life of ethnic groups in America is interpreted by people across the political spectrum as something special and precious; ethnic groups are portrayed as warm, open, and caring among themselves. They are seen as threatened, like an endangered species, by all the homogenizing pressures of American society—upward mobility, mass culture, rootlessness. The virtues of ethnic community in the past have become a yardstick to measure present-day communal emptiness, and, not coincidentally, an ideological weapon to fight reforms like racial integration.

Given the actual facts of ethnic community life in America, there is something obscene about politicians like Gerald Ford or Jimmy Carter celebrating "our precious ethnic heritage." The history of most ethnic groups in America, white as well as black, is appalling.

Most European peasants who migrated here had no consciousness when they came of being "ethnics." They identified themselves as members of a village or stetl; in America, unable to buy land and converted into an urban proletariat, they suddenly found their language, family patterns, even food habits, treated by the larger culture as signs of cultural inferiority.

Furthermore, the first generations of ethnics never experienced anything like the community "purity" these politicians speak of. The Lower East Side in 1910 was probably the most polyglot urban settlement in the world. Being an ethnic, any kind of ethnic, radically restricted where one could live as well as what one could work at; the ethnics found themselves jumbled together, among a mass of people who often could not stand each other's religions, understand each other's speech, or make sense of each other's customs.

That these tense, confused communities managed to work at all is the genuine tribute to be paid to the people who inhabited them. Celebrating ethnicity *per se* means celebrating the badges of cultural inferiority American society forced the agrarian immigrants to wear.

As the various ethnic urban groups gained a toehold in the American economy, they withdrew from each other. The pressures to make secure each ethnic group as a world unto itself were greatly increased by the coming of Southern blacks to Northern cities, for the latest migrants seemed to threaten to pull the Europeans back down into the chaotic world from which they had escaped. Granted that today the white ethnics' efforts to preserve ethnic homogeneity are more complicated and sympathetic than simple racial prejudice, nonetheless it is difficult to understand what a society has to celebrate when it forces people to act as if they were racists.

Finally, the issue of ethnic identity is a painful one between the generations in many ethnic families. People who have grown up in ethnic communities have often felt suffocated by them; when they leave, the old feel the ethnic culture is being abandoned because there is something unhealthy about it. This is more than a half-truth; in the extended families of many European ethnics, parental control of adult children is justified in the name of keeping up "tradition." The celebration of ethnicity ignores most of these realities.

Idealizing ethnicity fits into a pattern that antedates the large-scale arrivals of European peasants by more than a century. From the late 18th century on, concern about the eclipse or weakening of community is a constant theme in American writing. Madison, Tocqueville and Olmstead all worried about the decline of community. Many Progressives at the turn of the century feared ethnics because they feared that these outsiders threatened community life. By a perverse irony, modern-day ethnic consciousness is giving the myth of communal decline a new life.

The purpose of a myth of decline is not to revive the past but to create an attitude of resignation about the present. If what really matters has vanished, if community has broken down, then those left in the wake have some justification for feeling apathetic.

Americans are a peculiar people: economically aggressive, socially passive, not terribly interested in each other, convinced that the conditions under which people can live with some mutual concern are outside the bounds of practicality and that within these limits everyone has to take care of himself.

In celebrating as a precious heritage the horrors of the ethnic past, we give ourselves license to feel that the present is dead. It is not that we want to recover our real ethnic roots, but that we need to mourn the loss of them.

This is why the language of many of the ethnic revivalists is like the language of museum curators, talk of conservation, preservation, restoration. But a living tradition is not like a painting; it needs to be changed and retouched by each new generation. The ethnic revivalists speak as they do because they are obsessed with the idea that ethnicity is dying out. Whether ethnicity is in fact a living or a dead tradition in America is altogether another question; people need to believe it is something precious that is absent from their lives.

Jimmy Carter's so-called blunder in talking about the old-time virtues of ethnic purity may turn out to be a stroke of genius. Blacks, and Jews with good memories, may find the phrase chilling, but for those who have dimmed memory of their own ethnic pasts, or who have never had such a background, these words may strike a sympathetic chord, its root deep in the American experience.

The leader gives people an image of how much more decent things once were, and they want desperately to believe him. The ability to arouse their longing makes him a credible figure, more credible than the politicians who want to talk about what should be done now.

My own conviction is that if someone like Carter comes to power on these terms, it will be the beginning of a real and irreversible decline. Nostalgia is not a very good preparation for survival.

Richard Sennett, professor of sociology at New York University, is author of the forthcoming book, "The Fall of Public Man."

May 10, 1976

Of Ethnic Identity and Progressivism

To the Editor:

Richard Sennett (Op-Ed May 10) makes a strong and necessary case against romanticizing the past of American ethnicity. He is wise in warning against social and political programs based on nostalgia, ethnic or otherwise. Sennett is also correct when he intimates that when ethnicity does nothing more than protect turf against outsiders it is dangerous to a pluralistic society. Yet he is unnecessarily harsh and a bit cavalier when he fails to come to grips with the very human need to attach oneself through one's ethnicity to an authentic group identity.

One need not idealize ethnicity to identify positively with the struggles of millions of blacks, Hispanics, native Americans, Asian-Americans, Jews, Italians, Poles and other white ethnics to better define themselves and seek to restore a sense of community.

I wonder if Sennett, whose writing I deeply admire, would so easily cast a reactionary label on those who seek physical conservation and preservation. Why assume that a desire for "social conservation" must be backward-looking? Dr. Sennett's own studies show that there is a powerful need to make the shattered human ecological systems that relate to family, neighborhood and ethnicity work better. A movement which capitalizes on positive aspects of ethnicity and works to build coalitions among disparate groups might be a base for progressive politics.

We ought to counter the simplistic readiness of many American intellectuals to polarize such concepts as ethnicity and class and to cast advocates of ethnic pluralism and class analysis into enemy camps.

We have a responsibility in multi-ethnic America to teach and utilize the rich history, to enjoy the cultural differences and to respect the various group interests that emerge from ethnic identity. At the same time we must guard against group chauvinism and extreme separatism. The remaking of community based solely upon ethnicity is too limited a view, and the attacks on the concept of "ethnic purity" were highly justified, but why do we have to deny the continuing reality and power of the ethnic factor in order to maintain liberal credentials?

An America more responsive to its diversity may be a nation on the way to rediscovering itself.

IRVING M. LEVINE
New York, May 10, 1976

The writer heads the American Jewish Committee's Pluralism and Group Identity Institute.

May 22, 1976

The Search For Roots, A Pre-Haley Movement

By PHILIP NOBILE
and MAUREEN KENNEY

The roots movement, a growing subdivision of ethnic consciousness, predates "Roots." Alex Haley's splash merely rides on the long rolling wave of heritage-seeking among immigrant American children, that is, most Americans.

Two previous recent books took genealogical excursions similar to that of "Roots." Michael J. Arlen traced his family back to Armenia in "Journey to Ararat." Richard Gambino returned to Sicily for "Blood of My Blood." And Irving Howe's "World of Our Fathers" narrated the special experience of Jewish immigrants in New York City.

This literary trend continues, but the individual narrative has been superseded by several forthcoming how-to books. Examples are "Black Genealogy" by Charles L. Blockson and Ron Fry; "Finding Your Roots: How Every American Can Trace His Ancestors at Home and Abroad" by Jeane Eddy Westin; and "Finding Our Fathers: A Guidebook to Jewish Genealogy" by Dan Rottenberg.

Though there are no statistics on the extent of the roots movement, there are a number of reliable indicators of rising national interest. In 1974, for example, Congress passed the Ethnic Heritage Studies Program Act "to encourage greater understanding of the ethnic backgrounds and roots of all American citizens." The program has disbursed over $5.9 million in 140 separate grants. Last year's budget of $1.8

million has been increased to $2.3 million for 1977.

Most of the Ethnic Heritage bill funds have been funneled into universities. For example, in 1974, $10,000 went to Southern Illinois University for a project on the drama and theater of Baltic-American youth; $80,000 went to Brandeis University in Massachusetts for Jewish ethnic studies. And Duquesne University in Pennsylvania received $65,000 to devise an "ethnic heritage kit." Michael Novak, author of "The Rise of the Unmeltable Ethnics," says that the academic approach has legitimized the field of ethnic studies.

Two monthly Italo-American magazines—"I/Am" and "Identity"—began in late 1976. "I/Am," a slick full-color package emphasizing Italo-American personalities and problems, has an estimated 100,000 circulation after its fifth issue.

Even an airline is currently capitalizing on deepening feelings for the old country. A new series of sentimental radio and television commercials urges Americans to visit their land of national origin.

There is little doubt, however, that much current interest in family background does derive directly from the television version of "Roots." John La Corte, founder and director of the Genealogical Heraldic Institute of America, says that inquiries have more than doubled since the broadcasts. Mr. La Corte charges a minimum of $1,000 for a complete family tree and is already booked for the next two years.

Unfortunately, there is no central registry of facts equally useful to all Americans who want to trace their backgrounds. Individuals who want to do family research are advised by genealogists to start by looking at home. After locating as many names, dates, places, certificates and memorabilia as possible, they should then interview living relatives. Pedigree and family tree blanks, available at office supply stores, can be filled in as far as possible from living memory and available documentation such as family Bibles.

Once family sources are exhausted, there are other resources to turn to. Public libraries contain newspapers, telephone books, census records and, often, local genealogical collections. Church and county records, available for anybody's inspection, may also be helpful.

The next step may be to such institutions as the National Archives or the Library of Congress in Washington or the Mormon Church's genealogical library in Salt Lake City, where microfilmed documents on more than 30 million names lie. And there are thousands of local genealogical societies and libraries throughout the United States, such as the New York Genealogical and Biographical Society with its collection of more than 55,000 volumes. The final stage often involves a trip to the family's country of origin. The trail can be complicated, time-consuming and expensive.

Professional assistance is an alternative. Genealogical and heraldic societies which undertake family research often advertise in the classified section of telephone books. In addition, local historical societies and churches frequently know of individuals and institutions whose business it is to trace family roots.

The success of Alex Haley's investigation has encouraged the impression that with enough research, family roots are always there for the finding. But this is not always the case. Jewish documents, for example, were nearly obliterated in the Holocaust. And for Americans of Italian descent, who constitute one of the largest ethnic groups in the United States, family lineages are likely to be lost forever. Eighty-five percent of the Italians who emigrated to America came from southern Italy where, according to Dr. Gambino, director of Italian-American studies at Queens College in New York City, historical conditions precluded the keeping of accurate vital statistics. He notes that surnames were introduced as late as the 1830's there and that parents frequently failed to register the births of their sons in order to avoid conscription in the occupying armies. Although Italian consulates will forward birth and death inquiries to Italian town halls, Italy is often a dead end.

For another of America's largest immigrant groups, the Irish Government's genealogical office provides help for descendants. This office, which is scheduled to expand owing to soaring world-wide requests, receives 10,000 callers and several thousand letters a year. For $20 a detailed research report on family origins can be commissioned. About 1,000 of these reports are fulfilled annually.

For blacks, a successful African search is almost impossible. Alex Haley located his Gambian cousins by a great gamble. But his good fortune is not likely to be often repeated unless his Kinte Library Project, named for his African ancestor, proves successful. Since 1972, the Carnegie Corporation has spent $489,000 in start-up funds for a proposed repository of historical materials on black heritage and a resource center for scholars and interested persons. Although the library project has been delayed for eighteen months, further financing is now expected from Mr. Haley's royalties.

Philip Nobile is a contributing editor of Esquire. Maureen Kenney is a freelance writer living in New York.

February 27, 1977

ETHNIC GROUPS CLASH

THE MOB IN NEW-YORK.

Resistance to the Draft---Rioting and Bloodshed.

Conscription Offices Sacked and Burned.

Private Dwellings Pillaged and Fired.

AN ARMORY AND A HOTEL DESTROYED.

The initiation of the draft on Saturday in the Ninth Congressional District was characterized by so much order and good feeling as to well nigh dispel the forebodings of tumult and violence which many entertained in connection with the enforcement of the conscription in this City. Very few, then, were prepared for the riotous demonstrations which yesterday, from 10 in the morning until late at night, prevailed almost unchecked in our streets. The authorities had counted npon more or less resistance to this measure of the Government after the draft was completed, and the conscripts were required to take their place in the ranks, and at that time they would have been fully prepared to meet it; but no one anticipated resistance at so early a stage in the execution of the law, and, consequently, both the City and National authorities were totally unprepared to meet it. The abettors of the riot knew this, and in it they saw their opportunity. We say abettors of the riot, for it is abundantly manifest that the whole affair was concocted on Sunday last by a few wire-pullers, who, after they saw the ball fairly in motion yesterday morning prudently kept in the back ground. Proof of this is found in the fact that as early as 9 o'clock, some laborers employed by two or three railroad companies, and in the iron foundries on the eastern side of the City, formed in procession in the Twenty-second Ward, and visited the different workshops in the upper wards, where large numbers were employed, and compelled them, by threats in some instances, to cease their work. As the crowd augmented, their shouts and disorderly demonstrations became more formidable. The number of men who thus started out in their career of violence and blood, did not probably at first exceed three-score. Scarcely had two dozen names been called, when a crowd, numbering perhaps 500, suddenly made an irruption in front of the building, (corner of Third-avenue and Forty-sixth-street,) attacking it with clubs, stones, brickbats and other missiles. The upper part of the building was occupied by families, who were terrified beyond measure at the smashing of the windows, doors and furniture. Following these missiles, the mob rushed furiously into the office on the first floor, where the draft was going on, seizing the books, papers, records, lists, &c., all of which they destroyed, except those contained in a large iron safe. The drafting officers were set upon with stones and clubs, and, with the reporters for the Press and others, had to make a hasty exit through the rear. They did not escape scatheless, however, as one of the enrolling officers was struck a savage blow with a stone, which will probably result fatally, and several others were injured.

From the above it will be seen that the drawing by Provost-Marshal JENKINS did not commence punctually at 9 o'clock, as was intended. Intimations had been received that a riot was probable, and Acting Assistant Provost-Marshal-General NUGENT was applied to for a force which would be sufficient to preserve the peace. At 10 o'clock, however, no other response had been made to this application than the arrival of a dozen policemen, and Provost-Marshal JENKINS decided to resume the drawing. The wheel was placed prominently upon the table, the blindfolded man stood beside it, the man whose duty it was turn the wheel was ready, and Mr. JENKINS announced that the draft, which was begun on Saturday, would be concluded. At this time there were about two hundred persons present, and, during the twenty minutes before the riot was inaugurated, they freely made use of excited and threatening language. These ruffians did not hesitate at all about joining the main body of the rioters as soon as they arranged themselves before the building, and their exit was the signal for the attack, which commenced with a volley of stones. When the office had been cleared of the officers and other persons, many of the more excited of the rioters rushed in and played instant havoc with the machinery, and demolishing the furniture and papers. The books, lists, records, and blanks were dragged into the street, torn into fragments, and scattered everywhere with loud imprecations and savage yells. The men seemed to be excited beyond expression, and in their futile efforts to wrench open the iron safe, which contained the names of the drafted, gave themselves wholly to devilish rage and fury.

The destruction of the material in the office was hardly accomplished when smoke was discovered to be issuing from the rear of the room, and this evidence of the building being on fire was received with vociferous shouts, and other indications of delight. As the flames gradually increased, the passions of the mob grew deeper, and their yelling and brandishing of clubs, and threatening of everybody connected with the enforcement of the draft was more emphatic. Some of the crowd supposed that the Enrolling officers had secreted themselves in the upper part of the building, and notwithstanding the fact that women and children were known to occupy the upper floors, the cowardly wretches threw stones and other missiles into the windows.

BURNING OF THE ORPHAN ASYLUM FOR COLORED CHILDREN.

The Orphan Asylum for Colored Children was visited by the mob about 4 o'clock. This Institution is situated on Fifth-avenue, and the building, with the grounds and gardens adjoining, extended from Forty-third to Forty-fourth-street. Hundreds, and perhaps thousands of the rioters, the majority of whom were women and children, entered the premises, and in the most excited and violent manner they ransacked and plundered the building from cellar to garret. The building was located in the most pleasant and healthy portion of the City. It was purely a charitable institution. In it there are on an average 600 or 800 homeless colored orphans. The building was a large four-story one, with two wings of three stories each.

When it became evident that the crowd designed to destroy it, a flag of truce appeared on the walk opposite, and the principals of the establishment made an appeal to the excited populace, but in vain.

Here it was that Chief-Engineer DEOXER showed himself one of the bravest among the brave. After the entire building had been ransacked, and every article deemed worth carrying away had been taken—and this included even the little garments for the orphans, which were contributed by the benevolent ladies of this City—the premises were fired on the first floor. Mr. DECKER did all he could to prevent the flames from being kindled, but when he was overpowered by superior numbers, with his own hands he scattered the branches and effectually extinguished the flames. A second attempt was made, and this time in three different parts of the house. Again he succeeded, with the aid of a half a dozen of his men, in defeating the incendiaries. The mob became highly exasperated at his conduct, and threatened to take his life if he repeated the act. On the front steps of the building he stood up amid an infuriated and half-drunken mob of two thousand, and begged of them to do nothing so disgraceful to humanity as to burn a benevolent institution, which had for its object nothing but good. He said it would be a lasting disgrace to them and to the City of New-York.

These remarks seemed to have no good effect upon them, and meantime, the premises were again fired—this time in all parts of the house. Mr. DECKER, with his few brave men again extinguished the flames. This last act brought down upon him the vengeance of all who were bent on the destruction of the asylum, and but for the fact, that some firemen surrounded him, and boldly said that Mr. DECKER could not be taken except over their bodies, he would have been dispatched on the spot. The institution was destined to be burned, and after an hour and a half of labor on the part of the mob, it was in flames in all parts. Three or four persons were horribly bruised by the falling walls, but the names we could not ascertain. There is now scarcely one brick left upon another of the Orphan Asylum.

OUTRAGES UPON COLORED PERSONS.

Among the most cowardly features of the riot, and one which indicated its political *animus* aad the cunningly-devised cue that had been given to the rioters by the instigators of the outbreak, was the causeless and inhuman treatment of the negroes of the City. It seemed to be an understood thing throughout the City that the negroes should be attacked whereever found, whether they offered any provocation or not. As soon as one of these unfortunate people was spied, whether on a cart, a railroad car, or in the street,he was immediately set upon by a crowd of men and boys, and unless some man of pluck came to his rescue, or he was fortunate enough to escape into a building he was inhumanly beaten and perhaps killed. There were probably not less than a dozen negroes beaten to death in different parts of the City during the day. Among the most diabolical of these outrages that have come to our knowledge is that of a negro cartman living in Carmine-street. About 8 o'clock in the evening as he was coming out of the stable, after having put up his horses, he was attacked by a crowd of about 400 men and boys, who beat him wih clubs and paving-stones till he was lifeless, and then hung him to a tree opposite the burying-ground. Not being yet satisfied with their devilish work, they set fire to his clothes and danced and yelled and swore their horrid oaths around his burning corpse. The charred body of the poor victim was still still hanging upon the tree at a late hour last evening.

Early in the afternoon the proprietors of such saloons and other places of business as had negroes in their employ, were obliged to close up for fear that the rioters would destroy their premises. In most of them the negroes were compelled to remain over night, not daring to go home lest they be mobbed on the way.

July 14, 1863

FACTS AND INCIDENTS OF THE RIOT.

At a late hour on Tuesday night the mob, numbering 4,000 or 5,000, made an attack upon the clothing-store of Messrs. BROOKS BROTHERS, in Catharine-street, corner of Cherry. Sergeant FINNEY, of the Third Precinct, while in the discharge of his duty in endeavoring to protect the property of this establishment, was knocked down, beaten on the head and body with clubs, and afterward shot in the hand by a pistol by one of the rioters. He was subsequently conveyed to the Station-house, where his wounds were dressed. He is very severely injured, and no hopes are entertained of his recovery. Officer DANIEL FIELDS, of the same Precinct, was knocked down and brutally beaten about the head and face at the same time.

A man named JOHN MATZEL was shot and instantly killed. It is reported that he was one of the leaders of the mob, and that the ball which pierced his heart came from a revolver in the hands of one of the officers of the law. He was in the act of entering the clothing-store at the time he met his death.

Plunder seems to have been the sole object with the marauders in their attack upon the store of the Messrs. BROOKS. The fine ready-made clothing therein was tempting. Fortunately, the Police and the employes of the establishment successfully repelled the invaders before much property had been stolen. Three or four persons, whose names could not be ascertained, lost their lives at this place, and many others were badly injured.

An unoffending citizen named VAN R. FITCH, was quietly walking through Warren-street, about ten o'clock on Tuesday night, when he was knocked down and beaten in a shocking manner by some unknown parties. After the infliction of this outrage they left the man in the middle of the street, evidently believing him to be dead, and made good their escape. Mr. FITCH was picked up by some citizens and taken to a neighboring drug store, where his wounds were dressed. He was afterwards conveyed to his residence in Eleventh-street, in a dying condition.

THE MURDER OF COLORED PEOPLE IN THOMPSON AND SULLIVAN STREETS.

At a late hour on Tuesday night the mob made an attack upon the tenement houses, occupied by colored people, in Sullivan and Thompson streets. For three hours, and up to two o'clock yesterday morning there was what may be truly said to be a "reign of terror" throughout all that portion of the City. Several buildings were fired, and a large number of colored persons were beaten so badly that they lay insensible in the street for hours after. Two colored children at No. 59 Thompson-street were shot and instantly killed. Men, women and children, in large numbers flocked to the Eighth Precinct Station-house for protection. Over one hundred of them were there accommodated with temporary shelter.

July 16, 1863

An Appeal to the Irish Catholics from Archbishop Hughes.

In the present disturbed condition of the City, I will appeal not only to them, but to all persons who love God and revere the holy Catholic religion which they profess, to respect also the laws of man and the peace of society, to retire to their homes with as little delay as possible, and disconnect themselves from the seemingly deliberate intention to disturb the peace and social rights of the citizens of New-York. If they are Catholics, or of such of them as are Catholics, I ask, for God's sake—for the sake of their holy religion—for my sake, if they have any respect for the Episcopal authority—to dissolve their associations with reckless men, who have little regard either for Divine or human laws. †JOHN,

Archbishop of New-York.

July 16, 1863

The Mob Unmasked—Its Character and Spirit.

We advise everybody who wants to study the contortions of wizards who are being torn to pieces by the demons they have themselves evoked, to read carefully the articles of the anti-draft journals of this City upon the late riot. They raised the devil in order to confound and embarrass the Administration, and, as most of us expected, he behaved like a devil, and they are now tearing their hair, and calling on Heaven to witness that he horribly changed after they first made his acquaintance. They strove in vain to control him by a liberal use of euphemisms, just as the Irish peasantry seek to placate the fairies by calling them "good people," and showered on him such endearing epithets as "the laboring population," "the people," "the oppressed and outraged conscripts"—but in vain. He was not to be cheated into forgetfulness of his true character by any amount of caressing. They had introduced him into our streets, put a whole city at his mercy, and, true to his instincts, he rent and tore everybody within his reach, but especially the weakest, most helpless, most innocent and inoffensive.

The *Journal of Commerce* has drawn up an elaborate statement of the nature and causes of the riot, (which the *World* has adopted,) which maintains that while the feeling that led to the outbreak originally was one of opposition to the draft, the plundering and robbing was done by the regular professional thieves of the City, and that, therefore, the opponents of the draft are not fairly chargeable with it. Now we are satisfied that it would not be very difficult to show that, however true this may be of the mere robberies from the person committed on individuals in the streets, it is grossly inaccurate as regards the pillage of the numerous houses broken into by the rioters. The testimony as to the plundering, both in these and at the Colored Orphan Asylum, is unanimous as to the appearance and character of the persons engaged in it. Able-bodied laborers broke in the doors and stood guard, while the women and boys carried off the carpets and furniture and other valuables. There was no building sacked in which this did not take place, and it is mere waste of paper to ask the New-York public to believe that these were families of regular thieves, in which the fathers do the highway robbery and burglary business, while their wives and children attend to the petty larceny department. We venture to assert that among all the arrests made of persons accused of sharing in the pillage of Monday and Tuesday of last week, or found with stolen carpets and bedding in their possession, not six are regular thieves previously known to the Police. The persons, in short, who sacked houses, were seen by too many in broad daylight, and were too open in their dealings, to make any mystification as to their real character possible. They were *mainly*—for, of course, the regular thieves were not idle—neither more nor less than the poorest and most ignorant of the laborers and wives of laborers, who had been lashed into frenzy by the assurance of demagogues that they were about to be dragged unlawfully into the field, to be killed for the benefit of "the niggers;" and that the rich had so managed it that they were to pay three hundred dollars and stay at home quietly, while the poor men did all the fighting. From this impression, which has been carefully nursed and diffused by every Democratic journal in this City, there resulted intense hostility to the blacks, and a wild desire to be avenged on *property*; for, as a correspondent showed in these columns on Monday, the Irish have not naturally the slightest prejudice against color, and he might have added—or the slightest tinge of communistic hatred of the rich, in their composition.

We are far from accusing any of our cotemporaries of having instigated the riot knowing that it would result in or be attended with murder and pillage. We have not yet got to such a depth of degradation that even political partisans are utterly reckless as to the consequences of the commotions they excite. What the Copperhead orators and writers really desired—and we say unhesitatingly that they must either have desired it or their speeches and articles for the last four months on the subject of the draft and the war have been utterly meaningless—was that the resistance to the conscription should be organized, and strictly confined to that object—that no one should be harmed but the officers of the Government or the troops or police engaged in executing their orders, and that it should be so determined, that the Administration would be frightened into retreat. If half which the *World* and Messrs. Wood and Seymour have said, both of the nature of the Enrollment act, and of the intentions of the Government—and said in the bitterest and most inflammatory language—were true, such resistance would have been the duty of people less ignorant and excitable than the ruffians who burnt negro houses under the Dred Scott decision.

These agitators now pretend to be as much horrified as any of us at the turn things have taken, and beg of us to believe that the resistance to the draft and the pillage were two separate affairs, conducted by different parties. The public, however, knows better, and we know better. We have always denounced inflammatory addresses to the multitude, counseling and suggesting resistance to the law, because we believed what is now proved, that popular risings against the Government in such a society as ours, mean and will always mean *anarchy*, with its attendant horrors; that you cannot call anti-draft men into the streets of a large city, with pistols and clubs, and leave the thieves and murderers at home; and that if you excite a social convulsion for any purpose whatever, you must bear the responsibility of all the consequences of it. If you send men to burn the Provost-Marshal's office, you will be answerable before God and man for the burning of the houses adjoining it, which may follow it, and which in this instance did follow it. Our whole social and our whole political system is based on the understanding *that there is but one mode of getting rid of* the operation of unpopular laws, and that is by voting; and we denounce as novel, and as borrowed from the armory of European demagogues or malcontents, the threats and the suggestions of resistance by violence which have of late been so largely resorted to by the leaders of the opposition in this City. If we are to have peace, we must have the language of peace, and must not have even the possibility of a resort to armed force held up to the minds of the thousands now among us who have been bred in European notions of the relations between the Government and the governed. If they are to be taught by American journalists that they must rely on the same weapons for protection against the Government of the United States that they were accustomed in Europe to consider their only hope of deliverance, we shall all have to resort eventually to the same means of protection against their madness and folly, which now stands between the intelligence and property of European nations and the frantic violence of the "dangerous classes."

July 24, 1863

MINOR TOPICS.

—According to our San Francisco correspondent, the Irish are becoming dissatisfied with the presence of the Chinese in California. They got up a riot lately and drove off some twenty or thirty of the celestials, who were employed by the Pacific Mail Steamship Company, burning their houses, destroying their provisions and killing one of the obnoxious class. The Irish are impatient of the rivalry of any other foreign labor in the country. The Chinamen are among the most peaceable, orderly and useful of the laboring population in California, and have become so numerous that their labor is quite indispensable. They take no interest in politics, but attend strictly to their own private affairs. Having no votes, they receive very little attention from the politicians, and the Irish, who have votes, are allowed to treat them very much as they please. The ballot certainly is a very useful weapon is self-defense, and the Chinese immigrants will probably ere long avail themselves of it. They have the right to do so under the Civil Rights Bill.

March 18, 1867

Freedmen in the Indian Territory.

Of all the negroes to whom the war brought freedom, none, perhaps, were more miserable or oppressed than those of the Indian Territory. When emancipation came, it brought to them no change for the better. The Indians were, of course, compelled to release them from bondage, and nominally did so. But the reality of bondage remained. Their former masters became their enemies. There was no longer the motive to tend them as useful animals. Equality created hatred, born of jealousy, before impossible; and, while the nation gave to the negro a new social status, the Indian Territorial authority removed from him no legal or political disability whatever. Especially was this the case with those slaves which were owned by the *quasi* united tribes of Choctaws and Chickasaws. Manifestly this condition of affairs could not be allowed to continue.

An attempt was made in 1865, by a treaty, to secure some equality of local rights to these negroes. The two nations ceded "the leased district" to the United States, in consideration of which $300,000 was invested by the latter for the nations at five per cent. The principal was to be paid to them when they had given the negroes equal rights with themselves, and those who were residents forty acres each of the ceded land on the same terms as the Indians. If, however, any of the negroes desired to move out of the Territory, $100 was to be paid to each from the $300,000. If, further, the nations failed to fulfill these conditions within two years, the entire sum was to revert to the benefit of such negroes as wished to remove, those remaining having no interest therein. There were other provisions by which the ex-slaves were protected in respect to their labor and civil rights generally. With the execution of the treaty the matter ended. No subsequent action was taken by the two nations.

The negroes naturally complain of this, and demand some arrangement for their safety and welfare. They consider themselves full citizens of the nations with whom their lives have been passed, and, regarding the Territory as their proper home, prefer to remain in it. Conventions to urge these claims were, however, prevented by the Indians, who tore down the printed notices, and threatened the lives of any who should venture to attend; and they actually did arrest one colored man. Meetings were held in other localities in spite of these intimidations, and three delegates appointed to lay the case before Congress. But the delegates were too poor to go to Washington, and when Senator CARSON, of Arkansas, was able to leave his State for the purpose of representing them, Congress had adjourned.

The circumstances certainly entitle these poor freedmen to consideration by the Government. Their wrongs are felt all the more bitterly that many of them fought for the Union, and moreover that those who did were excluded from the treaty of 1865 by their absence on military duty, which enabled the Indians to call them non-residents. Whether what the negroes now seek can be wisely granted is another matter. Continued residence among the Choctaws and Chickasaws would subject them to worse cruelties after their forty acre allotments were secured than before. The general opinion of those who appear to understand the subject best, is that a country should be set apart for them in the leased district, where they could be all together, or in some other locality. At all events the question remains, whether these Indians are to oppress a loyal class of people, or whether the Government shall interfere and force a settlement which will insure peace, and protect the negroes in their homes and industries. There should be but one reply, and that is that justice must be done without further delay.

April 25, 1870

Is There to be a Riot on the 12th?

We have repeatedly maintained that foreign citizens have no right to introduce the discords of their own country into this. We can make a sufficient number of party quarrels of our own, and we can make them fast enough without being bored and annoyed by the hereditary feuds of foreigners. Processions which keep alive the quarrels of another country and another generation ought, therefore, to be discouraged, and we had half hoped that the good sense of Orangemen would have induced them to refrain from what they call a "demonstration" in the public streets on the 12th. The event which they commemorate was doubtless a very great one, and the man whom they honor is never likely to be coldly remembered by Protestants. It was owing to WILLIAM of Nassau that England was delivered from the thraldom of a monarch who thought that the Protestants ought to be swept from the face of the earth, who broke every law he had sworn to keep, who sent JEFFRIES on the Bloody Circuit, and who was at once the most bigoted, cruel, and cowardly of mankind. This was the last of the Papist Kings of England, and it was owing to WILLIAM that he was the last. Moreover, Protestants in Ireland might well exult over the victory gained by WILLIAM at the Boyne, for there was no form of injury which they had not sustained at the hands of the Catholics. TYRCONNEL's barbarities, sanctioned, if not ordered by JAMES II., were sufficient to sow the seeds of lasting enmity between Protestants and Catholics. All this is true, but it does not furnish any reason for reviving the animosities of the seventeenth century in the nineteenth, or for transplanting them from the banks of the Boyne to the greatest City of the New World. In a land of perfect civil and religious freedom it is worse than folly to fan into fresh life the flames of religious and political animosities which are dying out even in the countries where they originated. We have therefore advised the Orangemen to give up a procession which is at best an anachronism, and which is open to much graver objections on political grounds.

The line taken by the Roman Catholics in reference to the matter is, however, so outrageous that the Orangemen may well ask if the infamous intolerance of JAMES II. is really a thing of the past. At a "Convention of the Irish Societies," held last Friday evening, a delegate proposed that the Mayor should be requested to prohibit the Orange procession, and that if he declined, the Irish Catholics should turn out and settle the dispute "at once and forever, cutting down every Orangeman in the procession, and to be found in the City, with an emblem or insignia of Orangeism about him." On Saturday evening another meeting of Irish Catholics was held at Washington Hall, at which the crowd "gave expression to threats of annihilating the Orangemen." Everywhere the Irishmen are boasting in the same spirit. They will teach the Protestants, they say, to understand that the Catholics govern New-York, and that a Protestant procession cannot safely show its head in the streets. New-York belongs to the Catholics, and the sooner the public are made to understand that fact, the better.

This spirit will not surprise those who have watched the growth of intolerance on the part of Roman Catholics here. They are not satisfied with appropriating to themselves the lion's share of the money given to charitable institutions. They are simply resolved to set up a State Church here, and to drive Protestantism to take shelter in holes and corners. That is their "platform," and they mean to carry it out if they can. They are the first to complain of intolerance—even an article in the TIMES pleading for equal toleration for all sects arouses their virtuous indignation—while at the same time they, in accordance with the traditions of their Church, are the first to set examples of the worst kinds of intolerance. So far as right goes, the Orangemen are as much entitled to march in procession as the Papists. The Irish Catholics take possession of the City whenever they happen to want it, and the accomplished Mayor is ever ready to stand in front of the City Hall, and nod his head to the leaders as they pass, like a Chinese joss. The accomplished Mayor, and those less accomplished persons who are responsible with him for the peace of the City, have now had full and fair warning that the Catholics intend to cut down all men, women, and children who attempt to march in an Orange procession on the 12th. On the Mayor's accomplished head will rest no small part of the responsibility for any disturbance which takes place. The Police can preserve order if they are made to do it. But the Police generally are on the side of the meek and patient Catholics. Consequently, some persuasion will have to be used to induce them to do their duty. Superintendent KELSO must expect to pass a bad time on the 13th if the Catholics are permitted to carry out their threats on the 12th. And the Mayor will find something to occupy his attention besides those literary studies in which he is said to take delight, but in which his progress is almost as slow as that of TWEED, who cannot write a letter without suffering grievous agonies, or that of FISK, who is under the impression that it is proper to spell cart with a *k*, and to use a little *i* when writing in the first person.

July 10, 1871

THE RECENT RIOT.

Satisfactory Condition of Affairs—Restoration of Order—Resume of Events.

A greater contrast than that presented by New-York on yesterday and the day before has rarely been afforded. Wednesday was a day of horror, of consternation, of wild confusion; yesterday was one of peace, security and order. The vindication of the law had been so complete, the mob spirit had been so entirely eradicated by the action of the Police and militia that the City has not in years been so orderly and quiet as it has been since midnight of Wednesday. Streets and places which are usually filled at late hours of the night with disorderly characters, were deserted and quiet. Ruffians who are generally to be found at all hours defying the public peace, had slunk into their hiding holes. The change was marvelous and delightful.

Nowhere was it more marked than at Police Head-quarters. At 9 o'clock yesterday morning Superintendent KELSO and Gen. SHALER were convinced that order had been completely restored, and that no further outbreak need be feared. The few remnants of the force which remained on duty in the building were relieved, and the vast edifice in the early hours of yesterday was sepulchrally quiet and excessively dirty. The quiet continued throughout the day, but under the exertions of a large force of charwomen the filth had disappeared by nightfall, the furniture had been returned to its place, and there were no signs of the turmoil of the previous day remaining. During the morning Superintendent KELSO, Inspectors DILKS, WALLING and JAMESON, notwithstanding the exhaustive and continuous labors of the previous twenty-four hours, were yet on duty, and actively engaged in returning the Department to its routine duty. Fifteen hundred of the twenty-two hundred patrolmen attached to the force had been withdrawn from the precincts where their few remaining comrades had been constantly on duty, so that the entire force was as entirely exhausted as its chiefs. The men had earned the rest which was at last accorded them, not only by constant but effective service. Everywhere during the day of disturbance it was noticed that the Police invariably drove the rioters, no matter what was the disproportion of numbers. There was no flinching upon the part of any member of the Police except Patrolman O'GRADY, Sixth Precinct, who was so summarily stripped of his uniform by Superintendent KELSO, and the story in a morning paper that Capt. HELME showed cowardice is pronounced by Inspector WALLING to be "an infamous lie," as Capt. HELME was in his immediate view during the whole of the fight, and he knows that he bravely led his men throughout the conflict.

In actual incidents yesterday was almost a blank at head-quarters. The only affair that made any stir whatever was the removal of the rioter prisoners to the Tombs about 11 o'clock in the morning. They were congregated preparatory to being taken away in the rear room of the detectives, and so many men of ferocious and disagreeable appearance are rarely seen together. All were of unmistakable Irish nationality, all dressed in the coarse garb of the laboring classes, and many were stained with the dried blood which had oozed from their wounds inflicted by the detectives when they were beaten down in the streets on the previous day. Escorted by the detectives under Capt. IRVING, and thirty-five patrolmen under Capt. MCDONNELL, the sixty-two prisoners were marched to the Tombs by way of Houston-street, Broadway, and Franklin, and not to parade them to the general public, but to avoid the chance of an attempt at rescue by going through the Irish districts to the east of Broadway. When the Court-room was reached the prisoners were arraigned before Justice HOGAN, and all were temporarily committed for examination. At a later hour legal hairs being split in their behalf, nine of them were set at liberty for want of sufficient evidence of riotous acts. From these sixty-two prisoners, the detectives took, when they were arrested in the streets, seventy-eight deadly weapons, most of which were pistols, and those having these weapons invariably had several boxes of cartridges. Those not having fire-arms had swords, daggers, sheath-knives, and one fellow, with murder in his heart, had a trowel with all its edges ground sharp, thus converting it into a most terrible instrument of death.

July 14, 1871

RACE ANTIPATHIES.

A philosopher, desirous to discover the basis of race prejudice, would find an abundant field for investigation in the United States. Representatives of nearly all the races of mankind form a part of our homogeneous population. It is impossible that these strangers should not have their antagonisms. It is difficult to account for their antipathies. There appears to be no alien, however degraded he may be, who does not find some fellow-man whom he looks down upon with contempt. It is assumed that all foreigners are regarded with a certain assumption of superiority by the native American. If it were not for the fact that in most of our large cities the governing class is drawn from the Irish, we might be disposed to admit that there is a native American prejudice against the foreign element of our population which is imported from "The Gem of the Sea." But that statement will not bear a moment's examination in a city like New-York, in which it is possible for a party manager to put forth a ticket for county officers exclusively composed of foreigners, and the greater portion of which is distinctively Irish. It does not affect the case that the aforesaid political leader is, himself, an Irishman. As he has the reputation of being tremendously sagacious, we must assume that he would not have given us a ticket of foreigners to vote for unless he was certain that we would be glad to elect it for him.

More than once, criminal trials in this country have been made to hinge on race prejudice. In a murder trial, not long ago, in the City of Newark, N. J., the German element was arrayed against the defendant, who was American-born; and in a somewhat similar case, in another New-Jersey city, an Irish defendant was obliged to instruct his counsel to keep Germans from the jury-box. So far as representatives of these two nationalities are concerned, however, it is hardly likely that the most violent hater, on either side, would be willing to admit that he really considered any man of the other race as being below himself in the social or political scale. But how shall we account for the violent aversion with which the lower orders of the Irish regard the Italians? "A beggarly Italian," is the least objectionable phrase which the indignant Irish laborer finds to bestow upon the humble Italian who seeks an opportunity to earn his daily bread at as great a distance as possible from the haughty Celt. During the flush times, when the Ring ruled New-York, and when millions were recklessly spent in carrying on the public works, it was only by calling in the Police that small riots were crushed, after having been provoked by the determination of the Irish laborers to drive away the Italians who were employed on the works. And if the sons of Italy have finally been permitted to man the sea-going fleet of the Street-cleaning Bureau, (if we may apply such a misnomer to a bureau which does *not* clean the streets,) it is because the maritime Italian is considered only fit to unload garbage-scows and tempt the dangers of the deep and brave the attacks of the Gravesend Constabulary. In later years, we believe, Italians have been allowed to sweep the streets, but this at night and in regions where their traditional foes could not detect them. Between the Italian and the Irishman there is not that difference of religious faith which, too often, creates belligerency between races, and which has even separated Irishmen into two hostile factions.

The Italians, however, have at last found their hated foes in a deeper depth of obloquy than that to which they have themselves been consigned by their old enemies. The Chinese are our lowest race, and the Italians have found that out. The recent outbreak at one of the Colorado mines, promoted by the attempt of a contractor to employ Chinese, is a striking evidence of the fact that the Italians can vent upon another race the animosities which the Irish have poured upon them. The contractor having put eight Chinamen to work in the mines, the Italian laborers refused to allow them to remain, and they made a violent personal attack on the offending contractor, in order to emphasize their statements. To a New-Yorker, who recollects the meekness and peaceableness with which the Italian laborers bore the contumely heaped on them by the Irishmen who objected to their employment, there is something ludicrous in the belligerent attitude of the anti-Chinese Italians of Colorado; they, too, have found their inferiors.

All representatives of the white race look down upon the poor Chinaman. The fiercest enemy of the yellow race in this country is the Irishman. DENIS KEARNEY is the fittest type of that enmity. The Kearneyites, in the intensity of their hatred of the Chinaman, have apparently forgotten the once-despised "nagur," and, in his turn, the negro puts the Chinese far down below himself in the social and political scale. In California, where, as one would suppose, the degraded Digger Indian, the lowest form of man on this continent, would represent the social substratum, the proud aborigines look upon the Chinaman with inexpressible contempt. The Indian may be groveling in the filth and garbage of a vacant town-lot, happy to find an empty barrel in which to screen himself from the wind and rain; but if a Chinaman chance to pass that way, the haughty son of the soil throws after him an offensive epithet which he has picked up from the white man. The Indian's cur pursues, with canine instinct, the fleeing Asiatic. There may be a race yet lower than the Mongolian, but it has no representative in this Republic. How shall we keep the peace, if the mixed races should ever begin to fight it out among themselves? Clearly, we shall never have any more Know-nothing riots in this country. The day of Know-nothingism has gone by forever, let us hope. The riots of the more recent period have sprung from the animosities engendered between the Irish factions, and between the Irish and the Chinese.

November 28, 1879

GERMAN CATHOLIC WANTS.

MILWAUKEE, Aug. 17.—The approaching convention of German Catholics in Chicago on Sept. 6 has attracted widespread attention, and the report has been assiduously circulated that the German Catholics were discontented with their treatment at the hands of the Propaganda in Rome. As in all other sensational stories, there is a grain of truth in this one in that the German Catholics of America are making an effort to secure their just proportion of Bishops and Archbishops. The German Roman Catholic Central Society will meet at Chicago on Sept. 4, and two days later the General Congress will open its session. The arrangements for the latter have been made by a committee consisting of William Caspar, of Milwaukee; Father Lappert, of Covington, and Father Arends, of St. Louis.

The Germans of the provinces of St. Louis and Chicago forwarded their request to Rome some time ago—Father Aberle, a German-American priest, acting as messenger. He is now in the Eternal City, where he is being seconded by two German resident Cardinals. Archbishop Heiss, of the Province of Milwaukee, a Bavarian by birth, is the only German Archbishop in the United States. A reporter visited him at his residence, on Jackson-street to-day, and asked him, "Do the German Catholics and the German priests feel that they are entitled to a greater recognition in the appointments of Bishops and Archbishops?"

"There are some that undoubtedly feel that way," he replied, "but such matters cannot always be helped. We ought to have a few more, but then we are not so particular."

"What is the proportion of German Catholics in America?"

"There are about 8,000,000 Catholics in the United States, and of these 3,000,000 are Germans, but out of 12 Archbishops and 60 Bishops only one Archbishop, myself, and 11 Bishops are German. The Province of Baltimore is presided over by Cardinal Gibbons, who is of Irish descent. The majority of the six Bishops under him are Irish and one or two are British."

Bishop Becker had German Protestants for his ancestors, but he is not considered a German. At Boston they are either British or Irish, with one French Bishop and five Irish Bishops. The Chicago Province, the smallest in the United States, has Archbishop Feehan for its head. The Diocese of Alton, in this province, has been without a Bishop for a year, and it seems that Bishop Spaulding, of Peoria, an Englishman, cannot agree with Archbishop Feehan on some one to propose to the Propaganda."

"In Cincinnati, Archbishop Elder is British, and three of the nine Bishops are German—Bishop Dwenger, of Fort Wayne; Bishop Richter, of Grand Rapids, and Bishop Rademacher, of Nashville. Of the six Bishops in the Milwaukee province five are German and one Irish. The Germans are Bishops Katzer, Flasch, Vertin, Seidenbusch, and Marty. The latter is a Swiss, from Einsiedel. New-Orleans has a French Archbishop, Leroy three Irish Bishops, one Holland, and one French Bishop. In New-York we have the Irish Archbishop Corrigan, five Irish Bishops, one British, and one German Bishop. Archbishop Gross, of Oregon, is not a German, although he has a German name. I believe he is half Irish. His Bishops embrace one German, Junger, and three Flemish Belgians, who are of a German Netherland race, but not Germans as we understand the word. Archbishop Ryan and five Irish Bishops rule in Philadelphia, while the St. Louis province has Archbishop Kenryck, four Irish Bishops, and one German. At San Francisco there is Archbishop Reardon, an Irishman, and two Bishops of the same race. The Santa Fé province has a French Archbishop (Sailpoint) and two French Bishops. You see, the Germans are not very numerous, and if it wasn't for the Milwaukee province they wouldn't cut any figure whatever. Of course I cannot speak of any dissatisfaction among the Germans in my own province."

August 18, 1887

LABOR AND THE RACE PROBLEM.

A gleam of promise for the rational solution of the race problem comes from Western Pennsylvania. A leading firm of contractors in Pittsburg has just come to a decision to substitute negro labor for the cheap work of Italians, which it has hitherto employed. The Italians, the firm says, do not work well; they are inclined to indulge in quarrels and fights, and they are altogether too handy with the stiletto, when the slightest provocation for its use presents itself. A force of one hundred colored men has already been put to work by this firm, and the result has been so encouraging that the number is to be increased gradually until the Italian laborers have all been dispensed with. If the experiment proves successful, as there is now every indication that it will prove, other contractors will undoubtedly follow the lead of the Pittsburg firm, and the solution of the race problem in the South will have begun on a most favorable foundation.

The race problem, as it is called, is, after all, when reduced to first principles, simply a problem of making the negroes self-supporting citizens. The great trouble in the South has always been the idleness and consequent worthlessness of a large part of the negro population. The idle and vicious colored man is possibly no worse, essentially, than the idle and vicious white man, but when he lives in a community which but thirty years ago held to him the relations of master to slave it would be unnatural if the effect of these bad qualities were not magnified. Had the negroes as a class shown a willingness to work honestly for a livelihood immediately after their emancipation the race problem would have disappeared practically years ago; but their continued existence in the South as large consumers and small producers served to intensify the bitterness with which they were regarded by the dominant race, and a world of trouble, political as well as social, has been the result.

Possibly all the blame for his idleness and thriftlessness should not be too hastily thrust upon the freedman of a quarter of a century ago. No doubt his characteristics were due to a great extent to the condition in which he found himself and for which he was not responsible. After a lifetime of servitude, during which he had never been called on to attend to his own wants, he suddenly found himself thrown upon his own resources, with few friends to advise him in his ignorance. It was natural that he should be thriftless, logical that he should become vicious. But these conditions do not prevail to-day to any large extent. A new generation of negroes, born to freedom, has grown up in the South. It has received at least a smattering of education, and it recognizes the fact that it must work to live. In this generation lies the germ of the solution of the race problem.

If the negro is to work he must have work to do, and his claim to it is far superior to that of the Italian hordes who have been making such frequent raids into our country. The Italians who come to America are, as a rule, of the lowest class, and the majority of them are dangerous elements to incorporate into our society. Why should they be encouraged to come here by the bribe of plenty of work, when the negroes of our own land are without the means of earning a livelihood? That the negroes can and will do better work the Pittsburg contractors have found. Why should it not be offered to them rather than to undesirable foreign immigrants? The move now being made is in the right direction, and if it is extended, as it should be, thousands of colored men who are now idle will become self-sustaining members of scattered communities; the South will be gradually relieved of its surplus of negroes, without resorting to any foolish colonization scheme, and the relations between the races will adjust themselves equitably and to the mutual advantage of whites and blacks.

May 19, 1890

RUSSIAN JEWESSES MOBBED.

Their Employment in a Philadelphia Mill Resented by Americans.

PHILADELPHIA, Oct. 11.—Three young Russian Jewesses, Sarah Brakle, Rachel Seitzler, and Jennie Seitzler, of 786 South 6th Street started to work this morning in the combing department of Campbells' cotton mill, at Twenty-first Street and Washington Avenue. The mill employs about 500 hands, but these three girls are the first foreigners who have dared to join the ranks. The invasion of the foreign element angered the old employes, and all day long jeers and taunts were muttered against them. The Jewesses do not understand English, and paid no attention to the uncomplimentary remarks their presence called forth. At noon they were told in none too gentle tones that they must leave the mill or there would be trouble.

At 6 o'clock they left the factory for their homes, amid the taunts and jeers of the crowd. At Twentieth Street a demonstration was made to frighten the girls, but the result proved serious, as some thoughtless persons began pushing the Jewesses into the street, and a general scrimmage ensued. Blows were struck, and all three girls were frightened into hysterics. The police were notified. When they arrived the street was blocked, and it was with difficulty that they could force their way to the girls. All were taken to the station house, and the crowd dispersed. None of the girls was badly injured, but their clothing was torn, and an umbrella belonging to Rachel Seitzler was torn to pieces.

At 7 o'clock the girls were told to return to the scene of the disturbance, and Sergt. Ipe sent six policemen in citizens' clothes with them to make arrests if the girls were again molested. The crowd had dispersed, however, and no further violence was attempted. At the police station the girls declared they would return to work to-day, as they are laboring for the same pay as the Americans, and have an equal right to their positions. The crowd which attacked them was composed largely of American and Irish women, who declare that there will be trouble if Russian Jews are permitted to work in the mill.

October 12, 1897

STONE 2,000 PRAYING JEWS.

Bridge Laborers Fell Many at Worship at the Foot of Pike Street.

Some 2,000 Jews were attacked on the Pike Street Recreation Pier on the East River front yesterday afternoon by two-score Italian and Irish laborers. They were pelted with stones and several were hurt. The assault was kept up until the arrival of the police, when three of the assailants were arrested and locked up. There appeared to be no reason except hoodlumism for the attack on the Jews.

They had gone to the river front to observe the Tishla, an ancient ceremony which occurs on the second day of the Jewish New Year, and consists principally of the recitation of a prayer. They left the east side synagogues shortly before 4 o'clock in the afternoon, and had scarcely begun the ceremony at the meeting place when the assault took place.

The Manhattan Bridge, commonly known as No. 3, will have its terminal near Pike Street, and on the foundation structure already built there a skeleton bridge has been constructed to facilitate the work. Two freight cars are used on it to carry the material used in the construction. From this point a gang of bridge laborers hurled showers of stones upon the Jews.

A panic followed when other laborers cut off the retreat of the worshippers and stoned them from the approach to the pier. As one after another was struck and knocked down the confusion increased.

The cries for help were heard by the passengers on a Belt Line car three blocks away. Special Policeman Henry Kizowitz, who lives at 5 James Street, East New York, was on the car and ran to the pier. He drew his club and laid about him until the attacking party began to scatter. Before they all got away a dozen policemen, headed by Detectives Delaney and Wuschur of the Madison Street Station, arrived.

The assailants had started to run, but the detectives arrested two Italians who said that they were Frank Cotso, 20 years old, of 146th Street and Eighth Avenue, and Salvatore Mogocro, 34 years old, of 44 Oak Street. Policeman Louis Levy also caught Louis Rosso, 16 years old, of 88 James Street, after a chase of several blocks.

The more seriously injured are Jacob Baumm, 47 years old, of 520 East Houston Street, who had his arm cut and his nose broken, and Harris Brooks of 249 Cherry Street, whose jaw was dislocated. Others injured are Aaron Bomerantz of 9 Eldridge Street, Frank Libochoz of 40 Essex Street, Joseph Lubit of 126 East Broadway, and Joseph Rich, a coal dealer, at 240 Delancey Street.

An ambulance call was sent in to Gouverneur Hospital, and Dr. Smith drove to the pier and dressed the wounds of the injured.

October 2, 1905

Porto Rican Objects to Camp Plan.

To the Editor of The New York Times:

Reading your article under the heading "Texans Protest to President Against Negro Soldiers in the South," I learn that the War Department contemplates the encampment of the patriotic Porto Rican soldiers with the negroes in South Carolina. This plan is a great mistake and a revelation. It makes it clear that it will take long for us as a nation to understand the psychology of peoples of alien nationalities who come under our banner, and it teaches us Porto Ricans that our ultimate destiny is to possess one status in which we keep up the most cordial commercial and political relations, but as a distinct entity. R. M. DELGADO.

New York, Aug. 24, 1917.

August 27, 1917

TOPICS OF THE TIMES.

They Mustn't Import Their Quarrels.

Sunday's sanguinary little riot over in Newark was, in its essentials, no more reprehensible than the other like affairs scattered over our troubled land. It seems worse, however, for the participants, though in the United States, hacked and stabbed at one another in a quarrel relating to no American issue or controversy.

The opposing forces were believers in Fascismo and its opponents. Of course, any American has his right to an opinion as to the wisdom or the reverse of such a Government as MUSSOLINI'S, but, as we have neither hope nor fear of a similar rule, we all can view the state of Italy calmly, and anybody here who gets excited over it to the extent of shedding his neighbor's blood shows that he has imported a quarrel which ought to be settled where it originated.

Whether the Newark Fascisti or the Newark anti-Fascisti were the more to blame for what happened is as yet undecided. The antis had hired the hall, but as the meeting was for a discussion of Fascist ideas and principles it is not easily explicable why the exponents of them should have been unwelcome. Testimony varies widely as to whence the aggression came, but the war started so promptly that both sides must have been expecting it and ready. This is further shown by the fact that a large assortment of stilettos, razors and clubs was gathered from the floor of the hall after the hastily summoned policemen had made a desert and, correctly enough, called it peace.

Nobody was killed, but the wounded were numerous. This may or may not prove that the fighting was in desperate earnest, but evidently the earnestness wasn't quite so desperate as it would have been were the contestants of any one of several other races whose blood usually is held to be cooler than that in Italian veins.

August 18, 1925

PRO-NAZIS HECKLE BERNARD RIDDER

Boos and catcalls interrupted the speech of Bernard Ridder, publisher, when he warned delegates of the United Grman eSocieties last night in New Yorker Turn Halle, Lexington Avenue and Eighty-fifth Street, against pro-Nazi tendencies. His brother, Victor, also spoke.

A split developed in the united societies last week over the anti-Semitic issue. Several units of German Jews quit and last Friday night the entire board of officers resigned in a dispute as to whether the Swastika flag should be hoisted at the Sixty-ninth Regiment Armory on German Day, Oct. 29.

The meeting last night was called to give the arrangement committee an opportunity to work out details for German Day, but since the committee also had resigned, each unit of the society sent a representative. There were about fifty persons in the hall.

Warns of Racial Bias.

The Rev. William Popcke of Zion Lutheran Church, honorary president of the organization, was the only officer remaining after the wholesale resignations. As chairman of the meeting, he pleaded for peace. When contributions for rental of the armory were sought, the pro-Nazi element put up $150 annd the others only $2.

It was then that Bernard Ridder made his speech. He told the delegates that the "future of the United Societies lise in America" and that the disturbance had been caused "by those who have no right to be here." He warned the organization about being "carried away."

There were shouts of derision and booing. The Rev. Mr. Popcke restored order and the publisher continued.

"I tell the truth," he said, "though you may not like it. The moment you stir up racial hatred in this country you are lost. Opinion is this country will crush you, I ——"

Booing and catcalls broke out anew.

"Who will do it?" some one shouted. "Not the Jews."

The chairman, with the assistance of Heinz Spanknoedel, official representative of the Steel Helmets and the Friends of New Germany, pro-Nazi units, restored order after a time and Mr. Ridder continued.

Likens Agitators to Reds.

"The men who lead this racial agitation," he said, "are as certain to be driven out of the country as are the Communists and others. By raising the religious issue you are flying in the face of one of the most primitive emotions we have in this country."

Another uproar interrupted and the speaker quieted it by reminding the delegates that he had "suffered a great deal" because he defended Germany in the United States in the war years.

"For sixty years," he said, "the German Jews have been members of this society."

Mr. Spanknoedel interjected: "But they went out themselves last Monday."

"You blame them for getting out when they were attacked on a principle for which they, as Americans, have a right to fight?" Mr. Ridder asked.

"It is the Jews who stir things up, not we," said Mr. Spanknoedel.

He denied that the Steel Helmets annd the Friends of New Germany had any financial or political connection with the German Government. He also denied that the two units sought to import Nazi principles into the United States.

"I declare they do not suit the United States," he said. "They are for Germans only. The racial agitation comes from the Jewish side and there it remains."

The Rev. Mr. Popcke was authorized, however, to invite the German Jewish societies that withdrew to return and to make whatever steps he saw fit to restore unity. Another meeting of delegates will be held on Oct. 2 at Turn Halle.

September 26, 1933

Tactful Chinese Sign Fails to Save Laundry

The rioting in Harlem was not without a grim touch of humor here and there.

The proprietor of a Chinese laundry at 367 Lenox Avenue shared the fears of other store owners on the street as hoodlums raged along the sidewalks shattering windows with bricks and stones.

Then he noticed that Negro shopkeepers were painting on their display windows, in huge white letters, the word "Colored" to warn off the rioters. He adopted the idea and up went a sign: "Me colored too."

The window was smashed.

March 21, 1935

1,200 EXTRA POLICE ON 'WAR DUTY' HERE

Disorder Quelled in Harlem as Negroes Picket Italian Market—Two Hurt.

BROOKLYN PUPILS BATTLE

Boy Rivals With Ice Picks and Lead Pipes Dispersed—Many Detectives in Danger Zone.

About 1,000 uniformed policemen and from 200 to 300 detectives were ordered on special duty last night following the outbreak of trouble between Italian-Americans and Negroes in Harlem and Brooklyn. The precautions will be taken daily as long as the police authorities see danger of local violence as a result of the Italo-Ethiopian hostilities.

Disorder started in Harlem when Negroes began to picket the King Julius General Market, occupied by Italian butcher and vegetable shops, at 118th Street and Lenox Avenue about 2 o'clock yesterday afternoon. About eight Negroes and eight whites are employed at the market.

After picketing had gone on for more than an hour a Negro woman entered the market through the picket lines. A Negro picket followed her in and persuaded her to leave. When employes went outside to remonstrate with the pickets, a crowd of about 150 Italians and Negroes gathered.

They were exchanging insults and threats, and seemed likely to come to blows when the police arrived in response to a telephone call from the market.

Angered at Sight of Pistol.

Twenty-five patrolmen and fifteen detectives from the West 123d Street station, with three radio cars and an emergency truck, had little difficulty in dispersing the crowd, but Patrolman Chaimowitz drew his pistol and waved it over his head to make a group of Negroes move more quickly.

This gesture offended the Negroes and a short time later about 150 angry, grumbling men gathered in front of the West 123d Street police station. Police Captain George Mulholland ordered the Negroes to leave, but they held their ground, voicing protests against the action of Patrolman Chaimowitz. Captain Mulholland then ordered about forty-five patrolmen, who were just going out on their 4 P. M. tour, to disperse the crowd.

A Negro was arrested and a policeman was injured before this could be done. Most of the crowd fled, but Charles Linous, 33 years old, of 63 East 115th Street, refused to move from a stoop about 150 feet from the station, where he was waving the red, orange and green flag of Ethiopia.

The police said that he fought them off with the hard wooden flagstaff and struck Patrolman John J. Reilly a heavy blow on the right hand. Reilly was taken to the Harlem Hospital, where it was said some of the bones in his hand might be broken.

Linous also was injured in the mélée as he was dragged down from his stand, and was treated at the hospital for scalp wounds. He was then locked up in the West 123d Street station.

Fights at Brooklyn School.

The trouble in Brooklyn took place at Public School 178, on Dean Street, between Saratoga and Hopkinson Avenues, where 30 per cent of the 2,200 pupils are Negroes and another 30 per cent are Italians. On Wednesday afternoon, after school, a Negro boy beat an Italian lad in a fist fight. Shortly afterward the Italian returned to the school with a group of about ten boys, who beat four or five Negro boys who had been playing in the yard.

Feeling ran high when the students returned to school yesterday morning and a collection of weapons, including ice picks, sawed-off billiard cues, broom handles and lead pipes, was confiscated. There were threats of violence at the noon recess and both parents and the school authorities then asked the police for protection when school was dismissed at 3 o'clock.

About thirty policemen from the Liberty and Miller Avenue stations arrived at 2:45 o'clock with three radio cars. They seized weapons and dispersed a group of six Negro boys waiting outside the school.

After the pupils were dismissed, with appeals from the school authorities to go home, several hundred pupils and adults, both Negroes and Italians, gathered at Dean Street and Saratoga Avenue.

Their attitude was so threatening that the police sent for reinforcements and five radio cars and an emergency crew responded. After several fist fights among the schoolboys had been stopped the police dispersed the crowd.

Chief Inspector John J. Seery issued the orders for police mobilization over the departmental teletype system late yesterday afternoon after a conference in his office at police headquarters.

The orders specified that one squad of uniformed patrolmen, or one-tenth of the total patrol force, be held in reserve in each police station throughout the city, beginning at 8 o'clock last night and continuing until further notice. They also called for the transfer of detectives from quiet sectors to "danger zones" in Harlem, and in certain parts of the Bronx, Brooklyn and Queens.

These districts will be patrolled by detectives in fast automobiles, including radio cars.

The orders bring about a restoration of the old reserve system, which was abolished while Grover A. Whalen was Police Commissioner. Since then special emergency crews have been called upon for duty in handling crowds.

October 4, 1935

SONS OF ITALY SET UP GOOD-WILL BUREAU

Operated by Their Grand Lodge, It Will Combat Anti-Semitism

Dr. Santo Modica, Grand Master of the Sons of Italy Grand Lodge, Inc., has established a Bureau for Good-Will Between Italians and Jews in America to be operated by the lodge at its headquarters, 231 East Fourteenth Street, he announced yesterday. He made public a copy of a circular, describing the bureau, which has been sent to representatives of the organization's 200 lodges throughout New York State.

The circular declared that the resolution condemning anti-Semitism that was unanimously approved at the grand lodge's grand convention at Mount Vernon last December had been praised by Governor Lehman, Lieut. Gov. Poletti, Mayor La Guardia and "heads of various religious denominations, Catholics, Protestants, Jews, labor organizations, legislators, judges, municipal, State and Federal public officials." It continued:

"However, we have noted that anti-Semitism in Europe, unfortunately, has had a repercussion in America, particularly in the City and State of New York, causing a spirit of hatred and resentment between Italians and Jews that can only culminate in a daily struggle, with disastrous economic and moral effects on both sides."

The circular added that the grand lodge members are "mindful of our duty to stop such a movement, which is contrary to our traditions and American ideals" and would "do our utmost to revive their friendly relations and mutual cooperation for the purpose of securing civic and economic progress in this country."

The aim of his bureau, Dr. Modica said, was "to promote brotherhood and settle any dispute caused by discriminatory acts; to effectuate such, the spontaneous cooperation of the most outstanding Hebrew institutions is assured."

June 13, 1939

BOTH SIDES DEPLORE 'SMEAR' ON WILLKIE

Raising of 'German' Issue Is Denounced by His Backers as 'New Low' in Campaign

PAMPHLET IS REPUDIATED

Head of Negro Democrats Acts —Poletti Condemns Attacks Based on Ancestry

Republicans and Democrats here deplored yesterday as "vicious smear campaign tactics" the pamphlet issued Friday to campaign speakers by the Colored Division of the Democratic National Committee, which called attention to Wendell L. Willkie's German ancestry.

Bitterly characterized by Willkie campaign officials as "the most despicable campaign propaganda which has yet come into the Presidential campaign," the sections of the pamphlet dealing with the Republican Presidential nominee's ancestry were "repudiated" in Boston, according to the Associated Press, by Julian D. Rainey, Boston lawyer and chairman of the Colored Division of the Democratic National Committee.

Earlier in the day, Lieutenant Governor Charles Poletti, Democrat, speaking at the Columbus Day celebration in Columbus Circle, without specifically mentioning the pamphlet, held that "unjust attacks upon any one because of his antecedents should not be tolerated in America."

"Only a few days ago a candidate for the highest office within the gift of our people declared that a whispering campaign against him was being conducted because his ancestry was German," the Lieutenant Governor said. "Any such attacks deserve the condemnation of every right-thinking citizen of this country."

He made a similar statement in an address at the World's Fair.

Newspaper publication of a facsimile of the pamphlet being distributed by the Negro Democrats resulted in a storm of protest against the document from Willkie campaign leaders.

Alan Valentine, executive director of the National Committee of Democrats-for-Willkie, charged that "accredited leaders of Mr. Roosevelt's third-term campaign have descended to a new low in their attempts to smear Wendell Willkie and tie him up to Adolf Hitler."

Mr. Valentine placed the blame for the pamphlet squarely upon Charles Michelson, publicity director of the Democratic National Committee, and Edward J. Flynn, Democratic National Chairman, and contended that the President himself has made no attempt to condemn such tactics.

"If we judge correctly America's love of honesty and fair play, these latest drippings from poison pens will splash back at their authors," Mr. Valentine continued.

"These charges and slurs against Wendell Willkie, who enlisted the day of America's entry into the World War of 1917 and fought Germany then, who volunteered his services to defend Negro soldiers before court-martials in France, who fought the Ku Klux Klan in Indiana, make a new low in efforts to conceal the menace of the third-term 'grab.'"

Samuel F. Pryor Jr., vice chairman of the Republican National Committee and Willkie Eastern campaign manager, held that the issuance of the pamphlet indicated that "panic prevails in New Deal headquarters."

"I do not care to comment on as low a smear as came from the New Deal committee, but I'll wager any odds that no American Negro ever devised such a smear," Mr. Pryor said. "I believe the Negro race will resent such tactics. They already trust Mr. Willkie."

Francis E. Rivers, director of the Colored Division of the Republican National Committee, contended that the pamphlet represented the "last straw insulting and humiliating the colored race" and predicted that Negroes, in protest to such tactics, would overwhelmingly support Mr. Willkie on Election Day.

"I have received hundreds of telephone calls and telegrams from Negroes all over the country," Mr. Rivers said. "They insist that the foulness of this latest New Deal smear be taken off them immediately. The colored citizen has been so convinced by Mr. Willkie's forthright statements regarding his rights that these lying charges are powerless to affect the Negro's confidence in Mr. Willkie's sincerity and devotion to equal rights for all American citizens."

A similar statement was issued by Edwin F. Jaeckle, chairman of the Republican State Committee. Mr. Jaeckle denounced the pamphlet as "contemptible, and thoroughly in keeping with the New Deal's entire philosophy of government."

"Ever since Mr. Roosevelt took office, the entire New Deal machine has sought to array race against race, creed against creed, and color against color," Mr. Jaeckle contended.

George E. Wibecan, member of the National Planning Board, Negro Division, of the Republican National Committee and former Grand Exalted Ruler of the Colored Elks, bitterly assailed the pamphlet and contended that it was released in an attempt to counteract Mr. Willkie's "great reception" in Harlem last Tuesday night. He held that Negroes could not afford to "demean themselves through this type of vicious, unfair propaganda" and must repudiate it immediately.

Condemnation was also expressed by Erwin H. Klaus, president of the Roland German-American Democratic Society, with offices at 122 East Forty-second Street.

"The Colored Division of the Democratic National Committee certainly was ill advised in attempting to arouse discriminatory feelings against some Americans because of their national origin, since they, and rightfully so, oppose the discrimination colored people are shamefully subjected to," Mr. Klaus said.

"Roosevelt's re-election to us is of paramount importance to guide this country through the troubles of our time and make it instrumental in restoring world democracy. But we want to elect Roosevelt not because Willkie is a bad man but because Roosevelt is the better man."

Mr. Rainey, when questioned in Boston about the pamphlet issued by his division, said:

"I repudiate it, and please quote me on that. We did submit some material to Michelson for his approval, but none of it contained any reference to Willkie's ancestry."

October 13, 1940

Preferment for Negroes Is Sought by Board Here

Panel on Human Rights Cites Jobs and Housing —Plan Called Illegal

By SYDNEY H. SCHANBERG

The City Commission on Human Rights called yesterday for "preferential treatment" for Negroes to compensate "for the inequities of 100 years."

In a new policy statement, the commission said that all other methods of attacking racial bias in schools, housing and jobs had failed.

In the minds of some state officials, the proposal raised the question whether a city agency designed to eliminate discrimination was suggesting a plan that might discriminate against whites and thus possibly violate the state law against discrimination.

The law forbids discrimination in employment, housing and other areas because of race, creed, color or national origin.

Bernard Katzen, vice chairman of the State Commission for Human Rights, which administers the law, said:

"The concept of preferential treatment to compensate for the sins of the past is utterly inconsistent with equality of opportunity, and utterly illegal."

Mr. Katzen said the concept was self-defeating because it would antagonize whites and increase tension. He challenged Stanley H. Lowell, chairman of the city commission, to "stop talking in fancy words and riddles" and, instead, to work more vigorously within the law to "remove the evil road blocks" to equality.

Mr. Lowell, who is a lawer, insisted that there was nothing unlawful about preferential treatment for Negroes.

"But if it's found illegal in the courts," he said, "then the law must be changed. The protection of human rights needs the fist of government."

The commission was created in 1955 by a law passed by the City Council. It has a general mandate to "encourage equality and prevent discrimination." However, its powers of enforcement are limited.

The commission can investigate complaints and hold quasi judicial hearings, but for the most part, it depends on the Mayor and the City Council to put teeth into its findings and recommendations.

One exception is in the field of private housing, where the commission itself can seek a court injunction against a landlord.

Up to now, however, most of the commission's success, such as the agreement last summer by the White Castle hamburger chain to hire more Negroes, were gained through persuasion, reinforced by the pressure of public opinion.

'New Meaning' to Law

In its statement yesterday the commission said it was giving "new meaning" to the 1955 law, apparently in an effort to broaden its powers and to provide a reinterpretation of the law to sanction preferential treatment for Negroes.

The key section of the statement read:

"The Negro must receive recompense for the inequities of 100 years. This commission, therefore, urges preferential treatment to deal with the historic and existing exclusion pattern of our society."

At the same time,the commission took pains to make it clear that in terms of opening up new job opportunities, it was talking about "qualified" Negroes. The word was underlined in the text.

If the commission plan were successful, the statement continued, the preferential system should be necessary only "for a limited period, until the gap is closed."

Explaining what the system would mean in practice, Mr. Lowell said he would ask, among other things, that Negroes be moved to the top of waiting lists for apprenticeship programs in construction unions.

"If they had to wait their turn at this point, they'd be too old to qualify when they finally reached the top of the list," he said.

Another prime objective, Mr. Lowell said, would be more jobs for Negroes in the advertising field.

Mr. Lowell also asked that more money be spent on schools in Negro areas for special classes, such as remedial reading, to bring Negro children up to the educational levels of white youngsters.

Help 'Offset the Past'

This, the policy statement said, would "offset the inferior education they have received in the past."

Similar proposals have been made by Negro groups and leaders, including the Rev. Dr. Martin Luther King.

On the subject of jobs, Mr. Lowell stressed that he was not recommending the quota system, which he deplored as a "limitation on Progress." He also rejected "tokenism, the placing of two or three Negroes as a gesture toward integration."

Although the emphasis of the preferential plan was clearly on Negroes, the statement said it also applied to other low-income minority groups, "such as the Puerto Ricans, who have been subject to similar inequities."

Mr. Lowell made his remarks on a television interview program, WABC's "Page One," yesterday afternoon and in a conversation that followed the program. The policy statement was issued earlier in the day.

Would Step on Toes

When asked on the show Whether he believed that a lot of people might be angered by his plan, Mr. Lowell said: "I'm not concerned with stepping on toes."

It was also suggested that the plan might be considered just as undemocratic and unfair to whites as the Southern system of segregation is to Negroes.

The civil rights official said he thought the idea was completely fair because "Negroes have been deprived for 100 years" and this was the only way the scales could be balanced.

In fact, Mr. Lowell added, what he was asking for was not preferential but "realistic" treatment for Negroes.

On the credit side, Mr. Lowell noted, some strides toward giving special attention to Negroes have already been taken in New York City in the school system, in public housing and in municipal jobs.

The policy announcement, appraising the "old methods" of government civil rights agencies, said: "We have only to examine the educational lag of the Negro, the housing pattern in every Northern city and the job limitations of this one-tenth of our American citizenry to see that we have failed."

Mr. Lowell said that for the last 50 years, the civil rights movement had promoted "color blindness" — the belief that any reference to a man's ethnic origin is discriminatory.

"The time has come," he said, "when color consciousness is necessary and appropriate."

October 28, 1963

Bias on Mulberry Street

Community Charged With Hostility to Neighboring Minorities

TO THE EDITOR:

The response, the violent reaction and the attitude of Italian-Americans along Mulberry Street to the CORE pickets, Negroes and the civil rights movement in general was neither surprising nor unusual if one is familiar with their history in regard to other minority groups.

In fact, their behavior was consistent and clarifying to this observer. Being Chinese, and an American citizen (whether they like it or not), I can attest personally to the many years of discrimination, emotional and physical abuse and violent acts at their hands. I have also seen and heard of similar incidents to others of Chinese descent.

The word "Chink" and "Chinaman" was a daily greeting in school and in the street. We were told to "go back to China" or "Chinatown," where they said we "belong." We were forced to go to school (P.S. 130) in "their" neighborhood, and were shown, told and told again that we were not wanted. This despite the fact that the Chinese were as clean if not cleaner, as smart if not smarter and as friendly if not friendlier than the Italian-Americans. There can be no mistake that the basis of this feeling and attitude toward the Chinese was purely racial.

Before the Chinese it was the Jewish population in a nearby neighborhood, and now it is the Negroes and Puerto Ricans that are experiencing the venom of hate by this community.

Yet only a month ago at the James Center of the Children's Aid Society on Hester and Elizabeth Streets, speakers unanimously called upon their audience to welcome "strangers" into their community. Apparently the call was a mere whisper, and unheard or unheeded by most.

This community is in fact a "closed society" where the only American is a "white American" and their homeland the only one that one does not have to "go back to." This in spite of their own internal problems in the past as well as in the present.

No, this is not the "white backlash." Rather it is plain, unadulterated bigotry and discrimination.

New York, July 24, 1964. J. CHONG.

August 3, 1964

No Bias on Mulberry Street

TO THE EDITOR:

J. Chong on Aug. 3 complains of the abuses he suffered at the hands of "Italian-American" bigots in New York's Little Italy.

I submit that Mr. Chong would have a difficult time documenting his allegations of a pattern of discrimination and abuse. The fact of the matter is that the Italian and Chinese communities in this area have for years lived side by side with a surprisingly low rate of incidents and with a degree of mutual respect and friendship that is a model for other more prosperous areas of our city.

With respect to this community's reaction to the CORE pickets, the tenements of the Mulberry Street area are no newer than the tenements of Harlem or the brownstones of Bedford-Stuyvesant. The average income of the inhabitants is not high. The streets are narrow and overcrowded. By all outward appearances, Little Italy should be a slum. It is not, however.

Indeed, it is the one place in Manhattan where I would permit my wife or daughter to walk unescorted at any hour of the day or night without fear for their safety. Buildings and streets do not make slums. The principal ingredient of a slum is a shiftless and demoralized populace.

The Italian community of Mulberry Street spent two generations overcoming the same problems that the present minorities face and has made great progress by dint of its own hard work. Rather than turn new housing projects into slums, they have transformed a slum into a place where no man need fear to walk.

Who will criticize them for wishing to keep it that way?

LEONARD ABRUZZI.

Brooklyn, Aug. 4, 1964.

August 19, 1964

Grand Concourse: Hub of Bronx Is Undergoing Ethnic Changes

Transition Felt to Be Posing Threat to Stability of Area

By STEVEN V. ROBERTS

Hands behind his back, Rabbi Theodore Robinson swayed slightly, as Orthodox Jews do when they pray, and watched the traffic on the Grand Concourse. "The neighborhood is deteriorating, there's no getting away from it," he said with a shrug. "We had 140 families seven years ago and now we have 60. Even the big synagogues on the avenue are having trouble."

On 167th Street, Hyman Hans leaned on the white enamel counter of his kosher butcher shop and said: "Three shops closed around here last month. You couldn't buy a store on this block 10 years ago for any amount of money. Now they'd give it to you."

These are the marks of change on Grand Concourse, a broad, tree-shaded boulevard lined with sturdy brick apartments, which once meant status and prestige for Bronx Jews. From 161st Street, where it becomes residential, to Fordham Road, the avenue is in transition. The Concourse runs from 138th Street to Mosholu Parkway near 207th Street.

98% White in 1950

In the last decade, Negroes and Puerto Ricans have gradually moved north and west from the ghettos of the South Bronx and Morrisania, looking for better housing in the narrow, well-kept avenues parallel to the Concourse. A recent study by a social service agency predicted that the area, 98 per cent white in 1950, would be less than 50 per cent white in 1975.

The flight of the white middle class, almost imperceptible for years, has started to speed up. The builders of Co-op City, a 15,000-unit middle-income development planned for the northeast Bronx, have reported that a "sizable" number of their first 5,000 applications came from the Concourse area.

City officials fear that the area is on the edge of panic. Deeply troubled by the report on Co-op City, they are looking for some way to save the Concourse — not as a white preserve, but as a stable, integrated community.

Some things have already been done. Thirty building inspectors are thoroughly examining an area four blocks on either side of the Concourse from 138th Street to Burnside Avenue. The area from Tremont Avenue, just south of Burnside at 177th Street, to Fordham Road will receive city and Federal aid for rehabilitation and new construction under the low-rent public housing program.

But the complexity of the problem and increasing rate of change frustrates any simple solution. The transition of the Concourse began, in fact, not with a rejection of non-whites, but with a rejection by young white families of the world of their parents.

Vacancies Occurring

A typical attitude was expressed recently by Mrs. Rosalyn Hantman, as she watched her baby in Joyce Kilmer Park.

"There's not much play area here," she said, adding that she was moving to Ridgefield, N.J. "This isn't the perfect place to bring up children."

Her mother, Mrs. Hantman said, had grown up on Delancey Street on the Lower East Side. "To her the Bronx was a great step," said the young mother, "but I guess each generation wants something they can look ahead to."

For others, liberated by education and affluence from the narrow and exhausting lives of their parents, the relentlessly bourgeois character of the Bronx was just what they wanted to get away from.

The young have moved away. Many of the elderly—the area has twice the national average of older people—have died or are retiring to Florida. Vacancies have occurred.

"The landlords decided to accept Negroes rather than vacancies," said a real estate man.

View of Grand Concourse from 161st Street. Whites are leaving area and Negroes and Puerto Ricans are moving in.

City officials emphasize that they do not share the idea that Negroes inevitably cause a neighborhood to deteriorate. However, they say, the infusion of Negroes often produces reactions in a community that can lead to rapid decline.

"People become afraid simply because there is change," said one official, "and it is intensified if the change is also racial."

The slow exodus that opened good apartments to minority families has been accelerated by their presence. In certain spots a familiar pattern has begun to recur. Building managers have grown careless about maintenance. City services—police protection, sanitation, recreational facilities—have grown less reliable.

The migration again accelerates. The area becomes ripe for speculators, who have just begun to appear on the Concourse. "They've started to blockbust some buildings here," said an aide to Representative James H. Scheuer, who represents the area.

"When someone dies they move in an undesirable family, one on welfare with three kids in a bedroom. They cut services to the bone. Less stable families come in, sometimes several in one apartment so they can pay the rent, which of course goes up 15 per cent."

On the Concourse the problem is compounded by the nearness of slums to the east and south. The residents quickly say they like the new neighbors who can afford the rent in their buildings. But the black faces blur together.

As one resident said, "There is no wall between us and the slums." At night the streets are filled with children from the ghettos. Metal screens go up on store windows. Old people sit only on park benches near the street.

The New York Times July 21, 1966

The Concourse becomes residential at 161st Street.

The heart of the problem is fear of integration. And the people on the Concourse, who feel the fear in themselves and their neighbors, are pessimistic. "Let's face it," said one public school teacher, "for most of these parents a good school is a white school."

Others point to economics rooted in emotional and psychological attitudes. "It's too late to save the area," said a long-time resident. "We're not going to stop it. If Negroes and Puerto Ricans come in, there is no incentive to keep up the buildings. Maybe thats prejudice, maybe that's bigotry, but that's what happens."

To Negroes in the area such talk is deeply distressing. "It depends on what group of Negroes move in," said Anthony Hill, a young Air Force veteran who recently moved to the area. "If the whites won't move out when middle-income Negroes move in, property values won't go down. But if the whites do move, then the lower classes will start to move in, and then values will go down."

Government officials agree with Mr. Hill. Their job is to find some way to keep enough whites from moving so as to forestall the economic forces that produce slums.

More Public Services

Some, like Representative Scheuer, stress the need for more public services, making life more attractive and restoring confidence in the community. "We need a massive infusion of talents, money and services to provide the same integrated community facilities that will be built at Co-op City," said the Reform Democrat.

Bronx Borough President Herman Badillo emphasized the need for regular code enforcement, like the spot program recently started by the Buildings Department. He noted that the Federal Government recently rejected two code enforcement programs in areas east and west of the Concourse.

Previous attempts to upgrade declining neighborhoods have failed through lack of funds, manpower, and the authority to coordinate city services, Mr. Badillo said. This experience, he maintained, is an argument for a single agency with control over all phases of housing and development policy.

Morton L. Isler, head of the Community Renewal Program, an information-gathering division of the City Planning Commission, said that the city must tap Federal loan and grant programs and other means of financing the renovation of buildings. Code enforcement will not work, he asserted, if landlords cannot afford to bring buildings up to standard.

In addition, he said, "the city has to find ways to speed up the rate of Negro and Puerto Rican economic advancement." Otherwise, Mr. Isler said, there will not be enough middle-income members of these groups to move into the area and keep it sound.

What the city achieves in the Grand Concourse area is important, Mr. Isler said, because the buildings there must continue to provide good housing in the foreseeable future.

Of even greater importance, Mr. Isler added, was the fact that the battle against fear now starting there could provide valuable lessons for other struggles certain to erupt in other white areas facing an influx of Negroes and Puerto Ricans.

July 21, 1966

PRESIDENT URGES POLISH-AMERICANS TO SHUN BACKLASH

By TOM WICKER
Special to The New York Times

DOYLESTOWN, Pa., Oct. 16 —President Johnson, taking a last political fling before departing tomorrow for a tour of Asia, helped consecrate today a huge Roman Catholic shrine intended as a symbol of unity for Polish-Americans.

Touching several political bases before leaving the country on a trip that will take up most of the time remaining before the Nov. 8 election, Mr. Johnson was making an obvious bid for Polish-American support, but he also cautioned that ethnic group that it would betray its history by participating in a white backlash.

Invoking the spirit of Casimir Pulaski and Thaddeus Kosciusko, the Polish volunteers who fought in the American Revolution, the President reminded the dominantly Polish audience that General Kosciusko had bought Negro slaves in order to set them free.

'We Need That Spirit'

This, Mr. Johnson said, was the "true spirit of Polish-Americans."

"We need that spirit in America today—perhaps more than ever before," he said. "We need the spirit that says another man's dignity is more precious than life itself."

A crowd generously estimated by the Pennsylvania state police at more than 100,000 applauded Mr. Johnson frequently, if not always lustily. His remarks about equality and justice for Negroes appeared to be well-received.

Mr. Johnson spoke from a platform formed by a landing on the main entrance staircase of the concrete and glass shrine, which is nearly completed.

Overlooks Autumnal Vista

As he spoke, the President could look out over a broad expanse of the autumnal foliage of Bucks County.

A number of children in the audience were in Polish peasant attire, and a gaudily uniformed group from the Knights of Columbus lined up in front of the crowd.

Mr. Johnson's last day of activity before taking off for 17 days in Asia was not confined to his appearance here.

Mr. Johnson also issued a statement praising Congressional action to raise the level and extend the coverage of minimum wage provisions for the District of Columbia.

He signed numerous bills and appropriations, including the $58,067,472 appropriation for the Department of Defense.

Flies to Philadelphia

Early in the afternoon, Mr. Johnson flew to Willow Grove Naval Air Station near Philadelphia. From there he took a helicopter to a landing area near the Shrine of Our Lady of Czestochowa at Doylestown.

While Mr. Johnson did not appear here in the frank role of political campaigner, there was little doubt that the trip had election-year overtones.

Polish-Americans form large voting blocs in several important states—notably Pennsylvania, New York, Ohio, Michigan and Illinois.

It has also been among Polish-Americans that some of the strongest resistance to Negro social and economic gains—for instance, open housing—has been reported.

Last week, in Brooklyn, Mr. Johnson made a strong appeal for racial justice to a group of Italian-Americans. It is widely believed that white backlash against Negroes is most likely to occur among ethnic and minority groups, who themselves are not long removed from prejudice and economic disadvantage.

Mr. Johnson declared today that the nation needed "the spirit that says a man's skin shall not be a bar to his opportunity, any more than a man's name or a man's religion or a man's nationality."

He linked the nation's efforts to "expand the horizons" of its 20 million Negroes to its attempts to heed "the urgent pleas of others throughout the world," including "your friends and relatives in Poland."

He cited his pledge to build "bridges of friendship and trade and aid" to Eastern Europe, his appointment of John A. Gronouski to be Ambassador to Warsaw and his recent call for legislative authority to negotiate more favorable trade arrangements with Poland and other Eastern European nations.

As for the future, the President promised to work for closer cultural relations with Poland, for Polish cooperation in the satellite program, for the use of the American balance in Polish currency to benefit both countries, for liberalized travel regulations and for permission to send an American business mission to Poland.

Enthusiastic Applause

Perhaps the most enthusiastic applause came when Mr. Johnson, speaking of all the immigrant groups that had come to America, said that "most of all they brought a love of freedom and a respect for human dignity that was unsurpassed by any group in America."

The President was accompanied here by Mrs. Johnson, their daughter Lynda Bird, former Gov. David Lawrence of Pennsylvania and Farris Bryant, director of the Office of Emergency Planning.

As part of the ceremonies, the President was presented with a gold medallion of peace by the Rev. Michael Zembrzuski, the Vicar General of the Pauline Fathers, who conceived the idea for the shrine.

The shrine site is on an eminence called Beacon Hill 485 feet above sea level. It is two miles from Doylestown and about 30 miles north of metropolitan Philadelphia.

October 17, 1966

JEWS TROUBLED OVER NEGRO TIES

By IRVING SPIEGEL
Special to The New York Times

SAN FRANCISCO, July 4—The American Jewish community is confused and troubled over its relationship with Negroes.

A strain of ambivalence has crept into its long-time kinship with and support for the Negro's cause. It is perhaps even more uncertain, in the wake of the recent urban riots, about black attitudes toward Jews.

There is some Jewish backlash, whipped up by the slum explosions in which many Jewish merchants, generally small-business men, were looted and burned out. For some, it was the destruction of a business, the loss of a livelihood.

There is Jewish resentment over the anti-Israeli stance of black extremists who, in the parlance of the New Left, accuse the Jewish state of "Zionist imperialism" and "oppressions" against the Arabs.

There is Jewish bewilderment over the "step aside, whitey" thrust of black militants whose do-it-alone nationalism derides former Negro partnerships with white groups, among whom Jewish organizations were highly visible.

There is also, among the institutionalized elements of American Jewish life, a campaign of resistance to any trend toward Jewish disengagement from the Negro struggle. Its leaders, with few dissenters, say that the Jewish community has "a massive stake" in the outcome of the urban crisis.

It is impossible to give an accurate estimate of Jewish financial aid to Negro causes, but one reliable source, closely identified with fund-raising in behalf of the civil rights movement, remarked that "over the years it must have been in the multiplied millions."

There is no easy index to rank-and-file Jewish reaction to the new Negro militancy. But discussions with many of the 250 officials of Jewish intergroup relations agencies who attended a recent meeting here of the National Community Relations Advisory Council, and discussions with other Jewish organizational leaders in recent weeks, evoke the image of an American Jew soberly perplexed by the turn of events.

Civil Rights Supporter

Leonard Stein is an example. Mr. Stein is a small merchant in downtown Washington. He sells furniture and home appliances. Most of his customers are Negroes who buy on credit. His business employed—until the night it fell victim to an angry ghetto—a staff of eight, all Negroes.

Mr. Stein, who has headed B'nai B'rith's council in the Washington area, is a civil rights supporter. His contact with the ghetto has given him an insight into its desperate plight. It has troubled him and led him to become involved in community aid programs.

He says his relationships with customers and neighboring Negro tradesmen have always been excellent. He has had no personal fears about working in the ghetto.

When the riots came to Washington, his business premises were untouched while scores of establishments around his were looted and burned.

But two weeks later—after broad hints that it was coming—his business, too, was looted and burned.

Leonard Stein is not an angry man. He is frustrated and disheartened and, perhaps most of all, perplexed.

"I don't know why I was burned out," he said. "Because I'm a 'whitey'? Probably. Because I'm a Jew? That's hard to say. I was another phantom victim of an angry ghetto that exploded. Yet they didn't touch my place during the riots. Maybe I was picked out later by some of the extremists who figure they can exploit an anti-Semitic angle. But I really don't know. What happens next? I guess I don't know that either."

A Crucial Question

Mr. Stein's doubts and confusions point up a crucial question affecting Negro-Jewish relations today:

Should Jewish merchants and property owners, as well as Jewish welfare workers and schoolteachers, get out of the ghetto?

Dr. Judd L. Teller of New York, an authority on Jewish community life, says they should. He has recommended that Jewish community funds be used for the removal and relocation of "tens of thousands" of Jewish businesses—most of them "mama" and papa" stores—as part of an accommodation with black power.

Such proposals are criticized as "apartheid" by the American Jewish Congress and other groups. The Jewish Community Council in Washington, in a 12-point policy statement sent to Mayor Walter E. Washington after the recent riots, denounced such "false, simplistic solutions."

It asserted that both Negro and white racists were "trying to use the current ferment as a device to institute an apartheid system in America."

Nonetheless, in many cities there has been recognizable attrition in the number of Jewish-owned businesses in the ghetto.

Cleveland's black belt "kept its cool" following the murder of the Rev. Dr. Martin Luther King Jr. "because of the leavening influence of Mayor Carl Stokes," said Bennett Yanowitz of the city's Jewish Community Relations Committee. Jewish votes helped elect Mr. Stokes and he has "the respect of both whites and blacks," Mr. Yanowitz said.

But Cleveland had a ghetto eruption two years ago. Afterward, Jewish Community Relations officials met with Jewish merchants in the area and found that "surprisingly few felt any rancor," Mr. Yanowitz said. "But most were fearful," he continued, "and felt they were sitting on a powder keg. There has been a gradual drift to relocate elsewhere."

A study assessing the trend among Jewish merchants in nearly a score of cities following last year's riots showed that many "marginal" establishments went out of business, while others, some even well-established, were "seeking opportunities to get out on terms that would not be ruinous financially."

Jewish community relations specialists are all but unanimous in identifying Negro anti-Semitism as a small but highly vocal articulation among black extremists, unrepresentative of attitudes in the ghetto.

Backed by Polls

This is reinforced by a number of studies and polls conducted for the Anti-Defamation League of B'nai B'rith and the American Jewish Committee, which invariably showed a lower incidence of anti-Semitic sentiment among Negroes than among non-Jewish whites.

But Dr. Teller questions the implications of such studies. "The relative degrees of white and Negro anti-Semitism are immaterial," he says. "What is important is that circumstances make Negro anti-Semitism more explosive. Ten sticks of ignited dynamite are more dangerous than a ton of unlit dynamite."

"Black anti-Semites," said Harry Fleischman, race relations coordinator for the American Jewish Committee, "no more represent the Negro community than the Ku Klux Klan represents white America."

The drive of black power advocates to control the ghetto—and the eagerness of some Jewish merchants to flee—points up the irony of rising Negro-Jewish tensions despite a generation and longer in which the Jewish community has maintained a loose but effective alliance with its Negro counterpart in an integrated struggle on civil rights issues.

A Historical Accident

The confrontation in the ghetto today between the Jew as employer or landlord or merchant and the Negro as worker or tenant or customer is largely the result of historical accident. In the Brownsville section of Brooklyn, in Newark, Boston, Detroit and a score of other cities, Negroes have simply moved into urban areas that Jews, rising up the economic ladder, had left.

The indiscriminate nature of the recent burnings and lootings has left its residue of a "Jewish backlash."

Its extent is hard to measure and, as assayed by the Jewish community leaders who met here, differs from city to city, generally in relationship to the degree of white-Negro tensions in a particular area.

Some observers, among them Dr. Marvin Schick of Hunter College, a young leader in American Orthodox Judaism, find that the backlash is more pronounced among Jews who have not "made it" to suburbia and continue to live in the core city, in or near the ghetto, where they have regular contacts with Negro life.

"These people have had personal experience with Negro crime and juvenile delinquency," said Dr. Schick. "They may be wrong but they fear for their own safety."

Recently, the Jewish teacher has become a target of Negro militancy. Almost all of the 19 teachers and administrators barred from the Ocean Hill-Brownsville demonstration school district in Brooklyn are Jewish.

It is generally accepted that the board's desire is to have black children taught by black teachers. Yet the controversy points up an area of economic conflict between Negroes and Jews.

Jews, higher on the economic ladder, rarely compete with Negroes for the same jobs except in two service areas: teaching and welfare work. Many Jews work at these jobs in the ghetto and find themselves in conflict with Negroes who feel that since the work is with and for Negroes, the jobs should belong to blacks.

All told, however, Jewish backlash appears to be considerably less, proportionately, than backlash reactions in the white community as a whole. Moreover, its tone is different.

"It is situational rather than ideological," said Dr. Isaac Franck, executive vice-president of the Jewish Community Council of Greater Washington. He found that Jews were reacting to two situations—to the riots and to anti-Semitism voiced by the small number of black extremists.

The riots and their impact on Jewish merchants stimulated most of it. The verbalized anti-Semitism and anti-Israeli postures of black extremists added fuel.

One such outburst came last summer when the newsletter of the Student Nonviolent Coordinating Committee attacked Zionism, accused Israelis of committing atrocities against the Arabs and published anti-Semitic cartoons. Jewish organization leaders were shocked.

Another source of tension has been the opposition of many Negroes, particularly black separatists, to the leadership roles played by Jews in the civil rights movement.

Despite the active participation of many idealistic young Jews in the Negro struggle, a feeling has grown in the Jewish community that whites are no longer wanted.

Bayard Rustin, the Negro civil rights leader, has explained Negro behavior as a "love-hate syndrome." He defined the Negro community as one that had been kept "sociologically as children" by the white majority and was now maturing into a manhood "which Negroes are imposing on the white society."

In the growing-up process, said Mr. Rustin, "the desire to do it yourself becomes dominant."

'A Shrugging Withdrawal'

"When we Negroes have made it, we'll preach to our grandchildren the mythology of having done it all by ourselves," he said.

In the meantime, Mr. Rustin counseled his Jewish audience, Jews should understand the growing pains that compel the more strident elements in the Negro community to "hate those you love most."

"Where rank and file Jews have reacted negatively," says Isaiah Minkoff, executive vice chairman of the National Community Relations Advisory Council, "it has been less a matter of overt hostility and more a shrugging withdrawal. An attitude of indifference replaces their once active concerns for the Negro's struggle."

The institutionalized Jewish community is trying to combat such withdrawals as a denial of Jewish morality and conscience, as well as a surrender to Negro extremism. In a number of cities, local Jewish groups, often in concert with the Urban Coalition, have begun to divert funds and personnel from regular activities into poverty programs.

"We do ourselves a service," says Robert E. Segel of the Jewish Community Council of Metropolitan Boston, "ignoring imagined or real anti-Jewish manifestations among Negroes and concentrating on working for higher incomes for the poor, more adequate housing and quality education."

The fundamental problem in Negro-Jewish relations is, in the view of experienced Jewish community leaders, communication and understanding. There is, they all agree, a paramount need for Jews to be aware of not only the magnitude of the Negro's plight but also the reasons for his psychological reactions to it.

Dr. William A. Wexler, president of B'nai B'rith, put it this way in a recent address:

"We Jews never forget the martyred dead of centuries of persecution. Do we expect the Negro to forget and excuse 400 years of humiliation and discrimination? We Jews exult in our peoplehood. Do we deny the Negro his strivings for a peoplehood of new dignity, new values, new status and a new image?"

July 8, 1968

Question:

Is This Any Way for Nice Jewish Boys to Behave?

The Jewish Defense League answering a demand for reparations from synagogues.

Answer:

Maybe. Maybe there are times when there is no other way to get across to the extremist that the Jew is not quite the patsy some think he is.

Maybe there is only one way to get across a clear response to people who threaten seizure of synagogues and extortion of money. Maybe nice Jewish boys do not always get through to people who threaten to carry teachers out in pine boxes and to burn down merchants' stores.

Maybe some people and organizations are too nice. Maybe in times of crisis. Jewish boys should not be that nice. Maybe - just maybe - nice people build their own road to Auschwitz.

THE JEWISH DEFENSE LEAGUE

IS DEDICATED TO THE PROPOSITIONS THAT:

- **nice Jewish boys - or any nice boys - should not be forced out of their jobs by hoodlums.**

- nice Jewish boys - or any nice boys - should not be victims of quota systems and reverse discrimination in schools.
- nice Jewish boys - or any nice boys - should not become victims of totalitarian revolutionaries of the Radical Left.
- nice Jewish boys - or any nice boys - should not be forced out of their stores and see a lifetime of work destroyed by extremist thugs.
- nice Jewish boys - or any nice boys - should not be forced to pay a penny to extortionists for crimes they never committed.
- nice Jewish boys - or any nice boys - should not have to endure the potential rise of a Radical Right reaction which would destroy democracy.
- nice Jewish boys - or any nice boys - should not be victims of a do-nothing city, state or federal government.
- NICE JEWISH, CHRISTIAN, WHITE AND BLACK BOYS SHOULD CREATE A SOCIETY OF JUSTICE AND EQUALITY IN WHICH PEOPLE CAN GET BACK TO BEING NICE.

We Are Speaking of Jewish Survival!

We Are Speaking of The American Dream!

How Much Is Jewish Survival Worth To You?
How Much Are You Prepared To Give For It?

Gentlemen:

I am overjoyed at your work. I wish to help in any way I can. Enclosed is my contribution of

__ $10,000 __ $5,000 __ $1,000 __ other

Name ....

Address ....

Please mail to:
THE JEWISH DEFENSE LEAGUE
156 Fifth Avenue
New York, N. Y. 10010

I am overjoyed that your group has been formed.

— I would like to join. Enclosed is $10 for membership. ($3 for students)

— I also enclose an additional sum of .... toward your urgent needs.

Name............. Phone.........

Address ....

THE JEWISH DEFENSE LEAGUE
156 Fifth Avenue
New York, N. Y. 10010 Tel.: 989-6460

THE JEWISH DEFENSE LEAGUE

156 Fifth Ave., New York, N. Y. 10010 • Tel. 989-6460

MEIR KAHANE, National Director BERTRAM ZWEIBON, General Counsel

June 24, 1969

JEWISH COUNCIL ASSAILS QUOTAS

But 'Special Consideration' Is Urged for Negroes

By IRVING SPIEGEL
Special to The New York Times

PITTSBURGH, June 27 — A national assembly of human relations specialists, setting guidelines for Jewish community action on race relations and the urban crisis, today opposed "compensatory racial or ethnic quotas."

In adopting a policy statement, 250 Jewish communal leaders from 30 states favored "special consideration" for minority group members disadvantaged by past discrimination but emphasized their belief that the criterion "must be individual need, not race, religion or ethnicity."

This action was taken at the 25th annual conference of the National Jewish Community Relations Advisory Council, a coordinating body for nine national organizations and 82 local human relations agencies.

In opposing compensatory quotas, the council took issue with such black extremist demands as Negro college admissions based on population ratios—an issue that created the recent controversy at City College of New York—and similar "proportional quotas" in government and private employment.

Such a policy, the council declared, "would be inconsistent with the concept of equality of opportunity."

Special Help Backed

In its programmatic guidelines the council urged Jewish communities and agencies to support the right of Negroes and others "burdened by economic, social and vocational handicaps" to special education, job training, welfare aid, employment and other assistance.

The council also called for more effective action by government and private industry to correct "imbalances" in employment, education, housing and other areas evolving from past patterns of discrimination.

The policy statement affirmed that "the means employed to further Negro demands must be decided by Negroes."

But it added that "uncritical acceptance of Negro demands simply because they are voiced by Negroes is patronizing and condescending, as discriminatory in its way as uncritical rejection of Negro demands."

The council's declaration also welcomed movements in the black community that "foster group involvement, a common group culture, racial pride and a shared experience of a community controlling its own destiny—all necessary to full integration into society on a basis of equality with other groups."

Separatism Deplored

It deplored, however, proposals for "full and permanent" black separatism, rejecting the contention that white and black cultures are incompatible.

The council discussion on race relations followed a report yesterday by David Ginsburg of Washington, former executive director of the National Advisory Commission on Civil Disorders, (Kerner Commission).

Mr. Ginsberg assessed the urban crisis as "one year later—one year worse" since the commission issued its finding in March, 1968, that "the nation is moving toward two societies, one black and one white, separate and unequal."

Mr. Ginsburg said that the commission's recommendations for resolving racial imbalances in employment, housing, welfare and education "have neither been ignored nor implemented—they await the re-ordering of our national priorities."

June 28, 1969

ETHNIC ISOLATION IS SAID TO PERSIST

Tendency Called Factor in Slowing Racial Integration

A study of patterns of segregation in the metropolitan area has concluded that ethnic groups persist in their tendencies to isolate themselves from others, thus slowing the integration of whites with Negroes.

It had previously been thought that the inclination to isolation tended to decline as a group became accustomed to new surroundings and its socioeconomic status rose.

The study was conducted by Prof. Nathan Kantrowitz of Kent State University in Ohio, who is also a member of the University Seminar on Population and Social Change at Columbia University. It was financed by a grant from the National Institute of Mental Health to the Columbia University School of Social Work-Mobilization for Youth Research Project.

Censuses Analyzed

Professor Kantrowitz, who disclosed his findings in an article in The American Sociology Journal, based his conclusions on an analysis of censuses and a comparison of New York with 10 other cities. He found that segregation persisted in the second generation.

On a scale where total segregation is measured at 100 and no segregation at 0, he asserted that "the segregation index between the Norwegians and Swedes, 45.4, indicates a separation between the two Protestant Scandinavian populations which have partially intermarried and even have at least one community in common—the Bay Ridge neighborhood in Brooklyn."

Measuring the situations of 11 ethnic groups, along with Negroes and Puerto Ricans in the New York area, he found that 51.2 per cent of the population of southern European origin would have to be redistributed to achieve full integration with the northern European population.

He noted, for example, that the separation of the Catholic Irish from the Catholic Italians gave them a score of 45.5 nearly 40 years after the end of large-scale European migration.

"The strong prejudice against Negroes on the part of whites," Professor Kantrowitz wrote, "only compounds an existing separatism.

"For if Protestant Norwegians hesitate to integrate with Protestant Swedes, and Catholic Italians with Catholic Irish, then these groups are even less likely to accept Negro neighbors."

The indices of the separation of the Negro population from residents of European background vary from 79.7 to 88.4. For Puerto Ricans it is 66.

Professor Kantrowitz thinks this lower measure for Negro-Puerto Rican separation is largely because of the recent date of their migrations to New York.

July 20, 1969

Ethnic Group Ties to Blacks Sought

By BILL KOVACH

Special to The New York Times

CLEVELAND, Oct. 4—An effort organized by the Roman Catholic Church is under way here to direct a newly awakened feeling of cultural identity among white ethnic Americans into a coalition with the urban black community.

Although there have been gropings toward such cooperation elsewhere, the national pattern has been toward confrontation.

In city after city across the country, ethnic groups, reacting largely out of fear of the increasingly organized black population, have rejected the idea of a "melting pot." In most cases, attempts to assert themselves have placed ethnic group members in direct conflict with their black neighbors.

Mayor's Reaction

The growth of the trend toward polarization in a city such as this could be disastrous. Cleveland has 29 separate groups of white Americans of foreign stock that number more than 1,000 persons. These compete with a large black population (estimated at 47 per cent of the total) governed by a black Mayor, Carl B. Stokes. Mayor Stokes has reacted to the problem by assigning special aides to work with the ethnic communities. Community relations officers in areas of conflict meet almost nightly with community groups to develop programs of understanding.

The most far-reaching program, and one that could have national implication, is one called the Cleveland Plan, designed by an Ethnic Task Force of the Commission on Catholic Community Action.

'A Sleeping Giant'

In the introduction to the plan, Joseph F. Bauer, a Hungarian immigrant who headed the Ethnic Task Force, summed up the promise and peril of the situation:

"For whatever reason a sleeping giant is awakening and whether he awakens as a powerful ally in social progress or the biggest roadblock to social progress, depends entirely on how he is received.

"This plan is an attempt to enlist ethnic Americans, their press, churches, and other organizations in the fight against poverty and racial discrimination. It is an attempt to respond to the legitimate needs and aspirations of ethnic Americans and to develop greater communication and cooperation between ethnic and black Americans."

Specifically, the plan calls for organization of the ethnic groups largely along lines used by the Negroes in their drive for equal rights. The plan includes an ethnic council to seek out public funds and programs for low income ethnic people and a program to place ethnic group members on decision and policy-making boards of public and social agencies.

The ultimate goal would be to develop an ethnic leadership that could assure "that some of the brilliance which articulated black demands will have to be similarly developed to speak to and for lower-middle-class America."

"What we are saying, in effect," Mr. Bauer explained, "is that ethnic Americans, who feel lost, frustrated and abandoned, must be made secure enough in their identity that they can approach their black neighbors on an even footing—it's strange, but right now most of them feel unequal to such an approach."

The Ethnic Task Force is now awaiting a commitment for funding from the Roman Catholic Church, an important factor since the church is the strongest existing institution in ethnic communities. There is optimism based on the position of the Most Rev. William M. Cosgrove, Bishop of Cleveland, spelled out earlier in remarks to the Catholic Interracial Council.

"One of the major focuses of the Ethnic Task Force is to help ethnics to realize that they have many more economic and social interests with the black, brown and Appalachian white minority groups than they realize," the bishop said. "The differences which do exist are more fictional than factual and both groups are being victimized by those who profess to be interested in their welfare, but who in reality are allowing them to expend their energies in conflict with each other, instead of encouraging them to gather together to save our city."

Shared Problems

Mr. Bauer explained that in discussions conducted by the task force, some of which involved leaders of the black community, there was general agreement that they shared common social and economic problems that could be best tacked jointly. Both sides, however, have been conscious of the serious division between the two groups and of the obstacles to a coalition.

Meetings have been held through the city's Community Action Council on the local level between black and ethnic residents to emphasize their common problems and attempt to bring them together in joint cultural festivals and informal visits.

Proponents of the plan count heavily on the involvement of the Catholic Church, and support for such a move could be expected as well from the Mayor's office. One aide to Mayor Stokes said his office was "very interested" in the development and "would lend such an effort all the support and encouragement we could."

Although the interest of these two powerful groups is a hopeful sign, the effort is hampered by the disintegration of ethnic communities over the years.

Cleveland's huge ethnic population (an estimated 800,000 in the Greater Cleveland area) is largely made up of southern Europeans who came before and after World War II. Most of them settled in the inner city near the factories and steel mills.

Need for Self-Reliance

"When we came," said one Polish immigrant, "we knew we were coming to a country that expected us to take care of ourselves."

In a strange, alien land this acceptance of the need for self-reliance led to the organization of their own institutions to aid and protect them. Historically these institutions—usually centered on the church—were seen as a means toward assimilation. More recently, with the discovery that most cities contain large pockets of identifiable ethnic populations 50 years after their arrival, it has become clear that these institutions also served to isolate their people.

And, one observer said, "during the same time large numbers of ethnic Americans left the city for the suburb and their institutions—the fraternal clubs and welfare agencies—have largely disintegrated. The ethnics left behind at a time of increasing economic and social strain are finding it harder and harder to find a place in the city—to be accepted as unique people with unique desires."

In this environment the fears and concerns of the ethnic neighborhoods come out, as the task force study states, "in calls for law and order, the display of flags on their homes, cars and hard hats."

But, the task force suggests, the feelings are really of a group "for whom society has limited, if any, direct concern and little visible action."

Their hope is that the Cleveland plan will be a first step toward a satisfaction "of the most elementary human desire for recognition of the individual by his group and by his society."

October 5, 1970

On Ethnicity

By DANIEL P. MOYNIHAN

It would be getting on to thirty years ago that, as a freshman at C.C.N.Y., I was having lunch with a friend, a Rabbi's son from Brooklyn, who was engrossed in an article describing the admissions practices of well-known colleges and universities.

The proportion of Jewish students in each was somehow (probably inaccurately) calculated, and the institution was judged to be "tolerant" in that degree. My friend reported the more notable outcomes. An Ivy League university was found to be 5.4 per cent "tolerant." Something such. A fashionable girls' school turned out to be 2.2 per cent "tolerant." This was the general range of scores.

"Are we listed?" I asked. "What about that!" he exclaimed, with a sense of distinction both of us came to treasure. "We are!" "How do we do?" Deadpan now, he replied, "87 per cent 'tolerant.'"

The old place would not score anything so well today; but of course the reality is quite different and altogether welcome. Another ethnic succession is taking place in one of the great institutions of the city. This is the process that has given coherence and continuity to the life of the city from the outset.

With equal pleasure one learns that

C.C.N.Y. has now established separate departments of Afro-American, Asian, and Puerto Rican studies, which will evolve into a full-fledged School of Ethnic Studies, and also a Department of Jewish studies which will remain within the liberal arts faculty.

One gentle suggestion may be in order. A School of Ethnic Studies at C.C.N.Y. ought to provide for the study of all the ethnic groups of the city, large and small. Ethnicity is our history. It did not begin with the memory of the newest arrivals. Its marks are to be read in the substrata of every one of our major institutions. The Mayor is English, with a middle name that suggests some touch of German or Dutch. One deputy mayor is Italian, the other Irish. South Slavs, Indians, Nordics, Guatemalans, Israelis—name it—they are all here, part of the life of the city which has ever been thus.

Once more—gently—it should be understood that the preceding assertions, if accepted, have certain political meanings.

In 1963 Nathan Glazer and I published "Beyond the Melting Pot," a study of the major ethnic groups of New York. Liberalism was then much in thrall to the idea that ethnicity was vulgar—an affair of Tammany Hall and balanced tickets—and that the correct course of public policy was to eradicate ethnic identity as much as could be done by official decree. New York pioneered in laws that would forbid the author of my friend's magazine article from knowing anything about the race, color, creed, or national origin of, say, the students at C.C.N.Y. Glazer and I contended that whatever the laws, ethnicity persisted, and was seen by many as something of value. The book was not received with excessive generosity.

Last year we issued a new edition in which we surveyed events of the 1960's. This time it was we who were less than generous. We would have preferred the decade be tried over again. To be sure, the ethnic reality had come to be more widely perceived, but we felt it had also been seriously misinterpreted.

We posited two models of ethnic relations, for convenience termed Northern and Southern. In the North there are many groups; coalitions form and reform; positions in the pecking order shift; there is a rhetoric of civility and celebration. ("While we cannot yet establish that there was a Chinese member of Columbus' crew, it was of course the case that he navigated with a compass developed in China so that in every legitimate sense it may be claimed . . . .") In the Southern model there are two groups; no coalitions are possible; positions are fixed, one group being on top and the other on the bottom; there is a rhetoric of threat and abuse. We felt the latter model, with local variations, was the one being legitimized and encouraged in New York, and that while this served the interests of some individuals, it could be immensely destructive to the life of the city. We felt that the Northern model had been a social invention of some consequence, bringing stability to situations that seemed hopeless. By contrast, the Southern model is fundamentally instable.

Donald L. Horowitz has since offered a general theory of ethnic differentiation, distinguishing between vertical, that is to say, hierarchical systems, which would be our Southern model, and horizontal ones, which would correspond to the New York experience. The former is a caste situation: Winner and Loser. The latter may be thought of as a big poker game with many players and no permanent outcomes.

There will be consequences to the way we chose to understand the ethnic experience. To date most academic institutions have opted to study ethnicity as a form of caste. It can be that, and in many parts of the nation and of the world, it is. But there is another reality, which by and large has been ours in the North, and we should study it also. This is the style of ethnic relations that seeks accommodation. We ignore it at genuine peril.

Daniel P. Moynihan returned to Harvard this year after a spell as counselor to President Nixon.

May 2, 1971

Antipoverty Setup Seen Abetting Ethnic Conflict

By MARTIN ARNOLD

A Congressional subcommittee heard testimony yesterday that the city's antipoverty agencies had become mechanisms to foment ethnic conflict between blacks and Puerto Ricans and between blacks and Jews.

The House Subcommittee on Poverty and Manpower heard S. Elly Rosen, representing the New York Association of Jewish Community, Antipoverty and Municipal Employes, declare that Jews "are systematically denied participation in poverty programs," except in the Crown Heights section of Brooklyn.

Mr. Rosen said that where antipoverty elections were held to determine the make-up of community antipoverty corporations, poor Jews had been "forcibly prevented" from voting and had been actually beaten and otherwise abused. Asked who did this, he said usually blacks, who, he said, controlled most of the antipoverty programs in the city.

"The indignities suffered by the poor Jew in this city are enormous, horrendous and growing," Mr. Rosen said, noting that 800,000 Jews live below the poverty level in this city.

"The place one must go to is the New York City Council Against Poverty" to correct this situation, he said, adding:

"But after several rabbis were brutally beaten at such a meeting, had their heads broken and after witnessing meeting after meeting where people showed arms, where people came with masks against sitnk bombs and where resolutions were made strictly on a racial basis, who wants to bother with a kangaroo court, and who wants to complain against a beating only to be beaten again?"

In other testimony, David Billings 3d, a black, who is chairman of Council Against Poverty, said that "although Puerto Ricans are the major group in only three of these [26] poverty areas, 10 community corporations have a maority of Puerto Rican members on the board or a Puerto Rican as the executive director."

Long 'Oppression' Cited

He said the "inevitable struggle" between blacks and Puerto Ricans could only be eased by increasing Federal funds for antipoverty programs.

"Although recent labor statistics show that Puerto Ricans have the highest percentage of unemployment, blacks still feel that their oppression has over 400 years of history and no change in labor statistics changes the nature of white racist oppression against blacks as a matter of skin color," Mr. Billings said.

The subcommittee hearings, which lasted all day, heard testimony from, among others, Jule M. Sugarman, Human Resources Administrator; Dennis C. Gardner, executive director of the Morrisania Community Corporation, and Paul Cooper, director of the Brownsville Community Corporation.

The Congressman are conducting hearings on the city's antipoverty programs, but do not have specific legislation before them.

Representative Augustus Hawkins, Democrat of California, was chairman of the subcommittee. Representatives Herman Badillo, Charles B. Rangel and James H. Scheuer, Democrats of New York; Earl F. Landgrebe, Republican of Indiana, and Victory Veysey, Republican of California, also participated in the hearings, which will continue today at 9 A.M. in Room 305 of the the Federal Building, 26 Federal Plaza.

June 6, 1971

High Court Avoids Ruling On Quota for Law School

Refusal to Decide on Constitutionality of Preferential Treatment for Minorities Is Considered Only Postponement

By WARREN WEAVER Jr.

Special to The New York Times

WASHINGTON, April 23—

The Supreme Court, dividing 5 to 4, refused today to decide whether professional schools can constitutionally give preference in admissions to members of racial minorities at the expense of white applicants.

At stake, but never reached by the Justices, was the broader issue of "reverse discrimination," whether giving special treatment in education and employment to blacks unfairly deprives whites of their right to be judged on their own merits.

The narrow Court majority maintained in an unsigned opinion that there was no longer a live controversy requiring a decision because Marco DeFunis Jr., the student who originally brought the action to win admission to law school, will graduate from that school in June no matter what the Court might do.

This ruling, sharply attacked by the minority as "sidestepping," left completely unresolved one of the most controversial and emotional issues to come before the Court since the abortion cases.

Colleges and universities throughout the country must now wait for another case to be filed and work its way up through the court system—this one took nearly three years—before learning whether "affirmative action" programs to admit blacks with lower academic and test records are legal.

Voting to refuse to consider the case were the four men named to the Court by President Nixon: Chief Justice Warren B. Burger and Associate Justices Harry A. Blackmun, Lewis F. Powell Jr. and William H. Rehnquist—plus Associate Justice Potter Stewart, an Eisenhower appointee.

Charging that avoidance of the issue "clearly disserves the public interest" were the four dissenting Justices, all Democrats when they came to the Court: Associate Justices William O. Douglas, William J. Brennan Jr., Byron R. White and Thurgood Marshall.

In a long separate dissent, Justice Douglas said he believed the admission practices at issue were unconstitutional because they were based on race. He maintained that preferences for students from deprived backgrounds were valid if they were granted without respect to race.

The case turned aside by the Court (No. 73-235, DeFunis v. Odegard) had aroused widespread interest because of its potential effect on the legality of various quota systems currently used to help promote employment of blacks and other minorities as well as their access to higher education.

Some legal authorities favoring discretionary admissions policies had urged the Court to postpone a ruling on the issue until a more sharply defined case came before the Justices.

The Court action was brought by Mr. DeFunis after he was refused admission to the University of Washington Law School in 1971, on the grounds that three dozen successful applicants who were members of racial minorities had combined academic and aptitude ratings lower than his.

The trial court agreed with Mr. DeFunis's contention that this admissions system violated his constitutional right to equal protection of the laws. He was then admitted and remained enrolled after the Washington Supreme Court reversed the decision because Justice Douglas issued a stay pending final Supreme Court resolution of the case.

The law school maintained that it was not bound by the strict arithmetic of test scores and grades in deciding admissions and that a positive kind of quota benefiting a minority rather than restricting it was not unconstitutional.

Predictably, the Court's refusal to act satisfied almost no one. Mr. DeFunis declined to comment. The law school said no decision was preferable to a decision against its admissions policy. Organizations like the American Jewish Congress, which backed Mr. DeFunis, said they were disappointed.

The dispute split the liberal community into two angry and contentious segments, one arguing that each individual must be judged on his own merits and the other that long-standing racial discrimination demands some form of compensatory help for blacks and other minorities.

The division was particularly sharp among Jewish groups. Several of them joined the case on Mr. DeFunis's side, maintaining that quota systems that in the past deprived Jews of professional educations were equally objectionable when applied to help other different minorities.

Other Jewish organizations, however, including the National Council of Jewish Women and the Union of American Hebrew Congregations, defended the law school admissions policy on the general thesis that positive quotas designed to help the underprivileged were different from negative quotas designed to discriminate against them.

"All parties agree," the Supreme Court majority declared, "that DeFunis is now entitled to complete his legal studies at the University of Washington and to receive his degree from that institution.

"A determination by this Court of the legal issues tendered by the parties is no longer necessary to compel that result and would not serve to prevent it," the majority concluded.

The five Justices who refused to decide the case suggested that the responsibility was not theirs because if Mr. DeFunis had filed a class action on behalf of all law students denied admission, the controversy would have remained alive after his admission to the final quarter.

The majority alo conceded that they were probably only postponing a decision on the merits of a similar case.

"If the admissions procedures of the law school remain unchanged," they observed, "there is no reason to suppose that a subsequent case attacking those procedures will not come with relative speed to this Court, now that the Supreme Court of Washington has spoken."

The majority apparently assumed that the state Supreme Court would need less time on a second occasion to sustain the law school against another student denied admission.

Writing for the minority, Justice Brennan said that Mr. Defunis's objections to the school's admission policy would become "real, not fanciful" if he were forced to drop out by "illness, economic necessity, even academic failure" before graduating and had to reapply in the fall.

Mr. Brennan said he could "find no justification for the Court's straining to rid itself of this dispute. While we must be vigilant to require that litigants maintain a personal stake in the outcome of a controversy . . . there is no want of an adversary contest in this case."

The dissenters noted the broad range of national interest attracted by the DeFunis case, which promoted the filing of 26 friend-of-the-court briefs by interested and affected groups, and predicted that today's ruling was nothing more than a postponement.

"Few constitutional questions in recent history have stirred as much debate, and they will not disappear," Mr. Brennan said. "They must inevitably return to the Federal courts and ultimately again to this court."

In a relatively unusual move, none of the five Justices in the majority took credit for the authorship of the seven-page opinion. Ordinarily, many unsigned or "per curiam" opinions are only issued in relatively routine cases when the Court is unanimous or there is, perhaps, a lone dissenter.

In his separate dissent, Justice Douglas proposed that a new trial be held to consider, among other questions, whether the present national law school aptitude test should not be given to members of racial minorities because it measures "the dimensions and orientation of the organization man."

Mr. Douglas suggested that preferential treatment should be given to a promising law school applicant who was "a poor Appalachian white or a second-generation Chinese in San Francisco" just as readily as to other would-be lawyers with limited backgrounds.

Under the University of Washington system, special consideration was restricted to black, Chicano, American Indian or Filipino applicants.

April 24, 1974

TV 'SPOTS' SHIFT ETHNIC PATTERN

Drop 'Melting Pot' Theme and Focus on Reality

By LES BROWN

With the aim of easing the tensions fostered by the desegregation of schools, the Department of Health, Education and Welfare has financed a campaign of public service television "spots" that abandon the "melting pot" and "brotherhood" themes of previous race relations campaigns and, instead, concentrate on the problems and value of an ethnically mixed society.

Devised by experts on race relations and ethnicity from the University of Chicago, working with representatives of two major advertising agencies, the series of 30 spot announcements is being distributed this month to 500 commercial and public television stations.

The distribution is being handled by WTTW, the Chicago public television station, which produced the spots under a grant of $811,000 from the Federal agency.

Directed primarily at children of school age, the campaign makes a significant departure from the familiar "brotherhood" messages in recognizing that there are distinct thenic and racial differences among people, that the minorities have not "melted" into a homogeneous society, and that the differences can have positive value although they present problems.

The spots, which have been described as "sensitive and scrupulously honest" by William J. McCarter, general manager of WTTW, attempt to deal with the root causes of ethnic isolation.

Some of the announcements identify parentally induced prejudices, peer group pressures and the natural inclination of most human beings to fear the differences in others. In a recurring theme, children are urged to "be yourself."

'Value Reorientation'

"Television has given us one kind of message on desegregation, in its newsreel reports on riots in Boston over school busing," a spokesman for the project said. "This is another kind of message, a form of value reorientation."

As with all unsolicited public service material, the H.E.W.-funded spots may be televised at the stations' discretion, and the broad distribution of them carries no assurance that they will be televised in all communities.

H.E.W. was authorized to sponsor such a campaign under a provision of Title VII of the Emergency School Aid Act of 1972, which reads: "Congress finds that the process of eliminating or preventing minority group isolation and improving the quality of education for all children often involves the expenditure of additional funds to which local educational agencies do not have access."

WTTW organized a team of academics and advertising professionals to formulate the concepts and approaches. The task force on content was headed by Dr. Pastora San Juan Cafferty, director of the Latin Project and Ethnicity Study Group of the University of Chicago's School of Social Service Administration, and Laurence Hall, an expert in social welfare policy at the same school.

Ken Huchison of the Grey-North advertising agency enlisted the creative group to design the story board, and Leo Burnett Advertising conducted special consumer research on the spots.

October 8, 1974

Black and white in Catholic eyes

'The struggles for survival at the bottom of America are not over skin color, but jobs, turf and education.'

By Michael Novak

In Boston recently, Catholic mothers have said their rosaries in the streets to protest busing, rather like the "flower children" of 1967 protesting war. In Rosedale, Queens, some Catholic hoodlums have twice bombed the home of a brave black man from Trinidad, a despicable method favored by the New World Liberation Front. Protests, demon-

Michael Novak, author of "The Rise of the Unmeltable Ethnics," is executive director of Ethnic Millions Political Action Committee.

White ethnics and blacks have more reasons for unity than for division. If they should decide to form an alliance, the political consequences could turn out to be important, in terms of style as well as substance.

strations, riots and civil disobedience continue in the 70's—but not at the same places as in the 60's. They are likely to get worse.

These conflicts are usually presented by the media in terms of simple racism, viewed as they are against the common perception of urban blacks, their problems and their aspirations that has preoccupied opinion makers and politicians in this country for more than a decade. And yet the conflicts between urban Catholics and blacks are far more complicated, for they involve considerations of social class and clashes of genuine interests, as well as race. In fact, there is good reason to believe the proposition that affluent liberals who take a protective attitude towards the black underclass, with little regard for the white-ethnic lower class, are—at the risk of great social turmoil—participating in an unconscious, or perhaps conscious, conspiracy of the upper classes to prevent formation of an inner-city alliance that could function as a major political force.

Certainly, there is much to support the notion that blacks and white ethnics have more reasons for unity than for division. Robert Kennedy was the last politician who knew how to hold the trust of both groups, but another one could emerge (though it will not be his brother, whose record in bringing white ethnics and blacks together is abominable). If one does, the political consequences could be important—in terms of substance as well as style.

For example, no Brookings Institution is at present developing realizable policies on the undramatic issues of jobs and security, protection from migrant corporations, fair treatment from local banks, help in caring for the family's aged, a sense of public dignity in the media and the schools, a fair shot at representation in the higher echelons of power and status, a break in the stranglehold of the top 10 percent of the nation (in terms of education and income)—the "opinion leaders"—upon the definition of political issues. To understand the true importance of these issues, it is necessary to take a look at the Northern cities and their problems from a white-ethnic—which usually means a Catholic—point of view.

One hundred years ago, 90 percent of all black Americans lived in the South, and 90 percent of the nation's present Catholic population had not yet arrived from Europe. In the Northern cities, the 20th century has been largely the story of the fateful meeting of these two migrations, unprepared for each other, not linked through the mystic bond of slavery. In the last 20 years, we have learned a great deal about the black experience, and prejudice against blacks is no longer considered acceptable. But subtle forms of bigotry against Catholics are still respectable.

Catholic politics, highly personal and strictly accountable in its fashion, is called in pejorative terms "machine politics." Elite Protestant politics, highly impersonal and far vaster in wealth and power, is spoken of in terms of "conscience" and "reform." Catholic favoritism is called graft and nepotism; Protestant favoritism is forgiven as a result of an inevitable "old-boy network." At the top, there is no melting pot. On the boards of the foundations, among appointive judges, on boards of trustees of universities and in other places, white ethnics characteristically have as few token representatives as blacks. In Chicago's top 106 corporations, only 1.9 percent of 1,341 directors are Italian; 0.3 percent are Polish; 0.4 percent are black; 0.1 percent are Hispanic. In the Chicago police department today, at every level, Poles are almost as underrepresented as blacks. In Pittsburgh today, the foreign-stock families with incomes under $3,500 per year outnumber the black poor by more than 2 to 1. At one time or another, various groups have viewed Catholics as social antagonists. Sympathy for Catholic viewpoints requires from some an extra generosity.

One out of four Americans is Catholic in culture and tradition (not counting those who are no longer "practicing" Catholics). The Catholic population in Northern states is sometimes close to 50 percent, as in Massachusetts, and usually at least 30 or 40 percent. Moreover, in the Northern urban centers, practically the only whites left are Catholics and (in Chicago and Detroit) Appalachian migrants. Two out of every three persons in Detroit are black or Polish. Newark is divided by blacks, Hispanics and Italians. In recent decades, an increased urban migration of Hispanics, an influx of Cubans and the migration of many Catholics from the Midwest have raised the percentage of Catholics in Florida and Texas—usually thought of as Protestant states—to more than 20 percent. Catholics are especially highly represented in the 10 states, including California, with the largest electoral votes. Particularly in the North, urban history in the 20th century may have been more affected by Catholic cultures than by any other.

The black population of the North has increased since 1900 from under one million to over 10 million. The Catholic population there has increased from about 11 million to about 45 million. In all, Catholics in the U.S., North and South, number about 13 million Eastern Europeans, 10 million Italians, nine million Irish, seven million Germans, five million French-Canadians, and (here the figures are more uncertain) 15 million Hispanics.

One out of five Catholics lives in census tracts with blacks, compared to one out of 18 Anglo-Americans. Ethnics are already the most integrated white Americans.

As long ago as blacks were

slaves, many among these Catholics had themselves been serfs. Even today in Eastern Europe, the relatives of many of them live under an oppression more severe than in the days of the Czars, and in some respects as bad as any ever known to blacks. In the computation of human sufferings borne upon this planet, many do not consider themselves to have suffered less than blacks during the last 500 or 1,000 years. This is an important self-perception. Torture, imprisonment, grinding poverty, exclusion from education—these have all been visited on Catholic ethnics in Europe as well as on black slaves in America.

Once in America, the white ethnics often experienced the New World's own subtler forms of discrimination, if not oppression. By their names, accents and social links, Catholics have felt clearly identified and vulnerable before the powers-that-be in mills and mines, just as blacks have by their color. Whites still sort each other out by class, culture and kinship; some whites oppress others as they have blacks; many whites feel a burden of oppression, lack of status, and inequality. Blacks are not alone in suffering evil, in feeling self-doubt, in meeting prejudice. Slavic folksongs share with black melodies the melancholy of a long oppression. Nor are rebellion and resistance limited to blacks. Joe Namath and Andy Warhol, Boston's ROAR, Al Capone in Cicero, the long ethnic struggles in the coal fields, steel mills and auto factories, the building of Catholic "ward politics," the role of Slavic women, like Big Mary Septak in Pennsylvania, in pushing their men to rebellion—these suggest the range of Catholic populism, a study that much needs to be written. It is a populism that differs from the populism of Bryan, La Follette and Huey Long, as well as from that of King, Carmichael and Wilkins. A kind of self-hatred and passionate contempt for upper-class hypocrisy often suffuse it.

Blacks and Catholics had almost no cultural preparation for their fateful meeting in the Northern cities of America. They met not as slaveowners and slaves but as competitors in the grim and uncertain business of survival. About 1915, blacks were beginning to migrate North in sizable numbers. When, about 1919, blacks began to be used in large numbers as strikebreakers, as in Pittsburgh, a major change occurred in Catholic-black perceptions. From then on, the ethnic competition became serious, sometimes ugly and occasionally violent.

Many black leaders in the U.S. do not know the experiences of Catholic ethnic cultures, whether in Europe or in America. Hence, they often misjudge motives, interests and political behaviors. Without thinking about it very much, they often assume Protestant upper-class biases and put down the white ethnics. Thus, Stokely Carmichael took the epithet that upper-class nativists had long hurled at Catholic foreigners, "hunkie," (derived from "Hungarian" and connoting stupid, foreign, unworthy) and altered it to "honkie," applying it to all whites. Hearing it, white ethnics heard an old slur. The slur seems especially unfair; white ethnics are not the sole, or even the main, and certainly not the most powerful carriers of racism—a fact that has been publicly acknowledged by some blacks. Whitney Young was in 1971 the first black leader to issue a warning about the strategic danger of the public bias against white ethnic Catholics (and Jews). Recently, Robert Hill, research director of the Urban League, and Vernon Jordan, executive director, reviewed all the studies of white-ethnic attitudes toward blacks. Hill wrote: "How is it that, despite the overwhelming consensus of these studies, white ethnics are believed to be more racist than white Protestants? Generally, we have found that white Protestants throughout the nation are more likely to hold unfavorable racial attitudes than are white ethnics in similar size communities and regions." Young's Harris survey showed that 22 percent of Anglo-Americans favored racially separate schools, compared with 6 percent of the Irish, 5 per cent of the Italians and 16 percent of the Poles. Studies by Nathan Kantrowitz, Angus Campbell, Norman Nie and others—including a 1971 report in Scientific American by Paul B. Sheatsley and Andrew Greeley—consistently show the same patterns.

Furthermore, some issues—security in jobs and neighborhoods, for example—unite blacks and white ethnics, despite their real differences. Progressive majorities in the North have invariably depended on large majorities in white-ethnic and black neighborhoods together. Few white ethnics would deny that, generally, limiting comparisons solely to the North, Catholics are statistically doing better than blacks. But in terms of gross numbers, there are more white poor than black poor. What helps one can help the other, if programs are so designed.

Certain understandings are telegraphed between Protestants, including blacks, in the public language of such diverse figures as Lyndon Johnson, Sam Ervin, George Wallace, Albert Gore, Ralph Yarborough and also (in a different vein) of John Gardner, Elliot Richardson, Nelson Rockefeller and others. Catholics can only decipher them from the outside: the way America is spoken of as a "covenanted people," for example, and the interplay in Protestant consciousness between morality, guilt, reform and politics. Protestants somehow always manage to seem more high-minded and moral than Catholics, while not seeming to become less wealthy or less powerful in the process.

In the evangelical tradition, blacks seem to take up this politics of morality by second nature. Many describe racism as a moral failing, about which people should feel guilty and change their ways. (A different tactic would be to treat racism as an unconscious way of perceiving, and work on issues where coalition is possible and mutual loyalties can be built. Guilt divides; shared goals unite.) The first political move of many black politicians, in the media at least, seems to be to make their listeners feel guilty, in order to increase political leverage over them. Nativist Protestants (like Archie Bunker to his Polish son-in-law) used to call Catholics hunkies, dagos, bohunks, wops, micks and greasers; now others, including some blacks, call them racists, Fascists and pigs, terms not any the more affectionate. Or accurate.

Many ordinary Catholics, meanwhile, do judge blacks harshly and unfairly, reacting in some confusion to stereotypes of blacks promoted in the media. Bill Moyers asked a man in Rosedale why he wanted to keep blacks out. The reply: "If you really want to know, they're basically uncivilized. Wherever they go, the crime rate goes up, neighborhoods fall apart, whites have to leave. Well, we don't share their life-style and we're not going to live with them. Rosedale is the last white stronghold in this city and nobody's going to push us out. We're going to keep it crime-free, clean and white. If that's racism, make the most of it." A Ruthenian tool-and-die operator in Pittsburgh may have a grudging respect for the black worker who has worked beside him for several years and get along well with the black family across the street, in terms of wholly personal relations; but he may still dislike intensely the perception of blacks he gets from television and the papers. More to the point, he may hate the schemes of "social engineering" emanating from his hereditary enemies in the Protestant upper class, like busing or bank "redlining" or real-estate "blockbusting." In his eyes, he is being asked to pay the entire price for the injustices done to blacks—he who is living on the margin himself — while those who were enriched pay nothing. He also believes these schemes of forced intervention are both illusory and unfair. Persons of a more "modern" and "enlightened" mentality, by contrast, may have no relations with blacks as personal as his, and think of his resistance to blacks as a group as "prejudice."

The modern category of "prejudice," indeed, is based on an ideal of dubious merit. Instead of noticing differences between cultures and their imprint on personality, the "enlightened" or "nonprejudiced" person tries to see what all humans have "in common," a "universal brotherhood," the ways in which we are all, fundamentally, "the same."

In more highly educated and affluent circles, these attitudes are rather easy to maintain, or to fake, for at least two reasons. First, intimate competitive contact with persons of pronounced cultural difference is neither frequent nor disturbing. Money provides space, privacy, mobility and the ability to choose one's associations. Second, the persons one does meet tend to be highly cultivated and well-tutored in how to avoid unpleasant subjects; they tend to represent the best qualities of their own culture and to proceed in a manner wholly civil.

Cultural contact between the less privileged members of each group tends, by contrast, to be both deeper and more intimate when it is healthy, and rougher, less civil or even violent when conflicts arise. The Ruthenian tool-and-die worker may esteem his black colleague, drink with him in the bar after work, and talk with him as frankly as Archie Bunker with the Jeffersons. Yet his conflicts with blacks may be real. The bitter struggles for survival at the bottom of America are not merely irrational, and over skin color. They concern real goods: jobs, turf, real-estate values, education, culture.

What we call "lack of prejudice," therefore, is often rather more an aspect of privilege than a profound spiritual accomplishment. What we call "prejudice" is sometimes prejudice—that is, an irrational resistance to cultural difference, an attribution of inferiority based on group characteristics rather than on individual performance. On the other hand, it is sometimes also an accurate reading of actual economic and social competition. There *are* fewer jobs than there are job-seekers; there *are* neighborhoods fighting for stability and neighborhoods suddenly swamped by what Kenneth Clark has called a "tangle of pathologies"; there *are* skills structurally more highly developed in one culture than in another, skills of real economic and political effect in competition between groups. Those lower on the economic ladder do not have the money, mobility or freedom of choice to evade the bitter conflicts that result. Their resistance is not "irrational"; it concerns quite "rational" matters like their jobs, their schools, their mortgages.

Most Catholics, I believe, perceive at least three strata among blacks. (Among themselves, they also see the same strata.) First are the highly successful and prominent blacks, from Martin Luther King Jr. to Senator Edward Brooke, to the black athletes, professors, singers, actors and lawyers they see so often in the media (but seldom in real life). Second comes the black middle class, pious and disciplined and hard-working, whom they do see in their neighborhoods or meet in their schools — people very like themselves. Third comes the black underclass: that one-fourth or one-third of the black population that lives in poverty, joblessness and the widely publicized "tangle of pathologies."

It is this third group of blacks that dominates public consciousness, those whose "victimization" by society provides the rationale for special public assistance. Commentators conveniently forget that as much as one-fifth of the white-ethnic population is in similar economic and sometimes cultural desperation. Oakland Raider football player George Blanda grew up in the coal fields near Pittsburgh; he and his nine brothers and sisters never saw more than $2,000 a year from their father's work in the mines; even today the average income thereabouts is $6,000 a year. "They don't have to tell me what life is like in a ghetto," Blanda says. "I've heard 'dumb Polack' as often as they've heard 'nigger.' "

This misplaced emphasis and a host of other factors combine in the resentment—documented by writers from Robert Coles to Jimmy Breslin — working-class Catholics tend to feel over the alliance between rich, highly educated and powerful whites and the black underclass. It is as though an alien upper-class society has decreed that advances for the black underclass should come chiefly at the expense of working-class whites, while the privileged give up nothing at all.

The resentment and the resulting conflict with blacks is hardly mitigated by the fact that the black underclass, mostly newer to the impersonality and grimness of the Northern cities, is especially weak in an area to which both middle-class blacks and working-class Catholics give pre-eminent value: a strong, disciplined family life. According to the National Center for Health Statistics, 46 percent of all black births in 1973 were illegitimate, almost half of these (119,800) to girls between 15 and 19 years old. Without making moral judgments—for both Eastern and Southern Europeans have a sexual morality quite different from that of the Irish or Anglo-Saxons —one can estimate that the economic consequences are severe. Sons are normally socialized, both in white-ethnic and in strong black families, by the firm, even harsh, discipline of the father; and they are usually taught job skills in the home. By the time they graduate from high school, white-ethnic children have already had several types of job experience—the jobs usually being found for them by their fathers or their father's friends. Thus, youngsters without fathers suffer a severe economic handicap. Understandably, some members of the highly unemployed black underclass resent this situation, but to white ethnics, the valuing of strong family relationships is an important way to deal with economic adversity.

Many middle-class blacks also share the white ethnics' belief in the value of family life as well as their perception of the black underclass. Jesse Jackson, the Chicago black leader, has voiced it eloquently to Paul Cowan in The Village Voice. Like Catholic working people, they too believe passionately in discipline, hard work, education for their children and in a life of higher security, decency, peace and opportunity than they have known before. This black working class receives little or no attention from the black-militant leadership or from the press. It is considered "square," bourgeois, dull, not admirable.

Instead of trying to bring about an alliance between the black working class and the white working class, based on analogous patterns of attachment to church, school, family and hard work, too many social activists have for 15 years now lionized the black underclass and exulted in its differences from the black and white working class. This is a self-destructive political strategy.

Just as there are several different cultures among blacks so also are there several different cultural systems among Catholics. Symbolically, the Irish dominate the Catholic population, but in fact only 17 percent of American Catholics are Irish. Because they have spent the most time in America, and speak English as a native language, because of their educational and occupational success, their literary self-expression, their dominant roles in the church, in national politics and in other areas, the Irish are at the top of the

Catholic ladder. Still, those poor and working-class Irish locked in the old neighborhoods of Boston, Providence, Queens, Philadelphia and other Eastern cities retain the images of oppression they have experienced for 300 years in Ireland and in America. An insular people, they are less accustomed to welcoming foreign elements into their midst than some of the other Catholics. (In Western Pennsylvania, where I grew up, we Slavs and Italians were "foreigners." When we said "the Americans" we meant the Irish.) The high level of economic, forensic, managerial and political skills nourished in their culture for centuries has kept the Irish ahead of the Eastern European Catholics—the largest single family of Catholic cultures.

The Slavs, by and large, went to work in mines, mills and factories, working far more often for others and for weekly wages than, say, the Jews, the Greeks or the Italians, who with high frequency went into business for themselves. The American economic system tends to grant larger rewards to those who follow the latter rather than the former course. In Eastern Europe, the great symbol of winning liberty from serfdom was the ability to own one's own home. Even in the United States, home ownership remains the highest economic priority of Slavic families; higher, even, than education for their children.

The Poles vote more frequently than the other Slavs, and in higher proportion than any other ethnic group in America; but they have developed few highly placed political leaders. So far, the Slavs seem to lack the vivid charismatic and cosmopolitan intellectual and political leadership their numbers warrant. Edmund Muskie is really not very Polish and resisted identifying himself with white working - class immigrants; Mary Anne Krupsak and Barbara Mikulski may represent a new order.

The Italians, by contrast, rival the Irish in forensic and political skills. Sixty-two percent of the Italian immigrants settled within an arc from Providence to Philadelphia, and where they found the Irish dominating the Democratic Party, many of them began to take over the Republican Party. In Hollywood, the arts and the creative departments of advertising firms the Italians may be second to none in talent, visibility and success. In literature and scholarship, there may soon be a similar explosion. In the great corporations and in banking, they are making their presence known — but barely. Like the Eastern Europeans, the Italians tend to be a highly stable people; like the others, many move to the suburbs, but the old neighborhoods tend to live on longer than most.

More than other groups, they also seem willing to fight for their neighborhoods, to protect them, and to resist outside bullying or outside threats. Thus in Boston, when observers warn that South Boston may have been tough, but busing in the Italian North Ward may be even tougher, they are judging by a common folk wisdom. In New York, the word among other white-ethnic groups is "live next to the Italians, if you want to stay." Rosedale is largely Italian.

Conflicts such as that going on in Rosedale are hardly in the best interests of the city's future, but neither is a likely alternative—a massive exodus of whites from the old neighborhoods. In this sense, the cultural background that leads Italians and Slavs to prefer to stay in their old neighborhoods is one of the few remaining hopes of integrated cities in the North. Why is their willingness to stay not politically and economically reinforced?

First, it must be acknowledged, urban Catholics have often been betrayed by their own leaders. Disastrous Catholic "machines," like Rizzo's in Philadelphia, have won bluster but no real advance for their people; Catholic bishops, concentrating on narrow ecclesiastical issues, have ignored the vast social instability of the cities; Catholic intellectuals have been establishing ideological credentials as "conservative" or "liberal," with little attention to working - class sufferings; local Catholic political leadership, as on the Boston School Committee, has often been disgustingly poor.

Thus, Catholics must blame themselves for their political predicament. However, some share of the blame—and hope for change—rests with the larger, and far more powerful, Protestant Establishment. Banks, investment houses, corporate law firms and realtors — the institutions par excellence of the great Protestant Establishments in every Northern city—give the media beautiful ads about integration. In actual practice they destroy the family businesses, family investments and family morale of neighborhood after neighborhood. Banks "redline" changing neighborhoods; realtors "blockbust." Then the "tangle of pathologies"—undisciplined teen-agers (who commit 70 percent of all crime), other forms of crime, deterioration of housing, the bankruptcy of small businesses, the decline in services—moves in. The pundits then call those who flee "racists." The racism is in the Establishment. The institutions of the Establishment betray a neighborhood, drive people out, destroy their dreams and rob their savings. This is a scandal that cries out to heaven.

Black leaders have made a tragic mistake, strategically and tactically, in choosing white - ethnic neighborhoods as the target, the victim and the hostage of their desperation. Black leaders must know that the white-ethnic schools in Boston — or Philadelphia, or Detroit— are as bad as their own, that less money is often available to them, that the idealistic young teachers from Harvard and Columbia go first to black, hardly ever to white-ethnic schools. But some argue, "We will bus your children to our schools, as hostages to remind you of our plight." Why would black leaders choose their most potent voting allies as their hostages? It defies all rationality.

Black leaders have no closer bloc of political support than white-ethnic voters, who have provided majorities for every piece of progressive economic legislation and civil-rights legislation this nation has had these last 50 years. There is no progressive victory in Pennsylvania without large majorities in the white-ethnic wards of Pittsburgh and Philadelphia to overcome the Protestant, conservative, rural votes; the liberals of Bryn Mawr cannot carry even their own Main Line constituency. You cannot win Illinois for a progressive victory without huge white-ethnic majorities in Chicago; Michigan without Detroit; New York without Queens, the Bronx and Buffalo; Ohio without Cleveland and Toledo. In 1968, Wallace won 7.7 percent of the Catholic vote but 16 percent of the Protestant vote. White ethnics and black workers have the same economic interests and analogous cultural interests. Why divide them?

The Protestant and Jewish liberals who dominate the media are ignorant about the Catholic working class. Since they define the issues, they would save much grief if they would not divide working-class blacks from working-class Catholics and others, if they would praise the virtues of hard work and family economic wisdom by which parents give their children better opportunities than they received. Black and Catholic leaders share a similar responsibility. The destruction of urban ethnic neighborhoods is the No. 1 domestic tragedy of our time. This is the disease that will kill us all. This is the ecological abuse more fatal than dirty air.

"Redlining" and "blockbusting" must be stopped. There must be some plan of Federal insurance to protect the investments in homes and small businesses that hard-working people have made in their neighborhoods. These are more important social forms of savings than deposits in savings banks. When white-ethnic families can look ahead 20 or 30 years to reasonable prospects of financial security and neighborhood stability, they stay. They integrate. They have long since begun to be integrated. They do not love blacks, any more than Archie Bunker (*not* a white ethnic) loved the Jeffersons, but they do live next door, speak with the rough truthfulness of the working class, and—given no reason to fear for the future—live together decently.

There must also be greater honesty about the pathologies of the white and black underclass. The white-ethnic poor share the supposedly "black" pathologies—high drop-out rates, drugs, cynicism, an upsurge in broken families and economically unprepared youths. The white underclass of Boston described in George V. Higgins's novels is threatened by self-destruction. To expect suffering and wounded communities like these to assume, on top of their own, the even more grievous burdens of blacks is a measure of political foolishness that leaves one speechless.

I see on the horizon two possibilities. Either a political leader of the stature of Robert Kennedy will hold the trust of both black and white-ethnic workers and build a politics of human ecology designed to strengthen the cultural and family networks that make urban life faintly possible. Or the political foolhardiness of the sort chosen by the N.A.A.C.P. in Boston—a willingness to destroy a city in order to "save" it—will awaken the most enormous political and moral revulsion. Blacks and Catholics can go forward together. If they are pitted against each other, those on the top of the American ladder will rejoice in their own moral superiority—until turmoil and bitterness enshroud us all. ■

November 16, 1975

The Spirit of Education Present

By Richard Gambino

A managed-decline state of affairs exists today in higher education, partly the result of severe economic recession and inflation and a general disillusionment with the educational Establishment. It is also the result of great demographic changes.

The post-World War II "baby-boom" generation is much larger than the generations preceding it and following it. The United States birth rate during the 1930's was 2.1 children; in 1957 it was 3.7, while in 1975 it is only 1.8. During the period 1956-60, 4.3 million babies were born per year, while only three million are born annually now. The number of school-age children has dropped by two million since 1965, and will drop by another 2.5 million by 1985. The results are that the college-age population has shrunk and will continue to shrink.

Whatever the validity of managed decline as a demographic goal, in colleges and universities it is resulting in irreversible social injustice and even greater educational upset than experienced in the 1960's.

As decline is being managed in education, the large working and poor classes are being barred from opportunities. We are fast moving to a situation in which only the rich can send their children to college. With the disintegration of the educational ladder of class mobility, we face the prospect of our social classes freezing into castes in which people remain locked from birth to death.

A recent newspaper editorial concluded that "vague fears that the rising cost of tuition might squeeze out the lower middle class are now turning into concrete statistical reality." The nationwide college enrollment by students from families earning $10,000 to $15,000 per year dropped from 43 percent in 1970 to 36 percent in 1973.

Studies by the National Commission on Financing Post Secondary Education show that for every $100 of tuition increase there is a 2 percent drop in college enrollment for students from families earning $7,500 to $15,000 per year, and a 3 percent drop for students whose families have an income of less than $7,500.

We are talking about the large majority of the American people: Census Bureau reports place the present median annual income for families at under $9,000.

In addition, there is a high economic correlation between economic class, educational level, and ethnic background, as seen in 1972 census figures. In the 25-to-34-year age group, the ethnic group with families having the highest median income ($13,929), American Jews, has by far the highest percentage of college graduates (52.5 percent), while those in the lowest income group among whites, Hispanics ($7,595 per year), have the lowest percentage of college graduates in that age group, 5.3 percent. The correlation between income and higher education holds almost constantly as we identify the places of other groups in the class strata: Irish, Italian, Japanese, West Indian-Americans, etc.

Moreover, the rate of educational progress from one generation to another in each group is *directly* correlated to income—witness the fact that among Jews the rate of college graduates jumped 34.1 percentage points between generations (of Jews over 35 years old, only 18.4 percent are college graduates). Among Hispanics the rate increased by only 0.8 of a percentage point between generations (4.5 percent of Hispanics over 35 are college graduates).

The realities of financial aid given to students, inadequate and diminishing, will not alter the pattern much in the future. Scholarship monies are given mostly to blacks and to women of Jewish and Anglo-Saxon backgrounds.

This practice is also perpetuating the ethnic composition among college teachers and corporation executives in the United States. They are overwhelmingly Anglo-Saxon and Jewish males, with a sprinkling of women mostly from these two groups, and blacks of both sexes, being rapidly moved in. In terms of actual results, it is as if all others "need not apply."

In addition, decline in higher education is being managed mindlessly, following the almost complete take-over of colleges and universities in the last twenty years by administrators and other powers who are bureaucrats rather than educational philosophers.

Not having given much attention to questions of educational values and priorities in the period of vast growth just ended, bureaucracies are now making cutbacks by applying sterile formulas—"production" measured by non-educational criteria, and *in extremis* "across the board" cuts falling equally on good and fragile programs and nonsensical and hardy ones.

It is time to introduce equity and educational thinking into higher education. As we face a probably protracted time of low-growth or no-growth conditions in the United States, education is too important for the health of American democracy for us to permit it to become the privilege of the rich managed by mere managers.

Richard Gambino is associate professor of education at Queens College.

December 8, 1975

Affirmative Discrimination

Ethnic Inequality and Public Policy.
By Nathan Glazer.
248 pp. New York: Basic Books.
$10.95.

By WILLIAM V. SHANNON

After two centuries of slavery and another century of Jim Crow second-class citizenship, the court decisions and civil rights legislation of the 1950's and 60's finally destroyed segregation and overt racial discrimination—the barriers that had kept most black Americans, roughly 11 percent of the total population, out of the mainstream of national life.

"But freedom is not enough," President Johnson observed in his historic commencement address at Howard University in June 1965. "You do not take a person who, for years, has been hobbled by chains and liberate him, bring him up to the starting line of a race and then say, 'You are free to compete with all the others' and still justly believe that you have been completely fair.

"Thus it is not enough just to open the gates of opportunity. All our citizens must have the ability to walk through those gates.

"This is the next and the more profound stage of the battle for civil rights. We seek not just freedom but opportunity. We seek not just legal equality but human ability, not just equality as a right and a theory *but equality as a fact and equality as a result.*"

More than 10 years later, this promise remains unfulfilled. Can it ever be fulfilled? Is it the proper business of Government to go beyond the protection of individual rights and try to guarantee that any group, as a group, achieve equality in income, jobs, education and housing?

Harvard sociologist Nathan Glazer answers with a firm "no." He has written a tempered, factually argued, vigorous polemic against the predominant drift of public policy on racial issues over the past decade. He demonstrates that the Government effort to make good on President Johnson's promise "to fulfill these rights" (curiously, Glazer never mentions the Howard University speech) has led to an increasingly tangled skein of lawsuits, busing orders and racial quotas. There has been much juggling of statistics and tortured judicial reasoning.

The upshot, in Glazer's view, is that the chief beneficiaries of "affirmative action" have been the members of the black middle class and those poised to enter it, in short, the bright, ambitious people who would have gotten ahead anyway but whose progress toward better jobs and higher incomes has been accelerated by the Federal Government's prodding of private employers.

Affirmative action, Glazer argues, has failed to help the mass of working-class blacks. He makes clear that he has no principled opposition to preferential job treatment for blacks if he thought it would actually work: "For me, no consideration of principle—such as that merit should be rewarded, or that governmental programs should not discriminate on grounds of race or ethnic group—would stand in the way of a program of preferential hiring if it made some substantial progress in reducing the severe problems of the low-income black population and of the inner cities." In his view, lack of education, of useful skills and, sometimes, of good work habits and attitudes account for the relatively poor showing of many working-class blacks. Such handicaps cannot be remedied by preferential hiring.

My own view, as one who supports affirmative action in the employment field, is that despite the anomalies and absurdities Glazer highlights, it is healthy for the larger society to make a conscious effort to push blacks into positions of visible success and authority. All blacks, even the poorest inhabitants of the slums, benefit from the resultant gains in hope, self-image and sense of participation.

Moreover, although in the long run merit must be the controlling consideration, everyone knows there are many well-paid white collar jobs in our economy for which the criteria of competence are either so hard to define or so minimal that there are always enough white applicants whose hiring could be justified if the Federal Government did not intervene to make sure that blacks obtained their share of such jobs. Among the occupations that come to mind are sociologists and TV weathermen, receptionists and bank tellers, public relations assistants and journalists.

Glazer is more persuasive in his critique of affirmative action programs in housing and schools. With no help from the Federal Government except for the outlawing of restrictive covenants on real estate, blacks on their own have penetrated the suburbs, increasing their number from 2.2 to 3.7 million in 20 years. This natural process can be expected to continue. Plans to accelerate the pace or to mix population groups of widely varying incomes meet with insuperable practical obstacles in a society where people are so mobile and neighborhoods can change character so rapidly.

With regard to the aggressive effort of the Federal courts—notwithstanding the repeated admonitions of Congress—to realign school populations in order to achieve an ideal racial balance, Glazer is illuminating on the Sisyphean nature of this task. As Judge Garrity is discovering in Boston, it is extremely difficult, and probably impossible, to run a city school system from a judge's bench. The looked-for social gains remain elusive, while the costs to the city in "white fight" heighten racial antagonisms and severely disrupt the educational process.

Glazer argues that the promise of the Brown Decision of 1954 has been fulfilled; segregated schools no longer exist. In the North and South, what we have is the normal concentration of different races in different neighborhoods; this produces schools with a predominance of students of a particular race. The source of today's difficulty is the mistaken effort of the courts, beginning with the Charlotte-Mecklenburg Case in 1971, to alter the natural effect of this population concentration on our schools.

Glazer frames his attack on affirmative discrimination within a discussion of white ethnicity. He notes that affirmative action, while doing little for most blacks, has inadvertently legitimized white ethnic self-consciousness. "The saliency of ethnic identities has increased markedly since the middle sixties: since, specifically, the Negroes became blacks, and the dominant tone of black political rhetoric shifted from emphasizing 'we are like everyone else and want only integration,' to 'we are, of course, different from anyone else and want our proper share of power and wealth.'"

He believes that the sooner the blacks are recognized as just another ethnic bloc, making the usual efforts to get ahead in business and the professions, cutting the usual political deals and having no claims to special judicial preferment, the better off the nation will be. Glazer writes: "The judges should now stand back, and allow the forces of political democracy in a pluralist society to do their proper work."

Whether or not one agrees with this contention, "Affirmative Discrimination" is a book worth reading. Public issues only infrequently receive serious, sustained arguments of this high order. ■

William V. Shannon is a member of the editorial board of The New York Times.

February 8, 1976

Carter Issues an Apology On 'Ethnic Purity' Phrase

By CHRISTOPHER LYDON

Special to The New York Times

PHILADELPHIA, April 8— Jimmy Carter apologized today for using the phrase "ethnic purity" in his pledge two days ago to defend the stability of established neighborhoods. But he stuck to his original basic position, saying he would not "arbitrarily use Federal force" to change a neighborhood's ethnic character.

His apology notwithstanding, the "ethnic purity" reference and two others he had used in the same context—"black intrusion" and "alien groups"—constituted a haunting refrain throughout the Georgia Democrat's first full day of campaigning for the Pennsylvania primary on April 21.

Representative Andrew Young of Georgia, Mr. Carter's foremost advocate in black communities North and South, told the candidate for the Presidential nomination this morning that his phrasing was "a disaster for the campaign."

In a telephone interview from Washington, Mr. Young said, "Either he'll repent of it or it will cost him the nomination." At best, Mr. Young said, the former Georgia Governor's recovery will require time and an explicit policy statement on housing.

Senator Henry M. Jackson of Washington and Representative Morris K. Udall of Arizona, the other active candidates in the Democratic primary here, both seized today on Mr. Carter's remarks in Indiana Tuesday as the kind of mistake they had waited to exploit against the surprise early leader in the race for the nomination.

Senator Jackson said Mr. Carter's statements were "amazing," but did not explain why he thought so.

Mr. Udall, who has high hopes of winning votes in Philadelphia's large black community, said in a written statement: "Much worse than his ambiguity, which has become as much a trademark of Jimmy Carter as his grin, are some of the words and phrases he used to express himself on this issue. It is disturbing, very disturbing language to hear from a Presidential candidate, whatever he ends up saying he meant to say."

In an important policy shift, Mr. Carter announced today that he could now support the so-called Humphrey-Hawkins employment bill that organized labor and the Congressional Black Caucus have made their top priority Federal legislation.

The bill, sponsored by Senator Hubert H. Humphrey, Democrat of Minnesota, and Representative Augustus F. Hawkins, Democrat of California, would set a three-year goal of reducing unemployment to 3 percent and would guarantee Federal jobs, as necessary, to accomplish that end. Mr. Carter had heretofore made a point of emphasizing the cost to taxpayers of putting the unemployed to work. Today he said the bill has been amended sufficiently to make it acceptable, by adding emphasis to the development of jobs in the private sector, and by making 3 percent the goal of adult unemployment, not of the work force at large, which included teen-agers.

'Unfortunate Choice'

Mr. Carter used his first news conference of the day to volunteer his apology for "an unfortunate choice of words," referring specifically to his use of "ethnic purity."

"I don't think there are ethnically pure neighborhoods in this country," he said. The word "purity," he told a questioner, "bothers me, too."

Mr. Carter reiterated his support of Federal and state open-housing laws and for Government action to enforce equal opportunity in new housing built with Federal assistance.

He restated the position he took on Tuesday, saying, "I would not arbitrarily use Federal force to move people of a different ethnic background into a neighborhood just to change its character."

He acknowledged, at the same time, that he was attacking a kind of "arbitrary action" that had not been proposed or taken, saying, "I don't think the Congress or anybody else has advocated this."

When a reporter asked whether he was creating "a straw man, something that doesn't exist," Mr. Carter replied, "yes, that's correct." Yet, he insisted that he had not been trying to play politics with the overtones of his words.

"If the phrase had racial connotations," he said, "I've apologized, I hope, to the public, and I've already talked to my supporters."

Mr. Udall praised Mr. Carter for his swift apology, but he commented that the timing of Mr. Carter's earlier words was nonetheless "remarkable"—coming, Mr. Udall said, "when [Gov. George C.] Wallace [of Alabama] is leaving the race and Pennsylvania and Michigan are coming up."

Mr. Carter said he would sooner "withdraw from the race" than use "racist" appeals to win it. "My feelings are quite the opposite of that," he said. Blacks and others who know his record as Governor of Georgia will understand that his words on Tuesday were "careless," Mr. Carter said.

But he was also prepared to pay some political penalty. "If they don't try to to make political hay out of it," he said, referring to Mr. Udall and Mr. Jackson, "I would be surprised."

Mr. Young saw the danger of lasting damage, observing: "A lot of people who said 'You just can't trust a Southerner' are going to say, 'See, I told you so.'"

Mr. Young was torn today between defending Mr. Carter and denouncing his language. "This doesn't mean to me he's a racist," he said. "It means he made a terrible blunder that he's got to recover from."

"I just think it's an awful

phrase. I don't think he understood how loaded it is with Hitlerian connotations. My theme all along," Mr. Young continued, "has been that white liberals would eventually follow blacks to support him. But this gives them some reason not to. A lot of white liberals will hesitate, and blacks who don't know him personally will wait and see what he means."

"Those of us who do know him know he's had a good record on open housing," Mr. Young said. "He's kind of put himself into this trap. It wasn't Udall, Jackson or Humphrey. Nobody baited this for him. He's got to find a way to get out—and frankly I hope he does."

Many prominent Pennsylvania Democrats, both black and white, saw the "ethnic purity" controversy as a shaping event in their primary.

Chuck Stone, the black strategist and author, used his Philadelphia Daily News column this morning to call Mr. Carter, famous for his toothy smile, a "mandibulary phony."

He praised Mr. Carter as "the only candidate who campaigned aggressively for black votes in all the primaries," but this week, in Mr. Stone's analysis, "Carter quickly moved to reassure white America he was still their 'good ol' boy' and called for a new kind of segregation by homogeneity."

Some black leaders, meanwhile, were completely soothed by Mr. Carter's apology and explanation. Others reserved judgment.

"I was satisfied with the explanation he gave," said Ann Jordan, a member of the Democratic National Committee. "He said he would not use the powers of the Presidency to impose on a community low-cost housing in any setting. It's the press that's making this a major issue."

April 9, 1976

U.S. HIGH COURT BACKS USE OF RACIAL QUOTAS FOR VOTING DISTRICTS

By LESLEY OELSNER

Special to The New York Times

WASHINGTON, March 1—The Supreme Court ruled 7 to 1 today that a state drawing up a reapportionment plan may sometimes use racial quotas that are designed to assure that blacks and other non-whites have majorities in certain legislative districts.

The Court said that this type of race-conscious redistricting was constitutional, at least in some circumstances, when the state did it in an effort to comply with the Federal Voting Rights Act of 1965.

The Court acted in a case from New York involving a 1974 redistricting of state legislative districts. Among other things, it was designed to create non-white majorities of at least 65 percent in two State Senate Districts and two State Assembly Districts in Brooklyn.

The state devised this plan in an effort to comply with the Voting Rights Act, but Hasidic Jews in the Williamsburg section of Brooklyn challenged its constitutionality. The effect of today's decision, which upheld two lower-court rulings, was to leave the 1974 legislative lines intact.

Impact Hard to Gauge

The likely practical impact of the ruling and its long-term implications were impossible to gauge immediately, in part because the Justices differed among themselves on their reasoning, and there was no one statement of the Court's holding that was joined by a majority of the Justices. The ruling was thus the lowest common denominator of the various statements written or joined by the various Justices.

The ruling, however, seemed likely to have substantial significance.

Justice William J. Brennan Jr. said in a concurring opinion—in explaining why he was writing the opinion—that "this case carries us further down the road of race-centered remedial devices than we have heretofore traveled." This was accompanied, Justice Brennan added, "with the serious questions of fairness that attend such matters."

Jack Greenberg, director of the NAACP Legal Defense and Education Fund, which was on the winning side of the case, called the decision "a major and encouraging pronouncement on affirmative action with important implications in the areas of employment discrimination and education."

He also called it a civil rights "advance."

Deputy Solicitor General Lawrence G. Wallace, whose office also participated in arguing the winning side of the case, suggested that the ruling should not be interpreted as indicating how the Court might rule on some pending questions about affirmative action, such as the constitutionality of special minorities admissions programs at universities.

He said, however, that the New York case was significant, in that if the Court had ruled the other way, it would have "undermined" the effectiveness of the Voting Rights Act "a good deal."

The chief lawyer on the other side of the case, Nathan Lewin, said the ruling could be interpreted either quite narrowly, or broadly. The case involved a set of facts not likely to occur often and so the case could be viewed as limited, he said. However, he said, it also "gives more of an imprimatur, constitutionally," to the use of quotas.

Mr. Lewin contended that the Court had not adequately addressed the issues and problems raised by use of racial criteria, and said the Court in the voting rights area was in effect "abdicating" to the Justice Department, accepting Justice Department determinations on what kind of plans satisfied the Voting Rights Act.

White Writes Opinion

The opinions of the Justices in the majority—with Byron R. White writing the main opinion—appeared to seek to underplay the scope of the ruling.

The sole dissenter, however, Chief Justice Warren E. Burger, called the majority"s decision "racial gerrymandering" that "moves us one step farther away from a truly homogenous society."

Redistricting plans for areas that are subject to the act must be approved ahead of time either by the Attorney General, or by the Federal District Court in Washington. According to New York State's position in the case, state officials got the impression that a 65 percent non-white majority in those districts was necessary to obtain the approval of the Attorney General, who had raised questions about a prior redistricting proposal.

One effect of the redistricting, however, was to take the Hasidic Jewish community of Williamsburg out of a single Assembly District and a single State Senate District and place portions in two Assembly and two Senate Districts.

Representatives of the community filed suit in Federal Court challenging the redistricting, both under the 14th Amendment, which guarantees equal protection of the laws, and the 15th Amendment, which says that the right to vote may not be denied or abridged because of race or color.

The district court dismissed the complaint, and the United States Court of Appeals for the Second Circuit upheld the district court.

Technically, the Justices' holding was simply that New York, in the disputed 1974 redistricting, did not violate the constitutional rights of the redistricting"s challengers. Necessarily, however, by saying that this particular race-conscious plan was not unconstitutional, the Court was saying, as various statements of the Justices indicated, that race-conscious redistricting in general was a constitutional technique, at least in Voting Rights Act cases.

Justice White's opinion gave two independent grounds for ruling that New York did not violate the Constitution with its 1974 plan. Justices John Paul Stevens, W. J. Brennan and Harry A. Blackmun agreed regarding the first ground; Justice Stevens and Justice William H. Rehnquist agreed regarding the second.

Justice Potter Stewart wrote a separate opinion, joined by Lewis F. Powell Jr., that concurred in the final judgment for yet another reason.

Justice Thurgood Marshall did not participate in the case, apparently because of his past association with the National Association for the Advancement of Colored People, which was a party to the case.

The first ground relied on in the White opinion—the ground joined by a total of four Justices—involved the Voting Rights Act, which the Court in an earlier case had upheld as constitutional.

The opinion today said in effect that race-conscious remedies were permissible under the act in order to accomplish the goals of the act, and that the challengers here had yet proved that the state had done more than the Attorney General was authorized to require it to do under the act.

The second ground in Justice White's opinion, somewhat more far-reaching, was that the state was "free to do what it did" so long as it did not violate the Constitution, and that its actions did not violate it.

March 2, 1977

Suggested Reading

Barton, Joseph J. *Peasants and Strangers: Italians, Rumanians and Slovaks in an American City, 1890-1950*. Cambridge, Mass.: Harvard University Press, 1975.

Berthoff, Rowland Toppan. *British Immigrants in Industrial America, 1790-1950*. Cambridge, Mass.: Harvard University Press, 1953.

Billington, Ray Allen. *The Protestant Crusade, 1800-1860*. New York: Quadrangle Books, 1964.

Blegen, Theodore C. *Norwegian Migration to America, 1825-1860*. Northfield, Minn.: The Norwegian-American Historical Association, 1931.

Brown, Dee. *Bury My Heart at Wounded Knee, An Indian History of the American West*. New York: Holt, Rinehart & Winston, 1970.

Cortés, Carlos E., advisory ed. *The Chicano Heritage*. A reprint series of 55 books. New York: Arno Press, 1976.

Cortés, Carlos E., advisory ed. *The Mexican American*. A reprint series of 21 books. New York: Arno Press, 1974.

Cordasco, Francesco, advisory ed. *The Italian American Experience*. A reprint series of 39 books. New York: Arno Press, 1975.

Cordasco, Francesco, advisory ed. *The Puerto Rican Experience*. A reprint series of 33 books. New York: Arno Press, 1975.

Dinnerstein, Leonard and Frederic Cople Jaher. *Uncertain Americans*. New York: Oxford University Press, 1977.

Dolan, Jay P. *The Immigrant Church: New York's Irish and German Catholics, 1815-1865*. Baltimore, Md.: Johns Hopkins University Press, 1975.

Ellis, John Tracy. *American Catholicism*. Chicago, Ill.: University of Chicago Press, 1957.

Esslinger, Dean R. *Immigrants and the City: Ethnicity and Mobility in a Nineteenth Century Midwestern Community*. Port Washington, New York: Kennikat Press, 1975.

Fitzpatrick, Joseph P. *Puerto Rican Americans*. Englewood Cliffs, N.J.: Prentice-Hall, 1971.

Gambino, Richard. *Blood of My Blood*. New York: Doubleday, 1974.

Genovese, Eugene D. *Roll, Jordan, Roll: The World the Slaves Made*. New York: Pantheon, 1972.

Glazer, Nathan and Daniel P. Moynihan. *Beyond the Melting Pot*. Cambridge, Mass.: M.I.T. Press, 1963.

Gleason, Philip, ed. *Catholicism in America*. New York: Harper & Row, 1970.

Gordon, Milton M. *Assimilation in American Life*. New York: Oxford University Press, 1964.

Greeley, Andrew M. *Ethnicity in the United States*. New York: John Wiley, 1974.

Greeley, Andrew M. *That Most Distressful Nation*. New York: Quadrangle, 1973.

Greene, Victor, Oscar Handlin and John Appel, advisory eds. *The American Immigration Collection*, Series II. A reprint series of 33 books. New York: Arno Press, 1970.

Handlin, Oscar, Carl Wittke and John Appel, advisory eds. *The American Immigration Collection*, Series I. A reprint series of 41 books. New York: Arno Press, 1969.

Handlin, Oscar. *Boston's Immigrants: A Study in Acculturation*. New York: Atheneum, 1968.

Hawgood, John A. *The Tragedy of German-America*. New York: G.P. Putnam, 1940.

Higham, John. *Send These to Me: Jews and Other Immigrants in Urban America*. New York: Atheneum, 1975.

Higham, John. *Strangers in the Land: Patterns of American Nativism, 1860-1925*. New York: Atheneum, 1965.

Howe, Irving. *World of Our Fathers*. New York: Harcourt, Brace Jovanovich, 1976.

Hvidt, Kristian. *Flight to America: The Social Background of 300,000 Danish Emigrants*. New York: Academic Press, 1975.

Jordan, Winthrop D. *White Over Black*. Chapel Hill, N.C.: University of North Carolina Press, 1968.

Katz, William Loren, advisory ed. *The American Negro: History and Literature*. Two reprint series.

I. 44 books. II. 66 books. New York: Arno Press, 1968/1969.

Kessner, Thomas. *The Golden Door: Italian and Jewish Immigrant Mobility in New York City, 1880-1915*. New York: Oxford University Press, 1977.

Levitan, Sam, William B. Johnston and Robert Taggert. *Still a Dream: The Changing Status of Blacks Since 1960*. Cambridge, Mass.: Harvard University Press, 1975.

Lynch, Hollis R. *The Black Urban Condition: A Documentary History, 1866-1971*. New York: Thomas Y. Crowell, 1973.

McCaffrey, Lawrence J., advisory ed. *The Irish-Americans*. A reprint series of 42 books. New York: Arno Press, 1976.

McCaffrey, Lawrence J. advisory ed. *The Irish Diaspora in America*. Bloomington, Ind.: Indiana University Press, 1976.

Meier, Matt S. and Feliciano Rivera. *The Chicanos*. New York: Hill and Wang, 1972.

Metzker, Isaac. *A Bintel Brief*. New York: Doubleday, 1971.

Miller, Stuart Creighton. *The Unwelcome Immigrant: The American Image of the Chinese, 1785-1882*. Berkeley, Ca.: University of California Press, 1969.

Nash, Gary B. *Red, White and Black: The Peoples of Early America*. Englewood Cliffs, N.J.: Prentice-Hall, 1974.

Nelli, Humbert. *The Italians of Chicago: A Study in Ethnic Mobility*. New York: Oxford University Press, 1970.

Novak, Michael. *The Rise of the Unmeltable Ethnics*. New York: Macmillan, 1973.

Potter, George W. *To the Golden Door*. Westport, Conn.: Greenwood Press, 1974.

Prucha, Francis Paul. *American Indian Policy in Crisis: Christian Reformers and the Indian, 1865-1900*. Norman, Okla.: University of Oklahoma Press, 1976.

Rischin, Moses, advisory ed. *The Modern Jewish Experience*. A reprint series of 59 books. New York: Arno Press, 1975.

Rischin, Moses. *The Promised City: New York Jews, 1870-1914*. Cambridge, Mass.: Harvard University Press, 1962.

Saloutos, Theodore. *The Greeks in the United States*. Cambridge, Mass.: Harvard University Press, 1964.

Sanders, James W. *The Education of an Urban Minority: Catholics in Chicago, 1833-1965*. New York: Oxford University Press, 1977.

Satz, Ronald N. *American Indian Policy in the Jacksonian Era*. Lincoln, Neb.: University of Nebraska Press, 1975.

Taylor, Philip. *The Distant Magnet*. New York: Harper & Row, 1971.

Thomas, William I. and Florian Znaniccki. *The Polish Peasant in Europe and America*. New York: Alfred A. Knopf, 1927.

Washburn, Wilcomb E. *The Indian in America*. New York: Harper & Row, 1975.

Weglyn, Michi. *Years of Infamy: The Untold Story of America's Concentration Camps*. New York: William Morrow, 1976.

Index